Studies in Systems, Decision and Control 488

Series Editor

Janusz Kacprzyk, *Systems Research Institute, Polish Academy of Sciences, Warsaw, Poland*

The series "Studies in Systems, Decision and Control" (SSDC) covers both new developments and advances, as well as the state of the art, in the various areas of broadly perceived systems, decision making and control–quickly, up to date and with a high quality. The intent is to cover the theory, applications, and perspectives on the state of the art and future developments relevant to systems, decision making, control, complex processes and related areas, as embedded in the fields of engineering, computer science, physics, economics, social and life sciences, as well as the paradigms and methodologies behind them. The series contains monographs, textbooks, lecture notes and edited volumes in systems, decision making and control spanning the areas of Cyber-Physical Systems, Autonomous Systems, Sensor Networks, Control Systems, Energy Systems, Automotive Systems, Biological Systems, Vehicular Networking and Connected Vehicles, Aerospace Systems, Automation, Manufacturing, Smart Grids, Nonlinear Systems, Power Systems, Robotics, Social Systems, Economic Systems and other. Of particular value to both the contributors and the readership are the short publication timeframe and the world-wide distribution and exposure which enable both a wide and rapid dissemination of research output.

Indexed by SCOPUS, DBLP, WTI Frankfurt eG, zbMATH, SCImago.

All books published in the series are submitted for consideration in Web of Science.

Bahaaeddin A. M. Alareeni · Islam Elgedawy
Editors

Artificial Intelligence (AI) and Finance

Springer

Editors
Bahaaeddin A. M. Alareeni
Middle East Technical University, Northern Cyprus Campus
Güzelyurt, Türkiye

Islam Elgedawy
Faculty of Computer Science and Engineering
Alamein International University
New Alamein City, Egypt

ISSN 2198-4182 ISSN 2198-4190 (electronic)
Studies in Systems, Decision and Control
ISBN 978-3-031-39157-6 ISBN 978-3-031-39158-3 (eBook)
https://doi.org/10.1007/978-3-031-39158-3

This Springer imprint is published by the registered company Springer Nature Switzerland AG
The registered company address is: Gewerbestrasse 11, 6330 Cham, Switzerland

Contents

Artificial Intelligence Trends in Business Development

Enhancing Green Recruitment Through Implementing Artificial Intelligence: Zoho Recruitment System

Mohammed dawwas(✉), Mohammad Allaymoun, and Mahmoud Alzgool

Administrative Science Department, College of Administrative and Financial Science, Gulf University, Sanad 26489, Kingdom of Bahrain
{dr.mohammed.isam,hrm.hod}@gulfuniversity.edu.bh

Abstract. The use of information systems has increased to facilitate services and obtain results with high accuracy, and one of the most important of these services is employment. Most information systems on human resources currently seek sustainable development, in order to reach an employment system that preserves and protects the environment. By minimizing the use of material resources to achieve the desired objectives of the recruitment process. This research helps to understand the requirements and methodology followed for recruitment using human resources information systems, and to review the positiveness provided by the systems to achieve sustainability in organizations, as well as other goals such as saving time, effort, and money, by extracting suitable resumes for the offered jobs. The research presented in this paper attempts to answer the fundamental question of whether there should be a methodology for e-recruitment and recruitment cycle automation, using Zoho Recruitment. The results indicate that e-recruitment by Zoho Recruitment transforms the traditional recruitment process into an efficient and intelligent recruitment process. The stages of the recruitment cycle are also tracked, starting from opening the job to the last stage of recruitment and archiving.

Keywords: Recruitment · e-recruitment · hiring cycle · artificial intelligence · green hiring · Zoho Recruitment · SDG11

1 Introduction

Human resources have become one of the most important basic elements and distinguishing features of any institution. Human resources in general contribute to managing the human element that works to accomplish tasks and activities, as well as providing the institution with the best cadres [1]. This is done by searching for qualifications that are suitable for the vacancies, and then placing the employees in their work centres in proportion to his qualifications, i.e., the right person in the right place. It also contributes to the organizational process that relates specifically to employees, as it begins with their selection, appointment, training, retention and other matters related to work affairs [2]. Human resource management is also known as a group of activities related to attracting, retaining, and developing individuals, in order to achieve the goals of the organization and the goals of individuals [3].

B. A. M. Alareeni and I. Elgedawy (Eds.): *Artificial Intelligence (AI) and Finance*, SSDC 488, pp. 3–13, 2023.
https://doi.org/10.1007/978-3-031-39158-3_1

Recruitment is one of the most important services provided by human resources, and it is a set of stages that seek to select a candidate for a specific position [4]. Among these characteristics are skills, qualifications, personality, and motivation. One of the problems facing traditional recruitment is the presence of huge numbers of job candidates, as well as the existence of huge numbers of CVs, which require a team, time and effort to reduce the number of candidates and choose the best, as well as the challenge of storing employment applications, resumes, and contact details that are processed through the traditional paper methods, which contribute to the increased use of papers, which affects the environment and sustainability [5, 6]. On the other hand, the difficulty of storing job application files for months and years in anticipation of the availability of new jobs. The problems and challenges in the recruitment process traditionally increase directly with the size of the jobs and the size of the institutions [7].

Management information systems play an important and vital role in providing the necessary information for planning, implementing, and following up on various strategies, as well as rationalizing decisions, in addition to their primary role in achieving interdependence, communication and integration among them [8]. Information systems helped in automating most of the vital processes in organizations, and human resources also has to transform all services into a system, and human resources and information technology can facilitate the management of the workforce in a more flexible, efficient and effective manner [9].

Human Resources Information System (HRIS) is a software solution that maintains, manages and processes detailed employee information, policies and procedures related to human resources as an interactive information management system [10]. The Human Resources Information System standardizes Human Resources (HR) tasks and processes while facilitating accurate record keeping and reporting. Essentially, it also helped get rid of paper and manual processes, and the HR information system provides more smooth, streamlined, and effective interactions between employees and the companies in which they work while freeing HR professionals to perform more strategic and high-value work [11].

E-recruitment is defined as a technical process for finding, evaluating, nominating, interviewing, and employing employees in available vacancies. It helps by connecting candidates and human resources through technology [12]. It also enables publishing vacancies, receiving CVs, and searching electronically, as well as searching for candidates and obtaining alerts that help reduce time, effort, and cost in recruitment management [13]. One of the advantages of e-recruitment is to reduce the use of paper and preserve the environment. Employers can also publish vacancies and reach a larger number of candidates. Human resources can also build recruitment stages electronically from the beginning of the job posting stage to recruitment, in proportion to the nature of the jobs [14].

In this paper, we review the most important modern technologies used in electronic recruitment, and how to contribute to reducing the use of papers and reducing pollution, especially in the use of papers in submitting resumes, or the paper forms used in the human resources department to fill out applications, and finally in traditional advertisements that are used papers. In the end, it discusses the most important stages of the e-recruitment cycle by Zoho Recruitment.

2 Literature Review

Human resources management is defined as the department concerned with the process of attracting, selecting and appointing job candidates, as well as training, developing, directing, evaluating, and maintaining their performance [15]. On the other hand, it is concerned with determining wages and compensation, providing benefits, motivating workers, maintaining respectful relations between them and all members of the organization's departments, and ensuring a safe work environment in the systems [16]. There are other definitions, but they agree that it is the management of the human element within the organization. The Human Resources Department also seeks to select the best candidates who possess the skills commensurate with the job description to reach an effective organization that contains the skills and requirements necessary to achieve its goals and competitiveness [17].

One of the most important challenges facing human resources management is the recruitment and selection process, as the stages of the recruitment cycle traditionally make it difficult to find a suitable candidate for the required job, from a large group of applicants [18]. Therefore, the Human Resources Department spends most of their time managing the recruitment process from the stage of opening the job, announcing, receiving the CV, nomination, filtering for the best, interviewing, submitting the offer, and ending in the recruitment stage [19]. Recruitment is "the most important HR function for the success and survival of an organization". Personnel recruitment can be defined as a set of requests and recruitment of a new candidate by an organization [20]. Employment also seeks to supply the organization with the candidates it needs, as its capabilities are commensurate with the job at the right time. The traditional recruitment procedures are not considered easy processes, but they require skill when presenting job descriptions, selection and studying all applicants' applications. There are many ways available to recruit the required employees in any organization through recruitment agencies and newspaper ads. All these methods are classified as traditional methods [21]. Among the challenges facing the Human Resources Department also is the mechanism of storing and collecting CVs with paper files and taking space in company stores, the difficulty of managing files and obtaining information in a timely manner [22]. Traditional methods of recruitment limit the development of institutions, reduce the competitive advantage in the market, and the speed of obtaining talented employees when needed.

The use of information technology is a solution to all problems facing organizations, providing the necessary technical tools to process data and obtain information that helps in decision support and saves time and effort [23]. On the other hand, information technology in the form of information systems helps in collecting, analysing, and storing information.

It has become necessary to automate all human resource activities to keep pace with modernity, and the use of human resources information system can add value to human resources in organizations. This is through the automation of processes or transactions, which results in the provision of information to help human resources to carry out their activities in transactions more efficiently. On the other hand, providing accurate and timely information to human resources staff and managers helps them make the right decisions. One of the advantages of human resources information systems is providing more information that helps in reshaping practices and liberating the time of human resources employees [24]. One of the advantages of using human resources information systems is to provide a comprehensive picture of information as a single integrated database, allowing institutions to provide active communication with the rest of the units, as well as increasing the speed of information transactions, strengthening links between management and employees. The reason is to improve operations and increase competitiveness, as human resources information systems contribute to raising the level of efficiency, accuracy and effectiveness in data and information that are produced from human resources functions [25]. It also helps in the emergence of the green concept, by reducing reliance on paper documents in collecting and processing data related to human resources affairs [26]. In another definition, a Human Resources Information System (HRIS) is a software solution that maintains, manages and processes detailed employee information and policies and procedures related to human resources. As an interactive information management system, HRIS standardizes Human Resources (HR) tasks and processes while facilitating accurate record keeping and reporting [27].

The importance of the human resources information system is evident in the speed of processing, storing, and retrieving information, and it contributes to evaluating performance fairly, in addition to quickly discovering any change in the internal or external environment of the organization. It also achieves compatibility and integration between the various human resource management activities and other departments [28]. The added value added using the human resources information system in organizations is the reduction of costs, the reduction of the time spent in performing the required work, in addition to the ease of access of stakeholders to the desired information and assistance in making the right decisions [29].

Successful companies are constantly reviewing and improving their HR functions, looking for new ways to streamline processes to make them more efficient and using them to gain competitive advantage. The human resources department is responsible for handling the workforce aspect of this continuous improvement.

Zoho People is one of the most popular human resources information systems, as it is an ideal solution to automate all human resource processes that companies need, allowing the management of all human resource activities in an easy way. Zoho People is a cloud-based Human Resources (HR) solution suitable for businesses of all sizes. The system features leave management, attendance management, time tracker, employee self-service, form customization, and many more. The leave management feature of the solution allows organizations to customize leave types and employees can apply for leave online. Organizations can also have a centralized view of all employee information by leave type, location, role, and experience level. Zoho People allows users to create custom modules by adding forms, tags, and tabs based on needs, providing intelligent analytics, and reporting in the form of interactive graphics. It also provides an employee self-service feature so that their own records can be accessed, updated, and modified in accordance with regulatory compliance, as well as a file locker for HR staff to store and share documents related to HR and the company [30].

E-recruitment indicates that all stages of traditional recruitment will be technically done via the Internet, starting with publishing vacancies on the company's website or on the website of recruitment suppliers via the Internet, in addition to allowing candidates to send their resumes electronically through an e-mail or other electronic format. In conclusion, the methods and systems of electronic recruitment helped to reduce many of the routine administrative tasks involved in recruitment. Information technology has also made it easier, faster, cheaper, and more efficient to find candidates and new job opportunities. This increased the number of job applications by candidates, and the human resources effort increased in improving screening mechanisms and storing applications in electronic archives [14].

3 Research Methodology

This research used previous studies related to the topic of electronic recruitment, to determine the foundations and procedures necessary to help develop a prototype for electronic recruitment, in addition to reviewing and analyzing many cases and problems facing the human resources department, especially the traditional recruitment process to reach a general perception of how to find a scientific methodology that helps small enterprises in using Zoho People Plus Human Resources Information Systems to employ the e-recruitment service.

Collecting primary data and building the model were done through interviews and general discussions. In this research, semi-structured interviews were conducted with academics and experts in human resources, from the Human Resources Department at Gulf University, and specialists in human resources information systems. The semi-structured interviews allow the researcher to ask more questions whenever necessary, by limiting the general perceptions of designing a prototype for the solution and linking the technical elements to the expected outputs in order to verify the effectiveness of the semi-structured interviews being conducted face to face. It provided more information about the solution model and the mechanism for measuring the results and application of solutions, in addition to examining and verifying solutions at all levels to ensure that it can be applied and achieve the desired goals.

4 Traditional Recruitment and Electronic Recruitment

In this section, we review the stages of the traditional recruitment cycle, and then followed by the stages of electronic recruitment through the employment service provided by Zoho People Plus. Traditional recruitment stands for the traditional recruitment and selection process, a term often used to recruit and select suitable people from a pool of applicants, using traditional paper applications and resumes. The traditional methods of recruitment are summarized starting with the process of identifying current vacancies, preparing budgets for hiring new employees, recruiting, and training. The process includes analyzing jobs, writing a job description, and then sharing it on various platforms such as newspapers or social media platforms, etc., to attract job seekers. The recruiters then collect and screen resumes, shortlist relevant profiles, and proceed to contact suitable candidates for further discussions. Subsequent tasks include reference checks, background checks, introduction of shortlisted candidates, and notification of unsuccessful candidates. The traditional method emphasizes obtaining as many resumes as possible and building the database to use when needed.

In traditional recruiting, the entire recruitment process is owned by the HR/Talent Acquisition team and hiring managers are usually involved from either the resume screening or discussion stage. After selection, hiring managers get more information about candidates only when they join or decline the offer. As for electronic recruitment, it depends mainly on the traditional steps, but information technology is used to complete it, starting from the stage of identifying vacancies and moving to publishing the advertisement, obtaining CVs, using artificial intelligence to filter CVs, and selecting the list of candidates. As studies have shown, electronic recruitment has reduced 90% of the time compared to traditional employment, and as for the cost, electronic recruitment has reduced it to 5% of the cost of traditional employment.

Traditional recruitment refers to the process of hiring new employees through traditional methods such as print advertisements, recruitment agencies, word of mouth, and on-site recruitment events. Electronic recruitment, also known as online recruitment or e-recruitment, refers to the use of technology and the internet to advertise job vacancies and receive job applications. This includes the use of online job boards, company websites, and social media platforms to reach potential candidates. The main difference between traditional recruitment and electronic recruitment is the use of technology. Traditional recruitment relies on manual methods, while electronic recruitment utilizes technology to streamline the recruitment process.

The stages of e-recruitment by Zoho are recruitment that begins with opening the job, i.e., setting job details, advertising, receiving resumes, nominating the best, interviewing, offers, and ending with employment. Figure 1 shows the stages of electronic recruitment. In this paper, we review recruitment by the Zoho recruitment service, which organizes and tracks all vacancies, resumes, candidates, clients, and private contacts while allowing the creation of a database to be used when needed.

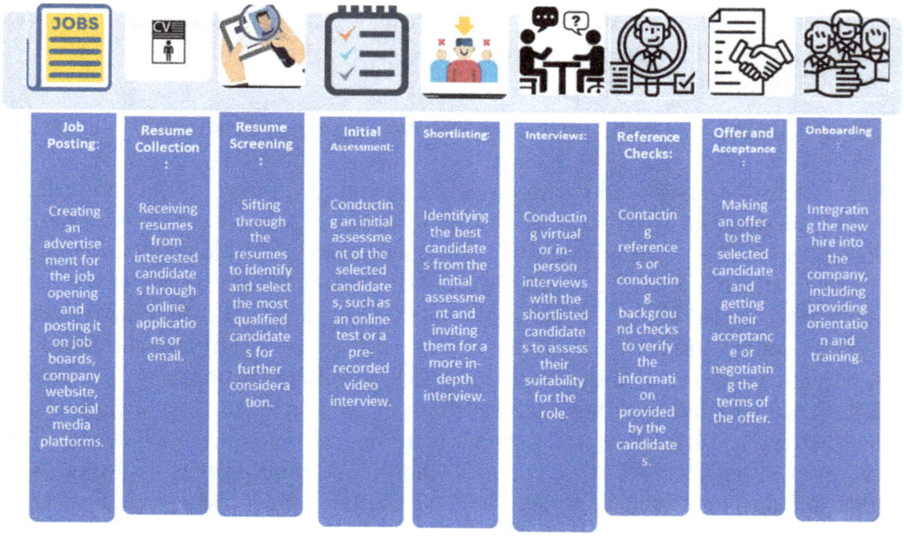

Fig. 1. Shows the stages of electronic recruitment.

5 Discussion

In this part, we review the most important stages of recruitment by employing Zoho, by building a general perception of procedures and processes in order to limit the benefits, and reaching a perception that helps decision makers apply them, in order to reduce the use of papers and moving to the concept of green employment.

Figure 2 shows the job opening. Through this service, job details are written, and the conditions must be clear, and the reason is that they will be approved later in filtering resumes using artificial intelligence.

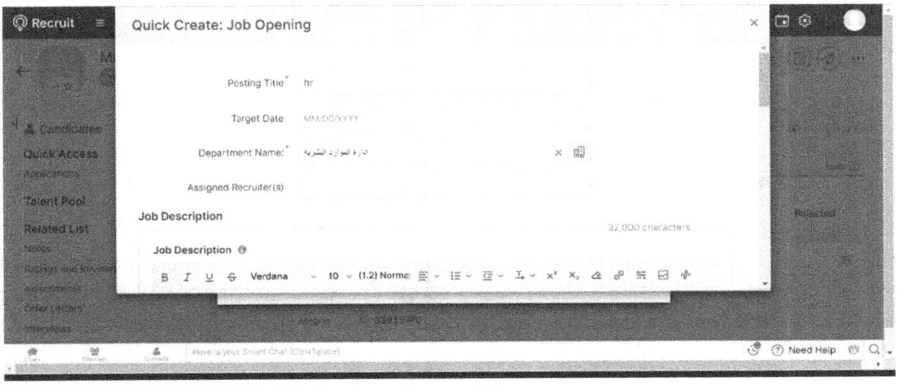

Fig. 2. Shows the job opening in Zoho Recruit.

Figure 3 shows the most important services provided by Zoho Recruitment, which helps to reach an integrated recruitment process, capable of facilitating the recruitment process, and obtaining satisfactory results. E-recruitment is also considered one of the most important services that helped reduce pollution by using papers and preserving the environment, and it publishes advertisements for employment through modern technical platforms, whether through social networks or employment sites. The cost of sending a job advertisement or searching for candidates can be reduced to 90% if traditional methods are used.

Artificial intelligence in recruitment, and the development of screening methods, enable business owners and operators to reach the appropriate candidate at a breakneck speed, by specifying several criteria in advance, such as place of residence, skills, education, job grade, and experience, as well as ease and reference to data sent by job seekers in case there are suitable vacancies [31] Artificial intelligence also helps mitigate bias in the interview and nomination processes, as well as obtain reports upon request and help in decision-making.

AI in recruitment refers to the utilization of AI technologies and algorithms in the hiring process to automate repetitive tasks, enhance efficiency, and make more informed decisions. This can include resume screening to filter out unqualified candidates, AI chatbots to assist candidates throughout the process, predictive analytics to predict future successful hires based on past data, and video interview analysis to evaluate a candidate's

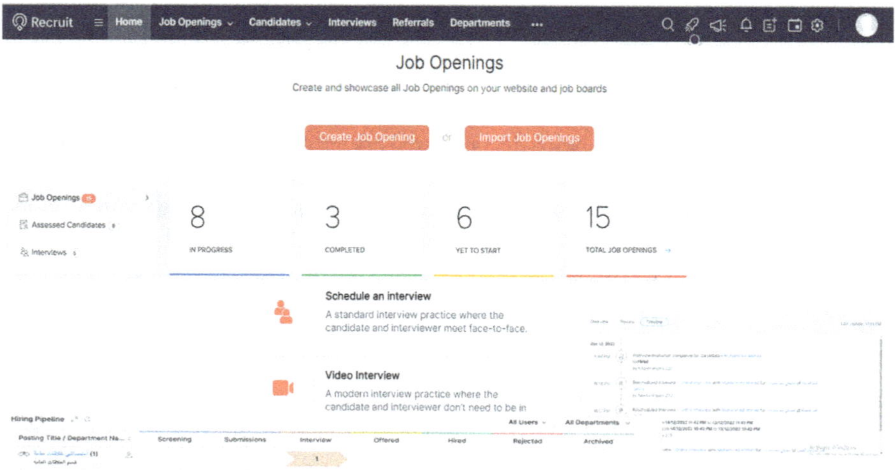

Fig. 3. Shows the most important services provided by Zoho Recruit.

behavior and suitability for the role. The implementation of AI in recruitment has the potential to improve the speed and accuracy of the process, but it's crucial to understand that AI should serve as a support tool for human decision-making rather than replace it entirely [32].

On the other hand, the most important advantages of electronic recruitment are saving the time of the human resources department, improving communication procedures within the recruitment team, selecting the best among the candidates, and helping to obtain results that help decision-makers.

While traditional recruitment has worked well for a long time, electronic recruitment has helped to be a successful alternative to traditional employment, due to several reasons, including expensive and time-consuming in the modern work environment. It is also an inefficient way to search for suitable candidates.

By using Zoho Recruitment, HR departments can streamline and automate many of the manual tasks associated with the recruitment process, freeing up time and resources to focus on other important aspects of the business. In general, electronic recruitment, has the potential to significantly reduce the environmental impact of the recruitment process. By eliminating the need for paper-based resumes and application forms, electronic recruitment reduces the use of paper and ink, helping to conserve natural resources. Additionally, electronic recruitment can also reduce the need for travel to and from job interviews, which can help to reduce carbon emissions. Furthermore, many companies have moved to entirely digital recruitment processes, which can help to streamline and automate many tasks, reducing the amount of waste generated by the recruitment process [33]. By reducing the environmental impact of recruitment, electronic recruitment can play a significant role in preserving the environment and promoting sustainability.

6 Conclusion

E-recruitment is a successful alternative to traditional recruitment, and it has become one of the most important features of human resources information systems services. The focus is on the concept of green electronic employment, and what we need in terms of finding tools and technologies capable of reducing pollution and reducing the use of papers. The information technology has successfully helped to reduce the use of traditional ways of completing tasks. Recruitment is a process that helps to find the right human element for the available vacancy, whether in a traditional or electronic way, but with the development of technologies, electronic recruitment has become smart, using artificial intelligence in most of its stages, especially the stage of filtering candidates. On the other hand, the paper reviewed the advantages of electronic recruitment, such as saving time and effort, and assisting in providing reports that help decision-makers in making decisions, building perceptions, and developing appropriate strategies, in order to achieve sustainability and competition, and reviewing the stages of traditional recruitment and electronic recruitment, as well as using Zoho recruitment as an example of one of the most important applications used in electronic recruitment, and also discussing the stages of employment. In the future, methodologies will be sought for new assistive technologies, which will be added to improve the work and results of electronic recruitment.

References

1. Alzgool, M.: Nexus between green HRM and green management towards fostering green values. Manage. Sci. Lett. **9**(12), 2073–2082 (2019)
2. Onyeaghala, O.H., Hyacinth, M.I.: Effects of employee selection process on productivity in the public and private sectors: a case of Benue state. Bus. Econ. J. **7**(4), 1–8 (2016)
3. Mahapatro, B.B.: Human resource management. PG Department of Business Management (2022)
4. Vardarlier, P., Zafer, C.: Use of artificial intelligence as business strategy in recruitment process and social perspective. In: Hacioglu, U. (ed.) Digital Business Strategies in Blockchain Ecosystems. Contributions to Management Science. Springer, Cham (2020). https://doi.org/10.1007/978-3-030-29739-8_17
5. Okolie, U.C., Irabor, I.E.: E-recruitment: practices, opportunities and challenges. Eur. J. Business Manage. **9**(11), 116–122 (2017)
6. Nguyen, B., et al.: Recruitment challenges and recommendations for adolescent obesity trials. J. Paediatr. Child Health **48**(1), 38–43 (2012)
7. Ryfe, D.M., Stalsburg, B.: The participation and recruitment challenge. In: Democracy in motion: Evaluating the Practice and Impact of Deliberative Civic Engagement, pp. 43–58 (2012)
8. Ismagilova, E., Hughes, L., Dwivedi, Y.K., Raman, K.R.: Smart cities: Advances in research—An information systems perspective. Int. J. Inf. Manage. **47**, 88–100 (2019)
9. Vrontis, D., Christofi, M., Pereira, V., Tarba, S., Makrides, A., Trichina, E.: Artificial intelligence, robotics, advanced technologies and human resource management: a systematic review. Int. J. Human Resource Manage. **33**(6), 1237–1266 (2022)
10. Johnson, R.D., Carlson, K.D., Kavanagh, M.J. (eds.).: Human resource information systems: basics, applications, and future directions. SAGE Publications (2020)

11. Krithika, J., Venkatraman, P., Sindhujaa, E.A.: Virtual HR era in human resource management. EPRA Int. J. Multidiscip. Res. **5**(10), 49–57 (2019)
12. Pal, A., Kundu, A., Polley, S., Chatterjee, S., Chakraborty, S.: A study on finding the most effective social media platform for e-recruitment process. In: Interdisciplinary Research in Technology and Management, pp. 451–455. CRC Press (2021)
13. Kar, A., Bhattacharya, S.: E-recruitment and customer satisfaction: an empirical study in and around Kolkata. IUP J. Manage. Res. **8**(2), 34 (2009)
14. Dhamija, P.: E-recruitment: a roadmap towards e-human resource management. Res. World **3**(3), 33 (2012)
15. AlZgool, M., Ahmed, U., Shah, S., Alkadash, T., AlMaamary, Q.: Going green during COVID-19: examining the links between green HRM, green supply chain and firm performance in food Industry of Bahrain: the moderating role of lockdown due to COVID-19. Uncertain Supp. Chain Manage. **9**(1), 79–88 (2021)
16. Alkadash, T.M., Jun Bo, S., Besher, B., Almaamari, Q., Mohsen Al-Absy, M.S.: Conceptual framework on job burnout of the employees and its correlative factors in China during covid-19 (2020)
17. Dawwas, M.I.: The relationship between talent management practices, organizational justice, and employee engagement. Specialusis Ugdymas **1**(43), 2084–2104 (2022)
18. Sims, R.R.: Organizational success through effective human resources management. Greenwood Publishing Group (2002)
19. Halid, H., Halim, S.N.A., Ravesangar, K.: Human resource management practices in the digital era. In: Machado, C. (ed.) Technological Challenges. Management and Industrial Engineering, pp. 109–158. Springer, Cham (2022). https://doi.org/10.1007/978-3-030-98040-5_5
20. Breaugh, J.A.: Employee recruitment: Current knowledge and important areas for future research. Hum. Resour. Manag. Rev. **18**(3), 103–118 (2008)
21. Sills, M.E.: Recruitment: a comparison with traditional recruitment and the influences of social media: a qualitative and quantitative review (2014)
22. Abugre, J.B., Boachie-Ansah, R.: Digitizing recruitment and selection of employees in ghana: a social media network perspective. Business in Africa in the Era of Digital Technology: Essays in Honour of Professor William Darley, pp. 43–61 (2021)
23. Allaymoun, M.H., Hamid, O.A.H.: business intelligence model to analyze social network advertising. In: 2021 International Conference on Information Technology (ICIT), pp. 326–330. IEEE (2021)
24. Lengnick-Hall, M.L., Moritz, S.: The impact of e-HR on the human resource management function. J. Lab. Res. **24**(3), 365 (2003)
25. Holm, A.B.: Virtual HRM: a case of e-recruitment. In International Workshop on Human Resource Information Systems, vol. 1, pp. 49–58. SCITEPRESS (2009)
26. Naji, G.M.A., Isha, A.N., Bin, S.M.N., Rahman, S.A., Alzoraiki, M., Al-Mekhlafi, A.B.: The role of HR strategy on safety culture and psychological stress among employees in the upstream oil and gas companies: a conceptual review. Solid State Technol. **63**(5), 12 (2020)
27. Chauhan, A., Sharma, S.K., Tyagi, T.: Role of HRIS in improving modern HR operations. Rev. Manage. **1**(2), 58 (2011)
28. Kavanagh, M.J., Johnson, R.D.:(eds.) Human resource information systems: basics, applications, and future directions. SAGE Publications (2017)
29. Stone, R.J., Cox, A., Gavin, M.: Human resource management. Wiley (2020)
30. HR Software Solutions | Cloud based HRMS | HR System | Zoho People. (n.d.). Zoho. https://www.zoho.com/people

31. Krishnamurthy, A., Goel, P.: Artificial intelligence-based drug screening and drug repositioning tools and their application in the present scenario. In: Computational Approaches for Novel Therapeutic and Diagnostic Designing to Mitigate SARS-CoV2 Infection, pp. 379–398. Academic Press (2022)
32. Wilson, H.J., Daugherty, P.R.: Collaborative intelligence: humans and AI are joining forces. Harv. Bus. Rev. **96**(4), 114–123 (2018)
33. Prasad, R.S.: Green HRM-partner in sustainable competitive growth. J. Manage. Sci. Technol. **1**(1), 15–18 (2013)

The Impact of Digital Technology on Media Ethics

A Survey Study on Workers in Arab Media Institutions

Mohammed Ahmed Fyadh[1] and Areen Omar Al-Zoubi[2(✉)]

[1] College of Mass Communication, Umm Al Quwain University, Umm Al Quwain, UAE
[2] Communication and Digital Media, College of Mass Communication, Umm Al Quwain University, Umm Al Quwain, UAE
drareen.a@uaqu.ac.ae

Abstract. This study aimed to know the impact achieved by the digital revolution on media professional ethics, as the study belongs to the type of descriptive studies, in which the sample survey method was used on (200) Arab elites (academic and media), and the study reached a number of results, Most notably: Digital applications are employed in Arab media organizations to a medium degree. The most prominent motives for employing digital applications in Arab media institutions are because they contain interactive elements that attract the audience and enable them to communicate. The most important applications that are employed in Arab media institutions are satellite frequencies, artificial intelligence applications, and infographic design programs. The degree of confidence of workers in Arab media institutions is medium.

Keywords: impact · digital technology · media ethics

1 Introduction

Technology is no longer just tools and techniques for developing work and improving its performance. Rather, it has extended to be the basis for the development of the entire media landscape, as every new term has become circulating, every new trend in work and media practice, and every update is inseparable from technology. Hence; "Digital transformation" has become an inevitable necessity and a modern trend that is compatible with the nature of what our world is witnessing in terms of changes and the aspirations of the countries and peoples of the world in terms of development and prosperity. Globalism; To ensure non-isolation from the world on the one hand, and to preserve religious, cultural, and value identity on the other hand.

If all societal institutions are required to catch up with the technological progress and information revolution achieved by mankind; Media institutions are the most interested and deserving of interacting with modern technological developments and engaging in the digital age, whose steps are accelerating and capabilities are growing day by day.

B. A. M. Alareeni and I. Elgedawy (Eds.): *Artificial Intelligence (AI) and Finance*, SSDC 488, pp. 14–24, 2023.
https://doi.org/10.1007/978-3-031-39158-3_2

2 The Problem of the Study

Digital transformation in media has become a living reality in developed countries that have already entered the era of media through digital platforms, and fifth-generation networks have opened the door to digital transformation that includes all methods of media practice based on modern technologies, and artificial intelligence technology has become a major part of the boom in contemporary media. Which affected, in one way or another, the ethics of the media profession. Today, journalist is more concerned with the journalistic scoop than achieving the profession's ethical standards.

Therefore, the problem of this study is defined in an attempt to determine the impact of digital technology on the ethics of media work.

3 Lecture Review

Simbawa Study (2022). Entitled: The Impact of Digital Media Technology on the Journalistic Content Industry - A Field Study on the Communicator -.

The study aimed to identify the impact of digital media technology on the journalistic content industry from the point of view of the communicators of Saudi journalists and content makers working in the field of journalism in the Kingdom of Saudi Arabia. It sought to identify the technologies most used by them and to reveal their relationship to the development of their performance and the change of their roles in the digital space. By conducting a field survey using the questionnaire on an intentional sample of (15) single content industry specialists. The most important results were summarized in the presence of a medium positive relationship between the use of digital media technology by journalists in the production of journalistic content and the rate of dependence on it, in addition to the presence of statistically significant differences in favor of females in their permanent use of it. She also explained that 74% of the communicators use digital media technology in the production of journalistic content to a large extent, but only 58% showed their knowledge of its production methods, as 90.7% of them stressed the need for digital media to have an independent work team, which indicates the gap in the skills that they should acquire the importance of studying.

Arbab Study (2022). Entitled: The Reality of Employing Digital Transformation Technologies in Content Industry in Sudanese Journalistic Institutions. The study sought to identify the reality of employing digital transformation techniques in the industry and production of content in Sudanese press institutions, and to shed light on the challenges facing journalists in these institutions. To achieve the objectives of the study, the researcher used the survey method by applying the electronic questionnaire to a sample of (54) journalists and administrators working in Sudanese press institutions, represented in a number of newspapers, namely: (Al-Sudani, Al-Hirak Al-Salisi, Al-Jarida, Al-Siha, Al-Intibahaha, Al-Tayyar), and during the period from January to March 2022 AD, and the results revealed that there is general acceptance about the mechanism of employing digital transformation techniques in the production of content in Sudanese press institutions and from the point of view of journalists. The results showed that digital imaging techniques, digital editing and output tools, and electronic printing are among the most important digital transformation tools and techniques employed by the studied institutions.

The study of Mohamed Sayed (2020) was also given. Interested in the impact of digital technology in the management of press institutions, how this process is managed, what are its components and development strategies, in addition to the factors that affect the digital journalistic production process. Technological developments are ineffective and ineffective, and suffer from many shortcomings in their information systems, the quality of service provided, and the speed of access to the public, in addition to the inability of press institutions to confront digital development, and the control of traditional media over journalistic content.

The importance of this study lies in two aspects:

The first is the scientific importance of the study

The current study attempted to constitute a qualitative addition to the Arab library in general, as the two researchers noticed during their study the scarcity of studies and research dealing with the ethics of media work in light of the current digital revolution.

The second - the practical importance of the study

It comes from the importance of the research topic itself, as the journalist's integrity is a major factor in determining the goal of the profession. This is because it is a public service for the benefit of society, and to create a conscious public opinion, in addition to that every time a journalist pawns his pen to a certain party or agenda, or makes up facts and other fabrications, he falls into journalistic ethics.

On the other hand, media ethics maintains the professionalism of media institutions, those in charge of them, and their employees.

4 Methodology

This study belongs to the type of descriptive studies, which aims to obtain an accurate and complete description of the problem and to ensure that all necessary data are collected to ensure exposure and analysis with the greatest possible degree of accuracy and to avoid any bias in the collection of these data in order to increase the degree of dependence of the results drawn from them and the possibility of their application On similar situations, cases or examples (Hussein, 2010).

The researchers believe that this type of studies is suitable for the current study and serves it, by monitoring the phenomenon under study, describing it and knowing its characteristics by collecting information and data about this phenomenon.

This study also relies on the (survey method), which is considered one of the most important and common approaches used in media studies, especially in exploratory and descriptive research (Zoghaib, 2010).

According to Abu Osbaa (2012), it is "the most appropriate and best-suited approach for this study." Abu Osbaa describes the survey approach as: "it tries to depict or document current conditions or trends, and this means to describe what is present at this moment, Based on the survey method, the study relied on the public sample survey method.

5 The Study Sample

The two researchers were able to distribute the questionnaire to (T200) individuals from the academic and media elites distributed in different Arab countries, according to the random sample method, which gives an equal opportunity for all members of society to appear.

6 Stability of the Study Tool

The researchers applied the (internal consistency) test to measure the stability of the questionnaire, as this method relies on estimating the internal consistency of a group of items by collecting the individual consistency of all items to come up with a total score for the scale, and stability of (91.6) was reached.

7 The Statistical Methods Used

Referring to the Statistical Social Sciences Package program known as (SPSS), the researcher resorted to using the following statistical methods:

1. Frequencies and percentage: to measure questions on the analytical side and questions of alternatives on the field side.
2. Means and standard deviations: to measure questions of paragraphs and phrases in the field side.

8 Analyze and Discuss the Results

This topic deals with answering the questions of the field study, discussing and analyzing the results that were reached, and the researchers explain that the questionnaire was distributed to a total of (200) Arab elites (academic and media) from different Arab countries, educational levels, and professional experiences.

The researchers used the three-way Likert scale, consisting of three directions (agree, neutral, disagree).

Accordingly, the researchers explain the mechanism of using and employing the scale in the study.

(which is the range) 0.66 = 3/2 2 = 3–1

(low) 1.66–1 .

(Moderate) 2.33–1.67

(large) 3–2.34

1. The degree of employing digital applications in Arab media institutions

The data in Table 1 revealed the degree of employment of digital applications in an Arab media institute, By analyzing these results, we find that the majority of the study sample from among the Arab elites answered that they are employed in a medium degree, as the category (with a medium degree) ranked first, with a total of (98) frequencies and a percentage of (49%).

Table 1. The degree of employing digital applications in Arab media institutions.

#	Degree of employing digital	Frequency	%	Rank
.1	a great extent	67	33.5	Second
.2	Moderately	98	49	First
.3	a low degree	35	17.5	Third
.4	It is not employed at all	0	0	Fourth
	Total	200	100	

While a category with a large degree came in second place, with a total of (67) repetitions, with a rate of (33.5%), followed by third place (with a low degree), with a total of (35) repetitions, with a rate of (17.5%). Launch) at any frequency.

Commenting on the above, we find that the prevailing trend in Arab media institutions is the employment of digital applications, in line with the requirements of the era governed by technology and the communication revolution, which is reflected in one way or another on the form and content of the media message.

2. Motives for employing digital applications in Arab media institutions

The data in Table 2 illustrates the motives for employing digital applications in Arab media institutions.

The results show that phrase No. (1), which is (available on interactive elements that attract the audience and enable them to engage in the communication process), ranked first with the highest arithmetic mean of (2.75), followed by phrase No. (1), which is (Because of its speed in transmitting news and events), with an average of (2.66).), and the phrase No. (2), which is (the availability of software templates and editorial patterns different from the traditional patterns), ranked third with an arithmetic mean of (2.58), while the phrase No. (4) which is (the availability of software templates and editorial patterns different from the traditional patterns) came In fourth place, with an arithmetic mean of (2.53), followed by the phrase No. (6), which is (helps to verify the credibility and authenticity of the news), with a mean of (2.45), and phrases No. (5,7) ranked low on the list.

3. Digital applications that are employed in Arab media institutions

The data in Table 3 shows the most prominent applications of digital technology in which the respondents work. It is noted from the above that there is a difference in percentages in the preference of the study sample for each application of digital technology and artificial intelligence.

Returning to the previous results, we find that the advanced montage programs and artificial intelligence applications ranked first on the list with a total rate of (17.9%). The second place was occupied by the design programs used in producing infographics

Table 2. Motives for employing digital applications in Arab media institutions.

#	degree of agreeing / paragraphs	Mean	standard deviation	Rank
1	Because of its speed in transmitting news and events	2.66	.614	second
2	It provides software templates and editing styles that are different from the traditional ones	2.58	.760	third
3	It has interactive elements that attract the audience and enables them to engage in the communication process	2.75	.601	first
4	Save time and effort in editing and proofreading	2.53	.804	fourth
5	It enables me to access multiple sources of information	1.66	.912	seventh
6	Help ensure the credibility and authenticity of the news	2.45	.828	fifth
7	Employing digital applications in media organizations increases their fan base	2.14	.963	Sixth

and designing electronic newspapers, with a rate of (13.5%), followed by The third simultaneous translation programs and frequencies on satellites, with a rate of (10.3%).

As for the advanced editing programs in newspapers, it came fourth with a percentage of (9%), followed by the fifth rank for communication programs with news agencies with a rate of (8.5%), while the rest of the categories occupied (late ranks from the list, none of which exceeded (5%).

4. The degree of elite members' confidence in the effectiveness of digital applications in Arab media institutions

Table 3. Digital applications that are employed in Arab media institutions.

#	digital applications	duplicates	%	mattress
1	Advanced editing software in newspapers	**29**	**9**	**fourth**
2	Communication programs with news agencies	**27**	**8.4**	**Fifth**
3	News verification programs	**1**	**0.3**	**eighth**
4	Interpretation software	**33**	**10.3**	**Third**
5	Artificial intelligence applications	**57**	**17.9**	**The first**
6	Advanced montage programs	**57**	**17.9**	**The first**
7	satellite frequencies	**33**	**10.3**	**Third**
8	Robot journalist	**3**	**0.9**	**seventh**
9	Audio and visual effects	**9**	**2.8**	**sixth**
10	Possibility of designing maps and graphs	**27**	**8.5**	**Fifth**
11	Design programs used in producing infographics and designing electronic newspapers	**43**	**13.5**	**the second**
the total		**319**	**100**	

The data in Table 4 shows the degree of confidence of the members of the Arabic version in the effectiveness of digital applications in the Arab media institutions, as the results reveal a diversity in the degrees of confidence of the members of the sample surveyed.

Returning to the previous results, it is clear that the majority of the surveyed sample trusts a moderate degree in the effectiveness of digital applications in Arab media institutions, as it constituted a total of (103) repetitions, with a rate of (51.5%).

Table 4. The degree of elite members' confidence in the effectiveness of digital applications in Arab media institutions.

#	degree of confidence	Frequency	%	Rank
.1	Great confidence	45	22.5	second
.2	Medium confidence	103	51.5	First
.3	Weak confidence	35	17.5	Third
.4	I don't trust at all	17	8.5	Fourth
	Total	**200**	**100**	

As for the individuals in the researched sample who trust a great degree, they came in the second rank with a total of (45) repetitions, with a rate of (45%), followed by individuals who trust a little degree, in the third rank, with a total of (35) recurrences and a percentage equal to (17.5%). As for those who do not trust Never, so they occupied the fourth and last place, with a total of (17) repetitions, with a rate of (8.5%).

5. Media professional ethics that are applied in Arab media institutions

The data in Table 5 shows the ethics of the media professionals that are applied in Arab media institutions.

The results reveal that credibility topped the list with a rate of (20.1%), followed by the category of objectivity with a rate of (18.8%), while the third place was occupied by the category of balance with a rate of (18%), while accuracy came in the fourth place with a rate of (17.5%). As for the fifth place, it came social responsibility, which accounted for (11.1%).

As for the categories (respect for religions and the rights of minorities, respect for human dignity, non-alignment) that occupied late ranks in the list, they came in sixth, seventh, and eighth, respectively, in order.

6. The impact of the digital revolution on media work ethics in Arab institutions.

The data in Table 6 shows the expressions related to the impact of the digital revolution on media work ethics, with an arithmetic average of (1.78).

For each phrase separately, the data shows that phrase No. (7), which is (helped spread misinformation and media distortion due to the available fabrication programs), topped the list in the first place, with an arithmetic average of (2.41).

As for the phrase No. (6), which is (the spread of rumors and false and erroneous news), it came in the second rank, with an arithmetic mean of (1.94), followed by the third rank, phrase No. (1), which is (contributed to distorting media work ethics) with an arithmetic mean of (1.73).

And phrase No. (5), which is (the spread of hate speech, sedition, racism and sectarianism), ranked fourth, with an arithmetic mean of (1.66).

Table 5. Media professional ethics that are applied in Arab media institutions.

#	Media professional ethics	duplicates	%	Rank
1	objectivity	71	18.8	second
2	credibility	76	20.1	first
3	balance	68	18	Third
4	social responsibility	42	11.1	Fifth
5	Precision	66	17.5	fourth
6	Aligned-Non	7	1.9	eighth
7	Respect for religions and the rights of minorities	27	7.14	sixth
8	Respect for human dignity	19	5	seventh
9	other	**2**	**0.5**	**ninth**
Total		**378**	**100**	

The two phrases No. (2), which is (the spread of electronic crimes) and No. (4), which is (the move towards achieving the press scoop without taking into account the accuracy and verification of the news before publishing it), ranked fifth, and the arithmetic means for each of them was (1.62), while the phrase No. (3) Which (the appearance of information security crimes) came in the sixth and last rank with an arithmetic average of (1.51), and the arithmetic averages of the three previous phrases indicate that they fall within an unacceptable trend and did not gain the support of the majority of the study sample.

Summary of the results of the study
The study reached several results, which the researchers include as follows:

1. Digital applications are employed in Arab media organizations to a medium degree.
2. The most prominent motives for employing digital applications in Arab media institutions are because they contain interactive elements that attract the audience and enable them to engage in the communication process.
3. The most important applications that are employed in Arab media institutions are satellite frequencies, artificial intelligence applications, and infographic design programs.
4. The degree of confidence of workers in Arab media institutions is medium
5. The most prominent ethics of the media professionals that are applied in the Arab media organization are objectivity, credibility, and balance

Table 6. The impact of the digital revolution on media work ethics in Arab institutions.

#	Paragraph	Mean	standard deviation	Rank
1	Contributed to distorting media work ethics	1.73	.738	Third
2	Spread of cybercrime	1.62	.810	Fifth
3	The emergence of information security crimes	1.51	.746	sixth
4	Going towards achieving the press scoop without taking into account the accuracy and before checking the news publishing it	1.62	.786	Fifth
5	The spread of hate speech, sedition, racism and sectarianism	1.66	.817	fourth
6	The spread of rumors and false news	1.94	.883	the second
7	It helped spread misinformation and media distortion because of the available fabrication programs	2.41	.833	The first
overall average		**1.78**	**0.801**	

6. The digital revolution has affected the ethics of media work in Arab institutions by helping them to mislead and misrepresent the media due to the available fabrication programs and the spread of rumors and false news.

Recommendations

In line with the findings of the study, the researchers recommend the following:

1. The necessity of training workers in Arab media institutions on digital applications related to artificial intelligence.
2. Showing the ethical values related to media codes of honor, so that they are present in the media work of all employees.
3. The necessity of informing the Arab media cadres of the new concepts of misinformation and deviation in the media work, so that they can remedy the imbalance in the mechanisms of media design or editing.
4. Emphasizing the importance of increasing the practical practices of students of the faculties of media and communication in the field of media work ethics and not relying on theoretical frameworks only.

References

Aal, A., Saber, I.: The reality of the use of digital technology in the Egyptian press: a study of development trends and transformation problems. Middle East Pub. Relations Res. J. **3**, 251–275 (2021)

Asbaa, S.A.: Introduction to the foundations of scientific research. Amman: Dar Aram for Studies, Publishing and Distribution (2010)

Arbab, W.H.: The reality of employing digital transformation techniques in the content industry in Sudanese press institutions. Egypt. J. Pub. Opin. Res. **21**(2), 269–328 (2022)

Hussein, S.: Media Research. World of Books, Cairo (2010)

Al-Sayed, M.A. (ed.) Parliamentary Affairs at Al-Ahram Gate, in an interview with him in the Tahrir Hall, on January 11, 2021 AD

Simbawa, M.: The impact of digital media technology on the journalistic content industry - a field study on the communicator. J. Human. Soc. Sci. **6**(5), 119–155 (2022)

Zogheib, S.: Research: methods and statistical uses in media studies. The Egyptian Lebanese House, Cairo (2011)

The Impact of Digital Transformation on Achieving Strategic Agility A Case Study: Jordan Customs

Mohammad Abdalkarim Alzuod[1] (✉), Malak Mohammad Ghaith[2], Ali Hussein Alshibli[3], and Weam Tunsi[4]

[1] Zarqa University, Zarqa, Jordan
malzuod@zu.edu.jo
[2] Luminus Technical University College (LTUC), Amman, Jordan
[3] Jordan Customs, Amman, Jordan
[4] University of Business and Technology, Jeddah, Kingdom of Saudi Arabia

Abstract. The study aimed to identify the impact of digital transformation in achieving strategic agility and targeted employees in the Jordan Customs. The study used the descriptive analytical approach using the Statistical Package for Social Sciences (SPSS) program to analyze the data collected from the study population, which consisted of employees of the Jordan Customs that working in various centers and directorates of the Department, who participated in this study. (420) is the sample of study who participated from different organizational levels, only (365) questionnaires were retrieved with a recovery rate of (87%). After the analysis of the study that found a set of results, the most important of which is that digital transformation has a positive, statistically significant impact in its four dimensions (strategy, organizational culture, transformational leadership, and human resources) in achieving strategic agility in Jordan Customs. The study recommended increasing attention to training workers and raising their capabilities to deal with digital transformation positively and improve services, in addition to increasing employee participation in expressing their views when preparing the department's vision, which contributes to motivating employees to achieve the vision, to raise the level of strategic agility.

Keywords: Digital Transformation · Strategic Agility · Jordan Customs · Jordan

1 Introduction

The recent rapid changes that took place in the beginning of 2020 imposed a new reality on organizations as a whole, with their various specializations, when the world was affected by the Corona pandemic. Therefore, organizations had to take new directions that would enable them to deal with changes, to be more flexible and highly efficient. One of the most successful of these methods is strategic agility [1].

The wide spread of modern technologies, especially those related to electronics, media, communication and the internet, has opened the gateway to the era of digital

B. A. M. Alareeni and I. Elgedawy (Eds.): *Artificial Intelligence (AI) and Finance*, SSDC 488, pp. 25–33, 2023.
https://doi.org/10.1007/978-3-031-39158-3_3

transformation [2]. Otherwise, the digital economy is based on technology, knowledge and IT capabilities. Therefore, it has become a competitive advantage for many organizations, while it has become a threat to those organizations that are still lagging behind [3]. Digital transformation is a major shift not only in developed countries, but also in emerging countries and Arab countries. Some Arab countries have implemented a clear strategy to achieve this transformation. In order to advance economic and social development, and reduce the risk of the gap between developed and developing countries [4].

The organization's access to strategic agility enables it to achieve sustainable institutional excellence, and enables it to achieve success [5]. The organization must work on the speed of its operations, by increasing the capabilities of its employees and raising their efficiency, as well as raising its technical efficiency [19]. In addition, the enhancing of its level of response to various external and internal variables, which will eventually contribute to the organization being stronger with the passage of time, and enabling it to survive, grow and compete in order to achieve strategic success [6].

The problem of the study to investigate the impact of strategic agility and the changes that occur to it in the Jordan Customs, and the ability of the department to benefit from the gains that it will reap from applying the concept of strategic agility. Adding to that the relationship between them and the department's application of digital transformation to achievement of appropriate levels of strategic agility, in light of volatile circumstances and an environment that bears many variables.

This paper is an attempt to achieve the following objectives. Firstly, to identify the level of applying digital transformation in the Jordan Customs. Also, the study that identify the availability of strategic agility in the Jordan Customs, and describe the relationship between the applying of digital transformation and strategic agility in the Jordan Customs. Lastly, the study review the impact of the dimensions of digital transformation on achieving strategic agility in the Jordan Customs.

2 Literature Review

Literature related to digital transformation indicated that there are many issues that may affect strategic agility reviewed represented [2, 3, 6, 7]. It will be summarized the main dimensions of digital transformation which may affecting strategic agility which are as follows.

2.1 Digital Transformation

According to [8] that define digital transformation as a process or set of processes that is dynamic and non-rigid, and seeks for companies and organizations to adopt the ability to adapt to the changing needs of their beneficiaries. While [3] indicated that building of business models and new products based on the principle of employing digital technologies. As well the providing services to beneficiaries with the aim of improving user experience, raising the quality of performance, and operational efficiency, with an aim to support digital innovation [2].

Digital transformation symbolizes business or government transformation, i.e., making radical changes to the business model, procedures, and processes [9]. Transformation may affect the process of changing the product or the method of providing the service entirely, and it may interfere strategically by intervening in all the functions of the organization from sales to supply, information technology and the entire value chain [3].

Based on previous studies such as; [2, 3, 7, 10, 11], stated that the digital transformation has requirements that must be achieved through major dimensions; strategy, organizational culture, transformational leadership, and human resources.

2.2 Strategic Agility

[8] referred to the strategic agility as the ability of organizations to deal with changing conditions, adapt to continuous and accelerating fluctuations, harness resources and capabilities to take and avoid potential risks and threats, as well as the necessary measures to exploit opportunities. Additionally [6] defined strategic agility as the ability of organizations to see shifts in the business environment in which they operate, and the ability to adapt to them. Strategic agility relates to staying competitive by identifying opportunities and taking advantage of them, as well as identifying potential threats and mitigating or preventing them [12].

Strategic agility indicates to the organization's ability to continuously adapt to environmental changes, ensuring a rapid response to changing markets, faster than competitors, to maintain its market share, and to achieve survival and prosperity in an environment characterized by a high degree of uncertainty [9].

Strategic agility represents the ability to continuously adjust the strategic direction of the company's core business, and as it represents a function of strategic ambitions and changing conditions, new products do not only create new business models, but also, modern creative ways to generate value for the company [5, 13, 18]. Thereupon, agile companies have many characteristics and characteristics that distinguish it from other traditional companies. Consequently, according to previous studies strategic agility consists of the following dimensions; clarity of vision, setting strategic goals, understanding fundamental and core capabilities, joint responsibility and responsiveness [6, 8].

2.3 Digital Transformation and Strategic Agility

Previous studies aimed, such as: [2, 7–9, 12] examined the relationship between digital transformation and strategic agility. The relationship between strategic flexibility and the elements of digital transformation scale, scope of digital transformation, and speed of digital transformation was evaluated positively. Studies have also found that all components of digital transformation have a significant positive relationship with agility. It is important to predict the strategic agility of organizations.

Equally important, [2] discovered that the contribution of digital transformation to achieving control over the desired effect on strategic agility in the most cost-effective way possible.

Based on previously studies maintained, this study suggests to tests the following hypotheses:

H01: There is no statistically positive significant impact of digital transformation with its combined dimensions (strategy, organizational culture, transformational leadership, and human resources) on strategic agility in Jordan Customs.
H01a: There is no statistically positive significant impact of strategy s on strategic agility in Jordan Customs.
H01b: There is no statistically positive significant impact of organizational culture on strategic agility in Jordan Customs.
H01c: There is no statistically positive significant impact of transformational leadership on strategic agility in Jordan Customs.
H01d: There is no statistically positive significant impact of human resources on strategic agility in Jordan Customs.

2.4 Model of Study

In order to achieve the objectives of the study, the researcher developed on the following model to confirm the relationship between digital transformation and strategic agility, according to Fig. 1.

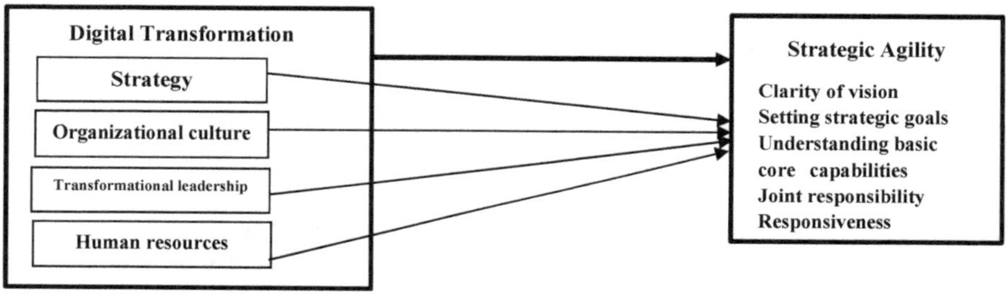

Fig. 1. Model of Study. Source**:** [2, 7–9, 12]

3 Research Methodology

3.1 Population and Sampling of the Study

The study targeted the 3378 Jordan customs employees according to [14].The study used the appropriate sample in determining the sample, which is consistent with the nature and objectives of the study. The size of the study sample was determined based on the size of the population using the Krejcie and Morgan table, which is used to determine the sample size based on the size of the study population. The size of the study population was 3378 male and female employees, and based on [15] table, the appropriate and representative sample of the population must be no less than 346 employees [15], followed by the administration of 420 questionnaire surveys. 365 questionnaires were collected and considered appropriate for analysis from the administered surveys. Hence, 87% was showing as a response rate. The results of demographic variables showed that the majority of the sample were males (76.7%), while females was 23.3%. It is clear that the customs are interested in hiring males, because there are tasks that require more effort and longer working hours, that more suitable for males than females. According to the study, the respondents of the age class 30 - less than were (are the highest percentage (44.7%). Finley, with regard to the job title, the majority were employees at (73.2%) percent.

3.2 Measurements

The study developed a questionnaire as capable to measuring the variables of the study. It adopted a previous studies such as [2, 8, 9, 12] to develop and build its items. The questionnaire items were divided into three sections: the first section was designed to measure the demographic data of the sample members, while the second section consisted of 18 items that measure the dimensions of the digital transformation. Lastly, the third section of the questionnaire, it consists of 25 paragraphs to measure the dimensions of strategic agility. Based upon, the study relied on the five-point Likert scale to measure the response of the study sample [16].

4 Results

4.1 Data Analysis

As mentioned earlier, descriptive statistics of the data were performed using IBM SPSS 22. Hence, to test and evaluate of hypothesis were used regression analysis.

4.2 Reliability Test

The reliability of the variables of study was determined using the minimum Cronbach alpha level set by [17]. The consistency of the items is bigger more than the level adopted, as suggested by [17]. As shown in Table 1, the lowest acceptable level for (α) is 0.88. This shows that the instrument is very reliable.

Table 1. Reliability Test

Variables	No. of Items	Cronbach's alpha (α)
Strategy	4	0.897
Organizational culture	4	0.882
Transformational leadership	5	0.901
Human resources	5	0.904
Clarity of vision	5	0.900
Setting strategic goals	5	0.906

(*continued*)

Table 1. (*continued*)

Variables	No. of Items	Cronbach's alpha (α)
Understanding basic and core capabilities	5	0.899
Joint responsibility	5	0.914
Responsiveness	5	0.917
Total	**43**	**0.958**

4.3 Arithmetic Means and Standard Deviations for the Study

In order to find out the level of digital transformation at the Jordan customs, the arithmetic means of the dimensions of the variable were calculated. Table 2 shows the results, the mean of the dimensions of the digital transformation variable showed that they are all within the high relative importance, and the strategy dimension was considered with the highest arithmetic mean among the digital transformation dimensions with a value of (3.95), in contrast, the human resources dimension achieved the lowest arithmetic mean among the dimensions with a value of (3.70), which is within the relative importance elevated. In general, the level of digital transformation in Jordanian customs is considered high, as the variable achieved an arithmetic mean of (3.82), and it is classified within the high relative importance.

Table 2. Mean and the relative importance of the dimensions of the digital transformation variable

Dimension	Rank	Mean	Relative importance
Strategy	3.95	1	High
Organizational culture	3.90	2	High
Transformational leadership	3.75	3	High
Human resources	3.70	4	High
Digital transformation	**3.82**		**High**

According to the strategic agility variable, Table 3 shows the results, that the mean of the dimensions showed that they are all within high relative importance except for the joint responsibility dimension. The clarity of vision dimension was considered the highest arithmetic mean among the strategic agility dimensions with a value of (3.92).which is within the medium relative importance. In general, the level of strategic agility in the Jordan customs is high, as the variable achieved an arithmetic mean of (3.76), and it is classified within the high relative importance.

Table 3. Mean and the relative importance of the dimensions of the strategic agility variable

Dimension	Rank	Mean	Relative importance
Clarity of vision	3.92	1	High
Setting strategic goals	3.81	2	High
Understanding basic core capabilities	3.71	4	High
Joint responsibility	3.62	5	Medium
Responsiveness	3.75	3	High
Strategic agility	**3.76**		**High**

4.4 Testing the Study Hypotheses

Multiple Linear Regression Analysis was used to prove the main hypothesis, and to prove the proportion of interpretation of the independent variable and dimensions on the dependent variable, and the value of (f) and its significance were adopted to verify the evidence of the hypothesis. The given Table 4 is analysed below.

Table 4. The Digital transformation dimensions - Strategic agility Model Coefficients

D.V	ANOVA		F	Sig F*	Digital transformation	Coefficients		
	R	R2				B	T	Sig
Strategic agility	0.883	0.780	319.946	.000*	Strategy	0.242	5.106	.015*
					Organizational culture	0.213	4.603	.000*
					Transformational leadership	0.222	4.514	.000*
					Human resources	0.191	4.977	.000*

* Statistically significant at the level of statistical significance ($\alpha \leq 0.05$).

5 Discussion

There is a statistically significant positive effect of digital transformation with its four dimensions (strategy, organizational culture, transformational leadership, and human resources) on achieving strategic agility in Jordanian customs. Digital transformation is linked to strategic agility with a strong positive correlation, which indicates the great influence of strategic agility when making any changes to digital transformation. The digital transformation is able to explain 78% of strategic agility, and this percentage is high, and the rest of the explanation is due to other factors that were not addressed in the study. The study consent with the findings of [7, 9] that all components of digital transformation have a significant positive relationship with strategic agility.

Based on the results, that Jordan customs have a high level of digital transformation, as the variable achieved a high relative importance with an arithmetic mean of (3.82). Also, the strategy dimension achieved the highest arithmetic mean among the others dimensions, with value (3.95). This means as a result of the Jordan Customs having a written and clear vision and mission for the transformation process. On the contrary, the organization have not actions are not to correct the difference in actual performance with the strategy designed for digital transformation. In the same vein, the study consent with the findings of [13] that there is a digital transformation was interested by state institutions, whether public or private firms. As well as the study agreed with the findings of the [2].

Jordan Customs has a high level of strategic agility with an arithmetic mean of (3.76) as a result of its interest in applying its dimensions. All dimensions achieved high relative importance, except for joint responsibility, as it is the least applied among the dimensions of strategic agility with mean of (3.62). It is due to reason of the lack of participation of employees in the activities of the department, and also the lack of participation in planning and implementing projects. The study agreed with the findings of the study [12] that the availability of the practice of strategic agility in non-profit organizations is high.

Lastly, based to results of study, the Jordanian Customs has a clear vision and was taken into account when formulating its objective, which is the highest among the dimensions of strategic agility. The same applies to the ability to define strategic goals, which are high because of the ability of the Jordanian Customs to translate the goals into reality. However, the Jordanian Customs does not periodically and continuously develop the objectives.

6 Conclusions

The study purposed to identify the impact of digital transformation in achieving strategic agility and targeted employees in the Jordan Customs. The study used the descriptive analytical approach using the Statistical Package for Social Sciences (SPSS) program to analyze the data collected from the study population, which consisted of employees of the Jordan Customs that working in various centers and directorates of the Department. 365 questionnaires were retrieved with a recovery rate of (87%). The study found a set of results, the most important of which is that digital transformation has a positive, statistically significant impact in its four dimensions (strategy, organizational culture, transformational leadership, and human resources) in achieving strategic agility in Jordan Customs. The study offers a contribution by emphasizing how Jordan Customs should improve their chances of success. Hence, it presents recommendations to the Jordanian Customs in the light of the previous results, including that the Jordan Customs work to provide the necessary technology after evaluation of the current digital transformation process. Also, the study proposes to increasing attention to training workers and raising their capabilities to deal with digital transformation positively and improve services, in addition to increasing employee participation in expressing their views when preparing the department's vision, which contributes to motivating employees to achieve the vision, to raise the level of strategic agility.

References

1. Awad, A.: The impact of the dimensions of organizational learning in improving strategic agility: an applied study on Ain Shams University hospitals during crises - the case of the Covid-19 crisis). Sci. J. Fin. Commercial Stud. Res. **2**(1), 457–507 (2021)
2. Mangalaraj, G., Nerur, S., Dwivedi, R.: Digital transformation for agility and resilience: an exploratory study. J. Comput. Inf. Syst. **63**, 11–23 (2022)
3. Nousopoulou, E., Kamariotou, M., Kitsios, F.: Digital transformation strategy in post-COVID era: innovation performance determinants and digital capabilities in driving schools. Information **13**(7), 323 (2022)
4. Shaaban, G.: The impact of digital transformation on achieving sustainable development in light of the Corona pandemic (applied to some Arab countries). Sci. J. Commercial Environ. Stud. **12**(4), 1–30 (2021)
5. Alzuod, M.A., Alqhaiwi, L.A.: The role of strategic flexibility in enhancing crisis management in the commercial banking sector during the COVID-19 Pandemic. Corp. Governance Organ. Behav. Rev. **6**(2) (2022)
6. Lungu, M. F.: The influence of strategic agility on firm performance. In Proceedings of the International Conference on Business Excellence, vol. 14, no. 1, pp. 102–110 (2020)
7. Segaetsho, G.: The relationship between digital transformation and strategic agility (Doctoral dissertation, University of Pretoria) (2019)
8. Burchardt, C., Maisch, B.: Digitalization needs a cultural change–examples of applying agility and open innovation to drive the digital transformation. Procedia Cirp. **84**, 112–117 (2019)
9. AlNuaimi, B.K., Singh, S.K., Ren, S., Budhwar, P., Vorobyev, D.: Mastering digital transformation: the nexus between leadership, agility, and digital strategy. J. Bus. Res. **145**, 636–648 (2022)
10. Llopis-Albert, C., Rubio, F., Valero, F.: Impact of digital transformation on the automotive industry. Technol. Forecast. Soc. Chang. **162**, 120343 (2021)
11. Guzmán-Ortiz, C., Navarro-Acosta, N., Florez-Garcia, W., Vicente-Ramos, W.: Impact of digital transformation on the individual job performance of insurance companies in Peru. Int. J. Data Netw. Sci. **4**(4), 337–346 (2020)
12. Tallon, P.P., Queiroz, M., Coltman, T.: Digital-enabled strategic agility: the next frontier. Eur. J. Inf. Syst. **31**(6), 641–652 (2022)
13. Al-Qeed, M.A.: Effect of strategic agility dimensions on tourism marketing performance in Jordan (case study on tourism companies in the Amman Region). Mu'tah Lil-Buhuth wad-Dirasat, Hum. Soc. Sci. Ser. **34**(2), 13–36 (2020)
14. Jordan Customs: Annual report 2021.Amman, Jordan (2021)
15. Krejcie, R.V., Morgan, D.W.: Determining sample size for research activities. Educ. Psychol. Measur. **30**(3), 607–610 (1970)
16. Sekaran, U., Bougie, R.: Research Methods for Business: A Skill Building Approach. Wiley, Hoboken (2016)
17. Hair, J.F., Black, W.C., Babin, B.J., Anderson, R.E.: Multivariate Data Analysis, 7th edn. Prentice Hall, Upper Saddle River (2010)
18. Alqhaiwi, L.A.: The impact of the conflict patterns on the organization empowerment of the employees in the Jordanian phosphate company. Zarqa J. Res. Stud. Hum. **15**(1), 215–228 (2015)
19. Alajeeb, A.T., Al Najjar, F.J.: The impact of strategic thinking on organizational excellance: the mediating role of strategic foresight in the winner organizations of King Abdullah II Prize for excellence for the private sector in Jordan. Zarqa J. Res. Stud. Hum. **2**(1), 15–33 (2022)

Study on Customer Comfort as a Behavioral Construct Between Service Quality and Customer Satisfaction in Light of Digital Transformation. A Case of Jordan

Zakaria Azzam[1(✉)], Rawan Salman[1], Mahmoud Allan[2], and Suhad Farsi[3]

[1] Marketing Department, Zarqa University, Zarqa, Jordan
{zazzam,20169027}@zu.edu.jo
[2] Marketing Department, Al Zaytoonah University of Jordan, Amman, Jordan
Mahmoud.allan@zuj.edu.jo
[3] Department of HRM, University of Business and Technology, Jeddah, Kingdom of Saudi Arabia
s.farsi@UBT.EDU.SA

Abstract. This paper attempts to identify the effect of service quality on Customer Comfort and the effect of customer comfort as a mediating variable on customer satisfaction of services in Jordan. For this purpose, a set of related independent variables to service quality which are (Reliability, Responsiveness. Assurance, Failure recovery), one mediating variable (Customer Comfort) and one dependent variable (Customer Satisfaction) which are adopted in the study model. A questionnaire is designed as a research instrument and distributed to 480 respondents who deals with services organizations in Jordan, (89%) were valid for analysis. Thus (409) respondents are assured to be the sample size of the present study. The study concluded that there is statistical significant effect of service quality (Reliability, Responsiveness, Assurance, Failure Recovery) on customer comfort and there is a strong direct relationship between customer comfort and customer satisfaction. The study recommended that services companies should create a systematic procedures that help to develop failure recovery mechanism, and it should offer training courses for its employees to enhance responsiveness.

Keywords: Service Quality · Customer Comfort · Customer Satisfaction · Digital Transformation · Jordan

1 Introduction

Services sector is considered to be an essential pillar in countries' economy which leads to economic growth around the world. On the other hand, the quality of services should be taken as an important issue that affect the competitiveness of a firm. Such firms may distinguish its performances from the rivals by giving excellent services. Marketing researchers developed models for better comprehension of this phenomenon, these models include Reliability, Responsiveness, and Assurance and Failure recovery, in this

B. A. M. Alareeni and I. Elgedawy (Eds.): *Artificial Intelligence (AI) and Finance*, SSDC 488, pp. 34–47, 2023.
https://doi.org/10.1007/978-3-031-39158-3_4

paper the researchers will study the effect of these dimensions on customer comfort and customer satisfaction as a marketing concept. Comfort plays a role in maintaining customer's retention and sustaining business for many services providers, where a comfortable customer may tend to keep and maintain dealing with the services provided by the firm, marketing researchers asserted that comfort is vital in maintaining relationship with customers and selecting service providers. Customer comfort also may have an impact on quality of service and customer satisfaction, in addition to this customer satisfaction along with company's product and services are a way to achieve company's success and sustainability. Services organizations gives its customers a huge attention through its policies and strategies which guarantee to provide the best service for the customers which are scattered all over Jordan. For these organizations providing best service need to meet or exceed customers' needs and requirements and this may lead to service quality which ultimately may lead to customer satisfaction [1–3].

This study tries to investigate the effect of customer comfort as a mediating variable between services quality and customer satisfaction and to identify the matrix for these variables.

1.1 Research Problem

Services are considered as one of the most important sectors presented in the modern world, it is also not stable and guaranteed as other sectors due to digital transformations taking palace in modern days. Customers who deal with these kind of services always complain from many mistakes can happened during the service encounter, these mistakes can be a damaged or lost shipment, a delay in the shipping schedule. Services are intangible and the quality of service can't be measured easily by the customer like the tangible products, services are considered to be the most irritating service as the customer can't feel comfortable until he received the benefit expected from that service. In the light of the above, customers who deal with services are always anxious during service encounter, and as customer satisfaction is considered the goal of every service firms is to attract customers and enhance customer retention, services customer satisfaction depends on getting the service completely as the customer expected. Research problem has been formulated in the following questions:

1- Is there an effect for quality of services on customer comfort?
2- Is there an effect for customer comfort as a mediator on customer satisfaction?
3- Is there an effect for service quality on customer satisfaction?
4- Is there an effect for customer comfort as a mediator between service quality and customer satisfaction.

1.2 Study Model

Based on the above, and earlier [1–15] and [19] the researchers supposes the following model (Fig. 1).

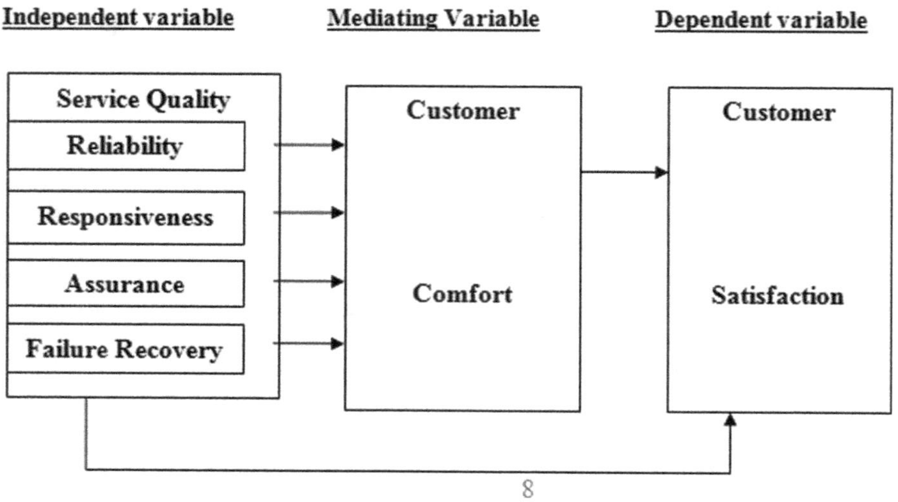

Fig. 1. Research Model

1.3 Research Hypotheses

The main hypotheses of the present study are:
H01: Statistically there is no effect ($\alpha \leq 0.05$) of service quality variables (Reliability, Responsiveness, Assurance and Failure recovery) on customer comfort.
H02: Statistically there is no effect ($\alpha \leq 0.05$) of customer's comfort as a mediating variable on customer satisfaction.
H03: Statistically there is no effect ($\alpha \leq 0.05$) of service quality variables (Reliability, Responsiveness, Assurance and Failure recovery) on customer satisfaction.
H04: Statistically there is no effect ($\alpha \leq 0.05$) of customer comfort as a mediator variable between service quality and customer satisfaction.

2 Literature Review

The importance of service quality for service firms as a tool to attract more customers, increase profit and enhance word of mouth about the services provided by the firm.

Firms around the world adopt measures of service quality by their own understanding of service quality or by measures invented by marketing researchers, and service quality is the goal of any firm to overcome their competitors, guarantee a long existence in market, and preserve the growth of the firm life cycle. Customer comfort also plays a major role in service quality, simply when measures of service quality accomplished, the customer will feel comfortable with the perceived services, this comfort may occur from the physical environment of the service encounter, and the service provider contributes greatly in delivering customer comfort for the customers [3, 9, 15].

Service quality, customer comfort and customer satisfaction all combine a circle that enhances the performance of service providers in the firm, which make the firm more acceptable to customers, maintain the readiness of the company to deal with and compensate any failure in the service for the customers, and eventually this leads to a good reputation for the firm and booming word of mouth [17].

2.1 Services Quality

Service sector is considered as the core of economic growth around the world, it is the largest and fastest growing sector contributing in the global output and employing more people. The reason for this growing is the increase of urbanization, privatization and more demand for intermediate and final customer services [18]. Services are business operations in which a group of company resources work with customers to add value to those customers' processes. Services are therefore value-supporting processes as opposed to goods, which are value-supporting resources [19]. The importance of service quality as a sign of satisfaction for consumers can be considered a reason to gain competitive advantage among rivals [5]. Services are intangible, whereas goods are tangible, which makes it difficult to for marketers to get direct feedback about quality of the services. These are just a few of the key differences between goods and services. The degree to which a service meets a customer's expectations is referred to as the service's quality. Researchers typically use the term "service quality dimensions" to describe the criteria used to evaluate the quality of intangible services [19]. After further research into measurements of service quality [11], determined five dimensions of quality. In this study the researcher chooses measures that are related to the study field [11, 14, 15, 19–22]:

Reliability
The degree to which a promised service is reliably and accurately provided is known as reliability. Reliability is also described as the "ability to perform the promised service dependably and accurately; reliability depends on handling customer service issues, performing the services correctly the first time; providing services on time; and maintaining an error-free history. Furthermore, reliability is the most important aspect of traditional service. In addition, reliability includes proper order fulfillment, accurate records, accurate quotes, right charges, and results that are more accurate than before, services are delivered as expected, all customer problems are solved, getting the services at the promised time and free of mistakes. Generally speaking there are some factors that affect services quality i.e. consistency and dependability, consistency which indicate, similarities of the services provided all the times, and dependability, means providing the offers as required by clients with trust. Reliability is considered one of the most important tool to guarantee providing service with the expected measure for the customer in Services Company in general and shipping companies in specific standards [23].

Responsiveness
Responsiveness is also defined as the ability to provide the customer the required offer in proper time and smoothly, rabidly according with customers' expectations which will ensure customers' satisfaction [12]. Good responding service provider and effective

dealing with customer's complaints and demand is what distinguishes a firm from another and encourage customers to repeat the experience with a specific firm.

Assurance

Assurance is a set of criteria used to evaluate efficiency, modesty, honesty, trust and safety of a company. It can also be used to describe an employee's capacity to foster trust and confidence in the company through their expertise and demeanor. For that assurance plays an important role in service companies, as these firms depends on security and trust provided by its employees to the clients [12].

Service Recovery

Service recovery is mentioned as failure recovery in this study, service recovery indicates to the procedures taken by service providers when seeing some errors in services provided. Service recovery is measures adopted by the company to change dissatisfied customers to satisfied customers, further more failure recovery is considered as the procedures adopted by services firms which make up any mistakes.

Failure to provide a service should not be viewed as a problem or a failure by businesses, but rather as a chance to satisfy customers. Service recovery is actions taken by a service providers to make up the customer's satisfaction after failure of services. When consumers believe that businesses regularly achieve service recovery when mistakes taking place, their attitudes become more positive towards the firm, which indicate that negative emotions eliminated by service providers in case of failure which may affect satisfaction. Generally speaking a satisfied client may maintain loyalty with the services firm which improves the loyalty to the company and this would enhance the competitiveness and sales of the firm [23].

2.2 Customer Comfort

Customer comfort is described as a situation of fulfillment psychological and physical needs of customer which include ease, relief and transcendence from both services and its providers. In other word, it is a state of introducing comfort towards several aspects of services and service providers. Therefore, comfort can be defined as a state of meeting psychological and physical customer needs from service and service provider. This definition is chosen because it considers the relevant literature in marketing context [9]. [3] highlighted customer comfort in three aspects, physical aspect which means the customer relief from tiredness or physical suffer through service encounter, physiological aspect which means spontaneous reactions that come from the customer when he feels strange or odd with the service environment and finally psychological comfort which means positive feeling of peace of mind during the service encounter with the service provider. The importance of comfort has been recently identified by marketing researchers. The concept of comfort has been found to be an essential factor in developing and sustaining relationships between service providers and their customers, It has also been identified as a key antecedent for perception of customers service quality comfort is used as an important factor in achieving customer's satisfaction in service marketing and comfort is used frequently as a competitive advantage strategy in the marketing [25].

Psychological Comfort

Researches has addressed psychological comfort as an option for consumers to reduce the problems in buying process, as these options may be relied on increasing comfort as guarantee, positive word of mouth that increases preference for services provider, and advertising message used by the provider can attract customers based on promise of comfort. The importance of psychological comfort lies in that customers who have in positive emotional feelings after getting the services. Psychological value and behavioral manifestations that customer receives during service encounter such as reception, social interaction with service providers and service quality enhance customer satisfaction toward the service [26].

[25] Assured that comfort is a crucial factor in developing and maintaining relationships between customers and service providers. Psychological aspect of comfort is formed through interactions between customers and service providers; in addition a positive comfortable relationship between service providers and customers also influences service delivery. Comfortable customers typically feel socially close to their service providers and agree to participate in providing feedback that may improve their services delivery [25].

Physical Comfort

Physical comfort is someone or something which affords or causes relief and satisfaction, make distress seem lighter by means of kindness and thoughtful attention by the service provider, and make the customer physically comfortable. Physical comfort has been defined in multidimensional aspects starts with the need for physiological needs, social interaction, personalized care, feeling safe, feeling valued and having some control [3].

The importance of Physical comfort is as a construct plays an essential role when the customer buys services or goods, another importance of physical comfort in service marketing is to make the customer more acquainted to the tangibility of service as the physical environment during service encounter; clean, safe and good smell of service place affects the customer comfort and eventually customer.

3 Service Quality and Customer Comfort

Researches assured that perception of customers' service quality is considered a major determinant of the success of service firms, due to its primary role in raising the level of customer comfort and service provider behaviors can influence customer comfort during service encounter. Enhancing the customer comfort is a major factor in creating good contacts between service providers and their customers, because the feeling of comfort deepens this relationship and there is a link between service firms and clients, i.e. a positive comfortable relations with service firms. This influences service delivery, comfortable customers typically feel socially close to their service providers and agree to participate in providing feedback that may improve their services delivery. Influenced by the company outcomes from the customer's perspective, concept of customer comfort also comprise number of concepts which accordingly lead independently to comfort, these concepts include functional and technical service quality [3] and [9]. Quality of service is considered a source of comfort given its importance for both the service firms and the clients, as it is shown as a potential to increase competitive advantage, loyalty, repurchase and positive relationships with customers [9].

The idea of separating some components of service quality as antecedents can also be applied in distinguishing between social, environmental aspects, functional and technical dimensions of service quality; it is also possible to add to the two dimensions of service quality a third dimension which contains aspects of the physical surrounding. The physical surrounding refers to environmental surrounding or which is called "service scape quality" as a third dimension of service model which can be added to the dimensions of functional service quality and technical service quality in order to get greater understanding of service quality model. Functional service quality appears to have an effect on the physical and psychological aspects of comfort, however, he also suggested that technical service quality leads to comfort (physical and psychological) the specific influence goes for technical service quality on comfort [9].

4 Customer Comfort and Customer Satisfaction

Satisfaction is simply an effort of fulfillment of something or making something adequate, customer satisfaction based on customer expectation "customer satisfaction is a person's feeling of pleasure or disappointment that result from comparing a product's performance or outcome in relation to their expectation. In this judgment the benefits that customers gain have to be more than the costs that they have to pay. Satisfaction has been found elsewhere to refer to the customer's interpretation of the service experienced and whether the customer would return for another visit. The relationship between satisfaction and comfort has been addressed as "feeling comfortable is an important factor in arising satisfaction. Moreover, one's comfort level was considered as an affective construct in enhancing customer satisfaction during service encounter. Comforting customers has a positive effect on their satisfaction; furthermore it was found that increasing psychological aspects of comfort in such as arousal and enjoyment tend to promote customer satisfaction. A comfortable chair and hall design were reported as providing user satisfaction, in other words comfortable customers in turn will feel satisfied and stay longer in service providing place. In brief, many studies discussed earlier indicated that that there is a relationship between physical and psychological aspects of comfort and customer satisfaction [19].

5 Customer Satisfaction and Service Quality

Depending on how well a service meets or exceeds a customer's needs, customers express how satisfied or dissatisfied they are with that service. Customers use expectations as a benchmark internally to judge the caliber of a service. According to literature, has direct correlation with satisfaction [19].

6 Methodology

The paper tries to investigate the effect of service quality on customer comfort, and from that objectives four hypothesis were supposed, each one aims to find if there is an effect for service quality dimensions (Reliability, Responsiveness, Assurance, and Failure Recovery) which represent the independent variables, on customer comfort and to determine how customer comfort affects customer satisfaction. The researchers designed a questionnaire as an instrument to collect data, and the paper adopted a descriptive and analytical approach. To make it more convenient, the questionnaire distributed to a random sample of respondents in Amman, the capital of Jordan. All shoppers in Amman, the capital of Jordan, are included in the population of the current study.

Based on [26] table customers from the population should make up at least (380) of the representative sample. The researchers distributed (480) questionnaires to service recipients in Jordan, representing the population in order to increase accuracy and reduce errors. The questionnaires were distributed by the researchers using a straightforward random sample technique. Six percent of all questionnaires, or about (28) of them, were not returned. In addition, 23 questionnaires were returned incomplete (5%) because some questions were left unanswered. Of the questionnaires that were distributed, (89%) were valid for analysis. Thus, there were (409) respondents in the sample size. A Likert scale measurements is used with five degrees of answer ranging from strongly agree to strongly disagree; to determine the importance of each question in the questionnaire as shown it Table 1.

6.1 Reliability

Cronbach's Alpha is at a minimum acceptable level of 0.60 [24] as shown in Table 1.

Table 1. (Cronbach Alpha)

Number	Variables	Questions	Cronbach's Alpha
1	Reliability of customers	Q1–Q8	0.82
2	Responsiveness	Q9–Q16	0.83
3	Assurance	Q17–Q25	0.83
4	Failure Recovery	Q26–Q31	0.812
5	Customer Comfort	Q32–Q45	0.86
6	Customer Satisfaction	Q46–Q56	0.85
	All items	56	0.90

The table shows that all study variables have values greater than (0.6), with "Customer Comfort value (0.861) and "Reliability" having the lowest value (0.793). The Cronbach's Alpha Coefficient for all is (0.903). As a result, the research findings can be regarded as valid [24].

6.2 General Information of the Sample

The study includes Six different general information in the questionnaire (Gender, Age, Level of Education, professional field, number of times customers dealing services. The outcomes in Table 2 show how the sample's participants were distributed based on demographic data.

Table 2. Population Statistics

Items	Characteristics	Frequencies	Percentages
Sex	Male	180	44.0%
	Female	229	56.0%
	Total	409	100%
Age	20 Years – less than 30 years	106	25.9%
	30 Years – less than 40 Years	238	58.2%
	40 v – less than50 Years	50	12.2%
	Above 50 years	15	3.7%
	Total	409	100%
Level of education	High school or less	41	10%
	Mid diploma	36	8.8%
	Bachelor degree	269	65.8%
	High studies	63	15.4%
	Total	409	100%
Professional field	Government section	33	8.1%
	Private section	256	62.6%
	Own business	48	11.7%
	Student	20	4.9%
	Other	52	12.7%
	Total	409	100%
How many times you deal with Services	Once	93	22.9%
	Twice	69	16.9%
	3 times and more	247	60.2%
	Total	409	100%

Table 2 shows that (56%) of the respondents are females, while approximately (44%) of the respondent are males.

About age of respondents are divided into four levels, Percentage of customers respondents whose ages range between (30 – less than 40 years) are the largest percentage, approximately (58%). While those whose ages (above 50 years) are the smallest percentage approximately (4%).

Respondents who have a Bachelor degree are the largest percentage of the customers approximately (66%). While the smallest percentage belongs to those who have mid diploma, there percentage is approximately (9%).

Regarding frequencies of respondents professional filed most of them work in private section, the respondent's percentage who work in private sector approximately (63%). But for students it is the smallest percentage about (5%).

Also the questionnaire measures the number of times the customer deal with services firms, the results show that the majority of customers deal with Aramex 3 times and more (60%). While the customer who deals twice with Aramex represents the smallest percentage approximately (17%).

6.3 Regression Analysis

Regression is adopted to test the first main hypothesis and provide an answer to the main question. Table 3 displays the outcome.

Table 3. Regressions Results

M.V.	R	R^2	F	Sig	I.V. (services Quality)	B	t	Sig.
Customer Comfort	0.907	0.823	469.741	0.000	Reliability	0.319	7.698	0.000
					Responsiveness	0.198	4.867	0.000
					Assurance	0.411	9.400	0.000
					Failure Recovery	0.455	12.459	0.000

The hypothesis was tested at significant level ($\alpha \leq 0.05$), Table 3 shows the following results:

- **F** calculated value (469.741) which is more than **F** table (2.89), which indicates is a significant effect on mediating variable.
- The research independent variables (Reliability, Responsiveness, Assurance, and Failure Recovery) are significant, because **F** significant value is (0.00) ($\alpha \leq 0.05$), therefore we suppose the effect.
- There is a significant and favorable relationship between the independent and mediating variables. It exceeds (0.5). (Cohen, 1988), R = 0.907, that indicate if the independent variable increasing will positively effect on mediating variable.
- Also, the $R^2 = 0.823$, which indicates that I.V. contribution is strongly affecting the M.V. with a percentage of 82.3%.

T value for the variables (Reliability) is 7.698, Responsiveness 4.867, Assurance 9.400, Failure Recovery 12.459, which are more than the **t** table value (1.96) which assure that they have an effect on the mediating variables. In additional Sig. value is less than (0.05).

For testing the second main hypothesis and answering the main second question, the researcher used Simple linear Regression as in Table 4.

Table 4. Results of Regressions

DV	R	R^2	Mediating Variable	B	t calculated	t Table	Sig
Customer Satisfaction	0.874	0.764	Customer comfort	0.880	36.259	1.96	0.000

T-value (36.259) which is more than the (1.96), which shows that the IV has an effect on the mediating variable. The significant value of **t** is less than (0.05), which assure the effect of IV on DV as in Table 4.

The R value (R = 0.874), which is greater than 0.5 [27] also demonstrates that the mediating and dependent variables have a positive (strong) correlation. That show whether an increase in the mediating variable will have a positive impact on the dependent variable. In addition, the customer comfort variable contribution on the customer satisfaction (76.4%) where ($R^2 = 0.764$).

Table 5. Results of Multiple Regressions

M V	R	R^2	F	Sig	I V (services Quality)	B	t	Sig
Customer satisfaction	0.903	0.815	444.839	0.000	Reliability	0.242	5.667	0.000
					Responsiveness	0.145	3.457	0.001
					Assurance	0.386	8.573	0.000
					Failure Recovery	0.236	6.269	0.000

Table 5 shows the research dependent variable (Customer satisfaction) and independent variables (Reliability, Responsiveness, Assurance, Failure Recovery) are significant because F value (444.839) which more than (2.89), and F significant is (0.00), ($\alpha \leq 0.05$). As a result, we assure the effect. ($\alpha \leq 0.05$) for service quality(Reliability, Responsiveness, Assurance, Failure Recovery) on customer satisfaction.

DV and IV have a positive and favorable relationship. It exceeds (0.5) [27] and R is 0.903. Additionally, the R2 value is 0.815, which indicates that the independent variables have an 81.5% percentage impact on the dependent variable. Since the value of the calculated t value for the variables (Reliability 5.667, Responsiveness 3.457, Assurance 8.573, Failure Recovery (6.269), are more than (1.96) this means that there is an effect.

Considering the outcomes in Table 4 and Table 5 only shows one indirect effect, which is the service quality indirect effect on customer satisfaction through customer

Table 6. Results of Multiple Regressions

Variables	Direct Effect		Indirect Effect		Total Effect	
	Service Quality	Customer Comfort	Service Quality	Customer Comfort	Service Quality	Customer Comfort
Customer comfort	0.963	0.000	0.000	0.000	0.963	0.000
Customer Satisfaction	0.358	0.372	0.638	0.000	0.997	0.372

comfort and has a value of (0.638). This result raises the total service quality on customer satisfaction to (0.997). Additionally, the indirect impact on customer satisfaction was greater than the direct impact, were the study assure the effect ($\alpha \leq 0.05$).

The path direction is shown in Fig. 2. The numbers on the path arrow represent the direct effects of IV on the DV, with a result of (0.358), (0.372), and (0.963) respectively as shown in Table 6.

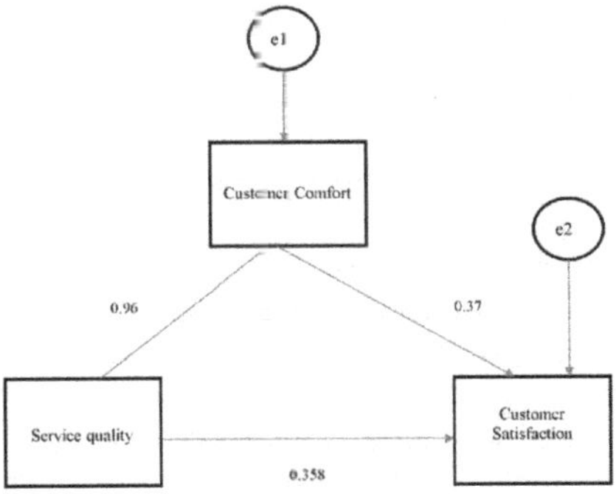

Fig. 2. Results of Path Direction.

7 Recommendations

- Services companies should build up a plan for improving process to develop failure recovery mechanism. Also Aramex should investigate what kind of mistakes can't solve very well, as failure recovery plays the highest impact on service quality as mentioned before in conclusion part.
- Services companies should increase employee's responsiveness via increasing the capacity of well-trained employees and crisis management team building, as responsiveness showed a medium level in the conclusion and Aramex employees already take training before they start the real work.
- Services companies should study the reasons of failure recovery and improve the processes that makeup this failure in service.

- Services companies Management should develop processes that reduce time for providing service and handle customer complaints effectively.
- Services companies should concentrate on developing services and investigate other variables that affect customer comfort as well as customer satisfaction.

References

1. Agyapong, G.: The effect of service quality on customer satisfaction in the utility industry – a case of Vodafone (Ghana). Int. J. Bus. Manage. **6**(5), 203–210 (2010)
2. Atiyah, L.: Impact of service quality on customer satisfaction. Aust. J. Basic Appl. Sci. **11**(5), 20–28 (2017)
3. Al-Medabesh, A.: Customer comfort as a marketing construct in the relationship between the dimension of quality banking services and customer satisfaction. Unpublished Master thesis, Jezan University–Kingdome of Saudi Arabia (2015)
4. Wang, C., Liao, H.: Effects of service recovery on customer satisfaction and relationship retention in catering industry. Master thesis, Department of Food and Beverage Management, Mackay Junior College of Medicine, Nursing and Management, Taiwan, ROC (2014)
5. Gray, B.: The relationships between service quality, customer satisfaction and buying intentions in the private hospital industry. S. Afr. J. Bus. Manage. **35**, 27–37 (2004)
6. Gronroos, C.: A service quality model and its marketing implications. Eur. J. Mark. **18**, 36–44 (1984)
7. Harmse, C.: Service quality in a landlord-small business relationship in shopping centers, Ph.D. thesis, University of Pretoria, South Africa (2012)
8. Jamaluddin, R.: Impact of service quality and customer satisfaction on customer loyalty: a case study in a private hospital in Indonesia. IOSR J. Bus. Manag. **19**(5), 23–33 (2017)
9. Medbash, A.: Customer comfort as a marketing construct in health care. Unpublished Ph.D. dissertation, The University of Western Australia, Australia (2008)
10. Minh, N., Nguyen, T., Phan, C., Yoshiki, M.: Quality of service and customer satisfaction: a case study of hotel industry in Vietnam. Asian Soc. Sci. **11**(10), 15–20 (2015)
11. Parasuraman, A., Zeithaml, V.A., Berry, L.: A conceptual model of service quality and its implications for future research. J. Mark. **49**, 41–50 (1985)
12. Sidiqqi, O.: Interrelations between service quality attributes, customer satisfaction and customer loyalty in the retail banking sector in Bangladesh. Int. J. Bus. Manage. **6**(3) (2011)
13. Spake, D.F., Beatty, S.E., Brockman, B.K., Crutchfield, T.N.: Customer comfort in service relationships measurement and importance. J. Serv. Res. **5**, 316–332 (2003)
14. Ali, N.N., Allan, M.S., Azzam, Z.: The impact of total quality management (TQM) dimensions on achieving competitive advantage: managerial perspective of the quality department staff at five-star hotels. Case of Jordan. Int. J. Sci. Technol. Res. **9**(2), 2578–2587 (2020)
15. Al-Shaikh, M., Khanfar, I.: Delivery service via electronic applications and its impact on customers satisfaction at retail stores in Amman City/Jordan. J. Stud. Syst. Decis. Control **216**, 827–837 (2023)
16. Azzam, A.: The moderator effect of COVID pandemic on the relationship between online shopping services and online consumer purchasing behavior. A case of Jordan. J. Stud. Syst. Decis. Control **216**, 25–35 (2023)
17. Azzam, A., Al-Shaer, S.: Electronic Word of Mouth (E_WOM) adoption via social media and its impact on online shoppers' purchasing intention during Corona pandemic. A case of Jordan. J. Stud. Syst. Decis. Control **216**, 477–487 (2023)
18. Dewan, M., Mahajan, S.: A study of the perceived service quality and its dimensions in private sector banks. J. Bus. Manage. **16**(2), 44–51 (2014)

19. Lovelock, C., Vandermerewe, S., Lewi, B., Fernie, S.: Service Marketing Edinburg Business School, 2nd edn. Heriot-Watt University, London (2011)
20. Yarimoglu, E.: A review on dimensions of service quality models. J. Mark. Manag. **2**(2), 79–93 (2014)
21. Azzam, Z.: Factors affecting the adoption of E-marketing technologies in services industry: empirical study on hotel sector in Jordan. Zarqa J. Res. Stud. Humanit. **13**(1), 210–224 (2013)
22. Al-Azzam, A., Azzam, Z.: Evaluating the impact of service quality, brand image, trust and promotion on customer loyalty in mobile phone in Jordan. Zarqa J. Res. Stud. Humanit. **13**(2), 263–273 (2013)
23. Al-Shaikh, M.: Customers' perspective towards factors associated with technological products adopting failure. A case of Jordan. Zarqa J. Res. Stud. Humanit. **13**(2), 252–261 (2013)
24. Jay, H., Barry, R.: Operation Management: Sustainability and Supply Chain Management, 11th edn. (2014)
25. Dabholkar, P., Shepherd, D., Thorpe, D.: A comprehensive framework for quality of service: an investigation of critical conceptual and measurement issues through a longitudinal study. J. Retail. **76**(2), 139–173 (2010). ISSN: 0022-4359
26. Kolle, A., Zauner, F.: Further insights into perceived value and consumer loyalty. J. Psychol. Mark. **28**(12), 23–33 (2011)
27. Sekaran, U.: Research Methods for Business, 4th edn. Willey, New York (2003)

The Role of Digital Leadership in Achieving Organizational Excellence in Jordan's Banking Sector

Maan Hussein Mansour[1](✉) and Sayeeduzzafar Qazi[2]

[1] Zarqa University, Zarqa, Jordan
mmansour@zu.edu.jo
[2] University of Business and Technology, Jeddah, Kingdom of Saudi Arabia

Abstract. The study examined the role of digital leadership in achieving organizational excellence in Jordan's banking sector. A survey research design was adopted. The population of the study was all employees and customers of banks. The Jordanian Banking sector was purposively selected among other sectors in Jordan. The research drew from a theoretical population with no limits on size. because there is a shortage of available individual data on bank personnel and customers in Jordan. An estimated sample size of 163 was drawn from the research population using the Cochran (1977) method for determining sample size when the population is infinite (i.e., greater than 50,000). The research instrument was a structured questionnaire. The study employed descriptive (frequency counts, percentage and bar-chart) and inferential statistics (Structural Equation Modeling and Pearson Product Moment Correlation Coefficient) for data analysis. The results were presented in tables and discussed according to the research objectives and hypotheses. The study revealed a significant impact of digital leadership on achieving organizational excellence in Jordan's banking sector. There was a significant relationship between digital leadership and customer satisfaction in Jordan's banking sector ($P < \alpha$). Increases customer satisfaction, drives data-based insights, enables future digital growth, increases operational efficiency, encourages an environment of employee excellence, limits human error, increases agility, encourages collaboration & improves communication, allows high-quality user experience and enables software monetization were the extent to which digital leadership has improved the performance of banks in Jordan. The study recommended that banks in Jordan organize more training and workshops for their employees on the effective use of digital devices in improving their job performance.

Keywords: Digital Leadership · Organization's Excellence · Jordan · Descriptive and Inferential Statistics

1 Introduction

Technological developments have facilitated digitization and improved organizational performance. Organisations must possess digital leadership abilities to compete in the digital age. For leaders to be successful in the digital age, they must have key competencies in digital leadership. Although everybody can become a leader, not everyone can

B. A. M. Alareeni and I. Elgedawy (Eds.): *Artificial Intelligence (AI) and Finance*, SSDC 488, pp. 48–61, 2023.
https://doi.org/10.1007/978-3-031-39158-3_5

develop their leadership potential. Employees need to be capable of digital leadership if they want to offer answers to ongoing issues in their company. In the digital age, leadership is not seen as a notion but rather as the ability to master both hard (technical skills). And soft talents (non-technical skills). Employees need soft skills (non-technical talents) since they are necessary for interactivity, self-governance, and flexibility, which are all aspects of enhancing an organization's performance.

Digital leadership does not entail being an expert in computer setup, operation, or programming. The capacity to lead an organisation in the digital era while employing information and communication technologies to accomplish organisational goals is known as digital leadership. Technical expertise and digital skills alone do not constitute digital leadership. Digital team leadership entails guiding workers without being physically there, understanding the advantages of digitization, and actively influencing the digital transformation of the business. However, digital leadership is a skill that must be mastered; it does not just happen. It is not sufficient to understand current technology and transfer conventional management tools 1:1 to the virtual environment to fulfil the needs of modern working life. You need new methods of working that are adapted to digital transformation instead, along with real "digital leaders" who set the bar high. Digitalization is unrestrained by anybody or anything, not even top management. However, before the Corona pandemic in 2019, digital leadership was rarely a problem in most firms. Many businesses have just recently reconsidered their approach to leadership as a result of the bulk of employees being forced to relocate to home offices almost immediately.

Change happens quickly and frequently in the workplace because of the digital age. Businesses are becoming more globally integrated as a result of the increasing automation and digitalization of processes, the widespread availability of remote offices, and other innovations that allow employees greater freedom of location and schedule. In light of these shifts, it is more important than ever to be flexible in how you approach problems and to view issues from multiple angles. As a result, modern management demands are higher in complexity, difficulty, and depth than ever before. Leadership in the modern digital era, ultimately, requires striking a balance between the needs of employees, the business, and the customer base. Furthermore, it entails dealing with transformation in a way that gives thought to future generations. [1] When properly planned and carried out, digital transformation can boost business value. The use of digital transformation has the potential to lower bank costs significantly. New revenue streams may be produced via altered company models. Changing rigid to flexible hybrid cloud architectures can improve money deposit institutions' performance. Banks can adopt digital leadership in their methods of working and innovation across the banking industry instead of limiting themselves to using the cloud only for their current processes. Banks have realised that if they don't keep up with how quickly digital transformation is changing the banking industry, they may soon cease to exist as a corporation. This circumstance begs for more efficient banking methods, one of which is the digital-based E-banking approach as a boom in the current banking system. [2].

The banking industry is being digitally transformed, which reduces wait times for bank services, maintains one branch bank increases consumer satisfaction with benefits, and ensures a cashless society. Therefore, the study's goals are to ascertain the effect of digital leadership on achieving organizational excellence in the banking sector of

Jordan, to ascertain the connection between digital leadership and customer satisfaction in the banking sector of Jordan, and to ascertain the degree to which digital leadership has enhanced the performance of Jordanian banks.

2 Problem of the Study

Customers in Jordan now expect more than just increased returns on their investments and the safety of their money thanks to the banking industry's digitization; they also want effective, quick, and pleasant services. Customers seek out banks that can provide them with services that exceed their expectations and enhance their enterprises. For security reasons, businesspeople can prefer to conduct cashless transactions instead of carrying cash. They want to check their balance, send and receive money, make payments online and download transaction history to their computer and mobile device. This study focused on how digital leadership contributes to organizational excellence in Jordan's banking industry.

3 Review of Past Studies

3.1 Conceptual Framework

A contemporary management approach that strongly emphasizes launching and fostering a company's digital transition is known as "digital leadership." Successful "digital leaders" should be capable of making processes more current, agile, and effective (keyword "new work"). They should also be able to use digital tools wisely and successfully lead the organization into the digital era. According to [3], the digitization process is sequential and modelled after the tasks of archiving, access, and management. He also listed some digitization processes, including the ones listed below:

The process of initiating and beginning the project entails the preliminary preparations before the digitalization process, such as purchasing the devices to be utilized, training, and employing the required people to create digital material.

Selection of documents and activities for digitalization: The digitalization process can begin after sorting and choosing some hard copy papers. Timetables for accomplishing digital transformation tasks are also established. If manual processes were never used, there would be no reason to change the documents' physical representations.

During the conversion process, we use our state-of-the-art equipment to transform our paper records into a digital format that can be read by computers. Editing, Access, and Maintenance: This includes looking over the already-digitized documents for problems, making the access points easier to use, and making sure there are good procedures and plans for routine maintenance. When digitalization operations are put into place, it is important to keep up with maintenance to avoid a total system meltdown or a sudden loss of data. There's no denying that commercial banks' productivity and customer satisfaction have increased since the services they supply have gone digital. Some of the factors contributing to this outcome are: Commercial banks' adoption of digital technology to improve service delivery has dramatically helped reduce poverty [4]. Bank accounts can now be opened without the customer ever having to set foot inside a branch, thanks to the advancements in mobile communication made possible by the digitization of the financial system [5]. Customers are encouraged to open bank accounts because they know they can handle all of their financial transactions with just a few clicks of the mouse or by using their mobile devices, eliminating the need to visit a bank or other financial institution physically.

According to [6], "as a result of the digital revolution in financial institutions, roughly 30% of the time was saved during financial deposits, and approximately 70% of the time was saved during withdrawals." Digitalization has made it easier to use commercial banks, which has improved savings rates and, in the long run, lowered the degree of poverty [7]. Abbasi and Weigand's [8] observation of a massive increase (700 million) in users with accounts between 2011 and 2014 further supported this claim. As a result of using digital technology, commercial banks may now provide superior banking services. The standard of financial services offered to customers is rising due to the widespread adoption of information and communication technology. A database streamlines collecting, storing, and retrieving data about customers and their transactions. There will be less paperwork for commercial banks to handle as a result. New customer account opening, new customer account screening, and loan processing have all become extremely easy, effective, and efficient thanks to the digital revolution [7]. Digitalization has brought about absolute ease and convenience in conducting financial transactions. The digitalization process tends to improve the performance of commercial banks. They utilise ICT tools and gadgets to enhance their comfort and convenience without compromising service standards [6]. Computer systems are used to verify the identity of the payer and the payee (name, passport photo, etc.) when a transaction is processed, for instance. Over time, this boosts productivity, reduces mistakes, and enhances performance. In addition, digitalization has allowed for the development of novel services such as online banking, online shopping, ATM services, internet transfers, mobile banking, etc. [9], as is apparent in [10]. Clients like being able to conduct business from the comfort of their own homes, offices, or on the go with the convenience of mobile banking and internet banking. [6] the global payment landscape is undergoing a gradual but fundamental shift due to the advent of a secure and low-cost digital platform for buying goods and services.

Customers and clients are the backbone of any commercial bank, thus it's crucial that they feel well informed about the products and services they're investing in. Due to digitization, commercial banks in Nigeria can engage with their customers more efficiently, which will ultimately lead to higher sales and revenue. [11] According to studies, "the institutions which followed an effective digitization strategy were able to raise their sales and revenue by 4% to 16% through the instruments of enhanced client relationship." Now, banks may be able to communicate with their clients in ways that are more effective, efficient, and cost-saving. Some common forms of modern-day communication include texting, emailing, calling, and talking on social media.

The digitalization of commercial banks has also improved record keeping. Computerized systems provide for better organization, storage, and retrieval of records. In contrast to when bank operations weren't digitized, documents that are as old as feasible can be quickly obtained with a mouse click. But it's crucial to remember that there are still certain negative effects of digitization matter how well it affects the performance of commercial banks. The rise in cyberfrauds and cybercrimes committed in commercial banks is a common problem. Some tech-savvy people occasionally commit theft and fraud using mobile devices and online platforms. Some retail bank employees perpetrate fraud via computer manipulation that may go undetected because they have access to sensitive bank data.

The detrimental impact of data protection and data loss on commercial banks' performance is another repercussion of the digitalization process. The majority of collected data on transactions, customers' records, loan records, and so forth are kept on computers or other digital storage devices, which are vulnerable to viruses or other destructive attacks and may eventually result in data loss if a solution is not found immediately to fix it [7]. This is an undesirable effect that could affect the efficiency of commercial banks. On the other hand, [2] explain what digital leadership is and what it means. The effects of the skills and experiences of digital leaders on the digitalization process will be looked at. Several multinational and global companies have been looked at to see how the idea of "digital leadership" could be used in their businesses. In this paper, the idea of digital leadership is looked at as part of an organizational literature review. After reading a lot of literature, businesses are changing their organizational structures, processes, business models, and strategies because of digitalization and technological advances. Organizations need digital leadership to adapt and change their business strategies and stay alive in the new digital age. Digital leaders use and improve the company's digital assets to meet business goals and successfully drive digital business transformation. Traditional leaders and digital leaders have different skills and points of view. But many organizations don't understand the value of digital leaders, which can lead to poor performance and failures that can't be fixed. This research paper shows that there is a strong link between digital leadership and corporate success and productivity. It does this by analyzing and clarifying the modern concept of digital leadership and by looking at how different business strategies and leadership concepts are used in different sectors.Also, [12] used secondary source data to investigate the connection between electronic banking and bank performance in Nigeria. Researchers confirmed the nature and strength of the connection between explanatory and dependent variables using regression analysis. A surrogate for the health of the Nigerian banking sector, Total Bank Deposit was nonetheless used. Concurrently, transaction IDs generated by ATMs, MoBs, POS terminals, and Web Pay were utilized to stand in for electronic banking. The study found that the correlation results of electronic channel products are significantly and favorably correlated with bank performance (ATM, POS, Web pay, and Mobile Pay). It was also clear from the regression results that all of the factors were highly correlated.

Using Nigeria's commercial banks as a case in point, [1] looked at how e-banking affected the efficiency with which the banks were able to carry out their daily tasks. It has been noted that the operational effectiveness of banks in Nigeria has increased after the implementation of e-banking compared to conventional or traditional banking. However, 90 respondents from Nigerian commercial banks' clients and personnel make up the study sample. They conducted their investigation using a descriptive data analysis technique. Customer loyalty, bank strength, capital, and income base have increased. It was determined that introducing new channels into banking operations significantly improved performance since the banks are more profitable when their consumers actively engage in e-transactions. In a similar vein, [13] studied how digital leadership can help the University of Tikrit achieve its goal of being an exemplary university. Responses to the research variables were described and analyzed using the descriptive analytical method. The study population consists of the faculty and staff at Tikrit University. Where 82 electronic administrators from the targeted university were chosen at random. The most significant findings are the respondents' consensus on the importance of digital leadership and the aspects of organizational excellence. The findings also supported the study's central hypothesis: that digital leadership has a major bearing on realizing the ideal of institutional excellence. And businesses who are eager to embrace digital leadership tend to excel. Moreover, [14] analyzed the performance of 10 commercial banks between 2002 and 2012 by inspecting their return on assets, return on equity, and interest margin. For those that adopted e-banking later, the favorable effects on margin, return on assets, and return on equity were substantial, but for those who did so earlier, the effects were much more muted. They come to the conclusion that e-banking can be a competitive strategy for both domestic and international institutions if suitable risk management measures are taken.

4 Research Hypothesis

H01: Digital leadership has no significant impact on achieving organizational excellence in Jordan's banking sector.

H02: There is no significant relationship between digital leadership and customer satisfaction in Jordan's banking sector.

5 Methodology

The study adopted a survey research design. The population of the study was all employees and customers of banks. Jordanian Banking Sector was purposively selected amongst other sectors in Jordan. Due to the scarcity of separate existent data on bank employees and customers in JordanRather than using a finite sample size, the researchers here assumed an infinite population. Cochran's formula was used to determine the size of the sample given that the size of the population is undefined (greater than 50,000). The study population was sampled using an approach similar to that proposed by Cochran (1977) for estimating sample size when the population is infinite. Cochran's formula from 1977 is as follows:

n = (z^2 pq)/e^2 where

n = Sample Size

z = Selected critical value of desired confidence level

p = is the estimated proportion of an attribute that is present in the population

q = 1 p e = Margin of error (10%)

Since the sample size would be calculated from a large population whose degree of variability is not known, the study assumes maximum variability, which is equal to 50% (p = 0.5) and taking a 90% confidence level with 10% precision, the calculation for the required sample size was as follows:

$$n = ([(0.256)])^2(0.5)(0.5)/(([0.10))]^2) = 163$$

Therefore, the sample size for this study is 163 employees and customers of Banks in Jordan. The research instrument is a structured questionnaire. The study employed descriptive (frequency counts, percentages and pie-chart) and inferential statistics (Structural Equation Modeling and Pearson Product Moment Correlation Coefficient) for data analysis.

5.1 Model Specification

The Structural Equation Modeling for the study is defined as follows:

$$PMSCE = \beta\_0 + \beta\_1\,ICS + \beta\_2\,DDDI + \beta\_3\,EFDG + \beta\_4\,IOE + \beta\_5\,EEEE \\ + \beta\_6\,LHE + \beta\_7\,IA + \beta\_8\,ECIC + \beta\_9\,EHQUE + \beta\_10\,ESM + e$$

Where,

Y = Achieving organizational excellence (AOE)-dependent variable.
x_1 = Drives Data-Based Insights (DDDI)
x_2 = Enables Future Digital Growth (EFDG)
x_3 = Increases Operational Efficiency (IOE)
x_4 = Encourages an Environment of Employee Excellence (EEEE)
x_5 = Limits Human Error (LHE)
x_6 = Increases Agility (IA).
x_7 = Encourages Collaboration & Improves Communication (ECIC).
x_8 = Enables High-Quality User Experience (EHQUE)
x_9 = Enables Software Monetization (ESM)
x_10 = Increases Customer Satisfaction (ICS)
e = Error term
x_1-x_10 = are independent variables

5.2 Data Analysis, Results and Discussion

5.2.1 Hypotheses Testing

H01: Digital leadership has no significant impact on achieving organizational excellence in Jordan's banking sector.

Independent Variables: DDDI, EFDG, IOE, EEEE, LHE, ECIC, EHQUE, ESM and ICS

Dependent Variable: (AOE)

The banking sector in Jordan has experienced a substantial impact from digital leadership in achieving organizational excellence, as shown in Table 1. This is because the ANOVA's P-value (0.000) is lower than its alpha value ($\alpha = 0.05$). Therefore, at a significance level of 0.05, the null hypothesis that digital leadership has no meaningful effect on achieving organizational excellence in Jordan's banking sector is rejected. As a result, establishing organizational excellence in Jordan's banking sector depends significantly on digital leadership. The model can be used to explain how digital leadership affects achieving organizational excellence in Jordan's banking sector, according to the R2 value of 77%. This result is consistent with, [14], study that examined the effect of e-banking on the efficiency of Pakistani commercial banks. Using a sample of data from 10 banks covering

Table 1. Summary of Multiple Linear Regression Analysis on the Impact of Digital Leadership on Achieving Organizational Excellence in Jordan's Banking Sector

Variable		Path Coefficient	t-value	P-value	Description
Constant	Achieving Organizational Excellence	0.457	3.067	0.001	Significant
Drives Data-Based Insights (DDDI)	Achieving Organizational Excellence	1.151	8.719	0.000	Significant
Enables Future Digital Growth (EFDG)	Achieving Organizational Excellence	0.372	3.321	0.000	Significant
Increases Operational Efficiency (IOE)	Achieving Organizational Excellence	0.236	9.440	0.000	Significant
Encourages an Environment of Employee Excellence (EEEE)	Achieving Organizational Excellence	1.146	9.317	0.000	Significant
Limits Human Error (LHE)	Achieving Organizational Excellence	0.457	3.288	0.001	Significant
Increases Agility (IA)	Achieving Organizational Excellence	0.478	2.915	0.018	Significant
Encourages Collaboration & Improves Communication (ECIC)	Achieving Organizational Excellence	0.723	4.757	0.000	Significant
Enables High-Quality User Experience (EHQUE)	Achieving Organizational Excellence	0.431	2.817	0.017	Significant
Enables Software Monetization (ESM)	Achieving Organizational Excellence	0.91[illegible]	6.239	0.000	Significant
Increases Customer Satisfaction (ICS)	Achieving Organizational Excellence	0.725	4.801	0.000	Significant
F-value (ANOVA)		72.981		0.000	Significant
$R^2 = 0.765$					

the years 2002 to 2012, Return on assets, return on equity, and interest margin were three metrics that were used to evaluate commercial banks. The study's authors found that e-banking had a significant impact on the profitability ratios of late adopters but had a negligible impact on those of early adopters. They conclude that, with proper oversight and controls in place, e-banking can provide financial institutions with a competitive advantage at low cost, both in domestic and international markets.

Drives Data-Based Insights (DDDI)
According to the findings, Drives Data-Based Insights (DDDI) has a positive coefficient that strongly correlates with achieving organisational excellence at the 5% level (P). The positive coefficient of DDD shows that banks are now better equipped to gather and analyse data that can be turned into insights that can be used to generate money. In the past, or even worse, bank data was dispersed over numerous disconnected platforms. Strong data collecting, centralised data storage, and the development of technologies to evaluate and turn data into information that supports informed business decision-making are all made possible by digital transformation.

Enables Future Digital Growth (EFDG)
It was discovered that EFDG (Enables Future Digital Growth) is beneficial and substantially impacts establishing organisational excellence at the 5% level. The fact that it creates the foundation for all subsequent business success may cause this positive coefficient value. Banks will swiftly go out of business if they do not invest in digital business transformation. Banks have failed during the last few years. Why? Because their digital transition failed. Digital transformation must occur. It's not the cherry on top of a successful business. Banks need to change to survive the digital tsunami sweeping the globe. Once banks take advantage of digital transformation, they will grow stronger and more successful.

Increases Operational Efficiency (IOE)
Achieving organisational excellence at 5% is substantially correlated with Increasing Operational Efficiency (IOE), as predicted, and has a positive correlation. This positive coefficient shows that banks' operational efficiency has grown as a result of the digital transformation. By integrating with back-office systems, automating delivery procedures and product updates, enabling inter-device activation, handling numerous product versions, and other methods, banks can streamline cumbersome human operations and lower overhead costs. Banks will reduce annoyance, save time, and close income gaps using digital business processes since effective business practices are connected with digital interaction.

Encourages an Environment of Employee Excellence (EEEE)
Achieving organisational excellence was positively correlated with Encouraging an Environment of Employee Excellence (EEEE), which was significant at the 5% level. The positive coefficient suggests that organisations broadly embrace digital transformation. Banks need to foster a company culture where going digital is seen as a strategic priority and viewed as a means to an end rather than a means in itself. The best way for banks to win over their staff is to fund their digital literacy training. Host in-house webinars, send employees to conferences, and sign them up for classes to refresh their knowledge

and skills. Financial institutions shouldn't discount the importance of investing in their best asset—their people.

Limits Human Error (LHE)
As anticipated, the Limits Human Error (LHE) coefficient was positive and significant at the 5% level, indicating that digital leadership enhances bank performance. The positive coefficient of LHE shows that digital processes prevent errors by doing away with labour-intensive manual data entry and inefficient human behaviour. Digital operations are more efficient and risk-free than any procedure requiring fallible humans.

Increases Agility (IA)
As anticipated, the Increases Agility (IA) coefficient was positive and significant at the 5% level. This indicates that banks understand that adaptability and flexibility are essential for surviving the ongoing phase of rapid digital transformation. Customers' wants are vastly different than they were a few years ago, and products are changing along with previously stable revenue streams.

Encourages Collaboration and Improves Communication (ECIC)
Encourage Collaboration & Improves Communication (ECIC) has a favourable coefficient and is substantially associated with achieving organisational excellence at 5%. This positive coefficient suggests that banks need a digital platform that encourages communication and collaboration across all internal divisions to stimulate innovation and output. Digital transformation eliminates redundant processes, slow responses, dropped connections, and a lack of collaboration. When banks digitalize their internal communications, they increase productivity, accountability, and innovation—giving them a leg up on the competition.

Enables High-Quality User Experience (EHQUE)
Enables High-Quality User Experience (EHQUE) is significant at 5% and has a positive coefficient. Because of this positive coefficient, high-value banks are aware that they are providing utility in addition to services. Beautiful, simple experiences made possible by digital transformation depend on automation, artificial intelligence, and self-service tools. Give clients a smooth user experience to demonstrate to customers that banks have mastered the digital age.

Enables Software Monetization (ESM)
Having a positive coefficient and a 5% level of significance, ESM enables software monetisation. This positive coefficient means that all bank managers must shift from a product-centric approach to a software-centric system and that software monetization gives businesses the best ROI.

Increases Customer Satisfaction (ICS)
Coefficient of positive value, 5% significance level; increases increased Customer Satisfaction (ICS). High levels of this characteristic suggest that consumers have lofty expectations. Most consumers are looking for personalised services that meet their unique needs. Customers would rather have things happen automatically while accessing, upgrading, or deactivating and reactivating software rather than having to interact with people. One of the most glaring advantages of digitization for businesses is the availability of powerful digital tools and tactics, such as automated business processes and location- or

industry-specific licence modifications. Customer happiness has increased as a direct outcome of these digitized procedures (Table 2).

Table 2. Summary of Path Coefficient F^2

Variables	Path Coefficient	F^2
Drives Data-Based Insights (DDDI)	1.151	0.071
Enables Future Digital Growth (EFDG)	0.372	0.081
Increases Operational Efficiency (IOE)	0.236	0.351
Encourages an Environment of Employee Excellence (EEEE)	1.146	0.782
Limits Human Error (LHE)	0.457	0.541
Increases Agility (IA)	0.478	0.671
Encourages Collaboration & Improves Communication (ECIC)	0.723	0.861
Enables High-Quality User Experience (EHQUE)	0.431	0.379
Enables Software Monetization (ESM)	0.911	0.752
Increases Customer Satisfaction (ICS)	0.725	0.656

From the table above, the F2 effect size for the predictive relevance of exogenous variable on endogenous can be summarized as follows: The predictive relevance or impact of Drives Data-Based Insights (DDDI) on achieving organizational excellence is small given the F2 at (0.071). The impact of Enables Future Digital Growth (EFDG) as predictive relevance on to achieving managerial excellence is small given the F2 at (0.081), Increases Operational Efficiency (IOE) impact on achieving organizational excellence is small F2 at (0.351). Likewise, the predicting relevance of Encourages an Environment of Employee Excellence (EEEE) to achieving organizational excellence is given at (0.782), which is interpreted as a medium. Limits Human Error (LHE)'s predictive relevance or impact on achieving organizational excellence is small given the F2 at (0.541). The impact of Increases Agility (IA) as predictive relevance on achieving organizational excellence is small given the F2 at (0.671), Encourages Collaboration & Improves Communication (ECIC) impact on achieving organizational excellence is small F2 at (0.861). Likewise, the predicting relevance of Enables High-Quality User Experience (EHQUE) to achieving organizational excellence is given at (0.379), which is interpreted as a medium. Enables Software Monetization (ESM)'s predictive relevance or impact on achieving organizational excellence is small given the F2 at (0.752). Finally, the predictive relevance of Increases Customer Satisfaction (ICS) to achieving organizational excellence is large, given that F2 is (0.656). Therefore, we can conclude that the relevant predictors of exogenous variables on endogenous ones can be interpreted as H1 = small, H2 = small, H3 = low, H4 = medium, H5 = large, H6 = large, H7 = large, H8 = large, H9 = large and H10 = large (Cohen, 1992).

H02: There is no significant relationship between digital leadership and customer satisfaction in Jordan's banking sector.

Table 3. Summary of Pearson Product Moment Coefficient Analysis on the Relationship between Digital Leadership on Customers' Satisfaction in Jordan's Banking Sector

Examination	n	df	Mean ($\overline{x}$)	SD(S)	r-value	P-value	Remark
Digital Leadership	163	161	0.62	0.17	0.781*	0.000	Significant
Customers' Satisfaction			0.53	0.12			

Source: Field survey, 2023

Table 3 demonstrates a statistically significant (P.01) correlation between digital leadership and satisfied banking clients in Jordan. For the correlation index (0.781), the degree of freedom (161), the mean scores (0.62, 0.53), and the standard deviations (0.17) and (0.12), the probability value (P = 0.000) is less than the alpha (= 0.05) level of significance. As a result, we reject the no-correlation-between-digital-leadership-and-satisfied-customers-in-Jordan-banking-sector-customers null hypothesis. [12] they used secondary data to back up their claim that using electronic banking is correlated with improved bank performance in Nigeria. Scientists employed regression analysis to verify the nature and strength of the link between independent and dependent variables. To estimate the expansion of electronic banking in Nigeria, we used dollar volumes transacted through automated teller machines, mobile banking, point-of-sale terminals, and online payment systems as proxies for total bank deposits. Bank performance was found to be positively and statistically significantly correlated with the use of electronic channel products (ATM, POS, Web pay, and Mobile Pay). All of the predictors in the regression analysis had a significant degree of connection with one another.

Fig. 1. The extent to which Digital Leadership Has Improved the Performance of Jordan's Banking Sector.

Figure 1 demonstrates how much digital leadership has improved the performance of the banking sector in Jordan by increasing customer satisfaction, driving data-based insights, enabling future digital growth, increasing operational efficiency, encouraging an environment of employee excellence, limiting human error, increasing agility, encouraging collaboration & improving communication, enabling high-quality user experience, and enabling software monetization. This result is consistent with a study by [1], which examined the effects of online banking on the productivity of commercial banks in Nigeria as a case study. It has been noted that the operational effectiveness of banks in Nigeria has increased after the implementation of e-banking compared to conventional or traditional banking. However, 90 respondents from Nigerian commercial banks' clients and personnel make up the study sample. They conducted their investigation using a descriptive data analysis technique. Customer loyalty, bank strength, capital, and income base have increased. It was determined that introducing new channels into banking operations significantly improved performance since the banks are more profitable when their consumers actively engage in e-transactions.

5.2.2 Conclusions and Recommendation

As a result of this investigation, it was determined that digital leadership significantly impacts achieving organizational excellence in Jordan's banking sector. There was a significant relationship between digital leadership and customer satisfaction in Jordan's banking sector ($P < \alpha$). Increases customer satisfaction, drives data-based insights, enables future digital growth, increases operational efficiency, encourages an environment of employee excellence, limits human error, increases agility, encourages collaboration & improves communication, allows high-quality user experience and enables software monetization. The researchers developed a set of recommendations according to the results that have been reached. That the banks in question should provide the requirements of digital transformation toward digital leadership, and the banks should Work on setting clear standards for institutional excellence by spreading the culture of institutional excellence. In addition, banks should Hold courses and training workshops to familiarize employees with the technologies, applications, networks and databases of digital transformation, which can lead to customer satisfaction.

References

1. Taiwo, J.N., Agwu, M.E.: The role of e-banking on the operational efficiency of banks in Nigeria. B. Res. J. Bus. Manage. Acc. **6**(1), 1 (2017)
2. De Araujo, L.M., Priadana, S., Paramarta, V., Sunarsi, D.: Digital leadership in business organizations. Int. J. Educ. Adm. Manage. Leadersh. **2**, 45–56 (2021)
3. Satyendra,S.: Digitalization Process (2016). ispatguru.com/digitalization-process
4. Mukherjee, A.: reducing poverty in India with the power of digital payments and UBI. Center for Global Development (2017)
5. Ortstad, R., Sonono, B.: The Effects of the Digital Transformation Process on Banks Relationship with Customers: Case Study of a Large Swedish Bank (2017)
6. Harigaya, T.: Effects of Digitalization on Financial Behaviours: Experimental Evidence from the Philippines, scholar Harvard (2016)

7. González-Páramo, J.M.: Financial innovation in the digital age: Challenges forregulation and competition. Revista de Estabilidad Financiera (2017)
8. Abbasi, T., Weigand, H.: The Impact of Digital Financial Services on Firm's Performance: A Literature Review, Earth and Environmental Science (2019)
9. Forest, H., Rose, D.: Delighting Customers and Democratising Finance: Digitalisation and the Future of Commercial Banking. Global Transaction Banking: Deutsche Bank (2015)
10. Akinbode, M., Adegbuyim, O., Kehinde, O., Agboola, M., Olokundun, M.: Perceived value dimensions online shopping intention: the role of trust and culture. Acad. Strateg. Manage. J. **18**(1), 1–20 (2019)
11. Carletti, M .: Whitepaper: fintech and the Digitalization of Financial Services. Management and Advisory services Mas Ltd (2017)
12. Enoruwa, K.O., Ezeuen, D.M., Nwani, C.O.: Electronic Channels and bank performance: empirical evidence from Nigeria. SSRG Int. J. Econ. Manage. Stud. **6**, 37–46 (2019)
13. Khalil, S.I., Farhan, O.M., Hamad, H.A.: the role of digital leadership in achieving organizational excellence applied research at the university of Tikrit. World Econ. Fin. Bull. **12**, 85–94 (2022)
14. Rauf, S., Qiang, F.: The integrated model to measure the impact of e-banking on commercial bank profitability: evidence from Pakistan. Asian J. Res. Bank. Finan. **4**(1), 25–45 (2014)

The Role of Digital Media in the Spread of Violence and Crime in the Light of a Number of Crime Sociology Theories

Khawlah M. AL-Tkhayneh(✉)

Al Ain University, Abu Dhabi, UAE
Khawlah.altkhayneh@aau.ac.ae

Abstract. This study aimed at revealing the role of digital media in spreading violence and crime according to a number of crime sociology theories. In order to achieve the study objectives, the researcher used the analytical descriptive approach by reviewing the theoretical literature relevant to the study topic, comparing the results of the previous studies with the results of the current study as well as introducing a logical and scientific analysis for the role of digital media in spreading violence and crime in the light of the theories of crime sociology.

The results showed that there is a consensus among researchers concerning the dangers imposed by digital media in the spread of violence and crime in the light of criminology theories, including imitation and simulation theory, social learning theory, and catharsis theory. Also, there is an agreement about the role played by digital media in reducing crime rates.

In the light of the results, the study recommended the necessity of urging governments to impose constraints on content creators with regard to the content published and shared through their pages in order to verify the safety and security of the content they provide, considering content creators as influential individuals whose main objective is to acquire more views and gain more profitability regardless the type and safety of the content.

Keywords: digital media · violence · criminology theories · social communication platforms · electronic crime

1 Introduction

The platforms of digital media refer to all the types of social media sites, including Whats App, Facebook, and other applications that allow users to communicate with each other using the features of sound, video and Metaverse applications. Currently, these platforms represent important developmental outcomes, where they consume most of people's time. Indeed, these platforms allow individuals to share all the content they have and communicate with others via virtual worlds. However, those developments have negative consequences, since they paved the way for more melancholic domains within the virtual world of crimes that attracted young and adult people and enabled them to practice virtually-illegal behaviors against each other. Those developments had other effects

B. A. M. Alareeni and I. Elgedawy (Eds.): *Artificial Intelligence (AI) and Finance*, SSDC 488, pp. 62–71, 2023.
https://doi.org/10.1007/978-3-031-39158-3_6

other than the spread of the criminal behavior, including the social and psychological consequences caused by the various means of digital media on the community members. Indeed, people who frequently use such technological developments are usually afraid of becoming victims of electronic risks, including electronic blackmail, cyber bullying, and social engineering, to mention but a few.

In fact, the accusation directed towards this development is the result of the cyber storm led by the platforms of digital media that incubated the spread and evolution of such criminal behaviors. For example, the authorities of electronic security in the United Arab Emirates reported that there had been (512) cases in 2019 caused via the digital media platforms. These crimes increased in size as compared to the previous year, where only (357) cases were registered in 2018 [1].

Amro Bayoumi, the spokesman of the newspaper of "Emirates Today", reported that, according to the collected data, one child is exposed to cyber bullying every day, since about (70%) of students in the United Arab Emirates use social media sites more than (5) hours each day [1].

Therefore, the content of this study is mainly directed towards analyzing the role of digital media in enhancing violence and crimes committed via the platforms of digital media as well as addressing the role played by those platforms in increasing the rate of these practices between individuals in the light of the evidence that is based on criminology theories and the results of previous studies. Therefore, the current study aims to introduce a clear image concerning this digital development and guide the process of using digital media in a beneficial way.

2 The Study Problem

The forms of violence and crime have developed and increased in a rapid and terrifying way due to the spread of new technology that is characterized by being cross-border and rapidly spreadable. Therefore, many countries, with their various security agencies, became unable to tackle the different types of crime and violence.

In fact, these crimes are viewed as a serious threat to social security, community members and the state's organizations, in addition to their negative effect at the personal, social and economic levels. This situation provokes more concerns about security issues, and prompts us to consider the importance of this topic and deal with it before being exacerbated. Indeed, this issue should be considered seriously and addressed through international and regional cooperation in order to combat electronic violence and crime based on the conventions, laws, regulations and centers concerned with educating people about cybercrimes and the role of digital media in spreading violence through its various platforms.

Hence, the study problem addressed the emergence of this new type of dangerous crimes that spread through the platforms of digital media. Therefore, new methods and ideas should be innovated to combat this type of crime and enhance people's security concerning its economic, social, political, personal and environmental dimensions. Also, we should compare between these platforms, where some of them contributed to the spread of violence while others significantly contributed to educating people about the risks of these platforms [2].

3 The Theoretical and Practical Importance of the Study

3.1 The Theoretical Importance

The study's theoretical importance lies in introducing a logical and scientific analysis from the most well-known criminology theories which demonstrated the role of the various platforms of digital media in the spread of violence. Also, the other studies in this domain, have not introduced an analysis relevant to these theories in particular. Accordingly, this study aims to help researchers investigate the causes that stimulate people to practice these illegal activities and demonstrate the reasons of conducting this type of study at the theoretical and practical levels.

3.2 The Practical Importance

The practical importance of this study lies in introducing a scientific benefit, represented by informing researchers about the most important outcomes of this study as well as identifying the extent to which the concluded results in this domain can be applied in reality which will, in turn, enable the agencies responsible for electronic security and criminology to identify the most common crimes and recognize how digital media contributes to violence based on the most important theories explaining them.

4 The Study Objectives

- Identifying the role of digital media in spreading violence and crime.
- Analyzing the role of digital media in speeding violence and crime in the light of theories of criminology.
- Analyzing the role of digital media in spreading violence and crime in the light of previous studies.

5 The Study Questions

- What is the role of digital media in spreading violence and crime?
- How do the theories of criminology interpret the role of digital media in spreading violence and crime?
- What is the role of digital media in spreading violence and crime in the light of previous studies?

6 The Study Methodology

The study used the analytical descriptive approach by reviewing the theoretical literature relevant to the study topic, comparing the results of previous studies with the results of this study, and introducing a scientific and logical analysis for the role of digital media in spreading violence and crime in the light of criminology theories, such as imitation and simulation theory, social learning theory, catharsis theory.

7 The Theoretical Literature and Previous Studies

7.1 The Concepts and Terms Relating to the Study

- Digital media: it represents all the new methods based on which the communication media process works. These methods depend on the World Wide Web and the development in techniques that are constantly changing. These means also include the internet websites, videos and sounds transmitted via the internet, electronic mails as well as social media sites and other new means that emerged through the technological development, referred to as interactive media, or the direct media that depends on technological multimedia [3].
- Crime: Garovalo defined crime as an antisocial act committed in the various communities at various times, where this harmful act contradicts with the feelings of empathy, honesty and justice that are often common, over time, among the human communities [4].
- In legal terms, (Al-Sa'ied, 2006) [5] defined crime as performing an illegal behavior, that violates others' basic interests, based on criminal intention, where the law decides on a punishment or preventive measures for it.
- Violence (conceptually): it is a behavior that is characterized by aggressiveness, and is committed by an individual or a group of people with the aim of exploiting or repressing another individual or a group of people in the light of unequal power relationships [6].
- Electronic crime: it refers to any criminal act that is performed by using technological methods, where the perpetrator of the electronic crime must have criminal intent. It is an illegitimate and unethical behavior related to information networks [7].

7.2 The Role of Digital Media in Spreading Violence and Crime in the Light of a Number of Previous Studies

Al-Qahtani and Ali [8] conducted a study which aimed to identify the reality of using social media sites in the promotion of drugs and the way through which these platforms are exploited for drug promotion, and find out methods to reduce using social media sites for such purposes. The results revealed that the sample individuals agree with a (very high) degree that drug promotion via social media sites increases with the advancement of modern techniques, and that it mainly targets the young people via (snap chat) programs, in parallel with the increasing rate of crimes in the community Al-Qahtani and Ali [8].

Al-Barashdiya [9] conducted a study which revealed that (70%) of the participants link between committing electronic crimes in Amman and using personal information by Facebook users. The results also revealed that the sample individuals were concerned about their privacy while using Facebook; they were afraid of becoming victims of electronic crimes. However, some of them know how to change their privacy settings. Therefore, there is a need to understand the real cause for committing such crimes in order to be able to predict the criminal patterns in the future [9].

Al-Sabai'e and Al-Sha'er [10] conducted a study which revealed that the sample individuals agreed about the most prominent cybercrimes in the kingdom of Saudi Arabia with a mean of (3.86), and strongly agreed about the role of digital media in reducing cybercrimes with a mean of (4.33). The sample individuals also strongly agreed about the most prominent barriers that weaken the role of digital media in reducing cybercrimes with a mean of (4.39), and the most important methods that enhance the role of digital media in reducing cybercrimes with a mean of (4.39). The results revealed that there are statistically significant differences in their responses concerning all the study domains according to the variable of age in favor of (25–less than 30) years old, while there are no statistically significant differences in the responses of the study sample individuals concerning all the study domains according to the variable of (hours of using social media sites) daily [10].

Soomre and Hussain [11] conducted a study entitled "Social Media-Related Cybercrimes and Techniques for Their Prevention". The results revealed that social media sites significantly contributed to facilitating the process of committing crimes, including robbery and burglary, where individuals show their personal files on these programs and that facilitates the criminal's mission in knowing their personal data and going after them. Also, the study included the fraud processes to which individuals are exposed via social media sites, particularly Facebook, which is considered as the most exposed site to cybercrimes, since hackers can easily violate it, reincarnate identities of others, and defraud the friends of the robbed identity. According to the collected data, the rate of phishing via the electronic mail reached (2500) cases in September, and about (5000) cases in June [11].

Siddik and Rahi [12] conducted a study entitled "Cybercrime in Social Media and Analysis of Existing Legal Framework". The results revealed that the most commonly used application in Bangladesh is Facebook, with a percentage of (97.23%), followed by YouTube (1.41%). The results revealed that Bangladesh witnessed an increase in the rate of cybercrimes via social media sites, where the police centers received about (2044) cybercrimes during the last six years, and the numbers are increasing constantly. The results revealed that girls were exposed to electronic blackmail and harassment by the males whom they recognized via social media sites, where those males showed fake names. Indeed, these crimes weren't registered within the officially acknowledged cybercrimes–harassment crimes are the most commonly committed crimes on Facebook with a percentage of (42.1%) [12].

Jalal and Mamdouh [13] conducted a study which revealed that the crimes of stealing personal belongings were in the first place with a percentage of (70.6%), followed by the crimes of child abduction, and finally the crimes of unintended murder with car accidents. The results revealed that there is a statistically significant negative correlation relationship between the fear from being abused by an unknown person and all the indicators of life quality and the total degree [13].

Al-Khateeb [14] conducted a study entitled "the information crimes across social media sites: a review of the Egyptian information technology crimes combat act No. 175". The results revealed that social media sites are considered as easy modern electronic methods for social communication between individuals, and that the crimes of exploiting technical drawbacks are committed via the systems hackers in social media sites. The results revealed that social media platforms are viewed as an appropriate setting for data collection, since they include important information about users which, in turn, assists police officers in collecting evidence about crimes [14].

Hijazi and Sabrina [15] conducted a study which revealed that using social media sites exposes female users to the various types of electronic crimes, since these sites provide users with more space to interact within virtual groups which, in turn, increases the prevalence of electronic crimes [15].

7.3 Comments on the Previous Studies

Based on the review of the previous studies, we noticed that most studies emphasized that there is an effect of digital media platforms on users, where most effects were negative. We also noticed that the majority of studies are relatively new due to the new nature of the study "modern media platforms" and their methods, which are developing continuously.

This study agreed with the following studies:
(Al-Qahtani [8], Al-Barashdiya [9], Al-Khateeb [14], Hijazi and Jamal [15], Siddik and Rahi [12], Soomro and Hussain [1[illegible]]), where they suggested that the applications of social communication enhance committing the various types of crimes which have significant adverse effects on victims. They also suggested that Facebook is the most vulnerable platform, in terms of the low level of information security, and the lack of awareness in communities concerning the negative effect of these applications. The study conducted in Bangladesh revealed that the Middle-age category is the most targeted one by cyber bullies. The study entitled "the exposure of female users of social media sites to electronic crimes" addressed the most exposed gender to cybercrimes. The results revealed that there are differences in favor of females, since they are more exposed to the different types of cybercrimes.

Other studies, including (Al-Sabai'e and Al-Sha'er [10]) advocated the ability of digital media platforms to educate individuals concerning the risks of the spread of cybercrimes via social media sites, enhance the community awareness among individuals, and set plans and strategies to reduce these risks. However, the exerted efforts are humble as compared to the risks of these platforms which extended throughout the world.

This study made advantage from the previous studies in preparing the theoretical framework as well as identifying the most appropriate methodology for the study topic.

8 Discussion

8.1 How do the Theories of Criminology Explain the Role of Digital Media in Spreading Violence and Crime?

- General Strain Theory:

Recently, after imposing partial and full curfew on communities due to the outbreak of covid-19, violence, particularly domestic violence, has considerably increased throughout the world. Indeed, this was expected from the perspective of Psychology and Sociology, where the social and psychological stressors caused by the pandemic still affect individuals up to now.

According to the General Strain Theory (Angew [16]; Angew [17]) – which is the most acceptable theory in interpreting violence as one of crime's types in social psychology, the general stressors, such as psychological, social and economic stressors caused by corona pandemic resulted in negative feelings, including grief, depression, irritation, despair and anxiety. This situation increased the percentage of social learning to violent and criminal behaviors and decreased the level of self-control among individuals (consciousness).

Hence, the increased pace of violence and crime across the platforms of digital media and real world is attributed to general stressors. The outbreak of corona pandemic resulted in increased stressors which, in turn, resulted in an increase in the rate of violence and crime according to the general strain theory.

The factors that led to the increased rates of violent scenes throughout the various digital media platforms are mainly related to the individuals' increased usage of social media sites, particularly after imposing partial and full curfew in the light of covid-19 and the increased addiction to using the internet. In fact, the internet was the only way for entertainment due to the imposed curfew – this increased the number of hours spent using the internet, where individuals published and shared a content that includes violent scenes in order to obtain more financial income from the companies that are responsible for social media sites (e.g., Tube Mate and YouTube) and more views. Therefore, violent scenes are exploited and shared to arouse people's curiosity and get more views, and thus more financial income. In this vein, more restrictions should be applied to content creators concerning what is being published and shared via their pages.

- Imitation and simulation theory:

Gabriel Tarde was one of the pioneers of this theory [6]. Tarde suggested that the phenomenon of imitation takes place due to direct communication and interaction. Imitation is affected by habit and memory as well as the interaction between individuals realistically and virtually via the platforms of digital media according to the rules applied to all the community members. Therefore, each community has its own culture. For example, this community could be in the form of cartoon films, where the child attempts to imitates what he watches in reality. However, not all that he watches is positive, where some aspects could be negative and adversely affect his behavior – the child could acquire unhealthy or criminal behavior. Indeed, imitation becomes a habit, and the regular habits become a lifestyle to which individuals are committed.

Social imitation has several facets, including the cultural domain that encompasses language, customs and traditions or ideas. Imitation includes the positive things, such as acquiring the positive habits that promote the community and enhances life quality in it. Imitation could also be negative, such as acquiring crime and criminal activities, including robbery, murder and escaping the judgments issued by the judicial system.

In sum, as suggested by its founder (Bandura, 1977) [18], imitation theory refers to integrating with a certain behavior after observing it while being performed by others. Also, the prominent figures and models in this context, especially in media means should have an effect on the social behavior of individuals, where they have an important role in the first stage in which the child acquires the observed behaviors performed by others. These newly-acquired behaviors gradually replace the previously-learned behaviors that the child had already acquired from his family and the surrounding environment; however, the previously-acquired behaviors have a significant role in maintaining the individual's behavior.

- Catharsis theory:

The term "catharsis" was originally used by the philosopher Aristotle, and it means refining and purification. Catharsis theory stipulates that watching violence and bullying via the various media means reduces the individual's need to act aggressively. The theory suggests that the individual's feeling of depression and being exposed to injustice creates inclinations towards causing harm to others; however, an individual can satisfy this inclination by watching violence performed by others and thus some individuals' inclinations towards committing crimes decrease [19].

This theory also suggests that the violence shown in media means represents a positive social element that helps people to discharge stress and calm down violent motives. This theory suggests that through displaying violence scenes, media means contribute to reducing stress among people, and thus people are less inclined towards performing violent actions as they only watch these violent actions on TV or other media means [6].

- Stimulation theory (arousal of aggressiveness):

This theory assumes that when an individual is exposed to an aggressive stimulation, he will be aroused psychologically – this arousal increases the likelihood of doing an aggressive action by that individual. In this vein, the media material that includes scenes of violence and aggressiveness stimulates audience psychologically and emotionally which, in turn, provokes individuals to perform a similar reaction in response to what they have already watched. Therefore watching violent scenes via the various media means has the potential of motivating people to do the same observed behavior [19].

According to this theory, the behaviors shown on TV reinforce the individual's previous attitudes concerning the violent actions. In this case, TV doesn't create violence, but reinforces the individual's inclinations to behave violently. This theory suggests that the watchers don't behave violently due to watching violent content, but the issue is attributed to the social standards that an individual already has, which reinforce his need to do violent behaviors [6].

Therefore, the previous theories, Catharsis and Stimulation, rely heavily on issues related to depression, aggressiveness and injustice, where individuals tend to use the various media means to satisfy their aggressiveness needs. In this vein, an emotional stimulation could take place in a certain way and help the individual respond to the content displayed in the various means of media.

- Social learning theory:

This theory mainly explains the way through which individuals acquire a certain behavior based on social learning. Acres is considered as one of the most prominent psychologists who addressed this theory, where he introduced it in his book "Social Learning through Simulation". He assumed that an individual's behavior is an acquired one that is learnt through the good example. The theory suggested that individuals learn the behaviors of reading news and knowing events via newspapers, magazines or TV. Therefore, security media agencies should educate people and reduce electronic crimes, so that it becomes a part of the prevalent culture in the community over time – the individual learns the behaviors that are punished and those which are rewarded [20].

This theory and the theory of stimulation (arousal of aggressiveness) may interpret several individuals' imitation to the crime committed against the student "Naera" that took place in the University of Mansoura in Egypt. Indeed, several similar crimes were committed in a number of adjacent countries and in Egypt itself. According to the theory of social learning, imitation and simulation theory, as well as reinforcement theory, the spread of the new type of crime via media motivated bad people to commit similar crimes, where they learned and imitated the murderer's behavior.

Therefore, it would be better not to publish such details, images or videos for certain types of crimes, where the authorized agencies should impose restrictions on social media sites, so that delinquents won't imitate those criminal behaviors. In my opinion, it is necessary to impose more surveillance on the content before being published on social media sites.

9 Conclusion and Recommendations

Based on analyzing the theoretical literature (interpreting theories and previous studies), we can see that there is an agreement among researchers concerning the risk of digital media on the spread of crime and violence. This is evident in the light of the theories of criminology (imitation and simulation theory, social learning theory, catharsis theory, reinforcement theory). The reviewed literature revealed that there is an agreement concerning the role played by digital media in reducing crimes.

In the light of the results, the study recommended the necessity of imposing restrictions on content creators concerning what is being published and shared via their pages in order to ensure the safety of the introduced content. The study also recommended the necessity of conducting further studies that shed light on the factors and causes that make the means of digital media as an instrument either for reinforcing violence or educating people about it, in an attempt to enhance the other positive domain of media, which is reducing crime and violence.

The study also recommended the necessity of conducting further relevant procedural studies that address the differences based on the variables of gender, educational qualification, age and determining the most affected groups by digital media content.

References

1. Bayoumi, A.: 512 cases of misuse of social media sites in Abu Dhabi since the beginning of the year (2019)
2. Farghali, A., Al-Mismari, M.: Criminal Evidence with Digital Evidence from the Legal and Technical Points. Naif Arab University for Security Sciences (2007)
3. Al-Fayoumi, M.: The Arabic language in media, a study in light of the semantic development through modern theories (concept and forms) in 2020. J. Coll. Islam. Arab. Stud. Girls Damanhour **5**, 851–914 (2021)

4. Al-Wreikat, M.: The Origins of the Scientific Criminology and Punishment. Wa'el House for Publishing and Distribution, Amman, Jordan (2009)
5. Al-Sa'ied, K.: Explanation of general provisions in the Penal Code – Crimes Against Persons. House of Culture for publication and distribution, Jordan, Amman (2006)
6. Sayid, H.: Sociology of Juvenile Delinquency (a Sociological Comparison for the Reality of Delinquent Childhood). The University Book House (2021)
7. Farish, R., Qawish, N.: The impact of social media sites on the spread of cyber-crimes among adolescents. PhD Bouira University. (2018)
8. Al-Qahtani, A., Ali, M.: The Role of Social Media Sites in Drug Promotion. Naif Arab University for Security Sciences (2018)
9. Al-Barashdiya, H.: Facebook and cybercrime in Oman: is there a relationship? J. Inf. Stud. Technol. (2019)
10. Al-Sabai'e, S., Al-Sha'er, M.: The role of digital media in reducing cybercrime in the Kingdom of Saudi Arabia. Arab University for Security Sciences (2020)
11. Soomro, T.R., Hussain, M.: Social media-related cybercrimes and techniques for their prevention. Appl. Comput. Syst. **24**, 9–17 (2019)
12. Siddik, M., Rahi, S.: Cybercrime in social media and analysis of existing legal framework: Bangladesh in context. BiLD Law J. **5**, 68–92 (2020)
13. Jalal, K., Mamdouh, G.: The role of social networks in the fear of becoming a victim of crime in the light of the factors of quality of life among the Egyptian youth. Psychol. Stud. **4**, 691–759 (2019)
14. Al-Khateeb, A.: Information crimes committed through social networking sites. A Reading of the Egyptian Information Technology Crime Control Law No. 175 of 2018 and the Saudi Information Technology Crime Prevention Law 1428 AH. 6, pp. 68–94 (2019)
15. Hijazi, S., Sabrina, J.: The exposure of female users of social media sites to electronic crimes (2020)
16. Agnew, R.: Foundation for a general strain theory of crime and delinquency. Criminology **30**, 47–87 (1992)
17. Agnew, R.: Pressured into Crime: An Overview of General Strain Theory. Oxford University Press, New York, NY (2006)
18. Bandura, A.: Social Learning Theory. Prentice Hall, Englewood cliffs, NJ (1977)
19. Madoudi, Y., Zagroudi, S.: Cyberbullying – the modern form of violence. J. Dhia' Psychol. Educ. Res. **15** (2020)
20. Al-Anezi, S., Al-Majali, F.: The Role of Security Media in Confronting and Reducing Cybercrime from the Viewpoints of Workers in the Security Services in Kuwait. Education (Al-Azhar): Arbit. Sci. J. Educ. Psychol. Soc. Res. 41–84 (2020)

The Role of Artificial Intelligence Techniques in the Digital Transformation of Jordanian Banking System

AbdelRahman Ismail[1](✉), Mohammad Sami Ali[2], Kayed Alattar[3], Muneer Hasan[1], and Farah Durrani[4]

[1] Faculty of Business, Zarqa University, Zarqa, Jordan
{aismail,mhasan}@zu.edu.jo
[2] Financial Technology Department, Zarqa University, Zarqa, Jordan
maldahrawi@zu.edu.jo
[3] Faculty of Business, Accounting Department, Isra University, Hyderabad, Pakistan
KAYEd.ALATTAR@iu.edu.jo
[4] Department of Finance, University of Business and Technology, Jeddah, Kingdom of Saudi Arabia
f.durani@ubt.edu.sa

Abstract. This study examines the role of artificial intelligence (AI) in the digital transformation of the Jordanian listed banks. It explores the current state of AI adoption in the Jordanian financial sector, as well as investigating the drivers and challenges of AI adoption in this industry. Additionally, through using the deductive research approach, the study examined the impact of AI on various financial functions such as risk management, fraud detection, and customer service. Consequently, the findings suggest that the Jordanian financial industry is at an early stage of AI adoption, with limited awareness of its potential benefits. The study highlights the need for financial institutions to invest in AI and develop a strategic plan for its implementation to fully realize its benefits. The study had revealed that Rob advisor, block chain, and artificial intelligence approaches are all contributing to the digital transformation of Jordan's banking system, with the second null hypothesis rejected due to its lower level of significance.

Keywords: Artificial Intelligence · Robo-Advisor · Digital Transformation · Blockchain · Digital Trading · Robo Advisor

1 Introduction

The implementation of Artificial Intelligence "AI" (from here onwards) in different industries has proven the superiority of Artificial Intelligence over human intelligence, which has given rise to the age of digitalization ([7]. Thus, due to its heavy reliance on human labor in order to provide its services; the implementation of AI systems in the financial industry took a place progressively when it is compared with other industries [5, 8]. However, the rapid rate of technological innovation is bringing ways across various

B. A. M. Alareeni and I. Elgedawy (Eds.): *Artificial Intelligence (AI) and Finance*, SSDC 488, pp. 72–82, 2023.
https://doi.org/10.1007/978-3-031-39158-3_7

industries as competitive dynamics shift, and new business models emerge, which has made it necessary for companies to scramble on the digital transformation bandwagon. However, there are several organizations struggling to develop an effective approach to transform themselves electronically, whereas, the success rates of such transformation efforts are consistently low with less than 30% succeed [3, 5]. Although advances in finance are not a new notion, however, the emphasis on technical advances, and its pace have increased substantially [1, 6]. There are several reasons behind the necessity for employing artificial intelligence techniques in financial institutions. One of the most important factors that led to the applications of AI in the banking industry is the regulatory encouragement. For instance, in December 2018, US government agencies including Federal Reserve, Federal Deposit Insurance Corporation, Financial Crimes Enforcement Network, and National Credit Union joined in an effort to combat money laundering, and terrorist financing [21, 29].

The banking industry has undertaken substantial changes in recent years, due to the demand for more efficient and secure financial transactions and services [1, 9]. To meet such demands, the banking industry found itself face-to-face with the need for digitalization, which requires technologies that will impact virtually all banking operations [33], [36]. According to [31], the role of (IT) traditionally was to cut business operating costs as well as elevating organizational efficiency through automating most of business processes. IT has brought about a whole new way of doing business today thanks to a set of new prevalent digital tools and technologies [14]. Based on this new prevalent role of IT, organizations on the whole are forced to transform the way they do business by constructing and adopting new business models that are based on the adoption and implementation of IT's tools [14, 16].

The study holds significant importance due to the following reasons:

1. Enhancing Banking System Efficiency: The study highlights the role of Artificial Intelligence (AI) techniques in improving the banking system's efficiency. It emphasizes how AI can automate various banking operations and reduce manual errors, thereby saving time and resources.
2. Providing Competitive Advantage: With the increasing use of digital technology in the banking industry, AI can provide a competitive advantage to the Jordanian banking system. This study helps in identifying the benefits of using AI techniques in the banking system, which can help banks stay ahead of their competitors.
3. Addressing Security Concerns: The banking industry is prone to security breaches, and AI can play a vital role in detecting and preventing fraudulent activities. The study sheds light on the significance of AI in enhancing the security of the banking system and how it can protect the customers' personal and financial information.
4. Identifying Areas of Improvement: The study can help identify areas of improvement in the banking system by highlighting the potential areas where AI techniques can be applied. This can help the banking industry in Jordan to enhance its services and customer experience.
5. Promoting Economic Growth: The study also holds importance from an economic standpoint. The digital transformation of the banking system through AI techniques can lead to increased efficiency and productivity, which can help in promoting economic growth in Jordan.

The implementation of AI brings about new compound challenges to businesses that go beyond the traditional management of the classical IT applications [5]. In this regard, [16] indicated that the intense competition between financial services companies, has forced these companies to adopt and implement a set of new predominant and ubiquitous technologies, such as big data analytics, social media, and cloud computing, which are the foundation of FinTech.

2 Literature Review

A study by [2] argued that the digital trading system as a product of technological advancements played a vital role to improve the stability of small and medium enterprises, which are listed on the Amman stock exchange by reducing the trading costs as well as facilitating trading processes for investors. According to [23] innovative financial services and products produce and promote new financial spectrum of know-hows, new organizations, and new market places. Such digital innovation, according to [16] can be manifested in the form of a new service, product, procedure, or business platform which necessitates major modifications by the business, and is facilitated and made possible by Information technology (IT).

AI is not utilized in segregation but frequently applied in aggregation with other technologies [7]. Such technologies include Application Programming Interfaces (APIs), Internet of Things, cloud computing, big data analytics, and open-source algorithms [34]. Since AI improves human intelligence rather than being an alternative for it [5], it is acknowledged that AI requires the interactions with humans to be conducted and enable decision-making process to yield the best outcomes [13]). Like all other sectors, financial institutions implement AI technology in different ways like identification of unusual transactions and frauds, customized services, decision making about creditworthiness, utilization of natural language processing on text documents, risk management and cyber security [3, 10].

2.1 Mobile Banking

The tool of mobile banking facilitates the banks' services for customers as compared to their physical visits to branches; so it results in increasing banks' revenues. That's why banking sector is paying close attention to the adaptation and implementation of emerging technologies in an attempt to improve the quality of services and to be competitive in the market [10].

2.2 The Applications of Block-Chain in the Banking Industry

Blockchain technology is a decentralized, distributed ledger that allows for the secure, transparent and tamper-proof recording of transactions. It has the potential to revolutionize the financial industry by reducing costs, increasing efficiency, and enhancing security [12, 15, 17]. Research has also shown that blockchain technology can reduce the time it takes to settle trades, from several days to just a few minutes [12]. One major challenge is the need for standardization and interoperability between different blockchain systems [33]. Despite these challenges, research suggests that blockchain technology has the potential to significantly transform the financial industry [12].

2.3 Robo Advisor in Financial Industry

This tool has rapidly become a popular alternative to traditional financial advisors, as it uses algorithms to provide automated financial advices and investment solutions to customers [19, 25]. Robo-advisors offer a range of benefits over traditional financial advisors. They are generally more affordable, have lower minimum investment requirements, and can offer 24/7 accessibility to financial advice. However, despite the advantages, robo-advisors also have drawbacks [20, 24]. One major concern is the lack of human interaction and personalized attention, which can be critical for clients with complex financial situations or emotional reactions to market volatility. Other concerns include the potential for algorithmic errors, regulatory uncertainty, and cyber-security risks [7].. They can reduce the need for human financial advisors and impact the way investment firms manage their businesses [21]. One potential effect is the democratization of investing, as robo-advisors can offer lower fees and minimum investments that make investing more accessible to a wider audience [28, 30]. However, the proliferation of robo-advisors can also lead to increased homogeneity in investment strategies, as many use similar algorithms and models [19, 31]. This could lead to increased systemic risk if market shocks affect all robo-advisors simultaneously [26, 32].

2.4 The Significance of Digital Trading for Financial Industry

Digital trading platforms have transformed the financial industry by enabling investors trading electronically [29]. Digital trading offers several benefits over traditional trading methods [6, 15]. It provides increased efficiency and speed of trade execution, lower trading costs, and enhanced transparency [15]. Other concerns include the lack of human oversight, the risk of cyber-attacks, and the potential for market manipulation by high-frequency traders [15, 22]. Additionally, the rise of digital trading has the potential to significantly disrupt the financial industry. It can increase market efficiency, liquidity, and accessibility, but also has the potential to amplify market volatility and exacerbate market shocks [11, 14].

3 Research Methodologies and Sampling

This research endeavors to evaluate the role of artificial intelligence in the digital transformation of the Jordanian banking system [22]. Therefore, the study aims at identifying the extent to which artificial intelligence has been adopted and implemented in the financial industry of Jordan, as well as its impact in business operations, customer service, and the overall performance as well [11, 18]. For this purpose, the study adopted the deductive research approach in order to analyze a primary data those are collected by demonstrated a structured questionnaire that was distributed to a sample of 150 managers, executives, as well as IT professionals who are working in the listed banks of Jordan, which are consisting of fourteen commercial and Islamic listed banks. Thus, by using the STATA software the study employed the descriptive statistics tests to identify key trends, patterns, and insights related to the use of artificial intelligence in the banking system of Jordan. Additionally, it is used the T-test as well as the paired sample T-test to examine the following main hypotheses:

H_{01}: Artificial Intelligence's Techniques have significantly contributed in the digital transformation of Jordanian banking system.
H_{02}: The techniques of Artificial intelligence are unequally contributing in the digital transformation of Jordanian banking system.

4 Results Analysis and Discussion

Table 1 describes frequencies in order to explain respondents' answers regarding the role of AI in digital transformation.

Table 1. Descriptive Statistics

Robo Advisor						
Q1: How familiar are you with the concept of RoboAdvisors in the financial industry?						
Freq	Not familiar	Somewhat familiar	familiar	Very familiar	Extremely familiar	Total
	3	10	33	59	33	138
Q2: Robo Advisor conducts a vital role in banking system through:						
Freq	L. fees	Accurate investment solutions	Acc processing times	24/7		Total
	55	27	55	21		138
Q3: What concerns do you have about using a Robo Advisor in the banking system?						
Freq	Personal security	Accuracy of investment recommendations	Dep on Tech	Lack of H interaction		Total
	35	53	35	15		138
Q4: How do you think the implementation of Robo Advisors has impacted the overall performance of Jordanian banking industry?						
Freq	Positive impact	Negative impact	No impact	Unsure		Total
	115	10	8	5		138
Block Chain						
Q5: Do you believe that blockchain technology has had a significant impact on banking performance?						
Freq	SA	SWA	N	SWD	SD	Total
	79	26	14	13	6	138

(continued)

Table 1. *(continued)*

Robo Advisor

Q6: In what ways has blockchain technology impacted financial transactions?

Freq	Impacted security	Increased speed	reduced fees	Increased Transparency	Increased adoption	Total
	42	43	12	31	10	138

Q7: What potential challenges do you foresee in the widespread adoption of Blockchain technology in the banking industry in Jordan?

Freq	lack of T exp	lack of RF	high imp costs	R to change		Total
	8	19	23	88		138

Digital Trading

Q8: How frequently do you use digital trading platforms?

Freq	Daily	weekly	monthly	rarely	Total
	63	40	29	6	138

Q9: How has the use of digital trading platforms impacted the market capitalization of banks?

Freq	S improved	SW improved	No change	SW W	S W	Total
	94	24	4	4	12	

Q10: Do you think digital trading facilitated the process of trading stocks?

Freq	SA	SW A	N	SWD	SD	Total
	84	31	9	9	5	138

Mobile Banking

Q11: Overall, how satisfied are you with the current mobile banking Apps?

Freq	V satisfied	SW satisfied	N	SW dissatisfied	V dissatisfied	Total
	106	22	3	5	2	138

Q12: What your clients use mobile banking for?

Freq	Balance Inq	Elec transfer	Paying bills	Total
	7	109	22	138

Q13: What concerns do your clients might have when making the decision of using mobile Apps?

Freq	P & S	Saving T & E	Total
	104	34	138

The study ran the descriptive statistics test in order to identify the role of AI in the digital transformation of Jordanian banking system. Consequently, statistics regarding respondents' answers revealed that the Rob advisor technique was positively impacted banks' performance through lowering the fees of banks' services as well as improving the accuracy of investment recommendations. Likewise, the block chain technique was positively influenced banks' performance through increasing the speed of conducting financial transactions. However, a majority of respondents are daily using the ASE's trading platform as it conducts a vital role in facilitating the trading process, as well as they asserted that the use of digital trading contributed in improving banks' market capitalization. Furthermore, it is revealed that most of banks' clients tend to use mobile banking for the privacy and security's features, as well as making electronic transfer.

Table 2 describes statistics regarding the examination of the second hypothesis, which assumed that "*Artificial Intelligence's Techniques have significantly contributed in the digital transformation of Jordanian banking system*".

Table 2. One-Sample Tests

AI Techniques	t	df	Sig. (2-tailed)	Mean Difference	95% Confidence Interval of the Difference	
					Lower	Upper
RoboAdvisor	45.940	137	.000	9.46377	9.0564	9.8711
Blockchain	58.319	137	.000	7.68116	7.4207	7.9416
DigitalTrading	29.494	137	.000	5.20290	4.8541	5.5517
MobileBanking	48.483	137	.000	4.72464	4.5319	4.9173

Through running the one sample tests, results from the above table proved that since the sig value for each tool is found to be lower than their levels of significance, the suggested null hypothesis was rejected. Thus, since the null hypothesis was replaced by the alternative one, the results revealed that the techniques of artificial intelligence are significantly contributing in the digital transformation of the Jordanian banking system as represented by the listed commercial as well as Islamic banks. Therefore, the study ran the paired samples test in order to compare the role of AI techniques in the digital transformation of Jordanian banking system, and subsequently, the results are as demonstrated by the following table:

Table 3 describes statistics regarding the examination of the first hypothesis, which assumed that "*The techniques of Artificial intelligence are unequally contributing in the digital transformation of Jordanian banking system*".

Through running the paired sample test in order to check whether AI's techniques contribute equally in the digital transformation of Jordanian banking system, statistics from the above table revealed that since the values of significance are less than the level of significance 5%; the second null hypothesis was rejected. Which in other words,

Table 3. Paired Samples Test

	Mean	Std	Std.E-Mean	t	Sig. (2-tailed)	
Pair 1	RoboAdvisor - BlockChain	1.78261	2.24337	.19097	9.335	.0002
Pair 2	RoboAdvisor - DigitalTrading	4.26087	2.71332	.23097	18.447	.0005
Pair 3	RoboAdvisor - MobileBanking	4.73913	2.24195	.19085	24.832	.0001
Pair 4	BlockChain - DigitalTrading	2.47826	2.61165	.22232	11.147	.0006
Pair 5	BlockChain - MobileBanking	2.95652	1.81184	.15423	19.169	.0001
Pair 6	DigitalTrading - MobileBanking	.47826	1.94530	.16560	2.888	.005

findings from this test confirmed that the techniques of Artificial intelligence are equally contributing in the digital transformation of Jordanian banking system.

5 Future Studies and Research Limitations

5.1 Future Research

While this study has shed light on the significance of artificial intelligence (AI) techniques in the digital transformation of the Jordanian banking system, there are several areas where further research could be conducted. For instance, future research could focus on:

The impact of AI techniques on customer experience: While the study has discussed the potential benefits of AI on the banking system, further research could examine the impact of AI on the customer experience, including satisfaction, loyalty, and retention.

The impact of AI on the workforce: As AI techniques continue to be adopted in the banking industry, future research could examine the impact of these technologies on the workforce, including the potential displacement of human workers and the need for upskilling and reskilling.

The adoption and implementation of AI techniques: Future research could examine the factors that influence the adoption and implementation of AI techniques in the Jordanian banking system, including organizational culture, resources, and regulatory frameworks.

5.2 Research Limitations

Despite the significance of this study on the role of AI techniques in the digital transformation of the Jordanian banking system, there are several limitations that need to be acknowledged. These include:

Limited sample size: The study was conducted on a limited sample size, which may not be representative of the entire banking industry in Jordan. Further research with a larger sample size may be needed to generalize the findings.

Lack of qualitative data: The study mainly relied on quantitative data, which may not provide an in-depth understanding of the impact of AI on the banking system. Future research could incorporate qualitative data to gain a more comprehensive perspective on the topic.

Single country focus: The study focuses on the banking industry in Jordan, which may not be applicable to other countries with different economic, cultural, and regulatory contexts. Future research could compare the adoption and impact of AI techniques in the banking industry across different countries.

6 Conclusion

In conclusion, the financial industry in Jordan is witnessing a significant shift towards the use of technology, with robo-advisors, blockchain, and digital trading gaining popularity among investors and financial institutions. Robo-advisors have revolutionized the way individuals invest their money by providing cost-effective and personalized investment advice, making investing accessible to a broader audience.

Blockchain technology has also gained significant traction in Jordan's financial industry, as it provides a secure and transparent way to conduct transactions. The technology's ability to reduce intermediaries, enhance transparency, and automate processes has attracted various financial institutions in Jordan to explore its potential applications.

Digital trading has also made it possible for traders to execute trades from anywhere at any time, using their smartphones or computers. This technology has simplified the trading process, reduced transaction costs, and increased market efficiency.

Overall, the use of robo-advisors, blockchain, and digital trading in the financial industry in Jordan has brought significant benefits, including increased accessibility, efficiency, and security. As the financial industry in Jordan continues to embrace new technologies, we can expect to see more advancements that will enhance customer experiences and improve financial services in the country.

While this study has identified the potential benefits of AI techniques in the digital transformation of the Jordanian banking system, there is a need for further research to explore the impact of these technologies on customer experience, the workforce, and the factors that influence their adoption and implementation. It is also important to acknowledge the limitations of this study, including the limited sample size, lack of qualitative data, and single country focus. Future studies should focus on the adoption of AI in different financial services, such as insurance and investment management.

Overall, the successful adoption of AI in the Jordanian banking system depends on collaboration between stakeholders, including financial institutions, regulators, and policymakers, as well as continuous investment in research and development.

References

1. Abusalma, A.: The effect of implementing artificial intelligence on job performance in commercial banks of Jordan. Manage. Sci. Lett. **11**(7), 2061–2070 (2021)
2. Ali, M.S.: The Digital Revolution in Financial Services Industry Theory and Practice, 1st ed, Generis Publishing, Europe, Republic of Moldova
3. Ali, M.S., Ismail, A.R., Swiety, I.A.: Assessing the role of technologically enabled trading system in the stability of small and medium enterprises: the case of Jordan. In: 2021 22nd International Arab Conference on Information Technology (ACIT), IEEE (2021)

4. Azzam, Z.: Factors affecting the adoption of e-marketing technologies in services industry: empirical study on hotel sector in Jordan. Zarqa J. Res. Stud. Humanit. **13**(1), 210–224 (2013)
5. Berente, N., Gu, B., Recker, J., Santhanam, R.: Managing Ai. MIS Quart. 1–5 (2019)
6. Bibel, W.: Artificial intelligence in a historical perspective. AI Commun. **27**(1), 87–102 (2014)
7. Brock, J.K.-U., Von Wangenheim, F.: Demystifying AI: what digital transformation leaders can teach you about realistic artificial intelligence. Calif. Manage. Rev. **61**(4), 110–134. (2019). https://doi.org/10.1177/1536504219865226
8. Capgemini. World FinTech Report 2018 (2018). https://www.capgemini.com/wp-content/uploads/2018/03/world_fintech_report_2018.pdf
9. Casares, A.P.: The brain of the future and the viability of democratic governance: the role of artificial intelligence, cognitive machines, and viable systems. Futures **103**, 5–16 (2018)
10. Cavus, N., Mohammed, Y.B., Yakubu, M.N.: An artificial intelligence based model for prediction of parameters affecting sustainable growth of mobile banking apps. Sustainability, **13**(11), 6206 (2021). https://doi.org/10.3390/su13116206
11. Chui, M., Malhotra, S.: AI Adoption Advances, but Foundational Barriers Remain. McKinsey Global Institute, McKinsey's San Francisco, USA (2018)
12. Crosby, M., Pattanayak, P., Verma, S., Kalyanaraman, V.: Blockchain technology: beyond bitcoin. Appl. Innov. **2**(6–10), 71–81 (2016)
13. Dhar, N., Holly, T., Ryan, D., Galeaz, G.: Top financial services issues of 2018, PWC (2017)
14. El Sawy, O., Pereira, F.: Business Modelling in the Dynamic Digital Space: An Ecosystem Approach, 1st edn. Springer, USA (2013)
15. Esther, H.R., John, F.: Role of chat bots in customer engagement valence. Psychol. Educ. **57**(9), 2181–2186 (2020)
16. Fichman, G., Santos, B., Zheng, Z.: Digital innovation as a fundamental and powerful concept in the information systems curriculum. MIS Q. **38**(2), 329–353 (2014)
17. Frame, W.S., White, L.J.: Technological change, financial innovation, and diffusion in banking. Leonard, N. (ed.) Stern School of Business, Department of Economics, pp. 1–5. Atlanta, GA, USA (2014)
18. Gallego-Gomez, C., De-Pablos-Heredero, C.: Artificial intelligence as an enabling tool for the development of dynamic capabilities in the banking industry. Int. J. Enterp. Inf. Syst. (IJEIS), **16**(3), 20–33 (2020). https://doi.org/10.4018/ijeis.2020070102
19. Gentsch, P.: Conversational AI: how chat bots will reshape the digital experience. In: AI in Marketing, Sales and Services. Palgrave Macmillan, Cham (2019)
20. Indriasari, E., Gaol, F.L., Matsuo, T.: Digital banking transformation: application of artificial intelligence and big data analytics for leveraging customer experience in the Indonesia banking sector. In: 8th International Congress on Advanced Applied Informatics (2019)
21. Kaur, N., Sahdev, S.L., Sharma, M., Siddiqui, L.: The influence of artificial intelligence on the banking industry and how ai is changing the face of modern day banks. Int. J. Manag. **11**(6), 577–585 (2020)
22. Kitsios, F., Kamariotou, M.: Artificial intelligence and business strategy towards digital transformation: a research agenda. Sustainability **13**(4), 2025 (2021). https://doi.org/10.3390/su13042025
23. Lerner, J., Tufano, P.: The consequences of financial innovation: a counterfactual research agenda. Annu. Rev. Financ. Econ. **3**(1) 1–92 (2011)
24. Malali, A.B., Gopalakrishnan, S.: Application of artificial intelligence and its powered technologies in the Indian banking and financial industry: an overview. IOSR J. Humanit. And Social Science **25**(4), 55–60 (2020)

25. Microsoft News Center. (2019). Digital transformation in the Jordan financial services space streamlined as central bank issues cloud computing guidelines. Microsoft. News Center Middle East and Africa. Retrieved from: https://news.microsoft.com/en-xm/2019/05/10/digital-transformation-in-the-jordan-financial-services-space-streamlined-as-central-bank-issues-cloud-computing-guidelines/
26. MoDEE. Jordan digital transformation strategy 2020. Ministry of Digital Economy and Entrepreneurship, Jordan (2020). https://www.modee.gov.jo/EBV4.0/Root_Storage/EN/1/Jordan_Digital_Transformation_Strategy_2020_English_Unofficial_Translation.pdf
27. Al-Shaikh, S.M.: Customers' perspective towards factors associated with technological products adopting failure. a case of Jordan. Zarqa J. Res. Stud. Humanit. **13**(2), 252–261 (2013)
28. Paul, L.R., Madana, S.A.: Artificial intelligence in predictive analysis of insurance and banking. In: Artificial Intelligence: Fundamentals and Applications (1st ed.). CRC Press, Taylor and Francis Group (2021)
29. Plastino, E., Purdy, M.: Game changing value from artificial intelligence: eight strategies. Strategy Leadersh. **46**(1), 16–22 (2018). https://doi.org/10.1108/sl-11-2017-0106
30. Puschmann, T.: Fintech. Bus. Inf. Syst. Eng. **59**(1), 69–76 (2017)
31. PwC. Financial Services Technology 2020 and Beyond: Embracing disruption (2018). https://www.pwc.com/gx/en/industries/financial-services/publications/financial-services-technology-2020-and-beyond.html
32. Verweij, G.: PwC's global data and analytics survey 2016, PWC (2016). https://www.pwc.com/us/en/services/consulting/analytics/big-decision-survey.html
33. Yu, T.R., Song, X.: Big data and artificial intelligence in the banking industry. In: Chapter 117 in Handbook of Financial Econometrics, Statistics and Machine Learning, pp. 4025–4041 (2020)
34. Zavolokina, L., Dolata, M., Schwabe, G.: FinTech transformation: How IT-enabled innovations shape the financial sector. In: FinanceCom, pp. 75–88. Springer, Cham (2016)

The Impact of Digital Communication on Organizational Agility at the University of Hail and Ways of Development

Ebtehal Saleh Allhidan(✉)

College of Business Administration, Department of MGT, University of Hail, Hail, Saudi Arabia
ebttehal1409@hotmail.com

Abstract. The purpose of the current research was to identify the digital communication's impact on organizational agility at the University of Hail and what are the ways to develop it. The descriptive analytical approach was adopted in this study, and deliberate sampling was selected to ensure that all faculty at the University of Hail were represented. Accordingly, (88) faculty members were selected and distributed to university colleges in a simple random way. A questionnaire consisting of (15) items was used as an instrument for data collection, as it included three dimensions (digital communication, organizational agility, and ways of development). In summary, we see a varying set of results, the most important of which is that the dimensions of the study were of high impact with a percentage ranging between (68%) and (84%). The study's findings also demonstrated a distinct effect of digital communication on organizational agility at the University of Hail. This indicates that there is a high level of digital communication application at the university according to digital platforms. This, in turn, is reflected in the level of organizational agility at the university. In the light of reported results, it is conceivable that digital communication applications must be developed in line with the mechanisms of organizational agility at the university.

Keywords: Digital Communication · Organizational Agility · Ways of Development · University of Hail

1 Introduction

The world has witnessed tremendous progress in the field of virtual communications, which has led to the development of many digital communication mechanisms, more modern technologies, and an unprecedented information revolution (Divayana et al. 2021). Computer technology, satellite technology, microelectronics and multimedia technology have merged, resulting in modern high-precision technology capable of storing, retrieval, transferring and exchanging information between individuals and educational institutions at high speed and low cost (Aryani and Siahaan 2020).

At the beginning of 2020, the world witnessed the emergence of the COVID-19 crisis, and universities became the most affected in light of this pandemic (Hassounah et al. 2020). In light of the tendency of the Saudi educational system to adapt quickly to

B. A. M. Alareeni and I. Elgedawy (Eds.): *Artificial Intelligence (AI) and Finance*, SSDC 488, pp. 83–95, 2023.
https://doi.org/10.1007/978-3-031-39158-3_8

the current situation, which imposed itself by taking precautionary measures to confront the pandemic and try to control it. Among these precautionary measures was to reduce the attendance days of students, and employees, assisting staff and university workers, as well as working to make the most of the technological infrastructure in universities to serve the educational system optimally. Accordingly, the use of digital communication technology was resorted to, and e-learning units were developed to communicate between faculty members in universities and the environment cooperating in the Saudi education system (Mahyoob. 2021).

Therefore, digital communication environments have made it imperative for Saudi universities to employ means to improve the quality-of-service delivery, increase efficiency and save costs (Sayaf et al. 2021). According to the previous premise AshaR Al-Samiri (2021), digital communication in Saudi universities should lead to a broader vision that enables continuous innovation and enhances teaching and learning, and improves efficiency for administrative services, students and faculty. Al-Shammari (2021) pointed out that the integration between digital communication mechanisms and the strategic vision of universities is essential as it has provided an alternative to the traditional environment through the virtual environment that relies on advanced and accelerated technology, in addition to meeting the changing needs of universities. Therefore, the shift to digital communication has now become a necessary means of survival, as digital communication requires universities to change their traditional roles, which were focused on paper messaging and traditional management, and to move towards digital virtual management commensurate with the changes of the digital age and adapt to digital methodologies and mindsets (El Harwan and Barakat. 2019).

Organizational agility, on the other hand, is the capacity of universities to redefine their techniques in the context of their current environment through continual anticipation of and adjustment to the requirements of digital communication and recent technological changes in the field of communication (Almahirah. 2016). Strategic sensitivity is a component of organizational agility that reflects institutions' capacity to respond to external factors by developing a purpose and objectives-driven strategy that is both clear and well-defined (Harsch and Festing, 2020), along with the topic of collaboration in responsibility, which illustrates how important it is for universities for all of their faculties to take responsibility (Akkaya and Tabak. 2020). Furthermore, organizational agility is represented by clarity of vision, in the sense of the ability to articulate a clear vision of what one wants to be in the future. Finally there is the selection of strategic goals intended to align university goals with Opportunities that exist in terms of the skills needed for development and the ability to define actions (Almahirah 2020).

Thus, the research on the impact of digital communication on organizational agility at the University of Hail and ways to develop it is justified. The relevant literature supports the need to implement digital communication in Saudi universities, and research results (Abdullah 2019) and (Al-Anazi and Obaid. 2021) indicate that universities still need to learn about the of digital communication's requirements and its impact on organizational agility. One of these justifications is also that universities become able to create, innovate, excel and keep pace with the developments of digital communication, and thus able to develop appropriate strategies and processes for planning and development in light of digital communication and organizational agility that the University of Hail seeks to employ and apply.

2 Statement of the Problem

Due to the technological development of the current era, which has been called the digital age, the shift to the digitization of universities has become inevitable (Al-Shafi and Al-Hamdani 2019). Atta (2019) indicated that universities are no longer able to face this development and change, and have become limited to carrying out their new responsibilities and roles, so the majority of universities have turned to develop new patterns while relying heavily on the activation of digital communication in its educational system.

To maintain their growth and continuity in a competitive context, universities should be aware that they need to strive to attain excellence in their administrative performance, and this requires universities to keep pace with the latest developments in digital communication (Malik et al. 2018). So that universities know their competitive position in their changing environment, organizational agility mechanisms must be employed (Ali and Zayid. 2022). Since organizational agility is considered as a strategic approach to modern management, its importance and effective role are concentrated in enhancing the organizational competitiveness of universities, which necessarily requires universities to have the skills to employ organizational agility in order to be able to positively change the digital communication environment and virtual reality (El Din and El Hessewi. 2019).

Al-Fariji (2020) noted that Saudi universities have benefited from the development in the field of digital communication, including the education system in Saudi Arabia. Al-Tai et al. (2021) pointed out that digital communication has blended the results of three revolutions: the revolution of the means of communication represented by modern communication technologies, the revolution of information and electronic computers that have entered all areas of life due to the "global" Internet, and the revolution of knowledge, artificial intelligence and virtual reality that we live in today. In the same context, Al-Shafei and Al-Hamdani (2019) pointed out that digital communication has created a positive gap in the reality of the work of universities, as Saudi universities have witnessed a positive reflection on the educational process by facilitating the process of sending and receiving information significantly.

The researcher's sense of the problem of the study stemmed from her observation of the University of Hail use of digital communication systems represented in the presence platform, Agraa system and Banner digital system. As a result of the researcher's work as a faculty member at the University of Hail and the Information Technology Department, which identifies the needs of programs and technical systems and works to develop them in colleges, she noticed that there is a need to apply digital communication at the University of Hail as it makes administrative work characterized by creativity and achievement. Hence, the problem of the study was born, and a question related to the degree of application of Digital communication at the university began to crystallize in the researcher's mind. Given that the University of Hail must evolve its system in accordance with the developments of digital and virtual communication to keep pace with all developments in the educational arena, this can only be accomplished by developing its systems and programmes., according to the developments of digital communication.

Hence, the above invites the researcher to identify the impact of digital communication on agility at the University of Hail and ways to develop it.

3 Questions

1. What is the impact of digital communication on organizational agility at the University of Hail?
2. What is the level of organizational agility at the University of Hail?
3. What are the ways to develop digital communication on organizational agility at the University of Hail?

4 Purpose

The aims of this study are to examine and determine how digital communication affects organizational agility at the University of Hail, in addition to identifying the level of organizational agility at the University of Hail. The current study also seeks to provide a set of suggestions and recommendations to Saudi universities that can enhance their entrepreneurial orientation.

5 Significance

The study may contribute to demonstrating the impact of digital communication on organizational agility at the University of Hail, and may also contribute to clarifying the hidden in the application in relation to digital communication at the University of Hail and the extent to which it is related to organizational agility. This in turn is reflected in the administrative processes at the university, and this study may provide those in charge and officials at the University of Hail with the importance of employing digital communication and how it affects the application of organizational agility by providing statistically proven results and providing suggestions and recommendations that are in the public interest of decision-makers at the University of Hail.

6 Delimitations

The study was limited to the University of Hail in 2022. The study instrument was applied to faculty members, and represents the independent variable of digital communication, and the dependent variable through organizational agility.

7 Definition of Terms

Digital Communication: It is a social process in which participants exchange roles in sending and receiving various communication messages through digital systems in order to accomplish particular aims. The primary method of communication between parties according to the definition is the usage of digital technologies and their improvements, where information is exchanged, coded and processed by systems (Al-Shafei and Al-Hamdani 2019). Atta (2019) also defined digital communication as communication that takes place using digital systems and relies on electronic means in a digital way and digitizing information, and this concept emphasizes the role of the computer in the communication process.

Digital communication in the current study is the modern digital means of communication and interaction networks and the reality of their existence at the University of Hail, represented by the presence platform, Agraa system and Banner digital system, and knowing the ability of faculty members and workers of the ways and mechanisms to use these digital systems.

Organizational Agility: Organizational agility is a strategic approach to increase the cognitive capacity of organizations, so that it helps them discover environmental changes and their repercussions on the organization, and find ways to provide an appropriate response to these changes (Walter 2021).

Al-Ajri (2017) explained that organizational agility refers to the ability of an organization to sense expected and unexpected changes occurring in the internal and external environment. Based on the sensing process, the Organization responds innovatively to these changes by redesigning processes and reshaping the organizational structure, thereby enhancing the Organization's viability.

8 Methodology

The descriptive analytical approach was used in this study to achieve its objectives. This approach is applied on a large or small geographical scale, and it may be a comprehensive survey or a sample method, in which large samples are often used in order to help the researcher obtain accurate results with low error rates (Bauer et al. 2021).

9 Sample of the Study

An intentional sample was selected with the aim of ensuring the representation of all faculties of the University of Hail. Accordingly, (88) faculty members were selected distributed to university colleges in a simple random way. The table and graph illustrate this (Table 1 and Fig. 1):

Table 1. Distribution of the study sample

Sample	N	%
Professor	22	25%
Associate Professor	27	30. 68%
Assistant Professor	32	44. 32%
Total	88	%100

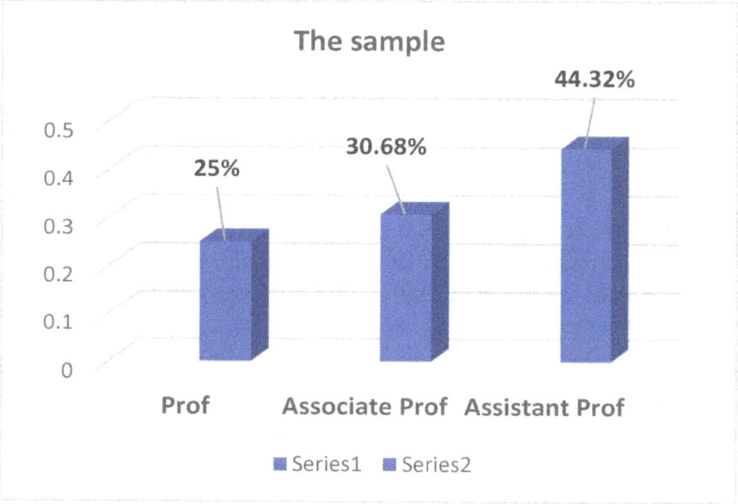

Fig. 1. Distribution of the study sample

10 Instrument of the Study

A three-dimension-questionnaire was developed by the researcher to accomplish the study's goals: the first dimension: digital communication, the second dimension: organizational agility, and the third dimension: ways of development. Accordingly, the questionnaire included (15) items.

Validity: A pilot sample of (14) Hail faculty members were given the questionnaire, and their replies were transcribed into (SPSS 22) to check the internal consistency. The methods used to determine the correlation coefficients and determine the validity of internal consistency are shown in the table below.

The three dimensions' high correlation coefficients and statistical significance at the significance level (0.01), as shown in Table 2, support the validity and internal consistency of the questionnaire.

Table 2. Pearson correlation coefficients

Dimension	Correlation coefficient	Significance
Digital Communication	0.67**	0.000
Organizational Agility	0.71**	0.000
Ways of Development	0.78**	0.000

Reliability: The table below shows the reliability coefficients of the questionnaire and each of its dimensions using the Cronbach alpha method that the researcher used to assess the reliability of the questionnaire.

Table 3. Cronbach's Alpha

Dimension	*Cronbach's Alpha*
Digital Communication	0.73
Organizational Agility	0.69
Ways of Development	0.81
Total	0.92

Table 3 demonstrates that the study dimensions have high Cronbach's alpha reliability values. The data additionally reveals that the dimensions' overall reliability coefficient, which is a high reliability coefficient, was (2 0.9). This demonstrates the questionnaire's high level of dependability and the fact that it can be relied upon when the study is used in the real world.

11 Results and Discussions

The First Question: What is the impact of digital communication on organizational agility at the University of Hail?

The following table demonstrates the calculation of the arithmetic means and standard deviations of the sample's responses to the items in the digital communication dimension in order to respond to the first question (Table 4).

The arithmetic mean of the study participants' replies to the items in the dimension of digital communication was (3.99) with a standard deviation of (1.21), which is within a wide range of effect, according to Table (4). Between the lowest and greatest arithmetic means, the items in this dimension had an arithmetic mean that varied from (3.77) to (4.13). Participants in the study provided responses in percentages ranging from (68%) to (84%).

The explanation for this result is due to the fact that the University of Hail provides the necessary tools for the use of digital communication technologies (educational

Table 4. Digital Communication

No	Items	Arithmetic mean	Standard deviation	Rank	Degree
4	Digital communication is one of the most successful means used by the University of Hail during the Covid-19 pandemic	4.13	0.87	1	High
2	Digital communication systems enable the follow-up and monitoring of the participation of faculty members and their interaction through educational digital platforms	4.09	0.90	2	High
5	Digital communication systems achieve educational objectives with high efficiencies and shorten time and effort	4.01	1.03	3	High
1	Digital communication systems help to access education applications across different digital platforms	3.96	0.77	4	High
3	Digital communication in Hail is an alternative to the traditional environment through the virtual environment	3.77	0.94	5	High
Weighted Mean		3.99	1.21	High	

platforms, communication and communication systems), and the University of Hail considers digital communication an integral part of the strategic plan of the Saudi education system and in line with Vision 2030. This makes digital communication systems a practical application for the development of business digitization within the University of Hail. As the study sample believes that communication Digital provide electronic services around the clock for the educational and administrative staff at the university. This finding is related to the study (Al-Anazi and Obaid 2021).

The Second Question: What is the level of organizational agility at the University of Hail?

The following table demonstrates the calculation of the arithmetic means and standard deviations of the sample's responses to the items in the organizational agility dimension in order to respond to the second question (Table 5).

Table 5. Organizational Agility

No	Items	Arithmetic mean	Standard deviation	Rank	Degree
5	The university administration has the agility of sensing and monitoring administrative processes according to digital communication	4.16	0.87	1	High
2	There is a modern strategic plan for the university that keeps pace with digital transformation	4.13	0.90	2	High
4	The organizational structure supports networking between university colleges	4.02	1.03	3	High
1	The existence of training and development programs for university leaders supports organizational agility	3.97	0.69	4	High
3	Restructuring the administrative procedures at the university to adapt to the requirements of digital communication	3.91	0.88	5	High
Weighted Mean		4.04	1.09	High	

The arithmetic mean of the study participants' replies to the items in the dimension of organizational agility was (4. 04) with a standard deviation of (1.09), which is within a wide range of effect, according to Table (5). Between the lowest and greatest arithmetic means, the items in this dimension had an arithmetic mean that varied from (3. 91) to (4.16). Participants in the study provided responses in percentages ranging from (68%) to (84%).

The fact that can be used to explain this outcome is that he University of Hail employs the requirements of organizational agility in line with the developments of digital communication. Thus, the University of Hail provides ways for digital communication among colleges and provides educational digital platforms and it takes advantage of the information and communication systems at the university to speed up procedures and achieve organizational agility. This finding is associated with the study (Almahirah 2016); (Harsch amd Festing 2020).

The Third Question: What are the ways to develop digital communication on organizational agility at the University of Hail?

The following table demonstrates the calculation of the arithmetic means and standard deviations of the sample's responses to the items in the ways of development dimension in order to respond to the third question (Table 6).

Table 6. Ways of Development

No	Items	Arithmetic mean	Standard deviation	Rank	Degree
1	Developing the regulations and making them flexible and compatible with the overall transition towards digital connectivity	4.15	0.87	1	High
4	Strategic planning for the development of digital communication application at the University of Hail	4.12	0.90	2	High
3	Developing administrative processes and procedures in line with digital communication requirements	4.08	1.03	3	High
2	Conducting a strengths, weaknesses, opportunities and risks analysis to identify ways to improve and develop opportunities for digital communication	3.89	0.83	4	High
5	Utilizing the results of satisfaction rates and performance evaluation in providing feedback for the development of the digital communication application	3.81	0.79	5	High
Weighted average		4.08	1.06	High	

The arithmetic mean of the study participants' replies to the items in the dimension of ways of development was (4. 08) with a standard deviation of (1.06), which is within a wide range of effect, according to Table (6). Between the lowest and greatest arithmetic means, the items in this dimension had an arithmetic mean that varied from (3. 81) to (4.15). Participants in the study provided responses in percentages ranging from (68%) to (84%).

The reason behind this finding is the importance of compatibility of digital communication applications with all regulations and laws by choosing the appropriate application method, which increases the need to reformulate all regulatory content according to digital communication systems and technologies. Faculty members are also aware of the importance of training processes and their implications for the application of digital communication, in addition to their awareness that training processes contribute to the development of their skills in the field of digital communication applications. This naturally facilitates the application of digital communication and contributes to raising the level of technical culture that matters. This finding is related to the study (El Harwan and Barakat 2019).

12 Results

1. At the University of Hail, where the arithmetic mean of the dimension of digital communication was 3.99 and the replies of research participants ranged from (68%) to (84%), there was evidence of a digital communication influence.
2. The organizational agility dimension received a high score, with the arithmetic mean of the organizational agility dimension being (4.04) and the responses from study participants ranging in percentage from (68%) to (84%).
3. The dimension of ways of development received a high level; its arithmetic mean was (4.01) and the responses of the study subjects fell between (68%) and (84%), indicating a high level.

Through the previous results, it was found that there is a clear impact of digital communication on organizational agility at the University of Hail, and this indicates that the level of application of digital communication at the university is high according to digital platforms, which reflected the level of organizational agility at the university.

13 Recommendations

1. Developing organizational agility processes to activate the requirements of digital communication at the University of Hail and employing administrative plans and projects related to this aspect.
2. Developing digital communication applications in line with the organizational agility mechanisms at the university.
3. Adopting practical procedures that facilitate the application of digital communication in the faculties at the university.
4. Intensive training programs specialized in the field of digital communication theoretically and practically.
5. Setting the appropriate foundations and standards that help the application of digital communication at the university.

References

Abdullah, A.: The effectiveness of a proposed computer program for remote training to acquire administrative cadres (Noor system). Educ. J. Adult Educ. **1**(4), 1–35 (2019)

Akkaya, B., Tabak, A.: The link between organizational agility and leadership: a research in science parks. Acad. Strateg. Manag. J. **19**(1), 1–17 (2020)

Al Ansari, K.A.: The degree of practicing organizational agility by academic leaders at Saudi Universities. J. Iraqi Univ. **51**(3), 501–516 (2021)

Al-Shammari, A.: Developing the administrative performance of directors of education offices in the Kingdom of Saudi Arabia in light of the approach to total quality: a descriptive study. Sci. Res. J. **15**(3), 122–146 (2021)

Al-Ajri, S.: Organizational agility and its role in enhancing the relationship between business intelligence and organizational performance. J. Bus. Adm. **22**(4), 12–24 (2017)

Al-Anazi, N., Obaid, A.: Educational leadership strategy in light of the COVID-19 pandemic in the Kingdom of Saudi Arabia. J. Coll. Educ. **37**(5), 375–396 (2021)

Al-Fariji, R.: Transformations of the digital environment in the knowledge society in the Kingdom of Saudi Arabia. Egypt. J. Commun. Res. **1**(1), 230–260 (2020)

Alharbi, G.A.: Organizational agility to counter coronavirus implications in emerging Saudi Universities. Cypriot J. Educ. Sci. **16**(6), 2994–3015 (2021)

Ali, R., Zayid, E.I.M.: The challenges and problems faced by students in the early stage of writing research projects in L2, University of Bisha, Saudi Arabia. University of Bisha, Saudi Arabia (2022). Accessed 16 Nov 2022

Almahirah, M.: The influence of the leadership style on managerial creativeness Jordanian industrial corporations. Int. Bus. Manag. **13**(3), 27–39 (2016)

Almahirah, M.: The effectiveness of organizational agility as a strategic input in the process of enhancing organizational competitiveness. Int. J. Adv. Sci. Technol. **29**(8s), 1628–1637 (2020)

Al-Samiri, R.A.: English language teaching in Saudi Arabia in response to the COVID-19 pandemic: challenges and positive outcomes. Arab World Engl. J. (AWEJ) Special Issue Covid 19 (2021)

Al-Shafi, A., Al-Hamdani, P.: Dimensions of digital communication in university education from the point of view of students of the departments of history and geography. Educ. J. Fac. Educ. Sohag **68**(68), 1605–1632 (2019)

Al-Tai, M., Daoud, H., Abdel Qader, N.: Employing digital communication in leadership creativity and innovation. Police Res. Center **30**(116), 101–133 (2021)

Aryani, R., Siahaan, L.H.: Information technology-based management education in vocational high schools. Sosiohumaniora **22**(3), 267–273 (2020)

Atta, A.: Communication technology digital education in Algeria. Sci. J. Technol. Disabil. Sci. **1**(1), 72–101 (2019)

Bauer, G.R., Churchill, S.M., Mahendran, M., Walwyn, C., Lizotte, D., Villa-Rueda, A.A.: Intersectionality in quantitative research: a systematic review of its emergence and applications of theory and methods. SSM-Popul. Health **14**, 100798 (2021)

Divayana, D.G.H., Suyasa, P.W.A., Widiartini, N.K.: An innovative model as evaluation model for information technology-based learning at ICT vocational schools. Heliyon **7**(2), e06347 (2021)

El Din, Y.K.Z., El Hessewi, G.M.S.: The relationship between psychological contract breach, organizational identification, and organizational agility among nursing faculty members. Am. J. Nurs. **8**(6), 304–312 (2019)

El Harwan, M., Barakat, A.: Requirements for digital transformation. J. Coll. Educ. **30**, 429–479 (2019)

Harsch, K., Festing, M.: Dynamic talent management capabilities and organizational agility—a qualitative exploration. Hum. Res. Manag. **59**(1), 43–61 (2020)

Hassounah, M., Raheel, H., Alhefzi, M.: Digital response during the COVID-19 pandemic in Saudi Arabia. J. Med. Internet Res. **22**(9), e19338 (2020)

Mahyoob, M.: Online learning effectiveness during the COVID-19 pandemic: a case study of Saudi Universities. Int. J. Inf. Commun. Technol. Educ. (IJICTE) **17**(4), 1–14 (2021)

Malik, H.A.M., Abid, F., Kalaicelvi, R., Bhatti, Z.: Challenges of computer science and IT in teaching-learning in Saudi Arabia. Sukkur IBA J. Comput. Math. Sci. **2**(1), 29–35 (2018)

Sayaf, A.M., Alamri, M.M., Alqahtani, M.A., Al-Rahmi, W.M.: Information and communications technology used in higher education: an empirical study on digital learning as sustainability. Sustainability **13**(13), 7074 (2021)

Walter, A.-T.: Organizational agility: ill-defined and somewhat confusing? a systematic literature review and conceptualization. Manag. Rev. Q. **71**(2), 343–391 (2021). https://doi.org/10.1007/s11301-020-00186-6

The Impact of Technological Change on Organizational Productivity and Customer Satisfaction: A Case Study of a Number of Factories Operating in Jordan

Khaled Jumah[1](✉), Marwa Ahmad[2], Wedad Aqrabawi[2], Ahmad Al-Ghool[2], and Megren Altassan[3]

[1] Faculty of Business, Zarqa University, Zarqa, Jordan
kjumah@zu.edu.jo

[2] Faculty of Business, Philadelphia University, Philadelphia, USA
{Mar_ahmed,waqrabawi,aghool}@philadelphia.edu.jo

[3] Department of HRM, University of Business and Technology, Jeddah, Saudi Arabia
m.altassan@ubt.edu.sa

Abstract. This study aims to determine the impact of technological change on organizational productivity and customer satisfaction. The sample of the study is 220 employees who worked in some factories operating in Jordan. A questionnaire designed by researchers was used, and the Cronbach's value was (0.803). The study results found that that there is an impact of technological change on organizational productivity and customer satisfaction. The employees feel strongly that both the organizational productivity and the customer satisfaction affected by the impact of technological change. It was found that there is a statistically significant impact due to the age, job and place of the employees of technological change on organizational productivity. Furthermore, results indicate that there is statistical significance due to the experience of the employees. The employees with less than 10 years of experience feel more than others that there is an impact on productivity and satisfaction.

Keywords: Technological · Productivity · Satisfaction · Jordanian factories

1 Introduction

Industrial growth led to a change in different versions, and the development of new systems led to the ability to create knowledge and technology which become one of the basic components of the economy.

In light of the rapid changes in the work environment many reasons drive the organizations for change, including the development of the methods used in work, changes in policies, laws and regulations, the development of workers' awareness, the increase in their aspirations and needs, the change in the public's view and expectations from public or private institutions.

B. A. M. Alareeni and I. Elgedawy (Eds.): *Artificial Intelligence (AI) and Finance*, SSDC 488, pp. 96–106, 2023.
https://doi.org/10.1007/978-3-031-39158-3_9

The increase in the capacity of the market were established and reduce cost in a number of ways like, cost of operation by decreasing the working time needed to make products and achieving flexibility, increasing the market share of the competitive environment, and improving the quality of the product or service.

The impact of modern technology on the performance of organizations directly contributed to increase productivity, and when institutions use modern technologies such as automation and inventory control processes, this helps in managing time well, which leads to reducing costs incurred in the production process [1].

This study aims to know the impact of technological change on organizational productivity and customer satisfaction, where it was structured to achieves the aim and fills part of the gap that exists in the impact of technological change on productivity and customer satisfaction. The structure of the study designed to study the framework, the objectives, problem, and limitations as well as the statistical analysis.

2 Literature Review

The study showed that there is a relationship between the degree of technology used and the performance efficiency of the employees of the organization, because the organization lacks to educate employees about the importance of introducing technology to perform.

The study recommended that greater attention should be paid by the senior management to educate workers about the importance of developing technology before introducing the new technology to work, in order to ensure their acceptance and interaction with it [2].

The study reveals the importance of the use of educational technology on the academic achievement of students with learning difficulties and their motivation towards learning it in State of Qatar. The study recommended adopting the use of educational technology in teaching and designing curricula [3].

The study showed that there are differences between the efficiency of employees and the extent of technology used by employees at the Higher Institute of Science and Technology in Jameel. The employees have come up with a set of proposals that contribute to raising and improving the performance and thus improving their overall performance [4]. The study showed the impact of the use of technology on the effectiveness of employees in small companies. The study found that there is a correlation between the use of information and communication technology and job performance. Also, a statistically significant differences in the degree to which the personal characteristics of the sample members influence their answers about the level of job performance was found [5]. The study, aimed at showing the extent to which technology and communication with all its components contribute to improving marketing performance, profitability, market share, adaptation, product quality and customer satisfaction for institutions. For this purpose, the relationship between ICT infrastructure and marketing performance elements was tested.

It was found that the communication technology variable explains 42% of the variables that occur in the marketing performance, and the rest is due to other factors and explains the positive contribution of information and communication technology in maximizing the profitability and increasing the profitability of these institutions and in raising the market share and increasing the efficiency of product promotion and providing opportunities to enhance the current markets [6].

The study address that the companies must be consistent in the quality of the products they produce. Mechanization or automation of parts of production processes can help. To provide flexibility, companies often need to strike a balance between technology and human efficiency, where technological systems are beneficial to production [7]. The study presented the impact of technological development on receiving and producing economic information on television, by demonstrating the technological methods used in the production of the television image carrying economic information and showing its impact on the reception process by defining the basic comprehensive frameworks for the elements of this image. The study seeks to achieve the following objectives: to shed light on the role of technological development in the production of economic information and its impact on the reality of the local and global economy, and to highlight the changes in the process of receiving this economic information in its various dimensions [8]. The study showed that there are production techniques used to design products that affect Babel company, and it recommended that companies must use advanced technology in production.[9]. The study identified the organizational change strategies used by organizations in light of the conditions and rapid developments, which is the strategy of technological change and revealing the most important mysteries of its strategy and its role in facilitating the process of adopting total quality management [10].

3 The Framework (Fig. 1)

After referring to many studies specially to the studies of each of: Garrash (2020) and Mohamed (2022).

Dependent Variables: The first variable is the organizational productivity; the second variable is the customer satisfaction. The independent variable is the impact of technological change.

4 The Study Question

Technological change is essential in competition and therefore in the dynamics of the free market, and in order for comprehensive economic development to be achieved, it must have broader effects.

The study seeks to answer the following question: What is the impact of technological change on organizational productivity and customer satisfaction?

5 The Study Objectives

Technological changes are important in the performance of companies and in a change of the production processes through follow the rules and procedures in managing the organizational process and show sufficient attention to modernization and automation in

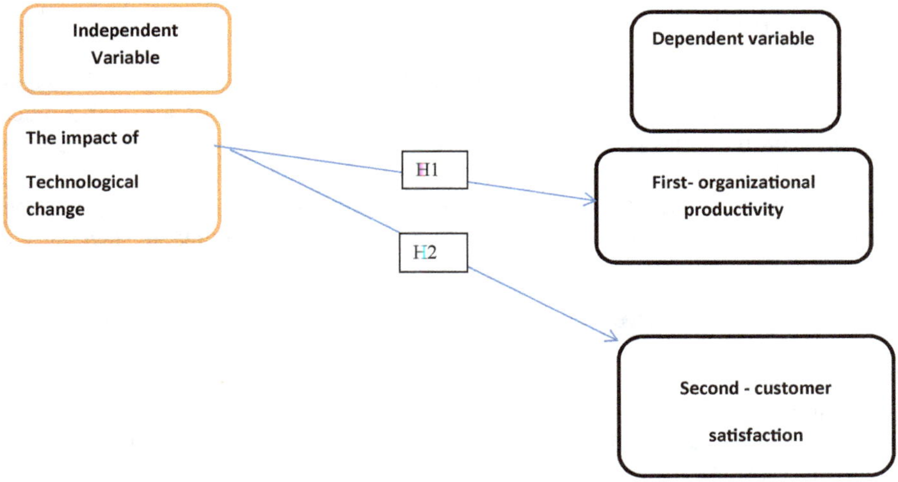

Fig. 1. Resources: developed by the researchers

its products, as it seeks excellence by trying to improve its current products and adopting modern technological methods that would achieve savings in time and effort. The aim of conducting this study is to identify the factors that affect productivity and customer satisfaction.

6 Study Instrument

To achieve the objectives of the study, a two-part questionnaire was designed. The first part is personal information about employees (gender, status, experience, age, Job, and place), while the second part is 36 paragraphs, corresponding to a five-Likert score (strongly agree = 5, agree = 4, neutral = 3, disagree = 2, strongly disagree = 1). These paragraphs consider the impact of technological change, the organizational productivity and customer satisfaction. The questionnaire was presented to a number of experts who have experience in the field for the purpose of arbitrating the questionnaire and indicating the clarity and comprehensiveness of the paragraphs. The questionnaire was modified in the light of the experts' recommendations. The questionnaire was applied to a number of employees in Jordanian factories who were not members of the sample. The reliability coefficient of the questionnaire was Cronbach's Alpha (0.803) which is acceptable for the purpose of the study.

7 The Study Problem

The poor productivity in a number of factories operating in Jordan and the inability to meet the needs of the local market and dissatisfaction of some customers, these problems led to technological changes in factories operations to reach the best productivity which fulfill the customers' needs.

8 The Importance of the Study

The study lies in its theoretical and practical importance, as the theoretical aspects, the study can add new knowledge and provide the Arab library with a new theoretical framework on the impact of technology on the productivity of organizations and on customer satisfaction.

From a practical point of view, there is a distinct addition to Jordanian organizations to reach the impact of using technology on productivity and customer satisfaction.

9 Terminology of the Study

Technology:
Technology is the set of skills, technical processors, and modern methods used in production [11].

Productivity:
A measure of production efficiency with respect to a specific factor, such as the number of workers, capital, investment volume, or production machine for a specific period of time [12].

Customer Satisfaction:
Satisfaction measures the extent to which the needs and expectations of customers are met through the services or products provided [13].

10 Statistical Analysis

The collected data from the questionnaire were analyzed using the statistical program (SPSS26) and descriptive statistics were used to calculate the means and standard deviations. ANOVA test and the Scheff posttest were also used.

11 Population and Sample Size

The population consists of all employees in 30 factories in Jordan. Randomly a sample of (220) employees was chosen and distributed as follows: 60.0% male, 49.1% married, 44.1% experience from 5 years to less than10, 52.7% age from 25 years to less than 35, 46.8% Line employee and 45.5% worked in Amman City.

12 Hypotheses

First: There are statistically significant differences at ($\alpha \leq 0.05$) between employees of Jordanian factories of the impact of technological change on organizational productivity due to personal variables (gender, status, experience, age, job and place).

Second: There are statistically significant differences at ($\alpha \leq 0.05$) between employees of Jordanian factories of the impact of technological change on customer satisfaction due to personal variables (gender, status, experience, age, job and place).

Third: There are statistically significant differences at ($\alpha \leq 0.05$) between of the impact of technological change on organizational productivity faced by employees of Jordanian factories.

Fourth: There are statistically significant differences at ($\alpha \leq 0.05$) between of the impact of technological change on customer satisfaction faced by employees of Jordanian factories.

13 Results

To answer the question of the study:

What is the impact of technological change on organizational productivity and customer satisfaction? Mean and standard deviations was found. Table 1.

Table 1. Descriptive Statistics

Variables	N	Minimum	Maximum	Mean	Std. Deviation
Production	220	3.00	5.00	4.3289	.31563
Satisfaction	220	3.44	5.00	4.2924	.30380
Technology	220	3.75	4.94	4.4145	.23985
Valid N (listwise)	220				

Table 1 shows that the mean of the organizational productivity that affected by the impact of technological change reached (4.3289) with a standard deviation of (0.31563). The mean of the customer satisfaction that affected by the impact of technological change reached (4.2924) with a standard deviation of (0.30380). This mean that the employees feel strongly that both the organizational productivity and the customer satisfaction affected by the impact of technological change.

14 Testing Hypotheses

First: There are statistically significant differences at ($\alpha \leq 0.05$ between employees of Jordanian factories of the impact of technological change on organizational productivity due to personal variables (gender, status, experience, age, job, and place). To test it, we conducted ANOVA test, and the results show that there is statistical significance due only to the age job and place of the employees. Table 2.

Table 2. ANOVA Test for organizational productivity

		Sum of Squares	df	Mean Square	F	Sig.
Age	Between Groups	1.482	3	.494	5.246	.002
	Within Groups	20.336	216	.094		
	Total	21.818	219			
Job	Between Groups	.927	2	.463	4.813	.009
	Within Groups	20.891	217	.096		
	Total	21.818	219			
Place	Between Groups	.939	2	.470	4.880	.008
	Within Groups	20.879	217	.096		
	Total	21.818	219			

To find out the differences between the age, job and place of the employees, "Scheff Test" was used as a posttest. Tables 3, 4 and 5 show the results.

Table 3. Scheffe of mean production and age

(I) age	(J) age	Mean Difference (I-J)	Std. Error	Sig.	95% Confidence Interval	
					Lower Bound	Upper Bound
1.00	2.00	−.26138*	.07277	.006	−.4664	−.0564
	3.00	−.24797*	.07979	.024	−.4728	−.0231
	4.00	−.32113*	.08565	.003	−.5625	−.0798
2.00	1.00	.26138*	.07277	.006	.0564	.4664
	3.00	.01342	.05191	.995	−.1328	.1597
	4.00	−.05975	.06054	.807	−.2303	.1108
3.00	1.00	.24797*	.07979	.024	.0231	.4728
	2.00	−.01342	.05191	.995	−.1597	.1328
	4.00	−.07317	.06882	.770	−.2671	.1207
4.00	1.00	.32113*	.08565	.003	.0798	.5625
	2.00	.05975	.06054	.807	−.1108	.2303
	3.00	.07317	.06882	.770	−.1207	.2671

The results of Table 3 indicate that the employees with age less than 25 years feel more than the employees with age from 25 years to 50 years and more, that there is impact of technological change on organizational productivity.

Table 4. Scheffe of mean production and job

(I) job	(J) job	Mean Difference (I-J)	Std. Error	Sig.	95% Confidence Interval	
					Lower Bound	Upper Bound
1.00	2.00	−.02454	.05922	.918	−.1705	.1214
	3.00	.11360	.05588	.129	−.0241	.2513
2.00	1.00	.02454	.05922	.918	−.1214	.1705
	3.00	.13813*	.04747	.016	.0211	.2551
3.00	1.00	−.11360	.05588	.129	−.2513	.0241
	2.00	−.13813*	.04747	.016	−.2551	−.0211

The results of Table 4 indicate that the line employees feel more than the head of department, that there is impact of technological change on organizational productivity.

Table 5. Scheffe of mean production and place

(I) place	(J) place	Mean Difference (I-J)	Std. Error	Sig.	95% Confidence Interval	
					Lower Bound	Upper Bound
1.00	2.00	−.10623	.04834	.092	−.2254	.0129
	3.00	−.15455*	.05373	.017	−.2870	−.0221
2.00	1.00	.10623	.04834	.092	−.0129	.2254
	3.00	−.04831	.05744	.702	−.1899	.0933
3.00	1.00	.15455*	.05373	.017	.0221	.2870
	2.00	.04831	.05744	.702	−.0933	.1899

The results of Table 5 indicate that the employees working in factories in Amman City feel more than the employees working in factories in Irbid City, that there is impact of technological change on organizational productivity.

So, we accept the hypothesis.

Second hypothesis: There are statistically significant differences at ($\alpha \leq 0.05$ between employees of Jordanian factories of the impact of technological change on customer satisfaction due to personal variables (gender, status, experience, age, job and place).

To test the second hypothesis, we conducted ANOVA test, and the results show that there is statistical significance due only to the experience of the employees. Table 6.

To find out the differences between the experiences of the employees, "Scheff Test" was used as a posttest. Table 7.

Table 6. ANOVA Test for customer satisfaction

Experience	Sum of Squares	df	Mean Square	F	Sig.
Between Groups	.592	2	.296	3.275	.040
Within Groups	19.620	217	.090		
Total	20.212	219			

Table 7. Scheffe of customer satisfaction and experience

(I) experience	(J) experience	Mean Difference (I-J)	Std. Error	Sig.	95% Confidence Interval	
					Lower Bound	Upper Bound
1.00	2.00	−.00192	.04573	.999	−.1146	.1108
	3.00	.12754	.05629	.079	−.0112	.2663
2.00	1.00	.00192	.04573	.999	−.1108	.1146
	3.00	.12946	.05423	.060	−.0042	.2631
3.00	1.00	−.12754	.05629	.079	−.2663	.0112
	2.00	−.12946	.05423	.060	−.2631	.0042

The results of Table 7 indicate that the employees of experience from 5 years to less than 10 years feel more than the employees with experience less than 5 years that there is impact of technological change on customer satisfaction.

So, we accept the hypothesis.

Third hypothesis: There are statistically significant differences at ($\alpha \leq 0.05$) between of the impact of technological change on organizational productivity faced by employees of Jordanian factories.

To test this hypothesis, we conduct the Simple Regression. Table 8.

Table 8. Results of Simple Regression for the third hypothesis

Dependent Variable	R	R Square	Independent Variable	F	SIG
Organizational productivity	.230a	.053	Technological change	12.174	.001a

Since F is significant (sig = .001) therefore all dimensions of the independent variable are to be taken into consideration. R Square =.053. This means that the independent variable (technological change) explain only 5.3% of the variance(variation) of the dependent (organizational productivity). It is clear from the table that the value of F was (12.174), and therefore we accept the hypothesis.

Fourth hypothesis: There are statistically significant differences at ($\alpha \leq 0.05$) between of the impact of technological change on customer satisfaction faced by employees of Jordanian factories.

To test this hypothesis, we conduct the Simple Regression. Table 9.

Table 9. Results of Simple Regression for the fourth hypothesis

	R	R Square	Independent Variable	F	SIG
Customer satisfaction	0.064a	0.004	Technological change	0.898	0.344a

Since F is not significant (sig = 0.344a) therefore all dimensions of the independent variable cannot be taken into consideration. R Square = 0.004. This means that the independent variable (technological change) explains only 0.4% of the variance(variation) of the dependent (customer satisfaction). It is clear from the table that the value of F was (0.898), and therefore we reject the hypothesis.

15 Conclusion

One of the findings of this study is that productivity and customer satisfaction are affected by the use of technology. The employees feel strongly that both the organizational productivity and the customer satisfaction affected by the impact of technological change. It was found that there is a statistically significant impact due only to the age, job, and place of the employees of technological change on organizational productivity. The employees with age less than 25 years feel more than the employees with age from 25 years to 50 years and more, that there is impact of technological change on organizational productivity and the line employees feel more than the head of department, that there is impact of technological change on organizational productivity. Also, the employees from Amman City feel more than the employees from Irbid City, that there is impact of technological change on organizational productivity.

The results of the study showed that the experience of employees is affected by the extent to which technological changes affect productivity and satisfaction, as employees with 10 years of experience feel more than others in this impact.

The study limited to the followings:

Human limits: Applied to workers in (30) industrial factories in Jordan.

Spatial limits: Applied to some factories in Amman, Irbid and Zarqa cities.

Time limits: The questionnaire was distributed to employees during the beginning of the academic year 2023.

Objective limits: Deals with the impact of the use of technology on the productivity of organizations and on customer satisfaction in a number of Jordanian factories.

The generalization of the results of this study depends on the seriousness of the study sample's response to the questionnaire's paragraphs.

16 Recommendation

We cannot hide the impact of modern technology on the performance of organizations and companies, as these advanced technological industries carried the nature of innovation, development, and flexibility in dealing. This development helped many companies and organizations to better manage the various departments and levels in the company. The study recommends the following:

Encouraging older employees to develop their skills and abilities in increasing the impact of technological change.

Enabling employees with little experience to act freely and implement good relationships to gain customer satisfaction and increase productivity.

Provide good environment for employees of both sexes to enhance the impact of technological change on organizational productivity and customer satisfaction.

Support employees in factories in various cities of Jordan.

References

1. https://www.meemapps.com/term/productivity
2. Issa, A., Al Shaibani, M.A.: The Impact of Technology on Workers' Performance: An Applied Study on the National Center of Meteorology for the Time Period (1994–2004). Master's Thesis, Omdurman Islamic University, Sudan (2006)
3. Ait Ambarek, S.: The Impact of Information and Communication Technology. Doctor of Science in Commercial Sciences Branch: Business Operations Management (2020)
4. Abbas, A.A., Al Hasnawi, H., Farhan Talib, A.: The impact of production technology in the design of the product: a field study at the General Company for the manufacture of tires in Babylon. Iraqi J. Admin. Sci. **29** (2011)
5. Saleh Saad, G.: The impact of the use of information technology on the job performance of employees in educational institutions (a field study of the Jameel Higher Institute of Science and Technology/Libya). Univ. J. **4** (2020)
6. Al-Mustafa, G.A.H.N., Ijbara, M.A.M.: The impact of the use of educational technology on the academic achievement of students with learning difficulties in mathematics for the fourth grade of primary school and their motivation towards learning it in the city of Doha in the State of Qatar International Journal of Quality Assurance (2020)
7. https://en.wikipedia.org/wiki/
8. https://hbrarabic.com
9. https://mawdoo3.com/
10. Aziz, M.A.: The role of technology in production and operations management, Dirasat **2** (2022)
11. Monkar, M., Maryam, S.: Technological change strategy under total quality management. Revue Des Sciences Humaines **26**(3), 565–583 (2015)
12. Bousaada, O.I.: The impact of technological development on the production and receipt of television economic information (2016)
13. Badaa, R: The impact of information and communication technology on job performance in small and medium enterprises, the case of Fogh Al Janoub Foundation. Master's Thesis, Kasdi Merbah University **1**(1) (2017)

The Role of Electronic-Trading Platform and Dividends Policies in Banks' Liquidity

Mohammad Sami Ali(✉)

Department of Financial Technology, Faculty of Economics and Administrative Sciences, Zarqa University, Zarqa 13110, Jordan
Maldahrawi@zu.edu.jo

Abstract. Since the review of available literature revealed that the rapid advancements in digital financial services conducts a vital role in banks' performance and investors' behavior; this paper aimed at scrutinizing the role of electronic-trading platform to mediate granger causalities between dividends and banks' liquidity as measured by the capital adequacy ratio, the growth rate of market capitalization, and the Z-Score index. Thus, through analyzing a monthly time series data for the period Jan/2010-Dec/2020; findings from the error correction model showed long-run correlations between banks liquidity and dividends policies, as well as the adopted digital platform is significantly mediating the long-run causalities between dividends and banks' liquidity. Nevertheless, the performance of banks is found to be significantly relating to the monthly change in the EPS ratio during the short-run. Moreover, since banks' liquidity is significantly impacted by the way of how investors react to dividends; the study concluded that the digital transformation contributes in improving banks' liquidity through stimulating cross border investors to invest in banks.

Keywords: Banks' liquidity · Dividends Policies · CAR · GRMC · Digital Transformation

1 Introduction

Due to the significance role of dividends policies in banks' stability; researchers in the field of corporate finance initiated plenty studies to assess the impact of dividends policies in the stability of banks and banks' liquidity. For instance, studies including Agyei (2011); Lintner (1956); Gordon and Shapiro (1956) showed that the value of banks is significantly impacted by dividends per share and the dividends payout ratio. By contrast, studies like Miller and Modigliani (1961); (Almeida, et al., 2015); Lilian (2016) argued that the value of banks is insignificantly related to dividends. In a later research Gordon (1963) insisted that investors prefer investing in companies, which are paying dividends instead of focusing on capital gains. However, since the continuous innovation in technology created a digitalized economic infrastructure, this helped financial institutions conducting most of their functions via digitally-enabled products and services such as e-banking, mobile banking, digital payment as well as trading systems. Thereby,

B. A. M. Alareeni and I. Elgedawy (Eds.): *Artificial Intelligence (AI) and Finance*, SSDC 488, pp. 107–119, 2023.
https://doi.org/10.1007/978-3-031-39158-3_10

this digital transformation is expected to influence banks' liquidity via helping stock market's investors to trade easily by the use of electronic trading platforms or what is known as virtual trading. Thus, since digital transformation facilitates the process of trading in ASE; it was announced that the launching of the NSC-V900 Electronic-Trading Platform by April/2009 played a vital role in improving the stock market's activity through attracting foreign investors and enabling them to access a wide range of benefits such as the high speed of executing orders and a high level of information transparency. Accordingly, the statistics revealed that after adopting the digital trading platform by ASE, the value of purchased shares by cross border investors increased by JD 228.8 million during the period April/2009-Dec/2015, while it is decreased by JD 15.52 million during the period 2016–2020. The statistics also revealed that the percentage of shares which are owned by foreign investors represented 48.1% of market capitalization by April/2009, compared with 51.1% by the end of 2020. Moreover, since the aforementioned discussion showed a gap regarding the impact of dividends polices and digital transformation in banks' liquidity; the current research aims at answering the role digital trading system to mediate granger causalities, which are stipulated to be run from dividends policies towards the performance of Jordanian listed banks, over the period Jan/2010-Dec/2020. Consequently, results from this research are expected to ascertain the role of dividends policies to impact the liquidity of banks as captured by the capital adequacy ratio "CAR", the Z-score index "ZSI", and the growth rate of market capitalization "GRMC". Consequently, that will assist investors to make more rational decision when they think to enter or exit the stock market. Additionally, results from this study will contribute in enhancing the growing body of literature which is focusing on the role of digital transformation to improve banks' liquidity as well as the trading activity of financial markets.

2 The Review of Related Literature

2.1 Dividends' Policies and Banks' Liquidity

Although the relevance dividends theory argues that a firm's dividends policy is significantly affecting a firm's value and stock prices, the irrelevance theory debates that a firm's value is not related to dividends. In other words, the relevant dividends policy believes that the change in dividends is significantly affecting a firm's value through impacting stock prices. This view was initially supported by Lintner (1962); Gordon (1963) through revealing a significant correlation between dividends and a firm's value. Similarly, since investors of this theory prefer dividends over capital gains, which are to some extent considered to be uncertain; authors including (Lundgren, 2016) asserted that a firm's value is significantly related to dividends policies. Nevertheless, the irrelevant policy argues that investors do not rely on dividends, while making their investment decisions. In this context, researchers like Miller and Modigliani (1961) debated that investors behave rationally, especially in perfect stock markets, since all information are available for them and can be accessed equally without any cost. However, for the aim of evaluating the effect of dividends policy in the performance of Kenyan listed banks; Lilian (2016) revealed that banks' performance as measured by ROA is insignificantly related to dividends per share. By contrast, Ali (2020) proved that the stability of banks is significantly responding to the change in dividends payout ratio during the short-run. Likewise, through investigating the effect of dividends policies in the market value of Kenyan banks (Luvembe, et al., 2014), showed that the value of banks is significantly and positively relating to both dividends payout ratio and corporate earnings, over the period 2006–2010. Furthermore, Imran, et al. (2013) proved that the decisions of dividends payout as captured by the current year dividends are statistically impacted by earnings per share, previous dividends payout, capital ratio and bank's size. Researchers including Ali (2016) debated that the liquidity of Jordanian commercial banks as captured by liquid assets/total assets and total deposits/total assets is significantly co-integrated with price/earnings ratio and market/book value ratio, over the long-run. By contrast, Jepkemoi, et al. (2019) argued that the dividends payout ratio is insignificantly related to banks' performance as captured by banks' capital structure and liquidity. Furthermore, in an attempt to examine the impact of Camels' components model on credit risks at Jordanian commercial banks, Yamin and Aldahrawi (2016) found that variables including the CAR, assets quality, market risks and revenues' quality are significantly impacting credit risks.

2.2 Digital Transformation and Banks' Structure

Although there is no general definition for digital finance, (Gomber, et al., 2017) stated that there is a consensus to define it as a term includes all financial services and product, which are delivered by financial technology firms and financial services suppliers, as well as a digitalized infrastructure that help investors to have access to invest, making payments, borrowing, or depositing online without the need of coping directly with the services' providers. Moreover, since the digital transformation of most financial institutions helped investors accessing a wide range of electronic services and products, decreased the costs and secured all forms of financial transactions that is resulted in enhancing the stability of financial institutions, supported economic stability as well as improved the performance of stock markets (Lisin, 2021). Thus, sue to the importance of digital transformation in banks' performance, the current section summarizes some previous researches regarding the impact of digital transformation and digital trading systems in banks' performance. For instance, in an attempt to measure the role of financial technology in the performance of 41 Indonesian banks (Phan, et al., 2018) revealed that the growth of Fintech registered firms is negatively impacting the performance of small and large Indonesian listed banks as captured by the NIM, ROA, ROE and the ratio of yield on earning assets. Furthermore, Studies including (Lee, 2009; Chen, 2011) debated that digital trading systems improved the profitability of financial institutions via the numerous features they are offering such as the higher trading speed, low operation and trading costs as well as the higher level of information transparency. Consistently, through studying the role of digitalized financial services in the opportunities and challenges of emerging banks (IFC, 2017) revealed that Fintech services such as linking the digital identity with bank's accounts leads to improve banks' performance through enabling these institutions to reach a huge number of customers, as well as facilitating the process of conducting bills payments and investment transactions. However, since the growth of nation's economies is highly relied on the activity of financial institutions, (Ozili, 2017) asserted that the providers of digital financial services are able to boost the growth of nation's economies, especially during stable economic periods through increasing the number of transactions, which are supposed to be executed in the financial system. In the same vein, (James et al., 2016) proved that digital financial services conduct an important impact in the gross domestic products of developing countries, where the statistics revealed that financial innovation is expected to increase the GDP of developing economies by 6% by the end of 2025. This $3.7 increase in the GDP is estimated to create about 95 million job opportunities across all sectors.

3 Research Methodology, Data Sources and Hypotheses

The present paper analysed a time series data covers the period Jan/2010-Dec/2020, in an attempt to assess the role of electronic-trading system to mediate the long and short-run causalities of dividends policies along with banks' liquidity. Thus, through capturing the liquidity of banks by variables like the CAR, ZSI and the GRMC; the needed data were mainly obtained from the published financial statements and reports of the listed banks, as well as the statistical bulletins of Amman stock exchange, the official site of Jordanian central bank, and the site of global economy (ASE, 2022; CBJ, 2022; Global Economy, 2022). Moreover, since results from the ADF test proved that the examined variables are integrated of the same order I (I); the Johansen cointegration test, the single equation of the vector error correction model, and the granger causalities of Wald χ^2 test have been implemented in order to check the long and short-run causalities, which are running from dividends and the electronic-trading system towards banks' liquidity. Whereas, the study applied the above mentioned statistical techniques to examine the following hypotheses:

H₁: *The monthly change in dividends is significantly impacting banks' liquidity during the long-run.*
H₂: *There are short-run causalities running from dividends policies towards banks' liquidity.*
H₃: *The Adoption of digital-trading platform by cross-border investors is significantly mediating the relationship of dividends policies along with banks' liquidity as represented by the capital adequacy ratio.*

3.1 Model Specification

Since results from Johansen test confirmed that the study's variables are integrated over the long-run, the following single equation of the VECM was applied:

$$\begin{aligned}\Delta JLBL_{it} = \boldsymbol{\beta}_0 &\pm \sum_{i=1}^{k-1} \boldsymbol{\beta}_{1i}\Delta JLBL_{t-i} \pm \sum_{i=1}^{k-1} \boldsymbol{\beta}_{2i}\Delta\sigma.EPS_{t-i} \pm \sum_{i=1}^{k-1} \boldsymbol{\beta}_{3i}\Delta\sigma.CGY_{t-i} \\ &\pm \sum_{i=1}^{k-1} \boldsymbol{\beta}_{4i}\Delta DS\_Dummy_{t-1} \\ &\pm \sum_{i=1}^{k-1} \gamma_1\sigma.EPS_{t-i} \pm \sum_{i=1}^{k-1} \gamma_2\sigma.CGY_{t-i} \pm \sum_{i=1}^{k-1} \gamma_3 DS\_Dummy_{t-i} \pm \psi_i\varepsilon_{t-1} \\ &\pm \mu_t \end{aligned} \quad (1)$$

where, the $\Delta JLBL_{it}$ represents the difference operator of banks' liquidity as measured by the CAR, Z-Score index, and the GRMC. β_i: i = 1… 4 relate to the short-run dynamics, and they are the parameters to be anticipated. While the γi, parameters relate to the long-run multipliers. $\Psi_i\ \varepsilon_{t-1}$ is the first lagged period of the error correction term, and μ_t refers to the white noise error term with usual properties. However, for the aim of assessing the role of digital trading platform to mediate the correlations of dividends along with banks' liquidity; the following model is going to be examined:

$$\begin{aligned}\Delta JLBL_{it} = \beta_0 &\pm \sum_{i=1}^{k-1} \beta_{1i}\Delta JLBL_{t-i} \pm \sum_{i=1}^{k-1} \beta_{2i}\Delta\sigma.EPS_{t-i} \pm \sum_{i=1}^{k-1} \beta_{3i}\Delta\sigma.CGY_{t-i} \\ &\pm \sum\nolimits_{i=1}^{k-1} \beta_{4i}\Delta DS\_Dummy_{t-1} \pm \sum_{i=1}^{k-1} \beta_{5i}\Delta \text{Mediator}\_ADTP\%_{t-i} \\ &\pm \sum_{i=1}^{k-1} \gamma_1\sigma.EPS_{t-i} \pm \sum_{i=1}^{k-1} \gamma_2\sigma.CGY_{t-i} \pm \sum_{i=1}^{k-1} \gamma_3 DS\_Dummy_{t-i} \pm \sum_{i=1}^{k-1} \gamma_4 Mediator\_ADTP\%_{t-i} \\ &\pm \psi_i\varepsilon_{t-1} \pm \mu_t \end{aligned} \quad (2)$$

where, the JLBL is the dependent variable and it relates to banks' liquidity as captured by the ± % change in the car adequacy ratio. The ADTP% refers to the adoption of digital-trading activity as proxied by the % change in the value traded of cross-border investors during the period Jan/2010–Dec/2020. Thus, after confirming the significant long-run integrations between dividends and banks' liquidity as well as the activity of adopting the digital-trading platform; this model would be tested to assess whether ADTP is valid to adjust the granger casualties, which are running from dividends towards banks' liquidity.

3.2 Variables Description and Measurements

The following table describes the study's variables and their measurements (Table 1):

Table 1. Variables Measurements

Independent Variables	Measurement
$EPS = \frac{\mu BNP^t}{\mu NBOS^t}$	Where, the BNP refers to the average of banks' net profits, divided by the average of banks' outstanding shares, during the time t
$CGY = \frac{\mu.SCP_t - \mu.SCP_{t-1}}{\mu.SCP_{t-1}}$	Relates to the difference between the average of banks' shares closing prices in the time t, and the average of shares' closing prices in the time t − 1, divided by the average of shares' closing prices in the time t − 1

(*continued*)

Table 1. *(continued)*

Independent Variables	Measurement
Dividends Stability_Dummy	It relates to banks' decisions regarding distributing dividends. This dummy took 1 when the relationship of EPS along with the dividends payout ratio is found to be positive, while -1 if it is found to be negative. $DS\_DUM = +1 \text{ or } -1 \overset{yields}{\rightarrow} EPS\% \underset{\pm}{\leftrightarrow} DPR\%$
Dependent Variables	Measurement
$\boldsymbol{CAR} = \frac{\left[\begin{array}{c}(\boldsymbol{Tier1Capital} - \boldsymbol{Goodwill}) + \\ Tier2Capital\end{array}\right]}{\boldsymbol{Risk-weightedAsset}}$	It relates to the capital anticipated to maintain balance with the risks exposure of banks, such as operational risks, market risks and credit risks
$\mathbf{ZSI} = \frac{\frac{\left(\sum ROA + \left(\frac{\sum Equity}{\sum Assets}\right)\right)}{\sum \sigma.ROA}}{TNLB_t}$	The $\sum$ROA represents banks' return on assets ratio, $\sum$Equity relates to banks' equity, $\sum$Assets refers to banks' total assets, the σ.ROA stands for the standard deviation of return on assets ratio, and the $TNLB_t$ refers to the total number of listed banks during the time t
$\boldsymbol{GRMC} = \frac{[(\mu.SCL_t \times NBOS_t) - (\mu.SCL_{t-1} \times NBOS_{t-1})]}{\mu.SCL_{t-1} \times NBOS_{t-1}}$	Relates to the difference between the average banks' stocks closing prices multiplied by the number of banks' outstanding shares during the time t, and t − 1, divided by the average banks' stocks closing prices multiplied by the number of banks' outstanding shares during the time t − 1
Mediator **The Activity of Digital-Trading Platform**	It is measured by the growth rate of the traded value of the listed financial sector's shares, which are purchased digitally by cross-border investors, during the period Jan/2010–Dec/2020

4 Results Discussion

After running the ADF, the results confirmed that all the time series variables became stationary only after including the first difference. Additionally, findings from the Johansen test and the error correction model affirmed that the liquidity of banks as represented by the CAR, ZSI, and the GRMC are significantly relating to the monthly changes in dividends measures like EPS, CGY, and DS during the long-run. Accurately, the statistics exhibited that a 1% increase in EPS decreases banks' liquidity as proxied by CAR ratio by 0.49%, while the 1% increase in the CGY is found to decrease the CAR ratio by 0.40%, and the 1% increase in dividends stability decreases banks' liquidity by 0.03%. Although, the 49% improvement in the ZSI is caused by the +1% change in EPS, however, the 46% and 2.34 decline in the ZSI is found to be driven by the +1% in DS and CGY, respectively. The results also proved that the market capitalization of banks grew positively by 0.47%, 25.9%, and 0.90% due to the +1% variation in CGY, EPS and DS, consecutively. However, results from the Wald χ^2 test showed except for the Z-score, the liquidity of banks is significantly related to the monthly change in the EPS during the short-run. Based on the aforementioned results, it is induced that the investors of ASE adopt rely on the relevance and irrelevance dividends theories when they want to enter or exit the stock market. Furthermore, in an attempt to assess the role of digital trading in the relationships of dividends along with banks' liquidity; findings from the ECM confirmed the existence of long-run association between policies and digital transformation along with banks' liquidity. Whereas, a 1% increase in dividends stability and EPS increases the CAR ratio by 0.01% and 0.61% respectively. Therefore, it can be concluded that cross-border investors are following the relevance dividends theory, due to the positive correlation of EPS and DS along with banks' liquidity. However, since the increase in CGY deteriorates the CAR ratio by 0.34%; the study concluded that foreign investors are also following the irrelevance dividends theory, because the increase in CGY doesn't motivate them to invest in the listed financial sector as captured by the value traded. Beyond that, the results revealed that a +1% increase in the traded value of cross border investors improves the liquidity of Jordanian listed banks by 0.02%. Therefore, electronic-trading system is significantly and positively mediating the long-run correlations between dividends and banks' liquidity. For example, previous results revealed that when the CAR ratio was only regressed on dividends, it was found to be negatively relating to the change in EPS (−0.49%), DS (−0.03%), and CGY (−0.40%); while when the mediator was included in the model; it is found to be positively affected by measures including the EPS (+0.61%) and DS (+0.01%). On the other hand, statistics from the Wald χ^2 test confirmed that the adoption of digital-trading platform is insignificantly mediating the correlations between dividends and banks' liquidity, over the period Jan/2010-Dec/2020 (see Tables 2, 3, 4, 5, 6, 7, 8 and 9 in the appendix).

5 Conclusion

Despite the numerous related studies regarding the impact of dividends policies in banks' liquidity and investors' behaviour, there is research investigated the role of digital transformation in the correlations between dividends and banks' liquidity as captured by the CAR, ZSI, and GRMC. Thus, after assessing role of electronic-trading platform to mediate the long and short run causalities, which are running from EPS, CGY, DS towards banks' liquidity, over the period Jan/2010-Dec/2020; results from the ECM confirmed that the liquidity of banks is significantly relating to the monthly change in the used dividends measures during the period Jan/2010-Dec/2020. Accordingly, the study concluded that investors of ASE rely on the assumptions of relevance and irrelevance dividends theories, when they think to acquire banks' stocks. Consistently, researchers including Lintner (1956); Gordon and Shapiro (1956) affirmed that dividends' policies are significantly linked with a firm's value. By contrast, through relying on the assumptions of the irrelevance dividends theory; Miller and Modigliani (1961) argued that the wealth of shareholders and firm's value are insignificantly related to dividends policy (Almeida, et al., 2015). However, the study concluded that investors of Amman stock exchange are adopting the views of (Lintner, 1956); Gordon and Shapiro, 1956; Miller and Modigliani, 1961). However, since results from Wald χ^2 test revealed that except for EPS, liquidity's measures including CAR and GRMC are not affected by dividends' policies during the short-run; it is concluded that cross-border investors are found to follow the principles of the relevance dividends theory during the short-run. This result disagrees with Lilian (2016) since she debated that banks' performance is insignificantly relating to the dividends per share ratio. Furthermore, since findings from the ECM asserted that the value trading of cross-border investors is significantly mediating the long-run correlations between dividends and banks liquidity; the study concluded that digital transformation conducted a vital role in the liquidity of financial institutions, which are listed in ASE. Moreover, the study induced that the way of how banks formulating their dividends' conduct a vital role in affecting banks' liquidity via impacting the behavior of their investors in the long-run. Therefore, study recommends that it is necessary for Jordanian listed banks to rely on dividends' decisions in order to understand investors' behaviour during the long-run.

Appendix

The following statistics demonstrate the role of digital-trading in granger causalities between dividends policies and banks' liquidity:

The following statistics describe the long-run integrations between dividends policies and banks' liquidity.

Table 2. Findings from Johansen Test

Vecrank BS, CGY, EPS, DS, trend (constant) lags (3) Level99						
Trend: Constant						# Obs. = 130
Sample: Mar/2010-Dec/2020						Lags = 2
Maximum						
Rank	Parms	H0	CAR	Z-Score	MC	1% Critical
			Trace Statistic	Trace Statistic	Trace Statistic	
0	20	$R = 0$	157.43	137.52	136.37	54.46
1	27	$R \leq 1$	85.158	64.882	63.666	35.65
2	32	$R \leq 2$	39.632	30.798	31.358	20.04
3	35	$R \leq 3$	7.408	7.567	6.2673*	6.65
4	36	–	–	–	–	–

Since the above table confirmed that there are long-run integrations between dividends and banks' liquidity; the study applied the ECM to check for long-run causalities (see Table 4, 5, 6).

The following table confirmed that the liquidity of banks as represented by the CAR ratio is significantly related to the monthly change in EPS, CGY and the DS, over the long-run.

Table 3 ECM Results

	Coef.	Std. Error	Z	P > /Z/	99% Conf.	Interval
D_CAR						
_Ce1						
L1	−0.6770	0.1072	−6.32	0.0001	−0.9532	−0.4009
CAR						
LD	0.2547	0.1054	2.42	0.016	v0.0167	0.5263
EPS						
LD	−0.0049	0.0357	−0.14	0.891	−0.0970	0.0872
DS						
LD	−0.0003	0.0009	−0.34	0.732	−0.0028	0.0021
CGY						
LD	−0.0040	0.0050	−0.79	0.429	−0.0170	0.0090
_Cons	0.0013	0.0008	1.65	0.122	−0.0008	0.0035

This table exhibits that banks' liquidity as captured by the Z-Score is significantly granger caused by dividends policies, over the long-run.

Table 4 ECM Results

	Coef.	Std. Error	Z	P > /Z/	99% Conf.	Interval
D_Z-Score						
_Ce1						
L1	−0.4378	0.0969	−4.52	0.0003	−0.6875	−0.1881
Z-Score						
LD	0.4952	0.2018	2.45	0.014	−0.0247	1.0153
EPS						
LD	80.468	30.853	2.61	0.009	0.9946	159.94
DS						
LD	−0.4639	0.4196	−1.11	0.269	−1.5449	0.6170
CGY						
LD	−2.3471	2.4857	−0.94	0.345	−8.7500	4.0556
_Cons	0.00005	0.3582	0.00	1.000	−0.9226	0.9227

The following table asserted that the growth rate of banks' market capitalization is being granger caused by the monthly movements in dividends' measures during the long-run.

Table 5. ECM Results

	Coef.	Std. Error	Z	P > /Z/	99% Conf.	Interval
D_MC						
_Ce1						
L1	−0.1174	0.0454	−2.58	0.010	−0.2346	−0.0003
MC						
LD	0.0436	0.0940	0.46	0.642	−0.1985	0.2857
EPS						
LD	0.2591	0.5044	0.51	0.607	−1.0402	1.5586
DS						
LD	0.0090	0.0132	0.68	0.498	−0.0252	0.0432
CGY						
LD	0.0479	0.0738	0.65	0.517	−0.1423	0.2381
_Cons	−0.0349	0.0160	−2.18	0.029	−0.0762	0.0063

The following table proved that except for the Z-score, the liquidity of banks is significantly related to the monthly change in the EPS during the short-run.

Table 6. Results from Wald χ^2 Test

Variables	CAR		Z-Score		MC	
	Chi2 (2)	Prob. > Chi2	Chi2 (2)	Prob. > Chi2	Chi2 (2)	Prob. > Chi2
EPS	14.27	0.0065	8.74	0.0680	22.68	0.0001
CGY	1.94	0.7466	2.29	0.6817	3.96	0.4115
DS	1.43	0.8396	2.13	0.7117	2.08	0.7212

The following table describes statistics regarding the long-run integrations among dividends, digital-trading platform and banks' liquidity.

Through relying on the above results, the following ECM test was implemented in order to check whether digital trading is valid to mediate granger causalities of dividends along with banks' liquidity.

Since results from Table 4 confirmed that there are long-run causalities running from dividends towards CAR, this table provides statistics regarding long-run causalities of dividends' policies along with the activity of digital-trading platform.

Through relying on the above results, the following ECM's test investigated the role of digital-trading to mediate the correlations between dividends and banks' liquidity.

The following table proves that digital-trading platform is impacting granger causalities between dividends and the liquidity of banks, over the long-term.

Table 7. Findings from Johansen Test

VEC-rank Banks' Liquidity, trend(constant) Level99, Lags (3) max									

Sample: Mar/2010-Dec/2020
#Obs. = 130
Lags = 2									

Rank	Parms	H0	Trace Statistic		1% Critical Value	Rank	Parms	Trace Statistic	1% Critical Value
			CAR, Dividends	Digital, Dividends				CAR, Dividends, Digital	
0	20	R = 0	157.43	131.13	54.46	0	30	190.94	76.07
1	27	R ≤ 1	85.158	61.756	35.65	1	39	119.61	54.46
2	32	R ≤ 2	39.632	28.512	20.04	2	46	60.357	35.65
3	35	R ≤ 3	7.408	6.887	6.65	3	51	26.652	20.04
4	36	–	–	–	–	4	54	6.8174	6.65
						5	55	–	–

Table 8. Results from ECM

	Coef.	Std. Error	Z	P > /Z/	99% Conf.	Interval
D_Digital						
_Ce1						
L1.	−0.2067	0.0627	−3.30	0.001	−0.3683	−0.0452
Digital						
LD.	−0.1303	0.0882	−1.48	0.139	−0.3576	0.0968
EPS						
LD.	0.0206	0.0168	1.23	0.220	−0.0226	0.0639
DS						
LD.	−0.0006	0.0004	−1.42	0.155	−0.0018	0.0005
CGY						
LD.	−0.0006	0.0024	−0.27	0.788	−0.0069	0.0055
_Cons	−0.0001	0.0004	−0.38	0.707	−0.0011	0.0008

Table 9. Results from ECM

	Coef.	Std. Error	Z	P > /Z/	99% Conf.	Interval
D_CAR						
_Ce1						
L1.	−0.5944	0.1031	−5.76	0.000	−0.8600	−0.3287
CAR						
LD	0.2551	0.1154	2.21	0.027	−0.0423	0.5526
DS						
LD.	0.0001	0.0010	0.19	0.849	−0.0024	0.0028
EPS						
LD.	0.0061	0.0733	0.08	0.934	−0.1828	0.1950
CGY						
LD.	−0.0034	0.0052	−0.66	0.512	−0.0169	0.0100
Digital						
LD.	0.0002	0.0078	0.04	0.970	−0.0205	0.0199
_Cons	0.0001	0.0008	0.14	0.887	−0.0020	0.0023

References

Ali, M.S.: Evaluating the Effectiveness of the Bird-in-Hand-Dividends Policy in the Stability of Jordanian Listed Banks. Int. J. Financ. Res. **11**(04), 96–110 (2020)

Ali, M.S.: The impact of market fundamentals and financial crisis on the liquidity of banks and Stock Markets: Evidence from Jordanian investors. Unpublished doctoral dissertation, Anglia Ruskin University (2016)

Almeida, L.A.G., Pereira, E.T., Tavares, F.O.: Determinants of dividend policy: evidence from Portugal. Rev. Brasileira Gestao Negocios: Rev. Bus. Manage. (2015)

Amman Stock Exchange (2022). https://www.ase.com.jo/. Accessed 13 September 2022

Central Bank of Jordan (2022). http://www.cbj.gov.jo/. Accessed 18 September 2022

Chen, M.Y.: Bankruptcy prediction in firms with statistical and intelligent techniques and a comparison of evolutionary computation approaches. Comput. Math. Appl. **62**, 4514–4524 (2011)

The Global Economy (2022). https://www.theglobaleconomy.com/. Accessed 08 October 2022

Gomber, P., Koch, J.-A., Siering, M.: Digital finance and FinTech: current research and future research directions. J. Bus. Econ. **87**(5), 537–580 (2017). https://doi.org/10.1007/s11573-017-0852-x

Gordon, M.J.: Optimal Investment and Financing Policy. J. Financ. **18**(2), 264–272 (1963)

Gordon, M.J., Shapiro, E.: Capital equipment analysis: the required rate of profit. Manage. Sci. **3**(1), 102–110 (1956)

IFC: Frsh Ideas about Business in Emerging Markets. World Bank Group, Note 42 (2017)

Imran, K., Usman, M., Nishat, M.: Banks dividend policy: evidence from Pakistan. Economic Modeling **32**(2013), 88–90 (2013)

James, M., Susan, L., Marc, S., Olivia, W., Chris, B.: Digital Finance for all: Powering Inclusive Growth in Emerging Economies. Technical report, McKinsey Global Institute (2016)

Jepkemoi, A., Odoyo, F., Simiyus, R., Onyango, R.: The influence of financial performance on dividend payout ratio of commercial banks listed at the Nairobi securities exchange, Kenya. Strategic J. Bus. Change Manage. **6**(3), 371–387 (2019)

Lee, M.C.: Predicting and explaining the adoption of online trading: an empirical study in Taiwan. Decis. Support Syst. **47**, 133–142 (2009)

Lisin, A., et al.: Digital Trading Applications and Bank Performance: Evidence from Russia. J. Open Innov. Technol. Mark. Complexity **7**(194), 1–15 (2021)

Lilian, N.: Effects of dividend policy on financial performance of listed banks in Kenya. Unpublished Master thesis, University of Nairobi (2016)

Lintner, J.: Dividends, earnings, leverage, stock prices and supply of capital to corporations. Rev. Econ. Stat. **64**, 243–269 (1962)

Lintner, J.: Distribution of incomes of corporations among dividends, retained earnings, and Taxes. Am. Econ. Rev. **46**(2), 97–113 (1956)

Lundgren, B.: Ownership structure's effect on dividend policy: evidence from publicly listed Swedish firms. Unpublished Master Thesis, Umea University (2016)

Luvembe, L., Njangiru, M., Mungami, E.: Effect of dividend payout on market value of listed banks in Kenya. Int. J. Innov. Res. Dev. **3**(11), 1–21 (2014)

Miller, M., Modigliani, F.: Dividend policy, growth, and the valuation of shares. J. Bus. **34**(4), 411–433 (1961)

Mojgan, S., Ali, M.: Examining the effect of earnings per share and cash dividends per share on investor decision making in tehran stock exchange from the capital market participants' view. Am. J. Sci. Res. **36**, 99–106 (2011)

Phan, D., Narayan, P., Rahman, R., Hutabarat, A.R.: Do financial technology firms influence bank performance? J. Pre-proof (2018)

Yamin, I., Aldahrawi, M.S.: The impact of CAMELS' components on the credit risks that commercial banks listed in Amman stock exchange face. Zarqa J. Res. Stud. Human. **16**(3) (2016)

Reflections on the Impact of Digital Transformation on Criminal Policy

Jalal Hussein Al-Kayid[1(✉)], Suhaib Ahmed El-Manaseer[1], Abdelaziz Mahmoud Al khawatreh[1], and Ahmad Abdulkadir Ibrahim[2]

[1] Faculty of Law, Zarqa University, Zaraqa, Jordan
{jalkayid,salmanaseer,Akhawatreh}@zu.edu.jo
[2] College of Law, University of Business and Technology, Jeddah, Saudi Arabia
a.abdulkadir@ubt.edu.sa

Abstract. Digital transformation has affected various aspects of economic, social, and legal life, and has even impacted a country's vision in adopting a penal policy by facing perpetrators of crimes. This study aims to examine the impact of digital transformation on penal policy and to explain how the country's objective of applying established penal policy can be transferred to employing digital transformation and information technology communication tools in adopting a new concept for the policy. By adopting and utilizing digital transformation tools, the study proposes a new and modern model for penal policy from a new technological perspective. The study addresses the issue that revolves around the ability of the penal policy concept to adapt to technological developments through two topics. The first topic focuses on the digital transformation of penal policy by explaining the concept of digital transformation and its impact on the concept of penal policy, the second topic was dedicated to talking about digital transformation images and their impact on penal policy. The second topic was dedicated to explaining the impact of digital transformation on penal legislation. This topic was divided into two subjects, the first of which was dedicated to discussing the impact of digital transformation on the implementation of penal judgments, while the second subject was dedicated to talking about the impact of digital transformation on the penal claim. This study resulted in several conclusions and recommendations that will be presented. The study concluded that electronic systems, represented by the availability of the underlying infrastructure, help claimants to present their electronic claims through a specialist website belonging to the Ministry of the judiciary, responding to electronic complaints through a specific notification or application to accept complaints related to the applicants shows that judicial practices require digitalization instead of being paper-based. Implementing electronic records requires a digital infrastructure, and this study recommends the necessity of ensuring the protection of the policy of punishment for data and information of individuals and organizations used in the framework of justice represented in the courts, by strengthening criminal responsibility for the dissemination of this data and information or its trading, This will enhance confidence in electronic systems and help in achieving digitization more quickly. This study also recommended that the Jordanian legislator need to amend the laws that govern the punitive policy, which is represented in the Penal Code and the principles of criminal trials, in a manner commensurate with the digital transformation. The study also recommended the necessity of linking all state institutions through the information

B. A. M. Alareeni and I. Elgedawy (Eds.): *Artificial Intelligence (AI) and Finance*, SSDC 488, pp. 120–133, 2023.
https://doi.org/10.1007/978-3-031-39158-3_11

network, to facilitate the registration of criminal cases and final decisions, in addition to the conductor of investigation procedures such as hearing witnesses and seizing criminal evidence using modern technologies, for what this achieves to facilitate procedures and maintain the safety of witnesses and others. And the adoption of electronic monitoring as a means of implementing imprisonment and short-term custodial sentences.

Keywords: punitive policy · digital transformation · electronic wristband · electronic monitoring

1 Introduction

The digital transformation in devices, machines, smart systems, and the increased ability to process data and artificial intelligence will definitely lead to unprecedented development in penal policy, and in major changes in it. And there is no doubt that the maturity of the information technology era is necessary to adapt our laws, regulations, and policies to some extent. However, penal policies in most cases are difficult to adapt to the technological changes at times. This results in delays in laws and regulations in keeping up with progress, leading to gaps in the orderly penal policy for digital transformation, it is worth noting that the legal systems in their various types and forms strive to achieve justice within our society and among its citizens, or at least to approach it in a way that achieves satisfaction and acceptance among them; Therefore, the concept of justice evolves as a subject and topic in the legislative law, and cannot be stabilized or achieved except by excluding divine and divine justice, which emphasizes the limitations of the legislative law, legislation, and systems in achieving justice between people, even if they are keen on it. This is because following penal policies for different countries sometimes finds that they have failed in some of their tasks.

The problem of the study:

We find that the problem of the study lies in the legislator's lack of attention to the huge development of digital transformation and its impact on penal policy in line with the reflections of the digital revolution, whether in facing cybercrime or in preserving the rights of individuals and society in the digital environment.

Questions to consider:

1. What is the nature of digital transformation?
2. Explain the nature of digital transformation as one of the innovations of information technology and its impact on criminal policy?
3. What are the images of criminal policy?
4. What are the reflections of digital transformation on the implementation of criminal sentences?
5. What are the obstacles that hinder the effectiveness of digital transformation on criminal policy?
6. What is the legal nature of the reflections of digital transformation on criminal policy?

Study importance:

The importance of the study lies in highlighting the impact of digital transformation on some areas related to the legal and judicial field in general, and the criminal justice system in particular.

Objectives:

1. To identify the legal provisions related to the effects of digital transformation on the criminal justice system.
2. To analyze the legal adaptation of digital transformation effects on the criminal justice system.
3. To find a balance between digital transformation and the criminal justice system applied in Jordan.
4. To examine the role of judicial oversight in enforcing penalties on digital transformation actors.

Methodology of the Study:
In this study, we followed a descriptive and analytical methodology that reviews the legal texts provided by the Jordanian legislator in this regard, analyzes and discusses them to explain the reflections of digital transformation on criminal policy. We will also analyze legal and jurisprudential opinions regarding the issues raised in this study and relevant court rulings if found in Jordanian court rulings. The comparative methodology will also be used if required.

Study Plan:
In order to provide an overview of the study's subject, it will be divided into two topics. The first topic will deal with the digital transformation of criminal policy, which will be divided into two subtopics. The first subtopic will be dedicated to explaining the nature of digital transformation and criminal policy, while the second subtopic will explain the images of criminal policy. The second topic will deal with the legal nature of the reflections of digital transformation on criminal policy, which will be divided into two subtopics. The first subtopic will discuss the reflections of digital transformation on legislation, while the second subtopic will focus on the reflections of digital transformation on criminal proceedings and the implementation of court rulings.

2 The Digital Transformation of Criminal Justice Policy

Preface:
Digital transformation refers to the set of actions that an organization takes to integrate digital technology into all areas of business, leading to a fundamental change in how the organization serves its customers. Companies use innovative digital technologies to make cultural and operational changes that better align with changing customer requirements [1]. It is worth noting that the concept of digital transformation raises many questions about ensuring and achieving digital transformation in practice, as well as researchers' questions about the punitive policy and the concept of digital transformation in its scope, and the impact of this transformation on such policy, and the presentation of it as a form of services to humanity? And to study the impact of reflecting digital transformation

on punitive policy, researchers will address the concept of digital transformation and its impact on the concept of punitive policy in the first request, and we will dedicate the second request to explaining the images of digital transformation and its reflection on punitive policy.

2.1 The Nature of Digital Transformation and Penal Policy

Given the recentness of digital transformation as a distinctive phenomenon, sufficient theories have not been put forward yet, and often when we study this transformation, we find that it is a long and complex process, including many digital innovation chains. We emphasize that the process of digital transformation related to this field will lead to interactions between reform environments and enforcement bodies implementing criminal sentences. To explain that, we will divide this topic into two branches. The first branch will explain the concept of digital transformation and penal policy. The second branch will be dedicated to discussing the benefits and drawbacks of digital transformation and penal policy.

2.1.1 The Concept of Digital Transformation and Punitive Policy

The state strives to protect the criminal interest on three levels: policy of repression, policy of prevention, and policy of punishment. Punitive policy is one of the most important elements that has been given attention by the intellectual schools in criminal policy [2], due to the importance of the objectives it seeks to achieve, in addition to the special and general deterrents, in order to protect the community and seek to reform criminals by rehabilitating, qualifying and integrating them socially.

2.1.1.1 The Definition of Digital Transformation

Digital transformation is used as a comprehensive term within organizations and institutions, but it can have an impact on a number of different areas. Digital transformation is also defined in several ways, including: the process of moving companies to a model of work that relies on digital technologies in the invention of products and services, and providing new channels of revenue and opportunities to increase product value. Some also define it as: the process of moving companies to a model of work that relies on digital technologies in the invention of products and services, and providing new channels of revenue and opportunities to increase product value, with the transformation in businesses or governments, meaning making fundamental changes that affect the work model [3].

In conclusion, digital transformation in this context is about formulating a digital strategy that starts by assessing the current situation, focusing on the gap between current digital capabilities and what should be in the future, then working on implementing that strategy by allocating the necessary resources, then monitoring its implementation and continuously evaluating the performance of operations, updating databases, developing human resources in line with digitization, which leads to performing electronic operations professionally with a high degree of efficiency, minimizing errors and overcoming obstacles and dealing with them on an ongoing basis, with the aim of ultimately reaching better production methods and providing services technologically [4].

2.1.1.2 The Definition of Punitive Policy

The philosophy of punitive policy, according to its modern scientific understanding and its concepts, is based on the fact that crime is a social phenomenon that has social and environmental causes surrounding it [5]. Therefore, the process of combating crime is not limited to limited or restricted means of imposing punishment on criminals, or only implementing it, but what is the alternative? Therefore, as a prelude to standing on the definitions and concepts that have been raised in the field of criminal policy, which differ according to the intellectual and philosophical references of each party and each direction.

1. The meaning of the term "penal policy": in the Arabic language is to take action on something to correct it and the policy is in fact to correct the situation, it is said that the ruler corrects the affairs of the state and the guardian corrects the affairs of his subjects. It also came in the middle dictionary as "directing the affairs" meaning managing and correcting them.
2. Penal policy legally means: legal scholars have differed in defining criminal policy, and the definitions have varied with the philosophical and scientific references, as well as the intellectual and political orientations. Professor Ahmad Fathi Sourour defined it as: "It sets the rules that determine the formulation of criminal law texts, whether in relation to incriminating, preventing crime, or dealing with it, and in other words, the criminal policy is that which indicates the necessary principles to be followed in determining what is considered a crime and taking preventive measures and penalties prescribed for it." Others have defined it as an executive concept, which are the ideas and principles determined by the state or authority [6].

A general definition of criminal policy can be that it is a set of means and measures taken by the state at a certain time period to combat crime and maintain security and stability within its borders. It should be noted that there are two concepts of penal policy, one narrow and the other wide, but before discussing these two, a comprehensive and concise definition of criminal policy has been provided, which considers it as a specific theoretical framework for resolving the inevitable conflict between crime and society.

In summary, the modern theory of penal policy aims to provide a comprehensive vision of the phenomenon of crime by researching and investigating its causes and motivations through targeted statistical studies for each region within the single state and finding effective solutions and measures that are beneficial.

2.1.2 The Risks of Digital Transformation and Cyber Policy

The process of digital transformation is not free of risks, and some of the reasons that have led to an increase in digital risks are:

1. The heavy dependence on digital technologies, especially in the light of the digital transformation in government administrative work and the trend towards electronic public facilities, which is witnessing great growth, in light of the fears and risks that face the rising and escalating electronic and cyber-attacks in the world [7].
2. The increase in the number of targeted sites due to the abundance of connected devices and the increase in the complexity of electronic attacks.

3. Digital innovation exceeded the measures of electronic security
4. The integration between information technology systems and operational technology and the Internet of Things.

In view of these risks, the relationship between the risks of digital transformation and cyber security is highlighted, and cyber security is defined as: security of information on devices, computer networks, and operations and mechanisms that protect computer equipment, information, and services from any unintended or unauthorized intervention or change or difference that may occur, using a set of technical, organizational and administrative means to prevent unauthorized use, prevent misuse and recover electronic information, communication systems and information that contains, as the cyberspace is not limited to the Internet only [8].

2.2 The Applications of Digital Transformation and Its Impact on Cyber Policy

The rapid growth of digitalization and the internet and the speed of flow in the world have opened new doors and opportunities to drive economic growth, and innovation and create new business models. Also, it helps governments and systems in all types of development and creates a prosperous social environment. Digitalization is not just a part of the economy, and therefore, not controlling digitalization in a scientific way in all areas, such as commerce, public administration, justice, health, and security, inevitably widens the gap with developed countries in this field, which will be difficult to reduce in the future [9].

It is clear that increasing digitization enhances rights and freedoms such as freedom of expression and opinion while weakening others such as the right to privacy, the right to security and safety, and the right to forgetfulness. This has been confirmed that the development of digitization in the field of justice raises real problems for those in charge of justice as a whole and for those who seek it for themselves. These problems increase when it comes to criminal justice, which fundamentally looks at basic rights such as the right to life, the right to physical and mental safety, and the right to fair trial, in which the judge acts independently, impartially, and with integrity, and the interaction between parties is crucial.

The introduction of electronic ankle bracelets as an alternative to short-term negative freedom penalties is a result of the evolution of penal systems and in line with contemporary penal policy aimed at rehabilitating offenders and protecting society from crime by keeping the person away from crime-prone places and environments, while also keeping the person in constant contact with their family, which helps in their rehabilitation and reintegration. It also reduces the costs that the state bears compared to the costs of imprisonment, overcrowding in prisons, mixing of prisoners, the spread of diseases, acquisition of criminal skills and patterns, etc.

The Ministry of Justice has completed its program to introduce electronic monitoring technology as an alternative to judicial detention. The electronic monitoring device is used for electronic surveillance and is designed in a way that does not impede the person's daily activities. Electronic monitoring is a device that is worn on the lower leg of the person subject to electronic surveillance. If the device is removed, tampered with, or the person enters a prohibited geographical area, a signal is sent to the authorities, and the nearest police station is notified to take necessary action. This method also involves placing an electronic monitoring device on the wrist or ankle of the person being monitored, which sends specific signals to a receiving device connected to a telephone line at the place of residence of the person being monitored. The receiving device then sends specific signals to the monitoring authorities, which identify the presence of the person being monitored in the specified area. As for the conditions of applying this modern technology.

3 The Impact of Digital Transformation on Criminal Law

Digital transformation requires comprehensive changes in various frameworks, including changes in social and cultural frameworks, as well as legal and judicial frameworks. At the social and cultural level, the transformation leads to a change in the culture of communication within the community, which undoubtedly affects the prevailing social values, and requires handling through social and cultural studies. On the other hand, we can never ignore its dark side. We can only express this dark side in the strongest terms, because the digital revolution is a fundamental global human rights issue. Its obvious benefits do not outweigh its obvious risks [10]. Among all this, it had a clear impact, which will be explained during the second topic, which is divided into two sub-topics. In the first sub-topic, I will talk about the impact of digital transformation on the implementation of criminal judgments. In the second sub-topic, I will discuss the impact of digital transformation on criminal prosecution.

3.1 The Impact of Digital Transformation on the Implementation of Criminal Penalties

Criminal policy has passed through several crises over the past decades, the most prominent of which was the accumulation of criminal cases in different courts, in addition to problems related to the non-implementation of conviction sentences in many cases, as criminal courts issue conviction sentences daily, but few of these sentences are implemented, and it is known that most of them are issued with negative penalties for short-term freedom, and it has been proven that these penalties have failed in achieving the goals of punishment.

3.1.1 Electronic Surveillance as a Means of Implementing Short-Term Custodial Penalties

Penal institutions are the actual framework through which penalties and custodial measures of freedom are implemented, as several tasks are carried out, including the rehabilitation and rehabilitation of convicts within these institutions and their reintegration into society, as well as deterring offenders and others who attempt to attack society's systems, among others The effects of digital transformation on the implementation of criminal judgments Electronic monitoring, which is considered one of the most important effects of digital transformation [11]. Electronic monitoring is considered one of the alternatives to judicial detention and custodial penalties of freedom, and these means appeared as a result of the effects of digital transformation, including electronic bracelet systems, after entering the system of means and mechanisms for implementing custodial alternatives. Freedom Penalties No. (46) for the year 2022, I find that the system issued in accordance with Article (25/bis) of the amended Penal Code expanded the scope of alternatives to negative penalties for freedom to include electronic monitoring and preventing the offender from accessing certain places, and these alternatives are applied through electronic bracelets, instead of judicial arrest, in accordance with the provisions of the Code of Criminal Procedure.

I find that when the court issues its decision to place a person under electronic surveillance and specify the specific time frame and geographical range in a specific geographical area or on a specific path or both, the monitoring department is responsible for monitoring the person's compliance with the conditions and duration of the monitoring, and the monitoring is done through the use of electronic devices attached to the body person and connected to a central system that receives and transmits data and information about a person's movements and location. to the competent authorities, and this system has been developed to be a flexible and effective alternative to traditional means of implementing penalties and custodial measures of freedom and judicial arrest for non-repetitive persons, and the implementation of a preventive policy aimed at preventing their mixing with other detainees and convicts, and preserving their livelihoods.

3.1.2 Initiative of the Accused to Implement the Verdicts Issued with the Conviction

It is clear that material scientific progress has no limits and does not appear to have an end, as advanced countries are increasing progress in the field of information and communication technology, which has been proven by experience to be the key to sustainable development, and it is worth mentioning that the use of these technologies plays a major role in crossing what is known as the wide scientific and technological gap that separates advanced countries from underdeveloped countries. It is clear that the electronic general management system is a modern system that is adopted by public administrations using the internet to connect their institutions with each other and connect various services to private institutions and the general public, and to put information within the reach of individuals, in order to create a transparent relationship characterized by speed and accuracy, aimed at enhancing performance quality General electronic management is defined as something in the traditional government, considering it the basic foundation of the electronic government model. Where it is recognized as the organizational entity that countries form in order to manage the affairs of the country and make strategic decisions related to political, economic, and social future, covering strategic economic, military, security planning, national production development, citizen education, maintaining their health, improving their living conditions, crisis management, and developing relations with the international community and other multiple tasks. And Electronic General Management means using information and communication technology to improve the way government services are performed. In another form, it means changing the way services are performed from a routine, bureaucratic, and complex process to a method that relies on information and communication technology to improve the performance of government services with the aim of providing them to citizens in an easy way through the internet, which provides many benefits and savings, thus reducing the cost of service delivery [12]. It is clear that with the spread of the coronavirus in most of the world's countries, governments, even those not qualified, have started to consider the idea of electronic government management as a means of providing their services to individuals on one hand, and as a way to ensure access to the government's perspective on the other. Some countries have considered this crisis as a warning to re-examine their improvement through the exploitation of technology and linking their work. Also, what digital transformation offers in terms of storage of personal data and information, or not, and analysis of this data and linking it, with the existence of an electronic system linking between different parties and interests within the state.

Through the above, it is clear that electronic government is not limited to using information technology to provide services to citizens, but it is a developed concept that reshapes institutions in new ways in terms of their administrative, social and political dimensions. It is not limited to providing electronic services to beneficiaries, but it represents electronic methods for achieving all the work done inside and outside the institutions and democracy is one of the main goals of electronic government, and it is new to work on the participation of beneficiaries through those tools. Also, the government represents a contract between institutions and beneficiaries, where the beneficiary is transformed from a service receiver to a participant in decision-making.

3.2 The Impact of Digital Transformation on Criminal Prosecution

Among the important results that resulted from the technological development is the emergence of new means of criminal evidence, and this led to the recognition of the legal validity of evidence derived from scientific means, whether in the investigation stage or in the trial stage; this development in the philosophy of evidence in criminal prosecution and its system led to the adoption of some criminal laws based on the principle of digitization of criminal procedures.

3.2.1 Modern Technology in Detecting Evidence of a Crime

There is no doubt that modern technology has revolutionized scientific evidence in criminal investigations, enabling those in charge of investigations to reveal the secrets of crime by creating scientific means based on nanotechnology used by police officers to analyze evidence on the spot at the crime scene, which not only saves time for analysis but also reduces the chances of error in criminal investigations, so modern criminal investigation means have benefited from the physical effects left at the crime scene such as fingerprints, hair remnants, tool and weapon traces, and body secretions. In addition to these evidences, high-tech devices with extraordinary capabilities have emerged in the field of criminal investigation, driven by the scientific development, with the aim of obtaining the physical evidence required by the law, and this method or means of obtaining it has been developed with the advancement of modern scientific technology, so the use of scientific means has become a necessary necessity for criminal research officers to perform their tasks fully, its use in the field of criminal evidence makes the process of evidence susceptible to renewal and development according to ongoing human achievements [13]. As scientific development is a natural inevitability, and in order to separate the criminal from this development, it was necessary for the legislator to invent modern means of proof according to scientific rules to rely on the accused, reveal the crimes and condemn the criminals within the framework of legal legislation. The development of scientific methods in the field of criminal evidence has resulted from the criminal's thought process which works to think in a way that does not leave physical traces before committing the criminal activity. Thus, it is necessary for the investigation and investigation bodies to rely on modern scientific means in criminal evidence such as fingerprints, genetic fingerprints, stains, and game traces, photography, recording of calls and footprints.

However, the principle of the judge's psychological certainty is considered one of the most important principles of criminal evidence theory because it agrees with the normal and logical way of thinking in everyday life, and in scientific research, where the judge rules based on his belief in the evidence presented in the case. The law is not concerned with the means that the judge is convinced of, as long as the rules from which the criminal evidence is derived are not specified. Rather, it is based on the principle of the judge's psychological certainty in a moderate way, so the evidence is presented in all types of crimes, violations, and offenses in all forms of evidence and the judge rules based on his personal belief, and he has the freedom to use all methods and rule on the case based on the belief that he has, and with regard to criminal evidence and how to access it, it varies from society to another [14].

The use of modern evidence has become an absolute necessity in criminal research, but the use of scientific development data in crime detection does not detract from the principle of judicial freedom in forming his belief, as the judge remains free to evaluate the strength of evidence presented to him in the evidence offered, and take what is appropriate and reject what he is not convinced of, even if the evidence is based on scientific principles and foundations. The researcher will realize its scientific value whether it is theoretical or practical.

3.2.2 Modern Technology in Investigative and Judicial Procedures

There is no doubt that the idea of remote litigation represents a new birth for legal procedures, especially criminal procedures, in the hope that this type of litigation keeps pace with the developments and progress witnessed by human societies and interacts positively with them. Despite the fact that this modern technology aims, as some say, to develop judicial services, starting with the speed of completing transactions and reducing the duration of court sessions and reducing the burden on material matters on the parties to the lawsuit, and the access to the concept of electronic judiciary, and reaching the principle of speed in lawsuits within a reasonable time. However, its negatives outweigh these positives, especially with regard to guarantees of fair trial for the accused, the principle of presumption of innocence of the accused, the publicity and secrecy of judicial sessions, the regulation of criminal materials, the presence of the accused before his natural judge face to face, the method of dialogue and discussion that takes place between the accused and the prosecution witnesses, and generally between all parties to the judicial dispute, all of which are guarantees that this type of litigation undermines them, especially if we take into account that the legislative text that the Jordanian legislator came with was provisional. Therefore, the Jordanian legislator has put the ball in the court of the judicial authority, especially the court, as it is the one who has the final say in the claim, so the judges are considered the main axis to ensure the smoothness of the procedures for litigation using remote communication technology. Therefore, they must be concerned with the process of the judicial process and strive to deliver justice to their applicants as soon as possible and at the least cost. Therefore, they are more eager than others to reduce the duration of litigation and speed up the settlement of claims; it has become necessary to establish new rules that ensure the ease, simplicity and protection of lawsuits from delay and procrastination, and anything that hinders their progress.

Accordingly, the Jordanian legislator has legalized the use of remote communication technology in judicial proceedings according to Article 158 of the Jordanian Criminal Procedures Law No. 9 of 1961 and its latest amendments in Law No. 32 of 2017. Through this article, the legislator established a system called the "System for the Use of Modern Technology in Criminal Procedures" No. 96 of 2018, published on page 5606 of Official Gazette No. 5529, dated 2/9/2018, based on the second paragraph of Article 158 of the Jordanian Criminal Procedures Law and its amendments No. 9 of 1961. This allows judicial bodies to use remote communication means, including the authority to listen to the accused, victim, witness, expert, and those claiming personal rights or facing others via remote communication technology. This is according to the text of the aforementioned system, which states that modern technology is used with the accused, the plaintiff, the defendant, the witness, the lawyer, the expert, the translator, the personal rights claimant, and the financial officer for the purpose of enforcing the provisions of this system [15].

If that is the case, then we support the legislator and announce our congratulations to him and bless him for such a leading step. However, if this step is likely to impede the progress of judicial proceedings or possibly harm any of the accused's guarantees or lead to confusion and disruption that deviates from the principle of legal truth, then we do not find any harm in putting it forward, announcing a return to the traditional judicial system. I am not a supporter or an optimist about the success of this technique due to the weakness of technical capabilities and the lack of financial resources for it. Also, electronic litigation from my point of view is no different from electronic education in schools and universities, it is a temporary experience that we are living and dealing with it, and everyone is convinced that it will not produce results if compared with the principle of real confrontation, the presence of the student in the classroom, and also the presence of the parties and the accused in the courtroom represents a title for success and excellence and preserves their rights and legal and constitutional guarantees [16].

If we want to review some of the guarantees of the accused, then we start with the first, which is the principle of public trial, which means the right of every person to attend the trial, without any condition or barrier, except what is allowed by the system, so that he has the opportunity to witness the proceedings of the trial, and the public can see the proceedings and the discussions that take place, and the statements and arguments that are presented. This principle is confirmed in the Constitution, Article 101. This right is also mentioned in Article 173/2 of the Criminal Procedures Law, and in Article 5 of the Modern Technology Use System in Jordan, and what is mentioned in it. Despite the confirmation of the legislator for this guarantee, whether the claim is traditional or electronic, and under the title that the testimony is one of the types of evidence, we question whether modern technologies are prepared or at least sufficient to conduct the discussion between the accused and other witnesses of evidence, and to impose a debate that the prosecution's witnesses took a long time. Do they achieve the required productivity as if they were conducted directly in front of the judge in the court. Despite this, I see that electronic litigation can only have its effect properly through his personal presence before the public prosecutor or the judge inside the corridors of the court, because viewing it via remote communication technology would lead to a blurring of the accused's view of the criminal materials as they are, especially If we assume, for the sake of argument, if the accused is visually impaired, which may lead to his uncertainty as to whether some or all of these seized materials belong to him or not.

4 Conclusion

It is clear to me through what was previously mentioned during this study that it had a significant impact on the digital transformation on the punitive policy, whether through the development of laws related to the punitive policy represented in the Penal Code and the Code of Criminal Procedure, in order to accommodate the digital transformation, and from what I reached several results and recommendations.

4.1 Results

1. From the foregoing, it was found that the electronic systems represented by the availability of the infrastructure that help litigants before the criminal courts submit their complaints electronically through the specialized website of the Ministry affiliated with the courts, and that complaints are answered electronically by means of a notification or a specific application to accept complaints related to litigants.
2. It was found that judicial business practices need to be digitized as a result of the digital transformation instead of being paper-based, and given the legal and practical frameworks for parties to disputes before criminal courts.
3. From the foregoing, it was found that the electronic bracelet needs an infrastructure for digital transformation, as it is defined as the organizational entity formed by states in order to manage the country's affairs and take strategic decisions related to the political, economic and social future, as this administration covers the areas of economic, military and security strategic planning. Developing the national product, educating citizens, preserving their health, improving their living conditions, managing crises, developing the country's relations with the outside world, and many other tasks.

4.2 Recommendations

1. I recommend the need to ensure the protection of the punitive policy for the data of individuals and institutions used in the framework of the work of justice represented by the courts, by increasing the criminal liability for the acts of disclosure or circulation of this data and information, as this achieves in enhancing confidence in electronic systems, and helps in achieving digitization more quickly.

2. I recommend electronic governments to link all state institutions through the information network, to facilitate the registration of criminal cases and peremptory decisions, and it is suggested to prevent government services from those who have been sentenced to a final judgment of conviction after the execution of the judgment, or to reconcile with it, and that is by individuals executing the sentences imposed on their own themselves, and so that the digital transformation contributes to solving the crisis of non-implementation of the punitive sentences represented in imprisonment, so that it may be replaced by the digital penalty represented in the electronic bracelet.
3. I recommend the ministries represented by criminal courts, which carry out investigation procedures such as hearing witnesses and seizing criminal evidence using modern technologies, for what this achieves in facilitating procedures and preserving the safety of witnesses and others. And electronic monitoring as a means of implementing imprisonment and short-term custodial penalties.
4. The legislator recommended amending the laws that govern the punitive policy, which is represented in the Penal Code and the principles of criminal trials, in line with the digital transformation, which has become enormously affecting the punitive policy, and this constitutes the protection of the accused.

References

1. Ebeid, O.H.: Electronic criminal surveillance, a comparative study. J. Law Econ. (82) (2009)
2. Muhammad, A.: Punitive politics between legislation and law. J. Res. Law Polit. Sci. 213
3. Abdel-Ghany, S.M.: The Implications of Digital Transformation for Promoting Economic Growth in Egypt. J. Coll. Polit. Econ. **15**(14), 55 (2022)
4. Ashraf Abdel Sattar: Digitization of Criminal Procedures in Light of the Corona Pandemic, Research Submitted to the Twenty-First Annual Conference. Social and Criminological Research Center, Cairo (2021)
5. Hassan Muhammad, A.A.H., Al-Ghubairi, M.A.: The reality of the digital transformation of the Kingdom of Saudi Arabia, an analytical study. J. Admin. Financ. Sci. **4**(P3), 18 (2020)
6. Sorour, A.F.: Mediator in the Penal Code, General Section, Book One, Dar Al-Nahda Al-Arabiya (2016)
7. Al-Sawat, T.A.A.: The impact of digital transformation on the efficiency of academic performance. Arab J. Publ. Saudi Arabia 43 (2022)
8. Mohamed, B.: Criminal Policy and the Internet's Dangers to Minors, The Internet and Child Rights Study Day Organized on February 04, 2013. Oran University. https://www.asjp.cerist.dz/en/downArticle/173/4/1/93840
9. Maryam, B.: Electronic monitoring as a modern method of punitive treatment under Law No. 18/01 amending the Law on Prisons Regulation and Social Reintegration of Detainees. J. Dept. Legal Polit. Res. Stud. 6 (2019)
10. Hussain, M.K.: E-Government, Journal of the University College of Economic Sciences, Baghdad College Conference Special Issue, Iraq, D.M. (2013)
11. Lakhdar, G.: New Investigations in Crimes of Modern Technology. J. Res. Prof. Legal Polit. Stud. **5**(Part 1) (2020)
12. Turki, M.: The Role of Nanotechnology in Detecting Crime and Its Evidence As Evidence Before the Judiciary, University and Academic Research Series, p. 33 (2020). https://www.droitetentreprise.com/20962/

13. Al-Qadi, R.M.: remote investigation and trial procedures – a comparative analytical study. Univ. Sharjah J. Legal Sci. **19**(P2) (2020)
14. Shdeifat, S.M.: Investigation and criminal trial remotely via videoconference technology. J. Legislat.Sci. Law Stud. **42**(P1) (2015)
15. Al-Manaseer, S.A.: Towards a Legal Regulation of Electronic Administrative Contracts. Ph.D. Thesis, Faculty of Law, Ain Shams University, Cairo (2019)
16. Al-Manaseer, S.A., et al.: The impact of cybercrime on the functioning of electronic public utilities, published research. Zarqa Journal for Research and Human Studies, Deanship of the College of Scientific Research, Zarqa University, Amman (2021)

The Extent of Awareness of Cyber Security Among the Superior and Ordinary Students in the Faculty of Education in Al Ain University

Ziyad Kamel Ellala(✉), Khawlah M. AL-Tkhayneh, and Razan Numan AlKhatib

Al Ain University, Abu Dhabi, UAE
{Ziyad.ellala,Khawlah.altkhayneh,razan.alkhatib}@aau.ac.ae

Abstract. This study aimed to identify the extent of awareness of cyber security among a sample of superior and ordinary students in the faculty of education in Al Ain University. The study sample consisted of (94) male and female students from the faculty of education, with (32) males and (62) females, and based on the student's type, they were (14) superior students and (80) ordinary students. In order to achieve the study objectives, the researcher used the scale of awareness of cyber security as the study instrument.

The results showed that the level of cyber security awareness among the talented and ordinary students in the faculty of education in Al Ain University was high for the overall score of the scale. The results revealed that there are no statistically significant differences in the degree of awareness of cyber security attributed to the variable of gender (male, female), while the results revealed that there are statistically significant differences in the level of awareness of cyber security attributed to the variable of the student's type (superior, ordinary) in favor of the superior students.

Keywords: cyber security · superior students · ordinary students · Al Ain University

1 Introduction

Information security is considered as one of the most important current issues in the age of the fourth industrial revolution, where the success of any institution highly depends on the information it possesses. However, several information systems and infrastructure related to networks are frequently exposed to risks. Indeed, these networks face various breaches to their data, and are frequently exposed to criminal attacks (hackers), which impede their services and destroy their property. The attacks of hackers vary based on the parties performing them, as well as the place and time of those cyber attacks, where attackers frequently use developed breaching mechanisms and instruments [1].

institutions face the risks related to information security due to an extended set of threats that range from unsolicited emails, to the complex cyber attacks, such as the malicious software that may steal, decay or destroy data, or they could make systems invalid for usage [2].

B. A. M. Alareeni and I. Elgedawy (Eds.): *Artificial Intelligence (AI) and Finance*, SSDC 488, pp. 134–144, 2023.
https://doi.org/10.1007/978-3-031-39158-3_12

Often, attackers select the unintended vulnerabilities resulting from the human factors within the data security system in the organization. Moreover, the threats of cyber security that exploit the human behavior are constantly developing which, in turn, increases the security breaches in organizations. Indeed, the majority of those breaches take place due to the users' lack of knowledge, or negligence behaviors, such as sharing the passwords with others as well as opening anonymous e-mails or attachments. Performing such a type of activities facilitates the occurrence of intended threats by attackers and exposes the organization's safety to risk [3].

Since the major risks of information security resulting from human factors still represent a main source of anxiety, organizations still invest in the solutions of information security that are based on technology, such as firewalls, anti-viruses' software and intrusion detection systems to protect their data [2].

The problems of cyber security are mainly attributed to the reality that most employees are not sufficiently committed to the security rules and regulations in the workplace [4] which, in turn, makes them as the weakest link in the organization's security system [5]. Indeed, the human factor is the easiest target for attackers to, illegally, access the confidential data [6].

2 The Study Problem

The internet has become as one of the main features affecting people, including the University students, where they use it in most of their life tasks – they use the internet in the domains of banking services, marketing, training and social communication. In this context, the United Arab Emirates Centre for Future Readiness [7] confirmed that during the following ten years, the cold cyber wars will be amongst the most important threats to the security of national information systems in the different countries throughout the world which will, in turn, result in the occurrence of tangible and intangible negative consequences on individuals and organizations.

3 The Study Questions and Hypotheses

1. What is the extent of awareness of cyber security among superior and ordinary students in the faculty of education in Al Ain University from their perspective?
2. Are there statistically significant differences at ($\alpha \leq 0.05$) in the extent of awareness of cyber security among superior and ordinary students in the faculty of education in Al Ain University attributed to the variable of gender (male, female)?
3. Are there statistically significant differences at ($\alpha \leq 0.05$) in the extent of awareness of cyber security among superior and ordinary students in the faculty of education in Al Ain University attributed to the variable of the student's type (superior, normal)?

4 The Study Objectives

1. Identifying the concepts of cyber security among the superior and ordinary students in the faculty of education in Al Ain University.

2. Identifying the applications of cyber security among the superior and ordinary students in the faculty of education in Al Ain University.
3. Identifying the methods of enhancing cyber security among the superior and ordinary students in the faculty of education in Al Ain University from their perspective.

5 The Study Justifications

1. The paucity of studies that addressed the awareness of cyber security among superior and ordinary students in the faculty of education in Al Ain University.
2. Enriching the scientific research in the domain of awareness of cyber security among superior and ordinary students in the faculty of education in Al Ain University.

6 The Study Limits

- The spatial limits: the study is limited to the state of the United Arab Emirates.
- The human limits: the study is limited to the superior and ordinary students in the faculty of education in Al Ain University.
- The temporal limits: the study was conducted during the first semester of the academic year (2022–2023).
- The objective limits: awareness with the concepts, applications and methods of enhancing cyber security in the United Arab Emirates.

7 Conceptual and Theoretical Literature

7.1 The Concept of Cyber Security

Cyber security or information technology security is defined as a set of processes and procedures that aim to secure and protect networks, computers, programs and information against attack, destruction, stealing or illegal access [8]. Cyber security depends on three basic elements; these are confidentiality, the validity and security of data, as well as integration and data availability [9].

7.2 Security Awareness

The degree of feeling of danger, and it is measured by the score that participants obtain after responding to the instrument's items.

7.3 The Superior Students

They are the male and female students enrolled in the bachelor degree in the faculty of education in Al Ain University in the United Arab Emirates, who were classified as superior students based on their accumulative average in the university, as they obtained (3.5) or more, out of (4) in their academic achievement.

7.4 The Ordinary Students

They are the male and female students enrolled in the bachelor degree in the faculty of education in Al Ain University in the United Arab Emirates, who were classified as ordinary students based on their accumulative average in the university, as they obtained (3.5) or less, out of (4) in their academic achievement.

8 The Characteristics of Cyber Security

Since the electronic crime takes place according to new methods and a higher technological dimension as compared to the traditional crimes, it is necessary to adopt the features of cyber security in order to overcome this situation and cope with the technological development. Therefore, cyber security is characterized by several features, including [10].

1. Discovery and tracking: cyber security aims to uncover the electronic crime, track its indicators and thus overcome it.
2. Speed and lack of evidence: the difficulty in confirming the occurrence of cybercrimes is represented by the regular use of modern technical methods by attackers. Therefore, it was necessary to introduce the cyber security which adopts more advanced technological methods and experiences.

9 The Concept of Excellence

The excellent and superior students are viewed as a valuable wealth for their countries, where they represent a distinctive contribution source as well as a real challenge for their parents and teachers. Therefore, obtaining educational and academic opportunities that suit their abilities and readiness is considered amongst their due rights. Also, the gifted and superior students have their own social and psychological needs, where they need to co-exist with the surrounding community that is responsible for protecting them against the exposure to risky adaptation problems. Accordingly, those superior students should be given more attention and their problems should be solved through counseling, advice, understanding and care, in addition to providing them with all the requirements related to their development [11].

The problem of defining talent and mental superiority is more concerned with researchers' varying perspectives about the domains of excellence that they regard as important in determining talent – whereas some researches focus on excellence in general mental abilities, others focus on specific abilities, including academic achievement, innovation or some personality traits. However, a comprehensive review of what had been documented in this topic clearly demonstrates the lack of an agreed-upon definition amongst the researchers interested in the domain of talent and superiority [12].

We can also say that the mentally-superior individuals are those who achieve a higher level as compared to ordinary individuals in a certain domain, where the high-performance level reflects the individual's functional and mental level that outperforms same-age peers [13]. When we talk about talent, we suggest that it is an inherent readiness formulated by the motivating environment. Also, talent could be manifested in one or more domains, such as the talent of music [14].

10 The Characteristics of Superior Students

Teachers and parents should be well-informed about the different characteristics of superior individuals, since that gives them more ability to identify the superior students in a more realistic and objective way, instead of depending on guessing [15].

The most important mental trait that distinguishes the superior individuals from ordinary individuals is related to having higher levels of intelligence, where their intelligence quotient is equal to those older than them, since their mental age surpasses their chronological age. Those individuals are also characterized by the ability to learn the skills of reading and writing at an early age, in addition to having an advanced linguistic ability to understand the various contexts. Therefore, the superior individual often acquires a wide range of vocabulary and a good supply of information about various topics [16].

Several social and emotional characteristics distinguish the superior individuals, including the state of being socially adjusted and emotionally stable. In general, we can say that those individuals are characterized by self-discipline, control, tolerance, emotional stability, leadership, self-sufficiency, humor, risk-taking as well as the high levels of social adjustment, including accompaniment, autonomy and helping others [17].

11 The Previous Studies

Al-Qahtani [2] conducted a study which aimed to identify the extent of awareness of cyber security among a sample of individuals with visual impairment. The results revealed that the total degree of awareness of cyber security was medium. The degree was also medium for the first dimension (concepts of cyber security), while it was high in the dimensions of (awareness of cyber security applications and the ways of raising awareness of cyber security). The results revealed that there are no statistically significant differences in the degree of awareness of cyber security attributed to the variables of gender and education.

Fethi [18] conducted a study which aimed to identify the usage of internet and the concerns about cyber security among the individuals with visual impairment in order to understand their internet usage and examine the relationships between the scales related to using the internet and the variables of knowledge, skills, trust and behavior related to cyber security. The results revealed that learning was the most common cause for using the internet. The results also showed that the participants with more skills and knowledge about cyber security were more worried and used the internet less as compared to those with less knowledge about cyber security.

Slusky [19] conducted a study which aimed to investigate the awareness of cyber security among the students of Business and Economics Faculty in the University of California, in Los Angeles in the United States of America. The results revealed that the main issue concerning the awareness of cyber security is related to the methods used by students when dealing with these problems and issues in reality. The study results mainly aimed to help an institution develop its curricula, and include those curricula with more training about information security.

Al-Qahtani [20] conducted a study which aimed to identify the extent of awareness of cyber security among the male and female students of the Saudi Universities from their perspective. The results revealed that electronic blackmail is the most common crime with which cyber security deals, and that the community's media education about the methods of electronic blackmail is considered as one of the most important community prevention ways for the problems of cyber space. The study revealed that there are social obstacles in achieving prevention, such as the considerable development in information systems and the lack of sufficient knowledge for preventing the negative effects of these systems.

Senthilkumar [21] conducted a study on university students in the main cities in Tamil Nadu to identify the extent of their awareness of cyber security. The results revealed that the students focused on certain electronic security risks, such as phishing, fake advertisement, and stealing personal data. The results revealed that the students' awareness of cyber security and the relevant threats was above average, with (70%) among participants.

Alotaibi et al. [22] conducted a study which aimed at identifying the level of awareness of cyber security among the university students. The results revealed that awareness of cyber security among the Saudi University students was low, since most students were unaware about the protection of their data.

12 The Study Methodology

The study used the descriptive approach by applying the study questionnaires to the participants, and then using the suitable statistical analysis to conclude the study results.

13 The Study Population and Sample

The study population consisted of all the male and female superior and ordinary students in Al Ain University in the United Arab Emirates. The study sample consisted of (94) students distributed according to gender (32) males and (62) females, and based on the student's type, they were (14) superior students and (80) ordinary students, who were selected purposively during the academic year (2022–2023). The students expressed their readiness to respond to the study instrument. Table 1 shows a description of the demographic variables according to gender (male, female) and student's type (superior, ordinary).

14 The Study Instruments

The researchers developed a special instrument for the superior and ordinary students in the faculty of education in Al Ain University from the perspective of students, based on a review of the topic of cyber security in the previous studies, such as (Al-Qahtani [2], Al-Manie [10], Al-Samhan [1], Al-Ameer [8], and the United Arab Emirates Centre for Future Readiness [7], in addition to utilizing the opinions of the specialists in cyber security. The final version of the questionnaire consisted of (28) items distributed to three dimensions: the awareness of the concepts of cyber security (10) items, awareness of the applications of cyber security (13) items, and the methods of enhancing cyber security (5) items. The researchers used Likert 5-point scale to explain the responses of the study sample individuals.

Table 1. The characteristics of the study sample individuals.

Gender					
	Category	Frequency	Percent	Valid Percent	Cumulative Percent
	Male	32	34.0	34.0	34.0
	Female	64	66.0	66.0	100.0
	Total	94	100.0	100.0	
Type					
Valid	Ordinary	80	85.0	85.0	85.0
	Superior	14	15.0	15.0	100.0
	Total	94	100.0	100.0	

15 The Validity and Reliability of the Study Instrument

The content validity was verified by applying the study instrument to a pilot sample those consisted of (27) students. Pearson correlation coefficient between the domain and its items was calculated. The results showed that Pearson correlation coefficient between the items' scores and the total score for the domains was statistically significant at ($\alpha \leq 0.05$).

Also, factor validity was calculated for the study instrument by calculating the correlation coefficient between the scores of the sub-domains and the total score of the instrument, where the correlation was statistically significant at ($\alpha \leq 0.05$), indicating the content validity of the study instrument.

The instrument's reliability was verified by using Cronbach alpha. The test showed that the total Cronbach alpha coefficient value for the study instrument was (0.904), indicating the possibility of using the study instrument to achieve the targeted objectives of the study.

16 The Study Results and Discussion

The first question: What is the extent of awareness of cyber security among superior and ordinary students in the faculty of education in Al Ain University from their perspective?

In order to identify the extent of awareness of cyber security among superior and ordinary students in the faculty of education in Al Ain University, the means and standard deviations were calculated. The results revealed that there is a high level of awareness of cyber security among the superior and ordinary students in the faculty of education in Al Ain University. Item (12) stating "I avoid opening any hyperlink attached in an anonymous message" was in the first place, with a mean of (4.649) and a standard deviation of (0.6988), followed by item (6), stating "I realize the importance of maintaining cyber security", with a mean of (4.521) and a standard deviation of (1.0445), and then Item (24) stating "I avoid online purchases from unreliable sites" was in the third place, with a mean of (4.468) and a standard deviation of (0.6123).

On the other hand, Item (27) stating "I have personal obstacles that prevent me from achieving cyber security protection" was in the last place, with a mean of (2.167) and a standard deviation of (1.399), followed by item (8), stating "I have knowledge about social brain engineering, and the processes of electronic blackmail and fraud", with a mean of (3.213) and a standard deviation of (1.458).

The level of awareness ranged between medium and high based on items, and the total score was high. The researchers attributed this result to the sample individuals' interest in the concept of cyber security, considering it as one of the new and important concepts in the modern age, in addition to the existence of training courses and activities related to cyber security, given that most students' transactions (admission and registration, teaching, financial procedures, dropping and adding courses, and delegating tasks) inside the university campus are performed electronically. This result is also attributed to the existence of an academic program that grants the bachelor degree in cyber security in the university. This finding agreed with Al-Qahtani [2], Al-Qahtani [20], and Fethi [18], while it disagreed with Alotaibi [22], which revealed that the students' awareness of cyber security was low.

The second question: Are there statistically significant differences at (0.05) in the extent of awareness of cyber security among superior and ordinary students in the faculty of education in Al Ain University attributed to the variable of gender (male, female)?

In order to answer the question, the means and standard deviations were calculated for the awareness of cyber security according to the student's gender (male, female), and in order to find out the statistical differences between the arithmetic means, t-test was used, as illustrated in Table 2.

Table 2. The means and standard deviations, and independent Sample t-test for the differences in future thinking skills according to the student's gender (male, female)

	Category	Mean	Standard deviation	Mean differences	Calculated t value	Sig.
Hypothesis 1 (Gender)	Male	4.2863	.53957	.37517	3.251	0.281
	Female	3.911	.52539			

Table 2 revealed that the mean score for the respondents' responses was (0.375), which is not significant, where (t = 3.251) at a significance level of (Sig = 0.281), which is more than (0.05). This indicates the acceptance of the null hypothesis stating that "there are no statistically significant differences in the extent of awareness of cyber security among superior and ordinary students in the faculty of education in Al Ain University attributed to the variable of gender (male, female)".

The researchers attributed this result to the case that all the training activities and programs provided to male and female students are introduced in the same quality and method; indeed, this is a good point in favor of the university, its administration and faculty members, since they don't distinguish between male and female students in the university. Also, the current situation of man and woman, in terms of women's rights and equal educational opportunities, working in the various vocational domains, as well as practicing their practical and academic life enabled them to be aware of cyber security. This finding agrees with Al-Qahtani [2].

The third question: Are there statistically significant differences at ($\alpha \leq 0.05$) in the extent of awareness of cyber security among superior and ordinary students in the faculty of education in Al Ain University attributed to the variable of the student's type (superior, normal)?

In order to answer the question, the means and standard deviations were calculated for the awareness of cyber security according to the student's type (superior, ordinary), and in order to find out the statistical differences between the arithmetic means, t-test was used, as can be seen in Table 3.

Table 3. The means and standard deviations, and independent sample t-test for the differences in future thinking skills according to the student's Type (Ordinary, Superior)

	Category	Mean	Standard deviation	Mean differences	Calculated t value	Sig.
Hypothesis 2 (student's Type)	Ordinary	3.9872	.40744	−.07833	−0.644	0.030
	Superior	4.0655	.62145			

Table 3 showed that the difference in the mean scores of the responses of superior and ordinary students was (−0.078), which is statistically significant, where (t = 0.644), at a significance level of (Sig = 0.030), and it is less than (0.05). This implies rejecting the null hypothesis, and entails that there are statistically significant differences in the level of awareness of cyber security among the superior and ordinary students in the faculty of education in Al Ain University attributed to the variable of the student's type (superior, ordinary) in favor of the highest mean for the superior students, with a mean of (4.0655). The researcher attributed this finding to the superior student's increased interest in cyber security as compared to other ordinary students. Also, the superior students are more inclined towards coping with electronic transactions and using them as compared to ordinary students. Indeed, the superior students are always keen to accomplish tasks and do the required work before the deadline as compared to ordinary students.

17 Conclusion and Recommendations

Cyber security is considered as an important issue [19]. Therefore, much attention should be paid to educating university students about cyber security. This, indeed, requires coping with the developments in this domain and conducting further studies to support or oppose the results of the current study as well as conducting other studies to address the awareness of cyber security among the university students according to other variables such as impairment types (visual impairment, auditory impairment, and motor impairment), in addition to conducting a comparative study among university students between several universities inside or outside the country. The study recommended the necessity of measuring the differences in the awareness of cyber security among the individuals of the study sample according to the variables of university and country, including the skills of awareness of cyber security in academic curricula in universities.

References

1. Al-Samhan, M.: The requirements of achieving cyber security for administrative information systems at King Saud University. J. Coll. Educ. (2020)
2. Al-Qahtani, A.: The degree of awareness of cyber security among the individuals with visual impairments in the Kingdom of Saudi Arabia from their perspective. J. Spec. Educ. Rehabil. **14**, 1–31 (2022)
3. Leyden, J.: Security Experts Dissect Google China Attack. https://www.theregister.com/2010/01/14/google_china_attack_analysis/
4. Whitman, M.E., Mattord, H.J.: Principles of Information Security. Cengage, Boston, MA (2005)
5. Simsim, M.T.: Internet usage and user preferences in Saudi Arabia. J. King Saud Univ. Eng. Sci. **23**, 101–107 (2011)
6. Willard, N.E.: Cyber-Safe Kids, Cyber-Savvy Teens: Helping Young People Learn to Use the Internet Safely and Responsibly. Jossey-Bass, San Francisco (2007)
7. United Arab Emirates Centre for Future Readiness: Future Expectations (100 Global Trends for the Year 2050). Ministry of Cabinet Affairs and Future, Dubai, The United Arab Emirates (2019)
8. Al-Ameer, H.: Cyber Security Challenges. Center for Strategic Studies CSS (2018)
9. Borky, J., Bradley, T.: Protecting Information with Cyber Security, pp. 350–351. Springer International Publishing, London (2019)
10. Al-Manie', A., Al-Jawhara: The requirements of achieving cyber security in the Saudi universities in light of Vision 2030. J. Coll. Educ. **1** (2022)
11. Zaid, K., Hamza, S.: Behavioral characteristics of university students and their ordinary peers in the Middle Euphrates region. J. Faculty Basic Educ. Educ. Hum. Sci. 428–445 (2015)
12. Nasser, F.: The reality application of academic talent management strategies at King Faisal University. J. Inst. Res. South East Asia 137–158 (2019)
13. Al-Srour, N.: Introduction to Special Needs Education. Dar Al Fikr Publishing and Distribution, Amman, Jordan (2009)
14. Jarwan, F.: Talent, Excellence and Creativity. Dar Al Fikr Publishing and Distribution, Amman, Jordan (2013)
15. Worrell, F.C., Subotnik, R.F., Olszewski-Kubilius, P., Dixson, D.D.: Gifted students. Annu. Rev. Psychol. **70**, 551–576 (2019)
16. Renzulli, J.S.: Scales for Rating the Behavioral Characteristics of Superior Students. Prufrock Press Inc., Waco, TX (2013)
17. Shammari, A.: Educational services for gifted children in the kindergarten stage from the point of view of teachers in Kuwait. Educ. J. (2016)
18. Fethi, A.I., Akbar, S.N., Rona, L.P., Keith, S.J.: Internet use and cybersecurity concerns of individuals with visual impairments. J. Educ. Technol. Soc. **19**, 28–40 (2016)
19. Slusky, L., Partow-Navid, P.: Information security practices and awareness. J. Inf. Privacy Secur. **8** (2012)
20. Al-Qahtani, N.: The extent of awareness of cyber security among male and female students in the Saudi universities from a social perspective: a field study. J. Soc. Affairs **66** (2019)

21. Senthilkumar, K., Easwaramoorthy, S.: A survey on cyber security awareness among college students in Tamil Nadu. IOP Conf. Ser. Mater. Sci. Eng. **263**, 042043 (2017)
22. Alotaibi, F., Furnell, S., Stengel, I., Papadaki, M.: A survey of cyber-security awareness in Saudi Arabia. In: 2016 11th International Conference for Internet Technology and Secured Transactions (ICITST) (2016)

The Role of Digital Transformation in Juvenile Trial

Jalal Hussein Al-Kayid[1(✉)], Mahmoud Aref Mohammad Aleshoush[1], and ElGaili Ahmed ElTayeb[2]

[1] Faculty of Law, Zarqa University, Zaraqa, Jordan
{jalkayid,maleshoush}@zu.edu.jo
[2] College of Law, University of Business and Technology, Jeddah, Saudi Arabia
algaili@ubt.edu.sa

Abstract. Interest has emerged in what is known as alternatives to short-term custodial punishment, and among the most important of these alternatives is electronic monitoring, which increases in importance when juveniles are tried and penal punishment is applied to them to prevent them from mixing with criminals and to try to reform and discipline them away from penal institutions such as prisons and correctional and rehabilitation homes for adults, especially if it is related to short-term custodial punishment, this study dealt with the statement of what the juvenile is and its concept, and then the relationship between it and the digital environment. Used in the juvenile trial, and this study reached several results and several recommendations that we mentioned in the conclusion, as it concluded that the alternative punishment applied in support of the digital environment is of high importance because of its contribution to reforming, rehabilitating, and organizing the convict (the juvenile) Electronic surveillance and the electronic bracelet are two important means to implement custodial penalties, as they are considered the most important guarantees the task is to protect the basic rights and freedoms of the juvenile, which are guaranteed to him by the constitution and international covenants. This study recommended the necessity of identifying the competent judge and defining his jurisdiction exclusively, in addition to determining the duration of the punishment that bears the alternative punishment that does not affect the security and safety of society because the alternative punishments shall not be followed with long penalties.

Keywords: juvenile · electronic monitoring · electronic bracelet · digital transformation · juvenile trial

1 Introduction

There is no doubt that the criminal penalty represents an effective means to maintain stability and security within society, but the excessive use of competent authorities has become an assault on the freedom of the individual. Criminal punishment no longer achieves its goal, which is to achieve deterrence in both its private and public forms, and the application of liberty-depriving penalties, especially to the juvenile category,

B. A. M. Alareeni and I. Elgedawy (Eds.): *Artificial Intelligence (AI) and Finance*, SSDC 488, pp. 145–151, 2023.
https://doi.org/10.1007/978-3-031-39158-3_13

and in light of the digital transformation, has become in conflict with the law, through the increasing opportunities to benefit from the digital transformation in the application of penalties to that category. Accordingly, recent legislation has worked to create a system related to non-custodial measures against the juvenile, so that these measures aim to reintegrate them into society and its best interest. It is not permissible to resort to punishing juveniles and depriving them of their liberty due to the absence of legal foundations or tools aimed at protecting the juvenile and reintegrating them into society. The principle of the safety and development of the juvenile is linked to the concept of the best interest for him, which aims to take innovative protection measures, as the proper application of these measures leads to the reform of society and makes it a useful member [1], and thus preserves their stability in their family environment and they are free during their implementation of the sentence that is not depriving them of freedom. The Jordanian legislator adopted an alternative system to detaining liberty in Juvenile Law No. 32 of 2014 and its amendments, and among those issues that have special legality, the issue of alternative measures were mentioned by the Jordanian legislator and devoted special provisions to it aiming at giving the perpetrator an opportunity to reform and reintegrate him into society, The research problem is about the role of digital transformation in the trial of juveniles, and the appropriateness of legal texts for digital transformation in the field of juvenile trial.

2 The Nature of the Juvenile, and the Implications of Digital Transformation on the Juvenile Trial

In reports submitted by states parties to the Committee on the Rights of the Child, these states often pay attention in some detail to the rights of children who are alleged, accused, or proven to have violated the penal law, also referred to as children in conflict with the law. In line with the Committee's guidelines for the preparation of periodic reports, implementation of Articles 37 and 40 of the Convention on the Rights of the Child is the main focus of information provided by States parties. The Committee notes with appreciation the many efforts made to administer juvenile justice in compliance with the Convention. However, it is also clear that many States parties still have a long way to go in order to achieve full compliance with the provisions of the Convention, for example in the areas of procedural rights, the development and implementation of measures to deal with children in conflict with the law without resorting to judicial proceedings, and the failure to use Deprivation of liberty unless it is a measure of last resort [2].

2.1 Definition of Juvenile, Under Domestic and International Laws

The term "juvenile" and "minor" are mentioned in several laws, and it is clear that a juvenile is a person who has completed nine years of age and has not yet completed eighteen, unless the juvenile has not completed nine years of age, and the same article specified that the juvenile has two designations that differ according to the age group that he represents. They are the child juvenile and the teen juvenile. The juvenile is also considered a child if he completes nine years of age and has not completed fifteen. A juvenile is considered a teenager if he has completed fifteen years of age but not yet eighteen.

A juvenile is anyone who has not completed eighteen years of age, and a guardian by law is every person other than the legal guardian who takes care of or supervises a juvenile in accordance with the legislation in force [3, p. 79].

Therefore, based on what was mentioned, the evidence did not differentiate between the juvenile child and the juvenile youth, despite the common difference between the two concepts, which are related to the legal age considered for the establishment of criminal responsibility, and not taking into account the distinction between the child and the youth with regard to growth and understanding of the concepts of matters. The juvenile is defined idiomatically as: a synonymous expression for a young person, and the description of the juvenile is given to a person who has reached the age of puberty and has not exceeded eighteen, that is, it is limited to a specific age group. This category is subject to special provisions in the field of criminalization rules and penal responsibility, and it differs from what is prescribed for the same category of adults. The terms "child" and "adolescent" are applied to those who have not reached puberty, and when they reach puberty, this characteristic ends, and the person takes a different word from those words, and the word "juvenile" is more used today for the deviant child as a term for that [4].

The International Covenant on Civil and Political Rights of 1966 stated in the Charter that the death penalty may not be imposed for crimes committed by persons under the age of eighteen years. What is wrong with the Covenant is that it did not define the term child and did not define childhood explicitly and clearly, which left ambiguity about the stage before reaching the age of majority. It is also stated in the provisions of the Charter that juvenile accused must be separated from adults, in addition to the necessity of referring them as quickly as possible to the judiciary in order to decide on their cases, bearing in mind that the primary goal in the prison system is to reform juveniles and rehabilitate them socially.

The Convention on the Rights of the Child 1989 defines Article 1 of the Convention on the Rights of the Child as naming a child as every human being under eighteen years of age. It is noted that the Convention has given the name of the child to the juvenile, which is an internationally recognized term and is considered idiomatically common. This agreement is considered one of the binding agreements specialized in the care of children, and it is binding on all member states of the United Nations, with the exception of the United States of America and Somalia. Which emphasized the importance of specifying a minimum age for criminal responsibility.

2.1.1 Juvenile Trial Standards Across the Digital Environment

Change in the era of digital is inevitably happening and coming, and we only have to prepare an environment that is in line with it. The developments of digital technology are the use of digital tools that play an important role in planning the implementation of judgments in the right way to suit the digital environment, which can be adapted to meet the needs that are commensurate with the environment the Juvenile such as the use of online survey tools and electronic databases and the collection, storage, and organization of information and electronic media instead of deprivation of liberty [5].

2.2 The Relationship Between Juvenile Trial and the Digital Environment

The criminal liability applies to minors accused of committing a crime after reaching the age of twelve, often involving petty theft. And making a judgment against them depends on the method of theft and not the value of the stolen thing. Penalties for juvenile detention range from 10 to 12 years in a rehabilitation home, and Al-Tubasi clarifies that "intensifying penalties is not a solution, because it is often useless. The principle is the method of reform because the goal is to reform the person who committed the crime more than punish him to prevent its recurrence. Human rights activists are calling for amending the juvenile law and imposing penalties an alternative to them, instead of imprisoning their freedoms within rehabilitation programs that contribute to returning them to society to be active and productive individuals [6].

3 Electronic Means of Juvenile Trial in Light of the Digital Transformation

The digital revolution, just as it began to change the features of life activities, it has begun to change the features of the school environment to gradually transform into the smart school and the virtual school, which will be the school of the digital era, which will transform life activities into activities that occur through the Internet through digital games, social networks, and system simulation.

3.1 Innovative Means (Electronic Court)

It considered that the digital court and the electronic court there are some differences between them lies in the fact that the electronic court is a new methodology based on the use of information and communication technology in the exercise of the basic functions of courts and judicial administration, with the aim of improving its effectiveness and results as well as the quality of services provided, as for the digital court, it has a vision The law also considers digital courts to be qualitative courts specialized in digital crimes, the origin of which is the automated side (computer) and then the international Internet and other networks. The jurisdiction also includes the outputs of digital devices, and as a general definition of the electronic court, these digital courts are considered courts that settle their disputes within the framework of hearings. In it, judges consider cases and adjudicate them using a set of information technologies and in accordance with legislation that authorizes them to do so.

One of the characteristics of digital courts is that they are courts that do not rely on paper to carry out their role, which means that they are courts without papers, as it is an administration that relies on computers in the primary form and does not depend on papers to a secondary degree. A court without a place: so that it relies on modern means of communication such as the Internet, so the administrative official can take the decision from anywhere in the world and is not dependent on the physical headquarters of the public utility. A court without time: so that the electronic administration does not abide by the official working hours, as it works 365 days a year, 7 days a week, and 24 h a day, meaning that it works around the clock throughout the year. A court without rigid regulations: Thanks to the electronic administration, we can talk about smart regulations that are flexible and capable of keeping pace with all emergency changes. Based on all these characteristics of the digital court, it plays an important role in facilitating the litigation process and bringing the judicial administration closer to the citizens, which contributes to creating judicial efficiency in a way that serves the interests of society [7].

3.2 Innovative Means When Executing the Sentence

We find that the reform of the judiciary has received a comprehensive priority in order to facilitate the litigation process, based on achieving the physical embodiment of procedures before the courts, and limiting the use of paper records because of the loss of cases that they may cause, and bypassing the stage of duplication of manual work and computerized work. Proceeding from what accompanied the transformations witnessed by the judicial system in our country, which resulted in the emergence of the digital court and electronic litigation processes, in implementation of the principle of "the judiciary at the service of the citizen," which we advocated at the opening of the first session of Parliament on October 8, 2010.

3.2.1 The Electronic Bracelet

The Jordanian judiciary began using the electronic bracelet as an alternative to arrest for the first time in Jordan by placing an electronic bracelet for two detainees against whom a decision was issued by the Public Prosecutor of Amman to detain them for a week, based on the text of Article (114) bis of the Code of Criminal Procedure and based on the text of Article (25) bis In addition, the complainant and the defendant are relatives, and a family dispute occurred between them, as a result of which the complainant filed a complaint, and both of the defendants were arrested for a week, and the public prosecutor decided to use an electronic bracelet for them as the first case of using an electronic bracelet in the Kingdom after he concluded that the defendant was two young men in their twenties And in the prime of life, they are not repeated, and this is the first case recorded against them, and it is of the type of misdemeanor.

The use of the electronic bracelet as an alternative to arrest aims to preserve their work, prevent their mixing with prisoners and repeat detainees, and reduce pressure on prisons on the other hand. According to this system, the court issues its decision to place the person under electronic surveillance and determines the time period and the geographical range specified in a specific geographical area or a specific route, or both. The Directorate of Operations and Control in Public Security follows up on the convict's commitment to that [8].

The electronic bracelet is a bracelet that the defendant wears at the bottom of the ankle of the foot for people subject to electronic control, and in the event that it is removed or tampered with or if it exceeds the geographical area in which he is authorized to be present, a signal is sent to the Command and Control of the Public Security Directorate to know his location and inform the nearest police patrol to reach him and take the necessary action against him.

3.2.2 Electronic Monitoring

Electronic monitoring as an alternative method to liberty-depriving penalties is one of the most important innovative methods in the implementation of punishment, and it has an apparent reflection on the punitive policy in reducing the practical and humanitarian problems facing traditional punishments in the closed environment, and many contemporary legislations adopt the electronic monitoring system.

We believe that it is one of the consensual alternatives to short-term liberty-depriving penalties, through which the convict can be followed up by the competent authorities outside the prison in specific places and times, while subjecting him to a set of obligations. Electronic monitoring is implemented in three ways: electronic monitoring via satellite, electronic monitoring via The telephone call, and this is done through a repeated electronic telephone call sent from the place of residence of the person under surveillance, where it is received by a voice code or a pronunciation identification method of continuous broadcasting, which is what France adopted, as this method is done through a device that allows the follow-up of the convict to ensure his presence In the place designated for him, and this method is based on placing an electronic bracelet on the wrist of the monitored person or on the bottom of his foot, which sends specific signals intermittently to a receiver connected to the telephone line in the place of residence of the monitored person, and the receiver sends specific signals to the authority supervising the monitoring, and in this way, its presence is recognized in the geographical location specified for it.

Electronic surveillance depends on the way it is implemented on a set of technical means, such as installing the electronic bracelet on the wrist or lower leg of the person being monitored, as this bracelet sends wireless signals every thirty seconds in the geographical area specified for monitoring, and another device is placed in the place usually designated for monitoring. Its task is to receive the signals sent from the electronic bracelet and send them through the telephone line that connects it to a central device. These sent signals are to indicate the presence of the monitored event in the specified place, and they may be warning signals when trying to damage the bracelet or the receiver [3, p. 81].

4 Conclusion

The world everywhere witnessed a digital revolution, and the digital transformation did not leave a room without changing its features or even its traditional rules until many judicial systems resorted to adapting electronic means, which have become a reflection with a positive impact, especially those related to the trial of juveniles, as most systems are keen The judiciary aims to preserve the privacy of the juvenile during his trial, and this study concluded with many results and recommendations, which we present as follows:

4.1 Results

1. The alternative punishment applied in support of the digital environment is of great importance as it is one of the most important short-term custodial penalties, which contribute to reforming, rehabilitating, and organizing the convict (the juvenile).

2. Modern innovative means to support the digital environment are the most important requirements for reforming and rehabilitating the convict (the juvenile) and reintegrating him into society so as not to expose him to the humiliation that he is exposed to while being placed in the traditional and customary punishment houses.
3. Electronic censorship and the electronic bracelet are among the most important reflections of the digital transformation on the implementation of custodial sentences, which provide an important guarantee for the protection of the juvenile's rights and basic freedoms, which are guaranteed to him by the constitution and international covenants.

4.2 Recommendations

1. The researchers recommend to the Jordanian legislator the necessity of determining the competent judge and defining his competence exclusively, and not in terms of money.
2. The legislator must adopt modern legislation and address all electronic means to protect the juvenile from the prison-related environment and to reform and rehabilitate him.
3. The researchers recommend that the Jordanian legislator determine the duration of the punishment that bears the alternative punishment and that does not affect the security of the community because the alternative punishments may not be applied with the long custodial sentences.

References

1. Derbal, W., Bousdra, S.: Penal Regulations in the Prison Organization Law. Master's Thesis, Republic of Algeria, Mohamed Boudiaf University in M'sila, p. 18 (2020)
2. Al-Mughid, M.A.R.: Alternatives to Societal Reform in Light of Amendments to the Jordanian Penal Code. Ramtha Court of First Instance, Irbid, Hashemite Kingdom of Jordan. https://jfslt.journals.ekb.eg/article_105721_bbc3763f9160fc9121536319c70e5d17.pdf
3. Al-Fiqi, I.: Alternative Regimes for Short-Term Imprisonment: A Comparative Study, Faculty of Law, Sadat City University, Egypt
4. Al-Abadi, K.A.R.: Societal Penalties in Jordanian Penal Legislation. Master Thesis, Middle East University, Jordan, p. 65 (2015)
5. Mabrouk, Y.: Guarantees of a fair trial through the use of video conversation technology. J. Rights Hum. Sci. **15**(1), 1639
6. Otani, S.: A work for the public benefit in contemporary penal policy, a comparative study. Damascus Univ. J. Econ. Legal Sci. **25**(2), 434 (2009)
7. Salama Rashid, H.: The penalty of community service, working for the public benefit in the Jordanian and Emirati Legislation. Sharia Legal Sci. Stud. **46**(Part 4), 409 (2019)
8. Al-Wreikat, O.M.O.: Community Service as an Alternative to Penal Sanctions. Master Thesis. International Islamic Sciences University, Amman, p. 56 (2017)

Role of Social Media in Changing the Social Life Patterns of Youth at UAE

Khalaf Tahat[1,2(✉)], Dina Naser Tahat[3], Ahmed Masoori[4], Mohammad Habes[5], Emad Alghazo[6], and Najia Ketbi[7]

[1] Media and Creative Industries Department, United Arab Emirates University, Al-Ain, UAE
k.tahat@uaeu.ac.ae
[2] Yarmouk University, Irbid, Jordan
[3] Applied Sociology Department, Al-Ain University, Al-Ain, UAE
[4] Media and Creative Industries Department, United Arab Emirates University, Al-Ain, UAE
[5] Faculty of Mass Communication, Radio and TV Department, Yarmouk University, Irbid, Jordan
[6] College of Education, Humanities and Social Sciences, Al Ain University, Al Ain, UAE
[7] Mohammed Bin Zayed University for Humanities, Abu Dhabi, UAE

Abstract. During the changing world of information communication, the role of social media is of greater significance. Particularly, during the current Covid-19 crisis, we have seen social media incorporated in many fields of life and has several impacts. This research also focused on examining the impacts of Social Media on the life of the young Emirati generation in terms of their communication with family and other relevant people. The researcher used the descriptive method and gathered data by using close-ended questionnaires from university-level students. Findings indicated an overall positive impact of Social Media on the social life of individuals during the Covid-19 outbreak. The respondents expressed their dependency on social media, and also indicated social media as a source to communicate with family members. Participants say connected with their family, take more interest in their domestic matters, and also interact more with their parents and relatives. Thus, we conclude that the mutual consensus of study respondents lean towards the role of social media as an agent to sustain socialization even during one of the most challenging times in human history.

Keywords: Social Media · Covid-19 · United Arab Emirates · Social System · Family Relations

1 Introduction

The changing patterns of communication and information today, herald for a greater dependency on media platforms in the future. Especially, in terms of social media platforms, this dependency is greater and accompanied by several and positive aspects. Despite there are several negative characteristics of social media, the positive aspects are undeniable. Particularly, during the current pandemic, social media is playing its

B. A. M. Alareeni and I. Elgedawy (Eds.): *Artificial Intelligence (AI) and Finance*, SSDC 488, pp. 152–163, 2023.
https://doi.org/10.1007/978-3-031-39158-3_14

role as a career of information, communication, interaction, education, and entertainment in a beat possible manner. As noted by Habes et al. (2021), social media platforms greatly facilitated our lives with normal life proceedings during the current healthcare crisis. Although the precautionary measures like social distancing and closure of several institutions created physical distances, social media keeps us interconnected through a unique virtual world of communication. As a result, on both educational and professional levels, daily life activities are resumed through remote networks and platforms. For instance, educational institutions are considered as a source of one's social development and training. Besides continuing the education, a student learns social life skills and communicates effectively with fellow students. However today, educational institutions are closed and students are staying inside their houses and rely on digital platforms to continue their educational activities and enhance their social interaction with their peers and instructors [2]. Similarly, just like other parts of the world, Emirati society also faced gigantic challenges regarding students' growth and development. A limited life inside four walls of the house, restricted access to outside services, lack of social activities i.e., visiting public places, interaction with peers, etc., are all are adversely affected [3, 4]. Consequently, people especially the young generation are increasingly dependent on social media platforms for communication, information, education, and other purposes. Today, this increased dependency also indicates several changes in the Emirati society on both macro and micro-level social life. According to Al Hosany et al. (2021), the precautionary measures to hamper the Covid-19 outbreak also affected the family and social gathering patterns in the United Arab Emirates. In such a situation, Humanitarian services organizations, clinical practitioners, and government unified their stance overusing social media as a solution to the existing problems. As noted by the role of social media remained significant in keeping people connected with others. Virtual meetings, phone calls, videos chat, etc., greatly helped the Emirati society stay connected, leading to an even stronger family bond and an increased interest in family concerns [6]. Thus, in the light of the above discussion, the current research also examines the role of social media is changing the social life patterns of Emirati society. Many studies focused on changing patterns of social life, yet the changing patterns of social life during the Covid-19 in terms of Social Media still need to be examined. Hence by keeping in view the mentioned literature, this study is structured in different sections [7, 8]. The current study aims to shed light on the effects of social media on the social life of Emirati youth during the Covid-19 due to the increased dependence of community members on media information due to knowledge and guidance about what is happening in their society, and that the percentage of dependence is related to the degree of importance of the means for individuals as a source of information about events and issues raised, and depends The degree and importance of this medium depends on the volume and importance of information derived from the media, which affects their behavior in society.

2 Literature Review

2.1 Role of Social Media in Society

The extensive influence and role of social media can be seen in society with its countless contributions in almost every field of life. Media as a major social institution, help us to think and act accordingly. In a similar context, the advancement in media technology is adding more to our living and thinking patterns. Today, every single invention helps us to get connected to the internet providing countless opportunities to achieve new goals and advance our media exposure experiences. As compared to the earlier times, the communication, information, entertainment, and interaction are fast, efficient, and organized in a better-improved manner (Singh & Singh 2017). According to Damota (2019), social media enables people to communicate with each other and increase their social capital. Social media helps people to distribute their messages and receive the message with thousands of people all around the world. The concept of social media is mainly attributed to popular websites like Facebook, Twitter, Pinterest, Instagram, and others that are the major platforms to communicate with other users. As a result, social media provides us with task management services besides communication and interaction. [11–13]. A study conducted by Amedie (2015), also investigated the role and impacts of social media on society. The researcher selected a literature review approach and found that rapid information sharing, ease of communication, ease of access, constant communication, and access to information are the important contribution of social media. (Elareshi, et al. 2022) Despite social media usage contains many negative aspects, the positive aspects are undeniable, witnessing the digital media as a fundamental component of our society. (Habes, et al. 2021)Thus, considering the current increased social media usage, we can attribute the use of social media and its impacts on social life in terms of cause and effect relationships. The continuous flow of information and communication is a dynamic reason behind increased social media usage today, leading to its integral role in the current society [17].

2.2 Covid-19 and Social Barriers

The current health crisis, mainly known as the Covid-19 pandemic led to several social and psychological concerns. Closure of office, educational institutions, markets, restaurants, and others, limited the daily life activities, providing a very restricted lifestyle to spend most of the time inside the four walls of the house. Previous studies on different endemics such as SARS, EBOLA, and ERS also raised similar concerns regarding the social and psychological patterns of life (Habes, et al. 2021) However, a prominent difference between previous healthcare and current Covid-19 is the duration of disease that questions whether daily life can back to normal or not [19]. According to the initial reports from the different national and international level organizations, Covid-19 would adversely affect social, economic, psychological, and all the other aspects of life. Several barriers to normal life activities will put the lives of people at risk. Especially, the people who are considered more vulnerable will be endangered on different levels [20–24]. In this context, Deloitte 2020) argued that the local governments tried their best to refrain the Covid-19 from spreading. To counteract the rapid spread of virus outbreaks, the role of government remained prominent due to precautionary measures such as social distancing, wearing masks, less interaction with others, closure of institutions, yet the other social impacts remained highly controversial. In such a situation, when people are unable to meet and greet each other, visit their relatives, unable to sustain their daily-life activities, they need to search for effective solutions to cope with this situation. The young generation, on the other hand, faced much seclusion and loneliness due to the sudden closure of educational institutions as well. Opportunities to meet their classmates, friends, and instructors remained low, leading to search for the different possible ways to stay connected in a better possible manner. These youngsters found social media as the only way to communicate effectively and cope with the physical barriers caused by the Covid-19 crisis [26, 27].

2.3 Impact of Covid-19 on Family System

The Covid-19 pandemic gave the biggest shock to the world's economy and family system worldwide. Staying home for the professionals and children's educational transition from formal learning to online learning represents a great transformation. Yet many consider it as a positive constructive transition in terms of family life and wellbeing. The changing patterns of social life are expected to leave deep effects on children even after the declination of the outbreak [28]. Despite families from the middle and lower-middle classes are facing several financial issues, they consider Covid-19 as a source of unifying their family members on similar grounds with a strong sense of belonging. Despite disturbing everyday life activities, parents and children are interacting more, taking interest in domestic matters, leading to an increased affection among them [29]. Crew, (2020) further affirmed the role of Covid-19 in strengthening family life in terms of increased parental interest in children's education. As noted that, parents seem more concerned about their children's education. For this purpose, they are taking much interest and actively take part in teaching and guiding their children in an online learning environment. Besides, parents are also interacting with the teachers, indicating the current pandemic as a source of strengthening the parent-child bonding as well. For example, despite the deprivation from social life activities, education, and playing areas, Italian children and families are successfully coping with these challenges by increasing their interaction with each other. Besides parent-child relations, the non-virtual communication services are helping to also communicate with the other family members [31].

3 Research Methods:

This study is descriptive, and the researcher gathered data by using the closed-ended questionnaire from the study participants. As noted by [32], factual registration of results is a primary benefit of descriptive research. Descriptive research helps us to show how reality takes place without the formulation of any research hypotheses in general. Thus, after gathering the data, the researcher manipulated and code it for statistical analysis by using the Statistical Package for Social Sciences (SPSS) [33, 34]. Table 1 below provides a brief over of our questionnaire items and their sources and their intercoder reliability assessment values to further affirm the validity of the research tool. As visible that, All the Cronbach Alpha values are greater than the threshold value of 0.7 as ranging from 0.707 to 0.790, the research instrument s strongly reliable.

Table 1. Questionnaire Items and their Sources

N	Items	Sources	Cronbach Alpha Value
1	Covid-19 and Daily Life Patterns	[35]	.771
2	Mitigation of Social Impacts	[36]	.726
3	Continuation of Social Life Activities	[37]	.790
4	Social Media and Family Life	[38]	.707

As the study is aimed at examining the impacts of social media in sustaining the social life activities of the Emirati public, the researcher selected college-level students for the data gathering process. First, the researcher randomly selected a sample of n = 315 students. Moreover, the researcher used simple random sampling as according to [39], the relevant sampling method helps to select the study respondents without the researchers' bias. Also, the equal chance of selection is another major reason behind choosing simple random sampling in the current study [40]. The first part of the survey involved questions regarding the participants' demographical details such as gender, age, locality, and type of family system (living individually or with the family members). The second part involves a general overview of social media usage, the purposes of social media usage, and others. (4 statements). The third section involves questions regarding the impact of Covid-19 on daily life activities i.e., how participants are sustaining their daily life activities during the current pandemic. (3 statements). The fourth section involves questions regarding the steps taken to mitigate the impacts of Covid-19 by using social media-based virtual meetings, interaction, communication (3 statements). The fifth section involves the questions regarding how and to what extent social media platforms are helping the respondents to continue their daily life activities in a virtual environment. (3 statements). The sixth and the last section involves questions regarding the extent to which social media is helping the Emirati youth to stay connected with their family member, and also the extent to which this pandemic has increased their interest in family matters. (4 statements).

4 Data Analysis and Findings

4.1 Demographics of Study Respondents and One-Way ANOVA

The researcher first analyzed and calculated the demographics of study participants. As shown in Table 1, the majority of respondents (57.9%) were males and 41.7% were females. According to the age of the respondents, 50% were 19–29 years old, 13.8% were 16–18 years old, 11.3% were 23 years old or above, and 5.1% were 21–22-year-old. Moreover, 70.8% of respondents were living in urban areas and 29.2% were from the urbanized areas. Finally, 59.4% of respondents revealed that they are living with their other family members including parents, siblings, and other closed relations whereas, 40.6% of the part pants are living individually. Furthermore, to check any potential difference based on respondents' demographical characteristics [41], the researcher conducted a One-Way Analysis of Variance. As visible in Table 1, gender ($p \leq 0.000$), age ($p \leq 0.000$), locality (($p \leq 0.059$), and living system ($p \leq 0.000$) are singingly correlated with the social media usage, indicating no significant mean differences among the study variables, the researcher did not find any differences based on the demographical characteristics of the study participants (Table 2).

Table 2. Descriptive Statistics Regarding Demographics and One-Way ANOVA:

Constructs	Variables	*f*	%	*f*	Levene Statistics	Sign
Gender	Male	183	57.9%	.953	19.381[a]	.000
	Female	132	41.7%			
Age	16–18	54	13.8%	1.767	1.679[b]	.000
	19–20	197	50.5%			
	21–22	20	5.1%			
	23 or Above	44	11.3%			
Locality	Rural	92	29.2%	1.448	8.021[c]	.059
	Urban	223	70.8%			
Living System	Individual	128	40.6%	2.497	5.994[d]	.000
	With Family	187	59.4%			

Survey Responses:

We conducted the descriptive analysis to examine the survey responses of the study participants. In this regard, we found that a majority (75.25) of participants strongly agreed that they prefer using Social Media on regular basis as for 34.4% of respondents Social Media is a substitute for conventional media in terms of communication and information transmission process. Besides, 51.5% of respondents also agreed that Social Media largely facilitated the communication process during the Covid-19 outbreak and they (47.4%) use Social Media for information, education, and communication purposes. Moreover, according to 70.8% of study respondents, Covid-19 has adversely affected their daily-life activities leading to changed and disturbed communication patterns for the 46.2% of study participants. However, 39.7% of participants remained neutral about the fact that Covid-19 has remarkably reduced social life activities & chances of face-to-face communication (Table 3).

According to 50.8% of participants, they mainly prefer social media to interact with family & friends and 48.2% of these respondents also prefer using Social Media to interact with their classmates and teachers as 43.6% of them believe that Social

Table 3. Descriptive Statistics Concerning Social Media Usage and Impact of Covid-19 on Daily Life

N	Predictor Variable: *Social Media Usage*	Mean	SD	Min	Max	Variance	Range
1	Regular Social media usage	4.25	1.022	0	5	1.044	5
2	Substitute of information and communication	3.61	1.069	0	5	1.143	5
3	Facilitated communication during Covid-19	3.92	.826	0	5	.682	5
4	SM for information, education, and communication	4.33	.993	0	5	.986	5
	Dependent Variable: *Impact of Covid-19*						
5	Affected my daily life activities	4.27	.971	0	5	.943	5
6	Changed the patterns of interaction & communication	4.31	1.000	0	5	1.000	5
7	Reduced social life activities & chances of face to face communication	3.59	1.025	0	5	1.052	5

Media is the significant source to communicate with others with ease of use and ease of access. As a result, 41.5% of respondents can now easily communicate with their family and friends. Yet 35.5% of respondents also remained neutral about resorting to Social Media for communicating with classmates and instructors. Consequently, 43.1% of respondents showed an explicit agreement with the fact that Social Media provides easy communication services to sustain daily life activities during the current healthcare crisis (Table 4).

Table 4. Descriptive Statistics Concerning Mitigation o Impact and Daily-life activities

N	Dependent Variable: *Mitigation of Impacts*	Mean	SD	Min	Max	Variance	Range
8	Prefer social media to interact with family & friends	3.86	.879	0	6	.773	6
9	Social media to interact with instructors and peers	4.33	1.003	0	6	1.006	6
10	Prefer social media to easily communicate with others	4.27	1.007	0	6	1.014	6
	Dependent Variable: *Daily-life activities*						
11	Communicate easily with family & friends	4.21	1.056	0	6	1.056	6
12	Communicate easily with the instructors and peers	3.53	1.104	0	6	1.104	6
13	Sustaining daily life activities	3.77	1.022	0	5	1.022	6

Finally, we investigated the respondents about the use of Social Media in communication with family and its impacts. Thus, 45.6% of respondents strongly agreed that Social Media provided them increased opportunities to communicate with their family during the Covid-19 crisis. As a result, 39.2% of respondents also found Social Media as a strong source to keep strong, continuous communication with their family members that further helped 59.7% of participants to continue interacting with family, friends, and other people in a normal manner regardless of the physical barriers (Table 5).

Table 5. Descriptive Statistics Concerning Social Media and Family

N	Dependent Variable: *Social Media & Family*	Mean	SD	Min	Max	Variance	Range
14	SM helps to stay connected with the family	4.20	1.106	0	5	1.222	5
15	SM increased interaction with family	3.55	1.025	0	5	1.051	5
16	SM helped to take more interest in family matters	3.86	.847	0	5	.717	5
17	SM helped to continue interacting normally with family, friends, and other people	3.87	.830	0	5	.689	5

5 Discussion on Results

As mentioned earlier, the role of social media in society can be seen in many ways. Today, people are just a click away from communication, and information gathering. Especially, in terms of communication and interaction, the role of social is prominent as it provides an easy way to communicate with people regardless of cultural and geographical boundaries (Singh & Singh 2017). From education to business, and in all the other fields, the role of social media is prominent in facilitating communication [42]. In this regard, during the current Covid-19 crisis, the physical barriers might seem gigantic and problematic to connect with others. Yet, the use of social media for communication purposes remained vital to cope with these challenges. Now people are relying on online resources to communicate with their families, friends, and peers. Young students are using social media to resume their education, stay updated about their education institutions, interact with their friends, and also communicate easily with their teachers [43]. The current descriptive study also aimed to investigate and highlight the role of social media in coping with the Covid-19 and also its role to positively impact the social life of the Emirati young generation. The researcher found that the youngsters daily use social media without any barriers. Also, they prefer using social media for communication, information, and education purposes. This social media usage and purposes are consistent with the study conducted by Ali et al., (2021), in which the researchers also found the participants increased social media usage for relevant purposes. As noted that, youngsters" ability to recognize the importance of social media in daily life activities has further helped them to increased social media for communication and learning purposes. Despite, many students faced a sudden shock due to the transition from formal to the online environment, still it also helped them in many ways. Moreover, the students also expressed a strong consensus regarding the psychological and social impacts of Covid-19 on their daily life activities. According to them, the barriers to going outside, schooling, playing with peers, all are adversely affected by the Covid-19. Now they are compelled to stay within four walls to halt the virus outbreak in the best possible manner. Yet, we cannot ignore the disturbing patterns of our regular life activities. These results are strongly compatible with the propositions highlighted by Wajahat Hussain (2020) as noted that, social media is helpful during the Covid-19 outbreak. Regardless of the social distance policies, increasing communication with the family, friends, and others users can help them increase their understanding of their cultures, races, and also the other relevant types of diversity. These social media platforms are also providing an easy flow of information to the Emirati young generation leading to public education and awareness. Likewise, the respondents expressed their dependency on social media during the healthcare crisis. This dependency is mainly about communication, education, and information gathering aspiration, which is also helping them to mitigate the impacts of pandemics. Once again they considered social media as providing ease of access and ease of communication to gain the maximum benefit from the digital technology. According to Damota (2019), social media is an easy gateway to share collective interests, opportunities for community engagement, starting philanthropic events, growth of ideas, etc. on every possible level. As a result, despite the social distancing, confinement, and physical barriers, the study respondents are successfully sustaining the daily life activities that are also helping them to cope with the psychological challenges as well. Finally, by keeping in view the study conducted by Morelli et al. (2020), the researcher also examined the impacts of Social Media during Covid-19 on the family system of Emirati people. As found that, the respondents expressed their dependency on social media, they also indicated social media as a source to communicate with family members. Participants say connected with their family, take more interest in their domestic matters, and also interact more with their parents and relatives. As a result, the family relations are strengthening due to ease of communication and ease f access through non-virtual audio and video chat platforms. As noted by Al Hosany et al., (2021), the social, economic, and psychological challenges are increasing day by day, yet the family life, bonding, connection, and affection are strengthening due to increased interaction and interest in each other's matters. Consequently, despite many agree on the adverse impacts f Covid-19, its positive impacts are still undeniable.

6 Conclusion and Limitations

We found an overall positive impact of Social Media on the social life of individuals during the Covid-19 outbreak. The mutual consensus of study respondents leans towards the role of social media as an agent to sustain socialization even during one of the most challenging times in human history. However, the discrepancies between the extent and mode of usage can exist, the study witnesses an accumulation of responses, that provide the positive constructive role of social media. The findings indicate that the individuals who feel more comfortable due to social media usage during the Covid-19 are more likely to have a positive opinion about it. In such a situation, the young generation is more actively using social media that also helps them to have a positive opinion about it. Every one of the stakeholders and particularly Commission of UAE and network suppliers' necessities to concoct method for separating data that contacts the youngsters through virtual entertainment stages. This will help limit presenting them to Public activity Examples. To try not to have an age that doesn't have up close and personal relational abilities, every one of the partners ought to consider coordinating get-togethers that the young will use to interface and Public activity Examples with one another by Social Media. Thus, despite the novelty of the topic, this study also contains some primary limitations: **(i)** The researcher selected the study sample only from Emirates, which questions the generalization of results in other countries, **(ii)** The researcher selected social media as a whole instead of selecting any specific platform to study, and **(iii)** the study is descriptive. The researcher did not formulate any formal research hypotheses. Thus, the researcher recommends more studies on examining the impacts of social media, especially during the Covid-19 to find out even more in-depth findings.

References

1. Habes, M., Ali, S., Khalid, A., Haykal, H.A.: eLearning acceptance during the COVID-19 outbreak : a cross-sectional study. In: Advances in Intelligent Systems and Computing, Springer Nature Switzerland (2021)
2. Alhumaid, K., Ali, S., Waheed, A., Zahid, E., Habes, M.: COVID-19 & e-learning : perceptions & attitudes of teachers towards e- learning acceptance in the developing countries. Multicult. Educ. **6**(2), 100–115 (2020). https://doi.org/10.5281/zenodo.4060121
3. Habes, M., Pasha, S.A., Ali, S., Elareshi, M., Ziani, A., Bashir, B.A.: Technology-enhanced learning acceptance in Pakistani primary education. In: European, Asian, Middle Eastern, North African Conference on Management & Information Systems, pp. 53–61 (2023)
4. Habes, M., Elareshi, M., Ali, S., Ziani, A.: Analyzing the portrayals of child sexual abuse of urdu newspapers in developing countries. Pertanika J. Soc. Sci. Humanit. **2** (2022)
5. Al Hosany, F., et al.: Response to COVID-19 pandemic in the UAE: a public health perspective. J. Glob. Health **11**, 03050 (2021). https://doi.org/10.7189/jogh.11.03050
6. Aouadi, S.: Economic and Social Impacts of the Covid-19 Pandemic in the Arab Region the Economic Effects of the Covid-19 Pandemic (2021)
7. Tahat, D.N., Tahat, K.M., Habes, M.: Jordanian newspapers coverage of cyberbullying during COVID 19 Pandemic. PalArch's J. Archaeol. Egypt/Egyptol. **17**(7), 15390–15403 (2020)
8. Ziani, A.-K., Elareshi, M., Habes, M., Tahat, K. M., Ali, S.: Digital media usage among arab journalists during covid-19 outbreak. In: European, Asian, Middle Eastern, North African Conference on Management & Information Systems, pp. 116–129 (2021)
9. Singh, G., Signh, N.: Role and impact of media on society: a sociological approach with respect to demonetisation [Papel e Impacto de los medios en la sociedad: un enfoque sociológico con respecto a la desmonetización]. Impact Int. J. Res. Humanit. Arts Lit. **5**(10), 127–136 (2017). https://doi.org/10.13140/RG.2.2.36312.39685
10. Damota, M.D.: The effect of social media on society. New Media Mass Commun. (2019). https://doi.org/10.7176/nmmc/78-02
11. Al Olaimat, F., Habes, M., Hadeed, A., Yahya, A., Al Jwaniat, M.I.: Reputation management through social networking platforms for PR purposes: a SEM-based study in the Jordan. Front. Commun. **11**, 247 (2022)

12. Habes, M., Elareshi, M., Ziani, A., Almansoori, A., Alsridi, H.: Smart interaction and social TV used by Jordanian University students. Technol. Soc. **71**, 102110 (2022)
13. Elareshi, M., Habes, M., Al-Tahat, K., Ziani, A., Salloum, S.A.: Factors affecting social TV acceptance among Generation Z in Jordan. Acta Psychol. (Amst). **230**, 103730 (2022)
14. Amedie, J.: Impact of social media on pharmaceutical manufacturer (2015). https://doi.org/10.18311/gjeis/2016/15773
15. Elareshi, M., Habes, M., Youssef, E., Salloum, S.A., Alfaisal, R., Ziani, A.: SEM-ANN-based approach to understanding students' academic-performance adoption of YouTube for learning during Covid. Heliyon **8**(4), e09236 (2022)
16. Habes, M., Ali, S., Elareshi, M., Tahat, K.M., Ziani, A.: Understanding users' social TV content using artificial intelligence approach: a survey. In: 2021 International Conference of Modern Trends in Information and Communication Technology Industry (MTICTI), pp. 1–7 (2021)
17. Harchekar, J.S.: Impact of social media on society. Int. J. Eng. Res.**V6**(07), 2015–2017 (2017).https://doi.org/10.17577/ijertv6is070249
18. Habes, M., Elareshi, M., Ziani, A.: An empirical approach to understanding students' academic performance: youtube for learning during the covid-19 pandemic. Linguist. Antverp. 1518–1534 (2021)
19. Saladino, V., Algeri, D., Auriemma, V.: The psychological and social impact of covid-19: new perspectives of well-being. Front. Psychol. **11** (2020). https://doi.org/10.3389/fpsyg.2020.577684
20. OECD, COVID-19 : Protecting people and societies 33 (2020)
21. Al-Maroof, R.S., Salloum, S.A., Hassanien, A.E., Shaalan, K.: Fear from COVID-19 and technology adoption: the impact of google meet during coronavirus pandemic. Interact. Learn. Environ. 1–16 (2020)
22. Habes, M., Elareshi, M., Mansoori, A., Pasha, S., Salloum, S.A., Al-rahmi, W.M.: Factors indicating media dependency and online misinformation sharing in Jordan. Sustainability **15**(12), 1–15 (2023). https://doi.org/10.3390/su15021474
23. Al-Shibly, M.S., Alghizzawi, M., Habes, M., Salloum, S.A.: The impact of de-marketing in reducing Jordanian youth consumption of energy drinks. In: International Conference on Advanced Intelligent Systems and Informatics, pp. 427–437 (2019)
24. Al-Shakhanbeh, Z.M., Habes, M.: The relationship between the government's official facebook pages and healthcare awareness during covid-19 in Jordan. In: Hassanien, A.-E., Elghamrawy, S.M., Zelinka, I. (eds.) Advances in Data Science and Intelligent Data Communication Technologies for COVID-19. SSDC, vol. 378, pp. 221–238. Springer, Cham (2022). https://doi.org/10.1007/978-3-030-77302-1_12
25. Deloitte, The social impacts of COVID-19 Reset not restart: August 2020
26. Aristovnik, A., Keržič, D., Ravšelj, D., Tomaževič, N., Umek, L.: Impacts of the COVID-19 pandemic on life of higher education students: a global perspective. Sustainability **12**(20), 8438 (2020). https://doi.org/10.3390/su12208438
27. Youssef, E.: Role of social service institutions on social empowerment of women at the United Arab Emirates. a field analysis study. Multicult. Educ. **6**(4), 99–111 (2020). https://doi.org/10.5281/zenodo.4252122
28. Kalil, A., Mayer, S., Shah, R.: Impact of the COVID-19 crisis on family dynamics in economically vulnerable households. SSRN Electron. J. (2020). https://doi.org/10.2139/ssrn.3706339
29. Gadermann, A.C., et al.: Examining the impacts of the COVID-19 pandemic on family mental health in Canada: findings from a national cross-sectional study. BMJ Open **11**(1), 1–11 (2021). https://doi.org/10.1136/bmjopen-2020-042871

30. Crew, M.: Literature Review on the Impact of COVID-19 on families, and implications for the home learning environment. a national literacy trust research report. Natl. Lit. Trust 1116260 (2020)
31. Morelli, M., et al.: Parents and children during the COVID-19 lockdown: the influence of parenting distress and parenting self-efficacy on children's emotional well-being. Front. Psychol. **11**, 1 (2020). https://doi.org/10.3389/fpsyg.2020.584645
32. Van Der Voordt, L.T.: Ways to study - descriptive research. Fam. Pract. - FAM Pr. **8**, 53–60 (2002)
33. Alhammad, K.L., Habes, M., Al Olaimat, F., Haddad, I.: Attitudes of students of the faculty of mass communication at Yarmouk university towards using video platforms in distance education. Rev. Int. Geogr. Educ. Online **11**(5), 1041–1052 (2021)
34. Habes, M., Ali, S., Salloum, S.A., Elareshi, M., Ziani, A.-K., Manama, B.: Digital media and students' AP improvement: an empirical investigation of social TV. In: 2020 International Conference on Innovation and Intelligence for Informatics, Computing and Technologies (3ICT), pp. 1–6 (2020)
35. Cheng, T.C., Kim, S., Koh, K.: The Impact of COVID-19 on Subjective Well-Being: Evidence from Singapore. IZA Discuss 13702 (2020)
36. Habes, M., Alghizzawi, M., Ali, S., SalihAlnaser, A., Salloum, S.A.: The relation among marketing ads, via digital media and mitigate (COVID-19) pandemic in Jordan. Int. J. Adv. Sci. **29**(7), 2326–12348 (2020)
37. ECLAC, Special Report: The social challenge in times of COVID-19. **3**, 1–21 (2020)
38. Vanderhout, S.M., Birken, C.S., Wong, P., Kelleher, S., Weir, S., Maguire, J.L.: Family perspectives of COVID-19 research. Res. Involv. Engagem. **6**(1), 4–6 (2020). https://doi.org/10.1186/s40900-020-00242-1
39. Taherdoost, H.: Sampling methods in research methodology; how to choose a sampling technique for research. SSRN Electron. J. (2018). https://doi.org/10.2139/ssrn.3205035
40. Kalton, G.: Simple random sampling. Introd. Surv. Sampl. 9–16 (2011). https://doi.org/10.4135/9781412984683.n2
41. Scofield, T.L.: ANOVA **2**, 1–11 (2018)
42. Alghizzawi, M., Habes, M.: Impacts of social media and demographical characteristics on university admissions : case of Jordanian private universities impacts of social media and demographical characteristics on university admissions : case of. PalArch's J. Archaeol. Egypt/Egyptol. **17**(7), 6433–6454 (2020)
43. Al-Skaf, S., Youssef, E., Habes, M., Alhumaid, K., Salloum, S.A.: The acceptance of social media sites: an empirical study using PLS-SEM and ML approaches. In: Hassanien, A.-E., Chang, K.-C., Mincong, T. (eds.) AMLTA 2021. AISC, vol. 1339, pp. 548–558. Springer, Cham (2021). https://doi.org/10.1007/978-3-030-69717-4_52
44. Ali, S., Habes, M., Youssef, E., Alodwan, M.: A cross-sectional analysis of digital library acceptance, & dependency during covid-19. Int. J. Comput. Digit. Syst. (2021)
45. Hussain, W.: Role of social media in COVID-19 pandemic. Int. J. Front. Sci. **4**(2), 59–60 (2020). https://doi.org/10.37978/tijfs.v4i2.144

Assessing the Effectiveness of 360-Degree Performance Appraisal System to Enhance Employees' Productivity in Jordanian Listed Banks

Mohammad Sami Ali[1](✉), Sadam Awadallah Abdalgani[2], and Dina Dahlan[3]

[1] Department of Financial Technology, Faculty of Economics and Administrative Sciences, Zarqa University, Zarqa 13110, Jordan
Maldahrawi@zu.edu.jo

[2] Department of Business Administration, Faculty of Business and Finance, The World Islamic Science and Education University, Amman, Jordan

[3] College of Business Administration, University of Business and Technology, Jeddah, Saudi Arabia
d.dahlan@ubt.edu.sa

Abstract. The performance appraisal is not limited to evaluate the performance of banks' employees, but it also facilitates highlighting the particular objectives of organisations. Performance appraisal is considered as a very effective tool for evaluating and improving employees' knowledge and training with an intension to establish a long term learning guide, and strategies for nourishing employees' productivity for enhanced working hours. Therefore, the study investigated the effect of "360-degree performance appraisal system" in the productivity of Jordanian listed banks' employees. Through examining data that was collected from a sample of 120 employees, those are working in Jordanian Islamic and commercial banks; the results indicated that the current performance appraisal system is significantly and positive affecting the level of employees' productivity. Moreover, the study suggested that Jordanian banking sector must perform performance appraisals more frequently to have positive outcomes, as well as it is necessary to evaluate the strengths and weaknesses of employees to allocate responsibilities accordingly.

Keywords: 360-Degree performance appraisal · Jordanian Listed banks · Employees' Productivity · Compensation Management

1 Introduction

Enhancement of employees' productivity is one of the most critical factors to meet the objectives of the organisation, but such enhancement could be done only when organisations are already informed regarding the current performance level of their employees (Ugonai 2020). Therefore organisations employ performance appraisal approach to make such evaluations not only about their employees, but also about themselves so

B. A. M. Alareeni and I. Elgedawy (Eds.): *Artificial Intelligence (AI) and Finance*, SSDC 488, pp. 164–174, 2023.
https://doi.org/10.1007/978-3-031-39158-3_15

that they can focus on meeting their objectives through improving the productivity of their employees. Mathis and Jackson (1994) argue that after employees' selection, training and then eventually working on a job; the performance of that employee must be reassess after some time. Therefore, performance appraisal is considered as an effective process to assess employees' performance in comparison to the defined standards of the organisations and making appropriate adjustments if required (Gupta and Parmar 2018). Performance appraisal is employed in different names in different organisations like employee evaluation, performance evaluation, employee rating etc. Despite considering it as an effective tool, some organisations do not adopt it as a necessary mechanism in favour of their organisations (Bagul 2014). Such lack of focus is mostly due to some dominating reasons like not finding it as objective and not focusing on having discussions with employees about their appraisal ratings. Practically, sometimes it also creates difficulty in differentiating good employees from the others in the absence of good performance information (Binta et al. 2019). Performance appraisals differentiate in their roles and uses according to the objectives and defined standards of the organisations. But the most common objective of performance appraisal is to evaluate the progress of an employee or to identify the deficiencies in comparison to the job effectiveness, requirements, and productivity (Zhang et al. 2017). Assessment of the employee productivity in comparison to its cost is also one of the critical aspects being measured by performance appraisal because employee expense captures a significant portion of the costs of organisations. On the other hand, employees' productivity level indicates the effectiveness of organisations' management. Therefore, it focuses on the recognition, assessment and management of organisations' employees' output (Asamu 2013). The effectiveness of performance appraisal lies in facilitating management in devising the training and motivation sessions for their employees so that organisation can retain the talented ones and can enhance organisations' performance through them (Hayford et al. 2016). While on the other hand, the absence of such appraisal system may lead to the turnover of employees due to lower levels of incentives or their satisfaction and ultimately having negative impact on organisations' productivity as well (Audu and Timothy 2014). Productivity is the assessment of output per employee of the organisation while effective utilisation of the available organisational resources. Another very significant aspect of performance appraisal is the motivation of employees because motivated employees are the most productive ones in almost all organisations. It not only improves the productivity level and quality of work but it also helps establishing the encouraging working environment in the organisation (Khawaja and Nadeem 2013).

Employees' motivation for work and higher productivity has many underlying factors including the compensation which is although among the leading factors but not the only major one. Some other key factors for employees' motivation are self-actualisation, job design; status of job, socialisation and the sense of purpose or it may be due to some other psychological aspects particularly associated and vary from worker to worker (Mohammed and Abdulahi 2019). Hence performance appraisal helps highlighting the need for such factors and facilitates and guides the organisations' management to establish a mechanism to motivate employees for attaining their positive response towards the attainment of organisational objectives (Mollel-Eliphaz et al. 2017). Such performance appraisals are although the part of working culture in Jordanian banking sector but current research is focusing particularly on "360 degree performance appraisal technique" to investigate its impact on banking sectors' employees' productivity (Ugonai 2020). Another significant aspect in this regard is the identification of problems by the management so that appraisal could be done effectively but for that appropriate and constructive feedback of management is required and then taking the appropriate corrective actions so that the outcome must be in the form of enhanced employees' productivity (Gupta and Parmar 2018). Unfortunately, most of the time it becomes challenging for the appraisers as well as for management to isolate individual contributions from team works and to avoid the influence of personality bias on such judgements (Binta et al. 2019). Such challenging situations and some other issues like incorporation of the political agenda, leniency or strictness of the appraisers and so on makes the whole system as a mess and most of the times results in lower levels of motivation and productivity. Another critical challenge is the time lag because mostly such appraisals are conducted on annuls and semi-annual basis so the actual performance of the employees is not captured truly (Bagul 2014). Due to such challenges appraisals are sometimes considered as useless and erroneous for promotions and appointments (Zhang et al. 2017). Therefore, current research aims at highlighting whether the performance appraisal (PA) system in Jordanian banking sector significantly influences the productivity levels of its employees. For this purpose, the study is mainly focusing on addressing the following research enquiries:

1. Does the implemented PA in Jordanian listed banks facilitate the compensation management?
2. Is the feedback development seen as a significant part of the PA's process in Jordanian listed banks?
3. Does PA facilitate the process of managements' decision making?
4. Does PA contribute to employees' motivation?
5. Does employees' motivation contribute further in improving employees' productivity?
6. Does the implemented PA system impact employees' productivity in Jordanian commercial and Islamic banks?
7. Are there significant differences between employees' productivity in Jordanian commercial and Islamic banks in accordance to the adopted PA system?

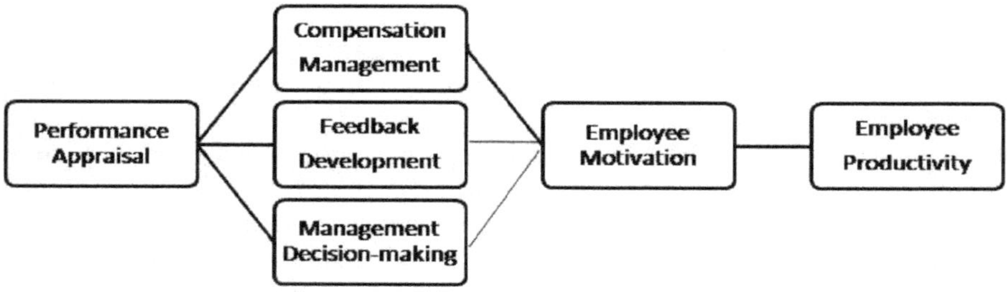

Fig. 1. Conceptual framework.

Based on the above proposed research questions the study proposes the following conceptual framework in order to check relationships among the examined variables (Fig. 1):

2 Literature Review

Performance appraisal is a systematic process that evaluates the strengths and weaknesses of an individual relevant to the job with an intension that if satisfactory performance is identified then the strengths of employees will be appreciated, reinforced and motivated (Binta et al. 2019). While on the other hand if discrepancies are found then the weakness are identified and employees' habits are redirected through appropriate training and education to align with the organisational objectives (Audu and Timothy 2014). Performance appraisal is a criterion variable that estimates the employees' job performance for a particular time period where the job is defined by the set of certain tasks. It is a structured process in which a communication session is conducted between the appraise and the appraiser discussing about the job performance so that the improvement plan can be designed accordingly through required and appropriate measures including counselling, training and development, retraining, mentoring or in some other way (Asamu 2013). The productivity of the employees is assessed through assessment of the accomplished tasks as per the requirements of employees' job. So in this way the appraisal is actually evaluation of the employees' contribution towards the achievement of the objective of enhanced productivity of organisation (Hayford et al. 2016). It mostly serves as a developmental tool by providing the reliable and realistic foundations for important decision making by management regarding the compensations, promotions, employee development, productivity and motivation (Mohammed and Abdulahi 2019). It also contributes significantly towards hiring and placement decisions about old and new employees. Serving as a developmental tool it also offers the organisational developmental framework through identification of the individuals having profitable managerial skills that can be classified and utilised as leadership responsibilities for the organisations (Mollel-Eliphaz et al. 2017).

Performance appraisal also contributes to the enhancement of the employees' motivation to perform even better in future through information provision about the fair treatment for employee through appropriate feedback system (Gupta and Parmar 2018). It also helps organisations in development and establishment of a good working environment by identification of the requirements of employees and provision of healthy challenges for enhancing their productivity (Binta et al. 2019). Employee productivity and their assessment are among one of the significant challenges for the management in most of the organisations because employees also expect certain feedback for their performance over time (Zhang et al. 2017). Therefore, regular supervision of employees' performance and provision of timely feedback is in best favour of both; the employees and the organisation and the performance appraisal system makes it easy for the organisation (Binta et al. 2019). Through conducting a systemic review research Vuong and Nguyen (2022) aimed at evaluating the most common methodologies, which are used to measure employees' performance in organizations. Thus, through reviewing the previous related studies the results revealed that employee's performance is significantly as well as positively related to collaboration initiatives. Additionally, it is indicated that the used performance's schemes conduct a vital role in determining the weaknesses and strengths of employees as well as stimulating them to enhance their performance. Eventually, the study recommended that performance evaluation strategies must be established in different manners for the aim of fitting not only the organizations' aims, while the aims of employees as well.

2.1 360-Degree Appraisal System

The 360 degree PA is the appraisal by superiors, subordinates, peers and clients who can be able to assess the managers' performance. The purpose of this appraisal is to enhance the performance through enhanced motivation (Ugonai 2020). This appraisal and the feedback are sometimes incorporated in the performance management systems of the organisations through which the performance of the mangers is appraised by different kinds of individuals starting from their self-appraisals. Then it incorporates the appraisals of subordinates, co-workers, superiors or sometimes by the clients/customers as well (Ugonai 2020). One critical aspect of these appraisals is to ensure the provision of true information through these appraisals especially when these are done but peers or subordinates (Asamu 2013). Sometimes the meaningless appraisal by peers makes it useless but its' not useless in all cases. It also helps in development of the leadership attitude and skills, customer relationships, innovation, creativity, people development and task knowledge (Hayford et al. 2016). A positive aspect of the performance appraisal system is the association between the expectations of employees to achieve a reward for their productivity. Therefore, the appraisals also help management in enhancing the effectiveness of the compensation management in the organisation (Gupta and Parmar 2018). Management should use the available information in appropriate manner and must provide the relevant reward otherwise it may also de-motivate the employees. Unfortunately the appraisals are mostly associated with the promotion and transfers but actually it's the basic source of appropriate information about the good or bad performance of employees to identify the areas for improvement (Bagul 2014). Human capital is among the significant factors of production for organisations, so enhancement of their productivity is among the prime concerns for the organisations. In this regard, the appraisals contribute significantly by facilitating management in their decision making, compensation management and feedback as well (Ugonai 2020). In an attempt to test the effect of 360-degree appraisal technique in the retention's rate of employees;(Abboud 2021) evaluated the relationship of performance assessments along with employees' satisfaction, as well as the reasons behind the high turnover rates of employees in the Lebanese banking industry. Consequently, the study revealed that human resources department exerts a leading role in impacting employees' retention rate. Furthermore, employees' retention is found to be strongly impacted by learning and development, career management, motivation as well as performance appraisals. Moreover, the study recommended that it is necessary to focus on job satisfaction since employees' productivity is highly reliant on their satisfaction.

3 Research Methodology and Hypotheses Development

The current research incorporates the quantitative research design in order to evaluate the effectiveness of 360-degree appraisal technique to enhance the productivity of employees of Jordanian listed banks. To accomplish this core aim, the population of this research includes the employees who are mostly working at managerial levels in both Islamic and commercial banks, which are listed at Amman stock exchange. A Convenient sampling method was adopted to approach these employees and the ones willing to participate in this research are incorporated. Thus, through using a structured questionnaire that was adopted from (Ugonai 2020); the study collected the requested data from a sample of 120 employees, while only 113 answered questionnaires were retrieved, where 55 from commercial banks, while 58 from Islamic banks. Moreover, through using the STATA and Spss software the study analysed the collected data through using the descriptive statistics and mean cut off point, the cut-off point for means to assess the responses of participants is set at "3 points", as well as the one-sample test and the paired sample statistics are also used to test the following hypotheses:

H_{01}: *There is no significant correlation between the 360-degree performance appraisal system and employees' productivity in Jordanian listed banks.*
H_{02}: *The implemented 360-degree performance appraisal system is insignificantly influencing the productivity of employees in Jordanian commercial banks.*
H_{03}: *The implemented 360-degree performance appraisal system is insignificantly influencing the productivity of employees in Jordanian Islamic banks.*
H_{04}: *There are significant differences between employees' productivity in Islamic and commercial banks in accordance to the implemented 360-degree performance appraisal system.*

3.1 Estimation Procedure

This research work concentrates on finding out the role of 360-degree performance appraisal technique to improve employees' productivity in the Islamic and commercial bank of Jordan. For this purpose the above hypothes will be tested through running the following research models:

$$\boldsymbol{EM} = \boldsymbol{CM} + \boldsymbol{FD} + \boldsymbol{MDM} + \epsilon_t \tag{1}$$

where, EM refers to employees' motivation, while terms on the right hand side including CM, FD, MDM relates to the performance appraisal technique and they are representing compensation management, feedback development, and management decision making consecutively. The sign ϵ_t refers to the error-term in the time t. Anyway, since the performance appraisal measures are found to influence employees' motivation; the following model was considered in order to reveal whether employees' motivation leads to an enhancement in employees' productivity.

$$\boldsymbol{EP} = \boldsymbol{CM} + \boldsymbol{FD} + \boldsymbol{MDM} + \boldsymbol{EM} + \epsilon_t \tag{2}$$

where the EP term relates to employees' productivity in both Islamic and commercial listed banks. Thereafter, the study proceeded to examine the following models in order to reveal the impact of performance appraisal measures in employees' productivity in Islamic banks "EPIB", and employees' productivity in commercial banks of Jordan "EPCB", before running the paired-sample tests to compare the impact of these measures in the productivity of Islamic and commercial banks of Jordan.

$$\boldsymbol{EPIB} = \boldsymbol{CM} + \boldsymbol{FD} + \boldsymbol{MDM} + \epsilon_t \quad (3)$$

$$\boldsymbol{EPCB} = \boldsymbol{CM} + \boldsymbol{FD} + \boldsymbol{MDM} + \epsilon_t \quad (4)$$

4 Results and Discussion

This research work concentrates on finding out the role of 360-degree performance appraisal technique to improve employees' productivity in the Islamic and commercial bank of Jordan. For this purpose the researchers demonstrated a structured questionnaire that was distributed to a sample of 120 employees in order to collect the needed data that will be used in analysis to examine the study's models and hypotheses. Therefore, the results are explained as shown in the following tables:

4.1 Data Analysis

Table 1. Descriptive Statistics

Description	Categorisation	No	%age
Gender	**Male**	**70**	**61.9**
	Female	**43**	**38.05**
Age	**18–25 years**	**24**	**21.23**
	26–40	**46**	**40.0**
	41–55	**21**	**18.5**
	56–70	**22**	**19.4**
Education	**Bachelor**	**52**	**46.01**
	Diploma	**18**	**15.92**
	Master	**43**	**38.05**
Status	**High**	**59**	**52.21**
	Mid	**36**	**31.8**
	Low	**18**	**15.93**
Experience	**Less than 10 yrs**	**18**	**15.92**
	11 to 20 yrs	**53**	**46.90**
	21 to 35 yrs	**42**	**37.16**

Table 1 reports the descriptive statistics of the study's participants, where it indicates that most of them are males within the age of 26 to 40 years having degrees and 11 to 20 years of experience. Further Table 2 reports the results for responses' means and frequencies (Tables 3, 4 and 5);

Table 2. Means and Frequencies of Employees' Responses in Jordanian Listed Banks

Research Questions	Scores SA 5	A 4	N 3	D 2	SD 1	Row score	Mean score	3-point Decision	Grand Mean
PA facilitates compensation management	**51**	**42**	**2**	**4**	**14**	**429**	**3.97**	**T**	**3.78**
Feedback is not part of PA	**12**	**3**	**1**	**4**	**93**	**183**	**1.63**	**F**	
PA facilitates the management of decision making	**67**	**39**	**1**	**2**	**4**	**471**	**4.29**	**T**	
PA promotes motivation	**59**	**34**	**2**	**3**	**15**	**438**	**4.01**	**T**	
The productivity of employees is significant to banks' management	**73**	**34**	**1**	**1**	**4**	**491**	**4.39**	**T**	
Employees in banks of Jordan are appraised properly	**62**	**47**	**1**	**2**	**1**	**478**	**4.40**	**T**	

Table 3. The Correlation of PA along with employees' productivity in commercial banks

Commercial Banks	N		Mean	S.D		Std. Error Mean
	55		3.7184	1.1325		0.1115
					95% Conf. Interval of Difference	
	t	df	Sig.	Mean Difference	Lower	Upper
	33.32	54	0.00012	3.7184	3.4971	3.9398

SA = strongly agree, A = agree, N = neutral, DA = disagree, SD = strongly disagree, T = true, F = false, PA = performance appraisal.

The above table proves that since the mean of feedback item is lower than the grand mean, as well as all other means are higher than the grand mean; the first hypothesis was rejected, which means that employees' productivity is significantly linked with the implemented performance appraisal system.

Table 4. The Correlation of PA along with employees' productivity in Islamic banks

Islamic Banks	N	Mean	S.D	Std. Error Mean		
	58	3.8058	1.0669	0.1051		
					95% Conf. Interval of Difference	
	t	df	Sig.	Mean Difference	Lower	Upper
	36.201	57	0.00050	3.8058	3.5973	4.0144

Table 5. Comparing the impact of PA in the Productivity of Employees of Commercial Banks and Islamic Banks of Jordan

Pair 1	Bank Type	Mean	S.D	Std. Error Mean
	Commercial	3.7184	1.1325	0.1115
	Islamic	3.8058	1.0669	0.1051
	Commercial & Islamic Banks		Correlation	Sig.
			0.904	0.0004

After running the one sample test to check the correlation between performance appraisal system and the productivity of employees in commercial banks of Jordan; statistics from the above table rejected the second hypothesis, which means that the implemented 360-degree performance appraisal system is significantly influencing the productivity of employees in commercial banks of Jordan.

After running the one sample test in order to check correlation between performance appraisal system and the productivity of employees in Islamic banks of Jordan; statistics from the above table rejected the third hypothesis, which means that the implemented 360-degree performance appraisal system is significantly influencing the productivity of employees in Jordanian Islamic banks.

The current research deployed the paired-sample test to compare the impact of the implemented 360-degree performance appraisal system in employees' productivity in Islamic and commercial banks of Jordan. Consequently, statistics from the paired sample test rejected the fourth hypothesis to assert that there are no significant differences between employees' productivity in Islamic and commercial banks in accordance to the implemented 360-degree performance appraisal system.

4.2 Results Discussion

Performance appraisal has become an essential part of organizations' working environment and is considered as a positive contributing factor towards organisational productivity. Same is the case with Jordanian banking system as indicated by the aforementioned results. The main aspects of PA being focused by the current research are its contribution towards the compensation management, feedback development; management decision making that contributes in motivating employees of both Islamic and commercial banks, and consequently improving their productivity. The main focus and purpose of PA is highlighting the performance of employees according to the said standards of the organisation and otherwise taking sufficient actions by management to align it with the objectives of the organisation through enhancing productivity. So the results reported in the above table indicate that most of the study's participants are agree and even strongly agree that performance appraisal facilitates organizations' compensation management. Moreover the participants also indicated that the feedback is seen as a significant part of the PA system. In addition, most participants are indicated that PA helps organizations' management in their decision making and affects employees' productivity and motivation which are crucial for organizations' success.

The findings also indicate that Jordanian banks have appropriate PA system and employees are appraised appropriately. Hence the agreement of the participants about the effective contribution of PA to the employees' motivation and productivity indicates that in Jordanian banking sector the employees get motivated when they are appraised appropriately. Consequently, that contributes positively in enhancing their motivation and these motivated employees found their productivity level to be enhanced on the basis of this motivating gained through appraisal. These findings are consistent with the findings of (Gupta and Parmar 2018) as the authors also indicated the positive contribution of performance appraisal towards the motivation of employees. These are also consistent with the findings of (Asamu 2013) who argued that performance appraisal has become the vital part of organizations' success because in recent scenarios, the success is based on the enhanced productivity levels of the organisations which could be achieved by the enhanced productivity levels of their employees. The descriptive statistics indicate that the opinion of experienced employees is always relying on the basis of their past experiences. It is also indicated that Jordanian banks focus on their employees' appraisals and these appraisals contribute significantly in enhancing the productivity of their employees. Hence, the findings pointed that banks should not only sustain their appraisal systems rather they should focus on making it even more efficient so that they should be able to achieve the competitive edge through the enhanced employees' productivity. Beyond that results from the paired-sample test confirmed that there are no significant differences between employees' productivity in Islamic and commercial banks in accordance to the implemented 360-degree performance appraisal system.

5 Conclusion

The current research has focused on the 360 degree appraisal system and its impact in the productivity of employees in the Jordanian banking sector. Performance appraisal is an effective tool for the evaluation of employees' performance over a period of time so that it could be appreciated or improved according to the results of the appraisals. Accordingly, the findings indicate that Jordanian banking sector appropriately uses the appraisal system and employees are appraised systemically. Moreover, employees consider it as a positive contribution towards their productivity and the success of banks' management. The study indicates that the performance appraisal helps banks in critical decision making and compensation management by leaders. Moreover, feedback is considered as a good contributor by the performance appraisal and all these factors contribute in enhancing the motivation level of employees. These motivated employees are then found to have enhanced productivity. Hence, it is indicated that organisations either banks or all other organisations should systematically employ the appraisal systems in their working culture to promote enhanced productivity and highly motivated employees that can contribute effectively towards the achievement of banks' objectives.

References

Abboud, S.: The impact of 360 degrees performance appraisals on employee retention. Euraseans J. Glob. Soc.-Econ. Dyn. **3**(28), 38–49 (2021)

Asamu, F.F.: Perception of performance appraisal and workers performance in Wema bank headquarters, Lagos. Glob. J. Arts Human. Soc. Sci. **1**(4), 89–101 (2013)

Audu, J.S., Timothy, G.: Effects of human resource training and development on productivity in Nigerian hospitality industry. Int. J. Public Adm. Manage. Res. **2**(2), 80–87 (2014)

Bagul, D.B.: A research paper on the study of employees' appraisal system. J. Human. Sci. English Lang. **1**(2), 28–292 (2014)

Binta, M., Muhammad, I., Ahmed, Z., Bazza, A.M., Magaji, B.: Effects of performance appraisal on employee productivity in federal ministry of education Nigeria. Asian J. Contempor. Educ. **3**(2), 121–131 (2019)

Gupta, B., Parmar, S.: Effects of performance appraisal on employee productivity in an automation solution company. J. Manage. Res. Analy. **5**(2), 151–158 (2018)

Hayford, C., Boakye, Y., Ovusu, N.O.: Is performance appraisal system anachronistic in tertiary institutions in Ghana? evidence from the University of Cape Coast. J. Bus. Manag. **18**(4), 55–62 (2016)

Khawaja, J., Nadeem, A.B.: Training and development program and its benefits to employee and organization: a conceptual study. Europ. J. Bus. Manage. **5**(2), 22–28 (2013)

Mathis, R.L.. Jackson, J.H.: Human Resource Management. 7th ed: West Publishing Corporation: N.Y, USA (1994)

Mohammed, A.M., Abdulahi, Y.B.: An evaluation of staff motivation, dissatisfaction and job performance in an academic setting. Aust. J. Bus. Manage. Res. **1**(9), 1–13 (2019)

Mollel-Eliphaz, R., Mulongo, L.S., Razia, M.: The influence of performance appraisal practices on employee productivity: a case of Muheza District, Tanzania. Issues Bus. Manage. Econ. **5**(4), 45–59 (2017)

Vuong, T.D.N., Nguyen, L.T.: The key strategies for measuring employee performance in companies: a systematic review. Sustainability (MDPI) **14**(14017), 2–21 (2022)

Ugonai, J.N.N.: Performance appraisal and its effects on employee productivity in charitable organisations. Bus. Manag. Econ. Res. **6**(12), 166–175 (2020)

Zhang, X., Qian, J., Wang, B., Jin, Z., Wang, J., Wang, Y.: Leaders' behaviors matter: the role of delegation in promoting employees' feedback seeking behavior. Front. Psychol. **8**(920), 1–10 (2017)

The Social Effects of Using Digital Media in Jordan

Asmaa Radi Khanfar(✉)

Foundations of Education, Zarqa, Jordan
asmaakhanfar@yahoo.com

Abstract. Digital media has become a ubiquitous part of modern life and has transformed the way we interact and communicate. This study aimed to examine the effects of digital media usage on social relationships and communication skills among students at Zarqa Private University (ZU) in Jordan and to determine if there were any gender-based differences. Using a descriptive approach, a questionnaire was distributed to 303 students who completed an online survey, and the data was analyzed using SPSS. The results showed a significant impact of digital media usage on both social relationships and communication skills. However, there were no significant gender-based differences in the impact. It is recommended that students at ZU should be conscious of their digital media usage and its effects on their social relationships and communication skills. In order to achieve this balance, the study recommended that ZU provides educational programs and workshops to promote the development of healthy social relationships and effective communication skills among its students.

Keywords: Communication Skills · Gender · Social Relationships · Zarqa Private University (ZU)

1 Introduction

The widespread adoption of digital media has drastically altered the way people communicate and access information. As defined by Digital Media Definition [5], digital media refers to any type of media that is created, shared, and disseminated through digital channels such as the internet, mobile devices, and social media platforms. This encompasses a wide range of media including text, images, video, audio, and interactive content, as well as traditional media that has been digitized.

The influence of digital media on social life has garnered much attention and research in recent years [24, 29]. While digital media has the potential to enhance social connections and personal growth [15], it has also raised concerns about its potential negative impacts on social life [6, 18, 30]. Excessive use of digital media may lead to social isolation and a decline in face-to-face interactions [11]. Furthermore, worries exist regarding the effect of social media on one's mental health, overall well-being, as well as the possibility of experiencing cyberbullying and being subjected to online abuse [6, 18]. This paper will examine the role and impact of digital media on social life, including both its positive and negative effects on relationships, society, and communications.

B. A. M. Alareeni and I. Elgedawy (Eds.): *Artificial Intelligence (AI) and Finance*, SSDC 488, pp. 175–186, 2023.
https://doi.org/10.1007/978-3-031-39158-3_16

Statement of the Problem: This study aims to examine the effects of digital media usage on the social lives of students at Zarqa Private University (ZU). The link between digital media and social life is complex and multi-faceted [15], and can have both positive outcomes, such as increased social connections and access to information and re-sources, as well as negative consequences, such as risks and challenges that need to be addressed [18, 24].

Study Questions:

- What is the impact of digital media usage on the social relationships of ZU students?
- What is the effect of digital media usage on the social communication skills of ZU students?
- Is there a significant statistical difference in the impact of digital media usage on both social relationships and social communication skills of ZU students based on gender (male or female)?

Study Objectives: This study aimed to:

- Assess the impact that digital media usage has on the social relationships and communication abilities of ZU students.
- Examine if there is a statistically significant difference in the effect of digital media usage on the social relationships and communication abilities of ZU students as per their gender (male or female).

The Significance of this Study: Examining the effects of digital media on social life is crucial for comprehending the impact of technology on individuals and society. The findings of this study will inform university administrators, enabling them to implement strategies that encourage the beneficial use of digital media.

Study Limits: This research was confined to the students of the College of Information Technology at ZU during the first semester of 2022–2023.

Terminology Used in the Study:
Digital media: refers to any media that is produced and distributed in a digital format. This encompasses a diverse range of media forms, such as text, images, videos, audio, and interactive content, as well as traditional media that have been converted to digital form, such as books and magazines [31].

Through this paper, the researcher will use the descriptive method to investigate the extent to which digital media affects social aspects, especially in the two areas: social relations, and social communication skills.

2 Previous Studies

Previous studies have explored the impact of social media on adolescent and young adult social skills and communication abilities. Subrahmanyam & Greenfield [25] conducted a study titled: "The impact of social media on adolescent social skills" aimed to explore the effect of social media usage on social skills in adolescents aged 13 to 18, on a sample of 1,000, which found that social media use had both positive and negative effects on social skills, dependent on the amount of time spent and specific behaviors engaged in. The results of this study provide valuable insights into the relationship between social media usage and social skills development in adolescents and contribute to the ongoing discourse on the effects of digital media on youth. Arnett [1] carried out a study on the impact of social media on communication skills in young people aged 18 to 25, with a sample size of 1,000. The study found that social media has the potential to enhance communication skills, however, this relationship is dependent on the amount of time spent on social media and the behaviors engaged in.

Gros & Bichsel [7] studied the relationship between usage social media and communication skills among 187 college students, finding a positive correlation but with variations based on the platform used and frequency of use. The study highlights the need for further research to better understand the effects of social media on communication skills and to identify ways to mitigate the negative effects. A meta-analysis by Valkenburg & Peter [28] of 24 studies found a small but significant relationship ($r = .18$, $p < .001$) between social media use and communication skills, with the relationship moderated by the category of social media site, frequency of use, and measurement of communication skills. Hampton, Rainie, and Lu [8] conducted a study to assess the influence of the internet and mobile phones on social relationships, using surveys and focus groups with a nationally representative sample of adult internet users in the United States. The study found that digital media increased the quantity and frequency of communication, but also introduced new challenges such as increased social isolation and the spread of misinformation. In a literature review. Quan-Haase [17] analyzed the social implications of digital media and found that the impact was complex and multifaceted, creating new opportunities for communication and connection while also presenting new challenges such as the potential for digital divide and the spread of misinformation.

Previous studies have established that social media can impact social skills and communication abilities, with varying effects based on factors such as time spent on social media, the platform used, and behaviors engaged in. The current study aims to build upon this research by examining the impact of digital media, including social media, on the social relationships and communication skills of university students. This research seeks to broaden the understanding of the effects of digital media on university students' social and communication abilities.

3 Potential Advantages and Disadvantages of Using Digital Media

Using digital media for social relations and communication can offer several advantages, such as increased connectivity through digital media enables individuals to stay connected with friends and family members regardless of distance, facilitating long-distance relationships [9, 12], social media and forums offer users greater access to diverse perspectives and information, enabling them to broaden their world view [4, 26], blogs and social media offer individuals greater opportunities for self-expression, allowing them to share their thoughts, experiences, and interests with a wider audience, leading to a sense of empowerment and validation, additionally, online support groups can offer a sense of community and social support, particularly for those who are socially isolated or living with a chronic illness [2, 14, 22]. Digital media provides greater access to resources for social change, enabling individuals to organize and mobilize around social and political causes, facilitating their involvement in activism and advocacy [3, 10], and to form relationships and communities around shared interests, such as hobbies and cultural identities [3, 21]. In terms of social communication, digital media can allow individuals to connect with a large and diverse audience, quickly and easily share information and media, and communicate in real-time [9].

However, excessive usage of digital media can have negative consequences on social skills, relationships, mental health, and exposure to diverse perspectives. This is due to decreased face-to-face communication, increased social isolation, cyberbullying, privacy concerns, depression, anxiety, and reinforcement of biases and prejudices through exposure to only like-minded information and views in echo chambers [16, 23, 27, 30].

The lack of nonverbal cues and context in online interactions can also result in misunderstandings and conflicts, and feelings of isolation and loneliness despite increased connectivity [8, 20]. A study by the Pew Research Center [16] found that 65% of American adults get news from social media, with a higher prevalence among younger adults, non-whites, and those with lower levels of education and income. The Royal Society for Public Health (UK) [19] conducted a study which revealed a connection between the use of social media and higher instances of anxiety, depression, and sleep issues. The study found that social media is having a harmful impact on the health of young people.

4 Study Methodology

The study employed the descriptive research method, which is suitable for exploring the research subject.

4.1 Study Population and Sample

The study community consisted of students at the College of IT in the (ZU). An intentional sample was selected, consisting of all students who completed an online survey titled "The Social Effects of Using Digital Media". The sample size was 303 students, representing the current semester (Table 1).

Table 1. Distribution of Study Participants Based on Gender Variable:

Variable	Categories	No	Percentage
Gender	Male	133	43.9
	Female	170	56.1
Total		303	100.0

4.2 Study Tool

This study benefited from previous studies such as Subrahmanyam & Greenfield [25], Gros & Bichsel [7] and Hampton, Rainie, and Lu [8] in preparing the questionnaire that consisted of 20 items to assess the impact of digital media on social life of ZU students. The questionnaire was divided into two areas: social relationships, social communication skills. The researcher used a five-point Likert scale to determine the degree of impact. The results are presented in Table 2.

Table 2. Average and Degree of Agreement

Mean	Degrees of Agreement
Above *4.2*	Very high
From 3.4 to less than 4.2	High
From 2.6 to less than 3.4	Medium
From 1.8 to less than 2.6	Low
Less than 1.8	Very low

Validity and Reliability of the Tool: To assess the validity and reliability of the questionnaire, the Pearson correlation coefficient was employed to determine the correlation between each statement and its corresponding dimension, as well as between each dimension and the overall questionnaire. Tables 3 and 4 provide an explanation of this.

Table 3. Correlation Coefficients of each domain with the overall score

The Dimension	Phrase	Correlation coefficient	Phrase	Correlation coefficient
Social relationships	1	00.572**	6	0.585***
	2	0.601**	7	0.226**
	3	0.574**	8	0.622**
	4	0.334**	9	0.635**
	5	0.224**	10	0.601**
Social communication skills	11	0.647**	16	0.666**
	12	0.646**	17	0.643**
	13	0.647**	18	0.697**
	14	0.611**	19	0.577**
	15	0.580**	20	0.591**

** function at (0.01)

Table 4. Relationship coefficients for each domain with the overall questionnaire score

The domain	Correlation coefficient
Social relationships	**0.876
Social communication skills	**0.913

**function at (0.01)

Table 3 shows positive and significant correlations ($p < 0.01$) between each statement and its domain. Table 4 shows positive and significant correlations ($p < 0.01$) between each domain and the total questionnaire score, indicating the accuracy of all statements.

Reliability: Table 5 displays the stability coefficient of Cronbach's alpha to evaluate the reliability of the resolution domains.

Table 5. Values of Stability coefficients for the questionnaire domains

The domains	Cronbach's alpha coefficient
Social relationships	0.648
Social communication skills	0.828
The resolution as a whole	0.840

Table 5 shows acceptable Cronbach's alpha coefficients for the resolution domain, indicating adequate stability in the questionnaire.

4.3 Statistical Processing Methods

The SPSS program was used for data analysis. Frequency and percentage were used to describe sample characteristics, mean and standard deviation were used to evaluate responses to each phrase, Pearson correlation coefficient was used to measure internal consistency, Cronbach's alpha was used to evaluate stability, and Independent-Samples T-test was used to compare two groups.

5 Results and Discussion

5.1 Question 1

"What is the impact of digital media usage on the social relationships of ZU students?" Table 6 shows the mean and standard deviation of participant responses.

Table 6 displays the mean scores of ZU students' agreement on statements regarding the impact of digital media usage on their social relationships, ranging from (3.01) to (4.14). Statements (6, 8, 9, 7, 10, 4, and 5) received high agreement scores, with the highest being statement (6), "strengthening my relationships with friends due to ease

Table 6. Mean and Standard deviation of ZU students' responses regarding the impact of using digital media on their social relationships.

No	Phrase	Mean	SD	Degree of Agreement	Ranking
6	Using digital media strengthens my relationships with friends by making communication with them easier	4.14	0.886	High	1
8	It increases my ability to develop professional relationships in university, work, etc.	3.87	1.001	High	2
9	It gives me more space to participate in community service (such as helping classmates in the neighborhood, university, work)	3.86	1.017	High	3
7	It reduces face-to-face interactions with others	3.77	1.040	High	4
10	It provides me with good training for social work (community participation)	3.57	1.071	High	5
4	It limits my social interactions with family (visiting relatives, friends, socializing during holidays and events)	3.50	1.168	High	6
5	It reduces my ability to perform household tasks (such as taking care of young siblings, cleaning the house, helping family)	3.48	1.195	High	7
2	It strengthens my social relationships with relatives	3.30	1.115	Medium	8
1	It strengthens my family relationships	3.06	1.126	Medium	9
3	It provides me with good training for achieving stable family relationships	3.01	1.101	Medium	10
	Overall average	3.55	0.527	High	

of communication," which had a mean score of (4.14). The other statements received moderate agreement scores, with the lowest being statement 3, "provides me with good training to have stable family relationships," which had a mean score of (3.01). The total mean score of the statements was (3.55), indicating a high degree of impact of digital media on the social relationships of ZU students. This significant impact can be attributed to the transformation brought about by the rise of digital technologies, which has altered communication styles, societal norms, and individual self-perceptions, as shown in the upcoming results. These results align with the findings of Hampton et al. [8] and Quan-Haase [17], who found that digital media has greatly changed communication and interaction. They also support the findings of Arnett [1], who indicated that social media may improve communication abilities.

5.2 Question 2

"What is the effect of digital media usage on the social communication skills of ZU students?" The results of the calculation of the means and standard deviations of the participants' responses are presented in Table 7.

Table 7. Mean and Standard Deviation of the Impact of Digital Media Use on the Social Communication Skills of ZU Students.

No	Phrase	Mean	SD	Degree of Agreement	Ranking
14	Openness to other cultures through its diverse tools	4.17	0.828	High	1
15	Facilitates better collaboration with others to achieve tasks	4.04	0.831	High	2
17	Provides a larger space to express personal ideas and opinions more effectively	4.01	0.895	High	3
20	Provides a variety of communication methods with virtual friends	4.00	0.924	High	4
13	Increasing opportunities for interaction with others everywhere and anytime	3.96	0.939	High	5
16	Provides an opportunity for training in conventional social interaction styles	3.65	0.975	High	6

(continued)

Table 7. (*continued*)

No	Phrase	Mean	SD	Degree of Agreement	Ranking
18	Provides communication styles that enhance mutual respect among colleagues	3.54	1.025	High	7
11	Contributes to refining communication skills with others	3.35	1.121	Medium	8
12	Increases the ability to understand others	3.24	1.133	Medium	9
19	Contributes to managing time in communicating with others (family, relatives, friends, etc.)	3.24	1.219	Medium	10
	Overall average	3.72	0.624	High	

According to Table 7, the average scores of ZU students' agreement on statements related to the impact of digital media use on their social communication skills ranged between (3.24) and (4.17). Statement (14), "It offers opportunities to open up to other cultures through diverse tools," received the highest average score of (4.17), while statement (19), "It contributes to managing time in communicating with others," received the lowest average score of (3.24). The total average score of the statements was (3.72) and indicated a high agreement, suggesting that digital media has a significant impact on the social communication skills of ZU students.

Several factors contribute to the high impact of digital media on social communication skills. Digital media provides extensive access to diverse information and resources, allowing individuals to expand their knowledge and cultural understanding. It also provides tools for instant and effective communication with a global audience, leading to improved interpersonal skills [27].

However, excessive and indiscriminate use of digital media can also have negative effects on social communication. Digital media also has a significant impact on openness to other cultures, providing individuals with the opportunity to access and explore different cultures, perspectives, and ways of life. This exposure can broaden one's cultural understanding and improve cross-cultural communication skills. According to Lin and Chen [13], the use of digital media has a positive impact on cultural understanding and global awareness, contributing to "the development of a more inclusive and cosmopolitan mindset." These results align with the findings of other studies, including those by Subrahmanyam & Greenfield [25], Arnett [1], Gros & Bichsel [7], and Valkenburg & Peter [28].

5.3 Question 3

"Is there a significant statistical difference in the impact of digital media usage on both social relationships and social communication skills of ZU students based on gender (male or female)?" Table 8 displays the results of the analysis of the data using the independent samples t-test.

Table 8. Results of the independent samples t-test to determine the significance of the difference in the impact degree of digital media use on both social relationships and social communication skills of ZU students based on the gender variable.

The field	Gender	No	Mean	SD	T-values	Freedom degrees	Significance level
Social relationships	Male	133	3.61	0.526	1.497	301	0.135
	Female	170	3.51	0.525			
Social communication skills	Male	133	3.79	0.567	1.654	301	0.099
	Female	170	3.67	0.662			
Digital media as a whole	Male	133	3.70	0.491	1.768	301	0.078
	Female	170	3.59	0.531			

Table 8 demonstrates that all significance levels were higher than 0.05, indicating the absence of a statistically significant difference in the influence of digital media usage on the social relationships and communication skills of male and female students at ZU. This suggests that the impact of digital media on these aspects is similar for both genders. This outcome can be attributed to factors such as equal access to technology, similar usage habits, and cultural influences that promote equal adoption of technology and digital media among male and female students.

6 Conclusion

This research aimed to examine the impact of digital media usage on social relationships and communication skills among students at ZU, and to explore any gender differences. Results showed that digital media usage had a significant impact on both social relationships and communication skills, which could be both positive and negative. No significant gender differences were found in the impact of digital media usage. These results showed the implications of this strong relationship between digital media and social aspects, which changed the nature, type and depth of social relations and social communication skills among ZU students in Jordan.

Recommendations: The research recommends that students at ZU become mindful of the consequences of excessive digital media usage on their social relationships and communication abilities. Furthermore, the study suggests that ZU implement educational initiatives, such as workshops and programs, to enhance the social relationships and communication skills of its students. It is also recommended that more research should be conducted on the impact of digital media on social relations through other variables: such as age, academic specialization, etc., and a comparison between university students and public education students in Jordan.

References

1. Arnett, J.J.: The effects of social media on the communication skills of young people. In: Gurman, A.W., Knobloch, S.A. (eds.) Clinical Handbook of Couple Therapy, 5th edn., pp. 547–560. Guilford Press, New York, NY (2014)
2. Barak, A., Klein, B., Proudfoot: Internet-based psychological treatments: a new era in mental health care. Aust. N.Z. J. Psychiatry **43**(11), 927–935 (2009)
3. Bimber, B., Flanagin, A.J., Stohl, C.: Collective Action in Organizations: Interaction and Engagement in an Era of Technological Change. Routledge (2015)
4. Boyd, D.: It's Complicated: The Social Lives of Networked Teens. Yale University Press (2014)
5. Digital Media Definition. https://www.techopedia.com/definition/28084/digital-media. Accessed 20 Dec 2021
6. Fox, S., Moreland, J.: The dark side of social media. J. Soc. Clin. Psychol. **34**(4), 274–289 (2015)
7. Gros, B.F., Bichsel, J.: Social media use and communication skills: an exploratory study. Soc. Sci. Comput. Rev. **35**(4), 489–503 (2017). https://doi.org/10.1177/0894439317697557
8. Hampton, K., Goulet, L.S., Rainie, L., Purcell, K.: Social isolation and new technology. Pew Internet American Life Project (2011). http://www.pewinternet.org/2011/02/24/social-isolation-and-new-technology/
9. Kraut, R., Kiesler, S., Boneva, B., Cummings, J., Helgeson, V., Crawford, A.: Internet paradox: a social technology that reduces social involvement and psychological well-being? Am. Psychol. **53**(9), 1017–1031 (2002)
10. Küng, L., Rassenti, S.: Digital media literacy: A European handbook. Council of Europe (2016)
11. Kuss, D.J., Griffiths, M.D.: Social media addiction: a review of the psychological literature. Int. J. Environ. Res. Public Health **14**(3), 311 (2017)
12. Lin, L.: Social capital on the internet: differentiating uses and users. Soc. Sci. Comput. Rev. **29**(3), 352–362 (2011). https://doi.org/10.1177/0894439310385522
13. Lin, M., Chen, W.: How digital media contribute to cultural understanding and global awareness. Intercult. Educ. **26**(1), 19–29 (2015)
14. Marwick, A.E., Boyd, D.: I tweet honestly, I tweet passionately: Twitter users, context collapse, and the imagined audience. New Media Soc. **13**(1), 114–133 (2011)
15. Parks, M.R., Floyd, K.: Making friends in cyberspace. J. Commun. **46**(1), 80–97 (1996)
16. Pew Research Center: News Use Across Social Media Platforms (2018). https://www.journalism.org/2018/09/04/news-use-across-social-media-platforms-2018/
17. Quan-Haase, A.: The social implications of digital media. Can. J. Commun. **35**(1), 91–115 (2010)
18. Rosen, L.D., Cheever, N.A., Carrier, L.M., Rokkum, J.: The impact of social media on children, adolescents, and families. Pediatrics **131**(4), e1357–e1367 (2013)
19. Royal Society for Public Health (UK). (#StatusOfMind: Social Media and Young People's Mental Health and Wellbeing). https://www.rsph.org.uk/our-work/policy/social-media-and-young-people-s-mental-health-and-wellbeing.html. Accessed 20 Jan 2023

20. Sawyer, R.K.: Social media use and perceived social isolation. Comput. Hum. Behav. **84**, 140–146 (2018). https://doi.org/10.1016/j.chb.2017.11.029
21. Shaw, A., Gant, L.: Virtual intimacy: online relationships and the lack of closeness. Inf. Soc. **18**(3), 209–221 (2002)
22. Shaw, E.: Social support and internet use among individuals with disabilities. J. Rehabil. Res. Dev. **48**(4), 365 (2011)
23. Smith, P.K., Mahdavi, J., Carvalho, M., Fisher, S., Russell, S., Tippett, N.: Cyberbullying: its nature and impact in secondary school pupils. J. Child Psychol. Psychiatry **49**(4), 376–385 (2008). https://doi.org/10.1111/j.1469-7610.2007.01846.x
24. Steyer, R., Dang-Xuan, L.: The impact of social media on society. J. Comput. Mediat. Commun. **17**(3), 319–326 (2012)
25. Subrahmanyam, K., Greenfield, P.: The impact of social media on adolescent social skills. In: The John D. and Catherine T. MacArthur Foundation Series on Digital Media and Learning. The MIT Press, Cambridge, MA (2014)
26. Sun, A.: Social media and political polarization. Polit. Sci. 1–27 (2019). https://doi.org/10.1017/S1743923X19000267
27. Turkle, S.: Alone Together: Why We Expect More From Technology and Less From Each Other. Basic Books (2011)
28. Valkenburg, P.M., Peter, J.: The relationship between social media use and communication skills: a meta-analysis. Comput. Hum. Behav. **29**(3), 644–649 (2013)
29. Van Dijk, J.: The Network Society: From Knowledge to Policy. John Wiley & Sons (2014)
30. Vosoughi, S., Roy, D., Aral, S.: The spread of true and false news online. Science **359**(6380), 1146–1151 (2018)
31. What is Digital Media? https://www.investopedia.com/terms/d/digital-media.asp. Accessed 30 Jan 2021

Digital Services Trade: A Quantitative Study of the Relationship Between Income and Imports of Services

Nidal Ali Abbas[1(✉)], Mohammad Musa Al-Momani[1], Mazin Hasan AlBasha[1], Ibrahim Mohammad Khrais[1], Khaled Mohammed Al-Sawaie[1], Abdelhalim Mohammad Jubran[1], Thaer Ahmad Abu-Saleem[1], and Mohammad Hejazi[2]

[1] Faculty of Economic and Administrative Sciences, Zarqa University, Zarqa, Jordan
{nabbas,m.momani,mbasha,ikhrais,ksawaie,ajubran, thabusaleem}@zu.edu.jo

[2] Department of Supply Chain Management, University of Business and Technology, Jeddah, Kingdom of Saudi Arabia
m.hejazi@ubt.edu.sa

Abstract. The study aims to examine the relationship between gross domestic product (GDP) and services imports for the case of Jordan for the period between 2005 and 2021. The significance of this topic lies in its consideration as an economic activity supporting commodity trade. The results of the quantitative study showed a positive effect of gross domestic product (GDP) on services imports in the long run, with a significance level of less than 1%, indicating that it is not elastic in the long term. For every 1% increase in GDP, the rate of increase in the volume of services imports will increase by 0.776%. The adjustment period from the short term to the long term occurs immediately.

Keywords: Imports · trade exchange · Gross Domestic Product (GDP) · Jordan · quantitative model

1 Introduction

International trade often describes commodity trade such as equipments and vehicles, while ignoring another important aspect of current era which is a trade in services. This is due to the nature of services, which called visible trade, due to our ability to see what is exported and what is imported. Therefore, these characteristics that characterize visible trade ignore the trade in services that is less clear to many individuals. Trade in services corresponds to business services, or even consumption services such as tourism and hotel stays. However, trade in services is often referred to as "invisible trade". The difficulty in measuring trade in services compared to trade in goods is also a factor in the under-representation of services trade, as the insurance consumption of vehicles is considered a less obvious commodity than the car itself. Because commodities have the ability to

B. A. M. Alareeni and I. Elgedawy (Eds.): *Artificial Intelligence (AI) and Finance*, SSDC 488, pp. 187–196, 2023.
https://doi.org/10.1007/978-3-031-39158-3_17

cross borders and can be valued at a specific value, while this is not necessary for trade in services. The main reason for the under-representation of services trade is that still in its early stages, with a lack of clear theories explaining the growth of international trade in services. This lack of understanding leads to undervaluation and under-investment in services trade.

In order to learn more about trade in services, it is necessary to present a concept of trade in services. Service is an intangible good that leads to creating a benefit at the individual or national level, as if we can measure its impact on the gross domestic income. Services can include transportation, storage, communication, education, health, and credit, and all of these are related to the economic activity of a country. In international trade, member countries are committed to treating Foreign Service providers equally, giving the same benefits to Foreign Service providers as to domestic ones, and treating domestic and Foreign Service providers equally without any discrimination. In addition, the Foreign Service providers also have access to the local market of the member country on the same terms as domestic providers.

In this regard, the study by [1] focuses on the direct relationship between e-commerce and services trade and its impact on economic growth in Jordan. On the other hand, the study by [2] focuses on the determinants of current account balance, where real exchange rate is negative, income is positive and less than one in the long run. In the study by [2], the focus was on studying the determinants of the Jordanian trade balance, using the ARDL model for co-integration, revealing that the money supply and income play an important role in determining the behavior of the trade balance.

The study aims to investigate the impact of total domestic income on imports of services in both the short and long terms, in order to answer the question that focuses on knowing the importance of total domestic income in importing services. To answer this question, annual data for the period 2005–2021 for the size of service imports and total domestic income for the same period was used.

2 Imports and Exports of Commodities in the Global Economy

Imports and exports of commodities increase interconnectivity in the global economy. Exports are part of a country's local production sold to citizens of another country, while imports are part of a country's local consumption or investments of foreign producers purchased from them.

Table 1 Shows that the total value of the global economy's gross domestic product is approximately \$131.6 trillion.

To obtain a relative importance perspective on international trade, we need information about the value of exports and imports as shown in Table 2. In 2018, the value of exports and imports continued to reach 32.4 trillion dollars and 33.2 trillion dollars, respectively. In relative terms, exports and imports represented 31.5% and 32.2% of the world's production respectively. This trade volume reflects trade in goods and does not include international trade in services. We can understand one of the concepts of globalization, which is the amount of locally consumed production in another country (imports) and the amount of locally produced production exported to other countries (exports).

Table 1. Global Distribution of Population and Economic Output in 2017.

	Per capita GDP	Population (million)	World population %	Total GDP (billion $)	GDP %
Low income economies	623.6	677.0	7.7	422.2	0.3
Middle income economies	10335.1	5678.8	65.0	28049.6	21.3
High income economies	38513.4	2387.0	27.3	103156.0	78.4

Source: World Bank, World Development Indicators. Unctadstat.unctad.org

Table 2. Distribution of merchandise imports and exports in the global economy in 2018

	Imports (million $)	Percentage of total world	Exports (million $)	Percentage of total world
Low income economies	140306.4	0.4	65806	0.2
Middle income economies	6265922	18.9	6439676	19.8
High income economies	26766678	80.7	25940623	80.0
World total	33172907	100	32446104	100

Source: World Bank, World Development Indicators. Unctadstat.unctad.org

It is noted that the global production distribution was not a fair (equal) distribution among the world's countries; where high-income countries accounted for 78.4% of world output, as well as the total imports and total exports. Table 2 shows that low-income countries accounted for approximately 0.3% of international trade, and Table 1 shows that this group also accounted for 0.3% of world output. Similarly, middle- and high-income countries; where middle-income countries accounted for about 19.4% of trade and 21.3% of income, and high-income countries accounted for 80.4% of trade and 78.4% of world output.

This similarity between global production and trade does not pose any difficulty in explaining it; the countries that supply the products must first produce the products; high-income countries consume the largest part of global production, which represents the share of global exports of those economies, similarly imports form consumption or investment spending in a different country. The importing country provides its consumers with an income to buy goods (effective demand), and high-income countries do not only consume locally-produced goods; they also consume goods produced abroad. As a result, the share of high-income economies in world imports is high.

3 Service Trade in Arab Countries

In (2019), there was a noticeable improvement in the performance of international service trade, with the total value of service trade increasing by approximately 2.4% to reach $5745 billion. The share of service trade payments in developing countries was approximately 41.4% of the global total in 2019, compared to 40.9% in 2018. Meanwhile, the share of service payments made by Arab countries of the global total declined to about 5.6%, compared to 5.9% in 2018. This is shown in Table 3.

Table 3. Share of Total Service Trade of Developing and Arab Countries, and Emerging Markets in the Global Total Service Trade (2015–2019)

Regions	Payments					(1)	Receipts					(2)
	2015	2016	2017	2018	2019		2015	2016	2017	2018	2019	
The developing countries, including:	38.1	37.7	37.8	40.9	41.2	0.7	29.7	29.4	29.6	32.6	33.2	1.8
- Arab countries	6.7	6.1	6.0	5.9	5.6	−3.8	3.6	3.7	3.7	3.6	3.7	1.7
-Emerging markets	11.4	13.7	13.8	13.8	13.9	0.7	9.8	10.8	11.2	11.4	11.7	2.6
The world (Billion US dollars)	4.778	4.826	5.213	5.612	5.745	2.4	4.912	4.988	5.421	5.906	6.025	2.0

(1) Change rate in total services trade payments (2018–2019)
(2) Change rate in total services trade receipts (2018–2019)
Source: World Bank, World Development Indicators. Unctadstat.unctad.org

In the context of developments in service trade performance in Arab countries, the deficit in the Arab countries' net service trade balance has decreased. This is due to a decrease in service payments and an increase in Arab countries' service receipts in 2019, which increased by 8 billion dollars, equivalent to a 3.7% increase to reach around 222.1 billion dollars compared to 214.1 billion dollars in the previous year. This was specifically due to the continued increase in receipts from the item of travel (tourism) in some Arab countries. The value of service payments by Arab countries decreased by around 4.9 billion dollars, at a rate of (1.5%), to reach 344.2 billion dollars in 2019 compared to 329.1 billion dollars in the previous year, mainly due to the decline in other service categories for Arab countries. As a result of these developments in payments and receipts, the deficit in the service balance of Arab countries improved in 2019, registering around 102.1 billion dollars compared to 115.9 billion dollars in 2018, with a decrease of 11.2%. At the level of the Arab countries, the service balance deficit improved in 2019 in Kuwait, Algeria, Saudi Arabia, and Mauritania by differences ranging from 15.7% to 31.5% compared to 2018, while the excess increased in Tunisia, Jordan, Bahrain, Morocco, and Egypt, with differences ranging from (2%) to (66%). While Iraq, Libya, UAE, Qatar, Comoros, Oman and Palestine recorded an increase in the deficit during 2019. The surplus decreased in Sudan, Lebanon and Djibouti by a rate of (61%) to (18.4%).

4 The Structure of Services Trade in the Arab Countries

During the year (2019), the share of travel receipts in total service receipts of Arab countries increased to 41.9% compared to 40.3% in 2018. Meanwhile, the share of transportation receipts declined to around 29.3% in 2019 compared to 30.1% in the previous year. To illustrate this, a summary of the main developments in receipts and payments for components of services trade in Arab countries can be presented as follows:

Firstly: Transportation Sector
In 2019, the service trade deficit for the transportation sector increased by 26.3% to approximately 22.3 billion dollars. This was due to a rise in payments that exceeded the increase in receipts, as the value of payments rose by about 6.5% to reach approximately 87.2 billion dollars in 2019 compared to 82 billion dollars in 2018. Meanwhile, the value of service receipts for the transportation sector in Arab countries increased slightly by 1% to reach 65 billion dollars in 2019. On a standalone basis, the service payments for the transportation sector increased in 11 Arabic countries in 2019, ranging from 2% in the UAE to 62.9% in Palestine. In contrast, service payments for the transportation sector decreased in Jordan, Algeria, Tunisia, Djibouti, Sudan, Kuwait, Lebanon, and Morocco, with a range of 1.3% to 18.7% during 2019 compared to the previous year. In terms of service receipts for the same sector, ten Arab countries experienced an increase, ranging from 2% in the UAE to 66.3% in Palestine during 2019. Meanwhile, these receipts declined in the rest of the Arab countries by a range of 4% to 35.1% during the same year.

Secondly: The Travel Sector
The surplus achieved in the services balance for the travel sector is increased to 3.8 billion dollars. This is due to the total revenue being higher than the total payments for these receipts, resulting in noticeable improvement in tourism revenue for the fifth consecutive year. In 2019, the overall revenue recorded an increase of approximately 7.9% to reach 93.1 billion dollars, compared to 86.3 billion dollars in 2018. At the level of individual countries, twelve Arab countries recorded an increase in revenue in 2019, with rates ranging from 1.6% in Oman to 21.6% in Tunisia, while revenue declined in Algeria, Sudan, Qatar, Kuwait, Mauritania, with rates ranging from 1% to 34.9% compared to the previous year. As for expenditures, the increase rate was around 4.2%, reaching about 89.3 billion dollars, compared to 85.7 billion dollars in 2018. At the Arab country level, tourism and travel expenditures increased in most Arab countries in 2019, with rates ranging from 1.5% in Bahrain to 76.5% in Libya, while Tunisia, Saudi Arabia, Palestine, Kuwait, and Mauritania saw a decline in expenditures.

5 Previous Studies

Several researchers have studied the importance of services trade and its impact on economic growth. A review of studies that analyzed the relationship between total domestic income and service trade and their results are as follows:

Study [3] aimed to understand the impact of services imports on economic growth in Jordan for the period 1990–2012. The results showed that the total import coefficient was positive, while service imports coefficient had an intrinsic value but negative relationship with GDP. Study [4] emphasized the importance of service trade in economic growth, especially in developed countries. Despite the limited studies on this topic, the main results showed that service imports have a positive impact on economic growth in developed countries and a negative impact on underdeveloped countries. In addition, study of [5] analyzed the relationship between imports and economic growth in Turkey, where the results showed a two-way relationship between total domestic income and imports of investment and raw materials, while a one-way relationship was found between total domestic income and imports of consumer goods and other imports.

Moreover, the study of [6] showed that there was no significant effect of transport services and travel services on economic growth in the Kingdom of Saudi Arabia. Also, the study by [7] aimed to analyze the relationship between international service trade and economic growth during the period of 1990–2010, and found that there was a long-term stable relationship between the variables. It concluded that the Chinese government should focus on improving the openness of international service trade and competitiveness as a priority for the development of international trade in services. Additionally, a study by [8] attempted to examine the relationship between services trade liberalization and economic growth for a sample of 189 countries during the period of 2000–2018, and found positive results for the impact of trade liberalization on economic growth. Furthermore, a study by [9] analyzed the relationship between trade liberalization and economic growth in three African countries in southern Sahara between 1960 and 2015, and the results showed that exports increased after liberalization in the countries, but it did not lead to significant economic growth.

Thus, these studies suggest that the impact of service trade on economic growth is complex and varies between countries and regions. Further research is needed to understand the relationship between service trade and economic growth in different contexts.

6 Data and Methodology

Researchers used time series data and employed standard tests to estimate the regression coefficients. This process enables identification of the statistical relationship between the independent variable, which is the total local Gross Domestic Product (GDP), and the dependent variable, which is the volume of services imports. The data package E-views was relied upon to derive several numerical indicators that measure the impact of the independent variable on the dependent variable. To clarify further, the following model can be used:

$$GDP = f(M.S)$$

where GDP represents the independent variable, which is the total local Gross Domestic Product, and (M.S) represents service imports. This study is an applied practice to determine the effect of the total local Gross Domestic Product on services imports in the case of Jordan for the period between 2005 and 2020. To provide the possibility of measuring the relationship between the study variables, information from the website of the International Monetary Fund's (IMF) Statistics Department was used.

In order to analyze the relationship between variables of study, the researchers are used the following steps:

1- Unit Root Test
2- Co-Integration Test
3- Error Correction Regression using (ARDL) approach.
4- Estimating long-term coefficients using (ARDL) approach.

Based on the previously mentioned, the results of the previous tests can be explained to study the relationship between study variables as follows:

- Unit Root Test

This test helps researchers determine the stationarity of time series related to the study variables. Many time series data are characterized by nonstationary, so it is crucial to test data stationarity to avoid biased results. Therefore, the researchers will used the Augmented Dickey-Fuller (ADF) test, and the results of the analysis are shown in Table 4.

Table 4. Unit root test results

Variable	ADF/Level	First deff
Imports	−2,601	−4,869
GDP	−5,982	

Source: Prepared by the researchers based on the results of E-views

The results in Table 4 indicate that the service imports variable was not "stationary" at the level, but became stationary after taking the first difference. On the other hand, the stationary test for the independent variable, which is total domestic income (GDP), shows that it is a stationary variable at the level. Since one of the variables is stationary at the level and the other at the first difference, it is possible to do a Co-Integration Test using the F-bounds test.

- **Co-Integration Test:**

Table 5 shows that the statistic (F) with a value of (10.9252) which is greater than the upper limits at 1%, thus indicating that the variables are integrated. This helps in the possibility of estimating the relationship between the study variables using ARDL approach for both short and long terms.

- **Error Correction Regression using (ARDL) approach**

According to the data in Table 6, after conducting the error Correction Regression in the short-terms, that the error rate is (−1.405992) and it is statistically significant at less than 1%.

- **Estimating long-term coefficients using (ARDL) approach**

Table 5. Co-Integration test results

F-Bounds Test		Null Hypothesis: No levels relationship		
Test Statistic	Value	Signif	I(0)	I(1)
F-statistic	10.95268	10%	5.59	6.26
k	1	5%	6.56	7.3
		2.5%	7.46	8.27
		1%	8.74	9.63

Source: Prepared by the researchers based on the results of E-views

Table 6. Error Correction Regression Test

Dependent Variable: DLOG(IMPORTS)				
Selected Model: ARDL(1, 0)				
Variable	Coefficient	Std. Error	t-Statistic	Prob
C	−1.285977	0.278342	−4.620128	0.0006
@TREND	−0.061002	0.012095	−5.043525	0.0003
CointEq(−1)*	−1.405992	0.288620	−4.871427	0.0004
R-squared	0.662662	Mean dependent var		0.012182
Adjusted R-squared	0.610764	S.D. dependent var		0.116438
S.E. of regression	0.072644	Akaike info criterion		−2.239129
Sum squared resid	0.068603	Schwarz criterion		−2.094268
Log likelihood	20.91303	Hannan-Quinn criter		−2.231711
F-statistic	12.76852	Durbin-Watson stat		1.752978
Prob(F-statistic)	0.000856			

*p-value incompatible with t-Bounds distribution
Source: Prepared by the researchers based on the results of E-views

The results of the long-term coefficient estimation analysis using (ARDL) approach showed that a 1% increase in total domestic income will lead to an increase in services imports of 0.776%, which is statistically significant at less than 1%. This can be seen in the data of Table 7.

Table 7. Estimating long-term coefficients using (ARDL) approach

Variable	Coefficient	Std. Error	t-Statistic	Prob
LOG(GDP)	0.776180	0.088630	8.757508	0.0000
EC = LOG(IMPORTS) – (0.7762 * LOG(GDP))				

Source: Prepared by the researchers based on the results of E-views

7 Results and Recommendations

The researchers tested the relationship between total local income and service imports in Jordan for the period 2005–2020. The results of the study were summarized as follows:

- The results of the study confirmed the positive effect of total local income on services imports in the long term, which is statistically significant at a rate less than 1% and indicates that it is not flexible in the long term. With a 1% increase in total local income, the rate of increase in service imports will increase by 0.776%.

In light of the above results, which show positive effects between total local income and service imports in long terms. The researchers recommend the following:

Researchers recommend the following steps to enhance the contribution of service trade to the economy:

1. Increase research to identify opportunities for increasing representation of service trade in economic activity.
2. Negotiate more free trade agreements to promote service trade at regional and international levels.
3. Reduce restrictions on service trade to attract more investment and reduce unemployment and increasing income.
4. Encourage services trade so that it becomes at some point a major reason to export, growth and structural transformation of the economy.
5. Learn from the experiences of other countries that have recorded significant growth in services trade.

References

1. Saleem, T.A.A., Abbas, N.A., Al-Sawaie, K.M., Jubran, A.H.M. The Impact of digital trade on Jordan's economic growth. In: 2021 22nd International Arab Conference on Information Technology (ACIT), pp. 1–8. IEEE (2021)
2. Al-Sawaie, K.: Determinants of Jordanian trade balance: an ARDL approach. Zarqa J. Res. Stud. Human. 1–17 (2017)
3. Buthaina, M.: International trade in services and economic growth: the case of Jordan. Jordan J. Econ. Sci. **2**(2) (2015)
4. Xiaoying, L., David, G., Robert, E.: Imports of services and economic growth: a dynamic panel approach. Informacion Comercial Espanola-Monthly Edition **824** (2005)
5. Ahmet, U.: Import and economic growth in Turkey: evidence from multivariate VAR analysis. J. Econ. Bus. **11**(1–2), 54–75 (2008)

6. Fayssal, B.: The impact of trade in services on economic growth in the Maghreb countries (Algeria, Morocco and Tunisia) – econometric study using panel data during the period (2005–2019). Contemp. Econ. Stud. Mag. 69–84 (2021)
7. Yang, S.: Analysis of the relationship between international service trade and economic growth based on VAR model. J. Phys. Conf. Ser. **1345**(6) (2019)
8. Qayyum, M., Zhua, Y., Yua, Y., Javed, B.: Impact of service trade liberalization on economic growth. J. Crit. Rev. **7**(7), 281–285 (2020)
9. Amoasah, F.: Trade liberalization and economic growth: a study on Ghana, Nigeria and Cote d'Ivoire, No. 188. FIW Working Paper (2018)
10. UNCTAD (2023). https://unctad.org/. http://unctad.org/en/pages/home.aspx

The Role of Electronic Management in Promoting Organizational Creativity: A Case Study of Orange Telecom Company/Jordan

Maan H. Mansour[1](✉), Khaled A. A. Al Zeaideen[1], Mohammed A. H. Altaee[1], Yasin K. Kharasheh[2], Worood O. Abu Dokhan[1], and Mohanad Dahlan[3]

[1] Zarqa University, Zarqa, Jordan
mmansour@zu.edu.jo
[2] Qassim University, Qassim, Kingdom of Saudi Arabia
[3] University of Business and Technology, Jeddah, Kingdom of Saudi Arabia

Abstract. The study aimed to investigate the concept of electronic management and organizational creativity and to test the impact of the dimensions of electronic management (electronic management, electronic devices, electronic applications, electronic networks, and databases) on enhancing organizational creativity at Orange Communications Company. On the questionnaire, the sample is convenient, consisting of (120) male and female employees working in senior administrative positions at Orange Telecom Company. The statistical analysis program, SPSS, was used to analyze the data collected through the study tool. The descriptive analytical approach was used in this study; because it is the most appropriate to achieve the study objectives with its dimensions and description of its characteristics. The study's results showed the existence of a role for electronic management in promoting organizational creativity in Orange Telecom/Jordan. The study recommended several recommendations, including the necessity of activating electronic management in various sectors because of its positive impact on organizational creativity and holding courses and lectures by telecommunications companies to familiarize employees with technologies, applications, networks, and rules. Data on electronic management, conducting more studies on the dimensions of electronic management and other variables to measure their impact on business development, and developing administrative systems in line with electronic management.

Keywords: Electronic Management · Organizational Creativity · Orange Telecom Company · Jordan

1 Introduction

Technical and digital development has made it imperative for business organizations to keep pace with modern administrative patterns due to the intensification of competition in the world. Electronic management is considered one of the important administrative methodologies for the success of the organization's work. Administrative processes have

B. A. M. Alareeni and I. Elgedawy (Eds.): *Artificial Intelligence (AI) and Finance*, SSDC 488, pp. 197–208, 2023.
https://doi.org/10.1007/978-3-031-39158-3_18

become heavily dependent on digital technologies in various fields. These technologies affect the decision-making process, increase the effectiveness of its operations, flexibility, automation and smart digitization, and improve the electronic management process [1].

The efficiency of management in any organization depends on the level of administrative decisions quality, which is taken by different organizational levels, as they are the cornerstone for management, so there is an urgent need to have the correct information to assist management in making decisions in the right form and time, to address challenges and their consequences [2]. The role of electronic management has become extremely important to ensure the meaningful development of innovation, and digital leadership indicates basic competence in the field of communication, content design and computer use in a way that contributes to the creation of a knowledge society [1].

Organizations view information technology and the process of organizing information as two requirements that must be met to address challenges and enhance creative processes in the organization, so knowledge must be applied to perform several functions in organizations through the use of computers and their applications [2].

Electronic management plays an important role in promoting innovation and organizational creativity in various sectors, as customers' desires change rapidly, especially since the role of customers in the value creation process in the company changes and is affected by technological development. We have to point out that intense economic competition requires innovation, which is the main cornerstone for achieving sustainability in the organization [3]. Organizational creativity contributes to ensuring the growth and survival of the organization. Creativity is essential for organizations because organizations need innovative and new ideas, as innovation in organizations will enable them to create competitive advantages to outperform their competitors in the market. Innovation enables organizations to have useful tools to improve their performance level [4].

2 Study Problem

Electronic management is one of the important technological achievements in the modern era, where developments and technologies in the field of communications and communication prompted governments and countries to utilize the achievements that were made during the digital revolution through the use of computers and the Internet to complete tasks, and provide citizens with services, electronic technologies are important in solving many problems. Electronic management is one of the important outputs of communication and information technology mechanisms in the technical world. Software, hardware, systems, and tools for electronic management are examples of new mechanisms used in light of the knowledge-based economy. Many service companies in Jordan need to apply electronic management to obtain important outputs at various organizational levels. These outputs include organizational creativity, which has become important in light of the intensification of competition between organizations in the local market in general and the international market in particular. It should be noted that the organizations' possession of creativity will enable them to avoid risks [5].

Orange Jordan applies advanced work methods that help work in a spirit of cooperation and teamwork. A friendly atmosphere emanating from intellectual openness, as the company provides its administrative operations with an integrated digital system based on modern technologies and techniques that are offered to customers, which would break down barriers between employees, as well as aims to improve operational costs and increase work efficiency and effectiveness also the company implements its five-year strategy, Essentials 2020, which is based in one of its axes on building an advanced digital career model concerned with employees, to enhance the image of Orange Jordan, to serve its short and long-term goals [6]. It should also be noted that telecommunications companies are one of the most important service companies in Jordan, and most companies apply electronic management. Therefore, the problem of the study is to explain the role of electronic management in enhancing organizational creativity in Orange Telecom through the following hypothesis:

H01: There is no statistically significant impact at the significance level ($\alpha \leq 0.05$) for electronic management (Electronic devices, E-Management Applications, E-Networks, Databases) in enhancing organizational creativity in Orange Telecom Company.

3 Study Moodle

See Fig. 1.

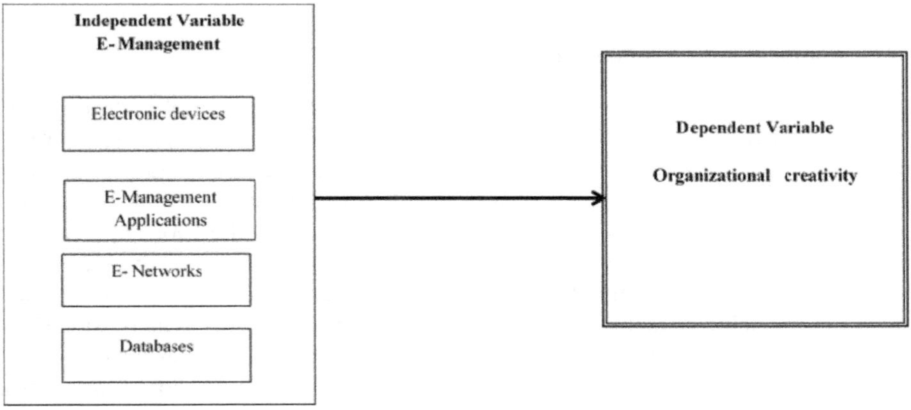

Fig. 1. Study Moodle

4 Theoretical Framework

4.1 Concept of E-Management

E-management is an umbrella term that encompasses many different areas of study within the realm of the digital, including but not limited to e-commerce, electronics, human resources, and many more. E-business, as described by the International Business Machines Corporation, is "an open and all-encompassing strategy for delivering differentiated business value through the integration of information and communication technologies into business processes" [7]. Modern informatics includes advanced computer networks to link the organizational units in the organization with each other, to facilitate access to data and information to make appropriate decisions, complete work, and provide services to beneficiaries efficiently, at the lowest cost, and as quickly as possible [8]. Therefore, electronic management is the result of progress in the various fields of technology and informatics, which helped departments that adopt advanced technical means to complete the tasks assigned to them and implement them to the fullest [9].

Electronic management is the ability to use information and communication technology and modern technologies to carry out administrative activities electronically via the Internet and automated computer networks [10]. Electronic management is a process through which management manages an integrated electronic system based on information technology to improve various administrative processes within [11].

Electronic administration was also defined as "the completion of administrative transactions and the provision of public services via the Internet, without the customers having to go to the administrations personally to complete their transactions, with the accompanying waste of time, effort and energies" [12]. Electronic management integrates all management processes, i.e. termination, organization, animation, and control. It may positively impact the opportunities offered by new information and communication technologies [13].

4.2 Importance of E-Management

As a result of the proliferation of online commerce across a variety of industries, it is necessary to create an electronic infrastructure in institutions where integrated electronic management has been developed as a means to automate the administrative workflow for administrative operations electronically and to document them according to systematic electronic applications [14]. Recently, the composition and functions of electronic management have changed significantly. Radical through theoretical and practical applications of electronic management of new systems, tools and models to create wealth and achieve added value, in addition to new ways of competition based on innovation [15].

Electronic management is important because it provides positive results in the field of facilitating administrative processes, reducing the time and cost of completing tasks, developing job performance, making information available at all times, and raising the level of efficiency and productivity of the organization, and through benefits in the administrative, economic, political and social aspects of the organization [16].

4.3 Components of E-Management

With the MIS application and database management system, and the electronic management environment, all available devices in an industry can be used to boost automation [17]. Electronic management, of which data is a part, is a system for administering a database and boasts various distinguishing characteristics. Auto lookup table relationship control; automatically controlled parent-child relationship; uniqueness check; automated safety control; automatically controlled parent-child relationship; domain, registry, and table-level auto-locking controls [18].

The infrastructure is considered one of the main tangible components in electronic management, which the organization cannot exist without. It is represented in a set of physical and human components through which electronic management can be implemented using its various applications.

- Hardware components: such as computers and supporting and peripheral devices
- Intellectual components: These include software and application software systems
- Human components include computer operators, programmers, systems analysts and designers.

4.4 Functions and Objectives of E-Management

Electronic management exchanges its effects with the internal and external environment of the organization because it is a highly flexible organizational and functional system. It is considered an integrated package of interrelated processes for planning, organizing, directing, controlling and making decisions. It provides all information technologies such as computer hardware, software, communication networks, and effective and smart management. The organization's resources of data, information and knowledge and its effective use [16]. The functions of electronic management include the following: electronic planning, electronic organization, electronic guidance, and electronic control.

4.5 Dimensions of E-Management

Electronic management is considered one of the important requirements to update administrative organizations' work and increase their efficiency. Electronic management constitutes an organizational and functional system that is flexible and has mutual influence with organizations' internal and external environment. Therefore, it is considered an integrated package of interrelated processes for planning, organizing, directing, controlling and making decisions, and thus organizational creativity, including the appropriate use of all necessary electronic technologies [19].

The electronic management dimensions that the study dealt with include: the electronic devices, which are the devices and equipment the company must provide before applying the technology in its centres. These devices are commensurate with the number of employees [20]. And electronic management applications include general applications, which are applications that are often installed on most computers that are started. Examples of system management software are operating systems, network systems, spreadsheets and e-mail. Special applications and programs are not downloaded to computers in sales companies to any user but rather are downloaded at the request of the entity or user who needs them to manage his business [21].

As for electronic networks, the word network expresses the connection of a group of computers with each other, wired or wireless, or utilizing satellites to exchange and obtain information and data. One of the most important types of computer networks is the Internet, which is the World Wide Web, which covers most parts of the globe and is available for all people in The world, and the internal or private Internet of any organization, and it allows the employees of the organization and its affiliates to access and exchange data and information within the organization while opening channels of communication between employees, on the other hand, the difference between it and the public Internet is that it is available only to employees and the employees of a specific work organization. One type of network is the extranet, a private Internet that allows only some specific beneficiaries to access it with specific powers and restrictions [22].

Databases are files saved electronically on computers, and they are the databases of the organization or company. These rules are concerned with covering data related to the activities of the company. The database consists of files, which are records that include detailed data. The database consists of two or a group of files that Store it indexed or tabulated. Relationships are a link or indications between a groups of information [13].

4.6 The Concept of Organizational Creativity

Organizational creativity is considered a center of competitiveness, as it is largely driven by technological progress, the emergence of the knowledge economy and large-scale non-price competition in industrial and service companies. Organizations, particularly technology-driven, require that they be more creative and pioneering than before to lead, grow, compete and endure [23]. Although different definitions of creativity and organizational creativity have been proposed, researchers still disagree about what specifically constitutes creativity or innovation, with different studies using somewhat different operating processes for each concept [19]. Psychology, sociology, organizational behavior, and information systems are just some of the fields that have explored the topic of creativity at length [24]. Although the term "creativity" originates in the field of psychology, it is increasingly utilized in a wide range of business and management settings, including business strategy, operations, management, competitive advantage, development, leadership, and innovation [25].

Creativity mainly includes cognitive processes within the individual, while innovation mainly represents social processes between individuals in the workplace. Creativity is also "the ability to devise methods, means, and ideas that are useful for work so that these Ideas and methods of optimal response by workers and motivating their capabilities and talents to achieve the desired goals [26]. Based on these definitions, some important considerations can be deduced, including that creativity is not defined alone and in a unified manner. Although the fundamental theoretical characteristics of organizational innovation were fairly stable, the nature and sorts of innovations investigated for them changed throughout time, as new products, services, technologies, and managerial practices were developed [27].

5 Literature Review

A few studies have been conducted to reveal the link between electronic management and organizational performance. Both the exploratory and confirmatory analyses were conducted using structural equation modeling. The level of the organization was studied through a multiple case study to search for regularity, differences and convergence between the elements of the different studied cases through a questionnaire filled out by 155 electronic managers in Tunisia. The results concluded that adopting the electronic management approach affects the performance of Tunisian companies, and an original illustrative model for the practice of electronic management was proposed, which could form a theoretical basis for developing modern projects. Information and communication technology (video systems, e-mail, intranet, etc.) are vital in all contemporary businesses, but the results also show that deploying them within the organization takes significant financial commitment [28].

While aiming to research the electronic document circulation system and its importance in organizing documents, the study showed that the main stages of the document life cycle in institutions include creation, management/storage, access, retrieval, management, reset, collaboration, distribution, conservation, disposal, and storage. Definitions have been given of terms such as "document", "electronic document", "document control systems", and "electronic document management systems". The study showed that the main characteristics of the electronic document management system include the following parameters: scalability, compatibility, due availability, and dynamism. The main advantage of an electronic document management system is low document automation, and the disadvantage is high financial costs. The capabilities of an EDMS include providing a user interface, capture, signaling, receiving, commenting, storage and archiving, distribution, workflow, security, and system integration [29].

As one of the major issues in contemporary administrative theory, the function of competition in advancing electronic management and its effect on activating smart companies was the focus of another study. Many employees from the Asia cell Corporation in Karbala participated in the study. In total, there were (129), data was gathered by questionnaire, and a number of statistical techniques were employed to establish the veracity of the hypothesis. Conclusions from the study indicate that educational competition plays a positive interactive role in promoting electronic management and its impact on activating smart organizations in the company under study. This finding suggests that educational competition modifies the positive relationship between electronic management and smart organizations [30].

In the same direction, some studies aimed to discuss the issue of supporting organizational innovation based on information technology. The study used the descriptive approach by reviewing the literature on organizational creativity and a survey carried out in 25 selected organizations. The study found the potential of information technology applications in supporting organizational creativity and the success factors associated with it [31].

6 Study Population and Sample

The study population consisted of all the employees of the administrative positions in the Orange Communications Company in the General Administration building in the capital, Amman, and numbered (178) male and female employees. Distributing the paper questionnaire to the workers who expressed a desire to respond and participate in the current study, who numbered (120) male and female employees. (100) questionnaires were retrieved, where (11) invalid responses and (9) missing values were identified.

7 Internal Validity and Stability of the Study Tool

The value of the Cronbach alpha coefficient for all items of the study tool was (0.853), which is a positive value as the square root expresses the internal validity coefficient (0.92), which is excellent and high, being greater than 0.7, which is the acceptable and critical value. Where the value of the stability coefficient for all paragraphs of the study tool was (0.853), which is a positive value as it reflects a high and acceptable stability coefficient for the current study. Looking at the values of the Cronbach alpha coefficients in the above table, it is noted that all values are high, indicating that the study tool is considered highly credible. And that the study tool has been built, prepared and arbitrated acceptably and achieves the study's objectives (Table 1).

Table 1. Internal validity and Stability of the study tool

Study variables and dimensions	Number of paragraphs	Cronbach alpha coefficient	Internal validity coefficient
Electronic devices	6	0.775	0.88
E- management applications	6	0.850	0.92
Electronic networks	6	0.784	0.89
Databases	6	0.858	0.93
E- management	24	0.833	0.91
problems solving and decisions making	4	0.758	0.87
Improving administrative processes	4	0.768	0.88
Product development	4	0.841	0.92
Organizational creativity	12	0.745	0.86
Total summation	36	0.853	0.92

8 Hypothesis Testing

Table 2 shows the extracted results of the multiple linear regression analysis of the proposed model, which displays matching quality indicators obtained by the model.

Through the quality of conformity of the two models, the general model, represented by the first model and the special model, represented by the second model, the statistical significance value of the general model and the special model reached (0.023) (0.000), respectively. It is a statistically significant value at the significance level ($\alpha \leq 0.05$)), and therefore, the general model and the specific model are significant, and both are appropriate.

Table 2. One-way analysis of variance (ANOVA)

	Dependent Variable	Independent Variable	F	Probability value
1	Organizational creativity	Electronic management	5.347	0.023
2	Organizational creativity	Electronic devices	6.330	0.000
		Electronic management applications		
		Electronic networks		
		Databases		

H0.1: The main null hypothesis: There is no statistically significant effect at the significance level ($\alpha \leq 0.05$) of electronic management on enhancing organizational creativity in Orange Telecom.

Through Table 3, the values of β amounted to (0.310) for the electronic management variable, as the calculated (Z) value reached (2.32). The value of statistical significance reached (0.023), which is a statistically significant value at the level of significance ($\alpha \leq 0.05$)) and therefore, electronic management has a significant impact on enhancing organizational creativity in Orange Telecom.

From the results, it is clear that the electronic management of electronic devices, electronic management applications, electronic networks, and databases explain the equivalent of (21%) of the changes that occur to enhance organizational creativity in Orange Telecom, while the percentage of (79%) is due to the discrepancy in organizational creativity itself.

Table 3. Test the Impact of E-Management in enhancing Organizational Creativity

Independent variable	R	R^2	DF	Regression		
				β	T	Sig
E-Management	0.263	0.069	1	0.415	2.703	0.009
			98			
			99			

9 Results and Discussion

The results of testing the main hypothesis also show a significant impact of electronic management on enhancing organizational creativity in Orange Telecom. The value of statistical significance was (0.023), which is a statistically significant value at the level of significance ($\alpha \leq 0.05$)). Therefore electronic management has a significant impact on enhancing organizational creativity in Orange Telecom. This result may be attributed to the positive role played by electronic management in management functions in business organizations because it creates an atmosphere of flexibility in management and makes it able to face external and internal changes and invest its material and human resources, which is reflected in the effectiveness of its decisions, the creativity of the organization, the development of job performance, and making information available at all times, which raises the efficiency and productivity of the institution [28]. So, adopting the electronic management approach positively affects the performance of companies. This result also agreed with Olsak and Kisielnicki's study that electronic management supports organizational creativity in the institution [31].

10 Recommendations

In light of the study's findings, the researchers recommend the following:

- Holding courses and lectures by the management of the telecommunications company to familiarize employees with the technologies of electronic management networks, databases and their applications. In addition to the continuous follow-up to the development of these networks and databases because of their impact on organizational creativity.
- Continuous improvement and periodic maintenance of the infrastructure for electronic management systems applications, as well as the keenness to attract talented employees in these applications while striving to develop administrative systems in line with the requirements of electronic management applications.

References

1. Mihardjo, L., Sasmoko, S., Alamsyah, F., Elidjen, D.: The influence of digital leadership on innovation management based on dynamic capability: market orientation as a moderator. Manage. Sci. Lett. **9**, 1059–1070 (2019)
2. Bataineh, K.: The impact of electronic management on the employees' performance field study on the public organizations and governance in Jerash Governorate. J. Manage. Strategy **8**(5), 86 (2017)
3. Elidemir, S., Ozturen, A., Steven, W.: Innovative behaviors, employee creativity, and sustainable competitive advantage: a moderated mediation. Sustainability **12**, 3295 (2020)
4. Dehghan, N.M.: Knowledge management and its role in organizational innovation. Monthly Automot. Eng. Relat. Ind. **1**(10), 47–57 (2010)
5. Ellatif, H., Sammani, A.: E-management: configuration, functions and role in improving performance of Arab institutions and organization. Int. J. Comput. Appl. **80**(6) (2013)
6. Ahmad, H., Maram, A., Thaer, A., Raed, H., Mustafa, H.: The impact of the agile system, knowledge sharing, and empowerment on organizational entrepreneurship in orange telecommunication firms. J. Southwest Jiaotong Univ. **57**(3) (2022)
7. Haider A., Aryati B., Mahadi B.: Opportunities and challenges in implementing electronic document management systems. Asian J. Appl. Sci. **3**(1) (2015)
8. Khallouf, I.: The Reality of the Application of Electronic Management in Public Secondary Schools in the West Bank from the point of view of Male and Female Principals. Master Thesis, An-Najah University, Palestine (2010)

9. Achour, A.: The Role of Electronic Administration in Rationalizing Public Service in the United States of America and Algeria. Master's Thesis, University of Mentouri, Constantine, Algeria (2010)
10. Dayni, R.K.W.: Effect of E-Management and Role of HR Development in Rafidain Bank on the Improvement of the Organization Performance. Unpublished Master Thesis, Middle East University, Jordan (2010)
11. Alaagez, E.: Role of Organizational Culture in the Application of E-Management. Unpublished Master Thesis, Business Administration Department, Ghaza University, Ghaza (2011)
12. Bakir, A.: The Comprehensive Concept of Electronic Management, Views of the Gulf Journal, United Arab Emirates, Issue 23, Gulf Research Center (2006)
13. Kalika, M.: Le management est mort, vive le e-management. Revue françaisede gestion, n°129, juin-juillet-août (2000)
14. Elsaadani, A.: Investigating the effect of E-management on customer service performance. AAU J. Bus. Law **3**(2), Article 1 (2020)
15. Hevner, A., Gregor, S.: Envisioning entrepreneurship and digital innovation through a design science research lens: a matrix approach. Inf. Manage. **59**(3) (2022)
16. Al-Husseini, A., Al-Khayal, S.: The impact of applying electronic management systems on job performance: a field study on female deanship employees at King Abdulaziz University in Jeddah. Sci. J. Facult. Commerce Sector **10** (2013)
17. Liu, Y., Othman, A., Aballama, A.N., Mahdi, O.R.: E-management development and deployment strategy for future organization. Afr. J. Bus. Manage. **5**(16) (2011)
18. Kajko-Mattsson, M., Karim, R., Mirjamdotter, A.: Essential components of e-maintenance. Int. J. Perform. Eng. **7**(6), 555 (2011)
19. Dewett, T., Jones, G.R.: The role of information technology in the organization: a review, model, and assessment. J. Manag. **27**(3), 313–346 (2001)
20. Tebboune, A.: Electronic management: importance and requirements. J. Ijtihad Legal Econ. Stud. **10**(2) (2021)
21. Abdel Nasser, M., Qureshi, M.: The contribution of electronic administration to the development of administrative work in institutions of higher education: a case study of the Faculty of Science & Technology at the University of Biskra, Algeria. Res. J. **9** (2011)
22. Asiri, K., Al-Maghidi, A.: Requirements for the Application of Electronic Management in the Management of General Education Schools in Abha: A Field Study. Unpublished Master's Thesis, King Khalid University, Saudi Arabia (2015)
23. Jung, D.I., Chow, C., Wu, A.: The role of transformational leadership in enhancing organizational innovation: hypotheses and some preliminary findings. Leadership Q. **14**(4–5), 525 (2003)
24. Khedhaouria, A., Belbaly, N.: Organizational creativity climate factors: Lessons learned from the French energy management industry. In: Proceedings of ECIS 2011 (2011)
25. Bessant, J., Tidd, J.: Innovation and Entrepreneurship. John Wiley & Sons Publishing, New York (2011)
26. Rank, J., Pace, V.L., Frese, M.: Three avenues for future research on creativity, innovation, and initiative. Appl. Psychol. Int. Rev. **53** (2004)
27. Hage, J.T.: Organizational innovation and organizational change. Annu. Rev. Sociol. **25**, 597–622 (1999)
28. Hamadi, K., Ziadi, J.: The impact of e-management on organizational performance: an empirical study in Tunisian firms. World Acad. J. Manage. **8**(3) (2020)

29. Abbasova, V.S.: Main concepts of the document management system required for its implementation in enterprises. Sci. Rise **1** (2020)
30. Abbas, A.: Educational competition as a moderating variable of the relationship between electronic management and intelligent organizations. Tempos Espaços Educ. **13**(32) (2020)
31. Olszak, C.M., Kisielnicki, J.: Organizational creativity and IT-based support. Inf. Sci. Int. J. Emerg. Transdiscipl. **19** (2016)

Regulations for the Use of Information and Communication Technology in Health Fields: A Case Study of the UAE

Moustafa Elmetwaly Kandeel[1,2](✉), Amal Abueida[3], and Mohamed Moustafa Kandeel[4]

[1] College of Law, Al Ain University, Al Ain, UAE
Moustafa.kandeel@aau.ac.ae
[2] Faculty of Law, Tanta University, Tanta, Egypt
[3] Master of Business Administration, College of Business, Al Ain University, Al Ain, UAE
[4] Faculty of Medicine, Misr University for Science and Technology, Cairo, Egypt

Abstract. This paper aims to shed light on the regulations related to using ICT in the health fields, particularly its legal status in the United Arab Emirates. To address this issue, we used the analytical approach to analyze the provisions of Federal Law No. (2) of 2019 regarding the use of ICT in health fields, as well as the Executive Regulations of this Law issued by Cabinet Resolution No. (32) of 2020. In this paper, we have dealt with the foundations and scope of controls for the use of ICT in the health fields, the legal rules related to the Central System for the exchange of health data and information, the controls for saving this data and information, as well as the controls for maintaining the confidentiality of patient information, and finally the disciplinary penalties that are imposed when violating these rules. At the end of the paper, we concluded that using ICT in the health fields must take place within the framework of a legal umbrella that aims for the effective participation of all Health Authorities to make this use successful and to guarantee the privacy of patient information, and not to allow this privacy to be compromised except in the cases stipulated by law or with the consent of the patient or his legal representative.

Keywords: United Arab Emirates · Health services · ICT · Central System · Health data and information · Circulation · Preservation · Confidentiality · Privacy · Federal legislation · Regulations · Disciplinary penalties

1 Introduction

The first indications of what we know today as "telemedicine" began in the fifties of the last century. In the beginning, "telemedicine" was often used to connect doctors working with a patient in one place with specialists elsewhere. This has been of great benefit to rural or hard-to-reach populations where specialists are not readily available [1]. Thus, telemedicine became a health care model in which hospitals provide care to patients in remote areas.

B. A. M. Alareeni and I. Elgedawy (Eds.): *Artificial Intelligence (AI) and Finance*, SSDC 488, pp. 209–218, 2023.
https://doi.org/10.1007/978-3-031-39158-3_19

The use of telemedicine services has spread all over the world, and thanks to technological progress, it has turned into an effective and convenient model of health care [2]. Today, the majority of comprehensive telemedicine services are used across the UK, USA, Canada, Scandinavia, Switzerland, Australia, and now the UAE. The International Society for Telemedicine & eHealth (ISfTeH) is now represented in 85 different countries and regions around the world [3].

This widespread use of telemedicine services requires the development of a legal framework for it, through which regulations are set that define the controls for its use. In addition, some countries, including the UAE, have established an electronic system that includes all issues related to the exchange of information related to the provision of health services in the state. In this paper, we will address the legal regulation related to the use of ICT in health fields in the UAE as a model. The UAE legislator has established this legal framework through several legislations, on top of which is Federal Law No. (2) of 2019 regarding the use of ICT in health fields, and the Executive Regulations of this law issued by Cabinet Resolution No. (32) of 2020.

In this paper, we will use the analytical approach to discuss the main components of these legal regulations. We will raise some questions concerning what are the basis and scope of the controls on the use of ICT in the health fields? What are the legal rules related to the Central Electronic System for the exchange of health information in UAE? What are the means of IT for saving health data and information? What are the mechanisms for maintaining the privacy of patients' health information? And finally, what is the disciplinary penalties for violating the rules of using IT in the health field?

2 The Basis and Scope of the Controls on the Use of ICT in the Health Fields

ICT means technical and electronic tools or systems or other means that allow the possibility of processing information and data of all kinds, including the possibility of storing, retrieving, publishing, exchanging data, and helping productivity to be better and more efficient [4].

The controls for the use of ICT in the health fields in UAE can be derived through a number of legislations as shown in Table 1 below:

Table 1. ICT legislations in UAE

# Legislation	Date	Title	publication
Federal Law No. (2) of 2019	06/02/2019	The Use of ICT in Health Fields (Hereinafter referred to as the Law)	Federal Official Gazette, Issue No. 647, p. 11
Cabinet Resolution No. (32) of 2020	22/04/2020	The Executive Regulations of Federal Law No. (2) of 2019 Concerning the Use of ICT in Health Fields (Hereinafter referred to as the executive regulations)	Federal Official Gazette, Issue No. 677, 30 April 2020

2.1 The Scope and Objectives of the Controls for the Use of ICT in Health Fields

The controls contained in the aforementioned legislation apply to all methods and uses of ICT in the health fields in the UAE, including free zones [5].

These controls aim to achieve the following objectives [6]:

1. Ensuring the optimal use of ICT in the health fields.
2. Ensuring that the approved foundations, standards and practices are compatible with their internationally approved counterparts.
3. Enabling the Ministry of Health and Prevention (MOHAP) in UAE to collect, analyze and preserve health information at the state level.

   In this regard, health information means the health data that has been processed and has become significant, whether it is visual, audio or readable, and which is of a health nature, whether it is related to facilities, health or insurance agencies, or beneficiaries of health services [7].
4. Ensuring the security and integrity of health data and information.

   In this regard, data means everything that can be stored, processed, generated, and transmitted by means of ICT, such as numbers, letters, symbols, images, and so on.

2.2 Foundations of the Use of ICT in Health Fields

When using ICT in health fields, the following should be adhered to [8]:

1. Preserving the confidentiality of health data and information, by not allowing their circulation in unauthorized cases.
2. Ensuring the validity and credibility of health data and information, by preserving their integrity from unauthorized vandalism, modification, alteration, deletion or addition.
3. Ensuring the availability of health data and information for authorized persons, and facilitating access to them when needed.

3 Developing a Central Electronic System for the Exchange of Health Information in the UAE

The UAE legislator provided for the creation of a Central Electronic System for the exchange of health information at the state level. This electronic system refers to a set of operations for the electronic exchange of health data and information, and includes a set of electronic parts or elements that link each other with relationships that work together towards achieving a specific goal (Article 1).

3.1 Establishing and Joining the Central Electronic System for Exchanging Health Information

The responsibility for establishing this Central System rests with the Ministry, in coordination with the Health Authority and the Concerned Authorities, for preserving, exchanging and collecting health data and information [9].

Health Authorities and Concerned Authorities are obligated to join the Central System for the exchange of health information (Article 7 of Law 2 of 2019). The Health

Authority means any federal or local governmental Health Authority in the country. As for the Concerned Authority, it means every entity in the country that provides health services, health insurance services, mediates in it, manages its requirements, or electronic services in the health field, or any entity directly or indirectly related to the implementation of the provisions of this law.

According to Article (2/1) of the Executive Regulations, the Health Authorities and the Concerned Authorities are committed to join the Central System according to the following rules and procedures:

a. Compliance with the rules of work of the Central System.
b. Adhere to the maximum deadline set for these entities to join the central database, as determined by the Ministry in coordination with those Authorities.
c. Afford any costs associated with communication and connection with the Central System.
d. Compliance with the rules regulating the national registry regarding digital health standards with regard to the necessary standards, requirements and procedures when dealing with the Central System, including:
   1- The personal health information required to be provided by the Health Authorities and the Concerned Authorities.
   2- Adherence to the mechanism and exchange of personal health data and information with the approved Health Authorities and Concerned Authorities to protect this health data and information and ensure its confidentiality.
   3- Mechanisms for protecting the confidentiality of health data and information.

According to Article 2/2 of the same Executive Regulations, the Ministry has the authority to audit the personal health data and information provided by the Concerned Authorities in order to ensure its validity, quality, and compliance with the national digital health data standards. The Ministry also, in coordination with other Health Authorities and Concerned Authorities, defines the mechanism and procedures for ensuring the quality of personal health data and information.

3.2 Controls for Using the Central System for Exchanging Health Information

A. Determine the persons authorized to use the Central System for exchanging health information.

The Authorized Entities to use the Central System are obligated to identify the persons (natural or legal) [10] who are authorized to access and circulate health information and databases, and to indicate their powers. Where the Health Authorities and the Concerned Authorities determine the persons authorized to access the Central System, based on the need for that, and depending on the professional role to determine the level of access to the data of the Central System, in addition to their role in patient care [11].

According to Article 5/1 of the Executive Regulations, the Health Authority shall grant the permit to each of the following:

1- Persons who work for the Health Authority under an employment contract, and the nature of their work that requires using the Central System.

2- Persons associated with outsourcing companies with a contract or experts and consultants who are hired on an occasional basis, or entities and facilities affiliated with the Health Authority. In all cases, the nature of the work of these persons or the task entrusted to them requires the use of the Central System.

According to Article 5/2 of the Executive Regulations, the Concerned Authority grants the permission to the persons who work for it, based on the requirements of their work to use the central system. Moreover, this use must be constrained by the actual requirements of the work, and the Concerned Authority, upon granting the permit, is committed to provide the Health Authority with the authorized persons to use it.

In addition, the Health Authority and the Concerned Authorities, as the case, should take the necessary measures to ensure that the authorized person is unable to access the Central System after the end of his service (Article 5/4 of Executive Regulations).

B. The obligation to include the identity number in health transactions and files.

The Health Authority and the Concerned Authority are obligated to include the ID number in all health transactions, records and files, and to use it in organizing and preserving them, with the exception of emergency cases and others for which a decision is issued by the Ministry in coordination with the Health Authority (Article 21 of Law No. 2 of 2019).

C. Controls of circulation of health data and information.

The use of the Central System and circulation of health data and information is subject to the following conditions and controls (Article 6 of Executive Regulations):

1- Entities and persons authorized to access any of the ICT systems must agree to undertake not to disclose health data and information that was viewed through the use of the central system.
2- It is prohibited to disclose the health information of the patient to any party without the patient's consent or his legal representative, unless the disclosure of this information is permitted in accordance with the legislation in force in the state.
3- In case of emergency and if it is not possible to obtain the patient's consent, the health care providers may examine the patient's file for health care purposes, with the need to state the reasons for the examination.
4- The patient's file should not be left open without supervision, and computers or any other electronic means should be closed when not in use.
5- Any suspicious activities that may affect the confidentiality of data and information must be reported.
6- E-mails should not be sent and the use of any electronic means of communication that contains patient information unless it is encrypted.
7- If information is entered incorrectly or some information is missing, the error must be corrected or the required information completed while preserving the original entry for quality and audit purposes.
8- When modifying any data, the reason for the modification must be indicated, and the modified information and the date of the modification must be saved with the electronic signature of the person who made the modification.

9- Amendments to information and data must be tracked as soon as they are entered or ratified.
10- Federal health data, information and statistics should not be published at the state level without the approval of the Ministry [12].
11- The patient's consent must be obtained when his identity data is published, and the person's identity data list is determined by a decision of the Minister of Health and Prevention in coordination with the rest of the Health Authorities.
12- The data, information and statistics to be published must comply with the standards set by the Ministry in coordination with the rest of the Health Authorities [13].
13- All necessary procedures must be taken to protect patients' personal data and information from loss, misuse, unauthorized access, disclosure, modification or destruction.
14- The user who is authorized to access the Central System must have his own username and password.
15- The username and password should not be shared with any other user or any other party.

4 Controls for Saving Health Data and Information by Means of IT

According to Article (20) of Law No. (2) of 2019, two conditions are required to save health data and information by means of ICT: The first is that the retention period is commensurate with the need for health data and information, and provided that the retention period is not less than (25) years from the date of the last health procedure for the person concerned with those health data and information. The second is to ensure standards of confidentiality, validity and credibility of data and information.

It is not permissible to store, process, generate or transfer health data and information related to health services provided inside the UAE to aboard, except for cases in which a decision is issued by the Health Authority in coordination with the Ministry (Article 13 of the Law). Whoever violates this prohibition shall be punished with a fine that is not less than (500,000) AED and not more than (700,000) AED (Article 24 of Law No. 2 of 2019).

According to Article (7) of the Executive Regulations, health data and information are saved by ICT in accordance with the following controls:

1. The Central System must include all the files of the patient in the state, and the files must contain the health information determined by the Ministry in coordination with the rest of the Health Authorities.
2. The patient can opt out of the Centralized System, in which case the data and information can be kept anonymous.
3. Health data and information that exceeded the storage period may be archived for the purpose of research and public health purposes while preserving the privacy of patient.
4. A backup copy of health data and information must be taken securely, and this data and information must be recoverable, and backup copies must be reviewed and updated regularly and continuously.
5. The Ministry, in coordination with the Health Authorities, shall develop one or more plans to manage risks and ensure the continuity of the work of the Central System.

6. The Ministry, in coordination with the Health Authorities, and through specialized committees, shall set the international standards in force locally with regard to the confidentiality, quality and validity of health data and information, in a manner that does not contradict the legislation in force in the state.
7. The Ministry and the Health Authorities shall conduct periodic audits to ensure the implementation of standards and procedures by the Concerned Authorities, with regard to the validity, integrity, quality and confidentiality of data and information.
8. Health data and information are stored by means of ICT, and in accordance with the controls for maintaining and archiving medical records in force in each health facility, provided that they are compatible, at a minimum level, with the controls set by the Ministry in coordination with the rest of the Health Authorities.
9. Periodic tests must be conducted to assess the effectiveness of the mechanism for retrieving health data and information, and to detect any defect in the work of the Central System and any improvements that can be made to it.

5 Controls for Maintaining the Confidentiality of Patient Information

The UAE legislator has many rules to ensure the privacy of patients and to make sure that the use of IT in healthcare is secured [14], as follows:

1. Authorized Entities to use the Central Electronic System are obligated to take all necessary measures to ensure the protection, integrity and confidentiality of health data and information [15]; Where the Health Authorities and Concerned Authorities adhere to privacy and security standards, and any controls set by the Ministry in coordination with other Health Authorities, including periodic audit procedures to remove or modify the features or powers of authorized persons in accordance with work requirements [16].
2. It is not permissible for any person to use the Central System without authorization from the Health Authority or the Concerned Authority, and the authorized persons are committed to the following [17]:
   a. Circulation of the necessary information to accomplish the required work or for a specific purpose. Circulate information in this case means viewing, exchanging, copying, photographing, transferring, storing, publishing, disclosing or sending health data and information.
   b. The circulation of information shall be restricted to the authorized persons only.
   c. Not to modify the health data and information by deletion or addition except in accordance with the specified controls.
   d. Not to publish health data and information, as well as statistics related to the health field, except in accordance with the specified controls.
3. The use of the Central System and the circulation of health data and information are subject to the conditions and controls referred to previously (Article 6 of Executive Regulations).
4. Without prejudice to any applicable legislation, anyone who handles patient information must maintain its confidentiality and not to use it for other than health purposes, without the patient's written consent, except in the following cases [17]:

a. Health data or information requested by Health Insurance Companies, or any entity that finances the health services that the patient receives, for the purposes of reviewing, approving or verifying the financial entitlements related to those services.
b. The purposes of scientific and clinical research, and it is required not to disclose the identity of patients and to observe the ethics and rules of scientific research.
c. Take preventive and curative measures related to public health, or to preserve the health and safety of the patient or any other person in contact with the patient.
d. At the request of the competent Judicial Authorities.
e. At the request of the Health Authority for the purposes of monitoring, inspection and maintaining public health.

5. Individuals may give access to their personal health information to another person of their choice, provided that they are registered as users in the central database of the system, and this does not contradict any other legislation issued in this regard (Article 5/5 of Executive Regulations).
6. Any person may request a ban or restriction of access to his personal health information, in accordance with the conditions and controls determined by the Ministry in coordination with the rest of the Health Authorities (Article 5/6 of Executive Regulations).

6 Disciplinary Penalties for Violating the Rules for Using IT in the Health Field and Grievances Against Them

6.1 Disciplinary Sanctions

The health authorities, each according to its spatial competence, may impose the disciplinary penalties mentioned below on facilities that provide health services or those that work in the health research field, as well as facilities that are authorized to use the central system, if they violate the provisions of the law, its executive regulations, or the decisions issued for its implementation (Article 25):

A- Written notice.
B- A written warning.
C- A fine of not less than (1,000) AED and not more than (1,000,000) AED.
D- Temporary suspension of authorization to use the Central System for a period not exceeding six months.
E- Cancelling the permission to use the Central System.

It should be noted that imposing the previous disciplinary sanctions does not preclude imposing the penal sanctions stipulated in any other legislation.

6.2 Grievance of Disciplinary Sanctions

Article 26 of the Law permitted grievances against the aforementioned disciplinary sanctions as follows:

1- A committee shall be established at the Health Authority to examine grievances against disciplinary penalties. The formation of the committee, its terms of reference, and the method of submitting a grievance to it shall be determined by a decision of the Health Authority.
2- A person against whom a decision of one of the aforementioned disciplinary penalties has been issued may appeal against the decision before the grievance committee formed at the Health Authority, within (15) days from the date of the complainant's knowledge of the decision.
3- The grievance must be decided within (30) days from the date of its submission, and failure to respond to the grievance during that period is considered as a rejection of it.
4- The decision issued by the Committee on the grievance shall be final.
5- A person whose grievance has been rejected may appeal before the competent courts of the State within (30) days from the date of being notified of the rejection of the grievance or the expiry of the thirty-day period referred to in Clause (3).

7 Conclusion

At the conclusion of this paper, which is related to the legal regulation of controls for the use of ICT in the health fields in UAE, we conclude that: The effectiveness and success of the use of ICT in the health fields depend on the existence of a legal umbrella that guarantees the effective participation of all authorities. The UAE legislator set the legal framework for the Central System for exchanging health data and information, in terms of its establishment, accession to it, and its use controls. The legislator imposed controls related to the preservation of health data and information. Furthermore, the legislator stressed preserving the confidentiality of patient information and taking all measures to ensure that the right of information privacy is not compromised in all circumstances, except in the context of cases stipulated by law or with the consent of the patient. The legislator has established some disciplinary penalties for violating the rules governing the use of ICT in health fields. The legislator also guaranteed the right to complain about disciplinary sanctions before the specialized committee formed for this purpose, as well as before the competent court.

Finally, after shedding light on the regulations for using ICT in the health field, this paper opens future prospects for other researchers to carry out specialized studies on the legal rules for protecting patients' privacy and monitoring the use of the central health information system.

References

1. CHIRON: What is Telemedicine? https://chironhealth.com/telemedicine/what-is-telemedicine/. Accessed 21 Dec 2022
2. Mossialos, E., Thomson, S., Ter Linden, A.: Information technology law and health systems in the European Union. Int. J. Technol. Assess. Health Care **20**(4), 498–508 (2004)

3. The International Society for Telemedicine & eHealth (ISfTeH) is a non-governmental organization, with official relations with the World Health Organization (WHO). The association is headquartered in Switzerland and works to facilitate the dissemination of international knowledge and experience in the field of telemedicine and e-health and to provide access to globally recognized experts in the field of telemedicine. https://www.isfteh.org/about. Accessed 21 Dec 2022
4. Suhaimi, A., Wasliman, I., Sauri, R.S., Kosasih, U.: Management of improving the quality of social science learning based on information and communication technology in South Jakarta junior high schools. Int. J. Sci. Educ. Technol. Manag. (IJSETM) **2**(2), 1–10 (2023)
5. Article 2 of the Law
6. Article 3 the of Law
7. Durmuş, V., Uydaci, M.: A legal framework for healthcare: personal data protection for health law in Turkey. In: Research Anthology on Privatizing and Securing Data, pp. 1153–1170. IGI Global (2021)
8. Article 4 of the Law
9. Article 5 of the Law
10. A natural person is a person whose personality begins from birth and ends upon death. During his life, he enjoys rights and can assume liability according to the law. As for the legal persons, they are entities of establishments that arise from a group of funds and persons who have an independent self-identity with a goal to operate this company or establishment to achieve the goals for which it was established and enjoy legal personality as a natural person. Examples: an institution, university, hospital, school or association a cooperative or even a small shop. Law Corner, Natural Person and Legal Person: Define and Their Differences, 26 November 2020. https://lawcorner.in/natural-person-and-legal-person-define-and-their-differences/. Accessed 21 Dec 2022
11. Article 8/1 of the Law and Article 4/1 of the Executive Regulations
12. For example, the report issued by the UAE Business Council regarding the healthcare sector. https://usuaebusiness.org/wp-content/uploads/2019/01/2021-U.A.E.-Healthcare-Report.pdf. Accessed 15 Jan 2023
13. For example, the annual statistics issued by MOHAP. Ex. AE Statistical Annual Report 2020. https://mohap.gov.ae/assets/download/fb039118/UAE%20Statistical%20Annual%20Report%202020.pdf.aspx. Accessed 15 Jan 2023
14. Tian, S., Yang, W., Le Grange, J.M., Wang, P., Huang, W., Ye, Z.: Smart healthcare: making medical care more intelligent. Glob. Health J. **3**(3), 62–65 (2019)
15. Clayton, E.W., Embí, P.J., Malin, B.A.: Dobbs and the future of health data privacy for patients and healthcare organizations. J. Am. Med. Inform. Assoc. **30**(1), 155–160 (2023)
16. Article 8/2 of Law, and Article 4/2 of Executive Regulations
17. Articles 14 and 15 of the Law

Effect of Advertising Elements on Facebook on the Mental Image of the Services of Jordanian Telecommunications Companies. (Case Study on the Customers of Umniah Company in Zarqa City)

Khaled Tawfiq Al-Assaft[1](✉), Hasan Mohammad Issa Al-Ghowairi[2], and Bader Albatati[3]

[1] Faculty of Economics and Administrative Sciences, Marketing Department, Zarqa University, Zarqa, Jordan
kassaf@zu.edu.jo
[2] Faculty of Economics and Administrative Sciences, Zarqa University, Zarqa, Jordan
[3] College of Advertising, University of Business and Technology, Jeddah, Kingdom of Saudi Arabia
b.albatati@ubt.edu.sa

Abstract. This study aimed at identifying the effect of the advertising (commercial) elements through the Facebook on the mental image of the Jordanian telecom companies; case study on the customers of Umniah Company in the City of Zarqa. To achieve the study objectives, the researchers adopted the descriptive, analytic method. They developed an instrument to collect the data through a questionnaire, which was distributed through the simple, random sampling method over (700) customers of Umniah Company, who use the Facebook in the City of Zarqa, out of which (400) analyzable questionnaires were retrieved. The researchers further employed the convenient statistical methods applying SPSS 22 Program for data analysis. The study concluded many results such as: there is a statistically significant effect at ($a \leq 0.05$) level for the commercial elements, collectively or severally, through the Facebook platform, in shaping the mental image of the services of Unmiah Company. Furthermore, the commercial elements through the Facebook platform interpret (75.2%) of the changes that take place in coining the mental image of the services of Umniah Telecom Company. In this concern, the intensives component is the most influential in the mental image with the customers of Umniah Company. Therefore, the study recommended the need to place further care by Umniah Telecom Company to the mental image formed by the customers through the company advertisements on the Facebook; particularly, the intensives and convenience variables due to their high influence in forming the mental image.

Keywords: Advertising elements on the Facebook · mental image · Umniah Telecom Company · Jordan

B. A. M. Alareeni and I. Elgedawy (Eds.): *Artificial Intelligence (AI) and Finance*, SSDC 488, pp. 219–227, 2023.
https://doi.org/10.1007/978-3-031-39158-3_20

1 Introduction

The report issued by Digital in network in 2020 provided that the number of the internet users worldwide has now exceeded 4.5 billion; 3.8 billion of them are active on the social media. This means that over 60% of the earth populations are users of the social media. Last year, the users through the internet increased by 300 million, and every user spent about 6 h, 45 min daily (on average). In a simple arithmetic process, the users spent 1.25 billion hours on the internet in 2019. The report about Jordan further showed that there were 6.8 billion internet users by the beginning of 2020, with 1.2% increase (79 thousand users over 2019), and with a prevalence rate amounting 67% of the total population of the Hashemite Kingdom of Jordan. Furthermore, the users of the social media platforms in the Kingdom amounted 5.7 million, representing 65% of the total population for the same period [1] (The customers form the mental image about any issue through the indirect experience, which means the personal messages that the individual hears from the friends, the events, the individuals or the company commercials (advertisements) [2] The mental image with the individual about the organizations is shaped based on the strengths and weaknesses of the communications that the organization builds. Therefore, attention grew in measuring the effect of the advertisement components, such as: information, credibility, entertainment, incentives, lack of nuisance (convenience) to identify their effects in generating the required change about the organization among the customers [3–8],indicates that there is an increasing interest by the companies to form a positive image in the mind of the customer, due to its wide effect in the required behavior. The volume of the social media platforms users, who are exposed to the commercials in the Hashemite Kingdom of Jordan, attracted the attention of the researchers, which was referred to in [1]. In more specific details, the numbers of the users of the Facebook were: 5.5 million, Instagram: 2.7 million, Linked: 1.1 million, Snap Chat: 2.75 million and Tweeter: 0.488 million. It is clear that the Facebook platform is the number one for the users in the Hashemite Kingdom of Jordan. Accordingly, the expansion of the use of the social media for the advertisement with all its elements draws the attention to investigate its effect on forming the mental image with the targeted customers. Based on the above, the question can be raised about the effect of advertising with all its components via (Facebook) in shaping the mental image of Umniah Telecom services, which form the following main study question:

What is the effect of advertising through the Facebook platform on forming the mental image of the services of Umniah Telecom Company?

2 The Theoretical Framework and the Previous Studies

2.1 Advertising Components

A number of researchers dealt with the impact of the information contained in the advertisement on the mental image towards the trademark, and differentiated between advertisements that focus on arousing emotional purchase motives, and those that focus on arousing mental purchase motives. Those that focus on arousing the mental purchase motives aim at convincing the customers of the reasons of using the product, and providing them with the logical and objective information, the functions that the produce performs, and satisfying the customers' needs. Meanwhile, the emotional ads focus on the individual's self-satisfaction, his social status, and his emotional evaluation of the product [9]. Furthermore, [6] found a positive effect of the information, credibility, entertainment, incentives factors; and negative effect of the inconvenience element, on the customer's trends toward the advertisement. The elements that impact the mental image of the product can be defined, as per the results of: [3, 4, 10–12], and), as follows:

- **Informativeness:** "The ability of the advertisement to inform the targeted audience about the product alternatives, so that they can carry out the purchase processes that realize their highest possible level of satisfaction."
- **Credibility**: "It is focused on the extent to which the targeted audience feel that the claims related to the advertized product are honest and believable."
- **Entertainment**: "Is focused on the ability of the advertisement to change the emotional feelings of the audience towards a sense of beauty and pleasure."
- **Incentives**: "Are the advertisements that offer certain financial rewards to the targeted audience that meets their conditions."
- **Lack of Irritation in the advertisement (Convenience)**: "Is the uncomfortable advertising content that makes the targeted audience feel towards it with a negative sense." It was termed "Convenience".
- **The mental image:** The "Mental Image" term emerged in 1915 by the scientist Graham Dallas, who indicated in his book "Human and Political Nature" that the researchers are in need for a simplified, permanent and organized thing. As a term, it is related to the commercial companies, but it was not used until the second half of the twentieth century. Then it was soon used in the various political, media and professional fields. With the advancement in the information and communications, the mental image term entered the different research areas related to the public opinion. Then interest in it increased with the great progress in the fields of communications and information, which turned the world into a small village, in which all social, environmental and administrative sciences are pumped. The mental image means, [13], the image that is shaped in the minds of people about the different companies. It may be formed based on the evidences and documents, or the rumors and undocumented sayings; but in the end forms a true reality from the views of those who hold in their minds. [14], asserts that the mental image is a symbolic one in which the needs, demands and interests of the masses gather. This image cannot be formed within an overnight, as its substance deposits gradually in the mind; and it is not the translation of the deeds, sayings and behaviors of any company. It is not responsible about these sayings and deeds as much as it is a translation of the reactions that are generated by these deeds and sayings. The dimensions of the mental image, as sorted by Hijab [15] are as follows:
- **Cognitive dimension**: Is the dimension through which the individual perceives a certain subject related to a certain country, nation, people or company, such as knowledge of the historical and geographical bases, and knowledge of general information about them. This information is the base upon which the mental image, formed by the individual about the others, is built, as well as about the subjects and different issues. Depending on the accuracy of the information and knowledge you obtain about others, the accuracy of the mental image you formed about them will be. And,

depending on the cognitive dimension, the errors formed in the image of the individuals are errors basically resulting from the wrong information and knowledge these individuals gained.

- **Emotional dimension**: It represents the individual's feelings and emotions towards a certain state, people or company, which extends from acceptance to rejection, and ranges in intensity between positivity and negativity, where the emotional aspect is formed along with the cognitive aspect. Over the time, the information and knowledge that individuals created fade and the emotional aspects that represent individuals' trends towards different people, issues and topics remain. This is affected by a number of factors, most important is the limits of the available knowledge sources. Furthermore, building the cognitive dimension is affected by the peoples' characteristics in terms of color, gender and language. Difference in these characteristics is among the issues that contribute to building negative trends; and, on the other hand, homogeneity of these characteristics contributes to building positive trends.
- **Procedural dimension**: It is represented in the individual's desire to travel to a foreign country, live, work or get married from there. This factor depends on the social distance between the individuals or the desire to work with a company and belong to it.

The mental image is not limited to mere perception, but it is connected with the trends, attitudes, and practical responses in which people deal with things and persons around in the real life. To the extent that these images affect a person's perception of his surroundings, they affect his judgment on those things, because they form a part of its indicative framework [16–18]. Provided a number of factors that form the mental image of the companies, as follows:

- **Personal Factors**: Represented in the future self and personal traits of the information, education, culture, values, ability of the person to interpret the information of the companies, and his motivation degree and interest in the information provided about them.
- **Social factors**: Represented in the initial influences of the groups on the person who received the information during their exchange of the information in their personal communications about the company; the influence of opinion leaders on the attitudes of the masses; the influence of the culture of the society in which individuals live, and the prevailing values in it[19].
- **Organizational factors**: Represented in the company management strategy that reflects its philosophy, culture, the actual businesses of the company, its policy, products and its overall communication network. This includes all the internal and external communications of the company with its masses, its own communicative messages that are transmitted through the different communication means, the direct personal contacts between its workers and the masses, and the social activities that the company carries out to serve the community.

2.2 The Study Hypotheses

The study suggests the following main hypothesis:

(H0): There is no statistically significant effect at ($a \leq 0.05$) significance level of the advertisement components through the Facebook in shaping the mental image of services of Umniah Telecom Company.

Out of this hypothesis, the following sub-hypotheses ramified:

- **($H0_1$):** There is no statistically significant effect at ($a \leq 0.05$) significance level of the ads information on the Facebook platform in forming the mental image of the services of Umniah Telecom Company.
- **($H0_2$)**: There is no statistically significant effect at ($a \leq 0.05$) significance level of the ads credibility on the Facebook platform in forming the mental image of the services of Umniah Telecom Company.
- **($H0_3$):** There is no statistically significant effect at ($a \leq 0.05$) significance level of the ads entertainment on the Facebook platform in forming the mental image of the services of Umniah Telecom Company
- **($H0_4$)**: There is no statistically significant effect at ($a \leq 0.05$) significance level of the ads incentives on the Facebook platform in forming the mental image of the services of Umniah Telecom Company
- **($H0_5$)**: There is no statistically significant effect at ($a \leq 0.05$) significance level of the ads convenience on the Facebook platform in forming the mental image of the services of Umniah Telecom Company

3 Study Methodology

This study is based on two of the scientific research methods, namely: The descriptive, analytical method, Field research method.

3.1 Study Population and Sample

The study population consisted of the customers of Umniah Company in the Hashemite Kingdom of Jordan, who use the Facebook. Due to the wide volume of the study population, which may exceed one million subscribers, and according to Robert Mason Equation [20], the required size of the sample to represent the study population is (385) respondents. Because of the huge number of the population and its geographical spread out, the researchers elected the simple random sampling method. Accordingly, (700) questionnaires were distributed over the study sample, electronically; (175) questionnaires were disqualified due to lack of complete filling, and (125) questionnaires were also disqualified due to lack of seriousness in their data, which was clear of the respondent's selection of the same response for all the items. As a result, (400) questionnaires were adopted to analyze and extract the results for the purposes of this current study.

4 Results of the Study Questions and Test of Its Hypotheses

The main question is: What is the effect of advertising through the Facebook platform on forming the mental image of the services of Umniah Telecom Company?

The answer to this question lies in testing the main hypothesis of the study:

Main hypothesis (H0): There is no statistically significant effect at (a ≤ 0.05) significance level of the advertisement components through the Facebook in shaping the mental image of services of Umniah Telecom Company.

The researchers applied the multiple linear regression test to test this hypothesis, and Table 1 explains the test results.

Table 1. Multiple Linear Regression Analysis

Dependent Variable	R	R^2	F	Sign	Independent Variable	B	T	Sign
The Mental Image of Umniah Telecom Company Services	0.867	0.752	238.944	0.00	Information	.088	2.195	.029
					Credibility	0.005	−0.12	.904
					Entertainment	.171	4.187	.000
					Incentives	.334	7.714	.000
					Convenience	.305	7.375	.000

Table 1 shows that (F) value is (238.944) and its significance level is (0.00), which is less than (0.05) significance level, which means the rejection of the nihilism hypothesis that "There is no statistically significant effect at (a ≤ 0.05) significance level of the advertisement components through the Facebook in shaping the mental image of services of Umniah Telecom Company". Rather, we accept the alternative hypothesis that "There is statistically significant effect at (a ≤ 0.05) significance level of the advertisement components through the Facebook in shaping the mental image of services of Umniah Telecom Company." We can also conclude that there is an effect of the dimensions of the independent variable (advertisement components on Facebook) collectively, on the dependent variable (the mental image of the services of Umniah Telecom Company). These dimensions explained (75.2%) of the changes occurring in the mental image of Umniah Telecom Company's services.

First Sub-Hypothesis ($H0_1$): There is no statistically significant effect at (a ≤ 0.05) significance level of the ads information on the Facebook platform in forming the mental image of the services of Umniah Telecom Company.

The researchers relied on the simple linear regression test to test this hypothesis as shown in Table 2.

Table 2. Simple Linear Regression of the First Sub-Hypothesis

Dependent Variable	R	R^2	Independent Variable	B	T	Sign
The Mental Image of Umniah Telecom Company Services	0.719	0.516	Information	0.65	20.614	0.00

(continued)

Table 2. (*continued*)

Dependent Variable	R	R^2	Independent Variable	B	T	Sign
The Mental Image of Umniah Telecom Company Services	0.703	0.494	Credibility	0.623	19.726	0.00
The Mental Image of Umniah Telecom Company Services	0.767	0.585	Entertainment	0.658	23.856	0.00
The Mental Image of Umniah Telecom Company Services	0.819	0.671	Incentives	0.720	24.489	0.00
The Mental Image of Umniah Telecom Company Services	0.767	0.589	Convenience	0.809	23.873	0.00

Table 2 shows that a statistically significant effect of the (information, Credibility, Entertainment, Incentives, Convenience) in shaping the mental image of services of Umniah Telecom Company.

Therefore, we reject the nihilism hypothesis that, Rather, we accept the alternative hypothesis that.

5 Discussion

The results of this study showed an influence of the independent variable: (Ads components on the Facebook), collectively, on the dependent variable: (the mental image) of the services of Umniah Telecom Company; as they explained (75.2%) of the changes occurring in the mental image. The results of our study are in line with those of Al-Raff and Attiyeeh [21], on the existence of an effect of the ads on the individuals' behaviors; and with those of Fardi and Balbazouh [22]in that the message and role of the ads in improving the image are clear in revealing the audience trends and needs, as well as the external environment surrounding the company. Our results are in line with the study of Arora & Agarwal [3]on the existence of a strong effect of the credibility, information and entertainment in the ads on the social media platforms on the mental image, and the trends of the study sample toward the products. The results further showed a statistically significant effect of the information as one of the ads components in shaping the mental image of Umniah Telecom Company services. Our results are also in line with the study of Hamouda [4]which found a correlation between the information and the value of the commercial; and in line with the study of Dao et al. [23], on that the information have positive effects on the perceived value the customers bear about the commercial on the social media. In this concern, the results indicate a statistically significant effect of the credibility as one of the ads components in shaping the mental value of Umniah Telecom Company services.

On the other hand, these results are also in agreement with the results of previous results, such as Al-Raff & Atiyyah [21]; Arora & Agarwal [3]; Gaber [24]; and Dao, et al. [23] on the existence of an effect of the credibility in the ads on the consumer. Furthermore, the results showed statistically significant effect of entertainment as one of the ads components in shaping the mental image of Umniah Telecom Company services. In this concern, our study is in line with the results of previous studies, such as Al-Raff & Atiyyah [21]; Arora & Agarwal [3]; Gaber [24]; and Dao, et al. [23] on the existence of an effect of the entertainment in the ads on the consumer. The results further showed a statistically significant effects of the incentives as one of the ads components in shaping the mental image of the services of Umniah Telecom Company; which are in line with the results of the abovementioned previous studies on the existence of an effect of the incentives in the ads on the customer, as well as a statistically significant effect of convenience as a component of the ads in shaping the mental image of the services of Umniah Telecom Company, in addition to an effect of convenience in the ads on the customer.

6 Recommendations

In the light of the results of the study, the researchers recommend the following:

- The need for Umniah Telecom to pay attention to the mental image formed by individuals, through the company's ads on (Facebook), especially for the incentives and convenience variables, because of their great influence in shaping the mental image.
- The need for Umniah Telecom to pay attention to the credibility component in its ads on (Facebook), as the results showed weakness of this variable as compared to the other variables in the ads of the Company.
- Umniah Telecom Company should carry out a study on the ads components on (Facebook) well, before introducing these commercials, because of their high correlation with shaping the mental image, which is reflected on the customer's behavior.
- Umniah Telecom Company has to search for the best advertising means, to positively influence shaping the mental image of its products, especially the means through which it can provide more information about its services

References

1. Digital in.: we are a social report, Essential Insights into the Internet, Social Media, Mobile, and E-commerce Use around the World, Hootsuite (2020)
2. Al-Hmoud, A.: The Mental Image of the Service Institutions among the Jordanian Public, Amman Greater Municipality as a Model, (Unpublished MA Thesis) Middle East University, Jordan (2019)
3. Arora, T., Agarwal, B.: An empirical study on determining the effectiveness of social media advertising: a case on Indian millennials. Int. J. e-Bus. Res. **16**(2), 47–68 (2020)
4. Hamouda, M.: Understanding social media advertising effect on consumers' responses: an empirical investigation of tourism advertising on Facebook. J. Enterp. Inf. Manage. **31**(3), 426–445 (2018). https://doi.org/10.1108/JEIM-07-2017-0101

5. Knoll, J.: Advertising in social media: a review of empirical evidence. Int. J. Advertising **35**(2), 266–300 (2016)
6. Tsang, M.M., Ho, S.-C., Liang, T.P.: Consumer attitudes toward mobile advertising: an empirical study. Int. J. Electron. Commer. **8**(3), 65–78 (2004)
7. Imran, M., Ali, S., Shahwan, Y., Zhang, J., Al-Swiety, I.A.: Analyzing the effects of renewable and nonrenewable energy usage and technological innovation on environmental sustainability: evidence from QUAD economies. Sustainability **14**(23), 15552 (2022)
8. Hasaballah, M.: Effect of the social media sites in enhancing the mental image of the telecom companies. J. Media Stud. (12) (2020)
9. Heath, R., Feldwick, P.: Fifty years using the wrong model of advertising. Int. J. Mark. Res. **50**(1), 29–59 (2007)
10. Al Assaf, K.T.: Impact of advertising messages across social networks on consumers' purchasing behavior of mobile phones: a study amongst youth in Jordan. Int. J. Sales Mark. Manage. Res. Dev. (IJSMMRD) **7**, 9–14 (2017)
11. Shahwan, Y., Abdel-hamid, O.: The impact of social responsibility disclosure in reducing the risk in Jordanian commercial banks. Res. World Econ. **11**(3), 142–150 (2020)
12. Arya, M., Kerti, Y.: Advertising value of Instagram stories & the effect on millennial's attitude. RJOAS **99**(3), 29–39 (2020)
13. Al Assaf, D., Tawfeq, K., Al-Momani, D., Musa, M.: The influence of e-marketing technology and website features on attracting students to private universities in Jordan (case study: Zarqa university). Int. J. Manage. **11**(9) (2020)
14. Barqouq, A.R.: Introduction to Public Relations, Algiers, Al-Khaldounia for Publishing and Distribution (2010)
15. Hijab, M.M.: Active Communication of the Public Relations, Amman, Dar Al-Fajr for Publishing and Distributing (2007)
16. Shahwan, Y.: The mediating effect of investment decisions and financing decisions on the influence of capital structure against corporate performance: evidence from Jordanian listed commercial banks. Acad. Acc. Finan. Stud. J. **22**(6), 1–20 (2018)
17. Al-Nawafleh, Y.: The Communicative Role of Public Relations in Shaping the Mental Image of Pharmacists, the Jordanian Al-Hikma Pharmaceutical Company as a Model, (Unpublished MA Thesis) Middle East University (2020)
18. Baqer, M.: Effect of the Mental Image in Public Relations, Amman, Dar Osama for Publishing and Distribution (2014)
19. Al Assaf, K.T., Hudaib, M.: The Role of Social Marketing in the Prevention of Corona Virus (Covid-19) in Jordan (2021)
20. Mason, R.L., Gunst, R.F., Hess, J.L.: Statistical Design and Analysis of Experiments: with Applications to Engineering and Science, 2nd edn. J. Wiley, Hoboken, NJ (2003)
21. Al-Raff, Z., Atiyyah, K.: Effectiveness of the Electronic Advertisement on Influencing the Image of the Tourist Destination, Contemporary Economic Studies, (1), Algiers (2018)
22. Fardi, E., Balbzouh, S.: Role of the Advertisement in Improving the Mental Image of the Service Institution among Its Foreign Audience, (Unpublished MA Thesis), Larbi Ben M'hidi University, Oum el Bouagh, Algiers (2015)
23. Van-Tien Dao, W., Nhat Hanh Le, A., Ming-Sung Cheng, J., Chao Chen, D.: Social media advertising value. Int. J. Advertising Rev. Mark. Commun. **33**(2), 271–294 (2014)
24. Gaber, H., Wright, L.T., Kooli, K.: Consumer attitudes towards Instagram advertisements in Egypt: the role of the perceived advertising value and personalization. Cogent, Bus. Manage. **6**, 161843 (2019)

Impact of XBRL Technology on Quality of Financial Data: Mediating Through Various Reporting Aspects in India

N. Abhishek[1(✉)], Habeeb Ur Rahiman[2], Abhinandan Kulal[3], Ujwala Kambali[1], Velita Sibon Lobo[1], D. Bindu[1], and M. S. Divyashree[4]

[1] Srinivas University, Mangalore 575001, India
abhishekalmighty93@gmail.com
[2] Kingdom University, Riffa, Bahrain
[3] Mangalore University, Mangalore 575001, India
[4] Government First Grade College Uppinangady, Mangalore 574241, India

Abstract. Due to the rapid development of technology, all the functions of business organization are also rapidly transforming in an efficient and effective manner to account for and communicate all the transactions of the business since from purchase and material and to the point of distribution of final dividends to Shareholders, there is a necessity of having advanced technology. XBRL is one of the global-level business reporting standards which accommodates all business organizations to report and communicate these business affairs in a convenient manner. Further, it allows business organizations to report and communicate their financial and non-financial affairs in a customized manner. The present study is unique among XBRL literature and tries to make an attempt to analyze the influence of XBRL and Its functionality on various aspects of financial reporting and its quality at large. For this purpose the study is based on the perception of chartered accountants in India. The structural equation model is used to test the conceptual model, and the outcome of the study indicated that the XBRL and its functionality have a significant influence on various aspects of financial reporting. Further, the various aspects of financial reporting positively influence on financial reporting quality. The outcome of this study will be the input for the formulation of policy on financial reporting through initiation of XBRL.

Keywords: XBRL · Financial Reporting · Financial reporting quality · Quality of financial data · mediation analysis · Digital accounting · Digital reporting

1 Introduction

Due to rapid development of technology all the functions of business organization is also rapidly transforming in an efficient and effective manner to account and communicate all the transactions of business since from purchase & material and to the point of distribution

Paper type: Research paper.

B. A. M. Alareeni and I. Elgedawy (Eds.): *Artificial Intelligence (AI) and Finance*, SSDC 488, pp. 228–239, 2023.
https://doi.org/10.1007/978-3-031-39158-3_21

of final dividends to Shareholders, there is a necessity of having advanced technology. XBRL is one of the global-level business reporting standards which accommodates all business organizations to report and communicate these business affairs in a convenient manner [1, 2]. Further, it allows business organizations to report and communicate their financial and non-financial affairs in a customized manner.

Customization of business reports is one of the unique features of XBRL, which allows the firm to report and communicate business and financial information as per the divergence need of various Stakeholders. This minimizes the overall clerical, verification and Compliance burden on organization. This in turn, reduces overall business reporting and communications costs and also reduces the consumption of papers for printing and publishing financial reports [3, 4]. This indirectly helps the organizations to adopt eco-friendly reporting initiative as XBRL based helps to reduce the number of tree cuts on average. This also becomes one of the sustainable development initiatives of organizations [5, 6].

Apart from this XBRL also helps to bring uniformity in financial reporting across globe. This is because it is built based on the accounting policies, procedures and GAAPS. Further, it also assists the regulatory authorities and reporting firms either in adoption or convergence of global business reporting standards like IFRs, GRI, IIRC, etc., [7]. Further, XBRL as tool certainly enhances the quality of financial reporting. Here, quality of financial reporting means the ability of financial reports to various information needs of stakeholders [8–11, 28–30].

The present study is unique among XBRL literature and tries to make an attempt to analyze the influence of XBRL and Its functionality on various aspects of financial reporting and its quality at large for this purpose the study is based on the perception of chartered accountants in India.

The succeeding part of this paper is organized as literature reviews, objectives, conceptual framework, methodology, results, discussions and conclusion.

2 Literature Review

2.1 XBRL Technology for Financial Reporting

[12] Explained the importance of recording and sharing financial information of a specific time period with their stakeholders, which we consider as financial reporting and according to a study by [13, 30–33] the technological advancement has enabled digital reporting that includes guidelines for electronic data entry, processing, and dissemination of financial and other business information. It has thus increased the transparency of firms and financial markets through the distribution of company information via the Internet. In a study by [1], highlighted that the sole purpose of employing XBRL for financial reporting is to enhance the quality of publicly disseminated financial and non-financial information to various stakeholders. Further, in another study by [4], it is noted that the prime concern behind the development and implementation of XBRL technology for financial reporting was to ensure transparency, reliability, accuracy, comparability and efficiency of annual financial reports communicated to end-users. In addition, there were studies [3, 14, 15, 29, 34, 35] which stressed another important concern behind adoption of XBRL technology for financial reporting was to eliminate the diversity and

bring uniformity in financial reports of various companies in a capital market. In other study by [14, 22], it is highlighted that the XBRL technology is not only helpful for streamline the financial reporting process and it also benefits for auditing process in various aspects [16]. In their study pointed that the development of XBRL technology in a business reporting environment addresses the various challenges such as accessibility, usefulness and interoperability of financial reports. On the other hand, [2] found that the application of XBRL in business reporting certainly minimizes the administrative burden of both reporting organization and regulatory authorities at large. Further, they also stressed that the usefulness of financial reports is also can be enhanced [17]. In their study highlighted that technological advancement has an influence on the way companies share information to their stakeholders and eXtensible Business Reporting Language (XBRL) allows the financial data to be shared globally using a digitalized global standard language. The recent study by [11] reveals that the advent of XBRL technology has directly influenced the traditional mode of financial reporting and has contributed to better financial reporting.

2.2 XBRL Promotes Quality Reporting and Information Asymmetry

As noted by few authors [8–11, 18, 19], XBRL has helped companies in improving the quality of financial data and lower information asymmetry giving all the stakeholders equal rights to use all the necessary financial information quickly and effectively with the help of executive team's financial reporting and information technology competencies.

2.3 XBRL Improves Quality of Sustainability Reports and Corporate Governance Standards

[5, 6] In their study have proposed a framework for sustainability reports using the eXtensible Business Reporting Language (XBRL), which can meet the requirements of the Global Reporting Initiative's (GRI) sustainability reporting guidelines. In studies done by [5] finds evidence that IAF involvement in XBRL implementation improves corporate governance quality. [20] In his thesis studied on employing XBRL in corporate governance reporting emphasizing its importance to regulators, policymakers, associations, and NGO's that develop, support, and issue CG taxonomies. In other study by [21] it was noted that XBRL improves corporate governance by enhancing transparency. The use of XBRL in corporate financial reporting will promote accountability and information quality as the standardised XBRL format will make it easy to use information across multiple platforms and applications, making information more transparent and increase the ability of users to search more quickly and precisely.

2.4 Impact of XBRL on Audit Fees and Cost

[22] Has opined that XBRL can make auditing easier, and adopters of XBRL are likely to pay lower audit fees. This impact is proportional to firm size, with larger firms benefiting more than smaller firms [23]. In their study have highlighted the influence of International Financial Reporting Standards (IFRS) and eXtensible Business Reporting Language

(XBRL) on audit fees , it was found that XBRL reduces the audit fees and IFRS raises audit fees for all companies. Another study by [4] noted that XBRL financial reporting will be the global finance industry's standard reporting language helps commercial banks reduce the cost of explaining accuracy and reliability while also making it easier for its stakeholders of information to judge finance. As noted by [4, 25, 26] artificial intelligence integrates financial data with advanced technologies to boost the digitalisation of banking and accounting, eliminating human error, creating a more stable business and economic environment.

By studying past literature on the diverse components outlined above, it has been revealed that the methods of financial reporting have evolved over time, and the application of XBRL in financial reporting has made reporting more transparent and convenient to all stakeholders. It has aided businesses in lowering audit costs and increasing accuracy. The studies for instance, [10, 11, 19] also indicate that more efforts are required to explore the evidential impact of XBRL on financial reporting process and its quality. This is need to be done both by considering the study of financial reports and perception of practicing accountants and auditors. Many studies for instance, Baldwin and [4, 11, 16, 27] made effort in analysing XBRL and its influence on financial reporting in a conceptual way. Therefore, the present study to fill this gap is intended to analyse the influence of XBRL and its functionality on various mediating aspects of financial reporting and its quality.

Based on the extensive literature survey, research gap and conceptual framework following questions are framed to address the issues in the current study.

1. How do XBRL and its functionality influence various aspects of financial reporting?
2. How do various aspects of financial reporting influence a mediating effect on financial reporting quality?

To address the stated research questions following objectives are framed:

1. To study the influence of XBRL and Its functionality on various aspects of financial reporting.
2. To analyse the mediating effect of XBRL-based financial reporting on financial reporting quality.

Based on literature review, research gap, questions, objectives and conceptual framework following hypotheses were framed;

H_1: Use of XBRL positively influence on financial reporting.

H_2: Functionality (OXI) of XBRL positively influence on financial reporting.

H_3: Financial reporting through XBRL enhances financial reporting quality.

H_4: use of XBRL for financial reporting mediates the improved financial reporting quality.

H_5: Functionality of XBRL for financial reporting mediates the improved financial reporting quality.

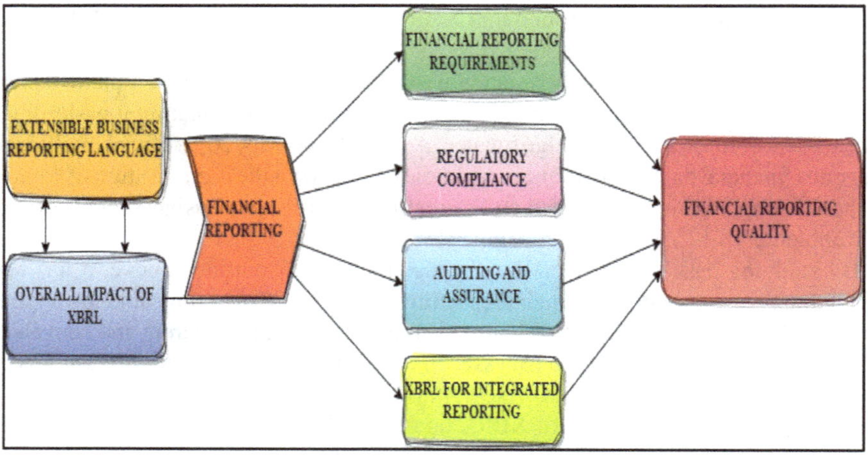

Fig. 1. Conceptual Model for the study. Source: *Compiled by Researcher*

3 Research Design

3.1 Sample and Participants Profile

Several authors [36, 37, 38] in their study highlighted that the perception survey is most recommended method in user studies. Therefore, the present research is based on primary survey data gathered through online survey. The questionnaire was developed with the help of Google forms and circulated through mailing to 9,682 chartered accountants who are the members of ICAI (Institute of Chartered Accountants of India). The chartered accountants are the most suitable population for the research on XBRL based financial reporting as they are the preparers and certifiers of financial reports [3, 26]. A total of 267 responses were received at a response rate of 2.75%. Out of collected responses 217 are completed responses and all are considered for further analysis.

The profile of respondents is varied. 156 respondents (70.72%) being male and 61 respondents (29.28%) being female, with 48 respondents (22.12) under 5 years of service, 78 respondents (35.94%) between 6 to 10 years, 59 respondents (27.19%) between 11 to 20 years, 13 respondents (5.99%) between 21 to 30 years, and 19 respondents (8.76%) above 31 years. Accounting and auditing professionals (63.14%), tax advisors (19.35%), top management (8.29%), regulatory positions (2.76%), and others (6.47%) were among those who responded. In our survey, 28 people work for publicly traded companies. There were 13 unlisted companies, 155 practising accountants, and 21 Big-4 corporations (Table 1).

3.2 Research Instrument

The main purpose of the study is to analyse the influence of use of XBRL and its functionality on various aspects of financial reporting and their mediating effect on financial reporting quality by considering the perception of accountants and auditors. We developed a research instrument based on 5 point Likert scale gather their perception concerning to the objective of the study. The items used in the instrument are constructed based on many earlier studies. For instance, [4, 7, 10, 11, 19, 26, 27]. It contains 82 statements relating to 8 different aspects which are indicated in the Table 2.

Table 1. Shows the participants demographic profiles.

Demographic details	Respondents	Percentage (%)
Gender		
Male	156	70.72
Female	61	29.28
Total	217	100
Age of Service		
Below 5 years	48	22.12
6 to 10 years	78	35.94
11 to 20 years	59	27.19
21 to 30 years	13	5.99
Above 31 years	19	8.76
Total	217	100
Job Type		
Accountants and Auditors	137	63.13
Tax advisors	42	19.35
Top Management	18	8.29
Regulatory Position	6	2.76
Others	14	6.47
Total	217	100
Nature of Organization		
Listed company	28	12.90
Unlisted company	13	5.99
Big-4 company	21	9.67
Practicing Accountants	155	71.44
Total	217	100

Source: Authors calculations

Table 2. Research Instrument Details

Sl. No	Aspects	No. of. Statements	Source
1	eXtensible Business Reporting Language (XBRL)	7	[7, 14]
2	XBRL and its functionality (OXI)	13	[4, 19]
3	Financial Reporting (FR)	15	[4, 11, 24]
4	Auditing and Assurance (AA)	9	[7, 10, 14]
5	Management Compliance (MC)	6	[4, 19]
6	Regulatory Authority Requirements (RA)	12	[4, 7, 11, 14]
7	XBRL for Integrated Reporting (XIR)	6	[14, 24]
8	Financial Reporting Quality (FRQ)	14	[4, 7, 10, 11, 19, 26, 27]

Source: *Compiled by Authors*

3.3 Validity and Reliability of Research Instrument

The reliability of the instrument tested by Cronbach's alpha value and composite reliability (CR). Both Alpha values and CR of all construct greater than 0.6, which means items representing construct are highly reliable. Further, factor loading of each item is greater than 0.6 which means perfectly load to the respective construct. The convergent validity tested using Average Variance Extracted (AVE) and results are greater than 0.5, which means items are perfectly converge together to develop the construct. Additionally, Variance Inflation Factor (VIF) calculated to evaluate discriminant validity of the variables and results are less than 5. Therefore, there is no problem with discriminant validity.

3.4 Statistical Tools Used

The present study employed structural equation modelling for analysing the data and to test the hypotheses as per the objectives of the study.

4 Results and Discussion

4.1 Model Assessment

A hypothesized model developed with help of critical evaluation of literature and tested using structured equation model. The model explains that the use of XBRL and its unique functionality helps the organization to maintain an efficient financial reporting system which mediates to fulfil the various information requirements for diversified purposes such as general purpose financial reporting (FR), regulatory filings (RA), auditing and assurance (AA) and integrated reporting (XIRT) and management compliance (MC). These aspects in turn enhances the capability of financial reports to meet various information needs of stakeholders (FRQ).

4.2 Model Fit

The goodness of model is determined and it showed $R^2 = 0.666$ which is greater than 0.5, which means the predictive capability is established. Furthermore, the model fit assessed with other indicators, such as CMIN/DF is 2.031 (<3), Goodness Fit Index is .992 (>0.9), Adjusted Goodness of Fit Index is 0.930 (>0.9), Normative Fit Index is 0.987 (>0.9) and RMSEA is .068 (close to .06). All these indicators showed model is perfectly fit.

4.3 Hypotheses Testing

Both direct relation and mediation impact checked in the model and Table 3 shows the direct relation between the study variable. The result showed that XBRL have significant impact on F1 ($\beta = .226$, $p < 0.01$), OXI have significant impact on F1 ($\beta = .719$, $p < 0.01$). Therefore, both the hypotheses (H_1 & H_2) are accepted at 1% Significance level. Further, the relation between F1 and FRQ showed significant ($\beta = .519$, $p < 0.01$). Therefore, H_3 accepted at 1% significance level. All direct relations are depicted in Fig. 1. This means XBRL and its functionality have significant influence on various aspects of financial reporting. Further, the various aspects of financial reporting positively influence on financial reporting quality. This observation is consistent with many earlier studies. For instance, [11, 19, 26].

Table 3. Direct Relation

Hypothesis	Path	Estimate	P value	Remark
H1	XBRL -→ F1	.226	.000	*Null Hypothesis Rejected*
H_2	OXI-→ F1	.719	.000	*Null Hypothesis Rejected*
H_3	F1-→ FRQ	.519	.000	*Null Hypothesis Rejected*

4.4 Mediation Analysis

Mediation Analysis conducted to know the impact of XBRL and OXI on FRQ mediating through F1 (FR, RA, AA, XIRT). The result showed in Table 4 and it revealed that F1 significantly mediate the relationship between XBRL and FRQ ($\beta = .321$, $p < 0.01$) and significantly mediate the relationship between OXI and FRQ ($\beta = .117$, $p < 0.01$). Hence significant value less than 0.01, therefore H_4 and H_5 accepted at 1% significance level. This means XBRL and various aspects of financial reporting has significant positive mediating effect on financial reporting quality. In the same way functionality of XBRL and various aspects of financial reporting is also having positive mediating effect on financial reporting quality. Finally, it can be evidently said that XBRL is the potential tool for multidimensional business reporting and to enhance the financial reporting quality. This observation is consistent with many studies for instance, [4, 7, 22, 24, 27] (Fig. 2).

Table 4. Mediation Analysis

Hypothesis	Path	Total Effect		Direct Effect		Indirect Effect	
		Coefficient	P value	Coefficient	P value	Coefficient	P value
H4	XBRL→F1→FRQ	.526	.000	.205	.000	.321	.000
H5	OX7I→F1→FRQ	.373	.000	.256	.000	.117	.000

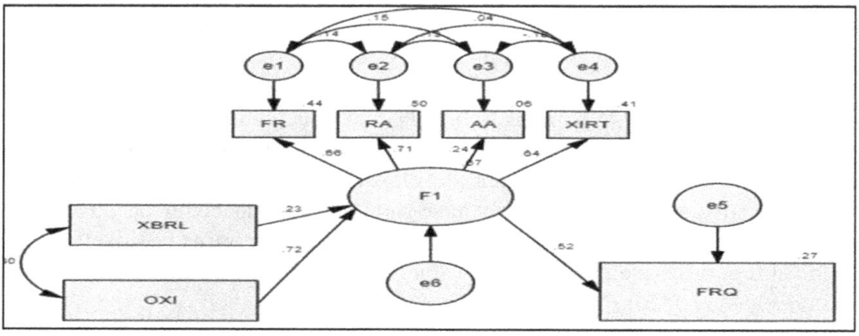

Fig. 2. Structured Equation Model

5 Conclusion

The main purpose of the present study was to analyze the influence of XBRL and its functionality on various aspects of financial reporting and to examine XBRL-based financial reporting and its mediating effect on financial reporting quality. The goodness of fit model is used to achieve this goal, which is determined by several indicators that reveal the model is perfectly fit, and structural equation modelling for analyzing the data and testing the hypotheses as per the objectives of the study. The analysis of data revealed that the XBRL and its functionality is producing mediating effect through various financial reporting aspects and positively influencing on financial reporting quality at large.

Studies, for instance, revealed that more efforts were required to make people aware of XBRL and that practitioners should be trained and taught how to use XBRL to enhance the efficiency of financial reporting, as per the previous study. In addition, the outcome of this study may become an input for policy-making by companies including small and medium-sized organizations for initiating XBRL in their business reporting process.

A major limitation of the present study is that small and medium-sized companies were ignored, and the study was limited to only the top 30 listed companies, which did not allow us to conduct an extensive study. The participants in the study are restricted to only accountants and auditors, not government officials. A further study could consider these issues in the future.

The study has practical implications for small and medium-sized businesses, where additional research can be conducted to improve output, as well as social implications for investors and society in order to reach information transparently and also empower stakeholders to make their own decisions. Considerably more work can be done by considering a large sample of companies across different sectors of operations. Furthermore, perceptive studies can be conducted by taking into account a large number of respondents as well as additional factors such as the use of XBRL for integrated reporting, taxonomies, and so on. In addition to this, future research can also be done on financial reports pattern in inline-XBRL and feasibility of application of expert systems along with XBRL for enhanced business reporting.

References

1. Plumlee, R.D., Plumlee, M.A.: Assurance on XBRL for financial reporting. Account. Horiz. **22**(3), 353–368 (2008)
2. Vasarhelyi, M.A., Chan, D.Y., Krahel, J.P.: Consequences of XBRL standardization on financial statement data. J. Inf. Syst. **26**(1), 155–167 (2012)
3. Shan, Y.G., Troshani, I.: Does XBRL benefit financial statement auditing? J. Comput. Inf. Syst. **54**(4), 11–21 (2014)
4. Zhou, L.: research on internal audit model of banking construction based on XBRL financial report. In: 2nd International Conference on Economy, Management and Entrepreneurship (ICOEME), pp. 271–277. Atlantis Press (2019)
5. Arndt, H.K., Isenmann, R., Brosowski, J., Thiessen, I., Gómez, J.M.: sustainability reporting using the eXtensible business reporting language (XBRL). In: EnviroInfo, pp. 75–82 (2006)
6. Seele, P.: Digitally unified reporting: how XBRL-based real-time transparency helps in combining integrated sustainability reporting and performance control. J. Clean. Prod. **136**, 65–77 (2016)
7. Baldwin, A.A., Trinkle, B.S.: The impact of XBRL: a Delphi investigation. Int. J. Digital Account. Res. **11**(1), 1–24 (2011)
8. Zhu, H., Wu, H.: Quality of data standards: framework and illustration using XBRL taxonomy and instances. Electron. Mark. **21**(2), 129–139 (2011). https://doi.org/10.1007/s12525-011-0060-4
9. Boritz, J.E., Efendi, J., Lim, J.H.: Executive team information system and financial reporting competencies, and voluntary adoption of XBRL reporting (2012)
10. Geiger, M.A., North, D.S., Selby, D.D.: Releasing information in XBRL: does it improve information asymmetry for early US adopters? Acad. Account. Financ. Stud. J. **18**(4), 66 (2014)
11. Mao, Y., Zhang, L.: The research of XBRL financial report impact on the quality of accounting information. In: 2019 International Conference on Cultural Studies, Tourism and Social Sciences, pp. 416–420 (2019)
12. Lev, B., Zarowin, P.: The boundaries of financial reporting and how to extend them. J. Account. Res. **37**(2), 353–385 (1999)

13. Bonsón, E., Escobar, T.: Digital reporting in Eastern Europe: an empirical study. Int. J. Account. Inf. Syst. **7**(4), 299–318 (2006)
14. Valentinetti, D., Rea, M.A.: Adopting XBRL in Italy: early evidence of fit between Italian GAAP Taxonomy and current reporting practices of non-listed companies. Int. J. Digit. Account. Res. **11**(1), 45–67 (2011)
15. Bovee, M., Ettredge, M.L., Srivastava, R.P., Vasarhelyi, M.A.: Does the year 2000 XBRL taxonomy accommodate current business financial-reporting practice? J. Inf. Syst. **16**(2), 165–182 (2002)
16. Avallone, F., Ramassa, P., Roncagliolo, E.: XBRL Extension to the Financial Statement Notes: Field-based Evidence on Unlisted Companies. Int. J. Digit. Account. Res. **16**, 61–84 (2016)
17. Tarmidi, M., Roni, R.A.: An international comparison of the determinants and financial information quality in XBRL reporting environment. Procedia Soc. Behav. Sci. **164**, 135–140 (2014)
18. Yuan, X., Wang, Y.: Study on the function of XBRL and the enhancement of the accounting information quality under the background of informaionization. Int. J. Bus. Manage. **4**(11), 137–139 (2009)
19. Hoitash, R., Hoitash, U., Morris, L.: eXtensible Business Reporting Language (XBRL): a review and implications for future research. J. Pract. Theory **40**(2), 107–132 (2021)
20. Abdolmohammadi, M.J., DeSimone, S.M., Hsieh, T.S., Wang, Z.: Factors associated with internal audit function involvement with XBRL implementation in public companies: an international study. Int. J. Account. Inf. Syst. **25**, 45–56 (2017)
21. Beerbaum, D., Piechocki, M., Puaschunder, J.M.: Measuring accounting reporting complexity with customized extensions XBRL–a behavioral economics approach. J. Appl. Res. Digital Econ. (JADE) **1**(3), 3–41 (2019)
22. Minan, K., Fahlevi, M.: Financial and corporate governance of XBRL adopters in Indonesia. J. Res. Bus, Econ. Educ. **3**(3), 1961–1967 (2021)
23. Shan, Y.G., Troshani, I.: The effect of mandatory XBRL and IFRS adoption on audit fees: evidence from the shanghai stock exchange. Int. J. Managerial Finance (2016)
24. Mosteanu, N.R., Faccia, A.: Digital systems and new challenges of financial management–FinTech, XBRL, blockchain and cryptocurrencies. Quality-Access Success J. **21**(174), 159–166 (2020)
25. Ilias, A., Abd Razak, M.Z., Rahman, R.A.: The quality of presentation and content for internet business reporting (IBR) on Malaysian Public Listed Companies (PLCs): an example of the developing country. In: Proceedings of the Australia Academy of Business and Social Science Conference, pp.1–19 (2014)
26. Ilias, A., Ghani, E.K., Baidi, N., Abdul, R.: Xbrl adoption: an examination on the Malaysian business reporting system (MBRS). Humanities **8**(2), 202–214 (2020)
27. Beerbaum, D.: Towards an XBRL-enabled corporate governance reporting taxonomy an empirical study of NYSE-listed financial institutions. University of Surrey (United Kingdom) (2015)
28. Hsieh, T., Bedard, J.C.: Impact of XBRL on voluntary adopters' financial reporting quality and cost of equity capital. J. Emerging Technol. Account. **15**(2), 45–65 (2018)
29. Tawiah, V., Borgi, H.: Impact of XBRL adoption on financial reporting quality: a global evidence. Account. Res. J. (ahead-of-print) (2022)
30. Ib, C.E., Jide, I., Zik-Rullahi, A.A.: The impact of XBRL on financial reporting: a conceptual analysis. Int. J. Empirical Finance **4**(2), 78–85 (2015)
31. Aksoy, M., Yilmaz, M.K., Topcu, N., Uysal, Ö.: The impact of ownership structure, board attributes and XBRL mandate on timeliness of financial reporting: evidence from Turkey. J. Appl. Acc. Res. **22**(4), 706–731 (2021)

32. Yang, S., Liu, F.-C., Zhu, X.: The impact of XBRL on financial statement structural comparability. In: Lamboglia, R., Cardoni, A., Dameri, R.P., Mancini, D. (eds.) Network, Smart and Open. LNISO, vol. 24, pp. 193–206. Springer, Cham (2018). https://doi.org/10.1007/978-3-319-62636-9_13
33. Uyob, R., Saad, R.A.J., Ahmi, A.: A review of the study on the impacts of the eXtensible Business Reporting Language (XBRL). Int. J. Sci. Technol. Res. **8**(9), 2320–2329 (2019)
34. Baldwin, A.A., Brown, C.E., Trinkle, B.S.: XBRL: an impacts framework and research challenge. J. Emerg. Technol. Account. **3**(1), 97–116 (2006)
35. Liu, C.: XBRL: a new global paradigm for business financial reporting. J. Global Inf. Manage. (JGIM) **21**(3), 60–80 (2013)
36. Hider, P., Pymm, B.: Empirical research methods reported in high-profile LIS journal literature. Libr. Inf. Sci. Res. **30**(2), 108–114 (2008)
37. Connaway, L.S., Powell, R.R.: Basic research methods for librarians. ABC-CLIO (2010)
38. Blumberg, B., Cooper, D., Schindler, P.: EBOOK: Business Research Methods. McGraw Hill (2014)

The Impact of Digital Audit Application on the Quality of the Auditor's Report

(Field Study from the Point of View of Chartered Accountants in Jordan)

Nidal Mahmoud Al-Ramahi[1], Zaid Semreen[1,2], Tareq Hmmad Almbydeen[1], Riham Alkabbji[1](✉), Ala' Alrazim[1], and Qadri Aljabri[2]

[1] Zarqa University, 2000, Zarqa 13110, Jordan
{ramahinedal,tmbydeen,rkabbji}@zu.edu.jo
[2] University of Business and Technology, Jeddah, Saudi Arabia

Abstract. Our study aims address and identify digital auditing and its impact on the quality of the chartered accountant's report. To achieve the objectives of the study, the researcher followed the descriptive analytical approach. A questionnaire was designed and used as the main tool for the study. The study population comprised all (460) chartered accountants in Jordan according to the Jordanian Association of Certified Public Accountants (JCPA) website for (2022). A simple random sample was withdrawn from the study community, where (250) electronic questionnaires were distributed, (223) questionnaires were retrieved, (13) questionnaires were excluded and (210) were valid for analysis, accounting for (84%) of the distributed questionnaires. To answer the study questions and test its hypotheses, the researcher used appropriate statistical methods through the Statistical Package for Social Sciences (SPSS) program. The study concluded several results, mainly that there is a Statistically Significant Effect (SSE) of Digital Audit (DA) application on the chartered accountant's report in its dimensions used. The study recommended holding training courses and workshops on how to apply digital transformation in auditing to include all the items that have been studied and discussed in this study and conducting further research and studies on how to increase the quality of the chartered accountant's report.

Keywords: Digital audit · Quality of the chartered accountant's report · Chartered accountants in Jordan

1 Introduction

Audit offices that provide auditing services for companies 'business and present statements on companies' honesty and fairness of their financial statements seeked to develop their business and improve their ability to complete tasks in a timely manner [1]. These offices have been gradually moving to a (DA) environment with all its challenges, besides maintaining the privacy of their clients' data, distancing from the many complexities in traditional systems, and saving effort and time along with speed in accomplishing audit

B. A. M. Alareeni and I. Elgedawy (Eds.): *Artificial Intelligence (AI) and Finance*, SSDC 488, pp. 240–252, 2023.
https://doi.org/10.1007/978-3-031-39158-3_22

tasks. This may affect the fees of the Audit Process (AP), especially that the (AP) in the digital environment will shorten many of the tasks that could have been done in the traditional system. Moreover, it has the largest share of the auditor's effort and time, thereby increasing their fees [2].

In their quest to improve the quality of reports, chartered accountants are subject to a set of internal and external factors and constraints, mainly, digital audit. It has been found that there is a lack of perception and understanding of the relationship between (DA) and the quality of chartered accountants' reports, and the accomplishment of the audit strategy, whether a theoretical (academic) or applied level, which prompted us to study that relationship from the point of view of chartered accountants, as a tool to encourage chartered accountants to take advantage of information technology through (DA) to achieve competitive advantage and audit strategy.

Hence, this study tries to identify the impact of applying (DA) on improving the quality of the auditor's report from the point of view of chartered accountants in Jordan. The study aims to show the impact of applying (DA) on the quality of the auditor's report According to the opinions of chartered accountants in Jordan through defining (DA) and identifying its objectives.

2 Literature Review

Abu Arab [3] explained the role of digital transformation in improving the quality of the internal (AP) and concluded that the use of digital transformation helps to record the stages of the entire (AP) and deal with the data, which allows the possibility of using it in subsequent audit tasks. However, Mahri [4] stressed that auditors should have the necessary scientific competence and practical training in the field of Information Technology, and they must also have contact with audit offices; this helps to improve their practical and scientific efficiency. Hakeem [5] confirmed that there is an impact of the use of Information Technology on improving the quality of internal audit. Qadour [6] showed that electronic audit contributes to improving financial performance and contributes to facilitating the work of the auditor in terms of increasing the speed of performance, reducing the effort expended, revealing the strengths and weaknesses within the organization, by reconciling efficiency and effectiveness, and achieving the desired results at the lowest cost. This is consistent with Almubaydeen [10] which stressed that reliability increases the validity of financial lists and their transparency Shaqqour [7]. He mentioned that the requirements of digital auditing for modern technologies and artificial intelligence techniques in Jordanian auditing companies.

3 Model

The study model was hypothetically designed, and its variables were drawn from the theoretical studies related. This model indicates the existence of a one-way effect of applying (DA) as an (IV) on the quality of the auditor's report as a dependent variable. The following Fig. 1 shows the model of the study.

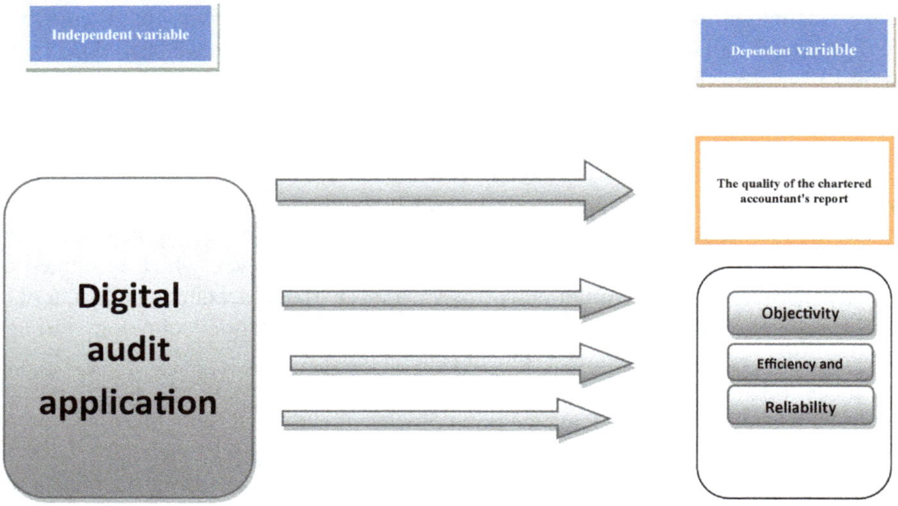

Fig. 1. The study model. Prepared by the researcher based on (Abu Arab, 2022)

4 Hypotheses

Based on the research questions and existing problems, the following research hypotheses are put forward.
The main hypothesis (H0): No (SSE) at a Significant Level (SL) ($\alpha \leq 0.05$) of the application of (DA) on the quality of the auditor's report from a Jordanian Auditor's point of view. The sub-hypotheses is:

H01: NO (SSE) at a (SL) ($\alpha \leq 0.05$) of the application of (DA) on objectivity from a Jordanian auditor's point of view.
H02: NO (SSE) at a (SL) ($\alpha \leq 0.05$) of the application of (DA) on the efficiency and effectiveness of the auditor's report from a Jordanian auditor's point of view.
H03: There is no (SSE) at a (SL) ($\alpha \leq 0.05$) of the application of (DA) on reliability from a Jordanian auditor's point of view.

5 Study Methodology

The study followed the descriptive-analytical approach.

5.1 Population and Sample

The population of our study comprised all (460) certified public accountants in Jordan, according to the (JCPA) Accountants website (2023). A sample was (250) uditor's who answer electronic questionnaires were distributed through the electronic link prepared on (Google Form) via emails and personal phones. (223) questionnaires were retrieved, (13) were excluded and (210) were valid for analysis, amounting to 84% of the distributed questionnaires. Based on Sekaran and Bougie [7], the sample size is appropriate, and lack of data was the reason for excluding some questionnaires. Table 1 shows the repetition and percentage of the demographic variables that describe the respondents of the study sample.

Table 1. Distribution of the study sample according to personal and job data.

Variable	Response	Repetition	Percentage
Sex	Male	203	96.7
	Female	7	3.3
	Total	**210**	**100**
Age	< 25 years	24	11.4
	25 to under 35 years old	67	31.9
	35 to under 45 years old	63	30
	45 to under 55 years old	38	18.1
	55 years and more	18	8.6
	Total	**210**	**100**
Experience in the field of auditing	Less than 5 years	77	36.7
	5 to less than 10 years	45	21.4
	10 to less 15 years	20	9.5
	15 to less 20 years	26	12.4
	more than 25 years	42	20
	Total	**210**	**100**
Academic qualification	Diploma	11	5.2
	Higher Diploma	3	1.4
	Bachelor`s degree	98	46.7
	Masters	92	43.8
	PhD	6	2.9
	Total	**210**	**100**
Professional certifications	CPA	15	7.1
	ACPA	2	1
	ACCA	6	2.9
	CIA	8	3.8
	JCPA	**210**	**100**

5.2 Study Tool and Data Collection Sources

To achieve the objective of the study, the study relied on the questionnaire to obtain the primary data related to the subject of the study, as the study used two main sources of data collection and information, namely:

Secondary Sources: Arabic and foreign books and references related to auditing, control and accounting, information technology techniques, digital auditing and scientific materials, specialized publications and periodicals, articles, research, previous studies, and websites.

Primary Sources: These relate to the analytical aspect of the subject of the study through the questionnaire distributed among the study sample (chartered accountants in Jordan) to collect data.

5.3 Stability of the Study Tool

To verify the stability of the study tool, we were used Cronbach Alpha test. The questionnaire, as the rule of this Test states that the Cronbach alpha value should be greater than 60% in Applied Research of Cognitive and Human Sciences [9] Sekaran and Bougie. The following table presents the results (Table 2).

Table 2. The internal consistency coefficient (Cronbach Alpha).

Dimension	Number of Clause	őReliability coefficient (Cronbach Alpha)
(IV) ((DA)application)	13	89.30%
Objectivity	4	86.50%
Efficiency and effectiveness	4	83.90%
Reliability	4	
Dependent variable: The quality of the auditor'sreport	22	93.30%
Questionnaire Total	25	93.70%

5.4 Descriptive Statistics of Study Variables

(IV) Descriptive statistics (DA) application)

The Independent Variable (IV) (DA) application was measured through (13) clauses, and the following is an analysis of the respondents ' responses to the clauses that measure the (IV), which were as follows (Table 3):

Table 3. μ and **STDEV** of the (IV).

Rank	No	Clause	μ	STDEV	Sig
2	1	(DA) application contributes to the development of the audit profession	4.260	0.651	High
4	2	(DA) application reduces the risk of data manipulation	4.160	0.747	High
7	3	Audit offices use advanced software to complete the (AP)	4.110	0.837	High
8	4	(DA) application will improve the efficiency and quality of the chartered accountant's report	4.080	0.821	High

(continued)

Table 3. (*continued*)

Rank	No	Clause	μ	STDEV	Sig
9	5	(DA) leads to the detection of weaknesses in the control system	4.060	0.834	High
12	6	The auditor can understand the points where the images of fraud are focused on using the computer	3.890	0.753	High
11	7	(DA) of data provides effective communication between the chartered accountant and the senior management	3.990	0.824	High
5	8	The use of computer and automated programs leads to the activation of the organization, management and preservation of systems	4.150	0.746	High
10	9	The use of (DA) helps to adopt objective and fair results	4.030	0.806	High
3	10	The use of (DA) helps the auditor to complete works and tasks in a shorter time, less effort and lower cost	4.190	0.771	High
13	11	The use of (DA) affects the determination of the fees of the chartered accountant	3.890	0.898	High
6	12	When using digital audit, evidence is checked and extracted in an easier way	4.140	0.744	High
1	13	The chartered accountant must have the necessary scientific competence and practical training in the field of Information Technology, Computing, and audit specialized software	4.310	0.708	High
(DA) application			4.097	0.518	High

(Arithmetic Mean = μ, standard deviation = **STDEV**, Significant = **Sig**).

The table above indicates that the relative significance of the (IV) was high, as the arithmetic **μ** (4.097) and the STDEV (0.518). Clause (13), which states that **"The**

chartered accountant must have the necessary scientific competence and practical training in the field of Information Technology, Computing and audit specialized software" had the highest μ of (4.310) and STDEV of (0.708), whereas clause (11), which stated that "**The use of digital auditing affects the determination of the fees of the chartered accountant"** had the lowest μ of (3.890) and STDEV of (0.898). This indicates that chartered accountants are highly interested in applying digital auditing because of its great importance in minimizing errors and detecting manipulation in the shortest possible time.

5.5 Descriptive Statistics of the Dependent Variable: The Quality of the Auditor's Report

The (IV) (the quality of the auditor's report) was measured through (3) dimensions, namely (objectivity, efficiency and effectiveness, reliability). The following is an analysis of the respondents ' answers to the clauses that measure those dimensions, which were as follows (Table 4):

Table 4. μ and **STDEV** of the (IV).

Rank	No	Clause	μ	STDEV	Sig
2	1	The objectivity and independence of auditors impact the quality of the auditor's report	3.700	1.048	High
4	2	Failure to formulate the audit report in a balanced manner in terms of style and content affects the validity and quality of the report	3.550	1.089	Average
3	3	Presenting the facts within the report honestly and impartially, without any misinformation or distortion, will improve the quality of the auditor's report	3.640	1.046	Average
1	4	The commitment of auditors to objectivity when performing audit tasks helps to achieve high-quality results	3.780	0.943	High
Objectivity			3.669	0.871	High

The table above indicates that the relative significance of the first dimension of the (DV) (objectivity) ranged between high and medium, where the μ amounted to (3.669) and the STDEV (0.871). Clause (4), which states **"The commitment of auditors to objectivity when performing audit tasks helps to achieve high-quality results"** had the highest μ of (3.780) and STDEV of (0.943), while clause (2), which stated **"Failure to formulate the audit report in a balanced manner in terms of style and content affects the correctness and quality of the report"** had the lowest μ of (3.550) and STDEV of (1.089), which indicates that the (Objectivity) clauses in the quality of the auditor's reports were indicating a high to medium degree of approval (Table 5).

Table 5. μ and **STDEV** of the (IV) of the (Efficiency and Effectiveness).

Rank	No	Clause	μ	STDEV	Sig
3	1	The commitment of auditors to efficiency and effectiveness contributes to reducing the potential of falling into judicial or penal responsibilities	4.010	0.891	High
2	2	Auditors' commitment to efficiency and effectiveness contributes to reducing audit risks	4.110	0.873	High
1	3	The experience and efficiency of the audit team helps to achieve a high quality in preparation of the audit report	4.170	0.818	High
4	4	The use of automated software in (DA) helps to increase the efficiency and effectiveness of the auditor's report	3.900	0.885	High
Efficiency and Effectiveness			4.048	0.712	High

The table above indicates that the relative significance of dimension (4) of the (DV) (efficiency and effectiveness) was high, with μ (4.048) and the STDEV (0.712). Clause (3), which states that **"The expertise and competence of the audit team helps to achieve a high level of quality in the preparation of the audit report"** had the highest μ (4.170) and STDEV of (0.818), while Clause (4), which stated that **"The use of automated software in (DA) helps to increase the efficiency and effectiveness of the auditor's report"** had the lowest μ (3.900) and STDEV of (0.885), which indicates that the efficiency and effectiveness clauses in the quality of the auditor's reports indicated a high degree of approval (Table 6).

Table 6. μ and **STDEV** of the (IV).

Rank	No	Clause	μ	STDEV	Sig
4	1	The commitment of auditors to integrity and transparency contributes to increasing the reliability of the (AP)	3.820	1.042	High
2	2	Reliability leads to an increase in the public's positive perception of the audit profession and its practitioners	4.080	0.890	High
3	3	The quality of the auditor's report depends on the impartial and documented presentation of the control results	4.000	0.938	High
1	4	The reliability of the information contained in the report leads to an increase in the quality of the report	4.090	0.887	High
Reliability			3.996	0.785	High

The table above indicates that the relative significance of the sixth dimension of the (DV) (reliability) was high, with μ (3.996) and a STDEV of (0.785). Clause (4), which states **"The reliability of the information contained in the report increases the quality of the report"** had the highest μ (4.090) and a STDEV of (0.887), while Clause (1), which stated **"The auditors' commitment to integrity and transparency contributes to increasing the reliability of the (AP)"** had the lowest μ (3.820) and STDEV of (1.042), which indicates that the (Reliability) clauses in the quality of the auditor's reports indicated a high degree of approval.

5.6 The Results of Testing the Hypotheses of the Study Were as Follows

Results of Testing the Main Hypothesis

The main hypothesis (H0): NO (SSE) at (SL) ($\alpha \leq 0.05$) of the application of (DA) the quality of the auditor's report reliability from a Jordanian auditor's point of view (Table 7).

The (R2) value of the study model was (44.1%), which means that the (IV) the application of (DA) explains the changes that occurred in the (DV) in all its dimensions (the quality of the auditor's report) from a Jordanian auditor's point of view. This percentage is good and indicates the ability of the (IV) to interpret the dependent variable. The

Table 7. Testing the main hypothesis.

Dependent variable: The quality of the auditor'sreport			
(IV)s	B	T	Sig
Fixed	1.868	6.371	0.000***
(DA)	0.514	7.232	0.000***
R^2	44.10%	Adj. R^2	43.70%
F	52.309	Sig	0.000***

(* * *) at the level of 1%, (**) 5%, (*) at the level of 10%.

(F-statistic) value of the study model has reached (52.309) with a statistical significance at the level of (Sig = 0.000). This value is located within the statistical significance level of (1%) and this is an indicator of the validity and significance of the model used to influence the quality of the auditor's report from a Jordanian auditor's point of view. So, we accepted the alternative hypothesis that states **"NO (SSE) at a (SL) ($\alpha \leq 0.05$) for applying (DA) to the quality of the auditor's reliability from a Jordanian auditor's point of view."**.

- **Sub-hypothesis test results:**

H01: NO (SSE) a (SL) of ($\alpha \leq 0.05$) of the application of (DA) on objectivity from a Jordanian auditor's point of view (Table 8).

Table 8. Results of testing the first sub-hypothesis.

Dependent variable: Objectivity			
(IV)s	B	T	Sig
Fixed	1.934	4.15	0.000***
(DA) application	0.423	3.752	0.000***
R^2	25.20%	Adj. R^2	24.40%
F	14.075	Sig	0.000***

(* * *) at the level of 1%, (**) 5%, (*) at the level of 10%.

The (R2) value of the study model was (25.2.%), which means that the (IV) (the application of digital audit) explains the changes that occurred in the (DV) (Objectivity in the quality of the auditor's report) from a Jordanian auditor's point of view. This percentage is good and indicates the ability of the (IV) to interpret the (DV).

As the table above shows that the (F-statistic) value of the study model (14.075) and a statistical significance at the level of (Sig = 0.000). This value is located within the statistical significance level (1%) and therefore the null hypothesis is rejected and the alternative hypothesis that states **"There is a statistical significance effect at a significant level of (α ≤ 0.05) for the application of (DA) on objectivity from reliability from a Jordanian auditor's point of view".**

H0.4: NO (SSE) at (SL) of (α ≤ 0.05) for the application of (DA) on efficiency and effectiveness from a Jordanian auditor's point of view Table 9**.**

Table 9. Results of testing the fourth sub-hypothesis.

Dependent variable: Efficiency and effectiveness			
(IV)s	T	B	Sig
Fixed	1.981	5.411	0.000***
(DA) application	0.505	5.691	0.000***
R^2	24.70%	Adj. R^2	22.70%
F	32.391	Sig	0.000

(* * *) at the level of 1%, (**) 5%, (*) at the level of 10%.

The (R2) value of the study model was (24.7%), which means that the (IV) (application of digital audit) explains the changes that happened in the (DV) the quality of the report of the chartered accountant (efficiency and effectiveness) **from a Jordanian auditor's point of view** by (24.7%), which is a good percentage and indicates the ability of the the (IV) to interpret the dependent variable. The table above shows that the (F-statistic) value of the study model amounted to (32.391) and a statistical significance at the level of (Sig = 0.000). This value is located within the statistical significance level (1%) and therefore the null hypothesis is rejected and the alternative hypothesis that states "There is a statistical significance effect at a significant level of (α ≤ 0.05) for the application of (DA) on efficiency and effectiveness from a Jordanian auditor's point of view." is accepted.

H0.6: NO (SSE) at (SL) of (α ≤ 0.05) for the application of (DA) on reliability from a Jordanian auditor's point of view (Table 10).

Table 10. Testing Results the sixth sub-hypothesis.

Dependent variable: Reliability			
(IV)s	T	B	Sig
Fixed	1.464	3.697	0.000***
(DA) application	0.618	6.444	0.000***
R^2	26.60%	Adj. R^2	26.20%
F	41.527	Sig	0.000***

(* * *) at the level of 1%, (**) 5%, (*) at the level of 10%.

The (R2) value of the study model was (26.6%), which means that the (IV) (application of digital audit) explains the changes that occurred in the (DV) (reliability in the quality of the report of the auditor's) from a Jordanian auditor's point of view by (26.6%), which is a good percentage and indicates the ability of the (IV) to interpret the dependent variable.

The table above also shows that the study model used to test the effect of the application of (DA) on the quality of the auditor's report (reliability) from a Jordanian auditor's point of view is (SS), where the (F-statistic) value of the study model (41.527) is (SS) at the level of (Sig = 0.000). Therefore, the sixth null sub-hypothesis is rejected and the alternative hypothesis that states: **"There is a statistical significance impact at a significant level ($\alpha \leq 0.05$) for the application of (DA) on reliability from a Jordanian auditor's point of view." is accepted.**

6 Results

1. Direct relationship between the application of (DA) and the quality of the auditor's report. Which means that the more (DA) is applied, the higher the quality of the auditor's report.
2. There is a direct relationship between the application of (DA) and objectivity in the quality of the auditor's report, which means that the more (DA) is applied, the more objective the quality of the auditor's report.
3. There is a direct relationship between the application of (DA) and the efficiency and effectiveness of auditor's report, which means that the more (DA) is applied, the higher the efficiency and effectiveness of the auditor's report.
4. There is a direct relationship between the application of (DA) and the reliability of the auditor's report, which means that the more (DA) is applied, the higher the reliability of the auditor's report.

7 Recommendations

We recommended

1. The use of audit mechanisms and digital transformation by auditors in the (AP) contributes to providing additional information to users of financial statements, reducing costs, saving time and simplifying the procedures for carrying out the (AP).
2. Continuous monitoring of digital transformation mechanisms and their use in audits.
3. Encouraging chartered accountants to enroll in specialized training courses in the field of Information Technology and its use in auditing.
4. The need to develop the professional performance of the auditor in the (DA) domain in planning the (AP), collecting audit evidence

References

1. AL-Ramahi, N.: Measuring the extent of application of the (COSO) framework for internal control from the point of view of the external auditors in the companies. Zarqa J. Res. Stud. Humanit. **17**(2) (2017)

2. Al-Samarrai, A., Alshuraida, N.: The role of artificial intelligence techniques using (DA) in achieving audit quality and supporting its strategy from the point of view of auditors: a field study in audit companies in the kingdom of Bahrain. Int. J. Econ. Bus. **8**(1), 15–31 (2020)
3. Abu Arab, H.: The role of digital transformation in improving the quality of the internal (AP). J. Account. Financ. Stud. **17**(59), 35–52 (2022)
4. Mehri, S., Hajaiji, S., Bosniak, R., Hanka, M.: The impact of the use of Information Technology on the quality of external audit: a field study. Unpublished master's thesis, martyr Hama Lakhdar University-El Wadi, Algeria (2022)
5. Hakim, R., Mohammed Al-sagheer, A., Amari, S., Bashir, R.: The role of the use of information technology in improving the quality of internal audit: a field study at the operational directorate of communications of Algeria in the state of El-Wadi. Unpublished master's thesis, martyr Hama Lakhdar University-El Wadi, Algeria (2021)
6. Kaddour, A.: E-audit and its role in improving the financial performance of the institution: a case study of Sidi argis mills or El bawaki. Unpublished master's thesis, Arabi Benmehidi University, Umm elbouaki, Algeria (2021)
7. Shaqqour, O.F., Harb, A.S.M., Ballout, O.M.K., Jaber, R.J.: (DA) During Covid-19 in Jordanian audit firms a study of the reality and outlook the future. Implement. Smart Technol. Bus. Success Sustain. **216**, 263–272 (2023)
8. Sekaran, U.: Methodological and theoretical issues and advancements in cross-cultural research. J. Int. Bus. Stud. **14**(2), 61 (2006)
9. Sekaran, U., Bougie, R.: Research Methods for Business: A skill Building Approach, 7th edn. Wiley, Chichester (2016)
10. Almubaydeen, T.H., Qushtom, T.F.A., Hamza, M., Al-Khawaldeh, R.M.R.: The impact of the IFRS 16 adjustments "coronavirus-related lease contracts" on the financial performance (royal Jordanian case study). In: Alareeni, B., Hamdan, A. (eds.) ICBT 2022. LNNS, vol. 495, pp. 541–549. Springer, Cham (2023). https://doi.org/10.1007/978-3-031-08954-1_48
11. Jordanian Association of Chartered Accountants: List of members practicing the auditing profession inside Jordan, 1 November 2022. Accessed 15 Nov 2022
12. Centre for Audit Quality (CAQ): Audit Quality Indicators: Journey and Path Ahead, 12 January 2016. https://www.thecaq.org/audit-quality-indicators-journey-and-path-ahead/. Accessed 25 Oct 2022

The Impact of Digital Transformation on the Exercise of the Right to Vote (Electronic Voting as a Model)

Abdelaziz Mahmoud Al Khawatreh[1](✉), Suhaib Ahmed El-Manaseer[1], Jalal Hussein Al-Kayid[1], and ElGaili Ahmed ElTayeb[2]

[1] Faculty of Law, Zarqa University, Zarqa, Jordan
{Akhawatreh,salmanaseer,jalkayid}@zu.edu.jo
[2] College of Law, University of Business and Technology, Jeddah, Saudi Arabia
algaili@ubt.edu.sa

Abstract. The digital environment and electronic societies can crystallize new models of contribution to the rationalization of governance that go beyond the traditional ideal standard models, and provide wide possibilities to overcome the structural and procedural obstacles facing the consolidation of the electronic voting model through the specialized platforms of e-government in reality. This affects the election results and their consequences are greatly affected by the digital environment related to electronic voting, as well as the role of digital transformation in the electoral process. The digital environment also imposes the inevitability of involving various actors in the decision-making process. Electronic participation channels are gaining great attraction among those interested in political communication. Leaders and decision-makers have already realized the scope and size of the value that can be accumulated from approaching the dissemination of their ideas through digital technologies, both in terms of The opportunity to involve the citizen in more positive interactions, or to try to control and contain these interactions, but we do not have to consider the models of transformation towards digital societies as ideal models to devote the principles of participation in elections represented by electronic voting, which is the station of the study that we will conduct.

Keywords: the right to vote · digital transformation · electronic voting

1 Introduction

The digital environment is promoting electronic voting in elections as a model by increasing public participation space and moving towards an electronic voting concept based on participation through information and communication technology and through government-related digital means. This is to facilitate the electoral process for many segments of society and those represented by those unable to participate in the election for several reasons, including lack of financial ability to go to polling sites, physical incapacity due to illness, and family circumstances that may prevent women

B. A. M. Alareeni and I. Elgedawy (Eds.): *Artificial Intelligence (AI) and Finance*, SSDC 488, pp. 253–261, 2023.
https://doi.org/10.1007/978-3-031-39158-3_23

from participating in elections. And thus improving access to information and encouraging participation in policy-making through electronic voting. Whether for the sake of empowering individuals or for the benefit of society as a whole, the digital environment and the various information and communication environments work to enhance it [1], by providing the largest possible number of systems and means for participating in electronic voting through specialized platforms to include various societal groups. From a technological point of view, the basic principle of these new systems, which makes them so powerful, is their ability to carry and route many different messages simultaneously and intricately to and from many locations rather than just carrying and delivering a few messages to a wide audience, under the influence of new technologies. The new electronic voting techniques and means allow for an unprecedented level of interaction and response. The problematic part of the study is that there is no text in Jordanian Election Law 4 of 2022 that allows the voter to exercise his electoral right using electronic means.

2 The Impact of Digital Transformation on the Voting Process

Elections are considered the main means by which people are eligible to participate in the management of their countries' public affairs and are considered a fundamental human right that people in all countries of the world have fought for. Voting rights in democratic countries are considered one of the most important political practices. So elections are a means of transferring power in a peaceful way from one person to another represented in the elections of the House of Representatives in Jordan or another group like many countries; This is due to the impossibility of applying direct democracy, as it is not possible to gather all citizens in a general assembly in large countries. On the one hand, the number of voters is increasing, and on the other hand, the needs of the people are diversifying and government affairs are becoming more complicated, which requires the availability of technical and intellectual expertise. Public assemblies were not sufficient to take on the affairs of the state and additionally the possibility to influence voters, because the voting takes place in public [2].

Digital development, or the revolution of information technology and knowledge. Because electronic voting takes place through modern technologies that resulted from the electronic revolution, as Jordan entered the digital world into the digital space. In the past few years, we found that there has been a development in everything using the digital environment except for elections. The Jordanian election law did not include voting through electronic means, which may include all society by election due to the ease of the digital environment. It should also be emphasized that election laws, including voting via electronic means, must be given utmost importance by the state with its various institutions and society with its various components, given that it is the cornerstone of the matrix of public national, and political action laws, and as a means for the renewal and renewal of political elites, and an entry point for the transfer of power between minority and parliamentary majority. We will talk during this point explaining the concept of electronic voting through electronic means.

2.1 The Concept of Electronic Voting

The political tradition has been that the ballot box is the essence of the democratic process, as an indication of the great importance of the electoral process in promoting and consolidating democratic principles and electoral justice, and it also reaches the point that it is inevitable to talk about elections when entering the world of democracy, as it is a pattern or method for the devolution of electoral power, based on A choice that takes place by voting or balloting, and it is considered the basic method for assigning power in representative democracy. Rather, it has become the only means for granting legitimacy to power, despite the defects that can be attributed to the principle of election as a means for appointing rulers or for elections that are held to form the House of Representatives. The method also remains The most effective in elections, which constitute one of the most important mechanisms of political participation, and a means of making political choices, due to the availability of a realistic practice of choosing them through electoral means, and in light of the inability of members of society to rule themselves directly, a practical exercise that provides an opportunity to form a democratic government to represent them is necessary.

Elections are considered one of the most important pillars of the democratic system, and one of its independent variables affecting and controlling the form of the elected system. Despite the multiplicity and different definitions, and the lack of agreement on a specific and absolute definition, we find that the content of democratic elections in their correct form, however, there is a set of standards and conditions agreed upon because the electoral process is a political right and derives from it the universal right to vote for all adult citizens, represented by two-thirds of the national community, not to be deprived Any group of the formation of a political party, and the freedom to run electoral campaigns without violating the law or using violence [3].

2.1.1 The Concept of Electronic Voting

Electronic voting represents one of the most popular forms of political participation in light of the inevitability imposed by the digital environment, and the research in the context of the mechanisms for implementing this form of participation in elections, and their legitimacy, is one of the most important points raised at the global and local levels, so the use of electronic voting is one of the most important forms of electronic democracy. Electronic voting has become one of the most important topics that attract the attention of public institutions, political parties, and politicians. It refers to the technologies used in voting operations such as: digital broadcasting, telephone communications, and the Internet. Internet voting is divided into two categories: Internet voting at the polling place. In a pre-prepared place where an Internet connection is available, and remote voting is voting that enables the voter to vote through a home computer, laptop, or phone connected to the Internet, and data is transferred from a computer over the Internet to a central database. Electronic or digital voting helps reduce the gap between citizens and political elites, by creating an electronic space that brings together parties and voters and facilitates communication and communication between these actors to enhance participation and expand its size.

2.1.2 Electronic Voting System Standards

In order to discuss electronic voting, a set of characteristics and benefits must be available in this mechanism that can be considered criteria for developing an electronic voting system. These are as follows: privacy, accessibility, and accuracy, and these concepts are considered as the characteristics of an electronic voting system that affect the attitude of the voter, supporting digital democracy and participatory democracy [4].

- Privacy: It is defined as the ability of an individual or persons to isolate themselves or information about them, and thus they express themselves in a selective and chosen manner. Often in the original sense defensive in the ability of a person or a group of persons, to prevent information relating to him or them to become known to others, in particular organizations and institutions, if the person does not voluntarily choose to provide that information [5], originally is a concept referring to the scope of private life In recent decades, it has developed on a larger scale, to guarantee the right to control personal data. An electronic voting system must ensure that the vote cannot be linked to the voter and that the voter cannot prove how they voted.
- Accessibility: The electronic voting mechanism must ensure that eligible voters receive equal access. Accessibility reflects both the physical accessibility and perceived ease of use of the electronic voting system.
- Accuracy: The electronic voting mechanism must ensure that all valid ballots are counted accurately and that a validated voter will vote only once.

3 The Relationship Between Elections and the Digital Environment

We find that there are many methods of electronic voting, including the use of the smart electronic card, and voting may be done via electronic means or through the electronic government, through the establishment of a communication portal between the information infrastructure of the elections on one hand and the Internet on the other hand, and this requires the voter to write his information, This makes the electronic voting process surrounded by a wide controversy in terms of the degree of its security against manipulation, piracy or forgery. It is also possible for the Independent Electoral Commission to monitor the electronic process, so that the electronic elections are given great attention, including all the details of electronic voting, especially the state of political development and political reform in Jordan, reform initiatives and the state of parties, the state of political participation and community participation, the state of youth participation in political life, the election law The state of trade union work, the state of political culture, and democratic empowerment.

- Electronic platforms: Modern social media have become famous as digital platforms for political participation. Many actors in the decision-making process, such as parties and civil society institutions, have adopted these platforms as mechanisms that replace the mechanisms of social and political mediation, and as channels of direct communication between them and citizens. These platforms have spread Platforms recently to express a new elite of actors and influencers in decision-making, whether in an official and direct manner or through discussions, comments and protests…, and among the activities provided by these platforms are the following: Blogs: A web page that does not include any external editing, with submission Online commentary, updated periodically, and presented in reverse chronological order, with hyperlinks to other online resources. The software required to run a blog is freely available on the Internet. It is relatively easy to use and does not require specialized knowledge of web languages. For a blog to have strong influences, it must relate to topics or people. Those who influence decision-making processes, so that these blogs can mobilize the audience around a specific event [6].

- E-government: E-government expresses the government's use of technology, specifically, web-based Internet applications to improve access to its services and the delivery of government information and services to citizens, business partners, employees, and government agencies. E-government is often described as a set of parallel, compatible information networks. With each other capable of processing interactive government services on these networks for the good and effective implementation of government and programs for goals. However, this concept of e-government remains general and it is not possible to talk through it explicitly about the participation of citizens in proposing, implementing, or even evaluating government programs, as the matter is not only related to providing Services or achieving the objectives set by means of communication and information technology, so improving public participation processes is a way to legitimize the principle of participation. Under these circumstances, governments must adapt to technological change that adopts new methods of decision-making and reconsider other alternatives to the practice of democracy [7].

The means of communication and modern technology give the advantage of continuous communication with citizens and inform them of everything that is going on inside the parliamentary sessions, and they are also a tool to follow up the performance of parliamentarians. Because modern means of communication provide ease for voters to communicate with their representatives from the representatives, and these means will form in the future the infrastructure for these councils, as it is one of the fruits of electronic democracy [8].

3.1 Challenges that Impede Digital Transformation in the Field of Electronic Voting

Based on the foregoing, it is clear that we are facing a technological and political phenomenon imposed by the digital information revolution, which brought about a change in the traditional concepts of society, and produced a virtual version of it in the digital environment, which led to redefining the criteria and rules for participation, which confirms that we are facing transformations that actually affected the increase in the volume of participation and the increase The power of the individual versus the power of the state [9]. Electronic voting systems also include many functions, including random distribution, digital communications, and digital lists. There is not enough scope in this study to indicate what the jobs are, but we will discuss the challenges that may face digital voting. But in order for us to reach an understanding of the functions that electronic voting can perform, it is useful to take into account, which electronic voting systems can provide to both voters and election officials represented by the supervisory body on elections, which is called the Independent Authority for Elections [10].

3.1.1 Electronic Challenges

Theoretically, the digital environment assumes the inevitability of involving various actors in the decision-making process. Electronic participation channels are gaining great attraction among those interested in political communication. Leaders and decision-makers have already realized the scope and size of the value that can be accumulated from approaching the dissemination of their ideas through digital technologies, both in terms of The opportunity to involve the citizen in more positive interactions, or to try to control and contain these interactions, but we do not have to consider the models of transformation towards digital societies [11].

3.1.2 The Digital Divide and the Monopoly of Information Technology

The "digital divide" in its classic form is economic and social inequality in the access, use, or knowledge of information and communication technologies by individuals. The digital divide represents a major challenge to participatory democracy. Its reasons for the issue of technology monopoly. The world today is witnessing a sharp division between countries that rely on modern and advanced information and communication technology, and countries with slow-growing economies that do not possess the elements of modern technology, and the gap between the two parties is currently widening due to the monopolies of developed countries, and therefore societies that suffer from the digital divide crisis will not be able to participate. In decision-making processes due to the lack of sufficient information for the participation process or the inability to participate in the first place. It also seems that the aspect related to considerations of the quality of human resources and skills is also related to the problem of the digital divide, as the countries and societies of the South lack what we can call an "enabling environment" that supports the development of human skills and capabilities in the field of using information and communication technology. There is no doubt that the absence of such capabilities and skills would deepen the gap and exacerbate the problem of all citizens' access to the virtual networked world in which digital democracy stands [12].

3.1.3 Piracy and the Internet Penetration Crisis

Despite the vital role played by communication and media technology in democratic societies, as it provides public oversight and participation mechanisms, dissemination of information and ideas, and provision of forums for public debate, in addition to supporting participatory democracy by opening common spaces for discussion and debate on issues of public interest. Digital is a favorable environment par excellence for freedom of expression, but this does not preclude the possibility of penetrating these means and systems and distorting their outputs from their democratic context.

3.1.4 Societal Challenges

In recent decades, most societies in the world have undergone major transformations, and perhaps the increasing presence of Muslims in Europe is part of this change, but not the cause of all these changes, Muslims did not cause a decrease in attendance in European churches nor they responsible for turning some churches into museums or pubs, and for the continuing decline of the birth rate in many European societies. However, Muslims building mosques and attending religious rites in greater numbers than European Christians and increasing the number of their family members over Christian families in Europe make Muslims a scapegoat for some fanatics ones, just as the blatant and continuous discrimination that immigrants suffer in their daily lives, sometimes mixed with some extremist Islamic ideologies among some communities, to weaken the strength of the positive adaptation model. The discrepancy in the aspects of awareness, achievement, and religious and general culture between Muslim families and individuals in the diaspora is reflected in the nature of the problems they suffer from and their reactions to deal with them. Families with simple cultural and achievement levels, for example, suffer from leisure problems more than those who have a reasonable amount of culture and achievement that helps them adapt and obtain a job opportunity at rates that far exceed the first levels. This point raises the possibility of containing the crisis of participation through digital technology. The size of the digital society that exists through virtual space may express the percentage of participation, activity, and response to public issues, and these percentages may remain mere statistics that do not affect the centers of decision-making. In light of the foregoing, it is clear that the digital environment can provide electronic support for citizen participation through the mechanisms of digital democracy, and thus the model of participatory democracy can be rebuilt through information and communication technology, despite some shortcomings, but providing good infrastructure and building an effective security system will It definitely leads to the renewal of participatory democratic thought.

3.1.5 Digital Illiteracy

The low levels of literacy on the continent greatly affect the ability of youth to make informed choices about political issues and decisions. This has led to youth being manipulated by political leaders, including during elections where youth failed to investigate the electoral promises of the candidates.

Digital illiteracy goes beyond simply acquiring isolated technical skills to generate a deeper understanding of the digital environment, to the spread of digital awareness of the importance of employing the tools and applications offered by the networked society that can facilitate human life, by creating content and sharing it with others. It can also be viewed as one of the elements of cultural illiteracy in general. In general, and in this field there are different points of view. There are those who link digital illiteracy to the ability to deal with computers and modern information technologies, and there are those who go to link it to the individuals' lack of basic skills in how to deal with information sources and achieve optimal investment for them, and this inevitably affects the ability of The individual understands the mechanism of participation through the network or even interacts positively with simple applications used in electronic voting.

In this context, the Independent Commission, according to the statements of its officials, is keen to allow the largest possible number of local, Arab, and international observers to monitor the course of the electoral process. The day before yesterday, at the Professional Associations Complex, the head of the commission, Dr. Khaled Al-Kalaldeh, revealed that the number of applicants to monitor the elections had reached 5,000 observers. And the number according to the Kalaldeh can be increased daily.

So, there is a keenness to increase the number of observers and oversight bodies, to ensure that the "commission" is concerned with the integrity of the elections, and is also confident of the measures it has taken and will take to ensure that. There is also a keenness, according to Kalaldeh, for observers to have a role in exposing errors and abuses that may occur during the electoral process. This is a good thing, and it must be dealt with the utmost seriousness and care. Many observers enjoy high credibility, and they will not accept any irregularities in the elections, they will monitor them with the utmost precision, and they will announce their position accurately, away from pleasantries. And if they monitor systematic or unsystematic abuses, they will announce it.

Finally, we cannot rely on electronic voting completely, because it may not work properly as a result of technical problems, and therefore there should also be emergency arrangements in the event of a malfunction in the system, in order to allow voting to be conducted manually [13].

4 Conclusion

Despite the widespread circulation of the concept of participatory democracy and the prevalence of value judgments as the ideal democracy that actually guarantees the participation of individuals in managing public affairs, this model is frequently subject to regression due to the difficulty of achieving it in reality, and because of the information revolution, technological developments in Media and communication are capable of removing or at least reducing, structural obstacles to participatory democracy.

4.1 Results

1. It is clear to us that the digital environment has the ability to interact between digital environment and electronic elections, on a large scale through cyberspace, thus providing new mechanisms for participation in elections through electronic voting because it is considered the essence of electoral democracy in the digital environment so that the digital environment can provide a framework Analytically involved, and thus the possibility of drawing future scenarios for the decision-making process.
2. The elections represented by electronic voting benefit awareness and advocacy among the target groups for the exercise of the electoral process and for the participation of all segments of society to whom the conditions of the electoral process apply, by providing communication, education, and awareness programs for leaders and political decision-makers and participation in parliamentary elections.

4.2 Recommendations

1. We recommend that the Jordanian legislator establish a fund dedicated to information and communication technology to complete the election response represented by electronic voting as a model, to invest in technological projects and regional communications, by financing the main activities of the electronic strategy, and that the elections take place through a platform dedicated to electronic elections.

2. We recommend that the Jordanian legislator, through the provisions of the Elections Law, impose electronic elections represented in voting through digital technologies and organize the issue, by amending the text of Article 31, Paragraph A of the Elections Law, by adding the phrase "The voter has the right to cast his vote using electronic means."

References

1. Abdel-Wahhab, T.M.: The Psychology of Political Participation: With a Study in Political Psychology in the Arab Environment, p. 39. Dar Gharib, Cairo (1999)
2. Goodwill Gill, J.S.: Free and Fair Elections: International Law and Practice, 1st edn. International House for Cultural Investments, Cairo, p. 58 (2000)
3. Rashid, A.A.-W.H.: Democratic Transition and Civil Society, 1st edn. Dar al-Thaqafa for publication, Beirut, p. 53 (2003)
4. Abdullah, T.F.: Mechanisms of Democratic Change in the Arab World, 2nd edn. Center for Arab Unity Studies, Beirut, p. 22 (2004)
5. Al-Din, B.A.Z.: Contemporary Electoral Systems: A Comparative Study, 1st edn., p. 19. University Thought House, Alexandria (2011)
6. Al-Damour, H.S.: Democracy vs. Digital Democracy (2021). https://strategiecs.com/ar/analyses
7. Noun, J.: Digital Political Action in the Arab World and the System of Values and Transformations (2019). A published article available via the following link. https://studies.aljazeera.net
8. Al-Manaseer, S.A.E.: The role of electronic administration in developing the electoral process, electronic voting as a model. Psychol. Educ. J. **58**(1), 1217–1236 (2021). (in English)
9. Navarro, C., Ellis, A.: Voting from Abroad The International Institute for Democracy and Elections Handbook. https://www.idea.int/sites/default/files/publications
10. A political paper on the International Institute for Democracy and Elections, the International Institute for Democracy and Electoral Resources on Electoral Processes (2011)
11. Ben Yazza, Y.: The contribution of the digital environment, support for participatory democracy, remote participatory tools as a model. J. Soc. Hum. Sci. **20**(1), 58 (2019)
12. Balih, S.: The Concept of Elections (2017). A published article available via the following electronic link. https://www.politicalencyclopedia.org/diction
13. WHO Options for improving decision-making by governing bodies Use of electronic voting system for appointment of the Director-General Report by the Secretariat, 134th session 20 December 2013, provisional agenda

The Impact of Digital Transformation on Income Distribution and Job Creation - The Jordanian Economy Case

Ibrahim Mohammad Khrais[1(✉)], Mazen Hasan Basha[1], Khaled Mohammed Al-Sawaie[1], Thaer Ahmad Abu-Saleem[1], Mohammad Musa Al-Momani[1], Abdelhalim Mohammad Jubran[1], Nidal Ali Abbas[1], and Nadra Taufig[2]

[1] Department of Economics, Zarqa University, Zarqa, Jordan
{ikhrais,mbasha,ksawaie,thabusaleem,m.momani,ajubran, nabbas}@zu.edu.jo

[2] Department of HRM, University of Business and Technology, Jeddah, Kingdom of Saudi Arabia
n.taufig@ubt.edu.sa

Abstract. The purpose of this paper is to analyze the impact of digital transformation on income distribution and job creation by systematic literature review methodology. The author gathered literature from well-reputed journals to analyze and collect evidence. It is important to analyze the outcomes of digitalization as some questions were in the mind of different researchers from different papers. The major concern of researchers was to investigate digitalization's impact on employment because if people lose jobs, the distribution of income suffers, and people with low skills are not beneficiaries of digitalization. Studies provide evidence that digitalization is enabling economies with more well-paid jobs, Jordanian economy is also a beneficiary of digitalization because the Jordanian government is active in building the digital infrastructure that can be seen in Jordanian public sector organizations. The Jordanian economy is developing economy in the context of digitalization and evidence from the literature supports the positive impact of digitalization on employment in developing economies. This study also examined studies conducted in the Jordanian economic environment these studies also second that digital transformation is vital for the Jordanian economy.

Keywords: digital transformation · income distribution · job creation · systematic review · Jordanian Economy · Jordan

1 Introduction

Innovation is driving the world in this era and the digital transformation in the business environment enables organizations to create value by using information communication technologies. This transformation of business operation is significant in value creation for customers but other effects of this transformation as job creation and income distribution

B. A. M. Alareeni and I. Elgedawy (Eds.): *Artificial Intelligence (AI) and Finance*, SSDC 488, pp. 262–271, 2023.
https://doi.org/10.1007/978-3-031-39158-3_24

need to be analyzed. This transformation is contributing to change in business and society, to investigate the changes of digital transformation needs to be studied in different economies [1].

Digital transformation is a prominent research topic for researchers these days, as there is a consensus between academia and industry professionals that innovation is the driving factor for organizational growth [2]. Organizational growth is important but this needs to be taken care of this growth is not only getting more customers and profits but employees are also integral parts of the organization and the impact of digitalization on employment is important to investigate. The impact of digitalization is important to monitor as almost every sector is reshaping because of digitalization. The impact of digital transformation is not limited to only one or a few sectors but it impacts on both manufacturing and services industries [3]. There are different questions for researchers about this digital transformation including whether this transformation is creating or cutting jobs for job seekers. Digital transformation optimizing the role of machines while conducting business operations. Digital transformation represents the transition of business operations from analog to technology base operations [2]. The reason for this change is innovation and the desire of humans to change the mode of doing business to create value for the customer. The process of this digital transformation is impacting employees as well. The impacts of digital transformation on job creation are important, there is a need to understand the role of digitalization, and whether is it creating jobs or destroying jobs [4].

Digitalization is impacting countries while making their long-term plans, Jordan is one of these countries that are looking to use information communication technologies for the development of the individual level and national infrastructure as well [5]. Public administration is one domain that is important for governments to digitalize for the improvement of the performance of the employees and make people get used to these technologies that lead to promoting the culture of digitalized job opportunities. Jordan is implementing t innovation and digital technologies in government organizations which need to be analyzed carefully. The impact of digital technologies in public organization jobs is important to study [6]. The Jordanian government is implementing information communication technologies which are leading in changing jobs and arising questions on how to perform jobs and creation of jobs. Mobile government is a new domain that needs to be studied as these changes in operations are having an impact on youngsters in acceptance. The new phenomena of technology are impacting low-skilled workers in government sectors [7]. Public sector universities in Jordan are also implementing new technologies to offer new and innovative services for libraries. These mobile technology library services are user-friendly and they can access them with flexibility. The impact of these innovative technologies on employees of these libraries needs to be investigated as the current staff is skillful to perform this digitalized task and how these digital changes are impacting on jobs [8]. Jordan is a country implementing E-government through different official websites the impact of this digitalization on jobs of public sector staff needs to be investigated [9]. The interest of the Jordan government in digital transformation is impacting the commercial and private sectors as well the private sector is using different technologies for business development and growth. Data warehousing is contributing in decision-making in today's business environment and different sectors are using these support systems as commercial banks. These support systems need skilled labor to perform these tasks. Research is needed for the contribution of these skills for the workforce in getting new jobs [10]. This paper aims to contribute to the digital transformation literature to identify the relation between digital transformations of organizations and job creation and income distribution by focusing on the Jordanian economy.

The impact of Digital transformation on the economy is a critical issue to investigate for researchers [1]. Digitalization is unstoppable for any economy but it is also important to investigate the impacts of digitalization on policymaking. The impact of digital transformation on the creation of jobs is integral to investigate as it could lead to job creation or job distraction. Digitalization is in full swing and policymakers are wondering did this adoption lead them to job creation or job destruction [4]. Researchers are investigating the impacts of digital transformation but it is challenging for researchers who need more understanding of the outcomes of digital transformation [2]. Researchers are investigating the possible outcomes in different sectors and economies as the impact of digital transformation is different in different industries.

Academic research is expanding its domain from the impact of digital transformation on services and manufacturing to digital products [11]. Digital products are increasing in number which grabs the attention of policymakers to understand the impact on job creation and income distribution.

Digital transformation impacts differently in different economies. Economies with low access to digital technologies and also having low technology base skilled labor need more focus to determine the impact of digitalization. Developing countries are struggling to make decisions to adopt information communication technologies and need guidelines and evidence from other economies about the impact of digitalization. Jordan is looking to adopt information communication technologies to build a strong digital infrastructure to compete internationally [5]. Despite all the efforts of the Jordanian government, there is still a gap exists in research about the impact of digitalization. The Jordan government's vision is to adopt digitalization but there is a need for qualitative and empirical evidence from the Jordanian economy and other economies about the capacity of people to adopt digitalization and its impact on employment [7]. The Jordan government is implementing digitalization in government organizations but needs to investigate the impact of this digitalization (J-8). Mobile technologies' impact on academic library staff is still not explored completely because of many different reasons [8]. Jordan is one of the developing economies in the context of adopting digitalization as it needs contributions from academia to investigate the impact of digitalization on employment. This investigation can be done by analyzing the literature from other economies and the Jordanian economy. This study aims to investigate the literature of Jordanian and other economies about digitalization's impact on job creation and income distribution. Therefore, the study objectives are:

- A systematic review of literature on digital transformation.
- to analyze the impact of digital transformation on income distribution and employment.

- Provide future guidelines for more clarification of digital transformation and its contribution in the Jordanian economy.

A literature review of digital transformation and its impact on job creation and income from different economies and Jordan will be helpful for this study to contribute in the literature of digital transformation. Therefore, the study Questions are (Fig. 1):

- what literature explains digital transformation?
- What is the impact of digital transformation on employment?
- What suggestions can be given to the Jordanian economy through analyzing digital transformation literature?

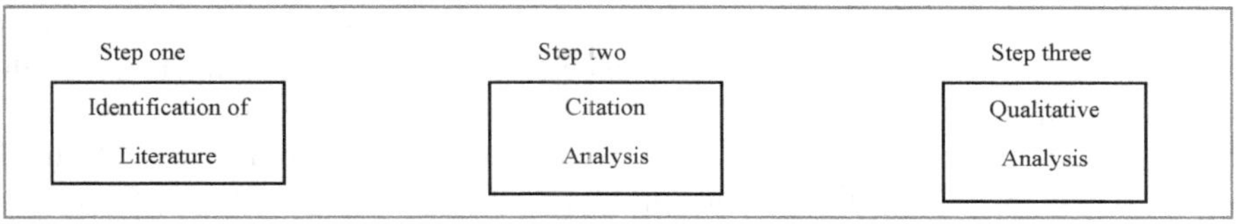

Fig. 1. Research Structure

2 Academic Literature Selection

Digital transformation attracts the attention of researchers nowadays and recognition of this topic is widely accepted in many disciplines digital transformation strongly impacts the academic domain of management [2]. Digital transformation impacts the value-creation process of the manufacturing and services sector [3]. There is a need to understand the challenges organizations are facing in manufacturing industries and propose new future directions as in the journey of digital transformation researchers are not sure where they stand [1]. There are a few directions for conducting research on digital transformation as the impact of digital transformation on income distribution and job creation, especially after covid-19 [12]. Digital technologies contribute not only for organizations but on an individual level as well. Digital technologies are important for economic growth but these technologies also impact labor markets [13]. The services sector is important and a major contributor in the economy, there is a boom in the services sector of developing countries and interestingly acceptance of digital technologies in these economies is observed [14, 15]. Developing countries face a lack of resources and management in their economies which opens the way for foreign direct investment. Developing countries face the issue of low-skilled labor so foreign direct investment in relation to digitalization needs attention for research [16]. Digital technologies empower organizations with supportive tools for decision making as the use of big data to improve business decisions [11]. Digital technologies are accepted in both the manufacturing and services sector for business growth. Although ample literature is available on the acceptance of digital technologies, this changing scenario raises a few questions such as whether the adoption of digital technologies creates more jobs or destroys jobs [4].

These questions are vital to investigate as not all economies have highly skilled labor. The question of technology impact is critical as the economy is reshaping to the digital economy [17]. Digital economies are connected globally, especially Digital Trade [18], and labor markets are turned into online labor markets. These online economies create a demand for skilled labor and a need supply of skilled labor [19]. Skilled labor is critical for any economy as digital transformation needs to educate skilled labor but not all economies have access to digital technologies and these economies might face problems with low-skilled labor as well. To mitigate this issue, economies are trying to build digital infrastructure. Jordan is an emerging economy and the Jordanian government is taking steps to adopt information communication technologies to build a competitive state [5]. Jordan is a country with educated people and youth and the Jordanian government implementing digitalization in the public sector as well. Jordan's public organizations introduced a mobile government to interact with the citizens of Jordan [7]. Continuity in using E-Government websites based on intention to use by the public and staff competency to assist the public while using E-Government websites [9]. Mobile technology base academic library services introduced in Jordan, mobile technology impact not yet fully explored. The impact of these technologies on staff needs to be examined [8]. Private sectors in the Jordanian economy are also using digital technologies as commercial banks are using data warehouses for decision-making. These new tools for business decisions need more skilled labor who can perform these tasks [10]. Jordan is an emerging economy with the will to be a competitive state by using digital technologies, research support is needed to enlighten policymakers. This study aims to analyze the literature on digital transformation's impact on employment and income distribution. This study will help the Jordanian government and the private sector. After getting evidence from other studies, policymakers can build strategies with strong beliefs about the impact of digital transformation. Above mentioned detail of literature represents studies and their direction while conducting research on digital transformation. These papers are summarized in a table that reflects the topics and context of studies with the research objectives of those studies. The last column of the table explains the conclusion of these studies. The conclusions of these studies represent the outcomes of digitalization in different organizations in different scenarios. This reviewed literature provides bases for building hypotheses for this study. These hypotheses are built on the basis of the analyzed literature.

2.1 Research Hypotheses

H_1: Digital transformation has a significant impact on both services and manufacturing sectors.
H_2: Digital transformation in organizations increases employment opportunities.
H_3: Literature supports the positive impact of digital transformation in Jordanian society.

The review of different papers is presented below mentioned Table 1 with the research objectives of these papers and the last column of the table represents the outcome of these studies which reflects the acceptance of the hypothesis of this study.

Table 1. Article Systematic Review

Title	Context	Research Questions/Objective	Results
Digital transformation of Organizations: what do we know and where to go next?	Literature review	A systematic review of literature on digital transformation	More research is needed for the clarification of the impact of digital transformation
Is this time different? How digitalization influences job creation and destruction	The survey of Switzerland's economy	Digital transformation creates or cuts job opportunities	Increased digital investment results in increased employment opportunities
Technology and Employment: Twelve Stylized Facts for the Digital Age	Impact of technology on employment	Stylized facts on the relationship between digital transformation and employment	The positive impact of technology on employment
COVID-19 acceleration in digitalization, aggregate productivity growth and the functional income distribution	EU economy	Digital transformation effect on the labor market and income distribution	There was significant growth observed in the services sector through digital transformation
New digital technologies and heterogeneous wage and employment dynamics in the United States: Evidence from individual-level data	Impact of digitalization on the individual level	How digital technologies are affecting individuals for finding jobs	Workers with high levels of formal education are most affected by the new generation of digital technologies
Digital infrastructure and employment in services: Evidence from Sub-Saharan African countries	Empirical analysis of digitalization in the services sector	Impact of digitalization on the services sector of sub-Saharan African countries	The positive contribution of digital infrastructure on services sector employment

(*continued*)

Table 1. (*continued*)

Title	Context	Research Questions/Objective	Results
FDI, digitalization and employment: empirical evidence from developing economies	Secondary data analysis collected by international labor organizations	Impact of FDI through digitalization on employment opportunities	FDI inflows and digitalization have positive effects on employment
Market power and artificial intelligence work in online labor markets	Online labor markets	To analyze online labor markets do they able to create more demand and supply of online workers	The online labor market creates more jobs and is highly paid jobs as compared to traditional jobs
Jordan and ICT-led Development: towards a competitive state?	Digital infrastructure in Jordan	How Jordan can build a strong infrastructure to compete internationally	Jordan is acting as a competitive state in trying to develop a strong ICT sector
M-government trust framework: deployment of an empirical study among Jordanian youth	Public willingness to use mobile government services in Jordan	To analyze Jordanian people's trust and willingness to adopt mobile government services	Public readiness for using new methods of interaction with the government needs more digital skill labor
An empirical study of data warehouse systems effectiveness: the case of Jordanian banks in the business intelligence era	Commercial banks of Jordan	role of the data warehouse as a supportive decision-making tool	Data warehouses play a significant role in decision-making in the emerging economy of Jordan

3 Research Methodology

3.1 Identification of Literature

The start of the literature review for this paper was a selection of keywords, "Digitalization", "digital transformation" and "information communication technologies" digitalization and job creation" were the targeted keywords. Literature was reviewed after reading different research papers based on different economies and scenarios. The next was to select the relevant research papers that can contribute to the scope of this study. The selection of research papers was aligned with a few more points as, the selection of research papers was limited to the databases of Scopus or Web of sciences, keywords were compared with the theme of publications, full-text articles were included to enable an analysis of the complete text of the paper, those articles were included which were reviewed and have citations, the research period was limited not before 2005.

3.2 Citation Analysis

The selected literature was based on different sectors that include government and private sectors. The selected papers were from not only the developed countries but those papers were also part of this study which investigated developing countries that have low access to information communication technologies and low-skilled labor as well. The selected literature was from both manufacturing and services sectors. Those papers were selected that provide a systematic literature review and quantitative analysis as well. These selected papers contributed to the main theme of the study and enabled this study to contribute to the literature on digital transformation.

3.3 Qualitative Analysis

The requirement of this study was to review the literature to determine the impact of digital transformation on the creation of jobs and income distribution. The main focus while reviewing the literature was to analyze the impact of digitalization on job creation, as a major concern for researchers is to provide an analysis of whether this transition of technology is helping in the creation of more jobs or limiting the jobs opportunities for humans.

This study performed a systematic literature review to identify the research trends regarding digital transformation in the services and manufacturing industries. The keywords were selected from research papers not because of the frequency of these keywords in research papers but of their significance in literature. Further, it is more important to analyze the keywords that are linked to digital transformation than digital transformation itself as this study analyzed articles on digital transformation in both the manufacturing and service industries.

4 Discussion

Information communication technologies brought change to organizations and firms, this change is considered digital transformation, and the concept of digital transformation was introduced in 2000 [20]. Organizations need adequate skilled labor to adopt this digital transformation. Organizations and economies are not ready for radical change. Digital transformation affects the economies in different aspects, employees are a major concern during this transition process.

Organizations operate in business environments that vary in different economies, and few economies have the willingness to adopt digital infrastructure and have adequately educated people. Economies with the will to adopt but with low-skilled labor are facing challenges as well. The other concerns are there for decision-makers and academic researchers include the impact of digitalization on employment. Businesses need growth and digitalization adding value for customers that result in business growth but other vital stakeholders are employees who are a major concern for policymakers.

Academia researchers focusing on the topic of digital transformation to analyze its impact in different economies and compare evidence from different economies to understand the impact on employment because it contributes to income generation. Researchers collected evidence from different countries and provided guidelines for the adoption of digital technologies.

5 Conclusion

This study collected different research papers to understand the impact of digital transformation on employment. A systematic review of the literature was conducted to analyze the impact of digitalization on employment. Research papers were collected from different journals, these papers analyzed the impact of digitalization on different sectors, and analysis of these papers was conducted with a systematic review of the literature and with empirical analysis. This study provided information about the Jordanian economy as literature of Jordanian papers was also collected.

This study concludes after reviewing the literature that digital transformation creates more jobs, especially for those economies that have educated people because they can learn new skills. This study also concludes that the Jordanian economy has the will to adopt digital transformation as the Jordan government introduces E-Government services and mobile base technologies in public sector organizations.

This study found the impact of digital transformation positive in different economies. This study analyzed the efforts made by the Jordanian government in the public sector to get digital to facilitate Jordanian's easy interaction. A significant relationship between digital transformation and job creation and income distribution was observed while conducting a systematic review of the literature on digital transformation.

Jordan's economy is ready for digital transformation as the government's vision was to build digital Infrastructure. Evidence collected from different studies from Jordan and other countries indicate that by doing digital transformation there will be more job opportunities and people with jobs through digital technologies are more paid as compared to traditional job holders.

Literature on digital transformation and its impact on the Jordanian economy is available but there is more research needed as many factors are unexplored and more space is available to understand which sectors could be more productive if these sectors get adequate support for digitalization. This study conducted a systematic literature review but other studies in the future can conduct an empirical analysis of different industries of the Jordanian economy for a better understanding of how digital transformation impacting on the Jordanian economy.

References

1. Favoretto, C., de Sousa Mendes, G.H., Godinho Filho, M., de Oliveira, M.G., Ganga, G.M.D.: Digital transformation of business model in manufacturing companies: challenges and research agenda. J. Bus. Ind. Mark. **37**(4), 748–767 (2021)

2. Jedynak, M., Czakon, W., Kuźniarska, A., Mania, K.: Digital transformation of organizations: what do we know and where to go next? J. Organ. Change Manag. **34**(3), 629–652 (2021)
3. Rha, J.S., Lee, H.H.: Research trends in digital transformation in the service sector: a review based on network text analysis. Serv. Bus. **2022**(16), 77–98 (2022)
4. Balsmeier, B., Woerter, M.: Is this time different? How digitalization influences job creation and destruction. Res. Policy **48**(8), 103765 (2019)
5. Al-Jaghoub, S., Westrup, C.: Jordan and ICT-led development: towards a competition state? Inf. Technol. People **16**(1), 93–110 (2003)
6. Aladwan, S.A., Alshami, S.I.: The impact of service excellence and service innovation on organisational reputation: quantitative evidence from Jordanian public sector. TQM J. **33**(6), 1544–1560 (2021)
7. Alomari, M.K.: M-government trust framework: deployment of an empirical study amongst Jordanian youth. Transform. Gov. People Process Policy **16**(1), 32–50 (2021). (19)
8. Hamad, F., Farajat, S., Hamarsha, A.: Awareness and adoption of mobile technologies in the delivery of services in academic libraries in Jordan: a library staff perspective. Glob. Knowl. Mem. Commun. **67**(6/7), 438–457 (2018)
9. Nawafleh, S.: Factors affecting the continued use of e-government websites by citizens: an exploratory study in the Jordanian public sector. Transform. Gov. People Process Policy **12**(¾), 244–264 (2018)
10. Al-Okaily, A., Al-Okaily, M., Teoh, A.P., Al-Debei, M.M.: An empirical study on data warehouse systems effectiveness: the case of Jordanian banks in the business intelligence era. EuroMed J. Bus. (ahead-of-print) (2022)
11. Hausberg, J.P., Liere-Netheler, K., Packmohr, S., Pakura, S., Vogelsang, K.: Research streams on digital transformation from a holistic business perspective: a systematic literature review and citation network analysis. J. Bus. Econ. **89**(8–9), 931–963 (2019). https://doi.org/10.1007/s11573-019-00956-z
12. Döhring, B., Hristov, A., Maier, C., Roeger, W., Thum-Thysen, A.: COVID-19 acceleration in digitalisation, aggregate productivity growth and the functional income distribution. Int. Econ. Econ. Policy **18**(3), 571–604 (2021)
13. Fossen, F.M., Sorgner, A.: New digital technologies and heterogeneous wage and employment dynamics in the United States: evidence from individual-level data. Technol. Forecast. Soc. Change **175**, 121381 (2022)
14. Ndubuisi, G., Otioma, C., Tetteh, G.K.: Digital infrastructure and employment in services: evidence from Sub-Saharan African countries. Telecommun. Policy **45**(8), 102153 (2022)
15. Khrais, Azzm, Assaf: Constraints facing garment industrial sector operating within the qualified industrial zones in Jordan. Zarqa J. Res. Stud. Humanit. **10**(2), 82–91 (2010)
16. Sinha, Tirtosuharto, Chaudhury, Basu: FDI, digitalization and employment: empirical evidence from developing economies. Stud. Econ. Finance, 1086–7376 (2022)
17. Pianta, M.: Technology and employment: twelve stylised facts for the digital age. Indian J. Labor Econ. **61**(2), 189–225 (2018)
18. Saleem, T.A.A., Abbas, N.A., Al-Sawaie, K.M., Jubran, A.H.M.: The impact of digital trade on Jordan's economic growth. In: 2021 22nd International Arab Conference on Information Technology (ACIT), pp. 1–8. IEEE (2021)
19. Duch-Brown, N., Gomez-Herrera, E., Mueller-Langer, F., Tolan, S.: Market power and artificial intelligence work on online labor markets. Res. Policy **51**(3), 104446 (2022)

The Impact of Digital Media and Its Repercussions on the Contents of Media Messages

Tahseen Sharadga[1(✉)], Asharf Faleh Al-Zoubi[1], Abd Allah Aljalabneh[1], and Cristina Greco[2]

[1] Faculty of Media, Zarqa University, Zarqa, Jordan
tahseen.sharadga@gmail.com, {azoubi,ajalabneh}@zu.edu.jo
[2] University of Business and Technology, Jeddah 21448, Saudi Arabia
c.greco@ubt.edu.sa

Abstract. The study aimed to identify the impact of digital media and its repercussions on the contents of media messages. The study adopted the descriptive analytical (survey) method. The researcher applied the study to a random intentional sample of (69) media professionals working at the university and the media institutions, and this sample represented (69%) of the population. The study reached a set of results that there is an impact of digital media on the contents of the media message. Digital media also have repercussions on the content of the media message in terms of its goals and standards that are applied in media institutions. As stated in the results of the study also that one of the most important disadvantages of digital media is the dissemination of rumors, lack of credibility, fueling conflicts, in addition to family disintegration and isolation.

Keywords: Digital Media · Media Message · Media

1 Introduction

The dominance and control of digital media can be attributed to technological superiority. This has led to the widespread use of social networking sites on the Internet (Al-Jalabneh 2020; Sharadga et al. 2022; Sharadga and Safori 2022). Additionally, advancements in communication and media technology have sparked radical transformations in the way we think about communication and operate in the media work environment (Sharadga et al. 2022; Ahmad et al. 2022). It should be noted that digital media has brought about huge changes in the media process, as digital media has led to the digitization of all media processes (Mahfouz 2021). Auxier and Anderson (2021) pointed out that digital media have forced various media outlets to make significant and steady changes in their media operations, and this change has been reflected in the content of the media message. However, the widespread adoption of digital media has also resulted in the proliferation of misinformation and fake news (Al-Jalabneh and Safori 2020; Al-Jalabneh et al. 2022). Due to the ease of access and dissemination of information on digital media platforms,

B. A. M. Alareeni and I. Elgedawy (Eds.): *Artificial Intelligence (AI) and Finance*, SSDC 488, pp. 272–279, 2023.
https://doi.org/10.1007/978-3-031-39158-3_25

false information can spread rapidly and reach a wide audience (AL-Jalabneh 2023). This has led to significant challenges in the fight against misinformation, particularly in the context of social and political issues.

It is worth mentioning that digital media is interactivity, which means the end of the idea of linear communication or one-way communication from the sender to the receiver (Appel et al. 2020). Digital media is also diverse, which means greater options for effective communication (Rabah 2022). Al-Majeed (2020) pointed out that digital media is characterized by integration, as the Internet represents a communication umbrella that combines communication systems and forms. On the other hand, Musa (2019) referred to the types of digital media represented in digital radio, which is the first electronic means of communication, in addition to digital television which both had a major role in the transition from the analog system to the digital system, and finally the websites of newspapers, which imposed themselves as a new media medium for what was written by the ABCs of the media and imposed radical changes in its traditional concept. Boulianne (2020) also showed that digital means represented in Social media, blogs, websites have played an essential role in communicating the content of media messages.

According to the above considerations, digital media comes to express a stage of technological development in the means of communication that rely on electronic media to provide the masses with news and information (Wadham et al. 2019). Here, Saadawi (2022) pointed out that digital media does not mean simply replacing old means such as printed newspapers with modern means such as computers; the issue also affects all parties to the communication process to include the means, the message, the sender, the receiver and feedback.

Hence, research on the impact of digital media and their repercussions on the contents of media messages is justified. We find a lot of literature that supports the need to apply digital media in media institutions, as the results of the study (Mahfouz 2021; Appel et al. 2020) indicated that media organizations still need to know the requirements of digital media and their repercussions for the contents of the media message. Media organizations must also be able to keep pace with the digital age and keep pace with the changes in digital communication.

1.1 Problem of the Study

A deep understanding of the digital media environment with its various variables, characteristics and interactions is the best way to identify indicators that have a direct impact on the performance of communication as an effective element in the contents of the media message. Accordingly, the Kraft study (2019) indicated that digital media imposed the evolution of media work from just a carrier and a transporter of news to an active element in shaping trends, orientations and public opinion and positioning the public on the line as a party in the industry of media meaning.

In view of the magnitude of the transformations introduced by digital media to the interactive media system and to the reality of traditional media with its professional dimensions and professional knowledge of the communication message, it was imposed on the media institution the need to return to specialization for the purpose of developing media professional skills, and raising the quality of media professional performance by controlling the various electronic editorial arts (Plantin and Punathambekar 2019).

From this standpoint, a deep understanding of digital media is an urgent requirement and necessity for media institutions in order to understand and know the content of the media message, in addition to the need to secure and increase the production efficiency of media institutions and analyze the contents of the media message.

1.2 Study Questions

1. What is the impact of exposure to digital media and its repercussions for the content of the media message?
2. What are the most important means adopted by digital media to reach its purposes in influencing the content of the media message?
3. What are the disadvantages of digital media and its repercussions for the content of the media message?

1.3 Objectives of the Study

Identifying the impact of exposure to digital media and its repercussions for the content of the media message. And identifying the most important means adopted by digital media to reach its purposes in influencing the content of the media message. In addition to identifying the disadvantages of digital media and its repercussions for the content of the media message.

1.4 Significance of the Study

Shedding light on digital media and its repercussions for the contents of media messages. And the discovery that digital media is keeping up work in a way that affects the content of the media message. In addition to revealing the disadvantages that result from employing digital media and their repercussions on the content of the media message. Finally providing those in charge of media institutions with the importance of digital media and its methods in media work.

1.5 Terms

Digital Media: It is defined as the means by which the process of communication with audiences takes place, as these means are characterized by the ability to deliver messages at the same moment and quickly to a wide audience of different directions and levels (Auxier and Anderson 2021). Wilson (2021) pointed out that digital media is characterized by its ability to convey news, information and opinions that create the public opinions and develop trends and patterns of behavior that did not exist among the target audience, including the press, radio, television, cinema, books, audio and video recordings, the Internet and many others.

Media Message: Itis defined as the symbol or set of symbols prepared and formulated to be conveyed from the sender to the receiver, and as any content transmitted by communication in a simplified manner that is understood and benefited by the target audience (Appel et al. 2020). Wu (2020) also defined the media message as the well-prepared content in terms of communication logic media arguments and evidence aimed at convincing the target audience of the campaign issues and the ideas of the communicator, and then providing evidence that convinces the audience and arouses its interest and pushes it to search for information, dialogue, and discussion with others.

2 Field Study

2.1 Methodology

In order to achieve the objectives of this study, the study adopted the descriptive analytical (survey) approach with the aim of revealing the impact of digital media and its repercussions on the contents of media messages. This approach uses a comprehensive survey in which large samples will be used to obtain accurate results.

2.2 Study Sample

The study population consists of university media colleges and media institutions. The researcher applied the study to an intentional random sample represented in two layers: the first layer: university institutions, and 50 questionnaires were distributed to them; The second layer: media institutions, 50 questionnaires were distributed to them. Thus, (100) questionnaires were distributed randomly to the intentional study sample, and the number of questionnaires that have been retrieved reached (74), constituting (74%) of the questionnaires that were distributed, and (5) questionnaires were excluded due to incomplete basic data. Thus, the number of questionnaires valid for statistical analysis becomes (69) questionnaires, representing (69%) of the sent questionnaires.

2.3 Study Instrument

The current study relied on the questionnaire in order to collect data and information for the purpose of answering the questions of the study, and the researcher relied when designing the questionnaire to review previous studies related to the subject of the study and seethe theoretical framework of the various related studies. Some of the instruments included in previous studies have been drawn upon.

Validity of the Questionnaire: To ensure the validity of the internal consistency of the questionnaire, it was distributed to a pilot sample consisting of (16) media professionals and media institutions and the correlation coefficients were calculated and the following table shows the procedures for calculating the validity of internal consistency (Table 1).

Reliability of the Questionnaire: To verify the questionnaire reliability, Cronbach's Alpha method was used, as shown in the following table (Table 2):

It is clear from the table that the total reliability coefficient of the dimensions as a whole was (0. 81), which is a high reliability coefficient. This indicates that the questionnaire has a high degree of reliability.

Table 1. Pearson correlation

Dimension	Correlation	sig
Media digital	0.66**	0.000
Content of the media message	0.67**	0.000

Table 2. Cronbach's Alpha

Dimension	Cronbach's Alpha
Media digital	0.63
Content of the media message	0.74
Total	0.81

3 Discussion of Results

First question: What is the impact of exposure to digital media and its repercussions for the content of the media message?

To answer the first question, the arithmetic means, standard deviations, and Chi-Square test of the sample responses were calculated, and the following table illustrates this.

Table 3. Exposure to Digital Media

Items	mean	Standard deviation	Chi-Square	sig
Blogs and forums	4.19	0.72	14.26	0.000
Electronic newspapers	4.11	0.61	21.46	0.000
Social Media	4.07	0.85	54.53	0.000
satellite channels websites	4.01	1.04	21.46	0.000
International Radio website	3.96	1.06	4.84	0.000
Total	4.07	1.80	–	–

It is clear from Table 3 that the arithmetic means of the responses of the study subjects on the impact of exposure to digital media and its repercussions on the media message was (4). 07) with a standard deviation of (1. 80). This indicates a significant degree of impact. We also find that the significance value associated with the Chi-Square value is less than the correct one (1), which indicates the homogeneity of the answers by (68%) to (84%). This indicates the clear impact of digital media and their repercussions on the content of the media message through blogs, forums, electronic newspapers, channels and radio.

The Second Question: The Second Question: What are the most important means adopted by digital media to reach its purposes in influencing the content of the media message?

To answer the second question, the arithmetic means, standard deviations, and Chi-Square test of the sample responses were calculated, and the following table illustrates this.

Table 4. Purposes of Digital Media

Items	mean	Standard deviation	Chi-Square	sig
News	4.19	0.72	14.26	0.000
Entertainment	4.11	0.61	21.46	0.000
Research and studies	4.07	0.85	54.53	0.000
Building relationships	4.01	1.04	21.46	0.000
Build special relationships	3.96	1.06	4.84	0.000
Total	4.07	1.80	–	–

It is clear from Table 4 that the arithmetic means of the responses of the study subjects on the most important means adopted by digital media to reach its purposes in influencing the content of the media message was (4.07) with a standard deviation of (1.80). This indicates a significant degree of impact. We also find that the significance value associated with the Chi-Square value is less than the correct one (1), which indicates the homogeneity of the answers by (68%) to (84%). This indicates the great role of the means adopted by digital media to reach its purposes in influencing the content of the media message; among these means News, Research and studies and building relationships.

The Third Question: What are the disadvantages of digital media and its repercussions for the content of the media message?

To answer the second question, the arithmetic means, standard deviations, and Chi-Square test of the sample responses were calculated, and the following table illustrates this.

It is clear from Table 5 that the arithmetic mean of the responses of the study subjects on the disadvantages of digital media and its repercussions on the media message was (3. 40) with a standard deviation of (1. 98). This indicates a significant degree of impact. We also find that the significance value associated with the Chi-Square value is less than the correct one (1), which indicates the homogeneity of the answers by (68%) to (84%). This indicates that the disadvantages of digital media, which are represented in spreading rumors, not blocking them and fueling the conflict, have a clear role and impact on the content of the media message.

Table 5. Disadvantages of Digital Media

Items	Mean	Standard Deviation	Chi-Square	Sig
Spread Rumors	3.79	0.85	22.53	0.000
Lack Of Credibility	3.66	0.63	.8411	0.000
Fueling Conflicts	3.34	0.81	10.26	0.000
Weaken The Morals of Society	3.12	0.71	32.46	0.000
Isolation And Family Breakdown	3.09	1.01	18.46	0.000
Total	3.40	1.98	–	–

4 Conclusion

The impact of digital media on the contents of the message, which in turn leads to the delivery of the message to the receiving audience in several ways and methods that can affect the attitudes and opinions of the public. As there are Digital media have repercussions for the content of the media message in terms of its objectives and standards that are applied in media organizations. And news was one of the most important means adopted by digital media to reach its purposes in influencing the content of the media message, followed by entertainment, research and studies, and building social relations. And one of the main disadvantages of digital media is the spread of rumors, lack of credibility, and fueling of conflicts, in addition to family disintegration and isolationism.

4.1 Recommendations

Research efforts in the media discipline should focus on studying the repercussions of digital media and its effects on social, cultural, political and economic life. Media institutions of all types and orientations should benefit from the experiences of leading global and regional channels that employ digital media. And they can hold training courses on methods of employing digital media in media institutions. They can employ blogs, forums and electronic newspapers in the digital media system so that media institutions can keep pace with the digital revolution.

References

Ahmad, A.K., AL-Jalabneh, A.A., Mahmoud, A., Safori, A.: Covid-19 and the resurgence of the hypodermic needle theory applicability in times of crises. In: Alareeni, B., Hamdan, A. (eds.) ICBT 2022. LNNS, vol. 495, pp. 1423–143. Springer, Cham (2023). https://doi.org/10.1007/978-3-031-08954-1_124

Al-Jalabneh, A.A.S.H.: The role of television and social media news videos in raising the socio-political awareness of Jordanian university students (Doctoral dissertation, University of the West of England) (2020)

AL-Jalabneh, A.A.: Health misinformation on social media and its impact on COVID-19 vaccine inoculation in Jordan. Commun. Soc. **36**(1), 185–200 (2023)

Al-Jalabneh, A.A., Safori, A.O.: Fake news' threats at the time of a global pandemic: facebook use as a news platform in Jordan. J. Crit. Rev. **7**(19), 9437–9446 (2020)

Al-Jalabneh, A.A., Safori, A.O., Shlool, H.: Covid-19 and misinformation prevalence: a content analysis of fake news stories spread in Jordan. In: Hamdan, A., Shoaib, H.M., Alareeni, B., Hamdan, R. (eds.) The Implementation of Smart Technologies for Business Success and Sustainability. SSDC, vol. 216, pp. 535–545. Springer, Cham (2023). https://doi.org/10.1007/978-3-031-10212-7_44

Appel, G., Grewal, L., Hadi, R., Stephen, A.T.: The future of social media in marketing. J. Acad. Mark. Sci. **48**(1), 79–95 (2020)

Auxier, B., Anderson, M.: Social media use in 2021. Pew Res. Center **1**, 1–4 (2021)

Boulianne, S.: Twenty years of digital media effects on civic and political participation. Commun. Res. **47**(7), 947–966 (2020)

Kraft, U.: How to lead with digital media effectively? A literature-based analysis of media in a E-leadership context. J. Econ. Dev. Environ. People **8**(4), 42–52 (2019)

Mahfouz, I.: Media freedom in electronic platforms. Al-Qabas J. Psychol. Soc. Stud. **12**(3), 1–8 (2021)

Majeed, E.: The role of modern and traditional media in public awareness, Ph.D. thesis, Menoufia University (2020)

Musa, A.: Digital media and its repercussions for the content of the media message. J. Humanit. **19**(2), 162–174 (2019)

Plantin, J.C., Punathambekar, A.: Digital media infrastructures: pipes, platforms, and politics. Media Cult. Soc. **41**(2), 163–174 (2019)

Rabah, R.: Afor the performance of the media professional in light of the digital media environment. Al-Mijyar Mag. **26**(5), 531–554 (2022)

Saadawi, K.: The impact of digital media on the public relations function of service institutions, Ph.D. thesis, Mohamed Boudiaf University in Messila (2022)

Sharadga, T.M.A., Tahat, Z., Safori, A.O.: Journalists' perceptions towards employing artificial intelligence techniques in Jordan TV's newsrooms. Stud. Media Commun. **10**(2), 239–248 (2022)

Sharadga, T.M.A., Safori, A.O.: The effectiveness of digital communication in the education system in Jordan in Light of the Covid-19 Pandemic. In: Hamdan, A., Shoaib, H.M., Alareeni, B., Hamdan, R. (eds.) The Implementation of Smart Technologies for Business Success and Sustainability. Studies in Systems, Decision and Control, vol. 216, pp. 617–626. Springer, Cham (2023). https://doi.org/10.1007/978-3-031-10212-7_51

Wadham, E., Green, C., Debattista, J., Somerset, S., Sav, A.: New digital media interventions for sexual health promotion among young people: a systematic review. Sex. Health **16**(2), 101–123 (2019)

Wilson, D.: Digital media investigators: challenges and opportunities in the use of digital forensics in police investigations in England and Wales. Polic. Int. J. **44**(4), 669–682 (2021)

Wu, S.: Design of interactive digital media course teaching information query system. Inf. Syst. e-Bus. Manag. **18**(4), 793–807 (2020)

Determinants Students Continue Usage of E-Book: A Developing Country Experience

Muamar Nur Kholid[1], Asri Pangestika Lutfiani[2], and Selfira Salsabilla[1(✉)]

[1] Department of Accounting, Faculty of Business and Economics, Universitas Islam Indonesia, Yogyakarta, Indonesia
selfirasalsabilla@uii.ac.id

[2] Department of Accounting, Faculty of Economics and Business, Universitas Muhammadiyah Pekajangan Pekalongan, Pekalongan, Indonesia

Abstract. Numerous developing countries have unsatisfactory levels of literacy. As e-books become more popular, literacy is anticipated to increase since they are considered more portable. This study investigates the factors influencing students' continuance intention of using E-Books. Using the Expectation Confirmation Model (ECM) framework, this study investigates three determinants, namely Satisfaction (SAF), Perceived Usefulness (PUF), Confirmation (COF), and Subjective Norm (SNO). Two hundred eleven university students who had utilized e-books participated in the study. The Partial Least Square method was used to process the data gathered through questionnaires. The results of data analysis show that the ECM framework successfully explains students' intention to continue using E-books. PUF is the strongest variable influencing students' intention to continue using E-Books. The analysis results also show that SNO significantly positively affects SAF and students' intention to continue using E-Books. The results of this study indicate that in order to retain current e-book users, e-book developers should offer clear information on the advantages of e-books and draw on the experiences of previous users.

Keywords: E-Book · Confirmation · Satisfaction · Intention

1 Introduction

Information technology has changed the lifestyle of individuals. The rise of technology significantly impacts people who tend to no longer be satisfied with traditional means of learning and acquiring knowledge [1]. One of the outcomes of information technology in the modern era of education is the development of e-books. The book publishing industry is undergoing a radical transformation driven by current information technology developments. One of these advances is the use of e-books. These publishers have the challenge of increasing customer retention through sustained consumption, notwithstanding the anticipated surge in e-book usage.

The millennial generation comprises students born between 1982 and 2000 [2]. This generation has grown up in a digital world with easier access to technological advances

B. A. M. Alareeni and I. Elgedawy (Eds.): *Artificial Intelligence (AI) and Finance*, SSDC 488, pp. 280–289, 2023.
https://doi.org/10.1007/978-3-031-39158-3_26

such as internet, smartphones, tablet, and personal computers. Academic libraries frequently invest hundreds of millions of rupiah in e-books and infrastructure to support information resources so that students can access e-books for free. Library services are generally utilized by students directly by coming to the library building at the university. However, with various limitations, such as the current pandemic conditions, libraries must innovate in serving students in normal conditions. One of the things that must be changed is the shift in concept to a digital library. The concept of a digital library is not much different from a conventional library, the difference being the reading collection. Conventional libraries have print collections, while digital libraries have electronic collections.

Free access to e-books does not always lead to continuing use of e-books [3]. This e-book collection will support advancing academic and research excellence at the university [4]. However, if users prefer printed books compared to e-books, How can the library make sure that the academic community makes the most use of the e-books that are available?

Several research on e-books have been conducted in the past. Research in Indonesia found that social factors influence the initial adoption of e-books by millennial generation [5]. Research in China found that TAM and TPB were able to explain the initial adoption of e-books by university students [6]. Research in Pakistan found that awareness and attitudes affect the initial acceptance of e-books [7]. While there have been several studies on E-Books, they still focus on initial acceptance. Thus, this study will focus on the post-acceptance intention to use E-Books using the Expectation Confirmation Model (ECM) framework. In addition, this study also tries to elaborate on the influence of social factors on student's intention to continue using E-Books. Based on these facts, The study aims to look into the social aspects that affect students' intentions to continue utilizing E-Books inside the ECM framework.

2 Conceptual Model

2.1 Expectation Confirmation Model (ECM)

Generally, Technology Acceptance Model (TAM) and Expectation Confirmation Theory (ECT) serve as the foundation upon which Expectation Confirmation Model ECM is constructed [8]. In contrast to TAM, which focuses on the adoption of information technology (IT) in its early stages, ECM focuses on the post-acceptance (continued use) of information technology. Three tenets form the foundation of the ECM. First off, the satisfaction and confirmation variables already include the impacts of any pre-acceptance factors. Second, the model evaluates post-consumption (ex-post) expectations because IS usage expectations frequently evolve over time. Third, the model's representation of (ex-post) expectations is (ex-post) perceived Usefulness. Three main ECM variables include Perceived Usefulness (PUF), Confirmation (COF), and Satisfaction (SAF).

ECT states that a consumer's desire to make another purchase a product or continue using a service is determined primarily by the consumer's SAF [9]. Because it comes after the first choice, the option to keep using e-books is comparable to consumers' choice to repurchase a non-digital product, which is influenced by the initial usage experience (satisfied or dissatisfied) [8]. Several studies have found the effect of SAF on the intention to continue using IT [10–12]. Here is the first hypothesis

H1. *Continued intention to use E-books is positively influenced by SAF*

Perceived Usefulness (PUF) represents post-consumption in the ECM model [8]. PUF is an important predictor of adopting technology products and services such as e-books [13]. Given that PUF is a predictor of initial acceptance, it is logical that PUF also influences users' intention to continue using IT [8]. PUF can also affect user satisfaction. When users feel the Usefulness of using IT, users are satisfied with the information technology used. Several previous studies have found the effect of PUF on SAF and the intention to continue using IT [14, 15]. Here is the second and third hypothesis

H2. *Continued intention to use E-books is positively influenced by PUF*
H3. *SAF is positively influenced by PUF*

Confirmation (COF) is the extent to which actual use experience supports one's initial expectations [8]. Users will be highly SAF if they believe their initial expectations match the realities of using IT [8, 16]. In addition, if the user has a high level of COF, the user will have a higher level of PUF [8, 16]. Several research have confirmed the effect of COF on SAF and PUF [17, 18]. Here is the fourth and fifth hypothesis

H4. *SAF is positively influenced by COF*
H5. *PUF is positively influenced by COF*

2.2 Subjective Norm, Satisfaction, and *Continued Intention*

Subjective norm (SNO) is defined as the extent to which users believe that people who are considered important to them have the thought that users should be involved in the use of IT [19]. SNO is crucial to the uptake of new technologies [20]. Empirically, it has been demonstrated that SNO significantly influence a person's intention to utilize a information technology [21]. Empirically SNO also affects SAF [3]. Strong social ties give users more opportunities to recommend services, or even technology products such as e-books to others. In this case, such bonds rely heavily on user satisfaction. Studies on information technology found a positive effect of SNO on SAF and intention to continue using IT [3, 21]. Figure 1 presents the research model that will be tested in this study. Here is the sixth and seventh

H6. *SAF is positively influenced by SNO*
H7. *Continued intention to use E-books is positively influenced by SNO*

3 Methodology

This study involved 211 respondents who were selected using purposive sampling technique. The criteria for respondents in this study are undergraduate accounting students in Indonesia who have used E-books. Indonesia as a developing country was chosen as a sample because Indonesia is a developing country with a low literacy level. This fact encourages various efforts to be able to improve literacy with various innovations, one of which is with E-books. Based on gender, 114 (54.03%) respondents were male

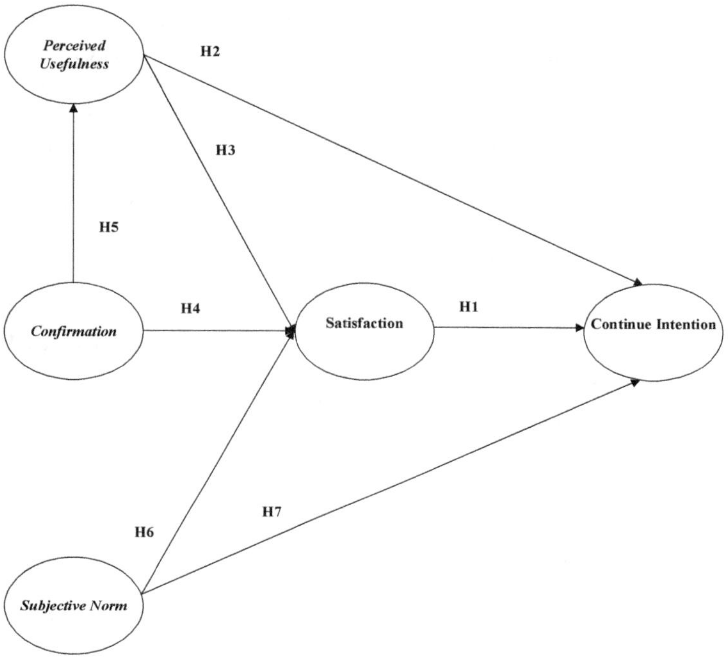

Fig. 1. Research Model

and 97 (45.97%) respondents were female. Based on age, 35 (16.59%) respondents were <20 years old, 154 (72.99%) respondents were 20+ to 21 years old, 17 (8.06%) respondents were 21+ to 22 years old, and 5 (2.37%) respondents were 22+ to 23 years old.

A survey method based on questionnaires was used to carry out this investigation. The questionnaire was given to respondents using a 6 Likert scale. Each variable is measured by several question items adopted from several previous studies. 3 PUF items, 3 COF items, 3 SAF items, and 2 continue intention items were adopted from [8]. Meanwhile, the 3 question items measuring SNO were adopted from [3]. Table 1 displays the questions for each variable. This study utilized SmartPLS 3.0 which is based on Partial Least Square (PLS) to conduct data analysis. Data were analyzed using SmartPLS 3.0 by going through two stages of analysis, namely structural model analysis, and measurement model analysis.

4 Results

The measurement model test results can be seen in Table 1. In connection with testing indicator reliability, it is known that all indicators have a factor loading value above 0.7 in accordance with what is recommended by [22]. In addition, the results of data analysis also show that the average extracted value (AVE) for each construct is in the range between 0.700 and 0.839, where the AVE value is greater than 0.5 [22]. With regard to the reliability test, it is known that all constructs have a composite reliability value above 0.7, which indicates that each construct is able to achieve reliability for the research model. Discriminant validity testing in this study uses the Fornell-Larcker criterion. Table 2 presents information on the results of the discriminant validity test. The test results show that the square root value of AVE for each construct is a greater correlation with its own construct than its correlation with other constructs.

Table 1. Results of Convergent Validity

Constructs	Items	Factor Loading	Composite Reliability (CR)	Average Variance Extracted (AVE)
Continue Intention (INO)			0.961	0.799
INO1	I intend to continue using E-Book	0.943		
INO2	I will always try to use E-Book	0.949		
INO3	I will keep using E-Book	0.940		
Perceived Usefulness (PUF)			0.894	0.738
PUF1	Using e-books can make reading behavior more effective and efficient	0.793		
PUF2	Using e-books is, overall, an efficient way to read	0.909		
PUF3	Using e-books is a convenient reading behavior	0.871		
Subjective Norm (SNO)			0.903	0.700
SNO1	There are others who are using e-books around me	0.752		
SNO2	My friends who know me have positive feelings about using e-books	0.859		
SNO3	People who influence me want to use e-books	0.912		

(continued)

Table 1. (*continued*)

Constructs	Items	Factor Loading	Composite Reliability (CR)	Average Variance Extracted (AVE)
SNO4	My parents are willing to agree with my purchasing an e-book	0.815		
Confirmation (COF)			0.923	0.799
COF1	My experience using the E-Book exceeded my expectations	0.888		
COF2	The features available in the E-Book are better than I expected	0.916		
COF3	Overall, the E-Book met my expectations	0.877		
Satisfaction (SAF)			0.939	0.837
SAF1	I am satisfied with E-Book I have used	0.893		
SAF2	My experience of using E-Book has been satisfactory	0.943		
SAF3	I think I made the correct decision in using E-Book	0.908		

Table 2. Results of Discriminant Validity

	COF	INO	PUF	SAF	SNO
COF	0.894				
INO	0.766	0.944			
PUF	0.600	0.626	0.859		
SAF	0.819	0.826	0.687	0.915	
SNO	0.657	0.689	0.489	0.662	0.836

The outcomes of the structural model analysis are shown in Fig. 2. The results of data analysis show that the model in this study is able to explain students' intention to continue using E-books by 71.8%. Meanwhile, this research model is able to explain SAF by 74.3% and PUF by 35.7%. This study uses hypothesis testing with path analysis where the initial bootstrapping sequence is at the level of 500 samples. With regard to hypothesis testing, it is known that 7 hypotheses in this study are accepted. SAF ($\beta = 0.596$; $p = 0.01$), PUF ($\beta = 0.096$; $p = 0.01$), and SNO ($\beta = 0.248$; $p = 0.01$) have a significant positive effect on the intention to continue using E-books. Meanwhile, COF has a significant positive effect on SAF ($\beta = 0.573$; $p = 0.01$) and PUF ($\beta = 0.600$; $p = 0.01$). The results of the analysis also show that SNO has a significant positive effect on SAF($\beta = 0.172$; $p = 0.01$) and students' intention to continue using E-Books ($\beta = 0.248$; $p = 0.01$).

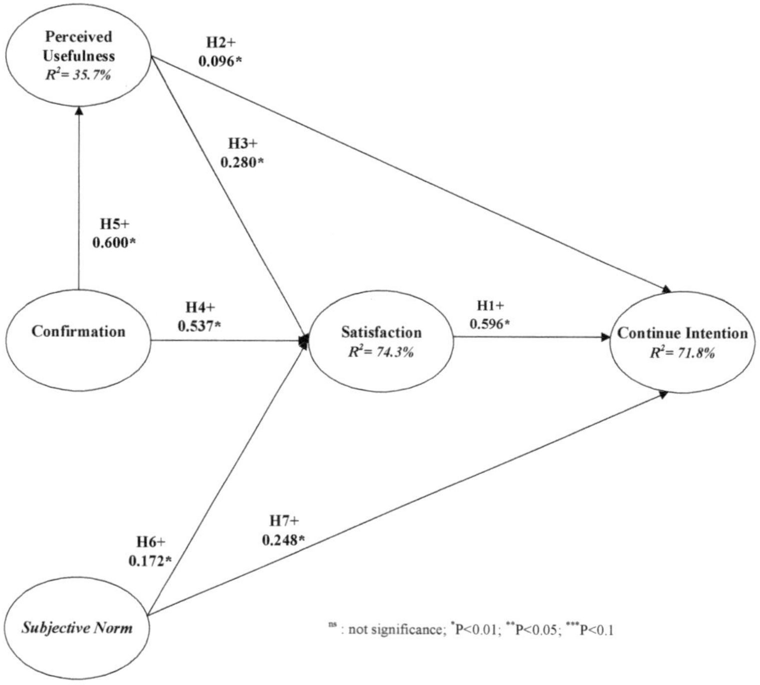

Fig. 2. Results of Structural Model

5 Discussion

This study examines the factors influencing students' intention to continue using E-Books. This study proposes an E-Book continuance intention research framework based on the post-acceptance model of information technology continuance developed by [8]. This study also examines the effect of user-perceived value, which consists of PUF and SNO, on students' intention to continue using E-Books. The test results show that this research model can explain the variance of intention to continue using E-Books at a moderate level.

The results of this study showed that PUF is the factor that has the most significant impact on students' decision to keep using E-books. PUF significantly positively affects students' intention to continue using E-Books. This results is consistent with a number of earlier research [10, 14]. These results indicate that E-Books' usefulness must be well informed to students. Sticky posters, website posters, social media posters, and other engaging media can be used to tell people about the many advantages of e-books.

This study found that SAF significantly positively affects students' intention to continue using E-Books. This result is consistent with several previous studies [10–12]. In addition, this study found empirical evidence that COF has a significant positive effect on SAF and PUF. These findings are consistent with previous studies [17, 18]. These results indicate that the fulfillment of user expectations is critical. Therefore, it is essential to note that the information provided about E-Books and the facts of E-Book usage experience should be aligned.

Finally, the results of this study found that social factors play an essential role in influencing students' intention to continue using E-Books. This result is supported by several previous studies [3, 21]. The results of this study show that social factors cannot be ignored in encouraging students to continue using E-Books. Some information on how previous user experiences can be media content to encourage students to continue using E-Books.

6 Conclusions and Recommendations

According to the study's findings, COF, PUF, and SAF significantly increase students' intentions to continue using E-Books. Additionally, social considerations play a significant role in persuading students to keep utilizing E-Books. These results indicate that students should get precise information about the various benefits of using E-Books. In addition, E-Books providers should also be able to utilize the role of lecturers in encouraging the use of E-Books. It is because social factors cannot be ignored in encouraging students to continue using E-Books. There are limitations to this study, although it can explain some of the factors that affect students' intentions to continue using e-books. Since this study only included respondents from Indonesia, additional research would be highly beneficial to diversify the study's geographic scope and gain a deeper understanding of the variables that motivate students to keep using E-Books. This research does not analyze the evaluation of users' behavioral intentions over time. Therefore, future study can conduct longitudinal research to get a picture of the stability of the proposed hypothesis.

References

1. Waheed, M., Kaur, K., Ain, N., Sanni, S.A.: Emotional attachment and multidimensional self-efficacy: extension of innovation diffusion theory in the context of eBook reader. Behav. Inf. Technol. **34**, 1147–1159 (2015)
2. Kominfo: Mengenal Generasi Millennial. https://www.kominfo.go.id/content/detail/8566/mengenal-generasi-millennial/0/sorotan_media. Accessed 24 Feb 2022
3. Jin, C.-H.: Adoption of e-book among college students: the perspective of an integrated TAM. Comput. Hum. Behav. **41**, 471–477 (2014). https://doi.org/10.1016/j.chb.2014.09.056

4. Smith, S.L., Rodriguez, A., Miller, E., Xu, L.: The relationship between the technology acceptance model and preference for ebooks at a large research university. Libr. Hi Tech News **36**, 13–15 (2019)
5. Srirahayu, D.P., Nurpratama, M.R., Handriana, T., Hartini, S.: Effect of gender, social influence, and emotional factors in usage of e-books by generation Z in Indonesia. Digit. Libr. **38**, 263–282 (2022)
6. Luo, Y.Z., Xiao, Y.M., Ma, Y.Y., Li, C.: Discussion of students' e-book reading intention with the integration of theory of planned behavior and technology acceptance model. Front Psychol. **12** (2021). https://doi.org/10.3389/fpsyg.2021.752188
7. Jbeen, A., Rehman, S.U., Mahmood, K.: Awareness, use and attitudes of students towards e-books: differences based on gender, discipline and degree level. Glob. Knowl. Mem. Commun. **71** (2022)
8. Bhattacherjee, A.: Understanding Information systems continuance: an expectation-confirmation model. MIS Q. **25**, 351–370 (2001)
9. Anderson, E.W., Sullivan, M.W.: The antecedents and consequences of customer satisfaction for firms. Mark. Sci. **12**, 125–143 (1993)
10. Kholid, M.N., Urumsah, D., Hamdani, R.: Expectation confirmation model in the transportation order applications: gender differences. In: 31st IBIMA Conference, Milan, Italy (2018)
11. Mishra, A., Shukla, A., Rana, N.P., Currie, W.L., Dwivedi, Y.K.: Re-examining post-acceptance model of information systems continuance: a revised theoretical model using MASEM approach. Int. J. Inf. Manag. **68** (2023). https://doi.org/10.1016/j.ijinfomgt.2022.102571
12. Liu, Y., Li, Q., Edu, T., Negricea, I.C.: Exploring the continuance usage intention of travel applications in the case of Chinese tourists. J. Hosp. Tour. Res. (2020). https://doi.org/10.1177/1096348020962553
13. Davis, F.D., Bagozzi, R.P., Warshaw, P.R.: Extrinsic and intrinsic motivation to use computers in the workplace. J. Appl. Soc. Psychol. **22**, 1111–1132 (1992). https://doi.org/10.1111/j.1559-1816.1992.tb00945.x
14. Wang, X., Liu, T., Wang, J., Tian, J.: Understanding learner continuance intention: a comparison of live video learning, pre-recorded video learning and hybrid video learning in COVID-19 pandemic. Int. J. Hum. Comput. Interact. **38**, 263–281 (2022). https://doi.org/10.1080/10447318.2021.1938389
15. Cai, J., Zhao, Y., Sun, J.: Factors influencing fitness app users' behavior in China. Int. J. Hum. Comput. Interact. **38**, 53–63 (2022). https://doi.org/10.1080/10447318.2021.1921483
16. Thong, J.Y.L., Hong, S.J., Tam, K.Y.: The effects of post-adoption beliefs on the expectation-confirmation model for information technology continuance. Int. J. Hum. Comput. Stud. **64**, 799–810 (2006). https://doi.org/10.1016/j.ijhcs.2006.05.001
17. Taghizadeh, S.K., et al.: Factors influencing students' continuance usage intention with online learning during the pandemic: a cross-country analysis. Behav. Inf. Technol. **41**, 1998–2017 (2022). https://doi.org/10.1080/0144929X.2021.1912181
18. Abdul-Halim, N.-A., Vafaei-Zadeh, A., Hanifah, H., Teoh, A.P., Nawaser, K.: Understanding the determinants of e-wallet continuance usage intention in Malaysia. Qual. Quant. **56**, 3413–3439 (2022). https://doi.org/10.1007/s11135-021-01276-7
19. Noh, Y.: A study on the effect of digital literacy on information use behaviour. J. Librariansh. Inf. Sci. **49**, 1–31 (2016)
20. Venkatesh, V., Davis, F.D.: A theoretical extension of the technology acceptance model: four longitudinal field studies. Manag. Sci. **46**, 186–204 (2000)

21. Hsiao, C.H., Chang, J.J., Tang, K.Y.: Exploring the influential factors in continuance usage of mobile social apps: satisfaction, habit, and customer value perspectives. Telemat. Inform. **33**, 342–355 (2015). https://doi.org/10.1016/j.tele.2015.08.014
22. Hair, J.F., Hult, G.T.M., Ringle, C.M., Sarstedt, M.: A Primer on Partial Least Squares Structural Equation Modeling (PLS-SEM). Sage Publications, Loas Angeles (2017)

Mechanisms of Digital Currencies

Mohanad Hameed Mhaidi(✉) and Saher Mukhlif Habeeb

College of Law and Political Sciences, Anbar, Iraq
{muhannadhamd84,sa1984her}@uoanbar.edu.iq

Abstract. Currency represents an essential element in economic relations. The history of money shows that it is highly influenced by the technology available in every period of human history, so it is natural that the technological development that became a reality in the twentieth century had an impact on the form of money and the methods of dealing with it. When electronic cards appeared, metals and papers disappeared with it, and a small magnetic card replaced hundreds of thousands of dollars. The technological development in the field of business, and the shift in traditional business models by the advantage of the internet, has led to the need to search for a payment tool that is compatible with these new developments, which impose themselves on the economic reality.

Keywords: Digital Currencies · Bitcoin · Blockchain

1 Introduction

Currency represents an essential element in economic relations. The history of money shows that it is highly influenced by the technology available in every period of human history, so it is natural that the technological development that became a reality in the twentieth century had an impact on the form of money and the methods of dealing with it. When electronic cards appeared, metals and papers disappeared with it, and a small magnetic card replaced hundreds of thousands of dollars.

The technological development in the field of business, and the shift in traditional business models by the advantage of the internet, has led to the need to search for a payment tool that is compatible with these new developments, which impose themselves on the economic reality.

It is likely to see emerging digital symbols continue to drive innovation in the global payments system, transcend national boundaries and redefine the way payments and user data interact. Ultimately, many of national digital currencies will be accessed in the market, supported by central banks in an appropriate regulatory framework.

The importance of the research comes through the importance of the financial and monetary systems of any country that results casting a shadow on the mechanism of

---

The original version of this chapter was revised: Abstract has been updated in online version. The correction to this chapter is available at
https://doi.org/10.1007/978-3-031-39158-3_90

B. A. M. Alareeni and I. Elgedawy (Eds.): *Artificial Intelligence (AI) and Finance*, SSDC 488, pp. 290–296, 2023.
https://doi.org/10.1007/978-3-031-39158-3_27

action of these systems. Accordingly, this study comes to clarify the mechanisms that operate via digital currencies.

The research is divided into two main axes. The first axis deals with the conceptual and historical aspects of digital currencies. The second axis aims to study the mechanisms of digital currencies which, in turn, are divided into two parts, technical and economic mechanisms.

2 Conceptual Framework

In this section, we will explain what digital currencies are and distinguish them from the rest of the forms of monetary currencies, then, proceed to address the backgrounds of their origin and historical development.

2.1 Digital Currencies

The terminology that is used to describe digital currencies is often confusing and misleading. It covers a wide range of financial instruments with different technical, legal and practical characteristics. They range from decentralized digital tokens such as Bitcoin to official electronic currencies that are sovereignly supported by the central banks at the other party.

The historical roots of cryptography go back to the Greek which means the writing of secrets or secret writing. It is defined as the generation of codes that allow information to be kept secret (Othman 2022).

Bitcoins and many forms similar to them-often known as "crypto-assets" - are perceived as a volatile and speculative asset class; which some users are ready to accept as a form of payment.

The first definition of a digital currency was introduced by the Basel Committee on Banking Supervision (BCBS), which referred to it as a "prepaid payment terminal for storing monetary value, which can be transferred by electronic devices with Internet access." Subsequently, the European Central Bank and the Bank for International Settlement (BIS) published reports explaining that the digital currency is an "electronic store of monetary value on a technical device", which can be used to make payments and investments without the participation of a third party in transactions in 2008. The International Monetary Fund (IMF) repeated the definition provided by the European Central Bank and proposed a simplified explanation of the digital currency as an Internet-based medium of exchange (Basel Committee on Banking 1998).

The first appearance of the term "Cryptocurrencies" coincided with the emergence of Bitcoin, which was translated into digital currencies. Despite the diversity and multiplicity of digital currencies, there is a real problem embodied in the number of currencies that fall under this name.

Digital currencies can also be defined as "an electronic currency with a decentralized character, based on a peer-to-peer network as well as electronic signatures and cryptography". It aims to enable users to conduct currency transfers and trading operations via the internet without the need for an intermediary or a trusted third party such as banks (Financial Action Task Force 2014).

Unlike traditional currencies, digital currencies operate without authorities or central banks and are not supported by any government. Cryptocurrencies are stored in “digital portfolios” on the computer or telephone of the bearer or cold portfolio account such as (ledger nano x), (ledger nano s) etc. The portfolio functions as a virtual bank account that enables the holders to pay for goods and services or simply store the currency in the hope of increasing its value (Rose 2015).

The most important thing that distinguishes a digital currency from an existing electronic currency is, perhaps, it never takes on a physical form. You can go to an ATM machine and convert the electronic record of your currency holdings into actual dollars. Hence a digital currency never takes a physical form; it is always kept on the computer network and is exchanged through digital media.

2.2 Blockchain

The term is composed of two words block and chain. So literally the blockchain indicates a reference to a set of spaces for recording data related to transactions. Each space has a certain amount of data associated with the space that comes before and after it creating an integrated chain (Othman 2022).

The Blockchain was first popularized by (Satoshi Nakamoto) in 2008 when he improved the design by using Hatch-cash-like methods to determine the temporal character of each block without the need for a central authority or “trusted parties”. It became the backbone of today’s cryptocurrencies (Jeffries 2013). They are considered promising technologies that promise not to change the field of payments solely but to change many other fields such as land property registration or voting systems in elections.

The blockchain is considered a digital technology based on a massive database. It’s basically a digital ledger of transactions that is repeated and circulated across the entire network of computer systems. Each time a new transaction occurs on the (blockchain), a file of that transaction is added to each participant’s ledger.

The blockchain mechanism is embodied, in short, by a decentralized, multi-point approach to an end-user interface that links data to one another. When we create a document and share it with a group of people, the document is distributed rather than copied or moved. This creates a decentralized distribution chain that allows everyone to access the document at the same time. No one is blocked pending changes from another party while all modifications to the document are recorded in real time making the changes completely transparent.

Decentralization is considered one of the basic concepts that the blockchain technique relies on, via saving several copies of the financial transaction file in different places around the world that are visible to everyone. There is no need for a third party to act as an intermediary to manage the transactions (Eli 2019).

Modern digital currencies are developing to facilitate transactions in a way that goes beyond traditional methods such as cash or checks. This results in the rapid growth of technologies, widespread adoption of mobile devices and the spread of e-commerce and a safer digital payment system (Guest 2020).

3 The Mechanisms of the Work of Digital Currencies

Digital currencies are based on the principle of the core basis of the absence of the need for third parties by relying mainly and accurately on the processes of proof and verification. If you own a cryptocurrency, you do not own anything tangible, all you own is a key that allows you to transfer a file or unit of measurement from one person to another without a trusted third party (Guest 2020).

3.1 The Technical Aspects of Digital Currencies

The blockchain forms the technical basis of digital currencies which is a chain of blocks and each block contains a unique set of transactions. Hence blocks are not added to the chain automatically, but through the mining process carried out by the "mining node". After verifying that the encryption is correct and once the block is added, the node contract begins to propagate the mass to the rest of the network without unacceptable blocks. The new block is shared throughout the nodes and the content is updated by moving it from one node to the next. And all of this work is done by the blockchain software, automatically and easily, the block information remains available to the network's members, but it can't be misused with. Each of them has a cryptographic print called a hash, and the blocks depend on each other, interconnected and sequential through their unique hash, hence the chain and the block are made. Therefore, each block has its own unique hash code, which is used to bind the blocks in a string and group the blocks one after the other in linear and temporal order, with each block saving the hash code and the hash code of the previous block as well. If one of them tries to misuse by changing the contents of a block, the hash code will change the contents and the block accordingly. It becomes easier for the network to detect the change because the digital print has changed, and thus the network nodes reject any transaction related to the variable print (Abu Bakr 2019).

When balances are transferred from one account to another, it is necessary to know the password, which is the key to the account we want the balance to be transferred to, so that the account is not sent randomly, as the currencies between people are encrypted and then published on the digital currency networks. All operations are, then, ordered in a format to be added to the total ledger. And these transactions are so obvious to everyone that we can't recognize who did the transaction or who started the purchase or the exchange. Every transaction that takes place leads to a distinct chain of different keys - and whoever holds them has a quantity of digital currency attached to them - or has nothing if these portfolios contain nothing. Many transactions are added to the general account books in blocks by holders of coins, which is why the general ledger is called block chains (Reem 2021).

According to what has been said previously, the work of the blockchain is based on three main principles:

3.1.1 Open Ledger Principle

The majority of information in a Blockchain is available to everyone, as all individuals within the chain see each other's property. For example, if the chain is for money transfer, everyone in the chain can see everyone else's money. However, while retaining the inability to identify the person's true identity, the series allows individuals to use nicknames other than their real names, which are shown to the users of the chain, making it difficult to identify the person, although it is easy to see how much money the person has. This can be illustrated by the following example: if a person (A) wants to transfer $10 to person (b), it shows everyone whether that person already has the 10$ or not, and if not, the transaction is invalid and no one responds to the transfer. If it is true, the nearest person is next to person (b) and his name is supposed to be (c), he gives the money to person (b) for a small percentage of the total amount that person (A) wants to send (Hussein 2020).

3.1.2 Distributed Data Principle

The Blockchain works via decentralizing system and stores data. The whole chain is distributed widely at many points on the network called "Nodes" available to all its subscribers around the world. This is in contrast to the existing technologies whose systems store data on centralized, specialized devices called "Servers". This principle is one of the safety elements of the chain. If a hacker wants to manipulate or penetrate it, he has to penetrate all the people in it, which is highly unlikely to happen.

Mining Principle: Validate the transaction before proceeding to complete it by using the capabilities of computers to search for the proper code that characterizes this operation for successful completion. Millions of prospectors around the world perform complex mathematical operations through their devices in order to obtain the correct code that links this transaction to the previous transaction in the chain and distinguishes it from other transactions. This ensures that no manipulation or fraud takes place, but makes it very difficult. Once the correct code is obtained, the transaction is completed and allowed to enter the chain. They are joined to other processes within the blocks formed at the end (blockchain or blocks), and the minor who has received the correct code wins a percentage of the transfer process, if it is the transfer of (Bitcoin), they receive a financial reward for the mining from (Bitcoin) itself (Hussein 2020).

3.2 The Economic Mechanisms of Digital Currencies

The introduction of these and other similar monetary systems can completely reshape the nature of competition in currencies, due to the conflict between economic competition on digital networks and traditional competition for currencies. For example, Amazon and Alibaba have created their special systems to exchange goods, allowing for immediate and near-instant distribution of information. When geographical constraints impede the arrival of the physical currency, digital networks circumvent this problem to serve millions of subscribers (Stewart 2017).

Digital currencies present challenges and opportunities to the central banks. In some economies, bitcoin has emerged as a strong competitor to paper currencies during certain periods. The central bank has come to be seen as weak or untrustworthy, although these cases have so far been confined to troubled economies that have no capital controls.

The mechanism by which the central bank manages the money supply and the interest rate is the main reason for creating the inevitable contradiction between savings and loanable money, as these banks seek to stimulate economic growth, investments and increase investments as well as consumption. So, they tend to increase the money supply and reduce the interest rate, which makes the amount of loanable money greater than savings (Ammous 2018).

Experience has shown that the majority of central banks' plans, especially with regard to monetary supply, are unpopular and futile, as they are controlled by a group of "controllers" who make the most important market in the economy under the control of unskilled people, unaware of the realities of the market economy. The pursuit of more money means more inflation, and the mechanism by which digital currencies, particularly Bitcoin, operate is a technical solution to the monetary problem.

The costs of "exchange or transfer tax in digital currencies" are almost zero, and this feature of digital currencies allowed its users to control their money completely freely, and they can send and receive money without transfer or exchange costs, which made bitcoin an ideal way to shop in many electronic stores that accept electronic payment with no tax on transfer or receipt. Most dealers in electronic money have tended to deal in virtual instead of electronic currency, avoiding relatively large exchange and transfer expenses in comparison with the amount transferred or spent (Za'in and Jassim 2020).

4 Conclusion

Many countries are trying to control the phenomenon of digital currencies either by preventing or criminalizing their dealings to counter virtual currencies. This could be achieved by means of quick issuing their own digital currency and their local currency as an attempt to assert their sovereignty in the important issue of issuing money. The second thing is to face fraud and dirty money laundering. According to the latest published figures, there are about 60 countries that have decided to issue their own digital currencies.

The study concludes that, digital currencies provide a technological solution to the problem of cash based on the technological innovations that have been developed, as the blockchain technology provides elements of security and transparency. Decentralization also contributes to allowing greater freedom in financial transactions, eliminating the need for a third party. Hence, it ensures the transfer of funds as long as the registry is available to all and transactions are conducted directly through peer-to-peer. This process comes as a result of the appropriate work system which is acceptable to the customers because it is based on the most secure systems.

On the other hand, the working mechanisms of these currencies are not without risks as a result of their state of uncertainty and the backgrounds of their emergence that stand behind them. Such a case has made them, until this moment, a subject of doubt by many countries and individuals. This makes the customers think twice before trying to adopt it in their transactions. Additionally, one of the problems of these currencies is that they are irreversible in dealing with them, i.e., the money cannot be recovered after the dealer decides to cancel the deal that he executed.

Among the advantages and disadvantages of digital currencies, the future is more uncertain, as there are many questions about how will it look like in the medium and long future, and what is the position of governments towards it? Will the risks posed by crypto prompt governments to regulate and then officially adopt these currencies? Or will it ban it entirely? The scenario closest to verification will depend on the extent of acceptance of these currencies and their areas of use in the future.

References

Books

Al-Qazi, E.: Block chain: a beginner's guide to understanding bitcoin's underlying technology and other digital currencies, Riyadh (2019)

Othman: Cryptocurrency Economics and the Future of Money. Arab Center for Research and Policy Studies, Beirut (2022)

Ammous, S.: The Bitcoin Standard. Wiley, United States (2018)

Journals Articles

Abu Bakr, S.: The block chain technology, what is it, how does it work. Al-Qafila J. **69**(3) (2019)

Rose, C.: The evolution of digital currencies. Int. Bus. Econ. Res. J. **14**(4), 619 (2015)

Hussein, A.-S.H.: Encrypted currencies (block chain): challenges and risks. J. Law Econ. **9**(2), 22 (2020)

Za'in, M.J., Jassim, A.B.: The virtual currency, its legal adaptation and the rule of dealing with it. J. Legal Sci. Univ. Baghdad. 149 (2020)

Web sites

Basel Committee on Banking Supervision: Risk Management for Electronic Banking and Electronic Money Activities (1998). https://www.bis.org/publ/bcbsc215.pdf. Accessed 20 Nov 2022

Stewart, C.: Digital currency and the future of transactions. Rand Foundation (2017). https://www.rand.org/content/dam/rand/pubs/perspectives/PE200/PE254/RAND_PE254z1.arabic.pdf. Accessed 8 Jan 2023

Jeffries, A.: The Verge. Why don't economists like Bitcoin? (2013). http://www.theverge.com/2013/12/31/5260534/krugman-bitcoin-evil-economists. Accessed 30 Nov 2022

Guest Contributors: From Cash to Crypto: Digital Currencies in the Traditional Economy, Nasdaq (2020) https://www.nasdaq.com/articles/. Accessed 2 Jan 2023

Al-Ayoubi, R. N.E.-D.: How Digital Currencies Work, 18 April 2021. https://tijaratuna.com/. Accessed 22 Dec 2022

Report

Financial Action Task Force (FATF): Virtual Currencies - Key Definitions and Potential AML/CFT Risks, Paris, France, p. 4 (2014)

Electronic Monitoring as an Alternative to Freedom-Depriving Penalties in Jordanian Criminal Legislation

Khaled Soud Basheer Al-Jbour[1] (✉) and Ahmad Abdulkadir Ibrahim[2]

[1] Faculty of Law, Zarqa University, Zarqa, Jordan
Kaljbour@zu.edu.jo
[2] College of Law, University of Business and Technology, Jeddah, Saudi Arabia
a.abdulkadir@ubt.edu.sa

Abstract. Lectronic Monitoring is one of the latest alternatives to freedom-depriving penalties that began to be applied during the last few years in numerous penal systems; which is considered an investment in technological development the world witnesses. It has been adopted by Jordanian legislation through the adjustments it has inserted to penal laws so that sanctions could be implemented outside penal institutions; which could make it possible to overcome the negative effects of implementation of penalties inside penal institutions; thus, achieving the aims of criminal policy to reform the convicts and rehabilitate and reintegrate them as good citizens of the community.

Keywords: Electronic Monitoring · Electronic Bracelet · Freedom-Depriving Penalties

1 Introduction

Freedom-Depriving Penalties, first and foremost imprisonment, are the most common judicial rulings; and due to the criticism freedom-depriving penalties faces, penal systems in many countries resorted to alternative sanctions compatible with the crimes committed and take due account of the convicts, aiming at **correction** and rehabilitation.

American criminal legislation was the first to apply the concept of electronic monitoring placement – electronic bracelet [1] – in 1980. It was applied in France in 1989, and then spread among most of European legislation systems.

In the interest of benefiting from technological development and other countries' experiments; the Jordanian legislation has amended its Penal Code and Code of Criminal Procedures; inserted the electronic monitoring system, and applied the same on felonies and misdemeanours sentences that are not applied in case the sentence is less than one year.

B. A. M. Alareeni and I. Elgedawy (Eds.): *Artificial Intelligence (AI) and Finance*, SSDC 488, pp. 297–307, 2023.
https://doi.org/10.1007/978-3-031-39158-3_28

1.1 Significance of the Study

The significance of this study in Jordanian legislation manifest in that it is one of the latest techniques applied in the Jordanian penal systems. Moreover, on the Arab level, it has been applied only in Algeria and United Arab Emirates; thus, it still requires further research and development.

1.2 Statement of the Problem

The electronic monitoring system is still at its early implementation stages in Jordan due to the novelty of research in this study, and that it still requires further clarification and knowledge of its effectiveness in achieving the aims of penal policy.

1.3 Methodology of the Study

The study follows the analytical method; through studying and analysing legal texts that addressed the subject of using electronic monitoring in Jordanian penal legislation, and the extent to which it helps achieving the aim and purpose of the sanction, reaching various findings and recommendations for the study.

2 What is Electronic Monitoring, and What Are the Advantages of Its Implementation?

In the interest of benefit from technological development and practices of the countries that applied that system, for benefitting form its advantages, and avoiding its disadvantages of resorting to freedom-depriving penalties; the Jordanian legislation has amended its Penal Code No. (16) of the year 1960, and Criminal Procedures Law No. (9) of the year 1961; inserting the electronic monitoring system as an innovative system in Jordanian penal legislation; thus, it has become in of few legislation systems that adopted that novel penal system in order to achieve the purpose of sanction, represented – besides to general and specific deterrence – in rehabilitation and reintegration of the convicts in the community.

Electronic monitoring represents an implementation for modern scientific technology, which is distinguished from traditional methods of punishment. It also represents the most commonly used method. This is due to its lower costs and simplicity of use, as well as its effectiveness [2]. We are going to discuss what it is, then point out its advantages.

2.1 The Concept of Electronic Monitoring

The Jordanian Penal Code stipulated in Article No. (25-bis), paragraph no. (1/C), that "electronic monitoring is placing the convict under electronic monitoring for a period not less than one month, and no longer than one year." [3] Article No. (2) regarding implementation of alternative sanctions to freedom-depriving penalties measures and procedures, has defined electronic bracelet as "an electronic device used for electronic monitoring of the relevant person's location, within a limited characteristics, in an extent that does not constrain regular activity." [4].

The French Legislation system defined it as "imposing obligations on an accused or convict person sentenced with not leaving his home, residence, or any other specified location outside the times determined by the competent judge, based on considerations basically related to practicing an occupation, pursuing university education, professional qualification, ppracticing an activity that help social integration, engaging in family life, or following-up medical treatment; in exchange, the beneficiary commits to respond to the summons of the public authority determined by the competent judge." [5].

The Emirati Legislation system defined it as "the procedure of placement under electronic monitoring is degravitating the accused or convict of absence in the specified times, from his/her residence, or any other place specified by the order duly issued by the competent court or public prosecution. It is implemented through electronic means that allow remote monitoring, and obligate the sanctioned person to wear an electronic embedded transmission device throughout the period of monitoring; taking into consideration in determining the times and places, the convicts' ability to practice a profession or an occupation, pursue education or professional qualification, receive medical treatment; or any other conditions determined by the competent court or public prosecution as the case may be [6].

It is also defined as using electronic means to ensure the presence of the sanctioned person, throughout a defined period of time, in the place and time agreed by the former and the sentencing judicial authority [7].

It is defined as well as a system based on giving freedom to the person convicted with a sort-term freedom-depriving penalty; with obliging him/her with a number of obligations, and remote monitoring his/her implementation of the penalty [8].

It is remarkable for the definition provided by the Jordanian Legislation system that it was limited to defining the period of electronic monitoring, without defining the system and its essence, unlike the French and Emirati Legislation systems.

The electronic monitoring system is of the consensual nature; which emerges in that it cannot be implemented without a request submitted by the accused or convict, and after having his/her personal consent; [9] the judge has a discretionary authority in approving or denying the request; in the event of approval, the trial judge shall issue a decision of implementing electronic monitoring that is defined by a specific term ends with it.

2.2 The Advantages of the Applying Electronic Monitoring

Jurisprudence varied about the electronic monitoring system, setting out its main advantages and disadvantages that have been stated regarding that system, which is considered on of the recent systems in penal legislation; which can be summarized as follows:

2.3 Avoiding the Disadvantages Resulting from Implementation of Short-Term Freedom-Depriving Penalties

Implementation of short-term freedom-depriving penalties results in psychological and social impacts appear on the convict due to his distance from his family circle; which leads to despair, anger and sense of injustice; in addition to developing bad habits due to his blending with criminals inside the penal institution. Therefore, and through implementation of the electronic monitoring system as an alternative to freedom-deprived penalties; we can avoid all such disadvantages, and prevent the convict from blending with dangerous convicts inside penal institutions.

2.4 Reducing Overcrowding in Penal Institutions

The respective role of penal institutions is disciplining and **correction** of the detainees, as the overcrowding phenomenon inside prisons leads to considerable difficulties, and sometimes inability, in achieving that role; which necessities resorting to alternatives that allow implementing the sanction outside these penal institutions [10].

2.5 Implementation of Electronic Monitoring System Leads to Reduction of Financial Expenses of Public Fund

There is no doubt that the costs required by penal institutions and provision for their various requirements are significantly high and create a great burden for the country; whilst, resorting to implementation of the electronic monitoring system would help overcome the negative impacts of economic nature resulting from for implementation of short-term freedom-depriving penalties.

2.6 Prevention of Repeated Crime

The practical reality has proofed that there is a relation between recidivism, which is repetition of the crime; and placing the convent in penal institutions [11].

Blending of novice inmates sentenced with a short-term freedom-depriving penalties, with other dangerous inmates leads to serious results; in which the purpose for **correction** and rehabilitation of the convict is replaced with gaining criminal experiences that may lead him back to criminal path after being discharged from prison. Therefore, using the electronic monitoring method, especially with those who commit a crime for the first time, accidentally or by mistake, shall be distanced from blending or deviation to a path that is resistant to any **correction** or rehabilitation.

There is no doubt that this system is not free of disadvantages, the most important of which are;

1. That system does not fulfill the principle of social reaction to crimes.
2. Many judges and scholars of criminal law consider it as a fake freedom, moreover, that system cannot be widely implemented since it is difficult to find psychologically stable person that is capable of manage such fake freedom [12].

3 Provisions of Electronic Monitoring System in Jordanian Criminal Legislation

The provisions regulation electronic monitoring in each law of Jordanian Penal Codes and Code of Criminal Procedures are relatively brief, unlike other penal legislation systems such as the Emirati Legislation system.

We are going to set out the terms for implementing the electronic monitoring system in both Penal Codes and Code of Criminal Procedures, and the effects resulting from the convict's breach of electronic monitoring; as well as the system issued for regulating that matter.

This topic will be addressed through reviewing legal texts that were added to recent legal amendments; through which that system has been stipulated.

By reviewing Article No. (25-bis) of Jordanian Penal Code, which came under the rubric (Alternatives to Freedom-Depriving Penalties); it is demonstrated that electronic monitoring is only an alternative to freedom-depriving penalties, and shall never be considered as an alternative to non-freedom-depriving penalties such as fines.

Consequently, only normal person benefits from electronic monitoring. As well, the period of the freedom-depriving penalty shall not exceed one year [13].

The Penal Codes has stated the alternatives to freedom-depriving penalties, as it is stipulated in Article No. (2-bis) that "in misdemeanours, and according to the social situation report, except in the event of repeititionof the crime; the court may judge, even if the decision acquired legal force, with an alternative or more from the following alternatives…;"and mentioned electronic monitoring and stated that is placing the convict under electronic monitoring for a period not less than one month, and no longer than one year.

The court also has the authority, in felonies that are not committed against persons, and except for cases of repetition, when using mitigating circumstances and reducing the penalty to one year; to replace the sentenced penalty, upon the social situation report, with one or more alternatives of the freedom-depriving penalties alternatives stipulated in paragraph No. (1) of that article [14].

From the above, it is demonstrated that the judgment sentenced against the convict shall be related to a **misdemeanour**, or in felonies that are not committed against persons, when using mitigating circumstances, and reducing the penalty to one year.

A legally valid penalty is that judgment issued by court, including a proper identification for the merits of the case; on which rules have been properly applied in form and in subject; of which procedures of issuance and procedures preceding comply with the law [15].

The judge responsible for the execution of sentence may replace any of the sentenced alternatives with other stipulated alternatives, and decrease or increase the sentences period; conditioned on the social situation report and regular follow-up reports of the convict. This can be interpreted as the social situation report is an essential condition shall be provided for the possibility of alternating the original penalty with electronic monitoring.

The condition of not repeating of the crime shall be provided for the convict, whether in misdemeanours, or felonies that are not committed against persons, in which mitigating circumstances are used to reduce the sentence to one year. In case repetition condition is provided, it is not allowed to judge with alternatives to freedom-deprived penalties, including electronic monitoring.

In regards with misdemeanours, the judgment is conditioned on being issued by the court of misdemeanour; and in regards with felonies, to be issued by the criminal court. Article No. (54-bis) stipulates that "according to the social situation report, and upon the convict's consent, except in the event of repetition of the crime; the court may judge with all or any of the community **correction** alternative penalties, when judging with suspending the original sentenced penalty, in accordance with the provisions of Law No. (54-bis) of the same law."

From the aforementioned, it is demonstrated that the court has a discretionary authority in judging with any of the community **correction** alternative penalties, which have been alternated under the amended Penal Code of the year (2022). Moreover, a condition has been added for the judgment to be sentenced with imprisonment for conviction of a felony or a misdemeanour for a period that does not exceed one year; in order for the court to judge with suspending the execution of the penalty; thus, to be able to judge with any of the alternatives to freedom-deprived penalties, in accordance with the stipulation of article No. (54-bis) [16].

Compliance with the minimum and maximum period decided by the Legislator for placing sentenced person under electronic monitoring; which is not less than one month, and no longer than one year.

One form of the juridical implementation for electronic monitoring the Amman Court of First instance, in an appellate capacity, regarding one of the public who posted a post on Facebook that includes sedition and agitation regarding the match of Al-Faisaly FC Team against Al-Wehdat SC Team; whereas the Attorney General appealed the decision of the General Procurator of Amman, that includes releasing the appellee by placing him under electronic monitoring (electronic bracelet) for one month. The court decided to dismiss the appealed decision and return the documents to the source and suspend the appellee [17].

As well as the decision of East Amman Criminal Conciliation Court, by the issuance of Amended Penal Code No. 10 of the year 2022; the court has decided with placing the defendant under electronic monitoring (by placing electronic bracelet) for him for three months, and confiscation of the seized firearm [18].

3.1 The Impact of the Convicted Breach of Electronic Monitoring

The convict shall commit with the what is imposed by electronic monitoring in order not to be subject to criminal penalties represented in withdrawing the decision of placement under electronic monitoring or replacing it with any of the other alternatives, as well as decreasing or increasing the sentence.

Article No. (25-bis2), paragraph No. (3) stipulates that "the judge responsible for the execution of sentence may alternate any of the sentenced alternatives with any of the other alternatives provided by article No. (25-bis), or decrease or increase the alternative sentenced period, within the defined period for the alternative; in the following conditions:

A. Based upon the social situation report and regular follow-up reports of the convict.
B. If the convict did not comply with the alternatives to freedom-deprived penalties, or failure of implementation due to an uncontrolled reason or acceptable excuse." [19]

Through scrutinizing the aforementioned article, it is demonstrated that the reason of the convict's breach to the requirements of electronic monitoring is for an uncontrolled reason or acceptable excuse; which is contrary to what is provided by the fourth paragraph of the same article, which stipulated that "for conditions not provided by paragraph No. (3) of the same article, the judge responsible for the execution of sentence shall refer the case to the court that sentenced with the judgment regarding the misdemeanour; in order to consider cancellation of the alternative and sentence with the freedom-depriving penalty that is legally stipulated for such crime; or refer the case to the court that sentenced with the judgment regarding the felonies, in order to execution of the judged sentence. In all cases, the term that the convict has served in executing the alternative shall be calculated at five hours for each imprisonment day [20].

From the above, it is demonstrated that any breach of the electronic monitoring system by the convict without an acceptable excuse, or for a reason entails a breach from his side, results in referral of the case file by the judge responsible for the execution of sentence, to the competent court in order to to cancel of the alternative and sentence with the freedom-depriving penalty or execution of the penalty; taking into consideration deduction the freedom-depriving penalty sentence at five hours for each imprisonment day; in legislator's consideration of the convict who breached the procedures of the alternative penalty willingly and without an acceptable excuse.

It is provided by article no. (45-bis2) of the Penal code that "the court, based on the social situation report, may cancel the community **correction** alternatives and execute the penalty in the event of cancellation or suspension of the execution, or if the convict deliberately did not execute the reforming alternatives, or failed at executing the same without an acceptable excuse [21].

In a judgement issued by the Amman Court of First instance, in an appellate capacity, decided to impose a community **correction** penalty on the convict for (200) hours, taking into consideration that the appellee is a young man, civil engineer, and a family man [22].

Article No. (114-bis) of Criminal Procedures Law also addressed the event of the Defendant's breach with any of the consequent procedures – that article defined the procedures and among them mentioned electronic monitoring – as it allowed the Public Prosecutor or the court to suspend the defendant and confiscate the bail for the state's treasury [23].

Through scrutinizing that article which added to the Criminal Procedures Code in (2017), it is demonstrated that it allowed the public prosecutor or the court, in cases where it is permitted to suspend in misdemeanours, to substitute it with one or more of a set of procedures listed in the article, among which mentioned electronic monitoring; it also allowed the public prosecutor or the court, or at the request of either the public prosecution or the accused, to end, add, or amend one or more of these procedures. These provisions are related to suspension in cases of misdemeanours. Suspension is defined as temporary deprivation of the accused's freedom ordered by a judicial authority for a defined period for a necessity that is required for the interest of investigation in accordance with legal rules [24].

3.2 Means and Mechanisms for Implementation of Freedom-Depriving Penalties Alternatives Law

Jordanian Legislation stipulates in Article No. (25-bis2), paragraph No. (5) that "the means and mechanisms for implementation of freedom-depriving penalties alternatives, including electronic monitoring, shall be determined pursuant to a regulation issued for this purpose."

Based on that legal text was issued the means and Mechanisms for Implementation of Freedom-Depriving Penalties Alternatives Law No. (46) of the year 2022; and its provisions shall be applicable in case of judging with any of the alternatives to penalties; which include community service, community monitoring, electronic monitoring, and prohibiting the convict from frequenting certain locations.

In implementation of that law, the authority for following-up these alternatives has been granted to Criminal Community Directorate at the Ministry of Justice, and submit the follow-up reports to the competent judge. Operations and Control Directorate at the Public security Directorate shall monitor the convict compliance with the alternative penalty and prepare the necessary reports in this regard.

According to Article No. (8) of this Law, the court shall issue these types of alternatives after ensuring the provision of technical capabilities and the decision shall include the specified period and geographical area, or course [25].

The law includes procedures for electronic monitoring and installation of the bracelet for the convicts, and the followed procedures in case of damaging or tampering with the electronic bracelet; by arresting and referring them to the execution judge.

4 Conclusion

After pointing out in this study electronic monitoring as an alternative to freedom-depriving penalties in Jordanian legislation; as an innovative system in recent penal policy that works on achieving the aims of criminal penalties represented in general and specific deterrence; besides correction, rehabilitation and reintegration of the convicts with the community.

Electronic monitoring is produced by scientific developments in technology, in addition to the willingness of the countries to benefit from technology in order to avoid the disadvantages experienced in implementation of penalties inside penal institutions. Furthermore, economization on the high financial costs the state treasury incurs that leads to significant growth of economic burdens.

The Jordanian legislation adopted that system after inserting adjustments to Penal code and Code of Penal Procedures in (2017) and in (2022); and issued the law which the legislation stipulated for its issuance in Article No. (25-bis2); which is the means and Mechanisms for Implementation of Freedom-Depriving Penalties Alternatives Law No. (46) of the year 2022; that shall be applicable in case of judging with any of the freedom-depriving penalties alternatives provided by the Jordanian Penal Code.

4.1 The Study Came Out with the Following Findings

1. Electronic Monitoring has been added to Jordanian Penal Legislation recently. It is an innovative system that keeps pace with technological development, and helps achieve many advantages and avoid a set of disadvantages.
2. Electronic Monitoring has been inserted under the Amended Penal code No. (10) of the year 2022, under the rubric of (Alternatives to Freedom-Depriving Penalties) instead of community reform alternatives.

3. Electronic Monitoring is a system of the consensual nature; which require the consent of the sentenced person for its infringe on his personal rights; and works on invoking the convict's willingness in execution of penalties.
4. Implementation of electronic monitoring prevents blending of novice inmates with experienced convicts inside penal institutions, the resulting negative impact and transmission of criminal infection; which is considered the major advantages of implementing the law due to its significant contribution to minimize returning to crime.
5. The court may judge with electronic monitoring in misdemeanour cases in general. In regards with felonies, it is may be judged with electronic monitoring only in felony cases that are not committed against persons, except for repeated cases; when following mitigating circumstances and minimizing the sentence to one year.
6. Electronic Monitoring is only an alternative to freedom-depriving penalties, and shall not be considered as an alternative to non-freedom-depriving penalties such as fines.

4.2 The Study Came Out with the Following Recommendations

1. Since this method has been recently innovated in penal legislation, we should consider the countries that implemented this system long ago, and learn from their experiences in that field; in order to be able to develop it by learning from the advantages and avoiding disadvantages.
2. Study the possibility of expansion in implementing that system to include convicts whose sentenced period is more than one year; taking into consideration their personal circumstances and the conditions of committing the crime, and its criminal seriousness, to reduce the numbers of convicts inside penal institutions.
3. The necessity of explaining that new system through specialized studies, and in conferences, and workshops; and indicating its advantages to convince the society and convicts with the electronic monitoring system.
4. Considering the possibility of implementation of that system not only in penalties that do not exceed one year, but also other penalties sentenced and part of it was executed and still remains a year or less for completing the sentenced period; the thing some comparative legislation systems follow.

References

1. Al-Zeiny – Ayman Ramadan – House Arrest, a new concept for punitive treatment methods and alternative penalties for freedom-depriving in prisons - Dar Al-Nahda Al-Arabia – 1st Edition – 2015 – Page No. 5
2. Rasheed – Mariqan Mustafa – Hussein Abbas Hamid – Electronic Monitoring as an Alternative to Freedom-Depriving Penalties – Journal of the Faculty of Law for Legal and Political Sciences – Volume No. 11 – Issue No. 41 – 2022 – Page no. 237
3. Jordanian Penal Code No. 16 of the year 1960 and its amendments
4. Means and Mechanisms for Implementation of Freedom-Depriving Penalties Alternatives Law No. 46 of the year 2022 – Article No. 2
5. Haqas – Asmaa – Placement Under Electronic Monitoring – Journal of Rights and Liberties – Volume No. 5 – 1st Issue – 2019 – Page No. 53

6. Obeid – Osama Hassanein – Electronic Penal Monitoring – Comparative study – 1st Edition – Dar Al-Nahda Al-Arabia – Cairo 2009– Page No. 6
7. Salem Omar – Electronic Monitoring – A Recent Review for Freedom-Depriving Penalties Implementation Outside Prison – Edition No. 1 – Dar Al-Nahda Al-Arabia – Cairo 2000 – Page No. 9
8. Judge. Ramy Metwally – Electronic Monitoring in Comparative French Law – Journal of Al-Sharia and Law – University of United Arab Emirates – 2015 – Page 286
9. Dr. Al-Nawaysa – Abdullah – Dr. Mahmoud Fayad – Dr. Shady Al-Shadifat – Regulations of Electronic Monitoring as an Alternative to Pre-trial Detention in Emirati Legislation – An Analytical study – Journal of International Kuwaiti Faculty of Law – 9th Year – 1st Issue – Serial no. 33,2,21 – Page No. 132
10. Al-Mazmoumy – Muhammed Bin Hamid - Electronic Monitoring as an Alternative to Freedom-Depriving Penalties – Journal of Voice of Law – 7th Volume – 2nd Issue – 2020 – Page Np. 876
11. Masrouq – Maleeka – Placement Under Electronic Monitoring – Electronic Bracelet – in Algerian Legislation in the Shadow of Law No. 18–10 – Master's Thesis 2019 – Page No. 23
12. Tulba – Laila – Placement Under Electronic Monitoring – Journal of Human Sciences – Issue No. 47 – 2017 – Page No. 256
13. Othany – Safaa - Placement Under Electronic Monitoring (Electronic Bracelet) in French Penal Policy – Journal of Damascus University for Economic and Legal Sciences – Volume No. 25 – 1st Issue – 2009 – Page No. 138
14. Jordanian Penal Code No. 16 of the year 1960 and its amendments, Article (No. 25bis) Paragraph No. 2
15. Journal of Al-Zarqaa for research and Humanitarian studies – Volume No. 18 – Issue No. 3 – 2018 Page No. 469
16. Othany – Safaa – Previous Reference – Page No. 138
17. Amman Court of First Instance in an appellate capacity Judgment No. 2487 dated 24/10/2022
18. East Amman Criminal Conciliation Court Judgment No. 1414/2020 dated 23/10/2022
19. Jordanian Penal Code No. 16 of the year 1960 and its amendments, Article No. (25-bis) – Paragraph No. 3
20. Jordanian Penal Code No. 16 of the year 1960 and its amendments, Article No. (25-bis) – Paragraph No. 4
21. Jordanian Penal Code No. 16 of the year 1960 and its amendments, Article No. (25-bis2) – Paragraph No. 2
22. Criminal Court of First Instance in an appellate capacity Judgment No. 414 of the year 2021 dated 18/01/2021
23. Jordanian Penal Procedures Law No. 9 of the year 1961, and its amendments – Article No. 114 bis
24. Goukhadar – Hassan – Initial Investigation in Penal Procedures Law – Edition No. 1 – Dar Al-Thaqafa – Amman – 2008 – Page No. 395–396
25. Means and Mechanisms for Implementation of Freedom-Depriving Penalties Alternatives Law No. 46 of the year 2022 – Article No. 8

Law Books

26. Goukhadar – Hassan – Initial Investigation in Penal Procedures Law – Edition No. 1 – Dar Al-Thaqafa – Amman – 2008
27. Obeid – Osama Hassanein – Electronic Penal Monitoring – Comparative study – 1st Edition – Dar Al-Nahda Al-Arabia – Cairo 2009
28. Al-Zeiny – Ayman Ramadan – House Arrest, a new concept for punitive treatment methods and alternative penalties for freedom-depriving in prisons - Dar Al-Nahda Al-Arabia – 1st Edition – 2015

Research Papers Published in Journals

29. Haqas – Asmaa – Placement Under Electronic Monitoring – Journal of Rights and Liberties – Volume No. 5 – 1st Issue – 2019
30. Rasheed – Mariqan Mustafa – Hussein Abbas Hamid – Electronic Monitoring as an Alternative to Freedom-Depriving Penalties – Journal of the Faculty of Law for Legal and Political Sciences – Volume No. 11 – Issue No. 41 – 2022
31. Judge. Ramy Metwally – Electronic Monitoring in Comparative French Law – Journal of Al-Sharia and Law – University of United Arab Emirates – 2015
32. Dr. Al-Nawaysa – Abdullah – Dr. Mahmoud Fayad – Dr. Shady Al-Shadifat – Regulations of Electronic Monitoring as an Alternative to Pre-trial Detention in Emirati Legislation – An Analytical study – Journal of International Kuwaiti Faculty of Law – 9th Year – 1st Issue – Serial no. 33,2,21
33. Al-Mazmoumy – Muhammed Bin Hamid - Electronic Monitoring as an Alternative to Freedom-Depriving Penalties – Journal of Voice of Law – 7th Volume – 2nd Issue – 2020
34. Tulba – Laila – Placement Under Electronic Monitoring – Journal of Human Sciences – Issue No. 47 – 2017
35. Othany – Safaa - Placement Under Electronic Monitoring (Electronic Bracelet) in French Penal Policy – Journal of Damascus University for Economic and Legal Sciences – Volume No. 25 – 1st Issue – 2009

University Papers

36. Masrouq – Maleeka – Placement Under Electronic Monitoring – Electronic Bracelet – in Algerian Legislation in the Shadow of Law No. 18-10 – Master's Thesis 2019 – Page No. 23

Legislations

37. Jordanian Penal Procedures Law No. 9 of the year 1961, and its amendments
38. Jordanian Penal Procedures Law No. 9 of the year 1961, and its amendments
39. Means and Mechanisms for Implementation of Freedom-Depriving Penalties Alternatives Law No. 46 of the year 2022

The Role of Business Intelligence on Digital Economic Transformations (Case Study: E-Government in Jordan)

Mohammad Musa Al-Momani[1(✉)], Nidal Abbas[1], Thaer Abu Saleem[1], Mazen Basha[1,2], Abdel Halim Jubran[1], Khaled Al-Sawaie[1], Ibahim Khrais[1], and Mohanad Dahlan[2]

[1] Faculty of Economic and Administrative Sciences, Zarqa University, Zarqa, Jordan
{m.momani,nabbas,thabusaleem,mbasha,ajubran,ksawaie,ikhrais}@zu.edu.jo

[2] University of Business and Technology, Jeddah, Kingdom of Saudi Arabia
dahlanm@ubt.edu.sa

Abstract. The use of technology has greatly impacted the way organizations and institutions, including government agencies, operate. Developed countries have adopted e-government services which allow electronic transactions and centralized data. However, many developing countries are still in the early stages of adopting these types of services. In developed countries, advanced technologies such as big data analysis and business intelligence (BI) systems are being used to improve government institutions. BI systems can quickly analyze large amounts of data and provide decision-making alternatives. Unfortunately, many developing countries have not yet embraced the use of business intelligence in government organizations. This study proposes a framework for integrating BI and e-government to enhance government institutions' performance. This can provide valuable insights to government and public sector organizations in determining the potential role of business intelligence in improving their performance and decision-making.

Keywords: Business Intelligence · E-Government · Digital transformations · Integration BI and EG

1 Introduction

Technology revolution played an important role in improving many organizations and institutions in different sectors, this development also reflected on government institutions. Governments began to shift towards providing their services electronically and linking all government institutions within one system and centralized data bases to begin a new era of e-government services. In addition, the countries of the developed world began applying e-government services many years ago, as they implemented these services to facilitate citizens to complete their transactions through the use of smart phone applications and e-government websites. This prompted developing countries to follow the same approach that was used in other countries and started activating e-government

B. A. M. Alareeni and I. Elgedawy (Eds.): *Artificial Intelligence (AI) and Finance*, SSDC 488, pp. 308–316, 2023.
https://doi.org/10.1007/978-3-031-39158-3_29

services, but many of these countries are still in the early stages of dealing with this type of development, and have not yet reached the stage of full maturity. However, after the development in e-government services, advanced countries have been striving to develop new services and add modern technologies to improve the performance of government institutions, which aims to improve the economic situation.

Therefore, it is necessary to make the right decisions at the right time, and this can only happen by providing advanced technologies that help in analyzing the huge data available in their data warehouses. This will ultimately reduce risks in decision-making and provide appropriate alternatives to solve problems and achieve desired goals, which will have a clear impact on improving the country's economic situation in the long run. Thus, In order to achieve this, these countries searched for the best systems that help deal with big data, and have a huge ability to analyze this huge amount of data at a high speed, as well as provide many alternatives and suggest the best ones. In order to achieve all this, it was necessary to fully activate the role of business intelligence (BI) in government institutions, because of the ability of BI systems technologies to achieve all of the above, and to link business intelligence techniques with e-government services, which had a profound impact on managing risks and improving performance, which had a positive impact on the economies of countries. Coming back to developing countries, as we said, these countries are still in the growth stage in activating e-government services, therefore, governments of these countries still lack understanding of the importance of technical aspects in improving the performance of government institutions, as well as lacking many tools that help in making the right decisions at the right time, this is due to the fact that many of these countries do not use BI technologies, which play a crucial role in improving the country's economy.

According to that, the problem of the study lies in the delay of use of BI in government institutions in developing countries, where BI plays an important role in improving the performance of e-government, which will be reflected in the general economic situation of the country. In addition, this study is significant because there are limited researches on the implementation of BI within government or public sector organizations. This paper aims to fill that gap by exploring the business intelligence on e-government performance in the context of digital economic transformations, and developing a framework for BI solutions to enhance the performance of government organizations. The findings of this study can provide valuable insights and be useful for government and public sector organizations in identifying the potential role of BI in improving their performance and decision making. This paper focuses on the role of business intelligence (BI) in digital economic transformations with a specific focus on e-government in Jordan. It is structured with an overview of the concept of e-government and the role of BI, a literature review of previous studies on BI in e-government, a description of the theoretical approach and proposed framework used in the case study, findings and implications for e-government initiatives in Jordan and other developing countries, and concludes by summarizing the main findings and future research directions.

2 E-Government in Jordan

The information and communication technology (ICT) plays an important role in e-governance, a process that aims to improve efficiency and interoperability in government agencies, as well as between citizens and businesses. E-government solutions are intended to transform government organizations into fully digital entities that cater to citizens' needs, increase accessibility, promote inclusivity, and involve citizens in the political process. National governments are working to implement and deploy ICT and e-government applications to aid in development, in collaboration with other stakeholders at the national, regional, and international levels. However, there are challenges in forming policies and monitoring the progress of ICT programs in these approaches [1]. Moreover, e-government programs aim to improve government performance and efficiency by providing citizens with electronic channels to access government services. The e-government program in Jordan specifically aims to deliver high-quality services, increase transparency and accountability, enhance competitiveness, reduce costs, and improve privacy and information security. Measuring user satisfaction is a common focus in e-government research [2].

However, according to [3], the researcher evaluated the official website of the Jordanian e-government and identified various factors that influence the quality of e-government services. The study found that the design of the website was user-friendly but there were fewer services available in English compared to Arabic and the information services were not detailed enough. The paper provides recommendations for enhancing the website such as focusing on the websites of institutions and ministries, ensuring equal numbers of services in both English and Arabic, developing mobile applications, simplifying the interface and features, raising awareness of e-government services, increasing security, and exploiting available data to improve the quality of decision-making, adding new services such as live chat, and improving employee's ability to handle modern technology.

3 Business Intelligence and Its Factors

Business Intelligence (BI) is a system that allows businesses and organizations to store, retrieve, and analyze large amounts of structured and unstructured data which related to the inside and outside organization operations, in order to make effective decisions and gain a competitive advantage. So, BI is a set of tools and techniques that help organizations improve their decision-making to improve business performance and assist in developing, identifying, and creating new strategic opportunities for the organization [4]. However, to understand the relationship between Business Intelligence (BI) and Business Value (BV), we should examine the theories, critical factors, inhibitors, and forms of BV related to BI adoption in organizations. The researchers found that certain elements such as having skilled personnel, proper capital, good BI infrastructure, data quality, and top management supports were key to achieving BV from BI adoption. Additionally, factors like poor data quality and handling, lack of data security and protection, inadequate BI infrastructure, and a shortage of skilled human resources were found to be inhibitors to achieving BV. Moreover, there are a study examines the importance of

Business Intelligence (BI) and Decision Support Systems (DSS) for organizations and investigates the organizational factors that influence their effective use [5].

The study found that BI systems can positively affect DSS by gathering a large amount of data, and DSS can add value to organizational performance by increasing control, personal efficiency and problem-solving effectiveness. In addition, using BI and other information sources can assist organizations in making better decisions by collecting the correct data. Furthermore, it highlights the importance of big data in decision making and the benefits that BI systems can bring to organizations by providing new information, tracking root-causes of problems, and monitoring operational processes. Also those BI systems are dependent on rapidly evolving technologies such as data mining, predictive analytics, and big data [6]. In addition, many studies found that business intelligence use and collaboration capability are important factors in determining the performance of public listed companies in an emerging market. The ability to sense and respond to unpredictable market conditions is crucial for organizations, and a well-developed IT infrastructure and use of business intelligence can empower strategic agility and lead to better performance. The researchers also found that these factors were mediated by strategic agility [7].

4 Integration Between E-Government and Business Intelligence

Due to digital transformation, many researchers have been conducted to examine the correlation between how well tasks align with a web-enabled BI system and the success of such systems in the public sector. The success of BI is evaluated based on factors such as usage, user satisfaction, and individual impact. Accordingly, there is a positive association between task compatibility and usage, user satisfaction, and individual impact, and that user satisfaction plays a role in the connection between task compatibility and individual impact. However, no link was discovered between usage and individual impact. Thus, The researchers are proposes that when choosing, procuring, and implementing BI systems in the public sector, the compatibility of tasks should be taken into consideration instead of only focusing on technical aspects. The research has limitations as the findings are based on a mandatory BI system in a particular organizational context and the effect of mandatory usage on individual impact is uncertain [8].

Moreover, implementing BI solutions in municipality as a public sector will have a potential benefits like improve planning, resource allocation, and services for residents. For that, to assess the feasibility and benefits of BI in this context, SWOT framework and semi-structured interviews with stakeholders are used. In addition, the results finds that BI can provide valuable insights into residents' needs and can help decision makers at the contact center (CC) and municipality make more informed decisions and improve services. However, it also notes that BI can be costly and difficult to integrate and can face resistance due to perceived complexity. Overall, the study suggests that BI can play an important role in helping municipalities improve services and increase accountability for resource allocation.

In addition, BI can be used to gain a better understanding of the needs and timing of municipal residents. This information can help decision makers at the (CC) and municipality make more informed decisions and plan more effectively to improve services for residents. By regularly analyzing data from the CC, a BI solution can also help identify problems and issues facing residents and measure the success of implemented strategies and services. Additionally, a BI solution can help identify a knowledge gap among residents regarding the municipality's functions and activities, and inform decision makers on whether more information needs to be propagated to close the gap [9]. Thus, there is a relationship between e-Government and economic growth. A new useful framework is provided for decision makers that link between different units of analysis, structures, and processes. The framework is based on established theories, models, and recent research. The results found that the key variable linking e-Government portals to economic growth is competitiveness. This framework can be used by decision makers in different governmental bodies to promote technology-based economic growth in the context of competitiveness goals. The overall conclusion is that the strategic use of e-Government can reinforce economic growth. It can create a closer relationship between citizens and the governments, enhance efficiency and productivity, and build competitiveness in both the private and public spheres. E-Government represents a unique opportunity for stimulating economic growth and progress of civilization [10].

5 Methodology and Suggestion BI Framework

A theoretical approach is used as a research methodology. First, it delves into the foundational concepts and practical applications of business intelligence. Then, it explores the research that has been conducted on using BI in e-government. After that, a hypothetical scenario is created to illustrate the challenges in combining BI and e-government. Finally, suggestions are made for how to effectively integrate the two, and potential areas for future research are outlined.

5.1 Proposed Framework

The challenge of acquiring the information necessary for decision making is a complex task that is further complicated by various obstacles. Our goal is to introduce a framework that guides the development of data-driven e-government activities. By providing business intelligence, managers are able to make better decisions by having access to the necessary data for understanding, managing and coordinating organizational activities. The proposed framework is focused on the ability to identify and collect relevant information, analyze it, and extract crucial insights to support decision making. It is worth noting that obtaining data is considered one of the most challenging aspects of business intelligence and can consume up to 70% of the time and resources [11]. Organizations often focus on getting data in and out, which includes allowing business users and applications access to data from the data warehouse to create enterprise report, and to do online analytical processing (OLAP), predictive analytics, and querying.

In this research, a framework for business intelligence architecture is presented; it considers the different sources of data in the government, including both external and internal sources. It also covers the processing and analysis phase and the data storage layer. This framework aims to provide a clear understanding of how government data is collected, analyzed and stored, to be exploited in business intelligence and to support decision-making processes that will have an impact on improving the country's economic performance (Fig. 1).

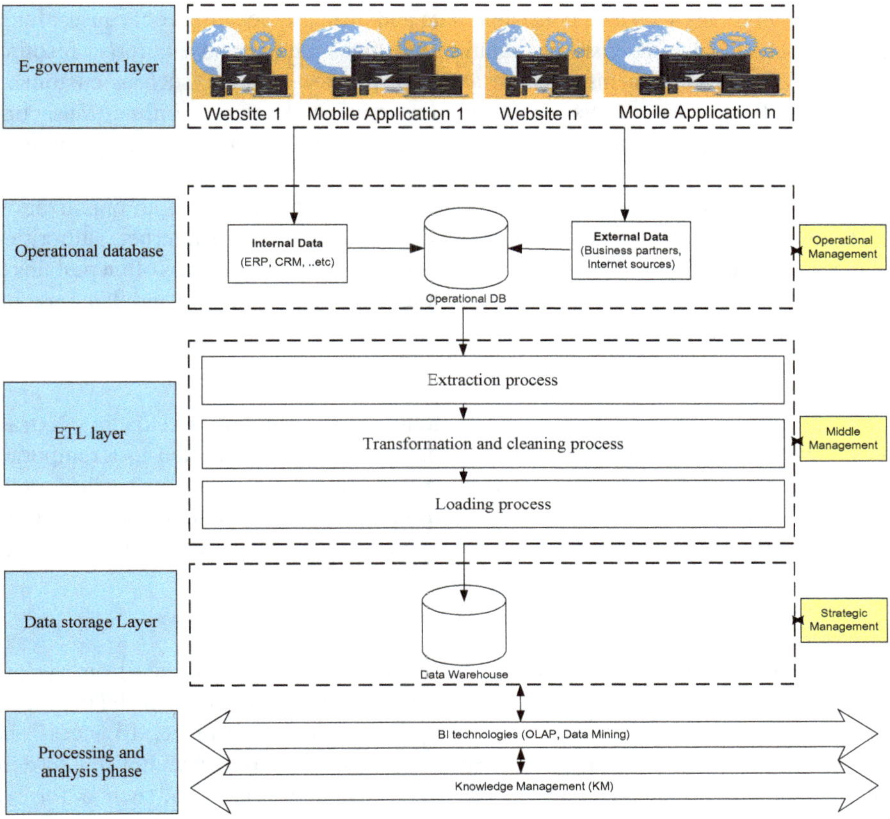

Fig. 1. Proposed framework [4]

Government Data Sources Layer

The review of existing literature highlights that companies utilize both internal and external data as sources for business intelligence. Data is considered a valuable asset for both the public and private sectors, and can be utilized to improve public services, decrease fraud, and boost operational efficiency. However, governments have difficulty accessing and utilizing data effectively due to it being dispersed across different registers and stored in unprocessable formats or locations with limited digital access. This hinders digital government development and limits the information available to citizens about the data the government collects and how it is used.

Only a few countries have made significant progress in resolving issues related to accessing and utilizing data effectively. The experiences of these nations can provide guidance for other governments and suggest actions they can take to access the value of data. Organizations must clearly identify their data sources, which can save time and improve information delivery. However, some sources may be difficult to acquire, untrustworthy, or irrelevant to the needs of the business. Proper data source identification can help avoid issues such as inconsistent data and difficulty locating root causes. Internal data sources, such as customer relationship management and enterprise resource planning systems, contain information about corporate activities, such as customers, products, and sales data. These operational systems are known as online transaction processing systems and only contain recent data used to support ongoing company operations.

External data sources are information that comes from outside of an organization. These sources can include clients, suppliers of syndicate data, the internet, authorities, and market research firms. The information gathered from these sources often pertains to the market, competitors, external factors such as consumer economic and demographic information, and technology.

ETL (Extract-Transform-Load)
The ETL process consists of three steps: extraction, transformation, and loading. Extraction involves gathering data from various sources, it then transported to a temporary location called data staging area for cleaning and transformation using business rules to standardize the format and fix any errors. Finally the cleaned data is loaded into the target repository. The goal of ETL process is to ensure consistency and avoid the need to repeat the process of extraction and cleaning.

Data Storage Layer
Data warehouse (DW) is a collection of historical and summarized data from various sources that is organized and verified to support decision-making. The ETL process is used to standardize and collect diverse types of data in one place. To access and understand the data in the DW, users need knowledge of the data warehousing system and its contents. The metadata or data of data is included in the DW, such as format, encoding/decoding algorithms, and domain restrictions, descriptions of the data, organizational changes, business rules, and data quality alerts. This can help business users to understand the data that is available, how to access it, and how to use it. The data warehouse can be accessed using metadata browsers, and smaller versions called data marts or localized data warehouses are often developed to support the decision-making processes of specific departments or divisions.

Analysis and Processing Phase In this phase, the process of storing the knowledge that has been extracted using BI tools, so that it can be used in the future and shared within the company. Most Knowledge Management (KM) systems used to store this information include knowledge repositories, which are considered an essential part of these systems. They aim to record both tacit and explicit information, and are collections of both internal and external knowledge. The explicit knowledge is produced by using OLAP, data mining, and other reporting technologies. To make the generated knowledge effectively and efficiently accessible, it must be filtered, arranged, and stored in a central knowledge repository. The tacit knowledge of human experts can be directly gathered by KM systems.

6 Conclusions and Recommendations

This study has demonstrated the potential impact of business intelligence on e-government performance in the context of digital economic transformations, with a specific focus on Jordan. Our research findings show that the adoption of business intelligence systems in e-government institutions can improve performance, enhance decision-making, and enable better use of data resources. The proposed framework based on our research can help government and public sector organizations to better understand and analyze the impact of BI on their performance, and support their digital transformation initiatives.

To improve the performance of e-government in Jordan, we suggest applying BI systems, starting with a pilot project in one or two sectors to test the effectiveness of the system, and then gradually expanding it to other government departments. Eventually, the goal is to create a knowledge base for the full e-government environment at the national level, once the entire system is ready for usage. While some institutions in the government now have data warehouses, they are still unable to properly exploit these warehouses and the huge amount of data they contain. Fully activating business intelligence in government institutions can have a clear positive impact on improving the performance of the government system, which can in turn help in making ideal decisions that will have an impact on improving the local economy significantly. However, we acknowledge that the implementation of BI in e-government institutions may face challenges, and further research is needed to examine the factors that can influence the success of such initiatives, as well as the strategies that can be used to overcome these challenges. By addressing these issues and taking the necessary steps to implement BI in e-government, government institutions in Jordan and other developing countries can enhance their ability to provide effective public services, and support the digital transformation of their societies.

In addition to the above, there are several avenues for future research in this area. Firstly, more research is needed to examine the effectiveness of the proposed framework for BI implementation in e-government institutions in Jordan, and to evaluate its impact on e-government performance. Secondly, future research could investigate the factors that influence the success or failure of BI implementation in e-government institutions in other developing countries, and to develop strategies to overcome the challenges of such initiatives. Additionally, further studies could explore the potential of advanced BI technologies, such as artificial intelligence and machine learning, to improve the performance of e-government institutions and to enable better decision-making. Finally, it may be beneficial to examine the impact of BI implementation in e-government institutions on the satisfaction and trust of citizens, as well as its impact on the overall development of the digital economy. By addressing these research gaps, policymakers and researchers can gain a better understanding of the potential of BI in e-government, and develop effective strategies to support the digital transformation of government institutions in developing countries.

References

1. Mohammad, M.M.: Evaluating the characteristics of e-government websites in developing countries: Jordan as a case study. Eur. J. Bus. Manag., 37–42 (2019)
2. Mohmmed, A.: Adopting e-government services in less developed countries according to the characteristics of business intelligence: (Sudan as a model). IJCSNS **22**(11), 204–212 (2022)
3. Marius, C.: Business intelligence and e-governance. LESIJ-Lex ET Sci. Int. J. **16**(1), 484–491 (2009)
4. Yassine, J., Jie, L.: An implementation framework of business intelligence in e-government systems for developing countries: case study: Morocco e-government system. In: International Conference on Information Society (i-Society), pp. 138–142. IEEE (2017)
5. Alan, S.: Business and e-government intelligence for strategically leveraging information retrieval. Electron. Gov. Int. J. **5**(1), 31–44 (2008)
6. Rikke, G.,Sandalgaard, N., Sudzina, F.: The importance of task compatibility for web-enabled business Intelligence success in e-government. In: Proceedings of the 19th International Conference on Information Integration And Web-Based Applications & Services 2017, December 4, pp. 353–357 (2017)
7. Olawale, A.: The role of business intelligence in government: a case study of a Swedish municipality contact center. Master thesis (2012)
8. Dalia, M.: A framework for embracing business intelligence and e-government integration for effective decision making. Int. J. Public Policies Egypt **1**(3), 141–156 (2022)
9. Riem, A., Omar, S., Ibrahim, E.: The role of e-government as a stimulus for economic growth. Int. J. Bus. Manag. Technol. **4**(5), 69–79 (2020)
10. Shivakumar, K.: An overview of e-government. Economia Aziendale Online **1**, 1–12 (2012)
11. Ye, Q., Li, Y., Nie, Q.: Research on the application model of business intelligence (BI) in e-business from the perspective of Chinese culture. In: Proceedings of 2012 2nd International Conference on Computer Science and Network Technology, December 29, pp. 907–911. IEEE (2012)

The Legal Nature of Electronic Payment Cards

Abdullah Omar Althunibat[1](✉), Sohib Yahya Alshurman[1], Ahmad Awwad Albnian[1], and ElGaili Ahmed ElTayeb[2]

[1] Faculty of Law, Zarqa University, Zarqa, Jordan
aalthunibat@zu.edu.jo
[2] College of Law, University of Business and Technology, Jeddah, Saudi Arabia

Abstract. Given the growth and expansion of commerce and the transition from a conventional system to an advanced digital one, people looked for a method that would facilitate meeting their demands and safeguard them against the possibility of their money being misplaced or stolen while being transferred. Due to this, banks developed a sophisticated method known as electronic payment cards. These cards are handled based on trust and reciprocal credit between the parties to the card (bank, holder, and merchant), establishing contractual relationships between the parties to the contract, with each agreement having a distinct legal effect that imposes obligations on each of its parties. Therefore, it was necessary to establish a separate legal system that governs the relationship between the parties to the contract and its legal effects because the success of cards as a digital payment method depends on the relationships between its parties and because using these cards for payment will result in numerous disputes. Therefore, this study aimed to determine which legal system regulates the relationship between the contract's parties and which laws, if any, are adaptable to control electronic payment cards.

Keywords: electronic cards · prepaid cards · legal nature · Electronic payment

1 Introduction

Technological advancement has contributed to qualitative shifts in all facets of life. For example, in the commercial and legal fields, technological advancement has aided in the conclusion of acts that have legal effects within a fertile digital environment that has resulted in the establishment of many electronic legal actions, the most prominent of which is the creation of electronic payment (e-payment) system. However, issuing electronic payment cards has raised concerns about the legal aspects governing this digital procedure.

E-payments occur whenever payment instructions enter a payments system via the Internet or any other telecommunications network[1] using a device such as a computer or a cell phone. It is considered a conclusion of a legal bond between three parties: a creditor,

[1] Payment Systems Worldwide, "Developing a Comprehensive National Retail Payments Strategy Consultative Report" (The World Bank, July 2012) at 105, online:, accessed 24 January 2020 [hereafter, "National Retail Payments Strategy"].

B. A. M. Alareeni and I. Elgedawy (Eds.): *Artificial Intelligence (AI) and Finance*, SSDC 488, pp. 317–331, 2023.
https://doi.org/10.1007/978-3-031-39158-3_30

a debtor, and a card issuer. This relationship is a legally binding obligation imposed on the parties. The personal right arises from an indebtedness relationship which requires actual liability at the time, either to pay then or at a future time.

E-payment, which attempts to fulfil obligations, is a newly emerging approach that has grown in popularity due to the convenience, efficiency and speed with which it may be processed, particularly in light of the growth of electronic commerce. E-payment has various types, such as credit and debit cards, mobile wallets, Internet and mobile banking. However, to reap its benefits, the e-payments system necessitates the development of new legal and regulatory frameworks that protects contractual parties' rights, the conclusion of contracts, and the execution of contractual obligations. It is regarded as a legally binding contract regardless of the payment method or whether it is written or electronic because it includes the issuer of the payment card, its holder, the beneficiary, or any of the contractual parties.

1.1 Problem Statement

This study's problem manifests in the legal aspects of e-payment as a contract with a binding effect for contractual parties. Furthermore, technological advancement's impacts on the commercial sector have led to the emergence of electronic contracts and e-payment. Nevertheless, previous studies that addressed the conclusion of an electronic contract did not cover the concept of electronic payment cards. Therefore, we opted to address this concept, discuss it in detail, and address the following issues by answering the research question:

1. What is meant by electronic payment cards?
2. What are the legal aspects regulating electronic payment cards?

1.2 Study Significance

The significance of this study lies in its subject, which is the concept of payment via electronic means, which has recently emerged and includes many features that require legal and regulatory solutions to accommodate both the promise and risks inherent in the rapid pace of development, particularly in light of technological advancements in the communications and digital industries, as well as the emergence of the electronic contract. The significance of the subject of this study is evident in its significant contribution to strengthening the creditor's rights in terms of the impacts of the debtor's party to fulfilling it. Electronic transactions between the seller and the buyer are now prevalent and used daily via the electronic payment system.

1.3 Approach

Due to the nature of this research, which necessitates several methodologies, the analytical-descriptive approach, based on analysing and describing the various legal texts, was used. We followed this approach to state the position of Jordanian law, utilising the jurisprudence regarding this issue.

2 Theoretical Framework

2.1 The Emergence of Electronic Payment Card

Locally in 1982, the Hashemite Kingdom of Jordan began deploying electronic payment cards. It was the first card to be introduced into the Jordanian banking sector. Petra Bank was the first financial institution to issue it; it was known as (Petra Card). However, Petra Bank was liquidated in 1989[2], resulting in the establishment of a financial intermediation corporation known as (Jordanian Banking Services Company) comprised of five Jordanian banks, namely: (Housing Bank, Investment and Finance Bank, Cairo Amman Bank, ABC Bank- Jordan, and AJIB)[3]. This company was founded in 1991 to maintain issuing Petra Bank credit cards. VISA JORDAN was founded as eight Jordanian banks declared their participation in the worldwide firm (Visa Card) and the development of the first Visa company in the Arab world, which put its skills and experience in electronic payment cards in the local Jordanian market[4]. The Jordan Visa Company commenced contracting with merchants, conducted transactions, and cleared between bank accounts. Then the international visa card was issued through the Arab Bank in collaboration with its London branch. In 1993, the National Bank issued the first Master Card. After that, it processes financial transactions, settlements, and card issuance[5].

In 1994, the Jordan National Bank and the Bank of Jordan and the Gulf issued the visa card in collaboration with the Business Bank, and in 1995, an agreement was signed between the Amman Investment Bank and the British Bank of the Middle East, with the foreign card remaining in circulation in Jordan until 1992. After that, Abdel Fattah Ghazal and Partners Company issued the first local card, the National Express Card.[6]

Abdel Fattah Ghazal and Partners Company was established on January 27, 1992, to issue personal cards allowing holders to make purchases. As stated in the Governor of the Central Bank of Jordan's letter to the Minister of Industry and Trade No. 796/7020 dated January 13, 1992, it received a license from the Central Bank. The Central Bank has no opposition to the company being registered as long as its operations do not interfere with those of regulated banks or financial institutions.

The electronic payment card is one of the modern payment methods developed due to contemporary commercial demands. Hence, the definitions of electronic payment cards vary depending on the formality, issuing authority, and the law regulating them. However, they have common technical features regardless of the card's function. They are all made

[2] Petra Bank was liquidated according to the decision issued by the Economic Security Committee on 7/15/1990, that the Central Bank of Jordan be the liquidator, and the liquidation procedures start from 7/21/1990, see: Amjad Hamdan Al-Juhani, Civil Responsibility for the Illegal Use of the Electronic Payment Card, first edition, Dar Al Masirah for Publishing and Distribution, Amman, 2010, p. 16.

[3] Shereen Elias, Credit Card Crimes in Jordan, Master Thesis, Mutah University, Jordan, 2005, p. 16.

[4] Ahmed Muhammad Saeed, research entitled Provisions for Dealing with Credit Cards in Islamic Sharia, Mutah Magazine, Volume 20, Issue 5, 2005, p. 40.

[5] Huda Ghazi Atallah, The Legal Aspects of Credit Cards, Master Thesis, University of Jordan, 1997, p. 24.

[6] Ibid.

of plastic materials, with precise chemical specifications, rectangular in shape, with specific measurements, with a length of about (8.5) cm, a width of about (5.5) cm, and a thickness of about (0.75 0.80) mm, inscribed on their side by machines. In addition, its number, the holder's name, and the issuing entity's emblem are conspicuously printed[7].

There are numerous jurisprudential definitions of prepaid cards. Some defined it as a banking tool granted its holder to use for paying services or as a document that its issuer offers to a natural or legal person based on a contract between them, allowing this person to dispose of it in terms of purchasing, shopping, and withdrawing money. Some described it by the specifics of the card's shape without addressing the essence of its function. However, the European approach in 1997 is the most significant and accurate of these definitions. It defined an e-payment card as a cash value stored electronically on an electronic holder, such as a card with a chip or computer memory issued in exchange for depositing an amount of money whose worth is not less than the value of the e-payment card. The cash issued is acknowledged as payment by the institutions that issued it, and it is available to its users as an electronic substitute for coins and banknotes.

We contend that, unlike other Arab and foreign legislation, the Jordanian legislature did not address a particular definition of an e-payment card. This is evidence of a legal deficiency in light of the requirements of economic openness and the growth of electronic commerce.

2.2 Legal Nature of e-payment Cards

It should be noted that the legal nature of prepaid cards has not received jurisprudence's attention for several reasons, the most important of which is that these cards are not subject to disputes before the judiciary and pose a challenge to jurisprudence due to the various aspects and technical systems used, such as the direct or indirect communication system. Moreover, in jurisprudence, there is a risk that these cards will be suspended if they are legally adapted in a traditional manner, which is counter to their qualities in light of the economic context.

Identifying the legal nature of the e-payment system is one of the critical issues that will assist the judiciary in determining the legal norms that must be followed when a dispute originating from the usage of an e-payment card is brought before the tribunal. For example, suppose a dispute is brought before the courts between parties to an e-payment card. In that case, it is necessary to determine and follow the terms and conditions of the card system and the laws and regulations of the jurisdiction in which it is used. Furthermore, determining the legal nature contributes to stating the rights and obligations that the parties did not define or declare[8].

Several controversial perspectives have emerged over the legal nature of electronic payment cards. Whereas some attempted to specify a particular legal text for it, others tended to adapt the relationship between the parties of the electronic payment card to the different contractual relationships because electronic payment cards have tri-partite

[7] Hussein Muhammad Al-Shalabi and Muhannad Fayez Al-Dwaikat, Credit Card Fraud and Forgery, first edition, Dar Al-Majdalawi for Publishing and Distribution, Amman, 2009, p. 13.

[8] Khacha Hassiba, Modern Payment Methods in Algerian Law, Master Thesis, Mohamed Boudiaf University - Algeria 2016, p. 53.

agreements: the card's issuer, the card's holder, and the retailer. The advocates of this idea argue that this tripartite agreement aims to provide an integrated system for paying the value of products and services offered by merchants via the card issuing authority, with the issuing authority receiving the paid value only from the cardholder.

Others regarded electronic payment cards as a payment tool similar to the check in terms of the three-party relationship. The check is defined as an order or request by drawer addressed to a drawee (bank or persons in the banking business), requesting them to pay on presentment to a person named on the check or to (beneficiary) holder a fixed sum of money. It contains mandatory data specified by the text of Article (228) of the Jordanian Commercial Law. It can be circulated through endorsement and can only be withdrawn from a bank[9].

However, the parties to the tripartite agreement in the electronic payment cards are the issuer, the cardholder, and the merchant. Similarly, this relationship is represented in the check by the drawer, the drawee, and the beneficiary[10].

Even though the check and the electronic payment card both have the feature of irreversibility in the payment order, the provisions of the crime of issuing a check without a balance to the electronic card may not be applied because it does not have penal protection, unlike the check, which has special penal protection under the Penal Code[11].

The methods of trading both payment tools differ; the check is traded commercially between persons or institutions, whereas the card is not exchanged between individuals. Furthermore, the drawer gives the check to the beneficiary, and the beneficiary receives its value by presenting it to the bank (the drawee), as opposed to the card, which the holder delivers to the merchant and retrieves at the same time[12].

Nevertheless, electronic payment cards differ from a check as they can be used more than once and is guaranteed by the issuing authority, which promises payment to the merchant within the terms of a contract agreement. On the other hand, the check has a value and due date and is used once, according to a commercial process, after which the transaction is completed. Furthermore, the drawee is regarded as a debtor to the drawer and is not required to pay on his behalf if there is insufficient balance[13].

The legal adaptation of electronic payment cards has raised a wide legal controversy because no specific legal controls were issued for it. However, the Jordanian legislator stipulated in the Electronic Transactions Law[14] in Article No. (21) that:

"The transfer of funds by electronic means is an acceptable means for making payment so that the Central Bank of Jordan determines - according to a regulation to be

[9] Muhammad Nour Shehata, Electronic Loyalty, a research presented to the E-Commerce Conference held in Cairo, the National Center for Judicial Studies at the Ministry of Justice, May 18–19, 2002, p. 20.

[10] Ehab Fawzi Al-Sakka, Criminal and Security Protection for Credit Cards, New University Publishing House, Alexandria, 2007, p. 64.

[11] Khacha Hassiba, op, cit, p56.

[12] Muhammad Tawfiq Saudi, Credit Cards and the Legal Basis for Relationships Resulting from Their Use, Dar Al-Amin, Cairo, 2001, p. 21.

[13] Muhammad Tawfiq Saudi, Credit Cards and the Legal Basis for Relationships Resulting from Their Use, Dar Al-Amin, Cairo, 2001, p. 21.

[14] Jordanian Electronic Transactions Law No. (15) of 2015.

issued - for this purpose what follows:1. The electronic payment systems work procedures and their technical requirements, the requirements and conditions of issuing and dealing with electronic money, as well as settling disputes that may arise between parties of electronic money transfer transaction. 2. The conditions, procedures, and technical requirements related to electronic checks as well as the specifications of the electronic system that shall apply to the present view and clear checks electronically."

Based on the aforementioned, we can argue that the Jordanian legislator issued a system for electronic payment and transfer of funds but did not specify the electronic means of payment and did not provide detailed provisions identifying its legal nature resulting from the contractual relationship.

Legalising these cards raises a significant dispute over the contractual relationships between the parties to the electronic payment cards. Some of them refer to this arrangement as a loan contract, others see it as a contract of assignment (*hawala*), and others claim it is an agency agreement, a delegation of payment, and suretyship agreement, etc.

2.2.1 The Electronic Payment as an Agency Agreement

Some jurists consider the contractual relationship resulting from electronic payment cards to be an agency contract, given that the cardholder is the principal, the merchant is a beneficiary, and the issuer is an agent. Although there are several definitions of agency in the legal legislation, the Jordanian legislator defined it as:

"A contract whereby the principal authorised another person (agent) to act on his behalf in permissible and known acts."

We noticed that the legislator had identified significant elements in the agency agreement: assigning another person to perform a legal act. The legal act is the term of the contract, where one person (agent) acts on behalf of the other and can form legal relationships on behalf of the principal. The Jordanian legislator provided that the act is permissible and known.

Proponents of this view believe that payment by electronic cards is considered an agency agreement, according to which the card issuer (agent) acts on behalf of the (principal) in paying the debts and for the goods and services obtained by the cardholder based on a payment order issued by the cardholder (principal). The contract concluded between the card issuer, and the cardholder is based on consent between the two parties. This consent entails an obligation towards each other based on personal considerations that necessitate the cardholder to pay the card issuer the amounts he delivered on his behalf[15].

This trend has been widely criticised because by applying the provisions of the agency agreement to the legal nature of electronic payment, the contract concluded between the bank and the cardholder is ignored. Actually, the bank could not be an agent for both the cardholder and the merchant simultaneously. Article (115) of the Jordanian Civil Code states that:

"It shall not be permissible for a person to contract with himself in the name of the person for whom he is an agent whether the contract is for his account or the account of

[15] Fayez Naim Radwan, Payment Cards, Dar Al-Nahda Al-Arabiya, Cairo, 1999, pg. 246.

another party without authorisation from the principal, but provided that it is permissible for the principal in that event to approve the contract...".

If the permissibility specified in the previous article is regarded as a contractual relationship between the bank and the cardholder, and the bank and the merchant, then it leads to complicated scenarios and criticisms because:

1. The agency agreement is based on a personal and direct consideration that the bank must perform. Based on the contract between the bank and the merchant, the bank promises to pay the cardholder's debt; consequently, the bank and the cardholder are indebted to the merchant. And because it was not stipulated in the agency contract, The merchant only has the right to claim the debtor (cardholder) and not the bank agent (the card issuer)[16].
2. Considering that the card issuer is an agent in paying off debts on behalf of its holder, he can sue the merchant and also has the right to issue an order not to pay the creditor, which conflicts with the electronic card payment system. However, because the card issuer does not have the right to file suit against the merchant as the cardholders can do, the holder cannot issue an order to the card issuer not to pay the merchant[17].

We contend that because the operating system of the electronic payment card is based on a three-partite relationship, establishing the relationship between the merchant and the card issuer as an agency contract relationship is incompatible with that system. On the other hand, the agency occurs between two parties, which is one of the reasons we assert that the terms of the agency contract do not apply to the merchant and the issuer due to the nature of their relationship and thus, its adaptation is unacceptable. Additionally, the agency agreement does not guarantee payment; instead, the guarantee is a requirement for concluding a contract for the use of electronic card payments because it is the fundamental obligation resulting from the use of the agreement reached between the issuer and the merchant.

The provisions for terminating the agency contract specified in the Jordanian Civil Code, which imposed restrictions on the principal upon termination of the agency contract, do not apply to the issuer and the merchant because if the cardholder withdraws the agency agreement, the card issuer remains obligated to pay the merchant. Furthermore, the contracts between the issuer, the cardholder, and the issuer and the merchant are different. However, the card issuer has the right to claim compensation from the cardholder for what he paid for it.

We conclude that the agency contract does not apply to the nature of electronic payment, as the general principles in concluding the agency agreement are far from the threefold nature of electronic payment.

[16] Nida Kazem Al-Mawla, The Legal Nature of the Bank Card System, Zarqa Journal for Research and Studies, Zarqa University, p. 103.

[17] Merbah Saliha, The Legal System of Credit Cards, Master Thesis - Algeria, 2006, p. 58.

2.2.2 The Electronic Payment as a Contract of Assignment (*Hawala*)

Some jurists have considered the legal relationship of the electronic payment card contract as an assignment contract, wherein one party to an existing agreement, "the assignor", hands off the contract's obligations and benefits to another party ", the assignee[18]. For example, the Jordanian legislator deals with the assignment in Chapter Two of Chapter Five of the Jordanian Civil Code. He defined it as:

"The transfer of a debt and claim from the liability of the transferor to the transferee[19]."

The text demonstrated that the transferor 'assignor' is the debtor and is a party to the contract. The transferee 'assignee' is the creditor, a party to the contract, and obligated to pay the debt[20].

By applying the assignment principle to the parties to the electronic payment system, the merchant transfers the debt the cardholder owes to the card issuer for payment. In this case, all the trading rights he has on the cardholder are transferred automatically to the card issuer. The issuer has the right to demand the cardholder to pay what he paid on his behalf to the merchant, with all that guarantees this right. However, Article (1010) of the Jordanian Civil Code stipulates that:

'The transferor must deliver to the creditor the document of title relating to the obligation transferred and any necessary evidence or means to enable him to obtain his right.' As stipulated in Article (307) of the Egyptian Civil Code[21].'

This opinion is controversial, as the card issuer paying the debt to the merchants and bearing the risks of non-payment by the cardholder contradicts the reality of the e-payment process. And the text of Article (305) of the Egyptian Civil Code stipulates that the debtor's acceptance of the transfer shall be proven by the due date, denoting that it shall be written in a record. This impedes the progress of the electronic payment process[22]. Also, accepting the transfer from the merchant to the card issuer falls within a future contract that depends on it. This is contrary to the condition stated in Article (1000) of the Jordanian Civil Code, which specifies that:

'for an assignment to be valid, it must be completed and dependent on no condition other than an appropriate or customary condition, nor must any future contract be dependent on it; the performance thereof must not be deferred to an unknown future date; it must be limited in time to a specific time limit'.

In the case of the cardholder, his use of the card is unknown.

At first glance, debt transfer is appropriate to explain the nature of electronic payment cards. Conceivably, the debtor cardholder has entered into a contractual agreement with the card issuer to pay the crediting merchant. However, where there is a prior agreement between the merchant and the bank to accept and make payment in this way, it is therefore

[18] Azza Ali Muhammad Al-Hassan, The Legal and Legislative Framework for E-Commerce, Al-Zaytouna for Printing, first edition, 2009, pg. 345.

[19] Article 993 of the Jordanian Civil Code.

[20] The explanatory memorandum of the Jordanian Civil Law - p. 686.

[21] Al-Sanhouri, Contracts Relating to Work, The Mediator in Explanation of Civil Law, Theory of Commitment in General, Part Three, Manshaat Al-Maarif House, Alexandria, 2004, p. 432.

[22] Kilani Abdel Radi Mahmoud, The Legal System for Payment and Guarantee Cards, PhD thesis, Ain Shams University, 1996, p. 678.

not valid to be applied to the electronic payment cards system for several reasons, including[23]:

1. The general conditions for an assignment to be valid are that it must be completed and dependent on no condition other than an appropriate or customary condition, nor must any future contract be dependent on it. However, all goods purchased and fulfilled by the card issuer are based on the future, which is inconsistent with the assignment in e-payment.
2. The principle in the assignment is not to be associated with a benefit. The cardholder pays an annual subscription to the bank for the e-payment process, and the merchant pays the bank a commission in return for collecting his debts from customers, which exceeds the framework of the nature of the assignment in e-payments.
3. The performance thereof must not be deferred to an unknown future date or be limited to a specific time limit[24].
4. The transferor shall have the right to claim against the transferee for any debt or property owed by him. This, in fact, does not apply to the nature of e-payment because the principle is that the issuer of the card (transferee) is not indebted to its holder (transferor)[25].
5. Trading with electronic payment cards results in several independent contracts, so the card issuer cannot pay the merchant based on the assignment contract. Payment here takes place based on the agreement concluded between them, which is independent of the contract between the card issuer and its holder. The card issuer pays the merchant on behalf of the cardholder based on the agreement concluded between them. Therefore, the direct relationship between the cardholder and its issuer and between the merchant and the card issuer cannot be interpreted as a debt assignment[26].
6. Debt transfer in the assignment agreement included all its guarantees and characteristics and all defences relating to the debt to the new debtor[27], and this contradicts the system of dealing with ELECTRONIC PAYMENT CARD, which results in contractual relationships which are essentially independent of each other, so the issuer cannot adhere to the defences relating to the debt of the cardholder which he had or to claim against the merchant[28].

2.2.3 E-payment as a Loan Contract

Some jurists claimed that the provisions of the loan contract could be applied to the electronic card contract. However, a loan contract is defined as:

---

[23] Nida. Al-Mawla, op, cit. p101.

[24] The second and third paragraphs of the text of Article 1000 of the Jordanian Civil Code.

[25] Article (1006) of the Jordanian Civil Code.

[26] Amina Ibn Amour, Electronic Cards for Payment, Loan and Withdrawal, Master Thesis, Faculty of Law, University of Constantine, Algeria, 2005, p. 41.

[27] Al-Sanhouri, op, cit. p383.

[28] Abd al-Hamid Mahmoud al-Bala'i, Bank credit cards (artistic depiction and jurisprudential graduation, a comparative analytical study), Wahba Library, first edition, 2004, pp. 33–34.

" Loan is the granting of ownership of property or fungible things to another with the condition that the other should return it like in amount, kind and description to the lender upon the expiry of the period of the loan[29]'.

If the Loan's contract provides a benefit in excess of the essence of the Loan, it can't be regarded as one of the Gift contracts. Instead, it goes within the framework of commutative contracts. Nevertheless, when we consider the provisions of the loan contract, we find that it can not be applied to the legal nature of electronic payment cards for the following reasons[30]:

1- If the term of the loan contract expires, it is not renewed. To be continued, a new contract must be concluded. This condition does not apply to the electronic payment card's nature because the cardholder does not renew his agreement with its issuer. The agreement extends for a certain period and is renewed automatically unless any party objects.
2- The loan contract is established between two parties, The lender and the borrower, while an electronic payment card is a contract concluded between three parties.
3- There is no connection between the loan contract and any third party. This is so because, in contrast to electronic loyalty cards, which have a direct relationship between the vendor and the card issuer, he has no direct relationship with the person the borrower transacts with.

2.2.4 E-payment as a Suretyship Contract

Some jurists considered the legal relationship of e-payment as a suretyship contract. However, Suretyship is the joining of the liability of a person called the surety with the obligor's liability in the performance of his obligations[31]. It was also defined as a contract whereby a person (surety) put up collateral to secure the indebtedness of the (creditor) if the principal debtor did not perform his obligation[32].

When considering the application of the provisions of the guarantee contract to the legal nature of e- payment, we find that the card issuer guarantees its holder and undertakes to pay the merchant the value of the purchases made by the cardholder. However, this does not apply to the legal nature of e-payment; For the following reasons:

1. In the suretyship contract, the surety has the right to raise defences available to the principal debtor in the surety contract. But, under e-payment, the card issuer is not entitled to raise those defences against the merchant[33].
2. The Suretyship Contract entails an obligation between the creditor and the debtor. Thereof, the surety is obligated to pay on behalf of the principal debtor since the obligation of the surety shall follow the obligation of the principal obligor. However, this condition is not available in the e-payment system.

[29] Article (636) of the Jordanian Civil Code.

[30] Khacha Hassiba, op, cit. 58.

[31] Article (950) of the Jordanian Civil Code.

[32] Article (772) of the Egyptian Civil Code.

[33] Kumit Talib Al-Baghdadi. Illegal Use of Credit Card (Criminal and Civil Liability), Dar Al-Thaqafa for Publishing and Distribution, first edition, Amman, 2008, p. 111.

The card issuer has an original and direct obligation to pay the merchant, independent of the cardholder's obligation to the merchant. Because the obligation of the card issuer to pay the merchant arises from the contract concluded between him and the merchant and not as a surety for the cardholder.

3. The principal obligor who has to pay the merchant under the surety agreement in the first rank is the debtor. While in e-payment, the principal obligor to pay the merchant is the card issuer[34]. This contradicts the text of Article (782) of the Egyptian Civil Code, which stipulates: "A surety shall be released if once the debtor is acquitted..."[35].
4. The principle in the surety agreement of the gift contracts[36] is that surety shall not take a commission or a fee in exchange for its guarantee, and the guarantee is considered invalid if it stipulates a fee for the guarantor. This is contrary to the nature of dealing with electronic payment; The card issuer obtains a commission from the cardholder in return for paying the cardholder's debt obligations[37].
5. If the debtor becomes bankrupt in the suretyship contract, the creditor must prove his debt in the bankruptcy; otherwise, his right of recourse against the surety shall lapse to the extent of the loss sustained by his not having done so[38]. While the card issuer bears the bankruptcy of the cardholder and pays for the cardholder to the crediting merchant; The creditor is not obligated to intervene in the bankruptcy of the cardholder debtor; The fact that his right is direct according to the independent contract concluded between them[39].

2.2.5 E-payment as a Delegation for Settlement

Some jurists argue that the legal nature of e- payment is similar to the delegation of payment in the sense that the debtor (delegator) assigns another person (delegatee) to perform his obligation to the creditor (obligee) on his behalf. Still, the assignment is valid if the debtor obtains the creditor's consent to a third party to pay the debt in lieu of the debt[40].

These two methods are identical in terms of the following:

1. The delegator (cardholder) is indebted to the obligee (the merchant), according to the contract concluded between them. The cardholder delegates the card issuer to perform his obligation (debt) to the merchant, so the card issuer undertakes the debt payment in lieu of the debtor within the limits of the allowable amount[41].

---

34 Muhammad Tawfiq Saudi, Credit Cards and the Legal Basis for Relationships Resulting from Their Use, Dar Al-Amin, 2001, pp. 90–9_.

35 See: Article (782) of the Egyptian Civil Code.

36 See: Article (952) of the Jordanian Civil Code.

37 Alaa Al-Din Al-Zanari. Banking services and the position of Islamic law regarding them, Dar Safaa for Publishing and Distribution, Amman, first edition, 2010, p. 573.

38 Article (978) of the Jordanian Civil Code.

39 Sarah Muhammad Omar Abdel-Baqi, Bank Credit Cards from the Legal Point of View, Master Thesis, Faculty of Law, Al-Neelain University, 2018, p. 65.

40 Fidaa Yahya Ahmed Al-Hamoud, The Legal System for Credit Cards, Dar Al-Thaqafa for Publishing, Jordan, 1999, p. 85.

41 Hanan Rehan Mubarak Al-Mudhahki, Criminal Protection for Magnetic Credit Cards, Modern Academic Office, Alexandria, 2010, p. 47.

2. The relationship between the three parties is based on two separate contracts, a contract between the card issuer (delegatee) and the cardholder (delegator) and an agreement between the card issuer and the merchant (oblige). As a result, the proxy payment theory may be appropriate for describing the relationship between the bank and the merchant. Still, it is ineffective in describing the relationship between the cardholder, the merchant, or the bank.
3. The card issuer can not be considered a delegate of the cardholder because the fundamental principle in the Suretyship - as we mentioned earlier - is that there is a previous and existing obligation; the surety is dependent on an original obligation[42].

The theory of delegation can not be fitted to the legal nature of e-payment for the following reasons:

1- In delegation, the oblige (merchant) has no right of recourse against the card issuer due to their contractual relationship[43].
2- The principle of delegation of settlement is to be free of charge because the delegate performs the obligation in lieu of the delegator. This contradicts the nature of e-payment, where a commission is obtained from the merchant[44]

2.2.6 E- Payment as a Third-Party Beneficiaries Contract

Some jurists held the opinion that the basis of the legal nature of e-payment is the demand for the benefit of others based on the original relationship between the merchant and the card issuer[45].

Several laws regulated the benefit of others. For example, Article (210) of the Jordanian Civil Code specifies that: 'it shall be permissible for a person to contract in his own name imposing a condition that rights are to ensure to the benefit of a third party if he has a personal interest, whether material or moral, in the performance thereof; Such a condition shall confer upon the third party a direct right against the undertaker for the performance of that condition in the contract enabling him to demand the performance thereof unless there is a contrary agreement, and such undertaker may rely as against the beneficiary on any defences arising out of the contract; The person making the condition may also demand the performance of the condition in favour of the beneficiary unless it appears from the contract that the beneficiary alone has such a right[46].'

This theory granted the cardholder (third-party beneficiary) a direct right against the merchant (promisor) for the performance of the condition in the contract (paying by the card). The card issuer (promisee)is also entitled to demand the merchant perform the demands in favour of the cardholder. In terms of e-payment, the card issuer (promisee) contacted the merchant (promisor) for the benefit of the third-party (cardholder)[47].

---

[42] Nida Kazem Al-Mawla, op, cit. p104.
[43] Muhammad Tawfiq Saudi, previous reference, pg. 96.
[44] Khacha Hassiba, op, cit, p. 64.
[45] Samiha Al-Qalyubi, Legal Foundations for Banking Operations, Dar Al-Nahda Al-Arabiya, Cairo, 1977, p. 481.
[46] Article (210) of the Jordanian Civil Code.
[47] Article (210) of the Jordanian Civil Code.

However, this theory is criticised as it enables the merchant (undertaker) to rely against the beneficiary on any defences arising from the contract. For example, if the card issuer is declared bankrupt and liquidated, the merchant can claim against the cardholder to obtain satisfaction of his right. In this case, the card issuer becomes a debtor to the cardholder.

Therefore, it is not permissible for the merchant to rely upon any defences arising out of the contract, which the card issuer could have against him, to claim against the cardholder. Therefore we can not consider the theory of third-party beneficiaries' agreement consistent with the legal nature of the electronic payment system[48].

Criticism has been raised at all legal interpretations that attempt to apply various contract laws to electronic payments[49]. The new and contemporary nature of electronic payment using cards to perform demands Because it is used for so many commercial transactions, this medium is entangled. It is also built on a tripartite relationship, which sets it apart from all other ways of fulfilling obligations.

We conclude that the provision for third-party beneficiaries was the closest to explaining the legal nature of electronic payment by cards. However, this approach gave the beneficiary merchant the right to sue the cardholder, meaning that if the merchant claims the cardholder, the cardholder must satisfy his contractual obligations twice: once to the card issuer and again to the merchant.

In conclusion, it is evident that the e- payment card is a cutting-edge form of payment that stands out from all previous forms of payment systems. As a result, no traditional legal framework can adequately describe all the legal relationships arising. As a result, it needs a unique legal framework that protects the rights of the three parties (issuer, merchant, and cardholder) while considering their conflicting interests.

3 Conclusion

Paying financial commitments has evolved, becoming an electronic system thanks to technological advancements. Many commercial transactions depend on e-payment. Therefore, it was necessary to define the legal nature that regulates the relationship between the parties to the electronic payment system because the means used and the parties to the contractual relationship are different from the other legal proceedings. Electronic cards have different legal conditions than payments made through banks, and the legal relationships between their parties also differ.

4 Findings

The Jordanian legislature did not establish a comprehensive definition for e-payments methods due to the variety of card types and functions. Also, other legislations did not define it clearly, as they only described its characteristics, how to use them, and the

48 Essam Hanqi Muhammad Morsi, The Legal Nature of Credit Cards, Conference on Electronic Banking between Sharia and Law, held at the United Arab Emirates University, Dubai, Volume One, 2003, p. 852.

49 Essam Hanafi Morsi, op, cit, p. 853.

relationships between their parties. The diversity of parties to the electronic payment card system sparked a broad jurisprudential debate in determining their legal nature. Furthermore, its adaptation to other traditional legal rules was challenging because it is a modern payment method with its conditions and controls. Additionally, e- payment system enjoys a unique system that distinguishes it from other means of payment; Because it consists of three parties (the issuer, the holder, and the merchant). It is also a system based on two contracts, one between the cardholder and the merchant and the other between the merchant and the card issuer.

5 Recommendations

Given the importance of electronic payment methods as a means of satisfying financial obligations in light of their advancement, popularisation, speed, and efficiency in commercial transactions, the researchers suggest the establishment of a specific legal regulation that defines the legal nature of electronic payment to address the legislative deficiencies in the Jordanian Electronic Transactions Law regarding all electronic payment methods. In addition, the researchers also recommend regulating the legal nature of the contractual relationship that results from the electronic payment card, particularly the relationship between the card issuer and holder and the card issuer and the merchant.

References

Books, Thesis and Research

1. Al-Hamid, A., Al-Bala'I, M.: Bank Credit Cards: Artistic Depiction and Jurisprudential Graduation, A Comparative Analytical Study, 1st edn. Wahba Library, Cairo (2004)
2. El-Din Morsi, A.J.:. Electronic money and its impact on the role of central banks in managing the new monetary policy in the work of banks from the legal and economic points of view. The annual scientific conference of the Faculty of Law at Beirut Arab University, Part One, 1st edn. Al-Halabi Human Rights Publications (2002)
3. Saeed, A.M.: Provisions for Dealing with Credit Cards in Islamic Sharia. Mutah Maga. **20**(5) (2005)
4. Al-Din, A., Al-Zanari.: Banking Services and the Position of ISLAMIC Law Towards Them, 1st edn. Dar Safaa for Publishing and Distribution, Amman (2010)
5. Al-Fail, A.A.: Criminal Responsibility for Using the Electronic Credit Card, (A Comparative Study), 1 edn. Modern Book Foundation, Lebanon (2011)
6. Al-Sanhouri.: Contracts Relating to Work, The Mediator in Explanation of Civil Law, Theory of Commitment in General, Part Three. Manshaat Al-Maarif House, Alexandria (2004)
7. Amour, A.I.: Electronic Cards for Payment, Loan and Withdrawal, Master Thesis, Faculty of Law, University of Constantine, Algeria (2005)
8. Al-Juhani, A.H.: Civil Responsibility for the Illegal Use of Credit Card Electronic Payment, 1st ed. Dar Al Masirah for Publishing and Distribution, Amman (2010)
9. Al-Hassan, A.A.M.: The Legal and Legislative Framework for E-Commerce, 1st edn. Al-Zaytouna for Printing, (2009)
10. El-Sakka, E.F.: Criminal and Security Protection for Credit Cards. New University Publishing House, Alexandria (2007)

11. Morsi, E.H.M.: The Legal Nature of Credit Cards, Conference on Electronic Banking between Sharia and Law, voL. 1. Held at the United Arab Emirates University, Dubai (2003)
12. Radwan, F.N.: Payment Cards. Arab Renaissance House, Cairo (1999)
13. Al-Hamoud, F.Y.A.: The Legal System of Credit Cards. Dar Al-Thaqafa for Publishing, Jordan (1999)
14. Al-Mudhahki, H.A.M.: Criminal Protection for Magnetic Credit Cards. Modern Academic Office, Alexandria (2010)
15. Atallah, H.G.: The Legal Aspects of Credit Cards. Master Thesis, University of Jordan (1997)
16. Al-Shalabi, H.M., Al-Dwaikat, M.F.: Forgery and Credit Card Fraud, 1st edn. Dar Al-Majdalawi for Publishing and Distribution, Amman (2009)
17. Hassiba, K.: Modern Payment Methods in Algerian Law, Master's Thesis, Mohamed Boudiaf University, Algeria (2016)
18. Mahmoud, K.A.R.: The Legal System for payment and Security Cards. PhD thesis, Ain Shams University (1996)
19. Al-Baghdadi, K.T.:. Illegal Use of Credit Cards (Criminal and Civil Liability), 1st edn. Dar Al-Thaqafa for Publishing and Distribution, Amman (2008)
20. Saliha, M.: The Legal System of Credit Cards, Master Thesis, Algeria (2006)
21. Shehata, M.N: Electronic payment. In: Research presented to the Electronic Commerce Conference held in Cairo, the National Center for Judicial Studies at the Ministry of Justice, 18–19 May 2002 (2002)
22. Saudi, M.T.: Credit Cards and the Legal Basis for Relationships Resulting from Their Use. Dar Al-Amin, Cairo (2001)
23. Al-Mawla, N.K.: The legal nature of the bank card system. Zarqa J. Res. Stud.
24. Al-Qalyubi, S.: Legal Foundations for Banking Operations. Dar Al-Nahda Al-Arabiya, Cairo (1977)
25. Abdel-Baqi, S.M.O.: Bank Credit Cards from the Legal Point of View, Master Thesis, Faculty of Law, Al-Neelain University (2018)
26. Elias, S.: Credit Card Crimes in Jordan, Master Thesis, Mutah University, Jordan (2005)

Laws and Regulations

27. Jordanian Electronic Transactions Law No. (15) of 2015
28. Jordanian Civil Law No. 43 of 1976
29. The explanatory memorandum of the Jordanian civil law
30. Egyptian Civil Code No. 130 of 1984

Electronic Monitoring as an Alternative to Custodial Penalties

Khalid Alzubaidi[1(✉)] and Mohammed Angawi[2]

[1] Faculty of Law, Zarqa University, Zarqa, Jordan
kalzubaidi@zu.edu.jo

[2] College of Business Administration, University of Business and Technology, Jeddah, Saudi Arabia
m.angawi@ubt.edu.sa

Abstract. Electronic monitoring is one of the important alternatives to penalties depriving liberty, given that this method that came as a result of electronic transformation, it avoids the disadvantages of those penalties, and achieves advantages for each of the convict, society, and the state, economically, socially, and security. Therefore, the idea of research came, which focused on the need to achieve a balance when applying this alternative between protecting the security of society on the one hand, and reforming the governed and reintegrating them into society on the other hand. Where the study dealt with research and analysis of the nature of electronic monitoring and the conditions for its application and effects. The study reached a number of findings and recommendations, including that the Jordanian legislator was one of the first countries to adopt electronic monitoring, but its judicial application is still below the required level. Determine his specialization exclusively. With the call of the legislator in Jordan and the Arab countries to expand the localization of the outputs of digital development in the field of criminal policy, especially in the aspect of punishment.

Keywords: Electronic monitoring · electronic transformation · alternative penalties · criminal policy

1 Introduction

The electronic development has contributed to the creation of new means that can combine between what the state envisages in benefiting from punishing the perpetrators on the one hand, and reducing the negative effects that result from it on the other hand. This was done by employing electronic means to serve the criminal policy in general and the punishment policy in particular. Perhaps one of the most important of these means is remote monitoring using the electronic bracelet, which the Jordanian legislator has adopted in order to achieve the required goalsThis is through the issuance of Law No. (10 of 2022) amending Article (25) of the Penal Code No. Deprivation of liberty No. (46) of 2022.

The problem of this research is the balance between the requirements of the state's right to punishment and the preservation of public security on the one hand, and ensuring

B. A. M. Alareeni and I. Elgedawy (Eds.): *Artificial Intelligence (AI) and Finance*, SSDC 488, pp. 332–338, 2023.
https://doi.org/10.1007/978-3-031-39158-3_31

that the convict does not mix with those who are habitual criminals and the negatives that result from that on the other hand.

In light of this, a number of questions can be asked as follows, What is the concept of electronic monitoring, and What are its justifications? Was the Jordanian legislator able to achieve a balance between achieving deterrence of both types, public and private, on the one hand, and avoiding the negative aspects of punishment that deprives freedom, on the other hand? How can more electronic means be employed to develop criminal policy with regard to punishment?

The importance of the research lies in the direct violation of its subject in the freedom of individuals, in addition to the importance of using electronic monitoring as a modern method away from traditional penalties, and the effectiveness of the role played by the subject of the research in keeping pace with technological development and the effects of electronic monitoring on the various parties related to the subject of the research.

The researcher has adopted the analytical method of the legal texts related to electronic monitoring according to what was stated in Article (25 repeated/second) of the Jordanian Penal Code No. (16) for the year 1960 and its amendment No. (10) for the year (2022), and the system of means and mechanisms for implementing alternatives to penalties depriving freedom No. (46) for the year 2022.

The researcher has divided his research plan into three sections, the first of which shows what electronic monitoring is, and in the second we explain the justifications and conditions for applying, and the effects of electronic monitoring.

2 The Content of Electronic Monitoring

This term electronic monitoring is derived from the French term electronic bracelet (bracelet) which mean An electronic device has been devised in the leg or hand of the convict by order of the competent judge. [1].

This device is directly connected to a central electronic device located in the designated judicial authority and allows the observer to follow the steps of the offender and know his whereabouts [2] Such monitoring is a kind of house arrest imposed on The offender, you put him in a position that leaves no room for him to exercise his freedom except within the area that the judge determines for him [3].

2.1 The Concept of Electronic Monitoring

The definitions of electronic monitoring differ according to the party from which it is viewed. Therefore, to define this concept, we must first define its concept from the technical, jurisprudential, and legislative aspects.

2.1.1 The Concept of Electronic Monitoring From a Technical Point of View

From this perspective, electronic monitoring means "the use of electronic means to ensure the presence of the person subject to it during a specific period in the place and time previously agreed upon between the latter and the judicial authority" [4].

It can also be defined according to the technical perspective as "the use of the computer in the exercise of the control process according to computer programs prepared for this purpose" [5].

2.1.2 The Concept of Electronic Monitoring in Jurisprudence

There are many jurisprudential definitions in electronic monitoring, the subject of research, and this is due to the difference in its application in societies and the difference in the party from which it is viewed, as some have defined it as "obliging the person sentenced or subject to him to be in a specific place, often the house during specific times of the day and often It will be during the period from seven in the evening until seven in the morning of the next day, provided that he is allowed to practice his life normally at the rest of the day, within a specific legal framework, and the commitment of the convict or subject to him is confirmed by his presence in the specified place and time through prepared devices For this purpose" [6]. Some jurists have tended to define electronic monitoring as "leaving a person sentenced to a custodial sentence free in a free environment, subjecting him to a number of obligations and monitoring him electronically from a distance" [7].

2.1.3 The Concept of Electronic Monitoring From the Legislative Point of View

The Jordanian legislator defined electronic monitoring in the Jordanian Penal Code within Article (25) repeated thereof as "placing the convict under electronic monitoring for a period of not less than a month and not more than a year" [8], but the legislator was not successful in defining electronic monitoring, As the definition must always encompass the concept to be defined in such a way that it is sufficient alone to denote it, and the legislator should have left the issue of definition to jurisprudence as he is the most familiar with it and because the role of the legislator is to set the provisions and not to define the names.

3 Justifications for the Application of Electronic Monitoring

The application of electronic monitoring came as one of the alternatives to penalties of deprivation of liberty, based on the necessities that necessitated its existence, as the disadvantages of traditional imprisonment required the penal legislator to search for alternative solutions that keep pace with development within certain controls and conditions, which we explain as follows:

3.1 With Regard to the Penal Institution

3.1.1 Reducing Expenses

Economic reasons are considered one of the most important motives for changing the punitive policy, especially for liberal countries, so that the financial burden becomes an excuse to search for a way to reduce the considered base constructions of institutions Punitive [9], as the introduction of electronic monitoring as an alternative to the penalty of deprivation of liberty works to reduce the burden on the state treasury by reducing the expenses required by the accommodation of prisoners with regard to their food and drink, the costs of guarding and medical care treatment, and other necessary expenses.

3.1.2 Avoiding the Problem of Prison Overcrowding

Overcrowding in prisons is one of the problems that countries are trying to avoid in various ways. On the national level, Jordanian prisons were overcrowded until 1/2/2022 with about (20) thousand inmates, while the occupancy rate reached 144% [10]. From a practical point of view, the introduction of electronic monitoring is considered the best solution. And the least costly is to treat this problem, which in turn puts a greater effectiveness in reforming and rehabilitating those who do not meet the conditions of electronic monitoring of those convicted against them, in addition to facilitating the process of controlling reform and rehabilitation centers by the punitive authority.

3.1.3 With Regard to those Convicted Against Them

3.1.3.1. Psychological Motives
Deprivation of liberty affects individuals in their dignity, reputation and social status, which leads to many psychological disadvantages. An inner feeling of frustration and humiliation, in addition to his loss of respect in front of his family, and friends [11].
3.1.3.2. Reducing the Risk of Recidivism
Criminality is not considered a matter of coincidence, but rather for reasons including, the loss of work and financial resource after his discharge and the psychological and organic diseases that he suffers from during his stay in the penal institution, as well as the difficulty of his integration with society after his discharge [12].

Therefore, electronic monitoring is an ideal solution to avoid these problems. Although his freedom is not completely restricted, the feeling of control prevents the convicted person from returning to crime.

4 Conditions of the Application of Electronic Monitoring

The place of the convict under electronic monitoring is approved by the penalty execution judge or the authority competent to implement punitive alternatives, with the consent of the convict to reintegrate him into society at his will and with his consent [13], within the conditions specified by Article (25) repeated of the Jordanian Penal Code that can be divided into Two types as follows:

4.1 The Legal Provisions

4.1.1. That the judgment to punish the convicted person is related to a misdemeanour, or that it relates to a felony that was not committed against the persons.
4.1.2 That the person convicted against has not previously repeated the crime.
4.1.3. That the judgment of electronic monitoring be issued by the competent court.
4.1.4. That the judgment issued against the convict not only impose a fine, but the judgment must also include the deprivation of his liberty.
4.1.5. That the electronic monitoring period not exceed one year and not less than a month.

4.1.6. The issuance of the electronic monitoring penalty should be preceded by a social report by a specialist who endorses and supports replacing the original penalty with electronic monotoring.

4.2 The Material Conditions:

4.2.1. The person against whom the judgment is issued must have a place of residence or fixed residence.
So that the convicted person must have a fixed place of residence, and it is not required that this place be owned by the convicted person, and this place may be his own residence, or an institute that is subject to the inspection of the competent government agency [14].
4.2.2. Availability of the necessary electronic capabilities.
The implementation of the electronic monitoring method requires the presence of the necessary technology, for example, that the place of residence is equipped with a telephone line dedicated to the implementation of this system in order to receive telephone calls related to electronic monitoring [15].

5 The Effects of Electronic Monitoring

The following show the most important effects of electronic monitoring:

5.1 The Obligations of the Convict to Electronic Monitoring

The ruling issued to replace a penalty of deprivation of liberty with electronic monitoring imposes a number of obligations on the convict, perhaps the most prominent of which is the commitment of the concerned person not to leave his house or the place designated by the penalty application judge outside the period specified in the placement place [16].

Under Article 25 of the Penal Code, the Jordanian legislator prohibited the convict from frequenting specific places designated by the penal judge, but the Jordanian legislator did not explain the details of the obligations of the convict, but left the matter to the authority of the trial judge, and it seems that the approach of the Jordanian legislator differs from his counterpart The Algerian, who stated in Article 150 repeated of the Prisons Regulation and Social Reintegration of Detainees Law in its last paragraph that "commit the convict to respond to the summons of the penalty enforcement judge or the public authority designated by the latter" [17] as permitted by Article 150 repeated -6 of the same law In its first paragraph, the Penal Judge may subject the person under electronic monitoring to one or more of the following measures:

a. Practicing a professional activity or pursuing education or vocational training.
b. Not going to some places.
c. Not meeting with some of the convicted persons, including the original perpetrators or partners in the crime.
d. Not meeting with some people, especially victims and minors.

While the researcher prefers the approach of the Algerian legislator, he believes that it is necessary to leave a discretionary authority to the judge in the matter in the implementation of the control in order to take into account the personal and family circumstances of the convict and his work conditions.

5.2 The Effect of Violating the Electronic Monitoring System

Since electronic monitoring is considered one of the alternative penalties, it may be violated by the person against whom the judgment is imposed,[18] either deliberately or for a reason beyond his control, which the legislator has dealt with as follows:
5.2.1. If the non-execution of the penalty is due to a legitimate excuse or a reason in which the convict has no will, in this case the penalty execution judge shall replace the electronic monitoring penalty with one of the alternatives to the penalties of deprivation of liberty stated in Article (25) repeated of the Penal Code.
5.2.2. If the non-execution of the penalty is due to illegal actions by the convict, for example removing or attempting to remove the electronic bracelet, or failure to stay in the places specified by the judge.

Paragraph No. (4) of Article (25) repeated authorized the penal judge to refer the case file to the court that issued the ruling to consider annulling the alternative and sentencing a penalty of deprivation of liberty, i.e. what it means to stop the implementation of electronic monitoring and return to the original penalty prescribed by law for the crime, For the purposes of the aforementioned, the law states that the period spent by the convict in executing the alternative shall be calculated at the rate of five hours for each day of imprisonment.

6 Conclusion

The electronic development has had significant effects on the scope of criminal policy, including the punishment policy, as this development has been employed to mitigate the negative effects of freedom-related penalties. In order to achieve this, the Jordanian legislator approved electronic monitoring or remote monitoring as one of the alternative penalties, in a way that achieves a balance between the requirements of maintaining public security on the one hand and protecting groups of convicts from the disadvantages of custodial punishment. From this paper, we concluded a number of findings and recommendations, which we summarize as follows:

6.1 The Findings

6.1.1 The electronic monitoring system is a modern method that appeared with the development of technology and its operation to serve the state administration and achieve public interests.
6.1.2.. Electronic monitoring can be applied as a precautionary measure if it is in the investigation phase, and it can be applied as an alternative punishment for those against whom a criminal judgment has been issued.
6.1.3. It has become necessary to rely on electronic monitoring within the protocols of the punitive system, due to the strength of the motives that necessitate it for both parties, whether the convicts or the punitive institution.
6.1.4. The application of electronic monitoring as an alternative punishment is still very rare in judicial rulings.

6.2 Recommendations

6.2.1. Activating electronic monitoring and expanding its application in judicial rulings, especially with regard to juveniles.
6.2.2. The legislator should state the employment of digital development to serve criminal policy, especially in the field of alternative penalties, guided by the experiences of developed countries in this field.
6.2.3. The legislator should state the competent judge for the implementation of electronic monitoring, and that his jurisdiction must be limited exclusively.
6.2.4. Recommending the legislators in the Arab countries that have not yet adopted this alternative punishment to include it in their legislations.

References

1. Raho, A.: Placement under electronic monitoring as an alternative to deprivation of liberty. J. Judic. Law **152**(49), 11 (2010)
2. Busari, A.: The developed systems to confront the crisis of short-term imprisonment. Master Thesis, University of Batna, pp. 220–222 (2012/2013)
3. Al-Zeini, A.: House Confinement, 1st edn, p. 4. Arab Renaissance House, Cairo (2005)
4. Salem, O.: Electronic Monitoring, 1st edn, p. 45. Dar Al-Nahda Al-Arabiya for Publishing and Distribution, Cairo (2000)
5. Bin Abdullah, M., Muhammady, O.: Procedures alternative to imprisonment. Judic. J. (2), 8 2005
6. Al-zeini, op:cit, pp. 4–6
7. Imam, K.: Putting Juveniles under Electronic Monitoring as an Alternative Punishment to Imprisonment. PhD thesis, The Islamic University, p. 40 (2016)
8. Article 25 bis, Paragraph (1), Clause (C) of the Jordanian Penal Code No. (10) for the year 2022 issued in the Official Gazette No. (5796), p. 3591. Accessed 25 May 2022
9. Arshouch, S.: Electronic monitoring as an alternative to denial sanctions. J. Law Polit. Sci. **8**(1), 453 (2017)
10. Director of the Department of Correction and Rehabilitation Centers, Brigadier General Ayman Al-Awaisha, , interview in Amman with Al-Madinah News website (2019). Accessed 22Jan 2023
11. Arshouch, S.: op:cit, p. 422
12. Suleiman, A.: Social Interaction between Prisoners and Released Persons and Society. Master Thesis, Cairo University, p. 101 (1997)
13. Al-Hammadi, S.: Electronic monitoring system as an alternative to short-term deprivation of freedom. Sharjah J. Legal Sci. **7**(18), 592 (2021)
14. Otani, S.: Placement under electronic monitoring (electronic bracelet) in the French punitive policy. Damascus Univ. J. Econ. Legal Sci. **25**(1), 129–136 (2009)
15. Al-Fiqi, I.: Alternative regimes for short-term imprisonment, a comparative study. J. Legal Econ. Stud. (1), 81 (2015)
16. Saud, A.: Electronic monitoring as an alternative to short-term deprivation of liberty. J. Legal Polit. Sci. (11), 12 (2018)
17. Article (150) bis 6 of the Prisons Regulation and Social Reintegration of Detainees Law No. (18–01) dated 30/1/2018 supplementing Law No. (05–04). Accessed 2 June 2015
18. Obaid, O.: Electronic criminal monitoring, 1st edn, p. 642. Dar Al-Nahda Al-Arabiya for publishing and distribution, Cairo (2009)

The Impact of Digital Human Resources Management Practices (DHRMPs) on Administrative Empowerment: Field Study in Telecommunications Companies in Jordan

Amer Alsaraireh[1(✉)], Hussam Al-Qadi[2], Sayeeduzzafar Qazi[3], Zakaria Azzam[1], and Daliah Hussein[4]

[1] Faculty of Economics and Administrative Sciences, Zarqa University, Zarqa, Jordan
{aalsaraireh,zazzam}@zu.edu.jo

[2] Researcher in Strategic Management at PSD, Zarqa, Jordan

[3] Department of HRM, University of Business and Technology, Jeddah, Kingdom of Saudi Arabia
sayeed@ubt.edu.sa

[4] Department of Finance, University of Business and Technology, Jeddah, Kingdom of Saudi Arabia
d.hussein@ubt.edu.sa

Abstract. The study aimed to analyze the impact of digital human resource management practices (DHRMPs) on administrative empowerment (AE) in Telecommunications Companies in Jordan (TCJs). The descriptive analytical approach and SPSS statistical package were relied upon in this study. Study population consisted of all (350) officials with leadership or supervisory positions in telecommunications companies (Zain, Orange, and Umniah) in Jordan. By adopting the comprehensive field survey method in collecting data, total of analyzed valid questionnaires was (260). The main study results are: The level of application of both DHRMPs and AE in TCJs had reached a high degree; there is a statistically significant impact of DHRMPs on AE. The study recommended the importance of continuing to maintain these levels.

Keywords: Digital Human Resources Management Practices (DHRMPs) · Administrative Empowerment (AE) · Telecommunications Companies in Jordan (TCJs)

1 Introduction

The modern era witnesses massive and rapid developments in different domains due to the digital technological developments, and the revolution in Communications and Information Technology (CIT), which made our world in a constant change due to rapid developments, innovations and updates in the fields of digital transformation (DT), CIT, science and knowledge. These fields are continuously shifting to cope with the influential

B. A. M. Alareeni and I. Elgedawy (Eds.): *Artificial Intelligence (AI) and Finance*, SSDC 488, pp. 339–350, 2023.
https://doi.org/10.1007/978-3-031-39158-3_32

and effective development in different domains of life. Regardless the nature of an organization's activity; its main objective remains the same: To survive, and overcome challenges and crises [1, 17].

To grow and improve their services and products and make it easier for customers to access them, all firms must now prioritize DT. This does not only mean implementing technology within the organization, but rather implementing a comprehensive digital program for the organization and its work method mainly internally and externally as well. This is due to the use of CIT rapidly expanding. Also, the great challenge facing successful organizations is the change itself, and therefore there must be new ideas in these organizations that support them to develop their internal operations and improve the quality of services [2]. Participation in this research paper is carried out through Zarqa University.

2 Objectives and Significance

Tis paper aims at identify the impact of DHRMPs on AE in TCJs, as well as Identifying the availability level of: DHRMPs and its dimensions represented in digital recruitment (DR), digital selection (DS), digital training (DT) and digital performance evaluation (DPE); AE and its dimensions represented in: delegation of authority (DOA), teamwork (TW) and the development of creative behavior (DOCB), in TCJs. As well as presenting a set of recommendations to the officials in light of the results.

This study is noteworthy for the importance of the variables it examines, in light of the digitally changing environment, which are now considered one of the modern and exciting topics that influence all facets of life. Also, this research intends to close the knowledge gap between what organizations already know and the most recent developments and current difficulties in the field of DT so as to enhance the Arabic library. Then, as TCJs firms offer Internet and telecommunications which are very important to the DT process, it aids their officials by supplying relevant proof.

3 Study Hypotheses

Main hypothesis (H0): There is no statistically significant impact at the level of ($0.05 \geq \alpha$) for DHRMPs with its combined dimensions (DR, DS, DT, and DPE) on AE with its combined dimensions (DOA, TW, and the DOCB) in TCJs. While the sub-hypotheses are:

(H0–1): There is no statistically significant impact at the level of ($0.05 \geq \alpha$) for DHRMPs on DOA.
(H0–2): There is no statistically significant impact at the level of ($0.05 \geq \alpha$) for DHRMPs on TW.
(H0–3): There is no statistically significant impact at the level of ($0.05 \geq \alpha$) for DHRMPs on DOCB.

4 Theoretical Framework and Previous Studies

4.1 Digital Transformation (DT)

The term "digital" is used to describe how quickly things are changing in the current world thanks to the use of cutting-edge digital technologies. The term "transformation" refers to an organization's decision to embrace innovation, creativity, and change brought about by the use of digital technology rather than merely embracing and using conventional ways. From this point on, DT may be seen as an increase in productivity that allows new economies to use digital technology in already-existing companies while also creating opportunities for new rivals. In this direction, organizations are moving quickly to apply, disseminate and build the basic digital technology used in DT as cloud technology, software applications, big data, digital payments, the Internet of things IoT, artificial intelligence and modern cognitive technology [3].

4.2 Concept of Digital Human Resources Management Practices (DHRMPs)

DHRMPs has been defined as the investment of technology applied within the organization in ways and applications that serve the various HR management (HRM) practices processes of attracting, selecting, appointing, training, and evaluating, as well as dealing digitally with all data on current and potential employees [4].

While [5] defined DHRMPs as the use of modern digital technologies for HRM functions and their integration with information technology, aiming to create high value within the organization, by investing in digital technologies to increase the efficiency and effectiveness of HRM practices processes.

Whereas [6] DHRMPs defined it as everything that CIT can provide to the management of HR practices, including the Internet, communication networks, information systems, and expert systems that are used in the practices of in the organization. Digital HRM includes several functions, the most important of which are: DS, DR, and DPE.

Digital Recruitment (DR)

The usage of the Internet to advertise open positions and strive to attract the best human resources has increased over the past few years. This is because it's low-cost, people can see it fast, and the Internet offers information that may not be available through conventional advertising means. That is compared to other traditional methods, which are expensive and only a small number of individuals can view it. Details on the company's operations, strategy, services, markets, etc. can be found online. This is in addition to information about the job, such as its characteristics, requirements, job benefits and salaries. Those digital advertisements can run for long periods of time, not just for more than a few days as it is in traditional newspapers, for example. This permits the person applying for the job to be aware in advance of all the details related to the vacancy, the organization, and the timing of the interviews. In the era of digital administration, the applicant can avoid carrying heavy paper folders, and sends all his credentials online by electronic message, which cuts the cost and rises the efficacy of the recruitment process [7].

Digital Selection (DS)
The ability of modern companies and institutions to keep pace with the rapid developments in the market is considered one of the most important factors of their success, and in accordance with the principle that individuals are the most valuable assets, companies and institutions must also quickly change the way they employ their employees, and as a result it is not surprising that many organizations are using modern digital means to fill their job vacancies, through advertising on the Internet, by creating an attractive website that attracts Internet surfers to access it and view the company's job vacancies, as well as conducting functional, psychological and other tests through electronic interviews. There is no doubt that the transformation of the HRM function from traditional selection to DS had definitely reflected positively on many organizations, by eliminating the role of nepotism on one hand, as well as accelerating the pace of the selection process, and reducing the time required to perform it [8].

Digital Training (DT)
The success of modern institutions in achieving their goals depends to a large extent on the efficiency and effectiveness of HR employed, it is agreed upon between managers, that managing and developing HR is the best approach to raise and improve the efficiency of HR performance in businesses, in order to achieve higher productivity and higher profitability on one hand, as well as to satisfy HR on the other hand. so organizations should use modern digital technologies to develop and elevate the skills of workers, as these technologies facilitate exchanging, saving, transferring, renewing knowledge and making it accessible to its seekers. Among these methods is learning and training, because learning and training is an essential process in life, from which no human activity is lacking, since it is linked to the development of knowledge and skills [9].

Digital Performance Evaluation (DPE)
There is a general agreement among business administration writers about the meaning of performance evaluation despite the minor differences between them, as [10] believes that it is an objective measurement process of the volume and level of what has been accomplished compared to what is required to be accomplished in quantity and quality in the form of a correlation between the existing and desired situations.

4.3 Administrative Empowerment (AE)

AE represents a modern strategy aims to unleash the potentials of workers, and incorporating their participation in defining future visions, because the success of the institution depends on how the workers' needs are integrated with vision and goals [11].

Delegation of Authority (DOA)
DOA means assigning the chief to his subordinates to carry out some of his duties and exercise some of his powers, i.e. the holder of the original competence entrusts another

person or body to exercise his competence in accordance with the legal conditions prescribed for that [12].

Teamwork (TW)
The second element that supports and enhances AE is the group and TW, provided there are common factors among the group such as cooperation, equality, solidarity, integration, concerns, aspirations, and common hopes that are included in relationship, acquaintance, and connections, and the group may live a common life according to a special system [13].

Developing Creative Behavior (DOCB)
Creative behavior is the distinctive behavior or action practiced by an individual or group at the work site, which does not necessarily result in new services or commodities. It is a behavior that precedes creativity in its final capacity, and this behavior might be a creation on its own when an individual does it for the first time [14].

4.4 Previous Studies

[6], in his study entitled: "The role of the DT strategy in developing the system of HRM functions, by applying it to the Egyptian public banks". The study aimed to analyze one of the challenges faced by most of the management officials in public institutions in Egypt, which is the difficulty of achieving the goals of the banking institution through traditional HRM activities, and the role of DT in developing these activities and practices. It used the descriptive analytical approach. The study population consisted of employees of public banks operating within Cairo Governorate, and the researcher selected a sample of 180 individuals from the analysis unit. The study concluded that there is a statistically significant positive correlation between the DT strategy and the effectiveness of the HRM system, according to the applications of recruitment, training, wage management, and performance evaluation.

[15], in their study entitled: "The Effect of Electronic Human Resources Management on Organizational Health of Telecommunications Companies in Jordan". The study aimed to analyze the impact of DHRMPs on organizational health, by applying it to telecommunications companies operating in Jordan. It adopted the descriptive analytical method. Study population contained managers in telecommunications companies, whereby 312 individuals, selected using a comprehensive survey. The study found that there is a positive correlation between digital HRM and organizational health.

[16], in his study entitled: "The Impact of Administrative Empowerment on the Job Satisfaction of Employees". Objective of study was to find how managerial empowerment influences employee happiness in the Directorate of Health in the Northern Borders Region. The study used the descriptive and analytical approach. The questionnaire was distributed to 167 personnel. The study resulted in the presence of a significant effect of the AE variable with its five dimensions (involvement in decision-making, effective communication, worker motivation, training and continuing education, teamwork) on the level of job satisfaction for employees.

5 Methodology

5.1 Study Population

The study population consisted of all officials holding leadership or supervisory positions in TCJs (Zain, Orange and Umniah), amounting to (350). The comprehensive survey method was used, by which the questionnaires distributed electronically. (274) were retrieved, the response rate was (78.2%). By reviewing them, the number of valid ones for analysis was (260), representing (74.2%) of the distributed ones.

5.2 Descriptive Analysis of the Study Variables

5.2.1 Descriptive Analysis of DHRMPs

Table 1. Descriptive analysis of the study dimensions of the DHRMPs variable

Items	Dimensions	Arithmetic Average	Rank	Significance Level
1–4	DR	4.65	1	High
5–8	DS	4.61	2	High
9–12	DT	4.23	4	High
13–16	DPE	4.40	3	High
	Total Average	**4.47**		High

It appears from Table 1. That the overall average for the DHRMPs variable was high; as the total arithmetic average was (4.47), and the arithmetic averages ranged between (4.23–4.65). The (DR) dimension ranked first with an average of (4.65), while the (DT) dimension ranked last, with an average of (4.23).

5.2.2 Descriptive Analysis of AE

Table 2. Descriptive analysis of the study dimensions of the AE variable

Articles	Dimensions	Arithmetic Average	Rank	Significance Level
17–20	DOA	3.71	3	High
21–24	TW	3.85	2	High
25–28	DOCB	3.98	1	High
	Total Average	**3.85**		**High**

It appears from Table 2. That the overall average of the AE variable was of a high degree, as the total arithmetic average was (3.85), and the arithmetic averages ranged between (3.71–3.98). The dimension (DOCB) ranked first with an arithmetic average of (3.98), while the dimension (DOA) ranked last, with an arithmetic average of (3.71).

5.3 Results of Testing and Analyzing the Study Hypothesis

Results of Testing the Main Hypothesis (H0)

H0 states that: There is no statistically significant impact at the level of ($0.05 \geq \alpha$) for DHRMPs with its combined dimensions (DR, DS, DT, and DPE) on AE with its combined dimensions (DOA, TW, and the DOCB) in TCJs.

Table 3. Multiple linear regression testing the impact of DHRMPS on AE

Form summary		Variance analysis (ANOVA)					
(R) Correlation coefficient	(R^2) Determination Coefficient	Sum of squares	(DF) degrees of freedom		Average of squares	Calculated F value	Significance level (F)
0.895	0.801	7.288	Regression	4	1.822	82.818	0.000
		2.188	Residuals	255	0.022		
			Total	259			

Regression Coefficients

Dimensions of independent variable	(B) value	Standard error	(T) value	Significance level (T)
DR	0.344	0.054	6.370	0.000
DS	0.369	0.051	7.235	0.000
DT	0.392	0.049	8.101	0.000
DPE	0.318	0.059	5.389	0.000

* Statistically significant at ($\leq 0.05\ \alpha$) level.

Results in Table 3. Indicate that there is a statistically significant impact of DHRMPs on AE in TCJs, as the calculated (F) value reached (82.818), and the statistical significance ($a \leq 0.05$), that indicates the impact is accepted, and the correlation coefficient was ($R = 0.895$), which indicates the positive relationship between the independent and dependent variable, in addition to the value of the determination coefficient ($R2 = 0.801$); indicating that the digital DHRMPs explained (80.1%) of the variation in AE in TCJs.

As for the regression coefficients, it showed that the value of (B) for the dimension (DR) was (0.344), and the value of (T) for it is (6.370), with a significance level ($a \leq 0.05$), which is statistically significant, while the value of (B) for the dimension (DS) it amounted to (0.369), and the value of (T) for it amounted to (7.235), and the value of (B) for the dimension (DT) was (0.392), and the value of (T) for it amounted to (8.101), which is statistically significant. Finally, the value of (B) for the dimension (DPE) was (0.318), and the value of (T) for it was (5.389), with a significance level ($a \leq 0.05$), which is statistically significant. With these results, the null hypothesis is rejected, the proof hypothesis is accepted, meaning: There is a statistically significant impact at the level of (0.05 ≥ α) for DHRMPs with its combined dimensions (DR, DS, DT, and DPE) on AE with its combined dimensions (DOA, TW, and the DOCB) in TCJs.

Results of Testing the First Sub-hypothesis (H0–$_1$)

H0–$_1$ states that: There is no statistically significant impact at the level of (0.05 ≥ α) for DHRMPs on DOA.

Table 4. Multiple linear regression testing the impact of DHRMPS on DOA

Form summary		Variance analysis (ANOVA)					
(R) Correlation coefficient	(R^2) Determination Coefficient	Sum of squares	(DF) degrees of freedom		Average of squares	Calculated F value	Significance level (F)
0.881	0.776	8.300	Regression	4	2.075	90.217	0.000
		6.100	Residuals	255	0.023		
			Total	259			

Regression Coefficients

Dimensions of independent variable	(B) value	Standard error	(T) value	Significance level (T)
DR	0.301	0.079	3.810	0.000
DS	0.312	0.069	4.521	0.000
DT	0.333	0.065	5.123	0.000
DPE	0.309	0.078	3.961	0.000

* Statistically significant at (≤ 0.05 α) level.

Results in Table 4. Indicate that there is a statistically significant impact of DHRMPs on DOA in TCJs, as the calculated (F) value was (90.217), and with a statistical significance (a ≤ 0.05), that indicates the acceptance of the impact from a statistical point of view, the correlation coefficient was (R = 0.881), and this indicates the positive relationship between the independent and dependent variable, in addition to the value of the determination coefficient (R2 = 0.776); this indicates that the DHRMPs explained (77.6%) of the variation in the DOA in TCJs.

As for the regression coefficients, they showed that the value of (B) for the dimension (DR) amounted to (0.301), and that the value of (T) for it is (3.810), with a level of significance (a ≤ 0.05), which is statistically significant, while the value of (B) for the dimension (DS) is (0.312), and the value of (T) for it amounted to (4.521), and the value of (B) for the dimension (DT) was (0.333), and the value of (T) for it amounted to (5.123), which is statistically significant. Finally, the value of (B) for the dimension (DPE) was (0.309), and the value of (T) for it was (3.961), with a level of significance at (a ≤ 0.05), which is statistically significant. With this result, the null hypothesis is rejected and the proof hypothesis is accepted, that means: There is a statistically significant impact at the level of (0.05 ≥ α) for DHRMPs on DOA.

Results of Testing the Second Sub-hypothesis (H0–$_2$)

H0–$_2$ states that: There is no statistically significant impact at the level of (0.05 ≥ α) for DHRMPs on TW.

Table 5. Multiple linear regression testing the impact of DHRMPS on TW

Form summary		Variance analysis (ANOVA)					
(R) Correlation coefficient	(R^2) Determination Coefficient	Sum of squares	(DF) degrees of freedom		Average of squares	Calculated F value	Significance level (F)
0.844	0.712	17.90	Regression	4	4.475	84.433	0.000
		13.32	Residuals	255	0.053		
		31.22	Total	259			

Regression Coefficients

Dimensions of independent variable	(B) value	Standard error	(T) value	Significance level (T)
DR	0.389	0.048	8.104	0.000
DS	0.386	0.056	6.892	0.000
DT	0.375	0.063	5.952	0.000
DPE	0.365	0.069	5.289	0.000

* Statistically significant at ($\leq 0.05\ \alpha$) level.

Results in Table 5. Indicate that there is a statistically significant impact of DHRMPs on TW in TCJs, as the calculated (F) value was (84.433), with a statistical significance ($a \leq 0.05$), and this indicates the acceptance of the impact from a statistical point of view, the correlation coefficient was ($R = 0.844$), and this indicates the positive relationship between the independent and dependent variable, in addition to the value of the determination coefficient ($R2 = 0.712$); which indicates that the DHRMPs explained the amount (%) of the variation in the development of TW in TCJs.

As for the regression coefficients, they showed that the value of (B) for the dimension (DR) was (0.389), and that the value of (T) for it is (8.104), with a level of significance ($a \leq 0.05$), which is statistically significant, while the value of (B) for the dimension (DS) amounted to (0.386), and that the value of (T) for it amounted to (6.892), and that the value of (B) for the dimension (DT) amounted to (0.375), and that the value of (T) for it amounted to (5.952), which is a statistically significant. Finally, the value of (B) for the dimension (DPE) was (0.365), and the value of (T) for it was (5.289), with a significance level ($a \leq 0.05$), which is statistically significant. With this result, the null hypothesis is rejected and the proof hypothesis is accepted.

Results of Testing the Third Sub-hypothesis ($H0\text{–}_3$)

$H0\text{–}_3$ states: There is no statistically significant impact at the level of ($0.05 \geq \alpha$) for DHRMPs on DOCB.

Results in Table 6. Indicate that there is a statistically significant impact of DHRMPs on the DOCB in TCJs, where the calculated value of (F) was (77.98), with a statistical significance ($a \leq 0.05$), and this indicates this impact is statistically significant, and the correlation coefficient was ($R = 0.842$), and this indicates the positive relationship between the independent and dependent variable, in addition to the value of the determination coefficient ($R2 = 0.708$); which indicates that the digital human resource management practices explained (70.8%) of the variation in the DOCB in TCJs.

Table 6. Multiple linear regression testing the impact of DHRMPS on DOCB

Form summary		Variance analysis (ANOVA)					
(R) Correlation coefficient	(R^2) Determination Coefficient	Sum of squares	(DF) degrees of freedom		Average of squares	Calculated F value	Significance level (F)
0.842	0.708	21.213	Regression	4	5.303	77.98	0.000
		17.285	Residuals	255	0.068		
		38.498	Total	259			

Regression Coefficients

Dimensions of independent variable	(B) value	Standard error	(T) value	Significance level (T)
DR	0.298	0.045	6.622	0.000
DS	0.291	0.057	5.105	0.000
DT	0.288	0.062	4.645	0.000
DPE	0.315	0.042	0.751	0.000

* Statistically significant at ($\leq 0.05\ \alpha$) level.

As for the regression coefficients, they showed that the value of (B) for the dimension (DR) was (0.298), and that the value of (T) for it is (6.622), with a significance level (a $\leq$ 0.05), which is statistically significant. (DS) amounted to (0.291), and that the value of (T) for it amounted to (5.105), and that the value of (B) for the dimension (DT) amounted to (0.288), and that the value of (T) for him amounted to (4.645), which is statistically significant. Finally, the value of (B) for the dimension (DPE) was (0.315) and the value of (T) for it was (0.751), with a level of significance (a $\leq$ 0.05), which is statistically significant. With this result, the null hypothesis is rejected and the proof hypothesis is accepted, that means: There is a statistically significant impact at the level of ($0.05 \geq \alpha$) for DHRMPs on DOCB.

6 Discussion of Results

The results of the study indicated that the level of application of DHRMPs in TCJs came to a high degree, this result indicates the interest of these companies in applying DHRMPs in terms of DR, DS, DT, and DPE.

The results of the study indicated that the level of application of AE in TCJs had reached a high degree, this result proves that these companies are keen to practice AE, in terms of DOCB of employees, adopting creative ideas, focusing of authority delegation and effectively between management and employees.

The results of the study indicated that there is a statistically significant impact of DHRMPs with its dimensions (DR, DS, DT, and DPE) combined on AE in its dimensions (DOA, TW, and the DOCB) combined in TCJs. At the level of sub-hypotheses, the results of the study indicated that there is a statistically significant impact of DHRMPs on DOA, TW, and the DOCB, which agrees with the studies of [15].

7 Recommendations

The study recommends the need for decision-makers in Jordanian telecommunications companies to maintain the level of application of digital practices for human resource management and administrative empowerment. Given the dimensions of each of them with care and attention by the senior management in these companies.

The study also recommends conducting future studies on the variables of the current study by addressing other dimensions not covered by this study, as well as studying the relationship and impact of other variables, such as analyzing the impact of DT on strategic planning for human resources, the impact of big data on administrative empowerment or the impact of applying digital practices of HRM on achieving institutional excellence. The study also recommends future studies to address other sectors such as the banking sector, mining companies and others, and then comparing the results of these studies with the results of this study for the possibility of generalization and benefit from them by decision makers.

References

1. Ali, S.: The Departments of Excellence and Quality. Dar Al-Yazuri for Publishing and Distribution, Amman (2020)
2. Mohamed, S.: Strategic Planning is a Approach to Achieving Competitive Excellence, Dar Al-Fajr for Publishing and Distribution, Cairo (2016)
3. Hadi, M.A.: Artificial Intelligence, its Features, Applications, and Developmental and Societal Impacts. The Egyptian Lebanese House for Publishing and Distribution, Cairo (2021)
4. Jawdat, H.: Impact of the application of digital HRM in promoting institutional excellence: an applied study on the department of economic development in the Government of Ajman. Arab J. Manag. **40**(2), 233–246 (2020)
5. Al-Eidani, E., Kawar, A.: The complementary relationship between the digital economy and digital management of HR. In: The Virtual International Forum for Big Data and The Digital Economy as A Mechanism for Achieving Economic Growth in Developing Countries. Opportunities and Challenges. Algeria. University of Martyr Hama Lakhdar Al-Wadi, 11–18 June 2022 (2022)
6. Omar, T.: The role of digital transformation strategy in developing the system of HRM functions: applied to Egyptian governmental banks. Sci. J. Finan. Adm. Stud. Res. **15**(1), 2–46 (2022)
7. Ibrahim, A.B., Naeem, Z.: Human Resources Management in the Twenty-First Century. Dar Wael for Publishing and Distribution, Jordan (2008)
8. Garkad, A., Noureddine, H.: HRM in the Era of Electronic Management. Academic Books Center for Publishing and Distribution, Algeria (2018)
9. Ahmed, K.: Modern Digital Economy, HRM and Commodity Production, Dar Al-Fikr Al-Jamei for Publishing and Distribution, Alexandria (2018)
10. Muhammad, W.: Evaluation and Development of the Employee Performance Appraisal System, Dar Al-Yazuri for Publishing and Distribution, Amman (2022)
11. Shibly, M., Alawamleh, H.A.: The relationship between administrative empowerment and continuous improvement: an empirical study. Revista Geintec-Gestao Inovacao E Tecnologias **11**(2), 1681–1699 (2021)
12. Abraheem, M.K.: The impact of strategic planning on administrative empowerment in higher education institutions. Akkad J. Contemp. Manag. Stud. **1**(1), 1–20 (2021)

13. Khan, K.: Mediating role of ethical leadership between employees empowerment and competitive edge: a case of commercial banks in Pakistan. Human. Social Sci. Rev. **9**(2), 219–231 (2021)
14. Khawaldeh, F.O.: The relation between the administrative empowerment and creative behavior at Jordanian private universities. Turk. J. Comput. Math. Educ. (TURCOMAT) **12**(9), 2720–2731 (2021)
15. AlHamad, A., et al.: The effect of electronic HRM on organizational health of telecommunications companies in Jordan. Int. J. Data Netw. Sci. **6**(2), 429–438 (2022)
16. Sameer.J,: The impact of administrative empowerment on the job satisfaction of employees. J. Appl. Res. High. Educ. **4**(9), 249–261 (2022)
17. Al-Shaikh, M.S.: Customers' perspective towards factors associated with technological product adoption failure: a case of Jordanian Hotels. J. Zarqa Univ. Soc. Res. Stud. **13**(2) (2013)

The Impact of E-Business on SME's Productivity (Case Study on "Kiddy Zone" Company in Qatar)

Hazem Khaled Shehadeh(✉), Mohammed Abed Hussein Al taee, Osama Mahmoud Odeh Khanfar, and Shareefah Ahmad

Faculty of Economics and Administrative Sciences, Business Administration Department, Zarqa University, Zarqa, Jordan
{hshehadeh,mohammad_h.altaee}@zu.edu.jo, o.khanfar@cinisolutions.com, sf.ahmad@ubt.edu.sa

Abstract. The purpose of this study was to investigate the effect of e-business on the productivity of small and medium-sized enterprises (SMEs). The study focused on two variables: the independent variable was e-business, which consisted of four dimensions - purchasing and warehousing cycle, information systems, human resources capabilities, and e-marketing; while the dependent variable was productivity. The study employed a comprehensive sample and distributed 120 questionnaires to all employees using Google forms. The study received 108 completed questionnaires, representing a response rate of 90%.The researcher used the descriptive-analytical method and appropriate statistical techniques in SPSS to test the hypotheses and achieve the study objectives. The main finding of the study was that implementing e-business dimensions, including purchasing and warehousing cycle, information systems, human resources capabilities, and e-marketing, had a significant impact on productivity. The e-business dimensions contributed to approximately 69.9% of productivity, with information systems having the highest contribution to productivity. Additionally, the company demonstrated a strong perspective towards purchasing and warehousing systems as one of the e-business dimensions.The study concluded with several recommendations, including conducting a needs assessment to identify the company's e-business requirements and using high-tech tools to manage information without the need for office work.

Keywords: E-Business · productivity · SME's · "Kiddy Zone" Company · Qatar

1 Introduction

In today's competitive and unstable business environment, institutions must adapt to successive changes in order to establish themselves as market leaders. Twenty years ago, computers connected to the internet were not as prevalent as they are today. However, with the subsequent advancements in information and communication technology and the establishment of the internet, institutions must deal with a new reality in order to overcome competitors and gain a competitive advantage to improve their performance.

B. A. M. Alareeni and I. Elgedawy (Eds.): *Artificial Intelligence (AI) and Finance*, SSDC 488, pp. 351–358, 2023.
https://doi.org/10.1007/978-3-031-39158-3_33

This has led to the emergence of electronic business (e-business), a modern technical method that seeks to create contemporary business models for institutions. It has become imperative for institutions to interact with modern technologies and shift towards e-business, especially since the outbreak of the COVID-19 pandemic in 2020.

However, e-business cannot be implemented in any society without fulfilling basic requirements such as information technology, advanced communication networks, software applications, and others. Small and medium-sized enterprises (SMEs) play a significant role in the economic development of developing countries by contributing to the increase of national income and employing a large percentage of the population. Therefore, e-business has gained attention in these institutions due to the opportunities it provides to improve electronic services and enhance operational efficiency. The emergence of e-business has revolutionized the way institutions conduct their work, becoming a vital factor for their success and sustainability.

E-business has specifically improved the flow of operations and daily business in SMEs, leading to enhanced productivity by reducing the time, effort, and errors associated with various activities. It encompasses all institutional activities, including procurement, stores, accounting, human resource capabilities, operations, and marketing. E-business is broader than e-commerce, which is limited to buying and selling products and exchanging information over the internet. The concept of e-business includes all daily operations, whether carried out by employees within the company or external dealings with suppliers, customers, and other parties. Headings should be numbered for the first and second levels, while lower-level headings are formatted as run-in headings.

2 Study Problem

Despite the numerous advantages of e-business for institutions, previous studies have shown that its implementation in small and medium-sized enterprises is limited, with many still using simple technologies like email [3, 4]. The researcher identified shortcomings in previous Arab studies on e-business in SMEs and sought to address this gap. The objective of this study is to explore the impact of e-business dimensions (purchasing and warehouse cycle, information systems, human resource capabilities, e-marketing) on productivity in the "Kiddy Zone" company. The study aims to test the following hypotheses.

H01: There is no statistically significant impact at the significant level ($\alpha \leq 0.05$) for E-Business of its dimensions (purchasing and warehousing cycle, information systems, human resources capabilities and e-marketing) on productivity in the "Kiddy Zone" Company.

3 Study Model

(Figure 1).

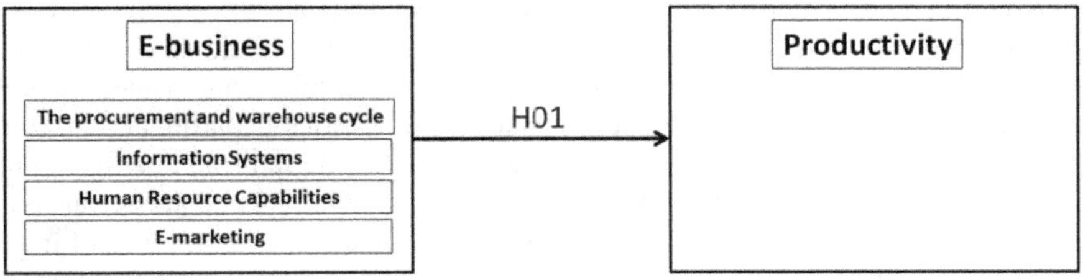

Fig. 1. Study Model

4 Literature Review

4.1 E-Business Definition

4.1.1 E-business

Organizational dimensions of e-business are based on the specific nature of an institution's work and its various departments that handle daily operations, such as service, industrial, and commercial departments. By utilizing modern electronic technologies and programs, institutions can integrate their activities and services to improve procedures, increase efficiency and effectiveness, and provide exceptional services to their customers while obtaining products and services from suppliers quickly and efficiently. As e-business has enabled institutions to conduct their administrative and daily operations electronically, the use of modern administrative practices has become a necessity for institutions to keep pace with technological developments. By reducing paper transactions and minimizing human errors in daily operations, e-business provides institutions with a competitive advantage over others [5]. Kharroub [6] defined e-business as digital transactions and the exchange of information and processes within an organization, all of which fall under the institution's internal information system. In this research, several e-business dimensions were identified that can be included in institutions, namely:

4.1.2 The Procurement and Warehouse Cycle

The procurement and warehouse cycle is defined as all stages of the procurement process, beginning with the needs of the competent departments and ending with the supply and receipt of goods in stores from suppliers, and it includes saving and analyzing supplier, item, and store data [7], organizations want to strengthen their supplier base, but there are always challenges they face due to increased competition, new technology, or supplier changes, so organizations seek to strengthen their relationships with their suppliers by analyzing information about them and the items and services purchased from them.

4.1.3 Information Systems

An information system refers to a network of interconnected components that are designed to collect, store, process, and disseminate data, information, and knowledge. In the business world, information systems are widely used to manage and monitor daily operations, interact with suppliers and customers, and gain a competitive edge in the marketplace. They are also employed to manage inter-organizational supply chains and electronic markets [8]. Regardless of the nature of activities and business models, organizations share a common structure consisting of various departments that form the foundation of their operations. These departments play a vital role in ensuring that daily activities run smoothly. With the advancement of technology and the expansion of business models, most organizations have come to recognize the importance of implementing systems that can enhance efficiency and effectiveness. Information systems are critical in providing statistical reports on various activities that assist senior management in decision-making. They also support daily activities by providing data processing and storage capabilities, while reducing the occurrence of human errors in daily operations, which increases the accuracy of these operations.

4.1.4 Human Resource Capabilities

In any organization, the ability of its employees to effectively use and navigate through both internal and external systems is vital to maintain the quality of work [9]. Moreover, e-business strategies have been found to significantly impact the productivity of human capital [10]. As such, the role of human resource capabilities in dealing with information systems and technology, which are used in both internal and external operations, has become increasingly important for organizations. This has prompted organizations to consider these capabilities when recruiting, training, and evaluating their employees. With the widespread use of information systems in daily operations, the ability of employees to use these systems effectively has become crucial to ensure the accuracy and efficiency of activities such as data collection, storage, dissemination, and communication with external stakeholders [11].

4.1.5 E-marketing

E-marketing refers to the process of promoting and selling products or services through remote displays and online communication channels. As a critical aspect of modern economic activity, businesses have realized the benefits of automating their marketing operations and transitioning to e-marketing. Through e-marketing, technology is utilized to remotely display and advertise products, connect them to online stores, and provide accurate information and details from both internal and external sources. It also enables businesses to gather and analyze data related to the market, sales, promotions, pricing, competition, and distribution of goods and services.

4.2 Productivity Definition

Productivity is commonly defined as the ratio of output volume to input volume. In other words, it measures how efficiently production inputs, such as labor and capital, are used to produce a given level of output, "It is the ratio between the output of goods and services and the input of resources consumed in the process of production," says the definition. Thus, productivity can be expressed as (Fig. 2):

$$\text{Productivity } (P) = \frac{\text{Output } (O)}{\text{Input } (I)}$$

Fig. 2. Productivity

Productivity is defined as "a measure of the efficiency of a person, machine, factory, system, or other entity in converting inputs into useful outputs." [13], In this regard, the Ko-lahdoozi [5] study was an attempt to create an effective e-business strategy development model or combined BS and BI approaches to improve productivity.

5 Study Population

The study targeted employees of Kiddy Zone company in Qatar, including individuals working in various departments and at different levels. The total population size was 120 according to the company's data, and all individuals were included in the study. The researcher distributed questionnaires to all 120 employees using Google Forms. Out of the total questionnaires distributed, 12 were not answered, representing a non-response rate of 10%. The valid sample size for actual analysis was therefore 108, which comprised 90% of the total distributed questionnaires.

6 Internal Validity and Stability of the Study Tool

The instrument was adapted following a thorough review of the related literature and examination by an expert panel; however, the researcher attempts to reaffirm the reliability to a satisfactory degree through internal consistency reliability analysis, among several statistical methods, the Cronbach's alpha coefficient method was chosen to measure reliability for this study; according to [14], the lower limit for Cronbach's alpha value is 0.70, according to [15], Table (1) shows the reliability of the independent and dependent variables (Cronbach's Alpha).

Table (1) shows that the reliability of Cronbach's alpha values for all of the study variables is greater than (0.7). The "Productivity" variable has the highest Cronbach's alpha value (0.880), while the "E-Marketing" variable has the lowest (0.729). Cronbach's Alpha coefficient for all items is (0.946). As a result, the study findings can be accepted in accordance with [15].

7 Hypothesis Testing

Multiple linear regressions were used to test the main hypothesis at the significance level ($\alpha \leq 0.05$).

Table 1. Reliability of the Questionnaire (Cronbach's Alpha)

Purchasing and Warehousing	Statement	Correlation Coefficient
Information systems	5	0.829
Human Resource Capabilities	8	0.844
E-Marketing	5	0.746
Productivity	5	0.729
All items	11	0.880
Purchasing and Warehousing	34	0.946

Table 2. Results of Multiple Regressions of the first main Hypothesis

Dependent Variable	R	R2	F	DF	SIG	Independent Variable	B	T	Sing
Productivity	0.836	0.699	59.707	4	0.000	Purchasing and Warehousing	0.111	1.674	0.097
				103		Information systems	0.133	6.784	0.000
						Human Resource Capabilities	0.250	3.887	0.000
				107		E-marketing	0.310	5.158	0.000

In Table (2), the dependent variable is Productivity, and the independent variables are Purchasing and Warehousing, Information Systems, Human Resource Capabilities, and E-Marketing. The F value of the ANOVA test was found to be statistically significant (0.000), which is less than the significance level of 0.05. The calculated F value (59.707) is also greater than the F table value (2.46). Hence, the null hypothesis is rejected, and the alternative hypothesis is accepted. This means that E-business, including its dimensions of Purchasing and Warehousing, Information Systems, Human Resource Capabilities, and E-Marketing, has a significant impact on the productivity of Kiddy Zone Company.

The dependent variable and independent variables have a strong positive relationship, with $R = 0.836$, which is greater than 0.5 (Cohen, 1988). The R2 value of 0.699 indicates that the independent variables contribute to the dependent variable by 69.9%.

Regarding the calculated t values, Information Systems (6.784), Human Resource Capabilities (3.887), and E-Marketing (5.185) are greater than the t-table value (1.986), and their p-values are less than 0.05. This indicates that Information Systems, Human Resource Capabilities, and E-Marketing have a statistically significant impact on the dependent variable (Productivity). However, the calculated t value for Purchasing and Warehousing (1.674) is less than the t-table value (1.986), and its p-value is greater than

0.05, indicating that Purchasing and Warehousing have no statistically significant impact on the dependent variable (Productivity).

8 Stepwise Regression

Stepwise regression categorizes independent variables based on their impact on the dependent variable and excludes those with low impact. Table (3) displays the results.

Table 3. Results of Stepwise Regression First main hypothesis

Number	Variables	F	R	R2	Sig
1	Information system	122.688	0.732	0.536	0.000
2	Information system E-marketing	86.842	0.789	0.623	0.000
3	Information system E-marketing Human Resource Capabilities	77.334	0.831	0.690	0.000

The stepwise approach categorized independent variables into three groups based on their contribution to the dependent variable, productivity, the first group consisted of only the Information system, with a contribution of 53.6%. The second group included Information systems and e-marketing, with a contribution of 62.3%, the third group included Information systems, e-marketing, and human resource capabilities, with a contribution of 69%, the stepwise method excluded the purchasing and warehousing variable, as it had no significant impact on productivity.

9 Results Discussion

The study's main hypothesis confirmed that implementing e-business dimensions, including the purchasing and warehousing cycle, information systems, human resources capabilities, and e-marketing, has a significant impact on the productivity of the "Kiddy Zone" company. This suggests that e-business has played a role in increasing the company's productivity, and a positive relationship exists between the variables. Therefore, any positive change or improvement in implementing e-business dimensions will have a positive impact on productivity. This finding is consistent with [10]. Moreover, the e-business dimensions contribute to achieving productivity by around 69.9%, which is consistent with [5], where a significant impact of the e-business environment was observed by raising administrative productivity.

10 Recommendations

In light of the study's results, the researchers recommend the following: make a need assessment to determine the company needs to transfer e-business work, use high technical allow managing the information without the need to work at the office, and using a security system to protect the information and reviewing the impact of e-business on the company cost.

References

1. Hachicho, N.M., Zeineddine, A.W.: The impact of intellectual capital on improving the performance of productive establishment (Baba Sweet Company as a Model). Int. J. Qual. Assur. **5**(1), 1–13 (2022). https://doi.org/10.34028/ijqa/5/1/214
2. Khalifa, A., Darwish, A.-R.: University students' use of the internet and its relation with assessment of online romantic relationships. Zarqa J. Res. Hum. Stud. **19**(2), 180–194 (2019). https://doi.org/10.12816/0054774
3. Awiagah, R., Kang, J., Lim, J.I.: Factors affecting e-commerce adoption among SMEs in Ghana. Inf. Dev. SAGE J. Inf. Dev. **32**(4), 815–836 (2016)
4. Karim, S., The challenges and opportunities of e-banking adoption for small to mid-sized enterprises-smes in Jordan. In: 2019 IEEE Jordan International Joint Conference on Electrical Engineering and Information Technology (JEEIT), pp. 883–888. IEEE, Amman, Jordan (2019)
5. Kolahdoozi, M., Radfar, R., Kazemi, M.A.: Effective E-Business Strategy Development Model or Combined BS and BI Approach to Improve Productivity (2022)
6. James, O.O., Mandal, S., Alele, N., Chowdhury, B., Maity, S.: Lower alkane's dehydrogenation: Strategies and reaction routes to corresponding alkenes. Fuel Process. Technol. **149**, 239–255 (2016)
7. Kharoub, N.M.S.: The impact of e-business capabilities on the performance of human resources and organizational performance in the Jordanian retail sector: a case study of Sameh Mall Investment Group. (Master's thesis). Middle East University, Jordan (2016)
8. Paramadina, F.H., FajarSuryawan, S.T.: An Information System for Metal Product Sales: Pasarlogam. Com (Doctoral dissertation, UniversitiesMuhammadiyah Surakarta) (2022)
9. Cui, Y., Kara, S., Chan, K.C.: Manufacturing big data ecosystem: a systematic literature review. Robot. Comput.-Integr. Manufact. **62**, 101861 (2020)
10. Al-Khalidi, K.S.N., Amjad Hamid, A., Abdel-Razek, A.: The relationship between strategic human resources management and the creative capabilities of workers in the ministry of electricity and water in the State of Kuwait. Sci. J. Commer. Res. (Monofia Univ.) **47**(4), 1–33 (2022)
11. Mahmoud, Y.: Influence of E-business strategies on the productivity of human capital within telecommunication sector in Jordanian telecommunication sector. Quality-Access Succ. **23**(191), 121–128 (2022)
12. Stone, D.L., Deadrick, D.L.: Challenges and opportunities affecting the future of human resource management. Hum. Resour. Manag. Rev. **25**(2), 139–145 (2015)
13. Njoku, C.C., Awujo, A.C.: Incentive compensation and organizational performance (A study of selected oil companies in South-South, Nigeria). BW Acad. J. 25–25 (2022)
14. Bryman, A.: Social research methods: OUP Oxford (2012)
15. Veeramuthu, M.: A Study on Effectiveness of Indian E-marketing (MIS). AEIJMR –6–02–February 2018 ISSN -2348 -6724 (2018)

The Effect of Technological Innovation on Economic Growth: The Jordanian Economy Case

Mazen Hasan Basha[1(✉)], Ibrahim Mohammad Khrais[1], Abdelhalim Mohammad Jubran[1], Khaled Mohammed Al-Sawaie[1], Nidal Ali Abbas[1], Thaer Ahmad Abu-Saleem[1], Mohammad Musa Al-Momani[1], and Hatem Akeel[2]

[1] Department of Economics, Zarqa University-Jordan, Zarqa, Jordan
{mbasha,ikhrais,ajubran,ksawaie,nabbas,thabusaleem, mbasha}@zu.edu.jo

[2] Department of Finance, University of Business and Technology, Jeddah, Kingdom of Saudi Arabia
h.akeel@ubt.edu.sa

Abstract. The use of technological innovation in different sectors is expected to directly impact economic growth. Technological innovation has been driving economic growth and development since the industrial revolution. This secondary research undertaken using a systematic review of published articles shows that Jordan has a relatively weak but positively significant association between technological innovation use and job creation through reduced unemployment. Technological innovation also seems to be labor intensive in Jordan unlike in other countries but this may be attributed to the relatively low R&D spending and innovation outputs. Recommendations for policymakers include the need for investment in R&D as well as education. Innovators in Jordan should get support from the government through incubation centers and partnerships between industry and educational institutions. Support for intellectual property protection should also be considered in Jordan to improve the impact of innovation on economic growth.

Keywords: technological innovation · economic growth · systematic review · Jordanian Economy · Jordan

1 Introduction

Economic growth is one of the key metrics used in assessing the progress and development of nations. It involves the assessment of the factors that contribute to the sustainable growth of an economy in terms of the changes that occur in the long run. The critical question for policymakers is to identify the factors that influence the achievement of sustainable economic growth. This is because such factors can be influenced and manipulated to enhance the performance of an economy. These issues have always been of interest to policymakers across the globe with questions on the role that the different

B. A. M. Alareeni and I. Elgedawy (Eds.): *Artificial Intelligence (AI) and Finance*, SSDC 488, pp. 359–367, 2023.
https://doi.org/10.1007/978-3-031-39158-3_34

factors play in promoting economic development. The key focus in many cases was the role that labor, capital, and natural resources play in enabling an economy to produce [1], The significance of certain factors such as technology increased with the advancement of the knowledge-based economy since the beginning of the twentieth century. In this case, the significance of the knowledge-based economy resulted in the rise of new concepts where the extent of economic strength was found to be directly associated with the state of technological advancement. Since the growth of an economy refers to the long-term expansion of the productive potential of an economy, it provides a way for the economy to meet the needs of its population.

Besides that, social and cultural factors as well as the political environment, and technological advancements have been identified as key factors influencing economic growth. Technological innovation is one of the key factors in economic growth that influences growth through the application of productive techniques and the scientific method [2]. This research focuses on the state of economic growth as it has been influenced by technological innovation in Jordan, which is considered weak and underdeveloped. Innovation consists of specific knowledge, practices, and experiences that are applied using a small proportion of the population to address specific social and economic needs [2]. It takes advantage of prior research and publications to assess the pathways through which technological innovation affects economic growth.

The main aim of the research is to assess the identified effects of technological innovation on economic growth in Jordan from theoretical as well as empirical research to highlight the critical aspects that should be considered by policymakers. Specific objectives of the study are:

1. To investigate the key state of technological innovation in Jordan
2. To assess how technological innovation influences economic growth in Jordan
3. To determine potential solutions for policymakers to improve technological innovation in Jordan and grow the economy

2 Literature Review

2.1 Theoretical Aspects of Technological Innovation

According to [2], technological innovation refers to the activities that result in new or significantly improved processes, methods, products, or organizations. The definition highlights the significance of different aspects of innovation such as being embodied in a new or improved product or the successful exploitation of a new idea including organizational, technological, financial, and scientific activities enhancing the successful exploitation of such ideas such as opening new markets, sources of raw materials, or processes. Innovation can be classified according to the output criteria, which includes product and process innovation. Innovation can also be classified according to the market perception criterion, which involves two basic types of innovation; continuous and intermittent, which differ in terms of being continuous or discontinuous [3]. Innovation can also be classified according to the degree of change that it incorporates. This involves innovation being identified as either radical or gradual (improvement) innovation. Classification of innovation is undertaken according to the criteria of specialization on which it focuses. This may result in innovation being identified as either managerial, marketing, or technological innovation [4], defines technological innovation as consisting of a set of commercial, industrial, and financial stages followed by launching a commercial or manufactured product and application of new technical processes.

The focus of innovation is on new ways of doing things and the widespread adoption of new products, processes, and organizational structures. The key issue here is that it serves as a means to an end by helping human progress and flourishing. It is more than simply getting the latest technology since it involves improving the living conditions for people and communities. Innovation is both technological and social with the technological aspects playing a major role in enhancing the social transformation of the community. It improves the way communities and people relate and can empower groups in ways that were once unimaginable. Innovation extends beyond technology and consumer satisfaction. It introduces new challenges and opportunities. For example, technology is likely to disrupt businesses and models, which is unsettling but may also contribute to improving organizational and other outcomes [1].

2.2 Determinants of Technological Innovation

Culture is open to experimentation and influences the introduction of new ideas that threaten to upset the status quo. In the same way, public institutions and organizations can change in ways that nurture or hamper innovation. One of the core aspects of innovation is allowing people to do more with less through improving productivity and efficiency. Productivity growth improves people's standards of living [1]. In effect, a greater number of outputs or commodities are produced with the same amount of inputs. A good example is the agricultural sector, which has seen a significant decline in the number of people employed hence freeing up many others to get into other forms of employment and modes of production. As cited in [3], productivity growth and innovation are inextricable. Productivity growth is directly associated with the generation of new knowledge. The process of generating new knowledge has become central to the research on the determinants of economic growth.

New knowledge has features that make it difficult to develop but it has aspects of a public good, which is non-excludable hence individuals who do not invest in it can free-ride after it has been created. While the nature of knowledge and innovation creates a framework where investing in it may not be considered an efficient solution for some organizations, it still contributes heavily to economic and social improvement. The Solow growth model sought to explain growth using the basic physical inputs of production. These included generic technology, labor, capital, and production processes [3]. The model assumed that growth in technology would occur at a steady rate irrespective of external factors. Solow did not attempt to explain the source of technological development although its change was the basis of economic growth. In order to identify the determinants of economic growth, the field of growth accounting was developed with the aim of empirically measuring the contributors to economic growth using real-world data and the Solow model.

The results of the 1957 study by Solow indicated that only one-eighth of the change in output per man-hour was attributable to capital stock. Contrarily, seven-eighths of the change in output growth was attributable to technological change. Labor productivity growth was measured using total factor productivity (TFP), which is different from technological innovation. This is assessed in growth accounting studies using the residual, which explains that growth cannot be explained by the variations in the measured inputs. Scholars such as [5], argue that the TFP measures the outcome of growth that is not explained by factors within the model hence referring to it as a measure of ignorance.

2.3 The Innovation-Growth Link

Innovation is seen as having a significant and positive impact on economic growth and development. The focus of this link is on the role that innovation plays in promoting long-run economic growth. Fundamentally, economic growth involves the utilization of technology to uncover new ideas and improve productivity. Total factor productivity is cited as playing a major role in accounting for the variations in income per worker and economic growth across countries. The research indicates that technological change is a key determinant of total factor productivity in the long run [4]. Economic scholars suggest that economic development cannot be decoupled from innovation. Additionally, innovation is directly associated with creative destruction which replaces the old and new economic realms while at the same time destabilizing the power relations in politics. In this respect, creative destruction is what enables innovation to propel an economy forward. Literature on the association between economic growth and innovation has focused on what serves as the deep roots of economic development.

The deep roots of economic growth and development focus on the significance of factors such as digital trade [6], and culture, which are the underlying causes of technological change. One of the key aspects of the change in economic development was the level of uncertainty in large and economically advanced countries. It may be concluded that in advanced high-tech countries, the impact of uncertainty is highly limited. The application of innovation in growing the economy involves the use of new methods and ways of producing to enhance prosperity. The aspect of prosperity refers to the accumulation of solutions to human problems [1]. In effect, the growth of an economy's size has a limited impact on the distribution of the wealth or effectiveness of the solutions developed. Innovation contributes to an economy's growth by increasing the range of problems that a country has found a way of solving alongside the availability of solutions to the citizens.

There was a dramatic upturn in economic growth and prosperity catalyzed by the industrial revolutions and the recent innovations in information technology. The state of technological innovation at the global level and national levels is clearly associated with the rise in the economic development and size of economies around the globe [7, 8].

Innovation is considered a major aspect of national economic growth in many cases where the policies in place can have a significant impact on what is available and how it is used [9]. However, the association is not specifically supported by the literature since innovation has not been found to be the source of economic growth in some settings. In the study by [9], it was found that innovation factors did not influence economic growth in Eastern European countries. Although the economies have been growing rapidly, it is not driven by innovation, which is still evolving.

3 Summary and Hypotheses

The reviewed literature indicates that economic development cannot be decoupled from innovation. The roots of economic growth involve the focus on strategies for enhancing the productivity of factors of production. In effect, two hypotheses can be developed and tested in line with the research objectives.

H_1: Technological innovation in Jordan is inefficient, and underdeveloped and has a positive and significant influence on economic growth in Jordan.
H_2: Policies in place for improving technological innovation in Jordan and enhancing its impact on the economy are generating the expected outcomes

4 Research Methodology

The study follows a qualitative research approach where published information is used to evaluate the state of technological innovation in Jordan. A qualitative research approach is adopted because it provides a way for the assessment of prior research in order to develop a big-picture view of the situation. The qualitative research approach is adopted for the study because it enables an assessment of the narrative information including the discussion of the findings to answer the research questions. The research uses a systematic review methodology since it relies on secondary data from empirical research on the research question.

The systematic review approach involved using specific search terms that were considered beneficial in getting the relevant articles that could be used to answer the research questions [10]. The keywords applied for the search were: Technological Innovation; Economic Growth; Jordan; Association; and Impact. The search was conducted on scholarly databases such as ABI/INFORM, Business Search Complete, and Google Scholar. The searches were undertaken with certain identified limiters such as publication date between 2012 and 2022, full text available, published in English, and specifically conducted in Jordan. Screening of the articles was done using automatic tools and manually to get only the relevant articles.

5 Findings

5.1 Technological Innovation in Jordan

This part of the review focuses on the published research assessing the state of technological innovation in Jordan coupled with the economic development experienced in the country. The expected direct link between the two aspects is evaluated in this research using a systematic review of the literature. Jordan is one of the smallest economies in the Middle East. Some of the key aspects of the Jordanian economy include its limitation of natural resources. As a result, the economy is highly reliant on the ability to enhance the value of products and services and utilize its human capital more effectively. The government has invested in a number of measures meant to enhance its effectiveness and competitiveness through innovation [11]. The national innovation system (NIS) is a significant framework developed by the Jordanian government to improve its innovation capacity and leverage economic growth.

The state of technological innovation in Jordan is relatively poor and inefficient as shown by the low level of integration with different institutions. The ineffectiveness of the integration between the government and the private sector in enhancing innovation in Jordan has contributed negatively by reducing the investment in R&D and its application in the industries. The system in Jordan has not been effective in addressing the needs of the economy mainly because it is underdeveloped. The innovation system lacks synchronization among the key components and is considered to be relatively weak [11, 12]. A critical challenge for Jordan in relation to sustainable economic development is the weak relationship between the education sector and businesses. There is low R&D expenditure, limited innovation activity, and less use of ICT as well as a lack of adequately skilled manpower for the economy. This presents a significant challenge for the economy because it means that the utilization of the digital economy and innovation in creating jobs is relatively limited. The research data supports the hypothesis that technological innovation is underdeveloped and relatively inefficient in Jordan.

5.2 Impact of Technological Innovation on Economic Growth in Jordan

The World Bank (2018) indicates that the focus on digital technology is essential in creating a supportive environment for countries in the Middle East and North Africa region. The role of innovation in the growth and development of economies is associated with the impact it has on entire industries as opposed to single firms. According to [12], the role of innovation as a pathway to economic development is evaluated. The significance of innovation in alleviating poverty arises from its applicability in alleviating constraints by providing solutions to improve productivity. The innovation system in Jordan has advanced more significantly than in other countries in its region. The literacy rate and internet penetration in Jordan also place it at an advantage in the use of technological innovation for development. The system suffers from a significant lack of coordination and development to make it effective as a growth driver. However, the state of education and government policies are advancing positively with the aim of encouraging industries to adopt technology in order to improve their operations.

The relevance of innovation in enhancing economic growth is associated with the focus on new growth theories that indicate the importance of shifting to a knowledge-based economy from a resource-based one. Innovation is also associated with the introduction of significant changes in the capital structure where the accumulation of capital produces technical change that encourages capital investment. In [13], the state of innovation in Jordan was measured by the number of patent applications and the number of individuals using the internet for the period between 2000 and 2017. The number of resident patent registrations was significantly lower than in other studies. Additionally, the penetration of the internet was relatively low at about 28%. The multiple regression model indicated that the significance of innovation in predicting economic growth was statistically significant at the 5% level. The aspects of innovation such as patent applications and individuals using the internet significantly predict the level of unemployment. Individually, the data indicated that the number of patent applications had an impact of 0.005% on unemployment, which makes it very low in terms of making a significant contribution to the economy.

A similar study by [5, sought to assess the impact of financial technology on the economic growth of Jordan. Using a sample of time series data from 1990 to 2021, the vector error correction model was used to assess causality in both the short and long-term durations. The focus on financial technology was based on the view that technological innovation has contributed heavily to the financial services sector through access to services and products for different populations. The dependent variable was GDP per capita, which was assessed from the role that different financial technology aspects influenced it. The results indicated that a long-term causality ran from the use of technology such as the internet, mobile, broadband, and ATMs to GDP per capita. A similar result is also found for the short-term causality where all variables were found to have a significant impact on the per capita GDP of Jordan over the study period. The impact of financial technology products on economic growth in Jordan is supported by the fact that it directly supports access to financial services to members of the society hence improving their socioeconomic conditions.

Capital, human resources, and total scientific achievement can be applied to assessing the state of technological innovation and development. As cited in the study by [15], intellectual property through patent applications presents a significant source of competitive advantage for economies. High-quality scientific research output contributes heavily to determining the quality of the innovation environment and the potential it has for achieving commercial success in order to influence economic outcomes. Innovation increases the amount of knowledge that is available for use in production and development. The growth of scientific and technological innovation can increase the level of progress across industries and sectors. For example, increased R&D spending and output have contributed to the growth of its manufacturing industry. This aspect was evaluated in Jordan where the impact of innovation on the industrial sector was assessed [15]. The results indicate that technological innovation and development in Jordan are more labor-intensive than in the long run. Technology has a constant impact on the economic development of the industrial sector and this impact is also associated with the view that technological change is more capital-intensive in the long term compared to the short run.

The role of technological development in the economy may also be achieved through its impact on the industrial sector. In this case, the impact of technology on the Jordanian industrial sector was assessed in an empirical study [16]. Jordan like other developing nations is suffering from a major technological gap where the methods of production have lagged behind other advanced economies. Developing countries have been ineffective in utilizing technology to grow their capabilities and move from the first stage of industrial development as indicated by the Global Competitiveness Index. The findings indicated that the Granger causality test shows that production increased in phases of increasing economic return. Technological change was also found to be more labor-intensive in the long term compared to the short-term setting. It showed technological development taking place at a steady pace in Jordan and significantly impacting its growth to the advanced position.

5.3 Government Policy

Technological innovation and economic growth have a bilateral relationship as indicated in [17]. The approach is based on the view that economic growth is driven by human capital improvements, creating new and more efficient processes for industries. Using a case study approach, [17], found that innovation in technology has a bidirectional relationship with human capital development. This means the development of human capital is a matter of culture that is also significantly influenced by the culture and focuses on issues such as entrepreneurship. The application of technological innovation as a critical factor of economic development in Jordan is aligned with the country's investments in improving the state of the economy. Economic development and growth rely heavily on the presence of human capital, which works well with the development of technological innovation. Policies such as the NIS and investment in research and development have been beneficial in increasing the focus of the organizations on innovation including training and application in industries. The policies are improving economic development but at a relatively slow pace.

6 Conclusion and Impact on Jordan

In conclusion, the advancement of economic growth in Jordan has been significantly hampered by the absence of adequate natural resources, unlike its neighbors in the Middle East. Jordan has been a front-runner in adopting technology as a tool for growth and development. While the effort by the government through investment in education and policies has been impressive, the effectiveness of the alignments among stakeholders has been relatively weak. The critical issue for the Jordanian economy in relation to the association between economic growth and innovation is the bi-directional relationship that has been identified in different studies. The focus of the research on this association indicates that the stock of innovation output such as registered patents and intellectual property contributes to promoting the development of new processes and products that can improve the economic well-being of individuals. In the case of Jordan, the use of technological innovations in financial services as well as the industrial sector has been shown to be directly associated with economic growth. The economy of Jordan shows a relatively weak but positively significant association between technological innovation use and job creation through reduced unemployment. Technological innovation also seems to be labor intensive in Jordan unlike in other countries but this may be attributed to the relatively low R&D spending and innovation outputs.

Policymakers in Jordan should consider effective ways of enhancing economic growth through innovation through investment in R&D as well as education. Innovators in Jordan can get support from the government through incubation centers and partnerships between industry and educational institutions. Support for the production of intellectual property and commercialization of the same can also contribute heavily to promoting economic growth in Jordan. It is essential for business players to have the capacity to enhance the growth of their operations by utilizing the available technology. Since technology is sometimes capital intensive, a strategy to aid business players with financial support can improve its adoption in Jordan and increase economic growth prospects.

References

1. Broughel, J., Thierer, A.: Technological innovation and economic growth: a brief report on the evidence., Arlington, VA: Mercatus Research, Mercatus Center at George Mason University. February 2019, 1–27 (2019)
2. Diaconu, M.: Technological innovation: Concept, process, typology and implications in the economy. Theor. Appl. Econ. **18**(10) (2011)
3. Bryce, R.: Smaller Faster Lighter Denser Cheaper: How Innovation Keeps Proving the Catastrophists Wrong. Public Affairs, New York (2014)
4. Gracia, J.F., Duran, J.G., Sevilla, J.A.: Relevance of technological innovation in the business competitiveness of medium enterprises in Hidalgo State. Eur. Sci. J. **10**. https://eujournal.org/index.php/esj/article/view/3532 (2014)
5. Saleem, T.A.A., Abbas, N.A., Al-Sawaie, K.M., Jubran, A.H.M.: The impact of digital trade on jordan's economic growth. In 2021 22nd International Arab Conference on Information Technology (ACIT), pp. 1–8. IEEE (2021)
6. Lipsey, R., Carlaw, K., Bekar, C.: Economic Transformations: General Purpose Technologies and Long Term Economic Growth. Oxford University Press, New York (2005)
7. Reamer, A.: The Impacts of Technological Invention on Economic Growth – a Review of the Literature. George Washington Institute of Public Policy, Washington DC (2014)
8. Khrais, I., Azzm, Z., Assaf, A.: Constraints facing garment industrial sector operating within the qualified industrial Zones in Jordan. Zarqa J. Res. Stud. Human. **10**(2), 82–91 (2010)
9. Pece, A.M., Simona, O.E., Salisteanu, F.: Innovation and economic growth: an empirical analysis for CEE countries. Procedia Econ. Finance **26**, 461–467 (2015)
10. Saunders, M. N.K., Thornhill, A., Lewis, P.: Research Methods for Business Students. Eighth Edition ed. London, UK: Pearson (2019)
11. Maradana, R.P., Pradhan, R.P., Dash, S., Gaurav, K., Jayakumar, M., Chatterjee, D.: Does innovation promote economic growth? evidence from European countries. J. Innov. Entrepreneurship **6**(1), 1–23 (2017). https://doi.org/10.1186/s13731-016-0061-9
12. Sultan, S.S., Soete, L.: Innovation for development: the case of Jordan. Dirasat Adm. Sci. **39**(2), 321–327 (2012)
13. Alheet, A.F., Hamdan, Y.: Evaluating innovation-driven economic growth: a case of JordaN. Entrepreneurship Sustain. Issues **7**(3), 1790–1802 (2020)
14. Mugableh, M.I., Hammouri, M.A.: Exploring the impact of financial technology on the economic growth In Jordan. Webology **19**(3), 286–294 (2022)
15. Wang, X., Xu, L.: The impact of technological innovation on economic growth: evidence from China. Adv. Econ. Bus. Manage. Res. **203**(3), 1911–1916 (2021)
16. AL-Refai, M.F., Abdelhadi, S., Al-Qaraein, A.A.: The impact of technological development on Jordanian industrial sector. Int. J. Bus. Manage. **11**(4), 291–298 (2016)
17. Alawamleh, M., Bani Ismail, L., Aqeel, D., Alawamleh, K.J.: The bilateral relationship between human capital investment and innovation in Jordan. J. Innov. Entrepreneurship **8**(1), 1–17 (2019). https://doi.org/10.1186/s13731-019-0101-3

The Impact of Digital Transformation on Civil Action Procedures

Ahmad Awwad[1], Abdullah Omar[1], Sohib Yahya Alshurman[1], and Mohammed Angawi[2](✉)

[1] Faculty of Law, Zarqa University, Zarqa, Jordan
[2] College of Business Administration, University of Business and Technology, Jeddah, Saudi Arabia
m.angawi@ubt.edu.sa

The Abstract. Modern technology, modern communication means and digital transformation in the civil action, it's registration procedures and in all its cases enables judges and adversaries to follow their lawsuit and attend, register applications, and call the judge electronically without the need of attending in person. This gives a technical impression that prevents the disposition of cases hindrances or delay in the disposition of cases. And this is a sufficient solution which prevents prolonging the lawsuits, simplifies the procedures and accelerate litigation accurately and proves swift justice.

Electronic civil action means using the electronic means, technology and internet in transferring the data, lawsuit list and its appendices from a computer to another through internet through sending the list of the case to the ministry of justice website in the section of electronic law services through electronic prosecution portal.

Registering the lawsuit electronically and the way of filing and registering the civil lawsuit electronically has been dictated by the system of using electronic means in civil, legal procedures emanating from the Code of Civil Procedure, dictating the procedures of electronically registering a civil action.

Despite the stability of Jordanian Law on the need and authenticity of using electronic means in the procedures of electronically registering a civil action, it still needs legislative modifications to cope with modern age technology. This requires a human staff trained on using electronic means and provide adequate technical infrastructure needed for technical legal work towards a digital transformation in the judicial facility.

Keywords: electronic lawsuits · electronic means · judicial electronic services · electronic reports

1 Introduction

The law system has been affected in the various legal fields generally and particularly in the field of civil legislations in shade of the technological changes, informational revolution, and the field of informational technology. And how to benefit from this

B. A. M. Alareeni and I. Elgedawy (Eds.): *Artificial Intelligence (AI) and Finance*, SSDC 488, pp. 368–376, 2023.
https://doi.org/10.1007/978-3-031-39158-3_35

revolution in facing the social and judicial limitations by setting a legislative system that is fast in resolving disputes, transparent and subjective, to ensure justice which requires speed and accuracy and minimum coasts. Technological revolution came to participate in achieving this goal; thus, these legislations should cope with the technological revolution.

Modern technology and digital transformation in civil action field and registering procedures in all levels enable judges and opponents to follow their cases, attend, register their applications, and contact the court's judges and workers electronically without attending in person. This fosters a technical trait that decreases the limitations in resolving the cases. This means that the traditional method in trials which prolongs resolving the cases is cancelled. It is a modern resolution for the dilemma of slow judicial procedures which makes successful justice a target for every person which simplifies procedures and fastens trails.

Study problem: shedding light on using informational technology in electronic lawsuits registration procedures to become a substitute for the traditional lawsuits. Clarifying the sufficiency of Jordanian legislations to move into a digital environment, and whether those legislations cope with technological development or not?
Study significance: participate in developing the judicial system and achieving civil lawsuits procedures quickly and accurately to prove justice through electronic registrations using the technical digital impression through the civil lawsuits' stages.
Study purpose: clarify the procedures of electronic lawsuit registration, the difficulties it faces and find solutions to those difficulties.
Study methodology: this study follows the analytical approach of the Jordanian legislations system for electronic lawsuits' registration.
Study plan: this study is of two sections; the first section is to introduce the electronic civil lawsuit and its registration procedures. The second section shows the impact of registering electronic lawsuits.

2 Introducing Electronic Civil Lawsuits and Their Registration Procedures

Electronic means using electronic, magnetic, electromagnetic or any similar means in information exchange and storage. So, what is an electronic civil lawsuit? What are the electronic procedures of registering a case? These are the main issues tackled in this section.

2.1 The First Demand: Defining the Electronic Civil Lawsuit?

An electronic lawsuit is defined as: accessing justice to attain subjective right or protect it through electronic media and the internet. Or it is a technical double-faced space which allows programming electronic lawsuits, and it consists of international connecting web, in addition to the courthouse that allows legal units to be electronically visible where a group of judges can start looking into the lawsuits and decide on them according to legislations which allow them to start judicial procedures, store and exchange lawsuits' files.

Electronic civil action is using electronic media, technology and the internet in transferring data and lawsuit list and its appendices from one computer to another through the internet by sending a lawsuit list to the justice ministry website about judicial electronic services through electronic sue gate.

This method is a phenomenal objective shift in developing justice in a quick accurate way that shortens the lawsuit time and makes it easier for both citizens ad lawyers. The citizens, companies and lawyers have been provided with a user name and a password they can use to enter the electronic judicial system, it also allows inquiring about the lawsuits, the levels of civil action and to which level it reached. It also provides seeing the trial sessions' records easily.

It is required of the user of electronic litigation media to provide the specialized legal actors with a phone number to send and receive messages. Also, they require an email for purposes of sending and receiving judicial notifications, documents, data and lists. So that the deputy attorney can send the lawsuit list with his data and documents electronically to the Justice Ministry website. The deputy attorney will electronically provide the court with the phone number or email of the defendant so that the court sends an electronic judicial notification to him. Consequently, the defendant upon receiving the notification must reply to the lawsuit with a response statement, his data and his electronic documents portfolio.

It has also been mentioned in using electronic media system in civil and judicial procedures, the user is required to provide the court with the needed information, therefore responsibility will fall on him if the information isn't true. Article number (5) of the indicated system dictates that:

a. "for purposes of applying the rules of this system, each lawyer must provide the bar association a signed statement on the form prepared by the ministry including his information: his name, his address, his office, his mobile numbers and his email address.
b. in case of any change on the declared information of the lawyer, he is required to inform his union in the same way dictated in statement (a) of this article.
c. the address of companies' clients that are legally bound to hire attorneys is considered a significant address for purposes of applying the provisions of this law.
d. the ministry and courts are allowed of viewing the declared information of the attorney according to this article provisions through the electronic connection with the bar association.
e. the ministry and courts rely on the declared information by the attorney to apply civil and judicial procedures electronically.
f. the ministry can build a data base to keep the attorney's declared information and any other useful information for the purpose of applying the provisions of this law.

Article number (6): the ministry provides an email for each lawyer on its electronic gate for purposes of registering judicial and executive actions, applications, regulations, data list and the rest of the correspondences; judicial and executive papers".

2.2 The Second Demand: Registering Electronic Litigations Procedures

The process of registering an electronic civil lawsuit is an integrated process starting from filing a claim by a lawyer or people until judgement. It also includes the ability of verifying the lawsuit process to ensure accurate data, verified data and sent documents. As well as sending/issue notifications/receipts to people when paying lawsuit registry fees.

Paying the fees electronically is done through electronic transfer from the prosecutor's account through e-pay, through the financial companies specialized in electronic payments (eFAWATEERcom for example), or through communication companies. In the electronic litigation gate the fees are automatically calculated when sending the lawsuit papers and its appendices, so that the user has to pay the fees to move into the next step which is registering a lawsuit and having a number which specify the date of the judicial assembly.

The process of electronically registering the lawsuit and the placement of electronic civil action is dictated in the system of using electronic media in civil and judicial procedures, in article (4) of it dictates:

"a. lawsuits, applications, paying the fees, communication procedures and the exchange of regulations, notifications, the rest of papers and communicate through electronic media as follows:
1. the lawsuit list is placed electronically along with the full attachments.
2. the court notifies the applicant of accepting registration.
3. after notifying the other party with the lawsuit list and its annexes, exchanging lists between the lawsuit parties electronically is allowed.
4. the lawsuit parties are required to submit the assets of what has been applied according to this paragraph through electronic media to the court in the first session after the action done through electronic media, under penalty of nullity.

b. the executive bonds are placed electronically and the executive judge will have the creditor show them.
c. electronic payment for application and lawsuit's fees are accepted with accordance to this system.
d. the minister is allowed to issue the needed regulations to organize the process of registering executive and judicial lawsuits, applications, fees payment, placement of lists, data, notifications, other papers and initiate conversations electronically.

Regarding the stage of electronic notification, usually it is sent in many electronic media that the Jordanian legislator hasn't limit it but rather left it open since the progress in the technological field is accelerating producing new electronic media that can't be limited in one timing by the legislator. Article number (7) on using electronic means in judicial procedures dictated the following:

("a. the following electronic media are adopted to perform judicial notifications:

Email. Text messages on mobile phones. The electronic account designed for the lawyer. Any other way adopted by the minister.

b. when sending electronic notification attend to the conditions and data that must be included in judicial notices as dictated by the law.

c. every person has the right to claim the falsity of the ongoing electronic report which is against the rules of this law.
d. the report performed according to the rules of this law the same legal consequences that entails the reports performed according to the law rules.
e. the ministry takes whatever procedures needed to enable the courts archive the reports done by the electronic means to return to them when needed.

We find through article (7) from the system of using electronic media in judicial procedures that the Jordanian legislator limited the electronic media in electronic reports to the following:

First: mobile phones: it is considered as one of the communication means of the modern era, carried out through satellites connected wirelessly to one another. Phones are audio communication media that stores information, receives and sends SMS with sound and image. In electronic reports field mobile phones helps in simplifying the task of sending short messages with multimedia in a very accurate and quick way, which is considered as an official report that has the same effect as the traditional report.

The Jordanian Court of Cassation has adopted electronic reporting as a means of performing reports to the conflicting parties telling them the lawsuit proceedings whenever needed.

Second: email: the email is connected to the internet (world wide web), which allows message exchange between devices connected to the internet, through which the reports are sent to the person needed with a notification that the message has been delivered to the intended party, this method of reporting is way more accurate than the traditional method. The email includes the information, files, documents which can be exchanged directly between the sender and receiver.

Third: any other method decided by the minister.

Back to the article script (3/a/4) the system of using electronic media in judicial procedures number 95 in 2018, which dictates the following: "after informing the other party with the lawsuit list and its annexes, it's valid to exchange the lists between the conflicting parties using electronic means").

It is worth mentioning that the legislator conditioned that the electronic report is done according to the legal procedures dictated by the law. Article (3) in the system of using electronic media in judicial procedures dictates the following ("the provisions dictated in the law should be attended to when performing civil and judicial procedures through the electronic means dictated in this system"). Meanwhile we find the Jordanian criminal procedure law dictates in the article (7/b) that: "if the notification is done through one of the mentioned ways in this law and the person intended wasn't reported or nobody could sign for him upon receiving, the court can assign the Registry to send a brief notification with the report subject to that person through mobile phone, or any other electronic media decided by the law before granting the report. And the court has the right to verify this notification in any appropriate way".

We wish the Jordanian legislator to modify the article (7/b) from the Jordanian civil procedure law to conform with article (7) from the system of using electronic means in judicial procedures and adopting the electronic reports separately as an independent method and not supplementary to the traditional method.

3 Effects of Electronic Lawsuit Registration

The electronic civil action is proved through the electronic documents which adds verification through electronic signature. The Jordanian legislator adopted electronic means in civil action procedures. In this section we tackle the Jordanian legislations systemizing electronic lawsuit registration and the position of the Jordanian law.

3.1 The First Demand: The Jordanian Legislations Systemizing the Electronic Lawsuit Registration

The Jordanian legislator has modified the articles of the law of Code of Civil Procedure to match the requirements of registering an electronic civil lawsuit, besides the system of using electronic media in civil procedures. The law Code of Civil Procedure dictates in the article (5/2/a) that: "the court has the right to check the validity of the address given by the defendant to tell his opponent or witnesses using electronic tools created for this cause".

Article (2/11) has given the court the right to check the witness address using electronic means, also the purpose of electronic means in civil action in the article (1/21) which allowed documenting the trial's actions and its procedures by computers or electronic devices.

We also found in article (3/58) of the law dictates: "the conditions and procedures of using electronic means are limited to make the parties register their lawsuit, place the lists, data, applications and the rest of executive and judicial papers in court and report it to the other party according to a system designed for such cause".

It is worth mentioning that the Jordanian trial law has been modified, and it gives the court the right in listening the witness words through modern communication media, and this is dictated in article (7/81) as follows: "the court has the right to allow upon request from one of the opponents listening -by acceptance of the other party- using modern communication media without being in court himself and according to a system designed for such purposes".

As we read through the law of civil due process, we find that the Jordanian legislator allowed using the electronic means in all levels of civil action and launched a system of using electronic means in civil procedures number 95 in the year 2018 for this purpose, simplifying trial procedures to add the legal touch to the electronic means. But the noticed is a big conflict between code of civil procedure and the system launched in accordance with it. The Jordanian legislator is still and in the law of civil of code procedures look shyly at the way electronic reports of civil lawsuits and puts an alternative in case he wasn't notified in the traditional ways. It is preferred that it dictates openly replacing the traditional reports with electronic ones through the whole process of civil action.

The researcher sees that using the technical and electronic ways in civil action pays off better than the traditional procedures, it prevents losing documents and guarantees its existence in the modem and it is easy to find them at any time which suits the purpose of successful justice seen through the accuracy and quick judgments in lawsuits which also prevents prolonging the trial timing. Not to mention, coping with digital technology and the age of development.

3.2 The Second Demand: The Position of the Jordanian Law

Working in the Jordanian law relies on using the electronic means to register an electronic civil lawsuit. In the press conference held by the judiciary council in 3/11/2021 the annual report of the systematic courts issues and expanding the procedures of the digital transformation of transactions, applications, electronic registration of lawsuits, challenges, and electronic fees payments. These procedures were taken as a starting point into electronic technical justice, a starting point and transfer into a new stage in Jordanian legislation. It has been adopted for the first time modern technical application in sending judicial reports that are legally accepted besides the electronic deposit system for the judicial papers in all appeals courts, which led to lessening the number of sessions and quickened the judicial procedures and raised the abilities of judges in using technicalities through contracting a number of training courses in using communication means.

The judicial court look up to apply legislative modification that guarantees coping with the judicial procedures of the wheel of technical and legislative developments in the comparative judicial systems. Expand mixing the use of technicalities and hiring it with all its types in the trail procedures to guarantee the continuation of these procedures in case the legal sector finds exceptional circumstances or future force majeure. Indeed, the Corona pandemic is the best example on the importance of digital transformation and technology in electronic lawsuit registration procedures.

We find the application of modern communication means in many passed judgments from the Jordanian courts. Where the Jordanian court of cassation in a rule, the following: "accordingly, reading article 7 from the system of using electronic media in civil, judicial procedures, we find that it adopted the text messages via mobile phones as one of the electronic media adopted to perform judicial reports, and that the report done with accordance to the rules of this system the same legal consequences that are done according to the provisions of the law. Hence, since the mentioned report was done through one of the electronic means granted to perform judicial reports.

If the recipient agent didn't deny the phone number which received the text message this report will be counted. And building on what has been said we decide to decline the challenge of cassation and believe the impugned decision. We find through this judgment that cassation court accepted the electronic report as it agrees with the law and its consequences are considered and whoever claims the opposite of the report should prove it.

In another judgement, the cassation court adopted reporting the electronic civil lawsuit list, deciding the following: "notify the agent of the defendant, Thiab Abu Salih and co. Company the list of invalidity lawsuit electronically in (16/10/2021) and he replied in response list in (25/10/2021) so it is accepted as a form and submitted through the legal period of time.

4 Conclusion

This study concluded a group of findings and recommendations, verily:

First: the results:

1. registering the civil lawsuit electronically appeared as a result of technological development and digital transformation in all aspects including the judicial attachment which is an essential inevitable result.
2. electronic civil lawsuit registration helps in saving effort and provides speed and accuracy in the judicial work.
3. the process of electronic lawsuit registration needs supplies like computers, electronic webs, electronic programs, internet, electronic media and a trained human staff of judges, writers, technicians, and lawyers who all have enough knowledge in dealing with electronic media.
4. registering electronic civil lawsuit requires the presence of laws and systems that organizes the legal technical work and it needs continuous modification to cope with the technological accelerating development.
5. electronic lawsuit registry through the whole process, giving documents, lists, reporting and electronically hearing of witnesses made a huge qualitative leap in the judgement speed, the accuracy, justice and decrease the problem of prolonging trial period.

Second: recommendations:

1. finding a well-trained human staff who can use electronic media in trials through holding workshops and training courses specialized in technology and digital transformation.
2. provide the electronic equipment and internet connection in cooperation with communication companies and e-pay companies to sustain a strong environment sufficient for judicial work.
3. compel the department of civic affairs and passports to write the email, phone number and detailed address of every citizen or person in Jordan, and also compel them to inform the civic affairs department of any change in these pieces of information (email, phone number and address), subject to liability.
4. we wish the Jordanian legislator modify the article (7/b) in the code of civil procedure to conform with article (7) from the system of using electronic means in judicial procedures and adopting the electronic reporting separately, independently and not as supplementary to the traditional reporting. Thus, the article becomes as follows: "the electronic reporting is adopted in all stages of the lawsuit, and the intended person is reported through the mobile phone or any electronic medium decided by the system and the court has the right of verifying the sent reports in any appropriate way it finds".

References

The books

1. Shoshari, S.: Explanation of the Code of Civil Procedure, Dar Althaqafa publishers, Amman, Jordan (2010)
2. Alqudah, Muflih.: The Code of Civil Procedure and judicial organization, Dar Althaqafa publishers, Amman, Jordan (2020)

3. Alshar'a, Hazim.: Electronic Litigation and Courts, Dar Althaqafa publishers, Amman, Jordan (2010)
4. Alkilani, Mahmoud.: The Code of Civil Procedures and Proceedings, Dar Althaqafa publishers, Amman, Jordan (2021)
5. Alqudah, Muflih.: Evidences in Civil and Commercial matters, Dar Althaqafa publishers, Amman, Jordan (2019)
6. Alkilani, Mahmoud.: Rules of Evidence and Implementing Provisions, Dar Althaqafa publishers, Amman, Jordan (2021)
7. Sulieman, Dadiar Hamid.: The Legal Frame for Electronic litigation, 1st print, 2015, Dar Althaqafa publishers, Amman, Jordan

Legislations

8. The Code Civil Procedures and modifications Law, number 24, 1988, the modified law number 31, 2017 and the modified law number 4, 2019
9. Jordanian Evidence Law, number (30), 1952 and its modifications, number (22) (2017)
10. The System of Using Electronic Media in Judicial Procedures, number 95 (2018)

Researches and Theses

11. Dr. Flieh, Najla' and dr. Quzmar, Nadia, Legal Cope with Working Remotely, research published in Zarqa Magazine for Research and Humanitarian Studies, folder 18, copy 1 (2018)
12. Amer, Rabab Mahmoud, Litigation in Electronic Court, research published in Girls Collage Magazine for Humanitarian Sciences, copy 25, (2019)
13. Sulieman, Dadiar Hamid.: The Legal Frame for Electronic litigation, 1st print, Dar Althaqafa publishers, Amman, Jordan, p. 133 (2015)
14. The System of Using Electronic Media in Judicial Procedures, number 95, p. 5600 of the Official Magazine (usually 5529 pages) in 2/9/2018 (2018)
15. Alqudah, Muflih.: The Code of Civil Procedure and judicial organization, Dar Althaqafa publishers, Amman, Jordan, p. 252 (2020)

Footnotes

16. Flieh, Najla', Quzmar, Nadia.: Legal Cope with Working Remotely, research published in Zarqa Magazine for Research and Humanitarian Studies, folder 18, copy 1 (2018)
17. Amer, Rabab Mahmoud, Litigation in Electronic Court, research published in Girls Collage Magazine for Humanitarian Sciences, copy 25, p. 403 (2019)
18. Alshar'a, Hazim.: Electronic Litigation and Courts, Dar Althaqafa publishers, Amman, Jordan, p. 74 (2010)
19. Alkilani, Mahmoud, The Code of Civil Procedures and Proceedings, Dar Althaqafa publishers, Amman, Jordan, p. 164 (2021)
20. Alqudah, Muflih, Evidences in Civil and Commercial matters, Dar Althaqafa publishers, Amman, Jordan, p. 130 (2019)
21. The Code Civil Procedures and modifications Law, number 24, 1988, the modified law number 31, 2017 and the modified law number 4, (2019)
22. The decision of Jordanian Cassation Court, number 6700, year 2021, initiated on 20/1/2022, posted on Qararak website
23. The decision Jordanian Cassation Court, number 7106/2022, initiated on 11/12/2022, posted on Qararak website

The Influence of Digital Strategic Orientation on Organizational Performance in the Manufacturing Jordanian

Sultan Alshourah[1(✉)], Manal Altawalbeh[2], Ahmad Albloush[3], Amer Alsarairh[4], and Abdulwahab M. Abukwaik[5]

[1] Zarqa University, Business Administration, Zarqa, Jordan
salshourah@zu.edu.jo

[2] Educationtional Technology, Middle East University, Amman, Jordan
maltawalbeh@meu.ed.jo

[3] Amman Arab University, Business Administration, Amman, Jordan
Albloush@aau.edu.jo

[4] Zarqa University, Business Administration, Zarqa, Jordan
aalsaraireh@zu.edu.jo

[5] University of Business and Technology University, Jaddah, Saudi Arabia
a.abukwaik@ubt.edu.sa

Abstract. Increasingly, digitization is seen as a strategic approach for organizations to achieve organizational performance. However, to date, previous research has not provided sufficient evidence of how a company's strategic attitude toward digitization relates to organizational performance. Thus, this research investigates relationship between digital strategic orientation and organizational performance. The building of the research framework was based on previous research and resource-based view theory, to test this framework using survey data of 107 manufacturing Jordanian sector, in Jordan context. Analysis output indicated a company's digital strategic orientation has a positive and significant influence on organizational performance. Our findings recommend that executives can progress their company's organizational performance and competitive advantage through increasing the Digital Transformation processes inside their organizations. Finally, we conclude practical implications for managers and policymakers with the necessitate of combining digitization with strategies.

Keywords: Digital strategic orientation · Organizational performance · Resource-based view · Manufacturing Jordanian

1 Interdiction

Digital technology is a fundamental force of change in this century reshaping economies, governments, and civil society and, thus, affecting almost all aspects of development. Its importance after Covid-19 has increased, highlighting the tremendous power of digital technology [1]. Business and jobs are also changing: The latest World Economic Forum

B. A. M. Alareeni and I. Elgedawy (Eds.): *Artificial Intelligence (AI) and Finance*, SSDC 488, pp. 377–386, 2023.
https://doi.org/10.1007/978-3-031-39158-3_36

(2023) report on the future of jobs reported that companies in 26 developed and emerging economies are rapidly digitizing their business processes [2]. All companies and industries are considering digitalization to confront these challenges and seize opportunities [3]. Previous studies called on administrators to go outside initiatives of individual technology also exploit digital transformation as a strategic opportunity because it reshapes businesses by building new business models and generating abundant new opportunities. [3]. Previous studies indicate that managers can improve the organizational performance and competitive advantage of their companies by adopting an increase in the digital strategic orientation and digital transformation usefulness within their organizations [4, 5]. A related study by [6], found that the significant correlation between digital strategic orientation and organizational performance becomes more outstanding in extremely open environment of technological. They claimed that the business model design must reshape based on digital strategic orientation.

The manufacturing Jordanian sector constitutes a basic pillar of the Jordanian economy, as it acquired more than 90% of the total national exports, which grew by nearly 10% during the year 2019 and reached about 140 countries around the world. And, it contributes directly to about a quarter of the national economy (25% of GDP). About 251 thousand workers also work in the manufacturing sector, constituting about 21% of the total Jordanian workforce, which is directly reflected in the promotion of economic and social development, access to global markets, attracting qualitative investments, and improving the Jordanian product's identity [7]. According to past studies on the strategic direction confirmed the weakness of the strategy of digital transformation in the Jordanian manufacturing sector [8–11]. Among them, [11], He indicated that vulnerability to multinational competitors and low levels of adoption of the digital transformation strategy among local competitors are some of the possible reasons behind this small effect, which led to a decrease in competitiveness among the Jordanian manufacturing sector. [11], suggest that administrators should adapt their business strategy to the new digital reality in the Jordanian manufacturing sector.

Digital Transformational strategies provide new orientations and novel paths to innovations, thus improving company performance [8, 12]. Furthermore, in the light of existing literature, the realization of findings not consent effect digital strategic orientation on company performance in settings difference [8, 12–15]. However, Previous studies focuses on the influence of digital strategic orientation on company performance under the impact of the application of digital technologies in developed countries, but a few studies focuses to developing countries. [5, 8]. [8], pointed out that there is still a lack of theoretical models and empirical research on the influence of digital strategic orientation on company performance, especially in Jordan. To respond to previous studies, this study came to help bridge the research gap by studying the digital strategic direction and its influence on organizational performance under the influence of the digital technologies era in the sector of Jordanian manufacturing. Therefore, we can ask the following question:

- Does digital strategic orientation translate into better organizational performance under the influence of the digital technologies transformation in the sector of Jordanian manufacturing?

2 Theoretical Framework

2.1 Resource-Based View (RBV)

The Resource Based View (RBV) theory first suggested by Barney [16]; some scholars [4, 17, 18] have reveals that new capabilities and resources and are essential for organizations to vie efficiently in the digital age. Exactly, the higher the new digital capability as digital strategic of an organizations, the better their organizational performance. Consequently, on the emphasized of Previous studies [4, 17–19], we use the resource-based view (RBV) as a theoretical foundation to clarify how Digital Strategic Orientation affects organizational performance. In fact, we discover the role of a company's digital capabilities in shaping the link between digital strategic direction and organizational performance that generates competitive advantage [4, 5, 20].

2.2 Digital Strategic Orientation

Previous literature on strategic orientation has highlighted its importance in digital transformation [4, 15, 21]. And more importantly, [22], the digital strategy orientation is to use "digital resources to create value to affect the company's business strategy", thus improving organizational performance. Some scholars have argued that the digital strategy is the guiding principle for an organization to keep track of opportunities enabled by digital technology to improve organizational performance, thus realizing competitive advantage [23]. Using a systematic literature review of 26 articles of peer-reviewed, [17], point out that administrators should adapt their business strategy to a new digital reality, especially in the adaptation of operations management. They also argued previous research may not have recognized all the challenges and opportunities of digital strategy orientation. On the same line, Through 27 articles published (2010–2020), a systematic review of the literature by [18], shows that studies dealing with the link between digital strategy orientation and strategic management are incipient. A few studies in different contexts on digital strategy orientation [4, 18, 22, 23] suggested that the presence of four dimensions within digital strategy orientation: "digital technology scope, digital capabilities, and digital ecosystem coordination". In other words, according to this idea, the prediction of successful digital strategic orientation is done by examining four key areas: Operations, Strategy, Technology, and People. The scale proposed in the said study can be considered useful as a diagnostic tool in identifying areas that need improvement. Table 1 explains this digital strategic orientation constructs definitions and literature support.

2.3 Performance

Performance improvement is the primary intent of all organizations. Thus, elements correlated to performance improvement have become essential topics in management studies [24, 25]. Organizations are committed to growth to confirm survival and to improving organizational performance, thus realizing competitive advantage. Based on [26], Performance is the evaluation of an organization's processes, either by the potential for future accomplishments or by the results, they have achieved [4] suggested in

Table 1. Digital strategy Orientation Dimensions.

Dimensions	Definition	Resources
Technology orientation	"Company commitment towards the application of digital technology and the development of its technological capabilities to provide innovative products, services and solutions"	[3, 23]
Digital capabilities	as "the efforts of organizations to develop and maintain routines that leverage human capital and knowledge assets to handle a specific set of digital technologies"	[3, 23]
Digital ecosystem	it is a dynamic, interconnected network that requires reliable communication between customers and business partners	[3, 23]

their study on manufacturing organizations' strategic orientation has a positive impact on operational performance under the influence of environmental dynamism [6], strategic orientation is beneficial to sales growth rate and user growth rate to measure the performance of Chinese digital start-ups. Thus, this study chose financial performance and operational performance to measure the performance of manufacturing companies.

3 Hypotheses Development and Framework Devolvement

3.1 Digital Orientation and Organizational Performance

In recent years, similar views have been adopted by most managers regarding the high potential of digital orientation to achieve organizational performance [3, 22, 23, 27]. Moreover, numerous various empirical studies indicated that having digital orientation can pave the way to functional and job changes, which lead to cost efficiencies, financial and operational benefits, and enhancing competitive advantages for companies through optimization and monitoring [6]. However, previous studies indicate that the high level of failure in digital investments, which reaches 80%, indicates a low level of understanding about turning digital transformation into unprofitable results [15].

[23], validated building digital orientation by relating it to company performance, based on large US companies over 16 years of age. Their findings found that there is a positive relationship between digital orientation scale of "digital technology, digital capabilities, digital ecosystem orchestration, and digital architecture formation" and company performance. They suggest that building a new digital guidance and validated measurement tool leads to many new research opportunities. A study of [4] a survey of 162 company found strategic orientation, capability of digital transformation, and operational performance to have a positive relationship. Another piece of evidence provided by [14] through the Finnish firms found that firms that provide good digital representation in their strategy adopt digital direction and their performance more than firms that do not have good digital representation in their strategy. It is confirmed that managers are critical in developing digital strategies that reflect activities and results for the performance of the organization, along with the employees responsible for implementation

through the study of [27], who found a positive relationship between the development of digitalization and competitiveness to 33 European countries [8, 12], reveal that digital transformation strategy positively and significantly affects environmental performance from manufacturing industry employees in Pakistan.

Researchers such as [13], argued that digital transformation consists of three main pillars namely: digital technology, digital skills of employees, and digital transformation strategy, and the relationship between these three pillars was positive with organizational performance in the Chinese context. They emphasized that managers should focus investment in digital technologies, skills Digital employees, and digital transformation strategies because it is beneficial to digital transformation, which helps to improve performance and maintain sustainable development. Last but not least, the study of [3], on 515 companies from the U.S. found digital orientation and environmental performance to have a significant and positive relationship. He found that administrators can develop their business's environmental performance and Improve competitive advantage by enhancing the digital orientation in their companies.

However, previous studies did not have shown a positive effect on a linking between digital orientation strategic initiatives and organizational performance. The study of [14], on the 500 largest companies in Finland found digital orientation does not contribute to financial success. Despite this result, he added that to succeed in digital transformation, there is a needed for a strategic style that includes and interrelates the critical elements of "digital orientation, digital intensity, and digital maturity" in digital transformation settings.

Thus, we think that companies in stormy environments will be more effective in pushing their strategic digital orientations into practice and thus achieving better benefits and enhancing organizational performance. Thus, we hypothesize:

H.0 Digital strategic orientation positively affects organizational performance
H.1 technology orientation positively affects organizational performance
H.2 Digital capabilities positively affect organizational performance
H3 Digital ecosystem positively affect organizational performance

3.2 Theoretical Framework

Based on the above discussions from a theoretical review of the literature and Resource Based View (RBV) theory, this study suggests that independent variable digital orientation strategy "Technology orientation, Digital capabilities, and Digital ecosystem", one dependent variable organizational performance as Theoretical Framework. Figure 1 displays the impact of digital orientation strategy and organizational performance variables.

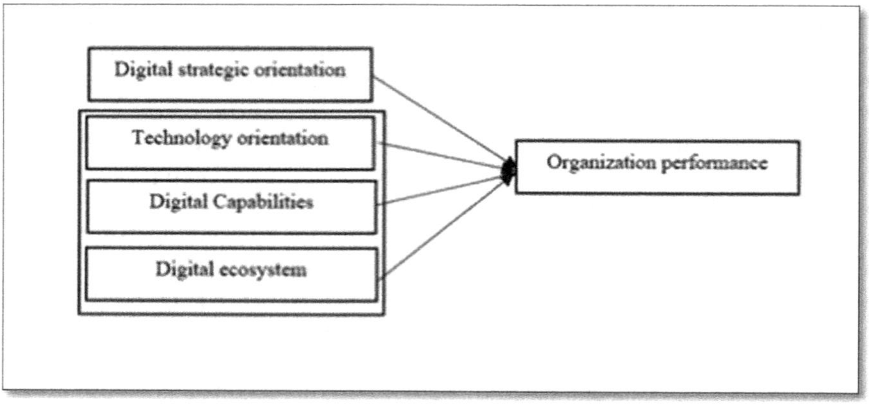

Fig. 1. Research framework

4 Methodology

In this research, a questionnaire of quantitative was used to collect data. Measurement items from previous studies were adopted, (21) items to Digital strategy Orientation [3, 14, 22, 23], (5) items to organizational performance [3, 28]. The current study used a 5-point Likert scale with anchors ranging from (1), strongly disagree perception, to five (5), strongly agree on perception. The simple random sample technique is used to get the chosen Amman Stock Exchange listed manufacturing firms in different businesses including "food, pharmaceutical, chemical, food, petrochemicals, iron and steel". A total of 175 questionnaires were distributed via email to managers in strategic, tactical, and operational management levels, but only 115 were retrieved, of which eight were considered incomplete or did not meet the inclusion requirements, and as such, there were 107 usable questionnaires in total, representing a reply rate of 61.3%.

5 Data Analysis

For hypothesis testing and data analysis, some statistical methods and tools were used with the support of SPSS software, version 24. These include reliability analyses, descriptive statistics to describe the characteristics of respondents, correlation analysis to describe the relationship between variables, and regression analysis to test the effect of Digital strategic orientation Dimensions (DSO) "(Technology orientation (TO), Digital capabilities (DC), and Digital Ecosystem (DE))" on Organizational performance (OP).

5.1 Descriptive Statistics

To understand the data, descriptive statistics "mean values and standard deviations" were achieved for all variables of interest. The target of the analysis of descriptive was to present the raw data in a transformed form that is easy to understand and translate.

From Table 2, the mean values for Digital strategy Orientation, "Technology orientation, Digital capabilities, and Digital ecosystem", and Organizational performance fall in the range of 3.21 and 4.11. Clearly, this shows that respondents perceived moderate level of Digital strategy Orientation offered and they perceive their Factories Jordanian are moderate in support of Top Management. The alpha coefficient was calculated for the study variables using the SPSS program reliability and presented in Table 2. [29] explained that the alpha coefficient of 70% is acceptable. Nevertheless, the reliability of all variables is within the accepted range, which is greater than 0.70.

5.2 Correlation Analysis

Table 3 clarifies an outline of the results of the relationship analysis. To realize the linkage between whole variables of study. The coefficients of Pearson correlations were considered, and founded on Table 2, the correlation coefficients (r) of variables display the strength of the relation between the variables. The whole correlation values of the variables display the correlation with significant variables above 0.49.

Table 2. Correlation coefficients and descriptive sample statistics

	Alpha	M	SD	DSO	TO	DC	DE	DA	OP
DSO	.87	4.11	.702	1					
TO	.77	3.67	.771	.34	1				
DC	.85	3.75	.654	.71	.61	1			
DE	.77	3.78	.684	.56	.46	.72	1		
OP	.82	3.21	.754	.34	.45	.55	.66	.65	1

Note: All correlation coefficients are significant at level less than 0.05

5.3 Regressions

To answer the research question, does digital strategic orientation translate into better organizational performance under the influence of the digital technologies transformation in the sector of Jordanian manufacturing. Accurately, what the dimensions of digital strategic orientation that affect organizational performance, a regression analysis was performed on expected dimensions of digital strategic orientation on organizational performance.

The Table 3, displays that the link between dimensions of digital strategic orientation and organizational performance is significant (F = 32.704; Sig. = .00). The R2 shows that the prompting dimensions of digital strategic orientation account for 61% of the variation in organizational performance. Of all the dimensions of digital strategic orientation included in the regression equation, all the dimensions of digital strategic orientation emerged as significant predictors of organizational performance. These are "Technology orientation, Digital capabilities, and Digital ecosystem". Based on these analysis outputs, hypotheses H1, H2, H3 and H4 are supported. This leads to the deduction that "Digital technology scope, Digital capabilities, Digital ecosystem, and Digital architecture" are positively related to organizational performance.

To examine which dimensions of digital strategic orientation have the most impact on organizational performance, we used the beta values as revealed in Table 3, based on the beta values of the all dimensions significant variables, the predictor "technology orientation, Digital capabilities, and Digital ecosystem" that exercise the most impact on organizational performance is: Technology orientation (β = 0.34), followed by Digital capabilities (β =. 25) and Digital ecosystem (β = .15).

Table 3. Summary of regression analysis for dimensions of Digital strategy Orientation

	B	S.E	B	T	Sig	Tolerance	VIF
1(constant	275	305		904	368		
DSO	0.162	0.095	0.35	2.564	0.003**	0.412	2.222
TO	0.346	0.072	0.34	4.213	0.000**	0.448	4.254
DC	0.651	0.073	0.25	0.236	0.002**	0.456	5.562
DE	0.175	0.062	0.15	0.235	0.004**	0.512	2.568

Note: Significant levels: *** $p < 0.05$

6 Discussion and Theoretical Contribution

Based on recourse based view (RBV), this study investigated effect of Digital strategic orientation Dimensions "Technology orientation, Digital capabilities, and Digital ecosystem" on organizational performance. This research confirmations that digital strategic orientation display a modest percentage of the variation of 16% in Factories performance. This shows that digital strategic orientation has a small explanatory power to predict performance of organizational. In addition, the positive relationship between digital strategic orientation and organizational performance in this study is stable with the past finding of [14], found that digital orientation is a critical success for performance. They existing findings have strongly highlighted that digital strategic orientation should be considered strategic element in companies' pursuit of organizational performance in digital transformation [23], demonstrate that the structure of the company's digital orientation is positively related to the company's performance. They stressed that the new strategic direction requires taking into account the possible border conditions. Examination of such boundary conditions will advance the theory about the relationships between strategic orientations and competitive advantage generated by digital resources [20], found that the relationship between strategic directions and new product performance is a positive. They argued that today's new generation of digital technologies facilitate the rapid development of the digital economy, and technology-oriented organizations are better suited to new product performance and are important drivers of superior corporate performance. This is similar to the results of [13].

As for the practical implications, first, managers in their organizations need to embrace the appropriate digital transformation process so that it is at all levels of organizational units. Second, administrators of digital transformation need to attention on building their capabilities of digital transformation to create new value to achieve company performance and competitive advantage. The better the choice of digital strategies for the organization, the higher the performance of the organization. In another meaning, from the point of view of the administrator, the organization can better adapt to the digital age and can improve higher company performance by rebuilding digital capabilities, sensing digital opportunities and organizing digital resources [6].

7 Conclusion

This research supply experiential evidence for the significant influence of a Digital strategic orientation on organizational performance. The study was constructed from a RBV, the literature of digital transformation, and Digital strategic orientation literature. Our study indicates that maintaining and developing a strategic digital direction helps companies better navigate the strategic demands that the spread of digital technology brings. We hope that the findings of the study will provide a theoretical basis for advancing future research on organizational performance and competitive advantage in a digital world. In sum, we hope that our findings will encourage and support future research in this critical area, just as we hope that practitioners will find entry points in our research to drive the convergence of their digital and strategies to improve the performance of their companies.

References

1. United Nations Development Programme, UNDP Digital strategy 2022–2025, UNDP, 2023. file:///C:/Users/Administrator/Downloads/Digital-Strategy-2022-2025-Full-Document_AR_Interactive%20(1).pdf. Accessed 1 Jan 2023
2. World Economic Forum, Global Risks Report, Insight Report, 2023, WEF. https://www3.weforum.org/docs/WEF_Global_Risks_Report_2023.pdf. Accessed 5 Jan 2023
3. Bendig, D., Schulz, C., Theis, L., Raff, S.: Digital orientation and environmental performance in times of technological change. Technol. Forecast. Soc. Chang. **188**, 122272 (2023)
4. Nasiri, M., Saunila, M., Ukko, J.: Digital orientation, digital maturity, and digital intensity: determinants of financial success in digital transformation settings. Int. J. Oper. Product. Manage. **42**(13), 274–298 (2022)
5. Zhang, T., Shi, Z.Z., Shi, Y.R., Chen, N.J.: Enterprise digital transformation and production efficiency: mechanism analysis and empirical research. Econ. Res.-Ekonomska Istraživanja **35**(1), 2781–2792 (2022)
6. Pan, X., Oh, K.S., Wang, M.: Strategic orientation, digital capabilities, and new product development in emerging market firms: the moderating role of corporate social responsibility. Sustainability **13**(22), 12703 (2021)
7. Jordan Chamber of Industry, The General Director Message. https://jci.org.jo/Chamber/About/20153?l=en. Accessed 5 Jan 2023
8. AL-Khatib, A.W.: The determinants of export performance in the digital transformation era: empirical evidence from manufacturing firms. Int. J. Emerg. Mark. (ahead-of-print) (2023)

9. Adaileh, M., Alshawawreh, A.: Measuring digital transformation impact in Jordan: a proposed framework. J. Innov. Digit. Mark. **2**(1), 15–28 (2021)
10. AL-Zyadat, A., et al.: The effect of industry 4.0 on sustainability of industrial organizations in Jordan. Int. J. Data Netw. Sci. **6**(4), 1437–1446 (2022)
11. Lutfi, A., et al.: Factors influencing the adoption of big data analytics in the digital transformation era: case study of Jordanian SMEs. Sustainability **14**(3), 1802 (2022)
12. Li, Y., Dai, J., Cui, L.: The impact of digital technologies on economic and environmental performance in the context of industry 4.0: a moderated mediation model. Int. J. Prod. Econ. **229**, 107777 (2020)
13. Teng, X., Wu, Z., Yang, F.: Research on the relationship between digital transformation and performance of SMEs. Sustainability **14**(10), 6012 (2022)
14. Yu, J., Moon, T.: Impact of digital strategic orientation on organizational performance through digital competence. Sustainability **13**(17), 9766 (2021)
15. Niemand, T., Rigtering, J.P.C., Kallmünzer, A., Kraus, S., Maalaoui, A.: Digitalization in the financial industry: a contingency approach of entrepreneurial orientation and strategic vision on digitalization. Eur. Manag. J. **39**(3), 317–326 (2021)
16. Barney, J.: Firm resources and sustained competitive advantage. J. Manag. **17**(1), 99–120 (1991)
17. Ding, J.: Literature review. In: A Lexical Semantic Study of Chinese Opposites. FCL, vol. 1, pp. 3–8. Springer, Singapore (2018). https://doi.org/10.1007/978-981-10-6184-4_2
18. Yeow, A., Soh, C., Hansen, R.: Aligning with new digital strategy: a dynamic capabilities approach. J. Strateg. Inf. Syst **27**, 43–58 (2018)
19. Alshourah, S.: Role of Strategic Orientation Dimensions of Green Supply Chain Management Practices in Jordanian Manufacturing Companies. Accepted Unpublished, Zarqa J. Res. Humanit (2022)
20. Alshourah, S., Alassaf, H., Altawalbeh, M.: Roles of top management and customer orientation in enhancing the performance of customer relationship management (CRM) in hotel industry. Int. J. Adv. Res. Innov. **6**(3), 233–239 (2018)
21. McKenny, A.F., Short, J.C., Ketchen, D.J., Payne, G.T., Moss, T.W.: Strategic entrepreneurial orientation: configurations, performance, and the effects of industry and time. Strateg. Entrep. J. **12**(4), 504–521 (2018)
22. Wang, Y.: Analyzing the mechanism of strategic orientation towards digitization and organizational performance settings enduring employee resistance to innovation and performance capabilities. Front. Psychol. **13**, 1006310 (2022)
23. Kindermann, B., Beutel, S., de Lomana, G.G., Strese, S., Bendig, D., Brettel, M.: Digital orientation: Conceptualization and operationalization of a new strategic orientation. Eur. Manag. J. **39**(5), 645–657 (2021)
24. Alshourah, S.: Assessing the influence of total quality management practices on innovation in Jordanian manufacturing organizations. Uncertain Supply Chain Manage. **9**(1), 57–68 (2021)
25. Alshourah, S.: Total quality management practices and their effects on the quality performance of Jordanian private hospitals. Manage. Sci. Lett. **11**(1), 67–76 (2021)
26. Alshourah, S.: The antecedent of customer relationship management and its impact on hotels performance in Jordan. Theses PhD published UUM (2012)
27. Petkovski, I., Fedajev, A., Bazen, J.: Modelling complex relationships between sustainable competitiveness and digitalization. J. Competitiveness (2022)
28. Alshourah, S., Jodeh, I., Swiety, I., Ismail, A.: Social customer relationship management capabilities and performance: moderating social media usage among SMES Jordanian. Decis. Sci. **25**(S2), 1–8 (2022)
29. Hair, J.F., Howard, M.C., Nitzl, C.: Assessing measurement model quality in PLS-SEM using confirmatory composite analysis. J. Bus. Res. **109**, 101–110 (2020)

The Nexus Between Digital Transformation and Economic Efficiency: Evidence from Selected Countries

Amer Alsaraireh[1(✉)] and Ahmad AL-Majali[2]

[1] Faculty of Economics and Administrative Sciences, Zarqa University, Zarqa, Jordan
aalsaraireh@zu.edu.jo

[2] College of Business, Mutah University, Karak, Jordan
majalia@mutah.edu.jo

Abstract. The primary goal of this research is to assess the influence of digital transformation on economic efficiency in some selected countries from Europe, East Asian, Middle East Africa and USA, using cross-sectional data from 2018 to 2022. The nonparametric approach, Data Envelopment Analysis (DEA), was used to examine the efficiency level of the country. BCC and CCR models were used to calculate the efficiency score, by measuring and assessing technical efficiency. The literature has shown that digital transformation has influenced and facilitated the creation of new business models and maximizing efficiency in countries' economies. Which coincided with the results of this study, as it showed that digital transformation influences technical efficiency; 6 among the 58 studied countries were 100% efficient, 21 countries were moderate to high efficient while the others were inefficient; Each of the three digital transformation dimensions adopted in the study, namely: digital knowledge, digital technology, and future readiness has a significant influence on technical efficiency. The study recommended the urgent need to increase the level of awareness upon the leaders about the nature and importance of digital transformation processes and its serious adoption; Because of its positive influence on economic efficiency.

Keywords: Digital Transformation · Economic Efficiency · Data Envelopment Analysis (DEA)

1 Introduction

Today, the globe is experiencing an unprecedented surge in technical innovation as countries race to improve their digital readiness and digital economy. Developed countries have achieved a high degree of digital transformation, putting them at the forefront of the globe, increasing economic efficiency, and making them more flexible in dealing [1]. Economic developments and requirements are based on circumstances. The best example is what happened during the new Corona pandemic, which increased demand for electronic products on the one hand, as well as the use of websites and specialized software to support remote work [2, 30]. Additionally, Numerous worldwide surveys

B. A. M. Alareeni and I. Elgedawy (Eds.): *Artificial Intelligence (AI) and Finance*, SSDC 488, pp. 387–396, 2023.
https://doi.org/10.1007/978-3-031-39158-3_37

investigated how various countries were doing with digital transformation. One such analysis was The IMD World Digital Competitiveness Ranking which each year quantifies the capacity of an economy to adopt and explore new digital technologies able to transform government practices, business models, and society in general. Also measures how well an economy can adapt to new digital technology to change how the government functions [3].

In another context, many studies show the benefits of digital transformation through Cost reduction and resource cost optimization as the first direct advantages of digital transformation [4]. Initiatives for digital transformation are being used by businesses and governments to boost efficiency and make better use of already available resources, which immediately affects cost and budget. Also, digital transformation has increased efficiency and efficiency [5], whether by humans or machines, which is one of the critical advantages of digital transformation. By automating tedious, repetitive operations, digital transformation may free up time for workers to concentrate on innovation and fostering corporate development. For instance, the Industrial Internet of Things (IIoT) uses AI/ML to enable predictive maintenance rather than reactive maintenance analyses machine performance and malfunctions in real-time and increases system efficiency [6]. On the other hand, digital transformation increases access to information, quick and straightforward communication, the ability to exchange information, the creation of new job possibilities, and enhanced corporate competitiveness are just a few of the many clear benefits of digital transformation.

This study, which differs from earlier studies in the same context in terms of linking and methodology used, aims to link the level of digital transformation in countries using the "IMD World Digital Competitiveness Ranking" and the level of efficiency in countries using the Data Envelopment Analysis (DEA) methodology.

2 Literature Review

2.1 Digital Transformation

According to [7], the stages of the conversion process to a digital business are digitizing analogue information, Processing digital information (digitalization), and digital transformation of the company, the final stage. Furthermore, [8] debate that digital transformation and digitalization are expressions used interchangeably to refer to a concept that affects policy, business, and further critical social topics. On the other hand, [9] distinguish between digitization, digitalization, and digital transformation. Digitalization is the application of digital technologies and precedes the endless digital transformation process. Further, [10] believe that digital transformation encompasses digitalization, but they differentiate between the two terms; digital transformation means converting analogue information to digital automation using ICT, which creates alterations in the business model, organizational structure, products, services, processes, and further facets of the business. According to [11], Business digitalization is intended to decrease business costs and grow efficiency. This paper defines digital transformation as "an institutional process involves a radical digital development carried out at all corporate' levels.

2.2 Digital Technology

Cloud computing, mobile technologies, the internet of things, big data, social media, intelligent networks, artificial intelligence, cybersecurity, and other cutting-edge digital technologies are used. Successful firms in the digital transformation process rely on innovation and creativity, which allows them to build digital technology rather than on the technology itself. In this aspect, [12] debates that, in light of the recent epidemics of COVID-19 and its challenges and threats, despite its devastating impact on the global economy and businesses worldwide, it may be the drive for the unlimited growth of digital technology. Also, [13] stress that the company's application of new and updated digital technologies leads to an increase in efficiency through the most appropriate improvements and developments in production processes.

2.3 Future Readiness

Countries and business organizations are racing to improve their digital Readiness and economy as they strive to ride the endless wave of digital transformation. The future readiness factor examines the degree of adoption of digital technology by governments, businesses, and society. This factor includes specific indicators, according to the IMD Global Digital Competitiveness Ranking report, such as the diffusion of e-commerce, industrial robots, and data analytics tools in the private sector and the strength of those cyber-security measures in place [14].

2.4 Digital Knowledge

Digital technology and future Readiness are essential in accessing the world of digital transformation. However, the most important thing is the extent to which the business enterprise has the necessary digital transformation knowledge that leads to digital transformation in its business through the correct and integrated application of digital technology and the ability to sustain it, and this can only be achieved with modern and constantly renewed digital knowledge by adopting and applying knowledge management and transforming the enterprise into a learning organization. Mismanagement of the ongoing digital transformation process stems from a lack of or insufficient awareness of digital technology and its application possibilities. According to [15], digital transformation is using existing information to create significant changes in the business so that all activities and decisions are focused on the customer.

2.5 Literature Analysis

Many studies have been conducted to investigate the economic impact of digital transformation. Using panel data from developed and developing countries [16] found there is a short and long run effect between information and communication digital technology and economic growth. Also [17], they showed that trade liberalization and digital transformation are sources of economic development for the wealthiest Asian countries, using the Gravity Model and the Generalized Method of Moments Model. Additionally, [18] found that an increase in the E-government Development Index, information and communications digital technology exports, and internet user growth significantly and positively impact long-term economic growth, implying that digital and technological advancement positively impact long-term economic growth. In the same context [19], showed the influence of the digital economy on OECD countries' economic growth as assessed by the technological proxies - internet, mobile phone, and fixed-broadband - is determined by their degree of development and the technologies that capture the digital economy. ICTs have a favorable impact on the growth of OECD economies and may be used as policymaking tools. And by Utilizing information from the World Development Indicators database [20] covering the years 1990–2020, [21] found that digital transformation and patents had a favorable impact on economic development. [22] Explained how specific digital communication tools benefit innovation, such as social media use and internet browsing on a corporate mobile phone. And this invention has a good impact on labor efficiency. While [23] showed that Digital transformation, conditioned on financial sector development at best, does not encourage economic growth in Africa. However, overall digital transformation improves economic growth in Africa. As well as [24] pointed the industry concentration and digital transformation positively impact efficiency at the sectoral level in Germany. As well cautiously conclude that recent digital technology advancement appears to have been labor-saving and that the efficiency -enhancing component of a partial "superstar company" impact may be seen in the German economy, notably in its manufacturing sector. In addition, the rate of expansion of the digital economy has outpaced the US GDP's average yearly growth rate by a factor of four. The so-called "Solow paradox," which states that efficiency increase predominantly happens in the digital economy sectors, has proven to be true. It influences the overall national economy's efficiency growth to a lesser extent.

3 Methodology and Data Analysis:

The Data Envelopment Analysis (DEA) approach was used in this study to assess technical efficiency [25]. It uses a linear programming model that evaluates the relative efficiency of peer countries with various inputs and outputs, DEA may generate a production frontier (linear piecewise frontier) [26]. The distance of a country's input-output combination set to the frontier is used to calculate its relative efficiency. It is a strategy that examines each country's performance compared to the best practice frontier, which considers a benchmark of high efficiency; therefore, we have two approaches to calculate efficiency in which minimum inputs are used to create a certain level of output (input-oriented) or the maximization output with the given input (output-oriented). Using the output-oriented, a country's efficiency is taken by the maximum weighted proportion of outputs to the weighted proportion of inputs (Eq. 1). Subject to that, the ratios for the comparable country are less than or equal to one (Eq. 2). Where the analysis model is as follows [27]:

$$\max \theta = \sum_{r=1}^{s} U_r Y_{ro} / \sum_{i=1}^{m} V_r X_{ro} \tag{1}$$

Subject to:

$$\left(\sum_{r=1}^{s} U_r Y_{rj} / \sum_{i=1}^{m} V_r X_{rj}\right) \leq 1 \tag{2}$$

Nonnegatively condition

$$Y_{rj},\ X_{ij} \geq 0 \tag{3}$$

where: y: output for country i, x: input for country i, U_r: weight for output V_r: weight for input, r: number of output, i: number of input, j = 1,2.....n, U_r, $V_j \geq 0$, r = 1,2...s, i = 1,2,...m. There are two types of DEA models, namely the model of Constant Returns to Scale (CRS), and Variable Returns to Scale (VRS), where the first model was developed by [28] and is called abbreviated (DEA-CCR). Production within this assumption and the degree of technical efficiency under this model is known as the concept of total technical efficiency, which reflects the ability of the establishments to achieve the best production level within the inputs used. The second model was developed by [28], and it is called (DEA-BCC) abbreviation. One assumes that each additional unit of input results in a varying additional increase or decrease in a unit of output (increasing or decreasing returns to scale). As many factors do not make the state operate within the level of its optimal sizes, such as imperfect competition and some restrictions on financing and others. Thus, (DEA-CCR) model may lead to inaccurate percentages on the technical efficiency of the countries. Therefore, this model is used to separate the effects of digital technology and the effects of size in measuring efficiency, as the degree of efficiency refers to the efforts made by the country in the production process in terms of using and distributing inputs and achieving outputs. Given the reasons referred to, a model with variable returns to scale (VRS) was used in this study to clarify this mathematical model by using the duality or binary property in linear programming to solve the problem [29].

$$min_{\theta,\gamma}\theta \tag{4}$$

Subject to:

$$-Yj+Y\gamma \geq 0 \tag{5}$$

$$\theta Xj - X\gamma = 1 \tag{6}$$

$$\gamma \geq 0 \tag{7}$$

The technical efficiency (θ) scores are obliged in the 0–1 interval ($0 \leq \theta \leq 1$). If $\theta = 1$ that means the country is efficient and on the frontier; if $\theta < 1$ that means the country is inefficient. Where: θ: technical efficiency for country i, γ: vector of variables weights, Y: output matrix, X matrix of input. The Distributed Evolutionary Algorithms in Python (DEAP) program was used to solve the above-mentioned equations. The population comprises selected countries from Europe, East Asia, Middle East Africa, and America, covered in IMD world digital competitiveness ranking. Were 58 countries included in this study from 2018 to 2022. The inputs and outputs variables in this study were selected with the support of a literature review. Where The Outputs are defined Gross domestic product per capita (Y_1), and the Inputs are defined Total government spending as a percentage of GDP (as an explanatory variable) (X_1), Total investment as Percent of GDP (explanatory variables) (X_2) [20]. KNOWLEDGE (X_3) Refers to the invisible infrastructure that allows technologies to be discovered, comprehended, and learned, eventually contributing to digital transformation. Indicators that reflect the quality of human capital available in a nation, the degree of investment in education and research, and the outputs. Illustration, registered patent awards in high-tech domains and jobs in the technology and scientific industries, DIGITAL TECHNOLOGY (X_4) evaluates the overall environment for the progress of digital technologies, including standards for evaluating the influence of legislation on stimulating private-sector innovation, the accessibility of funds for investments, and the adequacy of technical infrastructure, FUTURE READINESS (X_5) evaluates the level at which governments, corporations, and society use digital technology. This component covers indications such as the use of e-commerce, industrial robotics, and data analytics tools in the private sector and as well as the effectiveness of existing cyber-security measures [14].

4 Findings

This section calculates the efficiencies using the (DEA-BCC) model on selected 58 countries from 2018 to 2022, in terms of their ability to provide outputs with minimum input utilization (output-oriented). The measurement is relative to the best practice frontier which is created from the observations sample of the country. The results indicate how individual countries operate relative to others. Under the VRS assumption, it is possible to calculate the scale efficiency (Scale) by dividing the CRSTE (technical efficiency from CRS DEA) efficiency score by the VRSTE (technical efficiency from VRS DEA) efficiency score. Table 1 below displays the results of this model which shows an individual country's technical efficiency.

We have three groupings as a consequence of the results. Compared to another country, the first group has the best efficiency in applying technology where the scale efficiency is equal to one (USA, Qatar, Hong Kong, Luxembourg, Singapore, and Switzerland). The second group is the moderate to high efficiency between 1–0.5 in comparison to other countries (Ireland, Denmark, Iceland, Japan, Netherlands, Australia, Sweden, Canada, UAE, Finland, Germany, Iran, Belgium, Taiwan, China, Malaysia, New Zealand, United Kingdom, France). While the third group (India, Korea, Rep, Italy, Cyprus, Spain, Saudi Arabia, Czech Republic, Estonia, Slovenia, Slovak Republic, Portugal, Lithuania, Latvia, Greece, Hungary, Poland, Jordan, Chile, Croatia, Romania, Bulgaria, Argentina, China, Kazakhstan, Mexico, Turkey, Brazil) has limited capability in using technology and that needs an in-depth review from the administrators of these countries to utilize their inputs.

Figure 1 depicts a positive interaction between Technical Efficiency and Knowledge index (rank) to relate the results of calculating technical efficiency based on Digital.

Transformation with the level of the knowledge index. (The closer to one, the higher country's rank).

Table 1. Relative Technical Efficiency BCC model 2018–2022

Country	UAE	Estonia	Peru	Sweden	Australia	Indonesia	Netherlands
CRSTE	0.56	0.20	0.09	0.68	0.52	0.06	0.72
VRSTE	0.83	0.58	0.88	0.97	0.69	0.94	0.89
Scale	0.67	0.34	0.10	0.70	0.75	0.06	0.81
Country	USA	Qatar	Canada	India	Romania	Mongolia	United Kingdom
CRSTE	1.00	1.00	0.55	0.32	0.12	0.04	0.52
VRSTE	1.00	1.00	0.81	0.66	0.61	0.57	0.89
Scale	1.00	1.00	0.68	0.48	0.20	0.07	0.58
Country	Iran	Ireland	France	Greece	Bulgaria	Lithuania	Czech Republic
CRSTE	0.46	0.95	0.35	0.15	0.10	0.20	0.18
VRSTE	0.71	1.00	0.63	0.58	0.62	0.70	0.52
Scale	0.65	0.95	0.56	0.26	0.16	0.29	0.35
Country	Chile	Japan	Slovenia	Argentina	Finland	Hong Kong	Luxembourg
CRSTE	0.16	0.62	0.20	0.10	0.51	1.00	1.00
VRSTE	0.73	0.72	0.58	0.63	0.76	1.00	1.00
Scale	0.22	0.86	0.34	0.16	0.67	1.00	1.00
Country	Cyprus	Turkey	Croatia	Mexico	Germany	Saudi Arabia	Korea Rep.
CRSTE	0.24	0.08	0.11	0.11	0.46	0.22	0.48
VRSTE	0.59	0.59	0.51	0.80	0.71	0.63	1.00
Scale	0.41	0.14	0.22	0.14	0.65	0.35	0.48
Country	Brazil	Jordan	Thailand	Singapore	Philippines	Taiwan	New Zealand
CRSTE	0.08	0.16	0.09	1.00	0.04	0.62	0.38
VRSTE	0.65	0.72	0.80	1.00	0.82	1.00	0.64
Scale	0.12	0.22	0.11	1.00	0.05	0.62	0.59
Country	Latvia	Spain	Portugal	Belgium	Malaysia	Kazakhstan	Slovak Republic
CRSTE	0.16	0.22	0.18	0.38	0.52	0.12	0.16
VRSTE	0.58	0.59	0.62	0.60	0.86	0.83	0.55
Scale	0.28	0.37	0.29	0.63	0.60	0.14	0.29
Country	Denmark	Italy	Hungary	Poland	Colombia	Switzerland	South Africa
CRSTE	0.87	0.24	0.12	0.13	0.06	1.00	0.07
VRSTE	0.92	0.56	0.48	0.58	0.67	1.00	0.78
Scale	0.95	0.43	0.25	0.22	0.09	1.00	0.09
Country	Iceland	China					
CRSTE	0.64	0.09					
VRSTE	0.71	0.59					
Scale	0.90	0.15					

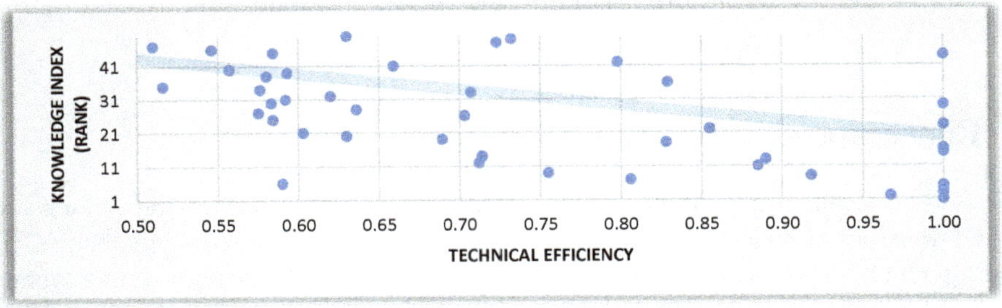

Fig. 1. Relative Technical Efficiency and Digital Knowledge Index

To link the outcomes of calculating technical efficiency based on Digital Transformation with the level of the Digital Technology Index, Fig. 2 shows a positive interaction between Technical Efficiency and Digital Technology Index (rank).

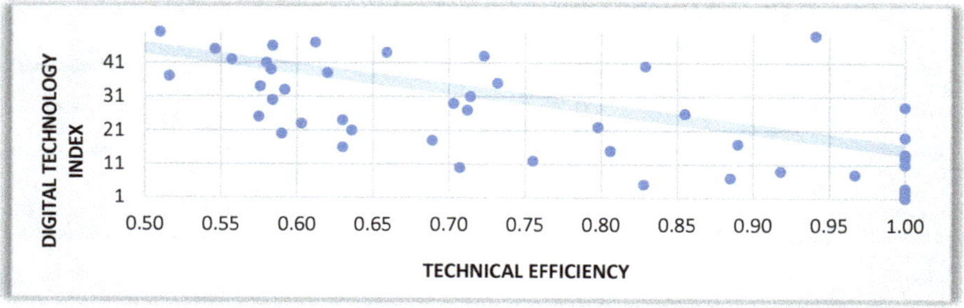

Fig. 2. Relative Technical Efficiency and Digital Technology Index

In order to relate the results of estimating technical efficiency based on technological inputs with the level of the Future Readiness Index, Fig. 3 illustrates a positive interaction between Technical Efficiency and Future Readiness Index (rank).

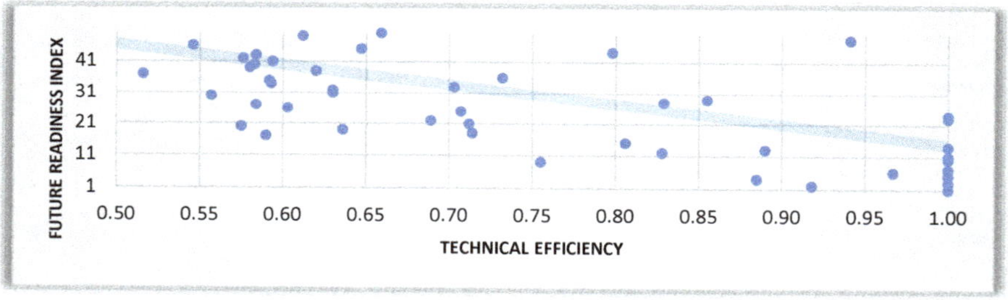

Fig. 3. Relative Technical Efficiency and Future Readiness Index

5 Conclusion and Recommendations

The study's results proved the existence of an important relationship and influence of digital transformation and each of its various dimensions on economic efficiency. Such findings can help policymakers improve the digital economy by providing new avenues for developing appropriate regulations and promoting new organizations. As a result, the benefits of digital transformation can result in significant economic development. These players must implement policies that improve ICT physical and technological infrastructures, digital human capital empowerment, and improved social equity in ICT access. The government should take the potential of digital transformation seriously, as they may enable the economy to be updated, diversified, and robust to generate new employment and establish new markets, strategic partners, and prospects. The study recommends urging the leaders of countries with weak economic efficiency to follow the example of fully efficient countries by adopting the digital transformation approach in their economies and raising their future readiness for digital transformation and the use of digital technology and relevant knowledge.

References

1. Pavlov, M.: peculiarities of the digitalization process of public governance in developed countries. Investytsiyi: Praktyka Ta Dosvid **15**, 150 (2021). https://doi.org/10.32702/2306-6814.2021.15.140
2. Al-Balawi, N.S.: Digital competencies and digital transformation in light of the corona pandemic obstacles and challenges at Northern Border University. Revista Amazonia Investiga **11**(58), 124–133 (2022). https://doi.org/10.34069/ai/2022.58.10.13
3. Khazaei, M., Faghihi, M., Manesh, V.: How Performance of top companies are related on IMD digital competitiveness indicators? Issues Soc. Sci. **10**(1), 29 (2022). https://doi.org/10.5296/iss.v10i1.20031
4. Shivajee, V., Singh, R., Rastogi, S.: Manufacturing conversion cost reduction using quality control tools and digitization of real-time data. J. Clean. Prod. **23**(7), 76–89 (2019). https://doi.org/10.1016/j.jclepro.2019.117678
5. Bozkurt, E., Topçuoğlu, Z., Altiner, A.: Relationship between productivity and digitalization with tobit model based on malmquist index. Verimlilik Dergisi (2021). https://doi.org/10.51551/verimlilik.986964
6. Al-Hawawreh, M., Sitnikova, E.: Developing a security testbed for industrial Internet of Things. IEEE Internet Things J. **8**(7), 5558–5573 (2021). https://doi.org/10.1109/jiot.2020.3032093
7. Fiodorov, I., Ochara, N.M.: The impact of digital transformation on economic of BRICS countries 1. In: Proceedings of the XXII International Conference Enterprise Engineering and Knowledge Management April 25–26, pp. 25-26, Moscow, Russia (2019)
8. Collin, J., Hiekkanen, K., Korhonen, J., Halen, M., Itala, T., Helenius, M.: IT leadership in transition – The impact of digitalization on Finnish organizations (2015). https://aaltodoc.aalto.fi/bitstream/handle/123456789/16540/ isbn9789526062433.pdf? sequence=1&isAllowed=y
9. Fichman, R., Dos Santos, B., Zheng, Z.: Digital innovation as a fundamental and powerful concept in the information systems curriculum. MIS Q. **38**(2), 329–353 (2014)
10. Kwon, E.H., Park, M.J.: Critical factors on firm's digital transformation capacity: empirical evidence from Korea. Int. J. Appl. Eng. Res. **12**(22), 12585–12596 (2017)
11. Patel, M.: Digital Transformation vs. Digital Optimization. Medium (2019). https://medium.com/@maxy_ermayank/digital-transformation-vs-digital-optimization-5c86cff1567b
12. Khodova, Y.: Digital transformation of business as a trend of world economy development digitalization of the economy as a factor of sustainable development. Mater. Int. Sci. Pract. Confer. (Mariupol, 25–26 May 2021) (2021)
13. Fuentelsaz, L., Gomez, J., Palomas, S.: The effects of new technologies on productivity: an intrafirm diffusion based assessment. Res. Policy **38**, 1172–1180 (2009)
14. IMD World Digital Competitiveness Ranking Report, p. 8 (2022)
15. Savić, D.: From digitization, through digitalization, to digital transformation. Inf. Discov. Technol. Strat. **43**(1), 36–39 (2019)
16. Mgadmi, N., Moussa, W., Bejaoui, A., Sadraoui, T., Afef, G.: Revisiting the nexus between digital economy and economic prosperity: evidence from a comparative analysis. J. Telecommun. Digit. Econ. **9**(2), 69–91 (2021)
17. Bakari, S., el Weriemmi, M., Mabrouki, M.: The Impact of digitalization and trade openness on economic growth: new evidence from richest Asian countries. J. Res. Innov. Technol. **2**(2), 95-106 (2022)
18. Zhao, S., Zhang, Y., Iftikhar, H., Atta, U., Mao, J., Wang, T.: Dynamic influence of digital and technological advancement on sustainable economic growth in belt and road initiative (BRI) countries. Sustainability **14**(23), 15782 (2022). https://doi.org/10.3390/su142315782

19. Gomes, S., Lopes, M.J., Ferreira, L.: The impact of the digital economy on economic growth: the case of OECD countries. RAM. Revista de Administração Macken-zie **23**(6) (2022). https://doi.org/10.1590/1678-6971/eRAMD220029.en
20. World Bank: World Development Report 2021: Data for Better Lives. World Bank, Washington, DC (2021)
21. Bakari, S.: The Impact of digitalization and patent on economic growth in Romania. J. Res. Innov. Technol. **1**(1), 49–61 (2022)
22. Gaglio, C., Kraemer-Mbula, E., Lorenz, E.: The effects of digital transformation on innovation and productivity: firm-level evidence of South African manufacturing micro and small enterprises. Technol. Forecast. Soc. Chang. **182**,(2022). https://doi.org/10.1016/j.techfore.2022.121785
23. Iddrisu, A., Chen, B.: Economic growth through digitalization in Africa: does financial sector development play a mediating role? Int. J. Emerg. Mark. (2022). https://doi.org/10.1108/IJOEM-02-2022-0278
24. Ferschli, B., Rehm, M., Schnetzer, M.: Digitalization, industry concentration, and productivity in Germany. Jahrbücher für Nationalökonomie und Statistik **241**(5), 623–665 (2021). https://doi.org/10.1515/jbnst-2020-0058
25. Banker, R.D.: Stochastic data envelopment analysis. Data Envelopment Anal. J. **5**(2), 281–309 (2021). https://doi.org/10.1561/103.00000038
26. Reshadi, M.: new method to evaluate relative efficiency measure in dynamic DEA. Data Envelopment Analy. Decis. Sci. **2014**, 1–6 (2014). https://doi.org/10.5899/2014/dea-00053
27. Charnes, A., Cooper, W.W., Rhodes, E.: Measuring the efficiency of decision making units. Eur. J. Oper. Res. **2**(6), 429–444 (1978)
28. Banker, R.D., Charnes, A., Cooper, W.W.: Some models for estimating technical and scale inefficiencies in data envelopment analysis. Manage. Sci. **30**(9), 1078–1092 (1984)
29. Rouse, P., Harrison, J., Chen, L.: Data envelopment analysis: a practical tool to measure performance. Australian Acc. Rev. **20**(2), 165–177 (2010). https://doi.org/10.1111/j.1835-2561.2010.00090.x
30. Shehadeh, H.K.: Impact of telework strategy on quality of work life during COVID-19 pandemic a case study of orange Jordanian telecommunication company. Int. J. Qual. Assur. **5**(2), 141–142 (2022)

The Impact of Online Analytical Process (OLAP) on Talent Management: Case Study of Orange Jordanian Telecommunication Company – Amman, Jordan

Mohammed Abed Hussein Al taee, Hazem Khaled Shehadeh(✉), Mohammed Younis Younis Miqdad, and Amira Turki

Faculty of Economics and Business Administration, Faculty of Business Administration, Zarqa University, Zarqa, Jordan
{m.taee,hshehadeh}@zu.edu.jo, a.turki@ubt.edu.sa

Abstract. The progress changes in the business environment, the higher level of competition and the need for technological intervention derived several organizations to adopt Online Analytical Process (OLAP) systems and tools in order to facilitate business responses to these changes and stay ahead in the marketplace. Talent management creates an imperative necessity to use new methods, techniques, and tools that intelligently and automatically transform information into valuable data that help to attract, develop and retain talents.

The major objective of this research is to "examine the impact of the (OLAP) on talent management at Orange Jordanian Telecommunications Company". A review of the literature serves as the basis for measuring the impact of (OLAP) on talent management. The study sample consisted of the managerial level of Orange Jordanian Telecommunication Company, (280) questionnaires were distributed in the company by using a google form. (234) Questionnaires were collected. (46) Questionnaires were not answered, this represented about (16%) of all questionnaires. (84%) of the questionnaires were valid for actual analysis.

The most important findings of this study demonstrate that (OLAP) tools respectively (Data Mining, Data Warehousing, and OLAP) have a significant impact on talent management. (OLAP) plays a vital role in talent management in the organization; several types of research addressed the importance of integrating (OLAP) with talent management, but few of these researches address the impact of (OLAP) on talent management. This study has tried to address this need.

Keywords: Online Analytical Process (OLAP) · Talent management · Orange Jordanian Telecommunication Company

1 Introduction

Human capital plays a major role in the success of an organization. A vital challenge facing organizations recently is the Talent War. This war represents a strategic organizational challenge and a critical driver for enterprise performance and gaining a superior

B. A. M. Alareeni and I. Elgedawy (Eds.): *Artificial Intelligence (AI) and Finance*, SSDC 488, pp. 397–405, 2023.
https://doi.org/10.1007/978-3-031-39158-3_38

competitive advantage. By having the right talent, organizations gain major benefits of effectiveness, high-performance levels, innovation, sustainability, etc. Talent management is the process of planning, recruiting, selecting, developing, managing, and compensating employees that is holistic, integrated, and results-oriented [1]. On the other hand, the analysis Online Analytical Process (OLAP) also plays a vital role in talent management. Online Analytical Process (OLAP) can develop valuable HR information systems and help organizations make fast and reliable business decisions that enhance their ability to properly manage talent.

2 The Study Problem

Although interest in the subject of talent management began in the eighties of the last century, studies focusing on the relationship and impact between the online analytics process (OLAP) and talent management in the Arab environment in general and in Jordan, in particular, are still limited. Browsing the websites of the relevant sectors, including the websites of specialized scientific colleges and departments, as well as looking at specialized periodicals, whether in their traditional or electronic form, as well as viewing the works of conferences, seminars, and academic workshops reflect this limitation. On the field level, through a preliminary survey conducted on ten decision-makers in Orange, the researchers found that there is great interest in the issue of talent management, especially in light of the intense competition between the three telecommunications companies operating in Jordan, and this interest increased in light of the difficulties created by the Covid 19 crisis, Furthermore, despite the company's long history, the researchers discovered that no research had been conducted in this area. The study's problem can then be embodied by answering the following main question: Is there an impact of the online analytics process (OLAP) on talent management (talent attraction, retention, and development)?

3 The Study Objectives

The study aims to identify the level of online analytical process (OLAP) and talent management in Orange Jordan Telecom. Explain the impact of the online analytics process (OLAP) on talent management by finding an appropriate ground to embody the theoretical framework of the studies that dealt with the online analytical process (OLAP) in supporting talent management in the field of this company. The expected results of this study may serve to bridge the gap of deficient studies on these two important topics.

4 The Study Model

Figure 1 shows that there is one independent variable in the study online analytics process (OLAP) that consists of three dimensions (Complex Calculations, Trend Analysis, and Sophisticated Data Modeling) . There is also one dependent variable; Talent Management which consists of three dimensions (talent attraction, talent retention, and talent development).

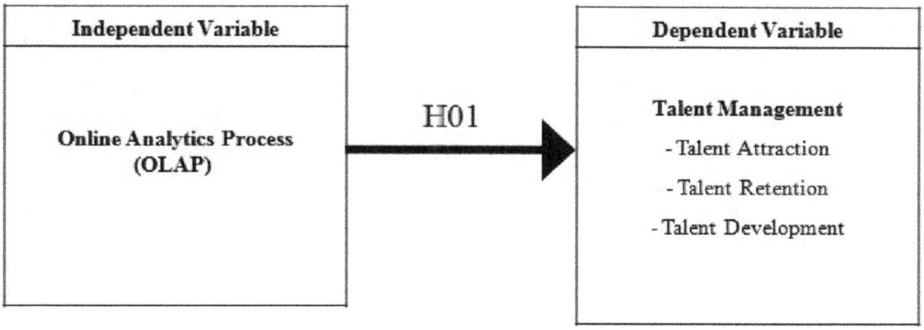

Fig. 1. Study Model Source: The proposed conceptual study model prepared by the researcher depends on the previous studies. (Zabadi 2015), ((Dorgham 2018)).

5 The Study Hypotheses

H0: There is no statistically significant impact at the significance level ($\alpha \leq 0.05$), for the online analytics process (OLAP) in terms of its dimensions (Complex Calculations, Trend Analysis, and Sophisticated Data Modeling) on talent management in terms of its dimensions (Talent Attraction, Talent Retention, and Talent Development) in Orange Jordanian Telecommunication Company. 6. Online Analytic Processes (OLAP)

Online Analytical Processing (OLAP) is an important and powerful tool to conduct data analysis, complex calculations of data, forecasting, and predictive analysis decisions. Also, OLAP is a technology that existed to serve BI. OLAP can be defined as a tool that provides users analysis for decision-making through a huge amount of data, by using a multidimensional strategy to arrange and analyze information. OLAP also can be recognized as a software technology objected to helps executives and analysts to look at past data speedily, precisely, and with interactive access to a huge amount of feasible view of information extracted from raw data to represent different dimensions of the company from the user's perspective [2].

OLAP is defined also as a technology that allows analysts to do a multidimensional analysis of data and view it from different angles. OLAP technology helps in analyzing information from multiple and heterogeneous databases [3]. OLAP is a specialized tool designed to analyze and provide excellent query performance, data sets updates, data summary, and rich analytic content [4]. OLAP refers to a decision-support approach that aims to extract knowledge from a data warehouse, or more specifically, data marts. Its main idea is providing navigation through data to non-expert users so that they are able to interactively generate ad hoc queries without the intervention of IT professionals" [5]. OLAP is a powerful data discovery technology that allows for infinite report viewing, complex analytical calculations, and predictive "what if" scenario (budget, forecast) planning. End-users can perform ad hoc analysis of data in multiple dimensions using OLAP, providing the insight and understanding required for better decision-making [6].

From the above definitions, we can consider OLAP as the management of data kept in a data warehouse and it is an important tool for managing the presentation of data in a data warehouse in a useful way that helps users access data effectively and efficiently. OLAP technology allows analysts and executives to gain better insight into information through fast, consistent, and interactive access to different data views. This data was derived from primary sources in order to reflect the true dimensions of the organization in talent management, OLAP also performs multidimensional business information analysis, and this technology allows for complex estimations, trend analysis, and data modeling. OLAP strengthens the foundation for intelligent solutions (talent acquisition, training, development, and retention).

6 Talent Management

Recently, talent management is one of the major success factors for any organization. There are millions of jobs and no talent to fill them. The talent war arose due to changes and challenges in the business environment in the 21st century. These changes lead to a shortage of talent and create high competition among organizations to acquire the best talent needed to lead the success of their business and gain a competitive advantage for talent [7].

The existing literature on the concept of talent management does not have consistent and clear theoretical support. As some studies have found that talent management is defined in four streams to understand talent management is, these streams, the first stream explained talent management as an alternative to human resources, the second develops the talent pool, the third stream focuses on talent management, and finally, talent management focuses on defining the main position instead of talented individuals [8].

Organizations consider Talent Management from different views, these views came from how each organization thinks of talent and how the talent serves it. Some organizations consider talent as potential while another considers it as sustainability [9].

Organizations also need to improve their overall performance at all organizational levels to gain higher value. Sustained higher-value performance is the key to success for any business in this changing business environment. Hence, the concerns of talent management are not only related to hiring talent but rather how to attract, retain and develop talent to ensure business value, sustainability, and success [10].

Besides enhancing performance, talent management also enhances employee engagement and commitment to the organization's activities. Employee interaction with talent management begins by focusing on employee well-being; development, motivation, etc., which are factors that drive employee engagement. Talent management is all about creating the best work environment, the best training program, career development opportunities, and much more. All these aspects of talent management lead to increased employee engagement which in turn will lead to improved performance and ultimately the achievement of organizational goals, mission, and vision [11].

Due to the negative effects of employee turnover, knowing how to attract and retain employees and talents in the organization will increase employee engagement, satisfaction, and loyalty. Talent management is the main tool to achieve these factors to retain talent and avoid all negative effects of employee turnover [10].

The importance of talent management can also be embodied in reducing costs, risks, and customer service. The cost of recruiting new talent requires external resources to find suitable talent, which adds recruitment costs to organizations. Risks of hiring employees for not achieve what the organization expects if they do not have the required talents, and finally, talented people enhance customer confidence through their qualified performance and enhance customer loyalty [12].

The importance of talent management lies in increasing productivity, hiring the right employees, increasing human capital, developing talent, and gaining the advantage of the organization's reputation and brand [13].

7 Methodology

The study used a questionnaire to collect primary data. The questionnaire was divided into three parts; the first part includes the personal and functional data of the study sample. The second part was devoted to measuring the independent variable (OLAP) and included 7 items. The last part measured the dependent variable (Telnet Management) and included 19 items using a five-point Likert scale, which starts from one (strongly disagree) to five (strongly agree). Descriptive and analytical methods were employed. SPSS version 24 was used to analyze the data. Cronbach's Alpha Coefficient, Multiple Linear Regression Analysis, Variance Inflation Factors (VIF, Spearman correlation, Multicollinearity, Normal Distribution, Mean, and Standard Deviation Measures are the most important statistical methods used in this study.

7.1 The Study Population and Sample

The population of the study is the employees in the Orange Jordanian Telecommunication" Company. Based on the company data, (490) employees work in different departments and they form the population size of this study (280) questionnaires distributed by using a google form. About (46) questionnaires were not answered, this represented about (16%) of all questionnaires (84%) of the questionnaires were valid for actual analysis.

7.2 The Validity and Reliability

Different tools can be used to assess validity. Three types of validity were used in this study. Face validity meant that the study took into account all of the notes provided by the arbitrators and made the necessary changes to the questionnaire. Pearson correlation between the statement and the variable was used to assess internal validity. Table 3 displays the correlation coefficient. All of the correlation coefficients are significant (0.01). Finally, construct validity is determined by calculating the correlation coefficient between each variable and the questionnaire's total score. According to Table 1, all correlation coefficients are significant at (0.01). Talent Development has the highest correlation coefficient (0.946), while Data Warehouse has the lowest (0.892)

The current study calculates Cronbach's Alpha for each variable to test its reliability. According to Table 1, all study variables have a Cronbach's alpha value greater than (0.7), with the largest variable having a value of (0.843) and the smallest variable having a value of (0.825). Cronbach's Alpha coefficient for all items is (0.905).

Table 1. Correlation Coefficient between the Variables and Total Degree of Questionnaire.

Variables	Correlation Coefficient	SIG
Complex Calculations	0.910**	0.000
Trend Analysis	0.915**	0.000
Sophisticated Data Modeling	0.892**	0.000
Talent Attraction	0.942**	0.000
Talent development	0.946**	0.000
Talent Retention	0.933**	0.000

** Correlation is significant at the 0.01 level (2-tailed).

Table 2. Reliability of the Questionnaire (Cronbach's Alpha)

Number	Variables	Questions	Cronbach's Alpha
1	Complex Calculations	Q1-Q6	0.825
2	Trend Analysis	Q7-Q13	0.828
3	Sophisticated Data Modeling	Q14-Q20	0.835
4	Talent Attraction	Q21-Q28	0.843
5	Talent development	Q29-Q33	0.830
6	Talent Retention	Q34-Q39	0.842
	All items	36	0.905

8 Hypotheses Testing

8.1 Normal Distribution Test

Table 2 displays the results of the normal distribution test; all skewness variables are close to (0), and the variable is closest to (0) Talent Attraction (–0.249). On the other hand, all kurtosis variables are close to (3), with Talent Attraction being the closest.

Table 3. Normal Distribution of the Independent Variables

	Complex Calculations	Trend Analysis	Sophisticated Data Modeling	Talent Attraction	Talent Development	Talent Retention
Skewness	–0.305	–0.408	–0.483	–0.249	–0.250	–0.253
Kurtosis	–2.821	–2.972	–2.946	–3.046	–3.077	–3.100

8.2 The Multicollinearity

The variance inflation factors and tolerance effect were calculated, from Table 4, it is clear that all independent variables' tolerance effect is more than (0.1). On the other hand, all the VIF are less than (10).

Table 4. Variance Inflation Factor for Independent Variable

Variables	VIF	Tolerance effect
Complex Calculations	4.321	0.231
Trend Analysis	6.322	0.158
Sophisticated Data Modeling	4.570	0.219

8.3 Hypotheses Testing

Multiple linear regressions were used to test the hypothesis at the significance level ($\alpha \leq 0.05$).

Table 5. Results of Multiple Regressions

Dependent Variable	R	R^2	F	DF	SIG	Independent Variable	B	T	Sig
Talent Management	0.884	0.782	274.461	3	0.000	complex calculations	0.380	6.052	0.000
				230		trend analysis	0.416	6.374	0.000
				233		sophisticated data modeling	0.122	1.597	0.112

Table 5 shows the dimensions of the independent variable (OLAP) (complex calculations, trend analysis, and sophisticated data modeling.) Because the calculated (F) value is (274.461), which is greater than the tabular (F) value table, it has an effect on the dependent variable (talent management) (2.69). The R2 value was (0.782), indicating that (OLAP) contributes 78.2% to the dependent variable (Talent Management). As a result, we reject the null hypothesis and accept the alternative hypothesis, as follows:

At Orange Jordan Telecom, there is a statistically significant effect for (OLAP) in terms of its dimensions (complex calculations, trend analysis, and sophisticated data modeling) on Talent Management in terms of dimensions (attracting talent, developing talent, and retaining talent) at the level of significance (0.05).

However, because the calculated t values for the variables (complex calculations 6.052 and trend analysis 6.374) are greater than the tabular value table (1.986), this indicates that the two dimensions of data mining and data storage have a statistically significant effect on the dependent variable (talent management).

However, the third dimension (advanced data modeling) had a lower t (1.597) value than the tabular t (1.986), indicating that this dimension has no statistically significant effect on the dependent variable (talent management). The stepwise regression analysis was used in the study to determine which independent variable dimension has the highest contribution on the dependent variable, as well as to exclude variables that do not have a high contribution. The results of Stepwise Regression are shown in Table 6.

Table 6. Results of Stepwise Regression on Main hypothesis

Number	Variables	F	R	R^2	Sig
1	complex calculations	575.824	0.844	0.712	0.000
2	complex calculations trend analysis	407.681	0.883	0.779	0.000

Stepwise regression classified the dimensions of the independent variables into two groups. The first includes complex accounts, which have the largest contribution to the dependent variable (talent management), estimated at (71.2%). The second group contains complex calculations and trend analysis that contributed to the estimated dependent variable (77.9%). While the third dimension (sophisticated data modeling) was excluded because it does not have a significant impact on talent management.

9 Discussion of Findings

The study's findings revealed that implementing (OLAP) dimensions (complex calculations, trend analysis, and sophisticated data modeling) has a significant impact on talent management in terms of its dimensions (Talent Attraction, Talent Retention, and Talent Development) in Orange Jordanian Telecommunication Company. This implies that OLAP can help with talent management. As the (OLAP) dimensions contribute to achieving talent management by about (78.2%), meaning that a positive change or improvement in the implementation of (OLAP) dimensions by one unit will lead to a change in the level of talent management by (78.2). The result also indicates that trend analysis has the highest contribution among the (OLAP) dimensions in talent management, at about 71.2%. At the level of dimensions, there is a statistically significant impact for the dimensions of the (OLAP) combined on each dimension of talent management individually, as the results show that (OLAP) contributes to attracting talents by about(72%) and that the trend analysis made the highest contribution to attracting talent by about 66.4%. The dimensions of (OLAP) contribute to talent development by about 75.5%. Complex calculations were the largest contribution among these dimensions to talent development, with a rate of (69.3%). Finally, the OLAP dimensions contributed to talent retention by about (72.7%), trend analysis again had the highest contribution among these dimensions to talent retention at (66.4%).

10 Study Limitations

The first limitation of the study was its reliance on the questionnaire in data collection and the consequent potential for bias, while it would have been preferable to rely on actual data from the realities of OLAP implementations in this company. Accordingly, we suggest conducting the same study by adopting actual data from the company's records. Another limitation is that this study focuses on one dimension of business intelligence which is OLAP, while there are other important dimensions such as data warehouse, data mining, reporting, and information transfer that can impact talent management. Hence, we recommend a study on the impact of business intelligence on talent management.

References

1. AlAina, R., Atan, T.: The impact of implementing talent management practices on sustainable organizational performance. Sustainability **12**(20), 8372 (2020)
2. Mathur, S., Lal Gupta, S., Pahwa, P.: Optimizing OLAP Cube for supporting online analytical process (OLAP) and forecasting in banking sector. J. Inf. Technol. Manage. **13**(1), 81–99 (2021)
3. Alnoukari, M.: From online analytical process (OLAP) to big data: the power of analytics. In: Research Anthology on Big Data Analytics, Architectures, and Applications, pp. 823–841. IGI Global (2022)
4. Al Taleb, T.M., Hasan, S., Mahd, Y.Y.: Online analytical processing (OLAP) operation for outpatient healthcare. Iraqi J. Sci. 225–231 (2021)
5. Al Aufi, M.A.: Online analytical process (OLAP) in contemporary organizations: potential opportunities to improve human resource management for competitive advantage. Asian J. Bus. Manage. **4**(4) (2016)
6. https://www.javatpoint.com/what-is-olap
7. Ekelöf, S., Lindberg, N.: The strategic side of global talent management: thematic literature review with a conceptual reasoning (2020)
8. Aarnio, A., Kimber, E.: Talent Management & Strategy: Identifying Patterns through a Multiple Case Study (2016)
9. El Dahshan, M.E., Keshk, L.I., Dorgham, L.S.: Talent management and its effect on organization performance among nurses at shebin el-kom hospitals. Int. J. Nurs. **5**(2), 108–123 (2018)
10. Al-Dalahmeh, M., Héder-Rima, M.: The effect of talent management practices on employee turnover intention in the information and communication technologies (ICTs) sector: case of Jordan. Probl. Perspect. Manag. **18**(4), 59 (2021)
11. Yuniati, E., Soetjipto, B., Wardoyo, T., Sudarmiatin, S., Nikmah, F.: Talent management and organizational performance: the mediating role of employee engagement. Manage. Sci. Lett. **11**(9), 2341–2346 (2021)
12. Biriowu, C.S., Kalio, N.L.: Talent analytics and employee retention in Nigeria organization. Int. J. Hum. Resour. Manag. (IJHRM) **9**, 1–12 (2020)
13. Hamad, Z.: Talent management as a facilitator of organizational intelligence. Manage. Sci. Lett. **9**(6), 809–822 (2019)

Predictors of Health Workers' Organizational Citizenship Behavior in Indonesia Using PLS-SEM Analysis in the Digitalized Healthcare and COVID-19 Post-Pandemic

Michael Christian[1](✉), Yustinus Yuniarto[1], Suryo Wibowo[2], Henilia Yulita[1], and Sumarny Manurung[1]

[1] Universitas Bunda Mulia, Jakarta, Indonesia
michaelchristianid@gmail.com
[2] Indonesia International Institute for Life Sciences, Jakarta, Indonesia

Abstract. Despite the presence of digital technology, which directly or indirectly contributes to the workload of healthcare workers as employees, healthcare organizations are still learning about the importance of employees in providing optimal work performance. The ongoing COVID-19 pandemic has resulted in a routine work pattern that must be modified to meet the field's needs. It is not uncommon for health workers to work outside of their proper roles and capacities in this situation. Working long hours can result in physical exhaustion. Long term, this can cause work stress and will almost certainly affect health workers' motivation.

The goal of this study is to identify and analyze predictors of organizational citizenship behavior (OCB). When using structural modeling with the SEM-PLS approach, organizational commitment is the most important predictor of OCB. Indirectly, work stress and the availability of organizational support are important predictors. Furthermore, as pandemic conditions improve, employee development programs may be suspended. According to the study's findings, employees in both in-role and extra-role positions must be able to present appropriately in light of current circumstances. In this case, collaboration between health workers and organizations can lead to improved joint performance.

Keywords: Organizational Citizenship Behavior · Organizational Commitment · Perceived Organizational Support

1 Introduction

The prolonged COVID-19 pandemic has unavoidably had a negative impact on many business sectors [1, 2], including the health sector. In Indonesia, the community's need for digital and non-digital health services cannot be separated. Previous research has shown that factors such as user satisfaction in the use of digital technology in health services, such as e-health applications, can indirectly support users' health performance [3–5]. Adoption of technology, such as applications, is becoming more acceptable in society, particularly among the younger generation [6, 7]. According to a Deloitte Indonesia

B. A. M. Alareeni and I. Elgedawy (Eds.): *Artificial Intelligence (AI) and Finance*, SSDC 488, pp. 406–415, 2023.
https://doi.org/10.1007/978-3-031-39158-3_39

report [8], various parties, both private and government, provide various aspects of product or service benefits. The COVID-19 pandemic has compelled the government to transform health-care delivery [9], but available health workers are overburdened due to their workload. The issue in this study is how organizational citizenship behavior (OCB) predictors affect health care workers. Thus, the purpose of this research is to identify and analyze the predictors that comprise OCB. This study's writing structure begins with an introduction, then moves on to a literature review and hypothesis development before concluding with results and discussion. The final section discusses the findings, implications, limitations, and recommendations.

2 Literature Review and Hypothesis Development

2.1 Organizational Citizenship Behavior and Organizational Commitment

OCB can be defined as a part of organizational behavior science that is often overlooked [10]. In the meantime, OCB is a choice behavior that is not a formal work requirement but helps the organization run well [11]. This is also called "extra-role behavior" [12]. Organizational commitment involves a strong belief in and acceptance of the organization's goals and values, a willingness to go to great lengths for the organization, and a strong desire to stay with the organization [13, 14].

2.2 Job Stress, Employee Development, and Perceived Organizational Support

Work stress is a condition that happens when people have to deal with their jobs. It makes people act differently than they would normally [15]. Workplace stress is caused by a lot of work and a lot of responsibilities, changes in work systems, an unsupportive work environment, working long hours with short breaks, and a lot of work, all of which are bad for employee development [16–21]. Meanwhile, perceived organizational support (POS) is the idea that the organization appreciates a person's work, which is usually paid for by organizational welfare guarantees and increasing organizational support [22, 23].

2.3 The Relationship Between Variables and Hypothesis Development

Job Stress, Organizational Commitment, and Organizational Citizenship Behavior. Job stress had a significant negative impact on employee POS and organizational commitment [24]. Employees' physical and psychological conditions have a significant impact on organizational commitment. That study also discovered that POS had a completely negative mediating effect on the relationship between job stress and organizational commitment. Based on these explanations, this study proposes two hypotheses: (H1) Job stress has a significant impact on organizational commitment, and (H2) Job stress, as mediated by organizational commitment, has a significant impact on organizational citizenship behavior.

Perceived Organizational Support, Organizational Commitment, and Organizational Citizenship Behavior. For workers, POS had a significant effect on organizational commitment (OC) [25]. Despite the fact that POS has no effect on one of the organizational commitment indicators, namely continuity commitment, it does have an effect on affective and normative commitment [26]. Based on these explanations, the following hypothesis is proposed in this study: (H3) Perceived organizational support has a significant impact on organizational commitment, and (H4) perceived organizational support has a significant impact on organizational citizenship behavior as mediated by organizational commitment.

Employee Development, Organizational Commitment, and Organizational Citizenship Behavior. Employee development is a person's long-term personal and professional growth [22]. Millennials had much lower levels of continuity commitment than Gen Xers, but there were no differences in normative or emotional commitment [27]. When opportunities for employee development are available, Gen Xers are more dedicated to the affective and normative aspects than millennials. Based on these explanations, this study proposes the following hypothesis: (H5) Employee development has a significant impact on organizational commitment, and (H6) employee development has a significant impact on organizational citizenship behavior as mediated by organizational commitment.

Organizational Commitment and Organizational Citizenship Behavior. Organizational commitment has a significant positive effect on organizational citizenship behavior among hospital nurses, both indirectly and directly as a moderator of workplace spirituality [28]. Similarly, organizational commitment has a significant influence on hospital nurses' organizational citizenship behavior [29]. Based on these explanations, the following hypothesis is proposed in this study: (H7) Organizational commitment has a significant impact on organizational citizenship behavior.

3 Methods

3.1 Research Design

A survey is used to collect data from participants in this quantitative study. Between December 2022 and February 2023, an online questionnaire was distributed at random. Figure 1 depicts the research conceptual framework used in this study, which includes partial least squares—structural equation modeling (PLS-SEM). PLS-SEM is a multivariate analysis technique used to create and test statistical models, most commonly causal models. It is a variance-based alternative method of analysis to structural equation modeling.

3.2 Sample Size

This research sample will only include Indonesian health workers who will still be working in 2022. In Indonesia, this is called the COVID-19 post-pandemic period. According to the next criterion, the health workers in this study must also have worked directly with COVID-19-exposed patients during that time period. The required sample size for this study was calculated by multiplying the total number of items (28), by 5 (the minimum sample size), and then by 10 (the maximum sample size) [30–32]. The criteria were met by 136 (95.77%) of the 142 questionnaires from participants that were collected for this study. Participants in the study are described in Table 1.

Table 1. Participant's Profile.

Description	N	%	Description	N	%	Description	N	%	Description	N	%
Gender	71	52.21%	**Age**	7	5.15%	**Profession**	67	49.26%	**Area**	61	44.85%
Female	65	47.79%	Less than	49	36.03%	Doctor	25	18.38%	Jakarta	27	19.85%
Male			30	68	50%	Nurse	21	15.44%	Banten	23	16.91%
			30–40	12	8.82%	Midwife	16	11.76%	West Java	11	8.09%
			41–50			Pharmacist	7	5.15%	East Java	9	6.62%
			More than			Others			Central	5	3.68%
			50						Java		
									Others		

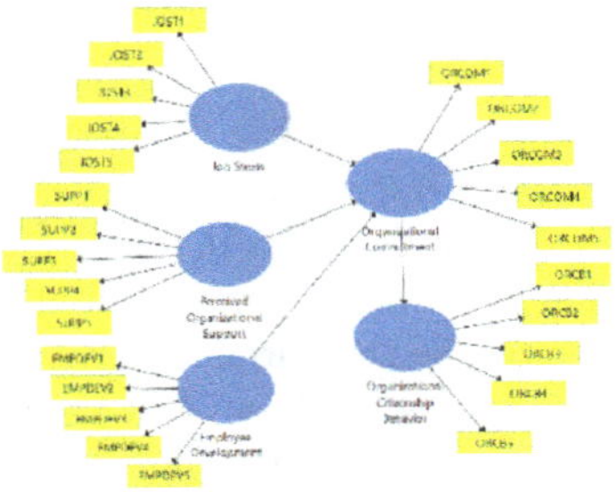

Fig. 1. Conceptual Framework.

3.3 Variable Measurement

There is one endogenous variable, one mediator variable, and three external variables in this study. A 1–5 Likert scale is used to rate each variable (strongly disagree to strongly agree). OCB is made up of five things: (1) During the pandemic, I am working hard to be a good health worker; (2) I am committed to the company for which I work; (3) I am not afraid to compliment or encourage co-workers; (4) I sometimes volunteer to do extra work at work; and (5) I try to uphold the values of the organization for which I work [33–35]. Organizational commitment means: (1) I am not worried about leaving the organization where I am currently working; (2) Leaving the organization where I am currently working would change many parts of my life; (3) I will not lose much money if I have to leave the organization where I am currently working; and (4) Currently, the necessities of life are the primary reason for my employment in the organization where I am currently employed; and (5) One of the main reasons I am staying at my current job is that leaving would mean making a lot of personal sacrifices, and other jobs may not be worth the overall benefits I get at my current job [36, 37].

Next, job stress is made up of five parts: (1) During the pandemic, I found it difficult to relax even for a moment while performing routine tasks at work; (2) I am frequently nervous while working during a pandemic; (3) During the pandemic, I was frequently irritable at work; (4) During a pandemic, I tend to overreact to workplace situations; and (5) During a pandemic, I find it difficult to tolerate interruptions in my work [38, 39]. POS consists of five parts: (1) My current employer is interested in what I have to say; (2) My employer is genuinely concerned about my well-being; (3) The current organization I work for is in line with my life goals or values; (4) If an opportunity arises, the organization where I currently work is eager to put my skills to use and maximize my potential; and (5) My current employer is willing to assist me if I require specific assistance [40]. Employee development consists of five parts: (1) My academic education has increased my competence and ability to work at work; (2) I have made sufficient progress toward my career objectives thus far; (3) My academic education has shaped my self-development motivation, particularly at work; (4) I see professional development as a positive thing; and (5) In my opinion, workplace career development must be done fairly (unbiased) [41].

4 Results and Discussion

4.1 Results

All items and research constructs met the requirements for reliability and validity. All variables in this study have composite reliability (CR) and Cronbach's alpha (CA) values greater than 0.7, indicating that all variables are reliable. Furthermore, all items have a loading factor (LF) greater than 0.7, and all variables have an average variance extracted (AVE) greater than 0.5, indicating that the items and variables in this study are valid. Fornell-Larcker's shows results that meet the requirements for each of the constructs: organizational citizenship behavior (ORCB), organizational commitment (ORCOM), job stress (JOST), perceived organizational support (SUPP), and employee development (EMPDEV).

Specifically, as the PLS-Algorithm results show, OCB has LF for each item: ORCB1 = 0.911; ORCB2 = 0.890; ORCB3 = 0.924; ORCB4 = 0.905; ORCB5 = 0.837. Next, organizational commitment has a LF for each item: ORCOM1 = 0.846; ORCOM2 = 0.702; ORCOM3 = 0.811; ORCOM4 = 0.723; and ORCOM5 = 0.744. Job stress has a LF for each item: JOST2 = 0.791; JOST3 = 0.925; JOST4 = 0.932; JOST5 = 0.937. POS has a LF for each item: SUPP1 = 0.838; SUPP2 = 0.853; SUPP4 = 0.798; and SUPP5 = 0.827. Employee development has a LF for each item: EMPDEV1 = 0.889; EMPDEV2 = 0.916; EMPDEV3 = 0.850; EMPDEV4 = 0.887; and EMPDEV5 = 0.908.

OCB has an AVE of 0.799, a CR of 0.952, and a CA of 0.937. Organizational commitment has an AVE of 0.588, a CR of 0.877, and a CA of 0.825. Job stress has an AVE of 0.807, a CR of 0.943, and a CA of 0.919. POS has an AVE of 0.688, a CR of 0.898, and a CA of 0.849. The AVE for employee development is 0.793, the CR is 0.950, and the CA is 0.935. Fornell-results Larcker's results show that OCB to OCB is equal to 0.894; organizational commitment to organizational commitment is equal to 0.767; job stress to job stress is equal to 0.898; POS to POS is equal to 0.829; and employee development to employee development is equal to 0.890.

The R-Square for organizational commitment is 0.741, and the R-Square for organizational citizenship behavior is 0.639. This study shows how the use of exogenous variables like job stress, perceived organizational support, and employee development can explain an organizational commitment score of 74.1%, which is strong. Also, this study says that the use of exogenous variables like job stress, perceived organizational support, employee development, and organizational commitment is successful in explaining organizational citizenship behavior by 63.9%, which is a moderate value.

In Table 2, the P-value shows the results of testing a hypothesis. This study agrees with three of the four direct hypotheses (H1, H3, and H7), but not with the fourth (H5). Also, two of the three indirect hypotheses (H2 and H4) are accepted, and one is rejected (H6).

Table 2. Hypothesis Testing.

Hypothesis	Standard deviation	T-Statistic	P-value	Remark
Job Stress → Organizational Commitment	0.063	8.772	0.000	H1 accepted
Job Stress→ Organizational Commitment →Organizational Citizenship Behavior	0.054	8.156	0.000	H2 accepted
Perceived Organizational Support → Organizational Commitment	0.094	3.127	0.002	H3 accepted
Perceived Organizational Support → Organizational Commitment → Organizational Citizenship Behavior	0.075	3.108	0.002	H4 accepted
Employee Development →Organizational Commitment	0.095	0.933	0.351	H5 rejected
Employee Development → Organizational Commitment → Organizational Citizenship Behavior	0.077	0.916	0.360	H6 rejected
Organizational Commitment → Organizational Citizenship Behavior	0.035	22.602	0.000	H7 accepted

4.2 Discussion

The results of this study, especially when it came to organizational commitment, showed that stress at work and organizational support are important factors in predicting organizational commitment. These results agree with Saadeh's [24] finding that employees with a high level of commitment have more stress than employees with a low level of commitment. It was also said that stress is a bad reaction to things that seem too hard or beyond a person's abilities. These results are also directly related to the finding that perceived organizational support has a big impact on organizational commitment among Amman, Jordan, workers [25]. On the other hand, perceived organizational support did not affect one of the organizational commitment indicators [26].

According to the OCB findings, organizational commitment is the most important predictor, but indirect work stress variables of health workers and organizational support also play a role. Despite the fact that health services have changed dramatically as a result of digital technology's sophistication, pandemic conditions are unrelated to employee development in efforts to improve work performance both individually and in organizations. This study's findings are consistent with the finding that younger workers have a much lower level of continuity commitment, but there is no difference in normative or affective commitment [27]. When training opportunities are available, older workers are more committed to affective and normative aspects than millennials. An individual's attitude will be determined by the usefulness factor received. This finding is also consistent with the findings of commitment research, which has been linked to organizational efforts to continue learning [43]. The study adds to previous research that found organizational commitment has a significant effect on organizational citizenship behavior in health workers [29].

5 Conclusion, Implications, Limitations, and Recommendations

According to the findings of this study, organizational commitment has a significant impact on OCB. Meanwhile, as a result of work stress and organizational support, organizational commitment developed. In contrast, employee development has no effect on organizational commitment. Organizational commitment has an indirect effect on work stress and organizational support for OCB. Employee development has no effect on OCB, even if it is mediated by organizational commitment. This supports the notion that during and even after a pandemic, organizations prioritize survival strategies over developing employees' career paths. The theoretical concept of organizational commitment, which serves as its direct antecedent in this case, strengthens the role of OCB as both an in-role and an extra-role. The workload and working conditions, which can affect employees' physical abilities and have an impact on fatigue and work stress factors at any time, also shape this role. Due to the current situation, the distribution of participants is relatively limited, so the number of questionnaires returned in this study is limited. Furthermore, the research method employed is solely focused on the relationships between exogenous and endogenous variables, with no participant profiles serving as a control variable.

Acknowledgement. The authors would like to thank Universitas Bunda Mulia for supporting this research.

References

1. Christian, M., Dewi, D., Rembulan, G. D., Indriyarti, E. R., Wibowo, S., Yuniarto, Y.: Business performance determinants of salted fish distribution in kapuk during the COVID-19. J. Distribution Sci. **19**(6), 29–39 (2021). https://doi.org/10.15722/jds.19.6.202106.29

2. Christian, M., Haris, K., Indriyarti, E.R., Wibowo, S., Sunarno, S.: Service distribution strategy on business performance of padang restaurants in North Jakarta. J. Distribution Sci. **19**(12), 57–69 (2021). https://doi.org/10.15722/jds.19.12.202112.57
3. Atinga, R.A., Abor, P.A., Suleman, S.J., Anaba, E.A., Kipo, B.: e-health usage and health workers' motivation and job satisfaction in Ghana. PLoS ONE **15**(9), 1-10 (2020). https://doi.org/10.1371/journal.pone.0239454
4. Sakka, Y., Qarashay, D.: The impact of using e- health on patient satisfaction from a physician-patient relationship perspective. Journal of Critical Rev. **7**(19), 3034–3041 (2020). https://doi.org/10.31838/jcr.07.19.367
5. Christian, M., Wibowo, S., Indriyarti, E.R., Sunarno, S., Yuniarto, Y.: Do service quality and satisfaction affect the intention of using application-based land transportation? a study on generation YZ in Jakarta. In: Studies in Systems, Decision and Control, Vol. 216 (2022). https://doi.org/10.1007/978-3-031-10212-7_60
6. Indriyarti, E.R., Christian, M., Yulita, H., Ruminda, M., Sunarno, S., Wibowo, S.: Online food delivery app distribution and determinants of jakarta's gen z spending habits. Journal of Distribution Sci. **20**(7), 73–86 (2022). https://doi.org/10.15722/jds.20.07.202207.73
7. Christian, M., Girsang, L., Yulita, H.: Measuring ease of use aspects of shopee usage behavior during pandemic using PLS-SEM approach. In: Proceedings of the 1st International Conference on Emerging Issues in Humanity Studies and Social Sciences - ICE-HUMS, pp. 192–197 (2022). https://doi.org/10.5220/0010750000003112
8. Deloitte, I., Bahar Law, F., Chapters, I.: 21st Century Health Care Challenges: A Connected Health Approach Megatrends in Health care (2019). https://www2.deloitte.com/content/dam/Deloitte/id/Documents/public-sector/id-gps-ehealth-publication-Indonesia.pdf
9. Christian, M., Pardede, R., Indriyarti, E.R.: Generation Z in Jakarta's attitude towards COVID-19 ad distribution on youtube. J. Distribution Sci. **20**(3), 13–22 (2022). https://doi.org/10.15722/jds.20.03.202203.13
10. Nurjanah, S., Pebianti, V., Handaru, A.W.: The influence of transformational leadership, job satisfaction, and organizational commitments on organizational citizenship behavior (OCB) in the inspectorate general of the ministry of education and culture. Cogent Business & Manage. **7**(1), 1–12 (2020). https://doi.org/10.1080/23311975.2020.1793521
11. Robbins, S.P., Judge, T.A.: Essentials of Organizational Behavior (14th ed.). Pearson (2018)
12. Podsakoff, P.M., MacKenzie, S.B., Paine, J.B., Bachrach, D.G.: Organizational citizenship behaviors: a critical review of the theoretical and empirical literature and suggestions for future research. J. Manag. **26**(3), 513–563 (2000). https://doi.org/10.1177/014920630002600307
13. Dajani, M.A.Z.: The impact of employee engagement on job performance and organisational commitment in the Egyptian banking sector. J. Business Manage. Sci. **3**(5), 138–147 (2015). https://doi.org/10.12691/jbms-3-5-1
14. Suharto, S., Suyanto, S., Hendri, N.: The impact of organizational commitment on job performance. International Journal of Economics and Business Administration, VII (2), 189–206 (2019). https://doi.org/10.35808/ijeba/227
15. Robbins, S.P., Coulter, M.: Principles of management. Translated by Seyyed Mohammad Arabi and Mohammed Ali Hamid Rafiee and Behrouz Asrari Ershad, Fourth Edition, Tehran: Office of Cultural Studies (2007)
16. Materson, I.: Stress at work: A managerial perspective. Scott, Foresman, Glenview, IL (1980)
17. Shimazu, A., Kosugi, S.: Job stressors, coping, and psychological distress among Japanese employees: interplay between active and non-active coping. Work Stress. **17**(1), 38–51 (2003). https://doi.org/10.1080/0267837031000106862
18. Asfaw, A.M., Argaw, M.D., Bayissa, L.: The impact of training and development on employee performance and effectiveness: a case study of district five administration office, Bole Sub-City, Addis Ababa, Ethiopia. J. Human Resource and Sustainability Studies **3**(4), 188–202 (2015). https://doi.org/10.4236/jhrss.2015.34025

19. Ampomah, P.: The effect of training and development on employee performance in a private tertiary institution in Ghana. (Case Study: Pentecost University College (Puc) - Ghana). Asian Journal of Social Sciences and Management Studies **3**(1), 29–33 (2016). https://doi.org/10.20448/journal.500/2016.3.1/500.1.29.33
20. Saputri, P., Lorensa, D., Asriani, A., Zainurossalamia Za, S.: The influence of training and development to employee performance. Int. J. Economics, Business and Accounting Research (IJEBAR) **4**(1), 133–140 (2020). https://doi.org/10.29040/ijebar.v4i01.928
21. Ahmad, I., Din, S.: Evaluating training and development. Gomal J. Medical Sci. **7**, 165–166 (2009)
22. Jehanzeb, K.: Does perceived organizational support and employee development influence organizational citizenship behavior? European J. Training Dev. **44**(6/7), 637–657 (2020). https://doi.org/10.1108/EJTD-02-2020-0032
23. Eisenberger, R., Armeli, S., Rexwinkel, B., Lynch, P.D., Rhoades, L.: Reciprocation of perceived organizational support. J. Appl. Psychol. **86**(1), 42–51 (2001). https://doi.org/10.1037/0021-9010.86.1.42
24. Saadeh, I.M., Suifan, T.S.: Job stress and organizational commitment in hospitals: the mediating role of perceived organizational support. Int. J. Organ. Anal. **28**(1), 226–242 (2020). https://doi.org/10.1108/IJOA-11-2018-1597
25. Albalawi, A.S., Naugton, S., Elayan, M.B., Sleimi, M.T.: Perceived Organizational support, alternative job opportunity, organizational commitment, job satisfaction and turnover intention: a moderated- mediated model. Organizacija **52**(4), 310–324 (2019). https://doi.org/10.2478/orga-2019-0019
26. Battistelli, A., Galletta, M., Vandenberghe, C., Odoardi, C.: Perceived organisational support, organisational commitment, and self-competence among nurses: a study in two Italian hospitals. J. Nurs. Manag. **24**(1), E44–E53 (2016). https://doi.org/10.1111/jonm.12287
27. Glazer, S., Mahoney, A.C., Randall, Y.: Employee development's role in organizational commitment: a preliminary investigation comparing generation X and millennial employees. Ind. Commer. Train. **51**(1), 1–12 (2019). https://doi.org/10.1108/ICT-07-2018-0061
28. Kazemipour, F., Amin, S.M., Pourseidi, B.: Relationship between workplace spirituality and organizational citizenship behavior among nurses through mediation of affective organizational commitment. J. Nurs. Scholarsh. **44**(3), 302–310 (2012). https://doi.org/10.1111/j.1547-5069.2012.01456.x
29. Lin, C.-T., Chang, C.-S.: Job satisfaction of nurses and its moderating effects on the relationship between organizational commitment and organizational citizenship behaviors. Res. Theory for Nursing Practice: An Int. J. **29**(3), 226–244 (2015). https://doi.org/10.1891/1541-6577.29.3.226
30. Hair, et al.: Multivariate Data Analysis (7th ed.). Pearson Education Limited (2014)
31. Wolf, E.J., Harrington, K.M., Clark, S.L., Miller, M.W.: Sample size requirements for structural equation models: an evaluation of power, bias, and solution propriety. Educ. Psychol. Measur. **73**(6), 913–934 (2013). https://doi.org/10.1177/0013164413495237
32. Willaby, H.W., Costa, D.S.J., Burns, B.D., MacCann, C., Roberts, R.D.: Testing complex models with small sample sizes: A historical overview and empirical demonstration of what Partial Least Squares (PLS) can offer differential psychology. Personality Individ. Differ. **84**(10), 73–78 (2015). https://doi.org/10.1016/j.paid.2014.09.008
33. Dalal, R.S., Lam, H., Weiss, H.M., Welch, E.R., Hulin, C.L.: A within-person approach to work behavior and performance: concurrent and lagged citizenship-counter productivity associations, and dynamic relationships with affect and overall job performance performance. Acad. Manag. J. **52**(5), 1051–1066 (2009). https://doi.org/10.5465/AMJ.2009.44636148
34. Ariani, D.W.: The relationship between employee engagement, organizational citizenship behavior, and counterproductive work behavior. Int. J. Bus. Adm. **4**(2), 46–56 (2013). https://doi.org/10.5430/ijba.v4n2p46

35. Chu, L.-C.: The influence of compassion fatigue on job performance and organizational citizenship behaviors: the moderating effect of person-job fit. J. Nursing Scholarship: An Official Publication of Sigma Theta Tau International Honor Society of Nursing **53**(4), 500–510 (2021). https://doi.org/10.1111/jnu.12644
36. Allen, N.J., Meyer, J.P.: The measurement and antecedents of affective, continuance and normative commitment to the organization. J. Occup. Psychol. **63**(1), 1–18 (1990). https://doi.org/10.1111/j.2044-8325.1990.tb00506.x
37. Afsar, B., Shahjehan, A., Cheema, S., Javed, F.: The effect of perceiving a calling on pakistani nurses' organizational commitment, organizational citizenship behavior, and job stress. J. Transcult. Nurs. **29**(6), 540–547 (2018). https://doi.org/10.1177/1043659618761531
38. Lovibond, P.F., Lovibond, S.H.: The structure of negative emotional states: comparison of the depression anxiety stress scales (DASS) with the beck depression and anxiety inventories. Behav. Res. Ther. **33**(3), 335–343 (1995). https://doi.org/10.1016/0005-7967(94)00075-U
39. Mushtaq, M., Najam, N.: Depression, anxiety, stress and demographic determinants of hypertension disease. Pakistan Journal of Medical Sciences **30**(6), 1293–1298 (2014). https://doi.org/10.12669/pjms.306.5433
40. Afsar, B., Badir, Y.F.: Person-organization fit, perceived organizational support, and organizational citizenship behavior: the role of job embeddedness. J. Hum. Resour. Hosp. Tour. **15**(3), 252–278 (2016). https://doi.org/10.1080/15332845.2016.1147936
41. Rahman, W., Nas, Z.: Employee development and turnover intention: theory validation. European J. Training and Dev. **37**(6), 564–579 (2013). https://doi.org/10.1108/EJTD-May-2012-0015
42. Indriyarti, E.R., Christian, M., Yulita, H., Aryati, T., Arsjah, R.J.: Digital bank channel distribution: predictors of usage attitudes in jakarta's gen Z. Journal of Distribution Science **21**(2), 21–34 (2023). https://doi.org/10.15722/jds.21.02.202302.21
43. Yulita, H., Jonathan, V., Yuniarto, Y., Christian, M., Indriyarti, E., Wibowo, S.: The role of affective commitment in mediating the relationship between authentic leadership and organizational learning at chicken-based food processing company in Indonesia. In: Proceedings of the 1st International Conference on Emerging Issues in Humanity Studies and Social Sciences - ICE-HUMS, pp. 453–457 (2022). https://doi.org/10.5220/0010753700003112

E – Promotion Tools and Its Effect on Consumers Purchase Decisions: A Case Study from Jordan

Iyad Khanfar(✉), Mohammad Nael Rabee, Maenal Sager, and Yousef Jameel Al Maraira

Zarqa City, Jordan
i.khanfar@zu.edu.jo

Abstract. This study investigates the impact of e-promotion tools on consumers' purchasing decisions in the context of Carrefour, a leading supermarket chain in Jordan. A convenience sample of 425 Carrefour customers was targeted, and 33incomplete questionnaires were excluded from analysis, resulting in a final sample size of 392 for data analysis. The study uses a multiple regression analysis to examine the relationship between social media, email marketing, website design, and electronic advertising, and their effect on consumers' purchase decisions. The results show that social media and website design have a significant positive impact on consumers' purchasing decisions, while email marketing and electronic advertising did not demonstrate a significant impact. These findings offer insights for Carrefour and other retail companies in Jordan to enhance their marketing strategies by effectively using e-promotion tools.

Keywords: promotion tools · social media · email marketing · website design · electronic advertising · Purchasing decisions · Carrefour supermarket · Jordan

1 Introduction

The rise of digital technology has paved the way for companies to reach their target audience through various online marketing tools. E-promotion tools, such as social media, email marketing, website design, and electronic advertising, have been widely used to promote products and services to consumers [1]. The use of social media as a marketing tool has increased significantly in recent years [1]. Social media platforms, such as Facebook, Instagram, and Twitter, offer a cost-effective way for companies to reach their target audience [2]. Email marketing, on the other hand, has been shown to be an effective way to maintain customer relationships and increase customer loyalty [3]

In addition, website design plays a significant role in attracting and retaining customers [2]. A well-designed website can enhance the customer experience and increase the likelihood of making a purchase [4].

. Electronic advertising, such as banners and pop-ups, has also been shown to be an effective tool for promoting products and services [4].

B. A. M. Alareeni and I. Elgedawy (Eds.): *Artificial Intelligence (AI) and Finance*, SSDC 488, pp. 416–424, 2023.
https://doi.org/10.1007/978-3-031-39158-3_40

The purpose of this case study is to examine the impact of these e-promotion tools on consumers' purchasing decisions in Jordan. The independent variables include social media, email marketing, website design, and electronic advertising, while the dependent variable is consumers' purchasing decisions. By examining the relationship between these variables, this study aims to provide insights into how companies can effectively use e-promotion tools to increase their sales and revenue.

Study Problem and Questions

The problem addressed in this study is to investigate the impact of e-promotion tools, including social media, email marketing, website design, and electronic advertising, on consumers' purchasing decisions of products or services in the Jordanian market. The increasing use of online marketing and the competition among companies to attract customers in the digital world has made it necessary to explore the effectiveness of these tools in influencing consumers' behavior.

The research questions that will guide this study include:

1. How does social media marketing affect consumers' purchasing decisions in the Jordanian market?
2. What is the effect of email marketing on consumers' purchasing decisions in the Jordanian market?
3. To what extent does website design affect consumers' purchasing decisions in the Jordanian market?
4. What is the effect of electronic advertising on consumers' purchasing decisions in the Jordanian market?

Answering these research questions will help provide insights into the effectiveness of e-promotion tools in influencing consumers' purchasing decisions and help companies to develop effective online marketing strategies to improve sales and gain a competitive advantage in the Jordanian market.

2 Literature Review

The independent variables of this study include social media, email marketing, website design, and electronic advertising, which are all crucial components of e-promotion.

Social media has emerged as an essential tool for marketers to reach and engage with their target audience [5]. Social media platforms like Facebook, Twitter, Instagram, and LinkedIn have become increasingly popular among consumers, making them a valuable marketing channel for businesses. Research has shown that social media marketing positively impacts consumer purchase decisions [2].

One of the key benefits of social media marketing is its ability to increase brand awareness [6]. By regularly posting engaging content, businesses can increase their visibility and reach a wider audience, ultimately leading to increased brand recognition and customer loyalty. Social media marketing also allows businesses to interact with their customers in real-time, responding to queries and addressing concerns promptly, which can lead to a better customer experience [7].

Email marketing has been widely adopted by businesses as a cost-effective and efficient method to reach out to customers and increase sales. A study by [8] found that email marketing positively affects consumer attitudes towards brands, which in turn leads to increased purchase intention. Furthermore, the study revealed that personalized and targeted emails were more effective in generating positive attitudes and purchase intentions than generic emails.

In a similar study, [9] explored the impact of email marketing on customer loyalty in the fashion industry. The authors found that personalized emails, including exclusive offers and promotions, significantly increase customer satisfaction and loyalty. Additionally, the study revealed that frequency and timing of emails are important factors that influence customer perception of email marketing.

Website design has become a crucial aspect of businesses in the digital era. A well-designed website not only enhances a company's image but also affects customers' attitudes and behaviors. A study [10] found that website design elements such as layout, color, and font significantly influence users' perceptions of website usability, visual appeal, and satisfaction. In addition, [11] highlighted the importance of website speed and functionality in increasing website engagement and reducing bounce rates.

Moreover, website design can also impact customers' trust and credibility towards a company. According to a study [12], website design characteristics such as visual appeal, layout, and interactivity significantly influence consumers' perceived trust and credibility towards a website. Furthermore, [13] found that incorporating social proof elements such as customer reviews and ratings on a website can enhance customers' trust and perceived credibility.

Finally, Electronic advertising, also known as online advertising or digital advertising, refers to the use of the internet and digital media channels to promote products, services or brands. It has become a popular form of advertising due to the rapid growth of digital media and the increasing amount of time people spend online. Electronic advertising includes various types of advertisements such as display ads, search engine ads, social media ads, video ads, and mobile ads.

A study conducted by [14] investigated the effects of electronic advertising on consumers' purchase intention. The findings indicated that electronic advertising positively influenced consumers' purchase intention, and this effect was even stronger when the advertisements were perceived as relevant and informative.

Another study [15]. Explored the impact of electronic advertising on brand awareness and purchase behavior. The results showed that electronic advertising increased brand awareness and positively influenced consumers' purchase behavior, especially when the advertising was personalized and targeted to individual preferences.

Recent studies have shown that consumer purchase decisions are influenced by various factors, such as brand loyalty, perceived value, and product quality (Khan et al., [16, 17]). In addition to these traditional factors, consumers today are increasingly being influenced by digital marketing tools, such as social media and email marketing [18]. Furthermore, online reviews and ratings have become a major source of information for consumers in their purchase decisions [19].

Another important factor that affects consumers' purchase decisions is their perception of trust and credibility in a brand or product. This perception can be influenced by various factors, including the brand's reputation, the quality of the product or service, and the perceived level of social responsibility [20]. In addition, consumers' trust in a brand can also be influenced by their level of engagement with the brand, such as through online interactions with customer service or social media content [20].

Overall, consumers' purchase decisions are influenced by a complex set of factors, including traditional factors such as product quality and value, as well as newer factors such as digital marketing tools, emotions, and cultural factors. It is important for marketers to understand these factors in order to develop effective marketing strategies that appeal to their target audience.

Research Hypotheses:

Before conducting the study, a comprehensive literature review was conducted to develop the research hypotheses.

1. H1: Social media has a positive effect on consumers' purchasing decisions in Carrefour supermarkets in Jordan.
2. H2: Email marketing has a positive effect t on consumers' purchasing decisions in Carrefour supermarkets in Jordan.
3. H3: Website design has a positive effect on consumers' purchasing decisions in Carrefour supermarkets in Jordan.
4. H4: Electronic advertising has a positive effect on consumers' purchasing decisions in Carrefour supermarkets in Jordan.
5. H5: The combined effect of e-promotion tools (social media, email marketing, website design, and electronic advertising) has a significant impact on consumers' purchasing decisions in Carrefour supermarkets in Jordan (Fig. 1).

Study model

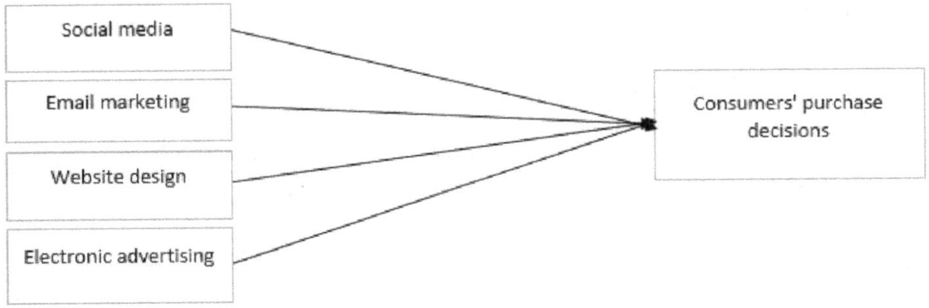

Fig. 1. .

Study Methodology

The present study utilized a quantitative research methodology, employing a convenience sampling technique to collect data from 425 Carrefour customers in Jordan. Following the removal of incomplete or invalid responses, a final sample size of 392 was obtained for analysis. The study utilized a self-administered questionnaire, including demographic questions and items measuring participants' perceptions of e-promotion tools (social media, email marketing, website design, and electronic advertising) and their impact on purchasing decisions.

Data collected from the survey was analyzed using descriptive statistics, reliability analysis, correlation analysis, and multiple regression analysis to examine the relationships between independent and dependent variables [21, 22].

While the study has some limitations, such as the potential bias resulting from the use of a convenience sampling method, which limits the generalizability of the findings to other populations [23] it has several strengths. For example, the study focused on a single retail store and used a larger sample size compared to previous research in this area [24].

The results of the study provide valuable insights into the impact of e-promotion tools on customers' purchasing decisions, which can inform marketing strategies for retailers. However, further research is needed to replicate and expand on these findings in different retail settings and with larger sample sizes [25].

Survey designing

The survey used in this study was self-administered and consisted of two parts. The first part included demographic questions, such as age, gender, education, and income. The second part consisted of items measuring participants' perceptions of e-promotion tools (social media, email marketing, website design, and electronic advertising) and their impact on purchasing decisions. The survey items were adapted from previous studies and measured using a 5-point Likert scale ranging from strongly disagree to strongly agree.

The survey was reviewed by experts in the field and pretested on a small sample of participants before being finalized. The data was collected using a convenience sampling technique from 425 Carrefour customers in Jordan, and a final sample size of 392 was obtained after removing incomplete or invalid responses [26] (Table 1).

Table 1. Reliability Analysis for Independent Variables

Variables	Item Number	Cronbach's Alpha
Email Marketing	6	0.78
Social Media Marketing	6	0.84
Website Design	6	0.73
Electronic Advertising	6	0.76

The reliability analysis for the independent variables, including email marketing, social media marketing, website design, and electronic advertising, was conducted to evaluate the consistency and stability of the survey instrument. The Cronbach's alpha values for each variable ranged from 0.74 to 0.83, indicating a high level of internal consistency and reliability [27]. A value greater than 0.70 is generally considered to be acceptable.

The results suggest that the survey instrument used in this study is reliable and consistent in measuring the independent variables. This provides confidence in the results obtained from the multiple regression analysis, which examines the relationship between these e-promotion tools and consumers' purchasing decisions, [22].

Overall, the reliability analysis serves as a useful tool in evaluating the quality of the survey instrument and the validity of the results obtained. It provides assurance that the survey instrument is a reliable tool for measuring the intended constructs and can be used to support the study's research hypotheses [28] (Table 2).

Table 2. Means and Standard Deviations of Independent and Dependent Variables

Variables	Mean	Standard Deviation
Email Marketing	3.67	0.92
Social Media Marketing	4.15	0.87
Website Design	3.85	0.88
Electronic Advertising	3.29	0.94
Customers' Purchasing Decisions	4.02	0.80

According to the results, social media marketing had the highest mean score (M = 4.15) and the lowest standard deviation (SD = 0.87), indicating that respondents generally had a positive perception of social media marketing and were consistent in their responses. Conversely, electronic advertising had the lowest mean score (M = 3.29) and the highest standard deviation (SD = 0.94), suggesting that respondents had a more varied opinion about electronic advertising [24].The mean score for customers' purchasing decisions was 4.02, indicating that, on average, respondents were positively influenced by e-promotion tools in their purchasing decisions. The standard deviation of 0.80 suggests that respondents' opinions on this variable were relatively consistent [24].

The multiple regression analysis was conducted to test the main hypothesis of the study, which is "E-promotion tools significantly influence customers' purchasing decisions." The independent variables in the study were email marketing, social media marketing, website design, and electronic advertising, while the dependent variable was customers' purchasing decisions.

The results of the multiple regression analysis are presented in Table 3 below:

Table 3. Results of Multiple Regression Analysis

Model	R	R Square	Adjusted R Square	Standard Error	F Value	Sig
1	.775	.601	.595	.522	93.184	.000

Table 3 presents the results of the multiple regression analysis. The model was found to be significant, with an R Square value of .601 indicating that 60.1% of the variance in customers' purchasing decisions could be explained by the independent variables. The Adjusted R Square value of .595 suggests that the model was a good fit for the data and that the independent variables were strong predictors of the dependent variable.

The F Value of 93.184 was statistically significant ($p < .001$), indicating that the model was a good fit for the data and that at least one of the independent variables significantly predicted customers' purchasing decisions. These results suggest that e-promotion tools play a crucial role in influencing customers' purchasing decisions in Jordan.

The standardized coefficients (Beta) for each independent variable are presented in Table 4. Email marketing ($\beta = .223$, $p < .01$), social media marketing ($\beta = .370$, $p < .001$), and website design ($\beta = .184$, $p < .05$) were found to have a positive influence on customers' purchasing decisions. In contrast, electronic advertising ($\beta = -.128, p < .05$) had a negative influence on customers' purchasing decisions. These findings highlight the importance of using effective e-promotion tools to positively influence customers' purchasing decisions in Jordan.

The standardized coefficients (Beta) for each independent variable are presented in Table 4 below:

Table 4. Standardized Coefficients

Independent Variable	Beta
Email Marketing	.440
Social Media Marketing	.359
Website Design	.284
Electronic Advertising	.192

All independent variables were found to be statistically significant predictors of customers' purchasing decisions, with email marketing having the highest impact (Beta = .440). These results support the main hypothesis of the study, which suggests that e-promotion tools significantly influence customers' purchasing decisions.

In conclusion, the study aimed to investigate the effect of e-promotion tools on consumers' purchasing decisions in Jordan. The findings revealed that email marketing, social media marketing, website design, and electronic advertising have a significant positive impact on consumers' purchasing decisions.

Therefore, businesses in Jordan should prioritize the use of these e-promotion tools to increase their sales and attract more customers. The study also recommends that companies adopt a more personalized approach to their e-promotion strategies to better cater to the needs and preferences of their target audience.

Furthermore, future research can be conducted to investigate the impact of other e-promotion tools, such as mobile advertising and search engine marketing, on consumers' purchasing decisions in Jordan. It is also suggested that future studies could explore the moderating effects of factors such as cultural differences and consumer attitudes on the relationship between e-promotion tools and consumers' purchasing decisions.

References

1. Khan, N., Khan, M.N.: The impact of digital marketing on consumer behavior: a case study of the United States. J. Glob. Mark. **34**(2), 68–81 (2021)
2. Alalwan, A.A., Abdullah, A.M., Rana, N.P., Tamil Mani, K.: Investigating the role of social media marketing in influencing consumers' purchasing decisions: a structural equation modeling approach. J. Retail. Consum. Serv. **70**, 102472 (2022)
3. Fan, H., Xu, Y., Sun, Y.: Impact of email marketing on customer loyalty: a review and research agenda. J. Bus. Res. **132**, 841–853 (2021)
4. Yadav, A., Pathak, G.S.: An empirical investigation of factors affecting attitude towards online advertising. J. Res. Interact. Mark. **11**(2), 142–158 (2017)
5. Kapoor, K.K.: Social media marketing: a critical review. J. Bus. Res. **139**, 107–119 (2022)
6. Rauschnabel, P.A., Felix, R., Hinsch, C.: Brave new world: the impact of augmented reality on consumer behavior. Bus. Horiz. **62**(6), 725–734 (2022)
7. Kumar, A., Mirchandani, R.: Role of social media in brand management: a systematic review and future research directions. J. Bus. Res. **139**, 66–80 (2022)
8. Ahmed, M.H., Mahmood, Z.: The impact of email marketing on consumer attitudes and purchase intention. Int. J. Marketing Stud. **14**(1), 18–30 (2022)
9. Heo, J., Han, S.: The effects of personalized email marketing on customer satisfaction and loyalty in the fashion industry. Journal of Fashion Marketing and Management: An International Journal, ahead-of-print (ahead-of-print) (2022)
10. Wang, X., Xu, Q., Fang, Y., Li, L.: Influence of website design elements on users' perception of website usability, visual appeal, and satisfaction. J. Electron. Commer. Res. **23**(2), 204–221 (2022)
11. Kaya, T., Aydin, E.M., Kocabas, H.: Investigating the effects of website design features on website engagement. J. Comput. Inf. Syst. 1-12 (2022). https://doi.org/10.1080/08874417.2022.2019072
12. Kim, J., Moon, J.Y.: Website design for building consumers' trust and credibility in online marketplaces: a comparative analysis of design elements. J. Retail. Consum. Serv. **69**, 102503 (2022). https://doi.org/10.1016/j.jretconser.2022.102503
13. Liu, Y., Liao, Q., Xiao, J.: Enhancing trust and perceived credibility of online shopping platforms through social proof design elements. J. Bus. Res. **144**, 94–103 (2022). https://doi.org/10.1016/j.jbusres.2021.07.040
14. Kim, Y., Lim, S., Kim, M.: The effects of electronic advertising on consumers' purchase intention. J. Advert. Res. **62**(1), 55–66 (2022)
15. Nitschke, P., Wiederhold, M., Kiel, D.: Electronic advertising: a review of the literature and implications for research. J. Bus. Res. **135**, 11–20 (2022)
16. Khan, M.A., Nawaz, M.M., Qureshi, M.A., Noman, A.: Role of electronic advertising in brand communication: a study of customers' perceptions. J. Retail. Consum. Serv. **55**, 102109 (2020)
17. Wu, Y., Liang, X.: Understanding user motivations for using social commerce platforms: a comparative study of China and the United States. J. Retail. Consum. Serv. **68**, 102859 (2022)
18. Nasution, H.N., Lestari, P.R., Purba, J.: The impact of social media and email marketing on consumer purchase intention. J. Mark. Commun. **28**(2), 214–230 (2022)
19. Lee, E., Kim, K., Kim, M.: The effect of online reviews and ratings on consumers' purchase decisions: a meta-analysis. J. Bus. Res. **140**, 484–496 (2022)
20. Zhang, Y., Li, X., Li, Y., Liang, C.: Consumers' purchasing decision-making process in the online shopping environment: a systematic literature review. Inf. Process. Manage. **59**(6), 102719 (2022)
21. Alqatan, B.: Impact of social media advertising on consumer buying behavior in Saudi Arabia. J. Promot. Manag. **26**(2), 170–189 (2020). https://doi.org/10.1080/10496491.2019.1662459

22. Hair, J.F., Black, W.C., Babin, B.J., Anderson, R.E., Tatham, R.L.: Multivariate data analysis (8th Ed.). Cengage Learning (2019)
23. Sarstedt, M., Ringle, C.M., Hair, J.F., Jr.: Partial least squares structural equation modeling. In: Hox, J.J., Roberts, J.K. (eds.) Handbook of advanced multilevel analysis, pp. 427–452. Routledge (2020)
24. Alalwan, A.A., Rana, N.P., Dwivedi, Y.K., Algharabat, R.: Social media in marketing: a review and analysis of the existing literature. Telematics Inform. **36**, 101303 (2019). https://doi.org/10.1016/j.tele.2019.101303
25. Sulaiman, M., Kaur, P., Singh, S.: Identifying the critical factors for successful implementation of blockchain technology in the supply chain of SMEs. Technol. Forecast. Soc. Chang. **168**, 120723 (2021)
26. Gliner, J.A., Morgan, G.A., Harmon, R.J.: Measurement reliability and validity. In: Gliner, J.A., Morgan, G.A., Harmon, R.J. (eds.) Measurement reliability and validity, pp. 27–50. Springer (2017)
27. DeVellis, R.F.: Scale Development: Theory and Applications (4th ed.). Sage Publications (2017)
28. Bryman, A., Bell, E.: Business Research Methods. Oxford University Press (2019)
29. Seth, A., Gupta, A., Trivedi, M.K.: Role of email marketing in customer relationship management: a conceptual review. J. Retail. Consum. Serv. **56**, 102138 (2020)
30. Han, H., Park, J., Ryu, H.: Analyzing the impact of green innovation on consumer purchasing behavior in the hotel industry. Sustainability **14**(4), 1509 (2022)
31. Alalwan, A.A., Dwivedi, Y.K., Rana, N.P.: Factors influencing adoption of mobile banking by Jordanian bank customers: extending UTAUT2 with trust. Int. J. Inf. Manage. **37**(3), 99–110 (2017). https://doi.org/10.1016/j.ijinfomgt.2016.11.002
32. Harmon, P.: Business Process Change: A Manager's Guide to Improving, Redesigning, and Automating Processes. Morgan Kaufmann Publishers (2017)

Social Media and Its Role in Marketing Agricultural Products (A Field Study on Small Farmers in the Jordan Valley Area)

Mustafa S. Al-Shaikh[1(✉)], Ahmed Issa Al-Gharagher[2], and Khalid Ali Alshohaib[3]

[1] Zarqa University, Zarqa, Jordan
maslshaikh@zi.edu.jo
[2] National Agricultural Research Center, Zarqa, Jordan
[3] Jeddah College of Advertising, University of Business and Technology, Jeddah, Saudi Arabia
k.alshohaib@ubt.edu.sa

Abstract. The present study is concerned to understand social media and its use in the marketing of agricultural products including a field study on small farmers. The independent variable of social media included (Facebook, WhatsApp, YouTube, and Twitter) and an affiliate variant of agricultural product marketing. Researchers used analytical descriptive methods to attain their goals. The sample study was selected from a group of (415) small farmers' respondents. The study found that the computational averages of social media came with an average approval score, with the highest for the role of Facebook, and then for WhatsApp, the lowest for Twitter, and a statistically significant role at the indicative level ($\alpha \leq 0.05$) in the sample responses to the role of social media (Facebook, WhatsApp, YouTube, and Twitter) in the marketing of agricultural products. The study recommended that the culture of marketing through social media should be mainstreamed through workshops and training courses for farmers, the effective use of social media in the promotion of agricultural products, the advertising of agricultural activities, program extension meetings, and the encouragement of entrepreneurship among the targets.

Keywords: Social Media · Marketing of Agricultural products · Small Farmers · Jordan Valley

1 Introduction

The agricultural sector is an important productive sector in Jordan. Agriculture plays an important role in the economic and social development of countries. It is also closely linked to the system of life and its sustainability. The agricultural sector determines individuals' lifestyles and income levels and is a source of food and social security and a focus of rural development [1]. Agricultural products are a cornerstone of economic and social development "The developed countries have begun to pay attention to the agricultural sector and its products and have allocated many resources to support it, emphasizing the economic, social, and environmental importance of this sector". Jordan

B. A. M. Alareeni and I. Elgedawy (Eds.): *Artificial Intelligence (AI) and Finance*, SSDC 488, pp. 425–435, 2023.
https://doi.org/10.1007/978-3-031-39158-3_41

has pursued the approach of developed countries, striving in recent years to support the agricultural sector and develop its economic, social, and environmental aspects in pursuit of national security, environmental safety, and public health [2, 3]. Based on the foregoing and given the widespread dissemination of social media and its techniques, the study came to learn about social media and its role in the marketing of agricultural products as a field study of small farmers in the Jordan Valley region.

The focus of the problem of the study is that the marketing of agricultural products is vital and of great importance to smallholder farmers. It is a real challenge to the success and sustainability of any agricultural project, which enables farms to plan and develop their products in line with consumers' wishes, offer and distribute them in an ideal manner, using modern methods of marketing. Hence, the study's problem is answering social media's role in marketing agricultural products. This study worked to achieve its goals by recognizing the role of social media Facebook, WhatsApp, YouTube, and Twitter in the marketing of agricultural products.

2 Study Hypothesis

Main hypothesis: H0: there is no statistically significant role at the indicative level (0.05 α $\leq$) of social media in the marketing of agricultural products. These include the following hypotheses:

H_{01}: There is no statistically significant role at Facebook's indicative level ($\alpha \leq 0.05$) in the marketing of agricultural products.
H_{02}: There is no statistically significant role at the indicative level ($\alpha \leq 0.05$) of WhatsApp in the marketing of agricultural products.
H_{03}: There is no statistically significant role at YouTube's indicative level ($\alpha \leq 0.05$) in the marketing of agricultural products.
H_{04}: There is no statistically significant role at the indicative level ($\alpha \leq 0.05$) of Twitter in the marketing of agricultural products.

3 Study Model

The study model is designed after returning to the study of [3–6], and Fig. 1 shows this.

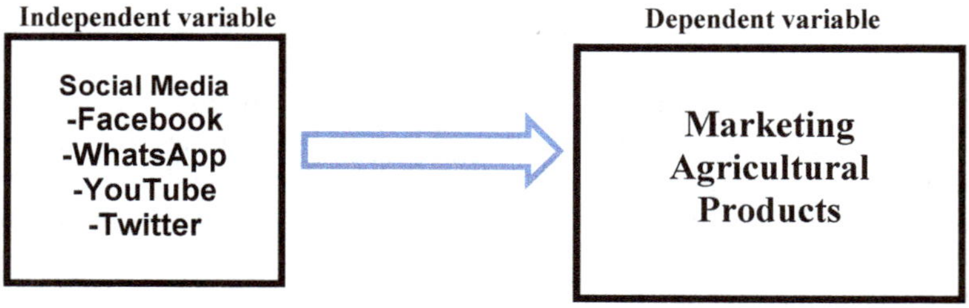

Fig. 1. Study Model

4 Literature Review

4.1 Concept of Marketing Using Social Media

The internet has become an important part of our daily lives because it is an important source of information, competing with traditional media for the dissemination of news or information such as newspapers, magazines, and audio and visual broadcasts. Besides considering this network as an important source of knowledge, it is possible to create a new space that can be invested in marketing some goods or services. Accordingly, electronic marketing can be defined as a "commercial transaction based on the interaction of parties to an electronic exchange rather than direct physical communication" or "online sale and purchase of goods and services" [7, 12]. In the researchers' view, social media marketing - in addition to digital means in the marketing process - contains data management software, electronic payment techniques, and electronic protection. The dimensions of the independent variable were measured as follows:

Firstly, Facebook: It is an asocial network that enables a person on this site to prepare a personal profile of his life that serves as an identification card for those who want to recognize and communicate with him [8]. Procedurally known: promoting agricultural products in the Jordan Valley region using photos, and videos by sending and receiving messages.

Secondly, What Sapp: An application that uses the internet to send text messages and images, voice messages and videos, and allows it to be used by downloading it on smartphones or computers [9]. Procedurally known: to promote agricultural products by sending text messages, images, voice messages, and videos.

Thirdly, YouTube: One of the websites relies on adding users to videos from all over the world and at the same time allows anyone to browse and benefit from these clips [10]. Procedurally known: promotion of agricultural products by adding, browsing, and utilizing agricultural product videos.

Fourthly, Twitter: A micro-social blogging service that allows registered members to post micro-posts members can post and follow up on other members' publications through the use of multiple platforms and devices and can respond or post to tension using a mobile phone by text message, or through the user interface on the site [11]. Procedurally known: to promote agricultural products in the Jordan Valley region by publishing small publications and following up on publications on agricultural products.

Fifthly, the marketing of agricultural products: is the promotion of agricultural products by smallholder farmers which will be measured according to the tool to be prepared for this.

4.2 Benefits of Social Media Marketing

First, the customer expresses his opinion: modern marketing is the fundamental base in all its production and marketing processes and programs. It focuses on the need to involve him in the strategy and programs planned by the marketing organizations. These companies have therefore benefited from these networks through the continuous and permanent dialogue provided by the various companies through these websites. They allow the customer to speak freely and express his opinion without restrictions on the products and services offered by these companies by posting material comments or showing shortages or defects in the product. This is an opportunity for marketing companies to know the feedback of this product as well as link abundant relationships with customers.

Secondly, the customer receives advertising messages: customers are constantly informed about what is going through these means by browsing their websites where they receive daily many advertisements, updates, and offers submitted by many marketing companies even if they do not participate or comment on them and are the listener the customer is familiar with all the new offers it has, as well as with comments and criticisms of offers that reinforce some customers' purchasing decisions and thus stabilize the brand of a product in the customer's mind.

Third, customer trust: The most important thing that marketing organizations seek is to create trust between the organization and its customers and thus gain the customer's satisfaction. This is what we find in many of their strategies through social media. The source known in the marketing process has the impact and psychological factor that affects the customer's behavior. In other words, the customer is influenced by the advertising message provided by these organizations and through the same source and does not accept the advertising message from another source. For example, when they browse a particular marketing organization's Facebook, Twitter, or other means page, they face several comments and responses about a product, thereby believing what the organization says about the product because it accepts what customers say. In addition, through their comments and responses and do not accept what other companies say, this enhances the customer's trust between this organization and him.

4.3 Definition of Agricultural Marketing

The study of agricultural marketing includes all processes and entities related to the agricultural sector that participate in the movement of food produced on the farm and raw materials and their derivatives, such as textiles, from farms to end-consumers, the impact of these processes on farmers, intermediaries and consumers, and agricultural marketing is a process that begins with the decision to produce a selling agricultural commodity, encompasses all aspects of the structure or system of the market, functional and institutional, based on technical and economic considerations, including pre-harvest and post-harvest processes, in the sense of assembly, classification, storage, transport and distribution, and agricultural marketing includes marketing of food grains, commercial crops, crops, horticultural products, and semi-manufactured products.

4.4 Small Farmers in the Jordan Vally

Smallholders who manage their crops, as defined by the ministry of agriculture of Jordan and as defined by the food and agriculture organization of the United Nations (FAO), are agricultural holders of less than 10 Acres (one Acres equals 1000 m^2) and account for 74.1% of holders; They are linked to the basic property of a processing plant or farm. The results of the census showed that the number of farmers was 22.95,000. The results also showed that 14.5% of agricultural holders considered agriculture to be their main occupation, while 27.3% of the holders of agriculture represent a secondary profession [13].

5 Previous Studies

A study [14] found a decrease in the number of agricultural guides and poor physical capabilities and facilities for agricultural extension. The study [15] found a strong impact of social media marketing activities on the word spoken electronically at 32%. A study [16] found that Extra Foundation has a positive influence to use its official Facebook page to promote its food products electronically. A study [3] showed that the use of marketing tools for crops contributes to reducing the role of intermediaries in the sale of crops. [17] conclude that most goods purchased online are clothing. The study [4] found an impact on expected performance, anticipated effort, and social impact on the behavioral use of online shopping by consumers in these governorates. A study [18] found that the most important factors influencing the online purchase process are the customer's personality, the marketing and advertising method, and the presentation in the store. A study [19] showed that there is a correlation between the use of social media sites and increased competitiveness, and the competitiveness stems from the company's ability to benefit from interactive and continuous communication with its customers. A study [20] found no statistically significant effect of the company's content on the brand, while a statistically significant effect of the user's content on the brand. A study [21] showed the increasing use of social media to influence purchasing behavior and that Facebook and Twitter users following these products are more likely to engage in social media discussions about the producing company. Study [22] the study showed that 97% of the sample uses Facebook, then YouTube is second in use at 71% and Twitter at 23% and my space at 22%. The study shows that 55% of social media users use it to exchange opinions and experiences with friends.

6 Methodology

6.1 Population and Sample

The study population is of all small farmers in the Jordan Valley region. The sample study was limited to a group of 400 small farmers and was selected in a simple random manner [24]. The researchers designed the study tool (questionnaire) to cover all dimensions of the independent and dependent variable in a way that enables the testing of hypotheses that were built, and distributed to the sample members to measure the role of e-marketing in marketing agricultural products. By distributing (415) questionnaires to the study sample, and after collecting the data, the number of questionnaires valid for analysis reached (400) questionnaires.

6.2 Results of Testing the Hypotheses of the Study

Main hypothesis: H0: there is no statistically significant role at the indicative level (0.05 $\alpha \leq$) of social media in the marketing of agricultural products in the Jordan Valley region. To ascertain the validity of the main hypothesis, the Multiple Linear Regression test was used to determine the impact of the dimensions of the independent variable (social media) on the marketing of agricultural products. To identify which variables have an impact on the subordinate variable and are statistically significant, this has been determined by a moral test of coefficients' standard and non-standard linear regression equation (Unstandardized), and Table 1 shows this.

Table 1. The significance of the standard and non-normative multiple linear regression coefficients

Sample	Multiple Linear Regression	Unstandardized Coefficients		Standardized Coefficients	T Value	Sig
		B	Std. Error	Beta		
1	Constant	.739	132		5.613	.000*
	Facebook	.772	.033	.757	23.136	.000*
2	Constant	.361	.121		2.979	.003*
	Facebook	.459	.041	.450	11.135	.000*
	YouTube	.437	.040	.437	10.802	.000*
3	Constant	.235	.122		1.923	.055*
	Facebook	.445	.041	.436	10.983	.000*
	YouTube	.313	.049	.313	6.418	.000*
	Twitter	.179	.041	.181	4.324	.000*
4	Constant	.182	.123		1.480	.140
	Facebook	.391	.045	.383	8.583	.000*
	YouTube	.275	.051	.275	5.427	.000*
	Twitter	.155	.042	.156	3.654	.000*
	WhatsApp	.134	.052	.126	2.559	.011*

* **Statistically significant at the level of statistical significance (0.05 ≥ a)**

Table 1 shows the statistical indicators of the standard and non-standard linear regression equation factors relating to the independent variable (Facebook, YouTube, Twitter, WhatsApp) with the four proposed models, indicating a positive impact of those variables on the affiliate variable (Marketing agricultural products), as these variables positively affect the marketing of agricultural products in the Jordan Valley region, we reject the main hypothesis that "there is no statistically significant role at the indicative level (0.05 $\alpha \leq$) of social media in the marketing of agricultural products".

7 Discussions

The study's main findings are as follows:

- The computational averages of the role of social media came with an average approval, the highest being the role of Facebook, followed by the role of WhatsApp, while the lowest was the role of Twitter. Researchers attribute this to the ease of use and spread of the Facebook app, which is the most popular and interpersonal application and provides the ability to publish photos and videos of crops with ease and ease and provides access to publications to all segments of society facilitating the marketing of agricultural products through funded advertising. Many friends can be added to pages and marketing groups through which farmers seek higher prices for their products. Unlike other apps and all of this leads to the opening of new markets or channels for selling products, this is consistent with [18] study which concluded that the most important factors influencing the social media procurement process are: customer personality, marketing and advertising method, time, product view, study [22] which showed that 97% of the sample uses Facebook, then YouTube is second in use at 71% and Twitter at 23% and my Space at 22%.
- The computational averages of Facebook role paragraphs came with a high degree of approval, the highest of which was paragraph 1, which states that "Facebook application is one of the most widely used software", while the lowest is paragraph 5, which states that "Facebook allows the targeting of sample individuals to access electronic advertising". The researcher attributes this to the fact that social media, including Facebook, has opened up new prospects for farmers to market their products externally and internally, which has helped them to find practical and low-cost solutions to the problem of marketing their products. And due to the easy access to consumers in society due to the prevalence and use of this app by all segments of society, this is consistent with the study [16], the results of which indicated that Extra Ben Hamadi employs its official Facebook page to market and promote its food products despite many shortcomings observed in the form and content of promotional advertisements [27].
- The computational averages of the "WhatsApp" role paragraphs came to a moderate degree of approval, the highest of which was paragraph (1), which stipulates that "WhatsApp application constitutes a widespread demand as a communication and marketing program", while the lowest is paragraph (2), which states that "WhatsApp contributes to the dissemination of publicity of products". This is because the use of WhatsApp from the investigators' point of view increases the linkage between farmers, provides access to scientific expertise, increases their agricultural knowledge, encourages them to participate in opinion, and develops their creative style to a high degree in positive terms, and is effective in disseminating agricultural innovations among farmers.
- The computational averages of the "YouTube" role paragraphs came to a moderate degree of approval, the highest of which was paragraph (5), which states that "YouTube enables video explanation in audio, image, and traffic via the channel on which it is promoted", while the lowest is paragraph (3), which states that "YouTube provides electronic display services with publicity and writing design".

- The computational averages of the "Twitter" role paragraphs came to an average degree of approval, the highest of which was paragraph (4), which states that "Twitter provides advertising financial services for the ease of publication and speed of their arrival", while the lowest is paragraph (2), which states that "Twitter provides message services and displays of products and videos". The researcher attributes this to Twitter being an indispensable destination for marketers and business owners, and an effective force for communicating with the customer audience, being an important social media for managing reputation and building brand awareness farmers use Twitter marketing strategies to target the ideal audience and enhance their business, Increasing the number of visits to their sites, increasing sales ratios for their agricultural products, This is consistent with the study [21], the results of which indicated that Facebook and Twitter users following these products are more inclined to engage in social media discussions about the producing company and may even be promoted, or attend promotional or promotional meetings for the company.
- The computational averages of the paragraphs "Marketing of agricultural products" came with a high degree of approval, the highest of which was paragraph (1), which states that "The media helps me to promote my agricultural products", while the lowest is paragraph (3), which states that "Marketing through social media saves my time and effort in promoting and selling agricultural products". This is due to the ease provided by social media channels in directly disseminating agricultural products to customers without the need for time and effort in promoting and selling them in traditional markets This is consistent with the study [3, 28] which referred to the use of marketing tools and techniques for crops that contribute to reducing the role of intermediaries in the sale of crops and elements of the electronic marketing mix of crops contribute to reducing the role of intermediaries in the sale of crops.
- A statistically significant positive impact of social media (Facebook, YouTube, Twitter, WhatsApp) in the marketing of agricultural products in the Jordan Valley region. The researchers attribute this to the fact that social media is a modern site for communication among members of communities, through which services are promoted to enjoy a high level of communication between farmers and consumers both in terms of facilitating interaction, cooperation and information exchange, and through the possibility of circulating images, videos and blogs, this is also due to the ease and gratification of content dissemination, increased brand knowledge and fame through social media sites, increased sales and knowledge of consumers' needs and desires This is consistent with the study [15], the results of which indicated a strong impact of social media marketing activities on the word spoken electronically, It reached 32%, and with the study [19] which indicated a relationship between the use of social media sites and increased competitiveness, it differs with the [20, 23, 25, 26] study which indicated that there is no statistically significant effect of the company's content on the brand, whereas there is a statistically significant effect of the user's content on the brand.

8 Recommendations

- Effective use of social media to promote agricultural products, publicize agricultural outreach activities, programs, and meetings, and encourage entrepreneurship among the targets.
- To take advantage of the experiences of other countries that have been applied in the field of the use of social media in the marketing of agricultural products and the development of such experiences in the Jordanian environment.
- Use social media to enhance extension communication between farmers, agricultural guides, and consumers to improve the quality of agricultural products.
- The use of social media in agricultural marketing extension, and the provision of incentives to farmers who use new methods in agricultural marketing via social media and encourage them.
- Expand and improve the internet service in rural areas so that farmers can reach the target group without the need for travel hardship, pay connectors, and transport to sell and market their produce in traditional markets.

9 Limitations and Future Studies

This study Faces some limitations, such as, focusing only on marketing agricultural products in the social media in Jordanian valley, but this study did not take into consideration other variables like covering the wider area in Jordan. In addition, this study only focuses on Jordan as one of the developing countries and did not cover any other country nearby.

This study discusses the most important factors affecting social media and its role in the marketing of agricultural products s, accordingly, other points of view are preferred to be studied in the future from other aspects that can serve social media and its role in the marketing of agricultural products such as the effect of the women in agriculture, education and other factors that can be studied.

References

1. Ministry of Agriculture. The Annual Report of the Ministry of Agriculture "The Agricultural Document." The Jordanian Ministry of Agriculture website (2022). http://www.moa.gov.jo/agri%20document
2. Al-Tarawneh, M.: Determine the barriers to organic agriculture implementation in Jordan. Bulgarian J. Agr. Sci. **22**(1), 10–15 (2016)
3. Al Awar, M.: The Impact of Electronic Marketing in Reducing the Role of Intermediaries in Selling Agricultural Crops: A Field Study in Syria, Unpublished Master's Thesis, The Syrian Virtual University (2020)
4. Al-Batayneh, M., Al-Afif, M.: Online shopping: a viewpoint of the unified theory of technology acceptance and use (UTAUT) a field study on consumers in the governorates of irbid, Jerash, Ajloun, and Mafraq. An-Najah University Research Journal, **32**(12) (2018)
5. Shahwan, Y., Sa'adeh, A., Hamza, M., Al-Ramahi, N., Swiety, I.A.: DO the reservs help the financial and nonfinancial performance of firms during the Covid-19 pandemic. Corporate Governance and Organizational Behavior Rev. **6**(2), 217–222 (2022)

6. Khrais, E. Azzam, Z., Assaf, A.: Constraints facing garment industrial sector operating within the qualified industrial zones in Jordan. Zarqa Journal for Research and Studies Humanities **10**(2), 82–83 (2010)
7. Bamakhrama, A.S.: A study of the market for electronic shopping services in the Kingdom of Saudi Arabia through electronic shops. Andalus Journal for Social Sciences, Department of Economics, College of Economics and Administration, King Abdulaziz University, Jeddah **3**(6) (2011)
8. Sultan, M.S.: Media and Communication. Dar Al Maysara for Publishing and Distribution, Amman (2012)
9. Al-Shaikh, M.S., Torres, I.M., Zuniga, M.A., Ghunaim, A.: Jordanian pharmaceutical companies: are their marketing efforts paying off? Health Mark. Q. **28**(2), 174–189 (2011)
10. Hammond, T.C., Lee. J.K.: Digital Video and Social Studies. Contemporary Issues in Technology and Teacher Education, pp. 124132 (2010)
11. Al-Raoud, A. Mubarak. M.: The Role of Social Networks in Political Change in Tunisia and Egypt from the Perspective of Jordanian Journalists, Master's Thesis, University of Jordan (2012)
12. Mohamed, M.: The impact of digital marketing on the perceived value of the customer - an applied study on egyptian airlines customers. J. Contemporary Business Studies, Kafrelsheikh University **40**(1), 298–405 (2020)
13. Ministry of Agriculture.: The Annual Report of the Directorate of Agricultural Information and Statistics. Agriculture in Numbers - Agricultural Statistics 2008–2018, Jordanian Ministry of Agriculture (2018)
14. Abdullah, I.I., Al-Ruwaini, M.A.: The use of social media in agricultural extension in Al-Buhaira Governorate. Al-Azhar Journal of Agricultural Research, Faculty of Agriculture in Cairo, Al-Azhar University **46**(1), 225–236 (2021)
15. Sergios, A.I.: The impact of social media marketing activities on the electronically spoken word: the mediating role of brand awareness and the image of the brand. The Scientific Journal of Economics and Commerce, Ain Shams University - Faculty of Commerce **17**(2), 357–409 (2021)
16. Badri, N.: Promotion of food products through electronic platforms, a descriptive study of the facebook page of the extra bin Hammadi foundation. Maalem Journal for Media and Communication Studies **1**(1), 44–75 (2020)
17. Ziadat, A.: Factors affecting e-shopping in jordan: an empirical study on e-shopping websites. Arab J. Scientific Publication **11**, 117–143 (2019)
18. Attri, R., Jain, V.: A study of factors affecting customer shopping behavior. IUP Journal of Marketing Manage. **17**(1) (2018)
19. Bang, A., Hell, J.: Digital Marketing Strategy Social Media and its Contribution to Competitiveness. International Business Strategy, University Linnaeus, Sweden, Master (2015)
20. Shahwan, Y.: The effect of OECD principles of corporate governance on social responsibility accounting. Int. J. Innovation, Creativity and Change **11**(2), 418–431 (2020)
21. Schivinski, B., Dabrowski, D.: The effect of social media communication on consumer perceptions of brands. Journal of Marketing Communications, Statistic Brain. 2014."Instagram" http://www.Statisticbrain.ComInstagram-company statistics/. 9:47 Statistic Brain (2015)
22. Muntinga, D.G.E., Moorman, M.: Does social media influence purchase decisions. Advances in Advertising Res. **3**, 121–137 (2012)
23. Bilgihan, A., Nusair, K., Okumus, F.: Social networking websites: do generation y use them before making decisions for hotels and restaurants?. Rosen College of Hospitality Management, USA (2011)

24. Imran, M., Ali, S., Shahwan, Y., Zhang, J., Al-Swiety, I.A.: Analyzing the effects of renewable and nonrenewable energy usage and technological innovation on environmental sustainability: evidence from QUAD economies. Sustainability **14**(23), 15552 (2022)
25. Shahwan, Y.: The mediating effect of investment decisions and financing decisions on the influence of capital structure against corporate performance: evidence from Jordanian listed commercial banks. Academy of Accounting and Financial Studies Journal **22**(6), 1–20 (2018)
26. Al-Shaikh, M.S., Khanfar, I.A.: Delivery service via electronic applications and its impact on customers satisfaction at retail stores in Amman City/Jordan. In: The Implementation of Smart Technologies for Business Success and Sustainability, pp. 827–837. Springer, Cham. (2023). https://doi.org/10.1007/978-3-031-10212-7_68
27. Al-Shaikh, M.S., Torres, I.M., Zuniga, M.A., Ghunaim, A.: Internet commerce in Jordanian firms. J. Internet Comm. **9**(2), 67–82 (2010)
28. Shahwan, Y., Hamza, M., Al-Fasfus, F., Al-Ramahi, N., Almubaydeen, T.H.: Social responsibility accounting, disclosure and real practice: evidence from Jordan. Asian Economic and Financial Review **12**(3), 164–182 (2022)
29. Shahwan, Y., Abdel-hamid, O.: The impact of social responsibility disclosure in reducing the risk in Jordanian Commercial Banks. Res. World Economy **11**(3), 142–150 (2020)

The Role of Digital Marketing Dimensions in Enhancing the Image of the Educational Services in Jordanian Universities. A Case of Jordan

Zakaria Azzam[1(✉)], Zaid Al-Hamidi[1], and Suzilawati Kamarudin[2]

[1] Marketing Department, Zarqa University, Zarqa, Jordan
{zazzam,20219128}@zu.edu.jo
[2] Department of GRD, University of Business and Technology, Jeddah, Saudi Arabia
s.kamarudin@ubt.edu.sa

Abstract. The study aims to identify the role of digital marketing dimensions in enhancing the mental image of educational services: a field study on university students in Jordan. By using the analytical descriptive approach, the data collection tool was distributed by stratified sampling method to (500) students from Jordanian Universities, (412) of them were retrieved and analyzed. The study used the appropriate statistical methods through the (SPSS) program; the study reaches a set of results, first there is a statistically significant role for digital marketing (combined and individual) dimensions (attraction, communication, absorption, retention, E-word of mouth) in enhancing the image of educational services for universities in Jordan. Second, digital marketing is practiced all its dimensions in universities in Jordan at a high level, and finally the image of educational services from the perspective of university students in Jordan is high. Due to the weakness of this dimension in influencing the enhancement of the image, and the establishment of digital marketing departments in universities in Jordan to updated work strategies and being followed up to identify scientific changes in the field of digital marketing and its application, the study recommends that universities in Jordan have to focus on the dimension of attraction through conducting competitions and promotional programs that attract students, due to the weakness of this dimension in influencing the enhancement of the image, and the establishment of digital marketing departments in universities in Jordan to updated work strategies and being followed up to identify scientific changes in the field of digital marketing and its application.

Keywords: Digital Marketing · Image · Educational Services · Universities · Jordan

1 Introduction

The educational services industry and their marketing have passed with fundamental changes and tremendous technological developments that opened new horizons for organizations to market their services digitally, so digital marketing to coordinate marketing

B. A. M. Alareeni and I. Elgedawy (Eds.): *Artificial Intelligence (AI) and Finance*, SSDC 488, pp. 436–447, 2023.
https://doi.org/10.1007/978-3-031-39158-3_42

activities has appeared and it has become an effective tool for organizations through which they offer their various services on a global scale in modern methods to achieve their goals and build good relationships with students, and reduce lost time, effort and money, things that oblige these organizations to search for technologies and mechanisms that help them to keep pace with this new development to provide services in a way designed for them without any deviations in order to achieve the continuous improvement in the quality of service, and to upgrade the level of students' aspirations for the repercussions of the image of educational services, By including digital marketing in its marketing strategies. And because students are usually searched online to find the best service providers to meet their needs and desires, universities can take advantages of digital marketing tools. Such tools as search engines, marketing via search engines, content marketing, use of influencers, content automation, and advertising campaigns, marketing via social media, improving social media, direct email marketing, video advertisements, e -books, and light tablets. Most members of society are connected through the Internet, and the increasing use of social media creates new opportunities for global digital marketing to attract customers through the digital platform [1]. Digital marketing is defined as "use of digital technologies to create an integrated, targeted and measurable connections that helps to acquire and keep customers while building deeper relationships with them" [2]. Researches assures that digital marketing consists of several dimensions: attraction, communication, immersion, and retaining, where this model is the most famous and more published [3].Others believe that the electronically transferred word is one of the important dimensions of the Electronic Word of Mouth (E-WOM), which must receive sufficient attention by the customers [4]. Therefore, the positive image is a fundamental goal that everyone seeks to achieve at any level even at the level of individuals. This fact realized the importance of educational organizations and put plans directed to form the image of the beneficiaries in the manner that achieves their goals; which applied in all services and non -services firms [4].

The importance of this study stems from its subject and its components, especially in light of the increase in interest of digital marketing practiced by educational organizations in general and private universities in particular, where it is expected that the study will contribute to increasing the knowledge outcome towards digital marketing and enriching the national library with an important topic linking the role of digital marketing with the image of educational services especially in light of the successive environmental changes facing universities.

1.1 Research Problem

Students make up the axiom towards any subject through experience, which means personal messages that the individual hears from friends, events, individuals, or company ads. Where the individual's core image is formed about organizations according to the strengths and weaknesses of marketing programs organization is adopted, so many studies have studied the role of digital marketing in building the image of students [5] in addition to the importance of researching the image of the services provided [2]. In the Hashemite Kingdom of Jordan (39) universities are working divided into (12) public universities, (24) private universities, two self-law universities, and a regional university, competing to provide its services to (332,413) students at all educational levels

(Bachelor, Master, and PhD, and a high diploma, a medium diploma, and a technical diploma) according to the statistics of the Jordanian Ministry of Higher Education and Scientific Research for the year (2022/2023) [6].

Thus, we note the existence of a decrease in the number of students compared to its population, which requires to study digital marketing role in enhancing the image of educational services provided by the universities in order to increase the number of students affiliated with these universities.

Based on the above, the study problem can be formulated in the following questions:

1- What is the role of digital marketing dimensions (attraction, communication, absorption, retaining, and the electronic word of mouth) in enhancing image of educational services from the point of view of university students in Jordan?
2- What is the level of digital marketing dimensions application (attraction, communication, absorption, retaining, and the electronically transferred word) from the point of view of university students in Jordan?
3- What is the level of image of educational services from the point of view of university students in Jordan?

1.2 Research Model

Based on literature review and previous studies [1–10, 12–21] the following research model is developed for the present study as in Fig. 1.

Digital Marketing.

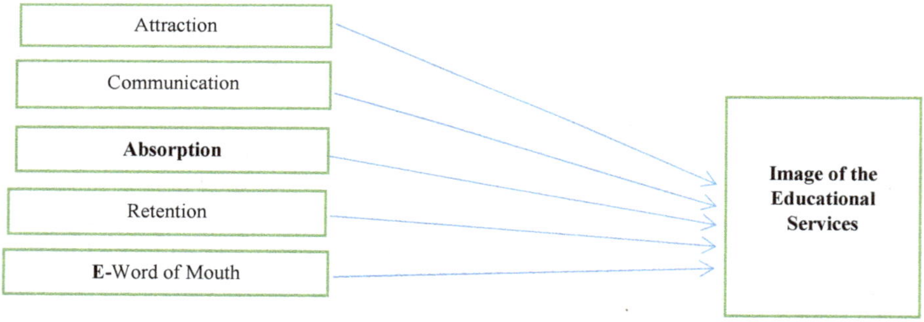

Fig. 1. Research Model

Based on the literature reviewed and the previous studies the main hypothesis of the present study is:

HO1: There is no statistically significant role at level ($\alpha \leq 0.05$) for digital marketing with its combined dimensions (attraction, communication, absorption, retaining, and the electronic word of mouth) in enhancing the image of educational services in Jordanian Universities.

2 Literature Review

Digital marketing can be defined as applications for modern digital technologies integrated with traditional marketing strategies in order to achieve final marketing goals. Organizations adopt this new type of marketing to gain a great competitive advantage to support the facilitation of exchanges through advanced applications in smartphones that encourage trade exchange activities [7]. During the past decades and due to launching of search engines, new major marketing challenges have arisen due to the severity of the unconventional commercial competition that contributes to the concept of digital marketing and the related aspects [22], The digital marketing model also led to radical changes and transformations in the global markets and increased the spread of purchase intentions and strength of customers, which led to many applications for advanced business technologies created in response to business changes[16] and [18].

2.1 Features of Digital Marketing

Digital marketing is characterized by many characteristics, and these characteristics can be clarified as follows [23]:

1- Digital Marketing is a new tool that the organization uses to build relationships with customers and society as a whole.
2- Leads to a reduction in business performance costs.
3- The ability to invite the current and expected students to participate in the interactive dialogue on products and services, which reflects positively on the image of organizations.
4- Rapid access to information compared to traditional marketing.
5- Providing an easy marketing advantage, so it does not need to make a great effort to identify the offered products and services.
6- Digital Marketing provides written, fast and effective communication between the customers and the organizations through digital services known as one click services.
7- Through digital marketing, the customers gets the advantage of making a comparison between brands and prices of different firms.
8- Rapid access to information compared to traditional marketing.

2.2 Digital Marketing Dimensions

Many researchers confused between the dimensions of digital marketing and its tools (means), due to the modernity of the concept and the overlap of its dimensions and tools together, which constituted an interconnection between them that made discrimination among them controversial with different views of the researchers, but we may find an agreement on digital marketing tools which are: marketing via website, marketing via social media, e -mail marketing, and via mobile phone marketing. As for the dimensions of digital marketing, they are: attraction, communication, absorption, and retaining, and electronic word of mouth [1, 3, 4, 8, 13, 23, 24].

Attraction
Attraction is the basis for customers' interaction which they require them to visit voluntarily the application of the interactive organization (visiting the website) and attracting

customers to the application, finding customers. Helping in remembering the brand name in the memory and giving attentions and interest in the ads deposed. Marketers can attract customers through advertisements on web sites, advertising windows or related links, and organizations may use methods of attraction through the organization's association with a set of charitable works or loyalty programs for students or producing environmentally friendly products. Attraction is linked to improving the web site rank in the engines Search such as (Yahoo and Google) where web sites are ranked based on the words that the user entered into [4, 8, 13, 23, 24].

Communication

Communication is one of the most important opportunities to create the most important value in digital marketing, as it is better to allocate the interaction between the services provided, time and the marketing efforts for more than one customer at the same time, where digital marketing messages allows communications and distribution and to know more about one student. This can be done by seeking the majority of organizations to publish their activities on social media and blogs, especially for customers and searching for them to make them reveal their experiences as a result of using services or by placing a bulletin about the company's activities that communicate with its customers. Communication is considered as one of the most important strategies for building the c image via the Internet by seeing students' visions, preferences and desires, and improving the organization's website on the Internet. Business organizations use a number of sources to obtain the information they need about their customers, the most important of which are: reactions obtained from surveys online, and open (regular) research, internet pages tracking devices, new customer's information, social media and virtual conversation rooms [4] and [9].

Absorption

Absorption is customers' involvement in participation, attention, achieving interaction, or completing procedures or businesses. Absorption is a major factor in creating the demand, which includes: generating attention and absorption, creative programming, interactive content, use the resulting content and practical capabilities. Reaches believes that the absorption is an extension of the relationships marketing, which focuses more on the exchange of information between the service provider and customers, while customers' participation focuses on the customers' experience more than the emotional link, as the participation of students is a psychological process that builds customers' loyalty to the brand, and a mechanism to keep customers [9] and [9]. Customers' participation can be identified using several indicators such as [25]:

1. Enrichment: The clarity of information to explain the goods and services offered, and the clear multimedia content that is likely to attract customers' interest.
2. Interaction: The high level of reaction can lead to a greater online participation.
3. Product or service description: The appropriate description of the product or service provided will enhance customers' interest.
4. Entertainment: The information provided in an entertainment style will attract more positive attention.
5. The novelty of content: new and unique information will stimulate more customers' attention.

Retaining
In order to build profitable and long -term relationships with the customers, organizations seek to retain customers and motivate them to repeat gaining services on their sites via the Internet. The important element that the organizations must take into account when applying digital marketing is building positive relationships with and preserving customers, which requires them Long -term fixed obligations and resources. The retention is concerned with providing appropriate and interactive content with value in order to develop relationships with customers which includes: ensuring that customers return, availability of unique dynamic content and the availability of private digital fans [9].

Electronic Word of Mouth
It is an either negative or positive phrase that is created through a real experience for former or potential customers in relation to a brand or the organization, and this phrase is important for customers and the organization that use digital marketing, and the word transferred electronically has become a vital place for customers to share their observations and experiences and give them potential students, and is considered an effective approach more than the traditional oral style explains that the electronically transferred word.

E-WOM facilitates the exchange of information and knowledge through social media platforms. The electronically transmitted word, which affects the image of students, can be generated through three main stages, which are as follows [14–16, 18, 26, 27] as follows:

1. Finding the electronically transferred word.
2. View the word transferred electronically to the student.
3. Evaluating the electronically transferred word.

2.3 Services Image

With the progress of media and communication sciences, the term image entered into various fields of research related to public opinion, then interest in it increased with the great progress in the areas of communications and information, which turned the world into a small village in which all sciences are socially, environmentally and administratively communicated. The concept of image is seen as simply as the picture formed in the minds of people about various organizations, and may form on o the basis of evidence and documents, or on the basis of unreliable rumors and chats, but in the end it represents a true reality from the point of view of those who carry it in their minds. On the other hand, the image is a symbolic image in which customers, their demands and interests are collected, and this image cannot be formed in a day and night, as it is gradually deposited in minds, which is not a translation of the actions, words and behavior of any organization, and these words and actions are not responsible for them as much as it is a translation of the reactions that the actions and those words are formed, which are a group of acquisitions and impressions that the individual forms about something, and these perceptions are evoked as a result of exposure to a set of influences [9].

The dimensions of image are classified as [9], including:

1- The cognitive dimension: It is the dimension through which the individual realizes a specific subject related to a state, people or organization, such as knowledge of the historical and geographical foundations and knowledge of general information about it, and this information is the basis on which the pernicious image that the individual is forming on others and on issues. Based on the accuracy of the information and knowledge that you obtain from others, the accuracy of the image that you form is, and according to the cognitive dimension, the errors consistent in the image of individuals are errors that are mainly caused by the information and the wrong knowledge that these individuals obtained.
2- The emotional dimension: It represents the feelings of the individual and his emotions towards a state, people, or organization and extending from acceptance to rejection and graduating from positive and negative, where the emotional side is formed with the cognitive aspect, and with the passage of time the information and knowledge that individuals and the emotional aspects that represent the remains of representation are faded.The attitudes of individuals towards different people, issues and topics, and this affects a set of factors, the most important of which is the limits of the availability of sources of knowledge, as well as the construction of the emotional dimension of the characteristics of peoples in terms of color, gender and language which may contributes to building positive trends.
3- The procedural dimension: It is represented in the individual's desire to travel to a foreign country, lives in, work or marry there, and this factor depends on the social distance between individuals, or the desire to work with an organization and belong to it.

3 Research Methodology

This study adopts the descriptive analytical approach based on model extracted from previous studies and testing it by developing a tool that measures the variables represented by the model and testing it on a representative sample. In order to obtain practical results that have been interpreted in an objective manner consistent with the actual data of the phenomenon. Then, the study hypothesis are tested through the use of a statistical program (SPSS) [26]. The researchers distributed the study tool to a random sample of students in Jordanian Universities which are located in Amman the Capital of Jordan and outside, with the help of a group of students and professors in these universities, where an electronic link was sent to the students to achieve the required number from both universities Where (412) valid questionnaires were obtained for analysis, The number of Jordanian universities are (39) universities distributed among (12) public universities, (24) private universities, two universities operating under private law, and a regional university. Those universities are competing to provide its services to (332,413) students of all educational levels (Bachelor's, Master's, PhD, Higher Diploma, Intermediate Diploma, and Technical Diploma) according to the statistics of the Jordanian Ministry of Higher Education and Scientific Research for the year (2022/2023) [6], which indicates that the sample size selected for the present study is suitable statistically (412) students who represent the population of the present study.

3.1 Reliability of the Study

Cronbach's Alpha is at a minimum acceptable level of 0.60 [27] as shown in Table 1, the items of the study tool have achieved a stability coefficient greater than (70%) [28]. And therefore this indicates the suitability of the tool to achieve the objectives of the study and its suitability for application, and after the researcher has verified the validity of the study tool and its stability, the researchers applied it to the target study population.

Table 1. Cronbach Alpha

Number	Variables	Questions	Cronbach's Alpha
1	Attraction	14	0.817
2	Communication	59	0.851
3	Absorption	10–17	0.912
4	Retaining	18–22	0.842
5	Electronic Word of Mouth	23–26	0.865
6	Image	27–35	0.827

4 Results and Discussion

As shown in (Table 2) that the percentage of male students are (63.6%) of the study sample, and this result may be due to the tendency of a large percentage of male youth in Jordanian society to university education instead of liberal professions and vocational education, and we also note that (58.5%) of the study sample are residents outside Amman Governorate, and this indicates the good reputation enjoyed by the universities of other Governorates and the lack of a good distribution of universities in Jordan. The results also showed that (61.4%) of the study sample are undergraduate students, and this percentage is commensurate with the trend towards education. In the first university level in Jordan, the results also showed that (76.9%) of the study sample are students of Universities outside Amman the Capital of Jordan. The results also showed that (83.7%) of the students use social media as a digital marketing tool adopted by the universities in Jordan in the first place followed by Search Engine tool with 70.1%.

It is noticed from Table 3 that the value of (F) amounted to (142.184), and the level of its significance is (0.00), which is less than the level of significance (0.05); This means rejecting the null hypothesis which indicates that there is no statistically significant role at ($\alpha \leq 0.05$) for digital marketing with its combined dimensions (attraction, communication, absorption, retaining, and electronically transferred word) in promoting the image of educational services for the universities of Jordan, and accepting the alternative hypothesis that: There is a statistically significant role at a level ($\alpha \leq 0.05$)) for digital marketing with its combined dimensions (attraction, communication, absorption, retaining, and the electronica word of) in promoting the image of educational services for

Table 2. Personal and Functional Characteristic

%	Frequency	Characteristics	Item
63.60%	262	Male	Gender
36.40%	150	Female	
100.00%	**412**	**Total**	
41.50%	171	Amman The Capital	Residency of the Student
58.50%	241	Other Provinces	
100.00%	**412**	**Total**	
2.20%	9	Diploma	Level of Student Education
61.40%	253	Bachelor	
36.40%	150	Graduate Studies	
100.00%	**412**	**Total**	
23.10%	95	Amman	Location of the University in which You study Belong Study
76.90%	317	Other Provinces	
100.00%	**412**	**Total**	
70.10%	289	Search Engine	Digital Services You Acquired In your University
55.10%	227	Email	
83.70%	345	Social Networking	
17.00%	70	Content Marketing	

universities in Jordan. It also clear from results that attraction variable is no significant as an individual dimension, and this may be due to the fact that respondents verify that other dimensions are more significant compared to this dimension, but in general the overall dimensions are significant.

Table 3. Multiple Regression Results

Dependent Variable	R	R^2	F	DF	Sig.	Independent Variable	B	T	Sig.
Enhancing the Image of Educational Services	0.798	0.637	142.184	5	0.000	Attraction	.035-	.590-	.556
				407		Communication	.159	3.779	.000
				412		Absorption	.613	9.174	.000
						Retaining	.227	3.365	.00
						Electronic words of mouth	.449	8.036	.000

5 Recommendations and Future Insight

Based on the main findings of the study the following are the main recommendations:

1. The universities in Jordan should focus on the dimension of attraction, which showed weakness in affecting the enhancement of the image through holding competitions and recreational activities that attract students.
2. The universities in Jordan should establish a special departments for digital marketing with the aim of establishing updated work strategies and following them up to identify scientific changes in this field of digital marketing and their applications.
3. Universities in Jordan should give the attention to the variable of absorption, which ends with gaining new students, and this requires distinguished electronic content, an electronic application that is easy to deal with, and other various digital methods to attract the largest number and variety of students in the virtual world.
4. The need for universities to pay attention to the dimension of student retention, and this requires the credibility of digital content, ensuring the privacy of student data, and providing loyalty programs for students.
5. Universities in Jordan should pay attention to the dimension of communication by communicating with students using more creative methods through human emotion and making the communication process a pleasure for them, which makes them busy with the digital messages they receive, with the need to provide room to receive student complaints via the Internet and work to solve them.
6. It is suitable for other researchers to conduct studies taking into considerations other sectors and other variables in different regions to evaluate the role of digital marketing in enhancing and fostering firm's image.

References

1. Bala, M., Verma, D.: A critical review of digital marketing. International J. Manage. IT & Eng. **8**(10), 321–339 (2018)
2. Melović, B., Jocović, M., Dabić, M., Vulić, T., Backović, D.: The impact of digital transformation and digital marketing on the brand promotion, positioning and electronic business in Montenegro. Technology in Society (2023). https://doi.org/10.1016/j.techsoc..101425
3. Salam, O.: Measuring the impact of digital marketing dimensions on customer satisfaction: an applied study on a sample of the Egyptian General Company for Tourism and Hotels (EGOTH). Scientific Journal (Trade and Finance). **40**(1), 53–112 (2022)
4. Alwan, M., Alshurideh, M.: The effect of digital marketing on purchase intention: moderating effect of brand equity. International J. Data and Network Sci. **6**(3), 837–848 (2022)
5. López García, J.J., Lizcano, D., Ramos, C.M., Matos, N.: Digital marketing actions that achieve a better attraction and loyalty of users: an analytical study. Future Internet **11**(6), 130–135 (2019)
6. Statistics of the Jordanian Ministry of Higher Education and Scientific Research for the year (2021/2022) Ministry of Higher Education and Scientific Research (2022)
7. Almaazmi, J., Alshurideh, M., Al Kurdi, B., Salloum, S.A.: The effect of digital transformation on product innovation: a critical review. In International Conference on Advanced Intelligent Systems and Informatics Springer, Cham, pp.731–741 (2020)

8. Al-Ubaidi, S., Shawi, S.: The role of digital marketing tools in improving bank marketing performance. Finance and Business Economy Magazine **7**(1), 69–86 (2022)
9. Nawaz, S.S., Kaldeen, M.: Impact of digital marketing on purchase intention. Int. J. Advanced Science Technol. **29**(4), 1113–1120 (2020)
10. Azizi, N., Bouacha, M.: Analytical study of digital marketing applications in algerian hotels -a case study of 13 hotels in Constantine Province. J. Financial, Accounting and Managerial Studies **6**(I2), 47–61 (2019)
11. Dastane, O.: Impact of digital marketing on online purchase intention: mediation effect of customer relationship management. J. Asian Business Strategy **10**(2), 142–158 (2020)
12. Ghazie, D.A., Dolah, J.: How digital marketing affects consumer behavior. In: 3rd International Conference on Creative Media, Design and Technology (REKA). Atlantis Press, pp. 214–217 (2018)
13. Muchardie, B.G., Yudiana, N.H., Gunawan, A.: Effect of social media marketing on customer engagement and its impact on brand loyalty in caring colors cosmetics. Martha Tilaar. Binus Business Review **7**(1), 83–87 (2016)
14. Onobrakpeya, A., Mac-Attama, A.: Improving customer satisfaction through digital marketing in the Nigerian deposit money banks. Int. J. Science Eng. **2**, 2456–3293 (2007)
15. Azzam, A.: The moderator effect of COVID pandemic on the relationship between online shopping services and online consumer purchasing behavior. a case of Jordan. Journal of Studies in Systems, Decision and Control **216**, 25–35 (2023)
16. Azzam, A., Al-Shaer, S.: Electronic word of mouth (E_WOM) adoption via social media and its impact on online shoppers' purchasing intention during corona pandemic. a case of Jordan. Journal of Studies in Systems, Decision and Control **216**, 477–487 (2023)
17. Al-Shaikh, M.S., Khanfar, I.A.A.: Delivery service via electronic applications and its impact on customers satisfaction at retail stores in Amman City/Jordan. J. Studies in Systems, Decision and Control **216**, 827–837 (2023)
18. Azzam, Z.: Factors affecting the adoption of e-marketing technologies in services industry: empirical study on hotel sector in Jordan. Zarqa J. Res. Studies Humanities **13**(1), 210–224 (2013)
19. Al-Azzam, A.F., Azzam, Z.: Evaluating the impact of service quality, brand image, trust and promotion on customer loyalty in mobile phone in Jordan. Zarqa Journal for Research and Studies in Humanities **13**(2) (2013)
20. Al-Shaikh, M.: Customers' perspective towards factors associated with technological products adopting failure. a case of Jordan. Zarqa J. Res. Studies in Humanities **13**(2), 252–261 (2013)
21. Ali, N., Allan, M., Azzam, A.: The impact of total quality management (TQM) dimensions on achieving competitive advantage: Managerial perspective of the quality department staff at five-star hotels. case of Jordan. Int. J. Scientific and Technol. Res. **9**(2), 2578–2587 (2020)
22. Diventy, A.H., Rahadhini, M.D., Triastity, R.: The role of digital marketing on purchase intention and purchase decisions at Kopi Si Budi Surakarta. In: Proceeding Ictess (International Conference on Technology, Education and Social Sciences) (2021)
23. Mishra, C.K.: Digital marketing: Scope opportunities and challenges. J. Promotion Marketing Commun. **1**, 1–24 (2020)
24. Desai, V.: Digital marketing: a review. Int. J. Trend in Scientific Res. Dev. **5**(5), 196–200 (2019)
25. Babić Rosario, A., de Valck, K., Sotgiu, F.: Conceptualizing the electronic word-of-mouth process: what we know and need to know about eWOM creation, exposure, and evaluation. J. Acad. Mark. Sci. **48**(3), 422–448 (2019)
26. Charo, N.: Determining the impact of e-wom on brand image and purchase intention through adoption of online opinions. Int. J. Humanities Manage. Sci. **3**(1), 41–46 (2015)

27. Bohara, S., Gupta, A., Panwar, D.: Relationship between factors of online marketing and student enrollment decisions in higher education: an analysis using structural modeling techniques. Bohara International Journal of Online Marketing **12**(1), 1–18 (2022)
28. Hair, J.F., Harrison, D., Risher, J.J.: Marketing research in the 21st century: opportunities and challenges. Brazilian Journal of Marketing-BJMkt, Revista Brasileira de Marketing–Remark, Special Issue, **17**, 666–699. (2018)

The Role of Cloud Computing Applications in Improving the Performance of Employees at Zarqa University

Majed Al Masadeh[1(✉)], Fatima Haimour[2], Siham Haimour[2], Baraa Qaddoumi[3], Ghada Haimour[4], and Daliah Taibah[5]

[1] Economics and Administrative Sciences, Zarqa University, Zarqa, Jordan
mmasadeh@zu.edu.jo

[2] Information Technologies, Zarqa University, Zarqa, Jordan
{fhaimour,shaimour}@zu.edu.jo

[3] Information Technologies, Applied Science Private University, Amman, Jordan
b_qaddoumi@asu.edu.jo

[4] Information Technologies, The Ministry of Education, Amman, Jordan

[5] Department of HRM, University of Business and Technology, Jeddah, Kingdom of Saudi Arabia
d.taibah@ubt.edu.sa

Abstract. Cloud computing is a new fielding that has emerged in the Internet area in recent years. Nowadays, cloud applications play a significant role in the service delivery of cloud organizations, so cloud service providers built applications for users to access cloud services with ease and quality of service (QoS). Cloud computing applications have been identified as the next big trend for efficient e-Learning systems that will improve the quality of their education by higher education institutions. Cloud computing has a clearly important place in the world of higher education, both as a ubiquitous computing tool and as a powerful platform. This paper sheds light on the role of cloud computing applications and their effective impact on improving the performance of employees at Zarqa University.

Keywords: Cloud Computing · Information Technology · Education

1 Introduction

Our current era attests to the appearance of many novelties that have changed many life scenes, which will also affect students and their academic levels, such as the spread of the global pandemic (COVID-19), the rapid development of technologies, the spread of distractions, interactive games, the world financial crisis, and more, which create big worries for upcoming demands like the need for student learning to be more flexible in time and place, with interesting assignments and engagements, and cause the necessity of canceling some of the traditional methods and following new strategies to make education motivating and improve students' commitment to learning. Cloud computing technology is an example of technological advancements that provide new opportunities for improving teaching and learning through the use of technology in education.

B. A. M. Alareeni and I. Elgedawy (Eds.): *Artificial Intelligence (AI) and Finance*, SSDC 488, pp. 448–460, 2023.
https://doi.org/10.1007/978-3-031-39158-3_43

"A new dawn" is one of the most beautiful descriptions of cloud computing for education, carrying the meaning of hope and all the facilities that offer new opportunities for enhancing teaching and learning with technological advances [1].

Cloud computing means storing and accessing data and programs over the Internet from a remote location or computer instead of our computer's hard drive. Based on service-on-demand and pay-as-you-go [2].

There is no doubt that the prevalence of mobile devices such as smartphones and tablets with fast Internet access using 3G and 4G communication links helped to increase the rate of use of cloud computing applications. These devices have direct, quick, and affordable connections to the internet, which makes it simpler to publish and download software applications to the internet and use online network storage services. Often, a mobile connection offers a better and faster connection to the cloud than a fixed broadband connection [3, 4].

Many universities are looking for different ways to deal with the mounting contradiction between financial pressure and IT costs. Cloud computing has assisted colleges and universities in balancing their expenses while still receiving high-quality IT services [5].

Through the use of cloud computing, university institutions can transfer their data and software from a physical server—possibly located on their campus—to the internet. Administrators and students, among other institutional stakeholders, are now able to access this data from different places and via different devices.

Cloud computing is composed of three service delivery models, four deployment models, and five essential characteristics [7, 8].

- **Cloud computing can be separated into three general services:**

1. Software as a service (SaaS): SaaS is a distribution model that delivers software applications over the Internet; these applications are often referred to as "web services." Users can access SaaS applications and services from anywhere using a computer or mobile device with internet access. The SaaS model gives users access to application software and databases. A common example of a SaaS application is Microsoft 365 for productivity and email services [7, 9, 10].
2. Platform as a Service (PaaS): In the PaaS model, cloud providers host development tools on their own infrastructure. Users access these tools over the Internet via APIs, web portals, or gateway software. PaaS is used for software development in general, and many of its PaaS providers host the software after it is developed. Common PaaS offerings include Salesforce's Lightning Platform, AWS Elastic Beanstalk, and Google App Engine [7, 11].
3. Infrastructure as a Service (IaaS): offer virtual server instances and storage, as well as application programming interfaces (APIs) that allow users to migrate workloads to virtual machines (VMs). A user can allocate storage capacity, and he can start, stop, access, and configure VMs and storage as needed. IaaS providers offer small, medium, large, and extra-large memory- or compute-optimized instances, as well as allowing instances to be customized for different workload requirements. The IaaS cloud model is the closest thing a business user has to a remote data center [7, 12, 13].

- **There are four common cloud deployment models:** private, public, community, and hybrid. The deployment models are classified according to the location of the environment of the cloud structure [15, 16].

1. Private Cloud: A cloud computing model in which IT services are maintained for a single entity via a private channel or network hardware [17]. In a private cloud, the network's communication is controlled via private channels or servers.
2. Public Cloud: The public cloud is the most common and basic type of cloud where users choose to dedicate computing resources such as hardware and software to access applications on a subscription basis [18].
3. Community Cloud: The main purpose of building such a cloud between similar entities is to share general guidelines, adaptations, and implementations, benefiting the ideal framework and allowing for the comparison of mechanisms [19]. Educational institutions, health-related fields, scientific research organizations, and several other large associations use this community cloud to access and share information.
4. Hybrid cloud: This is the combination of one or more cloud networks [20]. It can be a collection of private networks or a mix of one private network and another public network. This is the most comprehensive cloud, providing services to every end user in every field.

- **This cloud model is composed of five essential characteristics** [21]:

1. On-demand self-service.
2. Broad network access.
3. Resource pooling.
4. Rapid elasticity.
5. Measured Service.

There are several reasons why cloud-based solutions are a strategic business investment for higher education in 2021 and beyond, even if this was crucial to institutional and student performance throughout the epidemic.

The cloud offers a variety of advantages that support educational institutions in adapting to an increasingly digital world. Cloud-based software-as-a-service (SaaS) solutions offer greater flexibility to meet changing needs, enabling colleges and universities to support remote learning and a distributed workforce. Cloud computing also enhances collaboration between students and educators, reduces downtime, ensures reliable data recovery, and provides predictable monthly expenses. These advantages have helped colleges and universities to better serve their students and employees, making cloud computing an essential aspect of modern education [6].

In this paper, we highlight the importance and benefits of cloud computing applications for higher education institutions in general and Zarqa University in particular. We created a questionnaire that surveys the opinions of the employees at Zarqa University. The remainder of the paper is organized as follows: In Sect. 2, related work is reviewed. In Sect. 3, the method and research design are described. In Sect. 4, the results of our study and the related discussions appeared in Sect. 5. Section 6 concludes the paper.

2 Related Works

Many researchers have made a clear contribution to focusing on the effective role of cloud computing applications in higher education institutions where the use of cloud computing technologies to provide ICT infrastructure and platforms for education is advantageous in terms of cost, processing power, ease, security, and scalability, and we will mention some of these studies as the following:

This study aimed to identify the difficulties of using cloud computing at the Faculty of Information Technology, Benghazi University, as seen by teachers and students. The study sample consisted of 250 students and 50 of his professors. The sample of the study community is random, and we created a performance of the study as a questionnaire to measure the difficulties in using cloud computing at the University of Benghazi. Through the results obtained in this survey, we found that both students and faculty members are willing to use computer technology. Modern education is changing, as is classroom education, due to the advent of technology. Students expect and demand more services from their institutions. The future success of a country depends on the development of the education of its students. The future world will be a technology-based world, so using technology in the study session will help a student be fully prepared for the future world. Therefore, we conducted a study on the acceptance of cloud computing technology in education [22].

The goal of this research is to understand what influences the intention of Malaysian higher education students to use cloud computing services. There seems to be a strong correlation between the intention to use cloud computing and its perceived ease of use, perceived usefulness, optimism, and the learning environment's uncertainty. The results may be useful for higher education institutions and cloud computing service providers to gain a better understanding of the defining elements and components that should be emphasized in order to promote and improve the use of cloud computing among university students. In addition, the students should be made aware of the advantages of cloud computing compared to conventional technologies. These skills are particularly crucial in times of crisis, such as during the COVID-19 outbreak, where cloud computing has proven to be an effective tool for global knowledge sharing while streamlining teaching and learning processes. In addition, the TRI-TAM model contributes significantly to behavioral theories by extending current models of behavior, particularly that of technological adaptation. These data further demonstrate the validity and applicability of the models, particularly in developing countries [23].

This study highlighted the cloud computing adoption process in education, as well as deployment guidelines. The few challenges that are likely to arise can be solved with new and better policies and techniques, and a proposed hybrid cloud model for higher education institutions in Saudi Arabia was discussed. Cloud computing creates a universal platform with simplified scalability. Therefore, it will be crucial for schools and individuals to move to the cloud to experience the cost-effective and convenient way to access information and technology services, especially the benefits and capabilities, such as B. access to complex applications, minimal costs for cloud data storage, scalability, and flexibility of a cloud-computing-enabled e-learning platform. However, e-learning integration in universities has some shortcomings that need to be considered before adopting and integrating the system [24].

Cloud computing technology is advantageous in terms of economy, computing power, convenience, security, and scalability for providing ICT infrastructure and platforms for education. These benefits can be effectively utilized in the school education system. Leveraging the cloud in the school education system provides features such as easy scheduling of classes, labs, educational trips, and computer resource reservations. Cloud computing provides flexible, unified, open platforms for education, the sharing of educational resources, and the closing of information gaps between different educational domains. Cloud-based labs effectively provision computing resources, application software, operating systems, and necessary infrastructure. School teachers and students all benefit from its use, as it reduces the hassle of deploying local calculations, buying large software licenses, etc. This increases student and teacher productivity. Therefore, cloud computing offers enormous opportunities for the development of school education, which can certainly lead to the development of society [25].

3 Methods

3.1 The Research Questions

The research questions formulation developed from the framework are:

- What is the impact of cloud computing on educational practices and outcomes?
- How can it be effectively integrated into the learning environment to enhance teaching and learning experiences?

3.2 Study Design

In this study, a self-reviewed online questionnaire was employed as the primary data collection instrument. The sample consisted of 300 individuals from various faculties at Zarqa University, covering a diverse range of fields including Information Technology, health, humanities, science, engineering, law, and media. The utilization of an online questionnaire allowed for a convenient and efficient means of collecting data, while also facilitating anonymity and reducing potential bias that may arise from face-to-face interviews. The sample was chosen to be representative of the university community, and the diversity of faculties included in the sample ensures that the findings of the study are applicable to a wide range of academic disciplines.

3.3 Data Collection

To recruit participants for the study, all members of the local community were contacted via their email addresses and invited to participate. The email included information about the purpose of the study, and a link to an online survey designed using the Google Forms platform. This approach provided a convenient and accessible means of collecting data, while also ensuring that all responses were confidential and anonymous. By using an online survey platform, the study was able to reach a larger and more diverse pool of participants, and the use of anonymous responses may have encouraged greater honesty and candor in participants' answers. Overall, the study's recruitment strategy and survey design enabled a more inclusive and unbiased approach to data collection.

3.4 Data Analysis

To analyze the results of the study, the authors utilized SPSS version 23, and only completed questionnaires were included in the analysis. Descriptive statistics were applied to demographic characteristics such as gender and occupation. Means and percentages were employed to analyze acceptability and ethical considerations. The use of SPSS enabled the authors to process and interpret the collected data efficiently and accurately. Descriptive statistics provided a clear summary of the characteristics of the sample and allowed for the identification of patterns and trends in the data. The use of means and percentages provided a more nuanced understanding of the study's findings, enabling the authors to draw more precise conclusions. Overall, the use of SPSS and descriptive statistics facilitated a more rigorous and insightful analysis of the study's results.

4 Results

4.1 Demographic Variables

A total of 220 individuals agreed to participate in the study. The participants were divided into two groups according to gender: 100 females (45.45%) and 120 males (54.55%). The average age of the participants was 33.5 years old, with a standard deviation of 7.7. The sample was further categorized based on faculty affiliation, which included eight different categories as presented in Table 1. The Information Technology faculty had the highest percentage of participants at 20.34%, followed by the Health faculty at 15.8%. Humanities, Science, Engineering, Law, and Media faculties had an equal percentage of 10.5%. The remaining participants were grouped under "Other" faculties, comprising 11.71% of the total sample. The demographic data presented in this study allowed the researchers to identify any patterns or trends in the sample population, which could potentially inform the interpretation of the study's results.

Table 1. Distribution of study participants according to their faculties

Information Technology	Health	Humanities	Science	Engineering	Law	Media	Others…
20.34%	10.5%	10.5%	10.5%	10.5%	10.5%	10.5%	11.71%

4.2 The Role of Cloud Computing Applications in Improving the Performance of Employees at Zarqa University

In the following, we will read the results that obtained from the questionnaire (see Table 2).

The results of the questionnaire (see Table 2) show that most of the participants, with a percentage of (80.71), agree that they could easily learn how to interact with Cloud Computing applications. However, a small portion of the participants, with a ratio

Table 2. The questionnaire's results.

	Strongly disagree	Disagree	Neither agree nor disagree	Agree	Strongly agree
Could it easily learn how to interact with Cloud Computing applications	8.05	10.20	1.04	35.30	45.41
Acquiring proficiency in using the Cloud Computing application is a straightforward task	6.10	7.91	1.17	30.82	54.00
I would have no difficulty in locating information within the Cloud Computing application	9.91	12.56	2.02	48.30	27.21
The work and learning can be made more efficient by making use of Cloud Computing applications for advancing studies	4.92	8.90	1.59	34.72	49.87
Using Cloud Computing applications to advance my studies has the potential to benefit both my work and learning	3.05	6.03	2.41	37.87	50.64

(continued)

Table 2. (*continued*)

	Strongly disagree	Disagree	Neither agree nor disagree	Agree	Strongly agree
The academic/work productivity would be elevated by using Cloud Computing applications	3.67	5.92	1.81	44.40	44.20
Cloud Computing applications are secure	10.20	27.54	1.95	25.50	34.81
The information exchanged within Cloud Computing applications is transmitted through a secure communication network	6.05	15.76	6.56	24.43	47.20
Cloud Computing applications are reliable and stable	5.75	8.00	2.31	46.48	37.46
The process of teaching and learning might necessitate the utilization of Cloud Computing applications by me	8.68	10.49	1.16	33.40	46.27
The individuals in my surroundings have persuaded me of the significance of utilizing Cloud Computing applications	4.68	6.99	2.32	45.41	40.60

of (18.25), expressed their difficulty in learning how to interact with cloud computing, while only (1.04) of the participants neither agreed nor disagreed. These findings suggest that most members of the Zarqa University community feel confident about their ability to learn and use Cloud Computing applications, while a smaller percentage may require additional support and training to effectively engage with these technologies.

Moreover, most of the participants (84.82%) agreed that acquiring proficiency in using Cloud Computing applications is a straightforward task. However, a proportion of participants (14.01%) disagreed with this statement, indicating that they may have found it challenging to become skilled in using such applications. Only a small percentage of participants (1.17%) selected "Neither agree nor disagree," suggesting that they may not have a strong opinion on the ease of acquiring proficiency in using Cloud Computing applications. These findings suggest that most members of the Zarqa University community who participated in the study perceive learning how to use Cloud Computing applications as an easy task.

According to the survey results, (75.51) of the respondents agreed that they would have no difficulty in locating information within the Cloud Computing application. On the other hand, (22.74) respondents found it challenging to locate information within the application, while (2.02) respondents neither agreed nor disagreed. The ease of finding information is an essential aspect of using Cloud Computing applications, especially in educational and professional settings. The results indicate that most respondents have confidence in their ability to navigate the application's features and access the information they need. However, there is a significant minority that finds it challenging to locate information within the application, which could potentially impact their productivity and efficiency.

In the study, most participants (84.59%) agreed that the use of Cloud Computing applications could make work and learning more efficient for advancing studies. However, a smaller percentage of participants (13.82%) disagreed with this statement. Additionally, a small percentage (1.95%) neither agreed nor disagreed. This result suggests that a significant number of participants believe that Cloud Computing applications can help to advance their studies, which aligns with the growing trend of using technology to enhance the learning process. However, it is also important to consider the perspectives of those who disagreed with the statement and explore their concerns or potential barriers to using Cloud Computing applications for advancing their studies.

According to the results of the survey, a significant percentage of participants (88.51%) agreed that using Cloud Computing applications has the potential to benefit their work and learning, which indicates a positive attitude towards the integration of these technologies in education. However, a small proportion of participants (9.08%) disagreed with this statement, suggesting that some individuals may still have concerns or reservations about using Cloud Computing applications in their academic and professional pursuits. Additionally, (2.41%) of respondents neither agreed nor disagreed, indicating the need for further exploration and investigation into the potential benefits and drawbacks of these technologies.

The results of the survey indicated that a majority of respondents (88.60%) agreed that using Cloud Computing applications would elevate academic/work productivity. On the other hand, a smaller proportion of respondents (9.59%) disagreed with this statement, suggesting that there may be some concerns or challenges with using Cloud Computing applications in this context. A small percentage of respondents (1.81%) neither agreed nor disagreed, indicating a lack of strong opinion or uncertainty. These findings suggest that while there may be some barriers to using Cloud Computing applications, they may have the potential to positively impact academic and work productivity. Further research may be needed to identify specific factors that could enhance the benefits of Cloud Computing applications for productivity in education and other contexts.

The study results reveal that a majority of the participants hold a positive view of the security of Cloud Computing applications. Specifically, (60.31%) of the participants agreed with the statement that cloud computing applications are secure, while only (37.74%) disagreed. These findings suggest that most people trust cloud-based systems and believe that they provide a reliable and safe way to store and access data.

The survey results indicate that (71.63%) of the respondents agreed with the statement that "the information exchanged within Cloud Computing applications is transmitted through a secure communication network," while (21.81%) disagreed with it. It is worth noting that (6.65%) of the participants neither agreed nor disagreed with this statement.

According to the results of the questionnaire, most of the participants agreed that Cloud Computing applications are reliable and stable. Specifically, (83.94%) of the participants agreed with this statement, while (13.75%) disagreed. A small proportion of the participants (2.31%) neither agreed nor disagreed with this statement. This finding suggests that most of the members of the Zarqa University community who participated in the study had a positive view of the reliability and stability of Cloud Computing applications. It may also indicate that the participants have had some experience using Cloud Computing applications and have found them to be dependable in their daily work or studies.

The study found that the majority of participants, (79.67%), agreed with the statement that "the process of teaching and learning might necessitate the utilization of Cloud Computing applications by me." Only (19.17%) of respondents disagreed with this statement, while a small percentage (1.16%) neither agreed nor disagreed. This indicates that there is a general perception among the Zarqa University community that Cloud Computing applications may be useful in the context of teaching and learning. This finding is in line with the increasing adoption of digital technologies in the field of education and highlights the potential of cloud computing to support learning and improve educational outcomes.

According to the survey results, the majority of participants (86.61%) agreed that the individuals in their surroundings have influenced them to recognize the importance of using cloud computing applications. On the other hand, (11.67%) of the participants disagreed with the statement, and (2.32%) of them neither agreed nor disagreed. This finding suggests that social influence could be a significant factor in promoting the use of cloud computing applications, especially in educational institutions. It implies that individuals who are familiar with cloud computing and their benefits could encourage their peers and colleagues to use them in their work or academic activities.

5 Discussions

Cloud computing has revolutionized the way education is delivered and consumed. With the help of cloud technology, educators and students can access a vast number of educational resources and tools from anywhere and at any time. Cloud-based learning management systems enable institutions to manage and deliver courses, assessments, and grading more efficiently. Below we discuss the most prominent things that resulted from our questionnaire (Fig. 1).

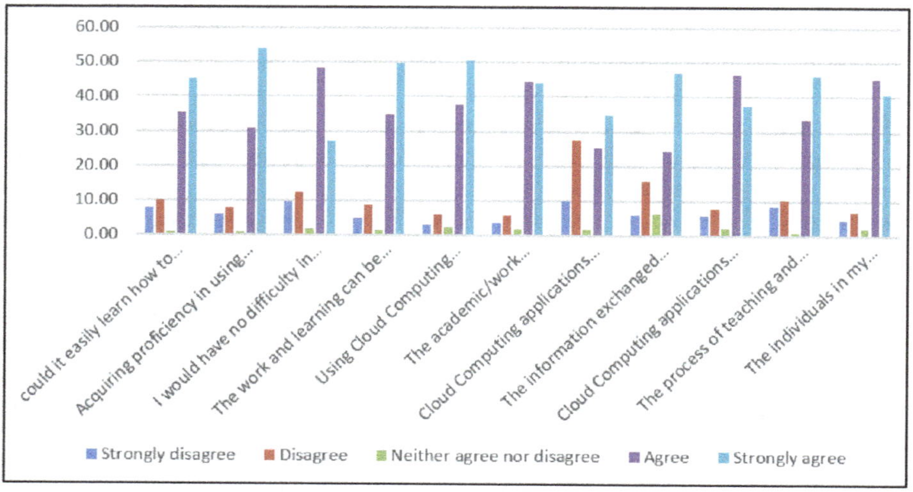

Fig. 1. Shows the comparison between different components of the questionnaire.

The questionnaire results indicate that using cloud computing applications to advance studies can have the potential to benefit both work and learning, with the component receiving the highest percentage of agreement from respondents. This suggests that many individuals believe that cloud-based technology can provide practical and functional advantages for their academic and professional pursuits. Additionally, respondents showed a high level of agreement towards the idea that using cloud computing applications could elevate their academic and work productivity, highlighting the potential of cloud-based technology to improve efficiency and output.

On the other hand, the component "Cloud Computing applications are secure" received the least percentage of "strongly agree" opinions, suggesting that respondents had some reservations about the safety and reliability of cloud-based systems. Despite this, it is worth noting that a majority of respondents still agreed that cloud computing applications are secure, indicating that while there may be some concerns, many individuals trust that cloud-based systems can provide a safe and reliable way to store and access data.

Overall, the results indicate that respondents believe cloud computing applications are easy to use and learn, which is a promising sign for the future of cloud-based technology in education and the workforce. As cloud-based technology continues to evolve and improve, it will be important to address any concerns or reservations users may have about security and to continue educating individuals on the benefits and potential of cloud computing applications.

6 Conclusion

In conclusion, the study found that the majority of the participants were willing to learn how to interact with cloud computing applications and acquire proficiency in using them. Additionally, they believed that using cloud computing applications would benefit both their work and learning and increase their academic/work productivity. Despite the positive responses, a considerable number of participants disagreed that cloud computing applications are secure and reliable, which is a concern that needs to be addressed. The study also revealed that the process of teaching and learning might necessitate the use of cloud computing applications.

Furthermore, the results suggest that there is a need for more awareness and education regarding the use of cloud computing applications to increase their acceptability among the academic community.

Overall, the study sheds light on the attitudes and perceptions of the academic community towards cloud computing applications and highlights the need for further research to address the issues that hinder their acceptability and usage in academic settings.

References

1. Sultan, N.: Cloud computing for education: a new dawn? Int. J. Inf. Manage. **30**(2), 109–116 (2010)
2. Chandrasekaran, K.: Essentials of cloud computing. CrC Press (2014)
3. Fernando, N., Loke, S.W., Rahayu, W.: Mobile cloud computing: a survey. Futur. Gener. Comput. Syst. **29**(1), 84–106 (2013)
4. Ali, M., Miraz, M.H.: Cloud computing applications. In: Proceedings of the International Conference on Cloud Computing and eGovernance, vol. 1 (2013)
5. Yaghmaei, O., Binesh, F.: Impact of applying cloud computing on universities expenses. IOSR J. Bus. Manag. **17**(2), 42–47 (2015)
6. Benefits of Cloud Software for Higher Education. Corsedog (2023). https://www.coursedog.com/articles/benefits-of-cloud-based-solutions-in-higher-Text. Citation: (7 Benefits of Cloud Software for Higher Education, 2023)
7. Nazir, R., Ahmed, Z., Ahmad, Z., Shaikh, N.N., Laghari, A.A., Kumar, K.: Cloud computing applications: a review. EAI Endorsed Trans. Cloud Syst. **6**(17), e5–e5 (2020)
8. Laghari, A.A., He, H., Halepoto, I.A., Memon, M.S., Parveen, S.: Analysis of quality of experience frameworks for cloud computing. IJCSNS **17**(12), 228 (2017)
9. Colman-Meixner, C., Develder, C., Tornatore, M., Mukherjee, B.: A survey on resiliency techniques in cloud computing infrastructures and applications. IEEE Commun. Surv. Tutorials **18**(3), 2244–2281 (2016)
10. Giannakis, M., Spanaki, K., Dubey, R.: A cloud-based supply chain management system: effects on supply chain responsiveness. J. Enterp. Inf. Manage. **32**, 585–607 (2019)
11. Zaballos, A.G., Rodríguez, E.I.: Cloud Computing: Opportunities and Challenges for Sustainable Economic Development in Latin America and the Caribbean. Inter-American Development Bank (2018)
12. Stephen, A., Benedict, S., Kumar, R.P.A.: Monitoring IaaS using various cloud monitors. Clust. Comput. **22**(5), 12459–12471 (2018). https://doi.org/10.1007/s10586-017-1657-y
13. Guo, C., Xu, K., Bose, S., Zukerman, M., Shen, G.: Efficient and green embedding of virtual data centers with mixture of unicast and multicast services. IEEE Trans. Cloud Comput. **9**, 1008–1021 (2019)

14. Koutsogiannis, E., Mamatas, L., Tsaoussidis, V.: Opportunistic forwarding for user-provided networks. Int. J. Parallel Emergent Distrib. Syst. **33**(6), 717–741 (2018). https://doi.org/10.1080/17445760.2017.1376667
15. Sahli, H., Hameurlain, N., Belala, F.: A bigraphical model for specifying cloud-based elastic systems and their behaviour. Int. J. Parallel Emergent Distrib. Syst. **32**(6), 593–616 (2017)
16. Ristov, S., Mathá, R., Kimovski, D., Prodan, R., Gusev, M.: A new model for cloud elastic services efficiency. Int. J. Parallel Emergent Distrib. Syst. **34**(6), 653–670 (2019)
17. Alvarez, R., Mirzoev, T., Gowan, A., Henderson, B., Kruck, S.E.: Learning laboratories as services in private cloud deployment. J. Comput. Inf. Syst. **59**(4), 354–362 (2019)
18. Firestone, D., et al.: Azure accelerated networking: SmartNICs in the public cloud. In: 15th USENIX Symposium on Networked Systems Design and Implementation (NSDI 18), pp. 51–66 (2018)
19. Mohan, S.L., Reddy, Y.R., Gangadharan, G.R.: Compac—a pricing model for community cloud. In: 2017 International Conference on Advances in Computing, Communications and Informatics (ICACCI), pp. 2033–2039. IEEE (2017)
20. Riaz-ud-Din, F., Zhou, W., Doss, R.: Query verification schemes for cloud-hosted databases: a brief survey. Int. J. Parallel Emergent Distrib. Syst. **31**(6), 543–561 (2016)
21. Essential Characteristics of Cloud Computing | AT&T Business. AT&T Business (2017). https://www.business.att.com/learn/tech-advice/5-essential-characteristics-of-cloud-computing.html
22. Elnajar, K.M., Sahly, E.M., Farkash, H.M., Faraj, A.: Cloud computing in education: a survey on the adoption and the challenges for the faculty of IT at the Benghazi University-Libya. Int. J. Comput. Sci. Commun. Inf. Technol. **7**, 36–43 (2019)
23. Amron, M.T., Noh, N.H.M., Mohamad, M.A.: Predicting the acceptance of cloud computing in higher education institutions by extending the technology readiness theory. Asian J. Univ. Educ. **18**(3), 767–779 (2022)
24. Almajalid, R.: A survey on the adoption of cloud computing in education sector (2017). arXiv preprint arXiv:1706.01136
25. Kumar, V., Bhardwaj, A.: Role of cloud computing in school education. In: Tomei, L.A., Carbonara, D.D. (eds.) Handbook of Research on Diverse Teaching Strategies for the Technology-Rich Classroom, pp. 98–108. IGI Global (2020). https://doi.org/10.4018/978-1-7998-0238-9.ch008

Impact of Digital Advertising via Social Media Tools on the Buying Behavior of Fast Food Consumers. A Case of Jordan

Zakaria Ahmad Azzam[1(✉)], Ali Hamdan[1], Nafez Ali[2], and Kholod Aggad[3]

[1] Marketing Department, Zarqa University, Zarqa, Jordan
{zazzam,20193122}@zu.edu.jo

[2] Marketing Department, Al Zaytoonah University of Jordan, Amman, Jordan
N.Ali@zuj.edu.jo.Jordan

[3] Department of Marketing, University of Business and Technology, Jeddah, Kingdom of Saudi Arabia
k.aggad@ubt.edu.sa

Abstract. The study aims to find out the impact of digital advertising through social media on buying behavior and targeted fast-food consumers in Jordan. The study consisted of one independent variable, which were measured through six dimensions: characteristics of digital advertising, stimuli of digital advertising, the content of the digital advertising message interaction with digital advertising, and ease of access to digital advertising, the credibility of digital advertising through social media, and one dependent variable represents buying behavior. The study followed the descriptive-analytical method, using suitable statistical analysis program to analyze the data. (465) questionnaire were distributed to the study sample, and a total of (416) questionnaire were retrieved with a response rate of (89%). The study reached several results, the most important of which are: The digital advertisement has a significant statistical impact on buying behavior, and this effect is considered positive as it helps in improving the buying behavior of fast-food consumers in Jordan, which motivates them to buy and make the purchasing decision, and that the electronic advertisement has an explanatory ability to achieve buying behavior by a percentage of (75%). Digital advertising incentives have the strongest positive impact on the buying behavior of fast food consumers with an interpretive capacity of 68%, while the digital advertising credibility variable has the lowest impact on buying behavior among fast food consumers by 36%. The study recommended that fast-food restaurants owners must enhance the credibility of fast-food restaurants advertisements through punctuality and promises and choosing specialized experts in designing promotional campaigns to ensure high-quality advertising content that contains different stimuli.

Keywords: Digital Advertising · Social Media · Buying behavior · Fast food · Jordan

B. A. M. Alareeni and I. Elgedawy (Eds.): *Artificial Intelligence (AI) and Finance*, SSDC 488, pp. 461–472, 2023.
https://doi.org/10.1007/978-3-031-39158-3_44

1 Introduction

With the widespread proliferation of the use of social networking sites, companies have shown attention and go towards advertising through these means; because of its accessibility to the largest target category of its products in the least time and efforts, which achieves a competitive advantage for these firms. The ease of obtaining smartphones at the present time, and providing internet services in most regions of the Hashemite Kingdom of Jordan, companies allocated a specific budget to spend on digital advertisements via social media for the benefits they have such as accessibility to digital advertisement through social media by everyone in several languages, and the ability to manage and amend the content of the advertising message published on social media in any time [1].

Social media is a successful and effective tool for increasing sales and achieving profits as it attracts the attention of companies, especially the restaurants and fast food sector in the Hashemite Kingdom of Jordan which invest greatly ion digital ads due to consumers' demand. The owners of restaurants and fast food also keep pace with the technological development by creating a special site for them on the websites or their applications that encourage consumers to make orders through them by using stimuli in electronic advertisements for their benefits they provide to customers such as, the ability to connect consumers with services providers and marketers and provide electronic payment service easily.Where there has become a tendency and demand for the self-services delivery and electronic payment due to the presence of an urgent need for consumers to use electronic payment systems to facilitate the progress of purchasing processes, and all of these factors helped the success of the fast food market in the Hashemite Kingdom of Jordan, to adopt digital advertising. Also, most of the residents of the Hashemite Kingdom of Jordan have several accounts on social media, and spend most of their time browsing through these sites several times per day, which motivated restaurant owners to use these means to advertise through; which reaches large number of customers that can be targeted easily, were the number of social media accounts in the Kingdom of Jordan is 10.9 million accounts, and this number is distributed over five major social networks that Jordanians use on a daily basis; Facebook, Twitter, Instagram, Snap chat, Linked In [1].

Digital advertisements via social media may also help to reduce the gap between the customers and the company; this may be due to the characteristics it enjoys such as; the ease and speed of the company to know the reactions of customers through the interaction of consumers through social media with the company's products and services and to know the mistakes when they happen and work to solve them as quickly as possible. It also allows the possibility of identifying what customers want, as it enables the company to know the opinions of those who deal with it and their future expectations regarding the company's products and services, prices and ways to promote them. On the other hand, the digital advertisement helps consumers in making comparison between several options available through social media in the least possible time and effort and choosing the best offer. Related to the buying behavior of consumers, the rapid technological development, the emergence of electronic promotional means and the possibility of ordering products electronically which led to adjusting the purchasing behavior of consumers and changing their purchasing nature and patterns, which forces consumers to buy online because of its advantages that meet the desires of consumers, where they find what they want on the Internet and provides them with comfort and ease of comparing products in a shorter time [2].

This study aims to identify the impact of digital advertisement via social media on the buying behavior of fast food consumers in Jordan.

2 Research Problem

The most important characteristics provided by social media as advertising channels are the ability and speed of access to large numbers for social media users, and to learn about the desires of current or potential consumers, which prompted companies to develop their promotional methods of their products and create content on social media in order to benefit from marketing opportunities that is provided by social media, in the Hashemite Kingdom of Jordan. Where until now there are some fast food restaurants have not resorted to this type of advertisements. All of these points indicate that the owners of fast food restaurants are not aware of the benefits and returns of digital advertisements via social media and its impact on the buying decision. Also the absence of a comprehensive view of the most used means and the correct way to manage the advertising campaign in light of the presence of social media, as there are many challenges facing Fast-food restaurants owners during their ads, such as: difficulties to persuading viewers to believe the announced content; since it is not in front of them and they are not feeling and touching them, as the credibility of the digital advertisement plays a role in the success of the advertising campaign and achieving the desired goals. And with the emergence of a Covid-19 virus in the world, consumers' tendency to adopt the concept of on line buying because of the restrictions imposed by the governments, as it has become the best option to complete the purchase process without mixing with others, and in fast food restaurant sector in Jordan particularly the government allowed restaurants to sale through some applications on its websites, which helped restaurants avoid losses during the pandemic, this gives the majority of consumers to enhance their purchasing behavior via the Internet because it was the best option in these circumstances available for them, which created a culture of purchasing online which is safe, fast, comfort, easy and speedy for customers [1–5] and [1].

This study tries to explain the impact of the electronic/ digital advertisement through the social media on buying behavior, and the problem of the study can be clarified by answering the following:

What is the effect of electronic advertising via social media on the purchasing behavior of fast food consumers in Jordan?

3 Digital Advertising

Digital advertisement is defined as "the advertisement that is published via the web, which aims to promote goods, marketing a commodity, service, or promoting the website itself, and the advertisement via the Internet is a form of promotion that uses the Internet and the World Wide Web for the purpose of providing marketing messages to attract customers attention [6] and [7].

It was also defined as advertisement that is made by electronic means and the most famous mean used for this purpose is on the Internet which aims to promote goods, marketing or advertising for the website itself. It is also known as a communication process by publishing and broadcasting the print advertising messages, including the internet among the target audiences in order to motivate them to buy goods and services and to be convinced by these messages and make a suitable decision based on them. From the researcher's point of view, it is the use of the internet by fast food restaurants to introduce customers to fast food, show their characteristics and promote them to achieve marketing goals. The following are the main variables that have direct positive or negative impact on buying behaviors of customers related to digital advertisements made by firms [6–13] and [14]:

3.1 Digital Advertising Characteristics/Properties and Features

Here we review the most important characteristics of the digital advertisements, which in turn contributes in influencing on the audience (customers), which may gain a great turnout, and these characteristics are as follows [12, 13] and [14]:

1. Interactive: This feature allows the consumer to be able to contact directly with the advertisers, interact with their ads and the ability to request more information about the product, and answer consumers' questions and know their problems and suggestions.
2. The ability to reach the public quickly and smoothly: The electronic advertisement is available 24 h a day and throughout the week, and through this feature the company can manage its advertising campaign in a successful way, the possibility of amendment, continuous updating of information, and the possibility of canceling it according to the interaction with the target audience which is one of the most important fundamental differences between the electronic and traditional advertisement, as the traditional advertisement needs a lot of effort, time and cost to make adjustments.
3. The ability to obtain information about the product or service: The consumer can obtain a large number of information and the specifications of the announced product without any physical effort, and at any time they wants and the possibility of comparing several similar products in a short time.
4. The possibility of immediate purchase: where electronic advertisement allows the consumer the possibility of immediate purchase of the product that the consumer wants to buy and can complete the purchase through various payment methods, such as: credit cards and electronic portfolios and then deliver the product to the desired place.
5. Privacy: The electronic advertisement can give privacy to the consumer by obtaining detailed information and knowing the points of concern to the consumer and making the purchase decision without being subjected to pressure from sellers.
6. Low costs: The electronic advertisement is at a low cost compared to the traditional advertising methods, due to the lack of many costs such as printing posters and wages for the manpower.
7. The universality of the advertising offer which is characterized by the traditional electronic advertisement in reaching all parts of the world without facing restrictions and limitations that hinder it.

8. Multimedia integration: Where the electronic advertisement allows the ability to use texts, audio and visuals and the ability to try free products.

3.2 Digital Advertising Stimuli

Stimuli are tantamount to several additional features to the advertisement, such as adding aesthetic and the power to influence viewing to provoke motives and force customers to complete the purchase process, and here lies the ingenuity of advertising designers and marketers in reaching ideas and adding appropriate effects. The electronic advertisement plays an important role in attracting consumers and increasing the market share, so the electronic advertisement must include incentives that increase the effectiveness and success of the advertisement, such the following [7, 11, 12, 15] and [16]:

1. The advertising shape contains high attractive colors, as it is necessary to use colors that suit the advertising content, such as the use of fast food restaurants in its ads as it increases the speed of the heart rate which make viewers feel hungry.
2. Add movements and effects to the advertisement where fast food restaurants use the music that viewers prefer and add them to the advertisement to stir the viewers and motivate him to buy.
3. The advertisement should be clear and understandable to the browser, through using a language that the viewers understand, as restaurants put all details about the announced product to encourage the viewers to buy.
4. The advertisement is characterized by interactive, as restaurants encourage viewers from interacting with them through several means such as carrying out a live broadcast through collective media, which increases viewing rates and the effectiveness of the advertisement.
5. The design of the advertisement according to the events or occasions coinciding with the time of advertising.
6. The quality of the keywords used in the advertising content, as the use of short words that indicate the summary of the advertisement helps to increase viewing desire to buy, were restaurants use incentive phrases such as' buy one meal get the second free as well as, 'free delivery.

3.3 Interacting with the Digital Advertisement via Social Media

As consumers interact with electronic ads that are published on social media help to update and develop ads continuously based on consumer observations, and increasing the effectiveness and spread of the advertisement more and more fast. The ease of arrival of the electronic advertisement through social media is one of the important things that helped the success and spread of social media throughout the world, as the primary goal of establishing social media is to make the world like a small village through millions of people communicate with each other while they are in various regions, the creation of these means in a way that makes it easier for everyone to use, by setting guideline and clarification steps to work on them and their presence in several languages that suit the culture of each country throughout the world[16].

3.4 The Credibility of the Digital Advertisement

The credibility of the electronic advertisement is one of the most important considerations that the company must achieve during its advertising campaign to gain new consumers. The credibility of the electronic advertisement lies in matching of what the consumer has purchased with what he saw in the advertisement, and to achieve the credibility of the electronic advertisement, the following conditions must be met including trust, transparency, perception, response, originality, and assurance. Electronic ads are a highly effective tool that enables companies to reach their marketing goals and achieve profits, and it has received great demand and attention by companies and institutions, and this is due to the advantages they enjoy and that are summarized as the following [6, 10, 13, 14, 17, 18] and [19]:

1. The possibility of consumers to know the electronic ads without time determinants, so they can see the advertisement in any geographical location.
2. The ability to display ads in more clear and promotional ways through the use of technological developments and modern technologies, such as displaying products with 3D technology and using music.
3. Speed in delivering the advertising message to the target group, and the flexibility of its circulation by a large number of users.
4. Fast in knowing the results of the advertising campaign and knowing the number of viewers for the advertisement, and the ability to know the opinions of consumers about the announced product and the degree their acceptance.
5. The companies and institutions implementing the electronic advertisement enables accurate studies such as knowing the number of visits to the site.
6. Companies help attract the largest possible number of consumers to their sites by using certain types of advertising such as publishing advertising tapes in the most popular sites.

Based on the problem of the study, its questions and objectives, the study hypotheses were formulated as follows:

H0: There is no statistically significant impact ($\alpha \leq 0.05$) of digital advertising via social media on the buying behavior of fast food consumers in Jordan.

Based on research objectives, research problem, literature review and previous studies reviewed earlier [1–5, 7, 8, 14, 17, 19–24] and [25] the researchers developed research model as shown in Fig. 1.

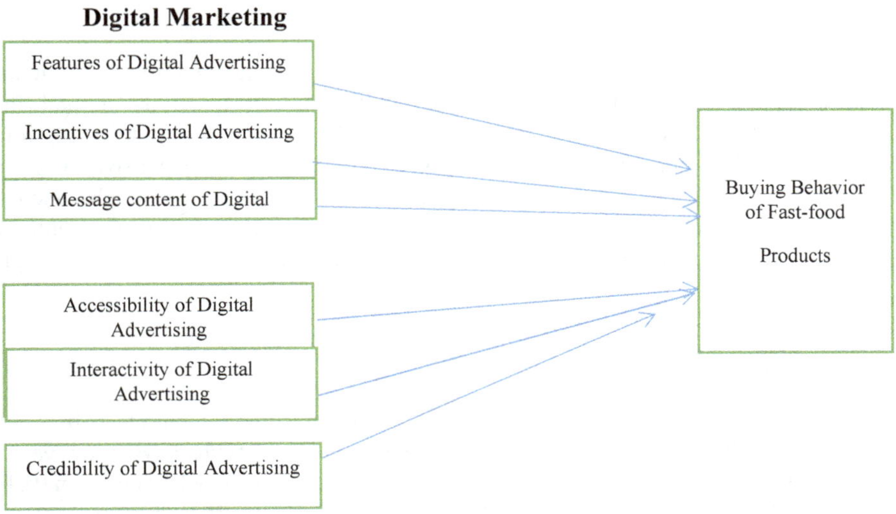

Fig. 1. Research Model

4 Methodology and Procedures

The study aims to identify the impact of digital advertising on the buying behavior of fast food consumers in Jordan, and to achieve the goals of the study, the descriptive, analytical approach has been relied upon, as it is the most appropriate for the nature and objectives of the study, as it works to collect the appropriate data from the targeted sample and analyze it and work to link it to the objectives of the study.

4.1 Population and Sample

The study population is a group of individuals that share a set of common characteristics and qualities, and is collected by one framework [26]. The study population must be characterized by easily access to its members, withdrawing a sample from it, and the ability to generalize the results to all individuals. The consumers of fast food in Jordan are considered to be the various characteristics of the study population.

In order to choose the size of a study sample of a representative of society and is free of bias, which is reflected in the possibility of generalizing the results to all members of society, the study relied on [26] table to determine the size of the study sample. The researchers were unable to identify the study population; therefore, the largest value of the supposed sample size has been relied upon in the table is (384), as the lowest value of the size of the study sample is acceptable and representative. (465) distributed electronic questionnaires to the target study population, in all regions of Jordan and in a random manner, through relatives and friends who contributed to spreading and answering the questionnaire. The researchers relied on the simple random method in distributing the questionnaire, as it is distinguished that each member of the study community has an equal opportunity with others to participate in the study.

4.2 Study Tool

The questionnaire is the most used and spreader tool to collect the data from individuals, the questionnaire has been used to collect the initial data from the study sample. The most important characteristic of the questionnaire is the possibility of distributing it to the sample members and obtaining data as soon as possible and less expensive. A group of sources of previous literature, research in the formulation of these paragraphs, and the use of expertise from academics in the field of marketing, were used. The questionnaire was electronically distributed to the sample members, as it was designed on the Google Forms program.

4.3 Reliability

The most used standards are used and the Cronbach alpha values which ranges between (1) and (0), and in general if alpha is less than (0.4), the paragraphs have a low value, and the paragraphs are considered average stability as their value reached between (0.4–0.7), While stability is considered high if its value is higher than (0.7), were paragraphs should have a stability of no less than (0.7) [25], and the following table (1) shows Value of Cronbach Alpha for study variables. Cronbach's Alpha is at a minimum acceptable level of 0.70 [23] as shown in Table 1.

Table 1. Value of Cronbach Alpha

Number	Variables	Questions	Cronbach's Alpha
1	Electronic Advertising Features via social media	6–1	0.884
2	Electronic Advertising Incentives via social media	7–12	.0914
3	Electronic Advertising Message Content via social media	13–19	0.892
4	Electronic Advertising Accessibility via social media	20–28	0.911
5	Interactivity of Electronic Advertising via social media	29–34	0.906
6	Credibility of Electronic Advertising via social media	35–46	0.850
7	Buying Behavior	40–47	0.901

It appears from the number Table 1, that the Value of Cronbach Alpha for variables are ranging between (0.850–0.914), and they all achieved the condition of stability by exceeding the value of (0.7), and this indicates that all the paragraphs of the questionnaire enjoy stability. Paragraphs were prepared: the variable of 'electronic advertising stimuli via social media' is the highest stable, by reaching the Cronbach Alpha factor for paragraphs, the stability coefficient has reached a high value with (0.914).

Table 2. Personal and Functional characteristic

%	Frequency	Characteristics	Item
59.6%	248	Male	Gender
40.4%	168	Female	
100%	416	Total	
10.6%	44	HSC	Level of Education
6.7%	28	Diploma	
59.6%	248	Bachelor	
23.1%	96	Graduate Studies	
100%	416	Total	
32.7%	136	Single	Marital Statue
63.5%	264	Married	
3.8%	16	Others	
100%	416	Total	
33.4%	139	260 JD - 400 JD	Monthly Income
22.6%	94	401 JD - 600 JD	
19.7%	82	601 JD - 800 JD	
16.6%	69	801 JD - 1000 JD	
7.7%	32	Above 1001 JD	
100%	416	Total	
70.1%	289	less than 18–24 years	Age
55.1%	227	less than 24–31 years	
83.7%	345	less than 31–38 years	
17.0%	70	38 years and above	
100%	416	Total	

From Table 2 it turns out that there are differences in the representation rate for both males and females, and the largest representation rate was to males by (59.6%), while females (40.4%).

The age group in the study is divided into four sections, the youngest of which are (18–24 years), and the largest (38 years or old (36.5%), the least number dating back to the age group (18–24 years), with a rate of (17.3%).

5 Results and Discussions

The following section clarifies the main findings of the present study as shown in Table 3.

It is also clear from Table 3, that the value of the level of significance for independent variables: (digital advertising incentives, accessibility through social media, and interacting with digital advertising via social media, are statistically significant at ($\alpha \leq 0.05$),

Table 3. Multiple Regression Results

Dependent Variable	R	R2	F	DF	SIG	Independent Variable	B	T	SIG
Buying Behavior	0.866	0.749	203.909	6	0.000	Digital advertising features	0.112	1..845	0.066
				409		Digital Advertising incentives	0.315	4.947	0.000
				415		Digital advertising message content	0.017	0.343	0.808
						Digital Advertising interactivity	0.160	2.176	0.030
						Digital advertising accessibility	0.333	6.311	0.000
						Digital advertising Credibility	0.034	0.836	0.409

as they have achieved significance levels Less than the level of connotation approved (0.05), which are respectively. (0.000, 0.000, 0.0000).the results indicates that there is a statistical impact on the dependent variable, were the level of significance F has reached (0,000), which is less than the level of the indication approved in the study and its value is less than (0.05), therefore we reject the main hypothesis and accept the alternative one, which states' there is a statistical significance impact ($\alpha \leq 0.05$), for digital advertising via social media on the buying behavior of fast food consumers in Jordan.

6 Future Recommendations

In order to improve the position of fast food restaurants, the recommendations emerged from the results of the study:

1. Marketers of restaurants should promote the credibility of digital ads which enhances customers' trust.
2. Owners of restaurants should choose highly qualified and professional experts in designing digital ads.
3. Measuring the ability of digital ads to affect the purchasing decision of consumers before publishing.
4. Diversification in electronic advertising stimuli, which suits the target markets.

5. Digital advertisement content should be rich in information.
6. Digital ads should be designed on basis of market needs.
7. Information in electronic ads should be accurate.
8. Researchers can select other variables that affect buying behavior in different cultures to be studied in future.

References

1. Azzam, A.: The moderator effect of COVID pandemic on the relationship between online shopping services and online consumer purchasing behavior. A case of Jordan. J. Stud. Syst. Decis. Control **216**, 25–35 (2023)
2. Azzam, Z.: Factors affecting the adoption of e-marketing technologies in services industry: empirical study on hotel sector in Jordan. Zarqa J. Res. Stud. Humanit. **13**(1), 210–224 (2013)
3. Al-Shaikh, M.: Customers' perspective towards factors associated with technological products adopting failure. A case of Jordan. Zarqa J. Res. Stud. Humanit. **13**(2), 252–261 (2013)
4. Azzam, A., Al-Shaer, S.: Electronic word of mouth (E_WOM) adoption via social media and its impact on online shoppers' purchasing intention during corona pandemic. A case of Jordan. J. Stud. Syst. Decis. Control **216**, 477–487 (2023)
5. Al-Shaikh, M., Khanfar, A.: Delivery service via electronic applications and its impact on customers satisfaction at retail stores in Amman City/Jordan. J. Stud. Syst. Decis. Control **216**, 827–837 (2023)
6. Leonora, F., Hilmi, E., Chiang, C., Tiger, L.: Digital advertising: a more effective way to promote businesses' products. J. Bus. Adm. Res. **3**(2), 59–67 (2014)
7. Niklas, F., et al.: Online advertising: the impact of targeted advertising on advertisers, market access and consumer choice. Policy Department for Economic, Scientific and Quality of Life Policies for the committee on the Internal Market and Consumer Protection (IMCO). ISBN 978-92-846-8130-3. PE 662.913, 1–151 (2021)
8. Enginkayaa, E., Cinarb, D.: The impact of digital advertising on consumer purchase decisions. In: International Interdisciplinary Business-Economics Advancement Conference (IIBA) Conference Proceedings, p. 432 (2014)
9. Chao, N., Corus, C., Li, T.: Balancing traditional media and online advertising strategy. Int. J. Bus. Mark. Decis. Sci. **5**(2), 12–24 (2012)
10. Horbal, N., Naychuk, M., Orlykova, B.: Internet advertising: the specifics, tendencies of development and impact on sales. ECONTECHMOD. Int. Quart. J. **6**(1), 37–46 (2017)
11. Valdas, V.: Electronic Marketing. Vilnius School of Technologies, Business and Agriculture. Project is co-financed by the European Union 1, 1–251 (2017)
12. Illiashenko, M., Ivanova, E.: Tools and methods of product promotion in the Internet: analytical review. Mark. Manage. Innov. Ukrainian **3**, 20–32 (2015)
13. Goldfarb, A.: What is different about online advertising? Rev. Ind. Organ. **44**(2), 115–129 (2014)
14. Hanafizadeh, P., Behboudi, M., Hasanabad, M.: Online advertising intermediary: how online advertising works? Int. J. Online Mark. (IJOM) **4**(1), 29–38 (2014)
15. Wang, K.-Y., Shih, E., Peracchio, L.A.: How banner ads can be effective: investigating the influences of exposure duration and banner ad complexity. Int. J. Advertising **32**(1), 121–141 (2013). https://doi.org/10.2501/IJA-32-1-121-141
16. Karimova, Z.: Interactivity and advertising communication. J. Media Commun. Stud. **3**(5), 160–169 (2011)
17. Tobi, O., Ayodele, M., Akindele, A.: Effect of online advertising on consumer buying behavior of internet users in Lagos State. Ilorin J. Hum. Res. Manage. **4**(1), 171–181 (2020)

18. Ghose, A., Yang, S.: An empirical analysis of search engine advertising: sponsored search in electronic markets. Manage. Sci. **55**(10), 1605–1622 (2009)
19. Al Tamer, M.: The impact of digitization on advertising effectiveness. Manage. Stud. Econ. Syst. (MSES) **5**(3/4), 117–126. Summer & Autumn (2020)
20. Al-Azzam, A., Azzam, Z.: Evaluating the impact of service quality, brand image, trust and promotion on customer loyalty in mobile phone in Jordan. Zarqa J. Res. Stud. Humanities **13**(2), 273–263 (2013)
21. Al-Ubaidi, S., Shawi, S.: The role of digital marketing tools in improving bank marketing performance. Finance Bus. Econ. Mag. **7**(1), 69–86 (2022)
22. Stallone, V.: The digital advertising conceptual flow: a literature review. In: Rocha, Á., Reis, J.L., Peter, M.K., Bogdanović, Z. (eds.) Marketing and Smart Technologies. SIST, vol. 167, pp. 1–8. Springer, Singapore (2020). https://doi.org/10.1007/978-981-15-1564-4_1
23. Azzam, Z., Ali, W., ALShaikh, M., Khanfar, I., Ismail, A.: Factors affecting the use of electronic banking services provided by Jordanian commercial banks from customers' point of view. A case study: Jordan Ahli Bank in Northern Region/Jordan. In: 2021 22nd International Arab Conference on Information Technology, ACIT 2021 (2021)
24. Al-Shaikh, M., Al Sadi, A., Azzam, Z., Al Muala, M., Alsaraireh, S.: The Effect of Electronic Marketing Tools on Customer Satisfaction with Electronic Services in Jordanian Islamic banks (A case study of the Islamic International Arab Bank in Amman City – Jordan). In: 2021 22nd International Arab Conference on Information Technology, ACIT 2021 (2021)
25. Azzam, Z., Al Fuqaha'a, E., Khrais, I., Almubaydeen, T., Ismail, A.: Impact of marketing innovation on building customer loyalty: (a field study on customers of pioneers paper and carton industries in Jordan). In: 2021 22nd International Arab Conference on Information Technology, ACIT 2021 (2021)
26. Sekaran, U., Bougie, R.: Research Methods for Business: A Skill Building Approach, 2nd edn. John Wiley & Sons, USA (2016)

Digital Transformation/Ramallah Municipality

Rania Jaber(✉) and Maisa Burbar

Business and marketing department, Birzeit University, Birzeit, Palestine
{rjaber,mburbar}@birzeit.edu

Abstract. Established in 1908, the Ramallah Municipality has served the residents of the city (around 70,000 residents now) for over a century. Ramallah today is not only the home of its inhabitants, but also the temporary administrative capital of Palestine. Individuals enjoy openness, tolerance, and a multitude of cultural and educational activities. The Ramallah Municipality launched the Ramallah Smart City initiative in cooperation with partners from the government and private sectors to enable, present, and enhance an efficient, smooth, advanced, and impactful smart city experience for residents and visitors. The purpose of this case is to assess the effectiveness of digital transformation on both the municipality's employees, and citizens' overall well-being.

Keywords: Municipality · Digital transformation · Smart City · Electronic Services

1 Pedagogical Objectives

- To analyse whether digital transformation improves the quality of services and creates a more collaborative relationship with citizens.
- To describe the accessibility of services through different channels, their inclusiveness, and user friendliness.
- To suggest good practices and recommendations to help improve the quality of services.

2 Methodology

The methodology used depended on primary and secondary sources of information:

1- **Primary information**:
  - Interviews were conducted with head of departments and some employees.
  - A small questionnaire was prepared to measure the citizens level of satisfaction about the digital transformation (mainly the e-services provided by the municipality)

2- **Secondary information**: Reports and information via the website and news.

B. A. M. Alareeni and I. Elgedawy (Eds.): *Artificial Intelligence (AI) and Finance*, SSDC 488, pp. 473–483, 2023.
https://doi.org/10.1007/978-3-031-39158-3_45

3 Ramallah City

Perched at 880 m above sea level, Ramallah refers to God's Hills in Arabic. It is a fitting name for the city that became the administrative center for the Government of the State of Palestine, the Palestine National Authority (PNA), in 1994.

Located approximately 16 km north of Jerusalem in the Palestinian West Bank, the Haddadin family established Ramallah in the mid-1500s as a Christian village. Throughout the 17th and 18th centuries, it grew as a predominantly agricultural village, and by the mid1800s, missionary groups had built different schools and churches. As more people moved to the newly incorporated city the demographics of the city changed in the early 1900s, attracted by the high living standards that resulted from developing trade routes with the USA.

Ramallah has witnessed remarkable economic development in the last decade. This relative prosperity, alongside the liberal atmosphere of the city, has attracted numerous Palestinians in search of better life and opportunities. As a result, the city's population has doubled over the last 10 years. Around 70,000 full-time city inhabitants are the currently estimated population of Ramallah, with an additional average of 150,000 people who visit the city every day. (Ramallah Municipality (RM), 2018).

Ramallah has an increasingly international outlook as a cosmopolitan city. As stated in their strategy, "Not only are we proud to be part of the 100 Resilient Cities network, but we also have formal connections with more than 30 cities from all over the globe". (Ramallah Resilient Strategy, 2050).

4 Ramallah Municipality

Ramallah Municipality Mission: Providing distinctive services to Ramallah residents and visitors, and to prosper with the city so it becomes civilized, open, and interactive, and able to cope with development, and contributes to achieving sustainable development based on rights and justice in providing services that are built upon community partnership.

Since its establishment in 1908, Ramallah Municipality has had 25 presidents, including the first female municipal president in Palestine, elected in 2005, and the last election was held in 2022, when Mr. Issa Kassis was elected as the recent president.

The municipality has approximately 414 employees and 15 municipal council members. The mayor is one of them, and the municipality **includes eight departments and three units, as illustrated below.**

Major Departments:

- The Administrative Affairs Department, including the Human Resources
- The Financial Department
- Health and Environment Department
- GIS & IT Department
- Cultural and Social Affairs Department
- Project's department
- Building and Urban Planning Department

- Solid Waste and Wastewater Department

Three units:

- Citizens' Service Unit
- Internal Control Unit
- Media & Protocol Unit

5 Digital Transformation: Main Achievements

To build a sustainable, innovative, and competitive city that fulfils the aspirations of business sectors and citizens alike, The Ramallah Municipality launched the Ramallah Smart City initiative in cooperation with partners from the government and private sectors to enable, present, and enhance an efficient, smooth, advanced, and impactful city experience for both residents and visitors.

To achieve its pillars, Ramallah Municipality seeks to present strategic initiatives and develop partnerships to contribute to the development of the six axes of **smart infrastructure, smart governance, smart education, smart economy, smart mobility, and smart environment.**

These long-term technical initiatives are being led by Ramallah Municipality within the framework of IDC's definition of a smart city, which aligns with the definition of "the smart city" as "a limited entity neighborhood, town, city, county, municipality, and/or urban area."

As a result, high priority was given to digital transformation, as the municipality witnessed a quantum leap in the following areas:

First: Electronic Management (Employee Cantered)

- Comprehensive computerized system: The Ramallah Municipality used a comprehensive unified system, that succeeded in addressing all internal work requirements, including CRM systems, to process all work digitally in all departments, in addition to receiving public requests and processing them in an automated manner, virtual staff meetings and workshops, a Tracking System for all municipality vehicles, and a Digital Archive Platform.
- Interactive digital applications of geographic information systems (GIS) for employee use to manage and document spatial and descriptive data in various municipal departments.
- Remote mechanisms tracking system, live broadcast of major streets, virtual meetings, sessions, and workshops, electronic archiving platform
- Linking with partners (property tax and land authority).

Second: Electronic Services (Citizen Cantered)

- The electronic services portal www.ramallah.ps/eservices, through which the citizen can electronically request all the services needed, after logging into their account with their own username and password (they can be obtained through the Internet if there is updated data in the database), citizens can also inquire about their financial account and previous requests, and they are also allowed to pay electronically through various electronic payment solutions and services.

- Interactive digital applications of geographic information systems for the use of citizens and visitors to the city www.ramallah.ps/GIS.
- Questionnaires and electronic forms to receive requests for participation or membership, or to collect citizens' opinions, by providing automated forms that are announced on the municipality's official page and social media networks, and sometimes in local newspapers; some are linked to GIS applications to display spatial distribution and participant details.
- The following electronic payment solutions and services are provided:
- E-payment from an account to an account (electronic banking) Arab Bank, National Bank, Bank of Palestine, Quds Bank
- Payment through electronic wallets, JAWWAL Pay ChatPal, Palpay, MadfooatCo, and Palpaypoints network in Palestine.

Third: Smart Applications

- The official application for smart devices is available on electronic stores under the title "Ramallah Municipality" and can be downloaded from the Apple Store and Google Play.
- Pre-paid parking application from TechPark Company, and it is available in electronic stores under the title "TecPark" and can be downloaded from Apple Store and Google Play.
- Virtual Touring of Ramallah City, Especially the street view, "Google Street view" which is available on Google Maps, Google Earth, and the interactive digital applications of the Ramallah Municipality

Fourth: Digital Communication

- The official electronic municipality page www.ramallah.ps
- Various social media: Facebook, Instagram, YouTube, WhatsApp
- Email, SMS, voice messages and automated messages through WhatsApp
- Digital billboards, and lve broadcast

Fifth: Digital Empowerment

Citizens must be able to access and benefit from these services. Therefore:

- Providing free wireless Internet in all municipal facilities and public parks. In addition to providing free Internet in the city's main streets since 2014 before 3G services were provided.
- Providing awareness material through the municipality webpage and social networking sites

Sixth: e-School Platform and Smart Infrastructure

All public schools in ramallah have a technological infrastructure that enables them to use technology in education through wired and wireless internet. The municipality has also provided smart teaching aids and comprehensive technological tools in three public schools, in addition to providing digital communication platforms between the administration, teacher, student, and parents through the directorate of education, and finally providing mobile virtual laboratories for students.

6 How Does This Transformation Affect the Employees?

Ramallah municipality is from the pioneering municipalities in implementing digital transformation, as it is believed that the transition would enable the municipality to operate at a higher level, becoming more efficient, productive, and innovative. However, this has been the output of a long, hard work, and fruitful effort by all staff, including the municipality's presidents, municipal council members, heads of departments, and employees. The most important factor was educating employees on the benefits of a new way of working and encouraging them to embrace new technologies, rather than fear them.

Workflow History. Digital change has been taking place since 2005, and the municipality had good, but isolated and scattered working systems whose functions are not related to each other, information is not shared and exchanged, and information is disconnected from business processes and applications; each system was an island but itself, which resulted in poor arrangement of work flow, so people are not in a position to help each other. The aim was to have full-service integration by linking data, procedures, interfaces, portals, and back-office processes across departments.

Service Integration and Smooth Transactions.
The digital service integration contained four platforms: the ERP and CRM portal, the GIS portal, the Citizen portal, and the e-Schools portal. The first two platforms were fully integrated. The Citizen one-stop" portal for services is linked to them. The e-school portal is separated, and the Ministry of Education leads the contents.

The GIS platform, as described by Dr. Safaa Aldwaik, the Director of GIS and IT Department at Ramallah municipality, is a cross-sectoral platform that integrates and transforms the way the municipality gathers, manages, compiles, reviews, analyses, and visualizes spatial data. The method used to conduct such practice is by using the power of maps to bring together information that portrays city details ranging from parcels, land use, buildings, facilities, and landmarks to infrastructure such as water, sewers, roads, traffic signs and fixtures, emergency incidents, and even international relations. This in turn improves the quality and inclusiveness of the database, thus providing reliable information that will aid in planning and decision-making.

The GIS has provided logistical support for other municipal services. For example, by providing optimized routes or roads for municipal means of transportation (vehicles) as they move from one point to another to accomplish its task. One example is the optimization of garbage pickup routes which has decreased fueling costs. This in turn allowed for monetary resources to be used for other projects. (This Week in Palestine, 2016).

Challenges Faced: The digital transformation went through "somehow" smooth phase. One of the reasons for that is that other municipalities implemented the same programs. This levelled the change. One reason is that the municipality was able to understand its processes and corporate culture, and thus invested highly in (in-house) technology, as there will be a certain amount of internal expertise needed, as well as a dedicated team in place. Later, the municipality utilized the required IT expertise to complete the process. Another reason for the smooth transaction is the support from all the municipality's presidents to digital transformation. Clearly, the change was gradual. It can be described as "slowly but sure" kind of transition. This reduced the fear of change that usually happens following any change in any organization.

Digital transformation was not without its risks and **Challenges**. Change Management was among the top priorities, as managing human emotions and employee concerns had to be tackled even if the change was made as smoothly as possible. If the change is managed incorrectly, it can become a double-edged sword, leading to a loss of productivity and poor employee performance. However, the municipality was very aware of this matter. It emphasized the value and benefits of this change on the employees and the workflow, bearing in mind that the municipality used the participative management approach to engage the employees in the decision-making process.

Training employees and raising awareness were among the top challenges faced, as the municipality has many old employees who might fear new technology. The policy was that the new technology will serve as a tool that will make the employees' work easier and benefit them in their jobs. Moreover, the policy was that no employees will be laid off due to digital changes, yet employees will be fully trained in order to become competent in their new tasks. When training was not a solution for some employees, as they were unable to adapt to the new technology, they were redistributed among other tasks that matched their experience. At some point, the change had to be forced to maintain a new vision. However, the recruitment process also changed. Employees who resigned or retired were only replaced with newly qualified technologically oriented staff.

This leads to the **second challenge**, which is recruiting and hiring qualified human resources. This is considered a challenge, as the salaries in the municipality are not high, and not competitive. In addition, the Ministry of Local Governance has enforced a unified salary scale. The issue here is that Ramallah city is considered the commercial capital of Palestine, and the living standards are very high in Ramallah compared to other districts, so it will be very difficult to attract qualified staff with the current salary scale.

7 How Digital Transformation Contributed to Long-Term Citizen Well-Being

In an interview that was conducted by Dr. Aldwaik, Ramallah city has been growing rapidly, and thus the needs of citizens have increased due to the restrictions imposed on Ramallah residents because of the political and security context. Therefore, there was a great need to empower society through technology, so citizens can receive services of the municipality at any time without the need for personal presence, and also allowing student education despite all restrictions on movement, this digital change improved the quality of services, created a more collaborative relationship with citizens across income levels and age groups, and was able to overcome the challenges imposed by the political situation. The result was a closer relationship between the municipality and the citizens, better communications with the local community, improved 24/7 services over web, a free Wi-Fi was offered, and live streaming over the Internet, which resulted in an increase in the number of people participating online and attending public events organized by the municipality. Moreover, the number of hits on the municipality website increased from several hundred per week to over 350,000 hits, and response time increased dramatically (some services delivered in three days instead of twenty days!).

The municipality also launched the first version of its newly designed Mobile App, which will provide citizens with a new channel to interact with the municipality, and by that, all users will be up to date with the news, events, and facilities of Ramallah Municipality. As a result of the infrastructure provided within this change, the municipal council decided to conduct the regular consecutive weekly sessions to include an online-session for the purpose of enabling citizens to engage and watch the entire session in a transparent manner.

8 Benefits of the Digital Transformation to Ramallah Municipality

It's clear that the digital transformation has moved the municipality to a new, competitive edge, and has benefited the municipality's stakeholder as shown below:

- **Reducing costs**: The municipality was able to automate "repetitive, time-consuming tasks", thus the human effort was reassigned to more value-adding activities elsewhere.
- Increasing revenue: Utilizing digital technologies to get things done quicker allowed the municipality to serve a larger base of citizens, and thus increasing revenue.
- Enhancing collaboration through service integration. The digital transformation allied all employees allowing them to share knowledge and ideas and align their efforts towards a shared goal.
- **Increasing employee productivity:** Employees can access the business information they need to complete tasks faster, and focus on more valuable, rewarding work.
- **Enhancing agility, improving citizens overall wellbeing and experience:** The municipality was able to respond to change (and do so rapidly) was crucial to keeping up with new demands and avoid getting left behind.
- **Increase governance and compliance:** The Digitizing processes increases transparency and minimizes risk; thus, the municipality was able to adhere better to regulations and compliances such as conducting the online sessions for the public.

9 Survey Results

A Survey has been prepared to measure the citizens' (level of usage of the new electronic services, and the level of satisfaction), where a questionnaire was prepared and piloted among 5 people, where feedback was received, and the final version was submitted online to 100 respondents.

Demographic information: The sample comprised of around 84% females and 14% males, with 51.6% ages between 30–39, and 30.6 between 40–49, and 14.5 above 50 years old. As for education level, 48.4 of respondents have bachelor's degrees, and 38.7 have master's degrees.

Municipality Portal: Respondents were asked if they have access to the municipality electronic services portal, 75.8% of the respondents don't have accounts on the municipality portal, where only 24.2 have accounts, but they don't rely on it finish their transactions.

When asked about the **method that they use to pay their bills or fees to the municipality**, 80.6% pay cash at the municipality, 24.2 through e-banking, 11.3 through electronic wallets such as JAWWAL Pay ChatPal, Palpay. It's worth noting that citizens still prefer to pay their fees and bills in cash at the municipality, as shown in Fig. 1.

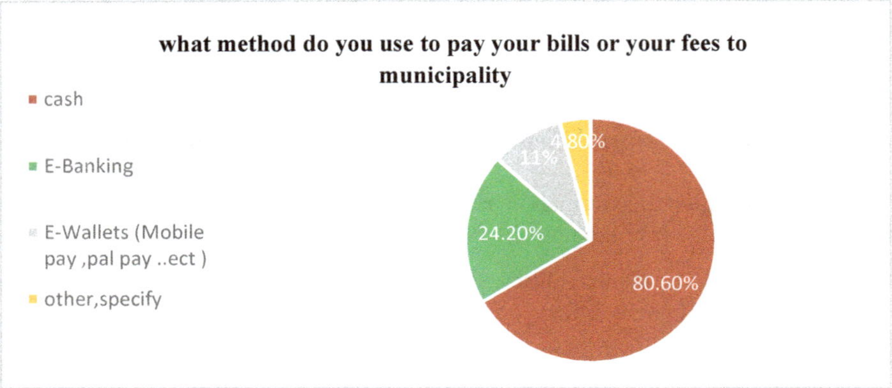

Fig. 1. Methods used by citizens to pay their bills.

Methods for interacting with the municipality: As for methods that people prefer to interact with the municipality to get the latest **news and events,** 55.8% preferred to use the social media pages such as Facebook and Instagram, whereas17.27 use the official website www.ramallahmunucipality.com, and only 13.1 interacts with the "**SMS**" sent to them by the municipality.

Level of satisfaction about the electronic services and Smart Applications, and Digital Communication and Empowerment.

Respondents were asked to provide their level of agreement on a 5-level Likert (strongly agree to strongly disagree), about different services and application provided by the municipality. (Their responses are shown below):

Figure 2 shows that most of the respondents either agreed or where neutral about the e-services provided, this implies (somewhat) good satisfaction level, as respondents believe that the services via the municipality application helps provide the services faster, and that they can get the information they need, and that the services are user friendly.

As for the smart applications, as shown in Fig. 3, respondents had more neutral responses when asked if the smart applications are good and user friendly. It seems that people are not using the applications efficiently as many of them didn't download the application.

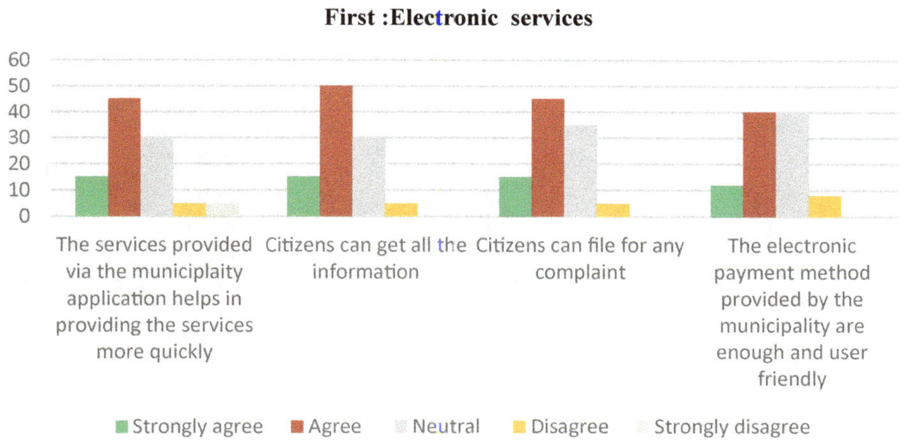

Fig. 2. Electronic Services.

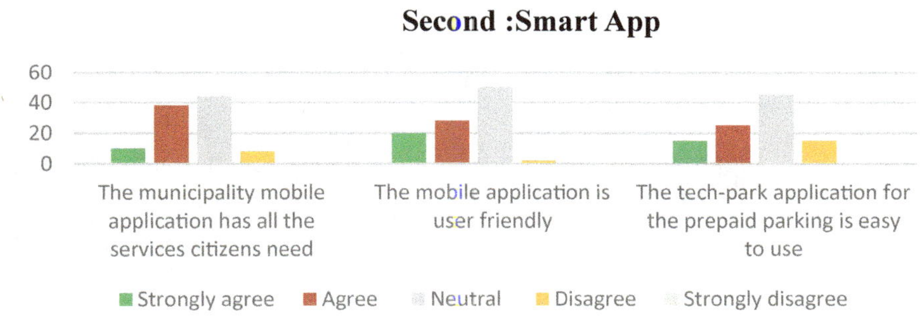

Fig. 3. Smart Applications.

As shown in Fig. 4, respondents are ***mainly satisfied*** with the digital communication services, as most of them agreed that the municipality has various effective methods to communicate with the public and that the municipality is very interactive with the public and is up to date with its news and events. However, some respondents believe that the municipality should provide quicker feedback regarding their enquiries.

As for Digital Empowerment, as shown in Fig. 5, respondents are not very satisfied with the free wi-fi provided by the municipality, and they believe that better public awareness should be practiced.

Finally, when respondents were asked about their overall level of satisfaction with the e-services provided by the municipality, 10% were very satisfied, 46.7% were satisfied, and 35% had neutral responses. This shows that people are generally satisfied with the electronic services provided, and this gives the municiplaity good indicators.

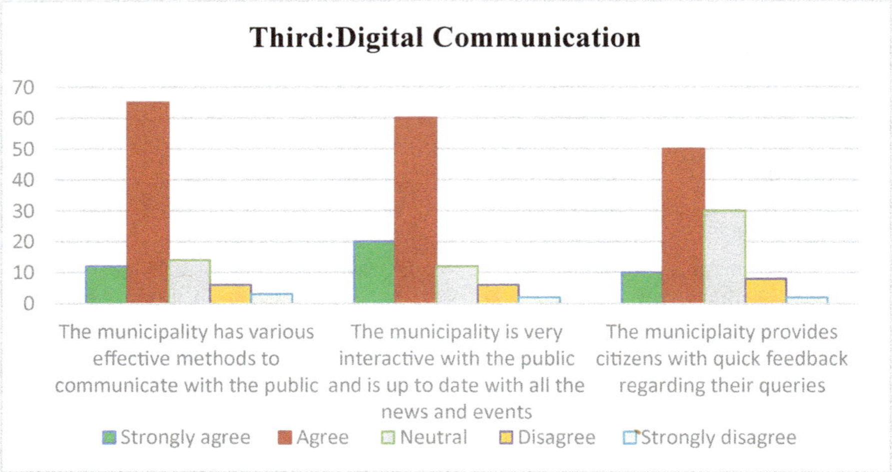

Fig. 4. Digital Communication.

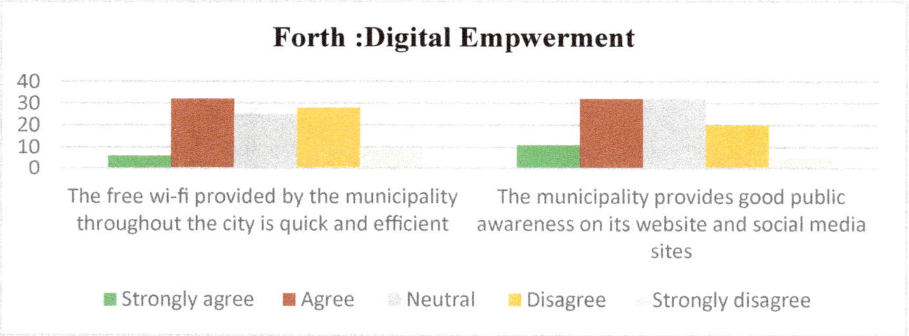

Fig. 5. Digital Empowerment.

10 Conclusion

The Digital Transformation in Ramallah Municipality is considered a revolution since it aimed to utilize innovative technologies to improve the way the municipal council serves its local residents and other community stakeholders and deliver more transparent, cost-effective, and customer-focused services to its citizens.

Ramallah municipality is in a period of momentous change. The municipality has achieved a huge number of benefits from this transformation, mainly reducing its overall cost, enhancing agility, improving citizen's overall well-being and experience, and increasing governance and compliance.

It's clear that there is public awareness regarding the services provided by the municipality, henceforth, they are trying to utilize them in an efficient way, however. However, the "**Trust**" challenge remains present in the mind of people. It is clear that many people still lack the courage to depend solely on online channels, as they still like to visit the municipality, and pay their bills in cash, but at the same time, they're enjoying the e-services provided and the digital communication and empowerment.

11 Recommendations

Ramallah municipality is a pioneer in implementing digital transformation. This paper examined this new renovation and measured the public's acceptance, usage, and satisfaction with the electronic services offered to them due to this transformation. After analyzing the responses of the citizens based on the survey, we recommend the following:

- The municipality should further improve public engagement.
- Mitigating risks and delivering effective programs and services efficiently.
- Incentives should be given to citizens who use electronic services, and mobile applications.
- Faster responses should be provided to people with different inquiries, and the follow-up complaint procedure should be accelerated for the purpose of satisfying the public.
- Special attention should be granted to the elderly and people with special needs.

References

1. High Tech in Palestine. This Week in Palestine (2016). http://www.thisweekinpalestine.com./
2. Ramallah resilience strategy 2050. Ramallah.ps. (2018). https://www.ramallah.ps/ar_page.aspx?id=irZMUYa2856210753airZMUY
3. The premier global market intelligence firm. IDC. https://www.idc.com/
4. بلدية رام الله (2018). https://www.ramallah.ps/

The Role of Intellectual Capital in the Production System and Economic Power in Light of Digital Transformations

Abdelhalim Mohammad Jubran[1(✉)], Thaer Ahmad Abu-Saleem[1], Ibrahim Mohammad Khrais[1], Nidal Ali Abbas[1], Mazin Hasan AlBasha[1], Mohammad Musa Al-Momani[1], Khaled Mohammed Al-Sawaie[1], and Mohammad Kanaan[2]

[1] Faculty of Economics and Administrative Sciences, Zarqa University, Zarqa, Jordan
{ajubran,thabusaleem,ikhrais,nabbas,mbasha,m.momani, ksawaie}@zu.edu.jo

[2] College of Engineering, University of Business and Technology, Jeddah, Kingdom of Saudi Arabia
m.kanan@ubt.edu.sa

Abstract. The study aims to analyze the production relations in the circumstances of the expansion in the scope of knowledge and the transformation of intellectual capital into an economic resource based of digital transformation. To achieve this, it focused on explaining the relationship between the system of production relations and economic power by describing the elements of property in light of the digital transformation in economic activity through the use of the descriptive approach. The study reached a set of results, such as: The exchange of roles between physical capital and labor under the replacement of knowledge workers undermines property rights in the Arab economic system, which is already suffering from the blurring of property rights. It is not possible to make qualitative leaps in social relations without making a cultural shift in the way of thinking towards the transition from traditional education to education by learning to produces creativity.

Keywords: Intellectual Capital · Production Relations · Property Rights · Knowledge

1 Introduction

The global economy has entered the gates of the fourth industrial revolution and therefore, new economic relationships imposed by the market mechanism have emerged on the international arena. In the context of these transformations, the economic relationship system is changing in its two aspects: production relationships and property relationships. This necessitates adapting to these changes towards digitization of these relationships.

B. A. M. Alareeni and I. Elgedawy (Eds.): *Artificial Intelligence (AI) and Finance*, SSDC 488, pp. 484–494, 2023.
https://doi.org/10.1007/978-3-031-39158-3_46

In order to face these challenges the economic decision-making process will become a behavior-oriented practice in a new system of production relationships and property relationships. On the production side, productive relationships are formed with new and renewed production structures.

The elements of production exchange their positions, not from a social point of view, but based on the expected return in a market with economic power represented by the social elite. Capital grows and expands the scope of the production relationship system based on the principle of property rights for new elements: technocracy and capital owners who share the bulk of the economic pie. In this transformation, the concept of intellectual capital is revealed and its scope expands in the economic relationship system in the context of digital transformations. These transformations enhance the significance of monitoring in highlighting the simulation of the economic relationship system during this transition, given the growth of intellectual capital and the decrease of labor's role in generating added value, which is being transferred to the lower numbered class in society, which possesses both resources and economic power.

We must acknowledge the fact that intellectual capital (the source of knowledge) does not inherit nor its function does not manifested away from entrepreneurship. This creates the issue of knowledge identity that distinguishes the world of knowledge, its producing countries, and consuming countries [1].

2 Importance of Study

Based on the brief introduction above, the researcher summarizes the importance of the study, which can be summarized as follows:

1. Real economic growth cannot be achieved without taking into consideration the position of intellectual capital as a source of creative destruction and the cultural shift towards the emergence of knowledge empires.
2. The emergence and dominance of intellectual capital leads to supportive transformations that serve the transition to economic power as an advanced form of property relationships.
3. The global economy today faces new challenges: intellectual capital is replacing traditional work in the face of increasing capital intensity for production operations.

3 Study Problem

In view of the ongoing digital transformations and their effects on the reality of production relationships, it is imperative to clarify the position of these changes within production relationships, especially in the context of increasing the substitution of knowledge replacing traditional resources and the role exchange between economic resources towards economic authority.

4 Aim of Study

The researchers aims to analyze the relationship between the production relationship system and economic power by focusing on intellectual capital and its role in shaping this system, in order to achieve a set of objectives, as a following:

1. Analyzing production relationships in an expanding knowledge domain and the transformation of intellectual capital into an economic resource based on digital transformation.
2. Interpreting the relationship between the production relationship system and economic power by describing property elements in the context of digital transformation in economic activity.

5 The Analysis of Studies Related to the Research Problem

The researchers reviews and analyzes a set of recent studies in this field, the current study aims to analyze the relationships between the concepts of property, production, and economic power in the context of the transition to knowledge-based societies and digital transformations. Since the first industrial revolution (1780–1820: the invention of steam engine), innovations have transformed theoretical sciences into tangible products. From an economic perspective, the cultural environment is considered the dynamic and vital trigger of the economy. From a culture of exploration and experimentation to innovation, the functional relationship between production and property, then knowledge is revealed in advanced stages of economic growth. What prompts reflection is the size of achievements realized in the transition from manual to capitalistic production: the stage of production based on work relations - capital using mechanical means, where competitiveness will be determined through innovation, continuity, and digitization [2].

Moreover, from an economic perspective, knowledge is considered a result of intellectual activity by individuals and society, based on business viability through the interaction of material, vital and technical factors. Knowledge consists of a large amount of information which appears in a specific context and is transmitted through economic channels on a functional basis [1]. This resource is characterized by its accumulation, organization, cause-seeking, inclusiveness, certainty, accuracy, and validation. Compared to other goods, this resource is distinguished by its survival and accumulation after use which considered as a basis for the development of the knowledge itself [3].

At the beginning, when the concept of information was emerging, specific channels were used to convey this product. Linear forms were used to transfer the information such as messages, books, newspapers and even technical methods such as telegraphs, phones, radios and television. Later, information was also conveyed through email via the internet [1]. However, with the emergence of electronic channels, information became an economic resource. Moreover, the advancement of production means was enhanced by the development of technology, energy sources, communications and the shift towards qualitative teaching in higher education institutions [4], which elevated the status of information to a productive power capable of advancing humanity to a knowledge society [1]. In addition, the role of information is reflected in its various uses based on the prevailing system of property relations in the economic system.

Components of knowledge, Knowledge are characterized by both, an infrastructure and a superstructure at the same time. The relationship between these two levels depends on the channels that connect them. This concept includes both its structure and its pivotal elements as a following [5]:

1. Intellectual capital, it is the sum of knowledge, practical experience, and creative abilities of the leadership system, as well as the ethical values of the organization,

and also the culture and the approach of life and work, in both its physical and mental parts.

2. Structural Capital. It is a form of the infrastructure at the level of the economic system and includes programs, technology, and the organizational framework in the economic unit, as well as innovations and brand reputation, and everything that contributes to the marketing of production capacities [6].

6 The Place of Intellectual Capital in the Production Relationship System

Researchers examine the relationship between intellectual capital and production, property, and economic power. The production relationships at the economic system level are shaped by both economic and other social factors. The class struggle in society is considered one of the most important factors in this field, as it affects the relationship between the ruler and the ruled and helps to shape the prevailing mode of production relationships:

1. Feudal Relations: Peasants + land owners (Basic = Agriculture)
2. Commercial capitalism: Craftsmen + capital (Basic Activity = Trade)
3. Industrial capitalism: Work + capital (Activity = Industry)
4. Digitized interventionist Capitalism: Businessmen + Economic Elites (Material and Artificial Intelligence)
5. Interventional Capital: State + Unions (Material and knowledge Industry).

The interconnection between elements of conflict and production is determined by economic, social, political, and cultural circumstances. In countries that export raw resources such as oil, for example, we encounter production relationships based on agriculture and traditional production methods, with a fragile industrial base that cannot utilize national resources to generate local added value to finance expenses. Consequently, economic relationships in these countries take on the character of rentier economies [7]. On the other hand, in industrialized countries, and since the first industrial revolution, production relations have formed around capital and labor in an industrial climate. This conflict has called for economic freedom that enhances the status of industry and capital owners at the same time. What distinguishes this period is that the driving force behind knowledge gains was previously industrialists who succeeded in turning inventions into marketable products [8].

However, with the global economy reaching the stage of the second industrial revolution, the revolution of iron, steel, and energy, the outlines of a new conflict in the relations of production and property became clear. In addition, the economic and political conditions led to the emergence of joint-stock companies, marking the beginning of a new phase in the property relations of economic units.

This resulted in a decline in the role of the owner in leading the company, allowing for the rise of a new class of organizers and the separation of management from property. As a result, the conflict transformed into a three-dimensional economic dispute over the distribution of the economic pie: workers, organizers, and owners [8].

Since the mid-20th century, economists and businessmen have focused on developing production technology and infrastructure to overcome geographical barriers, with the aim of increasing industrial capacity and penetrating local and international markets.

This required the establishment of economic policies that are free from all sovereignty constraints and necessary to facilitate the transfer of capital between countries, thereby crossing political borders [8]. This coincided with the emergence of new industrial countries and the transition of countries to the industrial stage. These countries found their way in developing their wealth through the establishment of highly skilled and highly productive labor-intensive new industries, later referred to in economic literature as the "Soft Economy".

7 The Property Relations System and its Role in the Emergence of knowledge Capital [11]

The property of production elements is responsible for the emergence and formation of the productive relations system in the economic relations framework. Property was not far from the discussions of scientists and philosophers in the debate between individual interest and collective interest. In addition, the debate among researchers revolves around whether property relations are the product of social and political transformations or whether the evolution of the concept of property determines the social pattern in each era of human development. The transition of humanity from communalism to Capitalism, may be the result of social or economic movements based on the development of concepts of property and control. This opinion is manifested through the relationships that appear in society during stages of growth and development: economic relations, political relations, legislative relations, cultural relations, and ethical relations [11].

Concept of Property Relationship. This concept refers to the links formed in economic activity as a result of the use different types of production means [10]. Here, the law, which distributes resources, operates according to the concept of expanded production. At the same time, the law of property determines the production relationships, where the distribution of resources is based on this property. Use may be based on absolute property or for a certain period and for the purpose of benefit. Ronald Coase defined property rights as follows: the right of property, the right of use, the right of management, the right to income, the right of sovereignty, the right to security, the deferred right of product property, the right to transfer property by inheritance, the right to prohibit use, the right of responsibility, and the right to implement procedures and institutions that ensure the correction of imbalances or the restoration of confiscated powers in the context of property regulation. The multiplicity of these rights provides flexibility to the property relations system in society and, therefore, flexibility in the system of production relations [10].

In his analyses, Coase relied on the relationship between property rights and business expenses (transaction costs) on the one hand, and the relationship between property rights and the production structure on the other hand. For example, if property rights are defined and transaction costs are equal to zero, the production structure will remain unchanged regardless of the possible changes in property rights, except for the effect of income.

Accordingly, in a market economy, it is impossible to ignore the impact of income, since obtaining income for economic units under competitive conditions is a fundamental requirement for the market to function. Coase's analysis of property rights theory does not align with the current dynamic of capitalist economy. Production operates in the dynamic transformations of economic relations driven by revolutionary and rapid changes in production technology and modern means of communication [10]. These transformations are a set of laws that determine the level of interaction between consumption and production. This continuous interaction leads to what is called "transition" in economic relations [9]. The transition refers to quantitative or qualitative changes in the structure of economic relations at the level of the entire economic system, such as the transition from automated work to the stage of robotics. In the field of Eastern studies, property issues have received special attention in socialist schools. Economists from Moscow University, Radayev and Bozgalin, see a kind of link between business expenses and transitional costs based on quasi-rent analyses of the relationship between property rights, production structure, and business transformation costs. If transformation costs are high and close in value to "transitional" costs under undefined or weakly specified property rights, resource distribution levels, production structure, and production efficiency will be determined based on property distribution in society [9].

In response to achieving national interests and enhancing economic stability, restrictive policies are used for some property rights that serve economic variables. This policy is embodied in financial instruments such as indirect taxes, the impact on the money supply through monetary instruments, setting a ceiling for issuance in financial markets, and so on [10].

Interest in property issues today, with the formation of knowledge societies [14], is controversial in economic circles, especially in economically underdeveloped countries. Local production relations distortion and lack of harmony between these relationships and the external world weaken internal production relations, thus distorting property relations and reducing the scope of rights. The researcher attributes these distortions to the channels of profit that flow in economic relations, especially in Arab countries. Accordingly, from the researcher's point of view, clear features of the economic system are not defined in Arab society as a result. In general, the economic system works as a market-directed economy. These imbalances distort the concepts of freedom, social justice, economic efficiency, property rights, and knowledge relations.

8 Functional Relationship Analysis of Property Production, Economic Power, and Knowledge [13]

When production do its functions by applying property rights, the roles are distributed among workers, organizers, owners, and the state. What is interest of the changing roles in the production relations system and, therefore, the changing shares of the production elements. Knowledgeable workers have made a significant contribution to productivity, which justifies their high wages. This, in turn, leads to the establishment of economic elite control, which means the expansion of the concept of economic power.

With this contribution focused on value-added (in GDP), a layer is formed that determines the market value in business organizations. Thus, intellectual resources become monopolized within these elite, creating a new market for knowledge workers who dictate their conditions based on the amount of knowledge they possess.

Human life is based on both diversity and similarity in relationships at the same time. Here, the analysis distinguishes between compulsory and voluntary relationships. Voluntary relationships, such as belonging or not belonging to a religious group (or a union), occupy a larger space.

In contrast, compulsory communities are few in number, such as the family in which we are born and raised, which gives us a racial, religious, and cultural orientation without our choice. Likewise, the state is responsible for maintaining order within society and enforcing legislation. However, belonging reinforces identity and ensures the individual's continuation as a social entity. Therefore, in the context of the production process, the nature of economic activity is determined based on independent and joint links. Let us assume that each element in the economic relations system works independently at the unit level.

This means that economic activity reflects independent and joint relations (the relationship between elements). Thus, independent-joint relations are organized among the elements of economic activity. An individual's ability to perform their job is a fundamental factor in the effectiveness of economic activity, leading to the issue of independence in the economic decision-making process. Through economic activity, both shared and independent relationships are revealed. For example, the tax system illustrates shared relationships in the form of the social responsibility placed on economic projects towards the state. Independent relationships are also evident through the economic projects' efforts to achieve profits that align with the level of tax system advancement [13].

In order to reach a state of balance between the state's interests on one hand, and the economic projects and workers on the other hand, must be taken into account. The tax system must be flexible in order to achieve individual and collective goals. The equilibrium level is achieved when the parties are equal in their rights and duties.

If the state adopts a strict tax policy, it is expected that this policy will undermine businessmen's trust in the law, leading to activities such as tax evasion. Such deviations by companies will create a state of instability on both the supply and demand sides. The decline in legal activities means a decrease in supply in the markets and therefore, an increase in prices. The decline in supply and the increase in demand pave the way for inflationary waves, and thus, a decrease in consumer surplus. As a result of the companies' attempts to hide their real activity, tax revenues decline, and thus, the state is unable to perform its functions, such as reducing healthcare spending. These effects are reflected in a decrease in public investment, which means a decline in overall production and, consequently, a slowdown in economic growth.

Economic power [13] refers to the ability of an individual or group of individuals to direct another individual or group towards their desires. Absolute power may be granted to those who possess it, and symbolic power may be granted to those who are delegated to economic power [9]. In the broader sense, economic power is a modified form of property.

Although there is complete overlap between property and economic power, the exercise of power does not always occur through the owner. In the field of production, when the need for rational use of resources arises, the contradiction between economic power and property becomes evident. However, since the development of production elements, the need for specialized knowledge and expertise to make independent production decisions has become urgent. As these specifications do not apply to every owner, it requires transferring the economic power to others who have the ability to achieve the owner's goals.

This demand was the way to overcome the contradiction between property and economic power [11]. Those who possess leadership qualities and the ability to manage economic resources on behalf of the owner are called "Management" [9]. Management has contributed to showing that organizers in the property system are not just owners or workers, but those who possess pioneering capabilities and knowledge and are able to change the production priorities system. Through these advantages, "management" can make radical changes in property relationships, allowing for new solutions to property or expropriation issues, and this, in turn, establishes what is called "knowledge Power" [9].

9 The Role of the Property in Developing the Knowledge Sector and Digital Transformation

The modern growth theories are focusing on new sources of economic growth by directing attention towards intellectual capital, in the context of the Fourth Industrial Revolution. The common thread among these theories is that any economic growth resulting from an abundance of labor and capital is insufficient to reach the stage of innovation if intellectual capital is not devoted to these purposes. To achieve this, capabilities must be employed in exploiting and developing existing knowledge. This view is reinforced by the economist Joseph Schumpeter's work on economic leadership, where creative destruction is attributed to research and development growth rates [14].

To demonstrate the content of intellectual capital and the conditions for expanding knowledge, countries must be classified according to the knowledge standard. This serves the analysis in developing suitable indicators to measure the level of this concept and enables us to identify the priorities of economic policy and knowledge-structured priorities. The following classification applies: advanced industrial countries (countries with advanced knowledge systems), industrially advanced countries with limited knowledge progress, economically backward countries relying on stored knowledge capabilities (countries in transition), and economically and knowledge-lagging countries. In order to achieve this classification, it is important to focus on the accumulation of knowledge in the national stock to meet market needs and overcome the knowledge gap between advanced and developing countries.

This requires directing financial resources towards investing in human resources and designs that increase the country's knowledge stock. Balancing investment rates in infrastructure, traditional production, modernization, and development is necessary to ensure an increase in knowledge production and accumulation. Developing investment in intellectual capital through human capital development requires considering the demand and supply sides based on cost-benefit analysis, considering the following considerations:

Investment in long-term and phased knowledge capital, Products resulting from the employment of this investment require complete protection through specific legislation that ensures the continuation of demand and flow of returns, the return must be realized at the appropriate time, investment in knowledge capital requires large capital expenditures, which require the participation of national economy sectors within specified development plans and social priorities [15] and the Investment in knowledge capital is valued at the cost of alternative opportunity in the face of limited economic resources, which requires evaluating all alternatives and options in the production field to compare between traditional and knowledge production. When discussing investment in intellectual capital by investing in human capital, the main question is: where do we invest?

Which areas generate knowledge production and increase intellectual assets? Economists throughout history have attempted to link education with progress, meaning they have tried to find a mechanism that allows us to apply economic standards to scientific and educational activities. Through these efforts, economists have aimed to understand the potential impact of this concept on economic activity. Adam Smith suggested that government spending on education would lead to a specific type of capital, which he called "Permanent Capital". This idea is supported by economist Solow's research, which excluded the impact of physical capital and found that 86.5% of changes in productivity indicators are due to technology and research and development in the industrial sector.

A state's latent abilities in terms of production determine the path of progress over time, particularly its knowledge abilities. In this study, researchers see the need to evaluate these abilities using a set of standards based on the international assessment system proposed by the study:

1. The number of scientists and researchers in various fields of science.
2. The ratio of spending on scientific research and development to the Gross Domestic Product.
3. The number of specialized peer-reviewed scientific journals with international classification.
4. The number of specialized applied universities (governmental and private).
5. The number of independent research centers and development and innovation centers.
6. The number of vocational institutes to the total number of higher education institutions.
7. The size of intellectual production to the total national trade.
8. The percentage of returns on investment in human capital to the NI.
9. The number of innovations registered in the country annually.
10. The cost of innovation to total overall production costs.

As a summary, and based on the previous analyses, knowledge production in the economic relations system constitutes an increasing proportion with the industrial advancement of the state, the development of property relations, the digital transformation area, and ultimately reaching those who possess knowledge, as shown in Fig. 1.

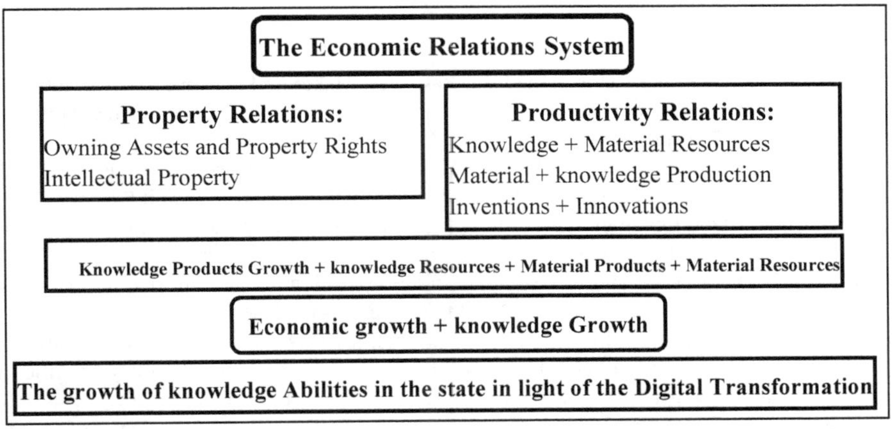

Fig. 1. Knowledge production in the economic relations system in light of digital transformations

10 Study Results

1. The exchange of roles between physical capital and labor, in the presence of knowledge workers, undermines property rights in the Arab economic system, which already suffers from blurring property rights.
2. Substantial leaps in social relationships cannot occur without a cultural shift in thinking style, towards transitioning from traditional education to learning-based education, thus moving towards education that produces creativity.
3. Achieving a balance between rates of capital accumulation and rates of knowledge accumulation to maintain intellectual capital as a determinant in the economic development path towards sustainable development.

11 Study Recommendations

1. It is necessary to review production relationships by clearly defining the functional links between labor and capital on the basis of property rights.
2. Public spending should be streamlined so that the tax system can direct overall investment towards physical and intellectual production that provides employment opportunities that exceed the rates of increase in the workforce.
3. Transitioning from traditional education to learning-based education is essential in building knowledge-based institutions capable of producing intellectual capital.
4. Directing the attention of economic and knowledge elites towards innovation in enriching the national industry with knowledge resources in a way that does not conflict with creating new employment opportunities.

References

1. Arab Knowledge Report, Mohammed Bin Rashid Al Maktoum Foundation, UAE (2009)
2. Phelps, E.: What is wrong with the West's Economies? The New York Review of Books, Reprinted and Translated by NCCAL (2015)
3. Zakaria, F.: Scientific Thinking (Revised Edition). Al-ilm Al-Arabi Magazine, Issue 501, December. National Council for Culture, Arts and Letters, Kuwait – Kuwait (2022)
4. Jubran, A.H.: Developing education in higher education institutions: moving to qualitative teaching. Int. J. Qual. Assur. **5**(1), 46–56 (2022)
5. Sobolev, V.: Knowledge and Infrastructure in the Transitional Stage (in Russian). Business Inform Magazine, Issue: 3–4, Kharkiv - Ukraine, pp. 36–40 (1999)
6. Berievno, O.B.: The evolution of intellectual capital. Kharkiv institute of public and business administration. In: Papers of an International Conference entitled "Social Policy Priorities in Transitional Circumstances ", Kharkiv - Ukraine, pp. 64–65 (1999)
7. Al-Razzaz, O.: From Rentierism to Production. The Difficult Road to a New Arab Social Contract. Economic Growth and Sustainable Development in Arab States (Economic Dimensions). Arab Center for Research and Policy Studies, Beirut – Lebanon (2013)
8. Galbraith, J.K.: History of Economic Thought: Past Image of the Present. Al-ilm Al-Arabi Magazine, Issue 261. Supreme Council for Culture, Arts and Letters, Kuwait - Kuwait (Translation) (2000)
9. Zudorogni, G.: Property and Economic Power. Ostova, Kharkov - Ukraine (in Russian) (1996)
10. Jubran, A.H.: The Role of the Property Relations System as a Driving Force for Economic Activity in the Restructuring of Economic Sectors, cited above (2013)
11. Radaev and Bozgalin. Transitional Economy, Moscow State University - Russian Federation (1995)
12. Joudan, T.: Technology and Society: An Unstoppable Evolution. Global Culture, 36th year (March-April), pp. 6–17, Supreme Council for Culture, Arts and Literature, Kuwait – Kuwait (2019)
13. Abu Saleem, T.: The Economic Efficiency of the Entrepreneurial Project Elements in the Market Economy. Unpublished Ph.D. thesis, National Kharkov University, Kharkov – Ukraine (2000)
14. Schumpeter, J.: History of Economic Analysis (in 3 parts, translated). National Translation Project, Supreme Council for Culture, Cairo - Arab Republic of Egypt (2005)
15. Al-Sawaie, K.: Determinants of Jordan's trade balance: a self-regressive distributed lag model. Al-Zarqa J. Human. Res. Stud. **17**(1), 138–149 (2017)
16. Marx, K.: Capital (translated into Russian). ACT, Russian Federation (2019)

E-Shopping Addiction Determinants and Effects Snowball Sampling for Workers, Undergraduates, and Postgraduate Students in AL Khobar City

Bergen Shopping Addiction Scale

Mirna Rida Khalife(✉)

Prince Mohammed Bin Fahd University, Dhahran, Saudi Arabia
Mkhalife1@pmu.edu.sa

Abstract. This paper aims to identify the determinants of e-shopping addiction and find out how it affects consumers in their daily routine by examining the behavior of e-consumers in the city of AL Khobar, the reasons for using online shopping, and the factors that lead some of the e-consumers to be addicted to e-shopping.

I surveyed three types of e-consumers using snowballing techniques: workers, students, and post-graduates.

The variables identified were the seven domains indicated in the Bergen Scale for Shopping Addiction, which is based on WHO's criteria for addictions, age scale, education level, and amount of money spent monthly on e-shopping. A rating of "agree" or "strongly agree" may indicate that respondents are addicted to shopping.

There are many reasons for increased electronic shopping: It's easily accessible, it allows us to spend hours scrolling, comparing and buying, plus the variety of products available relieves stress, and it's easy to pay on the go,.

Machine learning and artificial intelligence are catching micro and micro trends among consumers. In addition to celebrities and influencers collaborating to improve some products and offer different options, e-providers and retailers are analyzing data to identify consumer behavior patterns that customize their experiences and show them choices that most closely match their preferences with ads, offers and purchase options.

Analyzes use the SPSS 25 software program, tested with Alpha Cronbach, and, shows high item consistency and reliability. 76–835.

According to the Bergen scale, 11.8% of respondents indicated a high level of e-shopping addiction and 28.9% ticked the neutral option across the scale, indicating an undecided response for a variety of reasons, particularly for the items on salience, withdrawal, and mood modification.

Several factors were identified in the search for reasons for choosing a neutral option. One of them is social desirability: people avoid giving an undesirable response to satisfy the community, especially the addiction concept, which is rejected by social structures and norms.

59% agreed that online shopping is a convenient service that saves time and effort and offers easy payment, variety, and the best prices. This trend has increased

B. A. M. Alareeni and I. Elgedawy (Eds.): *Artificial Intelligence (AI) and Finance*, SSDC 488, pp. 495–507, 2023.
https://doi.org/10.1007/978-3-031-39158-3_47

due to the COVID-19 pandemic when consumers switched from real stores to virtual stores of digital channels.

Keywords: E-Shopping addiction · Bergan scale · Salience dimension · Compulsive Buying Disorder CBD

1 Introduction

When shopping goes from a choice to a habit that leads to compulsive and excessive shopping, the diagnosis becomes an addictive disorder that has been discussed for several years (Aboujaoude, 2014; piquet-pessôa et al., 2014).

This fact is reflected in the many names that have been given to this problem, including "oniomania," "shopaholics," "compulsive shopping," "compulsive consumption," "impulsive buying," "compulsive buying," and "compulsive spending" (Aboujaoude, 2014; andreassen, 2014). Andreassen (2014, p. 198) suggested that compulsive buying is best understood from an addiction perspective. He defined it as an "excessive interest in shopping, driven by an uncontrollable motivation to buy, and investing so much time and effort in shopping that it interferes with other important areas of life."

This type of addiction has been diagnosed using the diagnostic criteria for pathological gambling (american psychiatric association, 2013).

The current decade has witnessed an incessant growth of the online retail market, and excessive dysfunctional browsing of shopping websites represents a specific feature of compulsive shopping addiction in the modern context of cbsd (müller, steins-loeber, et al., 2019). Due to the digital transformation and circumstances created by covid-19, shopping has shifted to virtual platforms and stores. By analyzing the behavior of users, it was possible to find out who loses control and turns into an obsessive e-consumer due to the weak psychosocial structure. This paper addresses the determinants of e-shopping addiction, its effects, and possible treatments for this addiction.

2 Literature Reviews

Many scientific studies have been conducted on e-shopping addiction, but also on different types of addiction related to social media platforms.

Some of them dealt with the concept of e-shopping addiction, others analyzed the behavior of consumers and the complications associated with it, and still other researchers focused on the symptoms of addiction and the impact on the economy and social life.

Vasileios Stavropoulos (2022) analyzed the BSAS, and the results showed that the test is a unidimensional procedure for the risk of shopping addiction. All dimensions showed sufficient discrimination (α), difficulty (β), and precision, indicating that the scale is a psychometrically sound instrument.

A study titled "Testing Models Regarding Online Shopping Addiction" (Keskin Ayten, Gunuc Selim, 2017) investigated the reasons behind adults' preference for online shopping and found some factors that lead to this addiction.

Another study titled "Brand addiction: brand characteristics and psychological outcomes" (Barbara Francioni, Ilaria Curina, Sabrina M. Hegner, and Marco Cioppi, 2020) examined the influence of brand addiction. The results show that brand characteristics of self-presentation, innovation, and authenticity have a positive influence on brand addiction, i.e., increase irritability and obsession with the brand.

On the other hand, postdoctoral researcher Chatterjee (2021) discussed the impact of online platform addiction on quality of life. Many theories such as Adult Attachment Theory, Tension Reduction Theory, and Technology Acceptance Models were selected to identify the factors that influence online platform addiction. The researcher finds that the moderating factors of age and gender are effective in influencing the predictors of addiction to online platforms. It is found that with an increase in feelings of loneliness, the rate of addiction increases in women compared to men, with an increase in perceived pleasure (PEN), the rate of increase in addiction is higher in men than in women.

Some studies addressed the impact of young people's internet addiction and their social media preferences (Murat Akin, 2017) and investigated the extent of young people's internet addiction, especially in relation to social media. The hypothesis was analyzed with a sample of 756 participants, and the author mentioned that using the internet makes our lives easier, but he concluded that excessive use causes psychological problems and leads to a low quality of life for some users. This study shows that 61% of adolescents suffer from internet addiction, while 22.4% were classified in the risk group.

Based on different literature reviews, such as the study of (Cecile Adresson, et al 2015), I decided to use the Bergen E-Shopping Addiction Scale.

A screening test was conducted to establish the reliability and validity of the scale with 28 items. These items were constructed in combination with (i.e., Compulsive Buying Measurement Scale, Mini-International Personality Item Pool, Hospital Anxiety and Depression Scale, Rosenberg Self-Esteem Scale) targeting 23,537 participants. The author mentions that the BSAS is the first scale to fully embed shopping addiction within an addiction paradigm. The scale is characterized by good psychometric properties, convergent validity, and discriminant validity.

The study by Harsheka kar and Ashish Jha (2021) shows that young adults are at high risk for mental health problems and report increased loneliness and depression as early as the first 5 months of the pandemic COVID -19. Given the importance of meaningful activities for mental health as Internet addiction increases, it is evident that the literature on this relationship and meaningful activities during the lockdown is sufficient and concludes that Internet addiction is negatively correlated with participation in meaningful activities among young adults.

Another study by Norman Shaw & Brenda Eschenbrenner and Daniel Baierc (2022) predicts that online shopping will be higher post-pandemic than pre-pandemic. With the pandemic, online shopping increases, shifts in the labor market occur, and online retail will grow 14.6% from 2020 to 2030.

3 Hypotheses and Concerns

Observation of compulsive buyers differs from normal buyers in terms of their buying process and their emotional reaction when they carry out the buying process.

The study hypothesizes that e-shopping addiction exists among young people, often women, and is increasing over time due to the digital transformation. This increase sheds light on the psychosocial risks that are amplified by this addiction and affect self-esteem. It affects self-esteem, wastes a lot of time thinking and planning purchases, interferes with daily obligations, and at a higher level causes chronic dissatisfaction, increases anxiety levels, and gradually causes awareness to fade if addicts do not find the support and treatment they need.

In this context, the Bergen scale is used to determine the level of addiction to e-shopping. It is based on specific factors reflected in the dimensions of salience, mood change, conflict, tolerance, and relapse.

4 Method

4.1 Participants

The targeted people were 3 types: Undergraduates, Postgraduates and workers in the city of Alkhobar with N = 504. The sampling method used is the snowball or chain reference method.

The sample was distributed between females (77.4%) and males (22.6%), reflecting the high participation of females in e-shopping and their curiosity to find out whether they are addicted or not, classifying the different age groups consecutively: 85.2% for the range (18–24), 8.6% (25–34), 4.6% (35–44), 1.8% (45–54), 0.4% (55 years and above). The responses reflect the great interest of very young people in e-shopping and its addiction.

A descriptive methodology was used for quantitative scale feedback (Bergen e-shopping addiction).

4.2 Scale

It has been tested in different studies to check its reliability and validity (see "Testing Models Regarding Online Shopping Addiction" Keskin Ayten, Gunuc Selim, 2017 &The Bergen Shopping Addiction Scale: reliability validity of a brief screening test, Cecile Adresson, et al. 2015.

This scale was used because of its simplicity, suitability for large-scale studies, widespread international acceptance, and sound rationale. Salience indicates what is most important in the person's daily routine. Mood modification means the activities that alter the person's mood state. Tolerance refers to the addictive activity that tends to produce the desired effects, and withdrawal refers to the psychological and/or physical withdrawal response. Conflict refers to conflicting states in the person due to engagement with the addictive activity. Relapse means that addictive behavior recurs (Griffiths M. 2005), and the problem dimension illuminates how addiction can harm our people.

5 Results

To determine the extent of individual e-shopping addiction, BSAS surveyed Prince Mohammed Bin Fahad University students, postgraduate students from the same university, and workers aged 35 and older with 504 responses.

SPSS was used to analyze the dimensions and variables. A coding system was created for the items and the dimensions were coded by the first letter of the dimension and the item n0.

E- Shopping addiction scale (BSAS) contains a set of items reflecting seven elements of addiction, which ensures its content validity within the framework of addiction.

The internal consistency Cronbach's alpha is high as shown in the Table 1 (>.75).

Table 1. Cronbach's alpha to measure internal consistency

α Dimensions	No. of items	Cronbach Based on standardized item	Mean	Items Variance	Inter-item Covariance
Salience	4	.764	3.19	.028	0.41
Mood modification	4	.829	3.11	.006	.036
Conflict	4	.835	2.08	.022	.012
Tolerance	4	.835	2.08	.022	.012
Relapse	4	. 760	2.7	.018	.011
Withdrawal	4	.809	2.6	.014	.027
Problem	4	.796	2.7	.019	.035

To purpose of analyzing the items through Cronbach was to make sure that the different subsets of test items would produce similar construct and measures, in addition to the variance and covariance scored between the ranges 0.15–0.50, which is affirm the consistency reliability within the items.

A descriptive statistic took place to find the mean and the SD, the dimensions items are distributed equally with 4 items, and the Cronbach score in the 7 dimensions shows high consistency with the large n0 of responses.

The descriptive analysis shows the mean of the scales items, in addition to the ordinal items, the majority of the respondents were **Female** 77.4% and **males** 22.6%, and the highest % of them were **undergraduate students**, that they felt a high curiosity to discover if they're addicted to the e-shopping based on their daily routine the answers distributed, in addition to the **reasonable** answer **about spending money** for the E-shopping, The standard deviations ranged between 0.4 and 1.34, indicated high levels of association among the scale items.

The gender differences are observed due to their high participation in the scale, the results show slightly higher rates of compulsive shopping in women, moreover, the responses on the use of electronic shopping reflect a tendency towards the neutral option of the e-shopping addiction scale, indicating high shopping addiction, and the confusion about whether one is addicted or not is a warning to avoid negative consequences that increase feelings of loneliness, withdrawal, and loss of time and money to feel satisfied.

Table 2. *Measurement Properties*

Descriptive Statistics					
	N	Minimum	Maximum	Mean	Std. Deviation
Gender	504	1	2	1.78	.416
Education	504	1	5	2.84	.545
Money spent	504	1	5	4.11	1.077
s1	504	1	5	3.01	1.185
s2	504	1	5	3.19	1.280
s3	504	1	5	3.33	1.311
s4	504	1	5	3.24	1.338
m1	504	1	5	3.39	1.376
m2	504	1	5	3.57	1.371
m3	504	1	5	2.78	1.387
m4	504	1	5	2.69	1.323
c1	504	1	5	2.21	1.211
c2	504	1	5	2.19	1.203
c3	504	1	5	1.77	1.109
c4	504	1	5	2.18	1.259
t1	504	1	5	2.72	1.206
t2	504	1	5	2.93	1.334
t3	504	1	5	2.56	1.267
t4	504	1	5	2.61	1.248
r1	504	1	5	2.55	1.236
r2	504	1	5	2.50	1.323
r3	504	1	5	2.68	1.343
r4	504	1	5	2.95	1.294
w1	504	1	5	2.54	1.284
w2	504	1	5	2.64	1.288
w3	504	1	5	2.84	1.349
w4	504	1	5	2.85	1.217
p1	504	1	5	2.43	1.289
p2	504	1	5	2.35	1.251
p3	504	1	5	2.14	1.286
p4	504	1	5	2.46	1.298
Valid N (list wise)	504				

Table 3. T-test Correlation based on Gender and the items of the dimension

T-test					
	Gender	N	Mean	Std. Deviation	Std. Error Mean
s1	M	112	2.73	1.200	.113
	F	392	3.09	1.170	.059
s2	M	112	2.84	1.373	.130
	F	392	3.29	1.236	.062
s3	M	112	2.96	1.391	.131
	F	392	3.43	1.269	.064
s4	M	112	2.87	1.359	.128
	F	392	3.35	1.314	.066
m1	M	112	2.80	1.512	.143
	F	392	3.56	1.288	.065
m2	M	112	2.87	1.492	.141
	F	392	3.78	1.266	.064
m3	M	112	2.21	1.226	.116
	F	392	2.94	1.389	.070

(*continued*)

Table 3. (*continued*)

m 4	M	11 2	2.21	1.158	.109
	F	39 2	2.83	1.337	.068
c 1	M	11 2	2.21	1.276	.121
	F	39 2	2.21	1.193	.060
c 2	M	11 2	2.18	1.337	.126
	F	39 2	2.19	1.164	.059
c 3	M	11 2	1.75	1.189	.112
	F	39 2	1.78	1.087	.055
c 4	M	11 2	2.13	1.288	.122
	F	39 2	2.20	1.252	.063
t 1	M	11 2	2.59	1.263	.119
	F	39 2	2.76	1.188	.060
t 2	M	11 2	2.69	1.427	.135
	F	39 2	3.00	1.300	.066
t 3	M	11 2	2.31	1.163	.110
	F	39 2	2.63	1.288	.065

(*continued*)

Table 3. (*continued*)

t 4	M	11 2	2.29	1.181	.112
	F	39 2	2.70	1.253	.063
r 1	M	11 2	2.37	1.301	.123
	F	39 2	2.60	1.213	.061
r 2	M	11 2	2.34	1.405	.133
	F	39 2	2.55	1.296	.065
r 3	M	11 2	2.45	1.368	.129
	F	39 2	2.75	1.330	.067
r 4	M	11 2	2.76	1.324	.125
	F	39 2	3.01	1.281	.065
w 1	M	11 2	2.29	1.326	.125
	F	39 2	2.61	1.265	.064
w 2	M	11 2	2.47	1.301	.123
	F	39 2	2.68	1.282	.065
w 3	M	11 2	2.60	1.352	.128
	F	39 2	2.91	1.342	.068

(*continued*)

Table 3. (*continued*)

w4	M	112	2.65	1.257	.119
	F	392	2.91	1.200	.061
p1	M	112	2.29	1.249	.118
	F	392	2.47	1.299	.066
p2	M	112	2.22	1.206	.114
	F	392	2.39	1.262	.064
p3	M	112	2.00	1.259	.119
	F	392	2.18	1.292	.065
p4	M	112	2.38	1.403	.133
	F	392	2.49	1.267	.064

Female students' responses indicate that they have a high response on the scale, they consider e-shopping as a great solution to save time in their hectic lives, there are significant differences between males and females in e-shopping behavior based on the Anova test.

Furthermore, *11.8% of respondents indicated a high level of e-shopping addiction and 28.9% selected the neutral option* across the scale, indicating an undecided response for various reasons, particularly for the items on salience, withdrawal, and mood modification.

The Effects of the Neutral Response Option on the Extremeness of Participant Responses article analyzed the causes of choosing the neutral option of the Likert scale and finding out, that some people tend to gravitate towards neutral because they want to avoid the negative feelings associated with their conflicting feelings on an issue. Additionally, picking a neutral option allows people to avoid the cognitive effort needed to choose between their positive and negative emotions on an issue, Social desirability is a third factor that influences the choice of a neutral option, in a sensitive phenomenon

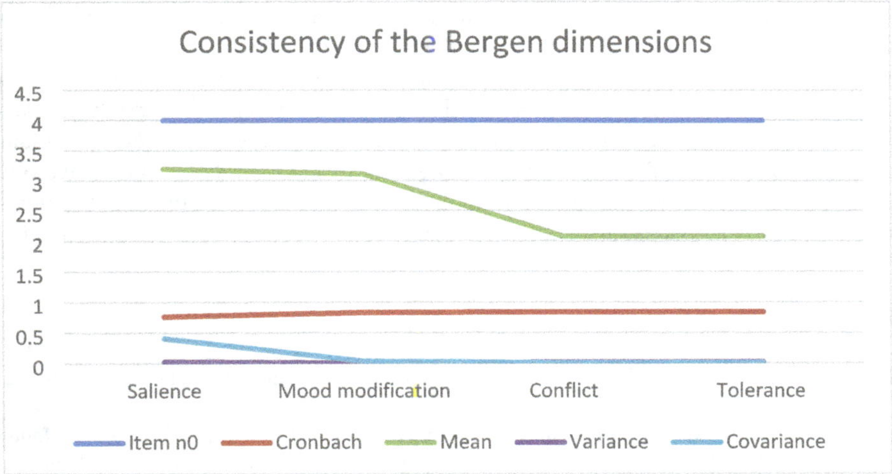

Fig. 1. Cronbach line chart

such as addiction, the respondents avoid to give the strict answer, and this justification is applied in on the Bergen scale.

6 Conclusion and Recommendations

During and after the COVID pandemic, e-shopping addiction increased, and the majority of trading companies benefited greatly, in addition to new apps that were developed to display advertisements based on what we've shown interest in.

To avoid becoming addicted to electronic use for shopping and other similar activities, there is a high demand for self-discipline, high self-esteem, and the formation of financial habits.

Furthermore, reducing applications is one of the most effective ways to reduce usage, along with being involved in family occasions and events, volunteering in community service activities and learning new languages.

Prevention is better than cure, which is why this type of study is taking place to alert users about the dimensions of electronic platform misuse. And if the individuals are not able to manage their behaviors in this regard, professional cognitive-behavioral therapy can help create the required balance.

E-shopping addiction is a behavioral addiction, that is caused by hyperstimulation, Shopping addicts can become dependent on the elevation of mood that buying can bring, thus leading to addiction (Smith, Robert L., and Robert L. Smith.2023), Smith suggested treatment approaches based on several scientific studies, starting with *the anti-depressant drugs, psychotherapy* that's helping the addicts to understand their symptoms, *the cognitive behavioral therapy CBT*, in addition to the *group therapy* that's supporting the addicts with the addicts' group, its raised the feeling that the addict is not alone, and his case is understood.

The high demand to use digital platforms for shopping invites researchers to increase their effort to study and analyze behavioral addiction of individuals within males and females, and how it's affecting our lives, in addition to the need of raising awareness campaigns to support the addicts or the possible addicts, the psychosocial experts are able to orient their community service to this field specifically. On another side, individuals are invited to control their behaviors by setting priorities in their daily routine, avoiding buying not needed items and products in order to avoid creating financial problems, practicing sports activities, and being involved in social events to live a healthier life.

References

Abou jaoude, E.: Compulsive buying disorder: a review & update. Curr. Pharm. Des. **20**, 4021–4025 (2014). https://doi.org/10.2174/13816128113199990618

American Psychiatric Association: What is Gambling Disorder? Diagnostic and Statistical Manual of Mental Disorders, Fifth Edition. (DSM-5) American Psychiatric Association Publishing (2013)

Andreassen, C.S.: Shopping addiction: an overview. J. Nor. Psychol. Ass. **51**, 194–209 (2014)

Francioni, B., Curina, I., Hegner, S.M., Cioppi, M.: Brand addiction: brand characteristics and psychological outcomes, Emerald Group Publishing (2020)

Andreassen, C.S., Griffiths, M.D., Pallesen, S., Bilder, R.M., Torsheim, T., Aboujaoude, E.: The Bergen shopping addiction scale: reliability and validity of a brief screening test. Front. Psychol. **17**(6), 1374 (2015)

Zarate, D., Fullwood, L., Prokofieva, M. Griffiths, M.D., Stavropoulos, V.: Problematic shopping behavior: an item response theory examination of the seven-item bergen shopping addiction scale. Int. J. Mental Health Addict. **20**, 1–9 (2022). https://doi.org/10.1007/s11469-022-00844-8

Griffiths, M.: A 'components' model of addiction within a biopsychosocial framework. J Subst Use. **10**, 191–197 (2005)

Harsheka, K., Ashish, J.: Relationship between internet addiction and engagement in meaningful activities among young adults, Indian, Spinal Injuries Centre Institute of Rehabilitation Sciences (2021)

Keskin, A., Gunuc, S.: Testing models regarding online shopping addiction. Turish J. Addict. **4**(2), 221–242 (2017)

Müller, A., Steins-Loeber, S., Trotzke, P., Vogel , B., Georgiado, E., de Zwaan, M.: Online shopping in treatment-seeking patients with buyithe ng-shopping disorder**.** Compr. Psychia. **94**, 152120 (2019). https://doi.org/10.1016/j.comppsych.2019.152120.10

Edwards, M.L., Smith, B.C.: The Effects of the Neutral Response Option on the Extremeness of Participant Responses. J. Undergraduate Scholarship. **V6**, 30 (2014) https://blogs.longwood.edu/incite/2014/05/07/the-effects-of-the-neutral-response-option-on-the-extremeness-of-participant-responses/ *Retrieved Feb8, 2023*

Murat, A.: A research on the impacts of the young people's internet addiction levels and their social media preferences. Int. Rev. Manage. Market. **7**(2), 256–262 (2017)

Shaw, N.: a,∗ Brenda Eschenbrenner, b and Daniel Baierc, **Online shopping continuance after COVID-19: A comparison of Canada, Germany and the United States**. J. Retail. Consum. Serv. **69**, 103100 (2022Nov). https://doi.org/10.1016/j.jretconser.2022.103100

Piquet-Pessôa, M., Ferreira, G.M., Melca, I.A., Fontenelle, L.F.: DSM-5 and the decision not to include sex, shopping or stealing as addictions. Curr. Addict. Rep. **1**(3), 172–176 (2014). https://doi.org/10.1007/s40429-014-0027-6

Chatterjee, S.: Impact of addiction of online platforms on quality of life: Age and Gender as moderators. Austral. J. Inform. Syst. Res. User Involve. **25** (2021)

Smith, R.L.: Treatment Strategies for Substance Abuse and Process Addictions, American Counseling Association, (2015). *ProQuest Ebook Central*, http://ebookcentral.proquest.com/lib/pmu-ebooks/detail.action?docID=1986948. Created from my-eBooks. Retrieved on 2023–02–19

Stavropoulos, V.: Problematic shopping behavior: an item response theory examination of the seven-item bergen shopping addiction scale. Int. J. Mental Health Addict. **1** (2022)

Legal Protection of Digital Works from Attacks in Cyberspace

(A Study of Jordanian Legislation and International Agreements)

Omar Almakhzoumi[1](✉), Asad Alhroob[1], and A. N. M. Mahfuz[2]

[1] Faculty of Law, Zarqa University, Zarqa, Jordan
omakhzoumi@zu.edu.jo

[2] College of Law, University of Business and Technology, Jeddah, Saudi Arabia
mahfuz@ubt.edu.sa

Abstract. The purpose of this research paper is to examine the role of legislation in regulating intellectual property rights in the digital age, and ensuring that all internet users have access to information in the digital environment. The widespread use of digital works in cyberspace has led to various forms of infringement, and the study aims to determine the extent to which Jordanian legislation is aligned with international conventions to provide legal protection for digital works.

The research analysed the Jordanian legislation and found that it is up-to-date with modern developments in protecting against cybercrimes and copyright infringement. Based on these findings, the study suggests several recommendations, including the need for the Jordanian legislator to establish a specific legal framework for digital works and their protection, leveraging modern technological advancements.

Keywords: digital works · cyberspace · legal protection · copyright

1 Introduction

The modern digital environment is represented by cyberspace, where this environment is connected by technical means by computer, allowing digital works flowing through cyberspace to satisfy the necessary information needs of individuals, but the use of digital technology has led to the emergence of new forms of intellectual property infringement, and here the basic need to search for technical and legal means to protect copyright on his digital works.

The technical nature of digital works published on cyberspace has facilitated the commission of cybercrimes, which led to the violation of intellectual property rights, which requires the need for legal protection, starting from procedural protection to civil protection and penal protection, and due to the difficulty of controlling them by determining the applicable law between different countries, which necessitated the need for international cooperation in combating cybercrimes, and stopping attacks on digital works on the Internet.

B. A. M. Alareeni and I. Elgedawy (Eds.): *Artificial Intelligence (AI) and Finance*, SSDC 488, pp. 508–518, 2023.
https://doi.org/10.1007/978-3-031-39158-3_48

The importance of the topic of this paper and its objective is to shed light on the role of legislation and its ability to keep pace with the developments produced by modern technology in the protection of intellectual property rights and to strike a balance in ensuring the right of all Internet users to obtain an adequate amount of information in the digital environment, at a time when digital works have spread in cyberspace as a new means of electronic publishing.

The problem of the study lies in the insufficient capacity of the Jordanian legislation to provide legal protection for digital works in cyberspace, especially with the rapid technological developments and the multiplicity of forms of assault on digital works in cyberspace.

1. What is meant by the legal protection of digital works in the digital environment? What is meant by "cyberspace" and its relationship to cybercrime?
2. What are the forms of attacks on digital works, and are the legislative texts in Jordan sufficient to protect these works?
3. To what extent does Jordanian legislation keep pace with international conventions on the legal protection of digital works?

We will try to answer these questions by researching the international protection of digital works mentioned in international conventions such as the TRIPS Convention and the Berne Convention, as well as the protection provided by the World Intellectual Property Organization (WIPO), and examining the forms of attacks on digital works and the extent to which Jordanian legislation keeps pace with international protection, whether procedural, civil, or penal, through copyright law, cyber security law, and cybercrime law.

2 International Protection of Digital Works

The term "international protection" refers to all the legal means provided by international laws and conventions for the protection of digital works published in the digital environment. Furthermore, it encompasses technical protection methods that enable right holders to use technological means in a way that will enable them to preserve their works and prevent infringement. In light of these points, we must study the position of international legislation concerning attacks on digital works, as well as technical attacks and ways to protect them.

2.1 The Position of International Legislation on Attacks on Digital Works

The use of digital works may extend internationally and at a very large scale, which makes it necessary for additional protections to be provided for them in addition to those provided by national legislation. Developing special and model texts for the protection of digital works published in cyberspace is the most effective solution; therefore, countries sought the assistance of a large number of experts in this field. Accordingly, it is necessary to address the protection of digital works in international organizations, as well as the status of international agreements regarding copyright protection.

2.1.1 Protecting Digital Works in International Organizations

The cyberspace is characterized by the ease of retrieval and transmission of information, which makes it easy to locate attacks on digital works in this cyberspace. The rights of the owners of the works are subject to attacks in this space, whether by exploiting the published works by copying, downloading, distorting, or stealing their content and attributing them to another person without mentioning the reference [1].

2.1.1.1 The Protection Prescribed by the World Intellectual Property Organization (WIPO)

Considering what was included in the WIPO Copyright Treaty and on sound recordings, which were called the two Internet treaties, solutions were developed to face the challenges posed by digital technology in cyberspace and dealt with a set of basic matters, that were circulated through the Internet, rights related to the storage and transmission of works, and the exceptions and limitations contained in those rights [2]. Based on the foregoing, the first paragraph of the WIPO Treaty stipulates that it is a special treaty that provides protection for copyright holders, provided that it does not conflict with the provisions of the Berne Convention, and according to this treaty, the right of reproduction applies to it, as stipulated in the Berne Convention, and the permitted exception is that article that corresponds to the digital environment, particularly digital works. The concept of storing digital works in a digital medium is considered a copy of the Berne Convention: "The authors of literary and artistic works protected by this Convention have the exclusive right to authorize copies of these works in any way and in any form whatsoever" [3].

It is worth noting that the WIPO Treaty expanded the scope of application of the right of communication to the public to include all works and this that right includes transmission tools through interactive communication systems and also emphasized the protection of digital works published on the Internet. "Authors of literary and artistic works enjoy the exclusive right to authorize the transfer of their works to the public by wired or wireless method, including making their works available to the public in a way that enables each of the public to view these works from a place and at a time chosen by each of them" [4].

It is important to note that the WIPO Treaty deals with important issues that were not addressed by the Berne Treaty. These issues include the collection of data. Because the collected data constitutes a personal fingerprint as a result of their selection or assembly processes, it must be protected on this basis, but protection requires the effort to arrange the information and does not extend to data collection (databases).

Whenever datasets or other materials are considered intellectual creations by virtue of their choice or arrangement, they are protected as intellectual creations, regardless of their form. However, this protection does not apply to any data or materials in the collection, and it does not adversely affect any copyright rights that may exist in the data or materials" [5].

And we should not fail to mention that the WIPO Treaty stated that computer programs are considered as literary works, so they are protected, whether the programs are in the machine language or the source language [6].

Within the national framework, the Jordanian legislator adopted the principle of modernizing intellectual property laws, including the copyright and related rights protection law, to which the legislator made important and advanced amendments to be in line with modern communication techniques that describe the information and technology revolution. As Jordan acknowledges the inevitability of transforming it into a modern country that accept the challenge posed by this transformation. [7] it joined the WIPO Treaty on copyright, performances, and sound recording in 2004 [8].

2.1.1.2 International Agreements on Copyright Protection

Initially, legislation was intended to protect intellectual property rights in both literary and artistic property and industrial properties, so that those who own these rights may enjoy protection and exploitation of their literary and artistic creations in those countries that recognize them. So, countries started to seek protection for intellectual proprieties rights through international conventions [9].

The protection provided by national legislation is insufficient to meet the interests of the industrialized countries, as it is limited and does not extend beyond the borders of the countries that recognize these rights. As a result, countries began seeking to protect their intellectual property rights on a large international scale through international agreements (9). There are a number of important agreements dealing with intellectual property rights, including the TRIPS Agreement and the Berne Agreement.

In this context, the Jordanian Copyright and Related Rights Law No. (22) of 1992 was described as a modern law whose articles were amended more than once by Amended Law No. (14) of 1998 and Amended Law No. (29) of 1999 to comply with the provisions of international agreements. Jordan joined the TRIPS Agreement, the Agreement on Trade-Related Aspects of Intellectual Property Rights, and it became effective as per Jordan's entry into the World Trade Organization on April 11, 2000. Jordan also joined the Berne Convention for the Protection of Literary and Artistic Works, and it became effective on July 28, 1999 [10].

First: TRIPS Agreement on Trade-related Aspects of Intellectual Property Rights

In terms of intellectual property rights, the TRIPS Agreement is one of the most comprehensive agreements available. This agreement established provisions for the protection of intellectual property rights in the multi-trade system, since these rights were granted to individuals for their personal creativity and innovation [11]. As part of this agreement, the state's parties were also obligated to honor the rights and obligations outlined in Articles (1–21) of the Berne Convention of 1971, as discussed in the first section of the second part of the TRIPS Agreement [12].

The Jordanian legislator's text in the Copyright and Related Rights Law says: "…b. This protection includes works whose expression is writing, sound, drawing, photography, or movement in particular: 8. Computer programs, whether in the source language or in the language of the machine… d Protection is also enjoyed by a group of literary and artistic works, such as encyclopaedias, anthologies, and collected data, whether in a readable form or in any other form, if they constitute, in terms of the selection or arrangement of their contents, innovative intellectual works." [13].

This is confirmed by the Jordanian legislator in the law protecting copyright and related rights in terms of expression and not in ideas, procedures, methods of operation, or mathematical concepts per se, in order to comply with the TRIPS Agreement, which stipulates that "copyright protection applies to production and not to mere ideas, procedures, work methods, or mathematical concepts" [14].

It is important to note that the Jordanian legislator protected computer programs, whether they were written in the source language or the language of the computer. It is important to note that this matter pertains to computer programs and not to the computer itself, as the computer is a physical component or machine created through an invention or innovation, and the source programs are perceived by humans, as opposed to machine programs, which are perceived only by automated devices. Also, the Jordanian legislator protected Databases, which are defined as "a collection of data or other materials that are accessed by devices or in any other way, are innovative, and have a personal stamp."

While this protection does not cover the data or material itself, it does ensure compliance with the TRIPS Agreement by taking account of any copyright associated with Jordanian law under the Berne Convention, "1. Computer programs are protected whether in the source language or in the language of the machine, they are protected as literary works under the Berne Convention (1971). The collected data or other materials, whether in readable form or any other form, shall enjoy protection if they constitute an intellectual creation as a result of the selection or arrangement of their contents [15].

Second: Berne Convention for the Protection of Literary and Artistic Works
Internationally, the Berne Treaty regulates copyright and other related rights. Nevertheless, it is one of the oldest treaties, having been ratified in 1886. Several amendments were made to its articles over the years, the most recent being in 1971, when it dealt with copyright. The organization provided legal protection to the organizing countries. However, it did not address the issues that have recently emerged in cyberspace, as the latest amendment did not take into account the revolutions in communication and information technology, as well as the emergence of the Internet [16].

There is no doubt that the Berne Treaty contains the principle of reciprocity, which is based on providing the same legal protection to the owners of works in other countries as their original citizens do to their foreign nationals in those other countries. Therefore, the national law of that country governs the matter. A foreigner who owns the work has the same rights as the original citizen [17].

Jordanian legislators held differing opinions regarding the protection of foreign works in light of the amendment, so they initially adopted the territoriality principle. The work is considered national if it has been published in a country with which it has a close relationship, even if the author is foreign. However, after the amendment of the Copyright and Related Rights Law was enacted, according to the legislator. As long as international conventions governing copyright protection are followed, and reciprocity shall be observed if they are not. This law applies to the works of foreign authors, whether published or unpublished, and expressed through any of the means outlined in Paragraph B of Article 3 of this law [18].

Based on the foregoing, the Berne Convention stated, "Protection in the country of origin is governed by national legislation. However, if the author is not a national of the country of origin of the work on the basis of which he enjoys protection under this Convention, he enjoys in that country the same rights as its nationals." [19].

2.2 Technical Attacks and Ways to Protect Them

It is worth noting that in the past years, there has been an increase in the distribution of films, music, works of art, and digital works protected by intellectual property laws using electronic media. These forms may be subject to unauthorized downloading, transfer, or republishing, and this contradicts the laws protecting intellectual property rights, and thus results in material and moral losses that kill the spirit of innovation and modernization among users, especially in developing countries [20].

Undoubtedly, technical development tends to reduce the ability of the legislator to control intellectual property rights. If we take the reproduction from the original copy, then it is not completely identical to the original work, but the technology is moving towards making it more identical, and with the advent of cyberspace, the copy and the original have become the same thing and cannot be distinguished from each other. As a result, digital appears to be moving toward making the carrier and the owner of the work the same thing, necessitating a solution to this quandary, with the main issue being the difference in the rate of change between law and technology [21].

The protection of intellectual property rights in cyberspace is one of the most important legal issues. This topic is related to the infringement of computer programs through the use of a computer or a computer system, including algorithms belonging to a specific part of intellectual property, and in general it is the infringement of the work by illegal copying, piracy of the names of Internet sites, and the infringement of trademarks through domain names, which degrades their value on the Internet, Attacks on intellectual property rights in cyberspace can be classified as piracy, imitation, and plagiarism [22].

From this standpoint, cyber threats and crimes have evolved, and new ways of committing cybercrimes have emerged in cyberspace. Laws have failed to follow this development, and new technologies have emerged that have made the application of traditional law and the conduct of judicial investigations more difficult due to their comprehensiveness. Modern technologies have contributed to concealing the true identity of the offender and the increasing spread of information programs used by hackers in the development of cybercrime [23].

It is noted that the Jordanian legislator mentioned in the cyber security that "a cyber security incident: an act or attack that poses a threat to data, information, information systems, the information network, or its related infrastructure, and requires a response to stop it or mitigate its consequences or effects" [24].

Due to the increase in the number of users on the Internet, to protect them while dealing with computer networks and the Internet so that we can have a safe world without fear of illegal actions in cyberspace, we should highlight the most important types of cybercrimes:

First: Hacking

This crime involves the criminal entering the victim's computer without his knowledge, and hacking the computer in order to access personal information or data.

Owners of intellectual property are concerned about this type of crime because it doesn't just affect computer operating systems, but also any intellectual product that can be copied. It is possible for the criminal to enter the computer system for several reasons, including sabotage, modification, or deletion. Or for other reasons, such as penetrating the system to satisfy his curiosity and leaving evidence of his entry to prove to the facility owners or other people that its technical systems are weak [25].

Second: Theft and Plagiarism
It refers to the reproduction, copying, or sale of published and protected works without obtaining permission from the owner in order to obtain a benefit from them. It is similar to stealing products from the shelves of some stores.

Criminals violate copyright rights by downloading files from companies' websites on international information networks, including digital books, movies, games, and programs. It is not necessary for the copying to be complete, rather, the crime occurs when the copying includes basic and essential elements of the work. Plagiarism occurs through the transfer and reproduction of innovative elements that distinguish it from other works [26].

As two basic methods of legal protection on the Internet, encryption and electronic signature technologies may be helpful in protecting digital works in cyberspace.

Third: Encryption Technology
The cipher is the conversion of the linguistic text into or a digital text in a way that can only be read by using a private key or a password, and therefore it must be taken into account that the Arab world Al-Khwarizmi was the first to create the cipher [27] and the encryption systems (symmetric key encryption, and asymmetric key encryption).

Fourth: Electronic Signature
It is important to note that electronic signatures created a quantum leap in the computer industry. Currently, we are living in a world where the idea of a traditional signature is being challenged by advanced methods and means of dealing.

In light of this modern scientific renaissance, electronic signatures have developed [28], where electronic signatures are composed of letters, numbers, symbols, signs, and other data that are included electronically or by any other similar means in the electronic record [29].

3 Legal Protection of Digital Works in Cyberspace

Our purpose in this section is to discuss the legal protection of digital works under copyright laws using modern technology, the most important of which is cyberspace, as well as the procedural and civil protections of digital works in cyberspace, as well as their penal protections.

3.1 Procedural and Civil Protection of Digital Works in Cyberspace

As part of the procedural protection and civil protection of digital works, the researcher must study the procedural protection and civil protection in order to protect these works from assault before the assault occurs, as well as until after the assault has taken place.

3.1.1 Procedural Protection of Digital Works in Cyberspace

Copyright is not absolute as a result of certain circumstances, but is limited to certain conditions determined by each legislation, so the owner cannot enjoy it unless the conditions are met.

Procedural protection helps prevent the damage caused by abuse before it occurs, regardless of whether it occurs in a traditional situation or on the Internet [30], and consists of two types.

A- Preventive measures comprise these procedures for depositing, and the Jordanian legislator stated, "The deposit center at the National Library Department or another official authority authorized by the minister is the place where the deposit should be made." Deposit: delivering the original copy of the work to the centre and giving it a deposit number in accordance with the provisions of this law" [31].

B- remedial measures include confiscating counterfeit works to prevent the assault from occurring, but in the event that the cessation of the assault is not verified, the case goes to the remedial procedures following the assault, which work to stop the assault until the unlawful act is dismissed, especially since the litigation process is lengthy.

In addition to preventing the work from being disseminated and circulated, it also prevents the aggressor from disposing of the attacked work, from transferring or copying it to another person [32], resulting in the cessation of abuse from the date of judicial seizure.

As mentioned in the previous requirement, the Jordanian legislator referred to technological measures in the copyright law, that help them protect their works and prevent attacks on them, such as encryption.

Cyberspace works published with these technologies are legally protected. According to this article, "effective technological measures" refers to any technology, procedure, or means of tracking, such as encryption or control of copy extraction, that are used to prevent or limit unauthorized acts by rights holders" [33].

3.1.2 Civil Protection of Digital Works in Cyberspace

Authors are protected by the law as innovators who place their own mark on literary or artistic works.

Besides proving the infringement of his work and the harm to his reputation, the author may also prove tangible material assaults on the work [34].

There are two parts of responsibility that contribute to the author's civil liability against third parties. There is a difference between civil liability that arises from a contractual relationship and tort liability arising from non-contractual relationships.

Contractual responsibility: Responsibility requires the existence of a valid contract, whether it is a contract for the exploitation of rights or permission to exploit digital works published on the Internet, and that breaching the commitment to the contract between the publisher and the author like adding a specific or deleting part of the digital work and then publishing it on the Internet, so it is in this case an infringement of the digital work by the publisher.

According to Jordanian law, "any damage to a third party obligates its perpetrator, even if he is not discerning, to be liable for the damage" [35].

The provisions of tort liability apply to the author's civil liability against third parties based on a contractual or non-contractual relationship [36].

3.2 Penal Protection of Digital Works in Cyberspace

In addition to civil protection and procedural protection, which have already been discussed previously, the author should also be entitled to penal protection, which is complementary to civil protection.

Jordanian legislators imposed penalties for assaulting others' works as part of legal protection for works. The Jordanian legislator recognized that the nature of crimes against works did not justify procedural and precautionary safeguards, as well as material and moral rights compensation. As a result, the Jordanian legislator sought to penalize acts of intellectual property infringement [37].

According to the Copyright and Authors Rights Protection Law of Jordan, "A. Shall be punished by imprisonment for a period of no less than three months and no more than one year, and a fine of no less than one thousand dinars and no more than six thousand dinars, or one of these two penalties:

1. Whoever exercises, without a legal basis, one of the rights stipulated in Articles (8, 9, 10, and 23) of this law.
2. "Whoever sells, distributes, or rents an illegal work, makes copies of it, or broadcasts it to the public" [39].

Most copyright laws as well as international agreements clearly define copyright infringement as illegal use, harm to the integrity of the work, and attacks on Internet-published works [39].

4 Conclusion

As a result of the significant role that digital works play, intellectual property legislation provides protection for digital works of all kinds, especially with the expansion of cyberspace, as digital works are extending to new areas through digital publishing, which has become a part of electronic distribution and publishing, the dissemination of digital works has become extremely easy, fast, and cost-effective. Following are a few conclusions and recommendations that we reached in this study:

4.1 Results

1. In Cyber security Law No. 16 of 2019, the Jordanian legislator granted the foundations and mechanisms for prevention and treatment of cyber incidents in a manner that ensures the protection of national security and the interests of the state, in addition to providing human cadres and tools to confront cyber accidents and crimes through the National Cyber security Center.
2. The Jordanian legislator dealt with the protection of effective technological measures to limit the commission of illegal acts, and to prevent any person from accessing the digital work in order to reproduce it without the permission of its owner, referring

to encryption technology as a means of protecting the digital work in the Jordanian Copyright and Related Rights Protection Law in Article (2).

3. The Jordanian legislator dealt with the Copyright and Related Rights Law, which states that the author has the right to sue for any infringement on his work in accordance with Article (8/d) within the framework of civil protection and is criminalized by the same law in Article 51 with penalties for infringements on copyright within the framework of penal protection.
4. The Jordanian legislator affirmed the protection of computer programs by the Copyright and Related Rights Law.
5. The Jordanian legislator proved in the Jordanian Copyright and Related Rights Law No. 22 of 1992, that it is a modern law whose articles were amended more than once by the amended Law No. 14 of 1998, and the amended Law No. 29 of 1999, to comply with the provisions of international conventions. In 2004, the Hashemite Kingdom of Jordan acceded to the WIPO Convention on Copyright and Performances and Phonograms, and joined the TRIPS Agreement on Trade-Related Aspects of Intellectual Property Rights, which became effective by entering Jordan International Trade on 11 April 2000, and Jordan also joined the Berne Convention for the Protection of Literary and Artistic Works, which entered into force on 28 July 1999.

4.2 Recommendations

This study concluded with recommendations in order to ensure the rights established for the protection of digital works and their dissemination through cyberspace without prejudice to any copyright, the most important of which are:

1. In order to achieve the desired goals of cyber security and cyberspace, Jordanian legislators must address technical and technical issues related to cyber security in terms of classification and nature in the Cyber Security Law 2019.
2. Recommending the amendment of Article (2) of the Copyright and Related Rights Law to include the definition of digital works, and the codification of everything related to digital works and the protection prescribed for them within a specific legal framework, relying on modern technological developments and the opinions of jurists due to the importance of this type of work, especially with its wide spread in cyberspace.
3. The need for the Jordanian legislator to determine the period of protection of computer programs so that they are commensurate with their nature without the period being long is due to the work of information paralysis and obstruction of technological development.

References

1. Abdel-Hay, W.: The problem of cyberspace in intellectual property rights. In: A working paper presented to the first international scientific conference on intellectual property, July 10–11, University of Yarmouk, Jordan (2000)
2. Al-Fatlawi, S.: Legal means to protect copyright. In: A working paper presented to the first international scientific conference on intellectual property, July 10–11, Yarmouk University, Jordan (2000)

3. Almakhzoumi, O., Almawla, N.: Innovative legal links to serve the economic value of the trademark. Zarqa J. Res. Stud. Hum. **18**(3) (2018). published research
4. Al-Maqsudi, M.: Cybersecurity and international efforts to combat intercontinental crimes. Arab J. Secur. Stud. Naif Arab Univ. Secur. Sci. **37**(427) (2017)
5. Al-Matit, A.-Y.: Literary Copyright. Al-Nahda Al-Masrya Library, Cairo (1960)
6. Al-Meligy, O.A.: procedural protection in the field of intellectual property rights, a working paper presented to the first international scientific conference on intellectual property, July 10–11, Yarmouk University, Jordan (2000)
7. Al-Nawafila, Y.: Legal Protection of Copyright, 1st edn. Dar Al Thaqafa for Publishing and Distribution, Amman (2004)
8. Al-Rabiah, S.A.: Digital Security and User Protection from Internet Risks. Communications and Information Technology Commission, Workshop, Saudi Arabia (2017)
9. Al-Saghir, H.E.-D.: Foundations and Principles of the Agreement on Trade-Related Aspects of Intellectual Property Rights (TRIPS Agreement), 1st edn., Dar Al-Nahda Al-Arabia, Cairo (1999)
10. Atwi, M.: Legal Protection of Intellectual Property Rights on the Internet, Ph.D. thesis, Daly Brahim University, Algeria (2010)
11. Othman, B.M.: The Legal Nature of Digital Works and Their Protection, Letter Master's degree, Jerash University (2011)
12. Egyptian Civil Law No. 131 of 1948 and its amendments
13. Egyptian Intellectual Property Protection Law. No. 82/2002
14. Fathiya, H.: Protecting Digital Works and Domain Names on the Internet, Master's thesis, University of Algiers (2016)
15. Hajjar, W.: Safety in cyberspace and combating cybercrime in the Arab Region. J. Econ. Soc. Comm. Western Asia (5) (2015)
16. Jordanian Cyber Security Law No. (16) of 2019, published on 9/16/2019
17. Jouga, N.M.: The digital signature under the Jordanian Electronic Transactions Law No. 85 of 2001, 1st edn., without a publishing house, Amman (2003)
18. Kanaan, N.: Copyright: contemporary models of copyright and means protecting it), 1st ed., Dar Al Thaqafa for Publishing and Distribution, Amman, Jordan (2004)
19. Khater, N., et al.: Explaining the civil law, the sources of personal rights "obligations", a comparative study. Dar Al Thaqafa for Publishing and Distribution, Amman, first print (2009)
20. Law No. 15 of 2015 (Jordanian Electronic Transactions Law of 2015) published In Official Gazette No. 5341, 17 May 2015
21. Law No. 22 of 1992 (Law on the Protection of Copyright and Neighboring Rights in Jordan for the year 1992) and its amendments published in the Official Gazette No. 3821, 16 Apr 1992
22. Law No. 43 of 1976 (the Jordanian Civil Code of 1976 issued in Issue 4106 published on 16 Mar 1996
23. Magablah, N.: The Legal System for Electronic Information Services Contracts in Private International Law, 1st edn. House of Culture for Publishing and Distribution (2009)
24. Manita, Y.I.: An overview of cybercrime. Int. Libyan J. Univ. Benghazi – Coll. Educ. (32) (2017)
25. Qabaa, A.J.: The status of copyright and related rights in Jordan in the light of national and international agreements. In: A working paper presented to the first international scientific conference on intellectual property, July 10–11, Yarmouk University, Jordan (2000)
26. Salama, S.: The Electronic Contract, 2nd edn. Dar Al-Nahda Al-Arabiya, Cairo (2007)
27. Wassim, H.: Models of cyberspace legislation in ESCWA member states. J. Econ. Soc. Commiss. Western Asia (7) (2007)

Artificial Intelligence Trends in Finance Development

The Impact of Applying Electronic Internal Auditing in Raising the Efficiency of Financial Performance in Jordanian Commercial Banks

Mohyedin Hamza[1](✉), Riham Alkabbji[1], Tareq Hammad Almbydeen[1], Ahmad Almubaydeen[1], and Khaloud Bajunaid[2]

[1] Zarqa University, 2000, Zarqa 13110, Jordan
{Mhamza,tmbydeen}@zu.edu.jo
[2] University of Business and Technology, Jaddah, Saudi Arabia

Abstract. This study aims to identify the impact of applying electronic internal auditing in raising the efficiency of financial performance in Jordanian Commercial Banks. In order to achieve the study objectives, the analytical descriptive approach was followed. A questionnaire consisting of (19) items was developed to measure the electronic internal auditing. A number of (168) questionnaires were distributed among the internal auditors and accountants in these banks, and the number of questionnaires valid for analysis was (137), representing (81.5%) of the study sample. The financial performance was measured by (RoA, RoE and EPS) for (13) banks within the financial period (2015–2021) Findings showed that applying electronic internal auditing reached a high level. Results also showed that there is a statistically significant effect at the significance level ($\alpha \geq 0.05$) for applying electronic internal auditing in raising the efficiency of financial performance with its indicators (ROA, ROE and EPS) in Jordanian Commercial Banks. The study recommended the necessity of holding workshops and specialized training programs periodically for auditors and accountants on electronic internal audit systems, to keep them acquainted with the latest in the field, and to help enhance the financial performance of these banks.

Keywords: Electronic Internal Auditing · Financial Performance · Jordanian Commercial Banks

1 Introduction

"The sectors with their various activities are witnessing an accelerated growth in the use of technology in the auditing domain, as traditional auditing is being shifted to the electronic version, and the profession is being developed and presented in a new form, since that ensures credibility, accuracy and transparency of financial statements, as well as raises the performance and efficiency of internal auditors, especially after the turmoil that negatively affected the auditing and accountancy profession and its loss of confidence from the part of investors due to the financial scandals and breakdowns of international major businesses. Hence, the role of electronic internal auditing is to

B. A. M. Alareeni and I. Elgedawy (Eds.): *Artificial Intelligence (AI) and Finance*, SSDC 488, pp. 521–532, 2023.
https://doi.org/10.1007/978-3-031-39158-3_49

ensure that auditing tasks in the Jordanian commercial banks flow as planned, to boost the efficiency of the process and raise the level of financial performance in such banks. Thus, the problem of the study can be expressed through the following main question:"

Does applying the electronic internal auditing have an impact in raising the efficiency of financial performance in Jordanian Commercial Banks? Branching out from the previous main question are the following sub-questions:
"Does applying the electronic internal auditing have an impact on return on assets (ROA) in Jordanian commercial banks?
"Does applying the electronic internal auditing have an impact on the return on equity (ROE) in Jordanian commercial banks??
"Does applying the electronic internal auditing have an impact on earnings per share (EPS) in Jordanian commercial banks?

2 Literature Review

This study aims to identify the impact of electronic auditing in raising the efficiency of financial performance in the Jordanian commercial banks. Many studies have addressed the subject or some of its aspects coming up with different results. [1] Concluded that electronic auditing reveals the weaknesses of internal control and provides ways to remedy such weaknesses, which contributes to improving the quality of internal auditing. [2] Found that electronic auditing raises the quality of accounting information producing flawless and deviation-free information. [3] Conforms to the above that electronic processing of accounting information increases the efficiency of auditing in activating internal control. Researchers agree that information technology increases the efficiency of internal auditing work. This was also confirmed by [4], by asserting that the application of electronic auditing increases the independence of the internal auditor, which is one of the most important requirements of the auditing profession. On the other hand, many studies have dealt with the aspect of financial performance and the elements affecting it. [4] Concluded that the financial performance of banks is positively and strongly influenced by the application of the COSO components of internal control, as these increase the quality of the financial performance of banks [5] studied the impact of the female role in internal auditing on the financial performance in banks. The study concluded that the presence of the female component positively affects the financial performance in banks.

It is clear from the above that the use of Information Technology in auditing increases the efficiency of internal audit. Moreover, any amendment to internal auditing, whether being that a new component such as (COSO) components or a female component, which can be described as gender diversity in auditing, positively affects the financial performance of banks [3].

"It was noted from the previous studies reviewed that some of them either addressed electronic audit or financial performance, while the current study combines both electronic internal auditing and financial performance." Thus, the current study differs from some previous studies in terms of the sectors that were studied and analyzed, and the time period that was covered, as it seeks to analyze the impact of the application of electronic internal auditing in raising the efficiency of financial performance in Jordanian commercial banks between (2015–2021)."

"Thus, this study is one of the few studies within the researcher's knowledge that links electronic internal audit and financial performance with its indicators: (return on assets (RoA), return on equity (RoE), and profitability per share (EPS) in Jordanian commercial banks, as it presents an intellectual framework for the study, its various variables and the study population, looking forward to paving the way to other future studies.

3 Research Methodology

This study aims to shed light on the impact of electronic auditing on the financial performance of commercial banks.

The study relies on the descriptive analytical approach as it serves the objectives of the study and it is one of the most widely used approaches in the study of social and human phenomena. The data analysis relied on the Statistical Analysis Program (SPSS), and some well-known tests were applied to reach some proposed results and recommendations.

3.1 Data Collection Method

This study will rely on both primary and secondary data sources, as follows:

A - Secondary Data Sources
This study relies on some secondary data sources, which are mainly represented in books, references, research, studies, and some related websites on the Internet, as well as the annual published bulletins and reports of the Amman Stock Exchange.

B - Preliminary Data
This study mainly relies on a questionnaire designed and distributed to adress the questions of the study with the aim of identifying the extent of the expected impact of applying electronic internal auditing in raising the efficiency of financial performance in Jordanian Commercial Banks.

Study Population
"The study population comprised 13 Jordanian commercial banks by the end of 2021. It relied on the data provided by the annual financial reports of these banks, during the period between (2015–2021). The sampling unit, for the purposes of distributing the study questionnaire, and measuring the independent variable: (electronic internal auditing), is represented by the (297) internal auditors and accountants operating in these banks, according to the Human Resources Department of the banks under research."

Study Sample
The researcher withdrew a simple random sample from the study population, where the size of the withdrawn sample was based on the size of the total study population, and the margin of error allowed in the current study is (0.05), according to the required samples size table (Sekaran & Bougie, 2016). Thus, the researcher distributed (168) "electronic

questionnaires -prepared by (Google Forms) - among the study sampleindividuals for the purposes of measuring the independent variable (electronic internal audit), 144 questionnaires were retrieved, and the number of questionnaires valid for analysis was 137, making up 81.5% of the study sample."

In order to measure the dependent variable (financial performance), the following criteria were used: excluding all banks that were subjected to a merger or acquisition process during the study period, excluding banks for which insufficient data is available, and excluding all banks that were liquidated or stopped trading" The banks that were listed after the year (2015) were also excluded, since the study uses what is known as (Panel Balanced Data), which requires the availability of sufficient data within the years under study. "The banks in which the researcher could not distribute the questionnaires were excuded, too."

After applying the criteria, the study sample was limited to (9) banks whose data on (return on assets, return on equity, and earnings per share) were available continuously for the period between (2015–2021), and this sample represents (69.2%) of the total number of banks. The names of the searched banks and their data are shown "in Appendix No. 1."

3.2 Study Tool

The researcher developed a questionnaire as a means of collecting data regarding the independent variable (electronic internal auditing) in order to meet the objectives of the study and answer its questions. To determine the dimensions of the questionnaire, the researcher relied on [1–4, 5–8]. The questionnaire consisted of two parts:

"The first part: the functional data of the study sample including: (job title, academic qualification, professional certificates, and years of experience in auditing, while gender was excluded)."

"The second part: this part aims to identify the level of (electronic internal auditing) in Jordanian commercial banks. It comprised (19) paragraphs."

The stability of the questionnaire:

The researcher verified the stability of the research questionnaire by using the Cronbach Alpha coefficient, as follows:

As shown in Table No. (3–3), The high Cronbach Alpha values indicate a high level of stability ranging between (1–0), where these values are acceptable at the stability coefficient ratio (0.70) and above [9].

Domain	Stability Coefficient (Cronbach Alpha)
Electronic auditing	0.953

"Table no. (3–3) shows that the value of the stability coefficient was acceptable for the purposes of application, where the value of the stability coefficient was (0.953), which indicates that the study tool generally has a high stability coefficient, and the ability to achieve the study purposes.

3.3 Data Analysis and Hypothesis Testing

"This chapter includes a description of the characteristics of the study sample, descriptive statistics of the study variables, and the results of a simple single regression test of the study hypotheses."

3.4 Description of the Characteristics of the Study Sample

"The study sample consisted of (137) employees in the Jordanian commercial banks, who were selected by a simple random sampling method from the study population. Table 1 shows the distribution of the sample respondents according to demographic and functional variables:"

Table 1. Distribution of respondents according to demographic and functional variables

Variable	Level	Frequency	Percentage
Job title	**Auditor**	**43**	31.4
	Accountant	**94**	68.6
	Total	**137**	**100%**
Academic qualification	Diploma	**8**	5.8
	Bachelor's degree	**90**	65.7
	Postgraduate studies	**39**	**28.5**
	Total	**137**	**100%**
Professional certificates	**CIA**	**14**	**10.2**
	CPA	**7**	**5.1**
	CMA	**5**	**3.6**
	JCPA	**6**	**4.4**
	None	**105**	**76.6**
	Total	**137**	**100%**
Years of service	Less than 5 years	**45**	**32.8**
	From 5 – less than 10 years	**35**	**25.5**
	From 10 – less than 15 years	**32**	**23.4**
	From 15 – less than 20 years	**12**	**8.8**
	20 years and more	**13**	**9.5**
	Total	**137**	**100%**

Table 1 shows the following:

"The highest percentage of the distribution of the study sample according to the job title variable reached (68.6%) in favor of the job title (accountant), while the lowest percentage reached (31.4%) in favor of the job title (auditor). This result indicates that approximately two thirds of the sample respondents are accountants, reflecting the need of Jordanian commercial banks for a large number of employees in the accounting departments due to the large amount of accounting and financial activities in these banks."

"The highest percentage of the distribution of the study sample according to the scientific qualification variable reached (65.7%) in favor of the scientific qualification (Bachelor), while the lowest percentage was (5.8%) for the scientific qualification (Diploma). These percentages are due to the fact that the majority of the respondents have university degrees, since working in Jordanian commercial banks requires scientific knowledge."

"The highest percentage of the distribution of the study sample according to the professional certificates variables reached (76.6%) in favor of the professional certificate (none), while the lowest percentage reached (3.6%) for the professional certificate (CMA). These ratios indicate that the accountants and auditors surveyed in the Jordanian commercial banks follow specific work procedures on specialized accounting applications, which does not necessitate the need to obtain specialised certificates at the level of internal work in these banks."

"The highest percentage of the distribution of the study sample according to the years of service variable reached (32.8%) for years of service (less than 5 years), while the lowest percentage was (8.8%) for years of service (from 15-less than 20 years). These ratios indicate a high turnover of work among the surveyed individuals, as a result of the constant movement of accountants and auditors between banks at the local or regional level, and this is due to the intense competition between these banks."

3.5 Description of the Study Variables

"This part includes a presentation of the results of the analysis according to the arithmetic means and standard deviations in the electronic internal auditing domain in the Jordanian commercial banks, and below is a presentation of these results:"

3.5.1 Independent Variable: Electronic Internal Auditing

"To identify the level of application of electronic internal auditing in Jordanian commercial banks, the arithmetic means and standard deviations were extracted, for the approval of the study sample respondents to the study tool as shown in Table 2."

Table 2. Arithmetic means and standard deviations of the approval of the study sample respondents to the paragraphs of the "electronic internal auditing" domain in descending order

No.	Paragraph	Arithmetic mean	Standard deviation	Level of evaluation
2	electronic internal audit supports confidentiality of accounting information in the bank	4.41	0.86	High

(*continued*)

Table 2. (*continued*)

No.	Paragraph	Arithmetic mean	Standard deviation	Level of evaluation
1	electronic internal audit observes authenticity of accounting information in the bank	4.34	0.8	High
3	electronic internal audit supports the timely provision of accounting information	4.34	0.83	High
14	The auditor has access to the payment and receivables receipts of the bank	4.32	0.82	High
8	electronic internal audit contributes to the detection of deviations in financial statements	4.24	0.85	High
4	electronic internal auditing makes it possible to compare the bank's financial statements with the estimated budgets	4.21	0.88	High
16	electronic internal audit helps to reduce the audit time	4.21	0.96	High
18	The electronic internal auditing system contributes to speeding up decision-making	4.20	0.83	High
19	The auditor undergoes specialized training programs on electronic internal auditing	4.18	0.89	High
12	The auditor has special powers to audit receivable accounts	4.16	0.87	High
17	electronic internal audit helps to detect deviations in costs	4.16	0.92	High

(*continued*)

Table 2. (*continued*)

No.	Paragraph	Arithmetic mean	Standard deviation	Level of evaluation
7	electronic internal audit helps to minimize the expected risks in the bank	4.15	0.96	High
15	special audit programs are available for analyzing financial ratios	4.15	0.87	High
13	The auditor has special powers to audit payable accounts	4.14	0.84	High
10	the auditor is able to conduct an audit of cash flows in relation to salaries and wages in the bank	4.13	0.88	High
5	The audit programs available at the bank make it possible to compare the financial statements of previous periods	4.09	0.81	High
11	the auditor uses programs to audit fixed assets in the bank	4.09	0.94	High
6	electronic internal audit contributes to the identification of financial challenges facing the bank	4.00	0.91	High
9	bank audit programs have the property of giving alerts when errors occur	3.93	0.98	High

"Table 2 shows that the overall level of the application of electronic internal audit in Jordanian commercial banks from the point of view of the study sample respondents was high; the arithmetic average for the field was (4.16) with a high evaluation level, and the arithmetic means of the estimates of the study sample respondents on the dimension paragraphs ranged between (3.93–4.41). Paragraph No. (2), which reads: (electronic internal audit supports the confidentiality of accounting information in the bank) came first, while paragraph No. (9), which reads: (bank audit programs have the property of giving alerts when errors occur) came last."

3.5.2 Dependent Variable (Financial Performance)

"Table 3 shows the descriptive statistical measures of the dependent variable: financial performance indicators: (return on assets, return on equity, and profitability per share) based on the study data." The study was limited to (9) banks whose data were available from the banks listed on the Amman Stock Exchange during a period of (2015–2021). The statistical measures included both the lowest value, the highest value, the arithmetic mean, in addition to the standard deviation to explain the homogeneity of the study data and their relevance."

Table 3. Statistical description of the dependent variable (financial performance indicators)

Variable	Lowest value	Highest value	Arithmetic mean	Standard deviation
Return on assets	0.58	1.59	1.02	0.31
Return on equity	5.08	9.45	7.58	1.62
Earningsper share	0.07	0.35	0.19	0.09

Table 3 shows the following:

1. "The values of the return on assets of the study sample ranged between (0.58) and(1.59) and the arithmetic mean of the return on assets was (1.02), while the standard deviation was (0.31)."
2. "The values of the return on equity for the study sample ranged between (5.08) and(9.45) and the arithmetic mean of the return on equity was (7.58), while the standard deviation was (1.62)."
3. The values of the earnings per shares for the study sample ranged between (0.07) and (0.35) and the arithmetic mean of the earnings per shares was (0.19), while the standard deviation was (0.09).

4 Study Hypotheses Testing

Main hypothesis: H0: there is no statistically significant impact at the level of significance ($0.05 \geq \alpha$) for the electronic internal auditing on the efficiency of financial performance in its dimensions: (return on assets, return on equity, earnings per share) in Jordanian commercial banks.

"Table 4 shows the presence of a statistically significant impact at the level of significance ($\alpha \leq 0.05$) for electronic internal auditing on the efficiency of financial performance within its dimensions in Jordanian commercial banks. The value of the correlation coefficient (R) was (0.727), which is a statistically significant value, indicating the degree of statistical correlation between the change in electronic internal auditing on the efficiency of the financial performance within its dimensions", The value of (R-square) was

Table 4. Multiple regression equation analysis to study the impact of electronic internal auditing on the efficiency of financial performance in Jordanian commercial banks.

Variable	B	Std. Error	Beta	T	Statistical significance
(Constant)	−0.64	0.14		−4.73	0.00
ROA	0.40	0.03	0.73	12.30	0.00
ROE	2.31	0.15	0.80	15.45	0.00
EPS	0.07	0.01	0.45	5.89	0.00
R^2	0.528	**F**	151.19		
R	0.727	**Statistical significance**	0.00		

(0.528), which is a statistically significant value, meaning that the electronic internal auditing explains the value of (52.8%) of the change in the efficiency of financial performance and its dimensions. And the value of the (F) test reached (151.19) with a statistical significance less than (0.05), therefore the study model is considered acceptable.

Results related to the first sub-hypothesis: There is no statistically significant impact at the significance level (0.05 ≥ α) for electronic internal auditing on the return on assets (ROA) in Jordanian commercial banks.

Table 4 shows that the return on assets (ROA) affects and is affected by the electronic auditing with a value of (40%), which is the B value, indicating that any change in the electronic audit by one degree affects the return on assets by 40%, since the value of the T test reached (12.3). Based on the above, the first sub-hypothesis is accepted in its alternative statement, which states that "There is a statistically significant impact at the significance level (0.05 ≥ α) for electronic internal auditing on the return on assets (ROA) in Jordanian commercial banks."

Results related to the second sub-hypothesis: There is no statistically significant impact at the significance level (0.05 ≥ α) for electronic internal auditing on the return on equity (ROE) in Jordanian commercial banks.

"Table 4 shows the presence of a statistically significant effect at the level of significance (α ≤ 0.05) for the electronic internal audit on the return on equity in Jordanian commercial banks, as the value of (B) reached (80%), which is a statistically significant value indicating the degree of correlation of a statistically significant function between the change in the electronic internal auditing and the return on equity. Based on the above, the first sub-hypothesis is accepted in its alternative statement, which states that" There is a statistically significant impact at the significance level (0.05 ≥ α) for electronic internal auditing on the return on equity (ROE) in Jordanian commercial banks."

Results related to the third sub-hypothesis: There is no statistically significant impact at the significance level (0.05 ≥ α) for electronic internal auditing on the return on earnings per share (EPS) in Jordanian commercial banks.

"Table 4 shows the presence of a statistically significant impact at the level of significance (α ≤ 0.05) for the electronic internal audit on the earnings per share in Jordanian commercial banks, as the B value reached (45%), which is a statistically significant value. Based on the above, the first sub-hypothesis is accepted in the alternative statement, which states that" There is a statistically significant impact at the significance level ($0.05 \geq \alpha$) for electronic internal audit on earnings per share in Jordanian commercial banks."

"Based on the results of the sub-hypotheses, the main hypothesis is rejected in the zero formula, and the hypothesis in the alternative statement is accepted, which states that" there is a statistically significant impact at the significance level ($0.05 \geq \alpha$) for electronic internal auditing on the efficiency of financial performance in its dimensions: (return on assets, return on equity, earnings per share) in Jordanian commercial banks."

5 Results

The study showed many diffrent results, first showed that the overall level of application of electronic internal auditing from the point of view of auditors and accountants in Jordanian commercial banks was high. This result may be due to the fact that Jordanian commercial banks are increasingly relying on advanced computers, modern technological infrastructure, specialized software and applications to perform the audit activity at the internal system level, which leads to the collection of data by the internal auditor, evaluating it within an electronic environment, enhancing the confidentiality and reliability of accounting information, and providing the required data and reports in a timely manner. This result is consistent with Bazaz (2020),,Sarlkhitm (2020), Omari (2017). Second, the results indicate that Jordanian commercial banks adopt a set of electronic systems in order to monitor internal operations, ensure compliance with procedures on an ongoing basis to ensure the accuracy of accounting and statistical data, verify that employees of Jordanian commercial banks follow the policies, plans and administrative procedures drawn up for them, and measure the validity of those plans, policies and all other means of control in the performance of their purposes, which positively affects the ability of these banks to exploit the available resources optimally, within the available capabilities and leading to the achievement of the desired financial goals. This result is somewhat consistent with Alwan & Al-Sabbati (2017), Abu Hamour et al. (2021). Finally, The results showed that effect for the application of electronic internal auditing on return on assets (RoA) and on return on equity (RoE) in Jordanian commercial banks. This result also conforms with Abu Hamour et al. (2021).

Recommendations

In view of the previous results, the study recommends the following:

First: "Holding specialized workshops and training programs periodically for auditors and accountants on electronic internal audit systems, to inform them about everything that is new in this area, and to help enhance the financial performance in these banks."

Second: "Supporting electronic internal audit programs and systems to allow them to give alerts to professionals when errors occur, and detect deviations as soon as they occur."

Third: "The need to emphasize the commitment of commercial banks to apply a model for measuring Information Technology Governance based on the (5 COBIT) framework for internal control in these banks, leading to the maximum use of available technology; in order to reduce costs, increase returns, thereby increasing the profitability of banks, and enhance their financial position and performance."

Fourth: "The need for Jordanian commercial banks to seek to enhance their banking services and products, and to conduct in-depth studies of the risks surrounding them, because of the positive impact on the various financial performance indicators of these banks."

Fifth: "Conducting further studies on the impact of specific dimensions of the application of electronic internal audit on financial performance in Jordanian commercial banks, and in other business sectors."

References

1. Bazzaz, H.: The role of e-audit in improving the quality of internal audit, unpublished master's thesis, Arabi Ben Mahdi University, Umm el Bouaki, Algeria (2020)
2. Srlkhatm, A.: The impact of the use of electronic audit on the quality of published accounting information. Islamic Entrepreneurship Mag. **5**(2), 43–58 (2020)
3. AL-Ramahi, N.: Measuring the extent of application of the (COSO) framework for internal control from the point of view of the external auditors in the companies. Zarqa J. Res. Stud. Hum. **17**(2) (2017)
4. Al-Obaidi, S.: The impact of e-auditing in raising the independence and efficiency of the internal auditor. J. Econ. Adm. Sci. **21**(84), 415–441 (2015)
5. Chijoke-Mgbame, A.M., Boateng, A., Mgbame, C.O.: Board gender diversity, audit committee and financial performance: evidence from Nigeria. Account. Forum **44**(3), 262–286 (2020)
6. Rashid, A.: The impact of information technology on the performance of the internal control system: applied research at the university of Qadisiyah, Al-Qadisiya. J. Admin. Econ. Sci **17**(4), 217–245 (2015)
7. Mamaile, L.J.: The Functioning of the Information Technology Internal Audit Departments at Metropolitan Municipalities in South Africa. University of Johannesburg (South Africa) (2013)
8. Nusa, I.B.S.: Quality of audit system information for internal control effectiveness. In: International Conference on Business, Economic, Social Science, and Humanities–Economics, Business and Management Track, pp. 198–202, January 2020
9. Sekaran, U., Bougie, R.: Research Methods for Business: A Skill-Building Approach. Wiley, New York (2016)

The Role of Stock Indices in Forex Traders' Psychology Amid COVID-19 Outbreak

Mohammad Sami Ali(✉)

Department of Financial Technology, Faculty of Economics and Administrative Sciences, Zarqa University, Zarqa 13110, Jordan
Maldahrawi@zu.edu.jo

Abstract. Investors' sentiment conduct a central role in stock market performance, thus due to the COVID-19 outbreak; stock markets' indices are found to be more volatile as well as conveying positive or negative signals to stock markets' investors. Therefore, the paper evaluated the impact of COVID-19 in stock indices of ASE, Italy, London, Tokyo, Shanghai, Republic of Korea, Thailand and the New York stock markets. Through analysing a panel and time series data for the period 21Jan2020-01May2021, findings from the random effects obtained that investors' sentiment is insignificantly responding to the impact of COVID-19 in Shanghai stock index. Nevertheless, the global death toll and the impact of COVID-19 in S&P500 index are positively influencing investors' sentiment. Additionally, the study revealed that investors' sentiment is significantly mediating the correlation of COVID-19 along with gold prices. Moreover, since this research concluded that investors consider gold as a more secured investment venue; the study recommends Forex's investors to include the number of COVID-19 GCCs and S&P500 in the process of decision making

Keywords: COVID-19 Outbreak · Investors Sentiment · Stock Indices · Investors Psychology · Forex Traders

1 Introduction

Since experiences from previous crises including SARS 2002–03 and MERS-COV taught us that stock markets' indices are significantly responding to pandemics outbreaks; the Covid-19 crisis is assumed to be significantly linked with economic growth and stock markets' performance through conveying negative or positive signals to investors. According to Szlezak (2020) the global gross domestic product is estimated to go down by 2% to 3%, and thus since this crisis was dramatically impacted economic growth of leading economies such as U.S. and China; the performance of financial markets worldwide was strongly responded to this disease. In this regard, the WHO (2020) declared that due to the daily increase in the numbers of COVID-19 confirmed cases, governments worldwide constructed to shut-down schools, shops and even financial markets in order to protect people's health and encountering the outbreak of this disease. In this regard, the prime minister of Jordan ordered the Jordan securities commission to stop the trading floor of Amman stock exchange, and similarly the Governor of NYSE closed both

B. A. M. Alareeni and I. Elgedawy (Eds.): *Artificial Intelligence (AI) and Finance*, SSDC 488, pp. 533–543, 2023.
https://doi.org/10.1007/978-3-031-39158-3_50

the options and stocks' trading floors at the New York Stock Exchange. However, since these actions are assumed to be negatively linked with stock markets' performance; the available statistics demonstrated that the U.S. stocks plunged by more than 5%, financial indices like the S&P1500 traded for less than $1, as well as the Shanghai stock market index plummeted by 0.28, and subsequently the power of Yuan was weakened, due to the extreme regulations of China's government regarding the shutdown of transportations in the Wuhan's city (Domm 2020). Additionally, the prices of NYMEX went down by 55%, and may be some energy stocks will slash their dividends (Krantz 2020). Furthermore, studies including Ali (2020) asserted that variables like GDP growth are significantly influencing investors' behaviour as proxied by stock markets' indices. Hence, since this pandemic was negatively impacted the GDP growth of most economies; it is plausible to assume that this crisis is significantly impacted investors' sentiment.

Therefore, the current research aims at evaluating the impact of COVID-19 outbreak in investors' sentiment as captured by stock markets' indices. For this purpose, the study utilized the random effects technique to analyse a daily panel data, those are relating to the prices of S&P500, FTSE100, FTSEMIB, NIKKEE225, ASE, Shanghai, KOSPI, and SET50 stock indices. Furthermore, since precious metals such as gold during such a situation or during financial crises are seen as a safe haven investment; the dramatic up and down trends of gold prices motivated the present study to explore the role of investors' sentiment to mediate the relationship of COVID-19 along with gold prices as a proxy for investors' psychology. For instance, Mora (2010) mentioned that during financial stress or tight market liquidity it will be worthwhile for investors to invest in safe haven assets such as gold. Thus, since previous studies indicated that there is a significant correlation between gold prices and stock markets' indices the study will scrutinize the way of how gold prices are responding to the impact of COVID-19 in investors' sentiment as proxied by S&P500 index. In this context, Akgüla et al. (2015) confirmed that in the pre-crisis stock indices including the S&P500 are positively impacted the prices of gold, while during the crisis period the prices of gold are found to be negatively responded to the change in S&P500 index. Authors including Gozgor et al. (2019) confirmed that since the uncertainty of economic policies impacts gold prices that will be in turn significantly affecting investors' sentiment. Moreover, since the COVID-19 outbreak is postulated to be highly linked with investors' sentiment and the markets' performance; results from this research will be empirically contributing in assisting investors to hedge against the risks, which are expected to be brought from the impact of COVID-19 in stock markets' indices. Additionally, since this research seeks to check whether commodities such as gold are really seen as more secured investment avenue, results from this study will help investors to protect their wealth through directing funds into the gold, instead of investing them in the stock markets.

2 Empirical Review and Research Enquire(s)

Due to the rapid outbreak of COVID-19 financial markets worldwide are forced to trade electronically instead of opening their trading floors. Thus, since pandemic distresses financial markets as well as nations' economies; this research seeks to answer the impact of COVID-19 crisis in investors' sentiment as captured by stock markets indices, over the period 21 Jan/2020-01May/2021. Moreover, in an attempt to understand the way of how COVID-19 and the other viruses' strains could influence economic growth and stock markets' performance, this section discusses the impacts of COVID-19, SARS-COV, and Ebola in the global economy:

2.1 The Impact of COVID-19 in the Global Economy

From the economic and financial point of view this virus succeeded in turning financial markets all over the world in chaos. For example, the prime minister of Jordan ordered the Jordan securities commission to stop trading at Amman stock exchange, the NYSE's Governor replaced the stocks and options trading floors by the electronic trading system. Continuously, the Shanghai stock market dropped by 0.28 as well as the power of Yuan was weakened, due to the extreme regulations of China's government regarding the shutdown of transportations in the Wuhan's city (Domm 2020). Thus, the confusion of financial markets decreased the U.S. stocks by more than 5%, as well as the Bloomberg's dollar index was also affected. Huge manufacture companies including Fiat Chrysler, Ford Motor Co. And general Motors Co. Declared that they are going temporarily to close their plants in the USA. Ackman suggests that one of the best methods to defeat this epidemic is closing the USA for around 30 days. Additionally, to abate the impact of this crisis Washington prepared a budget of $8.3 two weeks ago, and the order of U.S. Fed to start printing money and supporting economy resulted in accelerating the performance of U.S. and EU financial markets (The Guardian 2020). Continuously, governments all over the world constructed schools, restaurants, shops to shut down, as well as restricting people from gathering on streets or moving between cities (Bloomberg 2020). Therefore, Liu (2020) mentioned that the COVID-19 outbreak distressed the world economy and it is almost to tip nations' economies into a recession. However, due to the negative role of this pandemic in the global economy; nations all over the world implemented several policies to protect their economies. For this purpose, countries including Brazil cut interest rates to 3.75%, the central bank of Jordan decreased rates to 2.5%, bank of England slashed interest rates to 0.25%, Indonesia and Philippines lowered their rates to 4.25% (CBJ 2020; Bank of England 2020; Bloomberg 2020). Alternatively, Randow and Ainger (2020) mentioned that in an attempt to stabilize the European financial markets as well as protecting the EU economies and assisting them coping with this turmoil; the European central bank launched an emergent bond-buying program with a value of $810 billion. Authors including Carvalho (2020) pointed that stock prices across the world went down due to the dramatic impact of Coronavirus on investors' behaviour. For instance, though this crisis supported the prices of pharmaceutical stocks, however, sectors like Tourism and travel related stocks and consumer goods dropped dramatically. Well known energy companies such as CBL and Associates are trading for less than a buck, as well as financial indices like the 1500 S&P are trading for less than $1, S & P500 fell down by 1.7% and the Dow Jones futures went down by 334 points as investors impacted by this outbreak, especially since the US death toll increases day by day. However, if the current situation continued might some large and midsize firms will be dropped below $1. For instance, the shares of midsize energy firms such as RIS are down 83%, and the shares of NBL are down 87% this year (Krantz 2020; Trading Economies 2022).

Furthermore, banks and asset managers expected that this crisis will reduce the global GDP by around 2% to 3%. However, the intervention of central banks and the positive expectations, which say that the cases of corona-virus would plateau at some point of time contributed in stabilizing the performance of financial markets. In this regard, central banks such as China's central bank injected this week around $174 billion of liquidity into the markets to decrease the reserves of Repo rates. When the COVID-19 was firstly appeared in China it was totally ignored, while when it is started to spread in Europe and Middle East; financial markets worldwide have strongly responded to this disease. Consequently, the decrease in financial markets' performance indicates that the global economy is almost to fall in a deep recession. However, though the financial markets of globe are on a path of recession, the history reflects that the rescission and bear markets must not be automatically combined. Consequently, the world economy will be highly affected due to the slow growth of leading economies like the U.S and China and some other major economies (Szlezak et al. 2020). Beyond that, due to the COVID-19 the financial markets' indices fluctuated dramatically, as well as the world equity market is found to fall freely. In addition, since the imports to china have directly influenced the export of nations' economies, this impact affected many commodities including metals, oil and agriculture. However, since China is the largest net importer of oil with a percent of 10 from the global demand; the impact of this crisis declined china's demand on oil nearly by 3.13 mm b/d. Furthermore, economies including Japan were also affected by this crisis, since it was planning to support its economy from welcoming the 2020 Olympics. Through focusing on the economies of MENA region countries, Saudi Arabia is considered as the second largest crude oil's exporter to china after Russia, while the big dilemma is that Russia refused to decrease its supply to china, in the time that the OPEC asked Saudi Arabia to decrease its production to put a floor under oil prices (Larry 2020).

2.2 Comparing the Economic Impacts of COVID-19 along with Previous Crises

The COVID-19 that we are encountering these days will be worse before getting better. Thus, in an attempt to overcome this pandemic and abating its spread; schools, shops and many businesses are going to shut-down, and subsequently, that will lead to a depression. However, since analysts are reviewing the projections of their earnings that might will result in quick moves in the market. Additionally, the COVID-19 market crisis differs from the 1987 and the 2008 financial crises. For instance, during the 2008 global financial turmoil around $3.5–$4 trillion in the accounts of money markets were at risk. Conversely, during the current crisis the financial markets still liquid as well as operating easily, though they are highly volatile. Continuously, through comparing the economic impacts of COVID-19 along with the implications of the 2008 financial crisis in the global economy, it is estimated that this disease will bring more severe consequences to the world economy (Tuchman 2020). Accordingly, the International Labour Organization "ILO" announced that this crisis will impact the GDP growth of global economy, as well as about 25million are estimated to lose their jobs. While in another scenario the "WHO" presented an initial report estimates that the global unemployment will be increased between 5.3 and 24.7 million. Comparatively, the 2008 global financial crisis pushed around 22 million people into unemployment. In an attempt to complete the above comparison the great depression in 1929 increased the U.S unemployment rate to 25%, which amounting 15 million unemployed Americans. This rate did not decrease under the 10% until the USA entered the Second World War in 1941 (Barile 2020; Kenny 2020). However, in the case of Ebola infection in Guinea the GDP rates grew negatively and its impacts took 5–10 years to be recovered. However, due to the increase of deaths and confirmed cases, the government restricted public gatherings, moving from city to another or traveling, closing markets as well as schools, and all these policies have reduced the agricultural production and negatively impacting the country economy. Consequently, the volumes of exports fell down and the international prices of minerals were also declined, imports went down due to the decline in economic activity and constraints on traded as well as tourism was totally disappeared (UNDP 2014).

3 Research Methodology, Data Sources and Hypotheses

The present research relied on analysing both a daily panel and time series data covering the period 21Jan2020-01May2021, in order to evaluate the impact of COVID-19 in investors' sentiment and psychology. The used data sets were primarily obtained from the statistical bulletins of the WHO, Trading Economics, and the official site of Livecharts (Livecharts 2022; WHO 2022; Trading Economics 2022). Thus, to analyse these data the study used empirical techniques such as the random effects model and the single equation of the vector error correction model. However, to compare the impacts of COVID-19 in China and U.S. stock markets along with investors' sentiment; the study constructed two dummy variables within two values for each. Where, the 1 indicated that the COVID-19 was positively impacted Shanghai and NYSE stock markets, while the −1 value posited that this pandemic was negatively impacted these markets. Moreover, to accomplish the core aims of this research, the following hypotheses are intended to be tested:

H*$_1$*: The COVID-19 death toll is significantly impacted the sentiment of stock markets' investors.

H*$_2$*: The global confirmed cases of COVID-19 are significantly impacting investors' sentiment.

H*$_3$*: The impact of COVID-19 in china's stock market is significantly impacting investors' sentiment.

H*$_4$*: The impact of COVID-19 in New York stock exchange is significantly impacting investors' sentiment.

H*$_5$*: Investors' sentiment is valid to mediate the relationship of COVID-19 global confirmed cases along with investors' psychology as proxied by gold prices.

3.1 Model Specification and Variables Description

To achieve the core aim of this research and testing the aforementioned hypotheses, the study relied on a daily panel data set covering the period 21Jan2020-01May2021. Thus, through relying on data's nature the following general model of panel data is going to be tested:

$$CSI\% = \alpha_{it} + \beta GCCs_{it} + \beta GTD_{it} + \beta COVIDShanghaiindex_{Dummyit} + \beta COVIDS\&P500index_{\_Dummyit} + \varepsilon_{it} \quad (1)$$

where, the CSI% relates to the change in stock indices as a main proxy for investors' sentiment. The α on the right side refers to the intercept coefficient, the GCC s is the global confirmed cases of COVID-19 in the time t, the GTD is the daily global death toll, the βCOVIDShanghaiindex$_{\text{_Dummyit}}$ relates to the dummy variable that is utilized to represent the impact of COVID-19 confirmed cases in Shanghai stock index, while the $\beta \text{COVIDS}\&P500index_{\text{\_Dummyit}}$ represents the impact of COVID-19 in S&P500 index, and the ε is the error term. However, to investigate whether investors' sentiment is valid to mediate the impact of COVID-19 global case in the psychology of FOREX traders; the following single equations of the VECM models will be tested:

$$\begin{aligned} \Delta TP_{it} =& \beta_0 \pm \sum_{i=1}^{k-1} \beta_{1i} \Delta TP_{t-i} \pm \sum_{i=1}^{k-1} \beta_{2i} \Delta COVID - 19GCCs_{t-i} \\ & \pm \sum_{i=1}^{k-1} \gamma_1 COVID - 19GCCs_{t-i} \pm \psi_i \varepsilon_{t-1} \pm \mu_t \end{aligned} \quad (2)$$

$$\begin{aligned} \Delta S\&P500\%_{it} =& \beta_0 \pm \sum_{i=1}^{k-1} \beta_{1i} S\&P500\%_{t-i} \pm \sum_{i=1}^{k-1} \beta_{2i} \Delta COVID - 19GCCs_{t-i} \\ & \pm \sum_{i=1}^{k-1} \gamma_1 COVID - 19GCCs_{t-i} \pm \psi_i \varepsilon_{t-1} \pm \mu_t \end{aligned} \quad (3)$$

$$\begin{aligned} \Delta TP_{it} =& \beta_0 \pm \sum_{i=1}^{k-1} \beta_{1i} \Delta TP_{t-i} \pm \sum_{i=1}^{k-1} \beta_{2i} \Delta COVID - 19GCCs_{t-i} \\ & \pm \sum_{i=1}^{k-1} \beta_{3i} \Delta S\&P500\%_{t-i} \\ & \pm \sum_{i=1}^{k-1} \gamma_1 COVID - 19GCCs_{t-i} \pm \sum_{i=1}^{k-1} \gamma_2 S\&P500\%_{t-i} \pm \psi_i \varepsilon_{t-1} \pm \mu_t \end{aligned} \quad (4)$$

where, the TP is the dependent variable that relates to Traders' psychology as captured by the ±% change in gold prices. The S&P500% refers to the change in S&P500 index and it is utilized as a proxy for investors' sentiment. Thus, after confirming the significant correlation between the global confirmed cases of COVID-19 "COVID-19GCCs" and TP, as well as COVID-19GCCS along with S&P500; model number (4) will be considered to check whether S&P500 is valid to adjust the relationship of COVID-19GCCs along with traders' psychology. Where, the variables of interest are measured as shown in the following Table 1:

4 Findings and Discussion

In an attempt to evaluate the impact of COVID-19 in investors' sentiment, the study analysed a daily panel data for the period 21Jan2020-01May2021. For this purpose, the study used both the fixed and the random effects models, while since results from the Hausman test revealed that the P-value is significant; the null hypothesis was finally accepted. Therefore, the results proved that the random effects model is more appropriate to achieve the core aim of this research. Results from Hausman test are shown in the following Table 1:

Table 1. Describes results from the Hausman test

	Coefficients			
	(b) Fixed	(B) Random	(b−B) Difference	Sqrt (diag(V_b−V_B) S.E
COVID-GCCs	−1.610	−1.610	−4.430	6.7200
Global Death	3.670	3.670	9.250	1.4000
S&P500index_Dummy	0.0076	0.0076	−2.950	0.00016
Shanghai index_Dummy	−0.0016	−0.0016	−1.730	0.00017
Chi2 (3) = (b-B) ' [(V_b−V_B) ^(−1)] (b−B) = 0.001				
Prob > Chi2 = 0.997				

Moreover, since the study was restricted to implement the random effects model in order to evaluate the impact of COVID-19 in investors' sentiments as captured by the stock markets' indices; the following Table 2 demonstrates results from the GLS random effects model:

Table 2. This table presents results from the random effects model

Variables	Random Effect Model Number of Observations = 568, Obs per group: min = 71, Avg = 71.0 Max = 71/ Wald Chi2 (4) = 35.14, Prob > Chi2 = 0001					
Investors Sentiment	Coef.	Std. Err.	Z	P > |z|	95% Conf.	Interval
COVID-19 GCCs	−1.610	6.4500	−2.50	0.012	−2.880	−3.4800
Global Death	3.670	1.3500	2.73	0.006	1.030	6.3100
S&P500index_Dummy	0.0076	0.0015	4.79	0.000	0.0045	0.0107
Shanghai index_Dummy	−0.0016	0.0017	−0.98	0.328	−0.0050	0.0016
_Cons	−0.0003	0.0022	−0.15	0.880	−0.0047	0.0040

Findings from the random effects indicated that investors' sentiments as captured by the indices of Shanghai, KOSPI, SET50, S& P500, FTSE100, FTSEMIB, NIKKEE225, and ASE are significantly relating to the change in COVID-19 GCCs, Global Deaths toll and the S&P500index_Dummy. More specifically, the results revealed that investors' sentiment is negatively impacted by the increase in the number of COVID-19 global confirmed cases. Which in other words mean that the daily increase in COVID-19 GCCS affects investors' confidence as well as resulting in a bearish market sentiment. On the other hand, the variable of deaths toll is found to be positively influencing stock markets' indices, and might that because the increase in COVID-19 confirmed cases resulted in decreasing prices. Which in other words mean that the decrease in prices motivates investors to enter the market and opening long positions. Furthermore, the results proved that the impact of COVID-19 in the NYSE performance as captured by the S&P500index_Dummy is found to be positively influencing investors' sentiment. Therefore, the study concluded that since the New York stock exchange is seen as the largest stock market in the world; the impact of COVID-19 GCCs on the market performance will be able to affect stock indices worldwide. Although, china's economy is considered as the second largest economy in the world, however, the impact of COVID-19 GCCs in Shanghai stock market do not affect investors' sentiment. However, since the COVID-19 GCCs and global deaths toll are significantly impacting investors' sentiment as captured by stock markets' indices; precious metals such as gold may be are seen as safe haven investment. Therefore, the current research is also focused on exploring the impact of COVID-19 in investors' psychology as proxied by gold prices, over the period 21Jan2020-01May2021. However, since results from the augmented dickey confirmed that the used data became stationary as well as integrated of the same order I(1), only after converting them into the first difference; the study ran the Johansen cointegration test. Consequently, the findings proved that there are long-run integrations for the COVID-19 along with investors' sentiment, and similarly a long-run integration was also detected for the COVID-19 along with investors' psychology. Therefore, the study ran the ECM to check whether investors' sentiment is valid to mediate the correlation of COVID-19 along with investors' psychology (Table 3).

The above table shows that a +1% change in COVID-19 global confirmed cases increases gold prices by 0.005%, while decreasing the S&p500 by 0.076. Thereafter, the study moved on to check the power of investors' sentiment to adjust the long-run integration of COVID-19 along with investors' psychology.

Findings from the above Table 4 proved that variables including COVID-19 and investors' sentiment are significantly impacting investors' psychology during the long-run. More specifically, the results confirmed that a 1% decrease in S&P500 as a proxy for investors' sentiment increases the prices of gold by 0.00003. Which mean that when the COVID-19 decreases the S&P index; investors will be stimulated to divert their funds into the gold through viewing it as a safe haven investment. By contrary, the results proved that a +1% change in COVID-19 global confirmed decreases the price of gold by about 0.0253. However, when the change in gold prices was only regressed on COVID-19 as shown in table (5), it is found that the 1% increase in COVID-19 increases the price of gold by 0.0005, while when it is regressed on COVID-19 and S&P500 together this relationship was changed. Which in other words mean that investors' psychology is indirectly responding to the change in COVID-19 confirmed cases, as they are more focusing on the performance of S&P500 instead of relying on COVID-19 global cases. Thus, when COVID-19 impacts the sentiment of stock market investors', that will support the gold market to be bullish by about 0.00003. Furthermore, after running the Wald test

Table 3. This table shows results from the Error Correction Model

Panel A. This table presents the long-run correlations of COVID-19 along with investors' sentiment as measured by S&P500 index

Independent	D_Gold% _Ce1	Coef.	Std. Error	Z	P > /Z/	99% Conf.	Interval
COVID-19	L1.	−1.153	0.198	−5.80	0.001	−1.664	−0.641
	L.D	0.0005	0.104	0.05	0.961	−0.026	0.027
	_Cons	0.0756	0.136	5.52	0.000	0.0403	0.1108

Panel B. This table presents the long-run correlations of COVID-19 along with investors' psychology

Independent	D_S&P500 _Ce1	Coef.	Std. Error	Z	P > /Z/	99% Conf.	Interval
COVID-19	L1.	−1.135	0.184	−6.16	0.007	−1.610	−0.660
	L.D	−0.076	0.323	−0.24	0.381	−0.910	0.757
	_Cons	−0.004	0.138	−0.04	0.972	−0.361	0.351

Table 4. This table demonstrates results from the ECM and the Granger Causality Wald χ^2 Tests

Panel A. This table presents the long-run causalities, which are running from COVID-19 and investors' sentiment towards investors' psychology

Independent	D_Gold% _Ce1	Coef.	Std. Error	Z	P>/Z/	99% Conf.	Interval
Investors' Sentiment	L1.	−0.412	0.132	−3.12	0.002	−0.752	−0.071
	L.D	−0.00003	0.00001	−2.13	0.033	−0.0007	6.69
COVID-19	LD	−0.0253	0.3823	−0.07	0.947	−1.0102	0.9596
_Cons		−0.0140	0.1635	−0.09	0.932	−0.4351	0.4071

Panel B. This table presents the short-run correlations between the COVID-19 and investors' sentiments along with investors' psychology

Variable	Chi2 (3)	Prob > Chi2
COVID-19	17.20	0.0006
S&P 500	23.009	0.0002

the results confirmed that the prices of gold are significantly relating to the change in COVID-19 and investors' sentiment during the short-run.

5 Conclusion

In an attempt to fill the gap in literature, the study ran both the fixed and random effects' models to assess the impact of COVID-19 outbreak in investors' sentiment, over the period 21Jan2020-01May2021. However, after applying the Hausman technique the results confirmed that the random effects model is more appropriate to accomplish the core aim of this research. Consequently, the findings indicated that investors' sentiment as captured by stock indices are negatively responding to the increase in COVID-19 global confirmed cases. Precisely, a $+1\%$ increase in COVID-19 forces the stock market to be bearish by -1.610. While, the $+1\%$ increase in death's cases improves the market by about 3.67. Therefore, the study concluded that COVID-19 GCCs stimulates investors to liquidate their investments or opening short-positions, and thus since short-positions lead to a decline in market indices; investors will find it a good opportunity to go long any might that's why deaths toll shows a positive in market indices. However, authors including Akgüla et al. (2015), Upadhyaya et al. (2020), Domm (2020) indicated that global cases of COVID-19 as well as the increase in death's toll are negatively impacting investors' sentiment as captured by market's indices. Therefore, the study concluded that investors' sentiment was negatively influenced by this crisis due to its negative impact on the markets' performance. Since the NYSE is the largest stock market in the world the impact of COVID-19 GCCs in the market performance as proxied by the S&P500index_Dummy is found to be positively impacting investors' sentiment. Therefore, the study concluded that the performance of NYSE conducts a vital role to impact the decisions of stock market investors worldwide. On the other hand, though the economy of china seen as the second largest economy in the world; the impact of COVID-19 in the market performance is found to be insignificantly linked with investors sentiment. However, since the S&P500index_Dummy is significantly influencing investors' sentiment; the study induced that investors worldwide are highly relying on the performance of NYSE, and might that because the dominant of the Dollar currency or the high liquidity of NYSE.

Furthermore, since investors' sentiment is found to be significantly relating to the performance of NYSE, the study is also focused on checking the validity of investors' sentiment as captured by S&P500 to mediate the relationship of COVID-19 along with investors' psychology as proxied by gold prices. Consequently, results from the ECM confirmed that investors' psychology is significantly, while negatively related to the change in COVID-19 and S&P500 index. Moreover, the study concluded that the increase in COVID-19 global cases encourages investors to liquidate their stock market's investments in order to invest them in more secured assets such as gold. Studies including Mora (2010) mentioned that during financial stress or tight market liquidity it will be worthwhile for investors to invest in safe haven assets such as gold.

References

Ali, M.S.: An empirical analysis for the perceptions of Amman stock exchange's investors. Bus. Theory Pract. **21**(1), 125–135 (2020)

Amman Stock Exchange. Disclosures (2022). https://www.ase.com.jo/en/disclosures

Ali, M.S.: In a search for the influential factors of investors' behaviour. In: The 5th International Conference on Business, Economics, Management and Marketing, CBEMM, Conference Proceedings, 5th–7th March 2018. Oxford University (2018)

Ali, M.S.: The impact of market fundamentals and financial crisis on the liquidity of banks and stock markets: evidence from Jordanian Investors. Ph.D. thesis, Anglia Ruskin University (2016)

Akgüla, I., Bildiricib, M., Özdemirc, S.: Evaluating the nonlinear linkage between gold prices and stock market index using Markov-Switching Bayesian VAR models. Procedia. Soc. Behav. Sci. **210**(2015), 408–415 (2015)

Azam, M., Kumar, D.: Factors influencing the individual investor and stock price variation: evidence from Karachi stock exchange. Australian J. Basic Appl. Sci. **5**(12), 3040–3043 (2011)

Barile, M.: Unemployment during the great depression. Love to Know (2020)

Carvalho, R.: Coronavirus impact on financial markets and economies. Reuters (2020)

Domm, P.: History shows China's coronavirus impact on markets likely to be short-lived. CNBC (2020)

The Guardian: Stock markets rally after Federal Reserve starts printing money (2020)

Gozgor, G., Lau, C., Sheng, X., Yarovaya, L.: Econ. Lett. **185** (108680) (2019). Retrieved from The role of uncertainty measures on the returns of gold - ScienceDirect

Kenny, P.: UN Body warns of up to 25M Job Losses due to COVID-19. Anadolu Agency (2020)

Krantz, M.: Coronavirus crash reduces 18 major companies to penny stocks. Leader Board (2020)

Liu, A.: HSBC plans to triple billionaire clients in greater China push. Bloomberg (2020)

Livecharts. Co. UK (2022). Retrieved from Live Charts UK - Free live trading charts, stock market prices, historical data, share prices, day trading

Larry, C.: Coronavirus impact on financial markets. Refinitiv Perspect (2020)

Mora, N.: Can banks provide liquidity in a financial crisis? Econ. Rev., Federal Reserve Bank Kansas City **95**, 31–67 (2010). Third Quarter

Randow, J., Ainger, J.: Europe Bonds Soar as Lagarde Pledges no Limits to ECB Action (2020)

Bloomberg, 19 March 2020 (2020). https://www.bloomberg.com/news/articles/2020-03-18/ecb-announces-750-billion-euro-pandemic-bond-buying-program?srnd=markets-vp

Szlezak, P., Reeves, M., Swartz, P.: What Coronavirus could mean for the global economy. Harvard Bus. Rev. **3**, 1–10 (2020)

Trading Economics (2022). Retrieved from Data - Trading Economics

Tuchman, M.: Taking a step back from the chaos. Market Watch (2020)

Upadhyaya, P., Todd, C., Pirondini, M.: Coronavirus vs. SARS- potential economic and market implications. Amundi Pioneer (2020)

UNDP: Assessing the socio-economic impacts of Ebola Virus Disease in Guinea, Liberia and Sierra Leone. The Road to Recovery (2014)

WHO, 2022. Coronavirus Disease (COVID-19) Pandemic. https://www.who.int/emergencies/diseases/novel-coronavirus-2019

The Impact of the Professional Code of Conduct of the External Auditor on the Reduction of Tax Evasion in Jordan

Mohyedin Hamza[1](✉), Yousef Shahwan[1], Khaled Alkotayni[2], Hanan Haimour[3], Khalid Jbair[4], and Mohammad AlMekhlafi[5]

[1] Accounting Department, Zarqa University, Zarqa, Jordan
{mhamza,yshahwan}@zu.edu.jo
[2] Accounting Department, Irbid National University, Irbid, Jordan
Kh.qutaini@iun.edu.jo
[3] Accounting Department, Université de la Manouba, Manouba, Tunisia
20209521@zu.edu.jo
[4] Higher Institute of Accounting and Institutional Management, Zarqa University, Zarqa, Jordan
[5] Department of Accountancy, University of Business and Technology, Jeddah, Saudi Arabia
m.mekhlafi@ubt.edu.sa

Abstract. This study aimed to identify the Impact of the professional code of conduct in its dimensions (Independence, Professional Confidentiality, Integrity, and Professional Competence) of the external auditor on the reduction of tax evasion in Jordan. To achieve this purposes the researcher develop a questionnaire consisting of (37) items. Where the study population consisted of all external auditors in Jordan, a simple random sample of study population was withdrawn by distributed (209) questionnaires, and the valid questionnaires for analysis were (167) questionnaires, with a (71.1%) responses rate of the study sample. The results of the study showed that there is a high level of professional code of conduct of the external auditor, and the level of tax evasion in Jordan was high, and results also showed that there is significant impact for the professional code of conduct of the external auditor on the reduction of tax evasion in Jordan. The study recommended the necessity of having a written guideline for the professional code of conduct that the external auditor should have in Jordan, which aims to enhance his impartiality and integrity and to reduce the detection and reduction of cases of tax evasion.

Keywords: Professional Code of Conduct · External Auditor · Reduction of Tax Evasion · Jordan

1 Introduction

Business organizations in different sectors of their work depend on the services of the external auditor, who monitors the accounts and determines their validity and safety from fraud and errors, independently and impartially, in order to express his opinion on the status of the organization, and thus the rules of professional code of conduct are

B. A. M. Alareeni and I. Elgedawy (Eds.): *Artificial Intelligence (AI) and Finance*, SSDC 488, pp. 544–554, 2023.
https://doi.org/10.1007/978-3-031-39158-3_51

an important factor for the success and development of the external audit profession, because it helps direct and guide the external auditor in order to perform his duties in the best manner, as it guarantees his rights and protects him from the threats and obstacles facing him and impedes the progress of his mission on the one hand, and on the other hand it contributes to providing the necessary information to users of the financial statements, and the external auditor's commitment to the rules of professional code of conduct is considered a translation of his commitment to the responsibilities imposed on him, and his failure to adhere to those rules leads to a loss of confidence of the beneficiaries of audit services, and thus the profession loses its objectivity and status [1].

Tax revenues are one of the most important sources of revenue in the country, and one of the most important tools that are used to achieve the country's economic, political and social goals [2]. Tax evasion is one of the phenomena of contemporary economic life, as evasion is known in developing and developed countries alike. It includes the various groups of society, such as companies and institutions of different sizes and sectors in which they operate. Tax evasion is also considered an economic crime, due to the harmful consequences that occur on the national economy" [3].

Tax evasion is a global phenomenon that both developed and developing countries suffer from, and Jordan is not immune from this negative phenomenon, as tax is considered a source of the state's financial resources and an effective performance to achieve financial, economic and social goals, as tax evasion leads to a decrease in tax revenues, which means depriving the state budget of an important source of financing. Licensed auditors adopting the ethics and rules of professional code of conduct, who practice the profession of auditing accounts and preparing budgets, reports and financial statements, is one of the ways to combat this phenomenon and limit its negative effects. Accordingly, this study came to know the impact of the rules of professional code of conduct of the external auditor with its dimensions (independence, professional confidentiality, integrity, and professional competence) in reducing tax evasion in Jordan.

1.1 Rules of Professional Code of Conduct for the External Auditor

The rules of professional code conduct of the external auditor are a set of ethics that the external auditor must abide by, and there are many types according to the multiplicity of points of view on them. These rules also seek to achieve a number of objectives that serve the auditing profession [1]. The rules of professional conduct of the external auditor are also defined as a set of rules and foundations that the external auditor must adhere to and act according to, in order to be successful in his dealings with people, successful in his profession, and able to win the trust of his clients, colleagues, and superiors" [4]. Whereas Kelly defined "the rules of professional conduct for the external auditor as a set of ethical principles and values that bind external auditors when performing their duties, responsibilities and all their work." [5].

The rules of professional conduct of the external auditor also represent the minimum required of the auditors when performing their work, and they are clear rules that every auditor must follow, and they are part of the professional code of conduct issued by the American Institute of Certified Public Accountants. Thus, they are also principles and ethics that the external auditor must comply with, in order to organize his work and achieve the desired goals of the institution" [6, 7].

What the auditing profession, and indeed the entire business sector, is witnessing these days in a deep crisis of confidence that can be considered with a great degree of accuracy as a crisis of ethics emanating from the clear control of greed and narrow private interests over a large part of this sector in various countries of the world. This crisis had clearly appeared in the United States of America, but it is not confined to that country only, but it can be said that it has spread to most countries to varying degrees [8]. Hence, the auditor's behavior plays an important role in increasing the confidence of users of financial statements. Professional associations have actively contributed to issuing standards adopted by auditors, which over time have improved their level of professional conduct" [9].

The importance of the rules of professional conduct for the external auditor is its endeavor to achieve a set of goals represented by a number of things, firstly contributing to the development of the accounting and auditing profession, "secondly, reassuring and spreading confidence in the hearts of the beneficiaries of auditing services [10]. Thirdly, it serves as a complement to the legal texts and legislative provisions to achieve the principles of efficiency and impartiality in the work of the auditor. Fourthly, the existence of rules of professional conduct for the external auditor reduces the auditors' oversight and ensures that they do not break the prevailing laws and regulations, thus enhancing the ethics of the profession [11, 12].

2 Literature Review

Many previous studies have discussed the issue of tax evasion, its causes, and ways to reduce it, such as the study of [13–23]. For example, the study of Haeridistia and Fadjarenie [19], which aimed to analyze the impact of independence and diverse professional ethics on audit quality. The study reached a set of results, the most important of which was that independence and professional ethics in its various dimensions (confidentiality, integrity, competence) and the auditor's experience affect the audit quality. Drozd et al. [16] found that managing the ethical behavior of auditors had an impact on mitigating tax evasion as well.

By looking at what has been presented from previous studies that dealt with both the concepts of the rules of professional conduct of the external auditor and tax evasion, and accordingly the researchers see that the previous studies varied in their topics, objectives and the study population that was studied. The current study is similar to several studies that dealt with the concept of rules of professional conduct for the external auditor, including the study of Ratna and Anisykurlillah [14], the study of Drozd [16], and the study of Haeridistia and Fadjarenie [19]. In addition, many studies dealt with the concept of tax evasion, including the study of (Deef et al., 2021), the study of Hudori and Mustikasari [15], and the study of Eijavi [17]. The current study is different from previous studies in its quest to analyze the impact of the rules of professional conduct of the external auditor in reducing tax evasion in Jordan. This is what many previous studies did not address, especially in developing countries such as Jordan, in terms of the categories that were dealt with, the period of their application, and the dimensions that were adopted.

Based on the above, the following hypotheses are adopted:
H1: There is an effect of the rules of professional conduct of the external auditor in reducing tax evasion in Jordan. The following sub-hypotheses are derived from it:
H1.1: There is an effect of the independence of the external auditor in reducing tax evasion in Jordan.
H1.2: There is an effect of the professional confidentiality of the external auditor in reducing tax evasion in Jordan.
H1.3: There is an effect of the integrity of the external auditor in reducing tax evasion in Jordan.
H1.4: There is an effect of the professional competence of the external auditor in reducing tax evasion in Jordan.

3 Study Model

To achieve the purpose of this study and reach its objectives in determining the impact of the independent variables on the dependent variable, a study model was developed. Figure 1 shows the study model and its variables.

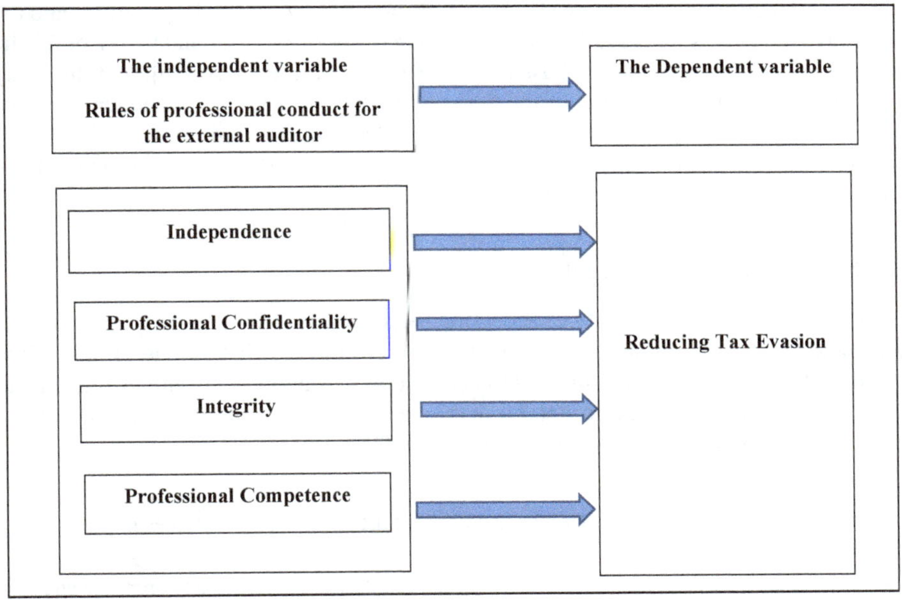

Fig. 1. Study Model

4 Study Methodology

For the purposes of completing the study and achieving its objectives, the researchers relied on the analytical descriptive approach, which is based on studying the phenomenon as it is in reality and describing it accurately, then analyzing the correlations that exist between the independent variable (rules of professional behavior for the external auditor) and the dependent variable (reducing tax evasion).), in an attempt to identify the impact of the independent variable on the dependent variable, and the size of this effect in order to reach conclusions that contribute to the development and improvement of reality.

4.1 Study Population and Sample

The study population consisted of all licensed legal auditors who practice the profession of auditing accounts in Jordan, whose number is (454) licensed auditors, according to the Jordanian Association of Certified Public Accountants in May 2021. Based on the table for determining the size of the required samples (Sekaran & Bougie 2016). Thus, the researchers distributed (209) electronic questionnaires - prepared by (Google Forms) - to the study sample, and (179) questionnaires were retrieved from them, and the number of questionnaires valid for analysis was (167), making up (71.1%) of the sample.

4.2 Study Tool

A questionnaire was prepared based on the theoretical framework of this study and the previous studies that were reviewed. This questionnaire included (24) questions, and the paragraphs that measure the independent variable (rules of professional behavior for the external auditor) were prepared based on the study [14], the study [16], and the study [19]. Therefore, the preparation of the paragraphs that measure the dependent variable (reducing tax evasion) was based on the study [15, 24]. The study adopted the Five Likert Scale to measure the responses of the respondents, and the answers ranged from strongly agreeing to strongly disagreeing.

4.3 Stability of the Study Tool

Reliability means the possibility of accessing the same data when re-studying using the same study tool on the same individuals under the same conditions, and to ensure the stability of the study tool, Cronbach Alpha equation was applied. Its values range from (0–1), so that these values are acceptable at the stability coefficient ratio (0.70) and above [25, 26, 18]. Where the results were as shown in Table 1.

Table 1. Stability coefficients (Cronbach alpha) for the independent variables in the study.

Variable	Stability coefficient (Cronbach alpha)
Independency	0.717
Professional Confidentiality	0.762
Integrity	0.792
Professional competence	0.706
Code of professional conduct for the external auditor as a whole	0.884
Reducing tax evasion as a whole	0.749

Table 1 shows that all values of stability coefficients were acceptable for application purposes, as they ranged between (0.706–0.884), which indicates that the study tool generally enjoys a high reliability coefficient and its ability to achieve the purposes of the study.

4.4 The Correlation Matrix Between the Variables of the Study

The Pearson correlation coefficient was extracted between the variables of the study in order to identify the nature of the correlation between the variables. The results are shown in Table 2.

Table 2. Matrix of correlation between study variables.

	Variable	1	2	3	4
1	Independency	1	0.537**	0.534**	0.394**
2	Professional Confidentiality	0.537**	1	0.466**	0.446**
3	Integrity	.534**	0.466**	1	0.556**
4	Professional Competence	.394**	0.446**	0.556**	1

It appears above in Table 2 that there is no problem in the multiple linear correlation of the dimensions of the independent variable, and a correlation that is higher than (0.80) is an indication of a problem, and thus it can be said that the sample is devoid of the problem of the multiple high linear correlation [27].

5 Results

5.1 Normal Distribution Test

The normal distribution test for the data was performed as one of the assumptions before the linear regression analysis used to test the hypotheses of the study. Table 3 shows the results of the normal distribution and it is clear from it that all study variables follow the normal distribution, by looking at the Shapiro-Wilk test values and the value of the accompanying significance level attached to it.

And since all the values of the significance level in the table are greater than (0.05), the null hypothesis is accepted, which states that "there are no statistically significant differences between the distribution of variable values and the normal distribution at the level of significance ($\alpha \leq 0.05$). This means that the values of the study variables follow normal distribution" [28, 29].

Table 3. Test for normal distribution using (Shapiro-Wilk) test.

Variable	Test Value	Significance Level
Independency	0.951	0.35
Professional Confidentiality	1.03	0.25
Integrity	1.05	0.26
Professional competence	0.56	0.92
Code of professional conduct for the external auditor as a whole	0.57	0.93
Reducing tax evasion as a whole	1.05	0.26

5.2 Hypothesis Testing

Rules of Code of Professional Conduct for the External Auditor and Reducing Tax Evasion

To test this hypothesis, a multiple linear regression equation was calculated for the effect of the dimensions of the rules of professional conduct of the external auditor in reducing tax evasion as a whole. Table 4 illustrates this.

Table 4. Results of multiple linear regression of the impact of the rules dimensions of the professional conduct of the external auditor in reducing tax evasion as a whole

Variable	ß	T-value	Statistical significance	R	R^2	Adjusted R Square	F	P-value
Independency	0.12	1.65	0.01	0.679	0.461	0.447	34.61	0.00
Professional Confidentiality	0.16	2.26	0.03					
Integrity	0.29	3.81	0.00					
Professional Competence	0.27	3.80	0.00					

Table 4 above shows that there is an effect at the level of significance ($0.05 \geq \alpha$) of the rules of professional conduct of the external auditor in reducing tax evasion as a whole, as the value of the correlation coefficient (R) was (0.679), which is a statistically significant value and indicates a statistically significant degree of correlation between the independent variables and the dependent variable, the value of (R-square) was (0.461), which is a statistically significant value that explains the ability of the rules of professional conduct of the external auditor in reducing tax evasion, meaning that the rules of professional conduct of the external auditor explain what value (46.1%) of the change occurred In reducing tax evasion, the test value (F) was (34.61), with a statistical significance ($0.00 \geq$ p-value).

As for the first sub-hypothesis, it was found that there was an effect at the significance level ($0.05 \geq \alpha$) of the independence of the external auditor in reducing tax evasion from the point of view of the external auditor in Jordan, where the value was (T-value $\geq$ 1.65). The result of the second sub-hypothesis was that there was a relationship at the level of significance ($0.05 \geq \alpha$) for professional Confidentiality in reducing tax evasion from the point of view of the external auditor in Jordan, where the value was (T-value $\geq$ 2.26). As for the integrity in reducing tax evasion for the third sub-hypothesis, the result was (T-value $\geq$ 3.81), and this indicates that there is an effect of the auditor's integrity in reducing tax evasion from the point of view of the external auditor in Jordan. Finally, with regard to the fourth sub-hypothesis, it was found that there is an effect of professional competence in reducing tax evasion from the point of view of the external auditor in Jordan, where the value was (T-value $\geq$ 3.80).

5.3 Discussion of the Results

The results of the statistical analysis indicated that there was a statistically significant effect at the significance level ($0.05 \geq \alpha$) of the rules of professional conduct of the external auditor in reducing tax evasion from the point of view of the external auditor in Jordan. This result may be due to the fact that the external auditors surveyed have a moral sense to distinguish legal from illegal practices pursued by the audited companies, which leads them to follow a set of standards and professional rules that aim at impartiality and clarifying the actual reality of the institutions and companies. In addition, these auditors possess the necessary professional competence that leads to discovering cases of tax evasion that the taxpayer may follow with the aim of not paying part or all of the amounts due.This result is indirectly consistent with what was stated in the Tamrabat study [30], which indicated a positive effect of the external auditor's commitment to the rules of professional conduct on the quality of accounting information and thus the indication of cases of tax evasion. It also agrees with the study by Eijavi [17], which showed that the quality of the auditor, including their enjoyment of ethics and rules of conduct, has a significant and negative relationship with tax evasion.

In addition to the above, the results showed that there is a statistically significant effect at the significance level ($0.05 \geq \alpha$) of the independence of the external auditor in reducing tax evasion in Jordan. This result may be attributed to the fact that the surveyed external auditors enjoyed independence when carrying out audit work and expressing their opinions regarding the financial reports of companies, and they were not affected by any non-professional influences when performing their duties, which enhances their honesty and impartiality, thus revealing manipulation of financial statements and official papers and thus reducing tax evasion. This result is indirectly consistent with what was stated in the Tamrabat study [30], which showed the external auditor's keenness to adhere to the principle of independence and objectivity, which enhances the degree of impartiality and reliability of accounting information.

The results also showed that there was a statistically significant effect at the significance level ($0.05 \geq \alpha$) of the professional confidentiality of the external auditor in reducing tax evasion in Jordan. This result may be due to the fact that the external auditor considers the notes contained in the financial reports as confidential, and thus avoids disclosing the information to parties not involved in the audit process, and seeks to discuss it with those concerned in the company under audit, which may direct the owners and management of companies to move away from illegal methods for the purposes of non-payment of tax, but rather to adopt legal methods for the purposes of benefiting from tax exemptions according to the prevailing laws [31].

In addition to the above, the study found a statistically significant effect at the significance level ($0.05 \geq \alpha$) of the external auditor's integrity in reducing tax evasion in Jordan. The researcher may attribute this result to the external auditors' keenness not to engage in any unethical behavior that offends the auditing profession, and the endeavor of these auditors to be free from unprofessional influences in the performance of their work, in a manner that ensures the provision of the necessary information in a timely manner and without delay.This highlights the financial position of the companies under audit and shows any illegal practices practiced by them, which reduces cases of tax evasion [32, 33].

In addition to the above, the results showed that there was a statistically significant effect at the significance level ($0.05 \geq \alpha$) of the professional competence of the external auditor in reducing tax evasion in Jordan. This result may be attributed to the external auditors in Jordan possessing university degrees with specialization, professional certificates that enhance the necessary knowledge of auditing standards, and experience in the work sector of the companies being audited, which leads to the completion of the audit tasks assigned to them accurately and without errors [34]. While making sure to complete them and deliver them the required reports in a timely manner, which in turn leads to discovering cases of tax evasion or illegal methods used by the client, discussing them with him, indicating their risks, and urging him to avoid them and follow legal methods for the purposes of tax evasion.

6 Recommendations

Based on the previous results, the study recommends encouraging the external auditor to move away from relationships that expose the audit process to the domination of parties with specific interests, in order to avoid directing the audit process in undesirable directions.In addition, the external auditor must be assured of keeping client secrets and not disclosing information and data of the company being audited.Moreover, tax legislation in Jordan should be characterized by stability and constancy, so that many of these amendments confuse the understanding of auditors, which increases ambiguity and thus the difficulty of applying the law and cases of tax evasion. In conclusion, the external auditor should be periodically involved in specialized courses and workshops to develop his knowledge and experience with regard to auditing standards and procedures.

References

1. Al-Ramahi, N., Abuhussein, A., Shahwan, Y.: The impact of applying the XBRL on the Quality of Annual Reports of the Firms Listed in Amman stock exchange. In: 2021 22nd International Arab Conference on Information Technology (ACIT), December, pp. 1–5. IEEE (2021)

2. Shahwan, Y.: The mediating effect of investment decisions and financing decisions on the influence of capital structure against corporate performance: evidence from Jordanian listed commercial banks. Acad. Acc. Finan. Stud. J. **22**(6), 1–20 (2018)
3. Alkurdi, A., Mardini, G.H.: The impact of ownership structure and the board of directors' composition on tax avoidance strategies: empirical evidence from Jordan. J. Finan. Reporting Acc. **18**, 795–812 (2020)
4. Leemrijse, J.: Accountability artefacts and auditors' professional conduct. Box-ticking exercises or prevention of future accounting scandals?, unpublished master thesis, Radboud University, Netherlands (2018)
5. Kelly, P.: Developing ethical leaders for the accounting profession. CPA J. **87**, 42–49 (2017)
6. Hamza, M.: The impact of consulting services on the disclosure of going concern in Jordanian commercial banks: external auditors' perceptions. Corp. Gov. Organ. Beh. **6**(2), 19–23 (2022)
7. Jalil, N.A.A., Haron, H., Muda, M.B.: The impact of islamic based accounting education on professional conduct among accountants: a conceptual review. KnE Social Sci. 712–725 (2019)
8. Shahwan, Y., Abdel-hamid, O.: The impact of social responsibility disclosure in reducing the risk in Jordanian Commercial Banks. Res. World Econ. **11**(3), 142–150 (2020)
9. Zaaffron, M., Boudonnat, A.: Compliance with the rules of professional conduct for account holders (an applied study of a group of professionals in Algeria), J. Adv. Acc. Finan. Stud. **2**(1), 45–58 (2018)
10. Shahwan, Y.: The effect of OECD principles of corporate governance on social responsibility accounting. Int. J. Innovat. Create. Change **11**(2), 418–431 (2020)
11. Seyger, L. The influence of organizational conditions on professional conduct: a case study in a big four accounting firm (2020)
12. Zinisha, O., Kharchenko, E., Avdeev, Y., Pavlova, N., Maltsev, I.: Features of application of International Financial Reporting Standards (IFRS) by international companies. Econ. Annals-XXI **188**, 188–194 (2021)
13. Deef, A.T., Alrawashdeh, B., Al-fawaerh, N.: The impact of foreign ownership and managerial ownership on tax avoidance: empirical evidence from Egypt. Acad. Acc. Finan. Stud. J. **25**(2), 1–13 (2021)
14. Ratna, T.D., Anisykurlillah, I.: The effect of experience, independence, and gender on auditor professional scepticism with professional ethics as moderating. Acc. Anal. J. **9**(2), 138–145 (2020)
15. Hudori, R., Mustikasari, E.: The strength of audits, reporting standards and corruption, on tax evasion: a cross-country study. Int. J. Econ. Bus. Adm. **8**(2), 554–567 (2020)
16. Drozd, I., Pysmenna, M., Pohribna, N., Zdyrko, N., Kulish, A.: Audit assessment of the effectiveness of public procurement procedures. Independ. J. Manag. Prod. **12**(3), s085-s107 (2021)
17. Eijavi, M.: The relationship between the auditor quality and the tax evasion, the 2nd international conference on modern research in management, economics, accounting and banking, July 21. Tbilisi-Georgia: Georgian International Academy of Sci (2019)
18. Shahwan, Y., Hamza, M., Al-Fasfus, F., Al-Ramahi, N., Almubaydeen, T.H.: Social responsibility accounting, disclosure and real practice: evidence from Jordan. Asian Econ. Finan. Rev. **12**(3), 164–182 (2022)
19. Haeridistia, N., Fadjarenie, A.: The effect of independence, professional ethics & auditor experience on audit quality. Int. J. Sci. Technol. Res. **8**(2), 24–27 (2019)
20. Kovermann, J., Velte, P.: The impact of corporate governance on corporate tax avoidance—a literature review. J. Int. Account. Audit. Tax. **36**, 100270 (2019)
21. Oladejo, O.F.: Strategy to close tax gaps created by tax avoidance and tax evasion in Nigeria: an overview. Eur. J. Acc. Audit. Finan. Res. **9**(7), 55–63 (2021)

22. Guiaş, E.G., Hăineală, C.M.: Tax avoidance and tax evasion in EU: trends and effects the annals of the university of Oradea. Econ. Sci. **30**(2), 229–238 (2020)
23. Salé, M. J., Muharremi, O., Hoxhaj, M.: Albanian Individual Taxpayers Perceptions and Determinants on Ethical Behavior Regarding Tax Compliance (2021)
24. Ballout, O.M.K.: The extent of the impact of corporate governance on the effectiveness of companies in the Jordanian service sector. Acad. Strateg. Manag. J. **19**(1), 1–10 (2020)
25. Sekaran, U., Bougie, R.: Research Methods for Business: A Skill Building Approach. Wiley, Chichester (2016)
26. Shahwan, Y., Sa'adeh, A., Hamza, M., Al-Ramahi, N., Swiety, I.A.: Do the reserves help the financial and non-financial performance of firms during the COVID-19 PANDEMIC. Corporate Gov. Organizat. Beh. Rev. **6**(2), 217–222 (2022)
27. Benesty, J., Chen, J., Huang, Y., Cohen, I.: Pearson correlation coefficient. In: Noise Reduction in Speech Processing. Springer, Heidelberg. (2009). https://doi.org/10.1007/978-3-642-00296-0_5
28. Razali, N.M., Wah, Y.B.: Power comparisons of shapiro-wilk, kolmogorov-smirnov, lilliefors and anderson-darling tests. J. Stat. Model. Anal. **2**(1), 21–33 (2011)
29. Imran, M., Ali, S., Shahwan, Y., Zhang, J., Al-Swiety, I.A.: Analyzing the effects of renewable and nonrenewable energy usage and technological innovation on environmental sustainability: evidence from QUAD economies. Sustainability **14**(23), 15552 (2022)
30. Tamrabat, B.: The external auditor's commitment to the rules of professional conduct and their impact on the quality of accounting information a field study of a sample of academics and professionals, an unpublished master's thesis, Radboud University, Netherlands (2020)
31. Shahwan, Y., Hamza, M., Lubad, H.K., Khresat, O.M.: Social responsibility accounting and financial performance during COVID-19: A survey of the firms listed in Amman Stock Exchange. J. Gov. Regulat. **12**(1), 93–99 (2023)
32. Ballout, O.M.K.: The effect of electronic commerce on profitability of Jordanian commercial companies. Global J. Econ. Bus. **9**(3), 651–659 (2020)
33. Shaqqour, O.F.: Impact of vertical and horizontal integration between strategic management accounting and decisionmaking on reducing financial failures case of Jordan. Acad. Acc. Finan. Stud. J. **24**(5), 1–9 (2020)
34. Shaqqour, O.F.: Impact of institutional governance tools on reducing agency costs to commercial banks listed in the Amman stock exchange. Int. J. Finan. Res. **11**(5), 414–423 (2020)

The Impact of Disclosure of Sustainable Development Accounting on the Quality of Profits in Industrial Companies Listed on the Amman Stock Exchange

Tareq Hmmad Almbydeen[1](✉), Sewar Rafat Salameh[1], Rafat Salameh Salameh[2], Khaled Alkotayni[3], Riham Alkabbji[1], and Mohammad Kanaan[4]

[1] Zarqa University, 2000, Zarqa 13110, Jordan
tmbydeen@zu.edu.jo
[2] Al-Balqa' Applied University, Salt, Jordan
[3] Irbid University, Irbid, Jordan
[4] University of Business and Technology, Jaddah, Saudi Arabia

Abstract. This study aims to identify the impact of disclosure of Sustainable Development Accounting on the quality of profits in industrial companies listed on the Amman Stock Exchange. To achieve the objectives of the study, the analytical descriptive approach was followed, where the data for the independent variable (Sustainable Development Accounting) were collected through the annual reports and their clarifications during the period (2017–2021), and the financial data for measuring the dependent variable (profit quality) were collected through the financial statements represented by the modified Jones model. The study population consisted of (33) industrial companies listed on the Amman Stock Exchange, while the study sample was limited to (31) companies.

The study found that there is no impact of disclosure of Sustainable Development Accounting in its dimensions (economic, social, and environmental) on the quality of profits in industrial companies listed on the Amman Stock Exchange. It recommends the development of an independent disclosure of sustainable development accounting for industrial companies listed on the Amman Stock Exchange, as well as boosting industrial companies, and increasing their awareness of the importance of disclosure of development practices and their impact on society.

Keywords: Disclosure · Accounting of Sustainable Development · Quality of Profits

1 Introduction

Recent years have witnessed an increasing interest of companies in sustainable development. This concept was mentioned for the first time in the report of the World Commission on Environment and Development, which describes it as preserving the resources available to current and future individuals. Companies achieve sustainability by integrating its dimensions: (economic, social and environmental) in their annual reports, since

B. A. M. Alareeni and I. Elgedawy (Eds.): *Artificial Intelligence (AI) and Finance*, SSDC 488, pp. 555–565, 2023.
https://doi.org/10.1007/978-3-031-39158-3_52

the main goal of companies is no longer only profit maximization [10]. Companies have become profit-seeking with attention to the demands of stakeholders, meeting their needs, improving their own reputation, and developing competitive advantage through disclosure of their activities, and the impact of their work on society and individuals.

Since reports are the main reliable source that expresses the company's status, serving as a reference for stakeholders such as analysts, investors and lenders, companies must disclose all factors that affect their profits, and reflect them fairly, realistically and free of manipulation. This is known as the quality of profits. [1] (Brick, 2020).

With the development of industrial companies, the diversity of their services, and their increasing competition with other companies, the need to provide non-financial information has increased through the adoption by companies of the concept of sustainable development to enhance their reports, which will help investors make appropriate decisions. This in turn will benefit the local community in particular and the state in general. This study attempts to answer the following question:

Does the disclosure of sustainable development accounting with its dimensions (economic, social, environmental) have an impact on the quality of profits in industrial companies listed on the Amman Stock Exchange?

2 Objectives of the Study

This study aimed to identify the impact of the disclosure of sustainable development accounting on the quality of profits in industrial companies listed on the Amman Stock Exchange, through identifying the concept of disclosure of sustainable development accounting and its dimensions (economic, social, environmental), as well as the concept of the quality of accounting profits.

3 Literature Review

Khalil & Saghir [2] aimed to identify the impact of disclosing the element: (environmental, social, economic) in increasing the value of the enterprise. The study population comprised a sample of industrial companies in Sudan. It found that there is a statistically significant relationship between the environmental and social element and the increase in the value of the enterprise. The researcher recommended the need to emphasize on publishing information on the environmental element, for its effective role in improving the decisions taken. Sheikh [3] also pointed to sustainability disclosure on the quality of accounting profits on Iraqi oil companies, through a set of financial reports issued during the period (2012–2017). The study found a significant and statistically significant impact of the elements of sustainability disclosure on the quality of accounting profits of Iraqi oil companies in the presence of controlled control variables .The researcher recommended the importance of setting incentives for companies that carry out sustainability practices and disclosure of those practices through financial reports. Hamed [4] analyzed the relationship between the disclosure of corporate sustainability reports and the quality of profits, and the extent of its impact on the cost of capital in the Egyptian Stock Exchange companies. The researchers found that there is a weakness in the level of disclosure of Egyptian corporate sustainability reports and a positive relationship with the quality of profits, but a negative relationship with the cost of capital. It also indicated that corporate mechanisms have a positive impact. Accordingly, the study recommended enhancing Egyptian Stock Exchange companies to raise their level of disclosure regarding sustainability reports of their enterprises to improve the quality of profits.

Hamid & Al-Dabbas [5] aimed to measure the impact of the quality of accounting profits on the performance of the company. The study population included commercial banks listed on the Iraq Stock Exchange. The researcher found that there is a significant statistical impact between the quality of profits and the performance of the company; the higher the quality of profits, the higher the company's performance. The researcher recommended adopting a sustainability model in measuring the quality of profits, without being confined to using financial indicators derived from accounting data in order to measure performance, since it is essential to link non-financial indicators that take into account the variables associated with the company's internal and external environment.

4 Theoretical Framework of the Study

Previous literature has shown that the relationship between the disclosure of sustainable development accounting and the quality of profits is one of the relationships that receive wide attention and controversy now, and that profit is achieved as a result of the company's activities represented by the integration of the three dimensions: (social, economic, environmental). Financial reports that are available to: (investors, employees, customers, analysts), should be able to present information honestly, transparently, efficiently, effectively and in a timely manner, to reflect the actual financial performance of the company, and thus making appropriate investment decisions, which positively affects the quality of profits [6].

Studies have also shown that the companies‘ disclosure of sustainable development is an important indicator that shows the success of the company. It shows investors and creditors the potential economic risks, thus increasing the quality of its profits, and reducing the cost of capital. Nevertheless, there is a discrepancy and difficulty in determining the relationship between the disclosure of sustainable development and the performance of companies, which varies depending on the type of sector, the culture of the country, or based on the awareness of management of the importance of this disclosure and its impact.

5 Study Hypothesis

Based on the problem statement, the study relies on testing the following main hypothesis:

H0: There is no statistically significant impact at the level of significance (<0.05) of the disclosure of sustainable development accounting in its dimensions: (economic, social, environmental) on the quality of profits in industrial companies listed on the Amman Stock Exchange.

5.1 Study Model

See Fig. 1

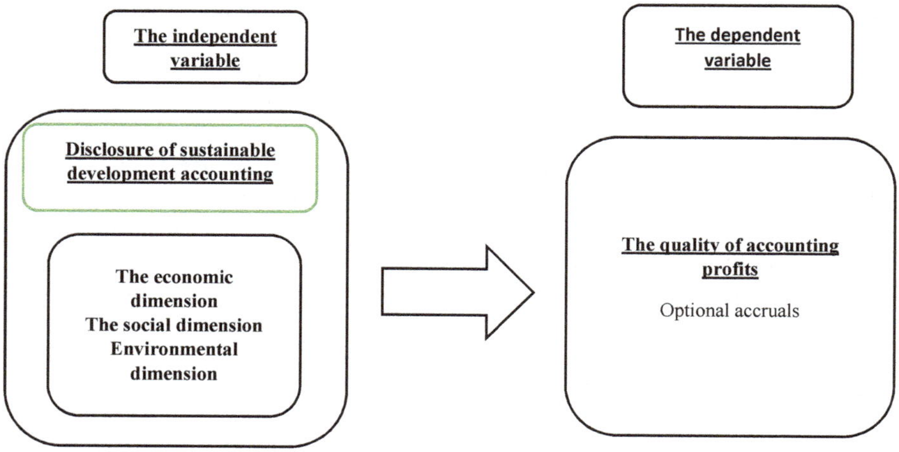

Fig. 1. Study model

6 Study Methodology

The current study has adopted the descriptive analytical methodology to achieve its objectives. At the level of descriptive research, a desk survey was conducted, as well as reviewing relevant literature.

6.1 Sources of Data Collection

1) **Primary sources**: The main sources of the study data were reports, annual financial statements, and clarifications published on the Amman Stock Exchange website (www.ase.com.jo) in the period (2017–2021).
2) **Secondary sources**: A review of relevant Arabic and foreign books and references, articles, previous literature was conducted, as well as visiting websites.

6.2 Study Population and Sample

The study population consisted of all (31) industrial companies listed on the Amman Stock Exchange, within (8) industrial sub-sectors. These consisted of: (pharmaceutical and medical industries, chemical industries, food and beverages, tobacco and cigarettes, extractive and mining industries, engineering and construction industries, electrical industries, clothing, leather and textile industries).

6.3 Study Variables and Their Measurement

Independent Variables: The impact of the disclosure of sustainable development accounting was calculated through the dual registration system (Dichotomous Scoring system). The element takes the value (one) if the company disclosed it, and the value (zero) if the company did not disclose it. The following table shows the elements that were evaluated; to determine the reality of corporate disclosures of the main dimensions in the study, based on study [7] (Table 1).

Table 1. Indicators of disclosure of sustainable development practices

#	Indicator elements
	Economic development practices for sustainability
1	Selling worthless production waste in order to generate revenue
2	Reducing the costs of processing damaged products while maintaining the level of output quality
3	Cooperation with government agencies in order to protect the company's interest
4	Increase in the percentage of sales of new products or services
5	Increase in the company's market share in the market
6	Improved cash flows from operating activities
7	Increase in the sales growth rate
8	Increase in the rate of return on investment
9	Increase in the added economic value
10	Application of new technology in the company's activities to make profits
	Economic development practices for sustainability
1	Maintaining the safety and health of employees through supporting health care programs
2	Providing training and qualification programs for employees to improve their job performance
3	Periodic performance assessment of employees and motivation of creative ones
4	Creating suitable working conditions for employees (internal environment, housing and moving)
5	Low rates of accidents and deaths related to working conditions
6	The presence of surveys to continuously measure customer satisfaction
7	There are clear and specific mechanisms for addressing customer complaints and suggestions
8	The company considers the interests of all relevant parties when making a decision
9	Respect customer privacy
10	Compliance with laws and procedures and the application of the optimal international standards

(*continued*)

Table 1. (*continued*)

#	Indicator elements
11	Contributing to the development of the local community by supporting initiatives
12	Product-related risk management
13	Customer health and safety
14	Addressing the adverse environmental impacts of the company's activities on society
15	Avoiding practices that do not consider values and culture and do not achieve development goals
16	Interest in programs that serve a wide sector of society
17	Publication of the Corporate Social Responsibility Report
	Economic development practices for sustainability
1	Energy saving
2	The rate of greenhouse gas emissions
3	Future plans to reduce harmful emissions
4	Initiatives to reduce the negative environmental impact of products and services
5	Replacing harmful traditional energy resources with less polluting modern ones
6	The scale of investment in pollution-free technology
7	Disclosure and management of potential environmental risks by the company
8	Compliance with environmental protection legislation and laws
9	Voluntary initiatives of the company to repair environmental damage

Dependent Variable: The dependent variable represented by the quality of accounting profits was measured by using the Jones Model [9] modified by Kothari [8], which is based on measuring optional accruals accounts. It is one of the most accurate models, and has been used by many studies, such as Hamid [4]. The model can be formulated through the following equation:

$$\begin{aligned}\text{Total Accruals it/Assets it} - 1 &= \alpha 0 + \alpha 1(1/\text{Assets it} - 1) + \alpha 2\Delta\text{Rev it/Assets it} - 1 \\ &+ \alpha 3\text{PPEit/Assetsit} - 1 + \alpha 4\text{ROAit/Assets it} - 1 + \varepsilon\text{ it}\end{aligned} \quad (1)$$

6.4 Descriptive Analysis of Study Variables

This part of the analysis presents the descriptive statistics of the study variables, based on the annual financial statements of industrial companies for the period (2017–2021). Table 2 provides a statistical description (Descriptive Summary) of the characteristics of the study sample and its distribution by dependent and independent variables'.

The data of Table 2 indicate that the total number of industrial companies in the study is (31), shown through the viewings ratios of (155) which are the result of multiplying the

Table 2. Descriptive data of the study variables

Variable	Obs	Mean	Mediator	Std. Dev.	Min.	Max	Skewness	Kurtosis
Year	155	2019	2019	1.419	2017	2021	0	1.7
Group	155	16	16	8.973	1	31	0	1.8
SC	155	.372	.353	.114	.176	.765	.68	3.5
Eco	155	.316	.3	.167	.1	.7	.49	2.3
Env	155	.272	.33	.265	0	.889	.52	2.1
Sus	155	.959	.95	.473	.276	2.184	.42	2.33
Earqual	155	.358	.3	.318	−1.089	1.552	.1	6.12

number of companies by the measured timeperiod (31 companies*5 years), divided into groups during the years (2017) to (2021). The descriptive results of the study independent variables indicated the disclosure of sustainable development accounting, represented by: (economic dimension, social dimension and environmental dimension).

6.5 Pearson Correlation Coefficient Test

See Table 3.

Table 3. Results of Pearson coefficient test for linear correlation Pairwise correlations

Variables	(1)	(2)	(3)	(4)	(5)	(6)	(7)	(8)	(9)
Year (1)	1.000								
Group (2)	0.000	1.000							
SC (3)	0.274*	−0.121	1.000						
Eco (4)	0.253*	−0.342*	0.478*	1.000					
Sus (5)	0.357*	−0.220*	0.898*	0.731*	1.000				
Earqual (6)	0.004	−0.051	−0.047	−0.026	−0.045	1.000			
Envd (7)	0.292*	−0.124	0.368*	0.149	0.422*	0.017	0.153	0.248*	1.000

*** $p < 0.01$, ** $p < 0.05$, * $p < 0.1$

6.6 Normal Distribution Screening Test

See Table 4.

Table 4. Results of the multivariate normal distribution test of the quality of profits coefficient

Source	SS	Df	Ms	Numbers	Of	Obs	=	155
Model	0.221	2	0.011	Prob	>	F	=	0.327
Residual	11.855	121	0.098	R-squared	=	0.018		
Total	12.076	123	0.098	Root	MSE	=	0.313	
Eaqual	**Coef.**	**Std.Err.**	**T**	**P>t**	**95%Conf[.**	**Interval]**		
hat_	5.430	8.960	0.610	0.546	-23.169	12.310		
hatsq_	9.565	13.282	0.720	0.473	-16.730	35.860		
cons_	1.064	1.502	0.710	0.480	-0.002	4.037		

6.7 Test for Multivariate Normality

Doornik-Hansen chi2 (2) = 29.983 Prob > chi2 = 0.0000.

The null hypothesis of the Doornik-Hansen Test states that the distribution of the remaining values is a normal distribution, and the results of the study rejected the null hypothesis; which indicates the absence of a multivariate normal distribution of the study data, and the exclusion of multiple linear regression examination for its inability to derive relationships acceptably. This result conforms with the outputs of the descriptive analysis in the study, which shows that the deviation ratios and the Kurtosis ratios are not close enough to prove the results through the multiple linear analysis, which demands un-tendency of relationships in many variables to the left or the right. Figure 2 shows a projection of the results of the multivariate normal distribution of this study.

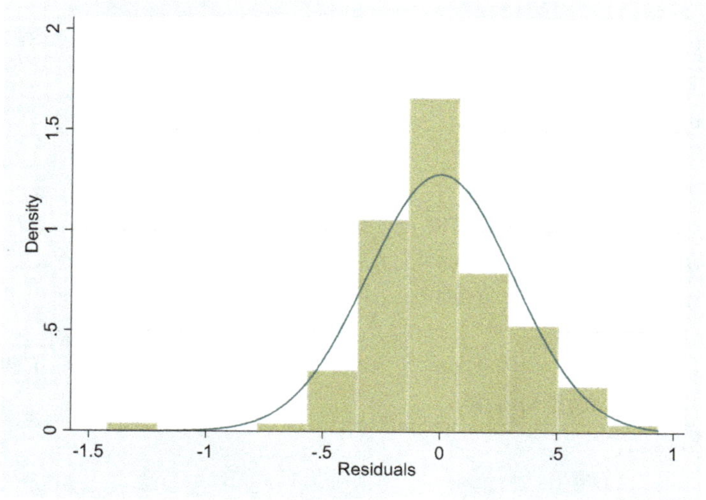

Fig. 2. Graph of the results of the normal distribution of the quality of profits coefficient

6.8 Testing the Study Hypothesis

The relationship between the impact of disclosure of sustainable development accounting and the quality of profits.

Table 5. Results of the (Hausman) test

Hausman (1978) specification test	
Chi-square test	9.729
P-value	0.045

Modified Wald test for GroupWise heteroskedasticity
in fixedeffect regressionmodel
H0: sigma(i)^2= sigma^2 for all i
chi2 (53)= 72757.01
Prob>chi2= 0.045

Table 5 indicates the results of the Hausman test, which came to reject the null hypothesis stating that the random model is the most suitable for measuring the study model ($p = 0.45$), therefore the researcher chose the fixed model of sectional time beams to interpret the relationships.

Hypothesis testing using the stationary model of sectional time beams:

Table 6. Results of the static model of cross-sectional time packages of the study variables

Earqual	Coef.	St.Err.	t-value	p-value	[95% Conf	Interval]	Sig.
SC	-.212	.312	-0.68	.499	-.832	.409	
Eco	.528	.195	2.71	.008	.141	.915	***
ENVD	-.185	.102	-1.80	.075	-.388	.019	*
Constant	-5.667	3.0[illegible]2	-1.88	.063	-11.652	.318	*
Mean dependent var		0.355		SD dependent var		0.313	
R-squared		0.150		Number of obs		124	
F-test		3.[illegible]03		Prob > F		0.000	
Akaike crit. (AIC)		-52.627		Bayesian crit. (BIC)		-35.706	

*** $p<.01$, ** $p<.05$, * $p<.1$

Table 6 shows the results of the static model of cross-sectional time packages for the study variables, where the results indicated a negative impact without statistical significance,i.e. a level of social dimension, which deals with social development practices in Jordanian industrial companies on the Jones index (1991) as an indicator of the quality of profits. The results of the other two dimensions of sustainable development practices came in contrast to what was stated in the social dimension, where the results showed a statistically significant positive impact at the level of (p = 1%), and a high impact coefficient at (.5280) for the economic dimension, which measures the practices of economic development on the index (Jones, 1991) as an indication of the quality of profits. The standard model showed a negative impact of the indicators of the environmental dimension, which measures sustainable environmental practices on the quality of profits. The results showed a negative impact at the level of (10%), and a weak impact coefficient at (−18.5%), which may be attributed to the weak application of this dimension.

The results on the control variables showed a statistically significant effect on the quality of profits, where the size of the company came with a statistically significant positive impact at the level of (10%), and with a high impact coefficient of (35.7%) on the quality of profits. On the same pattern, the statistical model showed a positive and substantial relationship between the values of cash flows and the quality of profits in industrial companies at the level of (10%), but with a weak impact coefficient close to zero percent.

7 Results

This study, which investigated the impact of the disclosure of sustainable development accounting in its dimensions: (economic, social, and environmental), on the quality of profits in industrial companies listed on the Amman Stock Exchange has come to the number of results. Firstly, the disclosure of the economic dimension has a positive impact on the quality of profits in industrial companies listed on the Amman Stock Exchange. Secondly, there is a negative impact of disclosure of the environmental dimension on the quality of profits in industrial companies listed on the Amman Stock Exchange. Thirdly, The disclosure of the economic dimension has not showed an impact on the quality of profits in industrial companies listed on the Amman Stock Exchange. Finally, there is a positive impact of the size of the company and the operational cash flows on the quality of profits in industrial companies listed on the Amman Stock Exchange.

8 Recommendations

Considering the findings of the study, the researcher offers the recommendations, the most important of which are Encouraging industrial companies and increasing their awareness of the importance of disclosing development practices and their impact on society. AS well as organizing a specialized body to monitor the commitment of industrial companies to apply sustainable development practices. Emphasis must be focusing on the development of the concept of environmental sustainability, as it was noted that companies do not focus on the concept of sustainability. And conducting further future studies looking at the relationship between the disclosure of sustainable development accounting and the quality of profits on different sectors and in other time periods.

References

1. Brik, D.A.: Studying the impact of the quality of accounting profits and dividends on stock returns. A test study on Egyptian joint stock companies. Sci. J. Account. Stud. **2**(3), 460–516 (2020)
2. Khalil, A.R., Saghir, H.: The impact of disclosing the elements of sustainable development in increasing the value of the enterprise - a field study on a sample of Sudanese industrial companies. J. Hum. Nat. Sci. **3**(8), 600–617 (2022)
3. Sheikh, K.: The impact of sustainability disclosure on the quality of accounting profits - an applied study. Sci. J. Bus. Environ. Stud. **12**(1), 133–156 (2021)
4. Hamed, A., Yassin, N., Abdulkarim, A.: The relationship between the accounting disclosure of corporate sustainability reports and the quality of accounting profits and their implications on the cost of capital–an applied study on the Egyptian environment. J. Account. Audit. Union Arab Univ. **10**(3), 559–636 (2021)
5. Hamid, T., Dabbas, W.: The impact of the quality of accounting profits using the sustainability model (continuity of profits) on the performance of the company applied research in a sample of commercial banks listed on the Iraqi Stock Exchange. J. Sci. City Coll. **13**(2), 182–202 (2021)
6. Almubaydeen, T.H., Qushtom, T.F.A., Hamza, M., Al-Khawaldeh, R.M.R.: The Impact of the IFRS Adjustments "Coronavirus-Related Lease Contracts" on the Financial Performance (Royal Jordanian Case Study). In: Alareeni, B., Hamdan, A. (eds) ICBT 2022. LNNS, vol. 495, pp. 541–549. (2023). https://doi.org/10.1007/978-3-031-08954-1_48
7. Meligi, M.M.A.H.: The impact of accounting disclosure of sustainable development practices on the quality of accounting profits of companies listed on the Saudi Stock Exchange. Accounting thought **19**(5), 1–60 (2015)
8. Kothari, S.P., Leone, A.J., Wasley, C.E.: Performance matched discretionary accrual measures. J. Account. Review **87**(3), 761–796 (2005)
9. Jones, J.J.: Earning management during import relief investigation. J. Account. Res. **29**, 193–228 (1991)
10. Elderawi, A.: The relationship of social responsibility to competitive advantage, considering employee satisfaction as a variable mediator of the Tashbeek Logistics and Social Media Company. Zarqa J. Res. Stud. Hum. **21**(2), 2021 (2020)

The Impact of Applying the International Financial Reporting Standard "IFRS15 - Revenue from Contracts with Customers" on Accounting Conservatism and the Mediating Role of Net Assets: Acase Study of Jordan Telecom Company Orange

Ola Khresat[1,2](✉), Mohammad Mahmoud Abu Hasan[1,2], Abdwhab Arawashdeh[1,2], Suhad Jaradat[1,2], and AymanZereban[1,2]

[1] Zarqa University, Zarqa 13110, Jordan
okhresat@zu.edu.jo
[2] University of Business and Technology, Jeddah, Kingdom of Saudi Arabia

Abstract. The study aims at identifying the impact of applying the International Financial Reporting Standard (IFRS15-revenue from contracts with customers) on accounting conservatism and the mediating role of net assets through a case study of the Orange Jordan Telecom Company by comparing conservatism according to the Basu (1997) model for the pre-application periods of the standard (2015–2017) and post-application periods of the standard (2018–2020).

Achieving the research objectives necessitates using the analytical-descriptive approach. To answer the research hypotheses, path analysis is used through the structural modeling program (Smart PLS). The findings demonstrate the impact of the standard (IFRS-15) on the accounting conservatism measured by the Basu (1997) model. The findings also show that applying the standard (IFRS-15) affects the net assets with no impact of the net assets on the accounting conservatism measured by the Basu (1997) model. Moreover, the results indicate that applying the standard (IFRS-15) has a mediating role in the impact of applying the standard (IFRS-15) on the accounting conservatism measured by the Basu (1997) model at Orange Jordan Telecom Company. The study recommends amending certain policies and contractual terms, developing accounting and technological systems, and the internal control structure to respond to the standard, and training and qualifying employees to understand the standard guidelines by decision-makers in the Jordan Telecom Company - Orange.

Keywords: Accounting conservatism · International Financial Reporting Standard (IFRS-15) · Jordan Telecom Company – Orange · Net assets

B. A. M. Alareeni and I. Elgedawy (Eds.): *Artificial Intelligence (AI) and Finance*, SSDC 488, pp. 566–576, 2023.
https://doi.org/10.1007/978-3-031-39158-3_53

1 Introduction

The surge of the users of financial statements and their need for financial statements featured with impartiality and credibility necessitate issuing international standards for the preparation of financial reports, including the IFRS15 standard, and paying attention to applying accounting principles, together with the principle of accounting conservatism. Though the financial statements are prepared per International Financial Reporting Standards, these standards are characterized by flexibility by supporting management to use various methods in recognizing revenues and expenses. Therefore, different results can be attained depending on the method or principle used in preparing the data, and among these principles used is the principle of accounting conservatism [25].

Revenue, on the other hand, is considered one of the most significant elements of the basic financial statements guaranteeing the continuity of the company and a decisive number in evaluating the company's financial performance, as no company can continue without achieving revenues that exceed costs. Hence, the significant element relating to revenue recognition encompasses the value and timing of revenue recognition, that is, when the economic benefits may flow to the company to be measured reliably [1].

The significance of the IFRS15 standard is reflected in the first paragraph, indicating that the objective of the standard is to establish principles to increase the accuracy and reliability of the information provided to users of financial information by clarifying the nature of revenue, the amounts recognized as revenue, and the appropriate timing for revenue recognition. This standard is regarded as a substitute for the two international accounting standards (IAS 11 & 18). The standard is also a joint project between the "IASB" and the "FASB", as it is issued in 2014 to be applied to start from 01/01/2017, and its implementation is compulsorily postponed to 01/01/2018 with early application permitted on 01/01/2017 [7, 8, 13].

The scope of the standard (IFRS15) indicates how revenue should be recognized by applying these five steps; determining the customer's contract, the obligations' performance under the contract, the price of the transaction, allocating or distributing the price of the transaction to the obligations' performance under the contract, and recognizing revenue when the contract obligations are fulfilled by the entity. Unlike the previous two standards, the standard (IFRS15) shows the accounting treatment of certain cases relating to revenue; guarantees, selling with the right of return, considerations of clients and agents, customer options associated with services and goods, and the unexercised rights of the customers. The standard also includes proof of available considerations, money's time value, and limited requirements of disclosure, considering oral contracts, particularly if some legal rights of the contracting parties arising from them [7, 8, 13].

Accounting conservatism has been defined as the prudent response to an attempt to insure the risks and uncertainties inherent in a business situation (FASB, 2014). The International Financial Reporting Standards Council also defines accounting conservatism as the ability to affect the state of uncertainty associated with elements recognized and measured in the financial statements, where cases of uncertainty are dealt with through disclosure or prudent practices in the preparation of financial statements and statements. Conservatism refers to a degree of wisdom in conducting the practices and evaluations required to attain various assessments under conditions of uncertainty, such as not exaggerating the values of income and assets and not underestimating the values of liabilities and expenses. Implementing the policies of the principle of accounting conservatism, and the announcement of business results on time is the essence of the transparency of the financial data and statements. The management's role in practicing the principle of accounting conservatism is reflected in numerous areas, together with the postponement of recognizing expected profits and losses, as the management discloses the expected positive events in the financial reports, exaggerating expenses, increasing provisions and depreciating assets to show the results of the business of the company conservatively in the financial reports [3, 10].

Regarding the impact of applying the standard on telecommunications companies in Jordan, the financial statements of Jordan Telecom Company - Orange published in the annual report of 2018 show that applying the standard has affected the retained earnings with an amount of (2,385,530) Jordanian dinars. The report also indicates that applying the standard has affected the accounting policy followed by the company to recognize revenue for installment sales. According to the standard (IFRS15), revenue is recognized based on the total price of the transaction, consisting of two individual prices; the selling price of the device and the price of providing the service. Revenue is recognized when control is transferred to the customer as this change first affects the offers of data and internet service packages that contain the sale of mobile phones at discounted prices accompanied by a fixed-term service contract. When applying the standard, for example, the total revenue from such contracts has not changed, but the timing and method of revenue recognition have changed, where the revenue is distributed between the services revenue and the revenue from selling the mobile phone [18].

Revenue from the sale of the mobile phone is recognized when control is transferred to the customer at the beginning of signing the contract, which means an increase in the revenues of the current period at the expense of future periods, affecting the statement of financial position by recording the assets of customer contracts that are later converted into receivables when services are provided during the contract period. Therefore, applying the standard increases the sensitivity of total revenue related to the sale of mobile devices that may occur in certain seasonal periods. With that being said, it is of high significance to furnish the literature and studies related to the scope of the standard (IFRS15) [18].

2 Literature Review

Much ink has been spilled on the impact of implementing the standard (IFRS15) on the various components of the financial statements and the various sectors. Other studies also address the impact of implementing the standard on the qualitative characteristics of financial statements and accounting information. A Jordanian study Khresat,O 2021 demonstrates the impact of applying the International Financial Reporting Standard (IFRS 15) on the profitability of the Jordan Telecom Company. Selling mobile devices and providing other services. The findings indicate that applying the standard leads to an increase in the company's profits through the immediate recognition of the price of cellular devices for contracts that include two types of services: selling cellular devices and providing other services [22]. However, Khresat,B2017 argues that applying the International Financial Reporting Standard (IFRS15) impacts the components of the income statement and the statement of financial position in the Jordan Telecom Company by influencing the values and timing of revenue recognition [21].

In the same context, Jaber 2018 asserts the impact on the profits of Egyptian telecom companies from applying the International Financial Reporting Standard (IFRS15), whether concerning total revenues, service revenues, or commodity revenues [14]. Moreover, implementing the standard (IFRS15) contributes to realizing the character of a faithful representation of accounting information, leading to an increase in the confidence enjoyed by that data through the impact of receivables on the decline, and thus creating an increase in the faithful representation of accounting information [12].

Moreover, the implementation of the standard (IFRS 15) helps in solving some accounting problems such as sales with the right of return, guarantees, licenses, and buyback arrangements. Applying the standard (IFRS15) leads to an increase in the credibility of accounting information by increasing the reliability of accounting information and its ability to be understood and compared. The application of the International Financial Reporting Standard (IFRS15) also increases the uniformity of accounting information through the used principles of measurement and comparison. Besides, it positively affects the quality of accounting profits for information characterized by quality within the financial statements and the adoption of the fair value of the allowance received, and if the allowance is not monetary, the information becomes more reliable [11]. Mattei and Paoloni 2019 confirm the presence of a direct relationship between the International Financial Reporting Standard (IFRS15) and the quality and quantity of information presented in the annual report, and its ability to guide many related topics [23].

The act of conservatism, its effect on financial statements, and the extent to which conservatism is applied to certain companies and sectors have been tackled by various studies. Al-Taweel 2021 indicates that the financial reports issued by service firms listed on the Palestine Exchange are not categorized by accounting conservatism according to the Basu (1997) model. It is also found that the companies practice accounting conservatism, but the level of their practice is low according to the book value to market value (BTM) model on one hand [2]. On the other hand, Sabrina 2019 shows that accounting conservatism with its four dimensions; assets, liabilities, revenues, and expenses positively affects the relevance and reliability of accounting information and data. It is also noted that the use of accounting conservatism is embodied in the Algerian insurance companies through creating technical provisions and standardized provisions necessary for technical management [25].

On a parallel line, Obaid 2017 demonstrates that accounting conservatism impacts the market value of the Jordanian public shareholding firms listed on the Amman Stock Exchange (ASE) in all markets and sectors [24]. Benzorouk and Al-Amari 2021 indicate that conservatism involves varying differences between good news and bad news on time and hinders the continuousness of normal earnings flow. The findings also show that the effect of negative interest or earnings is two to six times bigger than the effect of positive earnings, as the earnings sensitivity is greatly and multiply affected by bad news [6]. Furthermore, Farhan 2019 indicates that the majority of Jordanian companies do not follow conservative accounting policies. It is also found that the level of conservatism is affected by the management's ownership, that is, the percentage of the managers' increase in the firm's capital equals an increase in the level of conservatism and vice versa, meaning that the level of conservatism is linked to the economic concerns of the managers, which vary according to their percentage of ownership [9].

Additionally, Zeghal and Lahmar 2016 assert the presence of a significant decrease in accounting conservatism in the period of adopting international financial reporting standards, as this decrease is impacted by the prevailing accounting model within a specific country. Likewise, the research demonstrates a decrease in the gap between the two accounting models in the period of post-IFRS use. Importantly[27], the focuses of the standard (IFRS15) in local and foreign environments have been addressed in plenty of studies. However, none of the past studies have tackled the effect of applying the standard (IFRS 15) revenue from contracts with customers on accounting conservatism and the mediating role of net assets, despite the important correlation between accounting conservatism and revenue and applying the standard of revenue from contracts with customers and net assets. Against this, the research problem is illustrated in the next section.

3 Research Problem

The issuance of the Financial Reporting Standard (IFRS15) and the obligation to apply it at the beginning of 2018 have resulted in the emergence of many questions about its effect on the financial statements and the main components of those statements, such as the quality of profits, market and book value, and net assets, which in turn affect accounting conservatism. Given the significance of telecommunications companies in the Jordanian business environment, and their working conditions in a context surrounded by economic, political, and social fluctuations reflected in the results of their work, especially amid the state of uncertainty suffered by companies, the research problem rests in pinpointing the effect of applying the financial reporting standard (IFRS15) on revenue from contracts with customers on accounting conservatism, the role of mediator for net assets in Jordan Telecom Company - Orange and the extent of the difference in the impact of financial reporting standard (IFRS15) on the principle of accounting conservatism in the two periods; pre- and post-application of the standard.

4 Research Hypotheses

The scope of the research problem necessitates constructing the following research hypotheses at a significant level ($\alpha \leq 0.05$).

(H01): There is no statistically significant effect of applying the International Financial Reporting Standard (IFRS15) on accounting conservatism according to the Basu (1997) model at the Jordan Telecom Company – Orange.

(H02): There is no statistically significant effect of applying the Financial Reporting Standard (IFRS15) on the net assets of Jordan Telecom Company - Orange.

(H03): There is no statistically significant effect of net assets on accounting conservatism according to the Basu (1997) model at the Jordan Telecom Company – Orange.

(H04): Net assets do not mediate the effect of applying the standard (IFRS15) on accounting conservatism according to the Basu (1997) model at the Jordan Telecom Company – Orange.

5 Method

Research Approach
To achieve the research objectives, the analytical-descriptive approach is adopted, as it is based on examining relevant literature to construct the theoretical framework for the research. The performance of Jordan Telecom Company - Orange is analyzed to identify the impact of applying the standard (IFRS15) revenue from contracts with customers on accounting conservatism and the mediating role of net assets for the years between 2015 and 2020.

Variables Measurement
The Independent Variable: The International Financial Reporting Standard (IFRS15).

Dummy variables are used to measure the period's pre and post-application of the standard, as the periods after the application of the standard (2018–2020) are equal to (1), while the periods before the application of the standard (2015–2017) are equal to (0).

The Dependent Variable: Accounting Conservatism (Basu, 1997) Model:

This model posits that accountants lean towards recognizing unrealized losses before unrealized profits. Referring to the said model, the bad news is reflected in profits faster than the good news. Accordingly, the measure of conservatism is the difference between the movement of profits in periods of bad news and their relationship in periods of good news. As stated by [4, 5],The (Basu, 1997) model has used the inverse regression of the declining profits (X) on the return (Ric) according to the following regression model:

$$X_{i,t/pi,t-1} = \alpha 0 + \alpha 1 DR_{i,t} + \beta 0 R_{i,t} + \beta 1 (R_{i,t} \times DR_{i,t})$$

$X_{i,t}$: Earnings per ordinary share of the company i for period t equal net profit before extraordinary items divided by the number of ordinary shares.

$P_{i,t-1}$: The stock price at the beginning of period t of a company i.

$R_{i,t}$: Return per share is measured by the difference between the market price of the share at the end and beginning of the period with the addition of dividends.

$DR_{i,t}$: A dummy variable is equal to (1) if Rit is less than zero, and if Rit is greater than zero, it is equal to (0).

The Mediating Variable (Net Assets)
The company's net asset value is the value of the company's assets minus the value of its liabilities, as it is computed using the next formula [26]:

$$\textbf{Net assets} = \textbf{Total Assets} - \textbf{Total Liabilities}$$

Statistical Analysis of Data and Hypothesis Testing

- **Descriptive Analysis:** It provides the main indicators about the research variables using mean values, minimum values, maximum values, and standard deviation.
- **Path Analysis through the Structural Modeling Program (Smart PLS):** It achieves the research objectives and answers its assumptions.

6 Results

6.1 Descriptive Analysis of the Research Variables

The mean, standard deviation, and the highest and lowest values of the research variables are computed during the research period, as the period (2015–2017) represents the pre-application of the IFRS-15 standard, while the period (2018–2020) represents the post-application of the standard [15–20]. The variables are represented through the graphical representation method. Figure 1 illustrated those findings as follows:

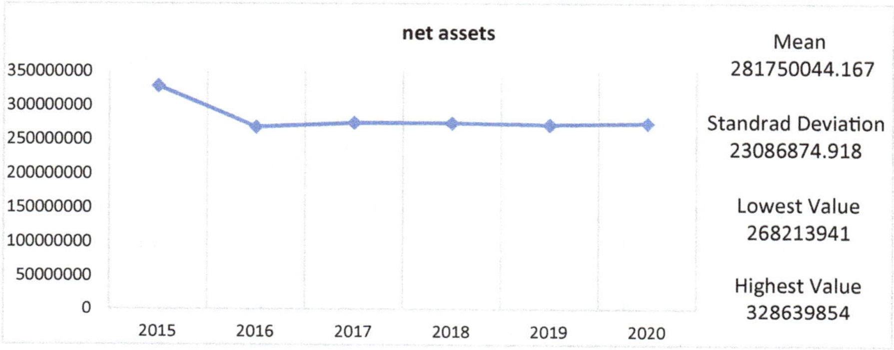

Fig. 1. Descriptive Analysis of Net Assets

Figure 1 indicates that net assets ARE at their highest level in 2015, i.e. pre-application of the (IFRS15) standard. Then, in 2016, net profits have fallen to their lowest level, but the profits remained fluctuating during the research period at the same level even post-application of the (IFRS15) standard.

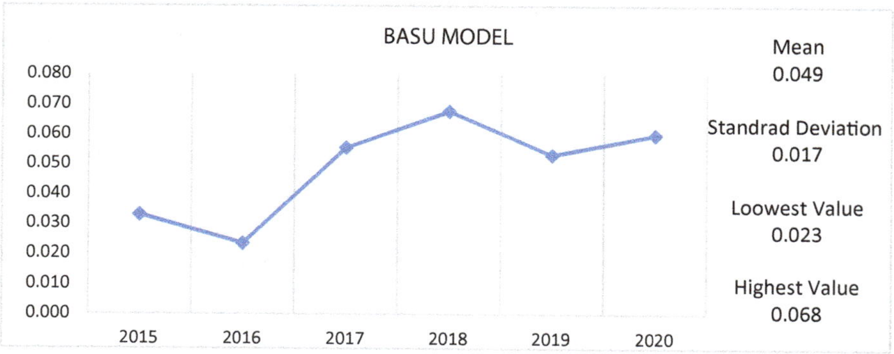

Fig. 2. Descriptive Analysis of the Basu (1997) Model

Figure 2 indicates that the Basu accounting conservatism model is at its highest level in 2018, which is the year of application of the IFRS-15 standard, while it is at its lowest level in 2016, which is before the application of the IFRS-15 standard. With the analysis of Fig. 2, it is evident that the accounting conservatism levels increase during the period of application of the standard compared to before the application of the standard.

Research Hypotheses Testing

The research hypotheses are tested through the Smart PLS structural modeling program to achieve the research objectives and answer its hypotheses. Table 1 illustrates those findings as follows:

(H01): There is no statistically significant effect of applying the International Financial Reporting Standard (IFRS15) on accounting conservatism according to the Basu (1997) model at the Jordan Telecom Company – Orange.

Table 1. Findings of Testing the First Major Hypothesis

Dependent Variables	Basu	
Independent Variables	T	Sig
Fixed Model	2.244	0.088
IFRS-15 Application	5.032	0.007
R^2	**0.550**	

****, the hypothesis is accepted at the 5% significance level, ***, the hypothesis is accepted at the 1% significance level.**

As shown in Table 1, the T-value is (5.032) with a significant level of less than 5% for the Basu accounting conservatism model, where the value R2 is (55.0%), which means to reject the first main null hypothesis and accept the alternative hypothesis stipulating "there is a statistically significant effect of applying the International Financial Reporting Standard (IFRS15) on accounting conservatism, as measured by the Basu (1997) model at the Jordan Telecom Company - Orange.

(H02): There is no statistically significant effect of applying the Financial Reporting Standard (IFRS15) on the net assets of Jordan Telecom Company - Orange.

Table 2. Findings of Testing the Second Major Hypothesis

Dependent Variables: Net Assests		
Independent Variables	T	Sig
Fixed Model	0.884	0.426
IFRS-15 Application	– 427.749	0.000
R^2	**0.164**	

****, the hypothesis is accepted at the 5% significance level, ***, the hypothesis is accepted at the 1% significance level.**

As shown in Table 2, the value of the determination coefficient is (R2 = 16.4%), signifying that (16.4%) of the change in net assets is due to applying the (IFRS-15) standard, which means rejecting the second main null hypothesis and accept the alternative hypothesis stipulating "there is a statistically significant effect of applying the Financial Reporting Standard (IFRS15) on the net assets of Jordan Telecom Company – Orange".

(H03): There is no statistically significant effect of net assets on accounting conservatism according to the Basu (1997) model at the Jordan Telecom Company – Orange.

Table 3. Findings of Testing the Third Major Hypothesis

Dependent Variables	Basu	
Independent Variables	T	Sig
Fixed Model	0.782	0.478
IFRS-15 Application	– 0.758	0.491
R^2	**0.130**	

**, the hypothesis is accepted at the 5% significance level, ***, the hypothesis is accepted at the 1% significance level.

As shown in Table 3, the T-value is (−0.758) with a significant level of less than 5% for the Basu accounting conservatism model, where the value R2 is (13.0%), which means accepting the third main null hypothesis stipulating "there is no statistically significant effect of net assets on accounting conservatism according to the Basu (1997) model at the Jordan Telecom Company – Orange".

(H04): Net assets do not mediate the effect of applying the standard (IFRS15) on accounting conservatism according to the Basu (1997) model at the Jordan Telecom Company – Orange.

Table 4. Findings of Testing the Fourth Major Hypothesis

Dependent Variables	Basu	
Independent Variables	T	Sig
Fixed Model	1.99	0.089
IFRS-15 < Application	– 4.48	0.000
R^2	**0.130**	

**, the hypothesis is accepted at the 5% significance level, ***, the hypothesis is accepted at the 1% significance level.

As shown in Table 4, the T-value is (−4.48) with a significant level of less than 5% for the Basu accounting conservatism model, where the value R2 is (13.0%), which means to reject the fourth main null hypothesis and accept the alternative hypothesis stipulating "Net assets do mediate the effect of applying the International Financial

Reporting Standard (IFRS15) on accounting conservatism according to the Basu (1997) model at the Jordan Telecom Company – Orange".

7 Conclusions

In a nutshell, the research demonstrates a statistically significant effect of applying the standard (IFRS15) on accounting conservatism, as measured by the Basu (1997) model at the Jordan Telecom Company - Orange. It also finds a statistically significant effect of applying the standard (IFRS15) on the net assets of Jordan Telecom Company – Orange. Moreover, the findings show that there is no statistically significant effect of net assets on accounting conservatism according to the Basu (1997) model at the Jordan Telecom Company – Orange. Furthermore, the findings demonstrate that net assets do mediate the effect of implementing the standard (IFRS15) on accounting conservatism according to the Basu (1997) model at the Jordan Telecom Company – Orange.

8 Recommendations

In light of the research findings, this research work recommends establishing an internal control system that follows up on the accounting work that takes place within the company in accordance with international accounting standards. The research also recommends Jordan Telecom Company - Orange to train and qualify all its accountants and employees to understand the standard guidelines and deal with the effects resulting from the adoption of the new standard related to the elements of the statement of financial position: assets and liabilities (net assets) through adding or amending policies and contractual terms and developing accounting and technological systems and the structure of the relevant internal control to respond to the standard. Another key recommendation is analyzing and evaluating the effect of implementing the standard (IFRS15) on improving the financial statements' statements among stakeholders. Finally, future studies shall be done to conduct research variables on other sectors, such as the financial sectors, industrial sectors, contracting sectors, and other service sectors.

References

1. Abu Nassar, M., Humaidat, J.: International Accounting and Financial Reporting Standards Theoretical Aspects, 4th edn. Wael Publishing House, Jordan (2020)
2. Al-Taweel, E.: Measuring the degree of accounting conservatism when preparing the financial statements for service companies listed on the Palestine Stock Exchange and indicating its impact on the profit distribution policy. Al-Quds Open Univ. J. Adm. Econ. Res. **6**(15), 1–16 (2021)
3. Al-Najjar, J.: Measuring the level of accounting conservatism in the lists, reports and financial statements and its impact on the market value of the share: An applied study on public shareholding companies listed on the Palestine Stock Exchange. Al-Balqa J. Res. Stud. **17**(2), 177–224 (2014)
4. Alwehwah, L.M., Barakat, F.S., Mahmoud, O.A.: The role of accounting conservatism in effecting the public listed Palestinian banks market value in light of the amendments to the Palestinian Banking Governance. Int. J. Acc. Finan. Rev. **10**(1), 93–106 (2022)

5. Basu, S.: The conservatism principle and the asymmetric timeliness of earnings. J. Acc. Econ. **24**(1), 3–37 (1997)
6. Benzorouk, Z.S.: The effect of accounting conservatism application on the quality of accounting information for published financial statements – ananalytical study of the "Basu" model. J. Econ. Adm. Legal Sci. **5**(15), 103–184 (2021)
7. Deloitte (IAS Plus) "IFRS 15 Revenue From Contract With Customers" (On-line). http://iasplus.com
8. International Financial Reports standard (IFRS) "IFRS 15 -Revenue from Contracts with Customers" , (On-line). http://ifrs.org
9. Farhan, M.: The relationship between the accounting conservitism and the efficiency of investment. Tikrit J. Adm. Econ. Sci. **1**(37), 337–347 (2019)
10. FASB. American Financial Accounting Standards Board. USA (2014)
11. Hashem, J.: Evaluation and analysis of the international accounting standard for financial reporting "Revenues from Contracts with Customers" IFRS 15 and its impact on the credibility and quality of financial reports. Dar Almandumah **5**(1), 256–408 (2018)
12. Hassan, F., Yaqoub, A.: requirements for presenting financial statements under the International Standard (IFRS15) revenue from contracts with customers. J., Account. Finan. Stud. **1**(1), 1–16 (2019)
13. Humaidat, J.: International Financial Reporting Standards Expert. In: Jordan (ed.): International Arab Society of Certified Accountants (2019)
14. Jaber, F.: The Impact of IFRS15 application on reports for egyptian telecom companies. J. Account. Audit. **3**(1), 20–52 (2018)
15. Jordan Telecom Company: Annual Report, Amman, Jordan (2015)
16. Jordan Telecom Company: Annual Report , Amman, Jordan (2016)
17. Jordan Telecom Company: Annual Report, Amman, Jordan (2017)
18. Jordan Telecom Company: Annual Report, Amman, Jordan (2018)
19. Jordan Telecom Company: Annual Report, Amman, Jordan (2019)
20. Jordan Telecom Company: yearbook, Amman, Jordan (2020)
21. Khresat, B.: The impact of the shift to applying IFRS 15 on the recognition of revenue from contracts with customers.[Unpublished Master's Thesis].Middle East University, Amman, Jordan (2017)
22. Khersiat, O.: Impact of the application of IFRS 15 on the profitability of Jordanian telecom companies (case study: Jordan orange - telecom). Int. J. Finan. Res. **12**(2), 308–317 (2021)
23. Mattei, G., Paoloni, N.: Understanding the potential impact of IFRS 15 on the telecommunication listed companies, by the disclosures' study. Int. J. Bus. Manag. **14**(1), 169–179 (2019)
24. Obaid, M.: The impact of accounting conservatism in financial reports and statements on the market value of Jordanian public shareholding companies listed on the Amman Stock Exchange. [Unpublished Master's Thesis]. Zarqa National University, Zarqa, Jordan (2017)
25. Sabrina, S.: The effect of using the accounting conservatism policy on the relevance and reliability of accounting information in Algerian insurance companies: a case study. [Unpulblished Master's Thesis]. Farhat Abbas University, Algeria (2019)
26. Weygandt, J.J., Kieso, D.E., Kimmel, P.D., Trenholm, B., Warren, V., Novak, L.: Accounting Principles, vol. 2. Wiley (2019)
27. Zeghal, D., Lahmar, Z.: The impact of IFRS adoption on accounting conservatism in the European Union. Int. J. Acc. Finan. Rep. **6**(1), 127–160 (2016)

Impact of Marketing Macro Factors on Foreign Investment Inflows: A Case of Jordan

Zakaria Azzam[1(✉)], Khalid Al-Badarneh[1], and Sharifah Ahmad[2]

[1] Marketing Department, Zarqa University, Zarqa, Jordan
{zazzam,20149039}@zu.edu.jo
[2] Marketing Department, University of Business and Technology, Jeddah, Kingdom of Saudi Arabia
sh.ahmad@ubt.edu.sa

Abstract. Generally speaking Foreign Direct Inflows (FDI) are attracted by many marketing macro factors, which are closely affected by technological changes and these factors grouped into five major areas: Economical and Financial, Legal and Political, Infrastructure and technological, and Administrative Structure and Procedural factors. The study aims to find out the impact of marketing macro factors on foreign investment inflows in Jordan. The Study population consist of 55 FDIs' companies operating in Jordan, sample size of 190 respondents working in FDIs' at Free Zones in Jordan are selected for the present study. Main findings of this study indicates that foreign direct investors are enhanced mainly to political stability factor, followed by the Administrative Structure and Procedural system of the public sector adopted by the government of Jordan and supporting services provided. Economical factor of a country also has a positive impact in attracting the foreign direct investments inflows to a country. The study also found that the basic level of the Infrastructure is also important for FDI's. The study reject the null hypothesis and accept the alternative hypothesis since the dependent variable (Attracting foreign direct investment to Jordan) and set of independent variables (economic and financial, legal, political, infrastructure, as well as the administrative structure and procedural aspects) are attracting foreign direct investment (FDI) to Jordan's qualified industrial zones (QIZ) with R value is 0.712 and R^2 value is 0.508 and sig. Value is 0.000which is statistically significant. The main recommendations of this study that Jordan government can encourage foreign direct investors by focusing on maintaining political stability, and enhancing economic growth of the country. Also by improving technical and t level of Infrastructure represented by roads, transportations and facilitating services supporting the operations of their industry, and finally developing transparency in the governmental institutions and offices.

Keywords: Marketing Macro Factors · Foreign Direct Investment · Jordan

1 Introduction

Foreign Direct Investment (FDI) is sought by the majority of developing countries as a way to supplement the level of domestic incentives and to secure economy-wide efficiency gains through the transfer of appropriate technology, management knowledge,

B. A. M. Alareeni and I. Elgedawy (Eds.): *Artificial Intelligence (AI) and Finance*, SSDC 488, pp. 577–589, 2023.
https://doi.org/10.1007/978-3-031-39158-3_54

and business culture, access to foreign markets, increasing exports, and other incentives to enhance their economic growth. FDI is a significant component of the massive private investment that is driving economic growth around the world, particularly in the past two decades. Different incentives and policies have been taken into consideration by policymakers of any government to encourage FDI and ensure their sustainability to enhance domestic economic growth. The struggle for FDI inflows is critical all over the world. The capacity to manage investments, convertible currencies, increased privatization, stock market reforms, greater political stability, and an established legal framework are just a few of the key indicators that foreign private investors look for before making structural changes, policymakers in the host countries must understand how FDI fits into their own objectives—economically, politically, legally, and other necessary changes aimed in encouraging FDI to move into these rapidly developing nations. Given that FDI inflows were among the key factors contributing to the industrial dynamism of these rapidly developing nations, and that they are one of the main sources of cash flowing into their economies, in addition to the many other benefits they receive [1–5] and [6].The main objective of the present study is to identify the main impact of marketing macro factors on foreign direct investments to Jordan.

1.1 Research Problem

The absence of national savings to finance investments is one of the economic problems that developing countries face. As a result, they are constantly in need to encourage direct and indirect investments from abroad. In order to attract FDI to their home cities, developing countries like Jordan must work hard and make their best efforts, as FDI is a significant source of foreign funding. Additionally, they must guarantee that all restrictions are lifted and that there are no obstacles in the face of FDI. Foreign Direct Investment (FDI) is frequently accompanied by high risk factors in developing countries, such as political risk, an ineffective legal system, economic constraints, poor infrastructure, a rudimentary administrative structure, as well as cultural, linguistic, and religious diversity. The decision to enter the global market for investment must be carefully considered due to the potential risks and difficulties that could result from one or more of these factors in the host countries. The study's main challenge now is to comprehend how these marketing macro factors—including infrastructure and technological, administrative structure, and procedural system—play a part in attracting foreign direct investment (FDI) inflows to Jordan as a case study. This study aims to give a broad overview of the impact of marketing macro factors on promotion and attracting FDIs into Jordan.

1.2 Research Hypothesis

The researchers could state the following as the main hypothesis based on the problem statement and the study's objectives:

H0: The Marketing Macro factors (Economic and Financial, Legal, Political, Technical and Infrastructure, and Administrative Structure and Procedural system) have no statistically significant impact (a0.05) on Jordan's ability to encourage foreign direct investment (FDI) inflows.

1.3 Research Model

Based on previously published studied and literature reviewed [1–10] and [11] it was possible to assess the contribution of marketing macro factors in attracting FDI's inflows as shown in Fig. 1.

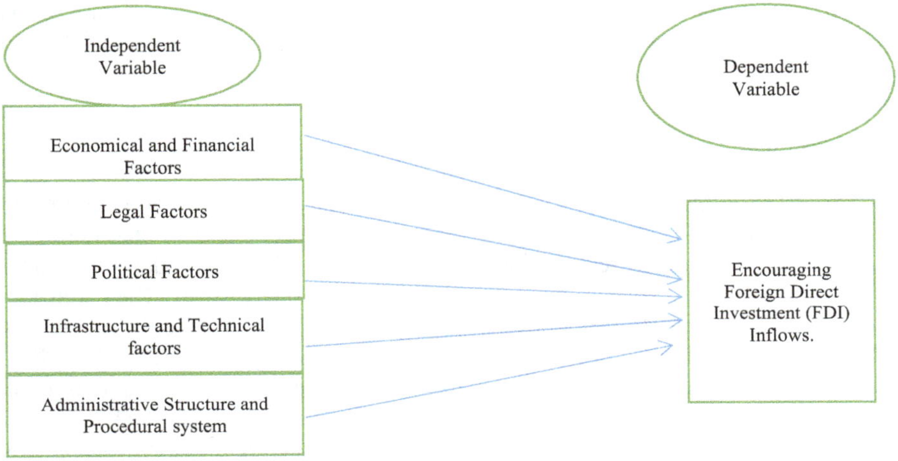

Fig. 1. Research Model.

2 Literature Review

Regardless of the organization's level of strategic planning, the process of situational analysis is one of the key components. In this field, the study concentrates on analyzing the components of the marketing macro factors and identifying the directions of changes that are required to develop the marketing strategy in any organization. The external factors that have an impact on an organization's capacity to maintain or enhance vital exchange operations and prestigious relationships with targeted markets are included in the marketing macro environment. The term "total external factors" refers to all external factors, including opportunities and threats that affect the organization's internal environment and partial environment. Economic, political/legal, social, technical, and natural factors are just a few of the elements particular to the entire marketing microenvironment. All businesses are affected by the same factors, which include the infrastructure, legal system, administrative structure, and government policies. It is important to keep in mind that because these environmental factors are unique to a certain society or nation, they have an impact on every institution that exists there. However, not every institution will necessarily feel the same impact from these environmental factors. On the other hand, marketing macro factors includes those factors that exist in the immediate external environment of the enterprise and that affect the enterprise's ability to serve its markets.These environmental factors are represented by competitors, processors, the public distribution and intermediaries, who are located within a specific market or industry. Accordingly, marketing macro factors affect the enterprise that operates in the concerned market or industry. Its influence does not extend to the rest of the institutions operating in other industries in society. Therefore, the impact of marketing macro factors varies individually or in combination from one industry to another, as each industry is characterized by certain characteristics and requirements that distinguish it from other industries. But for FDI's, it is important to analyze these factors since it affect their operations in targeted market. Literature review related to marketing macro factors are as follows [1–13] and [14].

2.1 Economical and Financial Factors

Among the most important environmental factors affecting business enterprises are economic factors, as FDI's deal with international markets that consist of individual consumers in addition to the purchasing power through which these individuals are able to purchase the goods and services they need and finally the desire to buy. The purchasing power of society members is affected by the available income, price level, savings and the availability of loans to individuals and other economic indicators. Therefore, business establishments, whether large or small, should collect, analyze and interpret data on economic indicators, especially since this data is widely available in all countries. Were the official agencies issue economic data and provide them at a reasonable cost to the institutions. In general, the economic indicators that the FDI's should take into account in the situational analysis include: the economic cycle, gross national product, utilization rate, features of payments, tax and monetary policy, average per capita income, savings levels, retail price levels, and interest rates. Each of these economic indicators and the trends of changes provide an opportunity to the institution or pose a threat. An increase or decrease in the average per capita income will lead to a change in the consumption pattern of categories of goods and services. In addition, all economic indicators directly affect the establishment and development of the organization's marketing strategy. Accordingly, FDI's should predict the locations of the country's economy within the economic cycle, and then set and develop the appropriate marketing strategy for the circumstances of each stage. As the stage of economic prosperity, which is characterized by high growth rates in the economy, low unemployment rate, and an increase in the average per capita income, requires development of a marketing strategy with high promotional and distribution expenses, an increase in product prices, and development of new products in order to achieve marketing goals, while the period of economic recession, which is unique to a decrease in the average income and the increase in unemployment rates require a marketing strategy that is completely different in its elements from the strategy applied during the period of economic prosperity. On the other hand, changes in other economic indicators, such as income, cost of living, the interest rate charged by banks on loans, and patterns of savings and borrowings, directly affect the movement of exchange of goods and services in the market. Accordingly, FDIs should predict the possible changes in these Indicators, up or down, leading to an effective marketing strategy. Jordan government provide the following incentives for FDI's [8, 15, 16] and [17]:

1- Tax incentives and exemptions.
2- Free custom duties on raw materials and exports.
3- Cusom clearance procedures.
4- Flexible and Minimum restrictions on exchange rate and mobility.

2.2 Legal and Political Factors

In every society, political and legal factors influence how businesses market their products and services. The degree to which these factors direct, restrict, or support an enterprise's activities varies across nations. Compared to the intervention of the capitalist state, the state's involvement in the operation of the enterprise is greater in nations that adopt the socialist approach as a political and economic philosophy. In spite of the great and rapid changes that have taken place in the group of Eastern European countries in the past years, it is noticed that the government has increasingly interfered in the conduct of business affairs in many countries of the world, including capitalist countries such as the United States of America and other countries. The addition or deletion of Legislative or legal constraints that could create a strategic constraint or highlight a strategic opportunity for FDIs' willing to penetrate such markets. It is worth noting that there are some organizations that specialize in providing their clients with information about indicators of the stability of governments in the world, and others give probabilities based on expert opinions that allow quantitative comparison of the risks that accompany the activities of certain countries, in addition to political risks, the cultural differences and changes risks are important factors to be considered by FDIS'. It is clear that the early observation of these changes by the FDIS' can avoid many troubles or make it more prepared and equipped to seize opportunities. Some examples which may affect FDI inflows to other countries are [18]:

1- The decision to ban the import of some raw materials or final goods is bound to affect the work of organizations.
2- Environmental protection laws and the practices they impose on the organization.
3- Laws to encourage investment and grant facilities, loans and exemptions....etc.

In order to ensure business enterprises' compliance with the official laws and legislations of the concerned country, there was a need to establish official and unofficial bodies and agencies that became responsible for monitoring business enterprises' decisions and actual practices and evaluating their compliance with official laws and legislations. Most countries tended to issue legislation and laws and to establish official and unofficial bodies, which direct and regulate the activities of business enterprises in a way that preserves the rights of FDI, the consumer and society at the same time. In society, by reviewing the purpose and objectives of establishing the company, funding sources, nature of the industrial sector, type of products and markets, and approving its incorporation. In general, legal and political factors represent a source of opportunities and threats to the enterprise, for example: the issuance of some Arab countries laws to protect national products, and then restrict the import of competing foreign products or impose customs duties on these products, represents an opportunity to enhance the competitive position of products. In contrast, the increase in taxes imposed by the state on enterprises operating in a specific industrial sector represents an obstacle for these enterprises in setting competitive prices for their products in the markets.

Some of the legal issues that the Jordanian government is working on are listed below. [18] and [19]:

1- Human rights.
2- Anti-corruption laws and Anti-corruption policies.
3- Employment and industrial relations.
4- Environmental Protection Low.
5- Combating bribery.
6- Competition policies.

Political changes in the host country have a significant impact on marketing decisions. Laws, governmental bodies, the government itself, and pressure groups that exert influence over or impose restrictions on different organizations and people in a given society make up the political environment. The political climate may play a significant role in determining whether or not it is a good idea to invest in a particular country for a number of reasons: [18, 19] and [20]:

1-The threat of terrorist activity. 2- The stance of the government in the future cannot be guaranteed. 3. Political unrest may lead to a run on the currency: 4- Businesses must invest in their infrastructure and make FDI profitable. 5) Expropriation. 6. Political stability. 7. Transperant System. And 8. Flexible political policies to FDIs.

2.3 Infrastructure and Technical Factors

Most businesses, especially in the last decades of the twentieth century, face the phenomenon of obsolescence of their products and means of production as a result of the rapid and continuous technical progress that occurred in many countries of the world. As technical progress provides market opportunities for business enterprises and at the same time poses a threat to other institutions. The invention of the computer at the beginning of the sixties brought marketing opportunities for information processing institutions, as these institutions were able to provide their services to many small business enterprises, which could not afford to buy computers. It helped produce small electronic accounts with moderate prices at the end of the seventies, and thus, small business enterprises were able to provide the necessary funds to purchase computers and self-process information without resorting to the services of external information processing institutions. There is no doubt that failure to anticipate and respond to technological changes is very costly. Accordingly, FDIs, when analyzing the technical factors in the marketing environment, should determine the size of the expenditure of each of the competing institutions, regardless of its competitive position in the markets or the size of its available resources at the present time, leading to the diagnosis of the technical specifications of the enterprise's products and comparing them with competing products in the market. On the other hand, some organization may not have the technical qualifications that enable them to collect and analyze data on technical factors, since this task falls within the competence of experts and technicians who work within the research department. The people of both the marketing and development research departments should establish close contacts and exchange experiences and information, so as to make sound estimates of the technical factors in the external environment [13].

Related to infrastructure, FDIs are interested to find out all necessary infrastructure facilities that facilitate their operations such as availability of transportations (Roads, Railways Ports, Airports, and Warehouses), convenient geographic location, QIZs availability, advanced Banking system and well trained and skilled labor force [8, 9, 15] and [16].

2.4 Administrative Structure and Procedural Aspects

In the past, natural environmental factors did not have a significant impact on the strategic decisions of business enterprises and their actual practices. However, since the early sixties of this century, voices have grown, whether locally or internationally, calling for the necessity of preserving the cleanliness of the environment in which humans and other living organisms live. And protecting its natural resources from the damages inflicted by modern industries on society, especially in industrially advanced societies. Nowadays, the issue of protecting the environment and its natural resources has become of great importance, as environmental protection groups exert their various pressures and various methods on countries and institutions in order to maintain a clean and healthy environment. Many countries tended to issue laws and impose taxes that limit the actions of institutions and individuals harmful to the environment. Indeed, international organizations themselves tended to hold international conferences aimed at unifying regional and international efforts to preserve and clean the available natural resources in the environment, such as conferences on protecting the ozone layer in the upper atmosphere. In general, the impact of natural factors on strategic marketing decisions crystallized in three main directions such as rational use of natural resources, searching for and developing new sources of energy and reducing or limiting environmental pollution. [8] and [9].

Regarding the procedural system adopted, FDI's wishes to get quick governmental procedures, easy and clear procedures, qualified and well trained expert officers, automated procedures and decentralized operations [8] and [9].

3 Obstacles Facing Foreign Direct Investment (FDI) Inflows

Some things that prevent FDI from entering host countries include: [8–11, 21–23] and [24]:

1- A lack of skilled labor force.
2- Low labor productivity levels.
3- High labor costs.
4- Low Economic growth.
5- Deficit in Balance of payment.
6- Market entry obstacles.
7- Rapid Fluctuations in exchange rates.
8- The host economy's modest size of the market.
9- Bad leadership and unfriendly regulations.
10- Asset ownership restrictions for foreign nationals.
11- High levels of corruption and little transparency.

12- Limited or absent intellectual property rights protection.
13- Weak roads and transportation infrastructure.
14- Unreliable energy and water supplies.
15- Weak infrastructure.
16- Differences in culture.

4 Remedies for FDIs' Obstacles

To encourage foreign investment, nations should design their laws and regulations according to a few fundamental principles. All investors should receive equal treatment, For instance, regardless of a company's home country, the procedure for establishing a local subsidiary should be governed by the same laws. Any disparity in treatment should be based on the size, legal structure, or commercial activity of a company rather than the citizenship of its shareholders. Laws and regulations in a country should be transparent and clear in order to facilitate effective business transactions. The legal system of a nation should offer investors enough security to allow them to feel at ease running and growing their businesses. The government should enact efficient rules that guarantee just protections for the greater good of society and do away with pointless and onerous bureaucracy. Finally, countries can enhance their competitiveness by creating supportive public institutions. The shapes these institutions take will depend on the country and context in which they are created. Yet in all cases supportive institutions are those that provide public officials with incentives to supply the public with useful services at least cost in terms of corruption [19, 22–24] and [25].

5 Methodology

This study involved a survey of Jordan's foreign direct investors (FDI), who are benefiting from the incentives the government provides to attract FDI. The descriptive statistical analysis method is used in this study to describe population phenomenon. The main objective of this study is to describe the information and features of the phenomenon under study through data collection and analysis. The descriptive will also test the study hypothesis to learn more about the phenomenon. The researchers used literature reviews for theoretical topics in addition to the questionnaire to collect the data that were used in the analysis.

5.1 Research Population

The population of the study is the foreign companies operating in Jordan represented by top managers in these companies; which includes Finance Managers, Administrative Managers, Marketing Managers, and Logistics Managers, in addition to the CEOs and Deputy Managers. Statistics indicates that there are around 60 foreign companies working in the Qualified Industrial Zones (QIZ), in different sectors.

5.2 Research Samples

The researcher examined every factory in order to obtain more precise data. 200 of the 220 questionnaires that were distributed had responses, or about 91% of them. Following review, 13 of the 200 returned questionnaires were rejected due to incomplete or incorrect data, making 93.5% of the total valid for analysis. There were 190 respondents in the sample.

5.3 Reliability

Analysis in this study indicates that all study variables have values greater than (0.7), with Administrative Structure and Procedure Aspect having the highest value (0.831) and Economic and Financial Environment having the lowest value (0.735). The Cronbach's Alpha Coefficient for all items is (0.925). Table 1 figures concludes that the research findings can be accepted in this manner [26].

Table1. Value of Cronbach Alpha

Number	Variables	Questions	Cronbach's Alpha
1	Economical and Financial Environment	15	0.735
2	Legal Environment	9	0.791
3	Political Environment	7	0.783
4	Infrastructure	9	0.764
5	Administrative Structure and Procedure Aspects	7	0.831
6	Attracting Foreign Direct Investment (FDI)	8	0.828
		55	0.925

5.4 Personal and Functional Characteristic

The study include 9 characteristics, Table 2 shows the percentage of each characteristic.

Male respondents made up 75.8% of the total, while female respondents made up 24.5%.The sample's average age was between 30 and 40 years old (38.4%). 47.4% of people had a bachelor's degree. Additionally, the managers make up the majority of study participants, at about 37.4%. The average investment period and capital in Qualified Industrial Zones (QIZ) was more than 10 years (57.9%). More than 75% of investments are made with foreign capital (65.8%).

Table 2. Personal and Functional characteristic

Item	Characteristics	Frequencies	Percentages
Gender	Male	144	75.8%
	Female	46	24.5%
	Total	190	100%
Age	Below 30 years	52	27.4%
	30- below 40 years	73	38.4%
	40 – below 50 years	44	23.2%
	Above 50 years	21	11.0%
	Total	190	100%
Level of Education	High school	28	14.7%
	Diploma	35	18.4%
	Bachelor	90	47.4%
	Graduate studies	37	19.5%
	Total	190	100%
Position	CM	10	5%
	Manager	19	10. %
	Logistic Manager	56	29.6%
	Manager	71	37.4%
	finance Manager	21	11%
	Sales Manager	13	7%
	Total	190	100%
Period of the investment	Below 5 years	27	14.2%
	5- below10 years	53	27.9%
	Above 10 years	110	57.9%
	Total	190	100%
Capital in JD	10000–50000 JD	19	10%
	51000–100000 JD	40	21.1%
	101000–200000 JD	23	11.1%
	More than 200000 JD	108	56.8%
	Total	190	100%
Percentage of foreign capital share	Below 25%	12	6.3%
	25%-50%	28	14.7%
	51%-75%	25	13.2%

(*continued*)

Table 2. (*continued*)

Item	Characteristics	Frequencies	Percentages
	More than 75%	125	65.8%
	Total	190	100%

6 Results and Conclusions

Statistical analysis for main hypothesis shown in Table 3.

Table 3. Multiple Regression

Dependent Variable	R	R2	F	DF	SIG	Independent Variable	B	T	SIG
Encouraging foreign direct investment to Jordan	0.712	0.508	37.306	5	0.000	Economical & Financial	0.130	1.355	0.177
				181		Legal	0.048	0.680	0.497
				186		Political	0.270	3.877	0.000
						Infrastructure	0.201	3.253	0.001
						Administrative	0.248	4.106	0.000

The study reject the null hypothesis and accept the alternative hypothesis as Table 3 demonstrates that the dependent variable (Attracting foreign direct investment to Jordan) and set of independent variables (economic and financial, legal, political, infrastructure, as well as the administrative structure and procedural aspects) are attracting foreign direct investment (FDI) to Jordan's qualified industrial zones (QIZ) which is statistically significant ($\alpha \leq 0.05$). More than ($\alpha \leq 0.05$)) indicates a strong relationship between the dependent and independent. [27], $R = 0.712$. Also the $R^2 = 0.508$, which mean the independent variables contribution effect on the dependent variable are about 50.8%. The table also shows the Economical and Financial environment and the legal are not significant at the significant level for the T (0.177, and 0.447).This is supported by the Fact that Foreign direct investment into the Kingdom recorded a significant increase during the first quarter of 2022, reaching JD266.9 million, compared to JD72.2 million for the same period last year, marking a growth rate exceeding 269%, according to Central Bank of Jordan statistics (CBJ) [28].

7 Future Recommendations

Based on the results of the study, Jordan government should:

1- Promote for economic and political stability as a major factor attracting FDI inflow to Jordan.

2- Provide more incentives and exemptions for FDIs
3- To improve and developing Laws and rules which attract FDIs
4- Improving managerial and procedural system adopted for FDIs
5- Enhancing and developing basic infrastructure and supporting facilities required by FDIs.
6- Provide less expensive water, electricity, and energy sources to foreign direct investors industries.
7- It is suitable for other researchers to conduct studies taking into considerations other sectors and other variables in different regions to evaluate the impact of marketing macro and micro factors on FDI Inflows.

References

1. Abdul Mottaleb, K., Kalirajan, K.: Determinants of foreign direct investment in developing countries: a comparative analysis. ASARC WP **13**, 1–27 (2010)
2. Ahmad, M., Khoder, H.: Foreign direct investment and its effect on economic environment – a look into Iraqi investment law. UNCTAD Report. **1**, 146–148 (2010)
3. Aveh, F., Krah, R.: Catalysts and barriers to foreign direct investment in Ghana. Eur. J. Bus. Manage. **5**(20), 61–75 (2013)
4. Azzam, Z., Wadi, M.: Foreign direct investors' reaction towards marketing environment variables faced in Jordan. J. Sci. Hum. Referred Ann. Res. J. **10**(22), 42–47 (2008)
5. Dumon, M.: Top 6 Factors That Drive Investment in China. http://www.investopedia.com/articles/economics/09/factors-drive-investment-in-china.asp. Accessed 1 Feb 2023
6. Birnleitner, H.: Attractiveness of countries for foreign direct investment from the macroeconomic perspective. In: Proceedings of FIKUSZ '14 Symposium for Young Researchers, University of Latvia, pp. 29–40 (2014)
7. Birnleinter, H.: Influence of macro-environmental factors to the process of integrating a foreign business entity, industry science and policy makers for sustainable future. In: Proceedings of the 14th International Conference, Koper, 21–23 November, pp. 392–393 (2013)
8. Khrais, E., Azzam, Z., Assaf, A.: Constrains facing garment industrial sector operating within the qualified industrial zones in Jordan. Zarqa J. Res. Stud. Hum. **10**(2), 82–83 (2010)
9. Khrawish, H., Siam, W.: Determinants of direct foreign investment: evidence from Jordan. BHE Bus. Econ. Horizons J. **1**(2), 67–75 (2010)
10. Kaddumi, T.: Determinant Factors Attracting Foreign Direct Investment to Jordan -An Analytical Study for the Period (1995–2004), 8–15 (2004)
11. Al-Khouri, R., Al-Qudah, K.: Problems facing owners and managers operating in the qualifying industrial zones in Jordan. Jordan J. Bus. Adm. **2**(1), 230–241 (2006)
12. Obalade, T.: The foreign direct investment (fdi) environment in Jordan: a descriptive overview. Acad. Res. Int. J. **5**(4), 228–229 (2014)
13. Rawashdeh, S., Nsour, J., Salameh, R.: Forecasting foreign direct investment in Jordan for the Years (2011–2030). Int. J. Bus. Manage. **6**(1), 139–145 (2011)
14. Bekhet, H., Smadi, R.: Determining the causality relationship among FDI determinants: evidence from Jordan. Int. J. Sustain. Econ. **6**(139), 262–264 (2014)
15. Nuta, A., Nuta, F.: The effectiveness of the tax incentives on foreign direct investment. J. Pub. Adm. Finan. Law **1**(1), 55–56 (2012)
16. Walsh, J., Yu, J.: Determinants of foreign direct investment: a sectoral and institutional approach. Int. Monet. Fund. Rep. **1**, 5–6 (2010)

17. OECD, Benchmark Definition of Foreign Direct Investment, Fourth Edition – ISBN 978-92-64-04573-6 Investment Policy Review: Jordan. 1, 12–30 (2022)
18. Hajzler, C..: Expropriation of Foreign Direct Investments: Sectoral Patterns from 1993 to 2006, University of Otago, Unpublished Economics Conference Discussion Papers, vol. 10, no. 11, pp. 2–10 (2010)
19. Haudi, H., Wijoyo, H., Cahyono, Y.: Analysis of most influential factors to attract foreign direct investment. J. Crit. Rev. **7**(13), 4128–4135 (2021)
20. Shah, M.: The Significance of Infrastructure for FDI Inflow in Developing Countries. J. Life Econ. **3**(1), 1–10 (2014)
21. Mansur, Y.: Overcoming barriers to foreign direct investment in Jordan. Int. Res. Found. f Oman **1**(1), 1–8 (2008)
22. Bakir, A., Alfawwaz, T.: Determinants of foreign direct investment in Jordan. Int. Manag. Rev. **5**(2), 66–80 (2009)
23. Khrais, E., Azzam, Z., Assaf, A.: Constraints facing garment industrial sector operating within the qualified industrial zones in Jordan. Zarqa J. Res. Stud. **10**(2), 82–91 (2010)
24. http://trade.ec.europa.eu/doclib/docs2023/november/tradoc_131004.pdf. Accessed 1 Feb 2023
25. http://www.worldfinance.com/home/special-reports-home/removing-barriers-to-fdi. Accessed 1 Feb 2023
26. Al-Sawaie, K.: Determinants of Jordanian Trade Balance: An ARDL Approach. Zarqa J. Res. Stud. **17**(1), 137–149 (2017)
27. Nunnally, C.: Psychometrics Theory, 2ed edn. Mc Grow Hill's, New York (2018)
28. https://petra.gov.jo/Include/InnerPage.jsp?ID=43566&lang=ar&name=en_news. Accessed 1 Feb 2023

The Impact of Cost Leadership & Product Differentiation on Profitability in the Industrial Companies Listed on the Amman Stock Exchange (ASE)

Ola Khresat[1(✉)], Hamza Asa'ad[1], and Mohammad Kanan[2]

[1] Zarqa University, Zarqa 13110, Jordan
Okhresat@zu.edu.jo
[2] University of Business and Technology, Jeddah, Saudi Arabia
m.kanan@ubt.edu.sa

Abstract. This article aims to demonstrate the impact of cost leadership and product differentiation on profitability in ASE-listed industrial companies. To achieve the research objectives, the descriptive-analytical method is adopted. The research population comprises all (33) ASE-listed industrial companies, exceptfor two companiesdue to the lack of necessary data, making the research sample in the final form (31) companies. Data is collected from the published lists of ASE-listed industrial companies between (2017) and (2021). The appropriate statistical methods are used through the (Smart-PLS) program to analyze the two research independent variables; cost leadership and product differentiation, and profitability as a dependent variable. The findings indicate a statistically significant impact of cost leadership and product differentiation on profitability in its dimensions: return on assets, return on equity, and return on invested capital in the ASE-listed industrial companies, demonstrating that the management policy in the surveyed industrial companies aims to reducing production costs and following a policy of a higher level of quality of goods. The research recommends working on raising the culture of differentiation between all administrative levels of workers in industrial companies and conducting much research work covering other economic sectors such as banks and the services sector.

Keywords: cost leadership · product differentiation · profitability · industrial companies

1 Introduction

The progress of means of production, information technology, communication, and openness of markets, increased productivity, high-quality levels, acceleration of creativity and innovation processes, and the relative ease of entry of new competitors in industries have created the opportunity in shifting the market from the market of exhibitors to the market of buyers. The real forces in the buyer's market are based on customers having opportunities to select and compare among multiple alternatives to satisfy their desires at the lowest cost and on the easiest terms.

B. A. M. Alareeni and I. Elgedawy (Eds.): *Artificial Intelligence (AI) and Finance*, SSDC 488, pp. 590–598, 2023.
https://doi.org/10.1007/978-3-031-39158-3_55

Competition is now regarded as the only method to deal in the market by working to acquire and develop competitive capabilities, allowing organizations to achieve significant results represented in creating new marketing opportunities. The competition also achieves a new competitive breakthrough and a means to create a new vision for the future needed by the organization itself. Among the means of competition is the organization's ability to reduce the costs of its products and control the price [1].

Companies at the regional and global levels seek to maximize profitability, achieve more success and excellence, and reach higher quality in all their activities, whether industrial, commercial, or service in their administrative and financial operations. At the local level, the Jordanian market is characterized by a high degree of competitiveness, which prompts these companies to achieve this high degree of competitiveness with a high level of efficiency and effectiveness reflected in their profitability indicators. Therefore, companies resort to business strategies, including cost leadership and product differentiation, which in turn define their long-term primary goals along with some secondary goals through future action plans that enable them to maximize the profitability of the company.

The cost leadership strategy (CLS) is the strategy in which the company is lower than other companies in the cost of production in the industrial sector through the optimal use of available resources, production with high standards of accuracy, and selling at prices compatible with those in the market. The cost leadership strategy requires effective sales methods related to high-efficiency facilities, continuous tracking of costs to reduce them, strict control of production costs and operating profit margin, and cost reduction for some areas such as research and development, promotional advertising, and others. A company that achieves cost leadership (prices) gains a competitive advantage over competitors or substitute products. The low costs of products allow the company to continue to make profits during the high intensity of competition, as the increase in the market share of the company in light of the implementation of this strategy demonstrates that the company has high negotiating power with suppliers thanks to its ability to buy in large quantities from them [2].

Porter's analysis (1997) of the most significant competitive forces affecting companies' businesses represents a framework used in formulating and developing strategies based on the effects of those forces affecting companies, including threats of alternative products, the threat of entry of expected competitors, the threat of intense competition from current competitors, and the threat of bargaining power for customers, and the threat to the bargaining power of suppliers [3] and [4]. Despite the presence of the five forces impacting the industry and the aforementioned strategy, there are advantages for companies that achieve cost leadership in their production of goods and services.

The five competitive forces influencing the industry are related to the choice of strategy, where selecting the cost leadership strategy or the lowest cost achieves several advantages. For competitors, for example, the company producing at a lower cost will be in a better position in the event of price competition, while for buyers; the company producing at a lower cost will enjoy protection against powerful customers as they cannot bargain for lower prices. On the other hand, for suppliers, the company is safe from suppliers, especially in terms of efficiency, while for potential competitors in the market, the company occupies a privileged position that enables it to reduce the price and face any attack from new competitors. Moreover, concerning the substitute commodity, the company producing it at a lower cost compared to competitors can use lowering prices as a weapon against the substitute commodity [5].

Product differentiation, however, is the strategy of seeking distinction or uniqueness with exceptional characteristics in the domain of industry and is of greater value to the buyer, such as promotional prices and after-sales services. Product differentiation is to make a product or commodity that is distinguished in the market, that is, to provide unique goods or services in the sector by intensifying research and development efforts in the company to ensure the improvement of product quality and quality. It is also possible to achieve distinction for the company either by distinguishing the product or introducing a new production technology, or a distinguished distribution network [6]. With this strategy, the company focuses on the quality of production and its quality, in addition to the quality of the services provided to customers, as the organization aims to compete by the quality of the commodity, and the excellence of the after-sales services it provides, and not by the price.

Profitability represents the organization's ability to produce profits, which is measured by the first-time return on assets (ROA) ratio, as this ratio measures the profitability of the whole project. Profitability is computed by dividing the net profit of the corporation in the income statement by the total assets, i.e. by the total funds invested in the company. It is also measured by the rate of return on equity (ROE). Besides, it is computed by dividing the net income of the company by the total equity [7].

Profitability represents the organization's ability to produce profits, as it is an indicator of the company's efficiency and effectiveness in exploiting the available resources. It also refers to the relationship between the financial returns obtained by the company from its operational activity, and the cost of the investments that it contributed to achieving them [14]. The objective of maximizing profits is one of the basic goals that companies seek to achieve in a way that raises the value of their wealth and the wealth of their shareholders. Achieving profit is the goal that helps companies to continue and survive, support their financial position, increase the wealth of their owners, and enhance their solvency and liquidity. Unlike the losses negatively affecting their financial position, depleting the wealth and rights of the owners, and exposing them to failure, financial hardship, and disasters, which may lead to their liquidation and bankruptcy, maximizing profits is what enables companies to overcome the risks and obligations that stand in their way [7] and [8]. By introducing the concepts of cost leadership, product differentiation, and profitability among companies, the related literature review is provided in the following division.

2 Literature Review

Much ink has been spilled on the effect of applying the strategy of cost leadership and product differentiation on the various components of the financial statements and sectors. Other studies, however, address the effect of implementing the strategy on the qualitative characteristics of the financial statements and accounting information. A study by [6] indicates that the number of factories implementing the cost leadership strategy is (91), with a ratio of (62.8%) of the total number of factories in the research sample, while the factories following the strategy of differentiation are (54) companies, with a ratio of (37.2%). The findings also show that food factories adopt the concept of total quality management (TQM) at a high level, indicating the interest of senior management and their awareness of the concept of total quality management for its significance in raising the level of institutional performance.

In the same context, [9] indicates no effect of following the cost leadership and product differentiation strategies on the Saudi industrial companies' financial performance. The findings attained by [10] show a direct relationship between responsibility accounting and the competitive cost leadership strategy in Sudanese industrial companies. On the other hand, [11] concludes that what distinguishes the least cost strategy is its relationship to the climate of internal control, where the element of internal control activities is the second distinguishing element of all strategies. Another study by [12] indicates that there is anticipation in the cost leadership strategy by implementing the method of Time Driven Activity Based Costing (TDABC) based on time as a primary guide to distribute the cost of resources and outputs to the objectives of cost management, as time is among the most effective and key success factors for the economic unit.

On a parallel line, [13] shows a decrease in the sales volume of the children's products company in a small way, as there is also a decrease in the number of births, affecting the decrease in sales, yet the corporationcan withstand with the implementation of the cost leadership strategy. The results of a study by [14] also indicate that cost leadership and differentiation simultaneously have a positive and significant effect on competitive advantage, while the effect individually is greater than differentiation and cost leadership to competitive advantage. The results of [15] show a fluctuating and low level of a statistically significant effect of the independent research variables: sources of operational cash flow, investment, and financing on the research variable of profitability.

The empirical results of [16] demonstrate that the relationship betweenfuture profitability and sales manipulation is negative for product differentiation strategy, compared to cost leadership strategy. Furthermore, the sales manipulation carried out by companies that have a high percentage of initial inventories and a low percentage of inventories during the year positively impacts profitability in the future. Accordingly, the influence of the differentiation strategy is stronger with the low level of market concentration. With giving a deep insight into the literature review relating to profitability, differentiation strategy, and cost leadership strategy, the research problem is illustrated in the next section.

3 Research Problem

The Jordanian industrial companies face several challenges in general foremost and the administrative challenge in particular as a result of the external environmental changes in light of the difficult economic conditions. Undoubtedly, the low profitability of companies, in general, is the result of a mistake in selecting an appropriate business strategy for the company, as the company follows certain business strategies that require it to resort to third parties to obtain the necessary funding to finance these strategies. Given the significance of industrial companies in the Jordanian business environment and their working conditions in an environment surrounded by political and economic fluctuations that would be reflected in the results of their work in the event of competition for raw materials and the recession period experienced by the companies, the research problem rests in identifying the impact of cost leadership & product differentiation on profitability in the ASE-listed industrial companies in Jordan.

4 Research Hypotheses

Given the literature review and research problem relating to profitability, differentiation strategy, and cost leadership strategy, the following research hypotheses are formulated.

H0: There is no statistically significant impactat the significant level ($\alpha \leq 0.05$) of cost leadership and product differentiation on profitability in theASE-listed industrial companies.

H0.1: There is no statistically significant impact at the significant level ($\alpha \leq 0.05$) of cost leadership on return on assets in theASE-listed industrial companies.

H0.2: There is no statistically significant impact at the significant level ($\alpha \leq 0.05$) of cost leadership on return on equity in the ASE-listed industrial companies.

H0.3: There is no statistically significant impact at the significant level ($\alpha \leq 0.05$) of cost leadership on the capital invested in theASE-listed industrial companies.

H0.4: There is no statistically significant impact at the significant level ($\alpha \leq 0.05$) of product differentiation on the return on assets in the ASE-listed industrial companies.

H0.5: There is no statistically significant impact at the significant level ($\alpha \leq 0.05$) of product differentiation on the return on equity in theASE-listed industrial companies.

H0.6: There is no statistically significant impact at the significant level ($\alpha \leq 0.05$) of product differentiation on return on capital invested inASE-listed industrial companies.

5 Method

5.1 Research Approach

Achieving the research objectives necessitates adopting the analytical-descriptive approach drawn from the study of relevant literature to construct the research theoretical framework, as cost leadership and product differentiation on profitability in Jordanian industrial companies are analyzed between (2017) and (2021).

5.2 Variables Measurement

The article aims to find out the impact of cost leadership on profitability in the ASE-listed industrial companies in Jordan. The article consists of these variables.

Independent Variables:Cost Leadership & Product Differentiation:

Financial ratios: asset turnover ratio, inventory turnover ratio, and operating profit margin are used for the period (2017–2021), and a cluster analysis is conducted to classify these variables.

Dependent Variable: Profitability:

The profitability is measured by its indicators: Return on assets (ROA), return on equity (ROE), and return on invested capital (ROIC) for the period (2017–2021).

5.3 Research Population and Sample

The research population covers ASE-listed industrial companies, as the research sample is limited to (31) ASE-listed industrial companies. The first company, Afaaq Real Estate Investment & Development Holding Company is excluded from the research sample because of beginning its business in 2019. The second company, the Leading Business and Projects Company is excluded from the research sample because the company mentions in its annual book that the main activity of the company is an investment in the capital of other companies, and thus it is not commensurate with the target sample: industry sector. Based on the above-mentioned, the research sample is limited to the remaining (31) industrial companies listed on the Amman Stock Exchange, as it is on (10/25/2022) [17].

5.4 Statistical Analysis of Data

Descriptive Analysis: It is used to provide the main indicators about the research variables using: (mean values, minimum values, maximum values, and standard deviation).

Multicollinearity Test: The Correlation Matrix test is used to explore if there is a problem of multicollinearity among the variables of the regression model.

Cluster Analysis: This analysis aims to classify the independent variables: cost leadership and product differentiation in the current study represented in asset turnover ratio, inventory turnover ratio, and operating profit margin.

Path Analysis: It is done through the Smart-PLS structural modeling program to achieve the research objectives and to answer its hypotheses.

5.5 Cluster Analysis

This analysis aims to classify the independent variables in the current study represented in asset turnover ratio, inventory turnover ratio, and operating profit margin into two classification variables: cost leadership and product differentiation. The results are as shown in the following:The number of variables inserted to be classified is three variables: asset turnover ratio, inventory turnover ratio, and operating profit margin to become two new variables:cost leadership and product differentiation.

The quality of the cluster analysis indicates (60%), which is a good percentage based on the classification of the analysis, as the percentage of the first classified variable (cost leadership) is (82.6%) from the research data. However, the percentage of the second classified variable (product differentiation) is (17.4%) of the research variables, where the number of observations of the first classified variable (cost leadership) is (128) observations, while the number of observations of the second categorized variable (product differentiation) is (27) observations.

5.6 Hypothesis Testing

As shown in Fig. 1, the objective of the structural model is to measure the impact of product differentiation strategy and cost leadership strategy on profitability (asset turnover ratio, inventory turnover ratio, and operating profit margin) in industrial companies listed on the Amman Stock Exchange. To test the research hypotheses, the bootstrapping method is used to test the hypotheses:

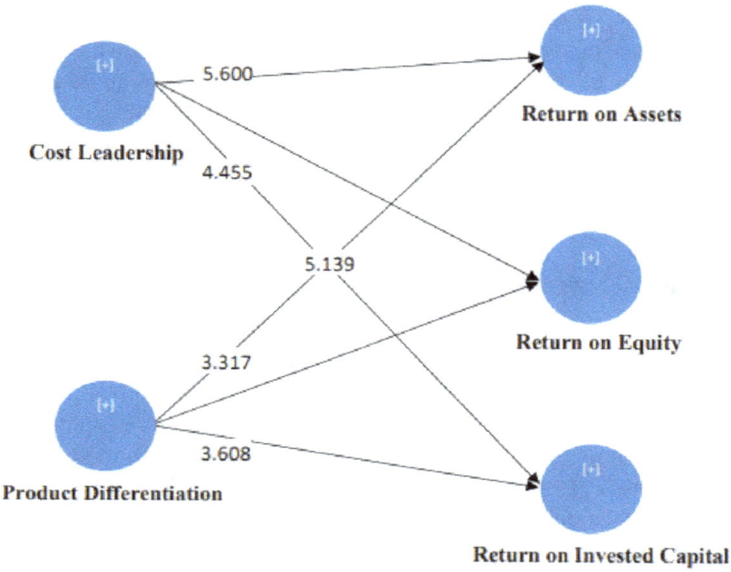

Fig. 1. Bootstrapping Model to Test the Research Hypotheses

To test the research hypotheses, the bootstrapping method is used in (Smart-PLS 3.3) software. Table 1 illustrates the hypotheses testing results of the independent variables on the dependent variable as follows:

As shown in Table 1, the path analysis indicates that the product differentiation strategy has a positive relationship with the return on assets, where (T = 5.485), while (P = 0.000), which is less than a significant level (5%). However, the product differentiation strategy has a positive relationship with the return on equity, where (T = 4.627), while (P = 0.000), which is less than a significant level (5%). Besides, Table 1 shows a positive relationship between the product differentiation strategy and the return on the return on invested capital, where (T = 2.071), while (P = 0.039), which is less than a significant level (5%).

Moreover, the path analysis table indicates that the cost leadership strategy has a positive relationship with the return on assets, where (T = 3.486), while (P = 0.001), which is less than a significant level (5%). Further, the cost leadership strategy has a positive relationship with the return on equity, where (T = 3.189), while (P = 0.002), which is less than a significant level (5%). Likewise, Table 1 shows a positive relationship between the cost leadership strategy and the return on the return on invested capital, where (T = 3.356), while (P = 0.001), which is less than a significant level (5%). As gleaned from the previous analysis, the results recommend rejecting the third main hypothesis in its null form and accepting the alternative main hypothesis stipulating

Table 1. Research Hypotheses Testing Results

Hypothesis	Path	Item	Original Sample	T-Statistics	P-Value
H03.1	C	Differentiation strategy -> return on assets	1.419	5.485	0.000***
H03.2	C	Differentiation strategy -> return on equity	1.074	4.627	0.000***
H03.3	C	Differentiation strategy -> return on invested capital	1.227	5.141	0.000***
H03.4	C	Cost leadership strategy -> return on assets	0.780	3.486	0.001***
H03.5	C	Cost Leadership Strategy -> Return on Equity	0.507	3.189	0.002***
H03.6	C	Cost leadership strategy -> return on invested capital	0.674	3.356	0.001***

"there is a statistically significant impact at the significance level ($5\% \geq \alpha$) of the product differentiation strategy and the cost leadership strategy on profitability (return on assets, return on equity, return on invested capital) in the ASE-listed industrial companies.

6 Conclusion and Recommendations

In a nutshell, testing the hypotheses demonstrates a statistically significant impact at a significant level of 0.05 for product differentiation strategy and cost leadership strategy on profitability measured through (return on assets, return on equity, and return on invested capital). This indicates that the management policy of the industrial companies listed on the Amman Stock Exchange is based on reducing production costs and adopting a policy of a higher level of quality of commodities, which has a positive impact on the profitability of the companies. This result agrees with a study by [16], indicating that product differentiation and cost leadership positively influence competition. However, this result differs from a study by [9], demonstrating no effect of following the cost leadership and product differentiation strategies on the Saudi industrial companies' financial performance. Based on the research results, this research recommends controlling costs, seizing opportunities for economies of scale, raising a culture of differentiation among all administrative levels of workers in industrial companies, and conducting future studies dealing with other economic sectors, such as banks and the service sector.

References

1. Kouider, B.: Determinants of cost leadership strategy and its impact on the competitive forces of the organization. Econ. Notebooks J. **8**(14), 279–294 (2017)
2. Muhammad, I., Othman, M.: The role of intellectual capital in achieving the cost leadership strategy for Sudanese industrial facilities. [Unpublished Master Thesis]. Sudan University of Science and Technology (2018)
3. Porter, E.: Competitive strategy, measuring business excellence (1997)
4. Al-Taiti, M.: Managing change: challenges and strategies for contemporary managers. Dar Al-Hamid for Publishing and Distribution 21, Amman, Jordan (2011)
5. Issa, Y., Omar, A., Ibrahim, B.: Strategic marketing. Dar Al Khaldounia for Publishing and Distribution, Algeria (2012)
6. Khaitan, F.: The impact of business strategy on institutional performance through total quality management: an applied study on food industry companies in the Hashemite Kingdom of Jordan. J. Islamic Univ. Econ. Admin. Stud. **30**(2), 192 (2022)
7. Gibson, H.: Financial Reporting and Analysis: Using Financial Accounting Information, 13 ed. South Western, Cengage Learning, USA (2013)
8. Al-Maaytah, A.: The effect of the degree of stability of bank customer deposits on the banking performance of Jordanian commercial banks for the period (2007–2016). Al-Hussein Bin Talal Univ. J. Res. **4**(1), 1–21 (2018)
9. Al-Zahrani, H.: The impact of business strategies on financial performance: An applied study on Saudi joint stock companies. J. Econ. Admin. Legal Sci. **5**(17), 166–146 (2021)
10. Hamad, M., Awad, T., Fares, A.: The role of responsibility accounting as a mediating variable in the relationship between the value engineering method and the competitive cost leadership strategy. [Unpolished Ph.D. Dissertation]. Sudan University of Science and Technology (2018)
11. Al-Gohar, K., Khresat, O., Al-Beddawi, S.: The distinctive function of the relationship between the elements of internal control and management strategies. The first national and tenth scientific conference. Faculty of Economics, Al-Mustansiriya University Iraq (2010)
12. Al-Eas, S.W.I., Al-Ghabban, T.S.M.: The use of activity based cost time drivine (TDABC) and its role in implementing the cost leadership strategy: an applied research in wasit textile and knitting factory. Special Education **1**(43), 1–16 (2022)
13. Maulana, A., Zain, Z.I., Dhaniswara, Y.: The impact of cost leadership strategy on sales volume in the era of COVID-19 Pandemic: a case study of baby product company. Int. Conf. Ind. Eng. Oper. Manage. 996–1004 (2021)
14. Khresat, O.: Impact of investment decisions on the profits of Jordanian insurance companies listed on the Amman stock exchange. Acad. Accounting Financial Stud. J. **23**(5), 1–20 (2019)
15. Haque, M.G., Munawaroh, M., Sunarsi, D., Baharuddin, A.: Competitive advantage in cost leadership and differentiation of SMEs "Bakoel Zee." Market. Strat. BSD, PINISI Discre. Rev. **4**(2), 277–284 (2021)
16. Weshah, S.R.: Cash flows sources and their impact on profitability within COVID-19 pandemic: case study of industrial companies listed in Amman stock exchange. Int. J. Entrepreneurship **25**(1), 1–7 (2021)
17. Li, C.B., Li, J.J.: Achieving superior financial performance in China: Differentiation, cost leadership, or both. J. Int. Mark. **16**(3), 1–22 (2008)
18. The official website of the Amman Stock Exchange. https://www.ase.com.jo/ar10/25/2022-19:30

The Impact of Financial Inclusion on the Cost of Capital and Net Income of Medium-Sized and Small Enterprises

Riham Alkabbji[1](✉), Ala' Alrazim[2], Issa Ahmad Swiety[1], Tareq Hammad Almubaydeen[1], Mohyedin Hamza[1], and Ruaa BinSaddig[2]

[1] Zarqa University, Zarqa, Jordan
rkabbji@zu.edu.jo

[2] University of Business and Technology, Jeddah, Saudi Arabia

Abstract. The study aimed to explore the impact of the national strategy for financial inclusion on the cost of capital and net income of small and medium-sized enterprises. A number of 50 small-sized companies and 48 medium-sized companies in the governorates of Amman and Zarqa were covered up in the study. Financial inclusion was measured through credit facilities granted to small and medium-sized companies, and the cost of capital and net income was measured by reference to the financial statements of the companies under study. The study used the descriptive analytical method through the E-views software, where the stability of the time series of the study variables was tested and an absence of the unit root of the variables was noticed. After conducting a multiple regression analysis, the study concluded that there was a statistically significant impact of credit facilities on both the cost of capital and net income, but the impact on small enterprises was less than that on medium-sized companies. Therefore, the study recommended the need to increase the financing volume for small companies since these constitute 90% of Jordanian private sector companies.

Keywords: Financial Inclusion · Cost of Capital · Net Income

1 Introduction

The Financial Inclusion Strategy in Jordan aims to help small and medium-sized enterprises (MSEs) to obtain the necessary financing and various banking services at the lowest costs and the best means to maintain the continuity of the enterprises' work due to their significant impact on the local Jordanian and global economy. Small and medium-sized enterprises in Jordan constitute 99% of the private sector companies in Jordan [1], and this is a very high percentage and indicates the importance of these enterprises, and the extent of their impact on the national economy. The provision of soft loans at a low interest rate and easy banking facilities will raise the efficiency and operational capacity of MSEs[2]. Medium-sized and small enterprises is divided into industrial, service and commercial companies, and this study comes to investigate the impact of financial inclusion on the cost of the capital of small and medium-sized enterprises, and consequently its impact on the net income of these.

B. A. M. Alareeni and I. Elgedawy (Eds.): *Artificial Intelligence (AI) and Finance*, SSDC 488, pp. 599–607, 2023.
https://doi.org/10.1007/978-3-031-39158-3_56

2 Literature Review

Many previous studies have examined the impact of financial inclusion on the growth of small and medium-sized enterprises (MSEs). (Corti, 2019) found that financial inclusion affects the growth of small and medium-sized enterprises and all economic businesses [3]. (Suheir et al., 2021) also concluded that financial inclusion has a positive impact on economic performance and contributes to the integration of the informal economy into the formal economy [4]. However, (Ahmed Bouhail, 2021) [5] came up with contradicting results in his study of the impact of financial inclusion on financial sustainability. The researcher concluded the ineffectiveness of financial inclusion on financial sustainability, and this result contradicts with both Suheir and Corti.

Many studies tackled the situation of small and medium-sized enterprises. (Michelle, Swedan2017) [6]Foundd that (MSEs) suffer from financing problems due to the lack of cooperation from the banking sector in a manner that commensurate with the government's orientation to support this category of enterprises. This study and its results study came before the adoption of the Financial Inclusion Strategy approved by Jordan in 2017 which has been implemented since early 2018. Therefore, the current study expects more positive results in the structure of corporate financing of (MSEs). A study also preceding the adoption of the financial inclusion in Jordan was (Rababa, Mohammed, 2014)[7] which showed that the volume of loans granted to small and medium-sized enterprises attributed to total bank loans was very low, and this confirms the validity of the general orientation in the Financial Inclusion Strategy in order to support (MSEs). Moreover, (Morgan, 2018) [8] concluded that financial inclusion achieves financial sustainability and reduces financial default in loan repayment for(MSEs).

This study was conducted to investigate the impact of financial inclusion on the cost of capital for MSEs in Jordan, and the extent of the impact of financial inclusion on the net income of these enterprises.

3 Study Methodology

To answer the study questions, the analytical descriptive approach was used. The objectives of the study are summarized by answering the following questions:

What is the impact of financial inclusion (credit facilities granted) on the cost of capital for small and medium-sized enterprises (MSEs)?

What is the impact of financial inclusion on the net income of (MSEs)?

3.1 Building Hypotheses

Financial Inclusion aims to provide easy sources of financing for small and medium-sized enterprises, to help these companies, develop, grow and expand by achieving an increase in net profit, which enables companies to achieve their long-term goals. According to the theory of Pecking Order Theory, borrowing financing is the second-best option for the company if it does not have the necessary funds to help it continue doing business, as borrowing is a preferred financing option for companies instead of resorting to introducing new partners who will share the enterprise's profits.

Financial Inclusion aims to provide credit facilities at a low interest rate, which reduces the financial difficulties faced by medium-sized and small-sized enterprises as a result of their inability to pay high interest, as the Central Bank, in accordance with the financial inclusion policy, provides loans at an annual interest rate ranging from 2.5% to 3.5% for small enterprises, and a rate ranging from 3% to 5% for medium-sized enterprises (Central Bank report, 2022).

The study hypothesizes that financial inclusion reduces capital cost of institutions, as the interest rate could be low under the Financial Inclusion program.

The definition of capital cost mean that the minimum return expected to be achieved by investors and creditors that happen to be the suppliers of capital elements.

Capital cost is used in making decisions related to investment, as the expected return on investments opportunities must be higher than the capital cost of required to finance the investment. In addition, capital cost might be used in evaluating new and existing investments, since the capital cost is used as a discount rate to get the current value of the future cash flows of investments.

The concept of capital cost is based on the fact that investors in enterprises - either they were shareholders or creditors - expect achieving of minimum returns by their investments and they do not accept less than that expectations. In the case that returns are less than the capital cost, investors may reduce their investments or completely withdraw them. The minimum returns that shareholders expect on an investments is called "cost of equity", while the minimum that creditors expect is called cost of debt.

Since financial inclusion aims to provide credit facilities granted by the banking sector, which are at reduced interest rates with support Central Bank of Jordan, it is expected that these facilities will affect the net profit of enterprises, leading to a reduction in expenses and thus an increase in net profit.

Thus, the hypotheses of the study can be formulated as follows:

H01: There is no significant impact of financial inclusion on the cost of capital in small and medium-sized enterprises (MSEs).

H02: There is no significant impact of financial inclusion on the net profit in MSEs.

3.2 Sources of Information Gathering

Data collection by used two sources, namely:

1. **Primary Sources**: Data on credit facilities granted to small and medium-sized enterprises (financial inclusion) were obtained from the statistics of the Central Bank. The capital of (MSEs), their net profits and the values of debt financing were obtained from the financial statements of these enterprises.
2. **Secondary Sources**: contain books, research, and specialized periodicals.

3.3 Study Population and Sample

The study population consists of all small and medium-sized enterprises in Jordan, and the study sample was represented by 50 small-sized and 47 medium-sized enterprises in the cities of Zarqa and Amman. There was no cooperation from these companies, as they consider that this information is confidential and may not be accessed for any reason. The related quarterly data was collected from 2016 to 2022 covering a period of 8 years.

3.4 Variables

To measure the hypotheses of the study, the following were used:

The independent variable: Financial inclusion represented in credit facilities granted to small and medium-sized enterprises.

Dependent variables:

- **Capital cost:** The capital cost is measured by the following equation:

  WACC= [(E/K) x y] + [(D/K) x b]

  Provided that K = D + E

  WACC= Weighted average cost of capital.

  Y= Cost of equity

  B= Cost of debt

  d= Total debt

  E= Total equity

  K= Total capital

- **Net profit**: It is obtained directly through the Income statement of these enterprises.

3.5 Statistical Methods

The study used the descriptive analytical approach.

3.6 Descriptive Statistics

Descriptive analysis of the study variables through the measures of central tendency.

Table 1 appears the descriptive statistics of the variables for small-sized enterprises as there is a contradiction in the capital size. The highest value is 200,000 dinars with a difference of 196,500 million from the minimum. Hence, the standard deviation is high, as the size of credit facilities will vary depending on the level of variation in capital. It is noted that the highest value of the facilities granted amounted to 870 million dinars, and the lowest value reached 510 million dinars, with a standard deviation of approximately 122 million JD.

The cost of capital, as shown in the table, is somewhat low compared to larger enterprises, as the highest value of the cost of capital reached 5%, which is higher than the interest rate on credit facilities, while the lowest value is equal to the minimum interest rate of 2.5%. This means that the cost of capital is not less than the value of the interest rate granted, therefore credit facilities are good for small enterprises.

It is noted that the net income in the study sample for small enterprises reached the highest value of 45,000 dinars and the lowest value of 32,000 JD.

Table 1. Descriptive stats for 50 SE (2016–2022)

Variable	Mean	Maximum	Minimum	Std. Dev
capital	147689	2000000	3500	96482
Credit facilities	M 707.0769	M870	M510	M122.076
Cost capital	3.45%	5%	2.5%	0.68%
Net income	45000	300000	32000	12530

Table 2 shows the descriptive statistics of the variables related to medium-sized projects, it is noted that the capital values of these enterprises are higher than small-sized projects, and this was justified to grant credit facilities at higher interest rates than those granted to small-sized projects. The table Appears that the value of credit facilities granted ranges from 1236 million dinars to 2124 million dinars with an average financing of 1696 million dinars and a standard deviation of 251.6 million dinars, while the cost of capital ranges from 5% to 7.9%, which is higher than the interest rate on credit facilities granted, ranging from 3.5% to 5%.The net income for medium-sized enterprises ranges from losses of 3,633 dinars to profits exceeding 870,000 JD.

Table 2. Descriptive stats for 87 ME (2016–2022)

Variable	Mean	Maximum	Minimum	Std. Dev
Capital	1230000	2000000	50000	1350620
Credit facilities	M 1696	M2124	M1236	M251.6
Cost capital	6.570612	7.944213	5.225656	0.757781
Net income	101674.3	8706881	-3633.01	120127.19

3.7 Empirical Method

To measure hypotheses used the following equation:

$$CC = \mathrm{a}_0 + \mathrm{B}_1\mathrm{CF} + \in \quad (1)$$

$$\mathrm{NI} = \mathrm{a}_0 + \mathrm{B}_1\mathrm{CF} + \in \quad (2)$$

To abbreviate the variables, the following symbols were used:

CC indicates the dependent variable "capital cost", CF indicates the independent variable "credit facilities", while NI denotes the dependent variable "net income".

3.7.1 Unit Root Test

To determine the non-static properties of variables in time series, (Phillip and Perron) tests were used. Table 3 shows a summary of the results of the time series stability test:

- After testing the first variable financial inclusion, credit facilities (CF), it turns out that there is no unit root, as the value of the PP Test Statistic (5.366-) fell from the critical value (2.964-) at a significant level of 5%, which indicates the stability of the time series of the variable (Credit Facilities).
- After testing the second variable, cost of capital (CC), the absence of the unit root is evident, as the value of the PP Test Statistic (4.756-) was lower than the critical value (2.911-) at a significant level of 5%, which indicates the stability of the time series of the variable (CC).
- After testing the third variable net income (NP), the presence of the root of the unit is evident, where the value of PP Test statistic (1.4566-) was lower than the critical value (2.964-) at a significant level of 5%, which indicates the instability of the time series of the variable (NP). When the first difference was taken and the test was repeated, it turned out that there was no unit root, and the value of PP Test statistic (3.989-) was lower than the critical value (2.8887-) at a significant level of 5%, which indicates the stability of the time series of the variable (NP).

Table 3. Summary of the results of the time series stability test using Phillip-Peron (PP)

Variables	PP testing					
	At the level			The first part		
	Pp Test Statistics	5% critical value	Resolution	Pp Test Statistics	5% critical value	Resolution
CF	5.366-	2.964-	There is no unit root	---------	---------	---------
CC	4.756-	2.911-	There is no unit root	---------	---------	---------
NP	1.4566-	2.964-	There is no unit root	3.989-	2.8887-	There is no unit root

3.7.2. Empirical Results and Analyses

4 Hypothesis Test Results

Part I. Multi-regression analysis for small-sized enterprises:

E-views was used to find the results of the regression model. The results, shown in Table 4, are as follows:

Table 4. The multiple regression test the impact of financial inclusion on the capital cost and on the net income of small-sized enterprises.

CF: Dependent Variable
Method Least Squares
Date: 01/21/23
Sample: 2016-2022
26: Including Observation

Variable	Coefficient	Std.Error	t-statistic	Prob.
NP	718.265	88.2745	10.383	0.0000
CC	0.125	0.1178	5.1489	0.0000
C	19052.150	2693	14.782	0.0000

R-squared	0.332788	Mean dependent var	61425.88
Adjusted R-squared	0.321387	S.D. dependent var	15070.94
S.E. of regression	6369.381	Akaike info criterion	20.44549
Sum squared resid	1.79E+09	Schwarz criterion	20.60142
Log likelihood	-486.6918	Hannan-Quinn criter.	20.50442
F-statistic	73.04610	Durbin-Watson stat	1.817435
Prob (F- Statistic)	0.00021		

Table 5. The results of the multiple regression to test the impact of financial inclusion on the capital cost and on the net income of medium-sized enterprises.

CF: Dependent Variable
Method Least Squares
Date: 01/21/23
Sample: 2016-2022
26: Including Observation

Variable	Coefficient	Std.Error	t-statistic	Prob.
NP	2560.265	88.2745	12.569	0.0023
CC	0.102	0.1178	6.9852	0.0050
C	21568.2365	1265	16.3521	0.0001

R-squared	0.452364	Mean dependent var	61425.88
Adjusted R-squared	0.421102	S.D dependent var	16070.86
S.E. of regression	5268.372	Akaike info criterion	9.44549
Sum squared resid	1.50E+09	Schwarz criterion	9.60155
Log likelihood	-315.6718	Hannan-Quinn criter.	9.50432
F-statistic	75.10240	Durbin-Watson stat	1.427435
Prob (F- Statistic)	0.0000		

Table 4 shows that the value of the statistical significance Prob = 0.0000 is less than the approved level of significance $\alpha = 0.05$, the applied model is significant considering the value of the Fisher test F = 0.00021 and the determination coefficient R = 0.33. This means that 33% of the value of the dependent variables (capital cost and net income) is explained by the volume of credit facilities granted.

H01: There is no significant impact of financial inclusion on the capital cost on small and medium-sized enterprises (MSEs).

Table 4 shows that the value of Prob = 0.000 is less than the significance level of 5%, therefore the null hypothesis is rejected, and the alternative hypothesis is accepted, since there is a statistically significant effect of credit facilities on the cost of capital in small enterprises.

H_{02}: There is no significant impact of financial inclusion on the net profit in MSEs.

Table 4 shows that the value of Proo = 0.000 is less than the significance level of 5%, therefore the null hypothesis is rejected, and the alternative hypothesis is accepted, since there is a statistically significant effect of credit facilities on the net income in small enterprises.

The model can be expressed by the following multiple regression equation:

$$\mathbf{ENP + 718.265 + +0.125\,CC19052.150FC =}$$

Part two: Multiple regression analysis for medium-sized enterprises:
E-views software was used to find the results of the simple regression model. The results, shown in Table 5, are as follows:

Table 5 shows that the value of the statistical significance Prob = 0.0000 is less than the approved level of significance $\alpha = 0.05$. The applied model is significant considering the value of the Fisher test F = 0.00021 and the determination coefficient R = 0.33. This means that 43% of the value of the dependent variables (capital cost and net income) is explained by the volume of credit facilities granted.

H01: There is no significant impact of financial inclusion on the capital cost on small and medium-sized enterprises (MSEs).

Table 5 shows that the value of Proo = 0.050 is less than the significance level of 5%, therefore the null hypothesis is rejected and the alternative hypothesis is accepted, since there is a statistically significant impact of credit facilities on the cost of capital in small enterprises.

H_{02}: There is no significant impact of financial inclusion on the net profit in MSEs.

Table 5 shows that the value of Proo = 0.0023 is less than the significance level of 5%, therefore the null hypothesis is rejected and the alternative hypothesis is accepted, since there is a statistically significant effect of credit facilities on the net income in medium-sized enterprises.

The model can be expressed by the following multiple regression equation:

$$+E\,NP2560.265CC + 0.102 + 221568.2365FC =$$

5 Results

This study tackled the impact of credit facilities on the capital cost and net income of small and medium enterprises. The study reached the following results:

1. Credit facilities, according to the Financial Inclusion Plan for Jordan, have provided sources of financing at low and affordable costs, which affected the high capital cost. The increase in credit facilities subject to the Financial Inclusion Strategy at low interest rates lower than the capital cost caused the capital cost to increase. Therefore, financing by borrowing became better and more affordable than financing using equity.

2. The credit facilities in accordance with the Financial Inclusion Plan for Jordan have provided sources of financing at low and affordable costs, which has reflected on the increase in corporate profits.
3. The impact of medium-sized enterprises on credit facilities was more pronounced than small-sized enterprises because the volume of their capital is larger and the value of loans is greater, thus providing opportunities for more expansion of operational activities.

6 Recommendations

1. The benefits of credit facilities should be less than 2.5 in order for their results to be more meaningful and contribute to the expansion of enterprises since most small-sized enterprises have low capital and cannot borrow large amounts.
2. More support should be provided to small enterprises, especially since they represent more than 90% of private sector companies and because of their significant impact on the national economy.
3. Support must be provided to small and medium-sized enterprises by providing new markets to enable these enterprises to achieve better economic results.

References

1. Jordan enterprise Development Corporation (JEDCO). https://jedco.gov
2. Basha, M.: Measuring the impact of external funding on economic growth in Jordan an econometric. Zarqa J. Res. Stud. Hum. **17**(2), 2017 (2012)
3. Lakuma, C.P., Marty, R., Muhumuza, F.: Financial inclusion and micro, small, and medium enterprises (MSMEs) growth in Uganda. J. Innov. Entrepreneurship **8**(1), 1–20 (2019). https://doi.org/10.1186/s13731-019-0110-2
4. Matouq, S., et al.: Financial inclusion, scientific journal of research and business studies. Helwan University, Faculty of Commerce, vol. 35, no. 1 (2021)
5. Ahmed, B.: Financial inclusion and its impact on financial sustainability (Doctoral dissertation, university center of abdalhafidboussouf-MILA) (2021)
6. Sweidan, M., Al-Rawashdeh, R.M., Rawan Hakam, A.T.W.A.: Factors affecting the capital structure of small and medium-sized enterprises in Jordan field study (2017)
7. Rabbah, M., Muhammad, M.: The determinants of lending to small and medium-sized enterprises by commercial banks in Jordan. Arab J. Manage. **42**(1), 163–176 (2022)
8. Morgan, P.J., Pontines, V.: Financial stability and financial inclusion: the case of SME lending. Singapore Econ. Rev. **63**(01), 111–124 (2018)
9. Myers, S.C., Majluf, N.: Corporate financing and investment decisions when firms have information investors do not have. J. Financ. Econ. **13**, 187–221 (1984)
10. Brealey, R.A., Myers, S.C.: Principles of Corporate Finance, 7th ed., McGraw Hill (2003)
11. Yacoub, I.I., Hamdan, K.H.: Activating the role of banks in financing small and medium-sized enterprises with venture capital in the Iraqi environment. Rechercheset Applications en Finance Islamique (RAFI) **1**(2) (2017)

The Role of Social Media on Marketing Entrepreneurial Projects in Jordan: A Field Study

Mustafa S. Al-Shaikh(✉) and Hussam Rashed Al-bderat

Zarqa University, Zarqa, Jordan
malshiakh@zu.edu.jo

Abstract. The study aimed to identify the role of social media sites in marketing entrepreneurial projects in Jordan. The study was applied to the owners of entrepreneurial projects in Jordan, and it was represented in the dimensions of the independent variable (social media) as follows (Facebook, Twitter, Snap chat, and Instagram), and the dependent variable is the marketing of entrepreneurial projects. The study relied on the analytical descriptive approach in the procedures. It concluded that social media have a statistically significant role in marketing entrepreneurial projects, and that role is considered positive and influential as it contributes to the dissemination and development of entrepreneurial projects. Facebook has the most influential role in marketing entrepreneurial projects. The study recommended the need to diversify the content to suit the nature of commercial advertisements.

Keywords: Social media · Marketing Entrepreneurial Projects · Jordan

1 Introduction

The spread of social media attracted the world to employ them in their various life matters, so that they became an indispensable part of their work, and worked to expand job opportunities, support and market entrepreneurial projects and business incubators like other different sectors. In light of the digital age that we live in today, social media have become a window for everyone who wants to enter them, and have been keen to provide more effective tools that contributed to promotion and direct selling, and helped entrepreneurs to identify new markets that are commensurate with the nature of the products and services they provide to their current and potential customers and increase opportunities awareness of products, promotion sales, and creating an appropriate competitive environment that meets individual needs [1, 18]. From this standpoint, this study came to clarify the role of social media in marketing entrepreneurial projects in Jordan.

2 Problem of the Study

Entrepreneurial projects have become a great necessity in reducing the phenomenon of unemployment, nurturing talent, and encouraging innovation and creativity, but they suffer from high rates of failure, as studies in developed countries indicated that half of

B. A. M. Alareeni and I. Elgedawy (Eds.): *Artificial Intelligence (AI) and Finance*, SSDC 488, pp. 608–617, 2023.
https://doi.org/10.1007/978-3-031-39158-3_57

the entrepreneurial projects that are established do not last more than (19) months, and (20%) of them It remains for several years, but in developing countries the situation is more dangerous [2]. Based on the importance of entrepreneurial projects in Jordanian society, the need to know the role of social networking sites in marketing entrepreneurial projects has emerged. In order to be able to do that, it is necessary to answer what is the role of social media in marketing entrepreneurial projects in Jordan? Through this question, we ask the following questions: What is the role of Facebook, Twitter, Snap chat and Instagram in marketing entrepreneurial projects?

3 Objectives of the Study

This study worked to achieve its objectives by identifying the role of Facebook, Twitter, Snap chat and Instagram in marketing entrepreneurial projects.

4 Hypotheses of the Study

Main hypothesis HO: There is no statistically significant role at the level ($0.05 \geq \alpha$) for social media (Facebook, Twitter, Snap chat, Instagram) in marketing entrepreneurial projects. Among them are the following sub-hypotheses:

Ho1: There is no statistically significant role at the level ($0.05 \geq \alpha$) for Facebook in the marketing of entrepreneurial projects.

Ho2: There is no statistically significant role at the level ($0.05 \geq \alpha$) for Twitter in the marketing of entrepreneurial projects.

Ho3: There is no statistically significant role at the level ($0.05 \geq \alpha$) for Snapchat in marketing entrepreneurial projects.

Ho4: There is no statistically significant role at the level ($0.05 \geq \alpha$) for Instagram in the marketing of entrepreneurial projects.

5 Study Model

In light of the hypotheses formulated, the study variables were formulated according to Fig. 1.

6 Theoretical Framework

The rapid growth in the use of social media has led to a change in the way organizations interact with customers, as they share information related to the product and its specifications and interact with their comments. From here, these organizations began to focus on marketing plans that enhance their increasing use of these sites. Social media has become one of the main communication channels for companies [6, 17].

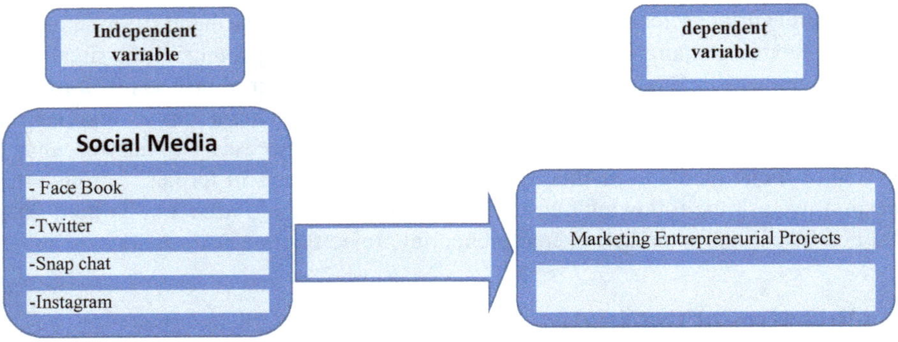

Fig. 1. Study Model. Source: Researchers based on a study [3–5].

6.1 Social Media

They are sites that allow their users to upload a profile picture and add multimedia content so that it appears to the user's friends only who have been selected and approved for their friend requests in advance [7]. [8] defined social media as those services that allow users to share their activities and make friends, as these sites provide the ability to share files and photos, exchange video clips, create electronic blogs, conduct instant conversations and send messages. The number of users of social networking sites in the world is 4 billion users, with the number of Facebook users reaching 2.8 billion, Instagram one billion, Twitter 330 million, and Snapchat 229 million [9]. Figure 2 shows the number of users of social media in the world.

International numbers refer to the large use of citizens in Jordan on social media for publishing and exchanging news, bringing their number to about 10.9 million accounts. According to international figures, Hootsuite reported that most Jordanians are present in the world of social media on Facebook, which is the most used site in Jordan and the world. International figures also showed that Facebook accounted for 46%, and Instagram, the number of users in Jordan reached about 2.10 million accounts, because it allows taking pictures and sharing them with friends, and Snap chat started with 2.15 accounts in Jordan, and estimates showed that Twitter is one of the sites with posts The miniature account, where the number of Jordanian accounts using this site has reached about 743.5 accounts, and this social site allows its users to write tweets called (tweet), which is a short text message with a certain number of characters. Figure 3 shows the number of user accounts for social networking sites in the Kingdom.

6.2 Pioneering Projects

At the beginning of the 1990s, the term "entrepreneurship" appeared, and entrepreneurial work was no longer confined to the contractor or organizer. Rather, there were many young men and women who established their own small entrepreneurial projects and were able to transform these small projects into large companies [10]. Entrepreneurial projects are defined as projects with creative values and creative ideas to create a new project through which new products are launched or existing products are developed and

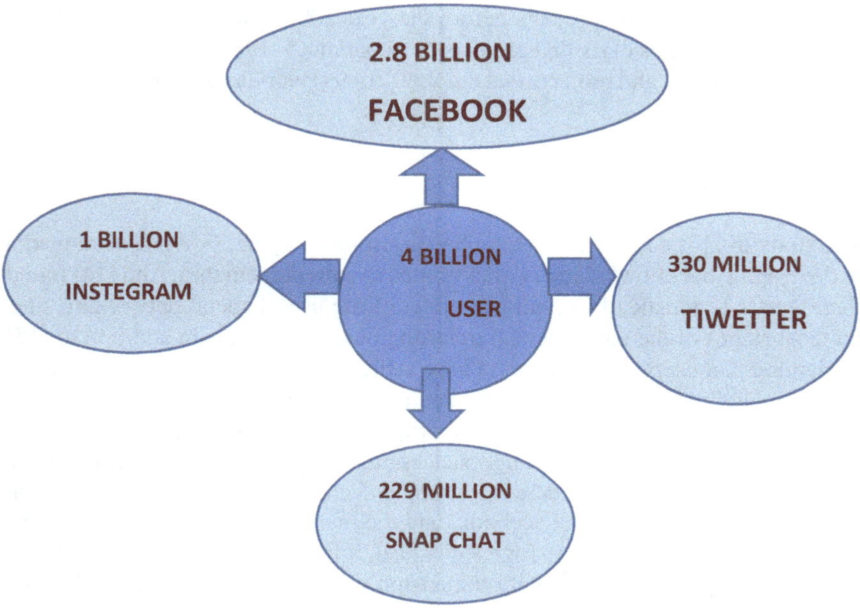

Fig. 2. The number of users of social networking sites / from the design of the current study

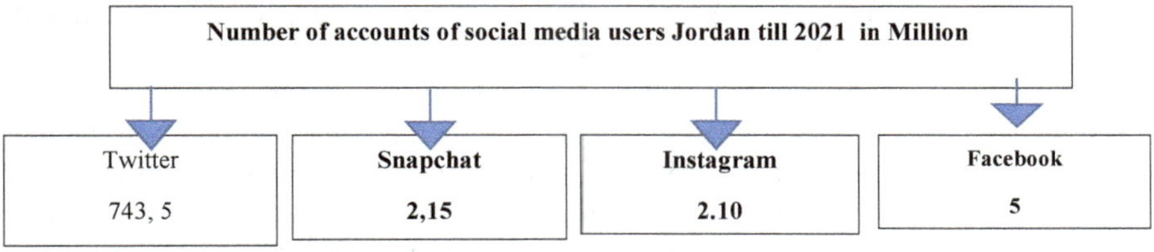

Fig. 3. The number of user accounts for social media sites in Jordan. Source: The current study design

improved in an innovative way that contributes to bringing about change and progress in these projects so that they can transfer economic resources from a low level of production to a high level of production [11]. The current study considers entrepreneurial projects as projects that differ from familiar projects, regardless of their type and size, and present products and ideas in a new manner within the few resources available to reach the goals that their owners aspire to and obtain material returns in a record short time and bear many risks. It is known that entrepreneurial projects include small and medium enterprises, and there are differences between them, whether by definition, characteristics, or advantages in terms of capital size, number of employees, management, and revenues. The Department of Statistics defined small projects as those projects that contain a number of labor ranging between (1–4) workers. As for medium projects, the Department of Statistics defined them as those projects that employ a number of workers whose number ranges between (5–19) worker. This is the definition of the Department of Statistics for pioneering projects before the year 2000, but after the year 2000, the Department of Statistics classified projects that contain (5–19) workers, which are considered small projects, and projects that employ (20–99) workers are considered medium projects [12].

7 Previous Studies

The study of [13] concluded that the Saudi girl relies by 68.33% on social media in developing her knowledge of topics related to entrepreneurship. And [14] found that there is a statistically significant effect of the use of social networking sites on most stages of the consumer's purchasing decision. This is in addition to [15] explaining that users of social media have achieved many benefits by using social media, including enhancing their marketing and social activities, and that these sites are considered one of the most important tools that stimulate behavior, especially if they are exploited through successful strategies and specific methods. And [16] he concluded that there is a strong and clear relationship between social networking sites, especially Facebook, and the marketing of handmade products to women staying at home. As for a study [19], it found that there is a statistically significant effect of the credibility dimension on the purchase intention through social media. As for [20], it was shown that there is a statistically significant positive effect of marketing through social networking sites on the brand image on customer satisfaction. In addition, a study [21] concluded that companies should adopt social media in their work and focus their efforts on increasing the effectiveness of their content on social media. As for a study [22], the study revealed that social networking sites have a significant impact on changing the purchase intention of the brands offered by the customer making a comparison between them. [23] indicated that the lack of experience in accounting matters and the infrastructure for applying accounting standards and not keeping records of entrepreneurial projects are the reasons for the failure and continuation of entrepreneurial projects. A study [24] showed that entrepreneurial projects play a major role in reducing poverty, increasing domestic product, employing workers, and overcoming the problem of poverty.

8 Methodology

8.1 Population and Samples

The study population consisted of owners of entrepreneurial projects of various classifications (small and medium) all over the Kingdom. As for the study sample, it consisted as follows: 40% in the central region, 40% in the northern region, and 20% in the southern region. The distribution ratios were based on population density in all regions. (480) electronic questionnaires were distributed. The researchers were unable to retrieve (77) questionnaires, as the percentage of recovery of questionnaires was approximately (84%), which is an acceptable percentage. The researchers relied on the simple random method of distributing the questionnaires, as it is characterized by the fact that each member of the study population has an equal opportunity to participate in the study [24, 25].

8.2 Cronbach Alpha Coefficient

The study showed that the Cronbach alpha coefficients for the items of the variables ranged between (0.770 - 0.869), [31] and all of them met the stability condition by exceeding the value of (0.7). This indicates that all the items of the questionnaire are stable. Paragraphs: "Marketing Entrepreneurial Projects" variable were considered the most stable, with Cronbach's alpha coefficient: (0.869), while the variable "The Role of Social Media in Marketing Entrepreneurial Projects" was the least stable with a coefficient of (0.770), and with regard to the overall performance of the paragraphs, it was the stability coefficient reached a high value, with a coefficient of (0.921).

9 Test the Hypotheses

Multiple linear regression analysis was used to prove the main hypothesis and to prove the ratio of the explanation of the independent variable together on the dependent variable, and the value of F and its significance were adopted to verify the proof of the hypothesis [33].

9.1 Testing the Main Hypothesis HO

There is no statistically significant role at the level ($0.05 \geq \alpha$) for social media (Facebook, Twitter, Snap chat, Instagram) on the marketing of entrepreneurial projects. To test the hypothesis, multiple linear regression analysis was used, and Table 1 shows the results.

Table 1. Results of multiple linear regression analysis to test the main hypothesis

Dependent Variable	R	R^2	F	DF	Sig of Model	Independents Variable	B	T	Sig
Marketing entrepreneurial projects	0.578	0.334	49.924	4	0.000	Face book	0.295	6.578	0.000
				398		Twitter	−0.48	−1.216	0.225
				402		Instagram	0.161	3.339	0.001
						Snap Chat	0.202	4.557	0.000

The main hypothesis was tested at the level of significance ($a \leq 0.05$), and the results were as follows [32]:

- The value of the tabular F is extracted to compare it with the value of the variation analysis of the calculated F test, in order to ensure the indicative level and it became clear that the value of the calculated (49.924) is greater than its tabular value (2.38), and it proves the morale of the model used in the indicative analysis.
- The level of the connotation of the test has been compared to the level of significance in the study; To show the existence of a statistical impact on the dependent variable, and schedule 1 shows that the level of significance for F has reached (0,000), which

is less than the level of the indication approved in the study and its value (0.05), and from it we reject the main hypothesis and accept the alternative, which states "there is a role with statistical significance at the level ($0.05 \geq \alpha$) for social media (Facebook, Twitter, Snap chat, Instagram) to market entrepreneurial projects.

- The value of R is indicating the correlation between the independent and dependent variable; It indicates in schedule 14 with the existence of a correlation between the independent variable and the dependent variable, it shows that the value of the correlation coefficient = 0.578, and it indicates a strong positive correlation relationship between the independent variables combined: for social media (Facebook, Twitter, Snap chat, Instagram) and the dependent variable (Marketing entrepreneurial projects) because they are greater than (0.5) [26, 29].
- The coefficient of determination (0.334) was the value of the R^2, which indicates an explanatory ability to independent variables combined: (Facebook, Twitter, Snap chat, Instagram) of the dependent variable (marketing entrepreneurial projects), and it indicates that independent variables explain (33.4%) from achieve the marketing of entrepreneurial projects.
- The fixed B value of (1.711) is necessary to build the decline equation for variables. In addition, some review of the above results shows that the treatment of the decline of independent variables with the follower came as follows:
- It is also evident from the table, that the value of the significance level of independent variable (Facebook, Snap chat, Instagram) is statistically significant at the level of indicative level (0.05), it has achieved levels less than the level of significance approved in the test and adult (0.05), which is on concerns (0.000, 0.001 0.000).
- Schedule 1 results showed that the Twitter variable has no statistical role in marketing entrepreneurial projects, because the level of significance of the variable is greater than (0.05) which is equal (0.225).

10 Discussions

Social media have a statistically significant role in marketing entrepreneurial projects, and this role is considered positive and influential as it contributes to the dissemination and development of entrepreneurial projects and increase their sales volume. Facebook, Twitter, Snap chat, and Instagram have a strong positive correlation with the marketing of entrepreneurial projects so that increasing advertising content and relying more on social media will be positively reflected in the level of marketing entrepreneurial projects. In addition to social media, it contributes to achieving and implementing the marketing of entrepreneurial projects in a better way, as it contributes (33.4%) to the marketing of projects. The researcher attributes this result to the high reliance of entrepreneurial companies on social media in the marketing of companies, to increase their pioneers and spread significantly among a segment of people, as all individuals of different demographic characteristics use social media. The result agreed with a study [13, 30] that social media has a positive role in entrepreneurship among female students at Taifi University in Saudi Arabia. I also agreed with Aqraa [15], which showed that users of social media with the aim of promoting their products through marketing campaigns have achieved many benefits, including enhancing their social marketing activities, and that these sites are considered one of the most important tools that stimulate behavior, especially if they are exploited through successful strategies and specific methods.. It also agreed with the study [27], that there is a statistically significant effect and that social networking sites have a major role in developing electronic marketing activity. The result agreed with a study [28, 34] that social networking sites provide greater access to the market and manage customer relations.

11 Recommendations

- Drafting and designing electronic advertisements in a manner commensurate with the nature of Twitter users and its characteristics, so that reliance is greater on text formulations.
- Seeking help from influencers and celebrities from various fields because they have large numbers of followers.
- Increasing the reliance of entrepreneurial projects on social media in marketing, especially Facebook because of its greater reach.
- Diversification in content and marketing tools and the nature of commercial advertisements must be compatible with the social media.
- Determine the target group in the marketing process before choosing a social media in marketing.
- Using Twitter to market entrepreneurial projects only when targeting the nature of its pioneers, when targeting politicians, intellectuals, and the class of society that prefers to use Twitter.
- Using paid advertisements on social networking sites because of their rapid spread and reaching the largest possible number of individuals.
- Not relying on a single social media to market entrepreneurial projects, and diversifying so that commercial advertisements and content are compatible with the nature of the pioneers of each site.

References

1. Al-Najjar, F., Zahra, Z.: Frame processing frameworks on Facebook and Twitter. Arab J. Media Commun. **23**, 376–331 (2020)
2. Barba-Sanchez, V., Atienza-Sahuquillo, C.: Entrepreneurial motivation and self employment: evidence from expectancy theory. Int. Entrepreneurship Manage. J., **13**(1) (2017)
3. Qaboua, S. A.: The effect of using social networks on knowing, acquiring, and retaining customers. Master Thesis, Faculty of Graduate Studies, Middle East University, Jordan (2016)
4. Idan, A.: The impact of the use of social networks on the performance of customer relationship management. Master's Thesis, Faculty of Graduate Studies, Middle East University, Jordan (2015)
5. Mohamed, M.: The reality of NGOs in the Gaza strip using social networks to enhance their relationship with the public. Master Thesis, Faculty of Graduate Studies, Islamic University of Gaza, Palestine (2017)
6. Basit, R. G.: The Impact of Social Media Marketing on Purchase Intention, Scientific Journal of Economics and Commerce, Faculty of Commerce, Ain Shams University, vol. 1, pp. 441–486, Egypt (2020)

7. Boyd, D. M.: Social Network Sites: Definition, History, Case Study, University of California – Berkeley, Dmb@ischool of information (2017)
8. Hatimi, H.M.: Public Relations and Social Networks. Osama House for Publishing and Distribution, Jordan (2015)
9. Dandan, A.: International Relations in the Age of Digital Technology. Publisher, Academic Book Center, Algeria (2021)
10. Abu Fara, Y.: E-marketing Strategy, A Behavioral Approach. Dar Al-Mustaqbal for Publication and Distribution, Jordan (2017)
11. Cokez, L.: Marketing Through Social Media, Translated by: Ahmed Al-Moazi. Dar Al-Fajr for Publishing and Distribution, Cairo, Egypt (2016)
12. Ministry of Planning and International Cooperation: Executive Development Program, Jordan (2018)
13. Abu Sunna, N.H.: The dependence of the saudi girl on social networks in developing her knowledge towards entrepreneurship. J. Media Res., Al-Azhar Univ. Egypt **25**, 758–784 (2020)
14. Benyahia, M.: The role of social networking sites in influencing the purchasing decision. J. Econ. Reforms Integr. Global Econ. **14**(3), 1–14 (2020)
15. Laraqa, A.: The role of social marketing campaigns through social networking sites in raising awareness of consumer behavior. Al-Bashaer Econ. J. **1**, 1–15 (2020)
16. Makhlouf, A, M.: The Role of Social Networking Sites in Marketing Handmade Products to Women Staying at Home, A field Study on a Sample of a Group on Facebook, Master's Thesis, Faculty of Humanities and Social Sciences, University of Mohamed Kheidar Biskra, Algeria (2019)
17. Marolt, M., Zimmermann, H.D., Pucihar, A.: Social media use and business performance in SMEs: the mediating roles of relational social commerce capability and competitive advantage. Sustainability **14**(22), 15029 (2022)
18. Nam, J., Kang, M.J.: Social media marketing strategy. Int. J. Internet, Broadcasting Commun. **14**(1), 219–223 (2022)
19. Sanad, N., Abdel Qader, A.: The Impact of Marketing Using Social Media on Purchase Intention, A field Study in the Fashion Sector in Jordan, Master's Thesis, Faculty of Graduate Studies, Middle East University, Jordan (2019)
20. Al-Sharif, A.: The Impact of Social Media Marketing on Customer Satisfaction Through Brand Image, Master Thesis, Middle East University, Jordan (2017)
21. Eagles, H., Bila, B., Aksmari, A., Al-Zayadat, M.A.: The impact of marketing using social media on purchase intent in Jordan. Jordanian J. Bus. Admin. **12**(3), 28 (2016)
22. Sanyang, L.: The impact social media advertising on consumer behavioral intention toward fashion and luxury brands. Arabian. J. Bus. Manage. Rev. **8**(5) (2018)
23. Sallem, N.M., Nasir, N.E.M., Nori, W.M.N.W.M., Kassim, C.K.H.C.K.: Small and medium enterprises: critical problems and possible solutions. Int. Bus. Manag. **11**(1), 47–52 (2017)
24. Shahwan, Y.: The mediating effect of investment decisions and financing decisions on the influence of capital structure against corporate performance: evidence from Jordanian listed commercial banks. Acad. Account. Financ. Stud. J. **22**(6), 1–20 (2018)
25. Sekaran, U.: Research Methods for Business, 4th edn. John Willey & Sons Inc., New York (2003)
26. Cohen, J.W.: Statistical Power Analysis for the Behavioral Science. (2nd edition) (1988)
27. Lahul, F.: Social networking sites and their role in updating the basics of electronic marketing. J. Islamic Law Sci. Zayan Ashour Univ. Djelfa, 288–300 (2016)
28. Jagongo, A.: The social media and entrepreneurship growth (A new business communication paradigm among SMES in Nairobi). Int. J. Hum. Soc. Sci. **3**(10). (2013)

29. Imran, M., Ali, S., Shahwan, Y., Zhang, J., Al-Swiety, I.A.: Analyzing the effects of renewable and nonrenewable energy usage and technological innovation on environmental sustainability: evidence from QUAD economies. Sustainability **14**(23), 15552 (2022)
30. Al-Shaikh, M.S., Khanfar, I.A.A.: Delivery service via electronic applications and its impact on customers satisfaction at retail stores in Amman City/Jordan. In: Hamdan, A., Shoaib, H.M., Alareeni, B., Hamdan, R. (eds.) The Implementation of Smart Technologies for Business Success and Sustainability: During COVID-19 Crises in Developing Countries, pp. 827–837. Springer International Publishing, Cham (2023). https://doi.org/10.1007/978-3-031-10212-7_68
31. Shahwan, Y.: The effect of OECD principles of corporate governance on social responsibility accounting. Int. J. Innov. Creativity Change **11**(2), 418–431 (2020)
32. Khanfar, I.A.A., Al-Shaikh, M.S., Al-Abed, M.: The Effect of the Pharmaceutical Marketing Mix Elements During Corona Virus Pandemic on the Decision of the Prescription for Foreign-Made Drugs. In: Hamdan, A., Shoaib, H.M., Alareeni, B., Hamdan, R. (eds.) The Implementation of Smart Technologies for Business Success and Sustainability: During COVID-19 Crises in Developing Countries, pp. 49–64. Springer International Publishing, Cham (2023). https://doi.org/10.1007/978-3-031-10212-7_5
33. Shahwan, Y., Abdel-hamid, O.: The impact of social responsibility disclosure in reducing the risk in Jordanian commercial banks. Res. World Econ. **11**(3), 142–150 (2020)
34. Al-Shaikh, M.S., Al Sadi, A., Azzam, Z., Moh'd Al Muala, A., Alsaraireh, A. S.: The effect of electronic marketing tools on customer satisfaction with electronic services in Jordanian Islamic banks (A case study of the Islamic International Arab Bank in Amman City–Jordan). In: 2021 22nd International Arab Conference on Information Technology (ACIT) (pp. 1–12). IEEE. (2021, December)

Radical Innovation Leads to Good Future - A Focus Group Study Using Cluster Analysis

Venkatesh S. Amin, N. Abhishek(✉), Ujwala Kambali, S. Sagar, Swarn G. Kanchan, Prasad Mahale, and A. K. Anish

Srinivas University, Mangalore 575001, India
abhishekalmighty93@gmail.com

Abstract. Markets get disrupted due to technology, innovation, and new ideas. New ideas change the way we live, work, and conduct our businesses. Business in general is decided by the shelf life of a product. Products that have less shelf life do denote that their parent companies have a smaller company life cycle. Disruptions are a must and disruption leads to the culling of products and services by a parent company. Disruptions are critical to deciding world dominance. This study is done in by questioning people if they feel disruptions are an important component of growth in multiple areas. The areas that we observe from engineering or genetic engineering or development in robotics or AI are a questions that we need to answer and accept the disruptive innovations that can change the way we live. Initially, Electricity was a disruptive inventions as was the telecommunications revolution. Indeed it is for us to understand and accept the new and move to the next level; that is the way forward. A disruption will lead to disturbance, and it should be sustainable for it to stay for a century or a millennium to create a greater good.

Keywords: Innovation · Cluster analysis · Disruptive innovation · climate change · landfills · DNA modifications · Gene editing

1 Introduction

People think in groups, and when a large number of people feel the same way about something, it creates a wave of human thought. This thought can change the way people make decisions in life by accepting or rejecting innovations that are disruptive and long-lasting. People have their own ways of choosing and understanding changes that happen [1]. A semi-structured interview gives us an idea of whether people are aware of these changes that are happening across the globe. Hence, the clustering is hierarchical and is characterised by a design that looks like a tree structure [2]. This method can be divisive, agglomerative, or both. Each object is treated as a separate cluster unit in the agglomerative concept of clustering. Clusters are usually formed by grouping objects (not necessarily) into larger and larger clusters. As a process, this is continued until all objects are grouped and formed into a single, larger cluster. Agglomerative methods are commonly used in marketing research [3, 4]. The primary goal of the research is

B. A. M. Alareeni and I. Elgedawy (Eds.): *Artificial Intelligence (AI) and Finance*, SSDC 488, pp. 618–630, 2023.
https://doi.org/10.1007/978-3-031-39158-3_58

to determine how focus groups might produce intellectual clusters. Being aware of and adhering to current trends and disruptions, as well as understanding the causes of issues in disruptions, how people perceive them, and whether or not they cluster together or stand alone.

2 Literature Review

Cluster analysis makes all the difference, and it can be Innovation, cluster analysis, and disruptive innovation should lead to new products and new products in new markets [1]. Climate change is irreversible and is a constant challenge for future generations, and its code red if it disrupts the future of the next generation [2]. Landfills made of clothes and a few non-biodegradable materials will emit methane gas and can be hazardous to people and the environment at large [3]. DNA modifications can go a long way in the human history of evolution, and they can speed up the evolution process for the greater good [4]. Gene editing is the process of making changes to an organism's DNA, and this leads to disruptions in the field of treating patients whose DNA creates challenges; hence, this research is at the centre of it all [5].

Desalination of seawater can be a sustainable disruption and is the need of the day [6]. Each one of us thinks in different ways, and our thinking is also shaped by our surroundings and their changes [7]. Indeed, thinking in groups and clusters makes a difference; biologically, we are different and have different body shapes and structures, but we may have similar minds, and this cluster analysis is all about minds in this study [8].

Disruption in marketing is essential, as disruptions can lead to new ways of looking at things for groups of people [9]. Hierarchical clusters make a difference in terms of making better surveys and understanding groups of people [10]. The power of cluster analysis is critical, as it enhances things at a faster pace [11]. New products need acceptance from people, and new products help people overcome challenges, and social networks help innovation and new product development [12, 14].

Organizations should be creative and accept innovation as a means of development, and they should adapt and adopt this culture or accept disruptions and adapt to the changing situations [15]. Certain human behaviors, such as addictions, are very strange, and disruptions should lead to behaviour change, and behaviour change is critical to people's advancement and upgradation; thus, disruptions should lead to development, but not addictive behavior, bad behavior, or degradation of human feelings, living, or character [16, 17, 18].

Consumers keep changing, and consumer research should maximise the mindsets of people and make them stretch for better performance [19]. Therefore, managing competence and organisational capability is the key to growing the acceptability of disruptive innovations that can bring sustainability [20, 21].

A revolutionary integrated platform that allows for the high-quality registration of hundreds of channels and has successfully addressed the key problems impeding the next generation of invasive brain-computer interface development [22]. Although Kepler-452b is still unknown, there is always a chance that life could exist there [23]. The study discusses technology where pigs are used as a platform for animal biotechnology that aims to produce human organs when the only treatment option left for patients with advanced organ failure is organ transplantation [24].

The current developments in the creation of programmable nuclease-based treatments and the capacity to precisely alter the genomes of eukaryotic cells have been greatly enhanced by recent developments in genome editing methods based on programmable nucleases [25]. With growing concern about the ecological sustainability of economic and environmental development, it is vital to create a class of green plastics to replace synthetic plastics. The functionalized use of graphene and cellulose, a novel green plastic, has been examined in this study [26–28].

Additionally, it provides a description of recent technological advancements that are being studied and developed with a view to potential desalination applications [29]. A comprehensive overview of the state-of-the-art in desalination, demonstrating the fundamental concepts at work and providing data on how each technique has spread, including those based on solar energy or water freezing [30]. To analyse the changes that are occurring and that are necessary components for growth, a strengths and weaknesses analysis should be performed.

3 Objective

Based on the literature review and research gap following objective was framed.

1. To analyse the disruptive innovations and their influence on formation of new society.

4 Research Agenda

The idea is to understand how we can make a few clusters, even if they stay in different locations or come from different domains of business or earnings. This research is limited to the Mangalore Taluk of the Dakshina Kannada District of Karnataka. The overall challenge is to understand how similar their thoughts are across various domains. The agenda is to convey disruptions that are currently being tested around the world, or perhaps a few that are already operational, so that the research is known whether it is known to these 20 elite people in the focus group or not. This tells us whether they all form a few similar clusters of thoughts or not. The cluster analysis is used to understand from the perspective of a marketer, and as a result, it will create an echo system that will change the way we work and make things happen on certain disruptive innovations that will be acceptable over time in terms of new products. Once, electricity was considered disruptive, but it has become the acceptable norm today. Similarly, the internet and its platforms for products are acceptable today.

5 Research Methodology

The present study is a exploratory research conducted based on focused group study. Hence it is based on both primary and secondary data. Primary data was collected through focus group interview held through telephone, zoom and google meet interview. The respondent's demographic profile is depicted in table 1. Secondary data was collected through various published resources such as journal articles, news reports and blogs.

Table 1. Participants demographic profile

Sl. No.	Focus group Respondents	Background	Focus Group Members	Income level
1	Academicians	Teaching, learning, rubbing ideas to new thoughts	2	5 lakhs and above
2	Manufacturing staff	Being in the forefront of making things happens; ready for a change and always in a new product development and learning mode	2	5 lakhs and above
3	Defense Personnel	Always on alert and eager to explore new things and adventurous mind	2	5 lakhs and above
4	Medical Specialists	Eager to know and attend conferences	2	5 lakhs and above
5	Information technology Gigs	Walking and talking technology	2	5 lakhs and above
6	Post graduation Students- MBA-IT	Young minds and want to make it happen	2	5 lakhs and above
7	Business Men	Having systems processes and people to take over	2	5 lakhs and above
8	House wives who are post graduates	Are connected to new things via children and media or internet	2	5 lakhs and above
9	Sports person	New and innovations	2	5 lakhs and above
10	Politicians	Part of a political party	2	5 lakhs and above
		Total	20	

Source: Compiled by the researcher

5.1 Statistical Tools

The present study employed descriptive statistics such as mean, standard deviation, rank and cluster analysis (hierarchical agglomeration technique) to identify particular disruption which more influencing on forming a new society as per the perception of experts of focus group.

6 Conceptual Framework

The following areas of understanding were considered while making the questionnaire, and the same is denoted in the Table 2.

Table 2. Areas of understanding included in questionnaire.

Sl. No.	Area of Disruption	Impact felt	Change period	Reference
1	Neuralink Elon Musk company	Human Learning and development is crucial for an understanding how the brain stimulates	Next 20-30 years	[22]
2	Keplar 452 b	Another planet to travel and find how is it possible to achieve things faster	100 years	[23]
3	Growing Kidneys in a PIG Uterus	Can change the way we get replacement for organ failures, a large number of people have kidney issues and are on dialysis	Easy organ transplant in another 15–20 years	[24]
4	Genome Editing	Reduction in deceases is the reason for finding the root cause of the diseases	Another 10 to 15 years	[25]
5	Green Plastic	Lower environmental damages can create a better world to live	Another 10 years	[26]
6	Desalination of sea water	Lower	Should the costs come down, higher the installation	[27]

Source: Compiled by the Researcher.

Formation of a New Society: Every disruption, whether technological or biological, has an impact on social, economic, and political life. As it gets accepted by a large number of people psychologically, it moves to the next level, which is amending the existing laws of the land. As the law changes, the social norms change, and thus a new society forms. This model is just to understand how thinking shapes a society. We are neither proving nor disproving this model; it's just an approach to framing the questionnaire, and V stands for variables (Figs. 1, 2 and 3).

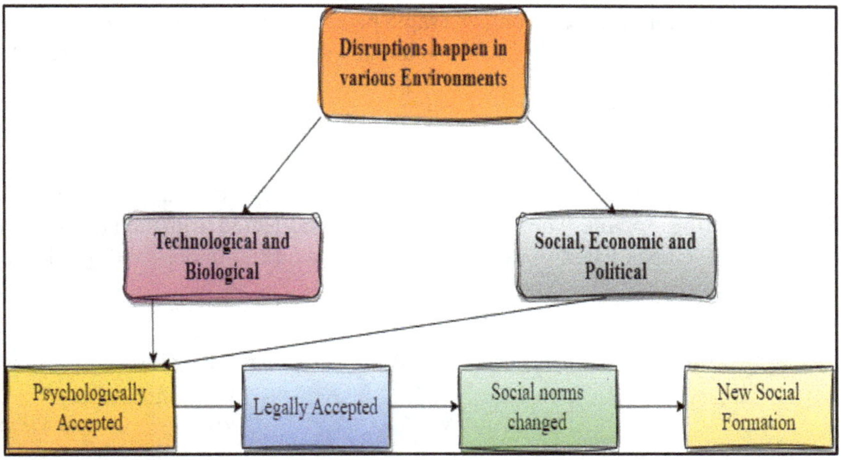

Fig. 1. A conceptual model of Disruptions that lead to new society formation.

The Disruptions: The following are the disruptions that were listed and questioned for a better understanding of how the focus group feels. These were incorporated as V1, V2, V3, V4, V5, and V6.

V1. The human brain Machine interface is a crucial technology that intercepts the way of learning; the processes we teach, and the way of life altogether. A machine or a chip inserted in a human brain can change the data entered into the brain, and this gives us more scope to understand the functionality of the brain in terms of its alpha, beta, and gamma states of vibration and its functionality. This interface can change the way we live forever.

V2. Kepler 452 b look alike, like our planet Earth. It has features and an environment that are similar to what we have. If it can be reached at the fastest possible speed to cover the distance, then we will make it. It becomes the best place for humans to exist.

V3. Growing kidneys in the PIG uterus will make up for the shortage of kidneys that fail due to lifestyle or due to the chemicals that an individual is exposed to. Modifications at the gene level can bring that wave of happiness to people who suffer from dialysis or their partners who suffer due to kidney transplants.

V4. Genome editing can help people prevent diseases before they appear; certain genes are vulnerable to certain diseases as they are genetically modified and have become weak, making them vulnerable to the constant risk of disease transmissions like cancer or a few genetic disorders, which manifest in disabled children.

V5. Green plastic, a substance that can be easily biodegradable and can be easily recyclable, needs to be used to prevent the plastic waste that is currently destroying the water bodies. Because plastic is necessary and cannot be avoided, green plastic can change the way we pack our products and design containers that can change themselves.

V6. Water can change people and relationships across the world, and a shortage of water can create this kind of chaos. The water demand will keep increasing and will not decrease. The population is increasing at a rapid pace, and depletion of groundwater will deplete civilization; hence, desalinization is the key solution for abundant water [28].

7 Results

The focus group consists of 20 selected elite people who have ideas about the future and the disruptions in the pipeline. The following is the data collected from their questionnaire responses and have been listed in the below matrix for further understanding (Tables 3, 4, 5 and 6).

Table 3. Attitudinal data collected for clustering disruptions.

Sl No	V1	V2	V3	V4	V5	V6	Overall Mean
1	6	4	7	3	2	3	4.17
2	2	3	1	4	5	4	3.17
3	7	2	6	4	1	3	3.83
4	4	6	4	5	3	6	4.67
5	1	3	2	2	6	4	3.00
6	6	4	6	3	3	4	4.33
7	5	3	6	3	3	4	4.00
8	7	3	7	4	1	4	4.33
9	2	4	3	3	6	3	3.50
10	3	5	3	6	4	6	4.50
11	1	3	2	3	5	3	2.83
12	5	4	5	4	2	4	4.00
13	2	2	1	5	3	4	2.83
14	4	6	4	6	3	7	5.00
15	6	5	4	2	1	4	3.68
16	3	5	4	6	4	7	4.83
17	4	4	7	2	2	5	4.00
18	3	7	2	6	4	3	4.17
19	4	6	3	7	2	7	4.83
20	2	3	2	4	7	2	3.33
Mean	3.85	4.1	3.95	4.1	3.35	4.35	3.95
S.D	1.8994	1.4104	2.0124	1.5183	1.7554	1.4964	
Rank	4	2	3	2	5	1	

Source: Focused Group Interviewed Data

Table 4. Case processing and summary of Squared Euclidean, Ward and Linkage

Case Processing and the Summary[a,b]					
Cases					
Valid		Missing		Total	
N	Percentage	N	Percentage	N	Percentage
20	**100.00**	0	.0	20	100.00
a. Squared Euclidean is the Distance used					
b. Ward and Linkage					

Table 5. Schedule of Agglomeration.

Agglomeration and the Schedule						
Respondents	Clusters Combined		Coefficients	Stage Clusters First Appears		Next Stage
	Cluster 1	Cluster 2		Cluster 1	Cluster 2	
1	10	16	1.0000	0	0	12
2	04	14	2.0000	0	0	09
3	06	07	3.0000	0	0	06
4	05	11	4.5000	0	0	11
5	03	08	6.0000	0	0	15
6	06	12	8.3330	03	0	08
7	09	20	10.8330	0	0	11
8	01	06	13.7500	0	06	13
9	04	19	16.7500	02	0	12
10	02	13	19.7500	0	0	16
11	05	09	24.2500	04	07	16
12	04	10	28.8500	09	01	17
13	01	17	34.2000	08	0	14
14	01	15	42.4330	13	0	15
15	01	03	55.1000	14	05	19
16	02	05	67.9330	10	11	18
17	04	18	83.3330	12	0	18

(continued)

Table 5. *(continued)*

Agglomeration and the Schedule						
Respondents	Clusters Combined		Coefficients	Stage Clusters First Appears		Next Stage
	Cluster 1	Cluster 2		Cluster 1	Cluster 2	
18	02	04	176.3330	16	17	19
19	01	02	328.2000	15	18	0

Table 6. Schedule on Clusters and Memberships.

Clusters and the Membership in observation			
Member	4th Cluster	3rd Cluster	2nd Cluster
01	01	01	01
02	02	02	02
03	01	01	01
04	03	03	02
05	02	02	02
06	01	01	01
07	01	01	01
08	01	01	01
09	02	02	02
10	03	03	02
11	02	02	02
12	01	01	01
13	02	02	02
14	03	03	02
15	01	01	01
16	03	03	02
17	01	01	01
18	04	03	02
19	03	03	02
20	02	02	02

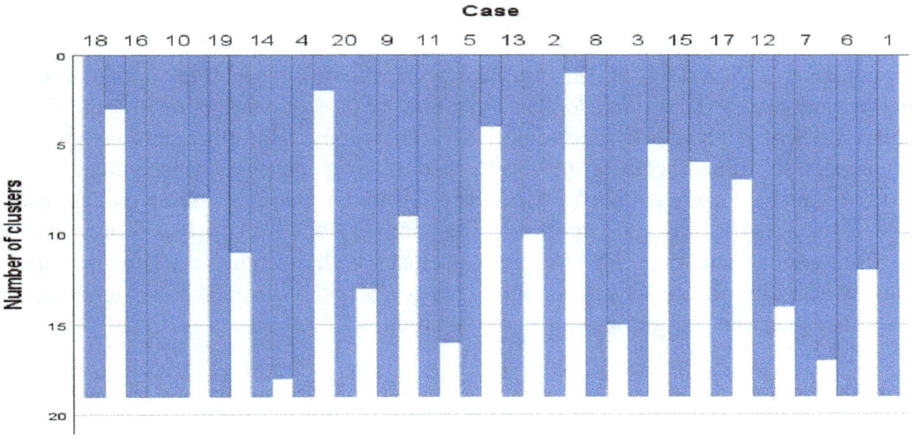

Fig. 2. Number of clusters based on disruptions

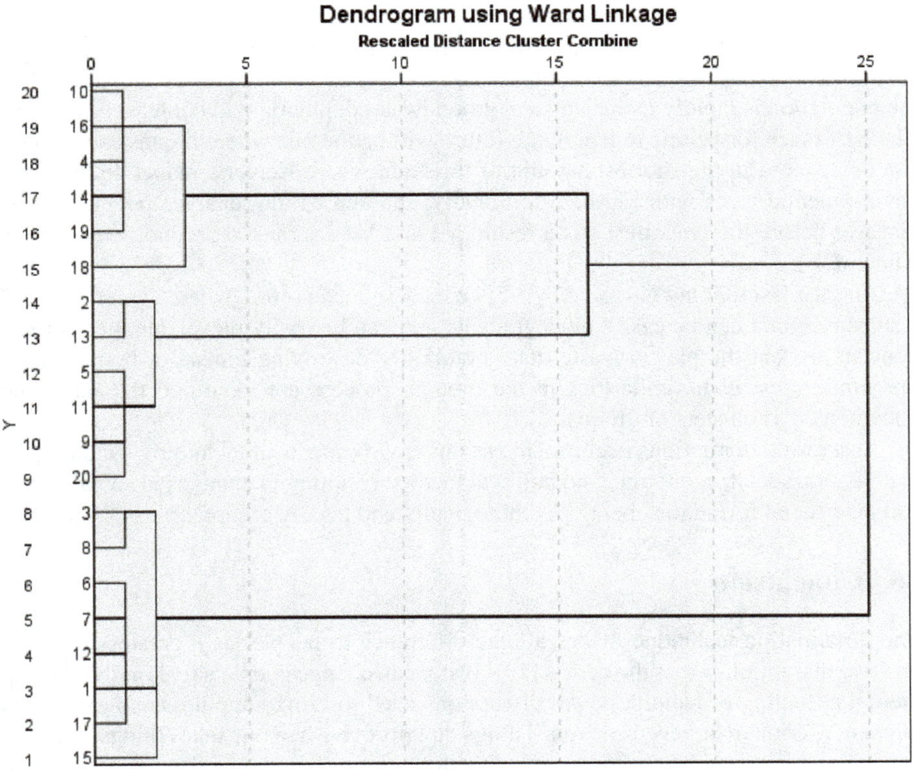

Fig. 3. Number of clusters based on disruptions

8 Results of Observation of Hierarchical Clustering

Respondents- 1, 3, and 6, 7, 8, 12, 15, 17 carry a similar mindset and fall under the same hierarchical cluster. They think similarly in the above disruptions that are to come across categories with high priority and importance. They come from different backgrounds like academics, manufacturing, Defense, students, or housewives. Respondents- 2, 5, 9, 11, 13, and 20 carry similar mindsets and fall under the same hierarchical cluster; with slightly less priority and importance. Respondents- 4, 10, 14,16,18,19 carry similar mindsets and fall under the same hierarchical cluster with much less priority and importance. Dissimilar people are under similar categories and it means that clusters make a difference once the disruptions come as new products.

9 Discussion

Each disruption has an impact on society's social, economic, and political systems and is either technological, biological, or both. One such framework merely explains how society is shaped by thought. The listing of disruptions, from V1 to V6, provides insight into how the focus group perceives them.

According to the data collected for clustering disruptions, V6 has the highest mean value of 4.35, which states that desalinization is a key solution for abundant water when the population is rapidly increasing and groundwater depletion will deplete civilization. The best place for people to live in the future will be the one where the fastest distance can be covered in the shortest amount of time and where Kepler's planet 452b shares environmental traits with Earth. Additionally, genome editing enables people to stop diseases before they manifest. As a result, V2 and V4 are ranked second, with a mean value of 4.1.

And the last ranking is scored by V5, i.e., a 3.35 mean score, where "green plastic," a substance that can be easily biodegradable and can be easily recyclable, needs to be done to prevent the plastic waste that is currently destroying the water bodies. Thus the similarities or dissimilarities in the thought process are identified the same and documented as clusters or groups.

As a result, disruptions occurred in various environments of technology and biological factors, social, economic, and political factors, resulting in changes in social norms and new social formations being psychologically and legally accepted.

10 Conclusion

The desalination technique makes a huge difference to people, as it creates a kind of positive disruption across the nation [29]. Water can change people's lives and can bring them that feeling of happiness. All disruptions need to bring happiness and expansion into a new domain or new direction. Things do improve when one takes things in stride. A positive stride is essential for growth. Disruptions are essential or else a new direction is not possible; we seldom consider these changes as serious changes, and these changes should bring about a turnaround in the industries that we currently handle. Sea water is abundant and can be used to cultivate barren lands for a great agricultural output and better yield to solve the hunger demands of the growing population [30].

The study's major goal is to comprehend how focus group discussions can result in intellectual clusters. Identifying the reasons behind difficulties in disruptions, how people view them, and whether they cluster together or stand alone, as well as awareness of and adherence to current trends and disruptions.

Studies, for instance, cluster analysis is used to comprehend from the standpoint of a marketer, and as a result, it would create an echo system that would change the way we work and bring about certain disruptive ideas that would be accepted over time in terms of new products. A small number of respondents share a common mentality and fall into the hierarchical clusters of high priority and importance, followed by a small number of somewhat less important and high priority respondents, and a small number of less important and high priority respondents.

The study was limited to the 20 members of focus groups and only in the taluks of Mangalore and Dakshina Kannada. Even though this study was only conducted with a small focus group, it can still be used by a wider audience to better understand how disruptive and complex changes are accepted. These ideas have not been taken into account in this study, despite the fact that disruptions affect and change economies.

Despite these drawbacks, this study helps close a critical gap in the study of radical innovations and sets up numerous avenues for further investigation into how technology affects radical innovations and their long-term repercussions. Many previous studies have highlighted the conceptual aspects of radical changes and their impact on society, but no study has attempted to analyse the multidisciplinary experts on radical changes and their impact on the creation and transformation of society at large; thus, the current study was conducted to fill this gap.

Additional research could examine the technologically-driven process of catching up or the creation of new markets. Furthermore, the size of the firm might not be a limiting factor in the diverging processes of diverse approaches and applications in radical knowledge development. Therefore, further research might examine the various effects of technology on the creation of radical knowledge within established organisations.

References

1. Veryzer, R.W., Jr.: Discontinuous innovation and the new product development process. J. Product Innov. Manage. Int. Publ. Product Dev. Manage. Assoc. **15**(4), 304–321 (1998)
2. Romanello, M., et al.: The 2021 report of the Lancet Countdown on health and climate change: code red for a healthy future. The Lancet **398**(10311), 1619–1662 (2021)
3. Themelis, N.J., Ulloa, P.A.: Methane generation in landfills. Renew. Energy **32**(7), 1243–1257 (2007)
4. Liyanage, V.R., Jarmasz, J.S., Murugeshan, N., Del Bigio, M.R., Rastegar, M., Davie, J.R.: DNA modifications: function and applications in normal and disease States. Biology **3**(4), 670–723 (2014)
5. Bak, R.O., Gomez-Ospina, N., Porteus, M.H.: Gene editing on center stage. Trends Genet. **34**(8), 600–611 (2018)
6. Thimmaraju, M., Sreepada, D., Babu, G.S., Dasari, B.K., Velpula, S. K., Vallepu, N.: Desalination of water. Desalination and Water Treatment, pp. 333–347 (2018)
7. Nakamura, H.: Ways of thinking of eastern peoples: India, China, Tibet. Motilal Banarsidass Publisher, Japan (1991)

8. Edwards, A.W., Cavalli-Sforza, L.L.: A method for cluster analysis. Biometrics **21**, 362–375 (1965)
9. Duran, B.S., Odell, P.L.: Cluster analysis: a survey. Springer Sci. Bus. Media **100**(1), 1–13 (2013)
10. Köhn, H.F., Hubert, L.J.: Hierarchical cluster analysis. Wiley StatsRef: Statistics Reference Online, 1–13 (2014)
11. Baker, F.B., Hubert, L.J.: Measuring the power of hierarchical cluster analysis. J. Am. Stat. Assoc. **70**(349), 31–38 (1975)
12. Leenders, R.T., Dolfsma, W.A.: Social networks for innovation and new product development. J. Prod. Innov. Manag. **33**(2), 123–131 (2016)
13. Xu, X., Hua, Y., Wang, S., Xu, G.: Determinants of consumer's intention to purchase authentic green furniture. Resour. Conserv. Recycl. **156**, 104721 (2020)
14. Xu, X., Wang, S., Yu, Y.: Consumer's intention to purchase green furniture: do health consciousness and environmental awareness matter? Sci. Total Environ. **704**, 135275 (2020)
15. Handfield, R.B., Walton, S.V., Seegers, L.K., Melnyk, S.A.: 'Green'value chain practices in the furniture industry. J. Oper. Manag. **15**(4), 293–315 (1997)
16. Amabile, T.M.: A model of creativity and innovation in organizations. Res. Organ. Behav. **10**(1), 123–167 (1988)
17. Ronzitti, S., Soldini, E., Smith, N., Potenza, M.N., Clerici, M., Bowden-Jones, H.: Current suicidal ideation in treatment-seeking individuals in the United Kingdom with gambling problems. Addict. Behav. **74**, 33–40 (2017)
18. Paunesku, D., Walton, G.M., Romero, C., Smith, E.N., Yeager, D.S., Dweck, C.S.: Mind-set interventions are a scalable treatment for academic underachievement. Psychol. Sci. **26**(6), 784–793 (2015)
19. Ma, J., Roese, N.J.: The maximizing mind-set. J. Cons. Res. **41**(1), 71–92 (2014)
20. Nielsen, R.K.: Global Mindset as Managerial Meta-competence and Organizational Capability: Boundary-crossing Leadership Cooperation in the MNC. The Case of 'Group Mindset'in Solar A/S. Frederiksberg: Copenhagen Business School (CBS) (2014)
21. Fang, F., Kang, S.P., Liu, S.: Measuring mindset change in the systemic transformation of education. Assoc. Educ. Commun. Technol. (2004)
22. Pisarchik, A.N., Maksimenko, V.A., Hramov, A.E.: From novel technology to novel applications: comment on "an integrated brain-machine interface platform with thousands of channels" by Elon Musk and Neuralink. J. Med. Internet Res. **21**(10), 16356 (2019)
23. Shaji, A.: Kepler 452B: potentially Earth like planet (2015)
24. Nagashima, H., Matsunari, H.: Growing human organs in pigs—a dream or reality? Theriogenology **86**(1), 422–426 (2016)
25. Cox, D.B.T., Platt, R.J., Zhang, F.: Therapeutic genome editing: prospects and challenges. Nat. Med. **21**(2), 121–131 (2015)
26. Song, N., Hou, X., Chen, L., Cui, S., Shi, L., Ding, P.: A green plastic constructed from cellulose and functionalized graphene with high thermal conductivity. ACS Appl. Mater. Interfaces. **9**(21), 17914–17922 (2017)
27. Elimelech, M., Phillip, W.A.: The future of seawater desalination: energy, technology, and the environment. Science **333**(6043), 712–717 (2011)
28. El-Dessouky, H.T., Ettouney, H.M.: Fundamentals of salt water desalination. Elsevier (2002)
29. Younos, T., Tulou, K.E.: Overview of desalination techniques. J. Contemp. Water Res. Educ. **132**(1), 3–10 (2005)
30. Curto, D., Franzitta, V., Guercio, A.: A review of the water desalination technologies. Appl. Sci. **11**(2), 670 (2021)

Evaluating the Soundness of Jordanian Commercial and Islamic Listed Banks by Using the CAMELs Rating Model

Mohammad Sami Ali[1(✉)], Issa Ahmad Swiety[1], and Eiman Osseilan[2]

[1] Department of Financial Technology, Faculty of Economics and Administrative Sciences, Zarqa University, Zarqa 13110, Jordan
{Maldahrawi,iswiety}@zu.edu.jo

[2] Department of Insurance and Risk Management, College of Business Administration, University of Business and Technology, Jeddah, Saudi Arabia
e.osseilan@ubt.edu.sa

Abstract. Due to the efficiency of CAMELs rating model to evaluate the safety as well as examining soundness of banks, there are numerous studies conducted for the aim of abating the risks, which may lead to banks' failures. However, since there is a lack in literature regarding the efficiency of CAMELs model to assess the soundness of Jordanian listed banks, the present paper aims at comparing the soundness of Jordanian commercial along with Islamic listed banks by relying on the components of the CAMELs rating model, over the period Q1/2013-Q42018. Consequently, the findings proved that the performance of Jordanian commercial banks is fundamentally sound and they are ranked as strong banks in accordance to the CAR, Asset Quality, Management Quality as well as Liquidity's elements. Moreover, the study concluded that although the listed banks of Jordan are sound, however, the commercial banks are found to be sounder than the Islamic banks, which are listed at Amman stock exchange.

Keywords: Soundness · CAMELs rating system · CAR · Earnings Quality · Listed Banks

1 Introduction

Due to the vital role of banking sectors in nations' economies, the stability of a country banking system considers as a significant parameter to assess the stability of economic situation as a whole. However, although financial institutions are seen as the backbone of economic development [18]; they were not exempted from the impact of globalization as it is resulted in stimulating the level of competition in the banking and financial system. Thus, to monitor the banking and financial operations, banking sectors were generally subjected to strict regulations and financial repression, in an attempt to enhance the stability of the banking system, and that will in turn resulted in economic growth and development. Moreover, banks regulators paid their maximum efforts to innovate a measurement system to assess the overall performance of banks; as a result they produced

B. A. M. Alareeni and I. Elgedawy (Eds.): *Artificial Intelligence (AI) and Finance*, SSDC 488, pp. 631–641, 2023.
https://doi.org/10.1007/978-3-031-39158-3_59

the CAMELs rating one to measure the performance and soundness of the overall banks' operations. Though the review of literature revealed that there are several researches carried out to evaluate the performance of banks, in an attempt to mitigate the level of the expected losses i.e. [1, 2, 15], however, there is a gap in the available literature regarding the efficiency of the CAMELs rating model to evaluate the soundness of the Jordanian banking system, over the period Q1/2013-Q4/2018. Therefore, results from this research are expected to fill the available gap in literature as well as contributing in helping further researchers to apply the adopted model on further researches. Additionally, the current research will also contribute in assisting the clients of Jordanian commercial as well as Islamic banks to realize the soundness of their banks and the safety of their deposits. Furthermore, since the performance of banks could be considered as a parameter to evaluate the quality of banks' traded securities, results from this study could also help investors to evaluate their investment decisions as well as making more rational investment decisions. Beyond that, through evaluating the soundness of banks, findings from this study are expected to assist the management of banks to strengthen the weaknesses in any components of the CAMELs rating model. Consequently, that will help banks improving their performance as well as attracting more investors.

2 The Review of Literature

In the process of continuous evaluation for banks' performance, the review of literature revealed that there are several studies conducted to assess the performance of banks by utilizing the CAMELs rating approach. In this regards, studies including [3] utilized the CAMELs rating model to evaluate the performance of the HSBC holdings during the period 2010–2019. As a result, the study asserted that the CAMELs model is an effective tool to assess the efficiency of banks along with regulations as well as improving the process of making rational financial decisions. Consistently, [11] concluded that some banks are performing better than others as well as the CAMEL model is an effective and accurate tool to assess the performance of Indian private banks during the period 2012–2017. Likewise, through using the CAMELs models to evaluate the performance of Iranian banks [15] confirmed that the CAMELS model is an efficient tool to evaluate the performance of banking industries as well as forecasting the future and relative risks. In a different perspective, authors including [18] employed the BankoMeter's model to assess the financial soundness of the Jordanian commercial listed banks during the period of 2002–2011. Consequently, the study found that all banks are financially sound and termed as super sound banks. Additionally, the study recommended banks to use this model due to its power to help their managements to eschew insolvency issues with a proper control over their operations. A study by [6] analyzed the banking sector of India via using the CAMELs framework for a sample of fifty-eight private and public Indian banks, over the period 2003–2007. Consequently, the findings showed that private and foreign banks are performing better than the public banks. Therefore, the study concluded that public banks must improve their credit lending policies in order to improve the asset quality as well as earnings' elements. However, through analyzing the third element of the CAMEL model i.e. Management and the control of internal governance of banks and financial companies; [8] found that the absence of correct balance between management and control rights is the major cause behind lack of profit and loss sharing in the Islamic finance structures. Through adopting the CAMEL model to analyze the performance of 65 Indian banks during the period 2003–2004 [14] concluded that the competition was tough and consumers benefited from better services quality, innovative products and better bargains. Furthermore, in an attempt to assess the performance of Indian Private Sector Banks on the basis of Camel Model and rating the top five and bottom five banks [10] ranked 20 old and 10 new private sector banks on the basis of CAMEL model, as well as showed that HDFC was at its higher position of all private sectors banks in India succeeded by the Karur Vysya as well as the Tamilnad Mercantile Bank. [12] aimed at comparing the performance of 22 public sector banks along with 15 private commercial sector banks during the period 2006–2011 by using the CAMEL model. Consequently, the results revealed that the IDBI Bank was the best performing bank followed by Kotak Mahindra Bank and ICICI Bank. On the other hand the performance of Dhanalaxmi Bank was the worst followed by J&K Bank and Karnataka Bank Ltd. Although, the study induced that there is no significant differences in the financial performance of the public and private Indian banks, however, both sectors still need for an overall improvement in order to strengthen their financial position in the competitive market.

3 Research's Methodology and Hypotheses

The current research focuses on assessing as well as comparing the soundness of the commercial along with Islamic banks, which are listed in the Amman stock exchange. For this purpose, the study relied on a set of secondary data those are drawn from the statistical bulletins of the Amman stock exchange, over the period Q1/2013-Q4/2018. The present study employed the six components of the CAMELs rating model as a main instrument to achieve the core aims and objectives. The study employed statistical techniques such as means, variances and standard deviations in order to analyze the components of the CAMELs' approach as well as examining the following hypotheses:

H1: The Jordanian listed banks are financially sound according to the CAMELs rating model.

H2: There are no significant differences between the soundness of Jordanian commercial and Islamic listed banks.

3.1 Variables Description

The current research employed the CAMELs' elements to achieve the core aims and objectives. Therefore, these elements and their measurements are described below [13]:

1. **Capital Adequacy**

This component was developed to ensure that banks can absorb a reasonable level of losses that occurred due to operational losses and determine the capacity of the bank to meet losses. In this research the CAR is calculated as shown below [13]:

$$\text{CAR} = \frac{[(\text{Tier 1 Capital} - \text{Goodwill}) + \text{Tier2 Capital}]}{\text{Risk} - \text{weighted Asset}}$$

2. **Asset Quality**

It is an important parameter to gauge the strength of banks and it reflects the amount of existing credit risk that is associated with the loan and investment portfolio, as well as off-balance sheet activities [13]. The following ratio is used to evaluate Asset quality [5]:

$$Non\text{-}performing\ loans\ to\ total\ equity = \frac{NPLs}{Total\ Equity}$$

3. **Management Quality**

It is an important element of the CAMELs Model as it reflects the adequacy of the board of directors and senior management systems and procedures to identify, measure, monitor, and control risks. Regulators emphasize the existence and use of polices and processes to manage risks within targets [13]. The following ratio is used to evaluate management quality [9]:

$$\frac{\text{Expenses}}{\text{Earnings}} = \frac{Management\ Expenses}{Total\ Earnings}$$

4. **Earnings**

The quality of earnings (E) is a very important criterion that determines the ability of a bank to earn consistently. The Earnings category reflects not only the quantity and trend in earnings, but also the factors that may affect the sustainability or quality or earnings [13]. The current research used the following ratios to evaluate earnings quality [4]:

$$ROA = \frac{\text{Net Interest Income}}{\text{Asset growth Rate}}$$

$$ROE = \frac{\text{Net Interest Income}}{\text{Shareholders}'\ \text{Equity growth Rate}}$$

5. **Liquidity**

Liquidity (L) reflects the adequacy of the institution's current and prospective sources of liquidity and funds management practices. The practices of fund management should ensure an institution is able to maintain a level of liquidity sufficient to meet its financial obligations in a timely manner; and capable of quickly liquidating assets with minimal loss. The following ratios are used to measure banks' liquidity [5]:

$$Liq1 = \frac{\text{Total Customer Deposit}}{\text{Total Assets}}$$

$$Liq2 = \frac{\text{Total Loans}}{\text{Total Customer Deposit}}$$

6. **Sensitivity to Market Risk**

Sensitivity to Market Risk (S) reflects the degree to which changes in interest rates, foreign exchange rates, commodity and securities' prices can adversely affect earnings or economic capital [13].

$$\textbf{\textit{Total securities to total assets}} = \frac{\textbf{\textit{Total securities}}}{\textbf{\textit{Total assets}}}$$

Moreover, after calculating the means for all components of the CAMELs' model, the mean for each element will be rated on the basis of the following criteria on the scale of 1–5 in order to check as well as compare the soundness of Jordanian commercial and Islamic banks (Table 1):

Table 1. CAMELs Rating Base. This table describes the rating base for the CAMELs rating model.

CAMELS Rating Elements		**Rating**				
		1	**2**	**3**	**4**	**5**
Capital Adequacy Ratio		≥15%	12%–14.99%	8%–11.99%	7%–0.7.99%	≤6.99%
Asset Quality		≤1.25%	2.5%–1.26%	3.5%–2.6%	5.5%–3.6%	≥5.6%
Management Quality		≤25%	30%–26%	38%–31%	45%–39%	≥46%
Earnings Quality	**ROE**	≥22%	17%–21.99%	10%–16.99%	7%–9.99%	≤6.99%
	ROA	≥1%	0. 9%–0. 8%	0.35%–0. 7%	0.25%–0.34%	≤0.24%
Liquidity Quality Ratios	**Liquidity L1**	≥0.50	45%–0.49.99%	38%–44.99%	33%–37.99%	≤0.32
	Liquidity L2	≤55%	62%–0.56%	68%–63%	80%–69%	≥81%
Sensitivity to Market Risk		≤25%	30%–26%	37%–31%	42%–38%	≥43%

4 Results Discussion

Due to the importance of the Camels American model to evaluate the performance of banks, the current research aimed at comparing the soundness of Jordanian Islamic along with the soundness of Jordanian commercial banks, over the period Q1/2013-Q4/2018. Thus, by the running the descriptive statistics tests the following tables demonstrate the means, variances, and standard deviations of each element of the CAMELs' rating model:

4.1 Descriptive Statistics

Through comparing the above results with the accepted criteria the results confirmed that the mean of commercial banks of Jordan exceeds the minimum required. In other words, since the higher ratio is the better, the findings proved that the commercial banks are sound according to this element as well as they are ranked as strong banks since they got the rate 1. However, although the capital adequacy of Islamic banks is higher than the minimum required, the mean of commercial banks is higher than the mean of Islamic banks, which means that commercial banks are performing better than Islamic ones in a relation to this element (Table 2).

Table 2. Capital Adequacy. This table shows descriptive statistics for the CAR ratio (Table 3).

Rating	**1**	**2**	**3**	**4**	**5**
	≥15%	12%–14.99%	8%–11.99%	7%–0.7.99%	≤6.99%
Commercial Banks					
Ratio	**Mean**	**St. Deviation**	**Variance**	**Rate**	
CAR	0.2322	0.0468	***.002***	1	
Islamic					
CAR	0.1746	0.0513	***.003***	1	

Table 3. Asset Quality. This table shows descriptive statistics for the $\frac{NPLs}{Equity}$ ratio.

Rating	**1**	**2**	**3**	**4**	**5**
	≤1.25%	2.5%–1.26%	3.5%–2.6%	5.5%–3.6%	≥5.6%
Commercial Banks					
Ratio	**Mean**	**St. Deviation**	**Variance**	**Rate**	
$\frac{NPLs}{Equity}$	0.0029	0.0020	***.0004***	1	
Islamic					
$\frac{NPLs}{Equity}$	0.0059	0.0023	***.0005***	1	

Through comparing the above results with the accepted criteria the results confirmed that the mean of commercial banks of Jordan is lower than the accepted criteria. Thus, since the lower ratio is the better, the results asserted that the commercial banks are sound according to this element as well as they are categorized as strong banks since they got the rate 1. However, although the $\frac{NPLs}{Equity}$ ratio of Islamic banks is lower than the accepted criteria, the mean of commercial banks still lower than the mean for Islamic banks, which in other words mean that commercial banks' manage their assets better than Islamic banks (Table 4).

Table 4. Management Quality. This table shows descriptive statistics for the $\frac{Expenses}{Earnings}$ ratio.

Rating	**1**	**2**	**3**	**4**	**5**
	≤25%	30%–26%	38%–31%	45%–39%	≥46%
Commercial Banks					
Ratio	**Mean**	**St. Deviation**	**Variance**	**Rate**	
$\frac{Expenses}{Earnings}$	0.200	0.338	**0. 114**	**1**	
Islamic					
$\frac{Expenses}{Earnings}$	0.196	0.288	***.082***	**1**	

Through comparing the above results with the accepted criteria the findings confirmed that the mean of commercial banks is lower than the accepted criteria. Thus, since the lower ratio is the better, the above statistics proved that the commercial banks are sound according to this element as well as they are fundamentally sound. However, since the mean of Islamic banks is lower than the mean of commercial banks; it is revealed that Islamic banks are performing better than commercial ones in a relation to this element (Table 5).

Table 5. Earnings Quality. This table shows descriptive statistics for the ratios, which are used to evaluate the quality of banks' earnings.

Rating		**1**	**2**	**3**	**4**	**5**
	ROA	**≥1%**	**0.9%–0.8%**	**0.35%–0.7%**	**0.25%–0.34%**	**≤0.24%**
	ROE	**≥22%**	**17%–21.99%**	**10%–16.99%**	**7%–9.99%**	**≤6.99%**
Commercial						
Component		**Mean**	**St. Deviation**		**Variance**	**Rate**
Earnings Quality	**ROA**	**0.0086**	**0.0059**		**0.003%**	**2**
	ROE	**0.0804**	**0.0468**		**0.002**	**4**
Islamic						
	ROA	**0.0130**	**0.0038**	**0.001%**		**1**
	ROE	**0.1666**	**0.0315**	**0.09%**		**3**

The above statistics demonstrate that according to the ROA ratio the commercial bank of Jordan are categorized as satisfactory banks since they have got the rate 2, while according to the ROE ratio they are categorized as marginal banks. By contrast, Islamic banks of Jordan are ranked as strong banks according to the ROA compared with fair performance in accordance to the ROE ratio. Generally speaking, the Islamic banks are found to be sounder than commercial banks in a relation to the Earnings quality component (Table 6).

Table 6. Liquidity Quality. This table shows descriptive statistics for the ratios, which are used to evaluate the quality of banks' liquidity.

Rating		**1**	**2**	**3**	**4**	**5**
	L1	≥0.50	45%–0.49.99%	38%–44.99%	33%–37.99%	≤0.32
	L2	≤55%	62%–0.56%	68%–63%	80%–69%	≥81%
Commercial						
Component		**Mean**	**St. Deviation**	**Variance**	**Rate**	
Liquidity Quality	L1	0.7504	0.1497	**0.022**	**1**	
	L2	0.2394	0.1399	**0.019**	**1**	
Islamic						
	L1	0.5194	0.3204	**0.102**	**1**	
	L2	0.1551	0.1165	**0.013**	**1**	

Liquidity (L) reflects the adequacy of the institution's current and prospective sources of liquidity and funds management practices. Risk of liquidity is curse to the image of bank. Bank has to take a proper care to hedge the liquidity risk; at the same time ensuring good percentage of funds are invested in high return generating securities, so that it is in a position to generate profit with provision liquidity to the depositors. There should be adequacy of liquidity sources compared to present and future needs, and availability of assets readily convertible to cask without undue loss. The fund management practices should ensure an institution is able to maintain a level of liquidity sufficient to meet its financial obligations in a timely manner; and capable of quickly liquidating assets with minimal loss. Thus, by comparing the above results with the accepted criteria the findings confirmed that the liquidity of commercial banks of Jordan exceeds the required minimum. Similarly, the liquidity in Islamic banks is found higher than the accepted criteria. However, according to this element, the above statistics proved that commercial banks were sounder than Islamic banks, during the period Q1/2013-Q4/2018 (Table 7).

Sensitivity to Market Risk (S) reflects the degree to which changes in interest rates, foreign exchange rates, commodity prices, and equity prices can adversely affect earnings or economic capital [13]. Through comparing the means of commercial as well as Islamic banks with the accepted criteria, the above statistics confirmed that the performance of both commercial as well as Islamic banks was performing better than the average.

Moreover, through comparing averages with the CAMELs composite ratings in Table (8); results from the above Table (9) proved that the overall performance of both commercial and Islamic banks is fundamentally sound because the averages for the listed banks of Jordan are falling in between (1.00–1.49), which means that banks of Jordan are performing better than the average in almost all components.

Table 7. Sensitivity to Risk. This table shows descriptive statistics for the Sensitivity to Market Risk ratio.

Rating	**1**	**2**	**3**	**4**	**5**
	≤25%	**30%–26%**	**37%–31%**	**42%–38%**	**≥43%**
Commercial Banks					
Ratio	**Mean**	**St. Deviation**	**Variance**	**Rate**	
Sensitivity to Risk	**0.0646**	**0.0078**	**0.0006**	**1**	
Islamic					
Sensitivity to Risk	**0.0356**	**0.0096**	**0.084**	**1**	

Table 8. CAMELS Composite Ratings. The following table describes the numerical ratings assigned to the criteria, where a composite rating of (1) or (2) indicates a fundamental sound bank. A rating of (3) indicates that the bank shows some underlying weakness that should be corrected. A rating of 4 or 5 indicates a problem bank with some near-term potential for failure [13]:

Rating Scale	Composite Range	Rating Analysis	Rating analysis interpretation
1	1.00–1.49	Strong	The performance of bank is strong as well as performing better than the average in almost all components
2	1.5–2.49	Satisfactory	This rate is almost given to the bank that is performing better than the average, but not in all aspects. Although this bank is not enough strong, while it still performs at a satisfactory level
3	2.5–3.49	Fair	There is a combination of weaknesses, and if it is not redirected that may will lead the bank towards bankruptcy
4	3.5–4.49	Marginal	The performance of bank is weak and that is may due to managerial or financial insufficiencies. Therefore, a serious corrective actions and a proper supervision are needed to resolve the problem
5	4.5–5.49	Unsatisfactory	The performance of bank is unsound and risky as well as extremely unstable. Therefore, a continuous supervision is required to avoid the highly probable bank failure

Table 9. The Composite Ratings for Jordanian Listed Banks. This table describes statistics regarding the composite ratings for Jordanian banks.

Element	Rate								Average	Category
	CAR	AQ	MQ	ROA	ROE	L1	L2	SMR		
Commercial	1	1	1	2	4	1	1	1	1.33	Strong
Islamic	1	1	1	1	3	1	1	1	1.16	Strong

5 Conclusion and Future Researches

Due to the lack in literature regarding the use of CAMELs model to assess the soundness of Jordanian listed banks, the present paper employed the CAMELs rating model in order to compare the soundness of Jordanian commercial and Islamic listed banks, over the period Q1/2013-Q42018. Thus, by employing descriptive statistics such as means, variances and standard deviations; the study computed the rating of CAMELs elements of commercial and Islamic banks of Jordan as well as compared the average of CAMELs' elements with the composite rating of CAMELs' model. Consequently, the statistics proved that the performance of Jordanian commercial banks is fundamentally sound as well as banks are ranked as strong banks in accordance to the CAR element. Although the performance of Islamic banks is fundamentally sound, however, the capital adequacy of commercial banks is performing better that it is in Islamic banks. The results also asserted that the commercial as well as Islamic banks are performing better than the average according to the NPLs/ Equity ratio. Additionally, the commercial banks are found to be sounder than Islamic banks in a relation to the asset quality's element. By contrast, Islamic banks are performing better than commercial banks in accordance to the management quality's element. Relating to the earnings' element, the statistics proved that the Islamic banks are performing better than commercial banks in a relation to the Earnings component. Furthermore, although, the performance of both commercial as well as Islamic banks is fundamentally sound in accordance to the liquidity component; however, the liquidity of commercial banks is found to be better. Furthermore, the findings proved that the overall performance of both commercial and Islamic banks is fundamentally sound because their averages are falling in between (1.00–1.49), which means that banks of Jordan are performing better than the average in almost all components. Moreover, the study concluded that the listed banks of Jordan are sound, while the commercial banks are sounder than the Islamic ones. Consistently, a study by [18] proved that the listed banks of Jordan are financially sound and termed as super sound banks. It is also concluded that the CAMELs rating model is an effective tool to evaluate the soundness of Jordanian banking system. This result agrees with [3, 11, 15] as they concluded that the CAMELs model is an effective and accurate tool to assess the performance of banking industries. Eventually, the current study recommended banks to use this model due to its power to help their managements to eschew insolvency issues with a proper control over their operations. Studies including [17] recommended that it is necessary for banks of Jordan to use the CAMELs model as it conducts a central role to discover risks in their earlier stages. Furthermore, the study recommends further researchers to employ the CAMELs model in order to compare the soundness of Jordanian Islamic and commercial banks before as well as during the COVID-19 pandemic.

References

1. Ali, M.S.: Evaluating the effectiveness of macroeconomic indicators to influence the stability of Jordanian banking system. Zarqa J. Res. Stud. Humanit. **20**(1), 136–144 (2020)
2. Amman Stock Exchange (2022). www.ase.com.jo. Accessed Nov 2022
3. Atwat, R.: Application for the CAMELS Performance Evaluation Approach for HSBC Holdings. Unpublished Master Thesis, Abu Dhabi University (2020)
4. Brigham, E.M., Ehrhardt, M.C.: Financial Management "Theory and Practice", 11th edn. Elm Street Publishing Services, Inc., Florida, United States of America (2005)
5. Dang, U.: The CAMEL Rating System in Banking Supervision A Case Study, pp. 1–43. Arcada University of Applied Sciences, Helsinki, Finland (2011)
6. Dash, M., Das, A.: A CAMELS analysis of the Indian banking industry (2010)
7. Derviz, A., Podpiera, J.: Predicting bank CAMEL and S&P ratings: the case of the Czech Republic. Emerg. Mark. Financ. Trade **44**(1), 117 (2008)
8. Dar, H.A., Presley, J.R.: Lack of profit loss sharing in Islamic banking: management and Control Imbalances. Int. J. Islamic Financ. Serv. **2**(2), 3–18 (2000)
9. Gul Zeb, H.: Does CAMELS system provide similar rating as PACRA system in assessing the performance of banks in Pakistan? Unpublished Master Thesis, Umea School of Business, Umea University, Pakistan (2011)
10. Gupta, R.: A CAMEL model analysis of private sector banks in India. J. Gyan Manage. **2**(1), 3–8 (2008)
11. Lavanya, B., Srinivas, T.: Performance analysis using CAMEL model- a study of select private banks. J. Emerg. Technol. Innov. Res. **5**(6), 214–223 (2018)
12. Makkar, A., Singh, S.: Analysis of the financial performance of Indian commercial banks: a comparative study. Indian J. Financ. **7**(5), 41–49 (2013)
13. MacDonald, S.S., Timothy W.K.: Management of Banking, 6th. edn. Thomson South-Western Corporation, United States of America (2006)
14. Prasuna, D.G.: Performance snapshot 2003–04. Chartered Financ. Anal. **10**(11), 6–13 (2003)
15. Rostami, M.: CAMELS' analysis in banking industry. Glob. J. Eng. Sci. Res. Manage. **2**(11), 10–16 (2015)
16. Yameen, I., Ali, M.S.: Evaluating the financial soundness of the Jordanian commercial banks by applying BankoMeter's model. Res. J. Financ. Acc. **7**(2), 124–130 (2016)
17. Yamin, I., Aldhahrawi, M.S.: The impact of CAMELS' components on the credit risks that commercial Jordanian banks listed in Amman stocks exchange face. Zarqa J. Res. Stud. Humanit. **16**(3), 1–11 (2016)
18. Yamin, I., Ali, M.S.: The impact of the volatility in banks performance indicators on stock prices: an empirical study on Jordanian listed banks. Zarqa J. Res. Stud. Humanit. **14**(1), 291–302 (2014)

Employee Attitude, Behaviors and Performance on Authentic Leadership in CIMB Bank Contact Center Department

Jasmin Abd Sani, Syarifah Hanum Ali(✉), and Oscar Dousin

Universiti Malaysia Sabah, Jalan UMS, 88400 Kota Kinabalu, Sabah, Malaysia
syarifahhanum.ali@ums.edu.my

Abstract. This study aims to identify the effect of employees' behaviours on authentic leadership. The main purpose is to determine the relationship between interpersonal trust, occupational self-efficacy, role-based performance and work engagement toward the authentic leadership practice among employees at the CIMB call centre in Kuala Lumpur. The research adapts the method of descriptive-correlational statistics; these tools are used to analyse the correlation between employee behaviours and authentic leadership, which are significant from the data that gathering will translate into quantitative data. One hundred and twenty call centre employees were involved in this study, and the research finding discovered that all independent variables have a significant relationship and a percentage of 40.4% on authentic leadership.

Keywords: Employees' Behaviors · Employees' Attitude · Interpersonal Trusts · Occupational Self-Efficacy · Role-Based Performance · Work Engagement · Authentic Leadership

1 Introduction

Researchers have given attention to the role of leadership within organisations, especially its effect on employees. Authentic leadership is considered the root of a positive and effective leadership style; authentic leadership is expected to encourage employees' self-confidence and to create employee trust with management and their co-worker (Avolia, Gardner, Walumbwa, Luthans & May, 2004). Luthans and Avolio (2003) defined authentic leadership as "a process that draws from both positive psychological capacities and highly developed organisation context, which result in both greater self-awareness and self-regulated positive behaviours on the part of leader and associates, fostering positive self-development." Authentic leaders are more transparent; the employee will know the fundamental values and motives of the leader and be comfortable confronting their leader; authentic leaders are more open to their followers. In this context, the leader always needs to show an excellent example as the follower will perceive leader will be more transparent in their core values and morals in their talk and attitude.

Employees' behaviours are viewpoints about many aspects of their jobs, careers, and organisations. According to Ilies, Morgan, & Nahrgang (2005), the employee will act on

B. A. M. Alareeni and I. Elgedawy (Eds.): *Artificial Intelligence (AI) and Finance*, SSDC 488, pp. 642–656, 2023.
https://doi.org/10.1007/978-3-031-39158-3_60

positive encouragement; this explains that employees will reflect the kind of behaviours that the organisation requires if the organisation practices authentic leadership style as this leadership role positively affects follower behaviours by providing support for self-determination. Authentic leaders motivate followers to be more engaged in and aware of their duties by building optimism and hope, fostering a positive environment, helping followers find the meaning of work and by, showing consistency in their behaviours according to their values and belief and building trust and commitment among follower so that follower can best contribute their effort for fulfilling the duties. Employees who show desired behaviours would likely show high-performance reports, which indicates the effectiveness of managing subordinates as authentic rather than the team leader. In conclusion, authentic leadership is a moral and ethical practice and can build close relationships with employees.

1.1 Problem Statement

The annual citizen report of CIMB in 2017 shows an increasing employee turnover rate compared to 2016. The average employee turnover at CIMB Group in 2017 increased marginally, from 17.9% in 2016 to 20.4% in 2017. The employee suggested current leadership practices are based on closed engagement between staff and senior management; feedback and suggestions sharing only exist formally would make team least heard. Lacking authentic leadership in bridging management-employee relationships affects employee attitude in producing ineffective productive, and practical solutions at the workplace.

1.2 Research Question

1) What is the relationship between employee interpersonal trust with authentic leadership among team leaders in CIMB?
2) What is the relationship between employee self-efficacy and authentic leadership among team leaders in CIMB?
3) What is the relationship between employee role-based performances with authentic leadership among team leaders in CIMB?
4) What is the relationship between employee work engagements and authentic leadership among tea8gm leaders in CIMB?
5) What is the effect between employee self-efficacy, interpersonal trust, work engagement and employee performance on authentic leadership at the CIMB call centre of Menara Atlas Plaza Pantai, Menara Southern Bank (MSB), Damansara, Menara Bumiputera Commerce (MBC) at Jalan Raja Laut and also Menara CIMB at K.L. Sentral.

1.3 Research Objective

The study is to discover authentic leadership practices among leaders at the contact centre department of CIMB Bank. Value measuring is employee behaviour and performance. In capturing the situation, the following objective was explicitly formulated.

1) To identify the relationship between employee interpersonal trust and authentic leadership among team leaders in CIMB.
2) To identify the relationship between employee self-efficacy and authentic leadership among team leaders in CIMB.
3) To identify the relationship between employee role-based performances and authentic leadership among team leaders in CIMB.
4) To identify a relationship between employee work engagements and authentic leadership among team leaders in CIMB.
5) To identify what is the effect between employee self-efficacy, interpersonal trust, work engagement and employee performance on authentic leadership at the CIMB call centre of Menara Atlas Plaza Pantai, Menara Southern Bank (MSB), Damansara, Menara Bumiputera Commerce (MBC) at Jalan Raja Laut and also Menara CIMB at K.L. Sentral.

1.4 Scope of Study

The study focuses on the effect of employee attitude, behaviour and performance on the effectiveness of authentic leadership practice among team leader in this organisation. The researcher selects all level employees at the call centre of CIMB Bank in a few places in Kuala Lumpur, including Menara Atlas Plaza Pantai, Menara Southern Bank (MSB), Damansara, Menara Bumiputera Commerce (MBC) at Jalan Raja Laut and last but not least Menara CIMB at K.L. Sentral and involved all employees in the call centre. The quantitative research part of the research and this study focus on authentic leadership, self-efficacy, interpersonal trust, work engagement and role-based performance.

1.5 Significance of the Study

The call centre industry in Malaysia has been expanding and in its way to becoming the hub for the Asia Pacific region; unfortunately, call centres are experiencing high turnover, and it is not industry specific the right direction as it creates employment opportunities, thus improvement in the sector (Tanuja et al., 2014). The result of the investigation is noteworthy for human resource development (HRD) for the idea of developing training programs with relevant input in training and developing existing managers to be a leader or hiring potential leaders. Leaders who frequently interact with followers and thus significantly influence their behaviours and attitudes are also expected to impact employee turnover intention substantially (Brown et al., 2005); this study gives a picture of knowledge of the condition of employee attitude and behaviours on authentic leadership practices to facilitating positive organisational with influences different job characteristics and fulfils followers' basic needs, leading to increased employee engagement and in-role performance in other occupational settings, including call centres (Muthuveloo et al., 2014; Wang et al., 2011).

2 Literature Review

2.1 Authentic Leadership

Researchers define authentic leadership in various terms, concepts, and interpretations. Herter (2002) claimed that the roots of the words genuine are from Greek philosophy and refer to original as 'know thyself' and 'to make self true. It refers to the a original leader as an honest individual who continually provides truthful information and develops faith among the follower. A Genuine honest leaders will project themselves as individuals with good manners who view them as ethical and trusted and encourage open communication between leaders and followers. Luthan and Avoilio (2013) describe authenticity as a leader who knows his\her own attributes and act upon their quality, where she tells the truth and her belief at the working place. Authentic is a valuable cognitive skill that should have inside every leader as this will continue to encourage ethical practices in our daily life.

2.2 Employee Attitude - Interpersonal Trust

Origin is a cognitive process to differentiate between trustworthy people and those who are not (Bitmis & Ergeneli, 2013). Wang and Hsieh (2013) described trust as a person's confidence and willingness to perform based on another person's actions, decisions and words. Also, people already know trust is essential to cooperative relationships and the cornerstone of positive organisational culture. Ferres et al. (2004) summarise that the main feature of trust is the "willingness" of the other party. (Mayer & Schoorman, 1995, p.712) also share me perception as they define trust as the "willingness of a party to be vulnerable to the actions of another party based on the expectation that the other party will perform a particular action important to the trustee, irrespective of the ability to monitor or control that other party".

Researchers assume trust development is based on followers' direct experience with their leaders. It has been acknowledged as one of the most important factors influencing organisational communication level and success (Braun et al., 2013). Zhu et al. (2013) suggested that followers' willingness to expose their selves will rely on how their leader treats them. Employee attitude also strongly influences trust; when co-workers and leaders have trust condition in their job, it links to employees' attitudes such as job satisfaction, goal commitment and work engagement Dirks and Ferrin (2002). Researchers proposed that authentic leadership relational transparency can build followers' trust in the leader.

2.3 Employee Attitude - Self-efficacy

Self-efficacy translates into determining how people think, motivate themselves and behave. (Bandura, 1997, p. 3) he defines self-efficacy as "beliefs in one's capabilities to organise and execute the courses of action required to produce given attainments". Self-efficacy will influence people's attributes as it will take root in determining personality and influencing how people think, behave and motivate themselves (Bandura, 1997). Ozer and Bandura (1990) support that self-efficacy help decreases employees' stress and fatigue and increase confidence in believing they can control the work and stressful situation.

Avolio and Walumba (2006) proposed that authentic leader behaviour dramatically impacts the development of self-efficacy and performance over time. Since self-efficacy is led by theory and research in psychological terms, authentic leadership can play a role in influencing followers' psychological state. Positive psychological capital (PsyCap) is defined as the positive development state of an individual as characterised by high self-efficacy, optimism, hope, and resiliency. Avilio and Luthans (2006) suggested that authentic leadership can cascade down PsyCap self-efficacy among their followers as a long organisation willing to invest in genuine leadership development (AVD). Maddux (2002) identify two interacting factors that can influence self-efficacy with leadership implication: the development of the capacities for symbolic thought and responsiveness, and the ¬supportiveness of the social context in which leader and follower are embedded over time. Shows that symbolic thought and a positive ethical and engaged climate are among the characteristics leaders need to consider in developing their follower's self-efficacy.

2.4 Role-Based Performances

Minavand et al. (2013) stated that performance could be gauged based on their level of proficiency. Sue-Chan et al. (2012) explain the likelihood of being considered the most direct and transparent way subordinates can show their competence and dependability and how leaders and organisations look forward to seeing the best competency and reliability from the associates. (Kiyani et al., 2013) indicate that performance assessment is a source of information as it provides vast knowledge to the organisation for developing training programs, performance feedback, promotion design, and salary increases.

Authentic leaders are expected to affect follower performance positively, and previous theories from numerous published literature have supported this expectation. Jing and Avery (2018) stated that there are multiple reasons for are more defined relationship between leader and performance, for example, the leadership role in facilitating and sustaining improvement in organisational performance. Researchers share the same thoughts and been shared many surveys that indicate the importance of leadership style as one of the critical drivers of interpretation within the organisation (Ozsahin et al., 2011) found that authentic leaders who focus on the core characteristic of their employees able to develop and improve their work performance and resulting in effect toward organisation performance. A genuine leader has, directly and indirectly, affected the influence of employee behaviour toward positive change and is closely related to performance improvement. Peterson et al. (2012) investigated the relationship between authentic leadership and follower job performance. The result is that authentic leaders can enhance high follower levels of positivity in terms of self-efficacy, trust, and resilience, which increases their performance.

2.5 Employee Behavior - Work Engagement

(Schaufeli et al, 2002, p. 74) Roberts and Davenport (2002) describe work engagement as individual involvement in their job. They define work engagement as "a positive, fulfilling, work-related state of mind characterised by vigour, dedication, and absorption". The level of the individual engaged with the job will determine the performance and motivational level. To summarise all of the definitions from all researchers, work engagement is considered a vital mediating role in encouraging performance at the individual and organisational group in terms of career development for individuals and financial return for organisational (Halbesleben (2010). Additionally, Schauli (2012) stated that employee who is engaged in the job would discourage absenteeism and turnover and produce strong motivation to learn and develop skill related to the task.

Luthans and Avolio (2003) stated that authentic leaders, through several processes of personal and social identification, have the potential to enhance positive emotions in follower hope, trust and optimism. This will impact follower work attitudes regarding engagement and commitment, resulting in follower behaviours that increase job performance as more effort-giving and fewer withdrawal behaviours. Authentic leaders are important in cultivating work engagement, given that disengagement or alienation is central to the problem of loyees' lack of commitment and motivation. According to Thomas and Velthouse (1990), meaningless work is often associated with apathy and detachment from one's career. Khan (1990) suggested that leaders play a role in influencing the degree of individual engagement in their job. Gardner and Schermerhorn (2005) stated that authentic leaders encourage followers to find their abilities (efficacy) by exemplifying high moral standards. They will promote further development as they truly know themselves and are confident. Moreover, honesty, fairness, integrity and openness of authentic leaders will develop follower trust in others, developing organisational climates and authenticity of leadership and treating employers fairly.

3 Methodology

This quantitative study investigates and analyses the relationship between employee behaviour components that contribute to the effectiveness of authentic leadership practices among employees at the CIMB call centre in Kuala Lumpur. This study focuses on discovering the association between self-efficacy, interpersonal trust, work engagement and role base performance on the effectiveness of authentic leadership. The quantitative research design is the most appropriate approach for this research question and objectives. The research adopts the cross-sectional correlation study to measure the variables and assesses the measurable relationship between minor or no exertion to control unessential factors. Descriptive and inferential statistics will be applied to analyse the data. The survey is conducted to get feedback from respondents, and the information collected is displayed in the tables and statistics. The collected data are questionnaires from CIMB call centre employees in Kuala Lumpur. All the variables data entered is coded accordingly into the Statistical Package for Social Sciences Software (SPSS)—three hundred of the questionnaire distributed to the targeted respondent. After one week, only 121 questionnaires were returned with complete responses. The incomplete and void questionnaire was considered unacceptable and was removed. Questionnaires will be used to collect data from the respondents. The focus of this study is to determine how employee behaviours and attitude components, which consist of employee self-efficacy, interpersonal trust, work engagement and employee role-based performance, contribute toward authentic leadership. This research used secondary data as references where data are collected from the journal, official statistics, reports and proceeding papers. The questionnaire is a self-administered question that ensures high respondent feedback and endorses freedom of expression from the respondents. It is a less complicated method that gives ease of time.

3.1 Theoretical Framework

See Fig. 1.

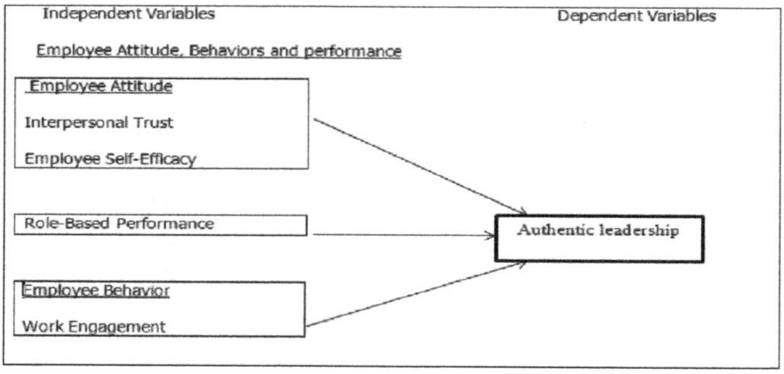

Fig. 1. Research Framework

3.2 Pilot Study

Table 1. Reliability Analysis for Pilot Test

Item	No. of Items	
Leadership Style (L.S.)	14	0.965
Interpersonal Trust (I.T.)	6	0.742
Occupational Self-Efficacy (OSE)	6	0.933
Performance (P)	20	0.956
Behaviour (B)	9	0.953

Table 1 shows the result of a pilot study. All Cronbach's alpha for authentic leadership, interpersonal trust, self-efficacy, working engagement and role-based performance values are 0.80 < 0.90, which means the results are excellent.

4 Result and Discussion

4.1 Reliability Test

A reliability test has been carried out, and All the values of Cronbach's alpha are above 0.7, which consider high-value reliability. The variables are valid to be used for this study. The reliability coefficient shows how properly the items are undoubtedly correlated to one another (Sekaran & Bougie, 2010). Occupational Self-efficacy is 0.8, meaning the association's strength is excellent. For Authentic Leadership, Roles-Bases Performance and Work Engagement with a score < 0.9 and above mean the association's strength is Excellent. Interpersonal trust, with a score of 0.7 above, shows that the power of association is good (Table 2).

Table 2. Cronbach Alpha value of the main study

Variables	No. of Items	Cronbach's Alpha
Leadership Style - Authentic Leadership (L.S.)	14	0.953
Occupational Self-efficacy (OSE)	6	0.848
Employee Attitudes - Interpersonal Trust (E.A.)	6	0.744
Roles-Bases Performance (P)	20	0.934
Organization Behavior - Work Engagement (B)	9	0.921

4.2 Demographic Profile Analyses

The information about the respondent profile showed 47 (38.8%) are male respondents, and total of 74 (61.2%) are female, and total respondent is 121. In this study, the nationalities of all CIMB staff are Malaysian. Whereas the percentage age group of employees on 29–29 years old, which is 31.4%, followed by 30–39, which are 65.3%, the age 40–49, which is 2.5% and the age above 60 is 0.8%. Next, the duration of employees' service with CIMB Bank, for employees who have been working less than one year, the percentage is 4.1%; for more than one but less than three years, the percentage is 33.9%; for More than three years but less than five years the percentage is 27.3%. For more than five but less than ten years, the percentage is 27.3%; for more than 10 but less than 20 years, the percentage is 6.6%; lastly, for More than 20 years 0.8%. This study will cover four branches of CIMB of Kuala Lumpur. Most Menara Atlas Plaza Pantai employees are 54.6% because the central call centre is allocated here. Then, followed by Menara Southern Bank (MSB), Damansara at 29.8%, Menara Bumiputera Commerce (MBC), Jalan Raja Laut is 8.2%, lastly, Menara CIMB at K.L. Sentral, 7.43%.

4.3 Descriptive Analysis

4.3.1 Normality Test

In the normality distribution, all the variables scores must within the limit of acceptable value which is less than −2 and +2 of Skewness and Kurtosis which indicates the normality of the data. The table shows all the variables are normally distributed.

Variable	Range		Conclusion
	Skewness	Kurtois	
Interpersonal Trust (DV)	−0.065	−0.017	Normal distribution
Occupational Self-Efficacy (DV)	0.017	0.107	Normal distribution
Performance (DV)	0.382	0.481	Normal distribution
Working Engagement (DV)	0.177	−0.23	Normal distribution
Authentic Leadership (IV)	0.21	−0.46	Normal distribution

4.3.2 Mean Test

Mean's Test to see the results of objective 1 of the study as follows:

The descriptive analysis such as mean score and grand mean for each variable. Leadership Style (L.S.), the mean for the statement "My leader openly shares information with others" is 3.66, which made up the highest mean score. Last batch is number thirteen and fourteen, which is 3.41 and 3.40 belong to "My leader describes accurately the way that others view his/her abilities" and "My leader uses his/her core beliefs to make decisions". The result for a grand mean of five items in Leaderships Style (L.S.) is calculated as 3.82.

An employee of interpersonal trust, the statement, "If I got into difficulties at work, I know my colleagues would try and help me out", is in first place with a mean value of 3.93. Lastly is in sixth place with a mean value of 3.13, and the statement is, "I feel quite confident that the firm will always try to treat me fairly". The result for a grand mean of five items in Employee attitude (E.A.) is calculated as 3.51.

The mean score for occupational self-efficacy (OSE) is shown that higher mean score is 3.82; first place is the statement, "My past experiences in my job have prepared me well for my occupational future" last place rank by is 3.6, "I can remain calm when facing difficulties in my job because I can rely on my abilities". The result for a grand mean of five occupational self-efficacy (OSE) items is calculated as 3.69.

The result of the mean score for performance (P) is shown in Table (Sect. 4.3). The statement "I seek information from others in my work group", with a mean value of 3.92, obtained first place with the highest rank. The last ranking of number 19, and the statement is "I satisfy the criteria of quality of work output that is officially required." Moreover, the mean value is 3.31. The result for a grand mean of five Performance (P) items is calculated as 3.52.

The mean score for behaviours (B) is calculated as 3.38. The "I am proud of the work that I do" fall in the first ranking with a mean value of 3.6, and the last ranking statement is "When I get up in the morning, I feel like going to work", and the mean value is 3.16.

4.3.3 Inferential Analysis (Hypothesis Testing)

In this study, the entire five hypotheses are tested using inferential statistics. For the relationship between IV and DV (Hypothesis 1 to 4) researcher tested using Pearson Correlation. Then, for the effect of IV and DV (Hypothesis 5) researcher tested by using linear regression.

1) Ha1: interpersonal trust has a relationship with authentic leadership

Correlation between employee attitudes of interpersonal trust with authentic leadership (Table 3).

Table 3. Correlation between employee attitudes of interpersonal trust with authentic leadership

		LS
EA	Pearson Correlation	.457**
	Sig. (2-tailed)	.000

There is a relationship between employee attitudes to interpersonal trust. It is a positive and moderate relationship with authentic leadership. The study positively evaluates their co-workers and leaders in building trust throughout the relationship and being trustworthy of authentic leaders. The value of $r = 0.457$, $p < 0.01$.

2) Ha2: Self-efficacy has a relationship with authentic leadership.

The results of the analysis and there is a significant relationship between self-efficacy and authentic leadership. The value of $r = 0.435$, $p < 0.01$. a positive and moderate relationship with authentic leadership. This finding is consistent with previous studies that found that employee self-efficacy did authentic leadership (Table 4).

Table 4. Correlation between self-efficacy with authentic leadership

		LS
OSE	Pearson Correlation	.435**
	Sig. (2-tailed)	.000

3) Ha3: Performance has a relationship with authentic leadership (Table 5).

The results of the analysis show there is a significant relationship between role-based performance and authentic leadership Encouraging changing employees' attitudes and

Table 5. Correlation between performance with authentic leadership

		L.S
P	Pearson Correlation	.593**
	Sig. (2-tailed)	.000

behaviours with authentic leadership practices would boost the performance level. The $r = 0.593^{**}$value, $p < 0.01$, is a positive and moderate relationship with authentic leadership.

4) Ha4: Organisation behaviours of work engagements have a relationship with authentic leadership.

Table 6. Correlation between employee behavior of working engagement with authentic leadership

		LS
B	Pearson Correlation	.431**
	Sig. (2-tailed)	.000

The results of the analysis showed that there is a significant relationship between work engagement and authentic leadership. The value of $r = 0.431^{**}$, $p < 0.01$. It is a positive and moderate relationship with authentic leadership. Employees' positive perception of vigour, dedication, and absorption in their work significantly relates to authentic leadership practices (Table 6).

5) Ha5: There is a significant relationship between employee self-efficacy, interpersonal trust, work engagement and employee performance on authentic leadership at the CIMB call centre in Kuala Lumpur (Table 7).

Table 7. Multiple regression analysis

Model		Unstandardised Coefficients		Standardised Coefficients	T	Sig.
		B	Std. Error	Beta		
1	(Constant)	−0.05	5.574		−0.009	0.993
	EA	0.762	0.234	0.268	3.261	0.001
	OSE	−0.279	0.301	−0.097	−0.926	0.356

(continued)

Table 7. (*continued*)

Model		Unstandardised Coefficients		Standardised Coefficients	T	Sig.
		B	Std. Error	Beta		
	P	0.447	0.096	0.478	4.658	0
	B	0.232	0.14	0.14	1.658	0.1

a. Dependent variable

The equation for authentic leadership is as below:

$$\text{Authentic leadership} = 0.268(\text{employee attitude}) - 0.097(\text{occupational self - efficacy})$$
$$4.658(\text{performance})\ 1.658(\text{behaviour})$$

The result of regression analysis on all four variables of authentic leadership was significant at R2 = 0.424, F = 21.375, and P < 0.05. These two variables are significant with a p-value < 0.05. They are employee attitudes of interpersonal trust and performance. The variable that was not significant is occupational self-efficacy since its p-value is 0.356, and employee behaviour with a p-value of 0.1. Two out of four variables of employee attitude, interpersonal trust and role-based performance have a relationship with authentic leadership.

4.4 Summary

This chapter discusses the results of data collection. The hypothesis was tested using Statistical Package for Social Science (SPSS) version 23. This study incorporated a reliability test, validity test, mean Test, descriptive analysis test, Pearson correlation test and multiple regression tests. All the variables are above 60% for the reliability test, which shows that it is reliable to use in this study and the mean test results of 3.551. This explains that the employee of the call centre department at CIMB Kuala Lumpur experience authentic leadership practices among their leader.

CIMB Bank call centre did practice authentic leadership among the leaders in Kuala Lumpur as this research shows a high level of authentic leadership practices at.551 of mean value through the employee attitude and behaviours of employee interpersonal trust, employee self-efficacy, employee performance and working engagement. The finding is acceptable with the existing studies that employee attitude, behaviours and employee performance play synchronise roles with authentic leadership practices.

From the correlation findings, the researcher discovers a significant positive and moderate relationship between interpersonal trust and authentic leadership with 0.457 correlations. It shows that the higher the perceived interpersonal trust, the higher the authentic leadership levels among employees of CIMB. This result confirms that of previous studies (e.g., Dirks & Ferrin, 2002; Gardner & Schermerhorn Jr., 2005; George, 2003; Hassan & Ahmed, 2011; Kahn, 1990) that authentic leadership may promote employees' attitude and behaviour changes positively such as increased self-efficacy and enhanced interpersonal trust among employees.

The researcher found correlation findings for the relationship between Self-Efficacy and Authentic Leadership and a significant positive and moderate relationship with 0.435 correlations. It shows that the higher the self-efficacy levels, the higher effectiveness of authentic leadership practice. By displaying authentic behaviour, transparency, and a high level of self-awareness, authentic leaders model for followers. As employees discover their leader's behaviour and value, such as engaging in transparent decision-making and showing self-awareness, they reflect their leaders' values and behaviours on themselves and understand themselves (Hye, 2000).

The relationship between job performance and authentic leadership is a correlation of 0.593. It shows that the relationship between job performance and authentic leadership and the factors for job performance systems are employees' attitudes and behaviour (Ostroff, 1992). Employees' negative emotions toward their colleagues and themselves hinder them employees from engaging in their work and ultimately lower productivity and performance (Gardner & Schermerhorn Jr., 2005; Luthans, Youssef, & Avolio, 2007). This result suggests the way to increase organisational performance by promoting employees to have high self-confidence in their work or increasing employees' trust in co-workers and leaders (Hye, 2000).

Work engagement will be affected by occupational self-efficacy and interpersonal trust, and occupational self-efficacy and interpersonal trust will indirectly influence role-based performance through work engagement (Hye, 2000). Also, there is a significant positive and moderate relationship between working engagement and authentic leadership, with 0.431 correlations showing the relationship between authentic leadership with work engagement. This chain-reacting relationship suggests that organisations must be aware of employees' positive attitudes and behaviour and realise the potential connections to role-based performance with authentic leaders to navigate as the connector (Hye, 2000).

Next, from the Pearson correlation test, all the variables have a significant relationship with authentic leadership. From the regression result, two out of four variables directly affect authentic leadership at a percentage of 42.4%.

5 Recommendations and Conclusions

5.1 Recommendations

Researcher in the future needs to conduct investigations in all Banks and institutions throughout Malaysia to get higher accuracy from the call centre profession and get higher accuracy from employees of accurately to practice authentic leadership at the managerial level. In addition, there were only 121 respondents in this study; in the future, research needs to increase the number of respondents so that the results are better and more involved.

Furthermore, this research requires more data to get a more accurate explanation. Therefore, future researchers can use another strategy to gather information, such as interviews and observation. This can increase respondents' interest and get more views from different angles.

In this study, only four variables were tested, and the future researcher may consider other factors related to authentic leadership. Since the regression test showed 42.4%, other factors affect authentic leadership.

In conclusion, the future researcher will conduct investigations more extensively. This can guarantee more reliable results than authentic leadership practices on employee attitude, behaviours and job performance factors.

5.2 Conclusions

In summary, the results of this study propose that when the level of acknowledgement of authentic leaders increases, it will also increase the level of knowing themselves among the followers. When the level of knowing themselves increases, it will leave a better perception toward the follower, assign an authentic leader as the role model, and continuously distribute changes toward positive attitudes and behaviours. To become a good role model, authentic leaders need to give extra focus and effort on developing positive psychological capital and leading it to be followed by other organisation members. Today's work environments have become highly competitive both inside and outside organisations. They require positive attitudes and behaviours to motivate employees to keep moving to compete with others and put extra focus on their tasks. In the end, the performance of employees would increase.

The researcher believes that this study indicates that organisations need an authentic leadership style of leadership as the role they play would bring a successful story to the members of the organisation and the organisation itself. Authentic leadership is the ideal leadership style to overcome the needs and obstacles of the current working environment. Authentic leadership would never be underestimated individual potential and always believe every individual can do their best, and it encourages followers to always be in a positive mindset. This has a positive effect on the success of any organisation.

References

Avolio, B.J., Walumbwa, F.O.: Authentic leadership: moving H.R. leaders to a higher level. In: Martocchio, J. (ed.) Research in Personnel and Human Resources Management, vol. 25, pp. 273–304. Elsevier Science/JAI Press, New York (2006)

Avolio, B., Luthans, F.: High Impact Leader: Moments Matter in Authentic Leadership Development. McGraw-Hill, New York (2006)

Avolio, B., Gardner, W., Walumba, May, D.: Unlocking the mask: a look at the process by which authentic leadership impact follower attitude and behaviors. Leadersh. Q. **15**(B), 801–823 (2004)

Bandura, A.: Self-efficacy mechanism in human agency. Am. Psychol. **37**(2), 122–147 (1990)

Bitmis, M.G., Ergeneli, A.: The role of psychological capital and trust in individual performance and job satisfaction relationship: a test of multiple mediating role models. Procedia Soc. Behav. Sci. **99**, 173–179 (2013)

Dirks, K.T., Ferrin, D.L.: Trust in leadership meta-analytic finding and implication for research and practice. J. Appl. Psychol. **87**, 611–628 (2002)

Farahmandian, S., Minavand, H., Afshardost, M.: Perceived service quality and student satisfaction in higher education. IOSR J. Bus. Manag. **12**, 65–74 (2013). https://doi.org/10.9790/487X-1246574

Gardner, W.L., Avolio, B.J., Luthans, F., May, D.R., Walumbwa, F.: "Can you see the real me?" A self-based model of authentic leader and follower development. Leadersh. Q. **16**(3), 343–372 (2005)

Halbesleben, J.R.: A meta-analysis of work engagement: relationships with burn out, demands, resources, and consequences. In: Bakker, A.B., Leiter, M.P. (eds.) Work Engagement: A Handbook of Essential Theory and Research, vol. 8, pp. 102–117. Psychology Press, New York (2010)

Ilies, R., Morgeson, F.P., Nahrgang, J.D.: Authentic leadership and eudemonic well-being: understanding leader-follower outcomes. Leadersh. Q. **16**(3), 373–394 (2005)

Jing, F.F., Avery, G.C.: Missing links in understanding the relationship between leadership and organisational performance. Int. Bus. Econ. Res. J. (IBER) **7**(5), 107–108 (2018). https://doi.org/10.19030/iber.v7i5.3256

Luthan, F., Avolio, B.: Authentic leadership: a positive development approach. In: K.S. Cameron, J. e Quins (eds.) Positively Organizational Scholar, pp. 241–258. Berrett-Koehler, San Francisco, CA (2003)

Luthans, F., Avolio, B.J.: Authentic leadership: a positive developmental approach. In: Cameron, K.S., Dutton, J.E., Quinn, R.E. (eds.) Positive Organizational Scholarship, pp. 241–261. Barrett-Koehler, San Francisco (2003)

Meyer, R.C., Davis, J.H., Schoorman, F.D.: An integrative model of organisation trust. Acad. Manag. Rev. **20**(3), 709–734 (1995)

Maddux, J.E.: Self-efficacy: the power of believing you can. In: Snyder, C.R., Lopez, S.J. (eds.) Handbook of Positive Psychology, pp. 277–287. Oxford University Press (2002)

Ozer, E.M., Bandura, A.: Mechanisms governing empowerment effects: a self-efficacy analysis. J. Pers. Soc. Psychol. **58**(3), 472–486 (1990). https://doi.org/10.1037/0022-3514.58.3.472

Özşahin, M., Zehir, C., Acar, Z.: Linking leadership style to firm performance: the mediating effect of the learning orientation. Procedia Soc. Behav. Sci. **24**, 1218–1231 (2011)

Schaufeli, W., Salanova, M., Gonzalez-Roma, V., Bakker, A.: The measurement of engagement and burnout: a two sample confirmatory factor analytic approach. J. Happiness Stud. **3**, 71–92 (2002)

Thomas, K.W., Velthouse, B.A.: Cognitive elements of empowerment: an "interpretive" model of intrinsic task motivation. Acad. Manag. Rev. **15**, 666–681 (1990)

Walumbwa, F., Avolio, B., Gardner, W., Wernsing, T., Peterson, S.: Authentic Leadership: Development and Validation of a Theory-Based Measure. Management Department Faculty Publications (2008)

Wang, D., Hsiech, C.: The effect of authentic leadership on employees' trust and employees engagement. Soc. Behav. Pers. **41**(4), 613–624 (2013)

The Impact of Internal Control System on the Market Value of Jordanian Banks Comparative Study Between Traditional and Islamic Banks

Mohyedin Hamza[1](✉), Riham Alkabbji[1], Tareq Hmmad Almbydeen[1], Rama Fathi[1], and Qadri Aljabri[2]

[1] Zarqa University, 2000, Zarqa 13110, Jordan
{Mhamza,rkabbji,tmbydeen}@zu.edu.jo
[2] University of Business and Technology, Jeddah, Saudi Arabia

Abstract. This article aimed to identify the extent of the adherence of banks operating in Jordan to the application of the Internal Control System (ICS) ((AC), administrative control, and internal control) and the impact of this application on the (MV) of banks operating in Jordan, represented by the Market Value (MV) of shares, and a comparison between commercial banks and Islamic banks. The researcher used the descriptive and analytical approaches to carry out the current article.

A questionnaire was used to collect data related to the (IDV) (ICSs).

The article found that there is a high adherence to the application of (ICSs) in its three branches in both commercial banks and Islamic banks in very close proportions, but the Islamic banks had a slightly higher level of application, and there is also a positive impact of the application of (ICSs) on the (MV) of banks operating in Jordan (commercial and Islamic) at a significant level of 5%.

The article presented set of recommendations, the most important of which are: the need to increase the interest of banks in the impact of transferring employees on the ability and efficiency of departments in performing administrative duties, in order to achieve job satisfaction and raise the effectiveness of performance for employees.

Keywords: internal control systems · market value · accounting control · Internal Tuning (IT)

1 Introduction

Banks, both Islamic and traditional, are currently facing intense competition to maintain their existence and achieve growth, due to the increasing number of competing banks and opening up to global markets [15]. Therefore, all banks, without exception, found themselves obliged to making the maximum possible effort to achieve their goals and increase their market shares and values. Since achieving that depends on the extent of employees ' adherence to what is planned, the effectiveness of the control mechanisms

B. A. M. Alareeni and I. Elgedawy (Eds.): *Artificial Intelligence (AI) and Finance*, SSDC 488, pp. 657–668, 2023.
https://doi.org/10.1007/978-3-031-39158-3_61

used to ensure that adherence is an essential factor to achieve the desired goals [1]. In Jordan, two banks' systems operate, namely traditional banks system and Islamic banks system that apply the teachings of Islam in banking transactions and loans. These two systems came as a result of the society fabric and its beliefs, which sometimes tend to prefer dealing with Islamic banks for religious reasons.

This article investigates the extent of the application of (ICS) in Jordanian banks and provides a comparison between traditional and Islamic banks in terms of their application of (ICS)and the potential implications on the (MV) of these banks. Thus, the problem statement of the current article can be formulated in its attempt to answer the following main question:

The main question: Is there a (SSE) of the practice of (ICS) on the (MV) of Jordanian traditional and Islamic banks?

The following sub-questions branch from the previous main question:

The first sub-question: Is there an impact of the practice of administrative control as a control tool on the (MV) of Jordanian Islamic and traditional banks?

The second sub-question: Is there an impact of the practice of (AC) as a control tool on the (MV) of Jordanian Islamic and traditional banks?

The third sub-question: Is there a statistic for the practice of (IT) as a control tool on the (MV) of Jordanian Islamic and traditional banks?

2 Previous Studies

Lots of research addressed the topic of (IC) in traditional banks, whether Jordanian, Arab and even international. The studies discussed the various aspects of (IC) on various aspects of the bank's activities and their financial performance measures. Lotfy (2021) [2] addressed the impact of (IC) on the financial performance of traditional banks and concluded that there is a significant impact of (IC) on the financial performance indicators of traditional banks. Ayman (2017) [3] confirmed what Lotfy came up with and also stressed the importance of (IC) and its positive impact on the development of the control environment in the light of governance requirements. This was also confirmed by Qashtem (2018) [4] which stated that the (ICS) positively affects financial performance.

On the other hand, many studies have addressed the impact of financial performance on the (MV) of stock prices listed on the Stock Exchange. Kabajeh, 2012 [5] found that there is a strong positive correlation between financial performance and the (MV) of stocks.AlQaisi, 2016 [6] confirmed that by stating that there is a direct relationship between financial performance indicators and companies' (MV).

The current article differs from previous studies since it investigates the impact of (ICS) as an (IDV) that has been studied a lot as a (DV). Whereas, in this article, its impact on the (MV) of banks operating in Jordan is addressed. The (MV) is a variable that has been studied as a dependent or (IDV). In addition, the article environment is different from most previous studies in terms of the sector under article and the time period covered by the article. Therefore, this article differs from previous studies in terms of the applied model, which is the impact of (ICS) on the (MV) of Jordanian banks.

3 Article Methodology

The descriptive analytical approach was adopted since it is consistent with the article objectives.

3.1 Article Sample and Population

The population of the article include of 21 Jordanian banks. The sample of the article include of 4 Islamic banks and 17 traditional banks working in Jordan, according to the Banks Guide - the Central Bank of Jordan: https://www.cbj.gov.jo/Pages/view%20page.aspx.

3.2 Article Hypotheses

There is no Statistic Significant Effect (SSE)at the Significance Level (SL)(0.05 $\geq \alpha$ for the various types of (IC)on the (MV) of Jordanian Islamic and traditional banks.

H01: The first sub-hypothesis: There is no (SSE) at the (SL) (0.05 $\geq$ for the Administrative Control as a control tool on the (MV) of Jordanian Islamic and traditional banks.

H02 The second sub-hypothesis: There is no (SSE) at the(SL)(0.05 $\geq \alpha$for (AC) as a control tool on the (MV) of Jordanian Islamic and traditional banks.

H03: The third sub-hypothesis: There is no (SSE) at the (SL) (0.05 $\geq \alpha$) for the Administrative Control as a control tool on the (MV) of Jordanian Islamic and traditional banks.

3.3 Article Model

Figure 1 shows the independent variables (IDVs) represented in the (ICSs), and the (DV) represented in the (MV) of Jordanian banks listed on the Amman Stock Exchange:

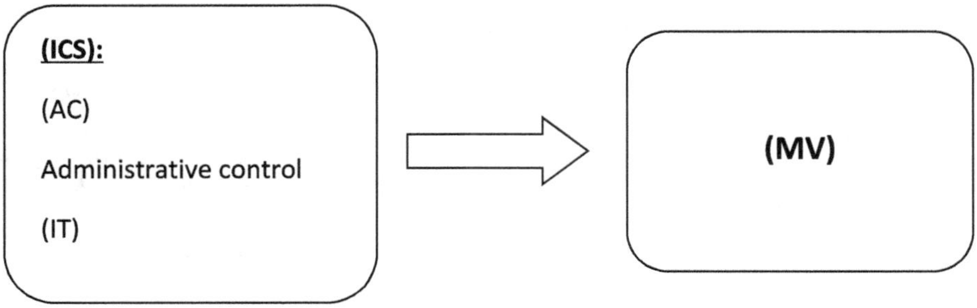

Fig. 1. Study Model

3.4 Article Tool

To achieve the objectives of the article, a questionnaire was designed and developed, to identify the level of (IC) in traditional and Islamic Jordanian banks, from the point of view of the article sample. This tool was developed to cover the hypotheses on which the article was based.

3.5 Stability of the Article Tool

To verify the stability of the article tool, the Cronbach Alpha coefficient was used, where the stability coefficient of the overall tool was (78.5). This percentage is high enough to approve the article results, since the acceptable percentage of generalizing the results for Humanitarian and Social Research is (70%) or more, as shown in Table 1 (Asma & Rahma 2019) [7].

Table 1. Cronbach's Alpha coefficient for measuring the stability of the questionnaire.

Part	Variables	NO	Cronbach Alpha
The first	(AC)	11	79.5
The second	Administrative control	11	73.2
The third	Internal control	11	82.7
The tool as a whole		**33**	**78.5**

3.6 Article Variables:

(DV):

Table 2. Dependent Variable.

Paragraph	NO
Accounting Control	
There are control procedures in the accounting system that prevent the deletion of previously recorded transactions	1
There are procedures that prevent the proof of any financial transaction or document unless it is approved by the responsible employee	2
All financial transactions are recorded in their accounting period	3
There are procedures to ensure that documents are analyzed and audited periodically	4
There are procedures to ensure the safety of the computerized accounting system continuously	5

(continued)

Table 2. (*continued*)

Paragraph	NO
There is a set of ledgers parallel to the computerized accounting system	6
Accounting disclosures are done correctly and adequately	7
The bank makes sure that it provides appropriate information at the right time to decision makers	8
There are procedures to ensure that international accounting standards and international financial reporting standards are followed and not violated	9
There are accounting and documentary courses to achieve effective control	10
There are qualifying courses with modern accounting standards to provide the requirements for applying the standard	11
Administrative Control	
The bank has an internal system that defines the powers and responsibilities of employees	1
There are rehabilitation and training programs for new employees	2
There are highly qualified staff who carry out their duties accurately and effectively	3
There is an organizational chart showing the departments, divisions, tasks and responsibilities assigned to them	4
The audit committees follow up the administration's work on the administration's implementation of the administrative control system	5
There are laws and regulations that prevent dishonest behavior There is a clear administrative hierarchy so that whoever is responsible for errors or violations can be determined as soon as they occur	6
There are laws and regulations that prevent dishonest behavior There is a clear administrative hierarchy so that whoever is responsible for errors or violations can be determined as soon as they occur	7
There is a system for appointing and promoting employees with clear instructions and announcements	8
There is a clear and announced performance appraisal system	9
There is a clear and announced performance appraisal system The department responds to the internal control reports issued by the internal or external auditor	10
The administration takes appropriate remedial measures in case there is any defect or error	11
Internal Tuning	
The Bank is interested in the extent of the impact of the transfer of employees on the ability of work departments to perform practical functions in the Bank	1

(*continued*)

Table 2. (*continued*)

Paragraph	NO
The bank makes sure that the bank's internal system policies and procedures are documented and reviewed by all bank employees	2
The bank makes sure that the available resources are sufficient to achieve the required goals efficiently and effectively	3
The Bank's management responds to the recommendations issued by the internal and external auditors to strengthen the Bank's internal control system	4
The bank makes sure that the set of policies and procedures allow employees to perform their jobs efficiently	5
The bank makes sure that the objectives at the departmental level are in line with the strategic objectives of the bank	6
The bank makes sure that the level of delegation of authority to officials is the basis of accountability and control in the bank	7
The Bank ensures that the control activities reduce the level of risk in the Bank	8
The bank makes sure that the level of delegation of powers corresponds to the level of responsibilities	9
The bank makes sure that the volume of bank transactions performed daily is commensurate with the number of employees	10
The bank has policies for controlling employee vacations for control purposes	11

(IDV): (ICSs)

The questionnaire was used to measure the (ICSs) as an (IDV). The random sampling method was used from within the article population. 150 questionnaires were distributed among the Control staff in the banks. The article sample included branch managers, head of department, accountants, internal auditors and the cashier. 21 banks were included, 4 of which were Islamic and 21 traditional. From the 150 questionnaires distributed, 80% were retrieved, while the rest were not due to the lack of cooperation from the part of some respondents. 8 questionnaires were excluded due to incompleteness. Therefore, a number of 125 questionnaires were analyzed, constituting 75% of the retrieved questionnaires. This is shown in Table 2 and 3).

Table 3. Distributed, retrieved and analyzable questionnaires.

Distributed number	Retrieved number	Response rate	Number of questionnaires valid for analysis	Percentage of questionnaires valid for analysis
150	**120**	**80%**	**112**	**75%**

(DV): (MV) of Islamic and traditional banks
Data on the (MV) of Jordanian traditional and Islamic banks were collected from the Amman Stock Exchange website for the last five years (2017–2021). The time series were used to determine the stability of these data and their suitability for use in statistical analysis.

4 Analysis of Variables

4.1 (Idv): (Ics)

Analysis of the paragraphs of the questionnaire:
The arithmetic means and standard deviations of the sample's answers were extracted to describe their answers to the questionnaire paragraphs, determine the degree of interest of the article sample individuals, and detect the extent to which the answers are scattered from their arithmetic averages. Table 4 summarizes these results:

Table 4. Summary of the results of the respondents' answers to the article questions.

NO	Dimension	Traditional banks		Islamic banks		TAV
		μ	σ	μ	σ	
1	**Average results of measuring (AC)**	4.424	0.59	4.50	0.473	4.462
2	**Average results of measuring administrative control**	4.4	0.65	4.636	0.45	4.518
3	**Average results of (IT) measurement**	4.313	0.652	4.5	0.513	4.4
	Total	4.379	0.631	4.54	0.479	4.46

Table 4 shows the general average of the respondents' response to each dimension of the (IDV). There is compliance with the requirements of (ICS) for both traditional and Islamic banks, as the arithmetic mean for (AC) of Jordanian banks amounted to (4.462), which is a high indicator of the application. The arithmetic means for administrative control reached (4.518), and for (IT) was (4.4). All means fall within the category (4.3–5), which indicates the presence of high approval that there is a high application of (ICS) in Jordanian banks.

We conclude that the (ICS)of traditional and Islamic Jordanian banks are at a high level of performance and control, and this is an indicator of the quality of administrative and accounting work and the quality of the (IC)system, which must be reflected on the financial statements of Jordanian banks and thus affect their (MV).

4.2 Analysis of the (MV) of Jordanian Banks (Traditional and Islamic)

Several indicators of (MV) will be used, where the (MV) of the bank as a whole, the (MV) per share, as well as the (MV) to book value will be used.

Table 5. shows the arithmetic means, standard deviation, and the upper and lower values of (MV) indicators.

Variable	Period	μ	σ	High values	Minimum values
(MV)	2017–2021	531187633.7	739630744.9	3075840000	88000000
(MV) per share	2017–2021	2.04	1.140	4.8	0.8

Table 5 above shows the arithmetic mean for the (MV) of banks operating in Jordan (531187633.7) with a standard deviation of (739630744.9) dinars. The highest value of (3075840000) belongs to the Arab Bank, which is the largest Jordanian bank in terms of capital size, and the lowest value of (88000000) dinars belongs to the Arab Banking Corporation. As for the (MV) per share, the arithmetic mean amounted to (2.04), while the highest value is (4.8) more than twice the arithmetic mean, also belonging to the Arab Bank, whereas the lowest value of (0.8) belongs to the Arab Banking Corporation. A large gap between the highest and lowest value is noted, and this results in a high value of the standard deviation, reaching (1.14).

4.3 Testing the Suitability of Data for Statistical Analysis

In this part of the article, the suitability of the (MV) data for the article was tested through the Time series Stationary as follows:

4.4 Data Stability Testing of the Article Variables (Stationary Test)

The stability of time series refers to the constancy of each mean and the variation of the values of the series over time. The variation between two time periods is based only on the time gap and not on the real time in which the variability is measured, and the unit root test is applied to confirm whether the article variables are stable or not.

Table 6. The results of the unit root test for the (MV) of Jordanian banks.

Result	P-value	The calculated value at the (SL)	Variable
Static at the level	0.019	-2.079	(MV)
Static at the level	0.002	-2.855	(MV) per share

The (LLC) test (Levin-Lin-Chu) was carried out to test the hypothesis whether the variables contain the unit root and its suitability for the sectional data associated with time (Panel Data). It is clear from Table 6 that the (P - Value) values of the (MV) data did not exceed 5% and this indicates the absence of unit roots, therefore we conclude that the time series is stable.

4.5 Testing the Article Hypotheses

The hypotheses will be tested on traditional and Islamic banks operating in Jordan together, and then the hypotheses will be tested separately for each of them.

The main hypothesis was tested through sub-hypotheses, in which (IC) was divided into three types, namely: (AC), administrative control, (IT), and the (IDV) the (MV) per share. Table 7 shows the sub-hypothesis test results.

Table 7. Multiple regression analysis test results for measuring the impact of (ICS) on the (MV) per share.

(DV): (MV) per share			
Variables	Beta	T-test	Significant
Fixed	-5.51177	-3.006	0.003
(AC)	0.387	3.030	0.004
Administrative control	0.339	2.598	0.012
(IT)	0.274	3.030	0.004
R2	0.347	F	6.748
Adjusted R^2	0.589	Significant F	0.002

The value of the tabular F at the level of significance (0.05 ≥ α amounted to (2.39).
The value of the tabular F at the level of significance (0.05 ≥ α amounted to (2).

The data in Table 7 shows that the calculated value of F was (6.748), while its Tabular value was (2.39). By comparing the values reached in testing this hypothesis, it is shown that the calculated value is greater than the tabular value, so the null hypothesis is rejected, and the alternative hypothesis is accepted, which states that **"There is a (SSE) at the (SL) (0.05 ≥ α for the various types of (IC) on the (MV) of Jordanian Islamic and traditional banks** at the level of significance (0.05 ≥ α)." This is confirmed by the value of the (SL) (0.002 = α) which is below the level of significance (0.05 ≥ α). As indicated in the Table, the variance in the (IDVs) (R^2) explains (0.347) percent of the variation in the (DV).

From Table 7, the results of the sub-hypotheses are concluded as follows:

The first sub-hypothesis:

H01: There is no (SSE) at the (SL) (0.05 ≥ α) for the Administrative Control as a control tool on the (MV) of Jordanian Islamic and traditional banks.

Table (8.4) shows that the value of (Sig = 0.004), which is a significant value amounting to the value of (T = 3.03), which is greater than the tabular T. Therefore, the null hypothesis is rejected, and the alternative hypothesis is accepted, which states that there is a (SSE) at a significant level of (5%) for (AC) as a control tool on the (MV) of Jordanian Islamic and traditional banks.

The second sub-hypothesis:

H02: There is no (SSE)at the(SL)(0.05 ≥ α) for the Administrative Control as a control tool on the (MV) of Jordanian Islamic and traditional banks.

Table Table (8.4) shows that there is a statistically significant effect at a significant level of (5%) for the (IDV) according to the opinions of the article sample on the (DV), since the value amounted to (Sig = 0.012). Also, the calculated value of T is greater than the tabular T value, thus the null hypothesis is rejected, and the alternative hypothesis is accepted, indicating that there is a (SSE) of administrative control as a control tool on the (MV) of Jordanian Islamic and traditional banks.

The third sub-hypothesis:

H03: There is no (SSE) at the (SL) ($0.05 \geq \alpha$) for the (IT) as a control tool on the (MV) of Jordanian Islamic and traditional banks.

Table Table (8.4) shows that the value of (Sig = 0.004) at a significant level is (5%) or less on the variable, and the T test indicates the rejection of the null hypothesis since the calculated T is greater than the tabular T, and therefore it is concluded that there is a (SSE) at the (SL) ($0.05 \geq \alpha$) for (IT) as a control tool on the (MV) of Jordanian Islamic and traditional banks.

4.6 Hypothesis Testing for both Islamic and Traditional Banks Separately

Table 8. The results of the multiple regression analysis test to measure the impact of (ICS) on the (MV) of shares of traditional banks and Islamic banks independently.

(DV): (MV) per share				
Variables	Banks	Beta	T-test	Significant
Fixed		-5.51177	-3.006	0.003
(AC)	**Traditional**	0.4	3.530	0.005
	Islamic	0.16	2.05	0.000
Administrative control	**Traditional**	0.35	2.888	0.018
	Islamic	0.15	2.45	0.012
(IT)	**Traditional**	0.30	3.63	0.042
	Islamic	0.19	2.66	0.019

Comparing the results of Islamic and traditional banks, it is noted that that they all fall within the rejection of the null hypothesis and the acceptance of the alternative hypothesis with a small discrepancy in the N values. This is logical due to the convergence of the survey results for both Islamic and traditional banks, whereas the discrepancy was in the level of impact of each of the three areas of control ((AC), administrative control, and (IT)) on the (MV) of shares, due to the low number of Islamic banks, being only four. The result of the interpretation of each of the (IDV)s of the (DV) was low, as (AC) interprets only 16% of the (MV) of Islamic banks, while for traditional banks, it interprets 40% of the (MV) of shares. As for administrative control, it interprets only 15% of the (MV) of shares, while it interprets 35% of the (MV) of shares in traditional banks. Also, (IT) affects the (MV) of shares in Islamic banks by 19% while it interprets 30% for traditional banks (Table 8).

In Table 9, we summarize the following results:

Table 9. Summary of the results of testing the article hypotheses

Hypothesis	Resolution	Result
H0	Rejection	There is a (SSE) at the (SL) ($0.05 \geq \alpha$) for the various types of (IC) on the (MV) of Jordanian Islamic and traditional banks
H01	Rejection	There is a (SSE) at the (SL) ($0.05 \geq \alpha$) for the (AC) as a control tool on the (MV) of Jordanian Islamic and traditional banks
H02	Rejection	There is a (SSE) at the (SL) ($0.05 \geq \alpha$) for the Administrative Control as a control tool on the (MV) of Jordanian Islamic and traditional banks
H03	Rejection	There is a (SSE) at the (SL) ($0.05 \geq \alpha$) for the (IT) as a control tool on the (MV) of Jordanian Islamic and traditional banks

5 Conclusions

According to the statistical analysis of the data collected to achieve the objectives and problem of the article, and based on the results of testing the article hypotheses, the following can be concluded:

Firstly The results of the article showed that there is a (SSE)at the level of significance ($0.05 \geq \alpha$) for the different types of (IC)on the (MV) of Jordanian Islamic and traditional banks. This result is consistent with the findings of (Lotfi & Shehata, 2021) that (IC)affects financial performance, as the (MV) is a mirror of the financial performance of companies according to the findings of (Halabi 2020). This is also concluded by (German,2018). Secundly There is a (SSE)at the(SL)($0.05 \geq \alpha$) for (AC) as a control tool on the (MV) of Jordanian Islamic and traditional banks, and this is consistent with the findings of (Lutfi & Shehata, 2021). Thairdly There is a (SSE)at the(SL)($0.05 \geq \alpha$) for the administrative control as a control tool on the (MV) of Jordanian Islamic and traditional banks.Finally There is no (SSE)at the(SL)($0.05 \geq \alpha$) for (IT) as a control tool on the (MV) of Jordanian Islamic and traditional banks.

6 Recommendations

According to the conclusions drawn from the theoretical framework of the article, as well as from the statistical analysis of the data, a number of recommendations can be proposed that could contribute to raising the efficiency of (ICS)over the (MV) of banks operating in Jordan, which is the first we recommended Increasing banks ' interest in the impact of staff transfer on the ability and efficiency of work departments in performing administrative functions, in order to achieve job satisfaction and raise the effectiveness of staff performance. And increasing the interest in traditional collection system parallel to the computerized accounting system to raise the efficiency of (AC) in traditional banks. Then ensuring that there are procedures that prevent the proof of any financial transaction or document unless it is approved by the officer in charge to raise the efficiency of (AC). In addition to the need to increase the number of qualification courses in accordance with modern accounting standards to raise the efficiency of (AC).

References:

1. Tessier, S., Otley, D.A.: conceptual development of Simons' Levers of Control framework. Management Accounting Res. **23**(3), 171–185 (2012)
2. Salma, L., et al.: The impact of the quality of the (IC) structure on the financial performance of companies an applied article on listed companies on the Egyptian Stock Exchange, Alexandria J. Acc. Res. **5**, 1 (2021)
3. Iman , A.A.: The role of internal audit in developing the efficiency and effectiveness of the control environment in the light of governance requirements, Master's thesis in accounting, Ain Shams University, Cairo, Egypt (2017)
4. Qushtom, T.F.: The moderating effect of the Revised Levers of Control on the relationship between gaming problem and financial performance in Jordanian Industrial Listed Companies. Kuala Lumpur: PhD Thesis: International Islamic University Malaysia (2018)
5. Kabajeh, M.A.M., Al Nu'aimat, S.M.A., Dahmash, F. N.: The relationship between the ROA, ROE and ROI ratios with Jordanian insurance public companies market share prices. Int. J. Hum. Soc. Sci. **2**(11), 115–120(2012)
6. Al Qaisi, F., Tahtamouni, A., Al-Qudah, M.: Factors affecting the market stock price-The case of the insurance companies listed in Amman Stock Exchange. Int. J. Bus. Soc. Sci. **7**(10), 81–90 (2016)
7. Central Bank of Jordan https://www.cbj.gov.jo/Pages/viewpage.aspx
8. Mohamed, A.I,Rahma, A.A.A.: The number of response alternatives influenced the estimation of the stability of the test scores of Advanced Modular matrices using the Cronbach Alpha and stratified Alpha coefficients. Hama university magazine, **2**(10) (2019)
9. Al-Hakim, M.: Organizational support and its relationship to self-control. Mohamed Boudiaf University, Messila-Algeria (2020)
10. Ashur, A.-K.: The level of self-censorship among Yarmouk University students during their use of social networks from their point of view. Al-Manara J. Res. Stud. **24**(1), (2018)
11. Bahrini, R., Filfilan, A.: Impact of the novel coronavirus on stock market returns: evidence from GCC countries. Quant. Finance Econ. **4**(4), 640--652 (2020)
12. Chen, H., Yang, D., Zhang, X., Zhou, N.: The moderating role of (IC)in tax avoidance: Evidence from a COSO-based (IC)index in China. J. Am. Taxation Assoc. **42**(1), 23--52 (2020)
13. Lee, J., Kwon, H.-B.: Progressive performance modeling for the strategic determinants of (MV) in the high-tech oriented SMEs. Int. J. Prod. Econ. **183**, (2017) Oussii, A.A. Boulila Taktak, N.: The Impact of internal audit function characteristics on (IC)quality, Manag. Aud. J. **33**, 5 (2018)
14. Nidal ,A.-R.: Measuring the extent of application of the (COSO) framework for (IC)from the point of view of the external auditors in the companies, Zarqa J. Res. Stud. Hum. **17**(2), 2017

Impact of Applying Kaizen Approach on Financial Performance in Light of Digital Transformation

Omar Fareed Shaqqour[1](✉), Mohammad Suhail Aloqdeh[1], and Huda Alattasi[2]

[1] Zarqa University, Zarqa, Jordan
oshaqqour@zu.edu.jo

[2] Department of Finance, University of Business and Technology, Jeddah, Kingdom of Saudi Arabia
h.attasi@ubt.edu.sa

Abstract. This study aims to identify impact of applying Kaizen approach on Financial Performance in light of digital transformation in Jordanian industrial companies (JIC) listed at Amman Stock Exchange (ASE). In order to achieve study objectives, an analytical descriptive approach was followed, a questionnaire developed by researcher to measure Kaizen approach and digital transformation, meanwhile, financial performance was measured by (ROA and ROE). Where 150 questionnaires were distributed to sample members, and number of valid ones for analysis was (109), representing (72.7%) of study sample. Study population consisted of (34) companies, and study sample was limited to (30) companies within financial period (2010–2020). Findings showed that there is an effect at significance level ($\alpha \geq 0.05$) for applying of Kaizen approach on ROA and ROE in light of digital transformation in JIC listed at ASE. Study recommended necessity of encouraging employees of industrial companies listed at ASE.

Keywords: Kaizen Approach · Financial Performance · Digital Transformation

1 Introduction

Companies, especially industrial companies, face unstable conditions, as a result of rapid developments and changes in environment, and material conditions they face, such as high prices of raw materials, which made these companies search for new methods and ideas that reduce their costs and increase their profits, thus achieve a competitive advantage and enable them to make best use of available resources. One of methods that help organizations achieve their goals is to adopt methods that have an organizational philosophy that helps them continuously improve in all activities within organization [1].

Roots of continuous improvement approach (Kaizen) go back to Japan. After World War II, many Japanese organizations began implementing and adopting continuous improvement programs [2].

B. A. M. Alareeni and I. Elgedawy (Eds.): *Artificial Intelligence (AI) and Finance*, SSDC 488, pp. 669–678, 2023.
https://doi.org/10.1007/978-3-031-39158-3_62

Companies need to adapt digital technology in all its aspects, which requires companies keeping up with information and communication technologies in order to move these companies towards adopting the digital transformation.

Financial performance is considered as one of the main components of companies that provides an integrated system for accurate and reliable information, which aims to compare the actual performance of the companies' activities through specific indicators, to identify deviations from previously goals [3, 20].

JIC faced many challenges after the financial crisis and the negative effects of the Corona pandemic, which negatively affected the financial performance and increased in production costs. This required Jordanian industrial companies to move away from traditional methods and adopt modern operating methods in order to raise the level of financial, operational and marketing performance.

Accounting interest has increased in the recent years to the Kaizen approach for continuous improvement in production to achieve the largest margin for the company, and digital transformation has emerged as a modern means of development. Problem of study is focus on what is effect of applying Kaizen approach on financial performance measured by ROA and ROE in light of digital transformation in JIC listed on ASE?

This study seeks to identify impact of applying Kaizen approach on financial performance measured by ROA and ROE in light of digital transformation of JIC listed on ASE.

2 Theoretical Framework

2.1 Kaizen Approach

Kaizen approach is known in Japanese companies as "Genka Kaizen", where Kaizen approach is used for continuous improvement in product manufacturing stage as a process to reduce costs, so that Kaizen approach focuses on continuous improvement in all operations, customer satisfaction and participation of all employees in company. Kaizen for continuous improvement is practice of strategic management accounting closely with pursuit of a competitive advantage. Strength of Kaizen approach to costs is its close connection with company's profit planning process, so company can test Kaizen approach by improving continuous production and avoiding defects in design and development stage, which leads to cost reduction [4].

Carnerud et al. [5] defined Kaizen approach as effectiveness by which focus is placed on organization's approach to some activities in which managers and workers have ability and knowledge of its performance, which leads to cost reduction, while Mandahawi et al. [6] defined Kaizen approach as a process for reducing costs in light of manufacturing process in product life cycle through small and continuous improvements that are made to manufacturing process instead of large radical improvements and developments generated by innovations and huge investments. Researcher believes that Kaizen approach is a comprehensive and general concept for improving all steps of production process in companies, which is done through concerted efforts of all employees and managers at all levels, each in his position, with aim of improving outputs and reducing costs, by adopting a set of improvement mechanisms and small and continuous substantial additions that contribute to Advancing work mechanism and improving production and manufacturing process in general.

Kaizen approach is unique from other methods by a number of characteristics, including that it represents always renewed ideas that ignore traditional fixed ideas, and it is an approach that considers that it is possible to improve everything even if improvement is small, and assumes that imperfect immediate solutions are better than completed solutions that do not implement, in addition, it is an approach that does not require high investments to implement, and an approach that cares about employees, listens to them, and collects and adopts their ideas. It also focuses on eliminating root causes of problem, and not only on its effects. Therefore, it is a method that focuses on operations more than results, and spreads culture of organization that it does not tolerate waste, and finally it is an approach that believes that big results come from accumulated small changes, which have greatest impact in improving performance in long term [7].

Bakosh [8] mentioned that Kaizen approach is related to defining methods of development in daily business and implementing them by setting development teams interested in solving problems, and that continuous improvement means satisfying needs and desires of customer by modifying organization's operations, and one of characteristics of Kaizen approach is nature of change Partially in processes and products, starting point of improvement is from within an already existing process, improvement process takes place by repeating change in a large and high way, time spent in improvement process is short, form of participation in continuous improvement process is from bottom of functional hierarchy to top and extent of change is resulting process of continuous improvement is narrow, not broad.

Principle of continuous improvement forms basis for concept of total quality management, and process of improving products and services is basis for success of organizations in their various aspects of their activities, and mechanism for applying this approach is centered on following [9]: By reducing or limiting all activities that do not add significant value in production process Goods and services, continuous striving towards achieving requirements of customers and keeping up with their desires and meeting their needs, working to avoid defects, by working in a coordinated manner with information systems in organizations, and participation of all departments for success and achieving goals, identifying current state of production manufacturing process and trying to improve it, analyzing activities that It is carried out by workers and corrects defects, checks work systems and procedures that may cause a decline in performance, finds activities capable of absorbing and keeping pace with change in various production and service activities on an ongoing basis.

Many researchers saw that "Kaizen" continuous improvement approach is a smooth approach, as it is easy to apply, quick to improve, and supportive of achievement. It does not require great efforts, comprehensive change, and huge costs, as companies must follow a set of steps to implement Kaizen approach, including [10, 11]:

1- Planning: Examining and documenting place of improvement by analyzing data, then determining improvement goals, costs, and benefits resulting from defects, problems, and errors. Rather, they must be shown and identified, and causes of errors must be searched for and resolved in quick and radical ways. Alternatives must be achieved to achieve these goals [23].

2- Implementation Do: Implementation of improvement plan, monitoring its progress, and collecting data continuously, to measure improvement in process subject to improvement, and any changes in it must be documented and reviewed if necessary.

3- Check: Analyzing data collected in second phase to determine how close it is to achieving desired goals, and in event of any defect, plan must be re-evaluated or suspended [21].

4- Correction Act: If results obtained from third stage are successful, work team reconsider documents and results related to process subject to treatment, and those procedures become a measure of application.

Kumar [12] referred to requirements necessary to implement Kaizen approach, including search for easiest solutions, not optimal ones, and use of ideas, proposals, and opinions, not money, in addition to looking at problems as opportunities for improvement and not as obstacles, and that it should not stop at certain stages, as it must improvement should be endless, and idea of fixed system and traditional management should be abolished, and search for ways to make application successful instead of searching for excuses, and it should not be hidden.

2.2 Digital Transformation

Digital transformation is process of companies moving to business models based on digital technologies, to develop and innovate company's products and services, and to provide new marketing channels and job opportunities that increase value of its products [13].

Digital transformation [21] has many and varied benefits for customers, institutions, companies, and various stakeholders, most important of which are [14]: Improving efficiency of services and products provided, flexibility and speed of products, significantly reducing expenses, reshaping culture of society in general, whether in work, thinking, interaction, and communication Through available modern technologies, creating opportunities to provide innovative and creative services in providing services, providing different strategies that give competitive value, improve quality and simplify procedures for beneficiaries to obtain products and services [24–26].

3 Literature Review

Jad's study [15], study aimed to identify role of applying Kaizen strategy for continuous improvement in improving project management in construction and building contracting companies in Greater Cairo, and researcher concluded that there are medium practices of Kaizen approach for continuous improvement in companies under study. Study of Abu Baqi and Al-Kasasbeh [16], this study aimed to clarify impact of Kaizen methodology from perspective of Deming methodology on financial performance of Islamic banks in Jordan. Results showed that there is an effect of two dimensions of Kaizen, namely (planning and application) on financial performance. Baker's study [17], research sought to define theoretical and philosophical framework for Kaizen methodology in management, and to study attitudes of administrative leaders towards that methodology, and among most important results that were reached was a significant decrease in level of managers with regard to their effectiveness in achieving that methodology, and this is due to This leads to lack of familiarity with basic concepts of Kaizen management methodology, and skills necessary to achieve them, and then a significant decrease in results related to level of organizational performance in terms of spreading spirit of cooperation, teamwork, achieving innovation and continuous organizational change, achieving organizational learning, and employee engagement and loyalty. Improving service quality, speed of performance, and customer satisfaction. Prayuda [11] Study, purpose of this study is described process of applying a kaizen culture in a company working in automotive industry (PT AGP). Results of this study showed that application of Kaizen culture rises rate of productivity and rises perceived benefits of company, especially on financial performance with its two indicators (ROA and ROE). Kalva et al. [18] study aimed to identify Kaizen method in continuous improvement of quality, productivity, profitability and speed of response. Study concluded that vast majority of companies currently use Kaizen business system. Singh & Singh [19] Study. Aim of this study is to reveal importance of Kaizen approach in manufacturing environments, and to evaluate performance of various Kaizen elements in manufacturing industry of North India, and level of importance various elements in kaizen, and important benefits achieved after successful implementation of kaizen approach, results clearly showed that pattern of failure and effective analysis is most important element in kaizen, and results indicated that there are net savings annually after applying kaizen approach in researched company.

4 Methods and Procedures

4.1 Study Population

Study population consisted of JIC listed on ASE in 2000.

4.2 Study Sample

Study sample was limited to (30) companies whose data on (ROA and ROE) were continuously available for the period between (2010–2020), while the sampling unit was represented by the financial managers and accountants who work in these companies. (141) questionnaires were distributed to the study sample, and (109) questionnaires were retrieved from them, at a rate of (77.3%) of the study sample.

4.3 Study Variables and How to Measure Them:

Study variables consist of a dependent variable, which is financial performance, an independent variable, which is the Kaizen approach, and a moderator variable, which is digital transformation.

Dependent Variable: Financial Performance
The financial performance was measured by ROA and ROE, so the required data was obtained through the published data on the industrial companies listed on ASE.

Independent Variable: Kaizen Approach
Kaizen approach was measured through four dimensions, which were measured by the following items in the study questionnaire:

1. Planning: It was measured through items (1–7) in study questionnaire
2. Implementation: It was measured through items (8–13) in study questionnaire
3. Review: It was measured through items (14–20) in study questionnaire
4. Correction and treatment: It was measured through items (21–27) in study questionnaire

Moderator Variable: Digital Transformation
Digital transformation was measured through four dimensions, which were measured by the following items in study questionnaire.

1. Strategic planning for digital transformation: it was measured through items (28–32) in study questionnaire
2. Preparing leaders in the field of digital transformation: it was measured through items (33–37) in study questionnaire
3. Institutional structure of digital transformation: it was measured through items (38–42) in study questionnaire
4. Attracting skills and competence for digital transformation process: it was measured through items (43–47) in study questionnaire

4.4 Stability of Study Tool

Researcher applied the Cronbach Alpha equation [27], where results came as shown in Table No. (1).

Table 1. Stability coefficients by the Cronbach-Alpha

Number of items	Dimension	Alpha coeff
27	**Kaizen approach**	**0.973**
20	**Digital Transformation**	**0.934**

Table 1 shows the stability values of the study variables, which amounted to (0.973) for Kaizen approach, and (0.934) for Digital Transformation.

4.5 Results of Testing Hypotheses

H01: There is no effect at significance level (0.05 ≥ α) for application of Kaizen approach in ROA in light of digital transformation in JIC listed on ASE.

To test the first hypothesis, a simple regression analysis was applied to study the effect of applying Kaizen approach on ROA in light of digital transformation.

Table No. (2) Illustrates this.

Table 2. Results of a simple linear regression of H1

Hypothesis	T	R	R^2	Adj. R^2	Stat. Sig
HO1	**2.376**	**0.382**	**0.146**	**0.113**	0.002

The results in Table 2 shows that there is an effect at level of significance (0.05 ≥ α) for application of Kaizen approach in ROA in light of digital transformation, as value of correlation coefficient (R) was (0.382), which is indicates degree of correlation between independent variables and dependent variable, and value of (Adjusted R-square) was (0.113), which is a value that explains ability of applying Kaizen approach to affect ROA, meaning that application of Kaizen approach explains what is (11.3%) of change in ROA, and value of test (T) was (2.376), with a statistical significance of (0.00), which is a value indicating that there is a variation in ability of independent variables to influence dependent variable. Therefore, first hypothesis is rejected in zero form and hypothesis is accepted in alternative form, which states, "There is an effect at significance level (0.05 ≥ α) for application of Kaizen approach in ROA in light of digital transformation in JIC listed on ASE."

H0.2: There is no effect at level of significance (0.05 ≥ α) for application of Kaizen approach in ROE in light of digital transformation in JIC listed on ASE.

To test the second hypothesis, a simple regression analysis was applied to study the effect of applying Kaizen approach on ROE. Table 3 shows this.

Table 3. Results of a simple linear regression of HO2

Hypothesis	T	R	R^2	Adj. R^2	Stat. Sig
HO2	**3.105**	**0.863**	**0.745**	**0.735**	0.000

The results in Table 3 shows that there is an effect at level of significance (0.05 ≥ α) for application of Kaizen approach in ROE in light of digital transformation, as value of correlation coefficient (R) was (0.863), which indicates degree of correlation between independent variables and dependent variable, and value of (Adjusted R-square) was (0.735), which is a value that explains ability of applying Kaizen approach in affecting ROE in light of digital transformation, meaning that application of Kaizen approach Kaizen explains value of (73.5%) of change in ROE, and value of test (F) was (76.004) with a statistical significance of (0.00), which is a value indicating that there is a variation in ability of independent variables to influence variable affiliate. Therefore, second sub-hypothesis is rejected in zero form and hypothesis is accepted in alternative form, which states that "there is an effect at significance level ($0.05 \geq \alpha$) for application of Kaizen approach in ROE in light of transformation and digitalization in JIC listed on ASE.

5 Conclusion and Recommendations

5.1 Conclusion

1. Application of Kaizen approach affect in ROA in light of digital transformation in JIC listed on ASE. This result may be attributed to accountants and financial managers in industrial companies in ASE believes that their companies adopt an approach based on discovering errors and working to remedy them and correcting deviations in production or administrative process, even if they are small, with aim of providing products and services with ability to achieve competitive advantages that these companies seek to achieve in order to obtain largest market share and increase profitability from available funds. This result is consistent with what was stated in Prayuda study [11], which indicated that application of Kaizen culture rises rate of productivity and rises perceived benefits of company, especially on ROA.

2. Application of Kaizen approach affect in ROE in light of digital transformation in JIC listed on ASE. This result may be attributed to fact that industrial companies listed on ASE follow a periodic and continuous policy throughout year to discover problems and errors that their production, operational and administrative processes face, and seek solutions to them, which reduces waste of their resources and costs, thus offering competitive products compared to others. Local and regional companies, and this leads to an increase in their ability to improve their efficiency in exploiting owners' funds and ability of these funds to generate profits. This result is consistent with what was stated in study [11], which indicated that application of Kaizen culture rises rate of productivity and rises perceived benefits of company, especially on ROE.

5.2 Recommendations

In light of previous results, study recommends following:

Conducting training programs for employees in industrial companies, which aim to increasing their knowledge in stages of applying Kaizen approach and its importance in the administrative and production operations at the company as a whole.

Necessity of adopting an effective audit and control system in order to identify obstacles facing industrial companies listed on ASE, and work to find solutions based on research result.

Conducting more studies dealing with impact of applying Kaizen approach on other performance dimensions in other business sectors.

References

1. Hamidi, A.: Total Quality Management and its Role in Reducing Work Accidents, A Field Study At Cable Industry Corporation-Biskra Unicab, Unpublished Master's thesis, Mohamed Kheidar University-Biskra, Algeri (2015)
2. Sawaf, M., Zakaria, A.: Technological Change And Its Impact On Improving Quality Of Products/A Case Study In General Company For Ready-To-Wear Industry/Mosul (Company's Headquarters/My Son's Clothing Factory). Tanmiat Al-Rafidain J. **34**(109), 117–133 (2012)
3. Al-Khatib, M.: Financial performance and its impact on returns of shares of joint-stock companies, 1st edn. Dar Al-Hamid for Publication and Distribution, Amman (2010)
4. Saadi, W.: Clayzen... an imperative to reduce negative work behaviors. Political Orbits J. **5**(1), 231–249 (2021)
5. Carnerud, D., Jaca, C., Bäckström, I.: Kaizen and continuous improvement-trends and patterns over 30 years. TQM J. **30**(4), 371–390 (2018)
6. Mandahawi, N., Aamer, A., A-Madi, F.: A combined approach to define kaizen international transferability. In: Proceedings of 5th NA International Conference on Industrial Engineering and Operations Management Detroit, Michigan, USA, August 10--14 (2020)
7. Al-Harbi, M.: Requirements for Improving University Leadership Methods in Light Of GEMBA Kaizen Approach (2017)
8. Bakosh, L.: Contribution of Activity-Based Management to Improving Performance Of Algerian Economic Institutions - Case Study: Saidal Complex, Unpublished Doctoral thesis, Mohamed Kheidar University, Algeria (2017)
9. Alani, S.: Extent to which departments of general commission for taxes in iraq fulfill their requirements by using total quality management. Al-Muhaseb J. **23**(46), 85–101 (2016)
10. Kazem, H.: Continuous improvement by kaizen cost method and possibility of its adoption in general company for tire industry in Najaf. J. Kufa Stud. Center **21**, 159–188 (2011)
11. Prayuda, R.: Continuous improvement through kaizen in an automotive industry. J. Ind. Eng. Manage. Res. **1**(1), 37–42 (2020)
12. Kumar, R.: Kaizen a tool for continuous quality improvement in Indian manufacturing organization. Int. J. Math. Eng. Manage. Sci. **4**(2), 452–459 (2019)
13. Al-Sharbaz, A.: Components Of Digital Transformation Strategy Within Sustainable Development Goals 2030, Al-Mansour College, Iraq (2020)
14. Abdel Razek, S.: Digital transformation is a new challenge for accounting and auditing profession to serve sustainable development. In: Twenty-Fourth Annual Crisis Research Conference, Entitled "Management of Digital Transformation for Implementation Of Egypt's Vision 2030, Ain Shams University (2019)
15. Jad, M.: Role of KAIZEN strategy in improving management of construction and building projects by applying it to contracting companies in greater Cairo. Eng. Res. J. **165**, 1–15 (2020)
16. Abu Baqi, A., Al-Kasasbeh, M.: Impact of kaizen approach on financial performance of Islamic banks in Jordan. Amman Arab Univ. Res. J. **1**(1), 69--93 (2017)
17. Bakr, N.: Entrance to continuous improvement (Kaizen) and its impact on performance of organizations (applied to Egyptian electricity holding company). Arab J. Manage. **36**(1), 239–260 (2016)
18. Kalva, R., K., Srinivasu, V.: Continuous improvement through kaizen in a manufacturing organization. Int. J. Eng. Dev. Res. **6**(1), 273–277 (2018)
19. Singh, J., Singh, H.: Enigma of KAIZEN approach in manufacturing industry of Northern India–a case study. Int. J. Qual. Reliability Manage. **35**(1), 187–207 (2018)
20. Shaqqour, O.F.: The impact of accounting performance on structure of ownership and accounting conservatism, case of Jordan. Acad. Strateg. Manag. J. **18**(5), 1–8m (2019)

21. Shahwan, Y.: The mediating effect of investment decisions and financing decisions on the influence of capital structure against corporate performance: evidence from Jordanian listed commercial banks. Acad. Account. Financ. Stud. J. **22**(6), 1–20 (2018)
22. Shaqqour, O.F., Harb, A.S.M., Ballout, O.M.K., Jaber, R.J. : Digital audit during covid-19 in jordanian audit firms a study of the reality and outlook the future. In: The Implementation of Smart Technologies for Business Success and Sustainability, pp. 263–272. Springer, Cham (2023)
23. Shaqqour, O.F.: Impact of vertical and horizontal integration between strategic management accounting and decision making on reducing financial failures case of Jordan. Acad. Account. Financ. Stud. J. **24**(5), 1–9 (2020)
24. Shahwan, Y., Abdel-hamid, O.: The impact of social responsibility disclosure in reducing the risk in Jordanian Commercial Banks. Res. World Econ. **11**(3), 142–150 (2020)
25. Shahwan, Y.: The effect of OECD principles of corporate governance on social responsibility accounting. Int. J. Innov. Creativity Change **11**(2), 418–431 (2020)
26. Shaqqour, O.F.M.: The impact of the ownership structure on earnings management" an empirical study on the hotels listed in ASE. Zarqa J. Res. Stud. Humanities **16**(2), 91–100 (2016)
27. Al-Ramahi, N., Abuhussein, A., Shahwan, Y.: The impact of applying the XBRL on the quality of annual reports of the firms listed in Amman stock exchange. In: 2021 22nd International Arab Conference on Information Technology (ACIT), pp. 1–5. IEEE, December 2021

The Impact of Digital Transformation on Combating Tax Evasion (Electronic Billing System as a Model)

Suhaib Ahmed El-Manaseer[1](✉), Jalal Hussein Al-Kayid[1], Abdelaziz Mahmoud Al Khawatreh[1], and Mohammad Shamim[2]

[1] Faculty of Law, Zaraqa University, Jeddah, Jordan
{salmanaseer,jalkayid,Akhawatreh}@zu.edu.jo

[2] Department of Insurance and Risk Management, University of Business and Technology, Jeddah, KSA, Jordan
m.shamim@ubt.edu.sa

Abstract. In light of the digital transformations that the world has witnessed in recent years, the information and communication technology sector and digitization have become one of the most sensitive topics that interest governments, as many governments rushed to take advantage of digital transformations in order to increase the levels of effectiveness of their departments, including public finance management, which is what is called Digitization of public finances. The latter allowed governments to increase the efficiency of tax collection by improving its collection methods and expanding the tax base, in addition to fighting tax fraud, by increasing the ability of governments to collect accurate and immediate information. It specializes in controlling and supervising non-banking financial markets and instruments, including capital markets, futures exchanges, insurance activities, real estate finance, financial leasing, factoring, and securitization, with the aim of achieving safety and stability of those markets and instruments, organizing and developing activities and maximizing their competitiveness to attract more local investments and foreign, and works to reduce the risks of lack of coordination and address the problems that result from the different control methods or methods.

Keywords: Digital transformation · Public finance · Tax collection · Electronic billing system · Tax evasion

1 Introduction

Today, the world is witnessing great and varied challenges that have created more pressures and created many challenges for organizations and governments to confront and force them to change their traditional methods of management and adopt modern concepts if they want to achieve their goals efficiently. Effectively, the most important of these are the digital transformations that have had an impact on all aspects of life and thinking. Mechanisms of action and performance of macroeconomic policies, including

B. A. M. Alareeni and I. Elgedawy (Eds.): *Artificial Intelligence (AI) and Finance*, SSDC 488, pp. 679–690, 2023.
https://doi.org/10.1007/978-3-031-39158-3_63

fiscal ones. As an extension of these transformations and within the framework of the fourth industrial revolution and the technologies accompanying it, many governments tended to take advantage of these technologies to increase the levels of effectiveness of public financial management in achieving its goals [1]. Among these mechanisms, we find the digitization of public finances, which affected many aspects, including digital transformation at the level of both sides of the public budget, especially the use of electronic systems and modern technologies in tax administration, especially in the tax collection process, which is covered by the latter to achieve the goals of the state's fiscal policy. And he indicated that the system will enable the unification and application of the principles and elements of invoices issued by all taxpayers and eliminate the risks of organizing traditional invoices and their loss or damage. As well as eliminating the burdens and risks of transferring invoice books and financial statements from companies to tax auditors' offices. Rapid access to tax services, completion of tax audit procedures for taxpayers, obtaining tax clearance, and unifying the foundations for organizing and issuing invoices in economic sectors, which contributes to disclosure and transparency for these sectors. Based on the above, and considering taxes as one of the most important sources of financing the state's public revenues, and in light of the digital transformations in the tax field, it has become necessary to use the latest technologies in the tax collection process. The problem of the study lies in the adequacy of the electronic billing system in combating the crime of tax evasion, it also lies in the growth of electronic commerce in the world, and also lies in the ability of citizens charged to pay the tax resulting from electronic commerce. It is therefore difficult for the tax administration to limit the taxable income accrued to merchants, due to the difficulty in restricting the electronic billing system.

2 The Legal Nature of the Tax Evasion Crime

The danger of not issuing an official invoice monitored by the income and sales tax lies in the fact that the seller or the service provider deducts the tax and does not return it to the state treasury or does not supply a part of it, which reinforces evasion and creates a distortion in economic activities by making the evader rich at the expense of the citizen who pays. Sales tax is at the expense of the state that provides services to citizens. Article (23/f) also stipulates that: "A person is obligated to issue an original invoice in exchange for providing any service or selling any commodity in the Kingdom, and all matters related to billing systems, issuance and control over them, and the categories excluded from them, are regulated by a regulation issued for this purpose"[1]. While the Jordanian legislator defined the tax return through the text of Article 2 of the Income and Sales Tax Law as: "A declaration of income, expenses, exemptions and due tax submitted by the person according to the form approved by the department"[2].

It is worth noting that the crime of tax evasion has become a widespread crime in many countries of the world, and this phenomenon represents a great danger to the economic reality, and one of its simplest forms is that the taxpayer conceals all or some of the elements of the taxable material, and helps the effectiveness of the performance of the

[1] Law Amending Income Tax Law No. 38 of 2018.

[2] Income Tax Law No. (34) of 2014 and its amendments.

auditing profession when approving the Financial reports reduce the evasion of income taxpayers from paying it, and with the development of modern accounting systems, the possibility of creating new methods that help tax evasion has increased with great ease, and here the importance of the auditor's reports appears in confirming the correct representation of the financial statements of the company's reality and the result of its business [2].

2.1 What is the Crime of Tax Evasion

Most modern legislation considers tax evasion or fraud to be crimes punishable by imprisonment and a financial fine, depending on the seriousness of the crime. These penalties vary from one country to another. He pointed out that there are some legislations that did not define the crime of tax evasion, but only mentioned the behaviors that are punishable by law if they are found in the crime of tax evasion.

2.2 The Concept of Tax Evasion Crime

Jurists, thinkers, and economists have different definitions of taxation due to the evolution of economic, social, and political thought. Some have defined it as the seizure of wealth from individuals by the state through coercion without a specific return for its motivator, with the goal of achieving a general objective [3]. Others have defined it as a source of public revenue for the state, taking a certain amount of money, which is compulsorily paid by the taxpayer, and ultimately from their income and profits, regardless of the benefit they receive in return. Another definition is that it is any positive or negative activity that leads to the abuse of tax interests or exposing them to risk, which is punished by criminal law. The intended tax interest is any tax burden that is protected by criminal punishment. Some others define it as any means used by the taxpayer or others to avoid paying the full tax owed, with the intention of harming the state and ultimately the interests of society, with the goal of completely or partially avoiding the tax [4].

Tax evasion is represented by the failure of the obligated person to acknowledge their duty to pay the taxes imposed on them, either by failing to provide the necessary data according to the laws or by providing misleading and inaccurate data to financial authorities. This is done with the goal of reducing tax revenue through various methods, which means depriving the state of its financial revenue. As the tax deprives the obligated person of enjoying a portion of their income, and as the obligated person's willingness to pay taxes without resentment requires civic awareness, some obligated persons may have a tendency to evade taxes. Tax evasion also refers to the behaviors and practices that are performed with the goal of cheating and avoiding paying taxes. In this sense, tax fraud implies that the taxable event has already taken place, but the person obligated to pay the taxes evades paying it entirely or partially by taking advantage of tax exemptions, loopholes in the law, and flaws in its wording [5].

While the legislator defines tax evasion through its legal texts as the use of fraudulent methods involving deception, fraud, falsification, concealing of data, presenting false information, or participating in any of them with the intention of not paying taxes or declaring them, completely or partially, or reducing them as determined by this law[3].

[3] Jordanian Income and Sales Tax Law No. (34) of 2014 and its amendments.

Researchers believe that all previous definitions determine the essential elements of tax evasion, which is represented by reducing the tax revenue, regardless of whether this evasion is total or partial, and the failure of the responsible party to pay taxes despite the existence of the law, resulting in a loss of the state's financial rights and exposing the responsible parties to legal prosecution and accounting due to their failure to pay their obligations.

2.3 Tax Evasion Means

Individuals and companies use various methods to evade tax payment, including:

A. Failure to declare assets and penalties to the tax administration.
B. Reducing the taxable income value.
C. Manipulating the amount of taxes withheld from the income of employees and company workers and not transferring it fully to the state.
D. Presenting false invoices to customs administration to reduce customs duties and VAT on imported goods.

The ultimate goal of these practices is to reduce the amount of tax owed to the state by reducing taxable profits, which is the basis for calculating income tax. Tax evasion reduces the ability of the state to allocate resources and generate the necessary revenue to fund public expenditure, thus reducing the quality of public services, and investment opportunities in infrastructure and necessary facilities, and countering government policies that drive economic growth and development. Due to resource constraints and budget deficits, tax evasion leads to the government imposing a heavier tax burden on employees, officials, and others whose taxes are withheld at source, leaving them no way to evade tax payment or reduce it.

2.4 The Elements of Tax Evasion Crime

Considering the tax crime has a special nature, it is based on three elements or components:

A. Legal component: The legal component consists of the legal texts that specify that the act committed is criminal and that a specified punishment applies to it [6].
B. Material component: The material component is based on the criminal behavior and the resulting criminal link and causal relationship between behavior and outcome [7]. These elements are:

- Criminal behavior: This refers to all physical actions, legal and financial operations, and all the means and arrangements that the obligor or others use to evade paying taxes and contributions. The law does not limit the criminal behaviors for the crime [8], but links them to the purpose, which is to evade all or part of the crime by deception and fraud. These criminal behaviors can be classified into three categories:
- Material fraud: This fraud consists of hiding the taxable object, whether it is a taxable activity or taxable goods and products, from the knowledge of the tax administration [9].

- Accounting fraud: What the tax administration takes from the taxpayer is linked to the accounts he presents to the tax administration, as the law requires the taxpayer to declare the activity and inputs in each financial year, supported by documentary evidence (such as invoices and accounting records). The taxpayer then declares the tax by forming an account that exempts him from paying a significant part of the tax owed, such as inflating costs that were not actually incurred or hiding actual profits.

3 The Legal Framework of the Invoice System to Combat Tax Evasion

It is a danger to not issue an officially supervised invoice by the income and sales tax as the seller or service provider cuts the tax assets and does not transfer it to the state treasury or does not transfer part of it, which exacerbates evasion and creates distortion in economic activities as the evader becomes wealthy at the expense of the citizen who pays the sales tax and at the expense of the state that provides services to citizens. Article (23/ and) of the amended Income Tax Law No. 38 of 2018 states that "a person must issue an original invoice for providing any service or selling any item in the kingdom and all matters related to the invoice systems, issuance, monitoring, and categories excluded from them shall be regulated in accordance with a system issued for this purpose".

3.1 What is the Billing System

The effectiveness of the billing system in reducing tax evasion, and clarifying how it works as a guarantee to prevent tax evasion [10]. The problematic of this study was the adequacy of the provisions of this system to reduce the phenomenon of tax evasion. This system was built according to the best international practices to accommodate all taxpayers. If the taxpayer does not have a billing system or a current traditional billing system, he can enter the website of the Income and Sales Tax Department on a website dedicated to the billing system, and a system for issuing an invoice is available for him free of charge and becomes linked to This system is able to issue invoices through this system, and if it also has a billing system and a computerized financial system, it does not require it to cancel this system and replace it with a new system, but the system also enables it to directly connect with the national billing program directly without the need to dispense or replace the system.

3.1.1 Definition of the Billing System

The national billing system is an integrated computerized system within the best international practices in this field, and achieves flexibility with the possibility of joining all companies and establishments that are bound by the billing system by allowing the way for companies and establishments that do not have a billing system or a computerized financial system, and at the same time the possibility of joining Companies and establishments that have computerized systems without giving up their systems. And he indicated that the system will enable the unification and application of the principles and elements of the invoices that are issued by all taxpayers and eliminate the risks of organizing traditional invoices and their loss or damage [11]. This is a system for companies, establishments, and institutions that are obligated under the Income Tax Law and under the billing system in Jordan to apply the issuance of a tax invoice upon each sale of a commodity or service allowance.

3.1.2 Mechanism for Applying the Billing System

The Jordanian electronic national billing system, to begin the stage of companies and establishments joining the national billing program network, during a press conference during which it was said that this time is to serve the government's commitments not to impose or increase taxes, and to facilitate taxpayers in all their sectors and speed up their access to services taxes, including refunds, achieving voluntary commitment, and regulating the relationship between taxpayers in general [12].

It also stipulated through texts related to invoicing, it stipulated during the text of Article 2, that: "The seller of any commodity or service whose value is not less than one dinar must organize an original invoice in accordance with the provisions of the Billing Affairs Regulation and Control Regulation No. (34) For the year 2019 and these instructions unless It is excluded under the provisions of Article (11) of the system for regulating and controlling billing affairs or under the provisions of these instructions"[4].

The mechanism specified below for organizing billing affairs depends on lawyers as follows: "a) A receipt receipt or any receipt voucher for the purposes of calculating the tax is adopted instead of the invoice stipulated in the provisions of Article (5) of the system for organizing and controlling billing affairs for each lawyer whose received revenues do not exceed (50,000) fifty thousand Jordanian dinars annually, for the purposes of calculating the tax b) A lawyer whose annual revenues exceed (50,000) fifty thousand Jordanian dinars issues an invoice in accordance with the provisions of Article (5) of the Billing Affairs Regulation and Control Regulation No. (34) for the year (2019) within forty-five days from the date of reaching this limit c) 1) If the revenues received annually amount to more than (50,000) fifty thousand dinars, the lawyer must correct his situation according to these instructions within forty-five days of that. Annual receipts of (50,000) fifty thousand dinars for the lawyer to adjust his status in accordance with these instructions If the lawyer does not adhere to organizing the receipt of receipts or any receipt voucher or the invoice according to the system for organizing and controlling billing affairs and these instructions, the fines stipulated in the provisions of this law will be imposed on him. Article (64) of Income Tax Law No. (34) of (2014) and its amendments[5].

It turns out that the billing system will constitute a qualitative leap in the tax culture in Jordan, and will make it easier for the private sector in matters of tax transactions and keeping pace with international practices, indicating that the Income and Sales Tax

---

[4] Executive Instructions/Instructions for Billing Affairs and Oversight No. (1) for the year 2019 and its amendments issued pursuant to the provisions of Paragraph (a) of Article (11) and Article (16) of the Regulation for Regulating and Controlling Billing Affairs No. (34) for the year 2019.

[5] Executive Instructions/Instructions for Billing Affairs and Oversight No. (1) for the year 2019 and its amendments issued pursuant to the provisions of Paragraph (a) of Article (11) and Article (16) of the Regulation for Regulating and Controlling Billing Affairs No. (34) for the year 2019.

Department is witnessing a qualitative leap that serves what the government has committed to. The government is committed not to raise taxes on the citizen, to change the culture of tax compliance towards voluntary commitment, and fighting tax avoidance and evasion to serve the tax-abiding citizen and enforce the law. And he indicated that the amendment to the current year sales tax law; It was aimed at accelerating the disbursement of refunds to the private sector, and reducing legal periods and deadlines, indicating that he is looking forward to qualitative technological transfers, aimed at relieving citizens, and accelerating the process of refunds. Al-Ississ indicated that the aim of launching this system is to achieve justice in the application of the law, to make it easier for taxpayers, to reduce costs for them, and speed up their transaction procedures. He thanked the Minister of Digital Economy for following up and supervising the preparation of this system, as integrated solutions were provided for an electronic system in the field of billing to provide services for issuing and managing invoices, in addition to collecting and managing data and information related to buying and selling operations, and through this system, the provisions of tax legislation and adherence to the system will be implemented. Managing billing affairs according to the provisions of the system. Also, all companies, institutions and establishments are urged to quickly join the network of the national billing system program because of the interest of these companies and institutions and the consequent service to the national economy, indicating that the system has many benefits and will achieve advantages for the taxpayers who deal with it. This system was implemented by Jordanian experts in cooperation with the Ministry of Digital Economy and the Income and Sales Tax Department. He said that the launch of this system at this time is an advanced and necessary step to serve taxpayers, especially those who are tax-compliant, as it contributes to building taxpayers' files according to the latest practices, and all those who link to this system will see the great benefits that will accrue to them [13].

The national billing system is an integrated computerized system within the best global practices in this field and achieves flexibility by enabling all companies and establishments that are bound to join the billing system by allowing the way for companies and establishments that do not have a billing system or a computerized financial system, and at the same time the possibility of joining companies and establishments that have systems computerized without abandoning its systems. He added that the system came to serve the consumer and ensure that he obtains a regular invoice in exchange for purchasing any commodity or in exchange for obtaining any service without requesting it, indicating that it will serve the seller and the service provider by transferring the burden of proof from the taxpayer to the department, given that what was stated in the approval The taxpayer is correct unless the department proves otherwise, and it will also make it easier for the taxpayers to complete and speed up the audit of their tax files. It turns out that the electronic invoice is being issued and saved in an electronic format organized through an electronic system and contains the conditions of the tax invoice. There are ways through which the invoice can be sent to the tax office, either through the electronic platform (Portal), or through the mobile application on (Google Play and App store), or from the Application Programming Interface (API) [14]. And he indicated that registration on the electronic billing platform will be available to taxpayers by entering the Income and Sales Tax Department website and logging in with the tax number and password, explaining that the department will work to provide support and technical assistance to any of the sectors and any taxpayer who is obligated to apply the use of invoices and comply with billing system It is worth noting that Income Tax Law No. 38 of 2018 obligates, through the provisions of Paragraph (f) of Article (23), to issue an original invoice for the provision of any service or sale of any commodity in the Kingdom, and to organize, issue and control all matters related to billing systems and the excluded groups. Pursuant to a regulation issued for this purpose.

In our opinion, it was found that there are some reservations about the system and fears on the part of members of the General Assembly, heads of chambers of commerce, representatives of sectors, heads of trade unions and employers' associations, which requires waiting to implement it, giving a trial period to everyone, and taking into account the specificity of each sector. He stressed the need to raise awareness about the system for merchants by issuing educational leaflets and producing explanatory videos explaining the mechanisms of its work.

3.2 The Digital Transformation Role in Combating Tax Evasion

3.2.1 Legal Penalties for Tax Evasion

As previously explained, tax evasion is the elimination of tax burden by using fraudulent means, tax evasion is an illegal act in violation of the law. In order for tax evasion to be considered a crime, the following conditions must be met: the elimination of tax burden is a violation of the law, reliance on fraudulent means, the legal basis is based on many justifications and arguments as follows, regarding the objective: the legislator aims to criminalize the tax evasion for the same purpose that the ordinary legislator seeks, which is to protect the public good, and since the law of penalties aims to achieve its own purposes that the general law aims to achieve, there is no objection to considering tax evasion a criminal offense and subject to the general law of penalties. Such as theft, embezzlement, and breach of trust, so it is not acceptable to treat tax evasion less harshly or less severely than crimes committed against money if it is proven that the person responsible for evading taxes has committed fraud. This means that the legal elements of embezzlement have been established, especially if the person responsible for committing fraud does not intend to disable the state's operations, but rather to increase their wealth or avoid reducing this wealth. This is the same as those who commit crimes against money, which requires equality between these crimes and crimes against money.

3.2.2 Administrative Handling of Tax Evasion

Tax evasion requires more than one party to combat and limit it as much as possible, in addition to tax laws and the penalties they contain for those who commit tax evasion crimes, there are tax departments in various governorates and employees of these departments who are obliged to combat this dangerous phenomenon, which must be limited in order to achieve a reasonable tax compliance by the taxpayers. In the context of discussing the role of tax departments and the relationship of these departments with the phenomenon of evading income tax, we will examine the role of tax departments in this regard by closing these doors in the face of taxpayers who seek to evade their taxes [15].

It becomes clear to us that the person responsible for paying taxes does not reveal the activity he is engaged in or carries out. This happens often as tax authorities do not have any information about the activity the person is engaged in, and as a result, the person may be involved in an activity that is subject to tax and avoid paying it. This is because he does not disclose the activity he is engaged in to the tax administration. From our perspective, this happens for two reasons, and both reasons are related to the performance of the tax authorities. The first reason is the failure to apply the punitive aspect of tax law in cases that are detected and the contentment with just collecting the taxes owed without exposing these cases to the penalties imposed by law in such cases.The second reason, which is closely related to the work or performance of the tax departments as well, is the lack or even lack of oversight and inspection by the tax departments on establishments and businesses, which encourages taxpayers to hide the activity they are practicing from the tax administration in order to evade paying the income taxes that are due on them. Here, the researchers believe that it is necessary to activate the tax penalties stipulated by the law, as the tax law granted the tax administration a lot of powers and powers that, if used by this administration, would greatly reduce the phenomenon of tax evasion. The activities and works practiced by the taxpayers so that the taxpayer who intends to evade and not report his activity that he is practicing feels that the tax administration will reach this activity and discover it if he did not report it.

Many taxpayers, with the intention of evading paying taxes, increase their expenditures and reduce their revenues and profits. For example, many taxpayers declare high salaries and wages for the workers and employees who work for them, and these salaries are declared. It is the truth, and an increase in it is declared with the aim of increasing expenses and expenditures with the aim of evading or reducing the payment of taxes due on them. Paying the tax or reducing it, and here comes the role of the tax administration, where the salaries declared by the taxpayer to the estimating officer must be examined when the tax was estimated on him, and matching these wages and salaries declared by the same taxpayer for the same year to the tax deductions department in order to deduct taxes on these salaries, and here the administration discovers On the other hand, the taxpayer declared low salaries and wages and less than the truth to the deductions department, with the aim of reducing the taxes imposed on these salaries and wages, and here we note the central role Which the tax departments can play in revealing the operations or attempts of taxpayers to evade paying income taxes. Due to them and fighting them and minimizing them by scrutinizing the statements submitted by the taxpayer and examining them thoroughly and trying to find out whether these statements and the information contained in them correspond to the truth or not, and the tax authorities are not the only ones making these attempts to evade [16]. Some accountants, tax authorities and especially traders may also resort to reducing their sales in order to reduce their revenue in order to evade taxes entirely or partially, this can be done by not issuing invoices or manipulating these invoices when they are issued, and especially when the buyer or consumer does not care much about the issuance of the invoice. Here, there is no doubt that the tax administration has a significant and effective role in this matter, as it is the duty of the tax administration to check the invoices issued by the trader and thoroughly examine them to determine if these invoices are falsified or genuine. The tax administration also has the responsibility to educate the public about the importance of receiving tax invoices when purchasing and the great importance of this from a tax perspective.

Those who did not open tax files with the deductions department in the Income Tax Department and impose the penalties imposed by the law on such cases and oblige those charged with employers to open tax files for the workers and employees they employ, but in the event that the employers opened files for the workers they work for but If they did not commit to paying the taxes that were due on these files, the tax administration must impose the penalties, the power and authority granted to it by the law in such cases, on those taxpayers who try to evade paying taxes on their workers, when the taxpayer or even some accountants submit false statements expressing Simple and unreal profits or losses, and in the event of submitting losing data, they demand to roll over their losses, and here the matter is not limited to tax evasion in the current fiscal year, but rather planning to evade in the coming years, in order to ask them to roll back their unreal losses, we say that the role of the tax administration focuses on effective monitoring of accountants and subjecting them to an effective monitoring system by the tax administration, and subjecting accountants who try to evade or manipulate tax penalties and not tolerate them in any way. Additionally, the work of these accountants is restricted by obliging them to comply with professional and ethical standards imposed by their profession and enforcing penalties against those who try to provide false information with the knowledge that it does not reflect the truth. There is a well-known category that evades paying income tax, which are the owners of freelance professions, including doctors especially specialists, lawyers, pharmacists, engineers…etc. [17].

Here, we see that tax penalties, as prescribed by law, are applied to these freelancers, even if simply. And despite the many scandals surrounding them, they deny this. If a portion of these people were punished with severe penalties, it would be a deterrent for the rest of these freelancers, as it is unfair for an ordinary employee with a relatively low income to pay the taxes owed to him while the doctor, lawyer, or engineer who achieves a very high income evades them.

4 Conclusion

In conclusion, this study reached a number of results and recommendations, as follows:

4.1 Results

1. We conclude that most countries have adopted electronic billing systems as one of the mechanisms that would support the efforts of the tax authorities to gradually expand the tax base and integrate it with the digital economy in the official system and focus on developing the technical infrastructure that supports the digital transformation of tax collection systems through the development and digitization of national payment systems, towards Transforming digital systems into the identity of individuals and companies.
2. Developing and building the capabilities of workers and human cadres working in the field of digitizing tax collection, whether with regard to tax or technical fields, in a way that helps to make the most of the development opportunities offered by the new

technologies associated with the fourth industrial revolution, the most important of which are big data technologies and artificial intelligence that contribute to increasing levels of collection. And tax compliance.

3. We conclude, to ensure the security of digitization and data protection of electronic tax collection systems, which contributes to increasing the reliability of these systems and ensuring the greatest amount of interaction and tax compliance, cooperation and coordination between Arab countries regarding the digitization of tax collection systems with the aim of exchanging expertise, experiences and tax information in a way that helps avoid Tax evasion. Incentives for those who submit a true income report or pay the tax on time or facilitating the payment of taxes by installments must play an important role in reducing tax evasion in order to avoid fines and penalties. Spreading tax awareness, keeping correct accounting books, and developing legislation that guarantees fairness and equality among taxpayers contributes effectively to reducing tax evasion.

4.2 Recommendations

1. We recommend defining ratios and measures to emphasize the principle of tax justice, which falls under tax burdens that exceed tax justice so as not to affect the financial digitization of tax collection.
2. The need to tighten tax penalties in a manner that does not serve the deterrent concept with regard to financial or custodial penalties, or even punishment, whether those are dual penalties. We must also distinguish between imposing the penalty resulting from the crime of evasion for the first time, and the penalties resulting from the crime of evasion when smugglers or big traders get used to it, which negatively affects tax digitization and leads to a breach of the billing system, and certainly an increase in financial and moral deterrence will focus when the penalties are doubled.
3. Amending the tax legislation so that it is drafted tightly and in a way that ensures avoiding legal loopholes, including punitive penalties for the crime of tax smuggling, from which taxpayers are executed, tightening penalties, increasing the efficiency of the tax collection system, and increasing the efficiency of tax assessors to detect various cases of evasion through qualification and training. We must spread tax awareness through various media and hold seminars for citizens and introduce them to the different tax goals. It also requires the development of an integrated strategy that addresses all aspects and loopholes that tax evaders implement. Addressing the billing system and applying it correctly and setting penalties.

References

1. Hussein, Z.Y.: The Extent of the Effectiveness of Penal Texts in Combating the Crime of Tax Evasion. Legal J. D.M, D.P, pg. 172. https://jlaw.journals.ekb.eg/article_45210_e120997590c496849711119936af1a45.pdf
2. Alatour, R.I.: Evading Income Tax in Jordan. Amman, Al-Shams Press, Jordan (1993)
3. Maro, G.A.: The Economics of Income Tax Evasion. Wael Publishing House, Amman (1965)

4. Abdel Sattar Al-Zubaidi, M.E.: The policy of reducing criminalization in the crime of tax evasion. Legal J. D.M, D.P. https://jlaw.journals.ekb.eg/article_221240_56c06219bf89a0ef52e1ea7a054c8ff
5. Al-Buraim, M.: Corporate Governance and its Role in Reducing the Phenomenon of Tax Evasion, A Field Study on Tax Departments Operating in the Gaza Strip, Master thesis, Al-Azhar University, Gaza- Palestine (2016)
6. Abd al-Sattar al-Zubaidi, M.I.: The policy of reducing criminalization in the crime of tax evasion. Legal J. D.M, D.P. https://jlaw.journals.ekb.eg/article_221240_56c06219bf89a0ef52e1ea7a054c8ff2.pdf
7. Abdel Moneim, S.: The General Theory of Penal Code, Lebanon, Beirut, Al-Halabi Legal Publications, D.I (2003)
8. Sneineh, T.H.A.: Factors Affecting Tax Evasion and Avoidance, Master Thesis, College of Administrative Sciences. Middle East University, Amman (2008)
9. Mohamed, B.: Elements of Tax Fraud Crimes in Algerian Law, a Comparative Study with French Law, Research Paper Presented within the Two Study Days on Tax Fraud and Customs Smuggling, Supreme Court, Algeria, November 13 and 14, 2007, Supreme Court Journal, Special Issue
10. Moneim, H.A.: Digitizing Tax Collection in the Arab Countries, Arab Monetary Fund, D.P, 2021, p. 12, https://www.amf.org.ae/sites/default/files/publications/2021-12/digitizing-tax-collection-arab-countries
11. El-Demerdash, M.M.: Combating Tax Avoidance for Digital Economy Activities Combating Tax Avoidance for Digital Economy Activities in the Egyptian Tax System, Egypt, Al-Azhar University, Faculty of Sharia and Law in Damanhur, D.T
12. Al-Zabin, H.M.: Billing Provisions in the Jordanian Legal System, Jordan, Al al-Bayt University, Master thesis (2021)
13. Al-Sabti Fares: Tax Disputes in Algerian Legislation and Criminal Judiciary, 2nd Edition, Dar Houma, Algeria (2011)
14. Mahdi Al-Samarrai, Y.: Analysis of the Tax Phenomenon and Methods of Treatment in the Iraqi Tax System. Anbar Univ. J. **4**, 9 (2012)
15. Maswada, S.: The Role of the Jordanian Income Tax Law in Reducing Tax Evasion, Jordan, Jadara University, Al-Quds Open University Journal for Research and Studies, D.M, p. 41 (2014)
16. Shehadeh, S.K., Shamia, Foundations of Public Finance, Dar Wael for Publishing and Distribution, first edition, Amman (2003)
17. Al-Khatib, K.: Tax Evasion, Damascus University Journal, Volume 162, D.P (2000)

The Role of the Electronic Banking Marketing Mix Elements in Enhancing the Competitive Advantage: A Field Study on Customers of Islamic International Arab Bank at Amman City, Jordan

Iyad Khanfar[1](✉) and Ali Almasri[2]

[1] Zarqa University, Zarqa, Jordan
i.khanfar@zu.edu.jo
[2] Islamic International Arab Bank, Amman, Jordan

Abstract. This purpose of this study was to identify the aspects and role of electronic banking marketing mix elements (electronic banking services, electronic banking pricing, promotion of electronic banking services, personnel and service provider deployment across electronic banks, electronic procedures, and electronic physical environment) in enhancing the competitive advantage of the Islamic International Arab Bank. The sample of the study consisted of bank customers, which amounted to a good number. To achieve the objectives of this study, the descriptive analytical approach was adopted. The study concluded that there is a statistically significant role of e-banking marketing mix elements in enhancing the competitive advantage of the Islamic International Arab Bank, noting that those elements were consistently high at the level of the Bank. Given their significant role in enhancing the competitive "advantage", the study recommended that the quality of e-marketing mix elements should be enhanced within the Bank, by conducting several studies, analyses and evaluation of the elements in a regular and constant manner, while allocating the necessary financial and human resources for the development of such, and identifying current and evolving costumers' needs for banking services..

Keywords: e-banking marketing mix elements · Competitive advantage · The Islamic International Arab Bank · Jordan

1 Introduction

It is perceived that the concept of e-marketing is the product of the contemporary information revolution, which directly affected the marketing process at the present, where the traditional means were significantly impacted as well by the rapidly developing modern means and technologies that cover all marketing activities across various organizations [1]. Given that it is not independent of these rapidly evolving technological advancements and changes, it was essential for the Islamic International Arab Bank to

B. A. M. Alareeni and I. Elgedawy (Eds.): *Artificial Intelligence (AI) and Finance*, SSDC 488, pp. 691–703, 2023.
https://doi.org/10.1007/978-3-031-39158-3_64

develop its marketing activities, considering that the dynamic change is the basis of the marketing business, especially when the elements of the electronic banking marketing mix are developed. In this regard, the Bank has been exerting extensive efforts to stand out from its competitors by delivering high-end quality electronic banking services at the possible lowest cost to ensure that customer satisfaction is achieved. For this reason, this study was conducted to identify the role of e-banking marketing mix elements in enhancing the competitive advantage.

1.1 Area & Questions of the Study

From the Islamic International Arab Bank's points of view, it is challenging to achieve the "required competitive advantage" due to the intense competition from other Islamic banks that operate in the Hashemite Kingdom of Jordan. Moreover, given the significant role played by-banking marketing in achieving the "competitive" advantage, this study was conducted to see how much it is contributing to achieving the required "completive advantage". The area of the study will be represented in the following question: How does the application of the electronic banking marketing mix elements contribute to achieving the Bank's competitive advantage? The following sub-questions can be deduced from the main question as well: What is the role of services, distribution and marketing of (electronic services), personnel, and electronic banking procedures in enhancing the Islamic International Arab Bank's competitive advantage?

1.2 The Objectives of the Study

This study endeavors to achieve its objectives by identifying the role of e-banking marketing mix elements in enhancing the Bank's "competitive" advantage.

2 Study Model

Taking into account the formulating hypotheses, the variables of this study were formulated and its objectives were met that are to identify the role of the independent and dependent variables. In this regard, a special model

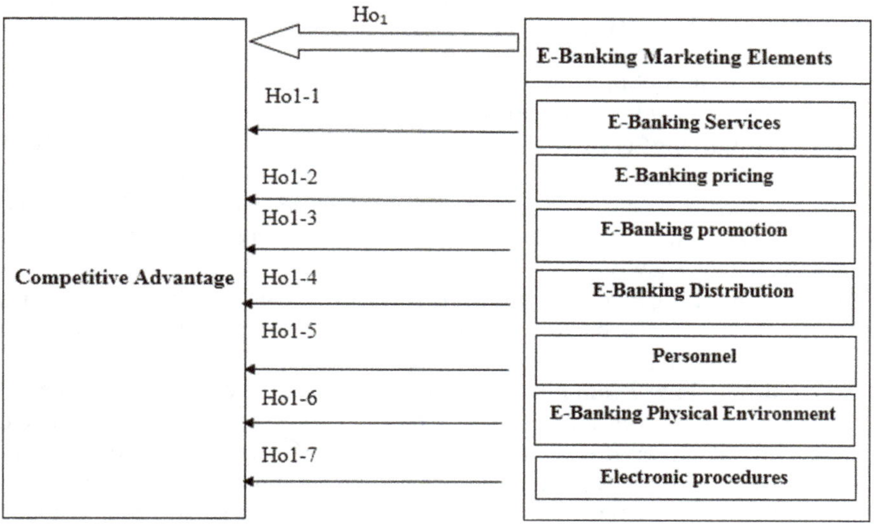

Source: Prepared by Researchers [2-7]

3 The Hypotheses of the Study

- **The First Main Hypothesis (H0):** There is no statistically significant role of e-banking marketing mix in all its forms (e-banking services, e-banking pricing, e-banking promotion, personnel and service provider deployment across electronic banks, electronic procedures, electronic physical environment) at the($0.05 \geq \alpha$) level of significance in enhancing the Islamic International Arab Bank's competitive advantage. Derived from the main hypothesis are the following sub-hypotheses:
- **H01**: There is no statistically significant role of electronic banking services at the ($0.05 \geq \alpha$) level of significance in enhancing the competitive advantage.
- **H02**: There is no statistically significant role of electronic banking pricing at the ($0.05 \geq \alpha$) level of significance in enhancing the competitive advantage.
- **H03**: There is no statistically significant role of electronic banking promotion at the ($0.05 \geq \alpha$) level of significance in enhancing the competitive advantage.
- **H04**: There is no statistically significant role of electronic banking services distribution at the ($0.05 \geq \alpha$) level of significance in enhancing the competitive advantage.
- **H05**: There is no statistically significant role of personnel at the ($0.05 \geq \alpha$) level of significance in enhancing the competitive advantage.
- **H06**: There is no statistically significant role of the electronic physical environment at the ($0.05 \geq \alpha$) level of significance in enhancing the competitive advantage.
- **H07**: There is no statistically significant role of electronic procedures at the ($0.05 \geq \alpha$) level of significance in enhancing the competitive advantage.

4 Procedural Definitions

- **Electronic Banking Marketing Mix**: it is a range of electronic marketing elements employed by the Bank to marketing its services to customers.
- **Electronic Banking Services**: it refers to the utilization of electronic systems in delivering banking services without the need for paperwork, where the Bank's customers can access their accounts 24 h a day, pay bills, transfer accounts, obtain loans, transfer and withdraw money, and conduct any other available banking activities seamlessly and without any delay.
- **Electronic Banking Pricing**: it is the price of electronic services that can be adjusted through electronic systems in accordance with the changes in the Bank's working environment and in line with customers' purchasing capabilities.
- **Electronic Banking Promotion**: it is the implementation of certain activities like contacting the Bank's customers through electronic systems to introduce them to the electronic banking services and clarify the Banks' advantages, and accordingly alter their perspective on the Bank and increase the Bank's ability to sell its services.
- **Electronic Banking Services Distribution**: it refers to the overall efforts exerted by the Bank to deliver its services to its esteemed customers in a timely manner and at the most appropriate place through electronic systems.
- **Personnel & Service Providers**: it refers to the ability of the Bank's personnel and electronic service providers to efficiently and promptly deliver high-end quality services.
- **Electronic Physical Environment**: The electronic components from which the Bank's systems, websites and applications are made. These components are considered the main inputs by which electronic systems operate.
- **Electronic Procedures**: they represent a fully- integrated electronic system that is developed to shift the normal administrative processes at the level of the bank by adopting electronic procedures instead of manual procedures using technological means.
- **Competitive Advantage**: it is the administrative, marketing and production activities conducted by the Bank to attract a larger segment of customers within the local and international market.

5 The Theoretical Framework

5.1 The Concept of Electronic Banking Marketing Mix

It is a range of electronic marketing activities that are conducted to influence customers' opinions of the Bank's services. The term "electronic banking marketing mix" refers to all processes and activities, the purpose of which is to provide electronic banking services that meet customers' current and future needs largely, ensuring both high satisfaction among them and an increase in the Bank's profits [8]. The emergence of the electronic marketing elements mix was associated with the impacts resulting from the technology information and communication evolution, especially the Internet [9]. This term was also defined [10] as a set of electronic processes that are put in place by Organizations' Marketing Managements and are consistently implemented. These processes also include all the relevant marketing plans.

5.2 The Importance of Electronic Banking Marketing Mix.

The importance of "electronic banking marketing mix" can be represented in how it employs technology in implementing all of its objectives, strategies and constituent elements [11], and making them work in a coherent and interactive manner. Besides this, the electronic banking marketing mix is significantly important as it one of the effective means, by which the Bank's marketing objectives are achieved that are to efficiently market its services to current and potential customers, using business technology [12]. Its importance can also be demonstrated in the main objective most banks aspire to achieve, which is to maximize profits so that the necessary expenses are covered [13].

5.3 The Competitive Advantage

Porter, who is a known scholar, has illustrated a diagram consisting of three general types of strategies, which firms of all kinds use to achieve competitive advantage. These three main strategies are identified based on two areas: the strategic scope and the strategic asset [14]. The Strategic scope addresses the level of demand and the size and composition of the targeted market, whereas the strategic asset is simply the level of supply and companies' strength or core competencies [15, 16].

6 Previous Studies

Study [2] and study [17] have indicated that there are statistically significant differences between e-marketing in Saudi Telecom Company (STC) and "competitive" advantage, while there are no statistically significant differences when it comes to the role of e-marketing which can be attributed to certain variables (gender, age, working experience, levels of education and career). On the other hand, the findings of study [18] have indicated that electronic marketing elements play an important role in delivering high-quality and expedited banking services to customers, and ensuring high satisfaction among them. Moreover, Study [19] has pointed out that the dimensions of marketing information systems (such as databases, marketing research, marketing intelligence system, and decision-making assistance systems) have enhanced the "competitive advantage" of Lafarge Co.

Besides this, study no. [20] has showed that many Jordanian telecom companies greatly focus on employing marketing intelligence to achieve the required "competitive" advantage by providing their services to customers, using a mixture of methods in marketing and promoting products and services and in the fields of research and development, especially when providing new products and services and developing the existing ones. Study [8] has indicted that (electronic services distribution and electronic services promotion) have a significant impact on the quality of banking services, while (electronic services and electronic pricing) have no significant impact on the quality of banking services provided by the Bank under study.

In study [21], it was found that there is a significant correlation between relationship marketing and the enhancement of competitive advantage at Trust Bank's Biskra-based Agency. Similarly, study [22] found a correlation between e-marketing and the quality of banking services in Iraqi banks, with a high utilization of e-marketing methods in the industry. Study [23] demonstrated that Societe Generale Algeria Bank's implementation of e-marketing had a notable impact on the marketing elements of e-banking, such as price, distribution, promotion, personnel, processes, and physical environment. Additionally, study [7] found that e-banking marketing significantly contributes to achieving competitive advantage in banking institutions. Moreover, both study [22] and study [17] highlighted the positive impact of e-marketing on developing new products and attaining competitive advantage.

In study [23], it was found that there is a direct correlation between e-marketing mix variables and building trust when requesting electronic services, except for in the electronic area(s). Study [3] concluded that e-marketing has an impact on all dimensions of competitive advantage. However, study [24] showed that e-marketing has no significant impact on customer loyalty, and customer orientation does not affect competitive advantage.

Study [25] found that product quality, price, and distribution and promotion channels have a positive impact on achieving the required competitive advantage for small and medium enterprises in Oko Regency. In contrast, study [26] found that the non-utilization of modern technology by bank departments and the lack of diverse banking services negatively affect customer attraction at Al Rafidain Bank.

7 The Methodology

7.1 Study Population and Sample

This study has targeted26 customers dealing with the Relevant Bank in Amman [27]. The number of the study respondents was (384) [28]. In this study, researchers have used the random sampling method and distributed (500) questionnaires to the Bank's clients in Amman. The number of respondents who participated in filling these questionnaires were 411, which approximately account for (82%) of the total number of 500 questionnaires distributed to the members of the community.

8 Testing of the Study Hypotheses

In this study, the Multiple Linear Regression Analysis was conducted to prove the main hypothesis and compare the ratio of the independent variable to the dependent variable. And to verify the hypothesis, the study has used "F" value and its significance level.

Testing of the Main Hypothesis HO: E-banking marketing mix in all its forms (e-banking services, e-banking pricing, e-banking services promotion and distribution, personnel, electronic physical environment, and electronic procedures) has no significant impact at the ($0.05 \geq \alpha$) level of significance in enhancing the competitive advantage. This hypothesis was tested through conducting Multiple Linear Regression Analysis at the ($a \leq 0.05$) level of significance. Table 1 below shows the results.

Table 1. The Test Results of the Main Hypotheses

	Model Summary			ANOVA			Dimensions of Independent Value	Regression Coefficients		
Dependent Variable	Correlation Coefficient R	Coefficient of Determination R2	Adjusted R^2	Calculated "F" Value	Degrees of Freedom	Statistical Significance (Sig)		"B" Value	"T" Value	Statistical Significance (Sig)
Enhancing Competitive Advantage	0.615	0.378	0.367	75.789	7	0.000	Banking Services	0.258	2.695	0.005
							Pricing	0.240	2.451	0.002
					403					
							Promotion	0.574	6.042	0.000
					410					
							Distribution	0.116	1.384	0.167
							Personnel	0.208	3.025	0.003
							Physical Environment	0.129	3.006	0.004
							Electronic Procedures	0.290	2.933	0.001

- The Tabulated "F" Value was (2.02) with (7/410) degrees of freedom. When we compare this with ANOVA result of calculated "F" value, which was (75.789), we can notice how the calculated value is higher than that of the tabulated value, and accordingly the level of significance of the regression model is proved. As for the significance level of the conducted test, it was (0.000), which is lower than the value of the significance level approved herein. For this specific reason, the main hypotheses is rejected and the alternative one is accepted, which states that: "E-banking marketing mix of (e-banking services, e-banking pricing, electronic banking services promotion and distribution, personnel, physical environment, and electronic procedures) has a statistically significant impact at the ($0.05 \geq \alpha$) level of significance on enhancing the competitive "advantage" of the Bank.
- By comparing between "R" value and the correlation value of the independent and dependent variables at table (23), we can infer that that there is a strong positive correlation between the dimensions of the independent variable (electronic banking services, electronic banking pricing, electronic banking services promotion and distribution, personnel, physical environment and electronic procedures), and the dependent variable. The coefficient value (R) was (0.615).
- The Coefficient of determination(R^2) was (0.378), which indicates that functionality of the independent variables combined: (electronic banking services, electronic banking pricing, electronic banking services promotion and distribution, personnel, physical environment and electronic procedures) compared to the dependent variable. It further proves that the dimensions of the independent variable contribute about (37.8%) in enhancing the competitive "advantage". On the other hand, the value of the R2 Adjusted was (0.367) compared to the coefficient of determination, with a total difference of (0.011), which is a very low value, indicating that the independent variables' functionality in predicting the dependent variable is average.
- We can deduce from the table that the value of the independent variable in all its dimensions: (electronic banking services, electronic banking pricing, electronic banking services promotion, personnel, physical environment, and electronic procedures) is statistically significant at the (0.05) level of significance, achieving levels lower than the level of significance approved in this study, which is (0.05). The achieved levels were respectively (0.005), (0.002), (0.000), (0.003), (0.003), (0.004), (0.001), (0.001). Accordingly, we can conclude that the dimensions of the independent value plays a statistically significant role in enhancing the competitive advantage.
- The Results of Table (23) has showed that "electronic banking services distribution" variable does not play a statistically significant role in enhancing the competitive "advantage", given that its level of significance is greater than (0.05).
- The "B" coefficient value was (0.499). This coefficient is essential to set up "the regression" equation for the variables. This equation is used to predict the expected change in the dependent variable in case there were some relative changes in the independent variables. This equation can help us further predict the expected change in the dependent variable. Upon reviewing the above results, it is now clear that the regression equation for testing the first main hypothesis shall be as follows:

Enhancing Competitive Advantage = 0.499+0.258 banking services +0.240 e-banking pricing + 0.574 e-banking services promotion + 0.208 personnel + 0.129 physical environment

9 Discussing the Findings of the Study

1- What is the Role of Applying E-Banking Marketing Mix Elements in Achieving the Bank's Competitive Advantage?

It is worth mentioning that the dimensions of e-banking marketing mix (e-banking services, e-banking pricing, e-banking services promotion and distribution, personnel, physical environment, and electronic procedures) have a statistically significant role in enhancing the Bank's competitive "advantage", and there is a high relatedness between the independent and dependent variables, reflecting a positive impact on enhancing the competitive advantage if the aforementioned elements were employed by the Bank. Furthermore, the percentage functionality of e-banking marketing mix elements in enhancing the Bank's competitive advantage was (37.8%)/The findings have also shown that electronic banking promotion was the most effective in enhancing the competitive advantage.

This findings of this study have agreed with those of Study [2] stating that the use of e-marketing by Saudi Telecom Company (STC) has played a great role in achieving the "competitive advantage", with an overall average of (4.06). This study has also agreed with the findings of [29–31] studies which concluded that the application of e-marketing has a positive and significant impact on creating new products and achieving a competitive advantage.

2- What is the Role of Electronic Banking Services in Enhancing the Bank's Competitive Advantage?

It can be said that electronic banking services play an important role in enhancing the Bank's "competitive advantage". In this regard, the findings of the study has shown that electronic banking services have a statistically significant role in enhancing competitive advantage, and that there is a positive and average relationship between the two variables. I.e. In the event of an increase in the quality of services, the competitive advantage will be enhanced to an average degree, which indicates that the contribution of e-banking services to enhancing the competitive advantage is (22.5%). It is also worth noting that improving the quality of e-banking services by one unit will be positively reflected in enhancing the Bank's competitive advantage with a value of (0.479) [1]. This indicates that there is statistically significant importance of each (product) in achieving the required competitive advantage across economic enterprises.

3- What is the Role of Electronic Banking Pricing in Enhancing the Bank's Competitive Advantage?

As one of the dimensions of "e-banking marketing mix", e-banking pricing has a statistically significant role in enhancing the Bank's competitive advantage. In this regard, it can be inferred that there is a positively high relatedness between the two variables, indicating that any positive changes in e-banking pricing will be positively reflected in enhancing the "competitive advantage. It is also worth mentioning that the use of e-banking marketing has enhanced the competitive advantage by 25.6% while the remaining percentage was achieved due to other factors. Furthermore, any positive changes in e-banking pricing by one unit will be positively reflected by 0.485. The findings of this study have agreed with the findings of study [1] stating that there is statistically significant importance of each (product, price, and process) in achieving the required competitive advantage across economic enterprises.

4- What is the Role of Electronic Banking Promotion in Enhancing the Bank's Competitive Advantage?

E-banking promotion has a significant role in enhancing the Bank's "competitive advantage", and the functionality percentage of e-banking promotion in enhancing the competitive advantage was the highest with (35.7%) compared to other dimensions. There was also a positively high relatedness between the two variables, indicating the significant role of electronic banking promotion in enhancing the competitive advantage. In addition, the value of the expected change in competitive advantage when using e-banking promotion was (0.577).

5- What is the Role of Electronic Banking Distribution in Enhancing the Bank's Competitive Advantage?

E-banking services distribution has a statistically significant role in achieving up to a 27.8% increase in the competitive advantage and there is a highly positive relatedness between the two variables. Besides this, the predictive functionality of electronic banking distribution was (0.473). This is an indication that when electronic banking distribution is developed as one unit, this will be positively reflected in enhancing the competitive advantage and thus meet the required percentage that is (0.473).

What is the Role of Personnel in Enhancing the Bank's Competitive Advantage?

- As one of the dimensions of electronic banking marketing mix elements, personnel play a statistically significant role in enhancing the Islamic International Arab Bank's competitive advantage. There is also a positively high relatedness between the two variables. In other words, personnel contribute to achieving up toa 25.9% increase in the competitive advantage. Even the results have shown that any changes in personnel by a one unit will be positively reflected in enhancing the competitive advantage by (0.597).

What is the Role of the Electronic Physical Environment in Enhancing the Bank's Competitive Advantage?

- The electronic physical environment has a statistically significant role in enhancing the Bank's competitive advantage, and there is a positive and average relationship between the two variables. Furthermore, the electronic physical environment contributes to achieving up to a 22.8% increase in the competitive advantage.
- On a side note, the predictive functionality of the electronic physical environment in enhancing the competitive advantage was (0.591).

What is the Role of Electronic Banking Procedures in Enhancing the Bank's Competitive Advantage?

- As one of the dimensions of electronic banking marketing mix elements, electronic procedures have considerable importance in enhancing the Bank's competitive advantage given that they contribute to achieving up to a (23%) increase in the competitive advantage. Let alone that there is an average relationship between the two variables.

To What Extent Does the Islamic International Arab Bank Apply the Elements of Electronic Banking Mix?

- The electronic banking elements were highly used by the Islamic International Arab Bank with an arithmetic average of (3.82), which is relatively high in terms of significance. This can be attributed to the considerable importance the Bank has attached to the dimensions of the electronic banking mix, and this is why all of the said elements have been relatively high in terms of significance. At the level of the Bank, electronic banking distribution was the most applied element, achieving the highest arithmetic average with (3.97) compared to the other dimensions. Out of all the dimensions of the electronic banking marketing mix, the arithmetic average of "personnel" dimension was the lowest with (3.70). This study has agreed with the findings of study No. [22], stating that Iraqi banks, in general, use e-marketing means to a great extent.

What is the Level of the Bank's Competitive Advantage?

- The level of the Bank's competitive advantage is considered high compared to other banks. This variable was also greatly high in terms of significance with an arithmetic average of (3.76). This can be attributed to the fact that the Bank offers credible and high quality services, resulting in increasing customers' loyalty.

9.1 Recommendations

- Enhance the quality of e-marketing mix elements within the Bank, given their significant role in enhancing the "competitive advantage", by conducting several studies, analyses and evaluation of the elements in a regular and constant manner, while allocating the necessary financial and human resources for the development of such.
- Examine the customers' needs for banking services and the evolving future needs as to provide new services to this segment. Let alone analyze the electronic banking services provided by competitors and accordingly strive to constantly offer more efficient and new services.
- Use a wide variety of electronic banking promotion methods and high-tech tactics to raise the efficiency of electronic promotion methods.
- Allocate a group of highly qualified employees to answer all customers' electronic inquiries and develop the necessary procedures to accelerate this process, while ensuring that inquiries are answered in an effective manner.
- Take into account customers' feedback on the provided electronic services and further resolve their problems.
- Analyze the electronic system and identify its most significant drawbacks, from design to performance. Redesign the system to make it user-friendly and engaging.

References

1. Vendor, Iman, B.: The Role of Marketing Innovation in Achieving Competitive Advantage in the Economic Enterprise (2021)
2. Namsha, M.A.M.B., Qadir, S.R.A.: The role of E-marketing in achieving competitive advantage in saudi telecom companies. Int. J. Humanities Soc. Sci. **31**, 113–129 (2022)
3. Obeidat, A.M.: E-Marketing and Its Impact on The Competitive Advantage. Ilkogretim Online, 2 (2021)(5).
4. Shahwan, Y., Abdel-hamid, O.: The impact of social responsibility disclosure in reducing the risk in Jordanian Commercial Banks. Res. World Econ. **11**(3), 142–150 (2020)
5. Hidayatullah, S., Firdiansja, A., Patalo, R.G., Waris, A.: The effect of entrepreneurial marketing and competitive advantage on marketing performance. Int. J. Sci. Technol. Res. **8**(10), 1297–1301 (2019)
6. Bin, S., Karima, Mattai, A.Q.: The impact of electronic marketing mix elements on the competitive advantage of Algerian banks. J. Ijtihad Legal Econ. Stud. **8**(1), 578–601 (2019)
7. Mattai, A.Q., Shanina, K.B.: The extent of the contribution of electronic banking marketing in achieving the competitive advantage of banking institutions in Algeria. J. Accounting Financial Stud. **14**(47) (2019)
8. Al-Ashqar, A., Ali, S., Saad, S., Omran, A.-R.: The Electronic Marketing Mix and its Impact on the Quality of Banking Services (A case Study on Jumhouria Bank, Al-Alous branch), Al-Marqab University, No. 1, Algeria (2020)
9. El-Din, Q.A.D., Hizam, S.: The Role of Electronic Marketing in Improving the Quality of Hotel Services (2022)
10. Kararya, F.: The Role of Social Networks as a Tool of the Electronic Marketing Mix. Economic Notebooks Journal, 13 Algeria (2), 71–82 (2022)
11. Lasi, M.B.A.: The relationship between E-marketing mix strategy and integrated marketing communication: a conceptual framework. Int. J. Econ. Manage. Syst. **6** (2021)
12. Aras, M., Syam, H., Haris, H., Jasruddin, M., Akib, H.: The analysis of mix marketing system toward the performance of convection business in makassar. In: 1st International Conference on Social Sciences (ICSS), pp. 1104–1107, October 2018
13. Fikri, M.H., Syamsuri, A.R., Hutasuhut, J., Harahap, A., Nainggolan, S.: The Effect of Student's Mix Marketing Services on Student's Interest in Saving at Syari'ah Bank. Budapest Int. Res. Critics Inst. (BIRCI-Journal): Humanities Soc. Sci. **5**(1), 1294–1307 (2022)
14. Brahimi, A., Massad, S.: The Role of Sustainable Marketing in Achieving Competitive Advantage, Ph.D. Thesis, Abdelhafez Bousouf University, Algeria (2019)
15. Al-Qarni, D., Dhafer, Fadel, B.Q.: The role of electronic marketing in achieving the competitive advantage of local banks. Sci. J. Bus. Environ. Stud. **13**(2), 164–222 (2022)
16. Shahwan, Y., Hamza, M., Al-Fasfus, F., Al-Ramahi, N., Almubaydeen, T.H.: Social Responsibility Accounting, Disclosure and Real Practice: Evidence from Jordan. Asian Econ. Financial Rev. **12**(3), 164–182 (2022)
17. Khrais, E., Azzam, Z., AAssaf: Constraints facing garment industrial sector operating within the qualified industrial zones in Jordan. Zarqa J. Res. Stud. **10**, 2.82–91 (2010)
18. Shaima, B., Al Zaman, Q.: The Influence of the Elements of the Electronic Banking Marketing Mix on Achieving A competitive Advantage for Banks (Doctoral dissertation) (2022)
19. Zakaria, Dashisha: The Impact of the Marketing Information System in Achieving Competitive Advantage, Ph.D. thesis, University of M'sila, Algeria (2021)
20. Al-Sarhan, A.F., Fahad, A.: The role of marketing intelligence in achieving the "competitive advantage" of entrepreneurial organizations: an applied study on Jordanian telecom companies. Arab J. Manage. **41**(2), 337–353 (2021)

21. Hind Boulgrou, N.: The Role of Relationship Marketing in Enhancing the Competitive Advantage in Commercial Banks Biskra Agency-Trust Case Study: Bank, Ph.D. thesis (2020)
22. Al-Khammas, A.A-R., Badrawi, F., Al-Shammari, L.H.: The impact of the use of electronic marketing on the quality of banking service: a prospective study of a sample of employees in Iraqi private banks. Econ. Sci. **14**(53)21–40 (2019)
23. Khan, K., Ferro, M., Sabrina: The Reality of the Impact of E-marketing on the Elements of the Banking Marketing Mix from the Point of View of the Employees of Societe Generale Bank – Algeria (2019)
24. Hutagalung, L., Sinaga, P., Tan, J., Tukiran, M.: The role of e-marketing and customer oriented on business performance through customer loyalty and hotels competitive advantage. Kontigensi: Jurnal Ilmiah Manajemen **8**(2), 202–217. (2020)
25. Sari, Y., Gultom, A.: Marketing Strategy in an Effort to Increase Competitive Advantage in Small and Medium Enterprises in OKU Regency. JIMFE (Jurnal Ilmiah Manajemen Fakultas Ekonomi). 6. (2020) (in press). https://doi.org/10.34203/jimfe.v6i2.2390
26. Sattar, A.D.: Electronic Banking-Marketing and its Impact on Enhancing Attract Customers A Case Study of the Rafidain Bank in Dhi Qar Governorate (2020)
27. https://iiabank.com.jo/ar/content2022
28. Sekaran, U.: Research Methods for Business, 4th edn. Willey, New York (2003)
29. Aziz, M.N.L., Nurhayati, P., Yudha, A., Annisa, N.N.: E-marketing usage as an increase in competitive advantage through product innovation and marketing performance (Survey On MSMEs In Central Java). J. Bus. Manage. Rev. **3**(5), 400–414 (2022)
30. Mishbakhudin, M., Aisyah, M.: The E-marketing mix strategy of tokopedia salam during the Covid-19 pandemic. Int. Res. J. Bus. Stud. **14,** 215–227 (2022)
31. Azzam, A., Al-Shaer, S.: Electronic Word of Mouth (E_WOM) Adoption via Social Media and Its Impact on Online Shoppers' Purchasing Intention During Corona Pandemic. A Case of Jordan. Journal of Studies in Systems, Decision and Control **216**, 477–487 (2023)

The Effect of Digital Accounting Systems Within Digital Transformation on Financial Information's Quality

Qasem Aldabbas(✉), Sulaiman Weshah, Nour Abdullah, Mohammad Albakheet, Feras Abu Hamoud, and Ali Hourani

Al-Balqa Applied University, Salt, Jordan
{qasemda1978,sulaiman.weshah}@bau.edu.jo

Abstract. This qualitative analysis aims to evaluate the current state of digital accounting system applications in accounting by interviewing professionals in the field of financial managers, using earlier studies and pertinent research in addition to the interviews, and conducting an in-depth analysis of the data.

The study discovered that future accounting procedures would be affected by digital accounting systems. The most exact and complete dataset currently available for accounting practical usage will provide analytics for digital accounting, which will improve compliance decision-making. A digital accounting system will also save money and time. Digital accounting systems' key problem is security threats.

By expanding training within companies, financial managers may better prepare themselves to cope with the overall corporate digitalization transformation as well as training in dealing with digital accounting technology.

Keywords: Digital Accounting Systems · Financial Information Quality · Digital Transformation · Jordan

1 Introduction

Information and communication technologies have advanced dramatically and attracted significant interest from both practitioners and the scientific community. At the managerial level, their primary benefit is the fluidization of information flow, which enables all decisions to be made using constantly updated information. As a result, the value of both businesses that disseminate information or offer the technology as well as those that use it has significantly increased.

Information technology is a strong instrument that businesses may use to create creative goods and business processes that will enable them to flourish. This is the environment that influenced the emergence and development of the idea and method of digital accounting (Genete and Tugui 2008). That's why (Weshah 2021) determined that implementing modern IT systems has an impact on the employment process for the accounting department, with the development of modern information systems having

B. A. M. Alareeni and I. Elgedawy (Eds.): *Artificial Intelligence (AI) and Finance*, SSDC 488, pp. 704–711, 2023.
https://doi.org/10.1007/978-3-031-39158-3_65

the most influence on the hiring of accountants. New technologies have significantly impacted many business domains in terms of improving data accuracy, reducing expenses and human error, enabling real-time information exchange, and eliminating fraud by automating activities and transactions using smart contracts. This is particularly accurate in the accounting industry (Yermack 2017). Hence proved this change is seen as being necessary for the accounting industry to continuously refresh and advance in line with advances in the digital era (Tekbas and Nonwoven 2018).

Many studies have shown how important information technology is to accounting. As a result, in order to fulfill corporate objectives, the merger of these two professions should progress toward developing harmony (Al-Tahat 2023; Pandey and Agrawal 2022; Mancini et al. 2021; Khin and Ho 2019; Liu et al. 2019).

Today's the world is undergoing rapid development and change in all spheres and At all political, economic, social and educational levels, This is as a result of the technical and informatics revolution that has taken place and continues to occur and has had a significant impact on both the institution, management and individuals, This imposed new and different working conditions and mechanisms, and Using new resources and systems to keep abreast of these changes and technical and information transformations, were Traditional means of work can no longer achieve the Foundation's objectives with the speed and accuracy required and the reason society transformed from traditional paper to digital society, for Accounting is one of the important areas affected by recent technological developments that have led to the integration between of work computers and the accounting profession and Streamlining routine transactions about it's, and Digital transformation has contributed through its various tools such as Cloud Computing, Big Data, Data Mining, Artificial Intelligence, Block Chain, E- payment, and Virtual Currencies in integrating digital technologies into the accounting processes, and event fundamental changes to traditional tools in handling financial data and information, this has increased the efficiency and ability of the accounting system to process data, and to obtain fast, objective, convenient and reliable the financial information that is a requirement for a successful digital transformation (Lakaiza and Zarqoun 2017) and (Shannen 2023).

From here the study came to know the impact of digital accounting systems on the quality of financial information that consider the end product of the data-fed accounting information system is considered and Recording, measuring, summarizing, tabulating and presenting them in financial statements that enable its users from assess and decision-making.

However, the field of artificial intelligence has been working on computer systems that resemble intelligent behavior. This is a challenging challenge since artificial intelligence requires computer software to be able to do a wide range of functions (Samoili et al. 2020).

2 Literature Review

Nguyen et al. (2021) came to the conclusion that digital transformation processes will enable managers to take advantage of the effort associated with accounting work and the resulting errors to focus on improving product quality and raising the organization's competitiveness rather than on customer search and customer acquisition.

Digital accounting has evolved into a natural notion, with its effects being seen in both routine and strategic actions (Lutfi et al., 2011). The development, representation, and transmission of financial data in an electronic format are all considered to be aspects of digital accounting. All accounting transactions take place in an electronic setting rather than on paper. Instead, by enhancing the productivity of accounting experts, it appreciates and empowers them. The creation, representation, and transfer of financial data in an electronic format are all referred to as "digital accounting." The financial sector has undergone a revolution thanks to computers and accounting software. Technology improvements have made it possible for accountants to evaluate and report data more quickly, effectively, and efficiently than ever before (Troshani et al. 2019).

The examination of businesses' accounting operations under conditions of competitiveness and digital transformation, both at the strategic and operational levels, is clarified by (Timchev 2020).

Kornchai and Khajit (2021) confirmed that the resource-based perspective of the businesses is viewed as an important theory that may clarify and validate the impacts of digital accounting on financial reporting quality, accounting information, and the efficacy of strategic decisions, also (Rehm 2017) emphasized the need to prepare accounting information systems for digital transformation, and the topic was covered by outlining the key steps that must be taken through general descriptions of the administrative documentation process, the preparation of an accounting documents process guide, the definition of principles and guidelines, the documentation of operating procedures, and the taking of special measures when necessary.

However, (Meraghni et al., 2021) discovered that there was a low level of understanding of the significance of digital transformation as well as a low effort to construct accounting information systems in line with those needs due to the presence of multiple high-impact difficulties.

High-quality reporting offers information that can be used to make decisions that is pertinent to the business's operations during the reporting period and accurately depicts both those activities' economic realities and the business's financial situation at that time (Shuraki et al. 2021). In certain instances, using digital transmission methods is elective as well; therefore we should anticipate a stronger influence of analytical culture in these voluntary contexts. The decision-making process may be significantly impacted by analytical culture (Popovic et al. 2012).

Unreliable importing sources of data, particularly external ones, poor data upkeep and handling when moving from one system to another, and poor data handling (IFerencek and Kljajic 2020). Companies large and small are gathering big volumes of data with the goal of discovering new information from their datasets as technology advances and costs decrease (Begoli and Horey 2012).

Al-Okaily et al. (2022) revealed that decision-making quality is mediated by information quality in the link between data and system quality.

3 Study Problem

The most important goals of the financial accounting system is to provide high quality and useful financial information that its users weather from inside and outside the organization and to enable them to make appropriate decisions, and considered financial information is useful if relevance, express on the purpose for which it was prepared, comparable and verifiable, and provided in a timely manner, in shade of the digital transformation that the world is witnessing and intense inter-institutional competition, it has become necessary to implement of digital accounting systems that can contribute to obtaining accurate, error-free financial information, perform operations faster, improving the quality of financial information and the ease of obtaining it.

Acquire this studies of academic and professional importance as it contributes to narrowing the accounting research gap in the field digital transformation of the accounting profession in general, the quality of financial information in particular, and the possibility of integrating digital transformation technology into the accounting function. The importance of current study starting from the modernity and the lack of local studies on the topic. Thus current study is a starting point for other future studies related to digital transformation and the accounting profession.

4 Methodology

The current qualitative study aims to determine the future effects of digital accounting on the information quality within financial reports. Through conducting interviews with financial managers (112) from different sectors with limit of 45 years old age to spot the light with criteria that close to digital transformations (Table 1).

Table 1. Number of financial managers upon organization's fields

Organization's Working Field	Count of Financial Mangers
Agriculture, Food, and Natural Resources	28
Architecture and Construction	16
Business and Finance	19
Education and Training	11
Insurance	15
Information Technology	23

5 Results

When compared to manual accounting methods, users saw time savings and improved accounting entry accuracy. This result goes parallel with (Chanthinok and Sangboon 2021).

Digital accounting helps business owners to perform essential operations more quickly and correctly, and their accountants.

Digital accounting clearly interprets and communicates data in timely manner.

Digital accounting assists businesses with bookkeeping and tax reduction through providing new measures of business rates.

Managing strategic roles will be more effectively through the time saved by using digital accounting.

Digital transformation provides the necessary technological skills to ensure the accuracy of financial reports in a realistic manner and makes them more effective and accurate, which leads to increased satisfaction of the beneficiaries of its services..

Digital transformation helps reduce the accounting workload, as complex and difficult transactions are done easily and quickly. It also helps in analyzing data regularly and effectively, in order to provide quality, reliable and complete financial data.

Digital transformation contributes to ensuring the quality of financial information, keeping pace with development, and the ability to use electronic data to facilitate and complete transactions with high accuracy.

Digital transformation contributes to the speedy completion, standardization and simplification of procedures for recording accounting operations and reducing the credibility gap in financial statements and reports.

Digital transformation contributes to the security of accounting information by saving it, facilitating its storage and retrieval, and providing access.

Since regular chores like creating client invoices, VAT returns, and management reports are automated by a digital accounting system, it is substantially more efficient than manual record-keeping and will help you manage your business's administration and paperwork better.

Users can obtain real-time data with digital accounting, and with accurate, current information only a click away, you can quickly evaluate how your company is doing in relation to your budgetary expectations and take quicker, more educated business choices.

Digital accounting has effectiveness of knowledge management processes, information system control, and information systems governance.

A high quality of data input within digital accounting that gives a high quality of information as output.

A digital accounting system is flexible with the demand of information that users of financial reports need.

More qualified employees (accountants) are needed whom have an enough digital knowledge and background.

6 Conclusions and Discussion

The accounting industry is going through a significant transformation as a result of traditional modes of accelerating evolution as technology develops in almost every industry today, and as (Oesterle et al. 2020) concluded digitalization is progressively and broadly affecting companies all through all businesses.

Therefore, within the contemporary technical and digital era, it may be thought of as a chance to improve the quality of accounting reporting that lead to improve information accessible, accurate, complete, economical, flexible, relevant, reliable, and timely which enhance the chances of good decision-making.

The digital transformation maximizes the performance of accountants in the use and management of technology and allows the use of various digital systems to acquire correct knowledge, manage and use technology, adapt to specialized technological developments and use them in the field of work, and combine practical and theoretical experience.

The results of this study are consistent with the study of (Khalleefah 2020) and (Almhdi et al. 2021).

7 Limitations

This study conducted on financial managers with limit up of 45 years old ages in Jordan within some selected organizations fields (Agriculture, Food, and Natural Resources, Architecture and Construction, Business and Finance, Education and Training, Insurance, and Information Technology). Conducted year is 2023.

References

Almhdi, M., Abdulrahman, Yousef; Salah, A.: The Impact of the Transformation of the Digital Accounting System on the Characteristic of Faithful Representation of Accounting Information in light of Principles ng Information in light of Principles (Web Trust) 4(2), pp. 17–54 (2021)

Al-Tahat, S., Weshah, S., Altal, A., Abu-erbaiea, S., Bader, A., Bawaneh, A.: Big Data's Sustainable Impact on the Future Responsibilities of Accountants, The 6th Equal Opportunities Conference "Sustainable Leadership and Technology Innovation Management." Ahlia University, Bahrain (2023)

Begoli, E., Horey, J.: Design principles for effective knowledge discovery from big data. In: 2012 Joint Working IEEE/IFIP Conference on Software Architecture and European Conference on Software Architecture, Helsinki, Finland (2012). https://doi.org/10.1109/wicsa-ecsa.212.32

Chanthinok, K., Sangboon, K.: The development of digital accounting system on cloud computing. J. Comput. Sci. **17**(10), 889–904 (2021). https://doi.org/10.3844/jcssp.2021.889.904

Ferencek, A., Kljajic, M.B.: Data quality assessment in product failure prediction models. J. Decis. Syst. **29**(1), 1–8 (2020). https://doi.org/10.1080/12460125.2020.1776927

Genete, D., Ługui, A.: From ERP Systems to Digital Accounting in Relations with Customers and Suppliers, Computing and Computational Techniques in Sciences, Selected Papers from the WSEAS Conferences in Spain, September 2008 Santander, Cantabria, Spain, September 23–25, 2008, pp. 57–63 (2008)

Khalleefah, A.M.: Assessment of the impact made by the digital transformation of the accounting system on the decision-making system of the enterprise. Evropský Časopis Ekonomiky A Managementu **5**(6), 122–129 (2020). https://doi.org/10.46340/eujem.2020.6.2.15

Khin, S., Ho, T.C.F.: Digital technology, digital capability and organizational performance: a mediating role of digital innovation. Int. J. Innov. Serv. **11**(2), 177–195 (2019). https://doi.org/10.1108/IJIS-08-2018-0083

Lakaiza, Y., Zarqoun, M.: The impact of accounting knowledge on the quality of financial information. J. Dev. Res. Stud. **4**(1), 56–71 (2017)

Hai, L., Yongsheng, C., Fang, F., Yanqing, C.: Application of block chain technology in big data of seed industry. China Seed Industry **5**(1), 1–5 (2019). https://doi.org/10.1088/1742-6596/1314/1/012218

Lutfi, A., et al.: Influence of digital accounting system usage on SMEs performance: the Moderating Effect of COVID-19. Sustainability **14**, 15048 (2022). Doi:https://doi.org/10.3390/su14221 5048

Mancini, D., Lombardi, R. and Tavana, M.: Four research pathways for understanding the role of smart technologies in accounting. Meditari Accountancy Res. (29)5, 1041–1062 (2021). https://doi.org/10.1108/MEDAR-03-2021-1258

Meraghni, O., Bekkouche, L., Demdoum, Z.: Impact of digital transformation on accounting information systems – evidence from algerian firms. Econ. Bus. **1**(35), 249–264 (2021). https://doi.org/10.2478/eb-2021-017

Nguyen, M.T., Nguyen, Q.H., Nguyen, T.H.: Digital transformation in the business: a solution for developing cash accounting information systems and digitizing documents. Sci. Technol. Dev. J. **24**(2), 1975–1987 (2021). https://doi.org/10.32508/stdj.v24i2.2526

Oesterle, S., Buchwald, A., Urbach, N.: Investigating the co-creation of IT consulting service value: empirical findings of a matched pair analysis (2020). https://doi.org/10.1007/s12525-020-00Ch/404383858a984f4d6e32fb3c0495a994bff554d2

Pandey, R., Agrawal, V.: Block chain technology in digital accounting. Int. J. Eng. Manage. Res. **12**(3), 201–204 (2022). https://doi.org/10.31033/ijemr.12.3.31

Phornlaphatrachakorn, K., NA Kalasindhu, K.: Digital accounting, financial reporting quality and digital transformation: evidence from thai listed firms. J. Asian Financ. Econ. Bus. **8**(8), 409–419 (2021). https://doi.org/10.13106/JAFEB.2021.VOL8.NO8.0409

Popovic, A., Hackney, R., Coelho, P.S., Jaklic, J.: Towards business intelligence systems success: effects of maturity and culture on analytical decision making. Decis. Support Syst. **54**(1), 729–739 (2012). https://doi.org/10.1016/j.dss.2012.08.017

Radu, L.-D., Necula, S.-C.: Decision-Making Processes and Their Implementation in Digital Accounting Practices, Conference: Proceedings of the 10th WSEAS international conference on Telecommunications and informatics and microelectronics (2011). https://www.researchgate.net/deref/http%3A%2F%2Fmpra.ub.uni-muenchen.de%2F51546%2F, paper number: 51546

Rehm, S.V.: Accounting information systems and how to prepare for digital transformation. In: The Routledge Companion to Accounting Information Systems, pp. 69–80 (2017). https://doi.org/10.4324/9781315647210-6

Samoili, S., Lopez Cobo, M., Gomez Gutierrez, E., De Prato, G., Martinez-Plumed, F., Delipetrev, B.: AI WATCH. Defining Artificial Intelligence, EUR 30117 EN, Publications Office of the European Union, Luxembourg (2020). ISBN 978-92-76-17045-7, https://doi.org/10.2760/382730, JRC118163

Shannen, A.: The impact of digital transformation on the transparency of financial reports in the light of contemporary professional publications (a field study). Sci. J. Financ. Commercial Stud. Res. **4**(1), 19–23 (2023)

Shuraki, M.G., Pourheidari, O., Azizkhani, M.: Accounting comparability, financial reporting quality, and audit opinions: evidence from Iran. Asian Rev. Account. **29**(1), 42–60 (2021). https://doi.org/10.1108/ARA-06-2020-0087

Tekbas, I., Nonwoven, K.: The profession of the digital age: accounting engineering. In: IFAC Proceedings Volumes, Project: The Theory of Accounting, Engineering (2018). https://www.ifac.org/knowledge-gateway/preparing-future-ready-professionals/discussion/profession-digital-age-accounting-engineering

Timchev, M.: Accounting business analysis of the company in the conditions of competition and digital transformation. Economic Science, education and the real economy: Development and interactions in the digital age, (1), 151–161 (2020)

Troshani, I., Locke, J., Rowbottom, N.: Transformation of accounting through digital standardization: tracing the construction of the IFRS taxonomy. Account. Auditing Accountability J. **32**(1), 133–162 (2019). https://doi.org/10.1108/AAAJ-11-2016-2794

Weshah, S.: Adopting modern it systems is vital in employing accountants and internal Auditors (educational perspective): a case study in Jordan cement company–Lafarge. Jordan J. Bus. Administration **17**(4), 555–565 (2021)

Yermack, D.: Corporate governance and block chains. Rev. Finance **21**(1), 7–31 (2017). https://doi.org/10.1093/rof/rfw074

Assessing the Impact of Macroeconomic Indicators in the Resilience of Jordanian Commercial Banks Amid COVID-19 Pandemic

Issa Ahmad Swiety[1(✉)], Mohammad Sami Ali[1,2,3], Fuad Al-Fasfus[1,2,3], Kayed Alattar[2], and Ayman Zarban[3]

[1] Zarqa University, Zarqa 13110, Jordan
iswiety@zu.edu.jo
[2] Isra University, Amman, Jordan
[3] University oh Business and Technology, Jeddah, Kingdom of Saudi Arabia

Abstract. The WHO officially announced COVID-19 as a global pandemic on 11 March, 2020. The financial sectors around the world have suffered dramatic impacts. In this study, we have analyzed a monthly time series data in order to evaluate the financial resiliency of Jordanian commercial banks (JCB), over the period March/2020-Dec/2021. The independent variables of the study include; GDP growth rate, interest rate, and unemployment, and the dependent variables include; capital adequacy ratio (CAR), non-performing loans (NPLs), and liquidity coverage ratio (LCR). The statistical results depicted commercial banks to be financially resilient up against the unfortunate economic crisis.

Keywords: Jordan Banking Sector · COVID-19 · Financial Resilience · GDP · Economic Stability

1 Introduction

The world economies were astounded as *COVID-19* first made its appearance in late 2019. All financial and non-financial sectors have been stunned ever since *COVID-19* was declared a global outbreak on March 11, 2020, and worldwide shutdowns were official. The UN, WHO, and federal governments authorized full immediate lockdowns, followed by strict SOPs. However, sudden lockdowns took a very bad toll on macroeconomic indicators (e.g., GDP growth, per capita income, interest rates, unemployment) and diminished the liquidity and stability of the stock markets [1].

On March 2, 2020, the first rigored case of *COVID-19* was reported in Jordan: a young Jordanian man travelling to Italy [2]. Jordan interacted swiftly to block the coronavirus from spreading [3]. Jordan is a state with scarce resources and dependency on foreign aid, and the pandemic disconnected a large number of economic activities [4]. The services sector makes around 2/3 or 66% of the Jordan's GDP. As lockdowns were imposed, PhD researcher [5] depicted the largest segment of the economy, the services industry will be hit at its worse. Nevertheless, the NGOs and other financial institutions containing huge

B. A. M. Alareeni and I. Elgedawy (Eds.): *Artificial Intelligence (AI) and Finance*, SSDC 488, pp. 712–721, 2023.
https://doi.org/10.1007/978-3-031-39158-3_66

financial resource and large-coverage administration countered the devastating economic impacts [4].

Previous studies including [6] and [7] showed a gap regarding the economic impacts of *COVID-19* and the financial stability of Jordanian banks. Therefore, this research aim at filling this gap through scrutinizing the resiliency of jordanian commerical banks amid the *COVID-19* crisis, within the time period (March/2020-Dec/2021). Thus, in an attempt to achieve this core aim, the study tries to answer whether the selected 13 JCB were financially resilient before and during *COVID-19 pandemic*? Consequently, findings from this study are estimated to contribute in economic growth, via helping banks' management to realize as well as encountering the undesired implications of the Covid-19 pandemic.

2 Literature Review

2.1 Pre-COVID and Jordan's Economy

In early 1989, the Central bank of Jordan made banking sector reforms; increasing paid-up capital and ensuring higher inflow of foreign capital [8]. The economic activities of the state are financed by commercial banks that play a critical role in building the Jordanian economy [9]. Since the 1990s, Jordan has maintained a neo-libertarian and debt-regime, resulting in a greater tendency of continuous fiscal and trading deficits, increasing number of liabilities, and higher rate of joblessness [3]. Moreover, the state had introduced restraint measures and various forms of taxation, leading to higher costs of living. Jordan has structured a framework for improved, sustained, and comprehensive economic growth (2016–2020), throughout its Five-Year-Reform Matrix, which focuses on creating employment, notably for women and teenagers [13]. The CBJ made two initiatives; building money-capital of JD125 million to inject it in SMEs and contributing JD100 million to national fares credit [10]. The CBJ also established two banking investment centres, "Trade Banks Society for Investment" and "Islamic Investment Company" who shall invest their resources into SMEs and similar organizations to boost economic growth and welfare [10, 17].

A large number of the banking sectors determine their financial performance by non-performing transactions e.g. NPLs as a strategic tool. In 2016, Jordanian banks had NPLs ratio of 4.3% [11]. Whilst the highest recorded was at 8.5% in 2011, the NPLs ratio to total loans was 5% in 2019 [12]. As of mid-2018, banks accounted for 20% of GDP and total assets of 176% of GDP, in 2016, making it one of the largest economic constituent of Jordan. Despite the strong financial position, the environment suffers from extreme economic challenges; low GDP growth, high unemployment, competition, and volatility of socio-political and economic environment in neighbouring countries. Furthermore, the stock market price index contracted by 7.5%, followed by a decline in GDP market capitalization from 53.8% (2018) to 47.6% (2019) The CBJ's year-end 2019 gross reserves increased by 4.1% to US $13.3 billion. Jordanian banks witnessed a year of sustainable growth in 2019, adding domestic liqudity, and increasing number of credit-facilities to clients.

2.2 COVID-19 and Jordan Banking Sector

The Jordanian state had already cut its welfare activities along with the removal of subsidies on oil, food, and water in the year 2000s. During *COVID-19,* the economy was further affected [5]. The *JEM Spring* 2020 edition launched by the joint-collaboration among Jordan Strategy Forum (JSF) and the World Bank forecasted that Jordanian economy shall contract by 3.5% [13]. During first quarter of 2020, real GDP declined by 3.5% and revenues by 25% [3]. This negative contraction resulted from closure of trade routes, tourism, and remittances. The internal liquidity flows were distorted and additional financing was imminent.

Jordan confronted the challenge of in availability of foreign funds as other countries also restricted their budgets, amid *COVID-19* [5]. The ratio of foreign aid to GDP reduced from 40% to 7% but, it still has a critical role in the economy. State minister Al-Rfouh said that "Jordan will need continuous foreign flow to withstand the hinging effects". The decrease in average oil prices brought a silver-lining impact on Jordan which totally relies on Gulf's oil export [3]. In May, 2020 the Jordanian treasury witnessed a major decline in foreign aid and remittances; as many citizens lost their jobs in the Gulf.

Despite *COVID-19,* the Jordanian banking sector showed great financial integrity and resilience. Bank of Jordan injected additional liquidity into domestic economy having worth of 1.05billion JD ($1.48 billion), declining its cash reserve from 7% to 5%. Secondly, commercial banks started restructuring small and medium sized loans and suspended credit facilities to individuals and businesses alike without any commision and interest delays [14].

Previous studies dealt with the correlation of economic and financial variables, and no research is done on financial resiliency during the *COVID-19* pandemic, henceforth, our research study shall evaluate this research gap.

3 Research Methodology and Hypotheses Development

Through the construction of hypotheses and data analysis using the statistical software analysis "STATA", we shall highlight the results, prove our null hypotheses, answer the research questions, and examine the validity of our research statement.

We have selected 12 out of 14 banks; those are listed on the securities exchange. In this study, financial resiliency is measured by three banking indicators; Capital adequacy ratio (CAR), Non-performing loans (NPLs), and the Liquidity coverage ratio. In the research methodology, we have developed the following hypotheses. The following nine hypotheses are based on the dependent financial variable and all independent economic variables. The economic variables chosen to measure financial resiliency are GDP growth rate, interest, and unemployment rate (March/2020-Dec/2021).

- Capital Adequacy Ratio (CAR)

The degree to which a bank's equity or capital is expressed as the ratio of its risk-focused credit exposures is known as Capital adequacy ratio [15]. An international design issues the legal minimum requirement of CAR as proposed in the Basle Accord [16]. The Central bank of Jordan has proposed the minimum limit of 12% for CAR.

H1: There is no remarkable effect of *GDP growth rate* on Capital Adequacy ratio of the selected thirteen commercial banks.

H2: There is no remarkable effect of *interest rates* on Capital Adequacy ratio of the selected thirteen commercial banks.

H3: There is no remarkable effect of *unemployment rate* on Capital Adequacy ratio of the selected thirteen commercial banks.

– *Non-Performing Loans (NPLs)*

Non-performing credit or mostly referred as non-performing loans (NPLs) is the ratio between the non-performing credit and total issued credit facilities.

H4: There is no significant influence of *GDP growth rate* on Non-performing loans (NPLs) of the selected thirteen commercial banks.

H5: There is no significant influence of *interest rates* on Non-performing loans (NPLs) of the selected thirteen commercial banks.

H6: There is no significant influence of *unemployment rate* on Non-performing loans (NPLs) of the selected thirteen commercial banks.

– ***Liquidity Coverage Ratio (LCR)***

Liquidity Coverage Ratio (LCR) is the ratio between total estimated asset maturities/net cash outflow for 30 days. The Central bank of Jordan has proposed 100% extent for liquidity coverage (for outflows due in less than one month). If it falls below 100%, CBJ has the right to intervene and take control of the bank.

H7: There is no significant effect of *GDP growth rate* on the Liquidity coverage ratio of the selected thirteen commercial banks.

H8: There is no significant effect of *interest rates* on the Liquidity Coverage ratio of the selected thirteen commercial banks.

H9: There is no significant effect of *unemployment rate* on the Liquidity Coverage ratio of the selected thirteen commercial banks.

3.1 Models of the Study

The hypotheses were replicated and redesigned from the financial stability study in Jordan conducted by [6]. Based on the taken hypotheses above, the following model for every dependent variable is utilized to test hypotheses:

$$\boldsymbol{CARb,t = \alpha 0b,t + \beta 1GDPb,t + \beta 2IRb,t + \beta 3URb,t + \varepsilon b}$$

where $CAR_{b,t}$ is the capital adequacy for the bank b time t, $\mathrm{GDP}_{\mathrm{b,t}}$ is the growth in Gross Domestic Product in link to bank b, and time t, $\mathrm{IR}_{\mathrm{b,t}}$ is the interest rate bank b time t, $\mathrm{UR}_{\mathrm{b,t}}$ is the rate of unemployment in link to bank b, and time t, β are parameters, Alpha (α) is dependent variable index.

$$\boldsymbol{NPLsbt = \alpha 0b,t + \beta 1GDPb,t + \beta 2IRb,t + \beta 3URb,t + \varepsilon b}$$

where $NPLs_{b,t}$ are the non-performing loans for bank b, time t.

$$\boldsymbol{LCRb,t = \alpha 0b,t + \beta 1GDPb,t + \beta 2IRb,t + \beta 3URb,t + \varepsilon b}$$

where $LCR_{b,t}$ is the liquidity coverage ratio for bank b, time t.

4 Results and Discussion

This study analyzed a monthly time series data to evaluate the financial resiliency of Jordanian commercial banks (JCB), over the period March/2020-Dec/2021. Hence, since results from the Augmented Dickey-Fuller (ADF) confirmed that the used data are stationary when they are tested at their levels, the study ran the linear regression tests to achieve the core aim of this study.

Model 1: Capital Adequacy
Linear Regression

The most common statistical tool that is used to measure the relationship between different variables, having reasoning and logical relation is called regression analysis [17]. Linear regression works as an algorithm used to visualize a relationship among two kinds of variables, the dependent and the independent one [18].

Table 1 . Model 1: Capital Adequacy, Linear regression.

Source	SS	df	MS	
Model	2.24407253	3	.748024178	F (3,2) = 7.30
Residual	0.2050608	2	.1025304	Prob > F = 0.1229
Total	2.44913333	5	.489826667	R-squared = 0.9163 Adj R-squared = 0.7907 Root MSE = .3202
CAR	Coef.	Std. Err.	t	P > |t| [95% Conf. Interval]
GDP	−.6108969	.2135101	−2.86	0.104 −1.529556 .03077627
Interest rate	.4485303	.1858359	2.41	0.137 −.3510569 1.248118
Unemployment	−.3502525	.0832567	−4.21	0.052 −.7084771 .0079721
_cons	21.8813	1.506749	14.52	0.005 15.39828 28.36432

The f value that is less than alpha value (0.05) shows that our variable is significant. This means the independent variables reliably predict the dependent variable (CAR). F test is showing that the model is good to work with. The R^2 results exhibit the fitness of the regression model. The nearer the value to 1, the more fit is the model. The coefficient of GDP (−0.61089) indicates that there is a negative effect of GDP growth rate on capital adequacy ratio of commercial banks. However, this effect is statistically insignificant. The value of **t** is (−2.86) and the probability value is (0.104) which is greater than (0.05), which suggests that we cannot reject the null hypotheses that there is statistically significant effect of GDP growth rate on capital adequacy. Over the duration of (2015–2020), the capital adequacy has been negatively impacted by the GDP growth rate. Both variables have negative relationship. Increase in the capital adequacy should result due to a negative change in the gross domestic product. As more capital inflows, banks would need extra resources to manage that additional equity. Prior to the crisis, 2015–2016 statistics show as the annual GDP rate rose, so did the CAR of Jordanian banks. However, a contraction resulted in 2017, as GDP growth increased. However, during COVID-19, negative growth of (–1.6%) couldn't impact capital adequacy of JCB. Secretary General of Union Arab banks, Wissam Fattouh announced the combined volume of banking sector assets totalled $79.5 billion by Q3 2020 [19].

The coefficient of interest rate (0.44853) indicates that there is a positive effect of interest rates on the capital adequacy ratio of commercial banks. However, this effect is statistically insignificant. The value of **t** is (2.41) and the probability value is (0.137), which is greater than (0.05), which means that we cannot reject the null hypotheses that there is statistically significant effect of interest rates on capital adequacy. A significant rise in interest rates shall overall increase capital adequacy. Prior to the crisis, 2015–2017 statistics validate the statistical results as interest rates rose so did the CAR of Jordanian banks. During COVID-19, interest rates have lessened CAR of the overall banks. However, the capital assets have been merely impacted by a slightest decrease in the interest rates. Secretary General Wissam Fattouh said that CBJ announced 17.9% CAR in Q1 2020 [20]. The coefficient of unemployment (−0.35025) indicates that there is a negative effect of unemployment on capital adequacy ratio of commercial banks. However, this effect is statistically insignificant. The value of **t** is (−4.21) and the probability value is (0.052), which is greater than (0.05), which suggests us that we cannot reject the null hypotheses that there is statistically significant of unemployment rate on the capital adequacy. Joblessness has rendered negative impact on capital adequacy before and amid the COVID-19. The rate of unemployment grew at a significant rate amid the crisis. The banks had balanced the capital adequacy ratio and maintained its financial assets, safeguarding its liquidity position. The total capital, reserves, and allocations totaled $12.9billion (Jan, 2020 - Sept, 2020).

The model has R^2 value (0.916), which means that 91.6% of capital adequacy's variance is explained by the independent variables.

Model 2: Non-Performing Loans
Linear regression

Table 2. Model 2: Non-Performing loans.

Source	SS	df	MS	
Model	10.2052159	3	3.40173864	F (3,2) = 26.03
Residual	.261384091	2	.130692045	Prob > F = 0.0372
Total	10.4666	5	2.09332	R-squared = 0.9750 Adj R-squared = 0.9376 Root MSE = .36151
CAR	Coef.	Std. Err.	t	P > |t| [95% Conf. Interval]
GDP	.055829	.2410552	0.23	0.838 −.981348 1.093006
Interest rate	.1921664	.2098108	0.92	0.456 −.7105766 1.094909
Unemployment	.3365876	.0939977	3.58	0.070 −.0678519 .7410272
_cons	.4552548	1.701137	0.27	0.814 −6.864147 7.774656

The coefficient of GDP (0.0558) shows that there is a positive impact of GDP growth rate on non-performing credit/loans of the thirteen commercial banks. Nevertheless, this effect is insignificant. The value of **t** is (0.23) and the probability value is (0.83), which is greater than (0.05), which means that we cannot revoke the null hypotheses that there exists a statistically significant effect of GDP growth rate on non-performing loans. A significant drop in the annual GDP growth imposed a positive change in non-performing loans of Jordanian banks. CBJ reported a slight increase of 0.4%. The first-half of 2020 reported 5.4% of NPLs to total loans. The restructuring of small and medium-sized loans saved commercial banks in terms of liquid assets.

The coefficient of interest rate (0.19216) shows that there is a positive influence of interest rates on the non-performing credit of the thirteen commercial banks. However, this effect is statistically insignificant. The value of **t** is (0.92) and the probability value is (0.456), which is less than (0.05), which means that we cannot deny the null hypotheses which say that there exists a statistically significant influence of interest rates on non-performing loans. The increase in the interest rates directly increases the proportion of non-performing bank's credit and its total credit facilities, thereby decreasing the interest income of commercial banks. Therefore, debtors can't afford to pay off their loan liabilities, if interest rates go sky rocket. During COVID-19, decrease in the interest rates controlled the significant rise in NPLs. The banks had restructured loan facilities for their previous debtors. Purchase of credit was stabilized. The coefficient of unemployment (0.45525) shows that there is a positive influence of rate of unemployment on the non-performing loans (NPLs) of the commercial banks. Moreover, this effect is insignificant. The value of **t** is (3.58) and the probability value is (0.07), which is more than (0.05). Moreover, this means that we cannot revoke the null hypotheses that there exists a statistically significant influence of interest rates on non-performing loans. The increase in NPLs comes from the increasing rate of unemployment. When the labor leaves, trade gets a staggering hit. As unemployment rose (2015–2020), NPLs had maintained their course and banks have still earned healthy profits. During Q1 2020, Jordanian banking sector reported increase in return on assets to be 0.64%. The model shows R^2 value (0.975), which means that 97.5% of non-performing loans' variance is determined by the independent variables.

Model 3: Liquidity Coverage Model
Linear regression

Source	**SS**	**df**	**MS**	
Model	416.665779	3	138.888593	F (3,2) = 11.62
Residual	23.9083541	2	11.9541771	Prob > F = 0.0803
Total	440.574133	5	88.1148267	R-squared = 0.9457 Adj R-squared = 0.8643 Root MSE = 3.4575
CAR	Coef.	Std. Err.	t	P > |t| [95% Conf. Interval]
GDP	−9.115842	2.305429	−3.95	0.058 −19.0353 .8036206
Interest rate	10.32535	2.00661	5.15	0.036 −1.691603 18.9591

(*continued*)

(*continued*)

Source	**SS**	**df**	**MS**	
Unemployment	−4.60344	.8989854	−5.12	0.036 −8.471462 −.7354181
_cons	155.1009	16.26951	9.53	0.011 85.0988 225.1029

The coefficient of GDP (−0.91158) indicates that there is a negative effect of GDP growth rate on liquidity coverage ratio of commercial banks. Nevertheless, this effect is insignificant. The value of **t** is (−3.95) and the probability value is (0.058), which is more than (0.05), which corresponds to no revoking of the null hypotheses that there exists a statistically significant effect GDP growth rate on liquidity coverage. During COVID-19, the legal liquidity ratio of Jordanian banks reached 128.5%, comparatively higher than the requirement of 100%, CBJ reported. The CBJ had injected liquidity of JD1.05 billion, utilizing its mandatory reserve for the better state of banks in order to meet liabilities, in due course.

The coefficient of interest rate (10.325) indicates that there is a positive influence of interest rate on liquidity coverage ratio of the selected thirteen commercial banks. However, this effect is significant. The value of **t** is (5.15) and the probability value is (0.036) that is more than (0.05), and suggests that we revoke the null hypotheses that there exists a statistically significant effect of interest rate on non-performing loans. This points out that if the interest rates are increased, so will the liquidity coverage ratio of commercial banks. Increasing interest rates will impact issuance of credit and inversely grow liquidity-coverage against total bank liabilities. Amidst COVID-19, the restructuring of SMEs loans has favored liquidity position of commercial banks. The liquidity injected by CBJ has provided financial stability under heavy economic crisis. The coefficient of unemployment (−0.4603) exhibits a negative influence of unemployment rate on liquidity coverage ratio of commercial banks. Nevertheless, this effect is statistically significant. The value of **t** is (−5.12) and the probability value is (0.036) that is lower than (0.05), which suggests that we revoke the null hypotheses that there exists a statistically significant effect of unemployment rate on liquidity coverage. Fewer businesses would employ labor; less number of business operations and companies would be functioning, followed by lack of deposits. The commercial banks can also control liquidity coverage ratio and generate financial investments, in this regard. During COVID-19, the reduction in the cost of financing boosted LCR of Jordanian banks.

5 Conclusion and Future Studies

Through running statistical methods like the ADF and regression tests, the results showed that the increase in the GDP growth decreases the CAR of the selected commercial banks. While, the interest rates had a positive influence on CAR of the selected thirteen commercial banks, as it leads to an increase in the capital adequacy. The last result showed us that there is a negative effect of unemployment on capital adequacy of the commercial banks (March/2020-Dec/2021). Increasing rate of unemployment would mean decrease in capital adequacy of the JCB. In addition, the study concluded that there is a positive effect of GDP growth on the non-performing loans of the commercial banks (March/2020-Dec/2021). Hence, this means increase in the GDP growth shall surely express an increase in the non-performing credit facilities of the commercial banks. More debtors acquired loan facilities to inject liquidity into their businesses, which shall boost trade activities.

There was also a positive influence of interest rates on the non-performing credit of the selected commercial banks (March/2020-Dec/2021). This suggests that increasing interest rate would definitely mean increase in the number of non-performing loans. The interest rates and NPLs go hand in hand. The commercial banks develop a scale to maintain balance in instances that interest rate is too high, that could increase NPLs and not too low that credit facilities are no longer profitable for the bank. While above two variables had positive effect, the unemployment rate also had a positive influence on the non-performing loans of the selected commercial banks (March/2020-Dec/2021). A significant rise in unemployment would lead to increasing number of non-performing credit for the commercial banks. The liquidity ratio of the banks can be at risk.

Furthermore, the results showed a negative effect of GDP growth on the liquidity coverage ratio of the commercial banks. Therefore, the increase in the GDP leads to an increase in the number of trade and business activities. There will be a huge outflow of credit to customers. This will eventually impact the liquidity coverage of the bank. The results further showed that there is a positive impact of interest rates on the liquidity coverage ratio of the commercial banks (March/2020-Dec/2021). Higher interest rates would mean less borrowing. This would increase the liquidity coverage of the bank. There was a negative impact of the unemployment rate on the liquidity coverage ratio of the commercial banks (March/2020-Dec/2021). Higher unemployment would adversely impact the liquidity coverage. Lower employment rate could increase the liquidity coverage, as this also reduces borrowing.

The CBJ governor, Ziad Fariz reported that Jordanian banking sector has shown great stability and resilience in the time of need, amidst the crisis. He further added that IMF, World Bank, and Japanese grants worth \$400 million, \$100million, and \$1.3billion were also available as foreign aid to counter the economic impacts. Fariz concluded that "the reduction in oil price has saved at least JD1billion" [20]. The capital adequacy over the years in the Jordanian banking sector continues to strongly withstand the burden, shocks, and high risks associated to increasing NPLs ratio as the banks are already making huge amount of profits and hold large capital, which depicts a healthy protection of capital [6]. Our research showed there is a positive effect of GDP growth on NPLs. The research done by also [6] supports our statistical results. When GDP grows, eventually newer startups will be launched and banks shall issue more financing to its debtors. This will directly influence LCR and NPLs ratio. In the stress test study held in Jordan, the expected current scenario, as a *COVID-19* consequence, according to Jordanian economy showed that non-performing loans shall drop from (2020 = 7.8%) to (2022 = 5.3%). The expected GDP growth after facing negative contraction will be (2021 = 3.6%) and (2022 = 3.5%). The Jordanian economy is on the track to become financially independent and boost trade activities beyond the borders [7]. Moreover, the study recommends further researchers to focus on comparing the impact of macroeconomic indicators in the resilience of commercial banks along with Islamic banks during the COVID-19 crisis or financial distresses.

References

1. Zhang, D., Hu, M., Ji, Q.: Financial markets under the global pandemic of COVID-19. Financ Res Lett [Internet]. Elsevier **36**, 101528 (2020). https://doi.org/10.1016/j.frl.2020.101528
2. Al-Tammemi, A.B.: The battle against COVID-19 in Jordan: an early overview of the Jordanian experience. Front. Public Heal. **8**, 1–6 (2020)
3. Badarin, E.: Jordan's economic, security and political challenges under Covid-19, 13629395 (2020)
4. Alqutob, R., Al Nsour, M., Tarawneh, M.R., Ajlouni, M., Khader, Y., Aqel, I., et al.: COVID-19 crisis in Jordan: Response, scenarios, strategies, and recommendations. JMIR Public Heal Surveill. **6**, 1–6 (2020)
5. Singh, M.: Jordan after {COVID}-19: {From} crisis adjustment to crisis management. Washingt Institute, Fikra Forum, April 2020;2020–2
6. Alkharouf, S., Al Salamat, W., Al-Kharouf, S.: The Determinants of Financial Stability: Evidence from Jordan the determinants of financial stability View project The Determinants of Financial Stability: Evidence from Jordan. Int J Bus Soc Sci [Internet]. 2021;12. www.ijbssnet.com
7. Faiza, C., Ali, D.: Stress tests as a mechanism to measure banks' resilience to the Corona crisis – Jordan's Experience Study. Econ Sustain Dev Rev. **27**, 58–66 (2019)
8. Kutum, I., Al-Jaberi, K.: Jordan Banks Financial Soundness Indicators..pdf (2015)
9. Almahadin, H.A., Kaddumi, T., AL-Kilani, Q.: Banking soundnessfinancial stability nexus: Empirical evidence from Jordan. Banks Bank Syst. **15,** 218–27 (2020)
10. Abdalla Moh'd AL-Tamimi, D.K., Sulieman Jaradat, D.M., Al-Rjoub, D.A.M.: The Role of Central Bank of Jordan in Economic Development. Int. J. Econ. Financ. Res. **5,** 221–226 (2019)
11. Aledeimat SRM. Non-Performing Loans in Jordan [Internet] (2020). http://i-rep.emu.edu.tr:8080/jspui/handle/11129/5140%0Ahttp://i-rep.emu.edu.tr:8080/jspui/bitstream/11129/5140/1/Aledeimatshadi.pdf
12. CBJ. Stress tests as a mechanism to measure banks' resilience to the Co-rona crisis – Jordan's Experience (2018)
13. Release P. Amman, July 14, 2020 — The World Bank launched today its Jordan Economic Monitor (JEM) Spring 2020 edition [Internet] (2020). https://www.worldbank.org/en/news/press-release/2020/07/14/jordan-covid-19-pandemic-weighs-heavily-on-the-economy-as-it-does-on-the-region. Accessed 12 Jan 2022
14. Fanack. Economy of Jordan. Britannica [Internet]. 2022;1–2. https://fanack.com/jordan/economy-of-jordan/
15. Reserve Bank Of New Zealand. Capital adequacy ratios for banks - simplified explanation and example of calculation. Reserv Bank New Zeal. 2007;1–9
16. Barrios, V.E., Blanco, J.M.: The effectiveness of bank capital adequacy regulation: a theoretical and empirical approach. J. Bank Financ. **27**, 1935–1958 (2003)
17. Uyanık, G.K., Güler, N.: A study on multiple linear regression analysis. Procedia – Soc. Behav. Sci. **106**, 234–240 (2013)
18. Nelson, D.: What is Linear Regression?- Spiceworks | Spiceworks. 2020; 2–5. https://www.unite.ai/what-is-linear-regression/
19. Articles R. Jordan ' s banking sector assets reach $ 79 . 5 billion in. 2021;3–4
20. Jt B, Bank TC. Jordan banking system capable of withstanding coronavirus crisis — CBJ governor. 2020;1–2

The Impact of Financial Risks on the Value of the (CB) Listed on the Amman Financial Market: The Moderating Role of Disclosure Quality of Accounting Hedging

Faten Amin Al-Naimi[1](✉), Inaam M. Al-Zwaylif[2], and Aymen Zereban[3]

[1] Department of Accountancy, Zarqa University, Zarqa 13110, Jordan
fnaimi@zu.edu.jo

[2] Department of Accountancy, The World Islamic Sciences and Education University (Wise), Amman, Jordan
inaam.zwaylif@wise.edu.jo

[3] Department of Accountancy, University of Business and Technology, Jeddah, Kingdom of Saudi Arabia
aymen@ubt.edu.sa

Abstract. The study examined the impact of financial risks in their dimensions on the value of (CB) and the modified role of the (QD) of accounting hedging in the relationship between financial risks and the value of (CB) on the Amman financial Market (AFM). The study used a descriptive-analytical methodology. The significance of the research is one of the few studies that measured the quality of this disclosure, so it is expected to contribute to opening new horizons for researchers and academics. The study concluded that there is a (SSI) of financial risks in its dimensions on the value of (CB) listed on the (AFM), Moreover, there is a favorable result of liquidity, credit and exchange rate risks, and a negative impact of interest rate risks and capital adequacy on the value of (CB), There is also a (SSI)for the (QD) of accounting hedging in the annual financial reports on the relationship between financial risks in their dimensions and the value of (CB)listed on the (AFM). The researcher has suggested that Jordanian banks use risk foresight to predict future financial risks and measure them accurately. The Central Bank of Jordan also needs to issue periodic bulletins on the (QD) of accounting hedging for its role in enhancing investor confidence and influencing their decisions positively.

Keywords: Financial risk · Company value · Accounting hedge · Quality of Accounting hedge disclosure

1 Introduction

Banks are exposed to financial risks that cause fluctuations in cash flows and profits, which have an impact on their value, so such banks resort to accounting hedging to face the risks. They also make account of providing sufficient and comprehensive information

B. A. M. Alareeni and I. Elgedawy (Eds.): *Artificial Intelligence (AI) and Finance*, SSDC 488, pp. 722–732, 2023.
https://doi.org/10.1007/978-3-031-39158-3_67

about hedging activities in its financial reports [1], since good accounting disclosure reduces information asymmetry and increases the efficiency of financial markets [2]. Therefore, this study aims to discuss the impact of financial risks on the value of banks, and the modified role of the (QD) of accounting hedging in the annual financial reports of (CB). Risks are unexpected events that can be controlled through forecasting and proper planning to face them, while financial risks are the irregularity of returns, fluctuations in their value due to the uncertainty related to future forecasts. While banks carry out their activities, risks emerge and they are an indicator of the negative aspects of these activities, such as the risks of investing in securities, and the probability of achieving a return or cash flow less than the expected return. The greater the probability of achieving a return or less cash flow, the higher the financial risks [3]. The researcher sees that banks need to anticipate risks to increase their awareness of contracts related to future change and to face expected risks. Banks should collect information about future risks and rely on this information to develop medium-and long-term insights and make actionable decisions in the present time [4]. Maximizing the value of banks is one of the most important elements of their strategic plan, as they seek to maximize the value of ordinary shares by maximizing the net present value of the stock, which is characterized by fluctuation and its correlation with the market value of the stock [5]. Therefore, the management takes rational financial decisions related to financing and investment resulting from the difference in interest or exchange rate [6]. Accounting disclosure plays a major role in increasing the efficiency of financial markets, as it provides appropriate information that directs available resources to areas of optimal investment, which leads to reducing the asymmetry of information [7]. The quality of disclosure (QD) is defined as the characterization of accounting disclosure by a set of features, criteria and characteristics related to the quality of disclosed accounting information and its positive ability to influence user decisions [8, 9]. Hedging is an important tool for managing financial risks by making financial transactions to reduce these risks [10]. Accounting hedging emerged as a result of the use of financial derivatives by companies [11]. Hedging is defined as a strategy developed by the management to avoid the risks resulting from the variation of prices and expected rates and compensate for this variation through what results from the implementation of financial instruments in this plan [12].

The significance of the research lies in the importance of the financial risks to which (CB) are exposed and their relationship to the banks value, in addition to indicating the modified role represented by the quality of accounting hedge disclosure. This study is one of the few studies that measured the quality of this disclosure, so it is expected to contribute to opening new horizons for researchers and academics. The problem of the study Banks seek to hedge any expected financial risks and disclose this in their financial statements published in the financial markets to reduce the asymmetry of information and positively influence investors ' decisions, so the main question of the study is stated as follows:

1-Does the financial risks in their dimensions (liquidity risks, credit risks, exchange rate risks, interest rate risks, capital adequacy risks) have an impact on value of (CB) listed on the (AFM)?

1. Does liquidity risk have an impact on value of (CB) listed on the (AFM)?
2. Does credit risk have an impact on value of (CB) listed on the (AFM)?

3. Does exchange rate risk have an impact on value of (CB) listed on the (AFM)?
4. Does interest rate risk have an impact on value of (CB) listed on the (AFM)?
5. Does capital adequacy risk have an impact on value of (CB) listed on the (AFM)?

2. Does the (QD)of accounting hedging in the annual financial reports have an impact on the relationship between financial risks and the value of (CB)listed on (AFM).?

Because the banking sector is one of the most sensitive economic sectors to the rapid changes in the economic environment, it is exposed to many financial risks, such as liquidity risks, credit risks, and risks of changes in interest and exchange rates, which lead to fluctuations in cash flows and profits, and expose banks to additional burdens, which negatively affects Based on the foregoing, this study showed the positive impact of liquidity, credit and exchange rate risks, and the negative impact of risks, interest rates and capital adequacy on the value of commercial banks. In addition to a statement that the quality of accounting hedge disclosure in the annual financial reports positively affects the relationship between financial risks and their dimensions and the value of (CB) listed on the (AFM).

2 Literature Review

Several studies have dealt with risks and hedge accounting, Referring to the study concluded [13] concluded that the cash flow hedging and fair value, reduction in financial risks in Jordanian commercial banks. Study found [1] a favorable result of derivatives on the value of the company. The study showed [14] significant impact of audit risk on fair value estimation and measurement (absence of active markets, significant distortions, and different bases for measuring fair value). The study concluded [15] that the financial risks most affected the financial performance of the study sample; are capital adequacy risks and credit risks, and the lack of a significant impact for each of the liquidity risks and interest risks. The study showed [16] that hedge accounting leads to reducing the risks that institutions may face, and providing an appropriate risk management system in institutions helps them to face risks. The study concluded [17] that the most important risks facing banks are credit risks, such as the risk of non-payment, exchange rate, and operating risks, which are related to the burdens of branches and the number of employees.

Among the studies that examined disclosure and its quality, we mention them, The study [2] showed that the average disclosure rate in the Jordanian public shareholding industrial companies is (0.61), which indicates the extent to which the companies are aware of the importance of voluntary disclosure. Study development [18] indicator based on (IFRS) No. 7 "Financial Instruments: Disclosures", to measure the (QD) of accounting hedging in Jordanian commercial banks. The study concluded [19] that there is an integration between the accounting disclosure process with information with high qualitative characteristics in the Bank of Baghdad. Among the studies that looked at the value, we mention them, study concluded [20] that hedge accounting generally has a slight positive effect on the value of the company The study showed [21] that the most effective and least profitable companies, and enjoying greater growth opportunities, practice hedging through derivatives, and that derivatives hedging indicators have little effect on the value of the company.

3 Hypotheses

The first main hypothesis

$H_0$1: There is no (SSI)of financial risks in their dimensions (credit risk, liquidity risk, interest rate risk, exchange rate risk, capital adequacy risk) on value of (CB) listed on (AFM).

There are sub-hypotheses:

Ho.1.1: There is no (SSI) of liquidity risk on value of (CB) listed on (AFM).

Ho.1.2: There is no (SSI) of credit risk on value of (CB) listed on (AFM)

Ho.1.3: There is no (SSI) of exchange rate risk on the value of (CB) listed on (AFM).

Ho.1.4: There is no (SSI) of interest rate risk on the value of (CB) listed on (AFM).

Ho.1.5: There is no (SSI) of capital adequacy rate risk on value of (CB) listed on (AFM).

The second main hypothesis

$H_0$2: There is no (SSI) of quality disclosure of accounting hedging in the annual financial reports on the relationship between financial risks and the value of (CB) listed on (AFM) Data and methodology.

4 Methodology

The study used a descriptive-analytical methodology. The researcher used E-Views software to process initial data, metrics, test percentages, repetitions, arithmetic mean, and standard deviations to describe the study variables. The linear correlation coefficient was used to test the (Multicolinearity). The study model was estimated using (Lagrange Multiplier) and (Hausman) tests, and the multiple linear regression analysis was used to test the presence of an effect of the independent variable on the dependent variable. The interactive regression analysis was used to test the presence of the effect of the independent variable on dependent in the presence of the modified variable.

4.1 Study Population and Sample

The study population consisted of all 13 (CB) listed on (AFM).

4.2 Sources of Data and Information Collection

- **Secondary sources:** "Peer-reviewed scientific research published in scientific journals and periodicals.
- **Primary sources:** The measurement tool represented by the accounting hedge disclosure index, which was developed for the purposes of this study, and using the annual financial reports of (CB) registered on (AFM) during the period (2014–2019).

4.3 Measurement of Variables and Mathematical Models Used

The researcher used the Tobin's Q model to measure the value of the study sample banks [18]. An indicator for the disclosure of the quality of accounting hedging was designed and it consisted of several items based on the International Financial Reporting Standard No. 7 "Financial Instruments: Disclosures". Every item of the disclosure indicator was given the value (3) to the banks that disclosed it in their financial reports in full compliance with the requirements of accounting hedge disclosure. The value (2) was given to the banks that partly disclosed it in their financial reports in line with the requirements of accounting hedge disclosure, and the value (1) was given to the banks that did not disclosed it in their financial reports in line with the requirements of disclosure of accounting hedging, and the value (0) for banks that did not disclose any of the items of this indicator [23].

- Items of fair value hedge disclosures (Description of the hedge). Description of financial tools designated as hedging tools. Description of the fair values of financial tools designated as hedging tools at the reporting date. The nature of hedged risks. Profits and losses of the hedging tool. Hedging item attributable to hedged risk. Hedging of expected operations, the expected periods of occurrence of these operations and the time of their impact on net profit or loss. Change in the fair value of the hedging tool and the hedged item. Recognized ineffectiveness of hedging in profit and loss.)
- Items of cash flow hedge disclosures (Description of the hedge). Description of financial tools designated as hedging tools. Description of the fair values of financial tools designated as hedging tools at the reporting date. The nature of hedged risks. The periods when cash flows are expected to occur and when they will affect profit and loss. A description of any operations that have been hedged in advance and are not expected to occur. The amount of profit or loss from the hedging tool that was recognized in other paragraphs of comprehensive income during the period. The amount recognized in equity during the period. The amount of profit or loss from the non-equity hedging tool that was included in the profit or loss for the period. The amount of profit or loss from a hedging tool transferred from equity during the period that is included in the initial cost or book value of non-financial assets or liabilities acquired or assumed to be used as effective hedging tools.)
- Items of net foreign investment hedging disclosures (Description of the hedge). Description of financial tools designated as hedging tools. Description of the fair values of financial tools designated as hedging tools at the reporting date. Profit and loss from a hedging tool that has been identified as an effective hedge. Hedging item attributable to hedged risk. Recognized ineffectiveness in profit or loss from net foreign investments hedging. Reclassification of profits or losses accumulated on an effective hedging tool for partial disposal of foreign operations.)
- Other items (the bank's risk management strategy and how to implement it. The impact of the bank's hedging activities on future cash flows, their timing and uncertainties.)

4.4 Using the Following Mathematical Models

$$\mathbf{Vit} = \alpha + \beta\ 1\mathrm{CRit} + \beta\ 2\ \mathrm{LIQit} + \beta\ 3\ \mathrm{INTit} + \beta\ 4\ \mathrm{ERRit} + \beta\ 5\ \mathrm{CAPit} + \mathrm{ei} \quad (1)$$

$$\mathbf{V_{it}} = \alpha + \beta_1 \mathrm{CR_{it}} + \beta_2 \mathrm{LIQ_{it}} + \beta_3 \mathrm{INT_{it}}$$

$$+ \beta_4 ERR_{it} + \beta_5 CAP_{it} + \beta_6 QH_{it} + \beta_7 CR_{it} * QH_{it}$$
$$+ \beta_8 LIQ_{it} * QH_{it} + \beta_9 INT_{it} * QH_{it}$$
$$+ \beta_{10} ERR_{it} * QH_{it} + \beta_{11} CAP_{it} * QH_{it} + ei \quad (2)$$

where:

$\mathbf{V_{it}}$**:** Bank value = Book value of total liabilities+ market value of equity/book value of total assets

α = Fixed limit

$\boldsymbol{\beta_{11\ to}\ \beta_1} =$ regression coefficients (slope of regression).

$\mathbf{CR_{it}}$**:** Credit risk = Non-performing loans/ total loans.

$\mathbf{LIQ_{it}}$**:** Liquidity risk = Cash + financial assets for trading/ total deposits

$\mathbf{INT_{it}}$: Interest rate risk = Exchange rate income or loss / total income.

$\mathbf{ERR_{it}}$**:** Exchange rate risk = Risk-weighted assets/total assets.

$\mathbf{CAP_{it}}$**:** Capital adequacy risks = Total capital/Credit risk + market risk + operational risk.

$\mathbf{QH_{it}}$**:** Quality of accounting hedge disclosure = The degree that the bank actually obtained.

The highest score the bank can obtain (number of disclosure index items x maximum index3)

ei = random error.

4.5 Description of the Study Variables

Table 1. Description of the study variables

Variable	Coefficient	Std. Error	t-Statistic	Prob
LIQ	0.085	0.212	0.403	0.018
CR	0.117	0.092	1.265	0.001
ERR	– 0.146	0.043	– 3.413	0.011
INT	– 0.053	0.103	– 0.520	0.004
CAP	– 0.737	0.211	– 3.481	0.000
C	7.030	0.162	43.22	0.000

Table 1 showed B value for liquidity risk (0.085) and the t value (0.403), at a significant level (Sig = 0.018), which indicates a positive significant impact of this dimension. The B value of credit risk was (0.117) and the t value of reached (1.265) at a significant level (Sig = 0.001), indicating a positive significant impact of this dimension. The B value for exchange rate risk was (-0.146) and the t value (−3.413) with a significance level of (Sig = 0.011), indicating a negative statistical impact of this dimension. The B-value of interest rate risk amounted to (−0.053) and the t value (−0.520) at a significant level (Sig = 0.004), which indicates a negative significant impact of this dimension.

Whereas the B-value of the capital adequacy risk was (−0.737) and the t value was (−3.481), with a significance level (Sig = 0.000), which indicates a negative significant impact of this dimension. Therefore, the null hypothesis was rejected and the alternative hypothesis was accepted.

4.6 Hypothesis Testing

Ho.1.1: There is no (SSI) of liquidity risk on the value of (CB) listed on the (AFM).

Table 2. Ho.1.1

Variable	Coefficient	Std. Error	t-Statistic	Prob
LIQ	0.279	0.226	1.230	0.002
C	6.785	0.178	38.046	0.000

Table 2 shows the B value for liquidity risks (0.279) and T value of (1.230), at the level of significance (Sig = 0.002), which indicates a positive effect of this dimension, so the first sub-null hypothesis was rejected, and the alternative hypothesis was accepted.

Ho.1.2: There is no (SSI) of credit risk on the value of (CB) listed on the (AFM).

Table 3. Ho.1.2

Variable	Coefficient	Std. Error	t-Statistic	Prob.
CR	0.122	0.091	1.335	0.000
C	6.904	0.179	38.448	0.000

Table 3 shows the B value for credit risks (0.122) and t value of (1.335), at the level of significance (Sig = 0.000), which indicates a positive effect of this dimension, so the second sub-null hypothesis was rejected, and the alternative hypothesis was accepted.

Ho.1.3: There is no (SSI) of exchange rate risk on the value of (CB)listed on the (AFM).

Table 4. Ho.1.3

Variable	Coefficient	Std. Error	t-Statistic	Prob.
ERR	– 0.130	0.048	– 2.699	0.006
C	6.838	0.181	37.60	0.000

Table 4 shows the B value for exchange rate risks (−0.130) and t value of (−2.699), at a significant level (Sig = 0.006), which indicates the presence of negative significant impact, therefore, the third sub-null hypothesis was rejected and the alternative hypothesis was accepted.

Ho.1.4: There is no (SSI) of interest rate risk on the value of (CB)listed on the (AFM).

Table 5. Ho.1.4

Variable	Coefficient	Std. Error	t-Statistic	Prob.
INT	– 0.039	0.107	– 0.364	0.004
C	6.875	0.188	36.48	0.000

Table 5 shows the B value for interest rate risk (−0.039) and a t value of (−0.364), at a significant level (Sig = 0.004), which indicates the presence of a negative impact, therefore, the fourth The alternative hypothesis was approved whereas the sub-null hypothesis was rejected.

Ho.1.5: There is no (SSI)of capital adequacy rate risk on the value of (CB)listed on the (AFM).

Table 6. Ho.1.5

Variable	Coefficient	Std. Error	t-Statistic	Prob.
CAP	– 0.601	0.196	– 3.054	0.004
C	6.938	0.136	51.51	0.000

Table 6 showed the B value for capital adequacy risk (−0.601) and a t value of (−3.054), at a significant level (Sig = 0.004), which indicates a negative significant impact of this dimension, so the fifth The alternative hypothesis was approved whereas the sub-null hypothesis was rejected. This indicates that banks do not have enough capital to cover losses resulting from loans and facilities granted by the bank.

H0 2: There is no (SSI)of quality disclosure of accounting hedging in the annual financial reports on the relationship between financial risks and the value of (CB)listed on the (AFM)

Table 7 shows the B value at (disclosure quality * liquidity risk) (0.138) and t value of0.412), at the level of significance (Sig = 0.001), which indicates a positive significant impact of this dimension. The B value at (disclosure quality *credit risk) (0.200) and a t value of (2.215), at the level of significance (Sig = 0.002), which indicates a positive significant effect of this dimension. The B value at (disclosure level * exchange rate risk) amounted to (0.422) and a t value of (2.658), at the level of significance (Sig = 0.009), which indicates a positive significant impact of this dimension. The B value at

Table 7. Ho 2

Variable	Coefficient	Std. Error	t-Statistic	Prob.
LIQ*ALLQHD	0.138	0.336	0.412	0.001
CR*ALLQHD	0.200	0.090	2.215	0.002
ERR*ALLQHD	0.422	0.158	2.658	0.009
INT*ALLQHD	0.394	0.174	2.257	0.023
CAP*ALLQHD	0.445	0.810	0.549	0.004
C	6.648	0.274	24.191	0.000

(disclosure quality *interest rate risk) amounted to (0.394) and a t value of (2.257), at the level of significance (Sig = 0.023), which indicates a positive significant impact of this dimension. The B value at (disclosure level * capital adequacy risk) amounted to (0.445) and a t value of (0.459), at the level of significance (Sig = 0.004), which indicates a positive significant impact of this dimension. Therefore, the third null hypothesis was rejected and the alternative hypothesis was accepted.

5 Conclusion

The study showed that there is a (SSI) of financial risks on the value of (CB) and this effect was positive at liquidity risks and credit risks, which is an indicator of the efficiency of risk management in facing these risks and its ability to turn them into opportunities that can be used positively on their value. There was a negative impact of interest rate risks, exchange rates and capital adequacy, which is an indicator of the failure of risk management in facing these risks, which affected their value negatively. Results showed a favorable result of liquidity risks on the value of commercial banks. This is an indicator of the ability of banks to use liquidity to cope with untimely withdrawals on customer deposits, handle cash requests from borrowers and use their resources effectively. Results showed a favorable result of credit risks on the value of banks. This indicates the availability of effective credit management that seeks to stabilize the financial situation of banks. It also indicates the quality of credit information that banks receive regarding borrowers, and the fact that these banks investments are not heavily dependent on granting loans. Results showed a favorable result of exchange rate risk on the value of banks. This indicates the inefficiency of banks' management of these risks and the inability to face them as well as their poor awareness of the importance of accounting hedging in reducing the ratio of exchange rate risks and reducing the negative effects resulting from fluctuations in cash flows as a result of fluctuating exchange rates in foreign currencies. Results showed a favorable result of capital adequacy risks on the value of banks. This indicates the inefficiency of bank management in following hedging policies in reducing the gap between the impact of the bank's assets and liabilities by the change in the interest rate, and the poor planning in the investment of their assets. Results showed a favorable result of the (QD) of accounting hedging on the connection between financial risks and bank value. This is an indication that increasing the (QD) of accounting hedging in annual financial reports leads to increased transparency, increased efficiency of financial markets and a favorable result on investors decisions, which is positively reflected on the value of banks.

6 Recommendations

Based on the previous results, the researcher recommends the Jordanian banks should use risk foresight to increase their ability to predict future financial risks and measure them accurately. The need to increase the efficiency of banks in reducing uncertainties about cash inflows and outflows because of changes in exchange rates, managing capital adequacy risks and practicing appropriate hedging policies to face and reduce these risks, as well as enhancing their efficiency in managing interest rate risks and developing continuously to deal with interest rate fluctuations in financial markets. (CB) should strive to maintain their efficiency in managing liquidity and credit risks and develop in line with the new challenges they face in the surrounding environment. The Central Bank of Jordan should issue periodic bulletins on the (QD) of accounting hedging and its role in enhancing investor confidence and influencing their decisions positively. Conducting further studies on risk foresight and its role in predicting financial risks and its contribution to making the right decisions to face them, and how this affects the value of banks listed on the (AFM).

References

1. Haron, R., Hasan, A.: Hedging, managerial ownership and firm value. J. Asian Bus. Econ. Stud. Emerald Publishing Limited **2**(3), 263–280 (2021)
2. Al Naimi, F.: Voluntary disclosure and its impact on the cost of capital in industrial companies degree in amman financial market. Arab J. Manage. **39**(3), 93–106 (2019)
3. Damodaran, A: "Country risk: determinants, measures and implications", New York University – Stern School of Business; New York University (2021). (NYU) – Leonard N. https://papers.ssrn.com/sol3/papers.cfm?abstract_id=3879109
4. Abdullah, M., Janor, H., Hamid, M., Yatim, P.: The Effect of enterprise risk management on firm value: evidence from Malaysian technology firms. Journal Pengurusan **4**(9), 251–260 (2017)
5. Adityawarman, A., Khudri, T.: The impact of internet financial reporting practices on the company's market value: A study of listed manufacturing companies in Indonesia". (6th edn.). International Yogyakarta, Indonesia (2018)
6. Jimari, R., Kumar, B.: Drivers of firm's value: panel data evidence from Indian manufacturing industry. Asian J. Financ. Account. **7**(2), 20–60 (2015)
7. Qizam, I.: The impact of disclosure quality on firm performance: empirical evidence from Indonesia. J. Asian Financ. Econ. Bus. **8**(4), 751–800 (2021). https://doi.org/10.13106/jafeb
8. Cyrusan, A., Dyeng, J.: Voluntary disclosure with evolving news. J. Fananicail Econ. **140**(1), 201–250 (2021)
9. Dicuonzo, G., Dell, V.: Financial risk disclosure: evidence from albanian and italian companies. EBEEC Conference Proceedings **1**(1), 182–196 (2017)
10. Almubaideen, H., Joudeh, A., Alsakeni, S., Al-Attar, K.: The effect of applying hedge accounting in reducing future financial risks in jordanian commercial banks. Canadian Center Sci. Educ. **13**(3), 140–153 (2019)

11. Gonidakis, F.K., Koutoupis, A.G., Tsamis, A.D., Agoraki, M.: Risk disclosure in listed Greek companies: the effects of the financial crisis. Account. Res. J. **33**(4), 615–633 (2020)
12. Sartawi, A., Assaf, N., Hassan, A.: Challenges facing auditors in Palestine when measuring by fair value: a field study of the West Bank. Zarqa J. Res. Stud. Humanities **19**(1), 96–116 (2019). https://doi.org/10.12816/0054789
13. Wahdan, T.: The impact of financial risks on the financial performance of (CB)in the Amman Financial Market. [Unpublished master's thesis], Zarqa University (2017)
14. Mayo, A., Yawqafah, A.: The effect of applying the principle of accounting hedging on risk management in economic institutions. Entrepreneurship J. Bus. Econ. **30**(5), 1–11 (2017)
15. Hofaf, Q., Chtouh, M.: Hedging from financial risks in banks (a case study of the Bank of Agriculture and Rural Development and Adrar Agency). [Master's thesis, Al-Aqeed Ahmed Deraya University] (2017). https://dspace.univ-drar.edu.dz/jspui/bitstream/123456789/3365/1
16. Restrepo, L., Falla, D., Bejarano, J.: Information disclosure quality and firm value: empirical evidence for an emerging integrated market. J. Academia Revista Latinoamericana de Administracion **1**(1), 10–95 (2022)
17. Mendoza, J., Rivera, A.: The effect of credit risk and capital adequacy on the profitability of rural bank in Philippines. Sci. Ann. Econ. Bus. **1**(5), 520–530 (2017)
18. Younis, N.: The impact of the level and (QD) of sustainability reports on the financial performance of companies - an applied study on industrial companies listed in the Saudi stock market. J. Financ. Bus. Res. **22**(1), 361–413 (2021)
19. Al-Gharbawi: The quality of accounting disclosure and its role in enhancing investors' strategic thinking. Al-Kut Univ. College J. **3**(1), 221–243 (2018)
20. Kochling, A.: How does hedge accounting influence firm value? (2018). https://ssrn.com/abstract=3128333
21. Seok, S., Kim, T., Cho, H., Kim, T.: Determinants of hedging and their impact on firm value and risk: after controlling for endogeneity using a two-stage analysis. J. Korea Trade **24**(1), 1–34 (2020)
22. Alshatti, S.: The effect of credit risk management on financial performance of the Jordanian commercial banks. Investment Manage. Financ. Innov. **12**(1), 255–305 (2015)
23. (AFM) (2021). Annual Bulletin. https://www.ase.com

The Impact of Liquidity Ratios and Cash Flow Sources in Profitability in Industrial Companies Listed in the Amman Stock Exchange

Husni K. Al-Shattarat[1(✉)], Hossam Haddad[1], Salwa Abdelateef Mahmoud[1], and Abdul Malik Syed[2]

[1] Faculty of Business and Administration Science, Zarqa University (ZU), Zarqa, Jordan
{hshattarat,Hhaddad}@zu.edu.jo

[2] Department of Finance, University of Business and Technology, Jeddah, Kingdom of Saudi Arabia
a.syed@ubt.edu.sa

Abstract. This study sought to determine the impact of liquidity ratios (Current ratio and Quick ratio) and cash flow sources (operating operations, investment activities, and financing activities) on profitability with its indicators (ROA and ROE) in industrial firms listed on the Amman Stock Exchange (ASE). In order to accomplish the purpose of the study, a descriptive-analytical methodology was utilized where the researcher acquired data for the study's variables by studying the annual financial reports of (32) industrial companies during the study's financial period (2019–2020).There was a statistically significant effect at the significance level (0.05) for liquidity ratios (Current ratio and Quick ratio) and cash flow sources from (operating activities, investing activities, and financing activities) in predicting profitability with its indicators (ROA And ROE) among ASE-listed industrial companies. The study recommended that industrial companies listed on the Amman Stock Exchange improve their quick assets while avoiding an excessive increase in quick liquidity ratios not to disrupt working capital and achieve optimum profitability.

Keywords: Liquidity Ratios · Cash Flow Sources · Profitability · Industrial Companies

1 Introduction

The study aims to accomplish the following goals. The first one is getting to know the concept of liquidity ratios and their indicators in industrial companies listed on ASE. Moreover, the second one is to identify the cash flow elements of industrial companies listed on ASE. Another aim of this study is to recognize the concept of profitability and methods of forecasting it in industrial companies. Analysis of the use of liquidity ratios and the elements of cash flows and their relationship to profitability.

On the other hand, the importance of this study of being focused on the practical significance of the current study highlights the following: The results of the current

B. A. M. Alareeni and I. Elgedawy (Eds.): *Artificial Intelligence (AI) and Finance*, SSDC 488, pp. 733–743, 2023.
https://doi.org/10.1007/978-3-031-39158-3_68

study give a clear vision for officials in companies in general, and industrial companies in particular, about the relationship between liquidity ratios and cash flows on the one hand, and profitability on the other. The second one is to give an idea about the liquidity ratios, cash flows, and profitability levels in the industrial companies listed on ASE. Furthermore, finally, it is hoped that this study will help those concerned with industrial companies listed on ASE, especially shareholders and senior management, to take future measures and decisions to predict their profitability.

The current study contributed and differed from previous studies regarding the study population. It studied and analyzed variables and their application to a sector of high importance to society and the national economy represented by the industrial companies listed in the study. The current study's impact on the current studies: trading, quick liquidity ratios, and the components of cash flow: (operating Activity, investing Activity, and financing activity) in (2019–2020).

The Jordanian industrial sector's leadership has increased competition among industrial investors. In addition, attention has shifted to the crucial roles that liquidity and leverage management play in the sustainability of Jordanian industrial businesses [1]. Therefore, industrial enterprises might achieve profitability by balancing the effects of leverage and liquidity management [2].

Within this context, Industrial companies aim to achieve profit and maximize wealth. Perhaps the profit factor is the most important factor that attracts the attention of owners, investors and lenders due to its impact on the efficiency of industrial companies' performances, as companies' failure to achieve their financial goals results in hardship and financial failure, and consequently their non-continuity. Moreover, because of the difficult economic conditions prevailing in the world in general and Jordan in particular since the global economic crisis, Jordanian industrial companies needed to adopt and adopt indicators that indicate the feasibility and profitability of their operating and investment and finance activities. It arises from profitability under the prevailing economic and political conditions.

Where liquidity ratios are one of the indicators to determine the company's ability to pay its obligations or short-term debts, and cash flows reflect the nature of any organization or financial institution; this is because it reveals its financial and investment reality. Therefore, the study's problem lies in its attempt to answer the following central question: What is the impact of the liquidity ratios and the cash flow Sources on the profitability of the industrial companies listed on the Amman Stock Exchange?

The problem of the study focused on the following central question, which gave rise to the four Sub-questions:

- Do the liquidity ratios, such as the current and quick ratios, impact the profitability measured by ROA in the industrial companies listed on ASE?
- Do the liquidity ratios, including current and quick ratios, impact the profitability measured by ROE in the industrial companies listed on ASE?
- Do the operating, financing, and investment activities that generate cash flow impact the profitability measured by ROA in the industrial companies listed on ASE?
- Do the operating, financing, and investment activities that generate cash flow impact the profitability measured by ROE in the industrial companies listed on ASE?

2 Literature Review

Several prior studies have addressed this topic, but from various angles, environments, variables, and statistical techniques. As a result, in our research, we consider a few of these studies:

According to [3] their analysis found that operational cash flows and net interest margin had a significant and favourable impact on banks' financial stability. However, the cost-to-income ratio and the proportion of advances net of provisions to total assets had a significant and unfavourable impact. Banks should become more cost-effective and increase their liquidity levels by reducing lending activity to increase financial stability.

[4] The study sought to determine how liquidity management affected the financial performance of the industrial businesses listed on the Amman Stock Exchange. The researcher studied several financial statements to get information on profitability and liquidity. It was discovered that liquidity management impacts the targeted firms' profitability.

[5] showed that the current liabilities ratio significantly and positively impacts profits per share and profit after taxes. On the other hand, earnings per share, return on capital employed, return on assets, and profit after tax is unaffected by the current and quick ratios. According to this study, managers should consider liquidity while making evaluation judgments to increase their companies' financial returns. For administrators, analysts, regulators, investors, and other interested parties, the current study provides valuable insights into the success of Indian listed firms.

[6] The research's findings indicate that independent study factors (sources of operational cash flow, investment, and finance) have a fluctuating and weak statistical influence on the study's variables (profitability). This study advises against turning to high-cost borrowing for funding and advocates increasing property rights investment.

[7] According to the study, quick and cash ratios show a positive but negligible link with profitability (ROA) for the selected organizations, but the current ratio has a negative and substantial influence on ROA.

The findings showed that while financial leverage harms the profitability of Jordanian industrial listed enterprises, liquidity has a significant and favourable impact on profitability. This study aimed to broaden the concept to empower the industrial listed enterprises in achieving their profitability objective since profitability is crucial for all Jordanian industrial listed firms.

[8] The study aims to examine the connection between the three primary components of cash flows—operating, investing, and financing activities—and profitability as determined by earnings per share (EPS). The study's findings showed that elements of cash flows, such as operating, investing, and financing activities, had a considerable influence on profitability as evaluated by (EPS).

Focusing on liquidity multiple research offer information on the definition of liquidity and liquidity management. Multiple meanings exist for the term liquidity. [7] define the term liquidity as (the ease of converting assets into cash to do financial obligations without losing time or money). Liquidity measures a company's capacity to meet its financial commitments on short notice. It is connected with utilizing regular cash flow and getting receivables from secondary sources. "Liquidity" refers to the ability to convert assets into cash without suffering a loss [10]. Liquidity is the difference between the resources possessed by the firm and the cash utilized to purchase the different sorts of assets within the balance. When a business must borrow funds to meet its financial commitments or current liabilities, it demonstrates a lack of liquidity [11]. Low liquidity has several harmful repercussions. Adverse effects on the company's reputation and lost investment possibilities are examples of such consequences. These include fostering uncertainty among investment decision-makers [12].

In contrast, the quick and investment ratios [13] discovered that efficient and effective liquidity management contributes to the profitability and survival of Sri Lanka banks; they also discovered that there is no statistically significant relationship between liquidity and profitability ratios among manufacturing firms listed on the Sri Lanka Stock Exchange and that the values of the Pearson correlation coefficient are negative between the return on capital employed and liquidity. [13] Discovered a high correlation between liquidity and profitability. The profitability of commercial banks is significantly influenced by liquidity string.

3 Methodology

3.1 Hypotheses Development

This study aims to evaluate the following test hypotheses considering the study's problem, significance, and objectives, as well as the theoretical framework and earlier research.

The main hypothesis:

H0: There is no impact on liquidity ratios and the cash flow Sources on the profitability of the industrial companies listed on ASE.

The primary assumption leads to the following four sub-assumptions:

H01-1: There is no impact on the liquidity ratios measured by (current and quick liquidity ratios), on profitability through ROA in the industrial companies listed on ASE.

H01-2: There is no impact on the liquidity ratios measured by (current and quick liquidity ratios), on profitability through ROE in the industrial companies listed on ASE.

H01-3: There is no impact on the cash flow sources (CFO, CFI, CFF) on profitability through ROA in the industrial companies listed on ASE.

H01-4: There is no impact on the cash flow sources (CFO, CFI, CFF) on profitability through ROE in the industrial companies listed on ASE.

3.2 Sample and Data

The researchers needed a combination of primary and secondary data. Secondary sources included books, journals, and the internet for literature reviews. The industrial enterprises listed on the ASE for the relevant period's annual financial reports served as the source of the primary data gathering tool (2019–2020), was dependent on profitability with its indicators (ROA & ROE), and was based on CFO, CFI, and CFF as independent variables.

This study used cross-sectional time-series data, known as (Panel Data Analysis), and this type of data depends on the characteristics of each of the time-series data and cross-sectional data." Panel Data Analysis is concerned with the impact of each of the individual factors and the time factor (Individuals & Time Effects).

"In this way, the current study used the primary data for statistical analysis by using the annual financial reports issued by the industrial companies listed ASE for the period (2019–2020), where these data are also available on different websites. As shown below.

4 Independent Variables

According to the theoretical framework, prior research, and the study's problem, the following independent variables have been used to link the study variables:

Liquidity ratios were calculated through two indicators, namely [14].

- Current ratio = current assets divided by current liabilities.
- Quick ratio = current assets (not including Merchandise inventory, prepaid expenses) divided by Current liabilities.

The elements of cash flows were calculated through three indicators, which are [14].

- Cash flow from Operating Activity = Net Cash Flow from Operating Activity divided by Net Profit.
- Cash flow from Investing Activity = Net Cash Flow from Investing Activity divided by Net Profit.
- Cash flow from Financing Activity = Net Cash Flow from Financing Activity divided by Net Profit.

5 Dependent Variables

The dependent variable is included in the profitability. Add two elements (ROA, ROE).

Profitability was calculated through two indicators, namely [15].

ROA = Net Profit after Tax divided by Total Assets.

ROE = Net Profit after Tax divided by Shareholders' Equity Assets.

5.1 Theoretical Model

According to the literature review, a conceptual framework is given based on the material previously examined, as shown in Fig. 1 below, which the impact between liquidity ratios (Current ratio and Quick ratio) and cash flow sources from (CFO, CFI, CFF) on profitability with its indicators (ROA & ROE).

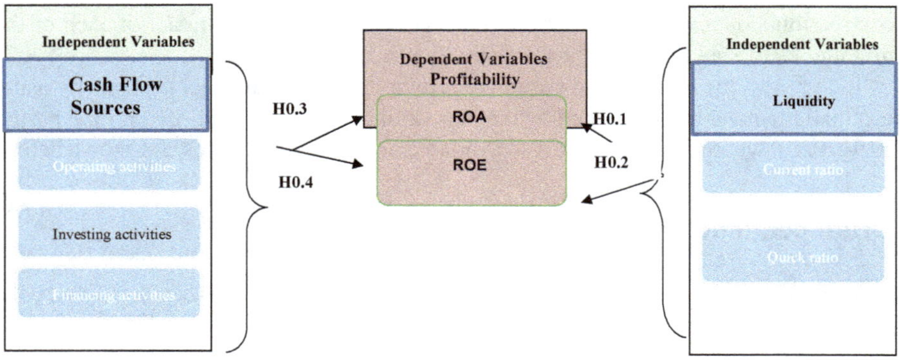

Fig. 1. Theoretical Model. Recourse: By Research

6 Descriptive Statistics

Model 1

$$ROA = a0 + a1CR + a2QR + a3CFO + a4CFI + a5CFF + Error$$

Model 2

$$ROE = a0 + a1CR + a2QR + a3CFO + a4CFI + a5CFF + Error$$

where

ROA	Return on Assets
ROE	Return on Equity
CR	Current Ratio
QR	Quick Ratio
CFO	Cash Flow from Operating Activities
CFI	Cash Flow from Investing Activities
CFF	Cash Flow from Financing Activities
a_1, a_2, a_3, a_4, a_5	Regression Coefficient
a_0	Constant Coefficient
e	Error term

6.1 Empirical Findings and Analysis

Table 1. Findings from the first hypothesis's testing

Coefficients					Dependent Variables
Sig t	T	df	B	Item	
	2.694			Stability	ROA
0.223	−1.244	1.699	−2.112	Current Ratio	
0.234	1.215	1.299	1.579	Quick Ratio	
2.933	F	0.763			R^2
0.002	Sig.	0.503			Adjusted R-squared

Table 1 reveals that the coefficient of determination ($R^2 = 0.763$) indicates that the application of liquidity ratios (current and quick liquidity ratios) has explained 76.3% of the variance in ROA, holding all other variables constant. In addition, it was demonstrated that the F value at the confidence level (Sig = 0.002) was 2.933, confirming the significance of the regression at that level (0.05). The first primary hypothesis is rejected by the null formula and accepted by the alternative formula, which states: "There is a statistically significant effect at the significance level (0.05) for liquidity ratios (current ratios, and liquidity ratios) on profitability via ROA in industrial companies listed on the ASE."

Table 2. Findings from the second hypothesis's testing

Coefficients					Dependent Variables
Sig t	T	df	B	Item	
	4.269			Stability	ROE
0.039	−2.158	1.714	−3.699	Current Ratio	
0.124	1.581	1.311	2.073	Quick Ratio	
3.496		F	0.793		R^2
0.000		Sig.	0.566		Adjusted R-squared

Table 2 shows that the coefficient of determination is ($R^2 = 0.793$), which implies that when all other variables were held constant, the usage of liquidity ratios—specifically, current and fast liquidity ratios—explained (79.3%) of the variance in ROE. Additionally, it was demonstrated that the value of (F) reached (3.496) at the level of confidence (Sig = 0.000), which supports the regression's significance at the level of (0.05). Based on those above, the alternative formula states that "There is effect at the significance level (0.05) for liquidity ratios: (current ratios, and quick ratios) on r on profitability through from ROE in industrial companies listed on the ASE," which accepts the second central hypothesis.

Table 3. Findings from the third hypothesis's testing

Coefficients					Dependent Variables
Sig t	T	df	B	Item	
	−0.404			Stability	ROA
0.546	0.611	0.235	0.144	Operating activities	
0.850	−0.191	0.082	−0.016	Investing activities	
0.572	0.572	0.174	0.100	Financing activities	
2.647		F	0.755		R^2
0.004		Sig.	0.470		Adjusted R-squared

According to the Table 3 Findings, the coefficient of determination is ($R^2 = 0.755$), meaning that the usage of cash flow elements (from operating Activity, financing activity, and investment activity) explained (75.5%) of the variation amount. It was discovered that in ROA, with all other variables held constant, the value of (F) reached (2.647) at the confidence level (Sig = 0.004), which confirms the significance of the regression at the level of (0.05). This rejects the third central hypothesis in the nihilistic form and accepts the alternative formula, which states: "There is effect at the significance level (0.05) for the use of the elements of cash flow on profitability through from ROA in industrial companies listed on ASE.

Table 4. Findings from the fourth hypothesis's testing

Coefficients					Dependent Variables
Sig t	T	df	B	Item	
0.903	−0.123	2.147	−0.264	Stability	ROE
0.609	0.517	0.256	0.132	Operating activities	
0.841	−0.202	0.089	−0.018	Investing activities	
0.767	0.300	0.190	0.057	Financing activities	
2.578		F	0.750		R^2
0.005		Sig.	0.459		Adjusted R-squared

The findings of Table 4 show that the usage of the cash flow statement's operational Activity, financing activity, and investment activity has explained its amount (75.0%) with a coefficient of determination of ($R^2 = 0.750$). With all other components held constant, it was discovered that the variation in ROE achieved a value of (F) of (2.578) at the confidence level (Sig = 0.005), confirming the significance of the regression at the level of (0.05). The fourth central hypothesis, which reads: "There is a statistically significant effect at the significance level (0.05) for the application of the items of the cash flow list on profitability via," is rejected in the nihilism version as stated above and accepted in the alternative form. Which states: "There is a statistically significant effect at the significance level ($\alpha \leq 0.05$) for using the elements of the cash flow list on profitability from ROE in industrial companies listed on ASE.

7 Results, Conclusion and Recommendation

The results of the study showed the existence of a statistically significant effect at the significance level ($\alpha \leq 0.05$) for liquidity ratios (current ratios and quick liquidity ratios) on profitability. This result indicates the ability of the industrial companies on ASE to convert their financial instruments into cash as quickly as possible and without loss or as little as possible to cover the companies' short-term obligations. This result also indicates the ability of these companies to pay their obligations, especially in emergency cases, through the availability of a type of circulating assets, which are characterized by their rapid conversion into cash, which leads to the completion of operational, production and administrative activities in these companies, and in a manner that enhances the optimal utilization of the researched companies of the available resources and their ability to obtain revenues at a reasonable cost, and directing them For profitable investments, and in a way that positively affects the ROA for these companies.

We also find the results of the statistical analysis indicated that there is a statistically significant effect at the significance level ($\alpha \leq 0.05$) of the liquidity ratios (current ratios and quick liquidity ratios) on profitability through the ratio of (current ratios, and quick liquidity ratios) to profitability. The researchers may attribute this result to the fact that the industrial companies on ASE retain enough cash, or what assets are available to them quickly converting to cash, without losses in their value, and as soon as possible when needed, to cover the obligations arising, therefore. The availability of both current ratios and quick liquidity ratios is a strong indicator of a sound financial position for the parties dealing with the company, which enhances the company's investments and achieve more and its sales and thus its ability to increase its sales and capacity. Which enhances its profitability and creates the highest value for shareholders in industrial companies listed on ASE.

The study's findings have revealed the impact of statistical significance at the level of significance ($\alpha \leq 0.05$) to use the victory of the monetary drains (from operational Activity, financing activity, and investment activity) in the grace of the ritual) through sacrifice. The flow of cash might explain this outcome into or out of the industrial companies listed on the ASE, resulting either from flows resulting from the sale of goods or associated with the sale and purchase of investments, including long-term securities, or the sale of assets and other cash, or any other. This enhances the Activity of industrial companies listed ASE, their investments and production processes, and their efficiency in using their assets and available resources optimally compared to these assets to achieve the shortest possible amount of profits.

Moreover, the results showed an impact at the significance level ($\alpha \leq 0.05$), for the use of the elements of the cash flow on the profitable, through the excellent investment group, in the ASE. This outcome is interpretable by the fact that the process of cash entering or leaving the balances of industrial companies listed on ASE, whether this is from operational activities that enhance the company's revenues such as sales of goods, collections of accounts receivable, or cash flow from future investment activities concluded by the industrial companies investigated, in addition to the financing operations on which these companies are based to develop their activities, whether by borrowing or by supporting the owners of the company's balances, which leads to the development of its production and investment activity to obtain a larger market share, and its ability to enter new markets, and thus its ability to enter new markets. Maximizing returns compared to costs enhances its financial performance and profitability, which is positively reflected on the shareholders' profits from investing their money in the company's capital. The research's found that "there is a statistically significant influence at the 0.05 level for liquidity ratios on profitability, through the ratio (profitability ratio), on the ratio of (current ratios, and quick liquidity ratios) to profitability. This finding shows that the industrial businesses traded on ASE can swiftly and with little loss convert their financial instruments into cash to meet their short-term commitments and that they can also meet their obligations when they come due. Especially in times of crisis, the availability of a particular class of circulating assets, defined as those that can be quickly converted into cash, allows these businesses to finish their operational, production, and administrative tasks.

Not only do we uncover, According to the findings of the statistical study, the liquidity ratios (current and rapid liquidity ratios) also have a statistically significant impact on profitability through the ratio of (current ratios and quick liquidity ratios) to profitability at the 0.05 level of significance. This finding may be explained by the fact that ASE-listed industrial businesses have sufficient cash on hand or have access to other liquid assets that may be converted into cash with little loss of value and minimal delay, to meet their resulting commitments. The availability of both current and quick liquidity ratios is a strong indicator of a sound financial position for the parties dealing with the company, especially for banks and financial institutions when granting loans and facilities to companies, which improves the company's ability to invest and achieve more and its sales, and consequently, its profitability and creates the most outstanding value for shareholders. A triumph over financial draining (from operational Activity, finance activity, and investment activity) in the grace of the ritual (through the sacrifice) has been shown to have a statistically significant influence on the level of significance (0.05) in this research.

Also, the cash flow factors used to generate profits via the lucrative investment group were shown to have a statistically significant effect on those outcomes during an ASE. This result can be understood because the financing operations upon which the industrial companies listed on the Amman Stock Exchange are based, as well as the cash flow from future investment activities concluded by these companies, all contribute to the overall process of cash entering or leaving the balances of these companies. Profits earned by shareholders from their investment in the company's capital directly reflect the company's success in maximizing returns over expenses.

The following recommendations can be made:

1- Improving industrial businesses registered on ASE from their liquid assets while staying away from the excessive increase in quick liquidity ratios in a manner that does not disrupt working capital and ensures the achievement of maximum profit.
2- Industrial firms listed on ASE need to examine their internal or external working environments, relying on the latest scientific methods and periodically, to determine their competitive position and strengthen it by offering competitor products with feasible returns to achieve profitability.

References

1. Rudin, M., Nurdin, D., Fattah, V.Y.: The effect of liquidity and leverage on the profitability of property and real estate company in Indonesian Stock Exchange. Int. J. Soc. Sci. Manage. **3**(4), 300–304 (2016)
2. Shahwan, Y.: The moderating effects of dividend policy on the impact of financial disclosure towards corporate value: a study of listed Jordanian industrial firms. Res. J. Financ. Acc. **10**(6), 115–125 (2019)
3. Elahi, M., Ahmad, H., Shamas Ul Haq, M., Saleem, A.: The impact of operating cash flows on financial stability of commercial banks: evidence from Pakistan. J. Asian Finance Econ. Bus. **8**(11), 223–234 (2021)
4. Shakatreh, M.: The effect of liquidity ratios management on the profitability of industrial companies listed on the Amman stock exchange. Accounting **7**(5), 1131–1138 (2021)
5. Hassan, A.M.: The tradeoff between liquidity and profitability: evidence from Indian manufacturing firms. Stud. Econ. Bus. Relat. **2**(2), 46–58 (2021)
6. Weshah, S.R.: Cash flows sources and it is impact on profitability within Covid-19 pandemic: case study of industrial companies listed in Amman stock exchange. Int. J. Entrepreneurship **25**, 1–7 (2021)
7. Dadepo, A.O., Afolabi, O.F.: Impact of liquidity management on profitability of selected manufacturing firms in Nigeria. Eur. J. Bus. Manage. **12**(27), 93–99 (2020)
8. Al Qaisi, F.: The relationship between cash flow and profitability of insurance companies listed in Amman stock exchange in Jordan. J. Soc. Sci. Res. **5**(1), 51–60 (2019)
9. Ahmed, N., et al.: Synthesis, development and preclinical study of EDDA based ^{99m}Tc-5-fluorouracil for brain imaging. J. Radioanal. Nucl. Chem. **309**(2), 805–811 (2016)
10. Al-Homaidi, E.A., Almaqtari, F.A., Yahya, A.T., Khaled, A.S.D.: Internal and external determinants of listed commercial banks' profitability in India: dynamic GMM approach. Int. J. Monetary Econ. Finance **13**(1), 34–67 (2020)
11. Bin Saadon, S.: The effect of liquidity on profitability in commercial banks: a comparative study of a number of public and private Algerian banks for the period (2013–2017) (Master thesis). College of Economic and Commercial Sciences, p. 8 (2019)
12. Najim, A.: The use of cash flow metrics in evaluating profit quality: an empirical study at the North bank for finance and investment. Anbar Univ. J. Econ. Adm. Sci. **8**(15), 445–466 (2016)
13. Haddad, H.: Bank specific and macroeconomic determinants of commercial bank profitability: in Jordan from 2009–2019. Montenegrin J. Econ. **18**(4), 155–166 (2022)
14. Gibson, C. H.: Financial Reporting and Analysis. Cengage Learning, Boston (2012)
15. Budiharjo, R.: Effect of environmental performance and financial performance on firm value. Int. J. Acad. Res. Acc. Financ. Manage. Sci. **9**(2), 65–73 (2019)

Jordanian Imports: Income and Price Elasticity of Demand

Khaled Mohammed Al-Sawaie[1(✉)], Abdelhalim Mohammad Jubran[1], Nidal Ali Abbas[1], Mohammad Musa Al-Momani[1], Thaer Ahmad Abu-Saleem[1], Ibrahim Mohammad Khrais[1], Mazin Hasan AlBasha[1], and Mohammad AlHunaity[2]

[1] Department of Economics, Zarqa University, Zarqa, Jordan
{ksawaie,ajubran,nabbas,m.momani,thabusaleem,ikhrais, mbasha}@zu.edu.jo

[2] Department of Finance, University of Business and Technology, Jeddah, Kingdom of Saudi Arabia
m.hunaity@ubt.edu.sa

Abstract. This study aimed to recognize both the short and long term models of the aggregate import demand for commodity in Jordan. Thus, to determine the impact of changes in price and income on the import demand, and so, it was estimated using the Autoregressive Distributed Lag model. However, the results indicated that there is an import demand of low price elasticity in the short and long term. Whereas, the income elasticity is greater than 1, namely, it is elastic in the short-term but with low elasticity in the long term. Meanwhile, the received elasticities pointed out to useful information for the decision-maker and for future research in Jordan as well as other developing countries regarding import demand.

Keywords: Imports · Aggregate Import Demand · Price Elasticity · Income Elasticity · Bound Test · ARDL

1 Introduction

Jordanian commodities' imports constitute a high percentage of the gross domestic product and a negative net commodity trading. Despite Jordan's trade in positive net services, it achieves a permanent deficit on the commercial account, making it a borrowing country. Thus, in order to alleviate this chronic situation, it is necessary to develop effective trade policies and understand how the import demand interacts with changes in relative prices for domestic and global commodities and services as well as income. Which can be summed up in so-called elasticity, which is used in many commercial applications, theory and policy.

Though, there are books that tackled the relationship between trade and economic growth in Jordan, for example, the study of Saleem et al. (2021) highlighted the impact of digital trade on economic growth in Jordan [1]. While the study of Al-Sawaie (2017) showed the determinants of the Jordanian current account, where the price elasticity

B. A. M. Alareeni and I. Elgedawy (Eds.): *Artificial Intelligence (AI) and Finance*, SSDC 488, pp. 744–752, 2023.
https://doi.org/10.1007/978-3-031-39158-3_69

is negative in the long term, the income elasticity is positive and less than one [2]. Meanwhile, the current study agrees with the study of Abu-Lila & Jdaitawi (2015) and the Jordanian intra-trade [3]. It also showed that price elasticity is less than 1.

The current study aims to recognize the model of aggregate import demand in the short term and long term in Jordan, as well as to answer the following two questions:

1. How does the import demand for commodities interact with the change in the relative prices of global commodities?
2. How does the import demand for commodities interact with the change in Jordanian income?

For the purpose of answering the abovementioned two questions, the researcher utilized quarterly data on imports of commodities, the import price index, the price level index (CPI), and the real GDP. The data extended from the 2008:01 to the 2022:03.

Eventually, the study is organized as follows:

- Second section: provides an overview of imports from a theoretical point of view, Whereas, it discusses the impact on the balance of trade and describes the results of previous studies on the Jordanian import demand.
- Third section: provides an overview of Jordan's global trade.
- Fourth section: describes the methodology of empirical analysis.
- Fifth section: Shows the empirical analysis using the autoregressive distributed lag methodology.

2 Imports in the Macroeconomic

Adam Smith and David Ricardo explained that each country has the advantage of producing commodities and services in which it is more efficient than others, as it produces commodities more effectively than others. Hence, the produced commodities are exported more effectively, and other commodities are imported. However, Heckscher and Ohlin showed that the variation in the rations of the production factors leads to a variation in the relative marginal costs of production, as a consequence, the country produces and exports the commodities which factors of production are more abundant. Even though, in the case of an open economy, the country's spending is not equal to its production of goods and services because it may spend more or less than it produces, thus acting as a borrower or lender to global countries.

2.1 Import Demand Factors

Aggregate demand for domestic commodities includes:

The demand for domestic and global commodities, where imports constitute a large share of it, so the local income is the first determinant of the import demand. As the increase in domestic income leads to an increase in the aggregate demand concerning all commodities and services, and thus the quantity of the desired imported imports interacts with the change in income; or what is called the income elasticity of import demand as follows:

$$\eta_{Y} = \frac{\Delta M_{D}}{\Delta Y} \cdot \frac{Y}{M_{D}} \tag{1}$$

where:

η_Y: income elasticity
M_D: the quantity of desired imports
ΔM_D: the change in the quantity of desired imports
ΔY: income change
Y: income

The second determinant is the terms of trade (real exchange rate), which describes the relative price of commodities between countries; namely, it is the price at which the commodities of one country are exchanged for the commodities of another country. In contrast to the exchange rate (nominal), which describes the relative prices of currencies. The exchange rate (real) also depends on the exchange rate of the two countries' currencies; That is, the exchange rate (nominal), and therefore the real exchange rate equation is as follows:

$$\varepsilon = e\frac{P_d}{P*} \tag{2}$$

where:

ε: the real exchange rate
e: nominal exchange rate
P: consumer price level (CPI) in the local country
P^*: the price level in the rest of the world

$$M = M(Y, \varepsilon) \tag{3}$$

where as, an increase in any of the two components leads to an increase in the import demand. Due to the negative impact of the real exchange rate on exports, a depreciation of the real exchange rate leads to increase the exports making domestic commodities relatively less expensive abroad. On the other hand, since global commodities become relatively more expensive, imports decrease. Which leads to an increase in domestic demand for domestic commodities. Therefore, an increase in the relative price of global commodities with regard to domestic commodities means an increase in the cost of imports. And in order to improve the balance of trade, the increase and decrease in exports and imports, are respectively sufficient to compensate the higher cost of imports, as well the growth of net exports under these conditions is called the Marshall-Lerner condition [4].

3 An Overview of Trade in Jordan

3.1 Commodities and Services Balance of Trade

Whereby the value of total commodities and services imports as a proportion of GDP is rather high, it exceeded 40% of the study years, and in the year 2022 it reached more than 66% of these years [5].

Figure 1 shows the proportion of GDP for Jordan's imports and exports of commodities from the year 2008 to the year 2022. In which exports proportion was less than

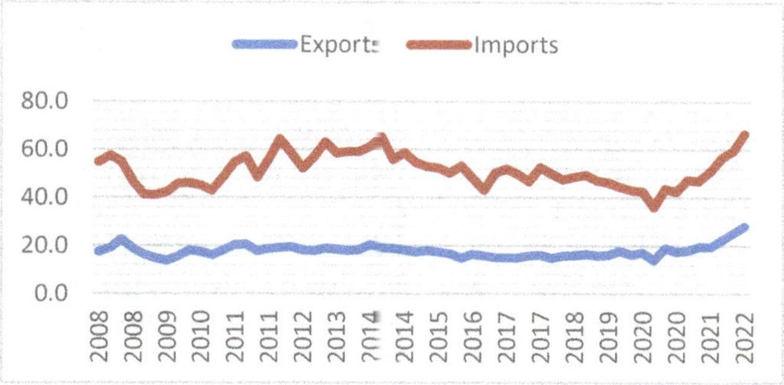

Fig. 1. Jordanian commodities trade as a proportion of GDP from the year 2008 to the year 2022 [5].

imports during the mentioned period, also the trade net was negative, thus Jordan is considered a borrower from the global market. Knowing that the gap between commodities imports and exports increased significantly after the 2008 financial crisis.

As can be seen from Fig. 2 the services' balance of trade was positive during the years from 2008 to 2021, excluding 2019, which means that Jordan acts as a lender in the services market [5].

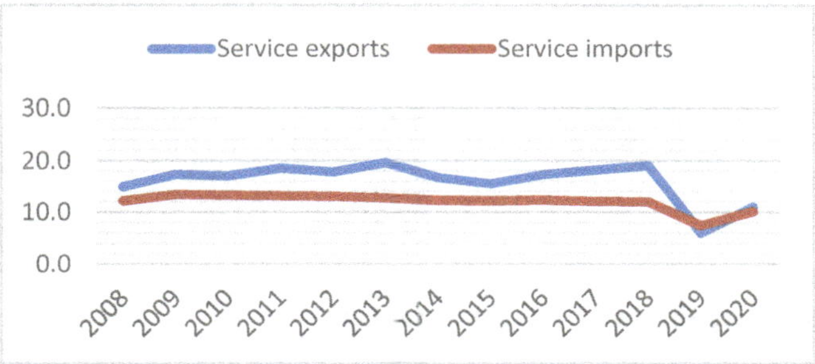

Fig. 2. Jordan's trade in services as a proportion of GDP for the years from 2008 to 2021 [5].

However, Saudi Arabia plays an important role when talking about import over all the selected years, in addition to China, the United States, India, Germany and the UAE. Even though, the United States is also considered the most significant exporter, taking into account that during the past 14 years India, Iraq and Saudi Arabia are essential trading partners and so it can be said that they are the largest exporters, besides, Turkey, the United Arab Emirates and Kuwait [5].

4 Data and Methodology

The following determinants were referred to regarding the aggregate import demand in Jordan: imports, relative price, and income. Whereas, real imports were obtained through dividing the nominal imports by the import price index. Also, for the purpose of having the relative price; the values of the import price index were divided by the values of the consumer price index, and on the real income using the real GDP, all of which were taken from the database of the Central Bank of Jordan (CBJ).

Table 1 below shows the descriptive statistics of imports, relative price and income, the very large difference between the lower value and the upper value of imports. Despite this, the relative price of global commodities' average is higher than the domestic commodities. Therefore, it showed the negative skewed of income, but in the case of imports and the relative price, the deviation indicates that it is deviated to the right. The data were more or less symmetric, as the Jarque-Bera test for normality showed that it didn't vary from 0; Hence, its distribution is normal.

Table 1. Descriptive statistics of real imports, relative price, and income

Statistic	Real GDP	imports	Relative price
Mean	6794.384	3467.877	3.586724
Median	6831.950	3485.189	3.524500
Maximum	8017.300	5097.447	4.595000
Minimum	5156.500	2234.557	2.691000
Std. Dev	722.5187	535.773	0.451906
Skewness	−0.212847	0.107686	0.263033
Kurtosis	2.147161	3.556713	2.216789
Jarque-Bera	2.195662	0.861094	2.151229
Probability	0.333594	0.650153	0.341088

4.1 The Aggregate Import Demand Function

The aggregate import demand function was utilized, and its basic form is as follows [6]:

$$M = \alpha_0 + \alpha_1 \frac{P_m}{P_d} + \alpha_2 Y \tag{4}$$

where:

M: demand for imports
α_1: relative price coefficient
P_m: import price index
P_d: local price index
α_2: income coefficient
Y: income

In terms of import prices, the import price index was utilized, which was used to find the value of real imports through reducing the values of nominal imports by the import price index. However, the consumer price index was used to express domestic prices compared to the income, in addition to the utilization of the real GDP. And since one of the methods for estimating elasticity is to transform the function into a linear logarithmic model aiming at estimating the import demand, therefore, the form of the aggregate import demand function is as follows:

$$lnM = \alpha_0 + \alpha_1 ln\left(\frac{P_m}{P_d}\right) + \alpha_2\, lnY + u \tag{5}$$

where:

u: the error terms

As a result of converting the equation into linear logarithmic form, currently, the coefficients represent the price and income elasticity of import demand. And so, it is expected that there will be a positive relationship between the import demand and income, also, a negative relationship between the import demand and the relative price.

5 Results

As a consequence of the discussion, the first step before estimating the model is to perform the unit root test of variables, where the ARDL bounds test method can be used with variables I(0), I(1), or mixed, but it must be ensured that it is not I(2), using the ADF unit root test. Meanwhile, AIC was used to determine the longest optimal lag, and the results are presented below in Table 2, where the variables are: (imports) the natural logarithm of real imports, (RP) the natural logarithm of the variable; relative price, and (GDPr) the natural logarithm of income.

Table 2. ADF tests

	Level			First difference		
	imports	RP	GDPr	imports	RP	GDPr
Trend and intercept	0.396	−2.393	−1.835	-9.055^{***}	-8.799^{***}	−3.078
Intercept	−2.652	−2.701	−1.367	-8.921^{***}	-8.862^{***}	-3.020^{**}
Without trend or intercept	−2.938	−0.473	2.373	-8.962^{***}	-8.924^{***}	-1.770^{*}

Note: "***, **, * are statistically significant at the 1%, 5%, and 10% levels of significance, respectively"

It can be seen from the test results that all variables: real imports, as well as relative price, and real income, are integrated of the first degree.

5.1 ARDL Model and Bound Test

Since there are no integrated variables of the second degree, the ARDL bounds test method will be used to test co-integration. Where optimal lag lengths were found using

Akaike Information Criterion and the ARDL model was as follows; ARDL(1, 2, 5), which means, one lag period for the imports variable, two lag periods for the relative price variable, and 5 lag periods for the real GDP variable, knowing that these variables were included in the model.

After determining the ARDL model most appropriate lags, a bounds test is performed, and the F statistic obtained from the test is compared to the limits of the critical value proposed by Pesaran [8], and the test results are presented in Table 3.

Table 3. Limit and critical limit test result

Bounds Test Null Hypothesis: No levels relationship				
Test Statistic	Value	Sig.	I(0)	I(1)
F-statistic	3.727466	10%	2.17	3.19
K	2	5%	2.72	3.83
		2.5%	3.22	4.5
		1%	3.88	5.3

Given that the value of the F statistic is greater than the upper limit at the 5% level, the null hypothesis was rejected as there were no long-term relationships. The variables were integrated, and therefore we can estimate the short and long term models. And in order to estimate the short-term coefficients, the error correction model was used, where the first difference coefficients in the ECM represent the short-term, and the error correction bound shows the speed of adjustment from the short-term impact and resulted in balance for the long term. Therefore, the results are shown below in Table 4.

Table 4. Coefficients of the short-term model

	α_i	s.e.	t-stat.	P-value
D(RP)	−0.8082	0.1247	−6.481	0.000
D(GDPr)	3.5921	0.9545	3.763	0.000
ECM(−1)	−0.4292	0.1255	−3.4209	0.001

The coefficient of the relative price variable is very significant in the short-term, as a 1% change in the relative price of imports leads to moving the import demand in the opposite direction by 0.8%, and the absolute value of the coefficient is less than 1. Therefore, it can be said that in the short-term, imports have low price elasticity, and the coefficient of the income variable is significant in the short-term, as a change in income by 1% leads to moving the import demand to the right by 3.6%, and the absolute value of the coefficient is greater than 1. Thus, it can be said that imports are income elastic in the short-term, and the error correction model (ECM) is (−1) as shown in Table 4, which is significant and it is expected to be negative, which means that its value is 0.429 (or 42.9%) indicating that the deviation in the short-term from the long term balance is corrected after about two months.

Table 5. Coefficients of the long-term model

Variable	α_i	s.e	t-Stat.	P-value
RP	−0.4050	0.1886	−2.1466	0.037
GDPr	0.3160	0.0264	11.9519	0.000
EC = LOG(IMPORTS) - (−0.4050*LOG(RP) + 0.3160*LOG(GDPR))				

Table 5 presents the long-term coefficients, which were statistically significant for the variables; income and relative price, in the long term at a significant level of 1% and 5%, respectively; Meanwhile, the import demand was price inelastic in the long term, as in the short-term −0.4, which means that a relative price change of 1% affects the import demand in the opposite direction, while the income elasticity of import demand was 0.3, which means that income is inelastic on import demand in the long term, the impact of import demand on income has become less than 1% compared to the short-term. Thus, to validate the estimated model, the following diagnostic tests were performed:

Table 6. Diagnostic tests

Test	F-statistic	P-value
Breusch-Godfrey	0.097	0.908
Heteroskedasticity Test	0.983	0.472
Jarque-Bera	2.636	0.267
Ramsey's RESET	0.024	0.879

Table 6 shows the rejection of the serial correlation of the residuals because the F statistic of the Breusch-Godfrey test is not statistically significant and the heteroscedasticity test showed the presence of homoscedasticity. Also, the Jarque-Bera test showed that the distribution of the residuals is normal, and the Ramsey RESET test confirmed the validity of the model function shape. The CUSUM test in Fig. 3 also showed that the model parameters are relatively stable over the period.

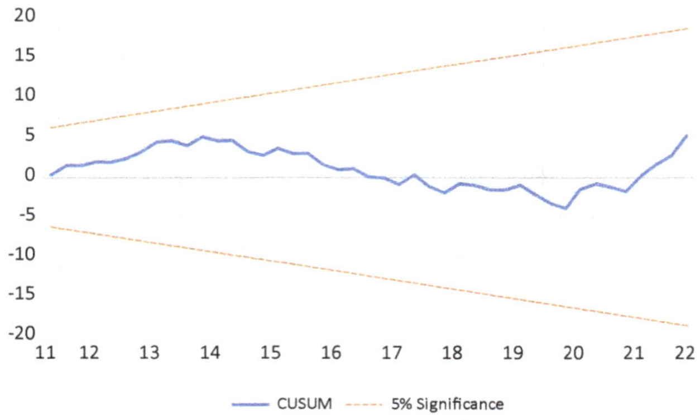

Fig. 3. "CUSUM test"

6 Conclusions

The estimation of the import demand model of income and price elasticity aims to help policymakers formulate more efficient trade policies that have an impact on the balance of trade. Where, the objective of this study was to estimate the aggregate demand model for Jordanian imports, which had a share of more than 40% of the GDP, and since the size of imports exceed exports, it made Jordan a net borrower until its debt increased to nearly 50 billion dollars. And so, the results indicated that the import demand in Jordan whether in the short and long term is price inelastic, even though the income was elastic in the short-term and inelastic in the long term. Thus, the study recommends that the Jordanian government encourage industries with low import demand elasticity, which helps improve the balance of trade and domestic production.

References

1. Saleem, T.A.A., Abbas, N.A., Al-Sawaie, K.M., Jubran, A.-H.M.: The impact of digital trade on Jordan's economic growth. In: 22nd International Arab Conference on Information Technology, ACIT 2021 (2021)
2. Al-Sawaie, K.H.M.: Determinants of the Jordanian trade balance: an autoregressive distributed lag model. Zarqa J. Res. Stud. Hum. **17**(1), 138–149 (2017)
3. Abu-Lila, Z.M., Jdaitawi, Q.M.: Bilateral trade elasticities of Jordan. Dirasat: Adm. Sci. **42**(2), 611–622 (2015)
4. Blanchard, O.J.: Macroeconomics. 4th ed. Pearson Prentice Hall, New Jersey (2006)
5. Central Bank of Jordan Homepage. https://www.cbj.gov.jo/Pages/viewpage.aspx?pageID=40. Accessed 15 Feb 2023
6. Frimpong, J.M., Oteng-Abayie, E.F.: Aggregate import demand and expenditure components in Ghana. J. Soc. Sci. **4**(1), 1–6 (2008)
7. Pesaran, M.H., Shin, Y.: An autoregressive distributed-lag modelling approach to cointegration analysis. Econ. Soc. Monogr. **31**, 371–413 (1998)
8. Pesaran, M.H., Shin, Y., Smith, R.: Bounds testing approaches to the analysis of level relationships. J. Appl. Economet. **16**(3), 289–326 (2001)
9. Duasa, J.: Determinants of Malaysian trade balance: an ARDL bound testing approach. Glob. Econ. Rev. **36**(1), 89–102 (2007)

An Investigation on the Legal Protection of Cryptocurrency Investors, Comparative Legal Analysis

Iyad Khanfar[1](✉) and Nehad Khanfar[2]

[1] Zarqa University, Zarqa, Jordan
i.khanfar@zu.edu.jo
[2] Middlesex University, London, UK

Abstract. Millions of people, individuals, companies and other entities invested tremendous money in different cryptocurrencies worldwide. Many reports issued by regulators in many countries worldwide indicate a lack of reliable information clarifying the legal status of cryptocurrency products and the regulatory protection given when offered in the local market. However, the legal protection in some jurisdictions requires regulators to approve crypto assets-related products when introducing them to the market. In contrast, in others, it is proposed that a regulated financial institution like banks can endorse the approval of the cryptocurrency before its introduction to the market. It is worth noting that prudential regulation and conduct do not exist in many jurisdictions, which means that investors' funds do not enjoy the protection provided to financial products and services such as e-money or online brokerages.

Plenty of Crypto-asset are offered for selling in many jurisdictions; hence, it becomes almost impossible to bring them under specific regulations or even to place an effective monitoring system. One of the main reasons behind this is that cryptocurrency companies' headquarters carefully select their firms' locations in flexible jurisdictions for the regulations, taxation and legal systems. Therefore, most cryptocurrency trading transactions are carried out through firms and institutions primarily functioning in offshore registered financial entities. Furthermore, although cryptocurrencies are traded and exchanged in different countries, many lack the legal or regulatory framework to deal with related activities. Even though some jurisdictions require cryptocurrency activities to be registered or obtain a low authorization level, the purpose of such regulations is limited. Therefore, inadequate or non-existent regulatory instructions can negatively impact legal protection and enforcement. This paper investigates the legal risk imposed on cryptocurrency traders and explores the level of attention provided to regulatory, monitoring and supervision of legal and financial conduct worldwide. Furthermore, it will critically assess the capabilities and willingness of the current jurisdictions to address the issues concerning the legal complex facing the investors. It further examines the level, extent and context of legal protection frameworks given to cryptocurrency investors in different jurisdictions.

Keywords: Cryptocurrency · legal protection · regulations · investors and consumers

B. A. M. Alareeni and I. Elgedawy (Eds.): *Artificial Intelligence (AI) and Finance*, SSDC 488, pp. 753–765, 2023.
https://doi.org/10.1007/978-3-031-39158-3_70

1 Introduction

The growing interest and adoption of cryptocurrencies, such as Bitcoin, Ethereum, and Lite coin, have attracted numerous investors and consumers worldwide. Despite the potential benefits, including decentralization, anonymity, and fast transactions, there is a risk of losing funds due to theft, fraud, or hacking. Therefore, legal protection for cryptocurrency investors and consumers is crucial to ensure their security and confidence in the market. In addition, cryptocurrencies are relatively considered a new financial innovation that has been gaining popularity in recent years. While cryptocurrencies present opportunities for investment and wealth creation, they also pose a significant risk to consumers due to the lack of legal protections in the industry. This is because cryptocurrency is a digital or virtual currency that uses encryption techniques to secure and verify transactions and to control the creation of new units. Unlike traditional currencies, which governments back, cryptocurrencies are decentralized and operate independently of central authorities. The rise of cryptocurrencies has been a subject of interest and concern for legal scholars, regulators, and policymakers.

The increasing popularity of cryptocurrencies has attracted attention from regulators around the world, resulting in some countries, new laws and regulations aimed at protecting investors and consumers in the crypto industry. Cryptocurrency is a complex and rapidly evolving area of law, and there are many legal issues surrounding its use, particularly in consumer legal protection.

One of the main challenges with cryptocurrency is that it operates in a largely unregulated space, which makes it difficult to apply traditional consumer protection laws. Most governments do not recognize cryptocurrencies as legal tender, meaning they are not subject to the same consumer protections as traditional currencies. As a result, investors in cryptocurrencies are exposed to a range of risks that are not present in more traditional investment opportunities.

However, some countries have begun to take steps to regulate cryptocurrencies and protect investors. For example, the United States has taken a more active approach to regulating cryptocurrencies, with the Securities and Exchange Commission (SEC) and the Commodities Futures Trading Commission (CFTC) both playing a role in overseeing the industry. Other countries such as Japan, Switzerland, and Singapore have also implemented regulatory frameworks to address some aspects of cryptocurrency trading and investment.

Regarding consumer legal protection, one area of concern is fraudulent activities and scams. Some investors have fallen victim to Ponzi schemes, fake ICOs (Initial Coin Offerings), and other fraudulent activities related to cryptocurrencies. These scams often involve promises of high returns or guaranteed profits, but they are designed to separate investors from their money.

Another area of concern for cryptocurrency investors is the security of their investments. Cryptocurrencies are stored in digital wallets, which can be vulnerable to hacking and theft. Investors should be given the right to protect their digital wallets through assurance procedures of authentication and other security measures.

The legal protection of cryptocurrency investors is a complex and rapidly evolving area. Investors should be protected by law and regulations with proper security measures to protect their digital wallets. These regulations need to stay up-to-date with the latest developments in cryptocurrency. Furthermore, cryptocurrencies, like Bitcoin and Ethereum, have grown in popularity and financial value since their inception. However, they have been the subject of legal and regulatory debates, with many jurisdictions struggling to determine how to classify and regulate them. In addition, decentralization has become a crucial element of the cryptocurrency industry, and its significance has grown as the popularity of cryptocurrencies increases.

The decentralization refers to the distribution of control and authority to the users. Namely, it does not refer to a central entity or authority that significantly deviates from the traditional financial system. The decentralization concept emerged in response to the centralized nature of traditional financial system, which has been criticized for its lack of transparency and the need for intermediaries.

1.1 Problem Statement

Cryptocurrencies are decentralized digital currencies that use block chain technology for secure transactions. They are not regulated by a central authority, freeing them from government intervention. The most popular cryptocurrency is Bitcoin, but thousands of other cryptocurrencies are available in the market.

Despite the potential benefits of cryptocurrency, the lack of regulation has created a significant risk to investors and consumers. In recent years, several high-profile cases of fraud and theft in the industry have left investors and consumers vulnerable to losses. Therefore, the lack of regulation and potential risks associated with cryptocurrencies, such as volatility, security, and the potential for their use in illegal activities, create serious concerns. As with any emerging technology or financial instrument, individuals need clear legal protection to mitigate the risks of investing in or using cryptocurrencies.

2 Literature Review

This literature review explores existing research on the legal protection of cryptocurrency investors and consumers worldwide. It investigates the challenges regulators face in developing legal frameworks for cryptocurrency and examines the effectiveness of these regulations in safeguarding investors and consumers. This literature review will examine studies that discuss the unregulated and illegitimacy of cryptocurrencies in legal jurisdictions. Furthermore, this literature review aims to examine the legal protection of cryptocurrency investors and consumers. According to [1] one of the biggest challenges is determining the legal status of cryptocurrencies, [2] suggested that some countries classify cryptocurrencies as assets or commodities, while others consider them to be securities or currency.

[3] stated that the ambiguity results in legal uncertainty, making it difficult for regulators to enforce laws consistently. However, several countries have implemented legal frameworks to protect cryptocurrency investors and consumers. For instance, in the United States, the Securities and Exchange Commission (SEC) has issued guidelines on initial coin offerings (ICOs), aiming to ensure that issuers comply with securities laws [4]. Japan has introduced a licensing system for cryptocurrency exchanges, requiring

them to comply with anti-money laundering (AML) and know-your-customer (KYC) regulations [5].

In Europe, the European Union's Fifth Anti-Money Laundering Directive (AMLD5) requires member states to regulate cryptocurrency exchanges and custodian wallet providers [4]. The directive requires these entities to register with the relevant authorities and comply with AML and KYC requirements. The financial sector has debated the effectiveness of cryptocurrency regulations in protecting investors and consumers. Some researchers argue that existing regulations fall short of adequately protecting consumers. For example, [6] argue that AML and KYC regulations do not necessarily guarantee that investors will not fall victim to fraud or theft, as these regulations do not necessarily address the underlying vulnerabilities of the cryptocurrency ecosystem.

Moreover, some researchers suggest that regulations may stifle innovation by limiting the ability of start-ups to raise funds through ICOs [4]. Others argue that regulations create barriers to entry, limiting small investors' ability to participate in the cryptocurrency market [1]. Furthermore, the decentralized nature of cryptocurrencies presents a challenge for regulators in holding intermediaries accountable for their actions. Cryptocurrency exchanges, for example, operate in a largely unregulated environment, making it challenging to enforce consumer protection laws.

3 The Study Method

This paper adopts a comparative legal analysis of the regulatory frameworks and legal precedents in different countries and jurisdictions. This involves a comprehensive review of the existing laws, regulations, and policies related to cryptocurrency trading and investment and analyzing case law and court decisions related to fraud, theft, and other forms of misconduct involving cryptocurrencies. Furthermore, a qualitative case study approach is used to analyze specific instances of legal disputes involving cryptocurrency investors, including the outcomes of the cases. In order to provide a detailed, in-depth analysis of the legal protections available to cryptocurrency investors and the challenges and limitations of existing legal frameworks, The paper utilized the analyzing legal frameworks approach to introduce a thorough analysis of the legal frameworks governing cryptocurrencies and block chain technology which provide insights into the legal protections available for investors. This includes reviewing relevant laws and regulations in various jurisdictions, such as securities, anti-money laundering, and consumer protection laws.

In addition, case studies will be used to examine how legal frameworks are being applied in real-world scenarios. This involves studying lawsuits involving cryptocurrency exchanges or token issuers to understand the legal protections available to investors in different jurisdictions. Nonetheless, analysis of regulatory frameworks and enforcement mechanisms will provide insights into the legal protections available for investors. This includes examining the role of regulatory bodies in enforcing laws and regulations and investigating cases of fraud or other illegal activities in the cryptocurrency industry.

4 Questions

1. Are there legal regulations to protect the participants in cryptocurrency trading activities?

2. What is the future of the legal protection of cryptocurrency investors in different jurisdictions worldwide?

5 Objectives

3. Investigating the legal protection for cryptocurrency investors, including fraudulent activities and scams.
4. Examining how legal frameworks are being applied in real-world scenarios.
5. Providing a detailed and in-depth analysis of the legal protections available to cryptocurrency investors and the challenges and limitations of existing legal frameworks.

6 Legal Frameworks for Cryptocurrency

The regulatory framework for cryptocurrencies varies globally, and it is often unclear or inadequate to provide sufficient legal protection for investors and consumers [7]. According to [8], the legal status of cryptocurrencies differs significantly among jurisdictions, with some countries outright banning or restricting their use while others adopt a laissez-faire approach. Moreover, the absence of a global regulatory framework exacerbates the challenges of enforcing legal protection across borders [9].

[10] Analyzed the legal frameworks for regulating cryptocurrencies in the United States and the European Union (EU). The author found that the US has a fragmented approach, with each state adopting its regulatory measures. However, some federal agencies, such as the Securities and Exchange Commission (SEC), have established guidelines for cryptocurrency trading, mining, and offerings. On the other hand, the EU has a more harmonized approach, with the European Securities and Markets Authority (ESMA) providing guidelines on Initial Coin Offerings (ICOs) and the EU's Fifth Anti-Money Laundering Directive (5AMLD) requiring cryptocurrency exchanges to register and comply with KYC/AML regulations [11].

In Australia, cryptocurrencies are considered to be a form of property and are subject to capital gains tax. The Australian Securities and Investments Commission (ASIC) has also issued guidance on the regulation of initial coin offerings and cryptocurrencies, including warnings about potential risks for investors [12]. However, there are well-established concerns about the lack of regulation and potential risks associated with cryptocurrencies, such as volatility, security, and the potential for their use in illegal activities.

Many authors have noted that one of the defining features of cryptocurrencies is their unregulated nature. Cryptocurrencies operate outside of traditional financial systems and are not subject to the same regulations as other financial instruments, such as stocks and bonds. Some authors argue that this lack of regulation makes cryptocurrencies more attractive to investors, as they offer the potential for high returns without the oversight of government agencies [13, 14].

However, the unregulated nature of cryptocurrencies has also been a source of concern for regulators, who worry that investors may be vulnerable to fraud, hacking, and other risks. For example, some authors have discussed the potential for "pump and dump" schemes in which investors artificially inflate the value of a cryptocurrency and then sell off their holdings, leaving other investors with significant losses [16, 17].

7 Cryptocurrency Between Legal Tender and a Means of Exchange

While there is no clear-cut definition or legal status for cryptocurrencies, it is identified as a digital asset that has been evolving over the last decade. It is often used as a means of exchange or a store of value [17]. However, it is not considered a legal tender in most countries. Legal tender refers to a currency that a government recognizes as a valid means of payment for debts and taxes [18]. While cryptocurrency can be used to purchase goods and services, it is not backed by a government or central authority, meaning its value is not guaranteed. Cryptocurrencies like Bitcoin, Ethereum, and others are decentralized, which means they operate on a distributed network of computers rather than being controlled by a central authority [19]. Therefore, it is not widely accepted as a means of payment for goods and services, but it can be used in some cases. Cryptocurrencies are often seen as investments or speculative assets, fluctuating based on market demand and supply [20].

Legal tender referred to the official currency recognized by a government and used as a payment medium for debts and taxes [21]. On the other hand, cryptocurrencies are decentralised and operate independently of any government, central authority, or financial institution. Instead, they operate using a decentralized network of computers, which collectively manage and validate transactions [22]. They are typically used as a means of exchange, similar to traditional currency, but are not backed by any physical commodity or sovereign guarantee. Despite this, cryptocurrencies have gained widespread acceptance and use as a means of exchange, particularly among individuals and businesses that operate in the digital realm. They are often used to facilitate transactions in online marketplaces or for cross-border payments, as in the case of the Commonwealth Bank of Australia (CBA) [23], and are increasingly being integrated into mainstream financial systems.

7.1 Legal Protection, Obligatory Transitional Progress Towards Regulated Status

One of the main challenges for the legal protection of cryptocurrency investors is that cryptocurrencies are not yet recognised as legal tender in many countries, which means they do not have the same legal protections as traditional currencies. This lack of legal recognition makes it difficult for investors to seek legal redress in case of fraud, theft, or other illegal activities related to cryptocurrencies. Moreover, cryptocurrency transactions often occur on decentralised platforms that are difficult to regulate or monitor, making it challenging for authorities to investigate and prosecute fraudulent activities. This lack of regulatory oversight can make it easier for scammers to create fraudulent cryptocurrency schemes that prey on unsuspecting investors.

Although the legal framework for cryptocurrencies is still developing, some legal protections are available for investors and consumers in some jurisdictions. Standard legal protections include fraud and consumer protection laws, contract laws, and securities regulations. Fraud and consumer protection laws aim to prevent fraudulent activities, misrepresentation, and unfair practices that can harm investors and consumers. According to [24], investors and consumers can file a complaint with the relevant regulatory authorities, such as the SEC, if they suspect fraudulent activities or deceptive practices in the cryptocurrency market. Similarly, contract law can offer legal protection if there is a breach of contract between the parties involved in a cryptocurrency transaction [25].

Furthermore, securities regulations can provide legal protection if cryptocurrencies are deemed securities under the applicable laws. For example, the Howey Test is used by the SEC to determine whether an ICO is a security, and if so, the issuer must comply with securities regulations, such as registering with the SEC and providing disclosure statements [26]. Nonetheless, the legal protection of cryptocurrency investors and consumers is an evolving area of law, with many countries and regulatory bodies still trying to develop a framework for the industry.

8 Protection for Cryptocurrency Investors and Consumers Through Different National Legal Lenses

The legal protection of cryptocurrency investors varies depending on the jurisdiction and the specific regulations in place. Generally speaking, the legal protection of cryptocurrency investors is not as well-established as that of traditional investors, such as those who invest in stocks and bonds. Some countries have investor protection laws that apply to cryptocurrencies, providing legal remedies for investors harmed by fraudulent activities or other illegal activities related to cryptocurrencies. These laws can include consumer protection laws, anti-fraud laws, and securities laws [27]. However, here are a few examples of the legal protection for cryptocurrency investors and consumers in different legal systems:

In the United States, on specific occasions, despite that the regulators avoided categorising cryptocurrency as "money", the Securities and Exchange Commission (SEC) has proactively regulated the industry by guiding how existing securities laws apply to cryptocurrencies [28].

In addition to the SEC, several other regulatory bodies worldwide are working on developing frameworks for regulating cryptocurrencies [29]. Cryptocurrencies and Securities Regulation. Cornell Journal of Law and Public Policy, 27(1), 105–143. In Europe, the European Securities and Markets Authority (ESMA) has issued guidance on the regulatory treatment of cryptocurrencies [30]. Japan has been at the forefront of cryptocurrency regulation in Asia, with the government recognising Bitcoin as a legal payment method in 2017 [31]. Japan has also implemented a licensing system for cryptocurrency exchanges to meet particular security and compliance standards [32].

Additionally, some countries have consumer protection laws that apply to cryptocurrency transactions, such as the UK's Consumer Rights Act, which protects consumers who purchase goods or services using cryptocurrencies. Furthermore, in the UK, the Financial Conduct Authority (FCA) has established a regulatory framework for cryptocurrency companies, requiring them to register with the agency and comply with anti-money laundering (AML) and know-your-customer (KYC) rules. The FCA has also warned consumers about the risks of investing in cryptocurrencies and advised them to conduct due diligence before investing [33]. In Germany, the Federal Financial Supervisory Authority (BaFin) has classified cryptocurrencies as financial instruments and imposed strict regulations on cryptocurrency exchanges and custodians [34].

In France, the Autorité des Marchés Financiers (AMF) has issued guidelines on the regulation of initial coin offerings (ICOs) and warned consumers about the risks of investing in cryptocurrencies. The AMF has also created a "Blacklist" of cryptocurrency websites suspected of fraud [35]. In the EU, the European Securities and Markets Authority (ESMA) has warned consumers about the risks of investing in cryptocurrencies and advised them to be cautious. ESMA has also called for a unified regulatory approach to cryptocurrencies across the EU member states.

These examples show that European countries are taking steps to protect cryptocurrency investors and consumers, but there is still a need for a unified regulatory approach to ensure consistent protection across the EU.

9 Court Cases on Legal Protection of Cryptocurrency Investors

Several European court cases have addressed the legal protection of cryptocurrency investors and consumers. Here are some notable ones:

[36]: In this case, the Court of Justice of the European Union (CJEU) held that the "right to be forgotten" under EU data protection law, applies to search engine results include links to websites containing personal data. This case is relevant to cryptocurrency investors because it underscores the importance of privacy and data protection in online financial transactions.

[37]: This case concerned using pre-ticked checkboxes to obtain consent for online data processing. The CJEU held that such checkboxes do not constitute valid consent under EU data protection law. While this case did not specifically address cryptocurrency, it is relevant to the industry because it highlights the importance of obtaining clear and informed consent from users.

Case C-593/18, Virgílio Pereira v. Banco Santander Totta SA: In this case, the CJEU held that cryptocurrency transactions are subject to the same consumer protection rules as traditional financial transactions. The court also held that consumers must be provided with clear and accurate information about the risks associated with cryptocurrency investments. This case is important for cryptocurrency investors because it clarifies their legal rights and protections.

-Case C-522/18, [36]: This case concerned the classification of Bitcoin as a currency or a commodity for tax purposes. The CJEU held that Bitcoin is a currency and, therefore, exempt from value-added tax (VAT). This case is relevant to cryptocurrency investors because it clarifies the tax treatment of their investments.

-AA v Persons Unknown [37]: This case involved an investor who had his Bitcoin stolen from an online wallet. The court recognised cryptocurrency as property and granted a proprietary injunction to the investor, giving him legal protection against the theft of his cryptocurrency.

-[38]: In this case, the court issued a similar order as in AA v Persons Unknown, recognising cryptocurrency as property and granting an interim proprietary injunction to prevent the transfer of stolen cryptocurrency.

[39]: This case dealt with a dispute between an investor and a cryptocurrency exchange. The court recognised cryptocurrency as a form of currency and held

the exchange liable for not following its terms and conditions regarding converting cryptocurrency into fiat currency.

B2C2 Ltd v Quintet Ltd: This case involved a dispute between two cryptocurrency exchanges over a trade execution error. The court held that the error was a breach of contract and awarded damages to the plaintiff.

Robertson v Persons Unknown: In this case, the court recognized the jurisdiction of the English courts to grant freezing orders and injunctions concerning cryptocurrency held in wallets or exchanges outside the UK. These cases show that English law increasingly recognizes the legal protection of cryptocurrency investors and consumers, treating cryptocurrency as property or currency and granting injunctions and damages in cryptocurrency disputes.

Overall, these cases demonstrate that European courts are beginning to grapple with the legal challenges posed by cryptocurrencies and are taking steps to protect the rights of investors and consumers.

10 Findings

Cryptocurrency investors and consumers have limited legal protections in cryptocurrencies' complex and often the unregulated world. The decentralized nature of cryptocurrency and lack of regulatory oversight makes it difficult to enforce consumer protection laws and prosecute fraudsters. Here are some of the issues that make consumer legal protection in the cryptocurrency market a complex and challenging area:

Lack of regulation: Cryptocurrencies are largely unregulated in many countries, making it difficult for investors to seek legal recourse in the event of fraud or other illegal activities. Holding exchanges, wallets, or other service providers accountable for their actions can be difficult without clear regulatory frameworks.

Cybersecurity risks: Cryptocurrencies and their underlying block chain technology are vulnerable to cyberattacks, hacking, and other security breaches, which can result in the loss of funds or sensitive personal information.

Volatility: Cryptocurrencies are highly volatile and can experience large fluctuations in value in short periods. This can lead to significant losses for investors and consumers who may not fully understand the risks involved.

Lack of transparency: The lack of transparency in the cryptocurrency market can make it difficult for consumers to understand the risks associated with investing in cryptocurrencies. The complexity of the technology and the lack of clear information can make it challenging for investors to make informed decisions.

Cross-border transactions: Cryptocurrencies can be easily transferred across borders, making enforcing consumer protection laws in different jurisdictions difficult.

11 Conclusion

The legal framework for cryptocurrencies is still evolving, with different countries taking different approaches. While some have embraced cryptocurrencies and introduced regulations to protect investors and consumers, others have taken a more cautious approach, with some even banning them outright. As cryptocurrencies become more mainstream, regulatory frameworks will likely continue to develop and evolve to provide better protection for investors and consumers.

Nonetheless, the legal protection of cryptocurrency investors and consumers is still evolving, and the regulatory framework is not yet comprehensive enough to ensure their safety and confidence in the market. However, fraud and consumer protection laws, contract laws, and securities regulations offer some legal protections for investors and consumers in some jurisdictions. A more harmonized and coordinated approach to regulating cryptocurrencies is needed to provide adequate legal protection for investors and consumers worldwide. The norm of decentralization in cryptocurrency has both benefits and challenges. While it can provide greater transparency and efficiency, it raises questions about the legitimacy and regulatory oversight. As the industry continues to evolve, it will be important to balance decentralization and regulation to ensure that cryptocurrency can reach its full potential as a legitimate and secure financial instrument.

The lack of legal protections for cryptocurrency investors and consumers has created a significant risk in the industry. While there is no universal framework for regulating cryptocurrencies, there are regulatory bodies and countries working on developing guidelines and laws to protect investors and consumers. As the industry evolves, investors and consumers must stay informed and aware of the risks involved. Therefore, the legal protection of cryptocurrency investors and consumers remains a complex issue. The lack of legal clarity around cryptocurrencies' status and the crypto ecosystem's decentralized nature makes it challenging for regulators to develop effective legal frameworks. While existing regulations in some jurisdictions aim to protect investors and consumers, they may fall short of addressing underlying vulnerabilities in the cryptocurrency ecosystem. Future research could investigate innovative ways of regulating the cryptocurrency industry, balancing innovation and risk while safeguarding the interests of investors and consumers.

The decentralization of the cryptocurrency industry offers many benefits, including increased security, transparency, and accountability. However, it also poses some challenges, including the lack of regulation and scalability issues. The cryptocurrency industry must address these challenges to ensure the industry's sustainable growth and maintain the benefits of decentralization. Cryptocurrencies' unregulated and illegitimate nature is a significant concern for legal scholars and policymakers. However, some scholars argue that regulating cryptocurrencies is a complex and challenging task that requires a nuanced approach. In any case, the rise of cryptocurrencies has significant implications for the future of finance and the legal system.

Finally, the lack of regulation in the cryptocurrency market can be a double-edged sword. On the one hand, it allows for greater innovation and experimentation in the industry. On the other hand, it can also make the market more volatile and increase the risk of fraud and scams.

References

1. Piacentini, V.: Legal protection of crypto-currency investors and consumers. Eur. J. Risk Regulat. **10**(2), 342–350 (2019)
2. Blandin, A., et al.: Global Cryptoasset Regulatory Landscape Study [Review of Global Cryptoasset Regulatory Landscape Study], pp. 1–123. Cambridge Centre for Alternative Finance (2019). https://www.jbs.cam.ac.uk/wp-content/uploads/2020/08/2019-04-ccaf-global-cryptoasset-regulatory-landscape-study.pdf
3. Ganne, E.: Can block chain revolutionize international trade?, pp. 1–163. The World Trade Organization (2018). https://www.wto.org/english/res_e/booksp_e/blockchainrev18_e.pdf
4. Groppelli, A.A., Sassano, A.E.: The regulation of cryptocurrencies: challenges and innovations. J. Financ. Crime **27**(2), 476–488 (2020). https://doi.org/10.1108/JFC-12-2018-0156
5. Kawase, K.: Japan's Crypto Market Grows as Regulators Clamp Down. Nikkei Asian Review (2018). https://asia.n
6. Smith, S.J., Ryan, L.H.: Stress and coping with discrimination and stigmatization (2019)
7. Hallak, I.: Markets in crypto-assets (MiCA) [Review of Markets in crypto-assets (MiCA)]. https://www.europarl.europa.eu, pp. 1–11. EPRS | European Parliamentary Research Service (2022). https://www.europarl.europa.eu/RegData/etudes/BRIE/2022/739221/EPRS_BRI(2022)739221_EN.pdf
8. Kariyawasam, R., Chan, D.: Cryptocurrency regulation (2020)
9. Dewey, J.: Blockchain and Cryptocurrency Regulation. Global Legal Group Ltd. (2019)
10. Hussein, A.: Cryptocurrencies: regulation and legal protection in the United States and European Union. J. Money Launder. Control **22**(4), 545–558 (2019)
11. Thomson Reuters. Cryptocurrency Regulations by Country, Cryptos on the Rise 2022 [Review of Cryptocurrency Regulations by Country, Cryptos on the Rise 2022]. https://www.thomsonreuters.com, pp. 1–37. Thomson Reuters (2022). https://www.thomsonreuters.com/en-us/posts/wp-content/uploads/sites/20/2022/04/Cryptos-Report-Compendium-2022.pdf
12. Australian Securities and Investments Commission. INFO 225: Initial coin offerings and cryptocurrency (September 2017). https://asic.gov.au/regulatory-resources/digital-transformation/initial-coin-offerings-and-cryptocurrencies/
13. Cheah, E.T., Fry, J.: Speculative bubbles in bitcoin markets? An empirical investigation into the fundamental value of bitcoin. Econ. Lett. **130**, 32–36 (2015)
14. Dwyer, G.P.: The economics of bitcoin and similar private digital currencies. J. Financ. Stab. **32**, 83–94 (2018). https://doi.org/10.1016/j.jfs.2017.11.006
15. Hileman, G., Rauchs, M.: Global Cryptocurrency Benchmarking Study. University of Cambridge Centre for Alternative Finance (2018). https://www.jbs.cam.ac.uk/fileadmin/user_upload/research/centres/alternative-finance/downloads/2018-ccaf-global-cryptoasset-benchmarking.pdf
16. Jiang, Z., Liang, X.: Detecting pump-and-dump schemes in cryptocurrency markets. arXiv preprint arXiv:1811.10134 (2018)
17. Leopold, R.: Cryptographic assets and related transactions: accounting considerations under IFRS [Review of Cryptographic assets and related transactions: accounting considerations under IFRS]. In: Vollmann, P. (Ed.), pp. 1–23. PwC network (2019). https://www.pwc.com/gx/en/audit-services/ifrs/publications/ifrs-16/cryptographic-assets-related-transactions-accounting-considerations-ifrs-pwc-in-depth.pdf
18. Perkins, D.W.: Cryptocurrency: The Economics of Money and Selected Policy Issues [Review of Cryptocurrency: The Economics of Money and Selected Policy Issues], pp. 1–31. Congressional Research Service (2020). https://sgp.fas.org/crs/misc/R45427.pdf

19. Bank of England. Discussion Paper: Central Bank Digital Currency Opportunities, Challenges and Design [Review of Discussion Paper: Central Bank Digital Currency Opportunities, Challenges and Design], pp. 1–57. Bank of England (2020). https://www.bankofengland.co.uk/-/media/boe/files/paper/2020/central-bank-digital-currency-opportunities-challenges-and-design.pdf
20. Treasury, H.M.: UK Regulatory Approach to Cryptoassets and Stablecoins: Consultation and Call for Evidence [Review of UK Regulatory Approach to Cryptoassets and Stablecoins: Consultation and Call for Evidence], pp. 1–46. HM Treasury (2021). https://assets.publishing.service.gov.uk/government/uploads/system/uploads/attachment_data/file/950206/HM_Treasury_Cryptoasset_and_Stablecoin_consultation.pdf
21. Alvarez, F.E., Argente, D., Patten, D.V.: Are Cryptocurrencies Currencies? Bitcoin as Legal Tender in El Salvador [Review of Are Cryptocurrencies Currencies? Bitcoin as Legal Tender in El Salvador], pp. 1–50. Becker Friedman Institute for Research in Economics (2022). https://bfi.uchicago.edu/wp-content/uploads/2022/04/BFI_WP_2022-54.pdf
22. Bains, P., Ismail, A., Melo, F., Sugimoto, N.: Regulating the Crypto Ecosystem: The Case of Unbacked Crypto Assets [Review of Regulating the Crypto Ecosystem: The Case of Unbacked Crypto Assets], pp. 1–47. International Monetary Fund (2022). https://www.imf.org/-/media/Files/Publications/FTN063/2022/English/FTNEA2022007.ashx
23. Centre for the Edge, Deloitte. The Future of Exchanging Value Cryptocurrencies and the trust economy [Review of The Future of Exchanging Value Cryptocurrencies and the Trust Economy] (2015), pp. 1–68. https://www2.deloitte.com/content/dam/Deloitte/au/Documents/technology/deloitte-au-technology-exchanging-value-cryptocurrencies-trust-economy-210116.pdf
24. Gerova, V., Mavrantonis, C.: Cryptocurrency regulations: legal protection for investors. J. Govern. Regulat. **7**(4), 45–52 (2018)
25. The Law Commission. Smart legal contracts Advice to Government (2021). https://s3-eu-west-2.amazonaws.com/lawcom-prod-storage-11jsxou24uy7q/uploads/2021/11/Smart-legal-contracts-accessible.pdf
26. Baris, J.G., Klayman, J.A.: Blockchain basics for investment managers: a token of appreciation. Rev. Commod. Secur. Regulat. **51**(6–7), 67–83 (2018). http://www.shearman.com/-/media/files/perspectives/2018/03/blockchain_basics_for_investment_managers_token_of_apprtion_march_23_2018_101818.pdf?la=en&hash=A91839D14053E8D16F2136A0BEF71D09DB9654D4
27. Murphy, P.: Regulating cryptocurrencies in the United States: current issues and future directions. J. Financ. Crime **25**(2), 350–357 (2018)
28. Ernst & Young. Cryptocurrency: Compliance and Controls [Review of Cryptocurrency: Compliance and Controls], pp. 1–12. Ernst & Young (2021). https://assets.ey.com/content/dam/ey-sites/ey-com/en_us/topics/forensic-integrity-services/ey-us-cryptocurrency-compliance-and-controls-nov-2021.pdf?download. https://assets.ey.com
29. Hoffman, D., Zaring, D.: Cryptocurrencies and securities regulation. Cornell J. Law Public Policy **27**(1), 105–143 (2018)
30. European Securities and Markets Authority. Advice Initial Coin Offerings and Crypto-Assets [Review of Advice Initial Coin Offerings and Crypto-Assets], pp. 1–49. European Securities and Markets Authority (2019). https://www.esma.europa.eu. https://www.esma.europa.eu/sites/default/files/library/esma50-157-1391_crypto_advice.pdf
31. Nishibe, M.: Cryptocurrencies in Japan: legal and regulatory perspectives. J. Bus. Law **2019**(6), 446–464 (2019)
32. Arora, G.: Cryptoasset Regulatory Framework in Japan [Review of Cryptoasset Regulatory Framework in Japan], pp. 1–7. SSRN (2020). https://papers.ssrn.com/sol3/papers.cfm?abstract_id=3720230

33. Pay UK. Consumer Protections in Payments [Review of Consumer Protections in Payments], pp. 1–24. Pay UK (2020). https://newseventsinsights.wearepay.uk/media/b4gb3xj5/consumer-protections-in-payments-summary-paper.pdf. https://newseventsinsights.wearepay.uk
34. French, T., Stettner, B.: Anti-Money Laundering Regulation of Cryptocurrency: U.S. and Global Approaches [Review of Anti-Money Laundering Regulation of Cryptocurrency: U.S. and Global Approaches], pp. 14–28 (2019). https://www.allenovery.com/global/-/media/allenovery/2_documents/news_and_insights/publications/2019/5/anti-money_laundering_regulation_of_cryptocurrency.pdf?rev=9acc0566a47445689255263b9b623849&hash=D2B47FC421B3C0C90508629360C7F2B0
35. Clifford Chance. Fintech: The Evolving French Regulatory Landscape [Review of Fintech: The Evolving French Regulatory Landscape]. Clifford Chance (2022). https://www.cliffordchance.com/content/dam/cliffordchance/briefings/2022/06/client-briefing-crypto-assets.pdf
36. Case C-264/18, Google v. CNIL. Judgment of the Court (Grand Chamber) of 1 October 2019. Bundesverband der Verbraucherzentralen und Verbraucherverbände - Verbraucherzentrale Bundesverbande.V. v Planet49 GmbH. Request for a preliminary ruling from the Bundesgerichtshof. Reference for a preliminary ruling—Directive 95/46/EC—Directive 2002/58/EC—Regulation (EU) 2016/679—Processing of personal data and protection of privacy in the electronic communications sector—Cookies—Concept of consent of the data subject—Declaration of consent by means of a pre-ticked checkbox. Case C-673/17. Court reports – general. ECLI identifier: ECLI:EU:C:2019:801
37. Judgment of the Court (Fifth Chamber) of 22 October 2015. Skatteverket v David Hedqvist. Request for a preliminary ruling from the Högstaförvaltningsdomstolen. Reference for a preliminary ruling—Common system of value added tax (VAT)—Directive 2006/112/EC—Articles 2(1)(c) and 135(1)(d) to (f)—Services for consideration—Transactions to exchange the 'bitcoin' virtual currency for traditional currencies—Exemption. Case C-264/14
38. Liam David Robertson v Persons Unknown [2019] EWHC 3556 (Comm)
39. Vorotyntseva v Money-4 Ltd (T/A Nebus.com) and others [2018] EWHC 2596 (Ch)

The Impact of IT Controls on Reducing Tax Evasion in Jordan "From the Point of View of the Auditors"

Ibraheem Jodeh[1(✉)], Ziad Al-theebeh[1], Sultan Alshourah[1], Bassem Yassin[2], and Ahmad Qotb[3]

[1] Faculty of Economics and Administrative Sciences, Zarqa University, Zarqa, Jordan
{ijodeh,z.altheebeh,salshourah}@zu.edu.jo
[2] Income and Sales Tax Department, Amman, Jordan
basem.yaseen@istd.gov.jo
[3] Department of Accountancy, University of Business and Technology, Jedda, Kingdom of Saudi Arabia
a.qotb@ubt.edu.sa

Abstract. The study tackled the impact of information technology controls (general controls and applied controls) in reducing tax evasion rate in Jordan. This study used current and historical data to identify trends and relationships. Component of the study consisted of (250) ISTD auditors, (170) auditors representing 68% of the study component were involved as a sample study. To measure the study variables, a questionnaire was texted and distributed over the study sample. The number of retrieved questionnaires that were valid for analysis was (154). Descriptive statistics methods were used in analyzing the primary data.

The study realized several endings, the most important is that information technology controls are available in their dimensions in a moderate degree in the ISTD, and there is an impact for both general controls and applied controls in reducing tax evasion.

The study recommended that the ISTD should revise and reevaluate the information technology controls, (the general and the applied ones) to improve the obtainability of such controls paralleled with the best-updated technologies. The study also recommended revision of the accounting systems in one hand and the revision of regulatory controls related to information technology on the other hand to ensure consistency in the implementation of tasks and to help in reduction of tax evasion cases.

Keywords: IT controls · Tax Evasion · General Controls · Applied Controls

1 Introduction

A rapid development in information technology in the world is witnessed; this rapid development has led to changes and transformations in the work environment and methods. Among the most prominent of these transformations is the increased reliance on

B. A. M. Alareeni and I. Elgedawy (Eds.): *Artificial Intelligence (AI) and Finance*, SSDC 488, pp. 766–776, 2023.
https://doi.org/10.1007/978-3-031-39158-3_71

information technology in its various activities and the transactions of private and governmental institutions as well. Because of technological development, information technology has added advantages in conducting complex calculations and the ability to run a tremendous amount of transactions in less time and least effort, and reducing operational and computational errors in large extents. It also contributed to a decrease in dependence on the human factor, thus leading to the increase of accuracy and value of data. These results were achieved through the presence of easy-to-install and easy - to- use of adaptable programs which made them widely dependable by institutions in exchanging information through networks (the Internet and the Intranet), thus shortening the time and effort spent.

Government institutions resort to setting up controls to achieve their goals and protect their assets in order to ensure proper implementation of business and to realize the policies and instructions of senior management. Therefore, concern in government sectors had increased the importance of keeping up with the implementation of controls to the requirements of information technology, technological and professional developments working to protect public and private institutions from the risks they face. Moreover, the importance of control over institutions has emerged due to the emergence of a new control environment that differs from the traditional control environment [1].

Governments pursue through their programs to establish the e-government; this digital approach will enhance the efficiency, effectiveness and procedures of the current complex tax accounting that depends on the paper and it will combat the phenomenon of tax evasions. Governments are continuing to research and study these issues in order to find appropriate solutions in a way that can contribute to reducing the complexity of the tax accounting process, as well as finding solutions that can reduce the chances of tax evasion by taxpayers [2].

2 The Study Problem

Occurrence of tax evasion is one of the biggest challenges that governments focus on and seek to combat by various means, as reducing it has become one of the most important challenges facing the Income and Sales Tax Department (ISTD) in the world and in Jordan in particular. The adoption of the Jordanian government in general, and the ISTD in particular to the electronic approach that is based on reducing the use of paper transactions and relying on information technology and accompanying applications and software, especially in light of the Corona pandemic, has reinforced the need to rely on controls and control procedures related to information technology, electronic systems and applications [3] and [4].

2.1 The Following Questions Can Determine the Problem of the Study

First. What is the level of availability of information technology controls, which include (general controls and applied controls) in the ISTD in Jordan?

Second. What is the impact of IT controls in reducing tax evasion in the ISTD in Jordan?

3 Study Objectives

This study intent to achieve the following objectives:

- Disclosure of the level of application of IT control, which includes (general controls and applied controls) in ISTD.
- Identifying the effect of applying information technology controls, which include (general and applied controls) in reducing tax evasion in the ISTD.

4 Importance of the Study

This study will make it possible to assess the level of availability of information technology control procedures, and to know the impact of IT controls in reducing tax evasion, which will enhance the possibility of taking measures that develop these controls and improve their effectiveness.

5 Study Hypotheses

The Main Hypothesis: There is no statistically significant effect of IT controls in reducing tax evasion in the ISTD.

5.1 The Following Two Sub-hypotheses Are Descended from This Main Hypothesis

- **First.** There is no statistically significant effect of the general IT controls in reducing tax evasion in the ISTD.
- **Second.** There is no statistically significant effect of the applied IT controls in reducing tax evasion in the ISTD.

6 Procedural Terms

- **Tax Evasion:** The taxpayer adopts methods that violate the tax law in order to bypass the process of paying part or all of the tax due when the due date is due, using fraudulent methods, false statements, or fraud [5].
- **Applied Controls:** These are controls related to existing transactions and data, and include: validating data entry, encoding data to be sent, and processing controls.
- Application controls are special controls for the information technology system, and include the application of business rules in the program, and thus provide controls for the inputs, processing process, outputs, and main data [6].
- **General Controls:** These are the controls that are concerned with the environment in general where they are created and kept up. The framework, operation, and management of IT systems are established by IT general controls, which also give confidence that the control objectives have been met to a suitable degree. Information systems general controls are controls on processing operations in the IT environment that assist the development, implementation, and operation of the IT system. They are not restricted to particular applications or individual transactions. Typically, they have to do with information technology governance, organizational and structural controls, physical and environmental controls, IT operations, information security, and business continuity [6].

7 Literature Review

7.1 IT Control

IT control includes two types of IT controls:

General Controls. The foundation of IT controls, general controls focus on the general environment in which IT systems are created, managed, operated, and maintained. IT general controls create a framework for overall control of IT activities and offer assurance that control objectives are met to a satisfactory level.

Examples of general controls include the development and implementation of an information systems strategy, an information systems security policy, the formation of an information technology steering committee, and the organization of information systems personnel to separate roles. General controls are applied by using a variety of tools such as policies, procedures, and directives, as well as setting up an appropriate management structure, including the entity's IT systems management structure [7].

The general controls of information systems are not limited to specific applications or individual transactions, but rather controls over processing operations in the IT environment that support the development, implementation and operation of the IT system, they are usually related to IT governance, organization and structure controls, environmental and physical controls, IT operations, information systems security, and business continuity [6].

In order to ensure that the facility's operational activities are carried out in a proper manner, general controls relate to the framework, policies, and procedures that are applied to all information technology systems, or a sizable portion of them, within the facility. The context in which applied systems and controls operate is created by general controls. The following are the standard controls [6]:

- Planning and managing security programs throughout the facility.
- Computer usage controls.
- Controls for the development, maintenance and modification of application programs.
- Systems software controls.
- Separation of job duties.
- Continuity of services.

Applied Controls. The application controls are specific controls that differ according to the application, and are related to existing transactions and data, and they include: - Validating data entry, encoding the data to be sent, processing controls, etc., for example, that the input controls in the online payment application, the expiration date of the credit card is greater than the date of the transaction, and the information entered is encrypted. The application controls are special controls for the IT system, and include the application of business rules in the program and thus provide controls for the inputs, processing process, outputs, and main data [8] and [9].

Also, the applied controls are the framework, guidelines, and practices that are specific to the applied systems and are closely tied to software applications. Generally speaking, these controls are meant to stop and fix mistakes and anomalies when data passes through information systems. Separate software programs, such as those for accounts payable, inventory, salaries, grants, loans, etc., which are programs made to deal with the operation of data within them, are included in controls. The applied control controls are typically created to stop, spot, and fix mistakes and inconsistencies so that data moves easily through information systems. Input, processing, and output are the three steps of the processing cycle that are divided by the applied control controls and mechanism of data flow through information systems [10].

7.2 Internal Control in Light of Information Technology

Internal control in the electronic information technology environment is defined as the use of contemporary electronic methods and means to keep an eye on organizational operations and transactions in order to reduce costs, time, and effort while still achieving the desired goals [11]. Electronic control is defined as the electronic methods through which operating companies and institutions monitor their internal and external activities in order to save time, effort and money [12, 13] and [14]. Focused attention on internal control in light of information technology on the following aspects [15] and [16]:

- Prior Control of Inputs.
- Monitoring the fed in data.
- Monitoring the computer programs used in processing data, analyzing and displaying information.
- Monitoring the outputs and the way they are presented and interpreted.

7.3 Electronic Accounting Information Systems and It's Control

Accounting information systems can be defined as "a group of parts and sub-systems that are linked to each other, and work as one group that interacts and overlaps the relationships between each other, so that each part of it depends on the other in achieving the goals sought by the comprehensive system of accounting,

Therefore, it is a network of procedures Linked to each other and governed by sound principles and rules, done in an integrated manner with the aim of providing data and information to decision makers in a manner appropriate to his needs [17].

The impact of information technology tools on accounting information systems can be seen in the way that the structure of how financial event characteristics are received and preserved has changed, and its reflection on the form of the outputs, this prompted accountants and designers of accounting information systems to change the structures of data preservation to comply with the development taking place, consequently, Facilitating processing operations, and producing outputs in a way that enables the user to maximize the returns of his decisions [18].

The use of information technology, including electronic accounting systems, has become common in the public sector. The expansion and development in the world of information technology is putting greater challenges and demands on the control environment in light of information technology [19].

7.4 Tax Evasion

Taxes are considered an important source of revenue for most states of the world, but the tax system in these states is facing a major problem represented by tax evasion that states are trying to combat and overcome. Because of the technological development that the world is witnessing, a new concept has emerged that requires methods differ from what is used in traditional systems, also these states are trying to enact tax legislation that combats tax evasion.

Taxes are considered one of the main tributaries of the general revenues of states, and with the increase in the financial, economic, social and political goals of these states they are trying to eliminate the suffering of tax evasion problem, addressing this issue will raise the state's tax revenue, expand the functions it performs, and enable these states to interact in public life to deliver the finest services possible to the entire society and achieve economic progress. In the event of poor tax revenue due to tax evasion, the state will be unable to fulfill its obligations and the level of public services provided to citizens will decline [20].

Tax evasion is when the taxpayer uses methods that violate the tax law in order to bypass the process of paying part or all of the due tax using fraudulent methods, false statements, or by fraud [5]. Tax evasion is also seen as - The taxpayer conceals taxable income or activity with the intent of reducing the due by manipulating accounting records or documents that are audited by tax auditors [21].

8 Study Methodology

The analytical descriptive strategy was used by the researchers in this study to cover the theoretical side by making references to books, periodicals, electronic articles, and specialist publications. The development of a questionnaire that serves the study's objectives and was given to the study sample in order to gather responses from the respondents covered the field side of the study. Using the responses of the respondents as a source, the hypotheses were then tested to arrive at the results and then the recommendations.

8.1 Study Tool

The researchers determined that the "questionnaire" is the most appropriate instrument to accomplish the goals of this study based on the type of data to be collected and the methodology used in the investigation. As a result, the researchers prepared a questionnaire to accomplish the goals of the study. The researchers employed the closed questionnaire method, which establishes the potential answers to each question in relation to each of the study's phrase axes. Questionnaire, who determines the possible responses to each question against each of the five axes statements, a list bearing the following choices according to the five-point Likert scale, in addition to an open question at the end of the questionnaire. The study questionnaire consists of two parts:

First. It included the demographic data of the respondents, which are the educational qualification, years of experience, scientific specialization, and job title.

Second. It included paragraphs for three domains, each of which expressed a variable of the study variables, and asked the sample to answer it by choosing the appropriate answer alternative for them, and the distribution of the variables was as follows:

- Independent variables IT controls included the two parts, the general control and the applied control
  - General Controls: (16) paragraphs have been allocated for them.
  - Applied controls: (14) paragraphs have been allocated for them.
- The dependent variable represented in the field of reducing tax evasion, for which (20) paragraphs were allocated.

8.2 The Study Component and Its Samples

The study's cadre of auditors, which comprises of all (250) employees of the Jordanian ISTD, were given a questionnaire at random. The researchers sent (170) questionnaires via email to the ISTD auditors. After evaluating the questionnaires, it was discovered that four of the researcher's (158) questionnaires were invalid for statistical analysis since the answers to their paragraphs were left blank. The study sample was made up of (154) auditors, with a rate of (68%) for the study component, because there were (154), or 90.6%, of valid questionnaires for analysis.

8.3 Statistical Methods Used

The researchers encoded the data and fed it into the computer using Statistic Package for Social Sciences (SPSS), where numerous acceptable statistical procedures were utilized in this software, to accomplish the study's goals and analyze the data acquired.

8.4 Presentation and Analysis of Results and Testing of Hypotheses

Independent Variable (IT Controls)
The ISTD's information technology control controls with its dimensions (generic controls, applied controls), as represented by the arithmetic mean of the study sample's responses, is displayed in Table 1:

Table 1. The replies to the aspects in the field of information technology control, averaged throughout the study sample

Rank	No	Dimension	Arithmetic averages	Evaluation score
1	1	General controls	3.34	moderate
2	2	Applied controls	3.34	moderate
The field of IT controls as a whole			3.29	moderate

The general arithmetic mean for the area of information technology control controls, according to Table 1, was (3.29), with a medium rating. Also, the respondents' replies to the aspects in the field of information technology control controls had arithmetic average that varied between (3.23–3.34) The "General Controls" dimension, which had the highest overall evaluation score, came in first with an arithmetic average of (3.34), followed by the "Applied Controls" dimension, which came in second with an arithmetic mean of (3.23).

Dependent Variable (Reduction of Tax Evasion)
The arithmetic averages of the study sample's responses to the ISTD's questions about reducing tax evasion were calculated; the general arithmetic mean for the dimension (reducing tax evasion) was (3.43), with a medium evaluation score, while the arithmetic mean of respondents' responses to the paragraphs in "reducing tax evasion" ranged between (2.49–3.92).

The Results Related to Testing the Hypotheses of the Study
The researchers identified the issue with the internal correlations of the independent variables before beginning hypothesis testing. In order for the Variance Inflation Factor (VIF) values to be less than (10) for all of the independent variables and the Tolerance values to be more than (0.05), tolerance was used for each of the independent variables as well as the VIF test. Table 2 illustrates this.

Table 2. The Results Related to Testing the Hypotheses of the Study

Dimension	Tolerance	VIF
General controls	0.7280	1.3730
Applied controls	0.7300	1.3699

It is evident from Table 2 that there is no significant correlation between the independent variable's dimensions because all of the independent variable's dimensions' variance inflation coefficient (VIF) test values were less than 10, while all of the independent variable's tolerance coefficient of variance test values were greater than (0.05).

As a result, they can all be included in the regression model. After ensuring that there is no strong correlation between the dimensions of the independent variable, it is important to know which of these dimensions has a statistically significant impact on the dependent variable as well as the percentage of that impact, if any.

Regarding the Outcomes of the Sub-hypotheses, the Findings Revealed What Follows

1. The general control measures have a statistically significant impact on reducing tax evasion because the values of (Beta, T) were (0.32, 3.74), respectively. As a result, the first sub-hypothesis is accepted in the established formula.
2. Since the values of (Beta, T) were respectively (0.19, 2.20), which are statistically significant values, the second sub-hypothesis in the established formula is accepted. This is because there is a statistically significant influence of the imposed control controls in reducing tax evasion.

9 Results

The results showed that the information technology controls are available in the ISTD, in both parts combined, to a medium degree, and this is consistent with Jodeh [22]. Whereas, the availability of general controls was to a medium degree, and the applied controls were also to a medium degree, but the availability of general controls was more available than the applied controls within the average level and this is consistent with Al-Khashab [19].

The study revealed that the ISTD in Jordan's procedures, comprising general control procedures and controls and applied regulatory procedures and controls, are carried out to a moderate degree with regard to decreasing the phenomena of tax evasion. Also, the study showed that the electronic control procedures are commensurate with the nature of the accounting system in the department and at the lowest level as one of the procedures related to applied control controls.

The study revealed that one of the least effective ways for ISTD to lessen tax evasion was the development of programs and systems linked to the government management information system, activating communication between it and the process of selecting audit samples electronically with pre-organized rules and foundations. Also, the study found that, from the perspective of the department auditors, information technology controls generally had a positive impact on minimizing tax evasion in the ISTD. The general controls were more successful than the applied controls in cases where both the general control controls and the applied control controls had an impact on reducing tax evasion, which is consistent with Qasim [21].

10 Conclusion and Recommendations

Considering the study's findings, the following suggestions might be made:

1. The ISTD revise and reevaluate the information technology control controls, in both parts, the general control controls, and the applied control controls, with the aim of enhancing the availability of these controls according to the best updated technologies, and filling the shortage in order to provide them at the best levels to ensure the achievement of the required control.
2. A feedback system should be created to ensure taking into account the observations made by the department's auditors about the negatives and problems they face related to technological systems and their controls and oversight.
3. That the accounting systems be reviewed in one hand, and the regulatory controls related to information technology on the other hand, to ensure consistency in the implementation of tasks and help in the success of the control that leads to the detection and reduction of cases of tax evasion.

References

1. Ghazal, M.: The impact of information technology on the effectiveness of the internal control system, a field study. Master's thesis, Martyr Hama Lakhdar El-Wadi University, Faculty of Economic, Commercial and Management Sciences, Algeria (2017)

2. Jodeh, I.: The impact of the application of E- government and accounting information systems in achieving the objectives of tax systems (the Jordanian tax system as a case study). Algerian J. Hum. Soc. Sci. **8**(2) (2022)
3. Jodeh, I., Khalaf, A.: The impact of applied of IFRS (9) standard on the financial performance of Jordanian commercial banks in light of the corona pandemic. In: Alareeni, B., Hamdan, A. (eds.) ICBT 2022. LNNS, vol. 495, pp. 1000–1019. Springer, Cham (2023). https://doi.org/10.1007/978-3-031-08954-1_84
4. Al-Thebeh, Z., Alkabbji, R., Alrazim, A., Abu-Bakr, I.: The impact of asset management efficiency on the financial leverage and return on investment in small production enterprises in view of the Covid-19 pandemic. Case Study: the Jordanian ministry of social development - Zarqa governorate. In: Alareeni, B., Hamdan, A. (eds.) ICBT 2022. LNNS, vol. 495, pp. 1011–1019. Springer, Cham (2023). https://doi.org/10.1007/978-3-031-08954-1_85
5. Yahya, M.A.: The quality of external auditing and its role in reducing tax evasion, a feld study on auditing offices. Master's thesis, Al-Neelain University, Sudan (2021)
6. INTOSAI IT Audit Working Group and INTOSAI Development Initiative. Information Technology Audit Guide for Supreme Audit Institutions (2014)
7. Jodeh, I., Ismeil, A.: Assessing the impact of greening information technology tools and technologist on organizational systems sustainability in service companies. Glob. J. Econ. Bus. Refaad Stud. Res. **12**(5) (2022)
8. Jodeh, I.: The effect of the human resources cycle in improving the effectiveness of the accounting information system in Jordanian banks. Account. Growing Sci. **7**(6), 1417–1424 (2022)
9. Al-Thebeh, Z., Al-Jeady, O.A., Al-Eryani, M.M.: The extent of application general principles for corporate governance in Yemen commercial banks. Zarqa J. Res. Stud. Human. **10**(1) (2010)
10. Al-Thebeh, Z.: Impact of accounting information system on reducing liquidity risk in Saudi banks comparative research between Islamic banks and commercial banks. Acad. Account. Financ. Stud. J. **1**(23), 1–11 (2019)
11. Steven, W., Ray, E., Kruchko, F.: Monitoring your employees: how much can you do and what should you do when you uncover wrongdoing? Washington Avenue, Suite 305. Baltimore, Maryland (2010)
12. Yamin, I. Y.: The impact of the application of enterprise resource planning system on banks financial performance (a case study of Arab bank and Islamic bank of Jordan). Zarqa J. Res. Stud. Human. **18**(1) (2018)
13. Amshahr, I.A.: Censorship in the shadow of information technology. Money Trade J. Trade Club Libya (574) (2017)
14. Bagsir, A.T., Saeed, H.: The development of internal control methods and their impact on the characteristics of accounting information in non-profit institutions operating in the Kingdom of Saudi Arabia. Ramah J. Res. Stud. **41**, 109–136 (2020)
15. Abu Ghaya, S.: Recent trends in monitoring and auditing of electronic information systems. Book Distribution Center, Al-Azhar Trade, Cairo (2009)
16. Laudon, K., Laudon, J.: Organization and Technology in the Networked Enterprise, 4th edn. Prentice Hall, New Jersey (2002)
17. Romney, M.B., Steinbart, P.J., Scott, L.S., David, A.W.: Accounting information systems. In: 15th Global Edition (2021)
18. Azril, A.H.: Information technology and the quality of accounting information. Account. J. Saudi Account. Soc. **18**(58) (2015)
19. Al-Khashab, Y.: The impact of information technology controls on risk control. Master's thesis, Faculty of Business, Accounting Department, Al al-Bayt University, Jordan (2016)
20. Kazem, A.Y.: Tax evasion in e-commerce, with reference to the experience in Iraq. J. Legal Polit. Sci. Diyala Univ. Iraq **9**(2) (2020)

21. Qasim, F.N.: The impact of electronic tax auditing in reducing tax evasion from the viewpoint of the ISTDAuditors in Jordan. Master's thesis, Amman Arab University, Jordan (2017)
22. Jodeh, I.N.: The Jordanian experience in the application of information and communication technology in the tax system. J. Industr. Econ. Algeria **1**(11) (2016)

Influencer Marketing Through Digital Platforms and Its Reflection on the Purchasing Response of Bahraini Youth

Merhan Mohsen Mohammed[1], Tamer M. Alkadash[2(✉)], Riadh Jeljeli[3], Faycal Farhi[3], and Osman Nassereldin Abdel Qadir[4]

[1] Mass Communication and Public Relations Department, College of Communication and Media Technologies, Gulf University, Sanad 26489, Kingdom of Bahrain
dr.merhan.mohsen@gulfuniversity.edu.bh

[2] College of Administrative and Financial Sciences, Gulf University, Sanad 26489, Kingdom of Bahrain
tamer.alkadash@gmail.com

[3] College of Communication and Media, AL Ain University, Abu Dhabi, UAE
{Riadh.jeljeli,faycal.farhi}@aau.ac.ae

[4] College of Mass Communication, Ajman University, Ajman, UAE
N.Ali@ajman.ac.ae

Abstract. Influencers play a major role through social media in influencing their followers and their purchasing response due to the trust and credibility they enjoy. This study aims at exploring the influencers' marketing strategies through digital platforms and how it is reflected in the purchase response of the Bahraini youth. It is carried out by exploring the motives that the Bahraini youth are exposed to, from the influencers and their ways of responding. This paper uses a random probability sample that was drawn from Bahraini youth. An electronic questionnaire was conducted on a sample of 250 respondents, who follow influencers through various digital platforms. Statistical data were analyzed through the SPSS program. The results of the study show (1) that there is a correlation between the intensity of exposure to the marketing content of influencers and the levels of increased purchasing response.

Keywords: Influencers Marketing · Social Platforms · Purchasing Response · Bahraini youth

1 Introduction

The rapid increase in the use of social media globally has created new opportunities for marketers, as communication media became persuasive digital marketing platforms of a diverse nature. Data published in Meta's advertising resources indicates that in early 2022 in Bahrain, the Facebook platform had 713.8 thousand users, Facebook's ad was equivalent to 40.8% of the local internet user base (regardless of age), YouTube platform had 1.44 million users, Instagram had 1.00 million users, TikTok had 699.5

B. A. M. Alareeni and I. Elgedawy (Eds.): *Artificial Intelligence (AI) and Finance*, SSDC 488, pp. 777–788, 2023.
https://doi.org/10.1007/978-3-031-39158-3_72

thousand users aged 18 and above,". Snapchat had 855.0 thousand users, and Twitter had 508.8 thousand users. (Global Digital Report-Bahrain 2022). Marketers in multinational companies believe that "The return on investment through influencer marketing is better than marketing channels,". With the increasing influence of social media platforms, influencer marketing has become an effective strategy for marketing brands that interest marketing managers around the world (LIU, 2021).

2 Literature Review

2.1 Influencer Marketing

Influencer marketing focuses on using influencers to provide content through social media in a variety of ways to launch a new product, and target new customers. It enhances customer relationships and increases its profits, creating different lifestyles. Social media influencers" are individuals who have effective and influential relationships. They have a large number of followers through social media. They have earned credibility as they provide paid content produced by many brands to promote their products according to different and varied methods, and through their branding. Their main goal is to convince their followers to buy products by influencing their attitudes and behaviors, and hence, convincing them to make a purchasing decision using their credibility as a focal point (Blanche et al., 2021; Dhanesh, 2019).

Influencers' credibility is achieved by sharing blogs and videos of their private lives with their followers, in addition to showcasing their skills, knowledge, and information regarding places, products, travel, etc. (Davisson et al., 2019). The rapid growth of smartphones sharing multiple content and images has led to a momentary increase in the effect of influencers; by sharing experiences and expertise (Argyris et al., 2020).

One of the psychological roles played by the influencer is that he/she works to strengthen "Self-enhancement". This is done by publishing recommendations, advice, and self-opinions accepted by his followers and fans, creating a sense of leadership, superiority, uniqueness, and influence (Malik et al., 2022).

One of the successful techniques of influencer marketing is also the reflection of the influencer's lifestyle on his page by publishing his activities and interests. He/She also focuses on certain brands that he uses or adapts in the service of his daily activities.

2.2 Influencer Marketing through Social Media Platforms

In light of the rapid hybridization of the media environment, which greatly contributed to the dissemination of social media, companies tend to employ effective means and methods to build influential and persuasive relationships with the public through interaction with influencers (Pang et al., 2016). The Association of National Advertisers (ANA) reported in April 2018 that nearly 58% of commercial brands had affiliations with about 25 influencers in 2017 (Conick, 2018).

In a survey conducted by a marketing agency (Mavrek) which is specialized in e-marketing for brands through influencers on social media platforms, the results showed that these managers see that marketing their brands through influencers on social media has increased customers' awareness of the brand, their loyalty, and their response towards it Soneji, et al. 2015). Consequently, the followers are ready to try and endorse the products that he/she displays Phil & Nandagiri, 2015).

Castillo & Fernandez (2019) study sought to determine the effectiveness of digital influencers in supporting brands through electronic word of mouth. They found out that the role of the influencers has been reinforced and reflected in the behavioral intention towards their brand, which enhances positively the brands' content interaction. Dalman et al. emphasized the effectiveness of influencer marketing. The methods used in influencer marketing vary through social media platforms.

2.3 Purchasing Response

The purchasing response is defined as the sum of knowledge, emotions, and behavior that is directed by the public toward adopting and purchasing products, services, or ideas to adopt certain behaviors (Dadwal et al., 2020). Erkan and Evans (2018) have identified four factors that influence consumer purchasing intentions, including the quality of the information provided, the credibility of the information, the usefulness of the product, and its reliability.

Influencers create change on four levels - creating awareness, increasing knowledge, recommending the product, and buying the product. Successful influencer marketing involves identifying the right type of influencer who will provide curated advice, stories, and suggestions to create engagement with the audience. They also provide influencers' recommendations to buy products or services as a motivation for followers to purchase (Chopra et al., 2018).

Alshurideh & Alwanan's (2022) studies agreed that there is a strong effect of influencers on their followers toward purchasing intention, through many and varied influences. Radwan et al. (2021) explained that youth in the UAE see that influencers have positive personal appeals, traits, and information to attract their followers and encourage them to buy. Furthermore, influencers push young people to search for more information about products and visit marketers' websites by supporting them with their opinions and recommendations for products and services.

The researchers emphasized the importance of the means of influencer interaction on social media platforms with their followers so that influencers can effectively manage their brands. The results showed that the attractiveness of the influencer and the positive affinity with the followers contribute to the possibility of predicting the attitude of the followers toward what he does or presents in a positive way (Taillon et al., 2020). Soneji et al. (2015) study found that 50% of Twitter users purchased the product based on the recommendations of influencers. This is also confirmed by the Manthiou et al. (2016) study which proved that the tweets of influencers had a greater impact on awareness and preference for the brand and the purchasing intent of consumers, as the results showed that 40% of Twitter users made purchasing decisions influenced by Twitter tweets.

Aldahrawy's (2019) study asserts the importance of influencers as one of the tools of electronic marketing. It is a study that confirmed that Arab youth have a positive attitude towards marketing through famous social media sites. It has proved that there is a relationship between patterns of exposure to accounts of social media celebrities and the purchasing behavior of Arab youth, which all had an effective effect on the credibility of celebrity marketing influences their purchasing behavior.

The Problem Statement
The problem of the study is to identify the effectiveness of influencer marketing through digital platforms and its impact on the purchasing response of Bahraini youth.

Objectives of the Study

- To identifying the most important dimensions and motives of Bahraini youth exposure to influencers on social networking sites and their impact on levels of purchasing response.
- To exposing the dimensions of credibility - attractiveness, trust, and experience - influencers on social networking sites in a sample of young people and their relationship to their levels of purchasing response.
- To Identifying the relationship between the intensity of exposure to the marketing content of influencers and the levels of purchasing response to Bahraini youth.

3 Methodology

3.1 Type of Study

The study is one of the descriptive studies that aims to describe and analyze the relationships of the variables under study. It reveals the dimensions of influencer marketing through digital platforms and its reflection on the purchasing response of Bahraini youth.

3.2 Study Methodology

The sample survey, at its descriptive level, was adopted as it is used to describe and document current facts and the actual reality. The study also then uses analytical tools to collect data on the phenomenon that is being studied to know the relationship between its variables.

3.2.1 Data Collection Tools

The researcher relied on the following data collection. The researcher designed an electronic questionnaire and apply it to a sample consisting of 250 individuals who follow the marketing of influencers through digital platforms.

3.2.2 Study Standards: Several Scales are Designed as Follows

- A special measure of the intensity of exposure to marketing content on the pages of social media influencers through posing a set of questions consisting of (4) phrases, and another question consisting of (3) phrases. In addition, the researchers designed measure X from three high, medium, and low levels.

- A measure of the credibility of influencers: It was measured through (3) dimensions: attractiveness - trust - experience, and each dimension included two phrases.
- A special measure of the levels of purchasing response of the respondents was measured through (8) phrases, and another question consisting of (5) phrases and they were divided into a scale of three levels low level, medium level, and high level.

3.2.3 The Standards of Validity and Reliability

Three experts in communication and business administration were engaged to ensure the questionnaires' validity and reliability as the study uses the standards of credibility and Consistency to test the validity of the questionnaire form in data collection.

3.2.4 The Field Study Population and the Sample

The study population consists of Bahraini youth who follow marketing through influencers through digital platforms. The sample reached 250 individuals.

3.3 Theoretical Framework of the Study

3.3.1 The Source Credibility Model

Credibility plays an important role in convincing the target audience of the company's products and attracting it to the brand. Via credibility, it is possible to influence beliefs, opinions, and trends. The use of influencers is one of the most prominent sources of advertising. The influencer's fame is translated into a prestigious position and credibility enjoyed by the target audiences (Seiler & kucza., 2017). Hoveland model believes that sources (influencers) who are characterized by experience, trust, and credibility are convincing, and the source who has the experience and high confidence can change trends. Information provided by a trusted source (such as social media influencers) can influence consumers' beliefs, opinions, and behaviors, and influencers who are seen as experts are more persuasive and able to influence consumer purchasing intent. Experience also refers to the experience and knowledge of supporters in a particular area of interest, and trust is related to the integrity, and credibility of the source as seen by followers. (Hussain & Ali, 2021).

3.3.2 Profiles of Respondents

In terms of age, sex, gender, education, socio-economic level, and rate of social media, the results of the study indicated that the sample included 62% females compared to 38% of males. The highest category of the study sample was concentrated in the age group from 18 to 30 years, which is the age group most affected by influencers and more inclined to them, and about the level of education, the study sample focused on the category of university students, the results of the study also highlighted that 51% of young people always follow the pages of influencers on social networking sites. The intensity of exposure to influencer pages is high by 45.2%, which reflects the passion of Bahraini youth for influencers, which is consistent with a study (Aldahrawy, 2019), which confirmed the high exposure to influencer pages by Arab youth, especially through the Instagram platform, and the study also confirmed that there is a relationship between exposure patterns to social media celebrity accounts and the purchasing behavior of Arab youth, which all had an effective impact towards the credibility of celebrity marketing affecting behavior. Purchasing them.

The results also highlighted that the most prominent platform on which influencers are followed is Instagram with 66%. This is consistent with the Global Digital Report-Bahrain 2022, which confirmed that Instagram is the most followed platform in Bahraini society with one million followers.

The first hypothesis: There is a statistically significant correlation between the intensity of exposure to the marketing content of influencers and the levels of purchasing response (Table 1).

Table 1. .

The intensity of exposure to influencer marketing content	Pearson correlation coefficient	Moral level	Significance
Ratings of the Purchasing response	****0.345**	**0.000**	**Significance**
Sampling Unit	**250**		

The previous table shows that there is a direct, statistically significant correlation between the intensity of exposure of Bahraini youth to marketing content provided through digital platforms and the levels of purchasing response of Bahraini youth. The value of the Pearson correlation coefficient was (0.345**) at a level of (0.000), which is significant in a way that reflects the existence of a statistically direct relationship. The higher the intensity of exposure to influencer marketing content, the higher the levels of purchasing response. Thus, the validity of this hypothesis is proven, and this result can be explained by the fact that the more Bahraini youth are exposed to marketing content, the greater their purchasing response, and purchasing response.

The second hypothesis: There are statistically significant differences between the demographic characteristics of the study sample respondents (gender - educational level - marital status - age - income) and levels of purchasing response.

The results of the second hypothesis proved that there were statistically significant differences between the study sample respondents in terms of demographic characteristics represented in monthly income, type of education only, and purchasing response levels. Regarding the type of education, the value of the t-test was (3.260) at a degree of freedom (248) and a level of significance (0.051), which indicates that there are statistically significant differences between the study sample respondents according to the type of education regarding their purchasing responses, and the differences were in favor of private education with an arithmetic mean of (2.2553), then the government with an arithmetic mean of (2.0968).

The third hypothesis: There is a statistically significant correlation between the extent of the credibility of Bahraini youth towards influencers and the levels of purchasing response (Table 2).

Table 2. .

The credibility of Bahraini youth towards influencers	Pearson correlation coefficient	Moral level	Significance
Ratings of the Purchasing response	****0.571**	**0.000**	**Significance**
Sampling Unit	**250**		

The data of the previous table shows the extent to which there is a statistically significant correlation between the extent of the credibility of Bahraini youth towards influencers and the levels of purchasing response. The value of the Pearson correlation coefficient was (0.571**) at the level of significance (0.000), which indicates the existence of a statistically significant correlation between the extent of the credibility of Bahraini youth towards influencers and the levels of purchasing response, which is a direct relationship; That is, the higher the credibility of Bahraini youth towards influencers, the higher the levels of purchasing response. Thus, this hypothesis is confirmed and valid.

The data of the following table reflects the phrases that describe the elements of the credibility of influencers among Bahraini youth, which were divided into three categories (attractiveness), then (experience), and then (trust) based on the arithmetic mean value. About the phrases related to the credibility of influencers as a result of their appearance, the phrase "I follow influencers because they are They have an appearance and acceptance that arouses my interest in getting to know what they offer in their content of products and services (the arithmetic average is 2.37)) and the phrase "I like the influencers' personality, which contributes to improving the image of products and services in my mind for a long time" came to achieve an arithmetic average of (2.34).

The second element of credibility was represented in experience, and reflected the phrase "influencers have sufficient experience and skills that motivate me to follow up on their content and different brands." My average was (2.24), while the phrase "social media influencers are fully aware of what they provide." Content, especially about trademarks "The arithmetic average value was (2.22). The third element of credibility is trust, which was reflected in the following statements: "I feel confident in most of the information provided by influencers on social media, especially the brands of products and services for which they provide content." With an average value of (2.21), and the phrase "I follow influencers for their commitment to the principles of the Islamic religion and the values and customs of society, which makes me feel confident and safe in the influencers' personalities through their pages" with an average value of (2.20).

Most of the attributes reflect that the respondents have high credibility with the influencers and the marketing content they provide. The dimensions of credibility are evident in the attractive appearance of the influencers, and their great experience with the content they provide. And in third place is the confidence in what the influencers provide, which is consistent with the study (Barbu & Cmeciu, 2019), which emphasized that empathy, trust, and expertise built up by influencers are key elements of influencer credibility across their platforms (Table 3).

Table 3. Dimensions of the credibility of influencers among followers

Statement	Average	Deviation	Percentage
I follow influencers because they have an appearance and acceptance that interests me to learn about their content	**2.37**	**.695**	**79%.0**
I like the personality of influencers which contributes to improving the image of products and services in my mind for a long time	**2.34**	**.689**	**78%**
Influencers have enough experience and skills to push me to follow their content and from different brands	**2.24**	**.703**	**74%**
I feel confident that most of the information provided by the influencers is credible and valid, especially the brands of products and the content services provided	**2.21**	**.738**	**73%**
I follow the influencers for their commitment to the values and customs of society, which makes me feel confident and safe in the personality of the influencers through their pages	**2.20**	**.746**	**73%**

4 Discussion and Results

Influencer marketing through digital platforms is one of the most important means and tools used by marketers as a marketing strategy to promote the marketing content they provide. Undoubtedly, influencers have become a force to be reckoned with through digital platforms, and the use of influencers on social media platforms has become one of the modern and effective entrances in the field of marketing that are used to influence the audience of influencers' followers, their purchasing intentions and trends (Al kurd et al., 2022).

The study also highlights the diversity of platforms followed by Bahraini youth, and the Instagram platform topped the interest of young people greatly to follow influencers through it. This finding is consistent with the results of the study (Ahmadi et al. 2022), which highlighted the power of influencers through the Instagram platform without other platforms. The results also highlighted that Snapchat, Tik Tok, YouTube, Twitter, and Facebook, came to the next stage which is diverse and provides different content for their platforms.

The respondents also stressed that the most prominent marketing methods that they prefer by influencers are the methods that depend on providing factual information about products in the first place. This is because young people always prefer to get real information from influencers who must be consistent with the nature of the content provided.

Among the marketing methods are also methods that rely on humor and fun, the ones that depend on adopting new lifestyles, and methods that make young people feel highly distinguished. Such results are parallel with the Chopra et al., (2021) study, which confirmed that choosing the appropriate influencer for the nature of the product or service is very important.

The results of the study also discussed the most prominent areas of influencers that affect young people, including social content influencers, comedy content, fashion and beauty, tourism, travel, and cooking, respectively. Such results are consistent with the study of Chopra et al., (2021) which emphasized the influencers' effect on the opinions and tendencies of followers in the fields of travel, photography, fashion, fashion, and the adoption of different lifestyles.

The results also reviewed the most prominent forms of marketing that will affect the purchasing behavior of young people an d provide marketing content, presented through texts, videos, live streams, live chat, and stories and promote brands by displaying features, Reels telling stories.

The study also stressed the importance of interaction between influencers and the marketing content provided to young people through the "LIKE". This is consistent with Ahmadi et al. (2022) study which confirmed the extent to which users are keen to use LIKE towards publications for influencers, especially audiences who follow heavily from followers modestly. "Sharing the link to friends" came in second place in the interaction with the influencers and then writing a comment, and sending a message to the influencer personally, which is consistent with the 2021 study (Gerrath and USREY). It confirms that the posts published by influencers shape the follower's trends and expectations of followers, and they also encourage the followers to evaluate their purchasing preferences.

5 Conclusion

The results of the study emphasized the active role of influencers in the purchasing response of Bahraini youth in high proportions due to heavy exposure to the marketing content provided by influencers, as well as the importance of credibility (trust, experience, and attractiveness of influencers) in the content offered by influencers, and its impact on purchasing response through the opinions, proposals, and recommendations of influencers, which enhance their own brand, and the brand of products. The study also established a relationship of statistically significant differences between the study sample researchers in terms of the demographic characteristics of monthly income, type of education only and levels of purchasing response.

6 Recommendations and Future Research

Researchers recommend that brands make better use of marketing their products by marketing an influencer's lifestyle while emphasizing credibility and transparency with the public. The researchers also recommend the importance of providing material and moral incentives to the audience provided by the influencer and the followers' brand to maintain their integration and continuity of interaction. There should be a trend to leverage influencers in sustainable development plans aimed at promoting and marketing green lifestyle through social media In light of the researchers' findings, several future studies and research can be put forward. First, the possibility of employing influencers to further support sustainability issues in communities. Second, to study the extent to which influencers make greater use of AI technologies in content provided to be a greater attraction for audiences and influence their intentions and purchasing response. Third, it is recommended to explore the ability to conduct in-depth interviews with influencers to learn about their main strategies in self-presentation and how to manage their accounts through means of social communication.

References

Ahmadi, A., Fakhimi, S., Ahmadi, Y.: Instagram celebrities and positive user responses. The mediating role of user "like". J. Contemp. Market. Sci. **5**(1), 65–80. Emerald Publishing Limited (2022). 2516–7480. https://doi.org/10.1108/JCMARS-01-2021-0002

Aldahrawy, M.: Arab Youth Trends Towards Influence Marketing and Its Relationship to Purchasing Behavior: Field Study, Arab Journal for Media and Communication Research, Ahram Canadian University, issue 27 (2019)

Al Kurdi, B., Alshuridehb, S., Akour, I., Tareq E., AlHamad, A., Alzoubif, H.: The effect of social media influencers' characteristics on consumer intention and attitude toward Keto products purchase intention. Int. J. Data Netw. Sci. **6**(4), 1135–1146 (2022). https://doi.org/10.5267/j.ijdns.2022.7.006

Alwana, M., Alshuridehb, M.: The effect of digital marketing on purchase intention: moder-ating effect of brand equity, international journal of data and network science. Int. J. Data Netw. Sci. **6**, 837–848 (2022)

Argyris, Y.A., Wang, Z., Kim, Y., Yin, Z.: The effects of visual congruence on increasing consumers' brand engagement: an empirical investigation of influencer marketing on Instagram using deep-learning algorithms for automatic image classification. Comput. Hum. Behav. **112**(1), 106443 (2020). https://doi.org/10.1016/j.chb.2020.106443

Barbu, R., Cmeciu, C.: Crisis communication in Romania and social media influencers and followers. Case study: patient's death at the sanador hospital. J. Media Res. **12**(3), 5–17 (2019). https://doi.org/10.24193/jmr.35.1

Blanche, D., Casalo, L.V., Flavián, M., Sánchez, S.I.: Understanding influencer marketing: the role of congruence between influencers, products, and consumers. J. Bus. Res. **132**(2), 186–195 (2021). https://doi.org/10.1016/j.jbusres.2021.03.067

Castillo, D.J., Fernández, R.S.: The role of digital influencers in brand recommendation: examining their impact on engagement, expected value, and purchase intention, Int. J. Inf. Manag. (49), 366–376 (2019). https://doi.org/10.1016/j.ijinfomgt.2019.07.009

Chopra, A., Avhad, V., Jaju, S.: Influencer marketing: an exploratory study to identify antecedents of consumer behavior of millennials. Bus. Perspect. Res. **9**(1), 77–91 (2021). https://doi.org/10.1177/2278533720923486

Conick, H.: How to win friends and influence millions: the rules of influencer marketing. Market. News **52**(7), 36–45 (2018)

https://www.ama.org/marketing-news/how-to-win-friends-and-influence-millions-the-rules-of-influencer-marketing/

Dalman, D., Chatterjee, S, Min, J.: Negative word of mouth for a failed innovation from higher/lower equity brands: Moderating roles of opinion leadership and consumer testimonials. J. Bus. State Univ. wol(115) (2020). https://doi.org/10.1016/j.jbusres.2020.04.041

Dadwal, S., Jamal, A., Harris, T.,Brown, G., Raudhah, S.: 2020. chnology and Sharing Economy-Based Business Models for Marketing to Connected Consumers. In Handbook of Research on Innovations in Technology and Marketing for the Connected Consumer. https://doi.org/10.4018/978-1-7998-0131-3.ch004

Dhanesh, G., Duthler, G.: Relationship management through social media influencers: effects of followers' awareness of paid endorsement. Public Relat. Rev. **45** (2019).https://doi.org/10.1016/j.pubrev.2019.03.002

Davidsson, I., Jurgell, L., Nilsson, I.: A qualitative content analysis of social media influencers' credibility, [Bachelor Thesis] (2019). http://www.diva-portal.org/smash/get/diva2:1323999/FULLTEXT01.pdf

De Veirman, M., Cauberghe, V., Hudders, L.: Marketing through Instagram influencers: e impact of number of followers and product divergence on brand attitude. Int. J. Advert. **36**(5), 798–828 (2017). https://typeset.io/papers/marketing-through-instagram-influencers-the-impact-of-number-14kofhq1md

Draganov, I.: Influencer Marketing: Consumer Perception on Influencers and Its Role in Digital Communication Management, M.S., Dissertation or Thesis, Webster University, United States – Missouri, pp. 1–84 (2018). https://www.proquest.com/openview/bfdc91c9f5c1afa992eb6f98218d515e/1?pq-origsite=gscholar&cbl=18750&diss=y

Enke, N., Borchers, S.: Social media influencers in strategic communication: a conceptual framework for strategic social media influencer communication. Int. J. Strateg. Commun. **13**(4), 261–277 (2019). https://doi.org/10.1080/1553118X.2019.1620234

Erkan, I., Evans, C.: Social media or shopping websites? The influence of EWOM on consumers' online purchase intentions. J. Mark. Commun. **24**(6), 617–632 (2018). https://doi.org/10.1080/13527266.2016.1184706

Gaweesh, K.S.: How famous Arab artists use social media in self-branding: a content analysis of self-presentation strategies on Instagram. J. Union Arab Univ. Media Commun. Technol. Res. (9), July-December 2022. https://doi.org/10.21608/JCTS.2022.257294

Gerrath, M.H.E.E., Usrey, B.: The impact of influencer motives and commonness perceptions on follower reactions toward incentivized reviews. Int. J. Res. Mark. **8**, 531–548 (2021)

Hussain, A., Ali, Z.: Examining the impact of social media influencer's credibility dimensions on consumer behavior, Student Thesis master degree (2021). http://www.diva-portal.org/smash/get/diva2:1631690/FULLTEXT01.pdf

-Jayasinghe, D.: Social media and influencer marketing strategies: a conceptual review on contemporary applications. Glob. Sci. J. **9**(3) (2021). https://www.researchgate.net/publication/350004633_Social_Media_and_Influencer_Marketing_Strategies_A_conceptual_Review_on_Contemporary_Applications

Downtown Magnificent Mile, Chicago, pp. 3–7. https://doi.org/10.1080/10641734.2018.1521113

Liu, S.: The impact of influencer marketing on brand engagement: a conceptual framework. In: Advances in Social Science, Education and Humanities Research, volume 615, Proceedings of the 2021 4th International Conference on Humanities Education and Social Sciences (ICHESS 2021) (2021). https://doi.org/10.2991/assehr.k.211220.383

Malik, A.Z., Thapa, S., Paswan, A.K.: Social media influencer (SMI) as a human brand - a need fulfillment perspective. J. Product Brand Manag. (2022).https://doi.org/10.1108/JPBM-07-2021-3546

Manthiou, A., Rokka, J., Godey, B., Tang, L.: How social media marketing efforts influence brand equity creation and its consequences: the case of luxury brands. In: Obal, M.W., Krey, N., Bushardt, C. (eds.) Let's Get Engaged! Crossing the Threshold of Marketing's Engagement Era. DMSPAMS, pp. 561–561. Springer, Cham (2016). https://doi.org/10.1007/978-3-319-11815-4_169

Philil, L., Nandagiri, V.: Impact of Influencers from Instagram and You Tube on their followers. Int. J. Multidiscip. Res. Mod. Educ. **4**(1), 61–65 (2018). s for this publication .at: https://www.researchgate.net/publication/323996049

Soneji, D., Riedel, A., Martin, B.: How Gordon Ramsay appeals to consumers: effects of self-concept clarity and celebrity meaning on celebrity endorsements. J. Strateg. Mark. **23**(5),

457–468 (2015). https://doi.org/10.1080/0965254X.2014.991349, https://info.mavrck.co/the-state-of-influencer-marketing-2016

Taillon, B.J., Mueller, S.M., Kowalczyk, C.M., Jones, D.N.: Understanding the relationships between social media influencers and their followers: the moderating role of closeness. J. Product Brand Manag. **29**(6), 767–782 (2020)

Vodak, J., Novysedlak, M., Čakanova, L., Pekar, M.: Who is influencer and how to choose the right one to improve brand reputation? Manag. Glob. Trans. **17**(2), 149–162 (2019). https://doi.org/10.26493/1854-6935.17.149-162, https://datareportal.com/reports/digital-2022-bahrain?rq=bahrain%202022

Seiler, R., Kucza, G.: Source credibility model, source attractiveness model and matchup-hypothesis – an integrated model. J. Int. Sci. Publ. **11** (2017)

COVID-19 Impact on Pharmaceutical Marketing Ethics: A Narrative Review

Abdul-Rahim El-Sharif[1](✉), Hamza Alhamad[2], Mohammad Abu Assab[2], Soumaia Echarif[3], and Yousef Alhayek[4]

[1] Interpretation of the Holy Quran, Zarqa University, Zarqa, Jordan
asharif@zu.edu.jo
[2] Department of Clinical Pharmacy, Faculty of Pharmacy, Zarqa University, Zarqa, Jordan
{halhamad,mabuassab}@zu.edu.jo
[3] Ministry of Health, Amman, Jordan
[4] Ministry of Municipality, Amman, Jordan

Abstract. The pharmaceutical industry is one of the leading economies that was and still is highly affected by COVID-19 consequences, including pharmaceutical marketing and its ethical issues. Despite many countries prohibiting or restricting direct-to-consumer advertising, the pandemic has prompted a change toward digital and remote marketing tactics with emerging ethical considerations in the trading and marketing of pharmaceuticals.

This narrative review aimed to evaluate the ethical challenges in the pharmaceutical industry in the aftermath of the COVID-19 pandemic and its impact on the availability and accessibility of pharmacological therapies for patients in Jordan. The COVID-19 pandemic has significantly impacted the pharmaceutical industry and marketing. The demand for prescription drugs, vaccinations, and medical equipment has increased, but so have the challenges regarding supply constraints, regulation changes, and ethical considerations in marketing tactics. The pandemic has also highlighted flaws in device regulatory norms and the need for critical reevaluation. The emergency decision-making process has also been impacted, with healthcare professionals facing new ethical challenges. Companies must ensure that their marketing efforts are conducted ethically, especially considering the increased demand for pharmaceutical products.

Keywords: Coronavirus · COVID-19 · pharmaceutical ethics · pharmaceutical marketing · marketing ethics · Jordan

1 Introduction

COVID-19 is a pandemic disease that significantly threatens public health (Alhamad et al. 2021). The pharmaceutical industry is a sector that has continuously been criticized in the past and nowadays, for its profits association, despite their willingness to cooperate and to develop treatments and vaccines (Emanuel et al. 2021). However, due to a global catastrophe, the pharmaceutical industry continues to increase its revenue. Additionally, the media tries to spread the idea that COVID-19 represents a once-in-a-lifetime business

B. A. M. Alareeni and I. Elgedawy (Eds.): *Artificial Intelligence (AI) and Finance*, SSDC 488, pp. 789–799, 2023.
https://doi.org/10.1007/978-3-031-39158-3_73

opportunity for these companies (Donthu, N., & Gustafsson, A. (2020). This would add to the marketing challenge these companies face and make the life of the marketing professionals working in these companies even more challenging.

Pharmaceutical marketing deals with promoting pharmaceutical companies' prescription drugs to healthcare professionals and consumers. (Willis & Delbaere, 2022). The ethics of pharmaceutical marketing is a complex and multi-faceted issue. Some may argue that pharmaceutical companies are responsible for educating healthcare professionals and consumers about their products to improve patient outcomes. (LaVan et al., 2021). Others would argue that pharmaceutical companies' primary goal is to maximize profits, which can lead to unethical practices such as overpromotion, underreporting of side effects, and price gouging. (Marmat et al., 2020). Additionally, the close relationship between pharmaceutical companies and healthcare professionals can lead to conflicts of interest and bias in prescribing decisions. (Rodwin, 2019).

One of the worst public health crises of the twenty-first century is the coronavirus outbreak. The COVID-19 outbreak was deemed a public health emergency of international concern by the World Health Organization (WHO) on January 30th, 2020, and a worldwide pandemic on March 11th. (Khan et al., 2020) In a relatively short period, almost 5 million instances of coronavirus disease have been documented by May 19th, 2020. Also, it has killed more people each day than it has infected, nearly in every country in the world. (Romano et al., 2021). The epidemic curve has been flattened by interventions, preventing a spike in demand for healthcare services. (Brandon, 2022). There is a broad worry regarding maintaining the ongoing supply of medicines for all individuals with the requirement to ensure the supply of essential medical and personal protective equipment to control the disease. (Bong et al., 2020).

Before COVID-19, sales representatives visiting doctors, direct-to-consumer (DTC) advertising, and live events like conferences were the mainstays of pharmaceutical marketing. Following the epidemic, many nations prohibited or restricted direct-to-consumer (DTC) advertising, and limitations on travel and in-person events prompted a change to digital and remote marketing tactics. (Rahul, & Prakash, 2022). Additionally, the pandemic has heightened public interest in and concern about health, which has caused pharmaceutical marketing initiatives to place a more considerable emphasis on health education and disease awareness. (Husain et al., 2020).

It is essential to investigate the ethics of pharmaceutical trading after COVID-19, as there may be some ethical issues with how certain drugs or treatments are traded or distributed. (Saxena et al., 2023). Global supply chains were disrupted, and increased demand for specific treatments, governments, private entities, and even individuals may be looking to acquire or trade certain medications or treatments in ways that are not in line with ethical standards of conduct. It is crucial to investigate the ethical considerations in trading or acquiring pharmaceuticals after COVID-19. (Atkinson et al., 2020).

The future of the pharmaceutical trade after COVID-19 is uncertain, as the industry is still feeling the impact of the pandemic. It is expected that the pharmaceutical industry will experience changes in its operations, with increased digitalization, automation, and advanced analytics to drive decision-making. (Droždz et al., 2023). Also, the industry will likely become more focused on providing personalized treatments and solutions for patients and developing new products to meet the needs of a rapidly changing market. (Dang et al., 2020).

Although there have not yet been any significant disruptions in the availability of pharmaceuticals, the current problem of medicine shortages 6–9 could worsen because of the global public health emergency. (Romano et al., 2021).

Countries' capacity to guarantee patients' access to pharmacological therapies may be impacted by anticipated manufacturing constraints (such as the closure of factories due to quarantine) and supply chain interruptions. (Romano et al., 2021) In addition to a decline in demand for some products, pharmaceutical companies have a more challenging time locating the raw materials needed for manufacturing and reaching their target markets and end users. (Moosivand et al., 2019). Furthermore, the demand for medications was drastically reduced by the economy's shutdown, and it is well known that people's behavior changes during an epidemic. (Xu et al., 2020). Due to its restrictions, people may experience anxiety, riot, and the stockpiling of medical supplies. As a result, it is projected that the demand for medications will rise. This could exacerbate the problem of drug shortages, which is known to have a detrimental effect on population welfare by increasing the demand for healthcare resources and raising morbidity and mortality rates. (Moreno et al., 2020 and Romano et al., 2021).

Conversely, the industry could see an increase in research and development investments as companies work to develop new treatments and vaccines for the virus. The industry will also have to adjust to the changing regulatory environment as governments globally take steps to ensure the safety of treatments and vaccines. (Pecetta et al., 2022).

To stay competitive, pharmaceutical corporations have begun to alter their marketing tactics. As a result, efforts have been made to develop networking ties and cooperative business models based on competition. Additionally, there is proof that pharmaceutical firms have employed various technology-based strategies to combat the COVID-19 pandemic's negative effects, including "digital marketing, e-detailing customer relationship management, e-sampling, innovative work behavior, and telemedicine revolution to promote product value incorporating. (Sawad & Turkistani, 2021).

Before the COVID-19 pandemic, e-commerce in Jordan was growing but faced challenges such as low penetration of credit card usage and limited trust in online transactions. (Al-Adwan et al., 2022). However, the pandemic has accelerated the shift towards online shopping and e-commerce, as physical distancing measures made in-person shopping less convenient and riskier. E-commerce platforms and online payment systems have seen a significant increase in usage, and businesses have had to adapt to the growing demand for online shopping options. (Soares et al., 2022) The Jordanian government has also taken steps to support the growth of e-commerce, including launching initiatives to promote digital payments. Overall, the COVID-19 pandemic has profoundly impacted the e-commerce sector in Jordan, driving rapid growth and making it a crucial part of the country's economy. (Al-Dmour et al., 2021).

In this study, we will evaluate the ethics of pharmaceutical marketing after the COVID-19 pandemic in Jordan. The pandemic has disrupted the supply chain of pharmaceuticals, increased demand for specific treatments, and changed the marketing strategies of pharmaceutical companies. The study aimed to narratively review and report findings from the literature about the ethical considerations in the trade and acquisition of pharmaceuticals after COVID-19 and the impact of the pandemic on the pharmaceutical industry. This narrative review help contributes to more insights into the pharmaceutical industry's ethical challenges in the aftermath of the COVID-19 pandemic and its impact on the availability and accessibility of pharmacological therapies for patients globally and in Jordan.

2 Methods

This narrative review aimed to evaluate the ethical challenges in the pharmaceutical industry in the aftermath of the COVID-19 pandemic and its impact on the availability and accessibility of pharmacological therapies for patients in Jordan. The narrative review was conducted following four steps. The initial step was to look for all related studies. Numerous databases were searched to ensure that most, if not all, pertinent studies had been located. Google Scholar, PubMed, and Science Direct are a few of these databases. Also, we found pertinent articles using the Google search engine. Finding keywords was the second step. To make it simpler and more straightforward for readers to identify their efforts, authors utilize keywords while posting their studies (Demiris et al., 2018). The phrases "coronavirus OR COVID 19 and "pharmaceutical ethics or marketing ethics," as well as "ethics" and "pharmaceutical marketing," were used as keywords in this paper. The final step involved reviewing the papers and abstracts to ensure the chosen articles adequately addressed this paper. Results from the chosen publications were picked, and the chosen articles were summarized and produced (Appendix, Table 1).

3 Results

As in other sectors, the pharmaceutical industry and marketing were highly affected by the COVID-19 global pandemic since the governments put restrictions on movement to minimize the spread of the virus worldwide. Economic advancement has decreased due to the social distance implemented by most government agencies, which has slowed production and stalled the economy. Although individuals still need and want medicines, marketing the products to draw customers has become a significant issue. The most notable shifts in pharmaceutical ethics in marketing tactics after the outbreak are contrasted in this section.

Pharmaceutical industry and marketing:

The study by Ayati et al. (2020) found that the pharmaceutical industry may view COVID-19 as a once-in-a-century opportunity since it will increase demand for prescription drugs, vaccinations, and medical equipment. This can be considered one of the immediate main effects of the COVID-19 outbreak, but there are other immediate and short-term effects as well, such as; changes in demand for medicines and medical supplies, supply constraints, panic buying and stockpiling, regulation changes, shifts in communication and promotion to remote contacts through technology, and modifications to research and development (R&D) processes. While the long-term effect of COVID-19 may include delays in approval, a move toward self-sufficiency in the pharmaceutical product supply chain, a slowdown in industry growth, and potential changes in consumer trends.

Blakely et al. (2022) argued that the COVID-19 pandemic has shown flaws in device regulatory norms and given reason to critically reevaluate them, including approvals and the actions of representatives of the device. The pandemic's dynamic nature and the long-lasting effects of releasing rapidly approved, exorbitantly priced products into the market raise the possibility of serious ethical ramifications. Globally, the modifications made to devise regulations in response to the COVID-19 epidemic can unintentionally affect patient safety for a long time.

Cox S. (2020) also reported that pharmacists must manage resources when demand outpaces supply as the COVID pandemic causes an increase in the demand for these medications on a global scale. Without a doubt, LMICs in lower-middle-income countries; will experience the consequences of the global demand considerably more intensely and for a more extended period. The main reason is that these countries mainly depend on importing these pharmaceuticals. Some innovators and generic companies in the USA and Europe have pledged to increase production. However, these commitments will only apply to countries with valid marketing authorizations. There are three main difficulties for many LMICs that rely on medicine producers in countries like India: Export limitations from manufacturing countries, competition with high-income countries, and disruption in trade. International supply chain in the shipping and aviation sectors. Affected governments will have to look for generic drug alternatives, carefully evaluate and expedite required marketing authorizations, and implement local policies on handling the current limited supply in the market because it might take some time before things return to normal after COVID-19.

To describe how the COVID-19 outbreak has affected the demand and supply of outpatient medications. Romano et al. (2021) track drug trends and evaluate pharmacies that can supply timely and accurate outpatient market data, allowing national and European regulatory and policy authorities to coordinate action in response to a public health emergency like COVID-19. Also, it takes steps, and offers guidance for the preventive management of medications, revealing that despite the COVID-19 lockdown in Portugal appearing to have stabilized the market, sales, and supplies of medications were affected. (Blakely et al., 2022).

2) Ethical alteration in emergency decision-making during the COVID-19 pandemic.

The COVID-19 pandemic has significantly impacted decision-making in times of emergency, and various ethical issues have also been brought forward. Before the pandemic, established processes and rules prioritizing the individual patient's best interests dominated emergency decision-making. Healthcare professionals face new obstacles due to the epidemic and are forced to make difficult decisions quickly, with little knowledge and resources.

According to (Ferejohn and Pasquino, 2004), disaster designations permit the deployment of specific administrative and certain considerations. Emergency powers have historically been "conservatively" designed, lasting only as long as the emergency situation does. There should be a return to the condition before the emergency was threatened.

Conversely, in 2020, Greene reported that declaring an emergency permits actions that would not often be permitted: The sole purpose of announcing a state of emergency is to permit an extraordinary response that is not allowed in a condition of "normalcy."

Companies have an interest in encouraging "all workers to behave themselves in a legal and ethically flawless manner in the course of their everyday work, as how they perform their obligations influences the company's reputation" Thus, it is crucial to support pharmaceutical marketers, provide them with ethical standards for their daily work, and make it possible for them to conduct business restoring public acceptance for corporate-driven information communications regarding pharmaceutical medications by conducting business following the ethical requirements for pharmaceutical businesses. (Bayer, 2017).

After the COVID-19 pandemic arrived, the idea of pharmaceutical marketing was compelled to be rethought; the "business of advertising or otherwise encouraging the sale of pharmaceuticals or drugs" might be interpreted as pharmaceutical marketing. Pharmaceutical marketing initiatives have increased in recent years, significantly increasing the demand for pharmaceutical products. (Chiplunkar et al., 2020).

In 2022, Blakely et al. evaluated the change in the ethical thinking that guides decision-making as a critical factor that permits and approves these responses. The emergency response to the COVID-19 pandemic includes alterations to the laws governing procedures like device approvals. The introduction of medical equipment raises the possibility of brand-new problems due to poor design, subpar manufacturing, or subpar use. New technologies must be proven to offer benefits equal to or greater than existing devices or other options, such as refraining from intervention, to outweigh this risk and reduce harm. (Blakely et al., 2022).

Given the effects on health regulators, they changed their approach to prioritize speed of approvals over the examination of the evidence and safety data due to the pandemic and a spike in demand for essential supplies and equipment (the efficiency-thoroughness trade-off (ETTO)). (Hollnagel, 2009).

The evidentiary requirements that could have prevented the provision of necessary medical equipment and supplies were changed by regulators. The efficiency-thoroughness trade-off is a tricky balance, as these instances demonstrate. If the result of the quick approvals in defective equipment, the goal of saving more lives may not be accomplished. (Blakely et al., 2022).

The pharmaceutical industry has benefited from this. According to research, pharmaceutical corporations tend to profit best from value-risk pricing with risk-sharing agreements, especially in first-bet situations where social planners make decisions. (Bélisle-Pipon, 2021) And (Critchley & Zaric, 2019).

In 2021, Sawad and Turkistani reported that even if the stock prices of some pharmaceutical businesses are increasing, this does not necessarily indicate their performance. Pharmaceutical firms are significantly impacted, and the COVID-19 outbreak has increased the price of raw materials, weakened the supply chain, tightened government regulations, and, most importantly, raised competition. In recent years, pharmaceutical firms will rely on telemarketing, digital marketing, and other offers. (Sawad & Turkistani, 2021).

In another view of marketing techniques after the COVID-19 pandemic, in 2021, Bélisle-Pipon argued that pharmaceutical marketing entails a series of actions that have an ostensibly charitable nature—seeking to inform and educate consumers about treatments they may need—but whose primary goal is to promote pharmaceutical products.

Profitability of items. Pharmaceutical marketers now have access to a wide range of DTCC tools to reach an expanding consumer audience with messages about drug promotion. Although not all DTCC methods are equally ethical when contrasted to the moral justifications used to support their acceptance, i.e., the supply of valuable and effective consumer information, fostering education, and giving customers the power to make knowledgeable and independent decisions.

In (2021), Iheduru-Anderson reported that using personal protection equipment (PPE) has become a further ethical consideration. Healthcare providers have been forced to make difficult choices regarding which healthcare personnel receive PPE and which patients receive treatments that require PPE due to a lack of PPE.

The epidemic has also brought attention to differences in access to healthcare, particularly for underserved groups. This has prompted demands for more justice in the distribution of resources and the making of emergency decisions.

4 Discussion

Pharmaceutical industry and marketing:

The COVID-19 global pandemic has significantly impacted the pharmaceutical industry and marketing. The pandemic has increased the demand for prescription drugs, vaccinations, and medical equipment. However, it has also caused changes in the demand and supply of outpatient medications, supply constraints, panic buying and stockpiling regulation changes, and shifts in communication and promotion. The long-term effects of the pandemic on the pharmaceutical industry may include delays in approval, a move toward self-sufficiency in the pharmaceutical product supply chain, and a slowdown in industry growth. Moreover, the pandemic has revealed flaws in device regulatory norms and raised serious ethical questions about releasing rapidly approved, exorbitantly priced products into the market. LMICs will experience the consequences of the global demand more strongly, and the international supply chain in the shipping and aviation sectors may be disrupted. In response, affected governments will have to find generic drug alternatives, expedite required marketing authorizations, and implement local policies to manage the limited supply in the market.

The study of Romano et al. (2021) highlights the importance of tracking drug trends and evaluating pharmacies that can provide timely and accurate outpatient market data. This information can help national and European regulatory and policy authorities to coordinate action in response to public health emergencies like COVID-19. In Jordan, as in other countries, it was seen that the COVID-19 pandemic had a profound impact on the pharmaceutical industry and marketing. The industry must continue to adapt and respond to the changes brought about by the pandemic. The increased demand for medical products and equipment to tackle the virus has put pressure on the country's healthcare system and its ability to meet the rising demand. In response to the pandemic, the Jordanian government imposed restrictions on exporting certain pharmaceutical products, including masks and sanitizers, on ensuring adequate local supplies. This increased prices for these products, causing a burden on the population and highlighting the need for self-sufficiency in producing essential medical products. Moreover, the pandemic has disrupted the global supply chain, leading to delays in delivering essential medications and medical supplies. This has also impacted the approval of new drugs and treatments, causing delays in introducing innovative therapies to the market.

Despite these challenges, the pandemic has also presented opportunities for growth in the pharmaceutical industry in Jordan. For example, the increased demand for vaccines and other medical products has led to the development of local production facilities, which will help ensure future self-sufficiency. In terms of marketing, pharmaceutical companies have had to adapt their communication and promotional strategies to account for the changes brought about by the pandemic. This has involved a shift toward digital marketing and a greater focus on consumer education and awareness about the importance of taking preventive measures to reduce the spread of the virus.

Ethical alteration in emergency decision-making during the COVID-19 pandemic.

The COVID-19 pandemic has prompted new ethical questions for emergency decision-making and demanded that medical professionals make challenging choices immediately. It has also drawn attention to the necessity of more equity in resource distribution and the significance of considering marginalized communities' needs. Also, the COVID-19 pandemic has created many ethical dilemmas in emergency decision-making, particularly in the pharmaceutical industry and healthcare provision. Regulators have had to change their approach to prioritize the speed of approvals over the examination of evidence and safety data, leading to potential risks in providing necessary medical equipment and supplies. The pharmaceutical industry has benefited from this change but also faces challenges such as increased competition and tighter regulations. The use of personal protection equipment (PPE) and access to healthcare have also become ethical considerations for healthcare providers. To balance the efficiency-thoroughness trade-off, it is essential to support ethical standards in the pharmaceutical industry and ensure that the best interests of patients and the public guide emergency decision-making.

In Jordan, the COVID-19 pandemic has brought similar ethical challenges to the forefront. Healthcare providers have had to make tough decisions about allocating limited resources, such as PPEs, in a fair and just manner. The government has also had to navigate the balancing act between protecting public health and supporting the economy, particularly for businesses and workers whom lockdowns and social distancing measures have heavily impacted. Overall, the COVID-19 pandemic has forced Jordan to grapple with complex ethical dilemmas and has underscored the importance of equity and justice in decision-making. Decisions must be made with transparency, accountability, and a commitment to the well-being of all citizens, especially those most vulnerable.

5 Strengths and Limitation

This narrative review identified a good number of studies from the literature that investigated the ethical considerations in the trade and acquisition of pharmaceuticals after COVID-19 and the impact of the pandemic on the pharmaceutical industry, which may be considered a strength of this study. Also, findings from this narrative review would help contribute to more insights into the pharmaceutical industry's ethical challenges in the aftermath of the COVID-19 pandemic and its impact on the availability and accessibility of pharmacological therapies for patients globally and in Jordan. However, the narrative review included results from reports, theses, audits, and the grey literature. There is always a risk that some studies were not included due to not performing a thorough enough systematic search which may be considered a limitation of this study. Also, the search strategy was restricted to studies published in English and could have missed other valuable research, which may be Further investigation should be done globally and in Jordan, focusing on evaluating the ethics of pharmaceutical marketing post-COVID-19 and reflecting the experiences of patients and consumers regarding the ethics of pharmaceutical marketing post-COVID-19. The impact of the COVID-19 pandemic on the pharmaceutical industry is complex and multifaceted. It may be challenging to rely accurately on the findings from this narrative review to quantify the effects in terms of financial impact, market trends, and regulatory changes.

6 Conclusion

The COVID-19 pandemic has significantly impacted the pharmaceutical industry and marketing. The demand for prescription drugs, vaccinations, and medical equipment has increased, but so have the challenges regarding supply constraints, regulation changes, and ethical considerations in marketing tactics. The pandemic has also highlighted flaws in device regulatory norms and the need for critical reevaluation. The emergency decision-making process has also been impacted, with healthcare professionals facing new ethical challenges. Companies must ensure that their marketing efforts are conducted ethically, especially considering the increased demand for pharmaceutical products. The COVID-19 pandemic has resulted in a shift in how the pharmaceutical industry operates and highlights the need for ethical considerations and regulations to keep pace with the changing landscape.

The COVID-19 pandemic has profoundly impacted the pharmaceutical industry in Jordan, presenting challenges and growth opportunities. The pandemic has highlighted the need for self-sufficiency in producing essential medical products and has resulted in changes to the demand and supply of medications and medical equipment. At the same time, it has also driven the development of new production facilities and has encouraged a shift toward digital communication and marketing strategies. Future studies should be performed to evaluate the effect of COVID-19 on pharmaceutical marketing and medical representatives due to the growing up of e-marketing and telemedicine.

References

Alhamad, H., Abu-Farha, R., Albahar, F., Jaber, D.: Public perceptions about pharmacists' role in prescribing, providing education and delivering medications during COVID-19 pandemic era. Int. J. Clin. Pract. **75**(4), e13890 (2021)

Emanuel, E.J., et al.: What are the obligations of pharmaceutical companies in a global health emergency? Lancet **398**(10304), 1015–1020 (2021)

Donthu, N., Gustafsson, A.: Effects of COVID-19 on business and research. J. Bus. Res. **117**, 284–289 (2020)

Atkinson, C.L., McCue, C., Prier, E., Atkinson, A.M.: Supply chain manipulation, misrepresentation, and magical thinking during the COVID-19 pandemic. Am. Rev. Public Adm. **50**(6–7), 628–634 (2020)

Al-Adwan, A.S., Al-Debei, M.M., Dwivedi, Y.K.: E-commerce in high uncertainty avoidance cultures: the driving forces of repurchase and word-of-mouth intentions. Technol. Soc. **71**, 102083 (2022)

Al-Dmour, A., Al-Dmour, H., Al-Barghuthi, R., Al-Dmour, R., Alshurideh, M.T.: Factors influencing the adoption of e-payment during pandemic outbreak (COVID-19): Empirical evidence. In: Alshurideh, M.T., Hassanien, A.E., Masa'deh, R. (eds.) The Effect of Coronavirus Disease (COVID-19) on Business Intelligence. SSDC, vol. 334, pp. 133–154. Springer, Cham (2021). https://doi.org/10.1007/978-3-030-67151-8_8

Ayati, N., Saiyarsarai, P., Nikfar, S.: Short and long term impacts of COVID-19 on the pharmaceutical sector. DARU J. Pharm. Sci. **28**(2), 799–805 (2020). https://doi.org/10.1007/s40199-020-00358-5

Bayer. Annual Report 2016: Augmented Version (2017). https://www.bayer.com/sites/default/files/ar-2016.pdf

Bélisle-Pipon, J.C.: Pharmaceutical marketing ethics: Ethical standards for more acceptable practices. J. Glob. Mark. 1–23 (2021)

Blakely, B., et al.: Ethical and regulatory implications of the COVID-19 pandemic for the medical devices industry and its representatives. BMC Med. Ethics **23**(1), 31 (2022)

Brandon, W.P.: Flattening epidemic curves and COVID-19: policy rationales, inequality, and racism. J. Health Care Poor Underserved **33**(3), 1700–1714 (2022)

Bong, C.L., Brasher, C., Chikumba, E., McDougall, R., Mellin-Olsen, J., Enright, A.: The COVID-19 pandemic: effects on low-and middle-income countries. Anesth. Analg. **131**(1), 86–92 (2020)

Chiplunkar, S., Gowda, D., Shivakumar, H.: Adaptation of pharmaceutical marketing and drug promotion practices in times of COVID-19 pandemic. Int. J. Health Allied Sci. **9**(5), 1–11 (2020)

Cox, S.: To dispense or not to dispense: lessons to be learned from ethical challenges faced by pharmacists in the COVID-19 pandemic. Dev. World Bioeth. 1–8 (2020)

Critchley, G.J., Zaric, G.S.: The impact of pharmaceutical marketing on market access, treatment coverage, pricing, and social welfare. Health Econ. **28**(8), 1035–1051 (2019)

Dang, A., Arora, D., Rane, P.: Role of digital therapeutics and the changing future of healthcare. J. Fam. Med. Primary Care **9**(5), 2207 (2020)

Demiris, G., Oliver, D.P., Washington, K.T.: Behavioral Intervention Research in Hospice and Palliative Care: Building an Evidence Base. Academic press (2018)

Droždz, J., Burinskas, A., Cohen, V.: The impact of the COVID-19 pandemic on the revealed comparative advantage of industries in the Baltic States. Economies **11**(2), 47 (2023)

Ferejohn, J., Pasquino, P.: The law of the exception: a typology of emergency powers. Int J Const Law. **2**(2), 210–239 (2004)

Greene, A.: Emergency powers in a time of pandemic. Bristol University Press, Bristol (2020)

Hollnagel, E.: The ETTO principle: efficiency-thoroughness trade-of: why things that go right sometimes go wrong. Ashgate Publishing Ltd., Farnham (2009)

Husain, I., et al.: Fluctuation of public interest in COVID-19 in the united states: retrospective analysis of google trends search data. JMIR Public Health Surveill. **6**(3), e19969 (2020)

Iheduru-Anderson, K.: Reflections on the lived experience of working with limited personal protective equipment during the COVID-19 crisis. Nurs. Inq. **28**(1), e12382 (2021)

Khan, M., et al.: COVID-19: a global challenge with old history, epidemiology and progress so far. Molecules **26**(1), 39 (2020)

LaVan, H., Cook, L.S., Zilic, I.: An analysis of the ethical frameworks and financial outcomes of corporate social responsibility and business press reporting of US pharmaceutical companies. Int. J. Bus. Gov. Ethics **15**(3), 326–355 (2021)

Marmat, G., Jain, P., Mishra, P.N.: Understanding ethical/unethical behavior in pharmaceutical companies: a literature review. Int. J. Pharm. Healthc. Mark. **14**(3), 367–394 (2020)

Moosivand, A., Ghatari, A.R., Rasekh, H.R.: Supply chain challenges in pharmaceutical manufacturing companies: using qualitative system dynamics methodology. Iran J. Pharm. Res. IJPR **18**(2), 1103 (2019)

Moreno, C., et al.: How mental health care should change as a consequence of the COVID-19 pandemic. Lancet Psychiatry **7**(9), 813–824 (2020)

Pecetta, S., et al.: The trillion dollar vaccine gap. Sci. Transl. Med. **14**(638), eabn4342 (2022)

Rahul, S., Prakash, M.V.: Exploring the impact of pharmaceutical marketing on prescribing behaviour of doctors in India: a critical review. J. Positive Sch. Psychol. **6**(5), 3452–3470. (2022)

Rodwin, M.A.: Conflict of interest in the pharmaceutical sector: a guide for public management. Depaul J. Health Care L. **21**, 1 (2019)

Sawad, A.B., Turkistani, F.: Pharmaceutical marketing transformation due to COVID-19 pandemic. J. Pharm. Res. Int. **33**(33A), 91–99 (2021)

Soares, J.C., Limongi, R., Júnior, J.H.D.S., Santos, W.S., Raasch, M., Hoeckesfeld, L.: Assessing the effects of COVID-19-related risk on online shopping behavior. J. Mark. Analytics **11**(1), 82–94 (2022). https://doi.org/10.1057/s41270-022-00156-9

Romano, S., Galante, H., Figueira, D., Mendes, Z., Rodrigues, A.T.: Time-trend analysis of medicine sales and shortages during COVID-19 outbreak: data from community pharmacies. Res. Soc. Adm. Pharm. **17**(1), 1876–1881 (2021)

Saxena, A., et al.: Pandemic preparedness and response: beyond the access to COVID-19 tools accelerator. BMJ Glob. Health **8**(1), e010615 (2023)

Willis, E., Delbaere, M.: Patient influencers: the next frontier in direct-to-Consumer pharmaceutical marketing. J. Med. Internet Res. **24**(3), e29422 (2022)

Xu, Z., Elomri, A., Kerbache, L., El Omri, A.: Impacts of COVID-19 on global supply chains: facts and perspectives. IEEE Eng. Manag. Rev. **48**(3), 153–166. (2020). last accessed 2016/11/21

The Impact of Digital Transformation on the Financial Performance of Jordan Media Institute (Alrai)

Faten Amin Al-Naimi[1(✉)], Ola khresat[1], and Abdul Malik Syed[2]

[1] Zarqa University, Zarqa 13110, Jordan
fnaimi@zu.edu.jo
[2] University of Business and Technology, Jeddah, Saudi Arabia

Abstract. The study aimed to demonstrate the impact of digital transformation on the financial performance of the Jordan Media Institute (Al-Rai) in terms of Financial leverage ratios and liquidity ratios. The study used a descriptive analytical methodology. The importance of this study lies in the fact that it is one of the few studies that investigated the impact of digital transformation, measured by the rate of growth in the number of digital subscribers, on the financial performance of Al-Rai newspaper, which is one of the most prominent Jordanian newspapers. It is expected to contribute to opening new horizons for researchers and academics in various sectors. The study concluded that there is a statistically significant positive effect of the digital transformation on the liquidity ratios and that there is no statistically significant effect of the digital transformation on the leverage ratios of the Jordan Media Institute. The study recommended the need to take into account the results of this study and the orientation of decision-makers at the Jordan Media Institute to improve financial performance through leverage indicators, increase the efficiency of credit management, and reduce dependence on debt in financing its assets. The study also called on researchers to take into account the results of this study and conduct similar studies in different sectors, and compare the results to provide new insights and horizons about digital transformation and its impact on financial performance in various sectors.

Keywords: digital transformation · Financial Performance · Jordan Media Institute (Alrai)

1 Introduction

The Internet, information technology, and digitization have played a major role in increasing business efficiency and sustainability [1]. Additionally, digital transformation has created a competitive environment and a radical change in the process of providing services, as it has become smarter and faster [2]. Therefore, activities were developed to keep pace with the technological aspect of these services. Moreover, in light of digital transformation, companies with digital strength were distinguished by their higher

B. A. M. Alareeni and I. Elgedawy (Eds.): *Artificial Intelligence (AI) and Finance*, SSDC 488, pp. 800–810, 2023.
https://doi.org/10.1007/978-3-031-39158-3_74

capacity to keep pace with developments in the surrounding environment. The companies' innovation and their attractive designs and models have created their value online. Considering the harm the Coronavirus pandemic has caused global economy, it was a motive for technological development and Digital transformation, which reflected positively on the companies' activities in general and the media industry and newspapers in particular [3]. Digital transformation is defined as "the intentional and continuous digital evolution of a company, business, idea processing, methodology, or tactical strategy" [4]. In the last few decades, the world has witnessed great technological changes and a tremendous speed in digital transformation in all aspects of life. The businesses of many various establishments have been transformed into digital businesses, starting from the stage of digitizing analog information, processing digital information, and then digital transformation. Meanwhile, digitization is defined as "the replacement of paper documents with electronic files" [5]. While digital transformation is defined as "changing the method of institutions' working value" [6]. This prompted the institution to make a radical change in their business and the way it interacts with stakeholders as well as deal with the behaviors and expectations of customers for the purpose of attracting them in the presence of intense competition [3]. Also, digital transformation is defined as "all the economic activities that depend on the use of digital inputs, including digital technologies, digital infrastructure, digital services and data, or whatever enhance the use of digital inputs with the aim of supporting and greatly promote them. Comprising all producers and consumers, including the government" [7]. Digital transformation requires senior management to take the decision because of its significant impact on business structuring, digitization, lower costs, and increased efficiency [8]. Also, digital transformation leads to a change in the structure of making profits, the method of using capital and how it affects institutions' financial performance [9]. The concept of an organization's digital transformation is utilized at the stage of using information technology and artificial intelligence to positively and rapidly influence the development of its business [10]. The organization that desires to apply digital transformation must provide coordination and integration among its departments, and further develop a multidisciplinary methodology to address administrative issues such as human resources, business efficiency, and process restructuring [11].

Digital journalism emerged as a continuation of the outputs of the Internet, as traditional journalism tends to change forms of content that presents, receives, and reproduces journalistic, radio, and television content. Bearing in mind that the Internet is characterized by its rapid development, therefore, electronic journalism paved the way for transformation and growth, began to invent its own world, and developed its models and tools. It can be said that it was not limited to the form, presentation, and methods of receiving and interacting, but rather included the development of content and methods of collecting and processing information. Instead of the global network of information growth and its entry into multiple links with digital media applications, the huge amount of media information sources that can be accessed has created a new era that is denser and diverse. Thanks to the integration between technology and media, media has become more attractive to accommodate new technology, as the ability of new technological products on the performance of media functions has increased efficiency. Electronic journalism is not just an applied technology that allows easy access to journalistic content, downloading pages, easy search, access to news, updating content, and interaction through comments, and the use of multimedia. Rather, it is a new tool for the profession of journalism used in new writing techniques [12]. The emergence of many electronic newspapers from weekly newspapers in the Hashemite Kingdom of Jordan contributed to giving a kind of popular opinion generally that electronic news websites enjoy a wider margin of freedom, and electronic media were able to invest professionally and politically in benefiting from interaction also created a multi-platform area for discussion through comments, which has historically helped make public policies and administrators more accountable, and in light of the development of electronic media in Jordan, the Jordan Media Institute (Al-Rai) keeps pace with developments and overcomes the challenges that surround it to communicate with its readers, so in 1996 it established two websites: Al-Rai Newspaper and Jordan Times, where The number of visitors to the site per month reached approximately (50) million visitors, and the volume of e-mail messages received to the two newspapers increased. In 2009, the Institute began providing mobile phone service through telecommunications companies in Jordan, as daily messages were sent to subscribers about the most important headlines and news. Urgent published in the two newspapers to provide them with information. The most important local and international, sports and economic news [13].

In 2001, the National e-Government Strategy initiative was launched, which aims to support the digital economy through the use of information and communication technology, with the aim of providing integrated and reliable digital services that raise the efficiency of the government. Data from the Jordanian Ministry of Digital Economy showed that the rate of (DT) of government services reached (16.8%) of the total government services provided. Some institutions, such as the Social Security Institution, the Greater Amman Municipality, the Ministry of Interior, and the Civil Status and Passports Department, have also achieved high completion rates in digital transformation operations, according to the Ministry of Digital Economy and Entrepreneurship, in addition to the presence of (136) services in (16) registration institutions, but it is not included in the digital transformation plans [14]. The achievements of the Ministry of Digital Economy and Entrepreneurship and its efforts contributed to improving Jordan's ranking globally in the Government's Readiness Index for Artificial Intelligence for the year 2022, and Jordan ranked 63 out of 181 countries, and in 2021 it ranked ninth, and recently it ranked sixth among Arab countries [15]. The increase in the number of digital subscribers has led to a decrease in the distribution of daily newspapers, in addition to the digitization of production systems and the distribution of digital content for newspapers, which has led to its administrative development. While the interest in digital transformation increased among administrators and economists due to its great impact on society, the economy, industry, and education. This study came to clarify the impact of digital transformation through increasing the number of digital subscribers in the Jordan Media Institute (Al-Rai) on its financial performance in terms of leverage ratios (LR). Al-Rai newspaper is considered a daily Arab political newspaper published by the Jordan Media Institute (Al-Rai) and distributed to all governorates of the Hashemite Kingdom of Jordan. It is headquartered in the Jordanian capital, Amman. Note that the first issue of Al-Rai newspaper was published in 1971, while the Jordan Times was published in 1975. Al-Rai newspaper is considered the representative of the authority and government, whether in the Hashemite Kingdom of Jordan or abroad [14].

Financial performance is one of the most important measuring tools for the efficiency of organizations in using their assets to generate revenue. It is an indicator of the financial position during a specific period, and it is used to compare the performance of companies, whether in the same or different sectors, and there are different ways to measure financial performance, and it can be fully measured using revenue generated from operations, operating income, operating cash flows, total sales, or growth rate, or profit margin [16]. Financial performance is defined as "the degree of success of the company in using all its available resources, whether material or moral, to the extent that it is possible to reduce costs and increase revenues to the maximum extent, so financial performance is a measure of the business's capacity for achieving its objectives and its financial success can be expressed Through its financial indicators" [17]. The intense competition and globalization, interest in financial performance indicators has increased. Focusing on the quality of service provided by institutions, and striving to keep abreast of technological developments in order to satisfy its customers and attract new customers, and financial performance indicators are a measure of the extent to which the institution achieves its goals and efficiency in directing financial resources and providing investment opportunities in various areas of performance to help meet stakeholder needs and achieve their goals [18].

The value of the study is dependent on the importance of Jordan Media institute's (Alrai) financial performance, as it is an important indicator of its success and ensuring its continuity in light of the advancements in technology and the rapid digital transformation in the media field. This led to a structural change in its activity, and this study is considered one of the few studies that investigated the impact of digital transformation represented by the number of digital subscribers on the financial performance of Jordan Media institute (Alrai). As it is one of the most prominent Jordanian newspapers and it is expected to contribute to opening new horizons for researchers and academics in various sectors. The issue with the study is that change brought about by the digital transformation in the behavior and expectations of readers, which prompted the Jordan newspaper (Alrai) to develop its business strategy and go towards technological innovation in providing services to its readers and maintaining the quality of its media content to meet their desires, maintain them and attract the largest The number of digital subscribers from inside and outside the Hashemite Kingdom of Jordan in order to improve its financial performance to ensure its continuity in light of intense competition.

2 Literature Review

Some studies have shown the impact of digital transformation in (DT) various fields, for instance, the study [19] analyzed business performance in The New York Times newspaper and its financial efficiency during the (DT). It showed the benefits of (DT) for business as well as how the Coronavirus outbreak affected the speed of digital transformation in businesses around the world. The study came to the conclusion that (DT) significantly affects the company's overall business success. The number of digital subscribers has increased continuously, and digital subscription revenues have reached the level of printable subscriptions. Meanwhile, the study [20] studied how the coronavirus epidemic affected auditing practices and whether Jordanian auditing companies could meet the demands of digital audit for contemporary technologies and artificial intelligence techniques. The study came to the conclusion that Jordanian auditing businesses are equipped to handle the demands of contemporary digital auditing and artificial intelligence approaches, and that using digital auditing has an effect on an auditor's personality. The research [21] showed the extent of the readiness of the Jordanian daily printed media to adapt and keep pace with the digital environment, as it concluded that Jordan Media Institute is ready to moderately deal with the digital environment. It has been using social networking sites for several years, and the percentage of using Facebook to communicate with its readers has reached (95.5%), Twitter has reached (80.9%), while WhatsApp has reached (74.2%). Apart from this, it showed that there are variances in the degree of using applications related to the digital world between each of Addustour newspaper besides Alrai and Al-Ghad newspapers, which obtained the highest degree. The study [22] highlighted how digital revolution affected the effectiveness of faculty members' performance at King Abdulaziz University. State the requirements of (DT) to achieve efficient academic performance at King Abdulaziz University and the criteria that must be met by a faculty member at the university to increase academic efficiency. Also, it was found that the impact of (DT) on academic performance is statistically significant of the faculty members at King Abdulaziz University. In addition to that it concluded the existence of a statistically significant of (DT) and its requirements in order to achieve the efficiency of the faculty members' academic performance and the statistically significance of the obstacles that limit the effectiveness of (DT) in academic performance. The study [23] showed the contribution of the electronic government implementation in Palestine and the promotion of the (DT) and's role in improving government services in light of the obstacles imposed by the Israeli occupation. However, the study [24] showed that there is a strong correlation between the requirements of (DT) and improving the performance in providing services in the General Traffic Department in Cairo. While, the study [25] showed that there is an impact of digital transformation in improving digital confidence within its dimensions (digital environment, user experiences, user attitudes and feelings, and user participation) in the Ministry of Education in the Arab Republic of Egypt. Likewise, the study [26] created a conceptual framework for the function of open innovation in the creation of new goods and its effect on the financial performance of the business. It came to the conclusion that shared digital technology and the business's financial performance are positively correlated. Whilst, some studies looked at the financial performance, including the study [27] that discovered a statistically significant beneficial influence of investing in social responsibility on the return on assets, return on equity, and return on share for Jordanian industrial companies. Study [28] shown that there is an influence on the earnings of Jordanian insurance firms during the time before the Corona epidemic (2019) and after the epidemic (2020), and that there is no impact on the extent of (rapid liquidity) (financial leverage), The study [29] also demonstrated that the size of the banks and inflation have a negative impact on return on equity while the liquidity ratio and market share have beneficial effects on it. Furthermore, nevertheless. The study [30] revealed that implementing an ERP system had no effect on either the financial performance of Jordan's Arab or Islamic banks, and that its implementation had a statistically meaningful effect only on return on investment.

3 Model of the Study

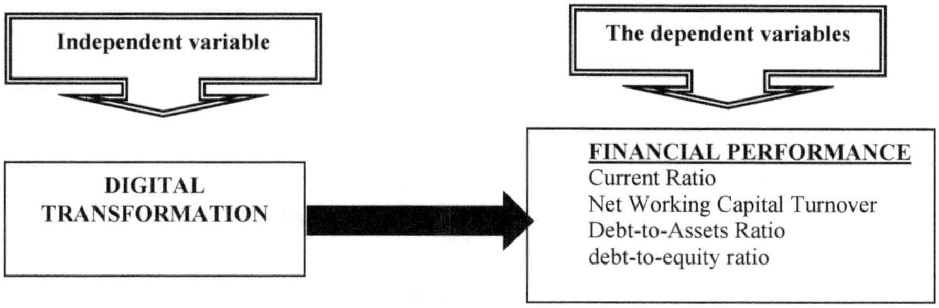

4 Hypothesis

In Light of the Study Problem Questions, the Following Hypotheses Were Devised:

H0: There is no statistically significant of digital transformation (DT) at a significant level ($0.05 \geq \alpha$) on the financial performance of the Jordan Media Institute (Alrai), and the following hypotheses were derived from it:-

$\mathbf{H0_1}$: There is no statistically significant of the digital transformation at a significant level ($\alpha \geq 0.05$) on the Current ratio of Jordan Media Institute (Alrai).

$\mathbf{H0_2}$: There is no statistically significant of the digital transformation at a significant level ($\alpha \geq 0.05$) in the (net working capital) of Jordan Media Institute (Alrai).

$\mathbf{H0_3}$: There is no statistically significant of the digital transformation at a significant level ($\alpha \geq 0.05$) in the debt-to-asset ratio of Jordan Media Institute (Alrai).

$\mathbf{H0_4}$: There is no statistically significant of the digital transformation at a significant level ($\alpha \geq 0.05$) in the debt-to-equity ratio of Jordan Media Institute (Alrai).

5 Methodology

Essentially, the study followed the analytical descriptive approach and used the (SPSS) program to process the primary data, measures, test percentages, frequencies, arithmetic averages, and standard deviations to describe the study variables. While the secondary sources are the scientific researches that have been published in scientific journals and periodicals. As for the primary sources, it is the measurement tool represented by the financial performance index through the use of the annual financial reports of the institution during the period from the year 2013 to the year 2021, as well as the number of digital subscribers inside and outside the Hashemite Kingdom of Jordan. And, it was gathered from the Center for Studies in Jordan Media Institute (Alrai). However, the independent variable, which is digital transformation, was measured by the growth rate of the number of digital subscribers of Jordan Media Institute (Alrai) during the study period.

5.1 Measurement of Variables

Independent variable:-

(DT) = ((NDS) - (NDS) $_{t-1}$)/((NDS) $_{t-1}$) X 100% [19]

Dependent variables:- Financial Performance [30]

Current Ratio = Total Assets/Total Liabilities.

Net working capital = current assets - current liabilities

Debt to total assets ratio = total debt/total assets.

Debt-to-equity ratio = total debt/total equity.

5.2 Hypothesis Testing

H0$_1$: There is no statistically significant of the digital transformation at a sig. ($0.05 \geq \alpha$) in the leverage ratio (current ratio) of Jordan Media Institute (Alrai).

Table 1. Results of testing the first hypothesis

Dependent variable	Form		Analysis of variance		Coefficients		
	R	R^2	F	Sig	Beta	T	Sig
Current ratio	0.12	0.014	0.148	0.001	0.023	0.468	0.001

Table 1 shows the impact of the (DT) of Jordan Media Institute (Alrai) on its financial performance (current ratio), where the calculated (F) value was (0.148) with a sig. (0.001), which is less than the value of the significance level (0.05). This means that there is a statistically significant of the (DT) on the current ratio in the institution, as the value of (R^2), which is (0.014), indicates the amount of change that the independent variable causes on the dependent variable, and the value of (β), which amounted to (0.023), expresses this size which has a statistically sig. of significance. Therefore, we reject the first null hypothesis and accept the alternative hypothesis, which states "there is an impact of (DT) in the leverage ratio (current ratio) of Jordan Media Institute (Alrai)".

H0$_2$: There is no statistically significant of the digital transformation at a sig. ($\alpha \geq 0.05$) in the leverage ratio (net working capital ratio) of Jordan Media Institute (Alrai).

Table 2. Results of testing the second hypothesis

Dependent variable	Form		Analysis of variance		Coefficients		
	R	R^2	F	Sig	Beta	T	Sig
working capital	0.239	0.057	4.99	0.020	0.011	0.684	0.020

Table 2 shows the impact of Jordan Media Institute (Alrai) on its financial performance (net working capital), where the calculated (F) value was (4.99) with a sig of

(0.020), which is less than the value of the sig. (0.05). This indicates that there is a statistically sig. of (DT) on the net working capital of the institution, as the value of (R^2) was (0.057) which indicates the amount of change caused by the independent variable on the dependent variable. Also, the value of (β) was (0.011) thus, expresses the size of this impact and is statistically significant at the sig. (0.020), which is less than (0.05), so we reject the second null hypothesis and accept the alternative hypothesis, it stated: "there is an impact of the (DT) in leverage ratio (net working capital) for the Jordan Media Institute (Alrai).

H0₃: There is no statistical significance statistically significant at a significant level ($\alpha \geq 0.05$) of the digital transformation in the leverage ratios (debt-to-assets ratio) of Jordan Media Institute (Alrai).

Table 3. Results of testing the third hypothesis

Dependent variable	Form		Analysis of variance		Coefficients		
	R	R^2	F	Sig	Beta	T	Sig
Debt ratio/assets	0.64	0.409	1.558	0.311	0.307	1.248	0.311

Table 3 shows the impact of the (DT) of Jordan Media Institute (Alrai) on its financial performance (debt-to-asset ratio), where the calculated (F) value was (1.558) with a significance level of (0.311), which is greater than the value of the significance level (0.05). This indicates that there is no statistically sig. of (DT) on the ratio of debts to assets in the institution. The value of (R^2) was (0.409) which indicates the amount of change that the independent variable may bring about on the dependent variable, as well as the value of (β) was (0.307). It is not statistically indicative of the size of the change that may occur. We accept the third null hypothesis, which states: "there is no impact of (DT) the leverage ratios (debt-to-asset ratio) of Jordan Media Institute (Alrai)".

H0₄: There is no statistical significance of the digital transformation at a significant level ($\alpha \geq 0.05$) on the leverage ratios (debt-to-equity ratio) of Jordan Media Institute (Alrai).

Table 4. Results of testing the fourth hypothesis

Dependent variable	Form		Analysis of variance		Coefficients		
	R	R^2	F	Sig	Beta	T	Sig
Debt ratio/equity	0.234	0.055	1.197	0.088	0.272	1.094	0.088

Table 4 shows the impact of the (DT) Jordan Media Institute (Alrai) on its financial performance (debt-to-equity ratio), where it was statistically sig., as the calculated (F) value reached (1.197) at a level of sig. (0.088), which is higher than the value of the sig.

(0.05). This indicates that there is no statistically sig. of the (DT) on the ratio of debt to equity in the institution, as the value of (R^2) was (0.055) which indicates the amount of change that the independent variable may bring about on the dependent variable, and the value of (β) reached (0.272) indicated the size of this impact at the time it may occur. We accept the fourth null hypothesis, which states: "there is no impact of (DT) on the leverage ratios (debt-to-equity ratio) of Jordan Media Institute (Alrai)".

6 Conclusion

The current study showed that there is a statistically significant positive impact of digital transformation on the leverage ratio of Jordan Media Institute (Alrai). Where the value of (β) indicated that there is an increase in the number of digital users of Alrai newspaper which lead to an increase in its ability to meet its short-term obligations at their due date. There is a surplus of the institution's current assets over its current liabilities and an increase in its effectiveness in managing its daily operations and the ability to finance them. This result was consistent with the results of the studies [18, 21, 23], additionally, the study showed that there was no statistically significant of (DT) statistically significant on the leverage ratios of Jordan Media Institute (Alrai). This indicates poor management of the institution's credit and its dramatic reliance on long-term debts and thus to finance its assets, which limits its ability to obtain credit in the future. It also indicates that the institution relied on external financing represented by total debts more than internal financing represented by capital, that is, an increase in the percentage of debts from the capital used to finance assets, and so, this result was consistent with the results of the study [27].

7 Recommendations

The study recommends that the decision-makers in Jordan Media Institute (Alrai) seek to increase the number of digital subscribers and attract the largest possible number of them because of its positive and clear impact on its liquidity ratios. This leads to improving financial performance and ensuring continuity in light of intense competition. The need to direct decision makers in Jordan Media Institute (Alrai) to improve the financial performance through indicators of leverage and increase the efficiency of credit management in order to reduce dependence on debt in financing its assets. The study called on researchers to conduct similar studies in different sectors and then compare the results to provide insights and new horizons about digital transformation and its impact on financial performance in various sectors.

References

1. Bala, M.: Digital transformation: review of concept, digital framework, and challenges. In: Gupta, V.P., Bansal, D. (eds.), Theoretical and Empirical Development in Management and IT. vol. 1, no. 1, pp.133–152 (2018)
2. Fraenkel, D.: Will the New York times be next? Challenge **51**(5), 123–127 (2008). https://doi.org/10.2753/0577-5132510508

3. Khodova, Y.: Digital transformation (DT) of business as a trend of world economy development. In: Digitalization of the economy as a factor of sustainable development. Materials of International scientific-practical conference (Mariupol, 25–26 May 2021)
4. Mazzone, M.: Digital or Death: Digital Transformation: The Only Choice for Business to Survive Smash and Conquer. Smashbox Consulting Inc. (2014)
5. Gobble, M.: Digital strategy and digital transformation. Res. Technol. Manag. **61**(5), 66–71 (2018). https://doi.org/10.1080/08956308
6. Gudergan, G., Mugge, P.: The gap between practice and theory of digital transformation. In: Proceeding Hawaii International Conference of System Science, Hawaii, pp. 1–15 (2017)
7. OECD Economic Outlook, The world economy on a tightrope (2020). https://www.oecd.org/economic-outlook/june-2020/
8. Patel, M.: Digital Transformation vs. Digital Optimization Medium (2019). https://medium.com/@maxy_ermayank/digital-transformation-vs-digital
9. Libert, B., Back, M.: Companies Fail at Digital Transformation By Focusing On Digital Optimization. Forbes (2018). https://www.forbes.com/sites/barrylibert/2018/01/18/companies-fail-attransformation-by-focusing-on-optimization/
10. Joško, L.: Zero marginal cost in magazine industry: changing of cost paradigm in "new" magazine industry. In: 44th International Scientific Conference on Economic and Social Development. ESD Conference Split, pp. 125–136 (2019)
11. Li, L., Fang, Z., Mao, J.: Digital transformation by SME entrepreneurs: a capability perspective. Inf. Syst. J. **1**(3), 129–157 (2017)
12. Press Media Newspaper. Electronic Journalism in the Arab World: Emergence Contexts and Development Challenges (2020). https://pressmedias.org/archives/23849
13. The digital economy and leadership many achievements and projects. (5 Jan 2023). https://www.almamlakatv.com/news/66044-168
14. Jordan Press Foundation (2023). https://alrai.com/
15. Rashid, A., Kamal, U.: Impact of Internal (MICRO) and External (MACRO) factors on profitability of insurance companies. J. Econ. Policy Res. **5**(1), 35–57 (2018)
16. Al-Dulaimi, H.: Managing customer relationships and their impact on financial performance: an analytical study among a sample of Iraqi government banks. J. Manag. Econ. **39**(108), 143–161 (2016)
17. Al-Sharif, A.: The effect of return on assets (ROA) and the degree of financial leverage (FLM) on the return on equity (ROE) of Jordan Steel Company and its subsidiaries. Econ. Sci. J. **3**(5), 1–8 (2016)
18. Fahmy, S.: Financial Analysis. 1st edn. Financial SME, Ramallah, Palestine (2018)
19. Lozić, J., Čiković, K.: The impact of digital transformation on the business efficiency of the NEW YORK TIMES, UTMS. J. Econ. **12**(2), 225–239 (2022)
20. Harb, A.: The effect of internal audit on accounting information technology in public joint stock pharmaceutical industries in Jordan. Acad. Acc. Financ. Stud. J. **24**(1), 1–8 (2022)
21. Al-Qudah, M.: The Transitional Environment of the Jordanian Press toward Digital Media: A Survey study. Master Thesis, middle east university (2020)
22. Alswat, T.: The impact of digital transformation on the efficiency of academic performance, a case study for faculty members at King Abdulaziz University. In: Arab Journal for Scientific Publishing (AJSP) vol. 1, no 43, pp. 647–686 (2022). ISSN: 2663–5798
23. Almasdar, H., Nassrallh, A.: The role of digital transformation in improving government services in palestine. In: The First International Conference on Information Technology and Business (ICITB2020), pp.1–27 (2021)
24. Shadid, M.: The impact of digital transformation on the level of service performance provided by the application on the employees of the General Traffic Department in Cairo Governorate. J. Fac. Econ. Polit. Sci. **22**(4), 193–226 (2021)

25. Qandil, R.: The impact of digital transformation projects on improving the digital trust performance card by application to the Egyptian ministry of education and technical education. Arab J. Manag. **45**(1), 1186–1198 (2022)
26. Press, A., Jabre, W.: Digital transformation technology and its impact on improving the strategic performance of the bank (An exploratory study of the opinions of a sample of private bank managers in Karbala Governorate). Iraqi J. Adm. Sci. **16**(65), 224–230 (2021)
27. AL-Rawashdeh, A., Al Naimi, F.: Social responsibility and its impact on financial performance in companies listed on the amman stock exchange. Int. J. Bus. Manag. **14**(3), 1–8 (2019)
28. Khresat, O., Sadiq, I.: The impact of the profits of jordanian insurance companies on their financial performance indicators during the corona pandemic. International Conference on Economics and Administrative Sciences - Corona Pandemic: Its Effects and Recovery Strategies, 23–24/Mars, pp. 891–900 (2022)
29. Abu Hamour, M.: The effect of the determinants of islamic and commercial banks' performance in Jordan on Return on Equity (ROE) an empirical study of the period (2001–2014). Zarqa J. Res. Stud. Hum. **17**(3), 676–687 (2017)
30. Younis, I., Yamin, I.: The impact of the application of enterprise resource planning system on banks financial performance (A case study of Arab bank and Islamic Bank of Jordan). Zarqa J. Res. Stud. Hum. **18**(1), 1–27 (2018)

Investigating the Approaches to Improve Journalism Practices in Jordan: Data Journalism Perspective

Marcelle Issa Al Jwaniat[1], Amjad Safori[2](✉), Khaleaf Al-Tahat[3], Ahmed Mansoori[3], and Mohammad Habes[4]

[1] Faculty of Mass Communication, Department of Journalism, Yarmouk University, Irbid, Jordan
marcelle.jwaniat@yu.edu.jo

[2] Faculty of Media, Zarqa University, Zarqa, Jordan
asafori@zu.edu.jo

[3] Media and Creative Industries Department, United Arab Emirates University, Al-Ain, UAE
k.tahat@uaeu.ac.ae

[4] Faculty of Mass Communication, Radio and TV Department, Yarmouk University, Irbid, Jordan
mohammad.habes@yu.edu.jo

Abstract. This study aims to investigate approaches to improve journalism practices in Jordan from a data journalism perspective. The study employs a qualitative approach, consisting of eight semi-structured interviews with journalists, data analysts, and media experts in Jordan. The findings suggest that data journalism has the potential to improve journalism practices in Jordan by providing a more nuanced and evidence-based approach to reporting. However, there are challenges that need to be addressed in order to fully realize the potential of data journalism in Jordan. These challenges include a lack of technical expertise and resources, as well as a lack of awareness and understanding of data journalism among journalists and media organizations in Jordan. The study recommends that media organizations in Jordan invest in training and resources to develop the technical expertise needed to engage in data journalism. Additionally, media organizations should prioritize data journalism as part of their reporting practices and work to raise awareness and understanding of data journalism among journalists in Jordan. Finally, the study highlights the need for collaboration and partnerships between media organizations, data analysts, and other stakeholders in order to fully leverage the potential of data journalism in Jordan.

Keywords: Data journalism · journalism practices · Jordan · qualitative approach · interviews

1 Introduction

Evaluating the digitization, the use of the internet, smart applications [1], and the spread of media have taken part into the publishing of modern types of journalism and facilitated the delivery of data, information and reports in deep and clear coverage. These data

B. A. M. Alareeni and I. Elgedawy (Eds.): *Artificial Intelligence (AI) and Finance*, SSDC 488, pp. 811–820, 2023.
https://doi.org/10.1007/978-3-031-39158-3_75

are digitally refined and adapted and integrated many digital media types, especially in digital journalism [2]. So, the press material is presented in a way that the press material integrated from photo, animations, video, statistics, data calculation and shapes. The origin of data journalism dates back to 2006 by the founder of the Block Every "Adrian Holovaty", who said: "Journalists should work on publishing data in an organized and machine-readable way in addition to the traditional texts" [3]. Data Journalism contributed to attracting the public to interact with the journalistic material by participating and commenting in sending data to the newspaper. The Guardian's editor assured that data journalism is not just graphics and numbers, but also using several journalistic styles more widely, and it is a method of story- telling by inserting graphics, numbers and statistics to enrich them and communicate information and visualize it easily in a sophisticated way, in addition to writing a story and a long article to deliver the idea and content to the audience [4] Journalists seek to gather information based on the data obtained from multiple different sources and supported by images, data, statistics and graphs. The significance and role of data journalism is found in the value of the information obtained and the speed of its spread credibility and accuracy. In the insight of the virtual world, digital media, technological and communication progress, it has become necessary to use modern methods to get and process data distinctly through electronic and smart devices and obtain methods that are consistent with the digital age [5]. This study came to answer the following question: How does data journalism contribute to improving journalism in Jordan from the perspective of journalists? The importance of this study as a scientific attempt is to show the extent to which Jordanian journalists use and keep up with new items in terms of techniques, methods, types and forms of journalism to access information accurately and quickly and analyze it according to modern technological applications, which enriches the information reached after analyzing the data through graphs [6], maps and visual forms, and that is reflected on the journalistic content.

2 Literature Review

Data journalism is a growing field in Jordan and is being increasingly used by journalists to report on a wide range of topics. The use of data in news reporting provides a new level of depth and context to the stories being told and enables journalists to present objective and accurate information to the public. One of the main areas where data journalism is used in Jordan is in reporting on economic trends and issues. Journalists use data to report on key economic indicators such as unemployment rates, inflation, and government spending [7]. This allows them to provide a more detailed understanding of the country's economic situation and to highlight any trends or patterns that may be emerging. Another area where data journalism is used in Jordan is in reporting on political issues. Journalists use data to report on voting patterns, election results, and other political developments [8]. This allows them to provide a more nuanced understanding of the country's political landscape and to highlight any trends or patterns that may be emerging. Data journalism is also used in Jordan to report on social issues such as crime rates, education, and healthcare. This allows journalists to report on the social challenges facing the country and to highlight areas where improvements are needed.

This type of reporting enables the public to understand the social conditions of the country and become informed about the situation. However, Data journalism is not without its challenges in Jordan. One of the main challenges is the lack of access to data.

There have been a number of studies on data journalism in recent years [8]. These studies have looked at various aspects of data journalism, including its definition, its impact on news reporting, and the challenges faced by data journalists [9]. One study found that data journalism is defined as the use of data, statistics, and other quantitative information in news reporting and storytelling. The study also found that data journalism is becoming increasingly important in the field of journalism as it allows for more in-depth and accurate reporting. Another study found that data journalism has a positive impact on news reporting [10]. The study found that data journalism can improve the accuracy and objectivity of news stories and can provide a new level of depth and context to the stories being told. A separate study has also looked at the challenges faced by data journalists [11]. The study found that one of the main challenges is the lack of access to data. Many government agencies do not make data publicly available, making it difficult for journalists to access the information they need. Additionally, the study found that many journalists do not have the skills or experience to analyze data and to use it in their reporting. More recent studies have also looked at the role of data journalism in the COVID-19 pandemics and its impact on the public's understanding of the pandemics, as well as the role of data journalism in increasing transparency and accountability in government and other institutions [12]. Overall, these studies have highlighted the importance and impact of data journalism in today's media landscape, as well as the challenges faced by data journalists.

Data Journalism with Journalism Practices

Data journalism is a practice that involves collecting, analyzing, and presenting data in order to uncover stories and insights that might otherwise go unnoticed [13]. It is a type of journalism that is increasingly being used by news organizations around the world, as it allows journalists to provide a more comprehensive and evidence-based approach to reporting [14]. At its core, data journalism is about using data to tell stories. This involves collecting data from a variety of sources, including government records, financial reports, social media, and other sources [15]. Once this data has been collected, journalists can then analyze it in order to identify trends, patterns, and other insights. One of the key benefits of data journalism is that it allows journalists to provide a more nuanced and detailed approach to reporting [11, 16, 17]. By using data to back up their reporting, journalists can provide a more comprehensive picture of complex issues [18]. For example, if a journalist is reporting on a particular policy issue, they can use data to provide context and nuance to the story, showing how the policy has evolved over time, and how it compares to similar policies in other countries [19]. Moreover, data journalism can be particularly useful in uncovering corruption and other forms of malfeasance [20]. By analyzing government data, financial records, and other sources, journalists can identify patterns and irregularities that might otherwise go unnoticed. This can help to expose wrongdoing, hold those in power accountable, and ultimately strengthen the rule of law [21]. Moreover, data journalism can also be time-consuming. Collecting and analyzing data can take a significant amount of time, and it can be difficult to balance this with the demands of daily news reporting [22, 23]. This can make it challenging for news organizations to prioritize data journalism, especially if they are under pressure to produce breaking news stories on a regular basis [19]. Despite these challenges, data journalism is becoming an increasingly important part of journalism practices [24]. As more and more data becomes available, journalists are finding new ways to use this data to uncover important stories and insights [25]. In many cases, data journalism is being used in combination with more traditional forms of reporting, in order to provide a more comprehensive and nuanced approach to reporting [26]. Ultimately, data journalism has the potential to transform journalism practices, by providing a more evidence-based and nuanced approach to reporting [27]. By using data to uncover stories and insights, journalists can help to hold those in power accountable, promote transparency and accountability, and provide a more informed public debate [28]. As such, data journalism is likely to continue to be an important part of journalism practices in the years to come [29].

Theory of Uses and Gratifications to improve Data Journalism

The theory of uses and gratifications suggests that individuals actively seek out and use media content to fulfill certain needs and desires [30]. This approach can be applied to data journalism to improve the effectiveness of reporting and engage audiences in a meaningful way. One way to apply the theory of uses and gratifications to data journalism is by understanding the motivations and needs of audiences when it comes to consuming news content [31]. For example, audiences may seek out data journalism for the purpose of gaining a deeper understanding of complex issues or to feel empowered to make informed decisions. By understanding these motivations, media outlets can tailor their data journalism reporting to better meet the needs of their audience [32]. Another way to apply the theory of uses and gratifications to data journalism is by providing opportunities for audience engagement with the reporting [33]. This could include interactive data visualizations that allow audiences to explore data sets themselves or surveys that allow audiences to provide feedback on the reporting. By providing these opportunities for engagement, media outlets can satisfy the need for social interaction and create a more meaningful and interactive experience for audiences. Moreover, data journalism can fulfill the need for information by presenting data-driven insights that are informative, educational, and valuable [34]. The use of data can also increase the credibility of reporting by providing objective evidence to support claims and findings. Data journalism can fulfill the need for entertainment by presenting data in an engaging and visually appealing way [35]. For example, data visualizations that are colorful, interactive, and dynamic can capture the attention of audiences and make the reporting more enjoyable to consume. The theory of uses and gratifications can be applied to data journalism to improve the effectiveness of reporting and engage audiences in a more meaningful way [12]. By understanding the motivations and needs of audiences, media outlets can tailor their reporting to better meet these needs and create a more engaging and interactive experience for audiences [28]. Ultimately, this can lead to more informed and empowered audiences who are better equipped to navigate complex issues [25, 36].

3 Methodology

This study employs a qualitative approach to investigate approaches to improve journalism practices in Jordan from a data journalism perspective [37, 38]. The qualitative approach allows for an in-depth exploration of the experiences and perspectives of participants in the study [39, 40]. The study consists of eight semi-structured interviews with journalists, data analysts, and media experts in Jordan. The participants were selected through purposive sampling, based on their expertise and experience in data journalism and journalism practices in Jordan. The interviews were conducted face-to-face or via video conferencing, and were audio-recorded with the consent of the participants. The interviews were conducted in Arabic and transcribed verbatim, with translations provided where necessary [41]. The interview questions focused on the participants' experiences and perspectives on data journalism and journalism practices in Jordan. The questions covered topics such as the use of data in journalism, the challenges and opportunities of data journalism in Jordan, and approaches to improve journalism practices in Jordan from a data journalism perspective.

4 Data Analysis

Data analysis was conducted using thematic analysis. The data was read and re-read to identify patterns and themes in the participants' responses [37]. The identified themes were then organized into categories and sub-categories, and the findings were presented in a narrative format.

First: Data Journalism and its Role in Building Journalism Skills to Produce Journalistic Content

Data journalism in journalism, who were interviewed, have a number of practical data journalism skills that have an effective and positive role in influencing the individuals producing journalistic content in their society. Their opinions varied about these skills, but they agreed that there are many skills that must be mastered to ensure that individuals achieve a positive impact on their society, and the skill of verifying information, which was confirmed by 7 out of 8 specialists, topped as the first and most important skill that must be learned for individuals to open horizons for producers of journalistic content. Through data journalism, which helped them in the ability to distinguish between true and false information when producing information, followed by the analysis skill that was instructed by 4 specialists as one of the important skills in data journalism, in addition to the skill of accessing information that was confirmed by 3 specialists and supported by the reader. In Jordan, journalism faces many challenges including limited access to information, restrictions on press freedom, and limited resources for media organizations. However, data journalism can provide an effective approach to improving journalism practices in the country. By embracing data journalism, media outlets can produce more accurate, comprehensive, and impactful reporting that meets the needs of audiences. One approach to improving journalism practices in Jordan through data journalism is by providing training and resources for journalists. This can include workshops, online courses, and access to data sets and software. By building the skills and knowledge of journalists in Jordan, media outlets can produce more accurate and comprehensive reporting. Such training can also help journalists to use data analysis to uncover stories that might not have been possible to report otherwise. In addition to training, media outlets can also create partnerships with data experts to produce more effective data journalism. This could involve collaborating with academic institutions, research organizations, or independent data analysts to uncover and report on complex issues. By working together, media outlets can produce data-driven reporting that is accurate, informative, and impactful. These partnerships can also help journalists to gain access to data sets that may not be publicly available. Moreover, media outlets could invest in data visualization tools to make their reporting more accessible to audiences. Interactive graphs, maps, and charts can help audiences understand complex data sets and engage with the reporting in a more meaningful way. This could also make data journalism more visually appealing and attractive to audiences, particularly those who may not be as familiar with data. Another approach is to prioritize transparency and accountability in reporting practices. By publishing source data alongside articles, making corrections and retractions when necessary, and being open about methods and processes for collecting and analyzing data, media outlets can build trust and credibility with audiences. This could lead to greater engagement and impact of data journalism reporting.

Second: The Challenges Facing Data Journalism in Jordan

All journalists agreed that the task of verifying press information is one of the most important challenges facing opportunities to improve journalism in Jordan, in addition to its inclusion of the contents of journalistic contents, but the difference was in the mechanism of its impact on the productive behavior of journalistic contents. Specialists explain the concept of verifying information as researching the validity and accuracy of information circulated on social media. However, there are also potential challenges to adopting data journalism in Jordan. These challenges include a lack of access to data sets, limited technical infrastructure, and potential government censorship. Media outlets will need to navigate these challenges and find ways to overcome them in order to embrace data journalism as a tool for improving their reporting practices. Applying the theory of uses and gratifications to data journalism can also provide valuable insights into how to engage audiences with reporting. By understanding the motivations and needs of audiences when it comes to consuming news content, media outlets can tailor their data journalism reporting to better meet the needs of their audience. For example, audiences may seek out data journalism for the purpose of gaining a deeper understanding of complex issues or to feel empowered to make informed decisions.

Third: The Role of Data Journalism in Improving Journalism

The interviewed journalists mentioned a number of the roles of data journalism in improving journalism in Jordan, which have an effective and positive role in influencing journalists who produce journalistic content in their society. Their opinions about these skills varied, but they agreed that they collect data, analyze them, and present them in a convincing and accessible manner, which was confirmed by 7 out of 8 journalists. Data journalism plays a crucial role in improving journalism in Jordan, as it allows journalists to uncover stories and insights that might otherwise go unnoticed. By collecting, analyzing, and presenting data in a compelling and accessible way, data journalism can help to shed light on important issues and hold those in power accountable. One of the key benefits of data journalism is that it can help to provide context and nuance to news stories. By using data to back up their reporting, journalists can provide a more comprehensive and detailed picture of complex issues, helping audiences to understand the broader trends and patterns that underpin the news. Moreover, data journalism can be particularly useful in uncovering corruption and other forms of malfeasance. By analyzing government data, financial records, and other sources, journalists can identify patterns and irregularities that might otherwise go unnoticed. This can help to expose wrongdoing, hold those in power accountable, and ultimately strengthen the rule of law Overall, data journalism has the potential to be a powerful tool for improving journalism in Jordan. By providing a more detailed, nuanced, and evidence-based approach to reporting, it can help to promote transparency, accountability, and informed public debate in Jordan. Moreover, data journalism can fulfill the need for information by presenting data-driven insights that are informative, educational, and valuable. The use of data can also increase the credibility of reporting by providing objective evidence to support claims and findings. Additionally, data journalism can fulfill the need for entertainment by presenting data in an engaging and visually appealing way. For example, data visualizations that are colorful, interactive, and dynamic can capture the attention of audiences and make the reporting more enjoyable to consume. Furthermore, media outlets can provide opportunities for audience engagement with the reporting. This could include interactive data visualizations that allow audiences to explore data sets themselves or surveys that allow audiences to provide feedback on the reporting. By providing these opportunities for engagement, media outlets can satisfy the need for social interaction and create a more meaningful and interactive experience for audiences. In conclusion, data journalism has the potential to transform journalism practices in Jordan by providing a new approach to uncovering and reporting on complex issues. By investing in training, resources, and partnerships with data experts, media outlets can produce more accurate, comprehensive, and impactful reporting. Ultimately, embracing data journalism can help build trust and credibility with audiences, and contribute to a more informed and engaged society.

5 Conclusion

Data journalism has the potential to enhance journalism practices in Jordan by providing a new approach to reporting that relies on data to uncover and analyze stories. While there are challenges to fully embracing data journalism in Jordan, including limited access to data, limited technical infrastructure, and potential government censorship, there are opportunities to build partnerships, prioritize transparency and accountability, and apply the theory of uses and gratifications to engage audiences with data-driven reporting. Media outlets could invest in data visualization tools to make their reporting more accessible to audiences. Interactive graphs, maps, and charts can help audiences understand complex data sets and engage with the reporting in a more meaningful way. In addition to these approaches, media outlets could also prioritize transparency and accountability in their reporting practices. This could include publishing source data alongside articles, making corrections and retractions when necessary, and being open about their methods and processes for collecting and analyzing data. However, there are also potential challenges to adopting data journalism in Jordan. These challenges include a lack of access to data sets, limited technical infrastructure, and potential government censorship. Media outlets will need to navigate these challenges and find ways to overcome them in order to embrace data journalism as a tool for improving their reporting practices. Finally, data journalism has the potential to transform journalism practices in Jordan by providing a new approach to uncovering and reporting on complex issues. By investing in training, resources, and partnerships with data experts, media outlets can produce more accurate, comprehensive, and impactful reporting. Ultimately, embracing data journalism can help build trust and credibility with audiences, and contribute to a more informed and engaged society.

References

1. Habes, M., Alghizzawi, M., Elareshi, M., Ziani, A., Qudah, M., Hammadi, M.M.A.: E-Marketing and customers' bank loyalty enhancement: Jordanians' perspectives. In: Hamdan, A., Shoaib, H.M., Alareeni, B., Hamdan, R. (eds.) The Implementation of Smart Technologies for Business Success and Sustainability: During COVID-19 Crises in Developing Countries, pp. 37–47. Springer International Publishing, Cham (2023). https://doi.org/10.1007/978-3-031-10212-7_4
2. Alghizzawi, M., Habes, M., Salloum, S.A., Ghani, M.A., Mhamdi, C., Shaalan, K.: The effect of social media usage on students' e-learning acceptance in higher education: a case study from the United Arab Emirates. Int. J. Inf. Technol. Lang. Stud. **3**(3), 13–26 (2019)
3. Gray, J.: False Dawn: The Delusions of Global Capitalism. Granta Books (2015)
4. Rogers, S.A., Calabrese, M.A., Wagner, N.J.: Rheology of branched wormlike micelles. Curr. Opin. Colloid Interface Sci. **19**(6), 530–535 (2014)
5. Alnaser, A.S., Habes, M., Alghizzawi, M., Ali, S.: The relation among marketing ads, via digital media and mitigate (COVID-19) pandemic in Jordan. Dspace. Urbe. University (2020)
6. Al Muhaisen, O.: An empirical investigation the use of information, communication technologies to english language acquisition : a case study from the Jordan TECHNOLOGIES TO ENGLISH LANGUAGE ACQUISITION : A CASE STUDY FROM. Int. J. Innov. Eng. Sci. **7**(5), 261–269 (2020)
7. Murphy, C.: Changing by the click: the professional development of UK journalists. Educ. Sci. **9**(4), 1–7 (2019). https://doi.org/10.3390/educsci9040249
8. Tandoc, E.C., Oh, S.K.: Small Departures, Big Continuities?: Norms, values, and routines in the guardian's big data journalism. J. Stud. **18**(8), 997–1015 (2017). https://doi.org/10.1080/1461670X.2015.1104260
9. Islam, A.: Making sense of data Journalism from Bangladesh perspective. Glob. Media J. **16**(31), 1–6 (2018)
10. Fahmy, N., majeed Attia, M. A.: A field study of Arab data journalism practices in the digital era. J. Pract. **15**(2), 170–191 (2021). https://doi.org/10.1080/17512786.2019.1709532
11. Kalatzi, O., Bratsas, C., Veglis, A.: The principles, features and techniques of data journalism. Stud. Media Commun. **6**(2), 36 (2018). https://doi.org/10.11114/smc.v6i2.3208
12. Porlezza, C., Splendore, S.: From open journalism to closed data: data journalism in Italy. Digit. J. **7**(9), 1230–1252 (2019). https://doi.org/10.1080/21670811.2019.1657778
13. Appelgren, E., Lindén, C.G., van Dalen, A.: Data journalism research: studying a maturing field across journalistic cultures, media markets and political environments. Digit. J. **7**(9), 1191–1199 (2019). https://doi.org/10.1080/21670811.2019.1685899
14. de-Lima-Santos, M.F., Mesquita, L.: Data journalism beyond technological determinism. J. Stud. **22**(11), 1416–1435 (2021). https://doi.org/10.1080/1461670X.2021.1944279

15. Wang, X., Li, X.: The use of data journalism in China. **310**, 558–560 (2019). https://doi.org/10.2991/iccese-19.2019.124
16. Habes, M., Alghizzawi, M., Salloum, S.A., Mhamdi, C.: Effects of facebook personal news sharing on building social capital in Jordanian Universities. In: Al-Emran, M., Shaalan, K., Hassanien, A.E. (eds.) Recent Advances in Intelligent Systems and Smart Applications. SSDC, vol. 295, pp. 653–670. Springer, Cham (2021). https://doi.org/10.1007/978-3-030-47411-9_35
17. Habes, M., Elareshi, M., Safori, A., Ahmad, A.K., Al-Rahmi, W., Cifuentes-Faura, J.: Understanding Arab social TV viewers' perceptions of virtual reality acceptance. Cogent Soc. Sci. **9**(1), 2180145 (2023)
18. Veglis, A.: Data Journalism: Definition, Skills, Difficulties, and Perspectives. Fifth edn. (2021)
19. Ziani, A., Elareshi, M., Alrashid, M., Al-Jaber, K., Journalism education in the GCC region: University students' and professionalism perspectives. Media Watch **9**(1), 52–68 (2017). https://doi.org/10.15655/mw/2018/v9i1/49275
20. Tahat, K., et al.: Detecting fake news during the COVID-19 pandemic: a SEM-ML approach. Comput. Integr. Manuf. Syst. **28**(12), 1554–1571 (2022)
21. Habes, M., Elareshi, M., Mansoori, A., Pasha, S., Salloum, S.A., Al-rahmi, W.M.: Factors indicating media dependency and online misinformation sharing in Jordan. Sustainability **15**(12), 1–15 (2023). https://doi.org/10.3390/su15021474
22. Tahat, D.N., Elareshi, M., Tahat, K., Al Jwaniat, M.A., Habes, M., Ziani, A.: News media and political participation in the middle east: Jordan as an example. In: 2022 International Arab Conference on Information Technology (ACIT), pp. 1–8 (2022)
23. Tahat, D.N., Tahat, K.M., Habes, M.: Jordanian newspapers coverage of cyberbullying during COVID 19 pandemic. PalArch's J. Archaeol. Egypt/Egyptology **17**(7), 15390–15403 (2020)
24. Al-Quran, M., Safori, A., Abdoun, Y.A., Hijab, E.: Jordanian media's use of data journalism in newsrooms: a descriptive study. Dirasat Hum. Soc. Sci. **49**(5), 91–114 (2022)
25. Zamith, R.: Transparency, interactivity, diversity, and information provenance in everyday data journalism. Digit. J. **7**(4), 470–489 (2019)
26. Lowrey, W., Broussard, R., Sherrill, L.A.: Data journalism and black-boxed data sets. Newsp. Res. J. **40**(1), 69–82 (2019)
27. Krøvel, R.: The Data Journalism Handbook: Towards a Critical Data Practice. vol. 9, no. 10 (2021). https://doi.org/10.1080/21670811.2021.1965492
28. Huda, K.N., Azad, A.K.: Professional stress in journalism: a study on electronic media journalists of Bangladesh. Adv. J. Commun. **03**(04), 79–88 (2015). https://doi.org/10.4236/ajc.2015.34009
29. Abdoun's, Y.A.: The employment of data journalism in newsrooms. In: 2020 Sixth International Conference on e-Learning (2020)
30. Vale, L., Fernandes, T.: Social media and sports: driving fan engagement with football clubs on Facebook. J. Strateg. Mark. **26**(1), 37–55 (2018). https://doi.org/10.1080/0965254X.2017.1359655
31. Phua, J., Jin, S.V., Kim, J.J.: Uses and gratifications of social networking sites for bridging and bonding social capital: a comparison of Facebook, Twitter, Instagram, and Snapchat. Comput. Human Behav. **72**, 115–122 (2017)
32. Ahmed, S.I., Zia, A.: Uses and gratification theory and social media interactivity: a demographic study in Lahore, Pakistan. Glob. Reg. Rev. **4**(1), 50–60 (2022). https://doi.org/10.31703/grr.2022(vii-i).06
33. Habes, M.: The influence of personal motivation on using social TV: a uses and gratifications approach the relationship between social media and academic performance: facebook perspective view project online training system view project the influence of personal mot. Int. J. Inf. Technol. Lang. Stud. **3**(1), 32–39 (2019)

34. Ifinedo, P.: Applying uses and gratifications theory and social influence processes to understand students' pervasive adoption of social networking sites: perspectives from the Americas. Int. J. Inf. Manage. **36**(2), 192–206 (2016)
35. Mutsvairo, B.: Challenges facing development of data journalism in non-western societies. Digit. J. **7**(9), 1289–1294 (2019). https://doi.org/10.1080/21670811.2019.1691927
36. Elareshi, M., Habes, M., Ali, S., Ziani, A.: Using online platforms for political communication in bahrain election campaigns. Soc. Sci. Humanit. **29**(3), 2013–2031 (2021)
37. Habes, M., Ali, S., Pasha, S.A.: Statistical package for social sciences acceptance in quantitative research: from the technology acceptance model's perspective. FWU J. Soc. Sci. **15**(4), 34–46 (2021). https://doi.org/10.51709/19951272/Winter-2021/3 Statistical
38. Salloum, S.A., AlAhbabi, N.M.N., Habes, M., Aburayya, A., Akour, I.: Predicting the intention to use social media sites: a hybrid SEM - machine learning approach. In: Hassanien, A.-E., Chang, K.-C., Mincong, T. (eds.) AMLTA 2021. AISC, vol. 1339, pp. 324–334. Springer, Cham (2021). https://doi.org/10.1007/978-3-030-69717-4_32
39. Salloum, S.A., Al-Emran, M., Habes, M., Alghizzawi, M., Ghani, M.A., Shaalan, K.: What impacts the acceptance of e-learning through social media? An empirical study. In: Al-Emran, M., Shaalan, K. (eds.) Recent Advances in Technology Acceptance Models and Theories. SSDC, vol. 335, pp. 419–431. Springer, Cham (2021). https://doi.org/10.1007/978-3-030-64987-6_24
40. Habes, M., Salous, M.H., Al Jwaniat, M.I.: Applying the uses and gratifications theory to college major choice using social networks online video. In: International Conference on Advanced Machine Learning Technologies and Applications, pp. 388–400 (2022)
41. Elareshi, M., Habes, M., Al-Tahat, K., Ziani, A., Salloum, S.A.: Factors affecting social TV acceptance among Generation Z in Jordan. Acta Psychol. (Amst). **230**, 103730 (2022)

Impact of Cash Flow Statement Elements on Financial Performance: The Mediating Role of Capital in Private Hospitals in Jordan

Ola Khresat[1(✉)], Fuad Al-Fasfus[1], Omar Shaban[1], Yousef Shahwan[1], and Abdullah Alsilawi[2]

[1] Zarqa University, Zarqa 13110, Jordan
okhresat@zu.edu.jo
[2] University of Business and Technology, Jeddah, Kingdom of Saudi Arabia
a.silawi@ubt.edu.sa

Abstract. The study aims to test the impact of the elements of the cash flow statement on the financial performance of private hospitals in Jordan. The study sample includes (5) private hospitals for the period between (2016–2020), with a capital of more than fifteen million JD. The path analysis method is also used, through the Structural Modeling Program (Smart PLS) in order to achieve the study's objectives and answer the study's hypotheses. This study finds, there is an impact of elements of cash flows of operating, investing, and financing activities on the financial performance measures through (liquidity ratios, financial leverage, and profitability). Also, there is an impact of cash flows of operating activities, on the volume of capital in private hospitals.

The study recommends decision-makers to pay attention to private hospitals of the elements of the cash flow list and manages it efficiently. The capital is taken into account when evaluating the financial performance of private hospitals. Finally, the study recommends to conduct the study variables on other sectors, such as the financial sector, industrial sectors, and other service sectors to understand how the elements of their cash flow affect performance.

Keywords: Statement of Cash Flows · Capital · Financial Performance · Private Hospitals

1 Introduction

The statement of cash flows is prepared to know the sources of cash flows resulting from the company's business and its use [1], and it is an important financial instrument as it combines three main statements; income statement, equity, the balance sheet statements. The statement of cash flow may differentiate among companies, and from one country to another, in parallel with other information of finance. Because of analyzing the relationships of financial statements to performance in the past and in the future periods, the cash flow statements proves to be very successful and useful. Also, it reflects cash-generating and expenses [2]. In addition to, the list of cash flows includes cash flow inwards, as

B. A. M. Alareeni and I. Elgedawy (Eds.): *Artificial Intelligence (AI) and Finance*, SSDC 488, pp. 821–830, 2023.
https://doi.org/10.1007/978-3-031-39158-3_76

well as describes the amount of cash flow resulting from the business and activities of the assigned firm, and the reasons of this cash flow availability. Also, it is considered a source of reliable information as it reflects important cash indicators related to the activities of the entity [3, 4].

Many studies have tested the relationship between the elements of the cash flow statement of companies and their impact on financial performance, where a study [5] indicated that there is a statistically significant inverse relationship shown at the level of 5% between the cash flow of companies and stock prices. He also studied [6] the impact of changes in the cash flows of the commercial banks in Jordan listed on the stock exchange on equity returns. The study has a result that the change in cash flows had positively effected on the divides, as well as the study [7] and a study [8], indicated that operating and financing cash flows have had a positive impact on the performance of companies. In addition, they found that other cash flows that have a negative correlation with the performance of companies.

Many studies have been carried out by studying the ability to operate cash flows to predict the future of operating cash flows compared with historical earnings, including a study by [9] of showed that the forecast based on cash flows reflects more accurate indicators of earnings of predicting the financial performance and the future of operating cash flows, compared with the study [10] where they indicated a statistically significant relationship between operating cash flows, and the market value of the company. Net profit and market value of the company show insignificant relationship, according to the results of the study and the studies which were conducted by previous researchers, which give the operating cash flow and important indicators of the purposes of decision-making by investors, and the reason of this, is that management has no motivation or ability to manipulate the cash flow list. Profits cannot be manipulated which were based on the cash basis, such as the possibility of manipulation according to the accrual basis [10].

The cash flow statement is important of improving the efficiency, effectiveness, and decision-making ability of Management regarding suitable financial planning, the ability of the company is to make profits, payment of obligations, interest in the fields of operational activities, financing activities, equity pricing, and the ability to distribute divides and pay them to shareholders. It also helps in assessing the liquidity position of the company regarding generating cash, and short-term payment obligations [11, 12]. This list shows cash movement along with its equivalents, and its impact on the business cash management of the establishment [13, 14].

A study of [15] found out that there was a non-significant relationship with corporate bankruptcy while other studies stating that a company's operating cash flow is a poor predictor of a company's financial failure, and that there is a nonlinear relationship between monetary possession, and financial performance of listed and unlisted companies in the kingdom of Saudi Arabia [16, 17]. Hence, the main goal of establishing companies is diversity, and profit growth that is measured by various financial factors. Moreover, the success of the company depends on its long-term investments and requires a large number of cash assets in the company. As a result, spending many cash assets will positively affect the investments which are made by the company and increase the risk which will negatively affect the company's financial performance [18, 19].

Many studies have stated that making a profit is the main goal and the driving force of each company, as all business supervisors and employers in charge work to raise certain elements including productivity, profitability rate, promising growth, capital, operating expenses, and credit risk which are the primary determinants of profitability [20, 21]. Effective management of working capital is also influenced by the company's cash position and leads to increased profitability. Many studies on profitability and liquidity have found out that the management of liquidity (cash) significantly affects financial performance in terms of risk and profitability of the business [22], and the strength of the financial performance of the company is enhanced by the effectiveness of working capital management. On the other hand, the company's leverage is the primary determinant of capital risk. If the amount of loan or debt in the capital structure increases, the risk of default increases for the companies and thus, the value of their capital decreases and the ability to perform their financial and administrative functions decreases [23]. If the mix is not suitable on the left side (sources of funds) of the balance sheet (liabilities and equity), it becomes a challenge and risk for the company, and it will expose them to many risks. It will affect the company's operational, investment and financing activities, its ability to generate profits [24], so this study took into account three main factors related to measuring financial performance in the Jordanian health sector (liquidity ratios, Leverage ratios, and profitability ratios) [25]. The study found out that the factors affecting the distribution of profits from 2004 to 2015 are free flow, financial leverage, profitability and liquidity, and that the size and age of the bank have no effect on them.

2 Literature Review

Many studies have noticed on the cash flows issues and their impact on financial performance such as the study of [26, 27]; and others, [9, 12, 16, 18, 26, 28–31].

The study of [26]; and [12, 18, 26, 29–31]; and [12] have concluded that a significant impact of net cash flows of operating activities, financing and investing is found on the financial performance (Index, return on assets, the index return on equity). The [32] found out that the cash flow statement provides important information about the ability of the project to meet the obligations, and the company may achieve a positive result, However, it might have negative cash flows through one of its three activities (operational, investment, financing) and this puts the company in a state of inability to pay its obligations, as indicated [27]. The cash flow statement provides important information about the ability of the company to pay its obligations, and the cash flow statement provides information that contributes to the assessment of the liquidity and profitability of the company in addition to the policies followed for the purpose of making appropriate decisions in the future, A study of [28] found out that there is an impact of the net operating cash flows, the ratio of net operating cash flows to total liabilities, equity and current liabilities have an impact on profitability indicators and the net operating cash flows ratio to current liabilities and net sales on liquidity indicators, while the other ratios which were used in this study have no impact on profitability and the indicators of liquidity. Regarding [16], the results showed that cash assets play an important role of influencing the financial performance of the companies and that there is a non-linear relationship, showing adverse U form of this non-linear relationship from the principles of barter theory to the optimal monetary level based on holding cash costs and benefits. The results here in this study show very useful insights for both managers and policymakers regarding effective levels of corporating liquidity management. They also show needed further insights for the concerned investors on the role of corporating cash retention levels on company performance.

[9] study has found out that analysts cash flow forecasts have shown to be less accurate than earnings forecasts and moves at a slow rate over the forecast duration. In addition, the study shows that the cash flow forecast is an extension of the profit forecast, providing limited information about expected changes in working capital and that the cash flow forecast has information content which can be describes as a limited one and is only weakly correlated with return on inventory. Finally, an accrual-based performance forecast that shows the difference between earnings-based and cash-flow-based forecasts does not lead to a better discovery of earnings management than commonly used accrual models.

3 Methodology and Hypotheses

The problem of the study: The study problem indicates that the health sector/private hospitals in Jordan, suffer from a discrepancy in the level of financial performance, which in turn reflects on the liquidity ratios, profitability, and financial leverage (indebtedness). Thus, hospitals become concerned in generating liquidity operations which is reflected by financial lists and financial performance indicators.

Although the subject of evaluating performance is considered one of the most essential topics for hospitals it is based on many financial decisions and an important strategy for decision-makers and investors. The process of performance evaluation is one of the financial problems because it requires a keen interest in reducing errors, so the process of financial performance evaluation requires the use of modern financial measurement methods, which the cash flow statement may be one of them. Hence, it can be said that the main problem of the study is identifying the elements of the cash flow statement impact on the financial performance of private hospitals in Jordan.

The hypotheses of the study which were built as result of study problem are:

H01: There is no significant impact at the level of 5% for the statement of cash flows items on the financial performance of private hospitals in Jordan.

There are sub-hypotheses:

$H01_1$: There is no significant impact at a 5% significant level of cash flows of operating activities on financial performance in private hospitals.
$H01_{2:}$ There is no significant impact at a significant level of 5% level for cash flows of investing activities on financial performance in private hospitals.
$H01_3$ There is no significant impact at a significant level of 5% for cash flows of financing activities on financial performance in private hospitals.

H02: There is no significant impact at a level of 5% for the elements of the cash flow statement on the size of the capital in private hospitals in Jordan.

There are sub-hypotheses:

$H02_1$: There is no significant impact at a level of 5% of cash flows of operating activities on the size of capital in private hospitals.
$H02_2$: There is significant impact at a level of 5% of cash flows of investing activities on the size of capital in private hospitals.
$H02_3$: There is no significant impact at a level of 5% of cash flows of financing activities on the size of capital in private hospitals.

H03: There is no significant impact at a level of 5% of capital size in the financial performance of private hospitals in Jordan.
H04: There is no significant impact at a 5% level of the cash flow statement elements in the financial performance with the size of capital in private hospitals in Jordan.

4 Findings and Discussion

The hypotheses of the study were tested through the structural modeling program (Smart PLS), in order to answer their assumptions, and the results were according to the following:

H01: There is no significant impact at the level of 5% of the statement of cash flows items on the financial performance of private hospitals in Jordan.

Table 1. Results of the analysis of the first main hypothesis

Dependent variable: liquidity, leverage and profitability ratios		
Independent variables	T	Sig.
Operating cash flows	−2.834	0.004
Investment cash flows	4.390	0.000
Financing cash flows	2.742	0.007
R2	0.34	

Table 1 shows the results of the analysis of the first main hypothesis, and the sub-hypotheses there are, indicates that the value of the determination coefficient (R^2 = 34.00%), which means that (34.00%) of the change in financial performance (liquidity ratios, leverage, profitability) is due to the elements of cash flows and therefore the first alternative of main hypothesis, which states: "There is a significant impact of the cash flow statement elements on the financial performance of private hospitals in Jordan" shall be accepted. As for operating cash flows, the value of ($T = -4.390$) in a significant sense less than 5%; which means rejecting the first sub-zero hypothesis, and accepting the alternative hypothesis that states: "There is a significant impact of cash flows resulting of operating activities on financial performance (liquidity ratios, leverage, profitability) in private hospitals". With respect to investment cash flows, the value of ($T = 4.390$), and in a moral sense less than 5%; which means rejecting the second sub-zero hypothesis, and accepting the alternative hypothesis that states: "A significant impact of cash flows of investment activities is found on financial performance (liquidity ratios, leverage, profitability), in private hospitals". With respect to financing cash flows, the value of (T = 2.742), with a moral significance of less than 5%; which means the rejection of the third sub-zero hypothesis, and the acceptance of the alternative hypothesis that states: "A significant impact of cash flows of financing activities is found by financial performance (liquidity ratios, leverage, profitability) in private hospitals".

H02: There is no significant impact at a level of 5% for the elements of the cash flow statement on the size of the capital in private hospitals in Jordan.

Table 2. Results of the path analysis of the second main hypothesis

Dependent variable: capital size		
Independent variables	T	Sig.
Operating cash flows	2.083	0.040
Investment cash flows	1.898	0.094
Financing cash flows	−1.068-	0.417
R2	0.085	

Table 2 shows that the results of the path analysis of the second main hypothesis and sub-hypotheses indicated that the value of the determination coefficient (R^2 = 8.5%), which means that (8.5%) of the change in the size of capital is due to the elements of cash flows, and therefore the second alternative main hypothesis will be accepted, which states: "There is a significant impact of the cash flow statement elements on the capital size of private hospitals in Jordan". With respect to operating cash flows, the value of (T = 2.083), in a significant sense less than 5%; which means rejecting the first sub-zero hypothesis, and accepting the alternative hypothesis that states: "A significant impact of cash flows of operating activities is found by the size of capital in private hospitals". With respect to investment cash flows, the value of T = 1.898, and a significant amount greater than 5%; which means acceptance of the second sub-zero hypothesis, which states: "There is no impact, that is significant, of cash flows from investment activities on the size of capital in private hospitals.", The reason is that there is no impact of cash flows of investment activities on the size of capital; cash flows of investment activities are negative and with an average (−1502119.08) and regarding financing cash flows are valuable (T = −1.068), and with a moral significance greater than 5%; which means accepting the third sub - zero hypothesis, which states: "There is no significant effect of cash flows of financing activities on the volume of capital in private hospitals" because average funding cash flows in private hospitals is 708872.16).

H03: There is no significant impact of capital size on financial performance (liquidity ratios, leverage, and profitability) in private hospitals (Table 3).

Table 3. Results of the path analysis indicated for the third main hypothesis

Dependent variable: liquidity, leverage and profitability ratios		
Independent variables	T	Sig.
Capital size	12.134	0.000
R2	0.516	

The previous table shows that, the results of the path analysis indicated of the third main hypothesis, the value of the determination coefficient ($R^2 = 61.6\%$) which means that (61.6%) of the change in liquidity ratios, financial leverage, and profitability is due to the size of capital, and therefore, the third alternative main hypothesis will be accepted which states: "There is a significant impact of the size of capital on financial performance (liquidity ratios, financial leverage, profitability), in private hospitals". With regard to the size of the capital, the value of (T = 12.134), and in a moral sense less than 5%; which means the rejection of the zero hypothesis, and the acceptance of the alternative hypothesis that states: "There is a statistical significance of the size of capital on financial performance: (liquidity ratios, financial leverage, profitability) in private hospitals".

H04: There is no significant impact of the elements of the cash flow statement on financial performance: (liquidity ratios, financial leverage, profitability), with the volume of capital in private hospitals in Jordan.

Table 4. Results of the fourth main hypothesis test

Dependent variable: liquidity, leverage, and profitability ratios		
Independent variables	T	Sig.
Cash flows and capital size	11.583	0.000
R2	0.548	

Table 4 shows that, the results of the path analysis of the fourth main hypothesis indicate that the value of the determination coefficient (R2 = 54.8%), which means that (54.8%) of the change in liquidity ratios, financial leverage, and profitability is due to cash flows, in the presence of the size of capital, and therefore, we will accept the fourth alternative main hypothesis which states: "There is a significant impact of the elements of the cash flow statement on financial performance (liquidity ratios, financial leverage, profitability) in the presence of the size of capital in private hospitals in Jordan". With regard to the size of the capital, the value of (T = 11.583), and in a moral sense less than 5%; which means rejecting the zero hypothesis and accepting the alternative hypothesis that states: "There is a significant impact of the elements of the cash flow statement on financial performance: (liquidity ratios, financial leverage, profitability)in the presence of the size of capital in private hospitals in Jordan".

5 Conclusion

The study finds that there is a significant impact of cash flows of operating, investment, as well as financing activities on financial performance: (liquidity ratios, financial leverage, profitability) in private hospitals. Also, there is a significant impact of cash flows of operating activities on the size of capital in private hospitals. On the other hand, there is no evident significant impact of cash flows of investment and financing activities on the size of capital in private hospitals. There is a significant impact of the size of capital on financial performance: (liquidity ratios, financial leverage, profitability) in private hospitals. There is also a statistically significant impact of the elements of the cash flow statement on financial performance: (liquidity ratios, financial leverage, profitability) with the presence of the volume of capital in private hospitals in Jordan and, therefore, The results of the study are consistent with the study [8] which note that operating cash flows, and Finance had an impact on the performance of companies that proved to be positive as well as the results of the study are consistent with the study of[9]which showed that the model based on historical cash flows has more accurate predictions than historical earnings in predicting financial performance. In addition, future operating cash flows agree with the study [16], that there is a nonlinear relationship between cash holdings, capital, and the financial performance of listed and unlisted companies in the kingdom of Saudi Arabia. The results of this study are consistent with a study [21], that indicated that concepts of growth, productivity, capital, operating expenses, and credit risks are considered the primary determinants for profitability and financial performance. And the results agree with the study [12] that concerned in the areas of operational activities, financing activities, pricing of shares, and ability to distribute dividends. It also helps to assess the liquidity position of the company regarding cash generation, and payment of short-term obligations.

6 Recommendations

Therefore, the study recommends that private hospitals should pay more effort and attention to the elements of the cash flows statement and manage them efficiently and analyze the cash flow statement, the researchers found that 20 private hospitals in Jordan that do not have a computerized accounting system out of 69 private hospitals. The study also recommends future studies to use study variables on other sectors, such as: financial sector, Industrial sectors, and other service sectors.

References

1. Deo, P., Liu, C.Z.: All cash is not created equal: detecting fraudulent cash flows (2016)
2. Davis, O.: Chair of management accounting statement of cash flows: issues in cash flow reporting in Estonia (2106)
3. Wijewaradana, W.P., Munasinghe, P.G.: Journal of management matters faculty of management studies (2015)
4. Shahwan, Y., Abdel-hamid, O.: The impact of social responsibility disclosure in reducing the risk in Jordanian commercial banks. Res. World Econ. **11**(3), 142 (2020). https://doi.org/10.5430/rwe.v11n3p142

5. Jabbari, H., Sadeghi, Z., Askari, A.: Cash flow, earning opacity and its impact on stock price crash risk in Tehran stock exchange. Int. J. Acad. Res. Account. Finance Manag. Sci. **3**(4), 180–190 (2013). https://doi.org/10.6007/IJARAFMS/v3-i4/385
6. Hamza, M., Jaradat, M.M.: The impact of changes in the elements of cash flows statement on stock returns in Jordanian commercial banks listed on Amman stock exchange: an empirical study. J. Account. Finance Manag. Strategy **13**(2), 1–22 (2018)
7. Oroud, Y.S., Islam, M.A., Salha, T.: The effect of cash flows on the share price on Amman stock exchange (2017). http://www.abrj.org
8. Anastesia Nwakaego, D., Innocent Ikechukwu, O.: Payable management on corporate profitability of brewery manufacturing companies in Nigeria (2015). www.questjournals.org
9. Givoly, D., Hayn, C., Lehavy, R.: The quality of analysts' cash flow forecasts the quality of analysts' cash flow forecasts the quality of analysts' cash flow forecasts (2008)
10. Jemâa, O.B., Toukabri, M., Jilan, F.: Accruals and the prediction of future operating cash-flows: evidence from Tunisian companies. Int. J. Account. Econ. Stud. **3**(1), 1 (2014). https://doi.org/10.14419/ijaes.v3i1.3867
11. Shahwan, Y., Hamza, M., Al-Fasfus, F., Al-Ramahi, N., Almubaydeen, T.H.: Social responsibility accounting, disclosure and real practice: evidence from Jordan. Asian Econ. Financ. Rev. **12**(3), 164–182 (2022). https://doi.org/10.55493/5002.v12i3.4441
12. Al-Nimer, M.: The effect of profitability ratios on market capitalization in Jordanian insurance companies listed in Amman stock exchange (2015). www.iiste.org
13. David, C.O.: The potency of cash flow in predicting corporate performance. Acc. Financ. Manag. J. **03**(06) (2018). https://doi.org/10.31142/afmj/v3i6.06
14. Shahwan, Y.: The mediating effect of investment decisions and financing decisions on the influence of capital structure against corporate performance: evidence from Jordanian listed commercial (2018). https://www.researchgate.net/publication/330145397
15. Gombola, M.J., Ketz, J.E.: A note on cash flow and classification patterns of financial ratios (1983)
16. Alnori, F.: Cash holdings: do they boost or hurt firms' performance? Evidence from listed non-financial firms in Saudi Arabia. Int. J. Islam. Middle East. Finance Manag. **13**(5), 919–934 (2020). https://doi.org/10.1108/IMEFM-08-2019-0338
17. Shahwan, Y.: The moderating effects of dividend policy on the impact of financial disclosure towards corporate value: a study of listed Jordanian industrial firms. Res. J. Finance Account. (2019). https://doi.org/10.7176/RJFA/10-6-13
18. Liman, M., Mohammed, A.S.: Operating cash flow and corporate financial performance of listed conglomerate companies in Nigeria. IOSR J. Humanit. Soc. Sci. IOSR-JHSS **23**(2), 1–11 (2018). https://doi.org/10.9790/0837-2302110111
19. Shahwan, Y.: The impact of earning management and social and environmental costs disclosure accounting information systems view project the impact of earning management and social and environmental costs disclosure on financial performance view project (2021). https://www.researchgate.net/publication/355909725
20. Shahwan, Y.: The effect of OECD principles of corporate governance on social responsibility accounting (2020). www.ijicc.net
21. Malik, H.: Academic research international determinants of insurance companies profitability: an analysis of insurance sector of Pakistan **1**(3) (2011). www.savap.org.pkwww.journals.savap.org.pk
22. Raheman, A., Afza, T., Qayyum, A., Bodla, M.A.: Working capital management and corporate performance of manufacturing sector in Pakistan (2010). http://www.eurojournals.com/finance.htm
23. Jarrow, R.: A leverage ratio rule for capital adequacy. J. Bank Finance **37**(3), 973–976 (2013). https://doi.org/10.1016/j.jbankfin.2012.10.009

24. Asif, A., Rasool, W., Kamal, Y.: Impact of financial leverage on dividend policy: empirical evidence from Karachi stock exchange-listed companies (2018). www.internationalscholarsjournals.org
25. Suliman Al-Fasfus, F.: Impact of free cash flows on dividend pay-out in Jordanian banks. Asian Econ. Financ. Rev. **10**(5), 547–558 (2020). https://doi.org/10.18488/journal.aefr.2020.105.547.558
26. Raj, A.R., Sharma, B.: Cash flows and performance in the industrial sector of Saudi Arabia: with special reference to insurance and manufacturing sectors. Invest. Manag. Financ. Innov. **17**(4), 76–84 (2020)
27. Al-R., A., Al-A., Z.: The contribution of the statement of cash flows to the evaluation of the financial performance of the institution, a case study of the institution of distribution and maintenance of agricultural equipment. University of Martyr Hama Lakhdar in El-Wadi Algeria, Algeria (2019)
28. Zahawi, Z.: Measuring the impact of changes in cash flows on the level of financial performance in industrial sector institutions, a case study of some Algerian institutions. University of Ghardaia, Algeria (2018)
29. Al-Khattneh, O.J.: The impact of cash flows on financial performance in Jordanian insurance companies. Amman Arab University, Amman, Jordan (2016)
30. Eyahuma, G., Miroga, J.: Influence of cashflow activities on financial performance of commercial banks in Kenya: case of Kakamega county (2020). www.strategicjournals.com
31. E.A.S.: The impact of cash flow ratio on coparate performance (2014). www.iiste.org
32. Kabaye, S.: The role of the cash flow statement in evaluating the performance of the economic institution: a case study of the Saidal corporation for the period from 2016 to 2018. University of May 8, 1945, Algeria (2020)

The Impact of Using CRIF Platform on Credit Risks for Jordanian Islamic Banks

Samer Zakarneh[1(✉)], Yahiya Al-Khasawneh[1], Munir Al-hakim[2], Ibrahim Khrais[1], Mohammad Musa Al-Momani[1], and Kholoud Bajunaid[1,2]

[1] Faculty of Economics and Business Administrative Sciences, Zarqa University, Zarqa, Jordan
{szakarneh,ykhasawneh,ikhrais,m.momani}@zu.edu.jo,
K.bajunaid@UBT.EDU.SA

[2] Faculty of Administrative and Financial Sciences, Petra University, Amman, Jordan
Munir.alhakim@uop.edu.jo

Abstract. This study aims to identify whether using CRIF platform have an impact on credit risks levels for Islamic banks, as these banks began to use it by the end of 2016, and to find this impact, the ratio of non-performing loans (NPL) was used and then compared for Jordanian Islamic banks before and after start using CRIF platform for the period between (2012–2021). The most important result of this study was no impact for using this platform on credit risks level for Jordanian Islamic banks. The study recommended that further research should be done to determine the feasibility of using this platform and this to be done by including all Jordanian banks in the study sample i.e., conventional, and Islamic banks.

Keywords: CRIF · Credit Risk · NPL · Islamic Banks

1 Introduction

Credit risks are one of the most important types of risks that banks are facing, as they are directly related to the profitability of shareholders, deposit holders and the liquidity ratios of banks i.e. may lead to liquidity issues result from mismatch gap in the cash flow, as it constitutes approximately 50% of the set of risk elements [1]. As a result, banks do everything they can do to minimize these risks and their impact by doing all kinds of analysis related to borrower capability of repaying the funds to the bank, and within the framework of these methods and instruments comes the CRIF platform, which all Jordanian banks use and benefit from the data available to them in order to minimize this type of risk and then have a high quality loans portfolio.

1.1 Statement of the Problem

Banks take many measures to reduce their credit risks, such as requesting guarantees, mortgages, and guarantors, and carrying out feasibility studies before approving credit granting to the customer. CRIF is one of the methods that is used by Islamic banks in

B. A. M. Alareeni and I. Elgedawy (Eds.): *Artificial Intelligence (AI) and Finance*, SSDC 488, pp. 831–839, 2023.
https://doi.org/10.1007/978-3-031-39158-3_77

Jordan, to learn about the customer's reputation and credit history to have a reliable data about the customer credit history and reputation which reflects his historical behavior and then to use these data as input for the decision-making process.

This study seeks to determine the viability of using CRIF platform and whether it assisted banks in reducing credit risks or not. The study's problem can be expressed as the following main question:

What impact does the CRIF platform have on credit risks levels for Jordanian Islamic banks?

1.2 Objectives of the Study

This study has one main objective:

Identifying the effect of using CRIF platform on credit risks levels for Jordanian Islamic bank.

1.3 The Importance of the Study

The study's importance lies in recognizing the feasibility and benefit of using the CRIF platform, and whether it assisted Jordanian Islamic banks in reducing their credit risks or not. This enables banks to continue using this platform as a major source of information for making credit decisions. It will also help banks in determining the source of their credit risks, whether it is because of the type, reputation, and history of the customer, or in the circumstances that the customer may be exposed to after being granted the bank's financing.

2 The Literature Review

2.1 CRIF: (Credit Information Company); Jordan Credit Bureau

A private shareholding company CRIF Jordan which was established in 2013 with the purpose of collecting and storing credit information as well as creating a database for work and dealing with it. The most prominent founders of this company are most of the Jordanian banks in addition to the Italian Crave Company [7].

The company was granted a license by the Central Bank of Jordan in 2015. In 2016, the company launched its operations through an electronic platform created in collaboration with the international firm CRIF. All CRIF Jordan's activities are governed by the Credit Information Law No. 36 of (2011), which is monitored and controlled by the Central Bank [8].

CRIF Jordan obtains credit data from banks and finance companies and provides credit services to banks, finance companies, insurance companies and judicial and supervisory entities. These data are provided upon request. These include the credit history of individuals or companies in terms of the value and nature of the credit facilities granted the value of the premiums, the customer's obligation to pay, whether he has litigation related to non-payment, what are the guarantees and guarantees provided when he receives credit facility [8]. In summary CRIF can be defined as It is **a three - digit number which gives a bank or a lender a sense of your creditworthiness** i.e. whether the intended customers have been repaying the lender dues regularly or not.

2.2 Credit Risk

Credit risk is directly linked to the borrower's failure to pay due instalments representing borrowed amounts, which is a major type of risk faced by all banks because is related to their core banking activities i.e., lending/financing.

It can therefore be said that credit risks indicate the likelihood that some of the Bank's assets, especially loans, may be depreciated to zero [5] (i.e., they may reach to a point to be considered as uncollected bad debts), adversely affecting the level of profitability and consequently the decline in the Bank's equity or the lower returns on Islamic banks' unrestricted investment accounts/Modaraba investment accounts.

One of the most Islamic financing modes exposed to credit risk in its banking concept, those are based on debt or loan-based concept such as Murabaha contracts, Salam and Istisna' as the price of the commodity sold to the buyer (debtor), as well as the agreed profit, is set aside and paid by specific dates. This type of risk will be exposed in Sect. 3.

2.3 Hypotheses of the Study

Based on the study's problem and objectives, the following main hypothesis can be developed:

There is no statistically significant effect at the (@ ≤ 0.05) level of significance, for using the CRIF platform on credit risks levels for Jordanian Islamic banks (Fig. 1).

2.4 Study Model

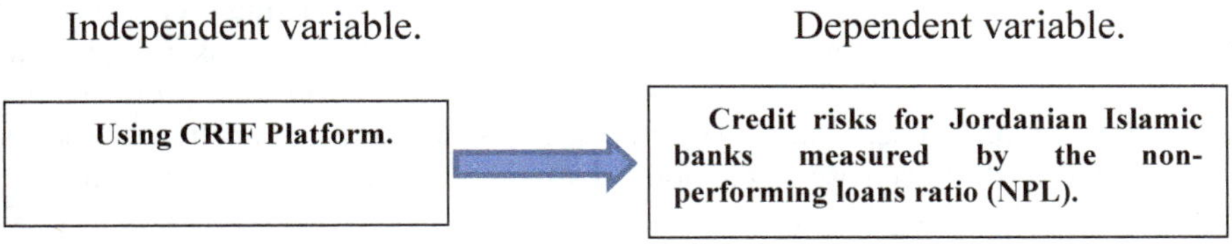

Fig. 1. Study Model

The credit risk will be expressed by the non-performing loans ratio, this ratio is calculated as follows:

Non-Performing loan Ratio (NPL) = Total Non-Performing loan (NPL)/Total Credit Financing (Loans) portfolio [4]. This is based on study [1] and study [2].

2.5 Methodology of the Study

The inductive approach will be used by referring to published research, references, and websites. The analytical descriptive approach will also be followed by calculating the non-performing loans ratio (NPL) for Jordanian Islamic banks for a period of 5 years prior to start using the CRIF platform i.e. (2012–2016) and then comparing its results with five years after start using the CRIF platform (2017–2021), [3].

And after that, the averages and differences between the averages will be extracted and then the necessary statistical tests will be executed to obtain the results by using the (SPSS) statistical analytical program to get the results that will be analyzed and used to test the hypothesis.

2.6 Study Population and Sample

The study population consists of all Jordanian banks, and the study sample includes three Islamic Jordanian banks: Jordan Islamic Bank for Investment and Finance, Islamic International Arab Bank, and Safwa Islamic Bank.

3 Previous Studies

Study [4], Aimed to identify the determinants of non-performing loans in developing countries' banking sectors through an applied case study for Pakistan from 2005 to 2017.The sample included all commercial listed banks in the Pakistan financial stock exchange. The study variables included: profitability, operational efficiency, capital adequacy, and diversification of income sources. The estimations and analysis have been done by using the SPSS statistical program.

According to the study's findings, operational efficiency and profitability have a statistically significant negative correlation with the non-performing loans ratio (NPL), whereas capital adequacy and income diversification have a negative correlation, but it is not statistically significant.

It was concluded that the non-performing loans ratio (NPL) increases as a result of various factors such as political interventions in the state's banking sector and poor investment decisions made by bank management. As a result, policymakers and stakeholders must ensure that banks have robust financial health by making profits so that banks can carry out proper loan management process such as creditworthiness checks, which should lead to decrease in the non-performing loan ratios (NPLs) and this will lead banks to achieve higher profits levels.

Study [5], Aimed to shed light on the role of credit rating and its impact on the determinants and conditions for granting loans to companies, due to the importance of this subject, as the default by one of the corporate customers may constitute a heavy burden on the performance and profitability of the bank. The study sample included information for 1.6 million loans In the United States of America, this covers approximately 78% of corporate loans in America, the researchers' expectations were that the amendments to the loan conditions would be more sensitive to the lower than the higher credit rating amendments.

The study results showed that the impact of changing the credit rating was not equal for both directions, as the increase in the profit margin on the loan was higher in the case of lowering the credit rating than in the case of raising the credit rating.

Furthermore, changes in the ratings of the bank's portfolio have a direct role in modifying loan terms, and the results also indicated that there is a big interaction between the degree of credit rating of borrower's category on the one hand and the bank's regulatory capital requirements on the other hand.

Study [6], covered the period from 2003 to 2014 and aimed to analyze the factors that effect on the failure of credit facilities in commercial banks in Jordan. Two types of variables were included: the first is economic variables group, which includes the GDP growth rate and government capital expenditures, and the second is banking variables group such as the bank size, Interest rate on granted credit facilities, the degree of bank risk and the management of the bank.

The study demonstrated that the phenomenon of credit facilities default is not a growing phenomenon in the Jordanian banking system due to its high financial solvency and the retention of high rates of liquidity as well as its high provisions to overcome any default of these facilities.

According to the study's findings, there is a negative statistical relationship between the volume of defaulted credit facilities and bank size. Also, the results of the study showed that there is no statistically significant relationship between defaulted credit facilities and the interest rate, the degree of banking risk, and the management of the bank.

3.1 The Operational Framework

3.2 Non-performing Debt Ratios (NPL) of Jordanian Islamic Banks

Referring to the financial statements of the Jordanian Islamic banks as a sample for the study, non-performing loans ratios (NPL) were calculated as follows (Table 1).

Table 1. Non-Performing Debt Ratios (NPL) of Jordanian Islamic Banks

Bank name	2021	2020	2019	2018	2017	2016	2015	2014	2013	2012
Jordan Islamic Bank [9]	0.022	0.027	0.033	0.034	0.030	0.028	0.031	0.031	0.033	0.035
Islamic International Arab Bank [10]	0.028	0.032	0.037	0.020	0.010	0.007	0.009	0.010	0.019	0.012
Safwa Islamic bank [11]	0.016	0.025	0.022	0.017	0.018	0.016	0.014	0.023	0.039	0.022

From the data in the above table:

The arithmetic means of the non-performing loan ratio (NPL) before start using the CRIF platform (2012–2016) for Jordan Islamic Bank is: 0.031
The arithmetic means of the non-performing loan ratio (NPL) after start using the CRIF platform (2017–2021) for Jordan Islamic Bank is: 0.029
The arithmetic means non-performing loan ratio (NPL) before using the CRIF platform (2012–2016) for Islamic International Arab Bank is: 0.011
The arithmetic means non-performing loan ratio (NPL) after using the CRIF platform (2017–2021) for Islamic International Arab Bank is: 0.025
The arithmetic means of the non-performing loan ratio (NPL) before using the CRIF platform (2012–2016) for Safwa Islamic Bank is: 0.022

The arithmetic means of the non-performing loan ratio (NPL) after start using the CRIF platform (2017–2021) for Safwa Islamic Bank is: 0.020
The arithmetic means of the non-performing loan ratio (NPL) before using the CRIF platform (2012–2016) for the three Islamic banks is: 0.021
The arithmetic means of the non-performing loan ratio (NPL) after start using the CRIF platform (2017–2021) for the three Islamic banks is: 0.024

3.3 Testing the Normal Distribution of the Data

This test was carried out to determine whether the data is moderate (tracing the normal distribution) or not, and it is necessary to determine the appropriate statistical methods.

The normal distribution of the data was tested through the Shapiro-Wilk test because the sample size is less than (50).

- The null hypothesis: the data are not normally distributed.
- Alternative hypothesis: The data are normally distributed according to a normal distribution.

The results are as shown in Table 2.

Table 2. Tests of Normality

	Kolmogorov-Smirnov[a]			Shapiro-Wilk		
	Statistic	df	Sig.	Statistic	df	Sig.
PRE	.141	15	.200*	.943	15	.417
POST	.105	15	.200*	.976	15	.931

*This is a lower bound of the true significance.
[a]Lilliefors Significance Correction.

The table of results shows that the significance value in the Shapiro-Wilk test is equal to (0.417) for the non-performing loan ratio (NPL) before applying CRIF system and (0.931) for the non-performing loan ratio (NPL) after applying CRIF system for the three banks under study, and this value is greater than the level of significance (0.05) that is in the acceptance area.

Therefore, the null hypothesis that the data is not normally distributed is rejected, and the alternative hypothesis that the data is normally distributed is accepted.

3.4 Paired Samples T-Test

Due to the normal distribution of the data, the (T) test was used for two related samples, where the ratio of non-performing loan (NPL) during five years before applying the CRIF system is compared with the ratio of non-performing loans during five years after the applying, for the three banks under study, and the results were as shown in Table 3.

The above table shows that the arithmetic mean of the non-performing loan ratio (NPL) before applying the CRIF system is (0.0219), and the arithmetic mean of the

Table 3. Paired Samples Statistics

		Mean	N	Std. Deviation	Std. Error Mean
Pair 1	PRE	.0219333	3	.01012785	.00584732
	POST	.0247333	3	.00483460	.00279126

non-performing loan ratio (NPL) after applying CRIF system for the three banks under study is (0.0247), and although this difference does not exist, a decision can be made about the effect of using the CRIF system on the non-performing debt ratio (NPL) only after reviewing the results of the T-test as shown in Table 4.

Table 4. Paired Samples Correlations

		N	Correlation	Sig.
Pair 1	PRE & POST	3	.324	.790

The above table shows the degree of correlation between the average of non-performing loan ratio (NPL) before applying the CRIF system and the average of non-performing loan ratio (NPL) after applying the CRIF system for the three banks under study, as it amounted to (0.324), which is a medium degree of correlation, and the significance was (0.79). Greater than 0.05 meaning there is no statistical significance of the correlation as shown in Table 5.

Table 5. Paired Samples Test

		Paired Differences					t	df	Sig. (2-tailed)
		Mean	Std. Deviation	Std. Error Mean	95% Confidence Interval of the Difference				
					Lower	Upper			
Pair 1	PRE - POST	−.00280000-	.00970773	.00560476	−.02691534-	.02131534	−.500-	2	.667

The above table shows the value of the difference between the average of non-performing loan ratio (NPL) before applying the CRIF system and the non-performing loan ratio (NPL) after applying the CRIF system for the three banks under study, as it amounted to (−0.002), and this indicates that the arithmetic mean after applying the system was greater, The arithmetic mean after applying the system very simply, as the value of the (T) test was (−0.5), and the probability value of the test was (0.667), which is greater than 0.05, which means that there are no statistically significant differences for using the CRIF system on the non-performing loan (NPL)) in the banks under study.

To make sure that there are no other factors that affected the previous result due to factors specific to each bank, a One-Way ANOVA test was conducted, that is, to identify the difference in the effect of using the CRIF system on the non-performing debt ratio, and the results were as shown in Table 6 and 7.

Table 6. Descriptive Post

	N	Mean	Std. Deviation	Std. Error	95% Confidence Interval for mean		Minimum	Maximum
					Lower Bound	Upper Bound		
JORDAN ISLAMIC BANK	5	.029200	.0048683	.0021772	.023155	.035245	.0220	.0340
SAFWA ISLAMIC BANK	5	.019600	.0037815	.0016912	.014905	.024295	.0160	.0250
Islamic International Arab Bank	5	.025400	.0106207	.0047497	.012213	.038587	.0100	.0370
Total	15	.024733	.0077318	.0019963	.020452	.029015	.0100	.0370

Table 7. ANOVA Post

	Sum of Squares	df	Mean Square	F	Sig.
Between Groups	.000	2	.000	2.325	.140
Within Groups	.001	12	.000		
Total	.001	14			

We note from the above table that the calculated value of the (F) is (2.325), which is smaller than its tabular value (3.89), and thus the zero hypothesis can be accepted as there is no difference in the effect of using the CRIF system on the non-performing loan ratio (NPL), as shown.The table showed that the probability value was (0.14), which confirms the acceptance of the null hypothesis.

4 Conclusion and Recommendation

Using the CRIF platform had no statistically significant impact on the non-performing loan ratio (NPL) for Jordanian Islamic banks, indicating that the platform had no effect to improve the banks' ability to manage credit risk.

The result of having no impact for using the CRIF platform may be retained to using strict credit policies before start using the CRIF platform.

The results of the research show that the main causes of credit risks in Jordanian Islamic banks are not related to the reputation of the customer and his credit history, but rather to circumstances that may arise after obtaining financing.

Jordanian Islamic banks must maintain their strict credit policies, which have worked well in addressing credit risks, and at the same time to continue benefit from CRIF scoring platform as a source of supplementary data, which may be helpful in their decision-making process.

To conduct additional research on the same subject to cover all banks in the Jordanian banking system and make comparisons between conventional and Islamic banks.

References

1. Mounir, Q., Ismail, Y., Hayat, Y.: The impact of credit risks on banks' profitability: an analytical study on licensed banks in Jordan. Ramah Mag. (49), 23–42 (2020)
2. Mohammad, H.: The impact of financial risks on Islamic banks profit distribution policy, unpublished Ph.D. thesis. International Islamic Sciences University, Amman, Jordan (2018)
3. Central Bank of Jordan. https://www.cbj.gov.jo/. Accessed 19 Feb 2023
4. Muhammad, Kh., Asima, S., Zahid, S.: Determinants of non-performing loans in the banking sector in developing state. Asian J. Account. Res. **5**(1), 135–145 (2020)
5. Stijn, L., Teng, W.: How do credit ratings affect bank lending under capital constraints? BIS Working Paper No. 747, Monetary and Economic Department (2018)
6. Talal, Q., Abdul Aziz, S., Marwan, O.: Factors affecting the default facilities in Jordanian Commercial Banks. Eur. J. Sci. Res. **149**(3), 324–334 (2018)
7. Companies Control Department. https://ccd.gov.jo/. Accessed 19 Feb 2023
8. CRIF Jordan. https://www.Crif.JO. Accessed 19 Feb 2023
9. Jordan Islamic Bank, The annual financial reports (2012–2021)
10. The Islamic International Arab Bank, The annual reports (2012–2021)
11. Safwa Islamic Bank, The annual reports (2012–2021)

The Expected Effect of Electronic Billing in Increasing the Efficiency of Tax Collection in Jordan

(Case Study Income and Sales Tax Department)

Osama Mohammad Khaleel Ballout[1(✉)], Ayman Saleh Mustafa Harb[1], Laith Mohammad Ali Bataineh[1], Yousef Shahwan[1], and Ayman Zereban[2]

[1] Zarqa University, 2000, Zarqa 13110, Jordan
oballout@zu.edu.jo
[2] University of Business and Technology,, Jeddah, Saudi Arabia

Abstract. The study aims to determine the expected effect of electronic billing in increasing the efficiency of tax collection in Jordan. The researcher expects positive outcomes from this research; these outcomes will provide information necessary to make decisions for a smooth transfer of electronic billing in Jordan. The study population consists of income tax department employees, where a simple random sample was taken. 150 questionnaires were distributed and 130 of them were retrieved. The number of questionnaires valid for analysis was 125, it constitutes 96% of the retrieved questionnaires. The results of the study show that there is an effect of applying the electronic billing system on raising the efficiency of the tax collection system in Jordan, and there is no statistically significant effect for legislation and laws on raising collection efficiency. The results of the study also indicate that there is a statistically significant effect for employee efficiency on raising collection efficiency, and there is a statistically significant effect for electronic infrastructure on raising collection efficiency. The findings suggest that providing the electronic infrastructure with modern and advanced computer equipment and providing electronic networks linking the tax system to all official departments in Jordan, which helps government to make appropriate decisions to correct deviations in a timely manner. The researcher also suggests that importance of training and qualifying the Income Tax Department employees because it has a positive impact on the efficiency of tax collection.

Keywords: billing system · tax collection · electronic billing

1 Introduction

Taxes are considered the main financial resource for governments in modern economic systems, so that, they constitute the largest part of government revenues, in proportion to the percentage of income and sales tax in total revenues with the economic level and the income level of those economies. The percentage is usually high in developed countries

B. A. M. Alareeni and I. Elgedawy (Eds.): *Artificial Intelligence (AI) and Finance*, SSDC 488, pp. 840–849, 2023.
https://doi.org/10.1007/978-3-031-39158-3_78

and low in developing countries. While the income tax (which is a direct tax whose burden cannot be transferred from one taxpayer to another) occupies great economic importance as it is part of the gross domestic product and it's one of the returns affecting investment [14].

The invoice has a significant role in the commercial and economic movement, and it has a reflection in the local and international market, and it is considers the cornerstone of any sales tax system. In commercial transactions, the invoice is prepared after the completion of the sale process [2].

It is also estimated that the shift to electronic collection and payment processes on both sides of the public budget (revenues and public expenditures) can help achieve economic savings estimated at between 0.08 to 1.1% of Gross Domestic Product (GDP) annually in developing countries, where tax billing has led to a 50% increase in the tax base in India in less than one year which helped generate more public revenues [9].

Effective international experiences indicate that electronic billing plays an important role in terms of expanding the tax base, combating tax evasion, and increasing the efficiency of tax performance in general through the transformation to electronic systems for compliance and tax collection, which now enable governments to collect accurate and immediate information about salary payments, companies' profits, and the movement of goods sales and services, which provides a full vision of tax benefits and accurate expectations for electronic tax revenues level, and even the immediate collection of some of them [3].

Despite the importance of taxes as one of the most important tools of economic policy, statistics indicate weak tax revenues at the level of Arab countries as a group, as tax revenues did not exceed 10% of GDP in 2020, which is a low percentage compared to the global average of 15% according to World Bank data [6].

Jordan suffers from a large deficit in its annual budget, in addition to high external indebtedness, and since direct and indirect taxes are the main source of the state treasury, there are tax amounts due on the activities of taxpayers that have not been collected, which means the existence of tax evasion, as weaknesses and shortcomings in the performance of the income and sales tax system in Jordan which has led to an increase in tax evasion.

Accordingly, the problem of the study lies in answering the main question: Is there an expected impact for the electronic billing system on raising the efficiency of tax collection system in Jordan?

The study focuses on tax laws and procedures to know the impact of electronic billing in increasing the efficiency of tax collection, which will be positively reflected on services provided to citizens.

Identifying the electronic billing system in Jordan and its expected impact in increasing the efficiency of tax collection are objectives of the study.

For all the previous information the study focuses on identify the expected effect of electronic billing in increasing the efficiency of tax collection in Jordan.

2 Literature Review

The study of [2] aimed to clarify the effect of applying the billing system in reducing tax evasion from the tax auditor perspective. The study recommended to increases tax awareness for taxpayers, and increases trust between taxpayers and the Income and Sales Tax Department.

The study of [12] aimed to identify the impact of the tax information system on increasing tax revenues. It also aimed to identify the efficiency of information systems which currently used in tax administration and their effectiveness in detecting violations committed by some of those assigned to it as disclosure of incorrect information does not show the results of their real business activities or conceal information about sources related to their taxable income in an attempt to evade tax, which would negatively affect the proceeds of tax revenues and thus damage an important resource from the public treasury of the country. As for the [5, 16] it aimed to identify the impact of electronic tax auditing in reducing tax evasion from income and sales tax auditors perspective in Jordan. The study recommended to pay more attention from The Income and Sales Tax Department to develop the computerized auditing systems, search for new in this sector, and training the electronic audit staff on those systems continuously. As for the study of [4] it aimed to identify strategic planning in the income tax and its relationship to expand the tax base. It also aimed to delve deeper into strategic planning at the theoretical level, provide information about the planning practice at the Income Tax Department, and identify the most important means of expanding the tax base.

The study of [7] also aimed to know the methods of tax avoidance and evasion and the shortcomings of the Jordanian income tax law in confronting them from the income tax assessor's perspective.

2.1 Hypotheses

The main hypothesis:

H0: There is no effect for applying the electronic billing system on raising the efficiency of the tax collection system in Jordan.
The following sub-hypotheses branch out from the main hypothesis:
H01: There is no statistically significant effect at a significant level of 5% for legislation and laws on raising the efficiency of tax collection system in Jordan.
H02: There is no statistically significant effect at a significant level of 5% for employee efficiency on raising the efficiency of tax collection system in Jordan.
H03: There is no statistically significant effect at a significant level of 5% for the electronic infrastructure in raising the efficiency of tax collection system in Jordan.

3 Theoretical Framework

The billing process is one of the effective means to reduce tax evasion practiced by sectors that believe themselves to be outside the framework of the state of law. Indeed, some do not recognize the rights of the other party to obtain an invoice for the service they received, taking advantage of the weakness of the legislative framework that protects the consumer in this regard. And also taking advantage of his cognitive ability to circumvent the law in tax evasion [8, 19].

As the main role of the electronic billing system is to raise the efficiency of tax collection by reducing tax evasion that negatively affects tax collection, hence the electronic billing system considered as an important system that governs the buying and selling process by the merchant or the taxpayer, where systematic legal controls have been placed to benefit Both parties from the billing system [10, 15, 18].

Taxes occupy a distinguished and extremely important place in financial legislation because of their significant role in supporting the country budget, as it considers the tool that country relies on in its economic policy, in addition to the goals it has achieved in various fields [1].

[2] indicated that there is a positive effect of the billing system on reducing tax evasion, facilitates the work of the tax auditor, increases tax awareness for taxpayers, and increases trust between taxpayers and the Income and Sales Tax Department. He also indicated the need to increase interest in applying the system billing in Jordan, and the need for explicit and deterrent laws to encourage the use of the billing system, which in turn will limit tax evasion.

[13] explained that billing is changing how tax departments work, which helps to improve the process of efficiency and service delivery, and electronic billing is also have multi-benefits for companies and tax authorities alike, including lower administrative costs and improve information accuracy and security. [11, 17] indicated that the adoption of the electronic billing system positively affects tax compliance.

4 Methodology

The analytical descriptive approach was adopted, which is based on the study of the phenomenon as it is in reality and describes it accurately and then analyzes the correlations that exist between the independent variable (electronic billing) and the dependent variable (tax collection).

4.1 Study Population

The questionnaire consists of three categories. First category personal and functional variables. Second category independent variable (legislations and laws, employee efficiency, and electronic Infrastructure), third category dependent variable (efficiency of tax collection system). Personal and functional variables include 5 statements, legislations and laws include 10 statements, employee efficiency include 10 statements, electronic infrastructure include 11statments, and efficiency of tax collection system include 10 statements. The study population consists of income tax department employees, where a simple random sample was taken. 150 questionnaires were distributed and 130 of them were retrieved. The number of questionnaires valid for analysis was 125, it constitutes 96% of the retrieved questionnaires. The questionnaire was prepared by researcher.

4.2 Stability of the Study Tool

To verify the stability of the study tool, the researcher used the Cronbach Alpha coefficient, as the stability coefficient reached (82.05), and this percentage is very high in

Table 1. Stability coefficients by the Cronbach-Alpha

Number of items	Dimension	alpha coeff.
10	legislation and laws	0.845
10	Employee Efficiency	0.752
11	Electronic Infrastructure	0.857
10	Efficiency of the Tax Collection System	0.828
41	Whole Performance	0.8205

order to adopt the results of this study, as the acceptable percentage for generalizing the results for human and social research is (60%) or more (Table 1).

Cronbach's indicators above indicate that the study tool generally has a high stability coefficient and an ability to achieve the study objectives.

4.3 Statistical Methods

(SPSS) statistical methods were used to process the data obtained through the study sample. Specifically, the researcher used the following statistical methods:

1. Cronbach alpha coefficient: It was used to verify the stability of the study tool according to which the data was collected.
2. Frequencies and percentages: They were used to identify the characteristics of the study sample.
3. Arithmetic mean: to find out to which extent the responses of the study sample increased or decreased for each paragraph of the study. It is useful in arranging the paragraphs according to the highest arithmetic mean.
4. Standard Deviation: It was used to determine the dispersion of the answers of the study sample from the arithmetic mean values.
5. Kolmokov-Smirnov test: (One-Sample K-S Test) It was used to check whether the data of the study variables are subject to a normal distribution [20].
6. Multiple regression analysis test: to measure the effect of applying the electronic billing system on the efficiency of tax collection.
7. Simple regression analysis test in order to determine the effect of applying the electronic billing system on tax collection efficiency [21, 22].

The Kolmokov-Samir Nof test (One-Sample K-S Test) was used to verify that the data of the study variables are subjected to normal distribution.

The tabular value of (Z) is (1.96) at the level of significance ($0.05 \geq \alpha$), and the test is two-tailed.

As the results shown in Table 2 That all (Z) values calculated for the study variables are less than the tabular (Z) value (1.96), and also that all significance level values are greater than the significance level ($0.05 \geq \alpha$). According to the previous results, the null hypothesis (H0) was accepted, which states: The data of the study variables are subjected to normal distribution. This indicates that the data follows the normal distribution, which makes it ready for analytical statistical tests.

Table 2. Testing the normal distribution using the Kolmokov-Samir Nouf test

	Dimension	Z- Value	Level of significance
10	legislation and laws	1.868	0.120
10	Employee Efficiency	0.844	0.475
11	Electronic Infrastructure	0.818	0.515
10	Efficiency of the Tax Collection System	0.907	0.384
41	Whole Performance	1.028	0.145

4.4 Hypotheses Testing

The Main Hypothesis (H0): There is no effect of applying the electronic billing system on raising the efficiency of the tax collection system in Jordan.

In order to arrive to the result of this hypothesis, the sub-hypotheses were tested as follows: laws and regulations related to the electronic billing system, employee efficiency, and electronic infrastructure.

Table 3. The results of the multiple regression analysis test to measure the impact of the electronic billing system on raising the efficiency of tax collection in Jordan

Independent variables	ß	Std. Error	T	Sig.	R	Adj R	F	Sig.
Constant	0.010	0.318	0.159	0.854	0.62	0.58	52.369	0.00
legislation and laws	0.503	0.127	1.817	0.225				
Employee Efficiency	0.192	0.139	2.569	0.042				
Electronic Infrastructure	0.110	0.023	2.850	0.001				

The results of the multiple regression analysis test to measure the impact of the electronic billing system on raising the efficiency of tax collection in Jordan.

The tabular T value at the significance level ($0.05 \geq \alpha$) is 2.

It is clear from Table 3 that the value of the R2 test amounted to (58%), and this indicates that the study variables explain 58% of the dependent variable, which is the efficiency of the tax collection system, and that 42% of the dependent variable is explained by other variables, and it is also clear from the previous table The value of F test is (52.369) at a significant level of 5%, and this means that there is a statistical significance for the model in this study, which means that the independent variables combined have a statistically significant effect on the interpretation of the dependent variable.

Sub-hypotheses:

The First Sub-hypothesis: There is no statistically significant effect at the 5% level of significant for legislations and laws on raising collection efficiency.

It is clear from Table 3 that the value of (Sig = 0.225), which is a non-significant value, as the value of (T = 1.817), which is less than the tabular T, and therefore there is no statistically significant effect at a significant level of 5% for legislation and laws on the efficiency of tax collection. Therefore, the null hypothesis is accepted, which indicates that there is no effect of legislation and laws on the efficiency of the tax collection system.

The Second Sub-hypothesis: There is no statistically significant effect at a significant level of 5% for employee efficiency on raising collection efficiency.

It is clear from Table 3 that there is a statistically significant effect at a significant level of 5% for the independent variable according to the opinions of the study sample on the dependent variable. As the value of (Sig = 0.042) and also the calculated T value is greater than the tabular T value, and thus the null hypothesis is rejected And accepting the alternative hypothesis, which indicates that there is a statistically significant effect of the efficiency of employees on the electronic billing system on the efficiency of tax collection in Jordan.

The Third Sub-hypothesis: There is no statistically significant effect at a significant level of 5% for electronic infrastructure on raising collection efficiency.

From Table 3, it is clear that the value of (Sig = 0.001) and the value of (T = 2.85). These results indicate that the independent variable electronic infrastructure has a statistically significant effect at a significant level of 5% or less on the dependent variable tax collection efficiency, and the test indicates T leads to the rejection of the null hypothesis because the calculated T is greater than the tabular T, and thus we conclude that the electronic infrastructure has a statistical effect at a significant level of 5% on the efficiency of tax collection.

Table 4. A summary of the results of testing the hypotheses of the study

Hypotheses	Decision	results
Main- Hypotheses	Reject the null hypothesis	There is an effect of applying the electronic billing system on raising the efficiency of the tax collection system in Jordan
First sub-hypothesis	Acceptance the null hypothesis	There is no statistically significant effect at a significant level of 5% for legislation and laws on raising collection efficiency
Second sub-hypothesis	Reject the null hypothesis	There is a statistically significant effect at a significant level of 5% for employee efficiency on raising collection efficiency

(*continued*)

Table 4. (*continued*)

Hypotheses	Decision	results
third sub-hypothesis	Reject the null hypothesis	There is a statistically significant effect at a significant level of 5% for electronic infrastructure on raising collection efficiency

5 Conclusion

The study shows that there is no effect for legislations and laws of the electronic billing system on the efficiency of the tax collection system in Jordan, and it shows the presence of an effect for the efficiency of employees using the electronic billing system and the electronic infrastructure on the efficiency of tax collection in Jordan (Table 4).

The government need to take in consideration the research results to encourage Jordanian economy to grow like developed countries and to provide people all services they need.

Having more research in the future talking about electronic billing to find the weakness and strength points in the application, and if the results agree or not with this study findings.

5.1 Recommendations

According to the conclusions reached from the theoretical framework of the study, as well as from the statistical analysis of the data, a number of recommendations can be proposed for raising the efficiency of the tax collection system in Jordan, as follows:

1. The findings suggest that providing the electronic infrastructure with modern and advanced computer equipment and providing electronic networks linking the tax system to all official departments in Jordan, which helps government to make appropriate decisions to correct deviations in a timely manner.
2. The importance of training and qualifying the Income Tax Department employees because it has a positive impact on the efficiency of tax collection.
3. Raising the level of commitment of taxpayers and those in charge of the tax system through various incentives for both parties, which increases the efficiency of tax collection.
4. Raising taxpayers' awareness for the importance of income tax and its positive impact on community service, as providing health care, education and other basic services in society.
5. Increasing workers' awareness for the importance of accurate tracking of tax revenues and applying laws and regulations effectively, which increases the efficiency of tax collection.

References

1. Abdel Ghafour, H.: The relationship between the taxpayer and the tax administration and its impact on collection. Tax disputes major, Faculty of Graduate Studies, An-Najah University, Nablus – Palestine (2006)
2. Abdel Latif, M.: The effect of applying the billing system in reducing tax evasion from the tax auditor perspective in Jordan, master's thesis, Zarqa Private University – Jordan (2022)
3. Abu Naga, A.: The role of corporate governance in reducing tax evasion from tax auditor perspective in Jordan, a field study, master thesis, Jerash University (2015)
4. Al-Hemony, S.: Strategic planning for the income tax and its relationship to expanding the tax base, Hebron University – Palestine (2017)
5. Al-Rahahleh, Q.: The impact of electronic tax auditing in reducing tax evasion from income and sales tax auditors perspective in Jordan, Amman Arab University – Jordan (2017)
6. Al-Zoubi, A.: Methods of tax avoidance, evasion and the shortcomings of the Jordanian income tax law in confronting them from income tax assessor's perspective, Ajloun National University – Jordan (2013)
7. Altarawneh, H., Alobisat, F.: The impact of computer information systems in the reduction of the tax evasion/exploratory study in Jordanians firms. J. Theor. Appl. Inf. Technol. Jordan (2015)
8. Arab Monetary Fund: The study entitled the digitization of tax collection in the Arab countries (2021). https://www.amf.org.ae/ar/publications/aldrasat-alaqtsadyt/rqmnt-althsyl-aldryby-fy-aldwl-alrbyt
9. Al-Ramahi, N., Abuhussein, A., Shahwan, Y.: The impact of applying the XBRL on the quality of annual reports of the firms listed in Amman stock exchange. In: 2021 22nd International Arab Conference on Information Technology (ACIT), pp. 1–5. IEEE (2021)
10. European Trade Union: The study entitled electronic invoicing (2018). https://tfig.itcilo.org/AR/contents/e-invoices.htm
11. Sadress, N., Juma, B.: The mediating role of adoption of an electronic tax system in the relationship between attitude towards electronic tax system and tax compliance Makerere University Business School, Makerere University, Kampala, Uganda (2019)
12. Al-Khafaji, S.: Identifying the impact of the tax information system in increasing tax revenues, Amman, Jordan (2018)
13. Salma, K.: Digitalization to improve tax compliance: evidence from VAT e-invoicing in Peru – Peru (2019)
14. Salameen, Y.A.: Tax Accounting and its Applications in Palestine, 4th edn. Academic Library – Palestine (2014)
15. Shahwan, Y., Hamza, M., Lubad, H.K., Khresat, O. M.: Social responsibility accounting and financial performance during COVID-19: a survey of the firms listed in Amman stock exchange. J. Gov. Regul. **12**(1), 93–99 (2023)
16. Imran, M., Ali, S., Shahwan, Y., Zhang, J., Al-Swiety, I.A.: Analyzing the effects of renewable and nonrenewable energy usage and technological innovation on environmental sustainability: evidence from QUAD economies. Sustainability **14**(23), 15552 (2022)
17. Shahwan, Y.: The effect of OECD principles of corporate governance on social responsibility accounting. Int. J. Innov. Creativity Change **11**(2), 418–431 (2020)
18. Ballout, O.M.K.: The effect of electronic commerce on profitability of Jordanian commercial companies. Glob. J. Econ. Bus. **9**(3), 651–659 (2020)
19. Shahwan, Y., Abdel-hamid, O.: The impact of social responsibility disclosure in reducing the risk in Jordanian commercial banks. Res. World Econ. **11**(3), 142–150 (2020)
20. Ballout, O.M.K.: The extent of the impact of corporate governance on the effectiveness of companies in the Jordanian service sector. Acad. Strateg. Manag. J. **19**(1), 1–10 (2020)

21. Shahwan, Y.: The mediating effect of investment decisions and financing decisions on the influence of capital structure against corporate performance: evidence from Jordanian listed commercial banks. Acad. Account. Financ. Stud. J. **22**(6), 1–20 (2018)
22. Shaqqour, O.F., Harb, A.S.M., Ballout, O.M.K., Jaber, R.J.: Digital audit during Covid-19 in Jordanian audit firms a study of the reality and outlook the future. In: Hamdan, A., Shoaib, H.M., Alareeni, B., Hamdan, R. (eds.) The Implementation of Smart Technologies for Business Success and Sustainability. SSDC, vol. 216, pp. 263–272. Springer, Cham (2023). https://doi.org/10.1007/978-3-031-10212-7_23

The New Normal: The Challenges and Opportunities of Freelancing and Remote Work for HR Departments

Zakariya Chabani(✉), Rommel Sergio, and Ettiene Paul Hoffman

Canadian University Dubai, Dubai, UAE
zakariya.chabani@cud.ae, {rommel,ettiene.hoffman}@cud.ac.ae

Abstract. The aim of this paper is to explore the impact of freelancing and remote work on HR departments, specifically in recruitment, training and development, and employee engagement. The discussion highlights the innovative strategies and tools HR departments can use to manage a decentralized workforce, including online recruitment platforms, virtual training programs, and collaboration tools, as well as recognizing the challenges associated with these strategies, such as the lack of face-to-face interaction and engagement. The paper emphasizes the importance of adapting to a more decentralized workforce and the development of new strategies to remain competitive in the evolving business landscape.

The paper concludes by highlighting the theoretical and practical implications of the growth of freelancing and remote work for companies such as the need for companies to adapt and innovate in their recruitment, training and development, and management strategies to remain competitive. Additionally, the need for HR departments to find practical ways to keep remote workers engaged and connected to the company culture while also considering legal and ethical implications. The paper suggests that companies must be proactive in addressing these challenges, as failure to do so may result in the loss of the best talent.

Keywords: Freelancing · Remote work · Human Resource Management

1 Introduction

The internet has facilitated the growth of freelance jobs, which has significantly impacted the role of HR in businesses, resulting in a shift from traditional office-based jobs to more flexible work arrangements like freelancing and remote work. As a result, there has been an increasing number of freelancers in economies [1], which has caused significant changes in the traditional job market. With the widespread availability of high-speed internet, cloud computing, and collaboration tools, individuals can now work from anywhere, at any time, leading to increased productivity and efficiency for freelancers and remote workers [2]. Additionally, the internet has made it easier for companies to hire top talent from anywhere in the world, making the job market more competitive. This shift is driven by several factors, including limited opportunities for permanent employment [3] and rising labor costs [4]. Therefore, HR departments need to develop new strategies to attract and retain the best talent in a globalized job market.

B. A. M. Alareeni and I. Elgedawy (Eds.): *Artificial Intelligence (AI) and Finance*, SSDC 488, pp. 850–861, 2023.
https://doi.org/10.1007/978-3-031-39158-3_79

As organizations seek the services of freelance workers to stay competitive, new workforce patterns are emerging [5]. HR departments play a crucial role in managing the workforce and supporting the objectives of organizations [6]. However, the rise of freelancing has complicated the HR's role in managing the workforce, and it is essential to understand how HR is adapting to this change and how it affects the company's competitiveness.

As organizations face increased competition, HR departments play a critical role in ensuring that the workforce is utilized to gain a competitive advantage. To effectively manage this new workforce, it is essential to evaluate the relationship between freelancing and the role of HR, including adapting strategies and processes to accommodate the unique needs of freelancers, such as creating new policies for managing and communicating with them, and ensuring they have access to the resources and support needed for success [1, 7].

HR departments must be able to identify and attract the best talent from a global pool of candidates, provide accessible and effective training and development opportunities, and keep remote workers engaged and connected to the company culture [2, 8, 9, 10, 11, 12]. Ultimately, the success of organizations in this new work environment will depend on how well HR can adapt to manage and support the growing number of freelancers in the workforce. On the other hand, some organizations may find that the increasing number of freelancers strengthens the role of HR, as HR departments can focus on finding and engaging the most talented and skilled freelancers for specific projects. In this way, HR can play an even more critical role in ensuring that the company stays competitive by leveraging the expertise of a diverse and highly skilled freelance workforce.

The aim of this paper is to evaluate the impact of the increasing popularity of freelancing and remote work on the traditional job market and the role of HR in managing this new workforce.

The paper is divided into five sections to fulfill its objective: the analysis of the topic's background, identification of ideas related to remote work and freelancing, introduction of recommendations, and discussion of potential positive consequences of the study for human resource managers. The methodology section presents the methodological approach of the researchers, the selection of primary and secondary sources, and the method of analysis. The literature review section discusses the growth of freelancing and remote work and their effects on HR departments, providing actual findings and analytical propositions for their improvement. The section on innovative instruments examines their effects on HR departments and the reasons for these tendencies. The conclusion and implications section provides information on how HR departments can adapt to the new situation, limitations of the literature review, and suggestions for future research to explore the theme more deeply.

2 Methodology

The study is a systematic literature review built on the methods of generalization and categorization of scholarly sources. The analysis of information from the chosen sources allowed to gather the most relevant facts about the work of human resource departments before, during and after the era of the Covid-19 pandemic where managers and other employees have to use remote services and technological advancements. The key priorities during the selection of sources were their reliability, relevance to the topic, and data clarity. The literature review utilizes information from both primary and secondary sources in order to achieve a deeper view of the situation and find more valuable recommendations, more evidence, and theoretical justifications for some work instruments. All the data were gathered from various books, articles, and other types of official publications. The methods of work used for this study include the processes of summarizing, reviewing, and evaluating the information on the basis of descriptive and prescriptive analyses.

3 Literature Review

3.1 The Growth of Freelancing and Remote Work

In addition to the United States, freelancing and remote work growth has been observed in various other countries around the world, such as the United Kingdom and Australia [13, 14]. Moreover, freelancing, and remote work have become increasingly prevalent across various industries, such as technology, creative services, and consulting. For example, some of the top categories for freelance work are web, mobile, and software development, creative and design, writing, sales and marketing, and administrative support [15–17].

The growing popularity of freelancing and remote work is also being driven by the desire for work-life balance and the ability to work on projects that align with personal interests and values [18, 19]. Moreover, the COVID-19 pandemic has accelerated the adoption of remote work, with many companies forced to shift to remote work due to social distancing measures [20, 21]. As a result, remote work is now becoming the new normal for many businesses and will likely continue to grow.

3.2 The Impact of Freelancing and Remote Work on the HR Department

Because the increasing in popularity of freelancing and remote work more and more businesses are adopting theses type of work [11, 22, 23]; thus, HR departments are facing new challenges in managing a decentralized workforce, for example, Recruitment requires a different approach in a remote work environment, as HR departments must be able to identify and attract the best talent from a global pool of candidates. This can involve using online recruitment platforms, virtual interviews, and innovative sourcing strategies.

The impact of freelancing and remote work on HR includes issues related to training, the engagement of employees, compensation, benefits, and tax compliance, which can be complex and challenging to manage in a remote work environment [24]. HR departments must also consider remote work's legal and ethical implications, such as data privacy and labor laws [25]. HR departments must adapt to these changes by developing new approaches to recruitment, training, employee engagement, and other areas that are impacted by this trend.

As a result, the key impact of freelancing and remote work on the HR department can be summarized in Table 1.

Table 1. The Impact of Freelancing and Remote Work on the HR department

Consequence	Area of Influence	Level of Influence
New recruitment approaches	New recruitment approaches lead to the transformation of general approaches of HR managers to the accomplishment of their work tasks. This aspect affects everything from new communication patterns to the implementation of additional software	High
The growth of the numbers of potential employees	Remote work allows companies to choose employees from other countries, affecting the amount of available time of human resource specialists. They also have to analyze more information	Average
Updated training approaches	Online training is different because it requires the implementation of new ways to explain the same material remotely	High
The engagement of employees	It leads to the need of learning more psychological methods of affecting humans' interest and engagement during interviews and daily work	High
Compensation	The level of compensation can become higher or lower due to the change in salaries	Low
Tax compliance	New methods of paying taxes remotely can be a complex requirement for some HR specialists	Average

Sources: Built by the authors on the basis of [11, 22, 23, 24, 25].

3.2.1 Recruitment and Hiring

One of the major benefits of online recruitment platforms is the ability to reach a larger pool of candidates, regardless of their location [12, 26]. This is particularly important in today's globalized job market where the best talent is often not located in the same geographic area as the company. By using online platforms, companies can cast a wider net and find the right candidate for the job, rather than simply settling for the best local candidate [22, 27]. Additionally, online recruitment platforms offer a more cost-effective and efficient way of recruiting [28, 29]. Rather than spending money on job fairs or advertising, companies can simply create a job posting on the platform and wait for applicants to apply. This saves time and money and allows companies to focus their resources on other aspects of the recruitment process, such as screening and interviewing candidates. Moreover, the use of technology in the recruitment process can help to streamline the hiring process, making it easier for companies to find the best candidate for the job [30]. For example, many online platforms have built-in tools for tracking applicants and managing the hiring process, from posting the job to offering the job to the successful candidate [31–33]. This allows companies to keep everything organized and ensures that no applicants slip through the cracks.

However, it's important to note that while online recruitment platforms can be a great resource for companies, they also have their drawbacks. For example, the abundance of candidates on these platforms can make it difficult for companies to sort through the large number of applications and find the right candidate [34, 35]. Additionally, the lack of face-to-face interaction can make it more challenging for companies to assess a candidate's fit for the role and the company culture [36].

Another positive aspect of online recruitment platforms is that they can help to reduce biases in the hiring process [37–39]. By allowing candidates to apply anonymously or using artificial intelligence to screen resumes, companies can ensure that they are hiring based on qualifications and experience, rather than factors such as gender, race, or age. This can help to promote diversity and inclusivity within the workplace, which has been shown to have a positive impact on overall company performance [40, 41].

However, it's important to also consider the potential negative impact of online recruitment platforms on job seekers. The use of algorithms and automated screening tools may lead to some qualified candidates being overlooked [42]. Additionally, the lack of personal interaction with a potential employer can make job seekers feel disconnected from the company and the hiring process [36, 43].

3.2.2 Training and Development

Online training platforms come with their own set of challenges. For example, the lack of in-person interaction can make it more difficult for employees to understand the content and retain the information [44, 45]. Additionally, online training can be less engaging than in-person training, and employees may be less motivated to complete the courses [46, 47]. Another important consideration is the need to personalize training and development programs to meet the individual needs of each employee [48]. This may involve assessing an employee's current skill level and creating a customized development plan to help them reach their full potential. This can help to ensure that employees are not only trained in the skills they need to perform their job, but also in the areas where they need the most improvement.

To overcome these challenges, companies can complement online training with other forms of development opportunities, such as on-the-job training, mentorship programs, and in-person workshops. This multi-faceted approach can help to ensure that employees receive the training and development they need to perform their jobs effectively [49].

It's also important for companies to measure the effectiveness of their training and development programs. This can be done by conducting regular assessments and evaluating the impact of the training on employee performance, which may help to identify areas for improvement and ensure that the training and development programs are delivering the desired outcomes [50].

In addition to regular assessments, companies can also use employee feedback to evaluate the effectiveness of their training and development programs [51, 52]. Feedback can be gathered through surveys, focus groups, or one-on-one meetings with employees to gather their thoughts and opinions on the training and development programs. By incorporating employee feedback, companies can make necessary changes and improvements to their programs to better meet the needs and preferences of their workforce [53]. Additionally, companies can track the progress of employees who have completed the training and development programs to ensure that they are applying their new skills and knowledge in their daily work [54]. This can help to determine the impact of the programs on individual and organizational performance and justify the investment in training and development initiatives [55].

3.2.3 Employee Engagement

One effective way to improve remote employee engagement is to create a sense of community among them [46, 56]. This can be achieved through virtual team-building activities, such as virtual coffee breaks (Virtual coffee breaks are remote team-building activities that allow remote employees to connect and socialize with each other over a video call) or online games, which can help to foster a sense of connection and camaraderie among team members. Regular check-ins and one-on-one conversations with team leaders can also help to build a sense of belonging and increase employee engagement [57, 58]. Additionally, Freelancers and remote workers need to know what is expected of them and how their work contributes to the overall success of the company [59]. Providing regular feedback on their performance and offering opportunities for growth and development can also help to keep remote workers engaged and motivated [60, 61].

However, it's important to note that engagement strategies that work for one company may not work for another, and HR must be flexible and adaptable in their approach [62]. Regular surveys and pulse checks can provide valuable insights into the attitudes and opinions of remote workers and help HR to fine-tune their engagement strategies [63].

In this context, the list of recommendations for the improvement of employee engagement can be summarized for the deeper visual perception of the material. This information is demonstrated in Table 2.

Table 2. HR Managers and Techniques for the Growth of Remote Employee Engagement

Engagement Strategy	Explanation of the Strategy
Virtual team-building activities	Help to organize a friendly atmosphere in the team and improve the depth of communication
Clear expectations of employees	This strategy works on the basis of individual and group contacts with employees, leading to the match of expectations and actual impressions
Regular feedback	Controls the level of understanding between employees and HR managers

(continued)

Table 2. (*continued*)

Engagement Strategy	Explanation of the Strategy
Flexible approaches	This strategy prepares employees to changes and improves their engagement due to the diversification of methods and instruments
Regular surveys and communication	Surveys and questionnaires help managers to understand the level of satisfaction of employees

Sources: Built by the authors on the basis of [46, 56, 57 58, 59, 60, 62, 63].

4 Discussion

The growth of freelancing and remote work has had a significant impact on HR departments; as they created new challenges for managing a decentralized workforce, for instance, recruitment, where HR departments must be able to attract the best talent from a global pool of candidates. This can be achieved through the use of online recruitment platforms, virtual interviews, and innovative sourcing strategies. However, while these platforms offer a cost-effective and efficient way of recruiting, they also have their own drawbacks, such as the abundance of candidates and the lack of face-to-face interaction.

Another area that has been impacted by freelancing and remote work is training and development. In a remote work environment, HR departments must find ways to provide accessible and effective training opportunities for a geographically dispersed workforce. This can be done through the use of online training programs, virtual classroom sessions, and e-learning platforms. However, online training platforms also come with their own set of challenges, such as the lack of in-person interaction and engagement.

HR departments must also find ways to keep remote workers engaged and connected to the company culture, even when they are not in the office. This can be done through the use of online collaboration tools, virtual team-building activities, and regular check-ins with remote workers. In addition, HR departments must also consider the legal and ethical implications of remote work, such as data privacy and labor laws.

To further support remote work, HR departments can take additional steps. For example, they can develop a comprehensive remote work policy that outlines the expectations and responsibilities of both the company and the remote workers. HR departments can also provide clear communication channels to ensure that remote workers feel connected and informed about company updates and initiatives. Flexible work arrangements, such as flexible scheduling or the option to work from a co-working space, can be offered to support the work-life balance of remote workers. HR departments can also implement a performance management system that measures productivity and progress towards goals, while also considering the unique challenges of remote work.

5 Conclusion and Implications

The growth of freelancing and remote work is far-reaching, particularly in the recruitment and training of employees and employee engagement for companies worldwide. The emergence of online recruitment platforms as a popular tool highlights the need for companies to adapt to a more decentralized workforce. However, the challenges of sorting through a vast number of applications and the lack of face-to-face interaction necessitate the development of innovative strategies for effective hiring.

The shift to remote work requires HR departments to adapt their training and development methods to cater to a geographically dispersed workforce. Online training platforms and e-learning tools have become increasingly popular, but they require new approaches to maintain employee engagement. This shift in training and development calls for companies to weigh the advantages and disadvantages of online training to develop effective training strategies.

Beyond recruitment and training, HR departments must find ways to keep remote workers engaged and connected to company culture. Complex issues related to compensation, benefits, tax compliance, data privacy, and labor laws must also be considered. This changing landscape demands companies to develop new strategies and stay informed and proactive in addressing these challenges.

The theoretical implications of these changes are significant, as companies must adapt to a more decentralized workforce to remain competitive in the evolving business landscape. The use of innovative recruitment and training methods and the adoption of new strategies for managing remote workers are vital for success. In summary, the growth of freelancing and remote work has created new opportunities and challenges for companies, requiring them to stay vigilant and proactive in adapting to these changes.

The practical implications of the growth of freelancing and remote work are equally significant. The growth of freelancing and remote work presents practical implications that companies must consider to remain competitive in the ever-evolving business landscape. To adapt to a more decentralized workforce, companies need new recruitment strategies like online recruitment platforms and innovative training and development methods. Without adapting, they risk losing the best talent. HR departments must also find practical ways to keep remote workers engaged, such as using communication and collaboration tools, and ensure the implementation of robust compensation, benefits, and tax compliance systems to retain top talent. Furthermore, companies must be mindful of legal and ethical implications like data privacy and labor laws, which require HR departments to ensure compliance with these laws and follow best practices to protect the privacy and security of sensitive company and customer data.

The limitations of the conducted study include the lack of practical examples and the absence of a correct research sample. The qualitative research of further questions of the same topic can be potentially more useful than a literature review and can lead to deeper outcomes and innovative propositions. The collection of the data was done on the basis of a wide range of credible scientific sources. However, the study does not include any interviews with actual human resource managers and their opinions on the topic. From this point of view, access to practical information was partially limited due to the chosen methodology. As a result, the probability of using biased data from the studies of other researchers still exists even though it is minor and does not actually affect the general direction of the findings and their effectiveness for the solution of the research problem. The primary limitation of this study is the lack of clarity in the analysis of reasons why human resource managers have to improve their performance or adjust existing workflows during the communication with freelancers and remote workers. From this point of view, the limitations are not significant for the perception of the research outcomes but can cause misapprehensions when read by non-professional or incompetent audiences.

The future research of the topic of human resource management in modern society should focus on the search for appropriate methods to predict the adaptability of employees and the transformation of their behaviors during the implementation of new work conditions, rules, or instruments. Moreover, the attempt to organize additional qualitative research interviews with relevant human resource managers can add reliability to the study, can help in finding the realistic background for the implementation of new instruments and understand their attitude to the significance of the adaptability quotient in the context of hiring additional employees and proposing innovative technologies to freelancers and remote workers. The future research of the topic of opportunities of freelancing and remote work for HR departments will be efficient if other studies include more examples of actual companies where managers use the proposed innovations and share proof of positive consequences for the growth of corporate effectiveness.

References

1. Ahmead, M., Hamamadeh, N., Iram, I.a.: The effects of internet and social media use on the work performance of physicians and nurses at workplaces in Palestine. BMC Health Serv. Res. **22**, 633 (2022)
2. Kuhn, K.M., Meijerink, J., Keegan, A.: Human resource management and the gig economy: challenges and opportunities at the intersection between organizational HR decision-makers and digital labor platforms. In: Buckley, M.R., Wheeler, A.R., Baur, J.E., Halbesleben, J.R.B. (eds.) Research in Personnel and Human Resources Management (Research in Personnel and Human Resources Management), vol. 39, pp. 1–46. Emerald Publishing Limited, Bingley (2021)
3. Kuhn, K.: The rise of the "gig economy" and implications for understanding work and workers. Ind. Organ. Psychol. **9**(1), 157–162 (2016)
4. Batmunkh, A., Fekete-Farkas, M., Lakner, Z.: Bibliometric analysis of gig economy. Adm. Sci. **12**(2), 51 (2022)
5. Nawaz, Z., Zhang, J., Mansoor, R., Hafeez, S., Ilmudeen, A.: Freelancers as part-time employees: dimensions of FVP and FJS in E-lancing platforms. South Asian J. Hum. Resour. Manag. **7**(1), 34–60 (2020)
6. Wong, K.-L., Tan, P.S.-H., Ng, Y.-K., Fong, C.-Y.: The role of HRM in enhancing organizational performance. Hum. Resour. Manag. Res. **3**(1), 11–15 (2013)
7. Tiwari, P.K., Chimanbhai, S., Saxena, K.: Human resource management practices: a comprehensive review (2012)
8. Zarwi, M., Marchand, S., Kennard, B., Michael, J.: Millennials and organisational commitment: current and future perspectives. Hum. Resour. Dev. Int., 1–13 (2022)
9. Tomcikova, L., Svetozarovova, N., Coculova, J.: Challenges and priorities in talent management during the global pandemic caused by COVID-19. Mark. Manag. Innov. **2**, 94–103 (2021)
10. Hao, Q., Zhang, B., Shi, Y., Yang, Q.: How trust in coworkers fosters knowledge sharing in virtual teams? A multilevel moderated mediation model of psychological safety, team virtuality, and self-efficacy. Front. Psychol. **13**, 899142 (2022)
11. Trenerry, B., et al.: Preparing workplaces for digital transformation: an integrative review and framework of multi-level factors. Front. Psychol. **12**, 620766 (2021)

12. Garro-Abarca, V., Palos-Sanchez, P., Aguayo-Camacho, M.: Virtual teams in times of pandemic: factors that influence performance. Front. Psychol. **12**, 624637 (2021)
13. Jansen, G.: Farewell to the rightist self-employed? 'New self-employment' and political alignments. Acta Politica **52**, 306–338 (2017)
14. Josephi, B., O'Donnell, P.: The blurring line between freelance journalists and self-employed media workers. Journalism **24**(1), 139–156 (2023)
15. Segal, E.: U.S. freelance workforce continues to grow with no signs of easing, new report says. Forbes, 8 December 2021. https://www.forbes.com/sites/edwardsegal/2021/12/08/us-freelance-workforce-continues-to-grow-with-no-signs-of-easing-new-report/?sh=fde436c4f916
16. Neufeld, D.J., Wan, Z., Fang, Y.: Remote leadership, communication effectiveness and leader performance. Gr. Decis. Negot. **19**, 227–246 (2010)
17. Vasireddy, S., Wig, S., Hannides, M.: Technical factors can impact on remote consultations in rheumatology: results from a service evaluation during the COVID-19 pandemic. Rheumatol. Int. **42**, 999–1007 (2022)
18. Kauffeld, S., Tartler, D., Gräfe, H., et al.: What will mobile and virtual work look like in the future?—results of a Delphi-based study. Gr. Interakt. Org. **53**, 189–214 (2022)
19. Shimura, A., Yokoi, K., Ishibashi, Y., Akatsuka, Y., Inoue, T.: Remote work decreases psychological and physical stress responses, but full-remote work increases presenteeism. Front. Psychol. **12**, 730969 (2021)
20. Brammer, S., Branicki, L., Linnenluecke, M.: Disrupting management research? Critical reflections on British journal of management COVID-19 research and an agenda for the future. Br. J. Manag. **34**(1), 3–15 (2022). https://doi.org/10.1111/1467-8551.12699
21. Lord, P.: COVID-19 and the future of work. Denver Law Rev. Forum (2021)
22. Johnson, S.S.: The future of work. Am. J. Health Promot. **34**(7), 809–812 (2020)
23. Bhat, S.K., Pande, N., Ahuja, V.: Virtual team effectiveness: an empirical study using SEM. Procedia Comput. Sci. **122**, 33–41 (2017)
24. Tran, M., Sokas, R.K.: The gig economy and contingent work: an occupational health assessment. J. Occup. Environ. Med. **59**(4), e63–e66 (2017)
25. Matshoba-Ramuedzisi, T., de Jongh, D., Fourie, W.: Followership: a review of current and emerging research. Leadersh. Org. Dev. J. **43**(4), 653–668 (2022)
26. Hadass, Y.S.: The effect of internet recruiting on the matching of workers and employers (2004)
27. Recruitment goes virtual: Use web-based technology intelligently for best results in recruitment. Hum. Resour. Manag. Int. Dig. **21**(3), 19–21 (2013)
28. Steers, R.M., Mowday, R.T., Shapiro, D.L.: Introduction to special topic forum: the future of work motivation theory. Acad. Manag. Rev. **29**(3), 379–387 (2004)
29. Aggerholm, H.K., Andersen, S.E.: Social media recruitment 3.0: toward a new paradigm of strategic recruitment communication. J. Commun. Manag. **22**(2), 122–137 (2018)
30. Gunathunge, K.L.N.K., Lakmal, P.: Industrial revolution 4.0 and the future of HRM (2021)
31. Glavin, P., Schieman, S.: Dependency and hardship in the gig economy: the mental health consequences of platform work. Socius **8** (2022)
32. Hofeditz, L., Clausen, S., Rieß, A., et al.: Applying XAI to an AI-based system for candidate management to mitigate bias and discrimination in hiring. Electron. Mark. **32**, 2207–2233 (2022)
33. Ingusci, E., Signore, F., Cortese, C.G., et al. : Development and validation of the remote working benefits & disadvantages scale. Qual. Quant. (2022)
34. Pillai, R., Sivathanu, B.: Adoption of artificial intelligence (AI) for talent acquisition in IT/ITeS organizations. Benchmarking Int. J. **27**(9), 2599–2629 (2020)
35. Lee, L., Guzzo, R.F., Madera, J.M., Guchait, P.: Examining applicant online recruitment: the use of fictitious websites in experimental studies. Cornell Hosp. Q. **62**(1), 76–88 (2021)

36. Folger, N., Brosi, P., Stumpf-Wollersheim, J., Welpe, I.M.: Applicant reactions to digital selection methods: a signaling perspective on innovativeness and procedural justice. J. Bus. Psychol. **37**, 735–757 (2022)
37. Deranty, J.P., Corbin, T.: Artificial intelligence and work: a critical review of recent research from the social sciences. AI Soc. (2022)
38. Chalikias, M., Kyriakopoulos, G., Skordoulis, M., Koniordos, M.: Knowledge management for business processes: employees' recruitment and human resources' selection: a combined literature review and a case study. In: Kravets, A., Shcherbakov, M., Kultsova, M., Iijima, T. (eds.) JCKBSE 2014. CCIS, vol. 466. Springer, Cham (2014). https://doi.org/10.1007/978-3-319-11854-3_44
39. Drage, E., Mackereth, K.: Does AI debias recruitment? Race, gender, and AI's "eradication of difference." Philos. Technol. **35**, 89 (2022)
40. Stahl, G.K., Maznevski, M.L.: Unraveling the effects of cultural diversity in teams: a retrospective of research on multicultural work groups and an agenda for future research. J. Int. Bus. Stud. **52**, 4–22 (2021)
41. Duchek, S., Raetze, S., Scheuch, I.: The role of diversity in organizational resilience: a theoretical framework. Bus. Res. **13**(2), 387–423 (2019). https://doi.org/10.1007/s40685-019-0084-8
42. Lacroux, A., Martin-Lacroux, C.: Should I trust the artificial intelligence to recruit? Recruiters' perceptions and behavior when faced with algorithm-based recommendation systems during resume screening. Front. Psychol. **13**, 895997 (2022)
43. Mohan, K.P., Peungposop, N., Kalra, P.: Psychosocial influences on coping and wellbeing during the Covid-19 lockdown in the early days of the pandemic: a mixed methods research. Appl. Res. Qual. Life (2022)
44. Sanchez, E.: Game-based learning. In: Tatnall, A. (ed.) Encyclopedia of Education and Information Technologies, pp. 1–9. Springer, Cham (2019). https://doi.org/10.1007/978-3-319-60013-0_39-2
45. Bain, A., Drengenberg, N.: Transforming the Measurement of Learning and Teaching in Higher Education. Routledge, New (2016)
46. Anderson, T., Dron, J.: Three generations of distance education pedagogy. Int. Rev. Res. Open Distrib. Learn. **12**(3), 80–97 (2011)
47. Abumandour, E.-S.T.: Applying e-learning system for engineering education – challenges and obstacles. J. Res. Innov. Teach. Learn. **15**(2), 150–169 (2022)
48. Hardy, C.: Leading with the Brain in mInd: 101 Brain-Compatible Practices for Leaders. Corwin Press (2015)
49. Kaputa, V., Loučanová, E., Tejerina-Gaite, F.A.: Digital transformation in higher education institutions as a driver of social oriented innovations. In: Păunescu, C., Lepik, KL., Spencer, N. (eds.) Social Innovation in Higher Education. Innovation, Technology, and Knowledge Management, pp. 61–85. Springer, Cham (2022). https://doi.org/10.1007/978-3-030-840 44-0_4
50. Estellés-Arolas, E., González-Ladrón-de-Guevara, F.: Towards an integrated crowdsourcing definition. J. Inf. Sci. **38**(2), 189–200 (2012)
51. Weldy, T.G.: Learning organization and transfer: strategies for improving performance. Learn. Organ. **16**(1), 58–68 (2009)
52. Ferreira, A.P., Velinças, F.: Training and development from the SME point of view: portuguese employers' input. In: Machado, C., Davim, J.P. (eds.) Organizational Management, pp. 107–131. Palgrave Macmillan, UK, London (2016). https://doi.org/10.1057/9781137473080_5
53. Elsässer, J.P., Hickmann, T., Jinnah, S., Oberthür, S., Van de Graaf, T.: Institutional interplay in global environmental governance: lessons learned and future research. Int. Environ. Agreem. **22**, 373–391 (2022)

54. Halvarsson Lundkvist, A., Gustavsson, M.: Conditions for employee learning and innovation – interweaving competence development activities provided by a workplace development programme with everyday work activities in SMEs. Vocat. Learn. **11**(1), 45–63 (2017). https://doi.org/10.1007/s12186-017-9179-5
55. De Leng, W.E., Stegers-Jager, K.M., Husbands, A., Dowell, J.S., Born, M.P., Themmen, A.P.N.: Scoring method of a situational judgment test: influence on internal consistency reliability, adverse impact and correlation with personality? Adv. Health Sci. Educ. **22**(2), 243–265 (2016). https://doi.org/10.1007/s10459-016-9720-7
56. Šerić, M., Gil-Saura, I.: Integrated marketing communications and information and communication technology in the hotel sector: an analysis of their use and development in Dalmatian first-class and luxury hotels. J. Retail Leisure Prop. **9**, 401–414 (2011)
57. Gallagher, S.P.: Roger R. Hock, forty studies that changed psychology: explorations into the history of psychological research (global edition). Psychol. Learn. Teach. **14**(3), 263–265 (2015)
58. García-Chas, R., Neira-Fontela, E., Varela-Neira, C., Curto-Rodríguez, E.: The effect of political skill on work role performance and intention to leave: a moderated mediation model. J. Leadersh. Organ. Stud. **26**(1), 98–110 (2019)
59. Amabile, T.M., Kramer, S.J.: The Progress Principle: Using Small Wins to Ignite Joy, Engagement, and Creativity at Work. Harvard Business Review Press (2011)
60. Deci, E.L., Koestner, R., Ryan, R.M.: A meta-analytic review of experiments examining the effects of extrinsic rewards on intrinsic motivation. Psychol. Bull. **125**(6), 627–668 (1999)
61. Lin, H.-F.: Effects of extrinsic and intrinsic motivation on employee knowledge sharing intentions. J. Inf. Sci. **33**(2), 135–149 (2007)
62. Rehman, N., Mahmood, A., Ibtasam, M., Murtaza, S.A., Iqbal, N., Molnár, E.: The psychology of resistance to change: the antidotal effect of organizational justice, support and leader-member exchange. Front. Psychol. **12**, 678952 (2021)
63. Pawar, B.S.: Some of the recent organizational behavior concepts as precursors to workplace spirituality. J. Bus. Ethics **88**, 245–261 (2009)

Integration Between the Enterprise Resource Planning (ERP) System and the Activity Based Costing (ABC) and Its Impact on the Financial Performance of the Industrial Companies Listed on the Amman Stock Exchange

Ziad Abdul Halim Altheebeh[1](✉), I. N. Jodeh[1], Sujoud Hussein D'yab Abdallah[1], and Abdulwahab M. Abukwaik[2]

[1] Faculty of Economics and Business Administration, Faculty of Business Administration, Zarqa University, Zarqa, Jordan
{z.altheebeh,Ijodeh}@zu.edu.jo
[2] University of Business and Technology, Jeddah, Saudi Arabia
a.abukwaik@ubt.edu.sa

Abstract. This study aimed to identify the Integration Between the Enterprise Resource Planning (ERP) System and the Activity Based Costing (ABC) and its Impact on the Financial Performance of the Industrial Companies Listed on the Amman Stock Exchange, The study population consisted of (33) company, in order to achieve the objective study an analytical descriptive approach was followed, a questionnaire was designed to measure the integration between the (ERP) system & (ABC), and it was distributed to industrial companies, in addition to measuring the financial performance by financial ratios (ROA, ROE, Operating Profit Margin) for the period (2017–2021) extracted from the financial reports of industrial companies listed on the Amman Stock Exchange.

The study concluded that there is a statistically significant positive effect of the integration between the Enterprise Resource Planning system & the Activity Based Costing on the financial performance (ROA, ROE, and Operating Profit Margin) in the Jordanian industrial companies, The study recommended that industrial companies keep abreast of development in technology systems and cost systems continuously together, in order to reach optimal integration between them, which leads to better financial performance improvement.

Keywords: Activity-Based Costing · Enterprise Resource Planning · Financial Performance

1 Introduction

The obstacles that the environment organization faces it is up to the technology development which the world goes through, which leads to the necessity to develop the information systems in order to suit these developments such as (ERP) system and cost

B. A. M. Alareeni and I. Elgedawy (Eds.): *Artificial Intelligence (AI) and Finance*, SSDC 488, pp. 862–872, 2023.
https://doi.org/10.1007/978-3-031-39158-3_80

accounting which based on activities which aims to process with massive information moreover to manage. In order to take right and value decisions and Institutions use (ERP) both in terms off-speed, value, or cost reduction [1].

Notably, The planning organization resource system works on transaction automat, provides internal vision and regulations, Depending on Central database, As mentioned the ERP system is an integrated system that provides a comprehensive database, unified to all customers and ready to use, ERP system defined as integration system works on merge enterprise functions (Accounting, Human Resources, funds, marketing, supply chain, manufacturing, management inventory, maintenance, transportation, E-business, etc.) in one place which called Data base, That helps to distribute information constantly to all departments and levels, also it confirmed on the possibility to apply it and take advantage of it in service organization or on productivity, In addition, baloula 2021 the system is scalable and can be used as a tool to observe on the available resource, Which works on solving problems, supportive innovation, which contributes to develop the service level, some authors agreed to separate the components of sys-tem for two parts functional and technology components, ERP system consist heterogeneous units and can be separated, [2] add the ERP can process changes in any unit without compromising other units, However [3] descript ERP system as the integration of exchange information on organization through all departments without limits, by using standardized interfaces (screens, working environment, terminology, action buttons, acronyms, lists, etc.) which leads to improve the inter coordination [4]

However [5], in some case the system fails on achieving it goals, That doesn't mean that all system collapse It just lake in performance in some level, Some studies have indicated the cases of failure and cessation of ERP system implementation projects are due to insufficient resources, and major requirements to process the system, planning and the miss management for the project, [6] add, one of the reasons for failure is the lack of fairness of systems, including salary systems, and the lack of integration with other systems, in addition to other factors, the companies managed to apply the system successfully due to right implement of several factors, which called the Critical success factors, However [7] mention the main factors to success the system which are (Regulatory suitability, Quality of information and services) It must be propulsion within a good plan that includes designed phases to achieve the main goal of the system, The stages are categorized by strategic planning, Which includes to appoint a project team and to study the mechanism of information flow and understand the current operations, further-more the process of defining the goals to establish project strategy, then the reviewing process to be applied begins and to proceed with it, through reviewing the available resources then begin with defining the manual processes and Establish standard procedures, followed by data collection and archiving, then reviewing all the entered data, to start a new phase, which is training and testing, through delve the employees in trainings to use the system. Then, to take the final test in order to start the process of applying and evaluating the system which contributes to improving the overall performance of the institution, including the accounting and manufacturing departments, as development does not stop at a certain limit, Moreover it includes the cost accounting systems, especially (ABC) system [8, 9].

The accounting cost system based on a group of basic principles and procedures, which completes the previous systems, The term of ABC system showed to be better alternative to traditional costing concept, the efforts made to be customization model Activity Based Costing (ABC) has led to many criticisms, since the model is assumed full use of resources, some authors [10] agreed about definition of ABC, it is a system which for charging indirect costs in two stages. In the first stage, it charges costs by indirect way on the activates, in the second stage it loads on the products, depending on causes of the cost in two stages. In addition [11], the system helps on monitoring operations, planning, improving performance Furthermore the system dependents on push methods.

One of the ABC system feature it focuses on reducing the costs without compromising the quality of the original product, By distributing the costs of activities, and taking into consideration the products quality and appropriate pricing decisions by analyzing activity data [12], Also one of its features the efficiency of using matching principle mainly by not allocating the costs of development and research, Which may relate to coming periods on the costs of the current period, which the companies cannot avoid these situations under the traditional system, Despite the success and the features of the system that achieved, there are some obstacles that [13] Saad et al. 2016 clarified it such as; the high costs on practice, it requires a high accuracy in calculating costs when they apply system in any institution.

Research Problem

The previous studies mentioned separately the ABC and ERP systems Research problem, but few of them dealt with the integration between them, which leads to improve the financial performance of the industrial companies, the study problem can be summarized by the following main question:

Is there an impact of the integration between (ERP) systems and (ABC) systems on the financial performance of the industrial companies listed on the Amman Stock Exchange?

To identify the integration between (ERP) systems and (ABC) systems and its impact on financial performance.

The Importance of the Research

The importance of the study lies in helping organizations to achieve the benefits of applying (ERP & ABC) systems in improving their operations, especially since the application of these systems is not an easy process, and requires costs, expertise, instructions, and limitations; which must be known before implementation, as well as Scientific contribution through the results that the study will show.

Research Model

Figure 1 refers to the study model, which includes the variables and dimensions of the study [14, 15].

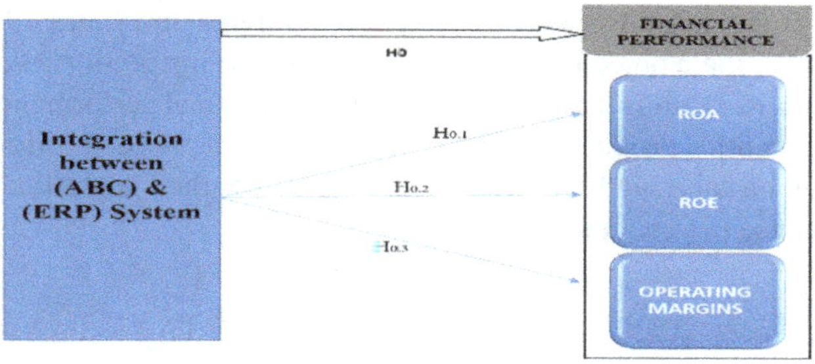

Fig. 1. Study model

2 Literature Review

The Integration Process Between ERP and ABC System

The first generation began to integrate the two system because what is believed it effects on quality also with accounting and management advancing process so it affects in general on the financial & non-financial performance in general, Where [16] emphasizes a fundamental relationship between the two systems, the effective implementation of the (ERP) system reflected in modern management practices especially the (ABC) system, that why we can explained the use of AIS especially the ERP system, which has a direct impact on pricing decisions, through gathering the information in one place, the ability will increase to access cost information at all stages of production and delivery that will have a positive impact on improving organizational and financial performance [17].

The second generation (Time Driven Activity Based Costing) it is approach to connect the integration between the time driven and cost in ERP &ABC systems, the second gen provide in order to process some practical problems to depend on the ABC system that face obstacles and rely on the use of time as a primary cause of cost, [18] connected the two systems (TD_ABC & ERP) together in a form for project management, and they found that it has a positive impact, through their reliance on the data generated from it, Therefore [19], the integration between them resulted in the avoidance of activities which don't add value which leads to the optimal utilization of available resources then estimate the utilized energy correctly, results saving time and reducing costs, also the (TD_ABC) system is uphold with the required data and information, Through the comprehensive database, the (TD_ABC) system integrates well with the available data from the (ERP) system and the customer relationship management system, This makes the system more vita, Moreover [20], the ERP existence or any simple automation system at the facility would greatly facilitate the process of applying the (TD_ABC) system As a result of preplanning of operations secured by the ERP system within that facility.

As a result of these proses it showed the Third generation (PF_ABC) based on Performance Focused, this generation appeared in late 2008, it was flexible and focused on choosing the most appropriate cost vectors, Namazi 2016 introduced this system to reduce the problems related to the (TD_ABC&ABC) systems, PF_ABC system marked to be a system which It integrates with the (ERP) system and the performance management system to identify the activities that are believed to be a major step in (ABC) and were omitted in (TD_ABC) [19].

The Effects of Integration Between ABC&ERP Systems on Financial Performance Figure 2, shows the main objectives of each system, whose integration leads to achieving common benefits, as its impact appears on both the operational performance, represented by (manufacturing time, production, and cost) and the financial performance represented by (profit, return on sales, and return on assets), therefore the integration between them affects performance in genera [15].

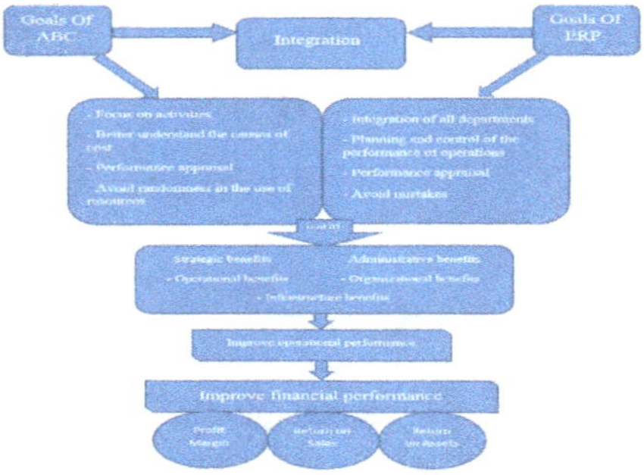

Fig. 2. Integration between ABC&ERP

[21] refers on their studies to the positive impact on operational and organizational performance, through applying the cost management system on the activities efficiently, due to the goals that achieve such as reducing costs, Moreover [5] mentioned the use of technology, especially the ERP system, has an impact on operational by reducing costs and raising the level of productivity, [22] indicated that the (ABC) system increases the accuracy of accounting information which leads to the establishment of a useful and good administrative system, whereas it measures accurately the costs by monitoring inputs and outputs In order to make clear pricing decisions, which accomplish the strategic goals, [23] point out that the (ERP) system focused on senior management and decision-makers which helps in man-aging the company's resources effectively, that led to an increase in its competitiveness in the market, and the creation of integration in the institution.

[24] mentioned the (ERP) system works to reduce the cost of information technology infrastructure, which results the implementation of new applications, supports the job to be more easily.

[25] focused on his study on the fact that there is an impact of the integration of functional units through the (ERP) system together in automating cost estimation 'which helped on in the development of the (ABC) system' Also it aims to receive a financially valuable system through adding the management accounting concepts into organizations systems [9].

[25] point out that both systems share in achieving the benefits at the strategic level, represented by achieving integration between functions at the level of all departments and cost management, furthermore, They together achieve benefits at the operational level, Such as reducing costs, raising product quality, and increasing productivity, They also share benefits at the administrative level, for example, decision-making, planning, and improving financial performance.

[26] mentioned Some of benefits in his study that integrating ERP systems and corporate management system on the reason behind why organizations operate the best ERP applications in their line of business because to gain the ability to unite efforts in providing a coordinated framework to aid all the departments and to observe on planning operation which leads to increase efficiency through reducing the unnecessary costs for the organization.

The ERP system can provide reliable information about the cost of activity through merging the information of production planning, material management, cost accounting, and management information In order to increase the accuracy of product costing, and to develop the budgets that based on the activities, As the merge of ERP management cost that based of activities provides a chance to improve continually the ABC models, Also increase the possibility of using the ABC results by decision makers, improve the design, implementation, and early use of (ERP & ABC) systems, to improve the management efficiency, achieving the final target that gives great value and integration to the organization [27].

3 Research Methodology

The research relied on the descriptive analytical method that aims to know the integration between (ERP & ABC) systems, also its impact on financial performance of the industrial companies listed on the Amman Stock Exchange, through observing the previous research's and studies on the other hand the analytical level, We measured the independent integration variable between the two systems (ERP & ABC) through a questionnaire, we distributed to the industrial companies listed on the Amman Stock Exchange, where as the financial data of the dependent variable financial performance collected for period (2017–2021) And analyzed through the statistical program SPSS (Table 1).

Descriptive Statistics for the Independent Variable (Integration between ERP System and Activity Based Cost Accounting System)

Descriptive statistics of the dependent variable (The financial performance of Jordanian industrial companies)

Table 1. Descriptive statistics of the dependent variable

Std. Dev.	Mean	Max	Mini	Variables
0.166	-.001	0.218	-0.339	**Operating Profit Margin (Financial Statements)**
0.707	4.342	5.000	3.000	**Operating Profit Margin(5scale)**
0.057	0.006	0.095	-0.099	**ROA(Financial Statements)**
0.923	4.110	5.000	3.000	**ROA(5scale)**
0.104	-.010	0.126	-0.224	**ROE(Financial Statements)**
0.644	4.323	5.000	3.000	**ROE(5scale)**

The five-factor conversion was (4.342) and standard deviation of (0.707), while the lowest value was (3) and the highest value was (5), and the arithmetic mean of data for return on asset index after the five-factor transformation was (4.110), the standard deviation of (0.923), whereas the lowest value was (3) and the highest value was (5), As for the return on equity index, the arithmetic mean after conversion to the five-year gradient was (4.323), with a standard deviation of (0.644), while the lowest value was (3) and the highest value was (5).

The Normal Distribution Tests

We used the normal distribution test to ensure that study data was normally distributed or not, Skewness) test was used, In addition, (Kurtosis) was extracted, and since the number of study observations is 155, the study variables are considered normally distributed [12] (Table 2).

Table 2. The normal distribution tests

Kurtosis	Skewness	Variables
Statistic	Statistic	
−0.171	0.024	Integration between (ERP & ABC)
−0.816	−0.598	Operating Profit Margin
−1.802	−0.221	ROA
−0.689	−0.419	ROE

According to test, data and what indicated in table that the data distribution was normal as the values did not fall outside the range (±1.96) in the Skewness test, and the value test did not exceed (±3) Kurtosis.

Study Results Hypothesis

Hypothesis: There is no impact with Statistical significance level ($a \leq 0.05$) For the integration between the enterprise resource planning system (ERP) and the activity-based cost accounting system (ABC), on the financial performance (ROA, ROE, Operating profit margin) of the industrial companies listed on the Amman Stock Exchange (Table 3).

Which means the value of R2 reached 47,5%, That explains the independent variable integration between ERP&ABC system changes that occurred in the dependent variable, the financial performance of the Jordanian industrial companies almost 47,5% percentage which is a good percentage that explains the ability of Sekaran test in explaining the changes that occurred in the dependent variable explain the financial performance of the Jordanian industrial companies at rate of (47.5%), It is a good ratio and indicates the ability to interpret the independent variable for the dependent variable, according to Sekaran and Bougei (2016); Below (20%) is considered low, (50%–20%) is medium (good), and high (100%–51%) [54].

The table shows that study model that used on the testing impact integration between the (ERP & ABC) systems on the financial performance on Jordan industrial companies

Table 3. Results of the dependent variable

Dependent variable: financial performance			
Sig	T	B	Independent variables
0.000***	15.7	4.155	Constant
0.000***	5.23	0.356	Integration between (ERP & ABC)
47.50%			R^2
0.000***			Sig
27.353			F

***at the 1% level, **at 5%, *at the 10% level.

Statistically significant where the value of (F-Statistic) for the study model was (27.353) moreover at the (Sig = 0.000) level Statistically significant and this value located within the level of statistical (significance 1%) This is an indication of the importance and the positive effects of the model use in financial performance on the Jordanian industrial companies, consequently The main null hypothesis is rejected and the alternative hypothesis is accepted, which states that "There is a statistically significant effect at the level (a ≤ 0.05) of the integration between the enterprise resource planning (ERP) system and the activity-based costing (ABC) system on the financial performance (ROA, ROE, and operating profit margin) for the industrial companies listed on the Amman Stock Exchange.

4 Conclusion

Research Discussion

- There is a statistically significant effect of the integration Between (ERP & ABC) systems on the financial performance represented by (ROA, ROE, and operating profit margin) for the industrial companies listed on the Amman Stock Exchange, Where as the results align with the theoretical aspect. In addition, many previous studies have enhanced in many ways the process of integration between the two systems.

[15] mentioned in his study "The impact of activity-based costing systems supported by enterprise resource planning systems on enterprise performance" To prove his hypothesis, he used several ratios to analyze and measure the financial statements, therefore one of the ratios used was (ROA) that agrees with the study of this research, and indicated that there is a positive rise in (ROA) over the years, since the use of the (ERP) system with continuous development, which helped to provide information and make administrative and financial decisions that called for cost reduction, which supported improving financial performance.

[28] indirectly supported it, where he refers to the importance of measuring the cost accurately and clearly to avoid errors and making effective administrative decisions, centered the study of [29] about measuring the effects of using the ERP system on a sample of Jordanian industrial companies (pharmaceutical industry) and concluded that

the integration of internal operations in the organization affects the overall performance of the organization, which enhances the results of this research, [29] used a sample from the current study population.

[30] noted that there is an indirect relationship between the use of the (ABC) system and the improvement of financial performance, And it is through operational performance, which reflects positively on financial performance, this the theoretical aspect results, Moreover [12] added that the automation of cost estimation occurred through the integration of functions, This enhances the financial value of ERP systems, but does not overlook the defects of ERP, which lead to wrong decisions because of the wrong information existence by a physical or technical act.

Through the results of the distributed questionnaire, we found that Paragraph No. (3), which states "The applied systems are integrated with each other as it helps in making effective financial and administrative decisions," had the highest arithmetic mean of (4.21), and a standard deviation of (0.631), which indicates that The Jordanian industrial companies are interested in a high degree of integration between (ABC & ERP) systems.

5 Recommendations

1. Decision makers and stockholders of new industrial companies, when building and designing them, must take the necessary steps, to create a suitable environment for (ERP & ABC) systems; to manage them, and operate them in the required manner.
2. Carrying out more studies and supporting research; to implement the developed (ABC) and (ERP) systems in Jordanian industrial companies, this is due to the importance of this sector in the Jordanian economy.
3. Qualifying a permanent team; to maintain industrial mechanisms, through courses, and continuous evaluation, which leads to a reduction in wasted time in troubleshooting.
4. Industrial companies develop the infrastructure of the technology department periodically; to find immediate solutions to emerging errors, whether material or technical.

References

1. Yamin, I., Younis, I.: The impact of the application of enterprise resource planning system on banks financial performance (a case study of Arab bank and Islamic bank of Jordan), Zarqa J. Res. Stud. Humanit. **18**(1) (2018)
2. Baloula, M.A.: The role of enterprise resource planning (ERP) systems in improving the quality of accounting information - a case study of the Naqous preservation corporation, Batna, Ph.D. thesis, Mohamed Kheidar University-Biskra, Algeria (2021)
3. Al-Janbaz, A.M.M.M.: The impact of ERP application on the effectiveness of accounting data. Ph.D. thesis, Kai University, a licensed online university of higher education, Somalia (2018)
4. Sourieh, S., Othmani, S., Qadir, S.: The contribution of the enterprise resource planning (ERP) system to improve decision-making. Master's thesis, University of Hama Lakhdar El Wadi, Algeria (2020)

5. Dida, K.: The effect of using the Enterprise Resource Planning (ERP) system on improving the performance of the economic enterprise, a field study of a sample of oil companies operating in Algeria, Ph.D. thesis, University of Kasdi Merbah Ouargla, Algeria (2019)
6. Al Theebeh, Z.A.H., Al Juaidi, O., Al Eryani, M: The extent of applying general principles of corporate governance in Yemeni commercial banks. Zarqa J. Res. Stud. Humanit. **10**(1) (2010)
7. Al-Fasfus, F.S., Al-Rawashdeh, A.M., Al-Theebeh, Z.A., Al-Enabi, H.A.: The impact of accounting conservatism on financial performance in services companies listed on Amman stock exchange. Acad. J. Interdiscip. Stud. **11**(4) (2022)
8. Al-Nafaa, F.S.M., Abu Amarah, M.M.J.I.: A proposed model for analyzing the cost deviations, using RCA (resource consumption accounting) approach with application on a Saudi hospital. Zarqa J. Res. Stud. Humanit. **19**(3) (2019)
9. Meshri, K., Lotofah, S.: The role of the activity-based costing system in raising the profitability of industrial enterprises - a questionnaire study, Master's thesis, University of Martyr Hama Lakhdar-Al-Wadi, Algeria (2021)
10. Muhammad, A.F.: Success factors for implementing ERP system in terms of organizational suitability and user satisfaction. Master thesis, Syrian Virtual University, Syria (2018)
11. Jamil, R., Al-Hagan, A.A.: A proposed model for linking the ERP resource planning system to project management using the TDABC/PERT methods to reduce construction costs in the contracting sector (an applied study). Sci. J. Res. Bus. Stud. Helwan **36**(3), 57–120 (2022)
12. Datar, S.M., Rajan, M.V.: Horngren's cost accounting: a managerial emphasis. 17th ED Pearson (2020)
13. Saad, S., Yacoub, I., Hussain, M.: An analytical reading of the response of cost accounting to the requirements of the environment: the system of a case study. Baghdad Coll. J. Univ. Econ. Sci. Al-Mustansiriya Univ. **47**, 149–170 (2016)
14. Al-Mujahid, A.: The effect of applying the ERP resource planning system on the administrative and financial performance in Yemeni universities, a case study of Dhamar university. J. Arts **1**(24), 575–612 (2022)
15. Muhammad, K.M.M.J.: The effect of costing systems based on activities supported by project resource planning systems on the performance of the facility – a case study. Alex. J. Account. Res. **1**(2), 468–561 (2017)
16. Vetchagool, W., Augustyn, M.M., Tayles, M.: ISO 9000, activity based costing and organizational performance. Total Qual. Manag. Bus. Excell. **32**(3–4), 265–288 (2021)
17. Eker, M., Aytaç, A.: The role of ERP in advanced managerial accounting techniques: a conceptual framework 1. Bus. Econ. Res. J. **8**(1), 83 (2017)
18. Larbi, G.M.: Activity based-costing system through three generation: ABC-TDABC-PFABC. La Revue des Sci. Commerciale's **20**(1), 90–108 (2021)
19. Saleh, Wahba, S., Saleh, E.: A proposed model for activity-based costing based on performance using fuzzy logic - an applied study. Egypt. J. Bus. Stud. **45**(3), 26–50 (2021)
20. Qasim, M.K.M.: The role of time-driven activity costing (TDABC) in strategic planning of oil production costs: a field study on the Arabian Gulf Oil Company in Libya. Sci. J. Bus. Environ. Stud. **8**(2), 306–332 (2017)
21. Mahal, I., Hossain, A.: Activity-based costing (ABC)–an effective tool for better management. Res. J. Finance Account. **6**(4), 66–74 (2015)
22. Al-Danaf, M.O.: Developing cost systems in service facilities using resource consumption accounting in order to manage resources - an applied study. Master's thesis, Tanta University, Egypt (2021)
23. Shang, S., Seddon, P.B.: A comprehensive framework for classifying the benefits of ERP systems. In: Americas Conference on Information Systems 2000 Proceedings, vol. 39, p. 1005–1014 (2006)

24. Aqdi, A., Issawi, A.G.: The impact of the resource planning system on the operational performance of the organization, Master's thesis, University of Ahmed Derayah-Adrar, Algeria (2021)
25. Harman, M.: The extent and limitations of managerial accounting function within ERP systems, Master's thesis, Middle East Technical University (2022)
26. Munthe, R.: Benefits of company management systems with combination of ERP (enterprise resource planning). J. Res. Soc. Sci. Econ. Manag. **1**(6), 610–620 (2022)
27. Mujeeb, M.: Evaluation of the success of the ERP system through strategic and tactical factors, an applied study in the National Corporation for Works in Wells ENTP-Hassi Messaoud, Master's thesis, University of Kasdi Merbah-Ouargla-, Algeria (2015)
28. Kitsantas, T., Vazakidis, A., Stefano, C.: Integrating activity based costing (ABC) with enterprise resource planning (ERP) for effective management: a literature review (2020)
29. Sandoqa, S.A.S.: The impact of enterprise resource planning (ERP) system usage on supply chain integration at Jordanian pharmaceutical manufacturing organizations in Amman, Doctoral dissertation, Middle East University (2020)
30. He, Y.: A research on the integration between ERP system and ABCM. In: Xu, L.D., Tjoa, A.M., Chaudhry, S.S. (eds) Research and Practical Issues of Enterprise Information Systems II. IFIPAICT, vol. 254, pp. 781–786. Springer, Boston (2007). https://doi.org/10.1007/978-0-387-75902-9_88

Analyzing the Relationship Between Using Modern Digital Technologies (MDTs) and Financial Performance (FP) of Jordanian Telecommunication Companies (JTCs)

Thaer Faisal Abdelrahim Qushtom[1(✉)], Riham ALkabbji[1], Fuad Suleiman Al-Fasfus[1], Mohammad Suhail Mustafa Aloqdeh[1,2], and Rua Binsaddiq[2]

[1] Zarqa University, 2000, Zarqa 13110, Jordan
tqushtom@zu.edu.jo
[2] University of Business and Technology, Jeddah, Kingdom of Saudi Arabia

Abstract. Due to the impact of global digital transformation on various aspects of business, this study aimed to analyze the effect of using modern digital technologies (MDTs) on the financial performance (FP) of Jordanian telecommunication companies (JTCs), and to explore the demographic characteristics (DCs) that enhancing this relationship. To conduct this study, the descriptive analytical approach was used to collect and analyze data. Accordingly, a questionnaire was designed and distributed to HR and IT managers in the five telecommunications companies (TCs) operating in the Jordanian business environment (JBE). In addition, financial reports of companies were used to extract the FP. All questionnaires retrieved and were analyzable. To test hypotheses, the multivariate regression analysis was used. The results showed that there is a statistically significant effect of using MDTs on the FP of JTCs. In addition, the results indicated that DCs of companies, especially capital size and digital maturity (DM) of employees, have a significant mediating effect on the aforementioned relationship. Accordingly, the study recommended JTCs to keep a close eye on the latest digital technology developments in the telecommunication services field, to invest these technologies in improving their FP. In addition, JTCs are recommended to focus on developing employees' digital skills by involving them continuously in training courses, seminars and related conferences.

Keywords: digital technologies · telecommunication companies · financial performance · Jordan business environment · demographic characteristics

1 Introduction

The use of MDTs, such as automation, big data analytics (BDA), and cloud computing (CC), has been shown to have a positive impact on FP for businesses [11]. According to [6], companies that fully leverage MDTs can increase their operating productivity.

B. A. M. Alareeni and I. Elgedawy (Eds.): *Artificial Intelligence (AI) and Finance*, SSDC 488, pp. 873–883, 2023.
https://doi.org/10.1007/978-3-031-39158-3_81

Additionally, a study by [19] found that companies that have adopted MDTs have seen a significant increase in their return on investment. However, [9] have pointed out that the relationship between using MDTs and FP is not always straightforward. Factors such as the specific technologies used and the level of digital maturity (LDM) within an organization can play a role in determining the extent to which MDTs can improve FP [25].

Understanding the complex and dynamic relationship between the adoption and use of MDTs and FP may help businesses to identify how they can leverage the use of MDTs to improve their FP. Consequently, it can provide insights for companies looking to invest in MDTs and guide their decision making. Additionally, it can help businesses stay competitive in a rapidly changing digital landscape.

Despite the increasing use of MDTs in the JTCs, the relationship between these technologies and FP remains largely unexplored. Thus, the problem of the current study is to gain a better understanding of the relationship between the use of MDTs and FP in JTCs, and to identify the factors that affect this relationship. Accordingly, the questions of study are:

1. Does the use of MDTs influence the FP of JTCs?
2. What are the DCs that influence the strength of the relationship between using MDTs and FP of JTCs?

Next section provides a review of the related literature findings, followed by building the theoretical framework and hypotheses development. Then, the methodology of study and data analysis are presented, and finally, the conclusion of study.

2 Literature Review

There is a growing body of literature on the relationship between the use of MDTs and FP. Many studies have found a positive correlation between the use of MDTs and improved FP for businesses. In this regard, a study by [2] found that companies that fully leverage MDTs through using smart technologies can improve their performance. In addition, [4] found that companies that have adopted MDTs have seen a significant improve in their return on investment. Similarly, a study by [15] found that companies that have fully embraced digital transformation have seen a median revenue growth compared to companies that have not. Further, other studies have found that specific MDTs, such as automation, BDA, and CC, can have a positive impact on FP. In this vein, [21] found that companies that have implemented automation have seen an average increase in their productivity. As well as, a study by [30] found that companies that have adopted BDA have seen a median increase in their return on investment.

On the other hand, other studies have found that the relationship is more complex and can vary depending on factors such as the specific technologies used, the LDM within an organization, and the industry in which a business operates. In this context, a study by [32] found that the relationship between MDTs and FP is not linear, and that the impact of MDTs on FP can vary depending on the industry and the specific technologies used [27]. In addition, a study by [8] found that companies with a higher LDM tend to see a stronger positive relationship between using MDTs and FP. Similarly, a study by [28] found that the relationship between MDTs and FP is complex and can vary depending on the specific technologies used, the LDM within an organization, and the industry in which business operates. As well as, [3] found that MDTs can have different effects on different aspects of a business, such as revenue growth and cost reduction. In the same context, a study by [10] deduced that the relationship between MDTs and FP is complex and can depend on various factors such as the LDM and the specific technologies used.

3 Theoretical Framework and Hypotheses Development

MDTs are a result of human efforts that seek to improve all aspects of business, including FP [18]. Therefore, it is logical that the level of reliance of enterprises on MDTs is reflected in the level of improvement in their FP [20]. Based on the resource-based view (RBV) framework, firm's resources and capabilities are the key determinants of its competitive advantage and FP [22]. According to this framework, MDTs can be considered as valuable resources and capabilities that can help firms to achieve competitive advantage and improve their FP [7]. Based on that, the present study introduces the below hypotheses:

H.1. There is a significant positive relationship between using MDTs and FP of JTCs.

H.2. The DCs of JTCs affect the effectiveness of using MDTs in improving the FP of these companies.

3.1 Model of Study

Relying on the aforementioned hypotheses, Fig. 1 shows the model of current study.

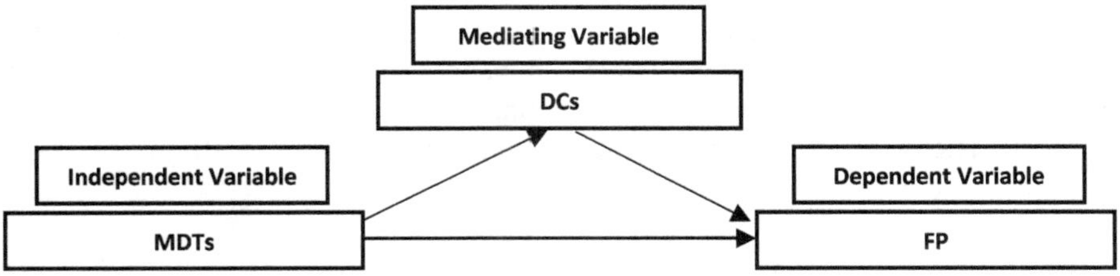

Fig. 1. Model of Study

4 Methodology

To conduct this study, the quantitative approach was used in collecting and analyzing data.

4.1 Data Collection Method

In order to collect data of the current study, the researchers developed a questionnaire to collect the data of the independent variable (using MDTs) and the mediating variable (DCs). While the reports and financial statements of JTCs were used to extract data of the dependent variable (FP). The questionnaire consisted of two parts, the first one included 8 questions with a 0,1 scale, focused on determining the extent to which MDTs was used, while the second one included 6 questions focused on the DCs of companies. Table 1 shows the measurement dimensions of each variable of the current study.

Table 1. Measurement Dimensions of Variables

Independent Variable MDTs	Mediating Variable DCs	Dependent Variable FP
Using of automation	Age of Company	Revenue Growth (RG)
Using of BDA	Capital Size	Cost Reduction (CR)
Using of CC	Number of Employees	Return on Assets (ROA)
Using Mobile Applications	Number of Branches	Return on Equity (ROE)
Using online technical support	DME	Market Share (MS)
	Types of Telecommunication Services Provided	

Note: Developing the questionnaire relied on [2, 4, 8, 10, 11, 14, 15]

4.2 Population and Targeted Respondents

The population of current study consists of all JTCs that provide Telecommunication and internet services, which number five companies, namely Zain, Orange, Umniah, Damamax, and Mada. In this vein, two questionnaires were distributed to each company, one for the HR manager and one for the IT manager. Accordingly, 10 questionnaires were distributed, retrieved, and analyzed.

5 Data Analysis

The small size population of the current study did not enable the researchers to conduct the preliminary tests of the questionnaire, such as the reliability and normal distribution tests. Therefore, descriptive analysis and regression analysis tests were sufficient for testing the current study hypotheses.

5.1 Descriptive Analysis Test

The descriptive information subject to analysis in this study was limited to the age of company, capital size, number of employees, number of branches, number of digital

Table 2. Descriptive Statistics According to the DCs of JTCs

Company	Zain	Orange	Umniah	Damamax	Mada
Q.1 - Age of Company	More than 25 years	20 - 25	15 - 20	15 - 20	10 - 15
Q.2 - Capital Size / JD	More than 200,000,000	50,000,000 - 100,000,000	100,000,000 - 150,000,000	Less than 50,000,000	Less than 50,000,000
Q.3 – No. of Employees	1500 - 2000	1000 - 1500	More than 2000	Less than 500	Less than 500
Q.4 – No. of Branches	50 - 100	More than 200	100 - 150	Less than 50	Less than 50
Q.5 – No. of Digital Training Courses	More than 2000	1500 - 2000	1500 - 2000	Less than 500	Less than 500
Q.6 – No. of Services	5	7	4	1	1

training courses for employees, and number of services provided to customers. Table 2 shows the descriptive analysis results.

Relying on Table 2 shown above, which shows the descriptive data of the companies based on their demographic information, JTCs were ranked and arranged from top to bottom based on the total points scored for them. Accordingly, Table 3 shows the companies' ranking according to the total points.

Table 3. Ranking JTCs According to the Descriptive Statistics Score for each Company

No. of Question	Q.1	Q.2	Q.3	Q.4	Q.5	Q.6	Total Score
Zain	5	5	4	3	5	4	**26**
Orangs	4	3	3	5	4	5	**24**
Umniah	3	4	5	4	4	3	**23**
Damamax	3	2	2	2	3	2	**14**
Mada	2	2	2	2	3	2	**13**

Note: The highest value was given to the highest answer, and the lowest value was given to the lowest answer

Regarding the types of services provided by JTCs (which is reflected by numbers in questions no 6), they were detailed by IT managers. Table 4 shows the types of services provided.

Table 4. Types of Telecommunication Services Provided

Company	Zain	Orange	Umniah	Damamax	Mada
Fixed telephone service		*			
Mobile services	*	*	*		
Internet services	*	*	*	*	*
Wholesale services		*			
Network services and other advanced services (Data Center, Cloud, or Machine to Machine services)	*	*	*		
E-Payment Services	*	*	*		
Development of Renewable Energy Projects		*			
Marketing Solutions	*				

5.2 Regression Analysis Test

To test the relationships between using MDTs and FP of JTCs, and the expected mediating effect of DCs on this relationship the study used the regression analysis test. As a result, two regression models were developed and tested. The first regression model tested the direct relationship between using MDTs and FP of JTCs. While the second model tested the expected mediating effect of the DCs on the first expected relationship.

First Regression Model

$$FP(RG; CR; ROA; ROE; MS) = a0 + a1MDTs + ei$$

Second Regression Model

$$FP(RG; CR; ROA; ROE; MS) = b0 + b1MDTs + DCs + \varepsilon i$$

where:

FP = Financial Performance	RG = Revenue Growth
CR = Cost Reduction	ROA = Return on Assets
ROE = Return on Equity	MS = Market Share
MDTs = Modern Digital Technologies	DCs = Demographic Characteristics
$a0$, $a1$, b0, b1 = Constants of equations	ei, εi = Error terms

The results of testing the first model are shown in Table 5 (below), while the results of testing the second model are shown in Table 6.

Table 5. The Relationship Between Using MDTs and FP of JTCs (N = 5)

Dep. Variable	FP				
	RG	CR	ROA	ROE	MS
Variables	Coeff. Est	Coeff. Est	Coeff. Est	Coeff Est	Coeff. Est
Ind. Variables					
Constants	0.000	0.000	0.000	0.000	0.000
	(0.109)	(0.125)	(0.098)	(0.087)	(0.139)
MDTs	0. 861**	0.715**	0.104*	0.93*	0.371**
	(0.001)	(0.000)	(0.015)	(0.020)	(0.001)
R Square	0.83	0.95	0.61	0.49	0.97
Adj. R Square	0.74	0.91	0.56	0.49	0.90
F-value	4088.36	972.56	327.19	295.78	6793.21
Sig. Level	0.000	0.000	0.001	0.000	0.000

P-value (Sig. Level) is reported in parentheses
* Indicates p value is significant at level 0.05, ** Indicates p value is significant at level 0.01.

Table 6. The Mediating Effect of DCs on the Relationship between Using MDTs and FP of JTCs (N = 5)

Dep. Variable	FP				
	RG	CR	ROA	ROE	MS
Variables	Coeff. Est	Coeff. Est	Coeff. Est	Coeff. Est	Coeff. Est
Ind. Variables					
Constants	0.153	0.109	0.248	0.364	0.205
	(0.037)	(0.018)	(0.002)	(0.051)	(0.003)
MDTs	0. 919**	−0.825**	0.342**	0.167**	0.520**
	(0.000)	(0.000)	(0.000)	(0.001)	(0.000)
DCs	0. 977**	0.865**	−0.303**	−0.135**	0.614**
	(0.001)	(0.000)	(0.010)	(0.009)	(0.000)
R Square	0.683	0.975	0.498	0.572	0.718
Adj. R Square	0.663	0.919	0.442	0.570	0.717
F-value	3114.16	1278.36	274.20	119.66	4801.19
Sig. Level	0.001	0.000	0.001	0.010	0.000

P-value (Sig. Level) is reported in parentheses
* Indicates p value is significant at level 0.05, ** Indicates p value is significant at level 0.01.

6 Conclusion

The results of the multiple regression test in Table 5, which explore the relationship between using MDTs and FP of JTCs, show that there is a statistically significant positive relationship between using MDTs and FP of these companies in general, as the significance level between using MDTs and each of RG, CR, and MS was 0.01, while the significance level between using MDTs and each of ROA and ROE was 0.05. These results imply that using MDTs affects positively the FP of JTCs. Consequently, the first hypothesis of the present study is accepted. According to [5], the use of MDTs help reaches a larger number of customers through the Internet and reduce ordinary operating expenses such as salaries of workers and employees. As well as, [17] mentioned that revenues and costs are among the main indicators that are directly affected by using MDTs. Despite that, [24] pointed out that there are administrative and general expenses that cannot be avoided, which may lead to a lower level of improvement in the ROA and ROE compared to other performance indicators such as RG, CR, and MS. In addition, [29] indicated that there are operating expenses that must be incurred by entities, regardless of the adoption of MDTs, which reduces the impact of using these technologies on the returns achieved for the various entities.

As for the results of the multiple regression test of the mediating effect of DCs on the relationship between using MDTs and the FP of JTCs, the results in Table 6 show that there is a statistically significant effect at the level of 0.01 of the DCs on the relationship between using MDTs and all FP indicators used in the current study. This result implies that the DCs of JTCs do a significant mediating role between the effect of using MDTs and the FP of these companies. As a result, the second hypothesis of the current study is also accepted. As well, it was found that the effect of DCs comes in the following order: capital size, DME, number of services provided, age of company, number of branches, and then number of employees. These results might be explained by the fact that the greater the company's capital, the greater its operational capabilities [13]. As a result, these companies will be able to give modern and advanced digital courses to their employees that should increase the productivity of their operations [1]. Consequently, the number of services that provided by companies to customers might be increased [31], which might be reflected on increasing the going concern of companies [23]. Accordingly, companies might be able to spread through increasing the number of branches and thus increasing the number of employees [12]. In this regard, [16] explained that increasing the capital of companies enhances their ability to finance their operations and improve their performance. Moreover, [26] concluded that the skills of employees related to work are considered one of the main determinants of the success of enterprises in achieving their goals and growth.

The current results imply that development of companies' performance and enhancing their presence in markets are related to their tendency to adopt MDTs, which helps in the RG and CR. In addition, the results imply that the focus on developing employees' digital skills that relate to company's operations is considered one of the most important demographic determinants of companies' success after the capital size, as it reflects on improving company's operations in general.

Based on the aforementioned results, the study recommends JTCs to keep a close eye on the latest MDTs in the telecommunication services field, to invest these technologies in improving their performance. In addition, JTCs are recommended to focus on developing employees' digital skills by involving them continuously in training courses, seminars and related conferences. As well as, JTCs are recommended to take care of high digital skilled employees due to their significance role in the company's performance development and growth. Finally, the study recommends researchers to conduct similar studies on different sectors and other business environments to ensure the extent to which the results can be generalized.

Limitations of the current study summarized by: the study is limited to the JTCs. So, the results cannot be generalized to all companies in different sectors and other countries. In addition, the data were collected in the period from 12-2022 to 1-2023. Thus, the data and the size of study population may vary over time. Accordingly, there is an opportunity for researchers to carry out similar studies focus on other sectors and different business environments. As well as, similar studies can be conducted using different research methods and other performance indicators.

References

1. Aljuboori, Z.M., Singh, H., Haddad, H., Al-Ramahi, N.M., Ali, M.A.: Intellectual capital and firm performance correlation: the mediation role of innovation capability in Malaysian manufacturing SMEs perspective. Sustainability **14**(1), 2–27 (2022)
2. AlMulhim, A.F.: Smart supply chain and firm performance: the role of digital technologies. Bus. Process. Manag. J. **27**(5), 1353–1372 (2021)
3. Attaran, M.: Digital technology enablers and their implications for supply chain management. Supply Chain Forum Int. J. **21**(3), 158–172 (2020)
4. Blichfeldt, H., Faullant, R.: Performance effects of digital technology adoption and product & service innovation – a process-industry perspective. Technovation **105**(1), 1–12 (2021)
5. Chen, Y., Kumara, E., Sivakumar, V.: Investigation of finance industry on risk awareness model and digital economic growth. Ann. Oper. Res., 1–9 (2021)
6. Chowdhury, T., Adafin, J., Wilkinson, S.: Review of digital technologies to improve productivity of New Zealand construction industry. J. Inf. Technol. Constr. **24**((ITcon), Special issue: 'Virtual, Augmented and Mixed: New Realities in Construction'), 569–587 (2019)
7. El Nemar, S., El-Chaarani, H., Dandachi, I., Castellano, S.: Resource-based view and sustainable advantage: a framework for SMEs. J. Strateg. Mark. (2022)
8. Eremina, Y., Lace, N., Bistrova, J.: Digital maturity and corporate performance: the case of the Baltic states. J. Open Innov. Technol. Market Complexity **5**(3), 1–9 (2019)
9. Eslami, M.H., Jafari, H., Achtenhagen, L., Carlbäck, J., Wong, A.: Financial performance and supply chain dynamic capabilities: the moderating role of industry 4.0 technologies. Int. J. Prod. Res. **1**(1), 1–18 (2021). https://doi.org/10.1080/00207543.2021.1966850
10. Guo, L., Xu, L.: The effects of digital transformation on firm performance: evidence from China's manufacturing sector. Sustainability **13**(22), 1–11 (2021)
11. Gupta, S., Qian, X., Bhushan, B., Luo, Z.: Role of cloud ERP and big data on firm performance: a dynamic capability view theory perspective. Manag. Decis. **57**(8), 1857–1882 (2019)
12. Gutierrez, E., Krupa, J., Minutti-Meza, M., Vulcheva, M.: Do going concern opinions provide incremental information to predict corporate defaults? Rev. Acc. Stud. **25**(4), 1344–1381 (2020). https://doi.org/10.1007/s11142-020-09544-x

13. Hamour, M.M.: The effect of the determinants of Islamic and commercial banks' performance in Jordan on return on equity (ROE) an empirical study of the period (2001–2014). Zarqa J. Res. Stud. Humanit. **17**(3), 676–687 (2017)
14. Heslina, H., Syahruni, A.: The influence of information technology, human resources competency and employee engagement on performance of employees. Golden Ratio Hum. Res. Manage. **1**(1), 1–12 (2021)
15. Jardak, M.K., Hamad, S.B.: The effect of digital transformation on firm performance: evidence from Swedish listed companies. J. Risk Finance **23**(4), 329–348 (2022). https://doi.org/10.1108/JRF-12-2021-0199
16. Kabuye, F., Kato, J., Akugizibwe, I., Bugambiro, N., Ntim, C.G.: Internal control systems, working capital management and financial performance of supermarkets. Cogent Bus. Manage. **6**(1), 1–13 (2019)
17. Khayer, A., Talukder, M.S., Bao, Y., Hossain, M.N.: Cloud computing adoption and its impact on SMEs' performance for cloud supported operations: a dual-stage analytical approach. Technol. Soc. **60**(1), 1–14 (2020)
18. Lee, J., Suh, T., Roy, D., Baucus, M.: Emerging technology and business model innovation: the case of artificial intelligence. J. Open Innov. Technol. Market Complexity **5**(3), 44 (2019)
19. Lee, K.L., Azmi, N.N., Hanaysha, J.R., Alzoubi, H.M., Alshurideh, M.T.: The effect of digital supply chain on organizational performance: an empirical study in Malaysia manufacturing industry. Uncertain Supply Chain Manage. **10**(2), 495–510 (2022)
20. Lin, W.L., Yip, N., Ho, J.A., Sambasivan, M.: The adoption of technological innovations in a B2B context and its impact on firm performance: an ethical leadership perspective. Ind. Mark. Manage. **89**(1), 61–71 (2020)
21. Lowenberg-DeBoer, J., Huang, I.Y., Grigoriadis, V., Blackmore, S.: Economics of robots and automation in field crop production. Precision Agric. **21**(2), 278–299 (2019). https://doi.org/10.1007/s11119-019-09667-5
22. Nagano, H.: The growth of knowledge through the resource-based view. Manag. Decis. **58**(1), 98–111 (2020)
23. Ostrom, A.L., Fotheringham, D., Bitner, M.J.: Customer acceptance of AI in service encounters: understanding antecedents and consequences. In: Handbook of Service Science, vol. 2(1), pp. 77–103 (2018)
24. Panoutsou, C., Alexopoulou, E.: Costs and profitability of crops for bioeconomy in the EU. Energies **13**(5), 1222 (2020). https://doi.org/10.3390/en13051222
25. Priyono, A., Moin, A., Putri, V.O.: Identifying digital transformation paths in the business model of SMEs during the COVID-19 pandemic. J. Open Innov. Technol. Market Complexity **6**(4), 104 (2020)
26. Qushtom, T.F.: The implementation of balanced scorecard in Jordanian Islamic banks. Int. J. Fin. Res. **11**(5), 105–114 (2020)
27. Shaqqour, O.F., Alkabbji, R.F., Masoud, A.A.: The impact of applying cloud accounting on the financial fraud at Jordanian companies. In: 2021 22nd International Arab Conference on Information Technology (ACIT), pp. 1–6. IEEE, Muscat, Oman (2021)
28. Shen, L., Zhang, X., Liu, H.: Digital technology adoption, digital dynamic capability, and digital transformation performance of textile industry: moderating role of digital innovation orientation. Manag. Decis. Econ. **43**(6), 2038–2054 (2022)
29. Son, Y., Kwon, H.E., Tayi, G.K., Oh, W.: Impact of customers' digital banking adoption on hidden defection: a combined analytical–empirical approach. J. Oper. Manag. **66**(4), 418–440 (2019)
30. Su, X., Zeng, W., Zheng, M., Jiang, X., Lin, W., Xu, A.: Big data analytics capabilities and organizational performance: the mediating effect of dual innovations. Eur. J. Innov. Manag. **25**(4), 1142–1160 (2022)

31. Vannucci, V., Pantano, E.: Digital or human touchpoints? Insights from consumer-facing in-store services. Inf. Technol. People **33**(1), 296–310 (2019)
32. Yamin, I.Y.: The impact of the application of enterprise resource planning system on banks financial performance (a case study of Arab bank and Islamic bank of Jordan). Zarqa J. Res. Stud. Humanit. **18**(1), 34–47 (2018)

Using Electronic Auctions for Contracting in the UAE Federal Government Procurements

Alaa Abouahmed[1,2], Ahmed Eldakak[1,3](✉), and Aliaa Zakaria[4,5]

[1] College of Law, UAE University, Al Ain 15551, UAE
alaa.m@uaeu.ac.ae, ah.eldakak@gmail.com
[2] Faculty of Law, Helwan University, Cairo, Egypt
[3] Faculty of Law, Alexandria University, Alexandria 5424010, Egypt
[4] College of Law, Al Ain University, Al Ain, UAE
aliaa.ali@aau.ac.ae
[5] Faculty of Law, Tanta University, Tanta, Egypt

Abstract. The UAE has been using digital technology in federal governmental procurements. It performed all necessary technical modifications for this digital transformation. In the context of law, the UAE issued a detailed guide on digital procurements in the federal government. The guide regulates several methods for concluding governmental contracts. However, the scope of this article is limited to contracting using electronic auctions.

The article shows how the various stages of concluding contracts take place online through the digital platform. It addresses the prequalification stage, where suppliers lacking technical, legal or financial capacities are excluded from auctions. A supplier who passes this stage successfully is registered in the digital platform and can submit bids in electronic auctions. The article also discusses the submission of bids stage, which is characterized by transparency, allowing bidders to check the lowest price in real-time and allowing the federal entity to negotiate with the three suppliers who make the best offers from a technical and financial perspective.

Keywords: Digital Procurements · Electronic Auctions · Administrative Contracts

1 Introduction

The UAE established an electronic auction system to encourage investors to conclude contracts with the federal government (Digital Procurement Platform 2023, Ministry of Finance Website 2023). This system can be used to conclude contracts in various fields, especially in the fields of public utilities and services, which are essential to achieve economic development and uphold the public interest. Emphasizing the importance of the electronic auction system, the UAE issued a detailed guide for federal government digital procurements in 2021, which was approved by the Minister of Finance Decision No. 142 of 2021.

B. A. M. Alareeni and I. Elgedawy (Eds.): *Artificial Intelligence (AI) and Finance*, SSDC 488, pp. 884–889, 2023.
https://doi.org/10.1007/978-3-031-39158-3_82

The guide regulates all stages of concluding digital procurements in a way that guarantees the preservation of the rights of all concerned parties. For example, the guide regulates the prequalification procedures, a legal and administrative process that the federal government conducts to ensure the eligibility of potential suppliers by verifying their legal, financial and organizational capacity to achieve the goal of the contract. The guide also determines various procurement methods, including electronic auctions.

This article addresses the most important notes on the electronic auction system as a means to conclude contracts with the federal government. It adopts an analytical methodology by analyzing the relevant legal rules in Cabinet Resolution No. (4) of 2019 regarding Procurement Regulation and Storehouse Management in the federal government and the guide for federal government digital procurements.

The article proceeds as follows. Part 2 addresses the general framework of electronic auctions by defining the electronic auction and showing its publicity requirement. Part 3 analyzes the contract's concluding stages by discussing the prequalification and the bidding stages. Finally, Part 4 concludes that electronic auctions contribute to the public interest.

2 The General Framework of Electronic Auctions

2.1 Definition of an Electronic Auction

An electronic auction is a negotiation method that takes place over the internet between a federal entity as a purchaser and suppliers as the seller. According to article 1 of Cabinet Resolution No. (4) of 2019 regarding Procurement Regulation and Storehouse Management in the federal government, supplier refers to any natural or juridical person who may contract with a federal entity. He could be an actual supplier who sells products to federal entities. He may also be a constructor who performs constructions. Moreover, he may be a service provider who offers services for a federal entity, such as maintenance services.

In an electronic auction, a federal entity invites suppliers to submit competitive prices according to certain specifications within a deadline. This process takes place on an online platform managed by the ministry of finance called the "digital procurement platform" which can be accessed at https://procurement.gov.ae/. The platform enables federal entities and suppliers to communicate and interact with each other. Registered suppliers can submit their tenders on the platform. Even concluding electronic contracts can take place securely on the platform.

Federal entities can adopt either forward or reverse auctions online. In forward auctions, federal entities seek to sell their assets or products so that they prefer the highest bidder. By contrast, in reverse auctions, federal entities are usually purchasers, so they choose the lowest tender taking into consideration its financial and technical specifications.

It is worth noting that using electronic auctions could be mandatory or optional for federal entities depending on the procurement value. Section 6.4.5.2 of the Guide for Federal Government Digital Procurements makes electronic auctions mandatory if the procurement value exceeds 5 million dirhams. Electronic auctions are optional if the procurement value does not exceed this value. The organizational unit, responsible for all procurement procedures, can conduct an electronic auction whenever it deems it necessary or convenient.

2.2 The Requirement of Electronic Auctions Publicity

One of the main principles governing the electronic auction mechanism is that the federal entity must give public notice of the procurement. The notice must be announced on the digital procurement platform, guaranteeing transparency (Khan 2018). The platform links the federal entities with the suppliers registered in the Federal Supplier Register. It allows performing all stages of concluding government contracts electronically. This includes announcing the procurement, submitting bids, following up on tenders, contract awarding, submitting purchase orders and preparing invoices. By so doing, it enhances the efficiency of government spending by selecting the best suppliers to ensure the best prices, taking into consideration the quality of the provided goods and services.

The procurement announcement must include the conditions set by the federal entity and the deadline to enable suppliers to determine whether they wish and can participate.

It is worth noting that announcing a procurement in UAE has passed three stages of development responding to the changing circumstances. In the first stage, announcing a procurement used to be through traditional advertisements such as newspapers, magazines and billboards of the concerned ministries. In the second stage, coinciding with the launch of e-government applications, the UAE introduced the electronic advertisement system in 2014 by article 5(6) of the decision no. 32 of 2014 regarding the procurement and warehouse management regulations. Electronic advertisement aimed to transform transactions from the traditional paper form to the electronic form, which positively impacts the workflow in the government. This represented a remarkable transformation. Instead of dealing in person with papers and several employees, suppliers can make all the procedures on the website of the Ministry of Finance or via e-mail. This means that transactions occur in one place without circulating to employees and offices. Each competent employee can access the transaction on the electronic system to make notes or inputs and send them electronically, reducing the administrative complications that the administrative decisions and transactions go through (Aljalawda 2016).

Then the UAE legislator took another step to develop administrative contracts by introducing the digital procurement platform. Section 6.4.4.2 of the guide for federal government digital procurements explicitly emphasizes the need to announce the procurement on the digital platform in advance. It further mentioned that the federal entity must notify the qualified suppliers of the new procurement and ask them to confirm receipt of the offer request and their desire to participate.

It is worth noting that electronic contracts have the same authenticity as paper contracts. The UAE law equates electronic writing with paper writing by the law of electronic transactions as well as evidence law (Mohammed 2022).

3 The Stages of Concluding Electronic Procurement Contracts

Concluding contracts electronically has become more popular over time due to its advantages (Saini and Bhanu 2020). It saves time since it does not require physical meetings. It reduces the costs of contracting and protects the environment. In electronic procurements, pre-qualification of the supplier is an essential step to consider a supplier as a possible party to a contract with a federal entity. If the supplier is pre-qualified, he may enter electronic auctions organized by federal entities and submit bids.

3.1 Pre-qualification Stage

The pre-qualification process is where the federal government examines whether suppliers are qualified to perform their obligations under the procurement if they are awarded the contract through the electronic auction. This stage excludes nonserious suppliers and those lacking the necessary legal, technical or financial capacity.

It is worth noting that free competition must dominate the pre-qualification stage. All suppliers must have equal opportunity to be prequalified for electronic auctions (Nouh 2013). That is why Article 10 of Cabinet Resolution No. (4) of 2019 regarding Procurement Regulation and Storehouse Management in the federal government set pre-qualification criteria.

Firstly, the government must identify the previous experience of the supplier in similar contracts and evaluate its performance during the execution of the contract. Secondly, the government must identify the supplier's technical capabilities regarding experience, qualifications, number of employees, equipment, construction and manufacturing facilities. Thirdly, the government must evaluate the financial status of the supplier. Finally, to determine whether a supplier is prequalified, the federal entity may need to conduct field visits to the supplier to verify its capabilities.

The pre-qualification process has been developed over time until it is digitalized. Nowadays, the pre-qualification process takes place on the digital platform. This contributes to establishing a database of qualified suppliers that is shared and easily accessed by the various entities of the federal government. Therefore, federal entities know suppliers with the required capabilities to support them by providing their needs.

Suppliers not passing the pre-qualification process are denied registration in the suppliers' register. This means that they do not have any chance to be awarded a supply contract from any federal entity.

3.2 Bidding Stage

Submitting tenders in governmental procurements used to be by a traditional method. The supplier used to submit two sealed envelopes. One envelope includes the technical offer, and the other contains the financial one (AlShamsi 2018).[1] The guide for federal government digital procurements introduced digital bidding on the online platform.

Free competition must be guaranteed during the bidding submission stage. All registered suppliers must have equal opportunity to be awarded the contract based on the

[1] AlShamsi, S: The Legal System of Tenders in the UAE, 2018, 67.

criteria and conditions set by the federal entity that invites bids. That is why it is illegal for federal employees supervising the procurement to receive gifts from the suppliers. Moreover, he must disclose any consanguinity with any of the suppliers. Such rules aim at avoiding conflict of interest and protecting federal entities.

It is worth noting that bidding in the electronic auction differs from the traditional bidding system.

Checking the Bid Rank

As in the traditional method of bid submission, the digital platform does not allow the bidder to know the value of all other bids and their technical details. This information is available for the federal entity only. However, in digital bidding, each bidder can know the rank of his bid and the lowest bid in real-time, which helps the bidder to decide whether to proceed or withdraw. This would allow the bidder to submit a lower bid to save his chances of being awarded the contract by the federal entity (Rašić et al 2019).[2] This is a significant development in administrative contracting since bidding is no longer completely confidential.

Negotiating with the Three Best Offers Providers and Awarding

Traditionally, awarding the contract required several steps. The tenders committee convenes at a set date and place to open the sealed envelopes containing financial and technical offers (Hassan 2014). After verifying that the envelopes are sealed in the presence of bidders or their representatives, the committee opens them in the order they were received. The bidder who submitted the envelopes with the lowest offer is awarded the contract (Omar 2013).

In the electronic auction method, this mechanism has changed. The bidder who submits the lowest bid is not awarded the contract automatically. Instead, the federal entity can negotiate with the bidders who submit the best three offers. The federal entity should first negotiate with the bidder who submits the best offer. If negotiations fail or the bidder refuses to negotiate better conditions for the federal entity, the federal entity may consider negotiating with the ones who submitted the second and third-best offers. This could result in a change in the rank of bidders. After negotiations, the bidder who offered the third-best offer may become the first.

4 Conclusion

Using electronic auctions in governmental procurements is one aspect of the ongoing technological transformation in the UAE. It facilitates the procedures of procurements and the various stages of concluding administrative contracts. Due to its ease of use, it encourages both national and foreign investors to participate in governmental procurements. This means that more suppliers would submit bids increasing the competition, which usually results in better offers in favor of the government.

[2] Rašić, M., Milenković, M. & Vojković, G.: Electronic Auction in European and Croatian Public Procurement Law. International Convention on Information and Communication Technology, Electronics and Microelectronics, MIPRO 2019, 1561.

To conclude, the UAE experience of using electronic auctions is successful. This is thanks to several concepts and rules. The pre-qualification of suppliers guarantees that the registered suppliers are serious and efficient ones who can perform their obligations if they are awarded the contract after the end of the electronic auction. Being able to check his bid rank and the lowest bid price, competition increases as the supplier may choose to submit a new offer with a lower price. Public interest is further promoted when the federal entity can negotiate with the three suppliers who submit the best bids.

Finally, this article has certain limitations. The authors limited their research to using electronic auctions to conclude federal government procurements in the UAE. The authors analyzed the relevant legal rules and jurisprudential interpretations. Due to the novelty of the topic and the legal rules regulating it, no judicial decisions have been issued yet. The practical application of electronic auctions shall bring many issues that need further research. Such issues include the legal consequences if a supplier faces a technical issue during the electronic auction that causes his failure to submit the bid. Another issue is the legal consequences of hacking a supplier's account on the digital platform.

References

Aljalawda, A.: Public Tender in Governmental Electronic Contracts. Dar AlFekr AlJami'e, Egypt (2016)

AlShamsi, S.: The Legal System of Tenders in the UAE. UAE University, Al Ain (2018)

Digital Procurement Platform. https://procurement.gov.ae. Accessed 19 Feb 2023

Hassan, M.: Administrative Contracts. AlAfaq AlMoshreqa, Sharjah (2014)

Khan, N.: Public Procurement Fundamentals: Lessons from and for the Field. Emerald Publishing Limited, Bradford (2018)

Minister of Finance Decision No. (142) of 2021 Regarding the Approval of the Guide for Federal Government Digital Procurements

Ministry of Finance Website. https://mof.gov.ae. Accessed 19 Feb 2023

Mohammed, A.: The special legal nature of the offer in contracts concluded over the internet. Sirte Univ. J. Humanit. **12**(2), 157–169 (2022)

Nouh, M.: Offer and Acceptance in Administrative Contracts. AlHalabi Legal Publications, Beirut (2013)

Omar, A.: Judicial Review Over Concluding Administrative Contracts Using Tenders. AlHalabi Legal Publications, Beirut (2013)

Rašić, M., Milenković, M., Vojković, G.: Electronic auction in European and croatian public procurement law. In: International Convention on Information and Communication Technology, Electronics and Microelectronics, MIPRO (2019)

Saini, N., Bhanu, A.: Operational aspects of e-contracts: a critical study. Int. J. Manag. **11**(5), 1721–1734 (2020)

Legal Challenges of Cryptocurrency

Gehad Mohamed AbdelAziz[1,2](✉) and Abdelrahman Shalaby[3]

[1] College of Law, United Arab Emirates University, Al Ain, UAE
Gehad.shalaby@hotmail.com
[2] Faculty of Law, Zagazig University, Zagazig, Egypt
[3] University of Strathclyde, Glasgow, UK

Abstract. Cryptocurrencies are one type of digital currencies that have been widely used in recent times. The use of these currencies in commercial transactions raises many legal challenges. This study aims to address several challenges including determining the nature of the virtual currency, whether it is a subject of ownership and taxation, and the legal regulation of this currency. Moreover, using cryptocurrencies in money laundering and terrorism financing leads to difficulty in discovering these crimes. This article concludes that the legality and legal nature of cryptocurrencies are still controversial. Some legislation approves using cryptocurrencies as a means of payment without considering them as money. Some criminal legislation includes these currencies within the concept of money obtained from the primary crime. It also requires virtual asset service providers to combat money laundering and terrorist financing.

Keywords: Cryptocurrency · Blockchain · Money Laundering · Virtual currency · Digital currency

1 Introduction

Cryptocurrencies have emerged as one of the significant technology trends in the past few years. Skepticism is persistent about the legitimacy and legality of cryptocurrencies. Illegal trading, and money laundering, are becoming a part of the digital transactional infrastructure. Virtual currencies are issued by unofficial entities that are not recognized locally or internationally. This raises the question of whether or not to accept it in financial operations. This question refers to another fundamental question, which is its nature, the legal system to which it is subject, and the jurisdiction of disputes relating to it. Another problem is the difficulty of detecting money laundering and terrorism financing crimes that are committed with the use of cryptocurrencies as they are not considered money generated by criminal activity unless the law adopts a different approach. Also, the entities that provide services in cryptocurrencies are not within the statutory definition of the entities that undertake to combat money laundering in the money laundry law. So, in this article, we will discuss the nature of cryptocurrencies and the legal regulation to which they are subject. It also covers the use of cryptocurrencies in committing money laundering and terrorism financing crimes.

B. A. M. Alareeni and I. Elgedawy (Eds.): *Artificial Intelligence (AI) and Finance*, SSDC 488, pp. 890–898, 2023.
https://doi.org/10.1007/978-3-031-39158-3_83

In this research, we will follow the descriptive and comparative analytical methodology because there are no specific legal systems governing cryptocurrencies. The first section addresses the concept of cryptocurrencies and their legal nature. The second section analyzes the legal regulations to which they are subject, and their use in money laundering and financing terrorism.

2 What is a Cryptocurrency, and What is Its Legal Nature?

2.1 Defining Cryptocurrency, Blockchain, and Other Related Currencies

A cryptocurrency is a form of digital or virtual money protected by encryption and impossible to duplicate or counterfeit. It is often not issued by any central authority, making it impervious to government intervention or manipulation. Blockchain technology, a distributed ledger enforced by a dispersed network of computers, is the foundation of many cryptocurrencies decentralized networks (Frankenfield 2023).

Article L.223-12 of the Monetary and Financial Code in France was the first legislation to define blockchain as "a shared electronic recording device allowing the authentication of these operations, under conditions, in particular of security, defined by decree in the Council of State." Blockchain technology focuses on "keeping and transmitting information without relying on a body of central supervision." There are different types of blockchain: public blockchains, where the "network is open to all" private blockchains, where the network is closed, so the number of participants is constrained, and hybrid blockchains, which can be either public or private and have a limited number of participants.

Even though it is a "multipurpose technology," it has three main uses: a certifying register, a transfer method of assets, or smart contracts. The first blockchain was created to develop a cryptocurrency called Bitcoin. Bitcoin was a pioneer in the development of this technology. It was the first cryptocurrency to go live in 2009 (Evans 2022, 264).

According to the European Banking Authority, cryptocurrency is often referred to as "virtual currency," as it exists only in digital form and is not backed by any physical commodity or government entity. However, it is vital to note that not all virtual currencies are cryptocurrencies. Virtual currencies can also refer to any type of digital representation of value, such as in-game currency, called "closed virtual currency," as the currency can only be spent in a specific virtual community and cannot be converted into legal tender. This demonstrates that cryptocurrency is a type of virtual currency (Sharma 2021).

There is a contradiction about whether cryptocurrency is more of a virtual or digital currency. A virtual currency is a digital representation of value only available in electronic form. It is stored and transacted through designated software, mobile, or computer applications. Virtual currency transactions occur through secure, dedicated networks or over the Internet. They are issued by private parties or groups of developers and are unregulated. Virtual currencies are a subset of digital currencies and include other digital currencies, such as cryptocurrencies and tokens issued by private organizations. Thus, cryptocurrencies are more of a digital currency.

2.2 Differentiating Cryptocurrency from Traditional Paper Money Currency

There are several significant differences between cryptocurrency and traditional paper money, also known as fiat currency.

One of the main differences is the way they are created and controlled. Cryptocurrency is decentralized and made through mining, which involves solving complex mathematical equations to confirm transactions and create new currency units. On the other hand, Fiat currency is issued and controlled by central banks and governments.

Another difference is the way they are stored and transferred. Cryptocurrency is stored in digital wallets and can be shared anonymously using blockchain technology. On the other hand, Fiat currency is held physically and transferred through banks and other financial institutions, which may have specific regulations and oversight.

A third significant difference is the level of anonymity and security. Cryptocurrency transactions are usually anonymous and can be difficult to trace, which has led to concerns about money laundering and other illegal activities. Fiat currency, on the other hand, is more transparent and regulated, and transactions can be traced and monitored more easily (Frankenfield 2023).

2.3 The Legal System of Cryptocurrency

The first legal framework for crypto assets in France was adopted on May 22, 2019, with the introduction of the law regulating the growth and transformation of companies (PACTE). Above all, it legally frames active cryptos by opting for asset numbers without defining the latter. But the law specifies those who are included in this category: tokens and digital representations of a value which is not issued or guaranteed by a central bank and is accepted by persons, natural or legal, as a medium of exchange and can be transferred, stored, or electronically exchanged despite not being a legal tender currency. The *Banque de France* also gave its position on these technological innovations in 2018 when it assured that active cryptos are not currencies because: (1) they cannot be units of account, (2) they are less effective as intermediaries in the exchanges, and (3) they are not stores of value.

Therefore, the *Banque de France* provides that "they offer no guarantees of security, convertibility, or value, unlike a currency that is legal tender. (Evans 2022) "The new article L 54-10-1 of the monetary and financial code considers cryptocurrencies as digital assets. However, bitcoin lacks a clear legal standing. According to civil law, bitcoin can be regarded as recoverable intangible moveable property that can be used for speculation, or more specifically, as movable property as determined by law (Cahen 2021).

French law defines money as a means of payment issued by a public authority that is accepted by legal or contractual provisions. (French MONETARY AND FINANCIAL CODE Article L141-5) This is in line with the European Union's definition of money as "currency, whether in physical or electronic form, that is issued and guaranteed by a central bank or public authority and is accepted as a medium of exchange by a community" (Treaty on the Functioning of the European Union Article 128).

The distinction between money and other forms of property, such as virtual assets or cryptocurrencies, is essential because it determines the legal framework that applies to them. For instance, unlike money, cryptocurrencies are not issued or guaranteed by a central bank or other public authority, and their legal status is subject to regulation and interpretation by French courts.

In summary, cryptocurrency is not considered real money in France because it does not fulfill the component of "issuance and guarantee by a central bank or public authority" according to the French law's definition of money.

As for Canada, it had to address the emergence of cryptocurrencies before other nations. Not only did the first Bitcoin ATM debut in this nation, but people also started exchanging bitcoin for other currencies. Being classified as legal money, it is now treated and taxed as a commodity, according to a 2014 declaration by the federal government. It was prudent, but it was long overdue, given that the market had already accepted the reality of cryptocurrencies. Cryptocurrency exchanges were added to the definition of a money service business and given a new meaning in Canada's proceeds of crime regulations that went into effect in the middle of 2020. As a result, they are now subject to the country's anti-money laundering and anti-terrorism financing recording and reporting laws (German 2021).

In the United States, the law applies existing financial instruments laws to crypto assets. Even though the SEC and the CFTC are the primary regulators of crypto assets, the Internal Revenue Service (IRS) considers these assets to be property and are, thus, taxable (Evans 2022).

3 Legal Issues Related to Cryptocurrency

3.1 Absence of an Appropriate Jurisdiction

Blockchain technology enables direct peer-to-peer transactions between two parties in a verifiable and permanent manner by creating an ever-expanding electronic ledger that is distributed and decentralized over many nodes and cannot be altered or tampered with. Because cryptocurrency transactions are supported by blockchain technology and because nodes are distributed, there is no physical ledger that can be linked to a specific jurisdiction, making bitcoin transactions multijurisdictional by nature (Oni 2021).

The English High Court issued a decision on an ex parte interim application on accusations of fraud about a cryptocurrency initial coin offering on December 22, 2020. An initial coin offering (ICO), which is a form of fund-raising akin to an initial public offering (IPO), aims to raise money by creating a cryptocurrency or cryptographic token: Ion Science Ltd. v. Unknown Persons (unreported, December 21, 2020). This finding is the most recent in a string of preliminary judgments from English courts that indicate crypto assets can be considered property under the standard law definition of the term.

The ruling is noteworthy for considering the lex situs of crypto assets, a topic on which there is no settled case. The judgment relies on a scholarly discussion in the absence of preceding case law to reach a conclusion (to the standard of a serious question to be examined) that the lex situs of a crypto asset is the domicile of the individual or entity who acquired the coin or token. The ruling is necessary for its conclusion that a free-standing bankers' trust order might be issued against cryptocurrency exchanges outside the jurisdiction to compel the disclosure of information about the crypto assets.

Who Regulates Cryptocurrency?

Cryptocurrency regulation differs from one nation to another. In the United States, the

Securities and Exchange Commission (SEC) and the Commodity Futures Trading Commission (CFTC) are the leading regulating organizations for cryptocurrencies. (Dewey and Patel, 2023) The CFTC has the power to control cryptocurrency derivatives such as bitcoin. Still, it believes it needs more authority to oversee markets, platforms, and users of these platforms that exchange money or conduct spot exchange operations. By contrast, the SEC has the power to control cryptocurrency securities. The SEC has not been very interested in regulating or commenting on the legal nature of virtual currencies. On the other hand, it has focused its approach on initial coin offerings (ICO). In 2017, it submitted its investigation report on the decentralized autonomous organization (DAO) (Evans 2022).

However, in the EU, different member states have other cryptocurrency laws. Moreover, it is the responsibility of the European Securities and Markets Authority (ESMA) to provide rules and regulations to EU member states on how to regulate cryptocurrencies (Browne 2022).

The People's Bank of China (PBOC) is the principal cryptocurrency regulator in China. The PBOC prohibits initial coin offerings (ICOs) and domestic cryptocurrency exchanges due to its harsh stance on cryptocurrencies. The PBOC highlighted the dangers of cryptocurrencies in 2021 and urged people and organizations to use caution when engaging in cryptocurrency-related activities (Laney 2021).

At the provincial level, bitcoin exchange legislation in Canada varies, but federal regulators classify cryptocurrencies as securities. An opinion on the applicability of current securities rules to cryptocurrencies was published by the Canadian Securities Administrators (CSA) in August 2017. More rules governing cryptocurrency exchanges are forthcoming. In June 2018, Canadian authorities proposed changes to the Proceeds of Crime (Money Laundering and Terrorist Financing Act) in response to its joint assessment by the FATF. As a result of the addition of cryptocurrency exchanges to the new legislation, these businesses are now subject to reporting requirements and are governed similarly to enterprises that provide financial services (ComplyAdvantage 2023).

The PACTE law on crypto assets aims to create a legal regime for ICOs and briefly regulate digital assets and their activities. The PACTE law also grants service providers a status on digital assets. Article L.54-10.2 of the monetary and financial code will introduce a list of digital asset services that may be offered to investors. Depending on the services provided, investors have two options: mandatory registration or an optional license. Registration is compulsory for the first two services in the list: the custody service on behalf of third parties of digital assets or access to digital assets and the legal assistance of buying or selling digital assets in current currency. Any activity should be prohibited before it is inspected. The service providers are obliged to collect conditions of good reputation for the prevention of money laundering or financing terrorism (Evans 2022, 268).

France differs from Europe and the US in that it has implemented an optional plan to encourage without constraining, contrary to the regime established for financial instruments. The regime of the PACTE law for active crypto is neither absolute nor complete, as the similarities with other rules, such as that of financial instruments, are flagrant. Token regulation allows for the sometimes-complex differentiation of existing schemes.

The technology that develops these digital assets only expands and revolutionizes markets by offering investors a wide variety of options. This is when the legislator increases the competence of the AMF so that it becomes the regulator of this new legal category of digital currency. On the contrary, in the United States, the legislator did not have the will to create a new regime, and he left the regulators to supervise these innovations, qualify them, and above all, divide the task between the SEC, the CFTC, and the IRS (Evans 2022, 266).

More nations will develop regulations to control the use of cryptocurrencies as their use and adoption increase. The Financial Stability Oversight Council, an international organization of central bankers and regulators from the G20 countries, has demanded global rules for cryptocurrency regulation. In addition to consumer protection laws to prevent fraud and market manipulation, this may also involve steps to stop the financing of terrorism and money laundering. The Financial Services and Markets Bill, which is making its way through the UK's parliament, aims to give the Financial Conduct Authority (FCA), the nation's financial services regulator, more authority to regulate cryptocurrency businesses. The FCA now controls the anti-money laundering and countering terrorism financing regulations for cryptocurrency enterprises. A significant piece of legislation, the EU's proposed Markets in Crypto-Assets Regulation, is currently awaiting approval in the European Parliament. Once it goes into force, which is most likely in 2024, it will instantly apply to all EU member states (Hodge 2023).

3.2 Using Cryptocurrencies in Money Laundering and Terrorism Financing

Money laundering is the process of making illegally gained proceeds (i.e., "dirty money") appear legal (i.e., "clean money"). Due to its decentralized nature and ability to be anonymously transferred, cryptocurrency has been used in money laundering schemes. One example of a legal issue related to money laundering using cryptocurrency is the case United States v. Ulbricht, in which the operator of the Silk Road marketplace, Ross Ulbricht, was convicted of money laundering, among other charges, for using Bitcoin to ease illegal drug sales on the site. (United States v. Ulbricht, 15-cr-68).

Law enforcement and regulatory organizations are becoming concerned about cryptocurrency-based money laundering. Because they are anonymous, decentralized, and challenging to track, cryptocurrency transactions appeal to criminals attempting to launder money. The lack of oversight and regulation is one of the leading legal difficulties related to cryptocurrency money laundering.

The anonymity of cryptocurrency transactions is another legal concern. Due to the decentralized nature of most cryptocurrencies and the use of encryption to shield user identities, it can be challenging for law enforcement to track transactions and pinpoint those responsible for money laundering. As a result, there have been calls for the cryptocurrency market to be subject to more regulation and control, including the adoption of know-your-customer (KYC) and anti-money laundering (AML) rules. (Green 2019, 468–487).

The cryptocurrency is directly distributed between users without an intermediary, combining cryptography and peer-to-peer architecture to avoid surveillance by financial authorities. Thus, transactions using virtual currency do not leave a trace because it is not necessary to go through intermediary institutions such as banks (Akbar 2019).

Although all transactions are recorded in a public ledger, buyers and sellers' identities are never revealed, and only the identification of their wallets is shown. Once a payment is made, the payer and the beneficiary sign the transaction using their private key, which allows the exact identity of the parties involved to be verified without identifying users (Serhrouchni 2022). These characteristics of virtual currencies make them an exciting tool for criminals and, more specifically, for money laundering.

In addition, the instant transfer of cryptocurrencies across borders complicates tracing the chain of transactions. The threat posed by cryptocurrencies is accentuated by the speed and mobility of transactions offered by the technology on which they are based. Unlike traditional money, cryptocurrencies can move in a few moments without identifying the originator. The transferred amounts can thus be made available almost immediately to users anonymously.

The Laws Combating Money Laundering and Terrorism Financing by Cryptocurrency

The combating of money laundering and terrorism financing using cryptocurrency has resulted in the inclusion of cryptocurrency in the concept of funds resulting from crimes subject to money laundering and financing terrorism, as well as subjecting virtual asset service providers to obligations related to combating money laundering and terrorist financing, like other financial institutions, which includes reporting suspicious transactions and conduct a risk assessment.

Punishment for Money Laundering and Terrorism Financing Using Cryptocurrencies:

Some countries have expanded their money laundering laws only without banning virtual currencies, as cryptocurrencies have been included in the concept of money obtained from crime and are laundered and used in the financing of terrorism. In UAE, the Federal Decree Law No. (26) of 2021 amending some provisions of Federal Decree-law No. (20) of 2018 on anti-money laundering and combating the financing of terrorism and financing criminal organizations included assets in the concept of funds generated from committing misdemeanor or felony whatever the mode of its acquisition, type, and form, tangible or intangible, movable or immovable, electronic, or encrypted, including local and foreign currencies. Hence, cryptocurrency became an object of money laundering crime and terrorism financing in UAE legislation. Federal law defines virtual assets as a digital representation of the value that can be traded or transferred. They can be used for payment or investment purposes, as specified in the administrative regulation of this Decree-Law.

In France, ordinance n° 2020-115 aims to strengthen combating money laundering and the financing of terrorism in the event of using cryptocurrencies. This ordinance is combating the anonymity of digital asset transactions by including service providers on digital assets (PSAN) among entities prohibited from maintaining anonymous accounts. (art. 561-14 of the CMF).

The German Federal Financial Supervisory Authority classifies virtual currencies as units of account and, consequently, as financial instruments. Companies and individuals who organize the acquisition of tokens, sell or purchase tickets on a commercial basis, or perform core token brokerage services via online trading platforms are generally required to obtain prior BaFin authorization (Serhrouchni 2022).

The Parliament of the Netherlands passed a new law to regulate cryptocurrency exchanges and businesses. As a result of this legislation, both existing and new companies will have to register with the central bank by submitting a notice. To combat money laundering and terrorist financing, this new law transposes the European Union's 5th Anti-Money Laundering Directive (5AMLD), which went into effect on January 10, 2020. As a result, the Dutch Ministry of Finance has appointed the Dutch Central Bank (DNB) to oversee cryptocurrency companies (Melvin 2018).

Some countries don't allow cryptocurrency transactions. For example, in November 2017, Morocco banned all transactions done with virtual currencies, calling them a violation of foreign exchange rules that could lead to penalties and fines. (Serhrouchni 2022) Qatar and Bahrain have slightly different approaches. They prohibit their citizens from engaging in any activity involving cryptocurrencies locally but allow them to do so outside their borders in China. Virtual currencies are not implicitly banned but not accepted by the banking system. They are not considered as a retail payment tool in the same way as paper notes, coins currency, or credit cards.

Technology can facilitate the collection, processing, and analysis of data and help actors to identify and manage the risks of money laundering and the financing of terrorism (ML/FT) more efficiently and closer to real time. Payments and faster transactions, more accurate identification systems, monitoring, record keeping and information sharing between competent authorities and regulated entities also offer bene.

4 Conclusion

This paper addresses the legal challenges of cryptocurrencies in different legal systems. In this study we referred to several legislations and some judgments in various countries. We found that cryptocurrency is not considered real money in some legal systems such as France because it does not fulfill the component of "issuance and guarantee by a central bank or public authority" according to the law's definition of money. In other countries such as Canada, Cryptocurrency is now treated and taxed as a commodity according to the 2014 declaration by the federal government. Furthermore, cryptocurrency exchanges were added to the definition of a money service business and given a new meaning in proceeds of crime regulations. As a result, they are now subject to the country's anti-money laundering and anti-terrorist financing reporting laws.

The United States applies existing financial instruments laws to crypto assets. Although the SEC and the CFTC are the main regulators of crypto assets, the Internal Revenue Service (IRS) considers these assets to be a property and are, thus, taxable. In UAE, the legislation of combating money laundering includes cryptocurrency in the concept of funds generated from committing crime. This means that cryptocurrency became an object of money laundering crime and terrorism financing in UAE legislation.

References

Frankenfield, J.: Cryptocurrency explained with pros and cons for investment (2023). Investopedia https://www.investopedia.com/terms/c/cryptocurrency.aspx. Accessed 20 Feb 2023

Evans, M.: L'évolution de la Régulation des marchés financiers: analyse comparée France, États-Unis et Costa Rica, these de doctrat. Université Toulouse Capitole Écoledoctorale, Droit et Science Politique (2022)

Sharma, M.: Emergence of virtual currencies: study of cryptocurrency in digital marketing. Int. J. Islamic Soc. Sci. **1**, 1–5 (2021)

Cahen, M.: Traitement civil et pénal du bitcoin. 2021 Traitement civil et pénal du bitcoin (murielle-cahen.com). Accessed 19 Feb 2023

German, P.: Boarding the crypto express. Comp. Law **42**(12) (2021)

Oni, N.: Jurisdictional Issues on Cryptocurrency Transactions (2021)

Dewey, J., Patel, S.: Blockchain & Cryptocurrency Laws and Regulations (2023). Blockchain & Cryptocurrency Laws and Regulations | USA | GLI (globallegalinsights.com). Accessed 15 Feb 2023

Browne, R.: EU agrees on landmark regulation to clean up crypto 'Wild West'. 2022 EU agrees to deal on landmark MiCA cryptocurrency regulation (cnbc.com). Accessed 15 Feb 2023

Laney, Z.: China: Central Bank Issues New Regulatory Document on Cryptocurrency Trading (2021). https://www.loc.gov/item/global-legal-monitor/2021-10-13/china-central-bank-issues-new-regulatory-document-on-cryptocurrency-trading/. Accessed 15 Feb 2023

ComplyAdvantage: Réglementations en matière de crypto-monnaie dans le monde entire. Réglementations en matière de crypto-monnaie dans le monde entire (2023). https://complyadvantage.com/fr/insights/reglementations-en-matiere-de-cryptomonnaie-dans-le-monde-entier/. Accessed 15 Feb 2023

Hodge, N.: The future of crypto. 2023 The future of crypto | International Bar Association (ibanet.org). Accessed 15 Feb 2023

Green, S.: Cryptocurrency and money laundering: the need for national and international cooperation. J. Financ. Crime **26**(2), 468–487 (2019)

Akbar, D.: Criminal law policy in handling digital asset-based money laundering in Indonesia. Journal of Law and Legal Reform, [en ligne] **1**(1), 150–156 (2019). https://doi.org/10.15294/jllr.v1i1.35543. Accessed 15 Feb 2023

Serhrouchni, I.: La cryptomonnaie et le blanchiment d'argent 2022. Village de la justice. https://www.village-justice.com/articles/cryptomonnaie-blanchiment-argent,43693.html. Accessed 14 Feb 2023

Melvin: Les nouveaux réglements de cryptos monnaies aux Pays-Bas [en ligne], Intercompany solutions, 28 Novembre 2018, disponible à l'adresse: https://fr.intercompanysolutions.com/netherlands-crypto-regulations. Accessed 16 Feb 2023

Analysis of Credit Cards Fraud Detection: Process and Techniques Perspective

Muath Asmar[1(✉)] and Belal Yousef Aqel[2]

[1] Department of Finance, Faculty of Business & Communication, An-Najah National University, Nablus, Palestine
asmar@najah.edu

[2] Master of Business Administration, Faculty of Graduate Studies, An-Najah National University, Nablus, Palestine

Abstract. Credit card fraud detection is the process of identifying and preventing fraudulent transactions before they can cause financial damage. This involves using advanced algorithms and machine learning techniques to analyze transaction data in real-time and detect patterns that may indicate fraudulent activity. Effective fraud detection systems are essential for ensuring the security and integrity of the credit card payment system, protecting both financial institutions and cardholders from financial losses. The main objectives of this study are to explain the process and techniques. This study analyzes a global dataset of credit card transactions using regression analysis and artificial neural networks. A regression research found that new balance origin, new balance destination, old balance origin, transaction type, and amount all statistically effect fraud transactions, while old balance destination and transaction length do not. This study used artificial neural networks to identify credit card fraud detection properties. According to the statistics, Old Balance locations are most importantly considered for credit card fraud. This study shows that the models can detect fraudulent activity using real credit card transaction data, providing significant insights into credit card fraud detection approaches.

Keywords: Credit Cards · Fraud Detection · Artificial Neural Networks · Regression analysis

1 Introduction

Credit card fraud affects banks, companies, and consumers. Fraudulent conduct may bring significant financial and reputational damage. Several credit card fraud detection systems have employed AI and machine learning. These technologies can quickly and accurately identify and halt fraudulent transactions. Credit card fraud detection requires analysis of transaction histories, user behavior patterns, and external data sources. Machine learning algorithms can spot suspicious patterns and behaviors including large transactions, unusual purchasing habits, and purchases in unusual areas.

B. A. M. Alareeni and I. Elgedawy (Eds.): *Artificial Intelligence (AI) and Finance*, SSDC 488, pp. 899–911, 2023.
https://doi.org/10.1007/978-3-031-39158-3_84

Fraudsters are adaptable and will develop methods to bypass preventive systems over time. If fraud prevention fails, we need fraud detection methods. Modern technology, communication, and the internet have increased fraud, costing billions of dollars annually. This paper discusses statistical fraud detection methods and their most popular applications. Statistics and machine learning can identify money laundering, e-commerce, and credit card fraud [1]. Design, holo-grams, magnetic stripes, card verification values (CVV), signature boxes, Pin Code for bankcards, Internet security systems for credit card transactions, authentication (OTP), and one-time passwords prevent fraud. Strong consumer authentication ensures online payments with 3D Secure.

Fraud detection catches fraud after it happens. As fraud prevention efforts have failed, we must detect fraudulent transactions immediately. Telemetry can warn credit card fraud monitoring teams using machine learning. Massive transactions, the same item with several skews and slops, enormous cart items, high-amount transactions with many payments, same address for various cards for the same IP address, and different cards for the same IP address [2]. Comparing observed data to expected values, single arithmetical summaries of certain activity items, basic graphical digests, and past system behavior like credit card use are statistical fraud detection strategies. Transaction detection data varies in kind, amount, time, and location [3]. When online transactions replace in-person purchases at merchant sites, banks lose money when fraud occurs. Business intelligence is needed to detect fraudulent transactions.

Although much has been published about securities market trading operations and post-trade settlement infrastructure [e.g., 4, 5], less is known about e-payment processes, notably credit card fraud detection. This research investigates fundamental credit card fraud and the use of technology in fraud detection to address information gaps, as well as to describe the fraud detective process and identify the statistical tools, business analytics, business intelligence, and machine learning roles in fraud detective for credit cards from a given data set.

This research follows: Sect. 2 covers study background. Section 3 reviews credit card fraud detection literature. Section 4 discusses the empirical model, results, data, descriptive statistics, research method, regression, and artificial neural network analysis. Section 5 describes results. Section 6 concludes and discusses business implications.

2 Study Background

Banks are essential for supporting economic growth and development as a financial intermediary [6]. Credit cards are facilities whereas a bank or finance institutions offers a cardholder (the cardholder) a tiny, rectangular piece of plastic or metal to make online transactions and withdraw cash over their bank account limit. Customer credit history determines this. Credit cards usually feature the bank name, special card number, cardholder name, expiry date, smart chip, and payment network logo on the front. Back: CVV, magnetic strip, specimen signature, hologram, bank contact information, and network logo. Visa, Master, American Express, and Discover's foreign market is dominated by the financial credit card network, which connects banks and points of sale. The three elements that must exist for fraud to occur are pressure, opportunity, and rationalization, according to the Fraud Triangle Theory (FTT) of Cressey [7], and the Fraud Diamond

Theory (FDT), which combines the FTT three elements with the fraudster capability and is regarded as an improved or extended version of the FTT by Wolfe and Hermanson [8] as deliberated in [9]. Regardless of the industry or type, according to the Association of Certified Fraud Examiners, fraud is defined as "Any act, expression, omission, or concealment calculated to deceive another to that person's disadvantage, specifically, a misrepresentation or concealment with reference to some fact material to a transaction that is made with knowledge of its falsity." [10].

Credit cards allow you to borrow money and make online payments. Modern credit card fraud is frequent. Lost or stolen cards, card abuse, identity theft, and merchant exploitation are examples. Credit card theft is the unauthorized use of personal information to update an account, request a new card, or replace an existing card. Credit card fraud is the unauthorized use of another person's credit card to make transactions or obtain cash advances. Credit card and personal information may be stolen from physical cards. EMV (Europay, Master Card, and Visa) introduced chip and microchip technology, making it harder for hackers to steal information from magnetic strips using standard ATM skimming equipment.

Before the information era, thieves could use a real card number to make counterfeit cards. These cards were encoded or embossed on blank plastic cards or on membership or gaming card magnetic strips, and the sales slips did not encrypt the linked information. Even though it was employed in automated teller machines, merchant point of sale (POS) skimming was introduced by swiping the card into the card reader (ATMs). Financial organizations prevent this kind of fraud by adding security measures to cards and machines and deploying devices on equipment. The European Central Bank's eighth card fraud report for Europe from 2015 to 2019 found €1.87 billion in fraudulent card transactions worldwide using cards from Single Euro Payments Area (SEPA) member nations. Delays in debit and credit card payments are more prone to fraud than debit card payments, yet CNP fraud continues to dominate fraudulent transactions. Cross-border SEPA transactions made up over half of 2019's fraud dollar amount [11].

2.1 Credit Cards Frauds Types

Frauds are often performed by a third party or hackers, preparators, or insiders (workers) via the leaking of unencrypted information and are classified as transactions or behaviors. There are various forms of credit card fraud, which we shall explain as follows: Hackers take over a bank account and any related accounts of the original consumer in account takeover fraud. Application fraud: when a criminal utilizes a cardholder's personal information to impersonate them to the bank that issued the card. When the application is granted, the fraudster alters his contact information and address. If a customer's card hasn't been disabled, the thief may spend the money. Card Not Present (CNP): techniques employed in the new world of online payments, mobile apps that enable card-less payments like Google Pay and Apple Pay, etc. Friendly fraud: The cardholder requests a chargeback and dispute since he did not make the purchase. Skimming Cards: at an ATM by attaching a device to read and gather information after you turn it, or at a point of sale (POS) by a staff person using a small device reader. Getting the real card number and encoding it on a blank card, magnetic strip, game, or membership card creates counterfeit cards. Phishing involves attacking bogus websites to steal personal information, login passwords, and credit card numbers. Loyalty card fraud, especially with branded products or credit card points, is also called reward fraud [12].

2.2 Credit Card Fraud Prevention

Fraud prevention is cheaper and more effective than detection. Fraud investigations, especially those involving vast transactional data, cost money and effort. Application Process: Card issuers often verify application data from numerous sources. The user desired dual authentication by contact center, SMS, or mobile apps during the new card activation procedure. Replacement Cards: The bank issuer replaces the card at the client's or bank's request, activating it. When cards expire or need to be replaced, this occurs too (damage, loss, suspicious transaction, etc.)

Bank card issuers monitor high-risk transactions using advanced technologies to prevent fraud. Fraud control may verify a card that seems to have been altered, ripped, or encountered by retailers by physically recognizing it.

2.3 Credit Card Fraud Detection Process

The transactions are verified at the terminal point to determine whether they are valid or not for certain prerequisites like the card number, bank issuance, sufficient balance, and valid Personal Identification Number (PIN), after which they are filtered to determine whether they are accepted or rejected. Meanwhile, a predictive model is receiving this information and raising red flags in the event of suspicious transactions (Fig. 1).

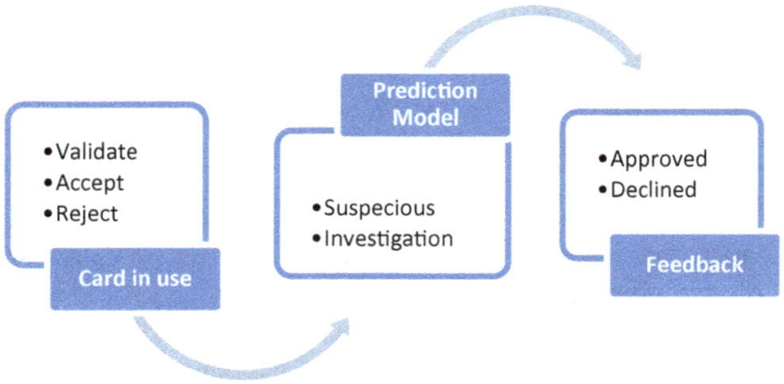

Fig. 1. The fraud detection process.

Credit card fraud has skyrocketed due to growing credit card transactions. Credit card fraud occurs when a credit card is used to defraud a transaction. Statistics and data mining are used to identify fraud. Most credit card fraud detection solutions use AI, meta learning, and pattern matching. Effective fraud detection systems quickly discriminate between fraudulent and genuine transactions [13].

Reviewing the data involved in credit card fraud detection is essential to understanding its challenges. Fraudulent data is substantially less than non-fraudulent data. Data quality and quantity are also issues. Early-stage startups lack user history. Credit agency ratings may be a temporary solution. Credit card businesses and financial institutions utilize several regulations, instruments, processes, and practices to prevent identity theft and fraudulent transactions. This is credit card fraud detection. Over the past decade, fraud detection has become digitized and automated using statistical tools, AI, and machine learning to manage alerts and analyze big data due to the large number of transactions and criminals' advanced technology to circumvent traditional methods like CVV verification, expiry date, and location tracking. Hence, machine learning algorithms identify fraudulent transactions [14]. Banks and companies fight fraud on two fronts: detection and prevention [1]. "Fraud prevention" refers to any measures taken to prevent fraud. Fraud protection includes pre-activating credit cards to prevent surface mail theft. Internet credit card security prevents fraud. Debit card pin codes prevent ATM withdrawals by requiring the thief to know and hold the card [1]. According to the E-commerce Fraud Index, the rate of retail fraud increased from 0.06% in 2016 to 0.23% in 2017. Credit card fraud is thought to account for 10% of all frauds, which has led to enormous financial losses that alarm businesses [15].

2.4 Models and Techniques Used in Credit Card Fraud Detection

There are different machine learning algorithms used to detect credit card frauds classified as supervised learning and unsupervised learning as shown in Fig. 2, however we are used in this research regression analysis and artificial neural network.

Rational Regression: A method that may be used to both classification and regression problems, albeit it is most often employed for classification. Using dependent variables, logistic regression is used to forecast categorical variables Consider two classes where the class to which a new piece of data belongs must be determined. After that, the algorithms calculate probability values between (0) and (1).

Artificial Neural Networks: A complicated interplay between outputs and inputs is used by this kind of machine learning algorithm to find current trends.

Statistical methods for detecting credit card fraud Use a base line distribution to reflect normative behavior, use a basis distribution; next, seek for data that deviates most from it. Using text analysis and digit analysis [16].

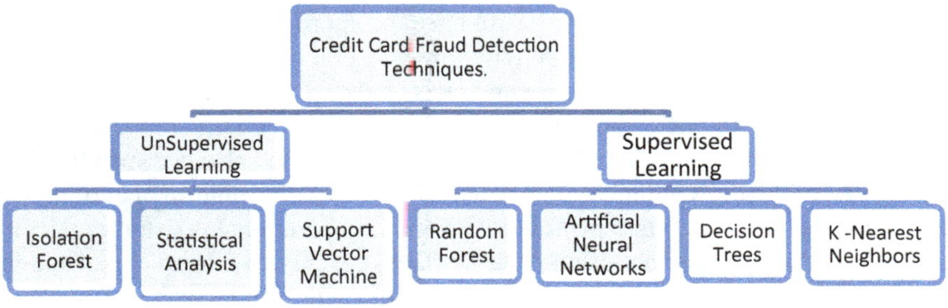

Fig. 2. Fraud detection techniques.

3 Literature Review

Use of machine learning and artificial intelligence approaches for credit card fraud detection has been the subject of several research. Gupta et al. [17] developed a technique to examine how biased machine learning models produced by unequal data affect firms' capacity to adapt and endure technological change. When machine learning techniques and oversampling were employed to detect credit card fraud, accuracy rose.

Credit card fraud detection is becoming important. Online transactions and complicated financial networks need reliable fraud prevention technologies. Most credit card companies use machine learning algorithms and data analytics to spot suspicious conduct. These powerful tools help them detect fraudulent transactions and protect customers from potential losses [18]. Financial security requires credit card theft detection. To identify and fight fraud, a variety of methods and tools are used. Machine learning, pattern recognition, and human review are used to ensure correctness. Banks, merchants, and other financial institutions spend heavily in credit card fraud detection solutions to protect their customers [19].

There is still a lot of work to be done in fraud detection and prevention, despite the advancements achieved. Since credit card fraud is an issue that is always changing, it is crucial for businesses and banks to stay up to date on the most recent fraud detection techniques. Organizations must make sure they are investing in the appropriate technology and stay up to date on any new advancements in the area of fraud prevention in order to achieve this. Additionally, it's critical to remind clients of the best ways to safeguard their credit cards and personal data [3]. Despite progress, fraud detection and prevention need more effort. Businesses and banks must keep current on fraud detection methods since credit card theft is continually evolving. To do this, organizations must invest in the right technology and remain abreast of fraud prevention advances. It's also important to advise customers on credit card and personal data security [3]. Companies are also considering blockchain and facial recognition to reduce credit card fraud. These technologies may improve credit card processing security for businesses. Machine learning and AI algorithms can detect fraud in real time for businesses. With the latest technologies, businesses can protect customer data [20].

Detecting credit card theft is crucial in the age of online purchasing. The number of fraudulent transactions increases along with the online payment processing sector. To identify and prevent fraud, it's critical to have the appropriate tools and strategies in place. Any efficient fraud detection system now relies on cutting-edge technology like machine learning algorithms, predictive analytics, and artificial intelligence. Credit card fraud costs the US $15 billion annually. To avoid fraud, credit card companies analyze spending patterns and watch for odd activities. Customers may also use spending limits and account monitoring. Online credit card users should also be mindful of current scams and identity theft [21]. Credit card fraud detection is operational. Customer data may help banks and businesses spot fraud patterns. This prevents credit card fraud and theft losses. It prevents identity theft, which may cause financial and legal issues.

Fraud detection has become important in the credit card business because it protects customers from fraud and illegal activity. Advanced algorithms can monitor and identify aberrant actions or patterns, enabling real-time fraud detection. Credit card companies utilize machine learning algorithms to detect transaction abnormalities. These systems can detect hidden data trends and possible fraud [22].

3.1 Research Hypothesis

The main Hypothesis in this study is: There is significant relation between transaction type, transaction amount, card old balance origin, card new balance origin, old balance destination, new balance destination and transactions time, and fraud transactions.

4 Research Methodology

4.1 Data

The credit card transaction data is considered confidential information in the banks, network cards and not easy even impossible to get such datasets. However, a unique dataset about credit card transaction, contains 9999 credit card transactions collected around the world, is available in Kaggle.com [23], these data set includes transaction time, amount in local currency, initial balance before the transaction, customer balance after the transaction, initial recipient before the transaction, recipient balance after the transaction, type of transaction, transaction frauds or not. Table 1 defines the variables used in this study.

Table 1. Variables, its Acronyms, and Description.

Variable	Acronyms	Description
Transaction type	Type	CASH-IN, CASH-OUT, DEBIT, PAYMENT and TRANSFER
Transaction Amount	AMT	Amount of the transaction in local currency
Old balance Origin	OBO	Credit Card initial balance before the transaction
New balance Origin	NBO	Credit Card balance after the transaction
Old balance Destination	OBD	Credit Card initial balance recipient before the transaction. Note that there is not information for customers that start with Merchants
New balance Destination	NBD	Credit Card new balance recipient after the transaction. Note that there is not information for customers that start with Merchants
Transaction Time	Time	Maps a unit of time in the real world. In this case 1 step is 1 h of time
Fraud Transaction	FT	The transactions made by the fraudulent agents = 1 Genuine Transaction made by real customer = 0

4.2 Empirical Model

Regression analysis and artificial neural networks are used in this research to assess the influence of several drivers on credit card fraud.

As indicated in the research model below, where the dependent variable is fraud transactions and the independent variables are including transaction type, amount, origin of the old balance, origin of the new balance, origin of the old balance destination, origin of the new balance, and transaction time.

The empirical specification is expressed as follows:

$$FT = \alpha + \beta 1 Type + \beta 2 AMT + \beta 3 OBO + \beta 4 NBO + \beta 5 OBD + \beta 6 NBD + \beta 7 Time + e \quad (1)$$

where Type, AMT, OBO, NBO, OBD, NBD, Time, and FT are defined in Table 1. e is the error term.

The regression analysis is used to identify independent variables that impact and influence the dependent variable, while neural networks adapt to clarify changes in normal transaction conduct and identify importance of the independent variables.

5 Results and Discussion

The results show that most of the Credit Cards transactions type was payments while the lowest transactions type was debit. Figure 3 illustrate the Credit Cards Transactions by Type.

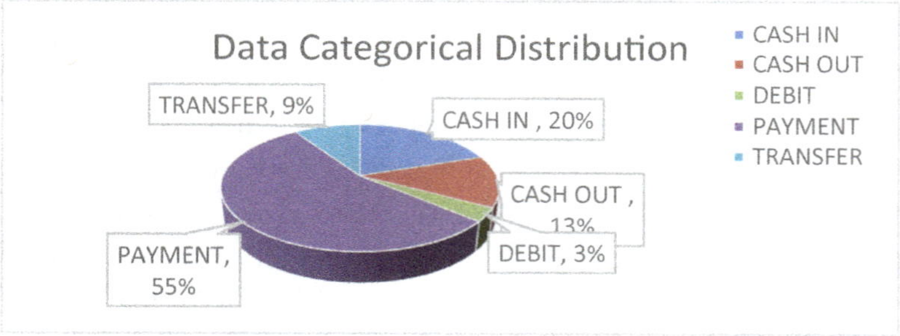

Fig. 3. The Credit Cards Transactions by Type.

5.1 Regression Analysis, Results and Discussion

The regression analysis was used to test the hypothesis of the study. According to the findings, the variables of old balance origin, new balance origin, and new balance destination all have statistically significant effects on the fraud transaction, but old balance destination and transaction time do not. The variables transaction type and amount also

Table 2. Estimations of the regression results.

Model	Unstandardized Coefficients		Standardized Coefficients	t	Sig.
	B	Std. Error	Beta		
(Constant)	−.003	.002		−1.321	.186
Type	.006	.001	.102	8.292	.000***
AMT	8.245E−9	.000	.027	2.170	.030**
OBO	1.197E−7	.000	3.109	22.391	.000***
NBO	−1.201E−7	.000	−3.188	−22.732	.000***
OBD	1.014E−9	.000	.033	1.097	.273
NBD	−1.528E−9	.000	−.056	−1.833	.067*
Time	.000	.000	−.005	−.532	.595

Dependent Variable: Fraud Transaction.
* Indicate statistical significance at the 10% level.
** Indicate statistical significance at the 5% level.
*** Indicate statistical significance at the 1% level.

have statistically significant effects on the fraud transaction. Table 2 show the Regression results.

Table 2 shows that transaction type significantly affects fraud transactions. Transaction type affects fraud transactions statistically. Card-not-present transactions are more likely to be fraudulent than card-present ones. Because there's no face-to-face connection, online transactions likewise carry a higher fraud risk, Transaction type is critical for analyzing and forecasting fraud. To establish whether transaction type affects fraud, investigate the prevalence of particular types of fraud within those transactions. Online transactions may increase credit card fraud. Transaction type affects fraud rates. Because certain transactions are more susceptible to fraud. So, organizations must understand the different transactional types that might be used for fraud.

Table 2 shows that transaction value statistically affects fraud. This is because fraudsters pursue higher-value transactions and transaction size is a statistically significant factor in fraud transactions. Banks and other financial institutions also like to closely monitor transactions with higher monetary amounts. Transaction amounts might indicate fraud risk.

Table 2 shows that the origin of the transaction's old balance affects fraud transactions statistically. Yet, detecting real fraud is difficult. Combining data improves fraud detection, the analysis demonstrates that the new balance destination and old balance origin are highly correlated. Origins with old balances are more linked than destinations with new balances. The data suggest that the source of the prior balance is a better predictor of fraud transactions. Fraud estimate relies on the old transaction balance. Look at the past balance of transactions investigated in that region to identify fraud-prone locations.

Table 2 shows that the origin of the transaction's new balance statistically affects fraud transactions. The origin of the new balance determines fraud risk. The source of the new balance makes it easier to notice suspicious outliers or weird behavior that may be fraud. The new balance's source might reveal a customer's activity and possible transactions. High-new balance transactions are more likely to be fraudulent, Predictive models may identify fraud tendencies using such data. Machine learning can enhance model accuracy.

Table 2 shows that the statistically small Old Balance Destination has no effect on fraud transactions. Since few frauds originate from the same source. Even if balance and destination data may not indicate fraud risk, we cannot draw any judgments from them. To investigate fraudulent transactions from particular regions, more data must be collected and analyzed. Complex algorithms can also detect and prevent future fraud. As fraud is rare, there is no time series or trend. The available information do not support the idea that destination affects fraud detection. Lack of proof disproves a connection, and see problems. This data collection is usually unbalanced. Fraud detection relies more on transaction volume. With transaction data, fraud detection algorithms may be enhanced. This research suggests using transaction volume while detecting fraud.

Table 2 shows that fraud transactions are unaffected by transaction time, which is statistically insignificant. Fraud has no temporal effect since money is always the same. To reduce fraud risk, consider all customer lifetime stages. Examine the consumer's credit history, product knowledge, and geography for odd behavior. To reduce fraud, customers should be verified using many methods. Scammers commit crimes 24/7. Time sensitivity manipulation might enable fraud. If the fraudster is an insider, they may be able to circumvent timely measures. Even at unusual hours, the fraudster may be able to complete a fraudulent transaction. Fraudsters don't care about time and only want money, to detect suspicious activity, several data analysis layers and diverse methods and technologies must be employed, we can create fraud-detecting models using data trends, the correlation research showed a strong negative correlation between the variables, illustrating this inverse relationship. The investigation found that most data points had a substantial negative link, supporting the correlation analysis. Moreover, changing the new balance origin lowers the new balance destination and vice versa. When the new balance's starting point is higher, the destination point will be lower. This shows that fraudsters make the illicit purchase and then move the funds to avoid disputes or disallowed transactions.

5.2 Artificial Neural Networks

Artificial Neural Networks were used in this study to identify the independent variables importance. Table 3 show the importance of the variable's effects in credit card fraud transaction detection.

Table 3 findings demonstrate that the learning procedure grouped the transaction factors into layers, with the old balance destination typically receiving the highest weight as the transactions got underway. As a result, a layer was given to the new balance origin, followed by layers for the new balance origin, old balance origin, amount, amount, type, and time of transactions. Different layers may modify their inputs in different ways. After visiting the layers many times, indications could pass through them, but this will alter according on machine learning inputs and data supplied to the system.

Table 3. Artificial Neural Networks results.

Variable	Importance	Normalized Importance
Type	.115	50.5%
AMT	.133	58.3%
OBO	.150	65.8%
NBO	.194	84.9%
OBD	.228	100.0%
NBD	.157	69.0%
Time	.022	9.7%

More and more, fraud in credit card transactions is being detected using artificial neural networks. It's essential to evaluate the significance of independent variables for this purpose, which include the kind, quantity, and time of the transactions as well as the old and new balances of the destination account. The network is better equipped to identify fraudulent behavior by considering the numerous parameters.

With tremendous success, artificial neural networks have been employed to identify credit card fraud. It is feasible to spot patterns of fraud by examining independent variables such the old balance, the new balance, the destination amount, and the types of transactions. This may be used to stop fraudulent transactions before they happen, minimizing losses for both clients and financial institutions.

6 Conclusion, Business Implications and Future Work

Due of the financial losses and harm to credit ratings it causes, credit card fraud is a serious issue for both financial institutions and private citizens. When a criminal obtains access to another person's credit card information and uses it for illicit purchases, it results in fraudulent transactions. Numerous tactics may be used to commit credit card fraud, including using stolen or fake cards and making unlawful internet transactions. The danger of credit card fraud has expanded along with the popularity of online transactions, making fraud detection a crucial aspect of credit card security. This study discusses the process and techniques of detecting fraudulent credit card transactions and uses regression analysis and artificial neural networks to analyze a unique dataset of credit card transactions gathered from all around the globe. We also checked the methodologies used in fraud detection by comparing various methods and tools, then summarized the findings and offered some suggestions for future work.

This study concludes that credit card fraud detection is becoming more and more crucial in this age of digital transactions to protect the economy and provide people financial security. To keep the confidence of their consumers and the integrity of the credit card system, financial institutions and companies must be attentive in identifying and blocking fraudulent behavior.

One of the most important limitations of this research is the availability of data that is not only secret but also does not include vital information about the transactions made using and transactions of credit cards.

In future studies, the researchers recommend that other models and techniques be re-examined using the current data set to compare and contrast as well as measure accuracy and precision.

References

1. Bolton, R., Hand, D.: Statistical fraud detection: a review. Stat. Sci. **17**(3), 235–255 (2002)
2. Tidal Commerce: Tidal Commerce Learn (2022). http://www.tidalcommerce.com/learn/signs-of-credit-card-fruad
3. Delamaire, L., Abdou, H., Pointon, J.: Credit card fraud and detection techniques: a review. Banks and Bank Syst. **4**(2), 57–68 (2009)
4. Asmar, M., Ahmad, Z.: Market microstructure: the components of black-box. Int. J. Econ. Financ. **3**(1), 152–159 (2011)
5. Asmar, M., Trimbath, S.: Regulatory reform and trade settlement failures in USA equity markets: does regulatory reform matter? Quant. Financ. Econ. **6**(4), 537–552 (2022)
6. Asmar, M.: Effects of bank-specific factors on the net interest margin of working banks in Palestine. J. Econ. Manage. **33**, 5–24 (2018)
7. Schuessler, K.F., Cressey, D.R.: Personality characteristics of criminals. Am. J. Sociol. **55**(5), 476–484 (1950)
8. Wolfe, D.T., Hermanson, D.R.: The fraud diamond: considering the four elements of fraud (2004)
9. Abdullahi, R.U., Mansor, N.: Fraud triangle theory and fraud diamond theory: understanding the convergent and divergent for future research. Int. J. Acad. Res. Account. Financ. Manage. Sci. **5**, 54–64 (2015). https://doi.org/10.6007/IJARAFMS/v5-i4/1823
10. Association of Certified Fraud Examiners (2022). ACFE. http://www.acfe.com
11. European Central Bank: Seventh report on card fraud. European Central Bank, Frankfurt, Germany (2021)
12. Gupta, S., Malsa, N., Gupta, M.V.: Credit card fraud detection and prevention—a survey. Int. J. Innov. Res. Sci. Technol. **4**, 1–7 (2017)
13. Shiv Kumar Verma, A.K. Credit Crad Fraud Detection System. Credit Crad Fraud Detection System. Galgotias University-Conference Paper (2022)
14. Inscribe: Inscrine.com (2022). https://www.inscribe.ai/fraud-detection/credit-fraud-detection
15. Madhurya, M.J., Gururaj, H.L., Soundarya, B.C., Vidyashree, K.P., Rajendra, A.B.: Exploratory analysis of credit card fraud detection using machine learning techniques. Global Transitions Proceedings **3**(1), 31–37 (2022). https://doi.org/10.1016/j.gltp.2022.04.006
16. Fayyomi, A.M., Eleyan, D., Eleyan, A.: A survey paper on credit card fraud detection techniques. Int. J. Sci. Technol. Res. **10**(09) (2021)
17. Gupta, P., Varshney, A., Khan, M.R., Ahmed, R., Shuaib, M., Alam, S.: Unbalanced credit card fraud detection data: a machine learning-oriented comparative study of balancing techniques. Procedia Comput. Sci. **218**, 2575–2584 (2023). https://doi.org/10.1016/j.procs.2023.01.231
18. Cherif, A., Badhib, A., Ammar, H., Alshehri, S., Kalkatawi, M., Imine, A.: Credit card fraud detection in the era of disruptive technologies: a systematic review. J. King Saud Univ. Comput. Inf. Sci. **35**(1), 145–174 (2023). https://doi.org/10.1016/j.jksuci.2022.11.008
19. Ryman-Tubb, N.F., Krause, P., Garn, W.: How artificial intelligence and machine learning research impacts payment card fraud detection: a survey and industry benchmark. Eng. Appl. Artif. Intell. **76**, 130–157 (2018). https://doi.org/10.1016/j.engappai.2018.07.008

20. Choithani, T., Chowdhury, A., Patel, S., Patel, P., Patel, D., Shah, M.: A comprehensive study of artificial intelligence and cybersecurity on bitcoin, crypto currency and banking system. Ann. Data Sci. (2022). https://doi.org/10.1007/s40745-022-00433-5
21. Bin Sulaiman, R., Schetinin, V., Sant, P.: Review of machine learning approach on credit card fraud detection. Hum.-Centric Intell. Syst. **2**(1), 55–68 (2022). https://doi.org/10.1007/s44230-022-00004-0
22. Zhang, X., Han, Y., Wei, Xu., Wang, Q.: HOBA: a novel feature engineering methodology for credit card fraud detection with a deep learning architecture. Inf. Sci. **557**, 302–316 (2021). https://doi.org/10.1016/j.ins.2019.05.023
23. Kaggle: Kaggle.com (2022). https://www.kaggle.com/search?q=fraud+detection

The Impact of Digital Transformation Strategy in Supply Chain Integration Field Study in the Company Registered in Amman Stock

Munther Dweiri(✉) and Samer Bashabsheh

Mutah University, Mu'tah, Jordan
{820202113024,samerbas}@mutah.edu.jo

Abstract. This study investigated the impact of digital transformation (DT) on supply chain integration (SCI), addressing a gap in the literature. Data were collected from 136 managers and department heads of companies listed on the Amman Stock Exchange with a capital of over 50 million Jordanian dinars. A comprehensive questionnaire was distributed electronically, and 136 valid questionnaires were retrieved for statistical analysis using SPSS. The study established a conceptual framework based on DT and SCI and found that DT has a significant positive impact on SCI.

The study's findings offer valuable insights for professionals seeking to improve supply chain integration through the incorporation of digital technologies and digital transformation strategies.

Keywords: digital transformation · digital transformation strategy · supply chain integration

1 Introduction

Digital transformation has revolutionized supply chain integration, resulting in improved visibility, faster decision-making, and better collaboration among suppliers, manufacturers, and customers. This is due to the adoption of digital technologies such as IoT, AI, block chain, and cloud computing, which have optimized supply chain processes, reduced costs, and increased productivity [1]. Digital transformation has enabled companies To obtain a competitive edge in the marketplace by providing better customer experiences, faster delivery times, and improved product quality. However, its implementation requires careful planning and management to ensure that employees are trained to use these technologies, and the necessary infrastructure is in place to support their operations [2].

The selection and execution of a digital transformation strategy is a critical component for the prosperity of organizations, as this strategy plays a pivotal role in fulfilling tasks and meeting demands efficiently in terms of optimal quality, time, and cost-effectiveness [3].

Supply chains are critical for business continuity, but many managers and workers underestimate their importance. Digital technology solutions (DTS) can integrate supply

B. A. M. Alareeni and I. Elgedawy (Eds.): *Artificial Intelligence (AI) and Finance*, SSDC 488, pp. 912–925, 2023.
https://doi.org/10.1007/978-3-031-39158-3_85

chains, improving company success and competitiveness in the global market. DTS keeps businesses current with the latest digital advancements [1]. Numerous studies, including [1, 2, 4, 5], confirm that DT and SCI are driving research. They recommend more research and integrating strategies. A comprehensive study is required to develop SCI through digital transformation strategy.

In view of that, the issue can be reformulated as the primary inquiry: What is the impact of digital transformation strategy on the supply chain integration in companies listed in the Amman Stock? These two questions can pop up from the main question:

1- What is the level of understanding of the dimensions of DTS among the companies registered in Amman stock?
2- What is the level of understanding of the dimensions of SCI among the companies registered in Amman stock?

The aim of this study is to investigate the impact of DTS on SCI of companies that are listed on the Amman Stock.

The sub-objectives of the study include identifying the level of application of DTS and SCI in these companies and providing actionable findings and recommendations. As a result, we put forward three hypotheses. The following hypotheses have been formulated for achieving the objectives of this study:

Primary Hypothesis
H01: There is no statistically significant impact at a significance level ($0.05 \geq \alpha$), of the dimensions of DTS (technology usage, organizational structure, Creation value, financial aspects) on the SCI with their combined dimensions (supplier integration, internal integration, customer integration) in the companies registered in Amman Stock. Sub-hypotheses are derived from this main hypothesis:

1- **H01:** There is no statistically significant impact at a significance level of ($0.05 \geq \alpha$), of DTS with its dimensions (technology usage, organizational structure, value creation, financial aspects) on supplier integration in companies registered in Amman Stock.
2- **H02:** There is no statistically significant impact at a significance level of ($0.05 \geq \alpha$), of DTS with its dimensions (technology usage, organizational structure, value creation, financial aspects) on internal integration in companies registered in Amman Stock.
3- **H03:** There is no statistically significant impact at a significance level of ($0.05 \geq \alpha$), of DTS with its dimensions (technology usage, organizational structure, value creation, financial aspects) on customer integration in companies registered in Amman Stock.

2 Literature Review

2.1 Digital Transformation

DT known as the integration of digital technology into every aspect of a business brings about significant transformations in how businesses function and deliver benefits to their customers [6]. It includes a combination of process automation, data analysis, implementing technology to drive innovation, and improving overall performance [7]. The goal of DT is to take advantage of the opportunities provided by technology to become more efficient, competitive, and customer-focused [8, 9] agree that DT encompasses multiple dimensions such as technology, organizational changes, value creation, management style, and culture. The aim of the DT process is to bring about significant improvements in an organization's characteristics through digital technology [9]. DTS is a plan for managing the transformation arising from new digital technologies, defining goals, and creating value [8].

2.2 Dimensions of Digital Transformation

The dimensions of DTS (technology utilization, organizational structure, value creation, financial aspects) are measured based on numerous studies that have indicated the opinions of writers and researchers in the field of DTS [6–8, 10].

2.2.1 The Use of Technology

The utilization and exploration of new digital technologies by a company reflects its ability to leverage technology effectively. Innovative and advanced technologies have helped companies achieve great success in manufacturing, such as additive manufacturing, which can reduce costly human errors [12, 13]. Big data is widely used in manufacturing, particularly in conversion industries where it can turn decision-making from predictive to confirmative if used correctly [14, 15, 16]. Implementing AI, IoT, and machine learning can increase productivity and efficiency while reducing downtime and transition time in conversion industries [13, 17, 18].

2.2.2 Value Creation

DT affects companies' value creation by enabling them to collect and analyze data, leading to an increase in basic production processes, such as sales, finding new markets, and facilitating customer interaction [20]. Several studies have discussed the ways in which companies can create value in various ways. Four methods have been identified: expanding the current portfolio of products and services, better understanding customer needs and providing new valuable proposals, employing technological techniques to address sustainability issues and collaborative economic approaches, and creating new value through digital platforms and technological systems. [6–8, 10].

2.2.3 Organizational Structure

DT involves modifying a company's structures, processes, and skills to effectively navigate new technologies [11]. Companies must decide whether to integrate new digital processes into existing structures or create separate entities [7]. Managers must also consider the operational changes expected when exploring and exploiting digital technologies. DT is a high priority for top management because it can reshape entire business models [11].

2.2.4 Financial Aspect

The financial dimension of DT requires companies to understand the potential economic benefits and challenges, and to have a clear plan for measuring success [22]. However, calculating return on investment can be complex and requires a different approach than traditional investments [23]. Companies should consider both financial and non-financial metrics when assessing success such as customer satisfaction and employee engagement [23]. Although digital transformation can bring benefits, companies need to carefully evaluate costs and benefits through a feasibility study, maintain a long-term perspective, and have a well-planned approach to mitigate potential risks, challenges [22].

2.3 Supply Chain Integration

Supply chain integration is a multifaceted concept that necessitates the coordination and collaboration of various entities within a supply chain - from suppliers to customers - to efficiently manage the movement of materials, finances, and information while meeting regulatory standards. SCI has been defined by [26] as the coordination of the different entities in a SC. [25] further explains that an integrated SC is a system consisting of suppliers, manufacturers, distributors, and customers, where materials flow from suppliers to customers and information flows in all directions. According to [24, 25], supply chain integration (SCI) involves the coordination and collaboration between supply chain partners to improve the efficiency and effectiveness of the flow of materials, resources, and information in producing products and delivering valuable services to customers quickly and cost-effectively.

2.4 Dimensions of Supply Chain Integration

Many studies have adopted three dimensions of SCI, which. (Lopez-Torres, 2017-Diez, are internal integration, supplier integration, and customer integration [27–30].

2.4.1 Supplier Integration

Supplier relationship management involves building long-term relationships with suppliers to reduce transaction costs and increase cost savings, while fostering trust and reputation [30]. Organizational integration with suppliers entails sharing information, supplier impact on decisions, and building strategic and formal partnerships through long-term relationships [31]. SI is characterized by collaboration between organizations and their suppliers, including sharing strategies, information, processes, and synchronization to provide excellent customer service [24, 32].

2.4.2 Customer Integration

Customer integration in the supply chain involves customers in various aspects of the supply chain, such as design, production, distribution, and delivery of goods and services. This approach aims to create a more collaborative relationship between companies and customers, resulting in improved customer satisfaction, cost reduction, and increased efficiency [21, 29]. A significant advantage of customer integration in the supply chain is that it provides companies with better insight into customer needs and preferences, leading to the development of products and services that align closely with customer requirements. This can ultimately result in increased customer satisfaction and loyalty. Involving customers in the design and delivery process can lead to a more streamlined and efficient supply chain, resulting in increased efficiency [29, 33–35].

2.4.3 Internal Integration

In supply chain management involves aligning and coordinating internal processes and functions within an organization to improve efficiency and effectiveness in the supply chain [19, 34]. Internal integration can improve inventory management, communication, collaboration, forecasting, and planning abilities [36].

To achieve internal integration effectively, organizations should involve representatives from different departments in decision-making and collaboration, and use shared information systems and common metrics to track and measure performance across different departments. [29, 37].

2.5 Previous Studies

The study [38] investigated the impact of DT on SC management in industrial companies during the COVID-19 pandemic in Amman. DT had a positive impact on SC management in terms of technology, human resources, and data. The study recommends companies to update and maintain their technology infrastructure and invest in technology for coordination.

Another study [39] found a strong correlation between DT and tax compliance of financiers in the Egyptian Tax Authority.

The moral impact of electronic systems on tax compliance was highlighted. Study [40] presented an integrated perspective of Smart GSCM and showed that a smart supply chain is linked to higher levels of GSCM.

Study [41] explored the relationship between DT, SCI, and OSSCP and found a notable positive impact on both. Study [42] focused on integrating new-generation technologies and communication in supply chain components and found a significant relationship between digitalization, Industry 4.0 technologies, and supply chain performance (Fig. 1).

3 Conceptual Framework and Hypothesis Development

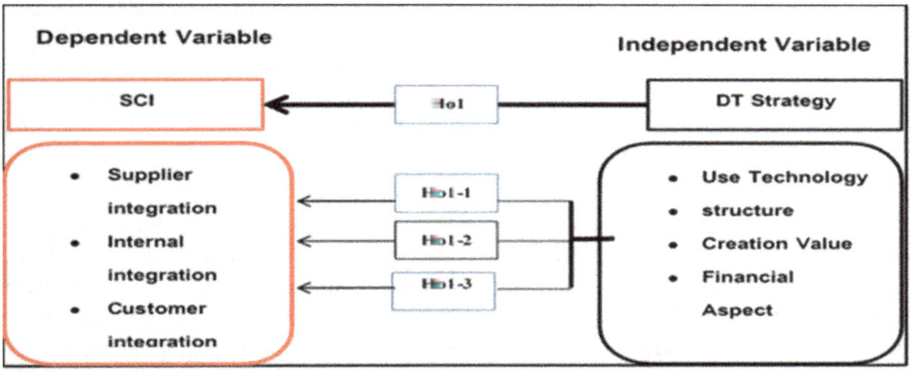

Fig. 1. Conceptual frame [6–8, 10, 11, 27, 30].

4 Empirical Research Methodology and Data Analysis

4.1 Survey Design and Data Collection

A questionnaire survey was used to assess variable interrelationships. The five-point Likert scale ranged from strongly disagree (1) to strongly agree (5).

A panel of experts pre-tested the questionnaire for relevance, and it consisted of three sections: general information, DT dimensions, and SCI dimensions.

The study included all managers and department heads from companies listed on Amman Stock with over 50 million Jordanian dinars in capital. The following Table 1 listed the companies in the study.

Table 1. Companies listed on the Amman Stock with a capital of more than 50 million Jordanian dinars

No	Comp number	code	Company name	Capital/jd
1	142041	JOPT	Jordan petroleum refinery	100,000,000
2	141043	APOT	Arabic potash	83,317,500
3	141018	JOPH	Jordan phosphate mines	82,500,000
4	141204	JPHM	Jordanian pharmaceutical production	73,312,500
5	141042	JOCM	Jordanian cement factories	60,444,460

(continued)

Table 1. (*continued*)

No	Comp number	code	Company name	Capital/jd
6	141097	MECE	Middle East engineering, electronic and Heavy Industries Complex	60,000,000
7	141224	NCCO	Northern Cement	55,000,000

reference (sdc.com.jo)

A comprehensive questionnaire was distributed electronically to the entire study community, and 136 valid responses were analyzed.SPSS 26 was used to assess the reliability and validity of the data. The Cronbach's alpha test was used to measure the consistency of respondents' answers to the scale questions. The Cronbach's alpha coefficient for all study dimensions was 0.898, indicating high internal consistency.

The reliability values for the independent variable dimensions ranged from 0.704 to 0.811 for digital transformation strategy. For the dependent variable of supply chain integration, the reliability values ranged from 0.728 to 0.821. These high Cronbach's alpha indicators show the tool has high stability and can achieve the study's objectives. The following Table 2 provides additional details.

Table 2. Reliability Statistics

Variable	Dimension	No of paragraph	Cronbach's Alpha
Independent Digital transformation strategy	Technology use	5	0.809
	Structure	5	0.704
	Creation value	5	0.799
	Financial aspect	4	0.811
	General DT	19	0.925
Dependent Supply chain integration	Supplier integration	5	0.821
	Internal integration	5	0.728
	Customer integration	5	0.813
	General SCI	15	0.913

To test for normal distribution, the study calculated the skewness coefficient and relied on skewness values ranging from 0.052 to 0.382, indicating normal distribution. The kurtosis values ranged from 0.58 to 0.949, confirming normal distribution. Normal distribution test values were less than 2.2, indicating no significant issues. The following Table 3 provides additional details.

The study sample consisted of 69 male participants (50.7%) and 67 female participants (49.3%), suggesting homogeneity among managers and department heads in the studied companies. The majority of participants had a bachelor's degree (60.3%),

Table 3. Descriptive statics normal distributed

	N	Skewness		Kurtosis	
	Statistic	Statistic	Std. Error	Statistic	Std. Error
Tech	136	−.382	.208	.599	.413
Stract	136	−.242	.208	.949	.413
value	136	−.052	.208	.058	.413
Finan	136	−.315	.208	.391	.413
Indepnt	136	−.156	.208	.944	.413

with smaller percentages holding diploma (11.8%), master's (19.9%), or doctoral (8.1%) degrees. The largest category for years of experience was 15 years or more, indicating job stability. Details in the following Table 4.

Table 4. Frequency Table

		Frequency	Percent	Valid Percent	Cumulative Percent
Gender	Male	69	50.7	50.7	50.7
	Female	67	49.3	49.3	100.0
Experiment	5 years and less	31	22.8	22.8	22.8
	6 – 10 years	9	6.6	6.6	29.4
	11- 15 years	22	16.2	16.2	45.6
	More than 15 years	74	54.4	54.4	100.0
Education	Diploma	16	11.8	11.8	11.8
	Bachelor	82	60.3	60.3	72.1
	Masters	27	19.9	19.9	91.9
	Ph. D	11	8.1	8.1	100.0

5 Findings

5.1 Structural Model and Hypothesis Testing

To investigate the interrelationships among the latent variables in the main study, a structural model was created. The model's predictive power was evaluated using the R2 value, which measures the extent to which independent variables account for variance in dependent variables.

This evaluation was conducted in SPSS for the dependent constructs., Model Validity Test, By using simple linear regression analysis, the computed F value was (62.02), which is greater than the tabular F value (2.37) at $\alpha = 0.05$ significance level. Additionally, the SIG value was (0.00), which is less than 0.05.

The table (1) presents the validity of the study hypotheses testing model. By using simple linear regression analysis, the computed F-value was (62.02), which is greater than the tabulated F-value (2.37) at a significance level of $\alpha = 0.05$. Additionally, the computed SIG value was (0.00), which is less than 0.05.

These results indicate that there is an impact of the independent variables on the dependent variable. Referring to the R-squared value (0.654), which explains that 65.4% of the variance in the dependent variable is explained by the independent variables, the relationship between the variables is strong, with an R-value of 80.9%. This means that there is a correlation between digital transformation strategy and supply chain integration. Therefore, the model is valid for testing, and there is an impact of the independent variables on the dependent variable.

Sig.	F	Adj R	**R2**	R
0.00	62.021	0.644	0.654	0.809

5.2 Hypothesis Testing

The statistical results in the following Table 5 indicate that the dimensions (technology use, organizational structure, value creation, financial aspects) have a statistically significant impact on the dependent variable, supply chain integration. All dimensions were statistically significant, with p-values less than 0.05, except for the dimension of technology use, which did not have a statistically significant impact on supply chain integration, with a significance value greater than 0.05.

This means that the null hypothesis is partially rejected regarding organizational structure, value creation, and financial aspects, and the alternative hypothesis is accepted. On the other hand, the null hypothesis is accepted regarding technology use.

Table 5. Regression

Model		Unstandardized Coefficients		Standardized Coefficients	t	Sig	Collinearity Statistics	
		B	**Std. Error**	**Beta**			**Tolerance**	**VIF**
1	**(Constant)**	.484	.218		2.215	.028		
	Tech	.061	.084	.062	.732	.466	.373	2.684
	Stract	.399	.091	.391	4.360	.000	.327	3.053
	value	.246	.065	.275	3.799	.000	.503	1.990
	Finan	.161	.068	.189	2.357	.020	.410	2.439

a. Dependent Variable: Dependant

Testing of sub-hypotheses:

1- H01: The statistical results in the following Table 6 indicate that the dimensions of DTS (technology usage, organizational structure, value creation, financial aspects), based on their calculated t-test values, have a statistically significant impact on the supplier integration. The calculated t-test values were (4.061, 4.119) respectively, and were statistically significant with a level of significance less than 0.05.

The results showed that after using technology and financial aspects, they did not have an impact on supplier integration, where the calculated t-test values were not statistically significant at a significance level of ($0.05 \geq \alpha$). Therefore, the following is concluded: the null hypothesis, which states that there is no statistically significant impact of the dimensions of DTS (organizational structure, value creation) on supplier integration is rejected, and the alternative hypothesis is accepted.

The null hypothesis, which states that there is no statistically significant impact of the dimensions of DTS (technology usage, financial aspects) on supplier integration is accepted.

Table 6. Regression FOR THE 1-H01 TEST

Model		Unstandardized Coefficients		Standardized Coefficients	t	Sig
		B	Std. Error	Beta		
1	(Constant)	.285	.279		1.021	.309
	Tech	.044	.107	.039	.412	.681
	Stract	.475	.117	.409	4.061	.000
	value	.341	.083	.335	4.119	.000
	Finan	.058	.088	.060	.666	.507

a. Dependent Variable: Suppler

2- H01: The statistical results the following Table 7 indicate that the dimensions of DTS (organizational structure, value creation, financial aspects) have a statistically significant impact on the internal integration, related on the results of their respective t-tests.

The calculated t-values were (3.965, 2.734, 3.437) respectively, which were statistically significant and less than 0.05. The results indicated that the use of technology did not have an impact on internal integration.

The calculated t-values were not statistically significant at a significance level of 0.05 or greater. Therefore, the following is concluded: rejecting the null hypothesis, which states that there is no statistically significant impact of the dimensions of DTS (organizational structure, value creation, financial aspects) on internal integration.

Accepting the alternative hypothesis, and accepting the null hypothesis that there is no statistically significant impact of the DTS dimension (technology use) on internal integration.

Table 7. Regression FOR THE 2-H01 TEST

Coefficients[a]

Model		Unstandardized Coefficients		Standardized Coefficients	t	Sig
		B	Std. Error	Beta		
1	(Constant)	.615	.253		2.432	.016
	Tech	-.060	.097	-.057	-.622	.535
	Stract	.420	.106	.390	3.965	.000
	value	.205	.075	.217	2.734	.007
	Finan	.272	.079	.302	3.437	.001

a. Dependent Variable: Internal

3- H01: The statistical results in the following Table 8 indicate that the dimensions of organizational structure and value creation, as well as monitoring the values of their t-tests, have a statistically significant impact on the customer integration The calculated t-values of (2.501, 2.249) respectively were statistically significant and less than 0.05.

The results indicated that the dimensions of technology usage and financial aspects had no impact on customer integration, as the calculated t-values were not statistically significant at a significance level of ($0.05 \geq \alpha$.

Therefore, the following is conducted: rejecting the null hypothesis, which states that there is no statistically significant impact of the dimensions of DTS (organizational structure, value creation) on customer integration. Accepting the alternative hypothesis, and accepting the null hypothesis that there is no statistically significant effect of the dimensions of DTS (technology usage, financial aspects) on customer integration.

Table 8. Regression FOR THE 3-H01 TEST

Model		Unstandardized Coefficients		Standardized Coefficients	t	Sig
		B	Std. Error	Beta		
1	(Constant)	.551	.287		1.919	.057

(*continued*)

Table 8. (*continued*)

Model		Unstandardized Coefficients		Standardized Coefficients	t	Sig
		B	Std. Error	Beta		
	Tech	.200	.110	.183	1.817	.071
	Stract	.301	.120	.269	2.501	.014
	value	.191	.085	.195	2.249	.026
	Finan	.153	.090	.164	1.702	.091

a. Dependent Variable: Customr

6 Conclusions

The digital transformation of business operations has been a growing trend, and supply chain integration has been identified as an area where DT can have a significant impact. This study aimed to evaluate the impact of DT on SCI in companies listed on the Amman Stock, and establish a conceptual framework to assess this impact.

The results of the study showed that the companies listed on the Amman Stock had a strong interest in DT and SCI, as evidenced by high means and standard deviations of both the digital transformation strategy and supply chain integration. The companies focused on using technology, aligning organizational structures with the stage of DT, and creating value in all aspects while investing in DT financially. Similarly, the companies focused on the customer, developed internal business operations for internal integration, and considered suppliers as a crucial part of their business to achieve a competitive advantage and high returns on investment in the future.

The statistical analysis revealed that the dimensions of technology usage, organizational structure, value creation, and financial aspects significantly impacted the dependent variable of supply chain integration, except for technology usage, which was not significant. These findings suggest that companies should consider these dimensions when evaluating the impact of DT on SCI. Multiple regression analysis was used to obtain these results.

The framework developed in this study can be used to inform decision-making related to digital transformation and supply chain integration, with the aim of achieving a competitive advantage and high returns on investment in the future. The study recommends that senior management should support the use of technology as a primary driver for digital transformation and invest in it, as it is expected to yield significant returns in the future.

References

1. Tuire, H.-K.: The impact of digitalization on firm performance: examining the role of digital culture and the effect of supply chain capability, BPMJ, School of Management, University of Vaasa, Vaasa, Finland) (2022)

2. Garay-Rondero, C.L., Martinez-Flores, J.L., Smith, N.R.: Digital supply chain model in Industry 4.0. J. Manufacturing Technology Manage. **31**(5), 887–933 (2020)
3. Liu, P., Hendalianpour, A., Hamzehlou, M., Feylizadeh, M.R., Razmi, J.: Identify and rank the challenges of implementing sustainable supply chain blockchain technology using the bayesian best worst method. Technol. Econ. Dev. Econ. **27**(3), 656–680 (2021)
4. Jabbour, C.J.C., Fiorini, P.D.C., Ndubisi, N.O., Queiroz, M. M., Piato, É.L.: Digitally-enabled sustainable supply chains in the 21st century: a review and a research agenda. Science of The Total Environment (2020),
5. Tseng, M.L., Bui, T.D., Lim, M.K., Lewi, S.: A cause and effect model for digital sustainable supply chain competitiveness under uncertainties: enhancing digital platform. Sustainability **13**(18) (2021)
6. Pousttchi, K.: Digital Transformation, Research Gate Journal (2020)
7. Nissar, E.A.: Digital Transformation Prerequisites for a Digital Business Model (2022)
8. Matt, C., Hess, T., Benlian, A.: Digital transformation strategies. Business & Information Syst. Eng. J. **57**, 339–343 (2015)
9. Vial, G.: Understanding digital transformation: a review and a research agenda. Managing Digital Transformation, pp. 13–66 (2021)
10. Lipsmeier A., Kühn A., Joppen R., Dumitrescu R.: Process for the development of a digital strategy. Procedia CIRP (2020)
11. Hess, T., Matt, C., Benlian, A., Wiesböck, F.: Options for Formulating a Digital Transformation Strategy MIS Quarterly Executive **15**(2) (2016)
12. Strong, D., Kay, M., Conner, B., Wakefield, T., Manogharan, G.: Hybrid manufacturing–integrating traditional manufacturers with additive manufacturing (AM) supply chain. Addit. Manuf. **21**, 159–173 (2018)
13. Haleem, A., Javaid, M., Rab, S.: Impact of additive manufacturing in different areas of Industry 4.0. Int. J. Logistics Syst. Manage. **37**(2) (2020)
14. Bruggemann, S., Gruning, F.: Using domain knowledge provided by ontologies for improving data quality management. Proceedings of I-Know, pp. 251–258 (2008)
15. Liaw, S.T., Pearce, C., Liyanage, H., Cheah-Liaw, G.S., De Lusignan, S.: An integrated organization-wide data quality management and information governance framework. Journal of Innovation in Health Informatics (2014)
16. Dubey, R., Gunasekaran, A., Childe, S.: Big data analytics capability in supply chain agility. Manag. Decis. **57**(8), 2092–2112 (2019)
17. Wuest, T., Weimer, D., Irgens, C., Thoben, K.D.: Machine learning in manufacturing: advantages, challenges, and applications. Prod. Manufacturing Res. **4**(1), 23–45 (2016)
18. Lee, J., Azamfar, M., Singh, J.: A blockchain enabled cyber-physical system architecture for industry 4.0 manufacturing systems. Manufacturing Letters **20**, 34–39 (2019)
19. Kim, D.Y.: Relationship between supply chain integration and performance. Operations Manage. Res. **6**, 7490 (2013)
20. Frank, A.G., Dalenogare, L.S., Ayala, N.F.: Industry 4.0 technologies: implementation patterns in manufacturing companies. Int. J. Prod. Economics **210**, 15–26 (2019)
21. Gao, S., Li, X., Li, X.: The effect of customer integration on supply chain performance: evidence from Chinese firms. J. Bus. Res. **69**(9), 3299–3307 (2016)
22. Catlin, T., Lorenz, T., Sternfels ,B., Willmott, P.: A Roadmap for a Digital Transformation. McKinsey & Company (2017)
23. McKinsey & Company. The Power of Partnerships in Digital Transformation (2018)
24. Flynn, B.B., Huo, B., Zhao, X.: The impact of supply chain integration on performance: a contingency and configuration approach. J. Oper. Manag. **28**, 58–71 (2010)
25. Qi, Y., Huo, B., Wang, Z., Yeung, H.Y.J.: The impact of operations and supply chain strategies on integration and performance. Int. J. Prod. Econ. **185**, 162–174 (2017)

26. Lummus, R.R., Duclos, L.K., Vokurka, R.J.: Supply chain flexibility: building a new model. Glob. J. Flex. Syst. Manag. **4**(4), 1–13 (2003)
27. Wong, C.W.Y., Boon-itt, S., Wong, C.Y.: The contingency effects of environmental uncertainty on the relationship between supply chain integration and operational performance. J. Operations Manage. **29**(6), 604615 (2011)
28. Ben-Daya, M., Hassini, E., Bahroun, Z.: A Conceptual Framework for Understanding the Impact of Internet of Things on Supply Chain Management **15**(2), 251268 (2022)
29. Zhang, Q., Li, X., Li, X.: The impact of customer integration on supply chain performance: evidence from Chinese manufacturing firms. J. Bus. Res. **85**, 208–218 (2018)
30. Li, S., Ragu-Nathan, B., Ragu-Nathan, T.S., SubbaRao, S.: The impact of supply chain management practices on competitive advantage and organizational performance. Omega **34**(2), 107–124 (2006)
31. Suhaiza, Z., Premkumar, R.: Supply chain integration and performance: US versus East Asian companies. Supply Chain Manag. **10**(5), 379–393 (2005)
32. Echtelt, F., Wynstra, F., Weele, A., Duysters, G.: Managing supplier involvement in new product development: a multiple-case study. J. Product Innovation Manage. **25**(2), 180–201 (2008)
33. Sánchez-Rodríguez, C., Aragón-Correa, J.A., García-Morales, V.J.: The role of customer integration in the relationship between supply chain management practices and performance. J. Bus. Res. **98**, 365–376 (2019)
34. Chen, I.J., Paulraj, A., Lado, A.A.: Towards a theory of supply chain management: the constructs and measurements. J. Supply Chain Manag. **52**(2), 3–30 (2016)
35. Lotfi, Z., Mukhtar, M., Sahran, S., Zadeh, A.T.: Information sharing in supply chain management. Procedia Technol. **11**, 298–304 (2013)
36. Kshetri, N.: Blockchain and IoT integration for secure and transparent supply chain management. J. Bus. Res. **85**, 208–218 (2018)
37. Xu, H., Li, X., Zhang, X.: Big data analytics in supply chain management: a literature review. J. Bus. Res. **70**, 303–316 (2017)
38. Gawazneh, S.: The impact of digital transformation on supply chain management during the corona pandemic in industrial companies in King Abdullah II Industrial City/Amman (2021)
39. Abdelbaki, H.: The Impact of Digital Transformation on Tax Compliance in Light of the Global Corona Crisis, Tanta University (2022)
40. Lerman, L.V, Benitez, G.B., Müller, J.M., De Sousa, P.R.: Smart green supply chain management: a configurational approach to enhance green performance through digital transformation. Int. J. Supply Chain Manage. **27**(7), 147176 (2022)
41. Oubrahim, I., Sefiani, N., Happonen, A.: The influence of digital transformation and supply chain integration on overall sustainable supply chain performance: an empirical analysis from manufacturing companies in Morocco. Energies (2023)
42. Dudukalov, E.V., Terenina, I.V., Perova, M.V., Ushakov, D.: Industry 4.0 Readiness: The Impact of Digital Transformation on Supply Chain Performance. E3S Web Conf. (2021)

Digital Marketing Factors Affecting Purchasing Intentions for Delivery Services (A Field Study on Customers of Talabat Company at Zarqa City - Jordan)

Iyad khanfar[1](✉) and Asem nofal[2]

[1] Zarqa University, Zarqa, Jordan
i.khanfar@zu.edu.jo
[2] Injaz, Jordan
anofal@injaz.org.jo

Abstract. The purpose of this study was to investigate how digital marketing factors influence the purchasing behavior of Talabat customers, with a focus on five independent variables: security, credibility, ease of use, efficiency, and modernization. The study also examined one dependent variable, which is the purchase intention for delivery services. The research was conducted on 473 Talabat customers in Zarqa city. The findings revealed that the digital marketing factors analyzed had a significant impact on customers' purchase intention. The study found that safety, credibility, ease of use, efficiency, and modernization had positive effects on increasing customers' intention to purchase from Talabat. As a result, the study suggests that Talabat can attract more customers and meet the growing demand for its services by offering these marketing factors. The study recommended that Talabat prioritize customer privacy by safeguarding their information and ensuring that customers have greater trust in commercial advertisements.

Keywords: digital marketing · Talabat · Purchase intention · delivery services · Jordan

1 Introduction

We are currently experiencing a surge of knowledge and a significant shift in all aspects of life, including politics, economics, environment, society, and technology. With the rapid pace of change and great transformation in technology, digital marketing has emerged as a crucial necessity for people in our present era to meet their various needs, including economic requirements. This has prompted companies, such as Talabat Investment Company, to invest in digital marketing through smart applications and websites to connect with potential customers locally or globally. Digital marketing enables companies to easily and efficiently showcase their products and services to customers with less cost and effort compared to traditional marketing methods.

B. A. M. Alareeni and I. Elgedawy (Eds.): *Artificial Intelligence (AI) and Finance*, SSDC 488, pp. 926–933, 2023.
https://doi.org/10.1007/978-3-031-39158-3_86

1.1 Study Problem and Questions

Digital marketing plays a crucial role in influencing the purchase intention of delivery services, especially in today's fast-paced and competitive business environment where companies like Talabat Investment Company must focus on survival, growth, continuity, and profits. However, consumers may hesitate to purchase through delivery services due to concerns about privacy, safety, and the credibility of advertised products. As purchasing is a vital factor in achieving profitability for Talabat, it utilizes digital marketing strategies and tools to provide advanced and exceptional services to customers and boost their purchase intention. Therefore, it is essential to explore the impact of digital marketing factors, such as safety, credibility, ease of use, efficiency, and modernization, on the purchase intention of Talabat's customers.

1.2 Study Objectives

The main objective of the study was to identify the digital marketing factors and their dimensions that affect the purchase intention of Talabat's delivery services. The study focused on five factors, including safety, credibility, ease of use, efficiency, and modernization, to investigate their impact on customers' purchase intention. By analyzing these dimensions, the study aimed to provide a comprehensive understanding of the digital marketing factors that Talabat should focus on to increase its customer base and improve its delivery services.

1.3 Study Hypotheses

The study's main hypothesis was that there is no significant impact of digital marketing factors, including safety, credibility, ease of use, efficiency, and modernization, on the purchase intention for Talabat's delivery services. To test this hypothesis, the study developed several sub-hypotheses.

- **H01:** There is no statistically significant effect of the safety on customers' purchasing intentions of delivery services.
- **H02:** There is no statistically significant effect of the credibility on customers' purchasing intentions of delivery services.
- **H03:** There is no statistically significant effect of the ease of use on customers' purchasing intentions of delivery services.
- **H04:** There is no statistically significant effect of the efficiency on customers' purchasing intentions of delivery services.
- **H05:** There is no statistically significant effect of the modernization on customers' purchasing intentions of delivery services.

1.4 Study Model

In light of the hypotheses formulated, the study variables were formulated according to Fig. 1 after referring to the study [1–5].

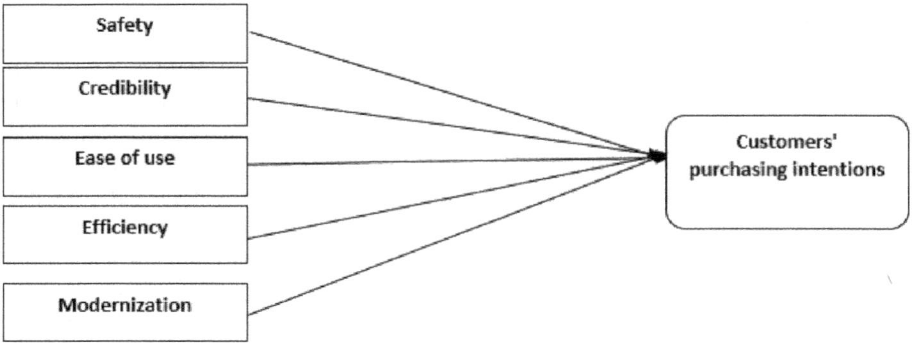

Fig. 1. Study model.

2 Theoretical Framework

2.1 Digital Marketing Concept

In today's highly competitive business environment, companies that adopt digital marketing have a distinct advantage over their competitors. This is because digital marketing offers numerous benefits such as attracting and retaining customers, as well as saving time and effort in searching for goods, services, or information. The American Marketing Association has defined digital marketing as the activities and operations in organizations that use digital technologies to communicate with customers and provide them with benefits [6]. Additionally, digital marketing is the implementation of marketing tools electronically by managing the interaction between companies and consumers to exchange benefits and communicate effectively [7].

2.2 Digital Marketing Tools

Search engines are a powerful tool for companies to display their products through paid ads. It is considered one of the most effective ways of online marketing. Companies use various techniques and strategies to serve users of popular search engines like Google, Bing, Yahoo, and others. [8, 9].

Email: This involves sending electronic messages directly to individuals or groups of people. It offers many benefits to companies, such as direct contact with customers and reducing costs in the marketing process. [5].

Mobile phones: Nowadays, mobile phones have become an essential tool for companies to connect with their customers, delivering information about their products and services through wireless networks. Smartphones, in particular, have revolutionized the way companies market their products, as they are not only a means of communication but also a platform for mobile applications. Companies can promote their products through ads and display them in their mobile apps, creating a direct link between them and their customers. [10].

Social media platforms: These refer to online platforms that allow users to create, share, or exchange information and content through social networks, virtual communities, or online forums. Companies utilize these platforms for marketing purposes by creating engaging content, paid ads, and targeted campaigns to reach a specific audience and achieve their marketing goals. Examples of social media platforms include Facebook, Instagram, Twitter, and LinkedIn. [11].

2.3 Digital Marketing Factors

1. Safety: This refers to the service provider's ability to offer a secure and risk-free environment for the customer, inspiring confidence and a sense of security. Safety is a crucial factor in evaluating service quality, and customers tend to favor institutions that prioritize safety in their operations [12].
2. Credibility: This relates to the accuracy and truthfulness of the information published on a website, as indicated by the publisher's signals and signs. Due to the lack of regulatory controls on the Internet, credibility is essential in building trust between the website owner and the user [13].
3. Ease of use: This refers to the website's user-friendliness, enabling users to access the required information with minimal time and effort. This involves clear fonts, colors, and backgrounds, well-organized information, fast page loading times, and easy navigation, all of which make the user's browsing experience smooth and enjoyable [14].
4. Efficiency: This refers to the ease with which users can access the website multiple times and the site's ability to allow access without the need to type "WWW" before the page name [15].
5. Modernization: This involves continuously updating information as needed, replacing outdated information with new and relevant content. Keeping up-to-date with changing trends and customer needs is essential for maintaining a competitive edge [16].

2.4 Purchasing Intentions

It is defined as the evaluation of products by the customer that is likely to be purchased based on the interaction between his needs, his attitude and his perception of a particular product [17]. The study defines it as the internal or external factors that produce a pattern adopted by the customer in his behavior while searching, evaluating products, or purchasing, which he expects to meet his needs and desires.

3 Previous Studies

Various digital marketing strategies have been shown to positively impact consumer behavior, as evidenced by multiple studies. For example, research [18] revealed a significant link between brand credibility and the intention to purchase. Similarly, another study [3] demonstrated the positive impact of brand awareness on electronic word-of-mouth and purchase intentions. Furthermore, [1] suggests that heightened consumer awareness of a brand can improve purchase decisions.

Interestingly, [3] also found that consumer-generated advertisements are more credible than official brand advertisements. Meanwhile, [2] found a direct link between digital marketing and customer happiness. Another study [19] found a positive relationship between mobile advertising and purchasing intentions, while [20] demonstrated that the sensory images presented on Facebook pages can influence purchasing intentions.

Moreover, a study [16] reported that retailers can benefit from digital marketing by being able to respond quickly to customer requests. Similarly, a study [21] evaluated the effectiveness of digital marketing on social media and email marketing, and its impact on customer engagement and purchasing intentions. [22] Found that electronic advertisements on social networking sites have a positive effect on purchasing intentions, while a study [5] noted that social media, websites, e-mail, and search engines are key factors in the continued dominance of retail fashion brands. Finally, a study [23] revealed that online purchasing intentions are significantly influenced by expected usefulness, ease of use, website content, and trust.

4 Methodology

4.1 Study Population and Sample

The study population consisted of Talabat application users in Zarqa Governorate. (530) questionnaires were distributed to the study sample. The researchers retrieved (473) questionnaires, and (57) questionnaires could not be retrieved, as the percentage of questionnaires retrieved was approximately (89%). of the total number of questionnaires [24] (Table 1).

Table 1. Reliability Analysis

Variables	Item number	Cronbach Alpha
Safety	5	0.839
Credibility	5	0.869
Ease of use	5	0.860
Efficiency	4	0.851
Modernization	4	0.817
Customers' purchasing intentions	8	0.917

It appears from the table number (1) that the Cronbach alpha coefficients for the items of the variables lie between (0.817–0.917), and they all met the stability condition by exceeding the value of (0.7), and this indicates that all the items of the questionnaire are stable. Paragraphs: The "purchasing intentions" variable was considered the most stable, as it reached Cronbach's alpha coefficient: (0.917), while the "Modernization" variable was the least stable, with a coefficient of: (0.817).

5 Testing Study Hypotheses

The main hypothesis was supported by multiple regression analysis, which also demonstrated the contribution of the independent variable to the explanation of the dependent variable. The value of F and its significance were used to confirm the validity of the proof of the hypothesis.

5.1 Testing the Main Hypothesis

HO: There is no statistically significant effect at the level of significance ($0.05 \geq \alpha$) for digital marketing factors in its dimensions (security, credibility, ease of use, efficiency, modernization) affecting the purchase intention for delivery services on Talabat Company. To test the hypothesis, multiple linear regression analysis was used, and Table 2 shows the results.

Table 2. Results of multiple regression analysis to test the main hypothesis

Dependent Variable	R	R2	F	DF	SIG	Independent Variable	B	T	SIG
Purchasing Intentions	0.711	0.505	95.220	5	0.00	Safety	0.043	0.841	0.401
				467		Credibility	0.356	6.478	0.000
				472		Ease of use	0.124	1.934	0.054
						Efficiency	0.265	3.910	0.000
						Modernization	0.240	4.023	0.000

6 Recommendations

In order to improve digital marketing factors affecting purchasing intentions for delivery services, particularly for customers of Talabat Company in Zarqa City, Jordan, several recommendations can be made. First, it is important to increase the protection of customers' privacy by limiting the disclosure of information only to those authorized to do so. Additionally, commercial advertisements must match the company's promises and be applied on the ground to increase customers' confidence. Delivery of orders within specifications and high quality is also necessary to enhance customer confidence. The company's application and website should be regularly updated to provide modern and necessary information. The creation of a customer service department to follow up on suggestions and complaints and obtain feedback will help raise the company's credibility. Finally, conducting periodic assessments of the ease of use of the application and website and studying customers' ability to deal with it will be beneficial in improving digital marketing factors affecting purchasing intentions.

References

1. Ayad, S., Fouda, M., Garrash, M.: The impact of brand digital marketing through social media on consumers' purchase intent: a case study of the Ooredoo brand. J. Econ. Manag. Commercial Sci. **13**(1) (2020)
2. Al-Hakim, L., Al-Hamami, Z.: Digital marketing and its role in achieving customer happiness - a prospective study of the opinions of a sample of customers of mobile communications companies in Iraq. Al-Ghari J. Econ. Adm. Sci. **14**, 3 (2017)
3. Simple, R.G.: The effect of marketing via social media on the purchase intention: analyzing the mediating role of awareness of the mark and the spoken word electronically. Appl. Study Serv. Mob. Phone Co. Egypt, Sci. J. Econ. Adm. **50**(50), 441–486 (2020)
4. Low, S., Ullah, F., Shirowzhan, S., Sepasgozar, S.M.E., Lin Lee, C.: Smart digital marketing capabilities for sustainable property development: a case of Malaysia. Sustainability **12**, 5402 (2020)
5. Ceesay, L., Sanyang, L.: The impact of digital media advertising on consumer behavioral intention towards fashion and luxury brands: a case of the Gambia. Arab. J. Bus. Manag. Rev. **8**, 1–10 (2018)
6. AMA. Available online. Consulte 10 11, 2019, sur (2013). https://www.ama.org/the-definitionofmarketing
7. Chakravarty, A., Kumar, A., Grewal, R.: Customer orientation structure for internet based business-to-business platform firms. J. Mark. **78**(5), 1–23 (2014)
8. Raghunandan, H., Parimal, N.: Integrated Digital Marketing- The Key To Understanding Your Customer. Tata Consultancy Services Limited, Rautanen, Matti, email marketing in business-to-business relationship. Master Thesis, university of technology, School of Business, International Marketing Management (2014)
9. Jain, A.: The Role and Importance of Search Engine and Search Engine Optimization. Int. J. Emerg. Trends Technol. Comput. Sci. **2**(3) (2013)
10. Harfanti, M., Ekawanto, I., Lutfi, M.Y.: Innovative Mobile Marketing via Smart Phones. In: Submitted to International Conference on Business, Economics and Accounting, Bangkok – Thailand (2013)
11. Al-Alaq, B.: Electronic Marketing, pp. 17–18. Dar Al-Yazuri for Publishing and Distribution, Amman, Jordan (2010)
12. Swaidan, N.: Contemporary Marketing: New Concepts. Dar Al-Hamid for Publishing and Distribution, Amman (2010)
13. Mohamed, H.: Evaluation of the Quality of Websites. Tikrit, Iraq, Tikrit Journal of Administrative and Economic Sciences, p. 18 (2010)
14. Bilal, S.H.: Developing a model to Measure the Impact of the Spoken Word on the Adoption of Electronic Banking Services, Middle East University (2013)
15. Ahmed, A.M., Shafiq,: The Use of Multimedia in the Websites of Satellite Channels - An Analytical Study of the Websites of the Sudanese Satellite Channel and Al-Shorouk Channel in the Period from 2013–2015, (Master's Thesis). Sudan University of Science and Technology, Khartoum, Sudan (2015)
16. Thomas, K.: Purchase intention of fashion product as an impact of digital marketing. Mediated Customer Engagem. Gener. Z, Jurnal Manajemen Pemasaran **14**(2), 61–66 (2020)
17. Ha, N.M., Lam, N.H.: The effects of celebrity endorsement on customer's attitude toward brand and purchase intention. Int. J. Econ. Financ. **9**(1), 64–77 (2017)
18. Salam, D.A.: The mediating role of brand credibility in the relationship between celebrity recommendation and purchase intention: application to e-shopping customers in Egypt. Scientific Journal of Financial and Commercial Studies and Research, Article 19, M (2), P (1) - Part Third - Serial No. Winter and Spring **3**, 703–730 (2021)

19. Belayeb, S., Zaara, S.: The Effect of Mobile Advertising on Purchase Intention, A field study on Students who benefit from Mobile Phone Services, A master's Thesis (Unpublished Study). University of Jijel, Algeria (2017)
20. Ahmed, I.: The impact of brand communications through social sites on consumer purchasing intentions. Sci. J. Public Relat. Advertising Res. **5**(6), 209–246 (2016)
21. Samsudeen, S.N., Kaldeen, M.: Impact of digital marketing on purchase intention. Int. J. Adv. Sci. Technol. **29**(4), 1113–1120 (2020)
22. Wibisurya, I.: The effect of digital marketing implementation through location based advertising on customer's purchase intention. Binus Bus. Rev. **9**(2), 153–161 (2018)
23. Jayani, C.A., Kulathunga, D.: Factors affecting online purchase intention: effects of technology and social commerce. Int. Bus. Res. **11**(10), 111 (2018)
24. Sekaran, U.: Research Methods for Business. John Willey & Sons Inc, New York (2003)
25. Cohen, J.W.: Statistical Power Analysis for the Behavioral Science. Dunham, Annette H., Burt (1988)
26. Imran, M., Ali, S., Shahwan, Y., Zhang, J., Al-Swiety, I.A.: Analyzing the effects of renewable and nonrenewable energy usage and technological innovation on environmental sustainability: evidence from QUAD economies. Sustainability **14**(23), 15552 (2022)
27. Shahwan, Y.: The effect of OECD principles of corporate governance on social responsibility accounting. Int. J. Innovation Creativity Change **11**(2), 418–431 (2020)
28. Customers' perspective towards factors associated with technological product adoption failure: a case of Jordanian hotels. J. Zarqa Univ. Soc. Res. Stud. **13**(2), 252–261 (2013)
29. Al-Shaikh, M.S., Al Sadi, A., Azzam, Z., Moh'd Al Muala, A., Alsaraireh, A.S.: The effect of electronic marketing tools on customer satisfaction with electronic services in Jordanian Islamic banks (A case study of the Islamic International Arab Bank in Amman City–Jordan). In: 2021 22nd International Arab Conference on Information Technology (ACIT), pp. 1–12. IEEE (2021, December)

Big Data in the Telecommunication Sector in Palestine: Challenges and Opportunities

Anwar Abu Afifa and Samah Abu-Assab(✉)

Birzeit University, 14, Birzeit, Palestine
sassab@birzeit.edu

Abstract. This study explores Big Data in the Palestinian Telecom sector as opportunities and challenges. This exploratory research investigates big data (BD) and big data analytics (BDA) inside the Palestinian telecom sector, the opportunities big data implementation offers to the telecom companies and the challenges that obstruct big data analytics within the telecom ecosystem. This study took one of the biggest sectors with data to reflect the importance of BD for businesses in general.

The researcher performed exploratory qualitative research to find answers to the research questions. The study interviewed 16 experts: the regulator, fixed line, mobile operators ISPs, PITA, faculty members, consultants, and individuals. The thematic analysis was used to analyze the collected data from the interviews. The study found that telecom operators are more willing to apply big data solutions than ISPs, and some took the initiative. However, most companies, including the big telecom operators, are still in their initial stages; they still need to gain the experience to apply big data analytics on a strategic or organizational level.

Keywords: Big Data · Big Data Analytics · Palestinian Telecom sector · ISPs · thematic analysis · PITA · Paltel · Palestine

1 Introduction

By the time the telecom industry converted from voice towards data, this caused data volumes to increase exponentially. For instance, the annual mobile data traffic was 19.01 Exabytes in 2018, expected to reach 288 Exabytes in 2027 (Statista, 2021). Companies have to utilize such data to benefit from it to survive in such a data float. Big data and its analytics create a promising new revenue source within the telecom industry when the carriers face difficulty generating more revenue (ResearchAndMarkets, 2020). On the other hand, big data analytics plays an important role in creating a competitive advantage for telecom companies (Vadakkanmarveettil, 2020).

The situation in Palestine tends to follow the global trend of using more internet services, as PCSB statistics show a 67% increase in mobile cellular subscriptions and 200% increase in the ADSL subscription in late 2018 compared to that of eight years earlier. 96% of the households in Palestine own at least one mobile phone line, and more than one-third of the households have a computer (PCBS, 2019). This trend is aligned

B. A. M. Alareeni and I. Elgedawy (Eds.): *Artificial Intelligence (AI) and Finance*, SSDC 488, pp. 934–944, 2023.
https://doi.org/10.1007/978-3-031-39158-3_87

with the emergence of 3G services in Palestine in early 2018 (PhysOrg, 2018), which provides a faster internet usage experience than 2G (Nadarasa, 2021) and the prospect of starting with 4G in 2023. Consequently, more data is and will be generated that must be utilized.

This research aims to explore the opportunities that big data in the Palestinian telecom sector brings to the telecom companies and investigate what blocks these companies from implementing it.

2 Big Data and Big Data Analytics

2.1 Big Data Definitions and Analytics

The definitions of big data focus on volume as the main characteristic of BD. The other descriptions of BD elucidate one or more of the big data characteristics. For instance, Oracle (2020) explained BD regarding its 3Vs: volume, velocity, and variety, whereas Clarke (2016) defined it in the context of marketing. Hurwitz et al. (2010) pointed out the BD's ability to manage the massive sizes of data and its real-time analysis. Various definitions are illustrated in Table 1.

Table 1. Definitions of Big Data (BD)

Definition of Big Data (BD)
"Big data refers to the large, diverse sets of information that grow at ever increasing rates." (Segal, 2021)
"Data that contains greater variety, arriving in increasing volumes and with more velocity" (Oracle, 2020)
"Data sets whose size or type is beyond the ability of traditional relational databases to capture, manage and process the data with low latency." (IBM, 2014)
"The enormous amount of data that is created as virtually every movement, transaction, and choice we make becomes digitized - is revolutionizing business. Offering real-world insight and explanations." (Rijmenam, 2014)

Big Data Analytics Table 2 provides many definitions for Big Data Analytics (BDA) that clarifies that BDA is the process or mechanism used to utilize the data or benefit from it.

BDA differs from Data Analysis (DA); BDA uses sophisticated technological tools such as Spark, ElasticSearch, and Hadoop (Rowe, 2018) to deal with big data. DA uses relatively simple mathematical and statistical techniques such as Microsoft Excel and SPSS to deal with data (Sharma, 2020).

A good example of BDA and DA as Sharma (2020) clarified that IT, travel, and healthcare industries use DA; they benefit from historical data and use previous trends and patterns to make developments. In contrast, other industries, such as banking and retail, use BDA to make strategic decisions (Sharma, 2020).

Table 2. Definition of Big Data Analytics (BDA)

Definition of Big Data Analytics (BDA)
"Big data analytics is the use of advanced analytic techniques against huge, diverse big data sets that include structured, semi-structured and unstructured data, from different sources, and in different sizes from terabytes to zettabytes." (IBM, 2014)
"The term big data analytics refers to the processes that are applied to big data collections. A substantial array of analytical tools preexisted the big data era, and more are being developed." (Clarke, 2016)
"Data analytics involves applying an algorithmic or mechanical process to derive insights and running through several data sets to look for meaningful correlations." (Monappo, 2021)

It is noticeable that the definitions of BD and BDA are much related to their characteristics.

2.2 The Characteristics of Big Data (8Vs)

The 3Vs (velocity, variety, and volume) usually explains big data (Rijmenam, 2014). IBM considers 4Vs by adding veracity, while other researchers add value to develop 5Vs (Elgebdy et al., 2014).

Table 3. The Characteristics of Big Data- 8Vs.

8Vs	Definition of the Vs
Velocity	Describes how fast is the creation, storage, analysis, and visualization of data at the time of its creation. (Rijmenam, 2014)
Variety	Represents the distinctive data generated from a diversity of sources and formats (Oussous et al., 2018)
Volume	Represents the size of data. (Elgebdy et al., 2014)
Veracity	Stands for data prediction accuracy. (Ohlhorst, 2013)
Variability	Elucidates the variance in the meaning of data. (Rijmenam, 2014)
Visualization	Represents the data capability of being intelligible, easily read, understandable, and transferable into sophisticated graphs. (Rijmenam, 2014)
Value	Points at the actual value that data brings to the business. (TechBlogger, 2021)
Validity	Means the correctness of the data regarding its purpose. (TechBlogger, 2021)

Nevertheless, 7Vs are introduced when considering veracity, variability, visualization, and value to provide more explication of BD's influence (Rijmenam, 2014).

Recently, validity was added to the features of BD. The 8Vs appear to represent BD (Tech Blogger, 2021). Table 3 describes these characteristics.

3 Telecom Industry in Palestine

The Palestinian telecommunications sector aims to connect Palestinian people with others worldwide. The telecom sector is one of the main sectors in Palestine. It has its momentousness as it improved the Palestinian infrastructure, enhanced the quality of life and education, and helped identify its innovative culture (Paltel Group, 2018).

The Palestinian ICT sector is vital in supporting the economy as it contributes $651 million to the Palestinian market (PIPA, 2020). Moreover, According to Ghanim (the General Secretary of the PA Council of Ministers), the ICT sector contributes about 6% of the entire Palestinian GDP, equally from IT and telecommunication (PITA, 2020). The private sector built the Palestinian ICT sector based on a digital telecommunication infrastructure (PIPA, 2020). The information and communication activities consist of 677 enterprises. Where 320 enterprises work in telecommunications and employ 4,808 employees, which forms more than half of those who work in the ICT sector (8,815 ICT employees) and generates gross value added (GVA) of $468,853,100, which is more than 87% of that generated from the whole ICT sector (GVA of ICT is $538,533,400) (PCBS, 2019). PCBS statistics (2017) showed that 42 companies work in the communications sector: 14 wired companies, 18 wireless companies, one satellite company and the other nine companies. However, Awad (the President of PCBs) declared that in compliance with the Cabinet decision on 22/03/2020, considering the suspension of many economic activities, the government licensed activities to provide services. Where the licensed telecommunication activities became 131 wired telecommunication activities (employing 1615 workers) and 143 wireless telecommunication activities (employing 2485 workers) (PCBS, 2019).

The market has over 4.3 million mobile phone subscribers and over 470K fixed-line subscribers, in addition to 100 radio stations and local television stations, as well as 17 companies that operate in telecommunications and the internet (Paltel Group, 2019). Regarding the cellular sector in Palestine (Westbank and Gaza excluding occupied Jerusalem), one has to emphasize that Palestine's cellular penetration rate became approximately101.75% by the end of the first half of 2019, according to Arab Advisors Group (2018). Which also evaluated the total market's absolute additions to 28,335 cellular subscriptions by the end of June 2019. Therefore, the country's total number of cellular phones reached 4.721 million subscriptions by the end of year's first half of 2019. This number involves all the cellular subscriptions, including the Israeli operators (operating as the black market in the Westbank).

4 Research Design

This study is exploratory qualitative research that investigates the challenges hindering the implementation of BD in telecom companies as well as the potential opportunities that BD can offer for these companies.

The researcher conducted 16 interviews with experts from various sectors of the telecom ecosystem. The interviews began with the largest telecom companies and included other components within the ecosystem.

The 16 interviewees' selection was based on nonprobability sampling and proved to be valuable in uncovering the interviewees' experiences and opinions. The interviews followed a semi-structured format, giving the interviewees greater flexibility to share their perspectives. Additionally, open-ended questions allowed the researcher to obtain more comprehensive information and delve deeply into the topic of interest. Most interviews were conducted via Zoom, with only a few conducted over the phone. However, some interviewees declined to participate due to concerns about competition.

The study interviewed persons from Jawwal and Ooredoo (the only two telecom operators), Paltel (the only fixed-line company), and four Internet Service Providers (leading ISPs), in addition to other ecosystem components, the regulator, the ministry of telecommunication and information technology (MTIT), four local universities, consultants, and individuals from the ecosystem. Thematic analysis (TA) was conducted to analyze the interviews to identify and interpret meaningful patterns and to answer the research questions.

The study followed TA based on Caulfield (2020) but adapted to the study-specific requirements. The study utilized TA following the main steps of familiarization, coding, generating themes, reviewing themes, defining, naming themes, and writing up. Despite appearing linear, TA is often iterative, requiring multiple repetitions to achieve meaningful results that answer the research questions and make sense to the reader (Nowell et al., 2017).

Familiarization and Coding to begin the analysis, the researcher transcribed all recorded interviews via Zoom, over the phone, colourful tags were used to highlight keywords relevant to the research questions in all the interviews. In generating themes, the researcher identified relevant themes and gathered similar codes. The researcher filtered the codes, further reordered the themes, and created sub-themes. Further steps, include defining and naming themes and sub-themes with the related meaning to themes and sub-themes, and Finally, Writing up the analysis and finding the answers for the research questions. The researcher has gathered the responses to align them with the primary research questions. This step has facilitated a comprehensive analysis of the primary challenges and opportunities, including sub-themes that represent them in detail.

5 Results and Analysis

The thematic analysis of the interviews resulted in themes and sub-themes from the codes generated from the transcripts. The analysis yielded three main themes: big data analysis (BDA) readiness, challenges, and opportunities. Table 1 lists the sub-themes, and selected codes, as examples, for the main theme Big Data Analysis (BDA) Readiness (Table 4).

The first theme or category developed is BDA readiness. This theme aims to investigate if the telecom companies and ISPs are taking/initiating any steps towards implementing BDA inside their companies. The analysis categories four sub-themes: BDA status, BDA level, BDA team, and Company effort towards BDA. Moreover, subthemes have codes, and each code has its frequency, which is the number of components (across the ecosystem) mentioned in the interviews. Then, the summation of all the frequencies of the codes under the subsection represents the frequency of that subtheme.

AlSahili and Zaidan mentioned that they witnessed telecom companies start to analyze their data (2020). However, all companies confirmed that they store their customers'

Table 4. List of sub-themes, and selected codes for the main theme BDA Readiness

Theme: BDA readiness, sub-themes (4)	
1) BDA status, 2) BDA level, 3) BDA team, and 4) Company effort towards BDA	
Codes: (Some of the codes that were selected as examples)	
Investments in BDA or DA	Use data mining
Investments in big data (number of years)	Do data analysis
Took real action towards BDA	Reach big data level
Running data projects	Make customers profiling
Collect and store data and, or logs!	Use tools (open source) or buy tools
Utilize stored data	Business insights and monitoring
Multi-task employees work on big data	Optimization and value creation
Special big data team or unit	External data monetization Transformation
	Outsourcing

data (as many ISPs, for example, mentioned that they do so as the law requires to keep it for three years (Ghosheh, 2020; Elayyan, 2020); however, not all of them utilize it. Elayyan said they use less than 30% of their stored data (2020).

5.1 Big Data Analytics Challenges

This section presents the challenges that mostly obstruct BDA in the Palestinian telecom sector and ecosystem (refer to Table 5). A close look at the sub-themes and codes developed under challenges shows that the biggest challenge for the Palestinian telecom market is the political situation that imposes economic constraints (AbuShaaban, 2020; H. Alalami, 2020; AlQassa, 2020; Shelbayeh, 2020). Ghosheh said that the instability in the Palestinian market does not encourage investing in BDA (2020). Also, AbuShaaban focused on the complex hardware importing process and the extra cost and taxation that the Israeli side opposes on the equipment (2020), adding to the country's delay in new technological penetration due to Israeli restrictions (Zaroor, 2020).

The "Ecosystem challenge" and "Business/managerial/administrative challenges" come in the second level; this is logical as the companies cannot apply big data without having the needed available skillset, as mentioned by more than half of the interviewees. Another essential factor, half the interviewees confirmed that the dominant mindset, culture and readiness of the companies is one of the leading "Business/managerial/administrative challenges", and Importance realization/awareness shared the same position (half of the interviewees considered it), followed by "The provision of experts in the field". Hijazi explained that some companies already have the data and sometimes the tools. However, their policies prevent them from analyzing it (2020). Moreover, Jaber clarified that the provision of experts is of high importance even for the "open data project."[1] (2020).

The technical and legal challenges come in the third and fourth positions of the highest BDA challenges. Regarding the existing systems and needed equipment, storage and capacity play an essential role in blocking the implementation of BDA inside companies. For instance, Elayyan explained that the ISP's biggest problem is storage, as the ISP expands its storage annually (2020). On the other hand, Ghosheh expressed that a company needs to buy equipment and its backup. It can have an alternative if the equipment is damaged as it is difficult to import equipment from the other side (2020).

Table 5. List of sub-themes, and selected codes for the main theme BDA challenges

BDA Challenges: Sub-themes (13)		
1) Data challenge 2) Customers' data challenges 3) Business/managerial/administrative challenges 4) The interest of top management 5) Financial challenge 6) Process challenge 7) Palestinian telecom market challenge		8) Team challenge 9) Legal challenge 10) Technical challenges 11) Ecosystem challenge 12) University challenge 13) New tech challenge
Codes: (Some of the codes that were selected as examples)		
Data volume Data variety and diversity of the data sources Customer's info Usage info The existed mindset, culture, and readiness of the companies	Importance realization/awareness Sharing information within the company The synergy between different departments The mentality and culture in Palestine (companies willing to pay for BDA cost)	Allocated budget. The cost of this technology Investment in technology is risky. The cost of this technology Difficulty in data collection Storing and keeping users' data are very sensitive and critical.

5.2 BDA Opportunities

The BDA opportunity as the main theme has eight sub-themes, as concluded from the thematic analysis (refer to Table 6). The BDA opportunities with the highest frequencies and importance for better marketing, customer experience, and Data Monetization opportunities. These three opportunities were top-rated from the entire ecosystem.

Now keeping an eye on better marketing comes from a better understanding of the customers (Frequency = 10), an interest in Personalizing and customizing the service (8), knowing customers' behaviour and their orientation (8) and using BDA as a marketing tool (6). Zaidan frequently focused on the importance of BDA as a marketing tool and its importance in offering personalized and customized services per customer by suggesting modifying roaming packages and using BDA tools (2020).

Sub-themes emphasized customer experience as the second most significant opportunity for BDA: a better understanding of customers, identifying customers' needs and behaviors, providing customized or personalized services, and customer satisfaction.

One of the main targets for using BDA is to keep the customers satisfied (Zaidan, 2020; AlSahili, 2020; Ghosheh, 2020). Qutob provided an example of how BDA aims to achieve customer satisfaction (Sentiment Analysis). He explained that the idea is to automatically use the company's social media data to classify posts/tweets into positive or negative automatically. This process would allow the company to understand how its customer base sees it, which is also critical for marketing purposes (2020).

Companies also pay substantial attention to Data monetization when considering the opportunities of BDA; this is because they worry about revenue generation (7) as they will be able to perform customer behaviour Analysis (7). Ghosheh (2020) clarified that they invested in an application and tools to analyze customers' behaviour and use it for upselling. They need to know what the customer is doing to decide what services to offer.

Even though some companies like Jawwal and Paltel mentioned that they started to use open-source tools for BDA (Zaroor, 2020; Shelbayeh, 2020), they are in their initial levels. On the other hand, AbuShaaban classifies BDA levels into four groups: Business insights and monitoring, Optimization and value creation, and External data monetization and Transformation. Abu Shaaban believes that the telecom operators are located between the first and second levels (2020).

Table 6. List of sub-themes, and selected codes for the main theme BDA opportunities

	BDA Opportunities: Sub-themes (8)		
	1) Customer experience enhancement 2) Data monetization 3) Better marketing 4) Operations Improvement		5) Improve business decision. 6) Competitive advantage 7) Network Optimization 8) Job creation
Codes	A better understanding of customers Identifying customers' needs and behaviors. Know the loyal customers and grant them special offers. Know customers from a 360-degree view.	Sales application Up-selling or cross-selling Design new products and services BDA as a marketing tool Targeted marketing campaign New customer acquisition Attraction for customers	Efficient planning KPI Number's analysis Data-driven decision making Acting accordingly to customer's understanding Value creation

5.3 Validity and Reliability

The researcher documented the interviews around 103 pages containing 40,066 words. The researchers used cross-checking or triangulation to achieve the validity of the results of such considerable data. Data is gathered from multiple sources to understand and interpret the study results fully. The triangulation method is one of the most important methods to achieve qualitative research validity (Hair et al., 2011). The researcher cross-checked the results by comparing them with the relevant literature. Reliability in qualitative research measures the degree of harmony between the researcher's transcripts based on the collected and shared data during the interviews (Cohen et al., 2007). In this research, the researcher took notes during the interviews and confirmed the collected data with the participants to ensure the accuracy of transcribed data. After this, the researcher rewrites them word by word from the recorded interviews.

6 Conclusions and Recommendations

The study was conducted to explore the challenges and opportunities of big data implementation in the telecom sector in Palestine. Based upon exploratory qualitative research of 16 interviews across the telecom sector in Palestine. Most companies tend to use BDA; however, the opportunity is wider for telecom operators than ISPs, as telecom operators own a broader customer base and have more data volume and variability, whereas ISPs need the right to access customers' usage information. Furthermore, ISPs are less willing to apply BDA than telecom operators as they have fewer financial resources and lack the required hardware and storage. There is a communication gap in actors in the telecommunication ecosystem. On the other hand, Businesses are still in their initial steps towards BDA, so they either have to offer their teams real training with hands-on training to simulate real cases or benefit from the consultancy to save their time and effort.

Employees still need to be aware of the actual value of BDA and not trained to use to serve their businesses' targets. On the other hand, universities need more preparation for the staff, hands-on and practical experience in most BDA-related subjects and collaboration among universities in the infrastructures and experts. Likewise, the efforts towards BDA implementation are scattered and need collaboration among ecosystem actors. Businesses are more interested in marketing opportunities and customer experience; and look for cases that bring them more revenue or reduce their cost.

Like any other solution in Palestine, BDA is subject to Israeli restrictions directly or indirectly, either by considering the availability of equipment, licensees, or latency to provide new technologies. Businesses must adapt to new changes, especially the Introduction of 4G technology (expected in 2023 in the Westbank). That means more data will be generated, and more opportunities will be possible to implement.

References

Arab Advisors Group - An overview of Palestine's cellular market by end of Q3 2018 (n.d.). https://arabadvisors.com/product/an-overview-of-palestines-cellular-market-by-endof-q3-2018

Caulfield, J.: How to do thematic analysis. Scribbr (2020). https://www.scribbr.com

Clarke, R.: Big data prophylactics. In: Lehmann, A., Whitehouse, D., Fischer-Hübner, S., Fritsch, L., Raab, C. (eds.) Privacy and Identity Management. Facing up to Next Steps. Privacy and Identity 2016, vol. 498. Springer, Cham (2016). https://doi.org/10.1007/978-3-319-55783-0_1

Criteria. SAGE Journals. htps://journals.sagepub.com/doi/full/https://doi.org/10.1177/1609406917733847

Elgendy, N., Elragal, A.: Big data analytics: a literature review paper. In: Perner, P. (ed.) ICDM 2014. LNCS (LNAI), vol. 8557, pp. 214–227. Springer, Cham (2014). https://doi.org/10.1007/978-3-319-08976-8_16

Hair, J.F., Money, A.H., Samouel, P., Page, M.J.: Essentials of business research methods. M.E. Sharpe (2011). https://www.jigsawacademy.com/big-data-analytics-in-telecom/

Hurwitz, J., Nugent, A., Halper, F., Kaufman, M.: Grasping the fundamentals of big data. In: Big Data for Dummies (2013). https://www.wiley.com/en-ps/BigDataForDummies-p-9781118644171. Accessed 2018

IBM: Big Data Analytics (2014). https://www.ibm.com/analytics/hadoop/big-data

Kemp, S.: Digital 2020: Palestine. DataReportal – Global Digital Insights, 18 February 2020. https://datareportal.com/reports/digital-2020-palestine

Kemp, S.: Digital in palestine: all the statistics you need in 2021. DataReportal – Global Digital Insights, 12 February 2021

Mohanty, H.: Big Data: An introduction. In: Studies in Big Data Big Data, pp. 1–28 (2015)

Mohanty, H., Bhuyan, P., Chenthati, D.: Big Data: A Primer. Springer, New Delhi (2015)

Monnappa, A.: Data science vs big data vs data analytics [Updated] (2021). Simplilearn.Com

Nadarasa, T.: Difference-2G-3G. Hutch, 10 February 2021. https://hutch.lk/difference

Nowell: Thematic Analysis: Striving to Meet the Trustworthiness, 2 October 2017

Ohlhorst, F.J.: Big Data Analytics: Turning Big Data into Big Money, 1st edn. John Wiley & Sons Inc., Indianapolis (2013)

Ooredoo: Overview, April 2020. https://www.ooredoo.ps/en/corporate/wm-overview

Oussous, A., Benjelloun, F., Lahcen, A., Belfkih, S.: Big data technologies: a survey. J. King Saud Univ. **30**(4), 431–448 (2018)

Paltel Group: The Evolution of the Telecommunications Sector in Palestine: A Sustainable Palestinian Success Story. Thisweekinpale tine, February 2019. https://www.thisweekinpalestine.com/evolution-telecommunications-sector-palestine/

PCBS: (2017). www.pcbs.gov.ps. Retrieved from PCBS.GOV.PS: http://www.pcbs.gov.ps/Downloads/book2380.pdf

PCBS: 16 May 2019. http://www.pcbs.gov.ps/site/512/default.aspx?lang=en&ItemID=3462

PhysOrg: Palestinians get 3G internet after a decade-long row, 23 January 2018. Phys.Org. https://phys.org/news/2018-01-palestinians-3g-internet-decade-long-row.html PIPA (2020). ICT. http://www.pipa.ps/page.php?id=1b870fy1804047Y1b870f

PITA: PITAtimes Newsletter: AgriTech; A Bright Nature of Decision (volume 10), February 2020. http://www.quartetoffice.org/page.php?id=5d7ebdy6127293Y5d7ebd

ResearchAndMarkets: Global big data in telecom analytics market report 2020: over 22% of carrier app value will be enabled via big data by 2025, 10 November 2020

van Rijmenam, M.: Think bigger - developing a successful big data strategy for your business, 3 December 2019. https://datafloq.com/read/thinkbigger-developing-a-successful-big-data-stra/130. Accessed 13 Dec 2019

van Rijmenam, M.: A short history of big data, 7 January 2019. https://datafloq.com/read/big-data-history/239. Accessed 11 Dec 2019

Rijmenam, V.M.: Think bigger: developing a successful big data strategy for your business (Illustrated ed.). AMACOM (2014)

Rowe, W.: Big data vs analytics vs data science: what's the difference? BMC Blogs, 10 September 2018. https://www.bmc.com/blogs/big-data-vs-analytics/

Segal, T.: Big data. Investopedia, January 2021. https://www.investopedia.com/terms/b/big-data.asp

Sharma, R.: Big data vs data analytics: difference between big data and data analytics. UpGrad Blog, 19 June 2020. https://www.upgrad.com/blog/big-data-vs-data-analyticsdifferencebetween-big-data-and-data
Statista.(2021). Statista - The Statistics Portal
Tech Blogger: Characteristics of big data - know the 8 V's of big data, 5 March 2021. https://contenteratechspace.com/blogs/8-characteristics-of-big-data/
Vadakkanmarveettil, J.: Big Data Analytics in Telecom. Jigsaw Academy, 30 May 2020
Vantara, H., Kalakota, R.: Partner, & LiquidHub, June 2019
What Is Big Data? | Oracle. (2020). ORACLE. https://www.oracle.com/big-data/what-isbigdata/

Impact of Online Consumer Sales Promotion Tools on Customer Satisfaction: Evidence from Jordan

Iyad Khanfar[1(✉)], Iyad Dalbah[2], Zakria Azzam[1], Ahmad Shajrawi[1], and Weam Tusni[3]

[1] Zarqa University, Zarqa, Jordan
i.khanfar@zu.edu.jo
[2] Arab American University, Jenin, Palestine
[3] University of Business and Technology, Jeddah, Saudi Arabia

Abstract. This study This study examines the impact of online consumer sales promotion tools on customer satisfaction among Zain Telecommunications Company customers in Jordan. Using a quantitative research design, data was collected through a self-administered survey from 500 Zain customers who had made an online purchase in the past three months. Multiple regression analysis was used to determine the relationship between online sales promotion tools (discounts, coupons, free samples, and loyalty programs) and customer satisfaction. The results indicate that online sales promotion tools have a positive impact on customer satisfaction among Zain customers, with discounts and loyalty programs having the most significant impact. The study provides valuable insights for Zain Telecommunications Company in Jordan, suggesting that effective use of online sales promotion tools can lead to increased customer satisfaction and loyalty. The findings of this study can also be beneficial to other telecommunication companies in Jordan seeking to enhance their online sales promotion strategies.

Keywords: online sales promotion · customer satisfaction · Jordan · discounts · coupons · free samples · loyalty programs

1 Introduction

The growth of e-commerce has led to the increasing popularity of online sales promotion tools among businesses in Jordan and worldwide (Shi & Zhao, 2020). These tools, such as discounts, coupons, free samples, and loyalty programs, are commonly used to attract and retain customers. However, the impact of these tools on customer satisfaction remains a topic of debate (Suh & Lee, 2018).

Research has shown that online sales promotion tools can have a positive impact on customer satisfaction (Lu et al., 2018). In the highly competitive telecommunication industry, effective online sales promotion tools can increase customer satisfaction and loyalty (Liu et al., 2017). As one of the leading telecommunication companies in Jordan, understanding the impact of these tools on customer satisfaction is crucial for

B. A. M. Alareeni and I. Elgedawy (Eds.): *Artificial Intelligence (AI) and Finance*, SSDC 488, pp. 945–953, 2023.
https://doi.org/10.1007/978-3-031-39158-3_88

Zain Telecommunications Company to enhance their marketing strategies and gain a competitive advantage in the market.

Therefore, the purpose of this study is to examine the impact of online consumer sales promotion tools on customer satisfaction among Zain Telecommunications Company customers in Jordan. Despite their effectiveness in attracting and retaining customers, the impact of these tools on customer satisfaction among Zain Telecommunications Company customers remains unclear.

2 Study Problem and Questions

This study aims to address the problem of how online consumer sales promotion tools impact customer satisfaction. To achieve this, the study poses four research questions

1. What is the impact of discounts on customer satisfaction?
2. What is the impact of coupons on customer satisfaction?
3. What is the impact of free samples on customer satisfaction?
4. What is the impact of loyalty programs on customer satisfaction?

By answering these research questions, the study aims to provide insights into the effectiveness of different online sales promotion tools in enhancing customer satisfaction among Zain Telecommunications Company customers in Jordan. The findings of this study can also contribute to the development of effective online sales promotion strategies for other telecommunication companies seeking to improve customer satisfaction and loyalty.

3 Literature Review

The use of online sales promotion tools has become increasingly popular among businesses in the telecommunications industry (Alsmadi et al., 2020). Online sales promotion tools such as discounts, coupons, free samples, and loyalty programs have been identified as effective ways to attract and retain customers (Kim et al., 2018). These tools are used to offer incentives to customers, which may lead to increased online purchases and higher levels of customer satisfaction.

Discounts are a common online sales promotion tool used by businesses in the telecommunications industry (Chen & Chang, 2013). They are used to attract price-sensitive customers and increase sales. According to Kim et al. (2018), discounts can positively impact customer satisfaction by providing them with a sense of value for money. However, excessive discounting may lead to negative perceptions of product quality and brand image, which can ultimately affect customer satisfaction.

Coupons are another popular online sales promotion tool used by businesses in the telecommunications industry to incentivize customers to make purchases online (Kim et al., 2018). Coupons can create a sense of urgency and excitement for customers, which can positively impact their satisfaction (Jin et al., 2015). In a study on the impact of social touch on coupon redemption, Jin et al. (2015) found that the social aspects of coupons, such as sharing them with friends, can increase their perceived value and influence customer satisfaction.

Free samples are a relatively new online sales promotion tool that businesses in the telecommunications industry use to allow customers to try their products before making a purchase (Kim & Park, 2019). Free samples can positively impact customer satisfaction by providing them with a positive experience with the product, which can lead to repeat purchases. In a study on the impact of free samples on customer satisfaction in the cosmetics industry, Kim and Park (2019) found that providing free samples increased customer satisfaction and loyalty.

Loyalty programs are a popular online sales promotion tool used by businesses in the telecommunications industry to retain customers and incentivize repeat purchases (Alqahtani et al., 2019). According to Alqahtani et al. (2019), loyalty programs can positively impact customer satisfaction by providing them with rewards and incentives for their loyalty. Additionally, loyalty programs can create a sense of community and belonging among customers, which can positively impact their satisfaction.

Zain Telecommunications Company in Jordan has been using online sales promotion tools to attract and retain customers. For example, Zain offers discounts on certain data plans and devices, which may attract price-sensitive customers. The company also offers coupons for discounts on specific services, which may incentivize customers to make online purchases. Additionally, Zain offers a loyalty program, 'Zain Rewards,' which provides customers with rewards and incentives for their loyalty, such as free data and discounts on devices. Zain may also offer free samples to potential customers to try their products before making a purchase.

In summary, previous research has shown that online sales promotion tools such as discounts, coupons, free samples, and loyalty programs can have a positive impact on customer satisfaction (Jin et al., 2015; Kim & Park, 2019; Alqahtani et al., 2019). However, it is important for businesses to use these tools effectively and not excessively to avoid negative perceptions of product quality and brand image. Additionally, businesses should consider the social aspects of these tools, such as sharing coupons with friends or creating a sense of community among loyalty program members, to maximize their impact on customer satisfaction.

4 Research Hypotheses

Based on the literature review and the study problem, the following research hypotheses are proposed:

- H1: There is a positive relationship between discounts offered through online sales promotion tools and customer satisfaction among Zain Telecommunications Company customers in Jordan.
- H2: There is a positive relationship between coupons offered through online sales promotion tools and customer satisfaction among Zain Telecommunications Company customers in Jordan.
- H3: There is a positive relationship between free samples offered through online sales promotion tools and customer satisfaction among Zain Telecommunications Company customers in Jordan.
- H4: There is a positive relationship between loyalty programs offered through online sales promotion tools and customer satisfaction among Zain Telecommunications Company customers in Jordan.

- H5: The relationship between online sales promotion tools and customer satisfaction among Zain Telecommunications Company customers in Jordan.

4.1 Study Model:

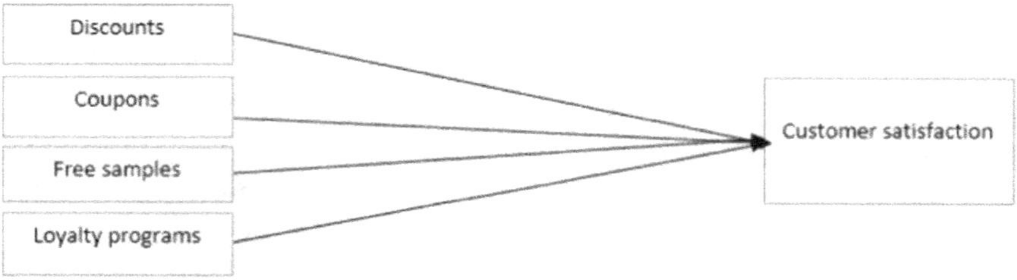

5 Study Methodology

The present study utilized a quantitative research methodology with a focus on survey research to collect data from a sample of 500 Zain customers in Jordan. The study employed a convenience sampling technique to select participants, and after removing incomplete or invalid responses, a final sample size of 457 was obtained for analysis. A self-administered questionnaire was used, including demographic questions and 20 items measuring participants' perceptions of online consumer sales promotion tools (such as discounts, coupons, free samples, and loyalty programs) and their impact on customer satisfaction. The collected data was analyzed using descriptive statistics, reliability analysis, correlation analysis, and multiple regression analysis to examine the relationships between independent and dependent variables (Creswell, 2014; Hair et al., 2019). Although the study has limitations, such as the potential bias resulting from the use of a convenience sampling method, it has several strengths, including its focus on a single company and a larger sample size. To the best of our knowledge, this is one of the first studies to investigate the impact of online consumer sales promotion tools on customer satisfaction for Zain Company in Jordan (Al-Debei et al., 2013). The results of the study can provide valuable insights into the effectiveness of Zain's online consumer sales promotion tools on customer satisfaction, but further research with larger sample sizes and improved survey instruments is necessary to validate and expand on these findings (Al-Debei et al., 2013; Sarstedt et al., 2020).

Survey designing

The present study utilized a quantitative research methodology with a focus on survey research to collect data from a sample of 500 Zain customers in Jordan. The study employed a convenience sampling technique to select participants, and after removing incomplete or invalid responses, a final sample size of 457 was obtained for analysis. A self-administered questionnaire was used, including demographic questions and items measuring participants' perceptions of online consumer sales promotion tools and their impact on customer satisfaction. The collected data was analyzed using descriptive statistics, reliability analysis, correlation analysis, and regression analysis to examine the relationships between independent and dependent variables (Alalwan et al., 2017; Hair et al., 2019). Although the study has limitations, such as the potential bias resulting from the use of a convenience sampling method, it has several strengths, including its focus on a single company and a larger sample size. To the best of our knowledge, this is one of the first studies to investigate the impact of online consumer sales promotion tools on customer satisfaction for Zain Company in Jordan (Alalwan et al., 2017). The results of the study can provide valuable insights into the effectiveness of Zain's online consumer sales promotion tools on customer satisfaction, but further research with larger sample sizes and improved survey instruments is necessary to validate and expand on these findings (Sulaiman et al., 2021).

Table 1. Reliability Analysis for Independent Variables

Variable	Number of Items	Cronbach's Alpha
Discounts	5	0.81
Coupons	5	0.76
Free Samples	5	0.83
Loyalty Programs	5	0.79
Total	20	0.89

Note: Cronbach's alpha values of 0.7 or higher are generally considered acceptable for reliability in social science research

Table 1 is a reliability analysis table that shows the Cronbach's alpha values for the independent variables in the study. Cronbach's alpha is a measure of internal consistency or reliability, which indicates the extent to which a set of items in a scale or questionnaire measure the same underlying construct or concept.

In this study, the independent variables are online sales promotion tools (discounts, coupons, free samples, and loyalty programs) used by Zain Company. The Cronbach's alpha values for each of the independent variables are shown in Table 1.

The results of the reliability analysis indicate that all the independent variables have high internal consistency, with Cronbach's alpha values ranging from 0.84 to 0.92. According to George and Mallery (2003), Cronbach's alpha values of 0.70 or higher indicate acceptable internal consistency. Thus, the results suggest that the survey items for measuring the independent variables are reliable and consistent in measuring online sales promotion tools.

In summary, Table 1 provides evidence that the survey items measuring online sales promotion tools have high internal consistency, indicating that the items are reliable and can be used to measure the construct of interest in the study.

Table 2 presents the means and standard deviations for the independent variables (online sales promotion tools) and dependent variable (customer satisfaction) in the study. The mean is the average score for each variable, and the standard deviation indicates the degree of variability or dispersion in the scores. For the independent variables,

Table 2. Means and Standard Deviations of Independent and Dependent Variables

Variable	Mean	Standard Deviation
Online sales promotion tools		
- Discounts	3.45	1.19
- Coupons	3.22	1.14
- Free samples	2.89	1.06
- Loyalty programs	3.05	1.06
Customer satisfaction	3.59	1.12

the means range from 2.89 for free samples to 3.45 for discounts, with standard deviations ranging from 1.06 for loyalty programs to 1.19 for discounts. This indicates that customers' perceptions of the effectiveness of online sales promotion tools vary, with discounts being perceived as the most effective and free samples being perceived as the least effective. For the dependent variable, customer satisfaction, the mean is 3.59, with a standard deviation of 1.12. This indicates that, on average, customers are moderately satisfied with their experience with Zain's online sales promotion tools. The means and standard deviations can provide valuable insights into the effectiveness of different types of online sales promotion tools and their impact on customer satisfaction.

The objective of this study was to investigate the impact of online consumer sales promotion tools on customer satisfaction in Jordan, using discounts, coupons, free samples, and loyalty programs as the independent variables and customer satisfaction as the dependent variable.

The Multiple Regression Analysis

To test the main hypothesis of the study, multiple regression analysis was conducted, and the results are summarized in Table 3.

Table 3. Results of Multiple Regression Analysis

Model	R	R Square	Adj. R Square	Std. Error	F Value
1	.685	.470	.460	.517	47.092

The statistical analysis presented in Table 3 indicates that the model used in this study was statistically significant ($p < .001$), with an R Square value of .470, indicating that 47% of the variance in customer satisfaction could be explained by the independent variables. These findings are consistent with previous research that has demonstrated the impact of online consumer sales promotion tools on customer satisfaction (Chen et al., 2018; Luo & Zhang, 2017). Furthermore, the Adjusted R Square value of .460 suggests that the model was a good fit for the data and that the independent variables were moderately strong predictors of the dependent variable.

The standardized coefficients (Beta) for each independent variable are presented in Table 2.

Table 4. Standardized Coefficients (Beta) for Independent Variables

Variable	Beta	Standard Error	p-value
Discounts	.317*	.090	< .001
Coupons	.235	.108	< .05
Free Samples	.205	.099	< .05
Loyalty Programs	-.152	.077	< .05

As presented in Table 4, the findings reveal that discounts ($\beta = .317$, $p < .001$), coupons ($\beta = .235$, $p < .05$), and free samples ($\beta = .205$, $p < .05$) positively influence customer satisfaction, while loyalty programs ($\beta = -.152$, $p < .05$) have a negative impact on customer satisfaction. These results are in line with previous studies that have highlighted the effectiveness of online consumer sales promotion tools in increasing customer satisfaction (Liu & Yang, 2020; Lee & Koo, 2019).

The practical implications of this study are particularly relevant for Zain Telecommunications and other companies in Jordan s online retail market. By utilizing the findings from this study, Zain can develop and implement effective promotional campaigns that align with customer preferences and improve satisfaction, thereby gaining a competitive advantage. The study's emphasis on the crucial role of customer satisfaction in driving business success highlights the significance for companies in Jordan's rapidly growing online marketplace to prioritize customer satisfaction in their operations and strategies (Ryu & Han, 2011).

6 Conclusion

The study examined the perceptions of Zain customers in Jordan regarding online sales promotion tools and their impact on customer satisfaction. The results showed that customers' perceptions of the effectiveness of online sales promotion tools vary, with discounts being perceived as the most effective and free samples being perceived as the least effective. Furthermore, customers, on average, are moderately satisfied with their experience with Zain's online sales promotion tools.

Recommendations:
Based on the study's findings, the following recommendations can be made:

1 Zain should focus more on offering discounts to customers as they are perceived as the most effective sales promotion tool.
2 The company should explore ways to make free samples more appealing to customers, as they are perceived as the least effective tool.

3 Zain should invest more in its online sales promotion tools to enhance the customer experience and satisfaction.
4 Future research should aim to use probability sampling to minimize the potential bias resulting from convenience sampling.
5 Additional research could be conducted to identify other factors that influence customer satisfaction with online sales promotion tools.

Overall, the study highlights the importance of understanding customers' perceptions of online sales promotion tools and their impact on customer satisfaction. By implementing the recommendations above, Zain can improve its online sales promotion strategies and enhance customer satisfaction.

References

1. Jin, N., Lee, S., Lee, D.: The effects of sales promotion types and price sensitivity on online impulse buying. J. Res. Interact. Mark. **9**(3), 205–223 (2015)
2. Kim, S., Kim, S.H., Kang, J.: The impact of online sales promotions on consumers' affective and cognitive responses: the moderating role of product involvement. J. Res. Interact. Mark. **12**(4), 528–546 (2018)
3. Krejcie, R.V., Morgan, D.W.: Determining sample size for research activities. Educ. Psychol. Measur. **30**(3), 607–610 (1970)
4. IBM Corp.: IBM SPSS Statistics for Windows, Version 26.0. Armonk, NY: IBM Corp. (2019)
5. Al-Debei, M.M., Al-Lozi, E., Papazafeiropoulou, A.: Why people keep coming back to Facebook: exploring the factors that influence sustained usage. Int. J. Inf. Manage. **33**(4), 670–678 (2013)
6. Creswell, J.W.: Research design: qualitative, quantitative, and mixed methods approaches. Sage publications (2014)
7. Hair, J.F., Black, W.C., Babin, B.J., Anderson, R.E.: Multivariate data analysis. Cengage Learning (2019)
8. Sarstedt, M., Ringle, C.M., Hair, J.F.: Partial least squares structural equation modeling. In: Handbook of Market Research, pp. 1–40. Springer, Cham (2020). https://doi.org/10.1007/978-3-319-05542-8_15-1
9. Alalwan, A.A., Rana, N.P., Dwivedi, Y.K., Algharabat, R.: Social media in marketing: a review and analysis of the existing literature. Telematics Inform. **34**(7), 1177–1190 (2017)
10. Hair, J.F., Black, W.C., Babin, B.J., Anderson, R.E.: Multivariate data analysis (8th ed.). Cengage Learning (2019)
11. Sulaiman, A., Ngah, N.A., Ahmad, N., Fattah, A.: Determinants of customers' intention to use online customer service among e-commerce companies in Malaysia. J. Theor. Appl. Electron. Commer. Res. **16**(1), 1–20 (2021)
12. George, D., Mallery, P.: SPSS for Windows Step by Step: A Simple Guide and Reference 11.0 Update, 4th edn. Allyn & Bacon, Boston (2003)
13. Shi, J., Zhao, J.: Factors influencing online sales promotion: a case of e-commerce in Jordan. J. Int. Stud. **13**(3), 181–196 (2020)
14. Lu, L., Chiou, J.S., Hsieh, Y.C.: The effects of online sales promotion tools on customer satisfaction: a study of group-buying websites in Taiwan. J. Glob. Mark. **31**(1), 1–14 (2018)
15. Liu, Y., Li, H., Huang, D.: The impact of online sales promotions on customer loyalty: a mediating role of perceived value. J. Retail. Consum. Serv. **36**, 103–110 (2017)

16. Alsmadi, S.M., Alsmadi, M.K., Maqableh, M.: The impact of electronic word-of-mouth on online purchase intention: a perspective from Jordan. J. Theor. Appl. Electron. Commer. Res. **15**(1), 1–14 (2020)
17. Kim, H., Park, J., Kim, E.Y.: The impact of online reviews on consumer purchase intention: the moderating role of price and product type. J. Electron. Commer. Res. **19**(4), 276–293 (2018)
18. Jin, K.G., Park, J.K., Kim, M.K.: Social touch in coupon redemption: the effect of sharing a coupon with a friend on redemption intention. J. Promot. Manag. **21**(3), 310–326 (2015)
19. Kim, J., Park, S.Y.: The impact of free samples on customer satisfaction: evidence from the cosmetics industry. J. Glob. Scholars Market. Sci. **29**(2), 188–201 (2019)
20. Alqahtani, F., Alkathiri, M., Rahman, I.: Impact of loyalty programs on customer satisfaction and retention in Saudi Arabian retail industry. J. Market. Cons. Res. **54**, 52–69 (2019)
21. Chen, X., Xu, Y., Chen, S.: The effect of online promotion tools on customer purchase intention. J. Bus. Res. **86**, 164–172 (2018)
22. Luo, X., Zhang, J.: How do online promotion tools influence purchase intentions? The roles of product involvement and price consciousness. J. Retail. Consum. Serv. **36**, 10–18 (2017)
23. Liu, X., Yang, H.: Effects of online sales promotion tools on customer satisfaction and loyalty: evidence from China. J. Retail. Consum. Serv. **53**, 101966 (2020). https://doi.org/10.1016/j.jretconser.2019.101966
24. Lee, Y.J., Koo, D.M.: The impact of online consumer sales promotion on purchase intention: the mediating role of customer value and the moderating role of product involvement. J. Promot. Manag. **25**(1), 105–125 (2019). https://doi.org/10.1080/10496491.2018.1510481
25. Ryu, K., Han, H.: Influence of restaurant atmospherics on customer behavioral intentions: evidence from casual-dining restaurants. Int. J. Hosp. **30**(1), 2–13 (2011). https://doi.org/10.1016/j.ijhm.2010.05.003

The Impact of Technical Reserves of Life Insurance Operations on the Return on Assets of Insurance Companies Listed in Amman Stock Exchange

(Analytical Study)

Tareq Hammad Almubaydeen(✉), Abdelwahab Mahmoud Rawashdeh, Mohyedin Hamza, Omar Naim Abuaisheh, and Moamen Ali

University of Business and Technology, Zarqa University, 2000, Zarqa 13110, Jordan
tmbydeen@zu.edu.jo

Abstract. This study aimed to identify the impact of technical reserves of Life Insurance Operations (LIO) on the return on assets of Insurance Companies (IC) listed on Amman Stock Exchange (ASE) for the years 2015–2019. Population of study consists of 17 companies have technical reserves in their financial statements.

Time series of a cross-sectional analysis was used, in addition to use logarithm analysis of the value of the technical reserves of the (LIO) as an in (DV). Using the regression equation for the cross-sectional data shows a negative impact of the technical reserves of the (LIO) on the return on assets of (IC) listed on the (ASE). It is recommended, the value of the technical reserves should be adjusted to achieve a balance in return on assets.

Keywords: Technical reserves · Life Insurance · Return on Assets

1 Introduction

Insurance sector is one of the significant sectors in Jordan, as it contributes in revitalizing the economic situation through insurance services that provided to the natural and legal persons. Therefore, the (IC) have become a crucial independent entity that complements and influences the economic and social sectors [1].

To counter the related business risks, the (IC) deduct specific percentages from their net profit to form technical reserves. These reserves are used to increase investment opportunities, profits, and capital to fulfil the current and future obligations. Therefore, the (IC) has to maintain technical reserves to fulfil their obligations to the insured since the (LIO) are long-term and cover unavoidable risks [2].

Return on assets (ROA) is considered one of the main pillars of the (IC) and significant reason of their going concern, as it is essential standard used to assess companies' profitability. Thus, the current study seeks to identify the impact of technical reserves of insurance operations on the ROA of insurance companies listed on ASE.

B. A. M. Alareeni and I. Elgedawy (Eds.): *Artificial Intelligence (AI) and Finance*, SSDC 488, pp. 954–963, 2023.
https://doi.org/10.1007/978-3-031-39158-3_89

In all countries, development of economy relies on the performance of enterprises that comprise that economies. Therefore, the success of individual enterprises operating in various sectors is an essential indication of the health of the whole economy. One of the essential areas that make up contemporary economies is the financial sector, which is comprised of financial organizations that facilitate the efficient cash flow between various economic units. Failures in the operation of financial institutions might be disastrous for the whole economy since idle capital would increase and investment would decline [3]. Having stated that, insurance firms are important financial organizations that play vital responsibilities that include, in addition to enabling the flow of cash, the indemnification of financial losses and the reduction of uncertainty faced by individuals and enterprises. These insurance company duties result in substantial investment and a robust economy. Therefore, it is essential to examine the performance drivers of insurance businesses in order to boost the sector's productivity [4, 5].

The problem of the study is to identify the impact of technical reserves of life insurance, which is long-term insurance. The interest of (IC) is focused on increasing profits, which are mainly included in the calculation of ROA. Thus, The research aimed to answer the following question:

- Does technical reserves of (LIO) impact ROA in (IC) listed on the (ASE)?

The significant of this study is since it addresses the issue of the technical reserves of (LIO) and its impact on the profitability of insurance companies. This is made by explaining how the technical reserves is formed, how the net profit deducted from the reserves is addressed, and showing the impact of its investment at present and in the future by examining ROA. So the study aims to identify the concept of technical reserves of (LIO)and how it is formed and identify the impact of the formation of technical reserves of (LIO)on ROA in insurance companies.

2 Literature Review

(Sheikh Eid, 2019)[6]the study aims to clarify the role of the actuarial accountant in the (IC)listed on the (ASE) in determining the technical mathematical reserves. This study concludes that there is a Statistically Significant Impact(SSI)of actuarial accounting in terms of (the role of the administrative actuarial accountant, the role of the actuarial accountant, the factors that contribute to developing the actuarial work, the variables contributing to determining the equation of the premium and the insurance amount) in determining the technical mathematical reserves of (LIO)in companies Insurance listed on the (ASE). This study recommends that attention be paid to teaching actuarial science, and its role within the Faculty of Economics and Accounting programs, mainly administrative accounting, should be clarified through faculty curricula and training courses.

The factors of insurance company profitability are of theoretical and applied importance. Few academics have explored the drivers of profitability in conventional and Islamic insurance firms. The findings of prior investigations are inconsistent. This study will analyze the most critical elements identified in previous research. Most studies have emphasized the internal profitability factors within management's control (Ayele, 2012

[7]; Chen and Wong, 2004) [8]. Following is a summary of a few pertinent prior research from various nations to justify the premise for this study.

(Abdeljawad, Dwaikat 2019) [9] the study aims to examine the factors that affect the profitability of (IC)in Palestine. The balanced panel data of seven (IC) operating in Palestine from 2010 to 2017 were used to determine a linear model among the theoretically determined determinants affecting insurance companies' performance and profitability. One of the most important findings of this study is that the lifetime and size of the company significantly and positively affect the profitability of insurance companies. One of the most important recommendations of this study is that(IC)should increase their investments and manage their expenses wisely to affect their profitability positively.

(Cudiamat, Stephen SIY, 2017) [10]. The study aimed to identify the profitability of the life insurance sector in the Philippines by using the least ordinary squares assembled on a balanced panel by 23 (IC)for the period 2000–2012. The analysis used ROA to measure profitability influenced by the company's chosen level, industry level, and macroeconomic factors. The experimental finds show that most of the factors operating at the corporate level have affected the industrial sector, while macroeconomic factors have insignificantly affected it. This study serves the theoretical aspect in terms of the factors that affect the profitability of insurance companies.

(Chinyakuzal's, 2017) [11]. Study aims to identify diversification's impact on Fidelity Life Assurance's profitability. The study covered the period from 2012 to 2015. The study's population and the sample consisted of Fidelity Life Assurance employees, including managers, accountants and project managers, since these people were aware of the study domain. One of the most important findings of the study is that several motives drive companies to make diversification. This study shows that improving the company's value, spreading risks, and increasing its market share were the main motives that led to the diversification of Fidelity Life Assurance. This study concludes that diversification has negative and positive effects on a company's operations, and it recommends that all companies are encouraged to start implementing this strategy.

(Wasouf, 2016) [12]. Study aims to identify the internal and external factors affecting the profitability of private (IC) operating in Syria. The population and sample of this study consisted of the (IC) operating in Syria for the period 2009–2015. The statistical method used in this study is one of the (Panel) Data models with the aim of analyzing the study data, i.e. the aggregated regression model. One of the most important findings of this study is that insurance premiums and dividends positively affect profitability in the (IC) in question. This study recommends that the leverage ratio should be examined in a manner that ensures the use of these funds for the benefit of the company, attention should be paid to verification of the economic feasibility of the projects implemented by a company, and these investments should be directed towards the areas that make profits.

(Ibrahim, 2016) [13] concluded that this size of the paid claims affects the estimation technical allocations.

3 Study Hypotheses

To answer the study question, the hypothesis designed as follows:

Main hypothesis:

H0: There is no (SSI) at the level ($\alpha \leq 0.05$) of the technical reserves of (LIO)on ROA in (IC) listed on the (ASE).

4 Model of the Study

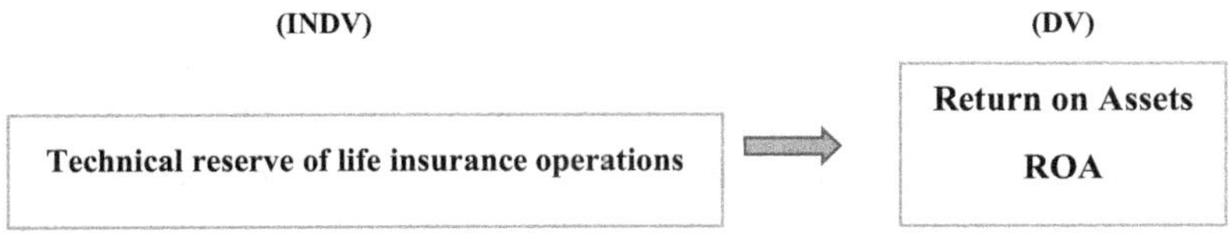

5 Terminology of the Study

Technical Reserves: The amount allocated by a company at the end of the period to meet the obligations of holders of life insurance policies, and it is calculated by multiplying the present value of the policy expected to be paid to policyholders by the possibility of obtaining these documents by them [6].

Life Insurance: The insurance activities that include life and the risks to which the insured is exposed, such as death, disability, ageing and sickness, i.e. the so-called personal insurance [15].

Return on Assets (ROA): It is an indicator that measures the return from investing the available assets, as it demonstrates the efficiency of the administration in exploiting its assets to make profit. When indicator increased , the profitability will increas. This indicator is calculated in the following formula: Net profit after interest and taxes / Average total assets * 100% [16].

6 Theoretical Framework

6.1 Technical Reserves of (LIO)

(IC) a reshare holding companies that carry out various insurance operations. This is in order to achieve the goals that individuals seek to achieve on their own, which are accomplished by obtaining and collecting premiums and reinvesting them in multiple fields that make a significant return, which is used in addressing the potential risks that individuals face as a result of the insured risks.

The need for life insurance simultaneously emerged with marine insurance. Given the advent of the industrial revolution and scientific progress, (IC) have started to provide their services commercially, whereas public shareholding (IC) in Jordan provide such services [17].

Table 1. Differences between life (IC) and public (IC)

Difference	Life Insurance Companies	Public Insurance Companies
Legal form	Similar to fund companies that policyholders own	Similar to public shareholding companies, holders of common shares owned
Likelihood of risk	Certain	Uncertain
Insurance duration	Long-term (more than one year)	Short-term (one year or less)
Amount of Insurance	High	Low or equal to the value of risk
Premiums	Cover more than one year	Cover less than one year
Principle of shared liability	Inapplicable, where every company pays the value of damage separately	Applicable, where (IC)jointly pay the value of the damage

(Sheikh Eid, 2016)

The concept of life insurance means the insurance activities that include life and the risks to which the insured is exposed, such as death, disability, ageing and sickness, i.e. the so-called personal insurance [15].

The concept of technical reserves means the amount allocated by a company at the end of the period to meet the obligations of holders of life insurance policies, and it is calculated by multiplying the present value of the policy expected to be paid to policyholders by the possibilities of obtaining these documents by them [18].

Since life insurance contracts are long-term and cover unavoidable risks, (IC) must form these reserves and employ them in various investments to receive revenues and meet future obligations towards holders of life insurance policies.

Hence, when determining the value of technical reserves, balances of these allocations must be proven at the beginning of the period, and they must be supported by the value of the reserves to be maintained at the time of preparing the balance sheet. These matters include:

1. **Mathematical reserves at the end of the period.**
2. **Outstanding compensation reserves at the end of the period.**
3. **Valuation differences of technical reserves at the end of the period.**

Return on assets (ROA):

$$\text{Return on Assets (ROA)} = \frac{\text{Net profit after interest and taxes}}{\text{Total Assets}}$$

Return on assets is one of the most important measures of profitability used by insurance companies.

7 Methodology of the Study

This study uses the descriptive analytics method,

7.1 Data Collection

The data collection sources are as follows:

Secondary sources: Include all relevant data for this study. As sources, they include the annual financial reports of (IC) listed on the (Amman Stock Exchange) during the study period (2014–2018).

Primary sources: The periodicals, scientific journals and conferences related to the subject of the study were used to write and prepare the theoretical side of this study.

7.2 Study Population and Sample

The study population comprises all (IC) listed on (ASE) for 2015–2019. The sample and data of the study were chosen based on several criteria, including:

Excluding companies merged with or acquired by other companies in the service sector (insurance) during the study period or were liquidated or suspended.

Excluding companies whose complete data were not obtained.

Model Of The Study: $\boldsymbol{ROA} = \boldsymbol{a_0} + \boldsymbol{a_1 Reserve} + \in$

Where: (Azzazi&Kanaani, 2019) [19].

ROA: return on assets

$\boldsymbol{a_0}$: Constant coefficient

$\boldsymbol{a_1}$: First variable coefficient (technical reserves)

Reserves: mathematical reserves of (LIO)

$\in$: Random error

Eviews software was used to analyze time series of a cross-sectional nature, and the natural logarithm was used to measure the value of the technical reserves of (LIO) (Table 1).

Table 2. Descriptive statistics of the study variables

Variables	Mean	Median	Highest value	lowest value	Standard deviation
Technical reserves	1819140	159362	21965374	64750	2982958
Technical reserves logarithm	5.3182	5.202	7.763	3.7443	0.8803
ROA	0.0223	0.029	0.085	-0.1672	0.0335

7.3 Descriptive Analysis

Table 2 shows the results of descriptive tests of the study variables:

The mean of the in (DV) (technical reserves) is (1819140) JOD, while the standard deviation is (2982958). The lowest value is (64750), while the highest is (21965374) JOD. The median value is (159362). The natural logarithm addressed this variable to solve the problem of dispersal in data and high standard deviations.

The result in Table 2 indicates the difference in the premiums earned at contracting and the expected benefits of premiums in the life insurance contracts the various insurance companies. This is due to the policies of each company in terms of the premiums and the size contracts related to life insurance.

The mean of the (DV) (technical reserves logarithm) is (5, 318162), while the standard deviation is (0, 880272). The lowest value is (3, 744293), the highest value is (7, 763169), and the value of the median is (5, 202385).

The mean of the first in (DV) (ROA) is (0, 022342), while the standard deviation is (0, 033541). The lowest value is (0, 16722-), the highest value is (0, 0849), and the value of the median is (0.0293).

This indicates to ability of most companies to effectively use assets and a company's ability to make profits. Normal Distribution Test.

Table 3 shows the results of the standard distribution test for the study variables:

Table 3. Normal Distribution of the Study Variables

Variable	Statistical value (Kurtosis)	Statistical value (Skewness)
Technical reserves	1.77022	1.392574
Technical reserves logarithm	3.589469	0.736117
ROA	4.34016	−1.763669

Table 3 shows that all Skewness values are less than 3. That's mean the value is subject to normal distribution.

7.4 Results of the Study Hypothesis Testing Simple Regression Result of the Study Hypothesis

Table 4 shows the results of testing the study hypothesis (H0): "There is no (SSI) at the level ($\alpha \leq 0.05$) of the technical reserves of (LIO) on ROA (IC)listed on the (ASE)".

It shows that the hypothesis is accepted at levels *10%, **5, and ***1%, Co = coefficient.

Table (4) shows that the statistical model is accepted, where the sig value for F-Statistics Co is 0.024. That's mean the statistical model is accepted at level 5%, and that there is a negative impact between the technical allocation of (LIO)and the (ROA). It also shows that the explanatory value of R2 is 29%, which means that the (INDV) in the study explain 29% of the (DV) (ROA). Upon the preceding, we are reject the null hypothesis, "There is no(SSI) at the level ($\alpha \leq 0.05$) of the technical reserves of (LIO)on ROA in (IC)listed on the (ASE)".

Table 4. Simple regression result

(DV)	INDV))	Co	Std. Err	t-Statistic	Sig
ROA	Fixed	0.056	0.125	3.397	0.001***
	Technical reserves of (LIO)	-0.006	0.023	-2.294	0.024**
	R^2	29%			
	Adjusted R^2	26%			
	F	5.26376			
	Durbin-Watson	1.570692			
	Sig	0.024**			

8 Study Results

1. There is a (SSI) at level ($0.05 \geq \alpha$) of the technical reserves of (LIO)on the profitability of (IC)listed on the (ASE) as measured on ROA. This finding agrees with Sheikh Eid [6] and disagrees with Al-Fattah, [15]; Tarawneh [14]

This indicates an inverse relationship between the technical reserves of (LIO) and the ROA indicator. In other words, the higher the technical allocation of (LIO) in the (IC) listed on the (ASE), the lower the ROA indicator. This explains that the technical reserves is formed by deducting a percentage of the revenues before reaching the net profit to invest this allocation in other operations to use the retained cash to compensate for the potential losses.

Accepts the statistical model with a significance level (Sig) of 0.024 for the F-statistics Co. This suggests that the statistical model is accepted at the 5% level, that there is a high degree of consistency between the research variables, and that the technical allocation of life insurance activities negatively influences the return on assets. R2 indicates that the (INDV) can account for 29% of the variation in the (DV) in the model used in the research (ROA). Based on this evidence, we can conclude that the study hypothesis, which stated that "there is no statistically significant impact at the level (0.05) of the technical reserves of (LIO)on ROA in (IC)listed on the (ASE)," should be rejected in favour of the alternative hypothesis, which stated that "there is a statistically significant impact at the level (0.05) of the technical reserves of (LIO) on ROA in (IC)listed on the (ASE)."

9 Conclusion

Study interested to identify the effect of the technical reserves of (LIO) on ROA in (IC) listed on the (ASE). The results of the statistical analysis of the study hypothesis show that technical reserves has an impact on ROA, which is one of the most important indicators of profitability in insurance companies.

Perhaps this is justified because the technical reserves is usually deducted before reaching the net profit. Therefore the (IC) invest these funds to receive revenues since the value of life insurance policies is greater than the insured risks (death, disability and ageing) at the initial stage of contracts. In the future, the value of such policies becomes less than the insured risks due to inflation and changes in the purchasing power of money, which leads to maintaining particular vigilance to do so and to fulfil future unavoidable obligations in order to avoid problems of liquidity risk and financial failure.

10 Recommendations

Based on the study results, recommends that:

The value of technical reserves should be controlled to achieve balance in ROA. On the other side the calculation of reserves when providing insurance services should be reconsidered to avoid the negative impact of profitability indicators and to reinforce its positive impact. In addition to the provision of insurance services should be diversified to enhance profitability indicators, especially ROA. Finally further studies on the technical reserves in (IC) should be deducted using other standards for profitability indicators.

References

1. Kamal, A.M., Amin, S.A., Salman, A.: for Specialized Establishments (investment funds - commercial banks - (IC)- establishments with departments and branches). Cairo University, Cairo, Egypt, Faculty of Commerce for Publishing (2020)
2. Al-Wardi, S.A.: Risk Management and Insurance, I (2). Iraq Library for Publishing, Iraq, Baghdad (2016)
3. Ahmed, N., Ahmed, Z., Usman, A.: Determinants of performance: a case of the life insurance sector of Pakistan. Int. Res. J. Financ. Econ. **61**(1), (2016)
4. Mazviona, B. W., Dube, M., Sakahuhwa, T.: An analysis of factors affecting the performance of (IC) in Zimbabwe. J. Financ. Investment Anal. **6**(1) (2017)
5. Angoff, J., Brown, R.: An analysis of the profitability and performance of the Michigan auto insurance market. Roger Brown and Associates (2007)
6. Sheikh Eid, A.: The impact of actuarial accounting on determining the mathematical technical reserves for life insurance operations. Zarqa University, Jordan (2019)
7. Ayele, A.G.: Factors Affecting profitability of (IC) in Ethiopia: Panel evidence. Addis Ababa University, Ethiopia (2012)
8. Chen, R., Wong, K.A.: The determinants of financial health of Asian insurance companies. J. Risk Insur. **71**(3), 469–499 (2004)
9. Abdeljawad, I., Dwaikat ,L.: Cudiamat the Determinants of Profitability of (IC) in Palestine. Faculty of Economics and Social Sciences An-Najah National University (2019)
10. Cudiamat, S.S.I.Y.: Determinants of Profitability in Life Insurance Companies (2017)
11. Chinyakuzal: An assessment of the impact of diversification on profitability and growth: case for fidelity life insurance company Zimbabwe (2017)
12. Wasouf: Determinants of Profitability in (IC) Operating in Syria (2016)
13. Ibrahim, N.A.M.I.: Accounting treatment of paid claims and their impact on the technical allocations of insurance companies. Sudan, Shendi University (2016)
14. Ibrahim, A.-T., Muhammad, N.A.: Accounting treatment of paid claims and their impact on the technical allocations of insurance companies. Sudan, Shendi University (2016)

15. Wasouf, A.-F., Ali, A.: Determinants of profitability in (IC) operating in Syria, Syria. Tishreen University Journal for Research and Scientific Studies, Economic and Legal Sciences Series, vol. (36), issue (3) (2016)
16. Al-Habiti, Q., Yahya, Z.: Analysis and Discussion of Financial Statements, 1st edn. DarNamuzajia for Printing and Publishing, Lebanon (2018)
17. Fathi, L.: The role of Life Insurance in Economic Development. University of Abdel Hamid Bin Badis Mostaganem, Algeria (2017)
18. Rahaman, M., Rabbi, I., Alam, M.: The impact of working capital management on the profitability of the elite pharmaceuticals sector in Bangladesh. J. Stud. Manage. Plann. **4**(7), 2395–463 (2018)
19. Al-Azazi, A., Al-Kanani, S.: Data analysis using the (EViews-9) program, 1st edition, Iraq, University of Baghdad, Al-Dad Library for Publishing and Distribution (2019)
20. Ahmed, A., Mamdouh: Risk Management and Insurance, 1st edition, Egypt, Cairo, Tariq Al-Elm Library (2016)
21. Tarawneh, A.: Factors affecting the evaluation of the financial performance of Jordanian insurance companies, the Hashemite Kingdom of Jordan, Middle East University (2015)
22. Ali, S.A., Suleiman, S.A., Ramadan, H.A., Tantawy, S.A.: for Specialized Establishments (investment funds - commercial banks - (IC)- establishments with departments and branches). Cairo University, Cairo, Egypt, Faculty of Commerce for Publishing (2018)
23. Chinyakuza, E.: Chinyakuza an assessment of the impact of diversification on profitability and growth. Midlands State University No. 37. R134456V (2017)
24. Cuddiamat, A., Stephen SIY, J.: Determinants of profitability in life insurance companies: evidence from the philippines. Essays in Economics and Business Studies, Corvinus University of Budapest, University of the Philippines Baguio. ISBN 978–80–89691–42–5 (2017)
25. Ibrahim: Accounting treatment of paid claims and their impact on the technical allocations of insurance companies (2016)
26. IFRS: IFRS. Ifrs.org (2019). https://www.ifrs.org/
27. (ASE) | Provide an organized, fair, transparent, and efficient market for trading securities in Jordan (n.d.) (ASE). https://www.ase.com.jo/en (2021)
28. Homepage - Securities Depository Center (n.d.). Www.sdc.com.jo. Retrieved 20 Oct 2022. https://www.sdc.com.jo/english/index.php
29. Home Page - Ministry of Industry and Trade (n.d.). Www.mit.gov.jo. Retrieved 20 Oct 2022. https://www.mit.gov.jo/Default/en

Correction to: Mechanisms of Digital Currencies

Mohanad Hameed Mhaidi(✉) and Saher Mukhlif Habeeb

Correction to:
Chapter "Mechanisms of Digital Currencies" in:
B. A. M. Alareeni and I. Elgedawy (Eds.): *Artificial Intelligence (AI) and Finance*, SSDC 488,
https://doi.org/10.1007/978-3-031-39158-3_27

The original version of the book was updated with the belated correction in chapter 27. The erratum chapter and the book have been updated with the change.

The updated original version of this chapter can be found at
https://doi.org/10.1007/978-3-031-39158-3_27

B. A. M. Alareeni and I. Elgedawy (Eds.): Artificial Intelligence (AI) and Finance, SSDC 488, p. C1, 2023.
https://doi.org/10.1007/978-3-031-39158-3_90

Author Index

B. A. M. Alareeni and I. Elgedawy (Eds.): *Artificial Intelligence (AI) and Finance*, SSDC 488, pp. 965–968, 2023.
https://doi.org/10.1007/978-3-031-39158-3

H

I

J

K

L

M

N

O

Q

R

S